# Thermodynamics and Heat Power

# Thermodynamics and Heat Power

## Ninth Edition

Irving Granet, Jorge Luis Alvarado, and
Maurice Bluestein

CRC Press
Taylor & Francis Group
Boca Raton  London  New York

CRC Press is an imprint of the
Taylor & Francis Group, an **informa** business

Ninth edition published 2021
by CRC Press
6000 Broken Sound Parkway NW, Suite 300, Boca Raton, FL 33487-2742

and by CRC Press
2 Park Square, Milton Park, Abingdon, Oxon, OX14 4RN

© 2021 Taylor & Francis Group, LLC

Seventh edition published by Prentice Hall 2003
Eighth edition published by CRC Press 2015

CRC Press is an imprint of Taylor & Francis Group, LLC

---

**Library of Congress Cataloging-in-Publication Data**

---

Names: Granet, Irving, author. | Alvarado, Jorge Luis, author. | Bluestein, Maurice, author.
Title: Thermodynamics and heat power / by Irving Granet, Jorge Luis Alvarado and Maurice Bluestein.
Description: Ninth edition. | Boca Raton, FL : CRC Press, [2021] | Includes bibliographical references and index.
Identifiers: LCCN 2020020516 (print) | LCCN 2020020517 (ebook) | ISBN 9780367280918 (hardback ; acid-free paper) | ISBN 9780429299629 (ebook)
Subjects: LCSH: Thermodynamics. | Heat engineering.
Classification: LCC TJ265 .G697 2021 (print) | LCC TJ265 (ebook) | DDC 621.042--dc23
LC record available at https://lccn.loc.gov/2020020516
LC ebook record available at https://lccn.loc.gov/2020020517

---

ISBN: 978-0-367-28091-8 (hbk)
ISBN: 978-0-429-29962-9 (ebk)

Typeset in Palatino
by Deanta Global Publishing Services, Chennai, India

*This book is dedicated to the memory of Irving Granet*

*This book is also dedicated to Gladys Vega, Solem Minjarez,*

*and to the memory of Luis Alvarado Torres*

# Contents

# *Preface*

The ninth edition of Thermodynamics and Heat Power includes a revised sequence of thermodynamics concepts, processes, and energy systems to enable learning outcomes in any introductory thermodynamics course. Offering a comprehensive approach to the analysis of energy conversion systems and methods, this text focuses on thermodynamics fundamentals such as the first and second laws of thermodynamics, and further explores concepts in the areas of engines and power plant analyses, refrigeration and air conditioning systems, and introductory heat transfer. It examines energy, heat, and work in relation to thermodynamics cycles, and explores the effects of fluid properties on system performance. The book emphasizes on practical energy systems, and incorporates step-by-step problems throughout each chapter to enable a better understanding of the fundamental concepts of thermodynamics.

The book introduces pertinent physics-based mathematical formulations and examples in an easily understandable manner to enable problem solving. The new edition contains extensive learning features within each chapter, including a review of applicable calculus concepts in a simple and straight manner. It uses everyday examples to foster a better understanding of thermal science and engineering concepts and includes examples and problems, and basic computational exercises for in-class and laboratory activities.

The ninth edition encompasses a wide range of topics pertinent and relevant in the teaching and learning of thermodynamics at the undergraduate level. This version covers the same topics covered in the eighth edition; however, the topics have been rearranged to enable adoption of the textbook by universities teaching undergraduate-level thermodynamics as part of engineering and engineering technology curricula. The content within each chapter has been revised in which basic concepts are explained systematically and graphically to facilitate learning. New examples and textbook problems include thermodynamic analysis of cooling towers and other energy-related systems such as microchannels and a variety of heat exchanger configurations. The new version also includes a variety of new pictures and diagrams to facilitate problem solving as well as new day-to-day examples to make sure students understand the relevance of thermodynamics in a range of applications. In conclusion, the new edition of *Thermodynamics and Heat Power* is suitable for undergraduate students in engineering and engineering technology.

I would like to acknowledge the immense contributions by the late Irving Granet and Maurice Bluestein over the past editions of this valuable textbook. My sincere appreciation also goes to the staff at Taylor & Francis for their help with this new edition: Jonathan Plant, Kyra Lindholm, and Bhavna Saxena. I thank my family, including Solem Minjarez and Gladys Vega.

**Jorge L. Alvarado**
*College Station, Texas*

# The Authors

**Maurice Bluestein** was a professor emeritus of Mechanical Engineering Technology at Indiana University–Purdue University Indianapolis. He taught for 19 years at the undergraduate and graduate levels, following a 25-year career in the biomedical engineering industry. His industrial experience included developing artificial limbs for the Veterans Affairs Department, designing waste management systems for the Apollo space mission, managing the clinical usage of the intra-aortic balloon pump as a cardiac assist device, and using ultrasound imaging to detect carotid artery blockages and to aid in the diagnosis of breast cancer. He received a Ph.D. degree in biomedical engineering from Northwestern University and MS and BS degrees in mechanical engineering from New York University and the City College of New York, respectively. He authored numerous scientific papers and is the co-developer of the Wind Chill Temperature Chart used by the weather services of the United States and Canada.

**Jorge L. Alvarado** is a professor in the Department of Engineering Technology and Industrial Distribution at Texas A&M University (TAMU). He holds a joint appointment in the Department of Mechanical Engineering at TAMU. He teaches courses in the areas of thermal sciences, fluid mechanics, and fluid power. Dr. Alvarado also conducts scholarly research activities in the areas of nanotechnology, micro-scale heat transfer, electronic cooling, phase change materials, thermal storage, bio-fuel combustion, and energy conservation and use of renewable energy in buildings. He received his BS degree in mechanical engineering (1991) from the University of Puerto Rico at Mayagüez; MS (2000) and Ph.D. (2004) degrees in mechanical engineering from the University of Illinois at Urbana-Champaign. He is also a registered professional engineer with more than 29 years of professional experience.

Dr. Alvarado has worked and collaborated with the U.S. Army Corps of Engineers, the University of Wisconsin at Madison, and Trinity University, among others. His publication record includes scholarly articles in journals such as *Applied Physics Letters*, *International Journal of Heat and Mass Transfer*, *Renewable Energy*, *Fuel*, *Energy* and *Numerical Heat Transfer*, among others. He is a member of the American Society of Mechanical Engineers (ASME), American Society of Thermal and Fluid Engineers (ASTFE), American Society for Engineering Education, and the American Physical Society (APS).

# *Symbols*

| | | Units | |
|---|---|---|---|
| **Symbol** | **Definition** | **British Engineering** | **SI** |
| $a$ | Acceleration | ft./s$^2$ | m/s$^2$ |
| $a, A$ | Area | ft.$^2$ | m$^2$ |
| $A, B, C, D, E$ | Constants | | |
| $c$ | Clearance | dimensionless | |
| $c$ | Specific heat | Btu/lb$_m$·R | kJ/kg·K |
| $C_d$ | Discharge coefficient | dimensionless | |
| $C_n$ | Specific heat of any process | Btu/lb$_m$·R | kJ/kg·K |
| $C_p$ | Specific heat at constant pressure | Btu/lb$_m$·R | kJ/kg·K |
| $C_p$ | Total specific heat at constant pressure | Btu/R | kJ/K |
| $c_v$ | Specific heat at constant volume | Btu/lb$_m$·R | kJ/kg·K |
| $Cv$ | Total specific heat at constant volume | Btu/R | kJ/K |
| $Cv$ | Velocity coefficient | dimensionless | |
| COP | Coefficient of performance | dimensionless | |
| $d, D$ | Diameter | ft. | m |
| $e$ | Base of natural logarithms | dimensionless | |
| $F$ | Force | lb$_f$ | N |
| $F_a$ | Geometric factor | dimensionless | |
| $F_e$ | Emissivity factor | Dimensionless | |
| $g$ | Acceleration of gravity | ft./s$^2$ | m/s$^2$ |
| $g_c$ | Gravitational constant | 32.174 ft.·lb$_m$/lb$_f$·s$^2$ | |
| Gr | Grashof number | dimensionless | |
| $H$ | Enthalpy | Btu | kJ |
| $h$ | Heat transfer coefficient | Btu/h·ft.$^2$·°F | kW/m$^2$·K |
| $h$ | Height | ft. | m |
| $h$ | Specific enthalpy | Btu/lb$_m$ | kJ/kg |
| $h_f$ | Specific enthalpy—saturated liquid | Btu/lb$_m$ | kJ/kg |
| $h_{fg}$ | Specific enthalpy of vaporization ($h_g - h_f$) | Btu/lb$_m$ | kJ/kg |
| $h_g$ | Specific enthalpy—saturated vapor | Btu/lb$_m$ | kJ/kg |
| $h_r$ | Heat transfer coefficient—radiation | Btu/h·ft.$^2$·°F | kW/m$^2$·K |
| $i$ | Current | amperes | amperes |
| $J$ | Mechanical equivalent of heat | 778 ft.·lb$_f$/Btu | |
| $k$ | $c_p/c_v$, heat capacity ratio | dimensionless | |
| $K$ | Proportionality constant | | |
| $k$ | Spring constant | lb$_f$/in. | N/m |
| $k$ | Thermal conductivity | Btu/h·ft.·°F | kW/$_m$·K |
| K.E. | Kinetic energy | ft.·lb$_f$/lb$_m$ | kJ/kg |
| $l, L$ | Length | ft. | m |
| $m$ | Mass | lb$_m$ | kg |
| $\dot{m}$ | Mass flow rate | lb$_m$/s | kg/s |
| mep | Mean effective pressure | lb$_f$/in.$^2$ | kPa |

*(Continued)*

| Symbol | Definition | Units | |
|---|---|---|---|
| | | **British Engineering** | **SI** |
| MW | Molecular weight | $lb_m/lb_m \cdot mol$ | $kg/kg \cdot mol$ |
| $n$ | Number of moles | dimensionless | |
| $n$ | Number of moles | mass/MW | mass/MW |
| $n$ | Number of particles | dimensionless | |
| $n$ | Polytropic exponent | dimensionless | |
| Nu | Nusselt number | dimensionless | |
| $p$ | Pressure | $lb_f/in.^2$ | kPa |
| $P_m$ | Mean effective pressure | $lb_f/in.^2$ | kPa |
| $P_m$ | Mixture pressure | $lb_f/in.^2$ | kPa |
| $P_r$ | Reduced pressure | dimensionless | |
| $P_r$ | Relative pressure | dimensionless | |
| P.E. | Potential energy | $ft. lb_f/lb_m$ | kJ/g |
| Pr | Prandtl number | dimensionless | |
| $Q$ | Heat interchange | Btu | kJ |
| $\dot{Q}$ | Heat transfer rate | Btu/h | kW |
| $q$ | Specific heat interchange | $Btu/lb_m$ | kJ/kg |
| $\dot{Q}_r$ | Radiant heat transfer rate | Btu/h | kW |
| $R$ | Electrical resistance | ohms | ohms |
| $r$ | Radius | ft. | m |
| $R$ | Universal gas constant | $ft. lb_f/lb_m \cdot R$ | $kJ/kg \cdot K$ |
| $r_c$ | Compression ratio | dimensionless | |
| $r_{c.o.}$ | Cutoff ratio | dimensionless | |
| $r_e$ | Expansion ratio | dimensionless | |
| $r_P$ | Pressure ratio | dimensionless | |
| $R_t$ | Thermal resistance | $°F \cdot h/Btu$ | °C/W |
| $Re$ | Reynolds number | dimensionless | |
| $s$ | Specific entropy | $Btu/lb_m \cdot R$ | $kJ/kg \cdot K$ |
| $S$ | Total entropy | Btu/R | kJ/K |
| $s_f$ | Specific entropy of saturated liquid | $Btu/lb_m \cdot R$ | $kJ/kg \cdot K$ |
| $s_{fg}$ | Specific entropy of vaporization $(s_g - s_f)$ | $Btu/lb_m \cdot R$ | $kJ/kg \cdot K$ |
| $s_g$ | Specific entropy of saturated vapor | $Btu/lb_m \cdot R$ | $kJ/kg \cdot K$ |
| $s_g$ | Specific gravity | dimensionless | |
| $t$ | Temperature | °F | °C |
| $T$ | Temperature, absolute | R | K |
| $t$ | Time | s (seconds) | s |
| $T_r$ | Reduced temperature | dimensionless | |
| $T_c$ | Critical temperature | R | K |
| $(\Delta t)_m$ | Logarithmic temperature difference | °F | °C |
| $U$ | Internal energy | Btu | kJ |
| $U$ | Overall heat transfer coefficient | $Btu/h \cdot ft.^2 \cdot °F$ | $kW/m^2 \cdot K$ |
| $u$ | Specific internal energy | $Btu/lb_m$ | kJ/kg |
| $u_f$ | Specific internal energy—saturated liquid | $Btu/lb_m$ | kJ/kg |
| $u_{fg}$ | Specific internal energy of vaporization $(u_g - u_f)$ | $Btu/lb_m$ | kJ/kg |
| $u_g$ | Specific internal energy—saturated liquid | $Btu/lb_m$ | kJ/kg |

*(Continued)*

| Symbol | Definition | British Engineering | SI |
|--------|-----------|---------------------|-----|
| | | **Units** | |
| $v$ | Specific volume | Ft.$^3$/lb$_m$ | m$^3$/kg |
| $V$ | Velocity | ft./s | m/s |
| $V$ | Volume | ft.$^3$ | m$^3$ |
| $\dot{V}$ | Volume flow rate | ft.$^3$/min | m$^3$/s |
| $v_c$ | Critical specific volume | ft.$^3$/lb$_m$ | m$^3$/kg |
| $v_f$ | Specific volume of saturated liquid | ft.$^3$/lb$_m$ | m$^3$/kg |
| $v_{fg}$ | Specific volume of vaporization ($v_g - v_f$) | ft.$^3$/lb$_m$ | m$^3$/kg |
| $v_g$ | Specific volume of saturated vapor | ft.$^3$/lb$_m$ | m$^3$/kg |
| $v_r$ | Reduced specific volume | dimensionless | |
| $v_r$ | Relative specific volume | dimensionless | |
| $W$ | Humidity ratio | dimensionless | |
| $\dot{W}$ | Power | Btu/min | kW |
| $W$ | Weight | lb$_f$ | kN |
| $w$ | Weight | lb$_f$ | N |
| $W$ | Work | ft. lb$_f$ | kJ |
| $w$ | Work per unit mass | ft. lb$_f$/lb$_m$ | kJ/kg |
| $x$ | Length | ft. | m |
| $x$ | Mole fraction | dimensionless | |
| $x$ | Quality | dimensionless | |
| $Z$ | Compressibility factor | dimensionless | |
| $z$ | Elevation above reference plane | ft. | m |
| $\alpha$ | Absorptivity | dimensionless | |
| $\Delta$ | Small change of variable | dimensionless | |
| $\varepsilon$ | Emissivity | dimensionless | |
| $\phi$ | Relative humidity | dimensionless | |
| $\gamma$ | Specific weight | lb$_f$/ft.$^3$ | kN/m$^3$ |
| $\eta$ | Efficiency | dimensionless | |
| $\eta_v$ | Volumetric efficiency | dimensionless | |
| $\mu$ | Viscosity | lb$_m$/ft.$^2$·h | N·s/m$^2$ |
| $\theta$ | A function of | dimensionless | |
| $\rho$ | Density | lb$_m$/ft.$^3$ | kg/m$^3$ |
| $\rho$ | Reflectivity | dimensionless | |
| $\sigma$ | Stefan–Boltzmann constant | Btu/h·ft.$^2$·R$^4$ | W/m$^2$·K$^4$ |
| $\tau$ | Transmissivity | dimensionless | |

*Note:* In addition to the symbols listed above, the following notation is used: Subscript $_m$ refers to the mixture property.

# 1

## Fundamental Concepts

### Learning Goals

After reading and studying the material in this chapter, you should be able to

1. Define thermodynamics as the study of energy and the conversion of energy from one form to another.
2. Understand the importance of thermodynamics in today's world.
3. Use the observable external characteristics that are known as properties to describe a system.
4. Establish and convert from one system of temperature measurement to another and understand the four methods of measuring temperature.
5. Use both the English and SI systems of units.
6. Use the elementary kinetic theory of gases to establish the concepts of pressure, temperature, density, specific weight, specific volume, and Avogadro's law.
7. Use the concept of pressure in both English and SI units (gauge and absolute pressure definitions are important ideas that are necessary for engineering applications).
8. Use the concept that fluids exert pressures that can be expressed in terms of the height and specific weight of the column of fluid.
9. Describe the various methods of measuring pressure and the methods used to calibrate pressure-measuring devices.

## 1.1 Introduction

*Thermodynamics* is the study of energy, heat, work, the properties of the media used, including fluids (liquid and gases), and the processes involved. Thermodynamics is also the study of the conversion of one form of energy into another. Because energy can be derived from electrical, chemical, nuclear, or other means, thermodynamics plays an important role in all branches of engineering, physics, chemistry, and the biological sciences.

In defining the word *thermodynamics*, we used the terms *energy*, *heat*, and *work*. It is necessary to examine these terms in detail, and this will be done in subsequent chapters. In this chapter, certain fundamental concepts are defined, and basic ideas are developed for future use.

Even in our modern age, which has seen changes in our understanding of how the world works, the basics of thermodynamics and its fundamental laws remain valid and always applicable. They apply to biological, chemical, electromagnetic, and mechanical systems. They apply to microscopic as well as macroscopic systems. In this textbook, the emphasis is on practical mechanical systems, including engines and heat-transfer devices.

The study of thermodynamics has never been so important as it is today. We live in an interconnected world where energy is a fundamental part of our lives. The cars that we drive, the computers we use, the appliances we rely on, and the homes and buildings we live in, all involve the use and transformation of energy for different purposes. Understanding the basics of thermodynamics and its fundamental laws is an important step in the design, analysis, and wise management of energy resources and systems.

## 1.2 Thermodynamic Systems

In physics, when studying the motion of a rigid body (i.e., a body that is not deformed or only slightly deformed by the forces acting on it), extensive use is made of *free-body diagrams*. Briefly, a free-body diagram is an outline of a body (or a portion of a body) showing *all* the external forces acting on it. A free-body diagram is one example of the concept of a system. As a general concept applicable to all situations, we can define a system as a grouping of matter taken in any convenient or arbitrary manner. We can consider a fixed amount of mass and follow it as it changes shape, volume, or position. In such a system, the mass will have a boundary that prevents any portion of mass from entering or leaving it; and this is called a *closed system*. However, it still permits energy (i.e., heat and/or work) to cross the specified or denoted boundary. Figure 1.1 shows a piston-cylinder device, which is an example of a closed system. As Figure 1.1 shows, energy in the form of heat ($Q$) or work ($W$) can cross its boundary; however, the mass ($m$) inside the system remains constant and cannot move across the boundary. Another example of a closed system is a closed container with fluid inside or a pressure vessel used to store air at high pressure.

On the other hand, we can choose as our system a region in space with a geographic boundary. Such a system permits mass to enter or leave the system across the boundary. It too allows for the movement of energy across the boundary. This is called an *open system*, as shown in Figure 1.2. The figure shows energy in the form of heat ($Q$) or work ($W$), and

**FIGURE 1.1**
Piston-cylinder device (closed system).

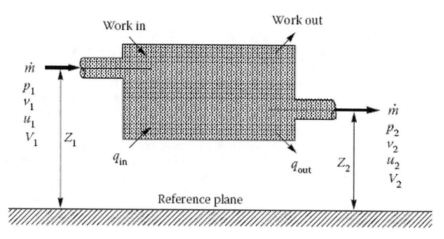

**FIGURE 1.2**
Open (steady flow) system.

mass ($m$) can cross its boundary. Moreover, any open system can be considered to be steady where the rates of energy ($Q$ or $W$) going in and out and mass going in and out are constant with respect to time. Another example of an open (steady flow) system is a gas turbine, in which hot gases flow through it steadily.

In either a closed or open system, a thermodynamic system will invariably have energy transferred from it or to it and can also have energy stored in it. From this definition, it can be noted that we have the ability to choose the grouping, but once having denoted the boundary of the system under consideration, we must take into account *all* the energies involved.

### 1.2.1 Application of System Concept

Thermodynamic equipment is best analyzed by rigorously applying the concept of a system. It can be applied to a volume in space, to a single body or a body of components, and even to a series of processes. To see how effective such a system analysis can be, consider an example of a well-insulated closed room with only a refrigerator in it, which in itself consists of a condenser, evaporator and compressor. There is an electrical outlet in the wall to power the compressor of the refrigerator. The room is so well insulated that no work or heat can pass through the walls except for electrical energy, which flows through the outlet to the electrically powered compressor. As is discussed later in the book, a refrigerator is designed to keep an internal compartment cold, while transferring absorbed heat to the surroundings, in this case the air in the room. If the refrigerator is plugged in and its door is left open, what will happen to the temperature of the air inside the room? Will it rise, fall, or remain constant? In this case, the room air is completely enclosed by the denoted boundaries of the closed system, which also includes the refrigerator. Since there is an inflow of energy to the system through the outlet and the refrigerator, and no other energy or mass change occurs; thus, the air temperature must rise as a result of the net inflow of energy into the system.

This concept should be remembered when considering alternative energy sources for transportation and other applications. This is relevant when considering energy substitutes for our dependence on foreign oil or other non-renewable energy sources. For instance, electric vehicles have been developed and ethanol has been added to gasoline as a secondary fuel; however, all too often such measures ignore the system cost for developing those energy sources. In the case of electric vehicles, one must consider the source of the electricity: batteries, coal-fired electric, natural gas plants, and nuclear power plants. As for ethanol, there is a significant energy cost in producing it since it requires planting and processing corn or other types of biomass.

Another proposed energy source has been hydrogen for fuel cells, which produce electricity and water as outputs in the presence of oxygen. While the benefits to the atmosphere and the efficiency of such devices are often stressed, the means of production of hydrogen and pure oxygen and their inherent costs are often ignored. Treating *all* components of production is system analysis at its best, and this should be kept in mind when considering different energy sources in energy systems.

## 1.3 Basic Thermodynamics Concepts and Properties

Let us consider a given system and then ask ourselves how we can distinguish changes that occur in the system. It is necessary to have characteristics that permit us to measure and evaluate system changes. If these characteristics do not change, we should be able to state that the system has not changed. Some of these measurements or quantifiable observations can be made on a system, which includes temperature, pressure, volume, and position, among others. These observable physical characteristics are called *properties*. When all the values of properties in a system are the same at two different times, we can say that we cannot distinguish any difference in the system at these times.

On the other hand, the properties and their values in a system enable us to uncover differences in the system after it has undergone a change. Therefore, the complete description of a system is given by its properties. The condition of the system, that is, its position, energy content, and so on, is called the *state* of the system. Thus, properties determine its state. Figure 1.3 shows three identical physical closed systems but at different states, given the different values of two properties, in this case, temperature and pressure. Therefore, as can be seen in the figure, the properties denote a state for a system.

It has already been noted that a given state of a system is reproduced when all its properties are the same. Because a given set of properties determines the state of a system, the state is reproduced regardless of the history or path the system may undergo to achieve the state. For example, consider a weight that is lifted vertically from one position to another. This weight can be brought to the same position by first lifting it vertically part of the way, then moving it horizontally to the right, then lifting it another part of the way, then moving it horizontally to the left, and, finally, lifting it vertically to the desired point. As this example illustrates, the state of the system at the end of the two processes is the same,

| m = 1 kg (air) | m = 1 kg (air) | m = 1 kg (air) |
| p = 100 kPa | p = 50 kPa | p = 10 kPa |
| T = 15 °C | T = 25 °C | T = 45 °C |

**FIGURE 1.3**
Example of different states in an identical system.

and the path the system took did not affect its state after the change had occurred. As we shall see in Chapter 2, a consequence of the foregoing is that the change in energy of a system between two given states is the same, regardless of the method of attaining the new or final state. In mathematical terminology, energy is a state function, not a path function; therefore, the difference in energy between the initial and final states are completely independent of the process path followed by the system. However, a fundamental question remains: How many properties are needed to define a state? In a simple compressible system—one in the absence of electrical, magnetic, gravitational, motion or surface tension effects—only two independent, *intensive properties* (explained below) are necessary to define a state.

In thermodynamics, properties are classified into two major categories: *extensive* and *intensive properties*. The properties that depend on the size and total mass of a system are termed *extensive properties*; that is, they depend on the extent or size of the system. An *intensive property* is independent of the size of the system. As Table 1.1 shows, pressure and temperature are examples of intensive properties since their values do not change with the size of a system under uniform boundary conditions.

In addition, several *intensive properties* such as density area also known as *specific properties* because they are given per unit mass or per defined mass within a system. Specific properties are always intensive properties. Figure 1.4 shows how the magnitude or value of each extensive property (i.e., $m$, $V$) depends on the initial or final size of the system, while intensive properties are completely independent of size or extent because their values are the same throughout the system.

As Table 1.1 shows, there are several fundamental thermodynamics variables that we should consider when designing and analyzing systems. For instance, properties such as temperature, mass, force, and pressure, which are used throughout this book. Therefore, it is necessary to have a good understanding of them. The following sections deal in detail with these properties.

**TABLE 1.1**

List of Representative Extensive and Intensive Properties

| Extensive Properties<br>Name, abbreviation (Units) | Intensive Properties<br>Name, abbreviation (Units) |
|---|---|
| Mass, $m$ (kg, lbm) | Temperature, $T$ (°C, °F, K, R) |
| Volume, $V$ (m³, liters, ft³, gallons) | Pressure, $P$ (kPa, Bar, Atm) |
| Energy, $E$, $W$, $Q$ (kJ, Btu) | Density, $\rho$ (kg/m³, lbm/ft.³) |
| | Specific volume, $v$ (m³/kg, ft.³/lb$_m$) |

**FIGURE 1.4**
Example of the effect of system size on extensive and intensive properties.

### 1.3.1 Temperature

The *temperature* of a system is a measure of the random motion or kinetic energy of the molecules within the system. Temperature is therefore also a measure of the thermal energy in a system (i.e., thermal energy is directly proportional to the kinetic energy of molecules within the system). It is different from *heat*, which is the transfer of thermal energy from one body or system to another. If there are different temperatures within the body (or bodies composing the system), the question arises as to how the temperature at a given location is measured and how this measurement is interpreted. Let us examine this question in detail, because similar questions will also have to be considered when other properties of a system are studied. In air at room pressure and temperature, there are approximately $2.7 \times 10^{19}$ molecules per cubic centimeter (molecules/cm$^3$). If we divide a cube whose dimensions are 1 cm on the side into smaller cubes, each of whose sides is 1/1000 of a centimeter, there will be approximately $2.7 \times 10^{10}$ molecules in each of the smaller cubes, still an extraordinarily large number. Although we speak of temperature at a point, we actually mean the average temperature of the molecules in the neighborhood of the point.

Let us now consider two volumes of inert gases not in contact with each other. By inert, we mean that the gases will not react chemically with each other. If the first volume is brought into contact with the second volume and left there until no observable change occurs in any physical property, the two volumes are said to be in *thermal equilibrium*. Figure 1.5 illustrates how such volumes reach thermal equilibrium after being in contact with each other for some finite amount of time.

If a third volume of inert gas is brought in contact between the first two volumes, and it is allowed to reach thermal equilibrium, it can be said that all the volumes are in thermal equilibrium with each other. Based on this discussion, the three volumes can also be stated to be at the same temperature after reaching thermal equilibrium, as illustrated in Figure 1.6.

This simple experiment can be repeated under the same conditions using solids, liquids, and gases, with the same result every time. The results of all these experiments are summarized and embodied in the *zeroth law of thermodynamics*, which states that

**FIGURE 1.5**
Two volumes reaching thermal equilibrium between each other.

**FIGURE 1.6**
Thermal equilibrium with a third volume or object.

two systems that have equal temperatures to a third system also have equal temperatures with each other. As an alternative definition of the zeroth law, we can say that if two bodies are each in thermal equilibrium with a third body, they are in thermal equilibrium with each other. The importance of this apparently obvious statement was recognized after the first law had already been given its name, and consequently it was called the zeroth law to denote that it precedes the first law. It should be noted that a thermometer measures only its own temperature, and for it to be an accurate indication of the temperature of a second system, the thermometer and the second system must be in mutual thermal equilibrium after being in contact with each other for a finite amount of time. As a consequence of the zeroth law, we can measure the temperatures of two bodies using a third body (a thermometer) without bringing the bodies in contact with each other.

The common scales of temperature are called the Fahrenheit and Celsius (centigrade) and they are defined using the ice point and boiling point of water at atmospheric pressure. The Fahrenheit scale was developed by Gabriel Fahrenheit in 1724, who wanted a scale that had a normal body temperature of 100; he came close, as the normal body temperature is 98.6°F. The Fahrenheit scale is used in the United States and a few much smaller countries. It is used with the English system units. The Celsius scale was developed by Anders Celsius in 1742. He wanted a scale that was much more numerically simpler. It is used in the metric system. In the Celsius temperature scale, the interval between the ice point and the boiling point is divided into 100 equal parts. In addition, as shown in Table 1.2, the ice point of Celsius is zero and the Fahrenheit ice point is 32.

The conversion from one scale to the other is directly derived from Table 1.2 and results in the following relations:

$$T[°C] = \frac{5}{9}\left(T[°F] - 32\right) \tag{1.1}$$

And

$$T[°F] = \frac{9}{5}\left(T[°C]\right) + 32 \tag{1.2}$$

The ability to extrapolate to temperatures below the ice point and above the boiling point of water and to interpolate in these regions is provided by the *International Scale of Temperature*. This agreed-on standard utilizes the boiling and melting points of different elements and establishes suitable interpolation formulas in the various temperature ranges between these elements. The data for these elements are given in Table 1.3.

**TABLE 1.2**

Defining Points for Temperature Scales

|  | °F | °C | K | R |
|---|---|---|---|---|
| Atmospheric boiling point | 212 | 100 | 373 | 672 |
| Ice point | 32 | 0 | 273 | 492 |
| Absolute zero | −460 | −273 | 0 | 0 |

**TABLE 1.3**

Temperature Data

| Element | Melting or Boiling Point at 1 atm | Temperature | |
|---------|-----------------------------------|-------------|------|
| | | °C | °F |
| Oxygen | Boiling | −182.97 | −297.35 |
| Sulfur | Boiling | 444.60 | 832.28 |
| Antimony | Melting | 630.50 | 1166.90 |
| Silver | Melting | 960.8 | 1761.4 |
| Gold | Melting | 1063.0 | 1945.4 |
| Water | Melting | 0 | 32 |
| | Boiling | 100 | 212 |

## ILLUSTRATIVE PROBLEM 1.1

Convert (a) 77°F to °C; (b) 10°C to °F; and (c) −40°F to °C.

**SOLUTION**

(a) Using Equation 1.1:

$$T[°C] = \frac{5}{9}\left(T[°F] - 32\right)$$

$$T = \frac{5}{9}(77 - 32)$$

$$T = 25°C$$

(b) Using Equation 1.2:

$$T[°F] = \frac{9}{5}\left(T[°C]\right) + 32$$

$$T = \frac{9}{5}(10) + 32$$

$$T = 50°F$$

(c) Using Equation 1.1:

$$T[°C] = \frac{5}{9}\left(T[°F] - 32\right)$$

$$T = \frac{5}{9}(-40 - 32)$$

$$T = -40°C$$

As the last problem shows, at −40°F, the temperature is also −40°C on the Celsius scale. In other words, at −40° both scales show the same numerical value for temperature.

By using the results of Illustrative Problem 1.1, it is possible to derive an alternative set of equations to convert from the Fahrenheit to the Celsius temperature scale. When this is done, we obtain

$$T[°F] = \frac{9}{5}\left(40 + T[°C]\right) - 40 \tag{1.3}$$

And

$$T[°C] = \frac{5}{9}\left(40 + T[°F]\right) - 40 \tag{1.4}$$

The resemblance of Equations 1.3 and 1.4 makes them relatively easy to remember and use.

Let us consider the case of a gas that is confined in a cylinder (with a constant cross-sectional area) by a piston that is free to move, as shown in Figure 1.7. If heat is now removed from the system, the gas inside will cool down, and the piston will move down as well; however, the pressure inside the cylinder will be constant due to the constant weight of the piston exerted on the gas inside. This procedure can be carried out for several gases, and if the volume is plotted as a function of temperature, we obtain a family of straight lines that intersect at zero volume, as seen in Figure 1.7. This unique temperature is known as the absolute zero temperature, and the accepted values on the Fahrenheit and Celsius temperature scales are −459.67° and −273.15°, respectively, with the values −460° and −273°, respectively, are used for most engineering calculations. It is also possible to define an absolute temperature scale that is independent of the properties of any substance, and we will consider this point later in this book.

The absolute temperature scale begins at absolute zero temperature; therefore, it is always positive, and more accurately represents the concept that temperature is a measure of the molecular activity in any substance. Thus, molecular activity (vibrations or motion) ceases at absolute zero temperature. In the English system of units, absolute temperature is given in Rankine. In the SI system, the absolute temperature is given in Kelvin. Thus, we define

$$T[R] = T[°F] + 460 \tag{1.5}$$

and

$$T[K] = T[°C] + 273 \tag{1.6}$$

**FIGURE 1.7**
Gas thermometer.

The relation among Rankine (R), degrees Fahrenheit (°F), Kelvin (K), and degrees Celsius (°C), is also shown in Table 1.2.

As noted earlier, the state of a system is uniquely determined by its properties. Thus, the accurate measurement of these properties is of great importance from both a theoretical and practical standpoint. Temperatures are measured in many ways, but in general, all the methods of measuring temperature can be categorized into four classes depending on the basic physical phenomena used to make the measurement:

1. Methods utilizing the expansion of gases, liquids, or solids
2. Methods utilizing the change in electrical resistance of an element
3. Methods utilizing the change in electrical potential of an element
4. Methods utilizing the optical changes of a sensor

The most common device used to measure temperature is the familiar *liquid-in-glass thermometer*, which consists of a reservoir of liquid and a long glass stem with a fine-line capillary. The operation of this type of thermometer is based on the coefficient of expansion of the liquid (usually mercury) being greater than the coefficient of expansion of the glass. For accurate measurements, these thermometers are calibrated by partial, total, or complete immersion in a suitable bath, as shown in Figure 1.8. If the thermometer is calibrated by one method but is used in a different way, it is necessary to make corrections to the readings for the difference in usage. Advantages of the liquid-in-glass thermometers include low cost, simplicity, good reliability, and long life.

**FIGURE 1.8**
Methods of calibrating thermometers.

Another device that is used to measure temperature or temperature differences depends on the expansion of materials and is called the *bimetallic element*. This element usually consists of two thin, flat strips placed side by side and welded together. The composite strip can be used flat or coiled into a helix or spiral. Changes in temperature cause the strip to change its curvature, and the motion produced can be used to move a pointer. The flat bimetallic strip is commonly used in room thermostats, where the motion of one end is used to close or open an electrical contact. The action of a bimetallic strip is shown in Figure 1.9.

Resistance thermometers (Figure 1.10) are commonly used in industry to measure process temperatures. The basic principle of this type of instrument is that the change in electrical resistance of a sensor due to a change in its temperature is easily measurable. The electrical resistivity of some metals increases very nearly in direct proportion to an increase in temperature. Thus, the measured change in resistance of a sensor can be converted to a temperature change. Metals used for the sensors include nickel, copper, and platinum. Because of their calibration stability, high-temperature coefficient, and moderate cost, nickel resistance units are normally recommended for temperature ranges between −100 and +500°F (−73.3 to 260°C). The resistance of the sensing element is usually measured by a *Wheatstone bridge* (shown schematically in Figure 1.11). The indicator is nulled to read zero by using the variable resistance $r_i$, where the various resistances are shown in Figure 1.11. Errors in using the Wheatstone bridge make the circuit shown in Figure 1.11

**FIGURE 1.9**
Bimetallic strip.

**FIGURE 1.10**
Industrial resistance thermometer. (Courtesy of American Chain and Cable Corp., York, PA.)

**FIGURE 1.11**
Wheatstone bridge.

unsatisfactory for highly accurate work. These errors are due to the contact resistance of the variable resistor, resistance changes in lead wires due to temperature gradients along them, and self-heating of the sensor due to the supply current. Modifications of the basic Wheatstone bridge have been made to compensate for and correct these faults. These circuits will be found in the references by interested students.

$$\frac{r_e}{r_b} = \frac{r_1}{r_2} \qquad (1.7)$$

Thermistors are also included as resistance elements. The name *thermistor* is derived from *thermally* sensitive *resistors*, because their resistance varies rapidly with temperature. Thermistors are included in the class of solids known as *semiconductors*, which have electrical conductivities between those of conductors and insulators. The advantages of the thermistor over the resistance thermometer and the thermocouple (to be discussed later) are as follows:

1. A temperature coefficient of resistance approximately ten times that of metals, with a correspondingly greater sensitivity to temperature change
2. A much higher resistivity than the metals so that small units may have high resistance, virtually eliminating the lead and contact resistance problem
3. No need for cold-end or lead material compensation, because the thermistor resistance is a function of its absolute temperature

For a limited temperature range, the thermistor combines all the best features of resistance thermometers and thermocouples and has greater sensitivity than either.

When two wires of different materials are joined at their ends, and their junctions are at different temperatures, a net thermal electromotive force (EMF) is generated that induces a net electric current. This is shown schematically in Figure 1.12a. Such a device is called a thermocouple, which is used as a thermometer by placing one junction in contact with

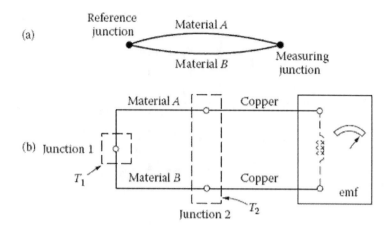

**FIGURE 1.12**
Elementary thermocouple circuit. (a) Basic thermocouple circuit. (b) Thermocouple measurement system. (b: Benedict, R.P.: *Fundamentals of Temperature, Pressure, and Flow Measurements.* 1969. Copyright Wiley-VCH Verlag GmbH & Co. KGaA. Reproduced with permission.)

the body whose temperature is to be measured and measuring the voltage produced at the other junction with a millivoltmeter, as shown schematically in Figure 1.12b. The practical application of a thermocouple as a temperature-measuring device in industry depends on three so-called laws or physical principles:

1. If each section of wire in the circuit is homogeneous, that is, if there is no change in composition or physical properties along its length, the EMF in the circuit depends only on the nature of the metals and the temperatures of the junctions.

2. If both of the junctions involving a particular homogeneous metal are at the same temperature, this metal makes no net contribution to the EMF. Thus, if the complete circuit consists of iron, constantan (60% copper, 40% nickel alloy), and copper, but both of the junctions involving copper are at the same temperature, we can consider the circuit as if it consisted entirely of iron and constantan, with only two junctions.

3. If all junctions of the circuit except one are held at a constant temperature, the EMF in the circuit will be a function of the temperature of the remaining junction and can be used to measure that temperature. It is customary to prepare tables giving this EMF as a function of temperature for the case where the reference junction (or junctions) is held at 0°C (32°F).

Figure 1.12b shows a thermocouple with two *continuous* dissimilar wires from the measuring function to the reference junction. From the reference junction, copper wires and the potentiometer (or millivoltmeter) complete the circuit. For the case of more than one thermocouple to be monitored, a circuit of the type shown in Figure 1.13 can be used. It is important to note that each thermocouple consists of two continuous wires between the measuring junction and the reference junction. Rather than use a circuit with multiple junctions, it is possible to use the circuit shown in Figure 1.14, which has a single reference junction. A typical industrial circuit using a potentiometer that is constructed to compensate automatically for the reference junction temperature is shown in Figure 1.15. Nowadays, portable or hand-held data acquisition (DAQ) systems are capable of converting

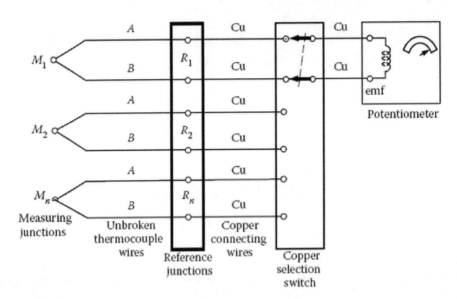

**FIGURE 1.13**
Ideal circuit when more than one thermocouple is involved.

the analog signal (i.e., current) produced by a thermocouple to a digital signal. This in turn allows for converted temperature readings to be fully integrated into a computer-based monitoring system.

Measurement of temperature can also be done non-invasively by measuring the level of radiation being emitted by objects. As a matter of fact, all bodies radiate energy at a rate proportional to the fourth power of their absolute temperature. This is based on the well-known Stefan–Boltzmann law, which is discussed in some detail in Chapter 11. For the moment, we concern ourselves with the optical pyrometer or infrared thermometers, which provides us with a method of converting this radiation to a temperature measurement. The optical pyrometer, shown schematically in Figure 1.16, consists of a telescope, a red glass filter and a small light bulb within it. In practice, the telescope is aimed at the body whose temperature is to be measured. The filament of the bulb appears black against the bright background being measured. The current through the bulb is varied by adjusting the rheostat until the brightness of the filament matches the brightness of the body being measured. By prior calibration, the reading of the ammeter is directly convertible to temperature. The advantage of this device is that no part of it is in contact with the body, and the optical pyrometer can be used to measure temperature above the melting points of either resistance thermometers or thermocouples. Currently, there are several commercially-available optical devices capable of converting radiation into temperature readings. These include infrared (IR) thermometers and IR cameras. For proper selection of non-invasive thermometers, it is important to understand the physical limits of the device by familiarizing oneself with its specifications and range of applicability. For most instances, previous knowledge of key optical properties (e.g., emissivity) of the surface of the object being measured is required.

A summary of commercially-available measuring devices is given in Table 1.4, showing their temperature ranges and characteristics.

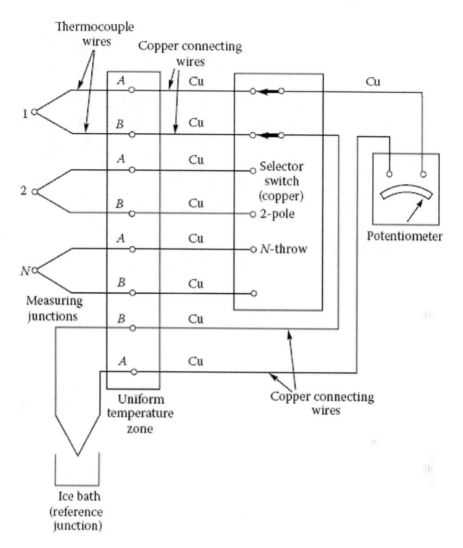

**FIGURE 1.14**
Single reference junction thermocouple circuit.

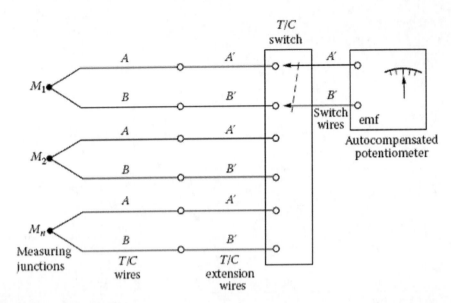

**FIGURE 1.15**
Typical industrial thermocouple circuit.

**FIGURE 1.16**
Schematic of the optical pyrometer.

**TABLE 1.4**

Summary of Information Regarding Commercially Available Temperature-Measuring Instruments or Systems

| Kind of Instrument | Glass Stem (Liquid-in-Glass) | | Filled Systems | | | | Thermocouples | | | |
|---|---|---|---|---|---|---|---|---|---|---|
| | Mercury | Other Liquids | Bimetal | Vapor Pressure | Liquid-Filled | Gas-Filled | Iron/Constantan (Type J) | Chromel/Alumel (Type K) | Copper/Constantan (Type T) | Platinum/Platinum-Rhodium (Types B, R, S) |
| Low temp limit | −38°F (−39°C) | −32°F (−196°C) (pentane) | −100°F (−73°C) | −127°F (−20 K) | −125°F (−87°C) | −452°F (4K) | About −320°F (−196°C) | About −300°F (−184°C) | −424°F (20 K) | Commonly 32°F (0°C) |
| High temp limit | 1100°F (593°C) | | 1000°F (538°C) | 600°F (316°C) | 1200°F (649°C) | 1000°F (538°C) | 1400°F (760°C) in oxidizing atm, 1800°F (982°C) in reducing atm, 2000°F (1093°C) spot readings | 2100°F to 2300°F (1149°C to 1260°C), 2400°F (1316°C) spot readings (not to be used in reducing atm) | 660°F (349°C) 930°F (499°C) spot readings | 2600°F to 2800°F (1427°C to 1538°C)3216°F (1769°C) spot readings (not to be used in reducing atm) |
| Remarks | Low price, reasonable accuracy, easily broken, short scale length | | Easier to read than liquid-in-glass, sometimes damaged by overheating | Normally has a nonlinear scale, has "cross ambient" effect, cheapest of the filled systems | Can be compensated accurately for ambient temperature variations | Suitable for wide ranges, linear scale, requires a large bulb | Thermoelectric power, microvolts per degree C | | | |
| | | | | | | | 64 at 760°C | 38 at 1100°C | 41 at 25°C | 12 at 1200°C (alloy 10% Rh) 14 at 1200°C (alloy 13% Rh) |
| | No auxiliary equipment required | | | | | | The couple most widely used in industry, poor reproducibility above 1600°F (871°C) | Chromel is 90 Ni, 10 Cr (chromel P); alumel is 94 Ni,) 3 Mn, 2 Al, 1 Si | Constantan is 57 Cu, 43 Ni | Expensive |
| | | | | | | | Can give a very rapid response, matched only by the radiation and optical pyrometers | | | |

*(Continued)*

**TABLE 1.4 (CONTINUED)**

Summary of Information Regarding Commercially Available Temperature-Measuring Instruments or Systems

| Kind of Instrument | Resistance Thermometers (RTD) | | | Thermistors | Radiation Pyrometer | Optical Pyrometer | Pyrometric Cones | Melting Pellets |
| --- | --- | --- | --- | --- | --- | --- | --- | --- |
| | Platinum | Copper | Nickel | | | | | |
| Low temp limit | −300°F (−184°C) | −220°F (−140°C) | −300°F (−184°C) | −76°F (−60°C) (most types) | About room temperature | 1400°F (760°C) (normally) | 1085°F (585°C) | 113°F (45°C) |
| High temp limit | 1400°F to 1800°F (760°C to 982°C) | 250°F (121°C) | 600°F (316°C) | 750°F (399°C) (glass coated) | As high as desired | As high as desired | 3659°F (2015°C) | 2500°F (1371°C) |
| Remarks | $\dfrac{(R_{100°c} - R_{o°c})/100}{R_{o°c}}$ | | | | Can operate a recorder and automatic controls | Cannot operate a recorder or automatic controls | Indication somewhat affected by heating rate | |
| | 0.0039 | 0.0043 | 0.0066 | −0.0098 (one type) | | | | |
| | May drift at the higher temperatures | | | High sensitivity, only fair stability | Absorbers of radiation, such as windows or dirt, can affect accuracy. Emissivity of source important. Very rapid response. | | A single cone or pellet can indicate only its rated temperature and is normally used only once. | |

Give greater accuracy than thermocouples, for the same investment. High sensitivity makes them suitable for narrow range instruments.

*Source:* Wolfe, H.C. (ed.), *Temperature, Its Measurement and Control in Science and Industry*, Reinhold Publishing, New York, 1955. With permission.

## 1.4 Force and Mass

### 1.4.1 English System

Force is very often defined in elementary physics texts as the push or pull exerted on a body. Although this definition serves to satisfy our daily experience, it is not satisfactory when dealing with the motion of bodies that are subjected to resultant forces that are not zero. The following paragraphs deal with the concept of force in a consistent manner, and we attempt to clarify the confusion concerning the units of force and mass.

The basis of much of the physical sciences rests on the work of Isaac Newton (1643–1727), English scientist from the late and early 17th and 18th centuries, respectively. For the present, two physical laws that are attributed to him will be used:

1. Law of universal gravitation
2. Second law of motion

The first of these, the *law of universal gravitation*, states simply that the force of attraction experienced by two bodies is proportional to the product of their masses and inversely proportional to the square of the distance (*d*) separating them. At this point, we define the *mass* of a body as the quantity of matter contained in the body. Thus, if the Earth is assumed to be spherical and its mass center is taken to be at its geometrical center, a body on the surface will experience a constant force due to the Earth's attraction. This force is given the name *weight*. Because the Earth is an oblate spheroid, a body at different locations at the surface will have slightly different weights. Also, the surface of the Earth is not smooth, so weight slightly depends on elevation. While a body's weight can change with altitude, its mass is constant (i.e., mass is directly related to the number of atoms present in a body).

So far, it would appear that the preceding concepts are both clear and relatively simple. In its simplest form, it can be stated that the weight of a body, in a given location, is proportional to its mass. By choosing the constant of proportionality to be unity, *the mass of an object in pounds at the Earth's surface may, for most practical purposes, be assumed to be numerically equal to the weight of the body*. The weight of a body can also be defined by *Newton's second law of motion* in terms of the fundamental units of length, time, and mass. This relation can be stated as follows:

$$F \propto ma \tag{1.8}$$

where *a* is the acceleration of a body due to gravity, *m* its mass, and *F* is its weight.

In the English system, the basic unit of mass is the slug; of distance, the foot; and of time, the second. The unit for acceleration is therefore ft./s². The derived unit of force, called the *pound*, is that amount of force, which will accelerate one slug at 1 ft./s². On the surface of the Earth, the acceleration of gravity is considered to be 32.174 ft./s², so a mass of one slug, per Equation 1.8, will weigh 32.174 pounds. The slug is an inconvenient term; it would be easier to express mass in pounds so that mass and weight could be expressed the same numerically using a familiar word. This is accomplished by the introduction of a proportionality constant in Equation 1.9, where $g_c$ is numerically equal to 32.174.

$$F = \frac{ma}{g_c} \tag{1.9}$$

It is in this constant of proportionality, $g_c$, that the confusion occurs, as well as in the ambiguous use of the word *pound*. To differentiate between the pound force and the pound mass, we will use the notation $lb_f$ and $lb_m$. Also, note that $g_c = 32.174 \ lb_m \cdot ft/lb_f s^2$. Use of $g_c$ provides a benefit in that at sea level, the number of pound mass is the same as the number

**TABLE 1.5**

Common Combinations of Units

| Length | Time | Mass | Force | $g_c$ |
|--------|------|------|-------|-------|
| Foot | Second | Pound | Pound | 32.174 ft./s² × pound mass/pound force |
| Foot | Hour | Pound | Pound | 4.17 × 10⁸ ft./h² × pound mass/pound force |
| Foot | Second | Pound | Poundal | 1.0 ft./s² × pound mass/poundal |
| Foot | Second | Slug | Pound | 1.0 ft./s² × slug/pound force |
| Centimeter | Second | Gram | Dyne | 1.0 cm/s² × gram/dyne |
| Meter | Second | Kilogram | Newton | 1.0 m/s² × kilogram/newton |
| Meter | Second | Kilogram | Kilogram | 9.81 m/s² × kilogram mass/kilogram force |
| Meter | Second | Gram | Gram | 981 cm/s² × gram mass/gram force |

of pound force. Thus, thermodynamic parameters can be stated as per pound regardless of whether it is a pound mass or a pound force. Table 1.5 lists some of the most common combinations of units that can be used.

To avoid the obvious confusion that the multitude of units can cause, the definition of pound mass and pound weight, as given in this section, is applied in conjunction with Equation 1.9 throughout this book. Note that $g_c$ is always 32.174 in the ft./s² system even if the acceleration of gravity changes, and remember that in the English system, which uses $lb_m$, any equation with mass must be divided by $g_c$. As a consequence of these considerations, weight and mass, at a location in which the local gravitational attraction is expressed as $g$, can be interrelated in the following manner:

$$w = \frac{mg}{g_c} \tag{1.10}$$

We also note that mass can be operationally defined as the inertia a body has in resisting acceleration. In the International System of Units (SI system), there is no ambiguity between force and mass, as explained in the next section.

---

**ILLUSTRATIVE PROBLEM 1.2**

A mass of 1 lb. weighs 1 lb. on the Earth. How much will it weigh on the moon? Assume that the diameter of the moon is 0.273 when the Earth's diameter is taken to be unity and that the mass of the moon is 0.0123 in relation to the Earth's mass.

**SOLUTION**

The force exerted on the mass by the moon will determine its "weight" on the moon. In general, the law of universal gravitation can be written as

$$F = K \frac{m_1 m_2}{d^2}$$

where $K$ is a proportionality constant. Using the subscripts $m$ for moon and $e$ for Earth, we have

$$F_e = \frac{K m_e m}{(r_e)^2}$$

because $r_e$, the radius of Earth, is the distance separating the mass center of the body $m$ on Earth and the mass center of Earth. Similarly,

$$F_m = \frac{Km_m m}{\left(r_e\right)^2}$$

$$\frac{F_e}{F_m} = \frac{m_e}{m_m}\frac{r_m}{r_e}^2 = \frac{1}{0.0123}(0.273)^2 = 6.06$$

Thus, a body on Earth feels a force (weight) approximately six times the force it would feel on the moon. The solution to the problem is that a mass of 1 lb. will weigh approximately 1/6 lb. on the moon.

## 1.4.2 SI System

For the engineer, the greater confusion has been the units for mass and weight. As we have noted, the literature abounds with units such as slugs, pounds, mass, pound force, poundal, kilogram force, kilogram mass, dyne, and so on. In the SI system, the base unit for *mass* (not weight or force) is the kilogram (kg), which is equal to the mass of the international standard kilogram located at the International Bureau of Weights and Measures. It is used to specify the quantity of matter (i.e., number of atomic particles) in a body. The mass of a body never varies, and it is independent of gravitational force.

The SI *derived* unit for force is the newton (N). The unit of force is defined from Newton's law of motion: force is equal to mass times acceleration ($F=ma$). By this definition, 1 N applied to a mass of 1 kg gives the mass an acceleration of 1 m/s² ($N=kg·m/s^2$). The newton is used in all combinations of units that include force, for example, pressure or stress ($N/m^2$), energy (N·m), and power ($N \cdot m/s = \dot{W}$). By this procedure, the unit of force is not related to gravity, as was the older kilogram force.

Table 1.6 gives the seven base units of the SI system. Several observations concerning this table should be noted. *The unit of length is the meter and a kilogram is a unit of mass, not weight.* Also, symbols are never pluralized and never written with a period, and uppercase and lowercase symbols *must* be used as shown *without exception*.

Table 1.7 gives the derived units with and without symbols often used in engineering. These derived units are formed by the algebraic combination of base and supplementary units. Note that where the name is derived from a person, the first letter of the symbol

**TABLE 1.6**

Base SI Units

| Quantity | Base SI Unit | Symbol |
|---|---|---|
| Length | Meter | m |
| Mass | Kilogram | kg |
| Time | Second | s |
| Electric current | Ampere | A |
| Thermodynamic temperature | Kelvin | K |
| Amount of substance | Mole | mol |
| Luminous intensity | Candela | cd |

**TABLE 1.7**

Derived SI Units

| Quantity | Name | Symbol | Formula | Expressed in Terms of Base Units |
|---|---|---|---|---|
| Acceleration | Acceleration | $m/s^2$ | $m/s^2$ | $m/s^2$ |
| Area | Square meter | $m^2$ | $m^2$ | $m^2$ |
| Density | Kilogram per cubic meter | $kg/m^3$ | $kg/m^3$ | $kg \cdot m^{-3}$ |
| Energy or work | Joule | J | N·m | $m^2 \cdot kg \cdot s^{-2}$ |
| Force | Newton | N | $m \cdot kg \cdot s^{-2}$ | $m \cdot kg \cdot s^{-2}$ |
| Moment | Newton meter | N·m | N·m | $m^2 \cdot kg \cdot s^{-2}$ |
| Moment of inertia of area | — | $m^4$ | $m^4$ | $m^4$ |
| Plane angle | Radian | rad | rad | Rad |
| Power | Watt | $\dot{W}$ | J/s | $m^2 \cdot kg \cdot s^{-3}$ |
| Pressure or stress | Pascal | Pa | $N/m^2$ | $N \cdot m^{-2}$ |
| Rotational frequency | Revolutions per second | rev/s | $s^{-1}$ | $s^{-1}$ |
| Temperature | Degree Celsius | °C | °C | °C |
| Torque (see "moment") | Newton meter | N·m | N·m | $m^2 \cdot kg \cdot s^{-2}$ |
| Velocity (speed) | Meter per second | m/s | m/s | $M \cdot s^{-1}$ |
| Volume | Cubic meter | $m^3$ | $m^3$ | $m^3$ |

**FIGURE 1.17**
Mass and force.

appears as a capital; for example, newton is N. Otherwise, the convention is to make the symbol lowercase.

As shown above, weight has been defined as a measure of gravitational force acting on a material object at a specified location. Thus, weight is a force that has both a mass component and an acceleration component (gravity). Gravitational forces vary by approximately 0.5% over the Earth's surface. For nonprecision measurements, these variations can normally be ignored. Thus, a constant mass has an approximate constant weight on the surface of the

Earth. The agreed standard value (standard acceleration) of gravity is 9.806650 m/s² (i.e., 9.81 m/s²). Figure 1.17 illustrates the difference between mass (kilogram) and force (newton).

The term *mass* or *unit mass* should be used to indicate only the quantity of matter in an object, and the old practice of using weight in such cases should be avoided in engineering and scientific practice. However, because the determination of an object's mass will be accomplished by the use of a weighing process, the common usage of the term *weight* instead of *mass* is expected to continue but should be avoided.

Based on the foregoing, Equation 1.10 can be written in SI units as

$$w = mg \tag{1.10a}$$

---

### ILLUSTRATIVE PROBLEM 1.3

A body has a mass of 5 kg. How much will it weigh on Earth?

#### SOLUTION

The weight of the body will be its mass multiplied by the local acceleration of gravity, that is,

$$w = mg$$

Thus,

$$w = 5 \text{ kg} \times 9.81 \frac{\text{m}}{\text{s}^2} = 49.05 \frac{\text{kg} \cdot \text{m}}{\text{s}^2}$$

and because 1 N = 1 kg·m/s²,

$$w = 49.05 \text{ N}$$

---

### ILLUSTRATIVE PROBLEM 1.4

A body has a mass of 10 kg. If the local gravitational acceleration is 9.5 m/s², determine (a) its weight and (b) its horizontal acceleration if it is acted on by a 10 N horizontal force.

#### SOLUTION

a. From Equation 1.10a,

$$w = (10 \text{ kg}) 9.5 \frac{\text{m}}{\text{s}^2} = 95.0 \frac{\text{kg} \cdot \text{m}}{\text{s}^2} = 95.0 \text{ N}$$

b. Because $F = ma$, $a = F/m$. Therefore,

$$a = \frac{10 \frac{\text{kg} \cdot \text{m}}{\text{s}^2}}{10 \text{ kg}} = 1 \frac{\text{m}}{\text{s}^2}$$

For the SI system to be universally understood, it is most important that the symbols for the SI units and the conventions governing their use, be adhered to strictly. Care should be taken to use the correct case for symbols, units, and their multiples (e.g., K for Kelvin, k for kilo, m for milli, M for mega). As noted earlier, unit names are never capitalized except at the beginning of a sentence. SI unit symbols derived from proper names are written with the first letter in uppercase; all other symbols are written in lowercase, for example, m (meter), s (second), K (Kelvin), Wb (Weber). Also, unit names form their plurals in the usual manner. Unit symbols are always written in singular form, for example, 350 megapascals or 350 MPa, 50 milligrams or 50 mg. Because the unit symbols are standardized, the symbols should always be used in preference to the unit names. Unit symbols are not followed by a period unless they occur at the end of a sentence, and the numerical value associated with a symbol should be separated from that symbol by a space, for example, 1.81 mm, not 1.81 mm. The period is used only as a decimal marker. Because the comma is used by some countries as a decimal marker, the SI system does not use the comma. A space is used to separate large numbers in groups of three. Thus, 3 807 747 and 0.030 704 254 indicate this type of grouping. Note that for numerical values less than 1, the decimal point is preceded by a zero. For a number of four digits, the space can be omitted.

In addition, certain style rules should be adhered to:

1. When a product is to be indicated, use a space between unit names (e.g., newton meter).
2. When a quotient is indicated, use the word *per* (e.g., meter per second).
3. When a product is indicated, use the square, cubic, and so on (e.g., square meter).
4. In designating the product of units, use a centered dot (e.g., N·s, kg·m).
5. For quotients, use a solidus (/) or a negative exponent (e.g., m/s or m·s$^{-1}$). The solidus (/) should not be repeated in the same expression unless ambiguity is avoided by the use of parentheses. Thus, one should use m/s$^2$ or m·s$^{-2}$ but not m/s/s; also, use m·kg/(s$^3$·A) or m·kg·s$^{-3}$A$^{-1}$ but not m·kg/s$^3$/A.

One of the most useful features of the older metric system and the current SI system is that multiples and submultiples of units are in terms of factors of 10. Thus, the prefixes given in Table 1.8 are used in conjunction with SI units to form names and symbols of multiples of SI units. Certain general rules apply to the use of these prefixes:

**TABLE 1.8**

Factors of 10 for SI Units

| Prefix | Symbol | Factor | |
|--------|--------|--------|---|
| Tera | T | $10^{12}$ | 1 000 000 000 000 |
| Giga | G | $10^9$ | 1 000 000 000 |
| Mega | M | $10^6$ | 1 000 000 |
| Kilo | k | $10^3$ | 1 000 |
| Hector | h | $10^2$ | 100 |
| Deka | da | $10^1$ | 10 |
| Deci | d | $10^{-1}$ | 0.1 |
| Centi | C | $10^{-2}$ | 0.01 |
| Milli | m | $10^{-3}$ | 0.001 |
| Micro | F | $10^{-6}$ | 0.000 001 |
| Nano | n | $10^{-9}$ | 0.000 000 001 |
| Pico | p | $10^{-12}$ | 0.000 000 000 001 |
| Femto | f | $10^{-15}$ | 0.000 000 000 000 001 |
| Atto | a | $10^{-18}$ | 0.000 000 000 000 000 001 |

1. The prefix becomes part of the name or symbol without separation (e.g., kilometer, megagram).
2. Compound prefixes should not be used; use GPa, not kMPa.
3. In calculations, use powers of 10 in place of prefixes.
4. Try to select a prefix so that the numerical value will fall between 0.1 and 1000. This rule may be disregarded when it is better to use the same multiple for all items. It is also recommended that prefixes representing 10 raised to a power that is a multiple of 3 be used (e.g., 100 mg, not 10 cg).
5. The prefix is combined with the unit to form a new unit, which can be provided with a positive or negative exponent. Therefore, $mm^3$ is $(10^{-3}\ m)^3$ or $10^{-9}\ m^3$.
6. Where possible, avoid the use of prefixes in the denominator of compound units. The exception to this rule is the prefix k in the base unit kg (kilogram).

There are certain units outside the SI system that may be used together with the SI units and their multiples. These are recognized by the International Committee for Weights and Measures as having to be retained because of their practical importance.

It is almost universally agreed that when a new language is to be learned, the student should be completely immersed and made to "think" in the new language. This technique has been proved most effective by the Berlitz language schools and the Ulpan method of language teaching. A classic joke about this is the American traveling in Europe who was amazed that 2-year-old children were able to speak "foreign" languages. In dealing with the SI system, students should not think in terms of customary units and then perform a mental conversion; it is better to learn to think in terms of the SI system, which will then become a second language. However, there will be times when it may be necessary to convert from customary US units to SI units. To facilitate such conversions, Table 1.9 gives some commonly used conversion factors.

---

### ILLUSTRATIVE PROBLEM 1.5

Table 1.9 lists the conversion factor for obtaining square meters from square feet as $9.290 \times 10^{-2}$. Starting with the definition that 1 in. equals 2.54 cm (0.0254 m), derive this conversion factor.

### SOLUTION

In solving this type of conversion problem, it must be kept in mind that units need to be consistent and that we can use the familiar rules of algebra to manipulate units.
Proceeding as noted, we have

$$1\,\text{in.} = 0.0265\,\text{m}$$

Or

$$1 = 0.0254\,\frac{\text{m}}{\text{in.}} \tag{a}$$

Because 1 ft. = 12 in.,

$$1 = 12 \frac{\text{in.}}{\text{ft.}} \tag{b}$$

If we now multiply Equation (a) by Equation (b), we obtain

$$1 = 0.0254 \frac{\text{m}}{\text{in.}} \times 12 \frac{\text{in.}}{\text{ft.}}$$

Because inches cancel, we have

$$1 = 0.0254 \times 12 \frac{\text{m}}{\text{ft.}}$$

We can now square both sides to obtain

$$(1)^2 = (0.0254 \times 12)^2 \frac{\text{m}^2}{\text{ft.}^2} = 9.290 \times 10^{-2} \frac{\text{m}^2}{\text{ft.}^2}$$

or the desired result,

$$1\text{ft.}^2 = 9.290 \times 10^{-2} \text{m}^2$$

which states that ft.$^2 \times 9.290 \times 10^{-2} = $ m$^2$.
The preceding can also be obtained in a single chain-type calculation as

$$0.0254 \frac{\text{m}}{\text{in.}} \times \frac{12 \text{ in.}^2}{\text{ft.}} = 9.290 \times 10^{-2} \frac{\text{m}^2}{\text{ft.}^2}$$

## 1.5 Elementary Kinetic Theory of Gases

The following derivation, based on elementary kinetic gas theory, is very useful and will serve later to introduce the theory of ideal or perfect gases. Imagine a cube, as shown in Figure 1.18, with sides *l*. In this cube, it will be assumed that there are *n* identical molecules (particles), each having a mass *m*. These particles are further assumed to be perfectly elastic spheres, each traveling with the same velocity. By *perfectly elastic*, it is implied that the collision of a molecule with another molecule or the walls of the container causes the molecule to rebound without loss of energy or momentum. The size of the molecule will be taken to be negligibly small compared to the dimensions of the system. If the velocity in the *x* direction is essentially constant over the length *l* and is denoted by $V_x$, the molecule shown in Figure 1.18 will travel a distance equal numerically to $V_x$ in a unit time. At this same time, the molecule will collide elastically with the walls (parallel to the *y–z* plane) a number of times equal to $V_x$ divided by *l*. This last statement has considered only the *x*-directed velocity of the molecule. It is also assumed that the molecules do not collide with each other.

**TABLE 1.9**

Conversions from Conventional to SI Units

| Multiply | By | To Obtain |
| --- | --- | --- |
| Atmospheres (760 torr) | $2.992 \times 10^1$ | Inches mercury (at 32°F) |
| Atmospheres | $1.033 \times 10^4$ | Kilograms/square meter[a] |
| Atmospheres | $1.013 \times 10^2$ | Kilopascals |
| Bars | $9.869 \times 10^{-1}$ | Atmospheres |
| Bars | $1.000 \times 10^2$ | Kilopascals |
| British thermal units (Btu) | $3.927 \times 10^{-4}$ | Horsepower-hours |
| British thermal units (Btu) | 1.056 | Kilojoules |
| British thermal units (Btu) | $2.928 \times 10^{-4}$ | Kilowatt-hours |
| British thermal units (Btu) | $1.221 \times 10^{-1}$ | Megawatt-days |
| Btu/hour-square foot | $3.153 \times 10^{-4}$ | Watts/square centimeter |
| Btu/hour-square foot-°F | $5.676 \times 10^{-4}$ | Watts/square centimeter-°C |
| Btu/minute | $2.356 \times 10^{-2}$ | Horsepower |
| Btu/minute | $1.757 \times 10^1$ | Watts |
| Calories | 4.1868 | Joules |
| Cubic feet | $2.832 \times 10^{-2}$ | Cubic meters |
| Cubic feet | $2.832 \times 10^1$ | Liters |
| Cubic feet/minute | $4.720 \times 10^{-4}$ | Cubic meters/second |
| Cubic meters | $8.107 \times 10^{-4}$ | Acre-feet |
| Cubic meters | $3.531 \times 10^1$ | Cubic feet |
| Cubic meters | $2.642 \times 10^2$ | Gallons (US) |
| Cubic meters/second | $2.119 \times 10^3$ | Cubic feet/minute |
| Cubic meters/second | $1.585 \times 10^4$ | Gallons/minute |
| Degrees Celsius | $(9/5)C + 32$ | Degrees Fahrenheit |
| Degrees Fahrenheit | $5/9(F - 32)$ | Degrees Celsius |
| Feet | $3.048 \times 10^{-1}$ | Meters |
| Feet of $H_2O$ (39.2°F) | $3.048 \times 10^2$ | Kilograms/square meter |
| Feet of $H_2O$ (39.2°F) | $4.335 \times 10^{-1}$ | Pounds force/square inch |
| Feet/second | $3.048 \times 10^{-1}$ | Meters/second |
| Foot-pound (force) | 1.356 | Joules |
| Foot-pounds (force)/minute | $2.260 \times 10^{-2}$ | Watts |
| Gallons | $3.785 \times 10^{-3}$ | Cubic meters |
| Gallons/minute | $6.309 \times 10^{-5}$ | Cubic meters/second |
| Horsepower | $4.244 \times 10^1$ | British thermal units/minute |
| Horsepower | $7.457 \times 10^{-1}$ | Kilowatts |
| Horsepower-hours | $2.545 \times 10^3$ | British thermal units |
| Horsepower-hours | $7.457 \times 10^{-1}$ | Kilowatt-hours |
| Inches of $H_2O$ (39.2°F) | $2.491 \times 10^{-1}$ | Kilopascals |
| Inches mercury (32°F) | $3.342 \times 10^{-2}$ | Atmospheres |
| Inches mercury (32°F) | $3.453 \times 10^2$ | Kilograms/square meter |
| Inches mercury (32°F) | 3.386 | Kilopascals |
| Inches mercury (32°F) | $4.912 \times 10^{-1}$ | Pounds force/square inch |
| Joules | $7.376 \times 10^{-1}$ | Foot-pounds (force) |

*(Continued)*

**TABLE 1.9 (CONTINUED)**

Conversions from Conventional to SI Units

| Multiply | By | To Obtain |
|---|---|---|
| Joules | 1.000 | Watt-seconds |
| Joules | $2.387 \times 10^{-1}$ | Calories |
| Kilograms | 2.205 | Pounds force |
| Kilograms | $1.102 \times 10^{-3}$ | Tons (short) |
| Kilograms/cubic meter | $6.243 \times 10^{-2}$ | Pounds force/cubic foot |
| Kilograms/square meter | $9.678 \times 10^{-5}$ | Atmospheres[a] |
| Kilograms/square meter | $3.281 \times 10^{-3}$ | Feet of $H_2O$ (at 39.2°F)[a] |
| Kilograms/square meter | $2.896 \times 10^{-3}$ | Inches mercury (32°F)[a] |
| Kilograms/square meter | $1.422 \times 10^{-3}$ | Pounds force/square in.[a] |
| Kilojoules | $9.471 \times 10^{-1}$ | British thermal units |
| Kilopascals | 4.015 | Inches $H_2O$ (at 39.2°F) |
| Kilopascals | $1.450 \times 10^{-1}$ | Pounds force/square inch |
| Kilopascals | $2.953 \times 10^{-1}$ | Inches mercury (32°F) |
| Kilopascals | $1.000 \times 10^{-2}$ | Bars |
| Kilopascals | $9.869 \times 10^{-3}$ | Atmospheres (760 torr) |
| Kilowatts | 1.341 | Horsepower |
| Kilowatt-hours | $3.413 \times 10^3$ | British thermal units |
| Kilowatt-hours | 1.341 | Horsepower-hours |
| Kilowatt-hours | $4.167 \times 10^{-5}$ | Megawatt-days |
| Liters | $3.531 \times 10^{-2}$ | Cubic feet |
| Megawatt-days | $8.189 \times 10^7$ | British thermal units |
| Megawatt-days | $2.400 \times 10^4$ | Kilowatt-hours |
| Meters | 3.281 | Feet |
| Newtons | $2.248 \times 10^{-1}$ | Pounds force |
| Pounds | $4.536 \times 10^{-1}$ | Kilograms |
| Pounds (force) | 4.448 | Newtons |
| Pounds mass/cubic feet | $1.602 \times 10^1$ | Kilograms/cubic meter |
| Pounds force/square inch | 2.307 | Feet of $H_2O$ (at 39.2°F) |
| Pounds force/square inch | 2.036 | Inches mercury (32°F) |
| Pounds force/square inch | $7.031 \times 10^2$ | Kilograms/square meter[a] |
| Pounds force/square inch | 6.895 | Kilopascals |
| Square feet | $9.290 \times 10^{-2}$ | Square meters |
| Square meters | $2.471 \times 10^{-4}$ | Acres |
| Square meters | $1.076 \times 10^1$ | Square feet |
| Tonnes (metric tons) | $2.205 \times 10^3$ | Pounds force |
| Tons (short) | $9.072 \times 10^2$ | Kilograms |
| Watts | $5.688 \times 10^{-2}$ | Btu/minute |
| Watts | $4.427 \times 10^1$ | Foot-pounds force/minute |
| Watt-seconds | 1.000 | Joules |
| Watts-square centimeter | $3.171 \times 10^3$ | Btu/hour-square foot |
| Watts/square centimeter-°C | $1.762 \times 10^3$ | Btu/hour-square foot-°F |

[a] Force units with $g_c = 9.81$.

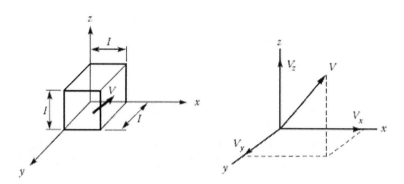

**FIGURE 1.18**
Kinetic theory derivation.

At this point, it will be necessary to involve the concepts of impulse and momentum for a rigid particle that has only translatory motion. On the basis of Newton's laws of motion for such a particle, it is possible to write Equation 1.9 as follows:

$$F = \frac{ma}{g_c} = \frac{m\Delta V}{g_c \Delta t}; \qquad F\Delta t = \frac{m\Delta V}{g_c} \tag{1.11}$$

(Equation 1.11 is written in English units.) The symbol $\Delta t$ is used to indicate a small interval of time, and $\Delta V$ is the change in velocity during this time interval. The term $F\Delta t$ is known as the impulse imparted to the particle by the force $F$ acting for the time $\Delta t$, and the term $(m\Delta V/g_c)$ is called the *momentum change* that the particle undergoes during this same time interval.

Let us consider the collision of a particle with a wall. As shown in Figure 1.19, the particle travels in the $x$ direction with a velocity $V_x$ prior to striking the wall. Because the particle has been assumed to be perfectly elastic, it will collide with the wall and rebound with a speed equal to $V_x$, but it will be traveling in a direction opposite to its initial direction. Thus, its velocity will be $-V_x$. The wall is stationary, and as a result of the impact, a reactive force $F$ is set up. Because the molecule has a momentum equal to $mV_x/g_c$ just before colliding with the wall and a momentum of $-mV_x/g_c$ after the collision, the change in momentum per molecule will be $2mV_x/g_c$. As stated earlier, the number of collisions with the walls in a unit of time will be $V_x/l$, and we must consider that the $n$ identical particles each undergo the same collision with the walls. The total number of collisions will therefore be $nV_x/l$.

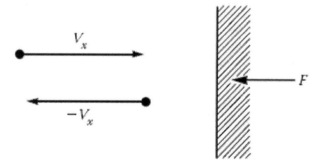

**FIGURE 1.19**
Particle collision with a wall.

Multiplying the total number of impacts per unit of time by the momentum change per impact yields the total change of momentum per unit of time for the system:

$$\frac{2mV_x}{g_c}\frac{nV_x}{l} \tag{1.12}$$

Equation 1.12 also expresses the *force* exerted on the wall. *Pressure* is the normal force exerted per unit surface area. For present purposes, it will be reasonable to assume that the force expressed by Equation 1.12 is distributed uniformly over the $y$–$z$ planes. Because these consist of two faces, each $l^2$ in area, the pressure exerted on the $y$–$z$ plane is

$$\frac{\dfrac{2mV_x}{g_c}\dfrac{nV_x}{l}}{2l^2} = \frac{mnV_x^2}{g_c l^3} \tag{1.13}$$

However, it was previously assumed that the molecule had the same velocity in each of the three component directions. Therefore, in the unit time interval under consideration, there is an equal probability that the particle will hit one of the three sets of parallel faces composing the cube. If this were not so, there would be unequal pressures on the various faces of the tube. Because the volume of the cube ($V$) equals $l^3$, Equation 1.14 follows from Equations 1.11 and 1.13 combined:

$$pV = \frac{1}{3}\frac{mnV_x^2}{g_c} \tag{1.14}$$

or

$$pV = \frac{1}{3}\left(mnV_x^2\right) \text{ (SI)} \tag{1.14a}$$

Equation 1.14 is used repeatedly later (in a slightly different form) and has special significance. However, for the present, only certain limited conclusions will be deduced from it. It should be remembered that this equation was derived on the basis of certain simplifying and therefore restrictive assumptions. In addition to the assumptions stated, it is important to note that the rotations and vibrations of the molecule have been neglected. Also, molecule collisions with other molecules have been ignored. This corresponds to assuming that the dimensions of the unit cube are small compared with the mean free path of the gas particles, that is, the distance molecules travel between molecule–molecule collisions. Although these assumptions and others inherent to this derivation lead to varying degrees of deviation of real gases from Equation 1.14, it is amazing that this equation can be used with a reasonable degree of accuracy for engineering purposes to describe the behavior of many gases over a wide range of conditions.

On the basis of experimental and theoretical investigations by Joule and others, it can be shown that for ideal gases, the pressure of these gases increases equally for equal temperature increments. This is equivalent to stating that the product $mV_x^2$ is a constant for all gases at a given temperature. As a consequence of this conclusion, Equation 1.14 yields *Avogadro's law*, which states that equal volumes of gases at the same temperature and pressure contain the same number of molecules (particles). It is also possible to deduce another concept from Equation 1.14, specifically, density. The product $mn$ found in the numerator of Equation 1.14 is the total mass of the molecules in the volume under consideration. The total mass divided by the total volume yields the property of *density* (ρ). The reciprocal of density is called *specific volume*. It is the total volume that a unit mass occupies under a given set of conditions. It is denoted by *v*.

We summarize some of the conclusions of the foregoing analysis (the first three conclusions are limited to gases and the last three are general definitions):

1. *Pressure* (which is defined as the normal force per unit area, i.e., $lb_f./in^2$, in the English system, or $kN/m^2$ in the SI system) results from the collisions of the molecules with the walls of the container and is a function of the number of impacts per unit time, the mass of the particle, and the velocity of the particle.

2. *Avogadro's law* states that equal volumes of gases at the same pressure and temperature contain the same number of particles.

3. Temperature can be taken to be a measure of the translational energy of the particles. In a more general sense, it can be said to be a measure of the molecular activity (and internal energy) of the gases.

4. *Density* is defined as the mass per unit volume (ρ).

5. The reciprocal of density is the volume per unit mass and is called *specific volume* (*v*, where $v = 1/\rho$).

6. *Specific weight* (γ) is defined as the weight per unit volume. Note that this term, which equals $pg/g_c$, does not exist in SI as a basic unit. However, in the SI system, specific weight consists of two units (kN and $m^3$), or $kN/m^3$.

7. *Specific gravity* (sg) is defined as the ratio of the density of a substance to the density of water at 4°C (i.e., $sg = \rho/\rho_w$). It can also be defined as the ratio of the specific weight of a substance to the specific weight of water at 4°C (i.e., $sg = \gamma/\gamma_w$). The density of water at 4°C is 1000 $kg/m^3$. In the older cgs (centimeter-gram-second) system, the density of water at 4°C is 1 $g/cm^3$, making specific gravity numerically equal to density in this system.

---

## 1.6 Pressure

We have seen that when a gas is confined in a container, molecules of the gas strike the sides of the container, and these collisions with the walls of the container cause the molecules of the gas to exert a force on the walls. When the component of the force that is perpendicular (normal) to the wall is divided by the area of the wall, the resulting normal force per unit area is called the *pressure*. To define the pressure at a point, it is necessary to consider the area in question to be shrinking steadily. Pressure at a point is defined to be the normal force per unit area in the limit as the area goes to zero. Mathematically,

$$p = \frac{\Delta F}{\Delta A} \lim(\Delta A \to 0) \tag{1.15}$$

**FIGURE 1.20**
The relation of a gauge to absolute pressure.

In common engineering units, pressure is expressed as pounds force per square inch ($lb_f/in.^2$) or pounds force per square foot ($lb_f/ft.^2$), which are usually abbreviated as psi and psf. In SI units, pressure is expressed in $N/m^2$, which is known as the *pascal* (Pa), or kilopascal ($kPa = kN/m^2$). The *bar* is $10^5$ Pa. These units are the ones used most frequently in the literature, but certain others are also used because of the manner in which the pressure measurements are made.

Most mechanical pressure gauges, such as the Bourdon gauge (see Figure 1.34), measure the pressure above local atmospheric pressure. This pressure is called *gauge pressure* and is measured in units of pounds force per square foot or pounds force per square inch, that is, psfg or psig. The relation of gauge pressure to absolute pressure is shown in Figure 1.20 and is

$$\text{absolute pressure} \left(p_{abs}\right) = \text{atmospheric pressure} \left(p_{atm}\right) + \text{gauge pressure} \left(p_g\right) \quad (1.16)$$

Absolute pressure is indicated as psia or psfa. It is usual to take the pressure of the atmosphere as being equal to 14.7 psia. However, in Europe, 1 "ata" has been defined and is used as 1 $kg/cm^2$ for convenience, but this unit is not 14.7 psia. It is more nearly 14.2 psia. Care should be exercised when using the unit of pressure that is expressed as ata. In SI, the equivalent of 14.696 (14.7) psia is 101.325 kPa. The terms *gauge* and *vacuum* are not used in SI, because all pressures are absolute in this system.

Referring again to Figure 1.20, it will be noted that pressure below atmospheric pressure is called a *vacuum*. By reference to this figure, we obtain the relation between absolute pressure and vacuum as

$$\text{absolute pressure} \left(p_{abs}\right) = \text{atmospheric pressure} \left(p_{atm}\right) - \text{vacuum} \left(p_{vac}\right) \quad (1.17)$$

---

**ILLUSTRATIVE PROBLEM 1.6**

Assume that a fluid whose specific weight is constant and equal to $\gamma$ $lb_f/ft.^3$ is placed in a uniform tube until its height in the tube is $h$ feet above the base of the column. Determine the pressure at the base of the column.

**SOLUTION**

From our definition of pressure as force divided by area, we must find the force being exerted on the base by the column of fluid. This force will equal the weight of the fluid, which in turn will equal the volume of fluid multiplied by the specific weight of the fluid.

Denoting the weight to be $w$, the cross-sectional area to be $A$, and the liquid height to be $h$, as shown in Figure 1.21, we have $w = \gamma V$ where $V$ is the volume of fluid. The volume of fluid is equal to $A \times h$, giving us

$$w = \gamma A h \tag{1.18}$$

which is the force on the base. Dividing $w$ by $A$ yields the pressure

$$p = \frac{\gamma A h}{A} = \gamma h \tag{1.19}$$

Because it is usual to measure distance up from the base, our result is

$$p = -\gamma h \tag{1.19a}$$

In English units,

$$p = \frac{-\rho g}{g_c} h \tag{1.20}$$

In SI units,

$$p = -\rho g h \tag{1.20a}$$

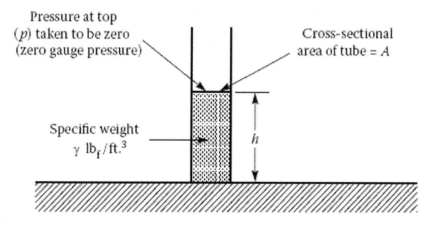

**FIGURE 1.21**
Illustrative Problem 1.6.

As Equations 1.19–1.20 show, static pressure increases with depth and decreases with elevation. Furthermore, static pressure only depends on the specific weight of the fluid and the height or depth of the liquid column. The term "static" is used if the fluid is standing still or where there is no fluid motion within the system.

---

### CALCULUS ENRICHMENT

Integration is a fundamental mathematical operation used to determine the antiderivative of a mathematical function. Once the antiderivative is obtained, it can be solved or evaluated within the range of the independent variable. In simpler terms, the evaluated antiderivative represents the area under the curve of a continuous mathematical function. In many applications in physics and other fields, knowing the magnitude of the area under the curve through integration is extremely valuable.

To better appreciate the usefulness of integrating a function and evaluating its antiderivative, let's consider a linear function, $y = x$, where y is a function of $x$ ($y = f(x)$), that's it, $y$ depends on the independent variable $x$, as follows:

$$y = x, \text{ where } y = f(x) \tag{a}$$

Now, let's integrate both sides, as follows:

$$\int y = \int x \, dx \tag{b}$$

After integrating, the following antiderivative, $Y(x)$ is obtained:

$$Y(x) = \frac{x^2}{2} \tag{c}$$

In most practical applications, it is important to evaluate or obtain the corresponding value of the antiderivative by taking into account the applicable range of the independent variable. For the antiderivative shown above, let's consider a range of $[x_1, x_2]$ for the independent variable $x$, as follows:

$$Y[x_1, x_2] = \int_{x_1}^{x_2} x \, dx \tag{d}$$

By evaluating the antiderivative, its value or magnitude is obtained as follows:

$$Y[x_1, x_2] = \frac{1}{2}\left(x_2^2 - x_1^2\right) \tag{e}$$

The example shows how the antiderivative is found and evaluated through integration. Later in the book, the reader will find how integration is used to solve a variety of thermodynamics problems.

## CALCULUS ENRICHMENT

The relation between the pressure and height of a fluid can be derived considering the following figure.

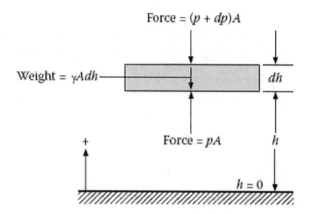

First, let's sum the forces in the vertical direction, noting that *up* is positive, as follows:

$$\text{Force}_{up} - \text{Weight} - \text{Force}_{down} = 0 \tag{a}$$

$$(p + dp)_{up} A - \gamma A dh - pA = 0 \tag{b}$$

By dividing the previous equation by $A$, the following equation is obtained:

$$(p + dp)_{up} - \gamma dh - p = 0 \tag{c}$$

By further simplifying the equation, the following relationship between differential pressure and differential height or elevation is obtained:

$$dp = -\gamma dh \tag{d}$$

where $\gamma$ is also $\rho g$
Thus,

$$dp = -\rho g dh \tag{e}$$

In order to solve Equation (a) or (b), it is necessary to know $p$ as a function of $h$. By integrating Equation (a) on both sides, we obtain:

$$\int dp = \int -\rho g dh \tag{f}$$

$$\Delta p = -\int \rho g dh \tag{g}$$

In general, we can integrate these equations as:

$$p(h) - p(0) = -\int_0^h \rho g \, dh \tag{h}$$

For the special case of a liquid having a constant density (i.e., $\rho$ = constant), Equation (c) can be integrated, starting at the surface, where $p = 0$, and noting that $h$ is measured positive upward. This gives us the result that

$$p = -\rho g h \tag{i}$$

Or

$$p = -\gamma h \tag{j}$$

Equations (d) and € are the same as Equations 1.19a and 1.20a. Equation e shows that pressure in a static fluid (i.e., fluid that's not moving) depends on the density or specific weight of the fluid, and the corresponding elevation or depth. No other factors come into play when determining static pressure in fluids with constant specific weight or density.

Let us consider the case of a fluid whose density is a linear function of pressure. We can write this in terms of the height as

$$\rho = \rho_0 (1 - \alpha h) \tag{f}$$

where $\rho_0$ corresponds to $h_0$ and $\alpha$ is the rate at which density decreases with elevation. Placing (f) into (c) gives us

$$p(h) - p(0) = -\int_0^h \rho_0 (1 - \alpha h) \, dh \tag{g}$$

Integrating

$$p(h) - p(0) = -\rho_0 h - \frac{\alpha \rho_0 h^2}{2} \bigg|_0^h \tag{h}$$

Simplifying

$$p(h) - p(0) = -\rho_0 h 1 - \frac{\alpha h}{2} \tag{i}$$

Equation (i) finds its use in meteorology, when the decrease of temperature with height is called the lapse rate. The lapse rate is related to Equations (c) and (i).

The most common fluid used to measure pressure differences, atmospheric pressure, and vacuum pressure is mercury. At approximately room temperature, the specific gravity (sg) of mercury is very nearly 13.6. It will usually be assumed that the specific gravity of mercury is 13.6 unless otherwise stated. The effect of temperature on the specific gravity of mercury is given in Figure 1.22.

**FIGURE 1.22**
Specific gravity of mercury.

---

**ILLUSTRATIVE PROBLEM 1.7**

A tube of glass open at the top has a 1-in. level of mercury in it. Assuming the specific gravity of mercury (Hg) to be 13.6, and determine the pressure on the base of the liquid column.

**SOLUTION**

To solve this problem, let us first convert the unit of weight of grams per cubic centimeter to pounds per cubic foot. In 1 lb$_f$, there is approximately 454 g. Also, there is 2.54 cm in 1 in. Thus, 1 g is 1/454 lb$_f$, and 1 ft. has $12 \times 2.54$ cm. To convert 1 g per cubic centimeter to pounds per cubic foot, we apply the following dimensional reasoning:

$$\frac{g}{cm^3} \times \frac{lb.}{g} \times \frac{cm^3}{ft.^3} = \frac{lb.}{ft.^3} = \gamma$$

Continuing yields

$$\frac{1g}{cm^3} \times \frac{1}{454}\frac{lb_f}{g} \times (2.54 \times 12)^3 \frac{cm^3}{ft.^3} = \frac{62.4\ lb_f}{ft.^3} = \gamma$$

Thus, the conversion factor from grams per cubic centimeter to pounds per cubic foot is 62.4. Students are strongly urged to check all computations for dimensional consistency when doing problems.

Applying Equation 1.19 gives us

$$p = \frac{1}{12}\ ft.(13.6 \times 62.4)\frac{lb_f}{ft.^3} = 70.72\frac{lb_f}{ft.^2}$$

In pounds force per square inch,

$$p \frac{\text{lb}_f}{\text{ft.}^2} \times \frac{\text{ft.}^2}{\text{in.}^2} \times \frac{1}{144} = \text{psi}$$

Therefore,

$$p = \frac{1}{2} \times 13.6 \times 62.4 \times \frac{1}{144} \frac{\text{ft.}^2}{\text{in.}^2} = 0.491 \text{ psi}$$

This is a gauge pressure. If the local atmospheric pressure is 14.7 psia,

$$p = 0.491 + 14.7 = 15.19 \text{ psia}$$

The value of 0.491 psi/in. Hg is a useful conversion factor.

---

### ILLUSTRATIVE PROBLEM 1.8

The density of mercury is 13.595 kg/m³. Determine the pressure at the base of a column of mercury that in 25.4 mm high.

### SOLUTION

Using Equation 1.20a, we obtain

$$p = -\rho g h = -13.595 \frac{\text{kg}}{\text{m}^3} \times 9.806 \frac{\text{m}}{\text{s}^2} \times -0.0254 \text{ m} = 3386.1 \text{ Pa} = 3.39 \text{ kPa}$$

*Note*: 1 psi ≃ 6895 Pa = 6.9 kPa. Therefore, 3386.1/6895 = 0.491 psi, which checks with Illustrative Problem 1.7.

---

In vacuum work, it is common to express the absolute pressure in a vacuum chamber in terms of millimeters of mercury. Thus, a vacuum may be expressed as $10^{-5}$ mm Hg. If this is expressed in pounds force per square inch, it would be equivalent to 0.000000193 psi, with the assumption that the density of mercury (for vacuum work) is 13.6 g/cm³. Another term, *torr*, is also found in the technical literature. A torr is defined as 1 mm Hg. Thus, $10^{-5}$ torr is the same as $10^{-5}$ mm Hg. Another unit of pressure used in vacuum work is the micron, $\mu$. A micron is defined as 1/1000 of 1 mm Hg, so $10^{-3}$ mm Hg is equal to 1 $\mu$.

---

### ILLUSTRATIVE PROBLEM 1.9

A mercury manometer (vacuum gauge) reads 26.5 in. of vacuum when the local barometer reads 30.0 in. Hg at standard temperature. Determine the absolute pressure in psia.

**SOLUTION**

$$p = (30.0 - 26.5) \text{ in. Hg absolute}$$

and from Illustrative Problem 1.7, 1 in. Hg exerts a pressure of 0.491 psi. Thus,

$$p = (30 - 26.5)(0.491) = 1.72 \text{ psia}$$

If the solution is desired in psfa, it is necessary to convert from square inches to square feet. Because there are 12 in. in 1 ft., there will be 144 in.$^2$ in 1 ft.$^2$. Thus, the conversion to psfa requires that psia be multiplied by 144. This conversion is often necessary to keep the dimensions of equations consistent.

---

**ILLUSTRATIVE PROBLEM 1.10**

A column of fluid is 10 m high. If its density is 2000 kg/m$^3$, determine the pressure at the base of the column if it is located in a local gravity of 9.6 m/s$^2$.

**SOLUTION**

Applying Equation 1.20a gives us

$$p = -\rho g h = -2000 \frac{\text{kg}}{\text{m}^3} \times 9.6 \frac{\text{m}}{\text{s}^2} \times -10 \text{ m} = 192 \text{ kPa}$$

The height is negative, because it is measured up from the base.

---

**ILLUSTRATIVE PROBLEM 1.11**

Using the data of Illustrative Problem 1.9, determine the absolute pressure in kPa.

**SOLUTION**

As before, $p = (30-26.5) = 3.5$ in. Hg absolute. Using "standard" temperature yields

$$\frac{3.5 \text{ in.} \times \dfrac{\text{ft.}}{12 \text{ in.}} \times (13.6 \times 62.4) \dfrac{\text{lb}_f}{\text{ft.}^3}}{12 \dfrac{\text{in.}^2}{\text{ft.}} \times 0.0254 \dfrac{\text{m}^2}{\text{in.}}} \times \frac{\text{kg}}{2.2 \text{ lb}_f} \times 9.806 \frac{\text{N}}{\text{kg}} = 11.875 \text{ kPa}$$

*Note*: $11.875 \times 1000/6895 = 1.72$ psia, which checks with Illustrative Problem 1.9.

In performing any measurement, it is necessary to have a standard of comparison in order to calibrate the measuring instrument. In the following paragraphs, five pressure standards currently used as the basis for all pressure-measurement work will be discussed. Table 1.10 summarizes these standards, the pressure range in which they are used, and their accuracy. Figure 1.23 shows the basic pressure-measurement concepts on which all pressure standards are based.

### 1.6.1 Dead-Weight Piston Gauge

The dead-weight free-piston gauge consists of an accurately machined piston inserted into a close-fitting cylinder. Masses of known weight are loaded on the top side of the free piston, and fluid pressure is applied to the bottom side until enough force is developed to keep the piston–weight combination in place, that is it, the weight of the piston–weight combination is compensated by the pressure-generated force, as follows:

$$\text{Force due to fluid pressure} - \text{Weight of piston \& weight combination} = 0 \quad \text{(1.21a)}$$

$$F_{\text{fluid}} - W = 0 \quad \text{(1.21b)}$$

Where $F_{\text{fluid}}$ is the force due to fluid pressure, as follows:

$$F_{\text{fluid}} = p_{\text{fluid}} A_{\text{piston}} \quad \text{(1.21c)}$$

By substituting Equation 1.21c into 1.21b, the following equation is obtained:

$$p_{\text{fluid}} A_{\text{piston}} - W = 0 \quad \text{(1.21d)}$$

By solving Equation 1.21d by $p_{\text{fluid}}$, the pressure due to the fluid or piston–weight combination can found as follows:

$$p_{\text{fluid}} = \frac{W}{A_p} \quad \text{(1.21e)}$$

where $W$ is the net weight of the piston–weight combination, and $A_p$ is the piston cross-sectional area. Figure 1.24 illustrates how a dead-weight and the piston generate a combined weight, which is counteracted by the pressure force on the fluid side of the system.

**TABLE 1.10**

Characteristics of Pressure Standards

| Type | Range | Accuracy |
|------|-------|----------|
| Dead-weight piston gauge | 0.01 to 10,000 psig | 0.01 to 0.05% of reading |
| Manometer | 0.01 to 100 psig | 0.02 to 0.2% of reading |
| Micro manometer | 0.0002 to 20 in. $H_2O$ | 1% of reading to 0.001 in. $H_2O$ |
| Barometer | 27 to 31 in. Hg | 0.001 to 0.03% of reading |
| McLeod gauge | 0.01 µm to 1 mm Hg | 0.5 to 3% of reading |
| Pressure Transducer | 0.5 psi to 10,000 psig | 0.05 to 0.5% of reading |

Pressure is the normal force $F$ (lb./in.$^2$, lb./ft.$^2$, dynes/cm$^2$) exerted on a unit area of a surface bounding a fluid.

Fluid pressure varies with depth, but is the same in all directions at a given depth.

Variation of fluid pressure with elevation is found by balancing the forces on a static-fluid element ($F_1$ is equal to $F_2$ plus the weight of the element). For a constant-density fluid, the pressure difference $p_2 - p_1$ is equal to the specific weight $\gamma$ times $(h_2 - h_1)$.

Pressure is independent of the shape and size of the vessel. The pressure difference between level 1 and level 2 is always $p_1 - p_2 = \gamma h$, where $\gamma$ is the specific weight of the constant-density fluid in the vessel.

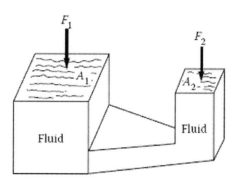

Constant pressure transmission in a confined fluid can be used to multiply force by the relative $p = F_1/A_1 = F_2/A_2$.

When the known weight is balanced, the gauge pressure is $p = W/A$; this is the basic principle of dead-weight testing.

**FIGURE 1.23**
Basic pressure-measurement concepts for static fluid applications.

**FIGURE 1.24**
Dead-weight piston gauge.

Corrections must be applied to the pressure obtained using the dead-weight piston gauge to obtain accurate fluid pressure. The two most important corrections concern air buoyancy and local gravity. The effective area of the dead-weight piston gauge is usually taken as the mean of the cylinder and piston areas, but temperature affects this dimension slightly. The effective area increases between 13 and 18 ppm (parts per million)/°F for commonly used materials, and a suitable correction for this effect may also be applied.

### 1.6.2 Manometer

We have already shown that the pressure at the base of a column of liquid is simply a function of the height of the liquid column and the specific weight of the liquid itself. Therefore, the height of a column of liquid of known specific weight can be and is used to measure pressure and pressure differences. Instruments that utilize this principle are known as manometers, and the study of these pressure-measuring devices is known as manometry. By properly arranging a manometer and selecting the fluid judiciously, it is possible to measure extremely small pressures, very large pressures, and pressure differences. A simple manometer is shown in Figure 1.25, where the right arm is exposed to the atmosphere while the left arm is connected to the unknown pressure. As shown, the

**FIGURE 1.25**
U-tube manometer.

fluid is depressed in the left arm and raised in the right arm until no unbalanced pressure forces remain. It has already been demonstrated that the pressure at a given level in either arm must be the same so that we can select any level as a reference and write a relation for the pressure (see Figure 1.23, static fluid example). Actually, it is much easier and more convenient to select the interface between the manometer fluid and the unknown fluid as a common reference level. In Figure 1.25, the pressure at elevation $AA$ is the same in both arms of the manometer, because static fluids cannot be at two different pressures if the fluid interfaces are at the same elevation or depth. Starting with the open manometer arm (right), we have atmospheric pressure $p_a$ acting on the fluid. As one proceeds down the arm, the pressure increases until we arrive at level $AA$, where the pressure is $p_a + \gamma h$. This pressure must equal the unknown pressure on the connected arm $p_u$.

Therefore,

$$p_u = p_a + \gamma h \tag{1.22}$$

or

$$p_u - p_a = \gamma h \tag{1.23}$$

Temperature and capillary effects must be considered for accurate pressure measurements. To minimize the effect of a variable meniscus, which can be caused by dirt, the method of approaching equilibrium, tube bore, and so on, the tubes are always tapped before reading, and the measured liquid height is always based on readings taken at the center or lowest point of the meniscus in each leg of the manometer. To reduce the capillary effect itself, the use of large-bore tubes (more than 3/8 in. diameter) is most effective.

To achieve greater accuracy and sensitivity in manometers, several different arrangements have been used. Perhaps the simplest of these is the inclined manometer. Consider a relatively large reservoir of liquid connected to a small-bore tube that makes an angle $\theta$ with the horizontal. The pressure or pressure differential to be measured is connected to the large reservoir; the inclined tube is open ended. Schematically, this is shown in Figure 1.26. The unknown pressure $p_u$ is given by

$$p_u = p_a + \gamma h \tag{1.24}$$

or

$$p_u - p_a + \gamma h \tag{1.25}$$

$$p_u - p_a + \gamma h' \sin\theta \tag{1.26}$$

Because $\theta$ is fixed, a scale placed along the tube can be calibrated to read directly in units of $h$ of a fluid. Usually, this is done by directly reading inches of water for $p_u - p_a$.

Another method of achieving greater sensitivity and accuracy is to use a manometer with more than one fluid. The manometer shown in Figure 1.27 can be used for this purpose by properly selecting the manometer fluid. Starting at level $A$, we have

$$P_A - h_A\gamma_1 - y\gamma_3 + h_B\gamma_2 = P_B \tag{1.27}$$

or

$$P_A - P_B = (-h_B\gamma_2 + h_A\gamma_1) = y\gamma_3 \tag{1.28}$$

**FIGURE 1.26**
Inclined manometer.

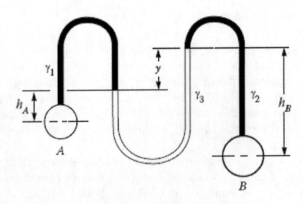

**FIGURE 1.27**
Three-fluid manometer.

In the usual case, the manometer is connected to different positions on the same pipe. Thus, $A$ and $B$ would be at the same level, and $\gamma_1$ can be taken to be equal to $\gamma_2$. Also, $h_B - h_A = y$. Thus,

$$P_A - P_B \approx y(\gamma_3 - \gamma_1) \quad \text{or} \quad y(\gamma_3 - \gamma_2) \tag{1.29}$$

Note that when you go down in a fluid column, you add $h\gamma$, and when you go up a fluid column, you subtract $h\gamma$. One can start at either end of the system and work around to the other end. Also, note that most manometer problems refer to the fluid's specific gravity rather than its specific weight. Thus $\gamma = sg \times 62.4$ lb./ft.$^3$ in the English system and $\gamma = sg \times 9.81$ kN/m$^3$ in SI units.

For small differences in $p_A - p_B$, it is apparent that the manometer fluid ($\gamma_3$) should have a specific weight very nearly equal to the specific weight of the fluid in the pipes. For large pressure differences, one can use a heavy fluid such as mercury to increase $y_3 - y_2$ and reduce the manometer reading.

---

### ILLUSTRATIVE PROBLEM 1.12

For the manometer shown in Figure 1.28, find the absolute pressure at point $A$ and the nominal pressure in the water pipe.

### SOLUTION

Starting at the top of the manometer, which is open to the atmosphere, $p = 100$ kPa. Going down the column of 1.2 m of 0.8 sg oil adds $(\gamma h)_{oil} = 0.8 \times 9.81 \times 1.2 = 9.418$ kPa, so the pressure at $A$ is 109.418 kPa. Continuing around the manometer until the water pipe is reached, we add 0.3 m of mercury, subtract 1.6 m of oil, and add 1.5 m of water. The net result is $109.418 + 13.6 \times 9.81 \times 0.3 - 0.8 \times 9.81 \times 1.6 + 1 \times 9.81 \times 1.5 = 151.6$ kPa.

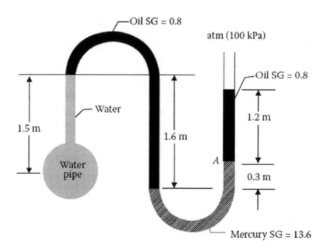

**FIGURE 1.28**
Illustrative Problem 1.12.

A word on the reference pressures employed in manometry is pertinent at this point in the discussion. If atmospheric pressure is used as a reference, the manometer yields gauge pressures. Because of the variability of air pressure, gauge pressures vary with time, altitude, latitude, and temperature. If, however, a vacuum is used as a reference, the manometer yields absolute pressures directly, and it may serve as a barometer. In any case, the absolute pressure is always equal to the sum of the gauge and ambient pressures (see Equation 1.16); by ambient pressure, we mean the pressure surrounding the gauge, which is usually atmospheric pressure.

### 1.6.3 Micromanometer

While the manometer is a useful and convenient tool for pressure measurements, it is unfortunately limited when making low-pressure measurements. To extend its usefulness in the low-pressure range, micromanometers have been developed that have extended the useful range of low-pressure manometer measurements to pressures as low as 0.0002 in. $H_2O$.

One type is the *Prandtl-type micromanometer,* in which capillary and meniscus errors are minimized by returning the meniscus of the manometer liquid to a null position before measuring the applied pressure difference. As shown in Figure 1.29, a reservoir, which forms one side of the manometer, is moved vertically to locate the null position. This position is reached when the meniscus falls within two closely scribed marks on the near-horizontal portion of the micromanometer tube. Either the reservoir or the inclined tube is then moved by a precision lead-screw arrangement to determine the micromanometer liquid displacement ($\Delta h$), which corresponds to the applied pressure difference. The Prandtl-type micromanometer is generally accepted as a pressure standard within a calibration uncertainty of 0.001 in. $H_2O$.

Another method for minimizing capillary and meniscus effects in manometry is to measure liquid displacements with micrometer heads fitted with adjustable, sharp index points. Figure 1.30 shows a manometer of this type; the micrometers are located in two connected, transparent containers. In some commercial micromanometers, contact with

**FIGURE 1.29**
Two variations of the Prandtl-type micromanometer.

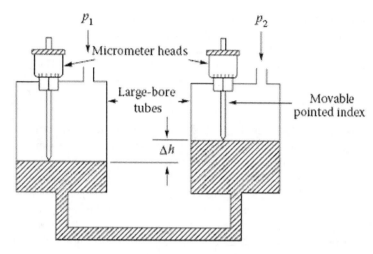

**FIGURE 1.30**
Micrometer-type manometer.

the surface of the manometric liquid may be sensed visually by dimpling the surface with the index point, or even by electrical contact. Micrometer-type micromanometers also serve as pressure standards within a calibration uncertainty of 0.001 in. $H_2O$.

An extremely sensitive, high-response micromanometer uses air as the working fluid and thus avoids all the capillary and meniscus effects usually encountered in liquid manometry. In this device, as shown in Figure 1.31, the reference pressure is mechanically amplified by centrifugal action in a rotating disk. The disk speed is adjusted until the amplified reference pressure just balances the unknown pressure. This null position is recognized by observing the lack of movement of minute oil droplets sprayed into a glass indicator tube. At balance, the air micromanometer yields the applied pressure difference as

$$\Delta p_{\text{micro}} = Kpn^2 \tag{1.30}$$

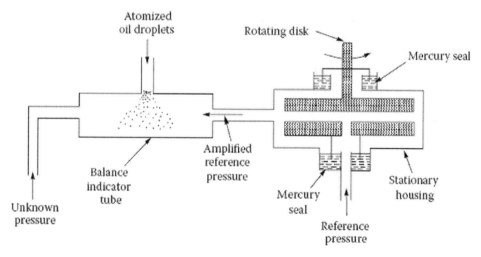

**FIGURE 1.31**
Air-type centrifugal micromanometer.

where $p$ is the reference air density, $n$ is the rotational speed of the disk, and $K$ is a constant that depends on disk radius and annular clearance between the disk and the housing. Measurements of pressure differences as small as 0.0002 in. $H_2O$ can be made with this type of micromanometer within an uncertainty of 1%.

### 1.6.4 Barometers

The reservoir or cistern barometer consists of a vacuum-reference mercury column immersed in a large-diameter, ambient-vented mercury column that serves as a reservoir. The most common cistern barometer in general use is the Fortin type, in which the height of the mercury surface in the cistern can be adjusted. The operation of this instrument can best be explained with reference to Figure 1.32.

The datum-adjusting screw is turned until the mercury in the cistern makes contact with the ivory index, at which point the mercury surface is aligned with zero on the instrument scale. Next, the indicated height of the mercury column in the glass tube is determined. The lower edge of a sighting ring is lined up with the top of the meniscus in the tube. A scale reading and a vernier reading are taken and combined to yield the indicated height at the barometer temperature.

Because atmospheric pressure on the mercury in the cistern is exactly balanced by the weight per unit area of the mercury column in the glass tube,

$$p_{\text{baro}} = \gamma_{\text{Hg}} h_{\text{to}} \tag{1.31}$$

The referenced specific weight of mercury, $\gamma_{\text{Hg}}$, depends on such factors as temperature and local gravity; the referenced height of mercury, $h_{\text{to}}$, depends on such factors as thermal expansion of the scale and the mercury.

Other factors may also contribute to the uncertainty of $h_{\text{to}}$. Proper illumination is essential to define the location of the crown of the meniscus. Precision meniscus sighting under optimum viewing conditions can approach $\pm0.001$ in. With proper lighting, contact between the ivory index and the mercury surface in the cistern can be detected to much better than $\pm0.001$ in.

To keep the uncertainty in $h_{\text{to}}$ within 0.01% ($\simeq0.003$ in. Hg), the mercury temperature must be taken within $\pm1°F$. Scale temperature need not be known to better than $\pm10°F$ for comparable accuracy. Uncertainties caused by nonequilibrium temperature conditions can be avoided by installing the barometer in a uniform temperature room.

The barometer tube must be vertically aligned for accurate pressure determination. This is accomplished by a separately supported ring encircling the cistern; adjustment screws control the horizontal position.

Depression of the mercury column in commercial barometers is accounted for in the initial calibration setting at the factory. The quality of the barometer is largely determined by the bore of the glass tube. Barometers with a bore of 1/4 in. are suitable for readings of about 0.01 in. Hg, whereas barometers with a bore of 1/2 in. are suitable for readings down to 0.002 in. Hg.

### 1.6.5 McLeod Gauge

The McLeod gauge is used in making low-pressure measurements. This instrument (shown in Figure 1.33) consists of glass tubing arranged so that a sample of gas at unknown pressure can be trapped and then isothermally compressed by a rising mercury column. This

**FIGURE 1.32**
Fortin barometer.

amplifies the unknown pressure and allows measurement by conventional manometric means. All the mercury is initially contained in the area below the cutoff level. The gauge is first exposed to the unknown gas pressure, $p_1$; the mercury is then raised in tube $A$ beyond the cutoff, trapping a gas sample of the initial volume $\bar{V}_1 = \bar{V} + ah_c$. The mercury is continuously forced upward until it reaches the zero level in the reference capillary B. At this time, the mercury in the measuring capillary C reaches a level $h$, where the gas sample is at its final volume, $V_2 = ah$, and at the amplified pressure, $p_2 = p_1 + h$. Based on the ideal gas law (discussed in Chapter 3), then,

$$p_1\bar{V}_1 = p_2\bar{V}_2 \tag{1.32}$$

**FIGURE 1.33**
McLeod gauge.

$$p_1 = \frac{ah^2}{\bar{V}_1 - ah} \tag{1.33}$$

If $ah \ll \bar{V}_1$, as is usually the case,

$$p_1 = \frac{ah^2}{\bar{V}_1} \tag{1.34}$$

The larger the volume ratio $(\bar{V}_1 / \bar{V}_2)$, the greater the amplified pressure $p_2$ and manometer reading $h$. Therefore, it is desirable that measuring tube $C$ have a small bore. Unfortunately, for tube bores less than 1 mm, the compression gain is offset by reading uncertainty caused by capillary effects.

Reference tube *B* is introduced to provide a meaningful zero for the measuring tube. If the zero is fixed, Equation 1.34 indicates that manometer indication *h* varies nonlinearly with initial pressure $p_1$. A McLeod gauge with an expanded scale at the lower pressures exhibits a higher sensitivity in this region. The McLeod pressure scale, once established, serves equally well for all the permanent gases (those whose critical pressure is appreciably less than room temperature).

There are no corrections to be applied to the McLeod gauge reading, but certain precautions should be taken. Moisture traps must be provided to avoid taking any condensable vapors into the gauge. Such vapors occupy a larger volume at the initial low pressures than they occupy in the liquid phase at the high reading pressures. Thus, the presence of condensable vapors always causes pressure readings to be too low. Capillary effects, while partially counterbalanced by using a reference capillary, can still introduce significant uncertainties, because the angle of contact between mercury and glass can vary by ±30°. Finally, because the McLeod gauge does not give continuous readings, steady-state conditions must prevail for the measurements to be useful.

In the earlier portions of this section, we discussed five pressure standards that can be used for either calibration or measurement of pressure in static systems. In the following, we discuss some common devices that are used for making measurements using an elastic element to convert fluid energy to mechanical energy or electrical energy. Such a device is known as a *pressure transducer*. Examples of mechanical pressure transducers with only elastic elements are dead-weight free-piston gauges, manometers, Bourdon gauges, bellows, and diaphragm gauges.

Electrical pressure transducers have an element that converts their displacement to an electrical signal. Active electrical transducers generate their own voltage or current output as a function of displacement. Passive transducers require an external signal. The piezoelectric pickup is an example of an active electrical transducer. Electric elements employed in passive electrical pressure transducers include strain gauges, slide-wire potentiometers, capacitance pickups, linear differential transformers, and variable-reluctance units.

In the Bourdon gauge, the elastic element is a small-volume tube that is fixed at one end but free at the other end to allow displacement under the deforming action of the pressure difference across the tube walls. In the most common model, shown in Figure 1.34, a tube with an oval cross-section is bent in a circular arc. Under pressure, the tube tends to become circular, with a subsequent increase in the radius of the arc. By an almost frictionless linkage, the free end of the tube rotates a pointer over a calibrated scale to give a mechanical indication of pressure.

The reference pressure in the case containing the Bourdon tube is usually atmospheric, so that the pointer indicates gauge pressures. Absolute pressures can be measured directly without evacuating the complete gauge casing by biasing a sensing Bourdon tube against a reference Bourdon tube, which is evacuated and sealed, as shown in Figure 1.35. Bourdon gases are available for a wide range of absolute gauge and differential pressure measurements within a calibration uncertainty of 0.1% of the reading.

Another common elastic element used in pressure transducers is the bellows, shown as a gauge element in Figure 1.36. In one arrangement, the pressure is applied to one side of a bellows, and the resulting deflection is partially counterbalanced by a spring. In a differential arrangement, one pressure is applied to the inside of one sealed bellows, and the pressure difference is indicated by a pointer.

A final elastic element to be mentioned because of its widespread use in pressure transducers is the diaphragm. One such arrangement is shown in Figure 1.37. Such elements may be flat, corrugated, or dished plates; the choice depends on the strength and amount

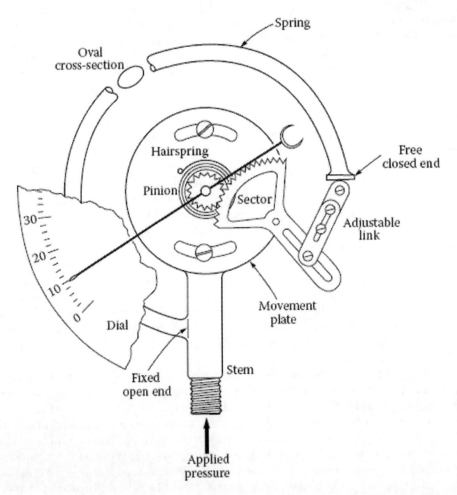

**FIGURE 1.34**
Bourdon gauge.

of deflection desired. In high-precision instruments, a pair of diaphragms are used back to back to form an elastic capsule. One pressure is applied to the inside of the capsule; the other pressure is external. The calibration of this differential transducer is relatively independent of pressure magnitude.

### 1.6.6 Pressure Transducer

Thus far, we have discussed a few mechanical pressure transducers. In many applications, it is more convenient to use transducer elements that depend on the change in the electrical parameters of the element as a function of the applied pressure. The only active electrical pressure transducer in common use is the piezoelectric transducer. The electrical pressure transducer is a device capable of measuring the pressure of gases or liquid, by generating a directly correlated analog signal when pressure is imposed on it. Most pressure transducers consist of a piezoelectric material that generates a small voltage when the imposed pressure force deforms the material slightly, as shown in Figure 1.38. The analog electrical signal generated by the device varies linearly with the imposed

Reference Bourdon tube sealed
at zero absolute pressure

Case open to
atmospheric
pressure

Pointer
movement

Sensing
Bourdon
tube

Pinion and
sector gauge
movement

Applied pressure

**FIGURE 1.35**
Bourdon gauge for absolute pressure measurement.

force, which in turn can be amplified and digitized so it can be read by computerized data acquisition systems.

Sound-pressure instrumentation makes extensive use of piezoelectric pickups in such forms as hollow cylinders and disks. Piezoelectric pressure transducers are also used in measuring rapidly fluctuating or transient pressures. In a recently introduced technique called electrocalibration, the transducer is calibrated by electric field excitation rather than by physical pressure. The most common passive electrical pressure transducers are the variable-resistance types.

The strain gauge is probably the most widely used pressure transducer element. Strain gauges operate on the principle that the electrical resistance of a wire varies with its length under load. In unbounded strain gauges, four wires run between electrically insulated pins located on a fixed frame and other pins located on a movable armature, as shown in Figure 1.39. The wires are installed under tension and form the legs of a bridge circuit. Under pressure, the elastic element (usually a diaphragm) displaces the armature, causing two of the wires to elongate while reducing the tension in the remaining two wires. The resistance change causes a bridge imbalance proportional to the applied pressure.

The bonded strain gauge takes the form of a fine wire filament set in cloth, paper, or plastic and fastened by a suitable cement to a flexible plate, which takes the load of the elastic element. This is shown in Figure 1.40. Two similar strain gauge elements are often

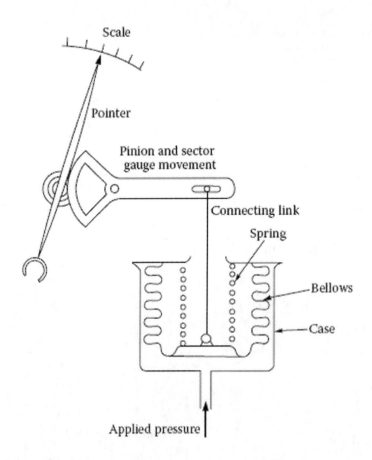

**FIGURE 1.36**
Bellows gauge.

connected in a bridge circuit to balance unavoidable temperature effects. The nominal bridge output impedance of most strain gauge pressure transducers is 350 ohms ($\Omega$), nominal excitation voltage is 10 V (alternating current [ac] or direct current [dc]), and natural frequency can be as high as 50 Hz. Transducer resolution is infinite, and the usual calibration uncertainty of such gauges is within 1% of full scale.

Many other forms of electrical pressure transducers are in use in industry, and a few of these will be discussed briefly. These elements fall under the categories of potentiometer, variable capacitance, linear variable differential transformer (LVDT), and variable-reluctance transducers. The potentiometer operates as a variable-resistance pressure transducer. In one arrangement, the elastic element is a helical Bourdon tube, while a precision, wire-wound potentiometer serves as the electric element. As pressure is applied to the open end of the Bourdon tube, it unwinds, causing the wiper (connected directly to the closed end of the tube) to move over the potentiometer.

In the variable-capacitance pressure transducer, the elastic element is usually a metal diaphragm that serves as one plate of a capacitor. Under an applied pressure, the diaphragm moves with respect to a fixed plate. By means of a suitable bridge circuit, the variation in capacitance can be measured and related to pressure by calibration.

The electric element in an LVDT is made up of three coils mounted in a common frame to form the device shown in Figure 1.41. A magnetic core centered in the coils is free to

**FIGURE 1.37**
Diaphragm-based pressure transducer.

**FIGURE 1.38**
Piezoelectric transducer for pressure measurement.

**FIGURE 1.39**
Typical unbonded strain gauge.

**FIGURE 1.40**
Typical bonded strain gauge.

**FIGURE 1.41**
Linear variable differential transformer.

be displaced by a bellows, Bourdon, or diaphragm elastic element. The center coil, the primary winding of the transformer, has an ac excitation voltage impressed across it. The two outside coils form the secondaries of the transformer. When the core is centered, the induced voltages in the two outer coils are equal and 180° out of phase; this represents the zero-pressure position. However, when the core is displaced by the action of an applied pressure, the voltage induced in one secondary increases while that in the other decreases. The output voltage difference varies essentially linearly with pressure for the small core displacements allowed in these transducers. The voltage difference is measured and related to the applied pressure by calibration.

The last elastic element to be discussed is the variable-reluctance pressure transducer. The basic element of the variable-reluctance pressure transducer is a movable magnetic vane in a magnetic field. In one type, the elastic element is a flat magnetic diaphragm located between two magnetic output coils. Displacement of the diaphragm changes the inductance ratio between the output coils and results in an output voltage proportional to pressure.

## 1.7 Review

After defining thermodynamics as the study of the conversion of one form of energy to another, we started our study by setting forth certain concepts concerning systems and properties. A *system* is any grouping of matter taken in a convenient or arbitrary manner, whereas a property is an observable characteristic that is determined by the state of a system and, in turn, aids in determining the state of a system. The condition of a system, described by its position, energy, and so on, is called the *state* of the system. We studied in detail the properties of temperature and pressure and the methods used to measure them. Because both the English and SI systems of units are used, we are required to understand how to use both systems of units. This caused us to spend some time in establishing and defining the basic rules and styles of these systems. Using elementary kinetic theory, we were able to quantify the concepts of pressure, temperature, density, specific weight, specific volume, and Avogadro's number.

Some of the material covered in this chapter may have been covered in other courses and may appear to be elementary. It cannot be emphasized strongly enough that this material must be fully mastered before proceeding further. You are urged to be sure of all the units used in any equation and to be equally sure of the meaning of all terms. Problems at the end of the chapters are included to develop a better level of understanding through the application of the text material. Reading and studying alone are not sufficient. The ability to apply the material studied indicates that you have really mastered it and are not just repeating material from memory.

## Key Terms

Terms used for the first time in this chapter are as follows:

**Avogadro's law:** Equal volumes of gases at the same temperature and pressure contain the same number of particles.

**Density:** Mass per unit of volume.

**Micron (μ):** $10^{-3}$ Mm of mercury, a unit of pressure used in vacuum systems.

**Pressure:** Normal force per unit of area.

**Property:** An observable characteristic of a state that is determined by the state and, in turn, aids in determining the state of the system. It is not dependent on the path or the means of attaining the state.

**Specific gravity:** The ratio of the density of a substance to the density of water at 4°C. It can also be defined as the ratio of the specific weight of a substance to the specific weight of water at 4°C. In the older cgs system, the density of water at 4°C is 1 $g/cm^3$, making specific gravity numerically equal to density in this system.

**Specific volume:** The reciprocal of density, that is, the volume per unit mass.

**Specific weight:** Weight per unit of volume.

**State:** The condition of a system that fixes the position and the energy stored in the system and is identified by the properties of the system.

**System:** A grouping of matter taken in any convenient or arbitrary manner.

**Temperature:** The temperature of a system is a measure of the random motion of the molecules of the system. The Celsius, Fahrenheit, Kelvin, and Rankine temperature scales represent the conventional and absolute temperature scales, respectively. Celsius and Kelvin are SI units, and Fahrenheit and Rankine are English units.

**Thermodynamics:** The study of energy, heat, work, the properties of the media employed, and the processes used; also, the study of the conversion of one form of energy to another.

**Torr:** A unit of pressure equal to 1 mm of Hg.

**Weight:** The force of attraction between a body and the Earth or another planet or moon.

**Zeroth law of thermodynamics:** When two bodies are in thermal equilibrium with a third body, they are in thermal equilibrium with one another.

## Equations Developed in This Chapter

| | | |
|---|---|---|
| Temperature conversion | $T[°C] = \dfrac{5}{9}\left(T[°F] - 32\right)$ | (1.1) |
| Temperature conversion | $T[°F] = \dfrac{9}{5}\left(T[°C]\right) + 32$ | (1.2) |
| Temperature conversion | $T[°F] = \dfrac{9}{5}\left(40 + T[°C]\right) - 40$ | (1.3) |
| Temperature conversion | $T[°C] = \dfrac{5}{9}\left(40 + T[°F]\right) - 40$ | (1.4) |
| Absolute temperature | $\text{Rankine} = T[R] = T[°F] + 460$ | (1.5) |
| Absolute temperature | $\text{Kelvin} = T[K] = T[°C] + 273$ | (1.6) |
| Force–mass relation | $F = \dfrac{ma}{g_c}$ | (1.9) |
| Weight–mass relation | $w = \dfrac{ma}{g_c}$ | (1.10) |
| Weight–mass relation (SI) | $w = mg$ | (1.10a) |
| Ideal gas relation | $pV = \dfrac{1}{3}\dfrac{mnV_x^2}{g_c}$ | (1.14) |
| Ideal gas relation (SI) | $pV = \dfrac{1}{3}mnV_x^2$ | (1.14a) |
| Absolute pressure | Absolute pressure, $p_a$ = atmospheric pressure, $p_{atm}$ + gauge pressure, $p_g$ | (1.16) |
| Absolute pressure | Absolute pressure = atmospheric pressure – vacuum | (1.17) |
| Pressure–height relation | $p = -\gamma h$ | (1.19a) |
| Pressure–height relation | $p = \dfrac{-\rho g}{g_c} h$ | (1.20) |
| Pressure–height relation | $p = -\rho g h$ | (1.20a) |

## QUESTIONS

1.1 Indicate which of the following are or are not properties: (a) location, (b) velocity, (c) time, (d) weight, or (e) size.

1.2 Would you expect that your weight would be (a) greater, (b) less, or (c) equal to your Earth weight on the planet Jupiter? Why?

1.3 State on which factor or factors the pressure at the bottom of a column of liquid depends: (a) its height, (b) the specific weight of the fluid, (c) the local acceleration of gravity, or (d) the cross-sectional area of the column.

1.4 What temperature does a thermometer measure?

1.5 What condition or conditions must exist if a thermometer is to measure a person's temperature accurately?

1.6    If a thermometer measures a temperature difference of 10°C, what is the temperature difference in K?

1.7    If a thermocouple measures a temperature difference of 5°F, what is the temperature difference in R?

1.8    Comment on the fact that commodities such as bread and meat are sold by the kilo (kilogram) in many countries of the world.

1.9    The unit of 100 is not a recommended unit in the SI system. Comment on whether it would or would not have been desirable to define the pascal as $N/cm^2$.

1.10   If an equation is not dimensionally consistent, is it necessarily incorrect? Why?

1.11   If the average velocity of the molecules in a box is doubled, would you expect that its pressure would (a) double, (b) stay the same, (c) be four times greater, or (d) be indeterminate?

1.12   What is temperature a measure of?

1.13   How are density and specific volume related?

1.14   A student states that the use of a digital readout will increase the accuracy of an instrument. Comment on this statement.

1.15   What must be done to establish the accuracy of any instrument?

## PROBLEMS

More difficult problems are preceded by an asterisk (*) in this chapter and in all following chapters. Unless indicated otherwise, use $g = 32.17$ ft./s² or 9.806 m/s², and use $p_{atm} = 14.696$ psia or 101.325 Pa. Also, use $g_c = 32.174$ $lb_m ft./lb_f s^2$.

### Problems Involving Conversion of Units

1.1    Convert a length of 15 ft. to meters.

1.2    Convert 3.85 in. to centimeters.

1.3    Convert 90.68 cm to inches.

1.4    Convert 5787 lb. to tons.

1.5    Convert 6.3 tn. to pounds.

1.6    Convert 35.4 in. to feet.

1.7    Convert 3.87 ft. to inches.

1.8    Convert 8.6 lb. to grams.

1.9    Convert 462 g to pounds.

1.10   Convert 2 td. 2 ft. 5 in. to centimeters.

1.11   Convert 550 mi./h to m/s.

1.12   Convert a flow rate of 2000 ft.³/min to L/s.

1.13   Convert 200 gal./h (gph) to m³/s.4

1.14   Convert 200 gph to L/s.

1.15   Convert 200 gph to L/min.

1.16   Convert 200 L/min to m³/s.

1.17   Convert 200 gpm to L/min.

1.18    Convert 50.7 cm³/s to m³/s.

1.19    Convert 861 gpm to L/min.

1.20    Convert 1.71 m³/s to L/min.

1.21    Convert 100 m³/s to gpm.

1.22    Convert 3.52 ft.³/s to gpm.

1.23    Convert 3.52 ft.³/s to m³/s.

1.24    Convert 150 gpm to m³/s.

1.25    Convert 6.52 m³/s to gpm.

1.26    Convert 3.65 m³/s to L/s.

1.27    Convert 1.2 ft.³/s to gpm.

1.28    Convert 4.8 ft.³/s to L/s.

1.29    Convert 101 kPa (kN/m²) to $lb_f$/in.².

1.30    A rectangle has the dimensions of 12 in. × 10 in. Calculate its area in square meters.

1.31    A rectangle has dimensions of 2 m × 3.5 m. Determine the area of the rectangle in square inches.

1.32    A box has the dimensions 12 in. × 12 in. × 6 in. How many cubic meters are there in the box?

1.33    A cylinder has a diameter of 10 cm and a height of 25 cm. Determine its volume in cubic meters.

1.34    A cylinder has a diameter of 5 in. and a height of 10 in. Determine its volume in cubic meters.

1.35    A cube is made 8 in. on a side. Calculate its volume in cubic meters.

1.36    An automobile engine has a displacement of 240 in.³. Determine its displacement in liters.

1.37    A car is claimed to obtain a fuel usage in Europe of 12 km/L. Determine its equivalent fuel usage in mi./gal.

1.38    A driveway requires 38 yd.³ of concrete. How many m³ of concrete does this correspond to?

1.39    A man weighs 175 lb. and stands 6 ft. 0 in. tall. What is his weight in newtons and his height in meters?

1.40    A barrel of oil contains 55 gal. If a gallon is a volume of 231 in.³, determine the volume of 15 bbl. in cubic meters.

*1.41    A gallon is a measure of volume that equals 231 in.³, so determine the number of liters in a gallon. Note that the liter is sometimes called the metric quart. Does your answer support this name?

1.42    A 1500-m race is often called the "metric mile." What part of a mile (5280 ft.) is the metric mile?

1.43    A car is operated at 50 mi./h. How many feet per second does this correspond to?

1.44    A car travels at 64 km/h. What is the speed of the car in mi./h?

1.45    A car weighs 2650 lb. and is 18 ft. long. What is its weight in kilonewtons and its length in meters?

1.46    A block of metal weighing 1200 lb. is hung from the end of a vertical rod, and it causes the rod to lengthen 0.001 in. Express these data in terms of newtons and meters.

1.47    Is one gigapascal (GPa) equal to (a) $10^6$ Pa, (b) $10^8$ Pa, (c) $10^3$ Pa, (d) $10^9$ Pa, or (e) none of these?

## Problems Involving Temperature

1.48    Convert 20°C, 40°C, and 60°C to equivalent degrees Fahrenheit.

1.49    Change 0°F, 10°F, and 50°F to equivalent degrees Celsius.

1.50    Convert 500 R, 500 K, and 650 R to degrees Celsius.

1.51    Derive a relation between degrees Rankine and degrees Kelvin, and based on the results, show that (°C + 273)1.8 = °F + 460.

*1.52   If a Fahrenheit temperature is twice a Celsius temperature, determine both temperatures.

*1.53   If a Celsius temperature is 2/3 of a Fahrenheit temperature, determine both temperatures.

*1.54   Fahrenheit and Celsius thermometers are used to measure the temperature of a fluid. If the Fahrenheit reading is 1.5 times the Celsius reading, what are both readings?

*1.55   An arbitrary temperature scale is proposed in which 20° is assigned to the ice point and 75° is assigned to the boiling point. Derive an equation relating this scale to the Celsius scale.

*1.56   For the temperature scale proposed in Problem 1.55, what temperature corresponds to absolute zero?

*1.57   A new thermometer scale on which the freezing point of water at atmospheric pressure would correspond to a marking of 200 and the boiling point of water at atmospheric pressure would correspond to a marking of minus 400 is proposed. What would the reading of this new thermometer be if a Fahrenheit thermometer placed in the same environment read 80°?

## Problems Involving Mass and Weight

1.58    A mass of 5 kg is placed on a planet whose gravitational force is ten times that of Earth. What does it weigh on this planet?

1.59    A mass of 10 kg weighs 90 N. Is the acceleration of gravity at this location equal to (a) $1/9$ m/s$^2$, (b) 9.0 m/s$^2$, (c) 90 m/s$^2$, (d) 10 m/s$^2$, or (e) none of these?

1.60    At a location on Earth where $g = 32.2$ ft./s$^2$, a body weighs 161 lb. At another location where $g = 32.0$, how much will the same body weight: (a) 161 lb., (b) 162 lb., (c) 160 lb., (d) 163 lb., or (e) none of these?

1.61    A body at mean sea level weighs 100 lb. ($g = g_c$). Estimate the weight of this body at an elevation of 7500 ft. Assume that the mean diameter of the Earth is 12,742 km.

*1.62   A mass of 100 kg weighs 980.6 N at sea level. Estimate the weight of this body at the top of a mountain 5 km high. Assume that the mean diameter of the Earth is 12,742 km.

*1.63  A balance-type scale is used to weigh a sample on the moon. If the value of $g$ is 1/6 of Earth's gravity and the "standard" weights (Earth weights) add up to 20 lb., what is the mass of the body?

*1.64  A balance-type scale is used to weigh a sample on the moon. If the value of $g$ is 1/6 of Earth's gravity and the "standard" weights (Earth weights) add up to 100 N, what is the mass of the body?

*1.65  Solve Problem 1.64 if a spring balance is used that was calibrated on Earth and reads 20 lb.

*1.66  Solve Problem 1.64 if a spring balance is used that was calibrated on Earth and reads 100 N.

*1.67  A mass of 100 kg is hung from a spring in a local gravitational field where $g = 9.806$ m/s², and the spring is found to deflect 25 mm. If the same mass is taken to a planet where $g = 5.412$ m/s², how much will the spring deflect if its deflection is directly proportional to the applied force?

*1.68  The mass of the planet Mars is 0.1069 relative to Earth, and its diameter is 0.523 relative to Earth. What is the weight of a pound of mass on Mars?

*1.69  A body having an unknown mass "weighs" 1 lb. on the planet Jupiter. Jupiter has a mass 318.35 times that of Earth, and its diameter is 10.97 times larger: (a) What will the body weight on Earth be? (b) What is the mass of the body?

*1.70  Solve Problem 1.68 for a mass of 10 kg.

*1.71  Solve Problem 1.69 for a weight of 10 Newtons on Jupiter.

## Problems Involving Newton's Second Law, $F = ma$

1.72  A force of 100 N acts horizontally on a 10-kg body. What is its horizontal acceleration?

1.73  An unbalanced force in pounds equal numerically to the weight of a body in pounds causes the body to accelerate. What is its acceleration?

1.74  If a body is accelerating at the rate of 5 m/s² when acted on by a net force of 10 N, determine its mass.

1.75  A body has a mass of 321.7 lb. and is accelerated at the rate of 5 ft./s². Determine the magnitude of the unbalanced force acting on it.

1.76  A mass of 2 kg has an acceleration of 1.2 m/s². Calculate the magnitude of the unbalanced force acting on it.

1.77  What is the acceleration of a body if a net force of 25 lb. acts on it if its mass is 160.8 lb.?

## Problems Involving Specific Weight, Specific Volume, and Specific Gravity

1.78  Determine the density and specific volume of the contents of a 10-ft.³ tank if the contents weigh 250 lb.

1.79  A tank contains 500 kg of a fluid. If the volume of the tank is 0.5 m³, what is the density of the fluid, and what is the specific volume?

1.80  Calculate the specific weight of a fluid if a gallon of it weighs 20 lb. Also, calculate its density.

1.81    The specific gravity of a fluid is 1.2. Calculate its density and its specific weight in SI units.

1.82    A fluid has a density of 0.90 g/cm$^3$. What is its specific weight in both the English and SI systems?

1.83    It is proposed by gasoline dealers to sell gasoline by the liter. If gasoline has a specific gravity of 0.85, what is the weight of 60 L of gasoline?

1.84    A household oil tank can hold 275 gal. of oil. If oil has a specific weight of 8800 N/m$^3$, how many pounds of oil will there be in a full tank?

1.85    A car has a fuel tank that holds 14 gal. of gasoline. If the gasoline has a specific gravity of 0.70, calculate the weight of gasoline in units of pounds.

1.86    An oil has a specific gravity of 0.8. Determine its specific weight, density, and specific volume in SI units.

1.87    An oil has a specific weight of 0.025 lb$_f$/in.$^3$. Calculate its specific gravity.

1.88    A can weighs 12 N when empty, 212 N when filled with water at 4°C, and 264 N when filled with an oil. Calculate the specific gravity of the oil.

1.89    The density of a liquid is 48.6 lb$_m$/ft.$^3$. Calculate the specific weight and the specific gravity of the liquid.

1.90    A shipping company requires that packages have a base perimeter plus the height not to exceed 108 in. By calculation, a student determines that the optimum package size should be an 18-in. square base and a 36-in. height. If the company further specifies that the weight must not exceed 70 lb$_f$, determine the maximum specific weight of the contents of the package.

1.91    A body has a specific gravity of 1.48 and a volume of $7.24 \times 10^{-4}$ m$^3$. What is the weight of the body?

1.92    Oil has a specific gravity of 0.83. What is the weight of a liter of this oil in SI units?

1.93    What is the weight in pounds of a gallon of oil that has a specific gravity of 0.86?

1.94    A cylinder 6 in. in diameter and 10 in. in height contains oil that has a density of 850 kg/m$^3$. Determine the weight of the oil in English units.

1.95    A liquid has a density of 1100 kg/m$^3$. Determine its specific gravity and its specific weight.

1.96    The fuel tank of a car holds 60 L of gasoline. Assuming that the gasoline has a specific gravity of 0.74, determine the weight of the gasoline in the tank.

1.97    If the density of a liquid is 780 kg/m$^3$, calculate its specific gravity and its specific weight.

1.98    A liquid has a specific weight of 200 lb$_f$/ft.$^3$. Calculate the volume needed to have a weight of 390 lb.

*1.99   In the older cgs system, the density in g/cm$^3$ is numerically equal to the specific gravity of the fluid. Show that this is indeed true.

## Problems Involving Pressure

1.100   A skin diver descends to a depth of 60 ft. in fresh water. What is the pressure on the diver's body? The specific weight of fresh water can be taken as 62.4 lbf/ft.$^3$.

1.101   A skin diver descends to a depth of 25 m in a salt lake where the density is 1026 kg/m$^3$. What is the pressure on the diver's body at this depth?

1.102   If a Bourdon gauge reads 25 psi when the atmosphere is 14.7 psia, what is the absolute pressure in pascals?

1.103   A column of fluid is 1 m high. The fluid has a density of 2500 kg/m$^3$. What is the pressure at the base of the column?

1.104   A column of fluid is 25 in. high. If the specific weight of the fluid is 60.0 lb$_f$/ft.$^3$, what is the pressure in psi at the base of the column?

1.105   Determine the density and specific volume of the contents of a 20-ft.$^3$ tank if the contents weigh 250 lb.

1.106   A tank contains 500 kg of a fluid. If the volume of the tank is 2.0 m$^3$, what is the density of the fluid, and what is the specific gravity?

*1.107   Convert 14.696 psia to 101.325 kPa.

1.108   A pressure gauge indicates 25 psi when the barometer is at a pressure equivalent to 14.5 psia. Compute the absolute pressure in psi and feet of mercury when the specific gravity of mercury is 13.0.

1.109   Same as Problem 1.108, but the barometer stands at 750 mm Hg and its specific gravity is 13.6.

1.110   A vacuum gauge reads 8 in. Hg when the atmospheric pressure is 29.0 in. Hg. If the specific gravity of mercury is 13.6, compute the absolute pressure in psi.

1.111   A vacuum gauge reads 10 in. Hg when the atmospheric pressure is 30 in. Hg. Assuming the density of mercury to be 13.595 kg/m$^3$, determine the pressure in pascals.

1.112   A U-tube mercury manometer, open on one end, is connected to a pressure source. If the difference in the mercury levels in the tube is 6.5 in., determine the unknown pressure in psfa.

*1.113   Two sources of pressure $M$ and $N$ are connected by a water-mercury differential gauge, as shown in Figure P1.113. What is the difference in pressure between $M$ and $N$ in psi?

**FIGURE P1.113**

*1.114  Determine the difference in pressure between $A$ and $B$ in Figure P1.114 if the specific weight of water is 62.4 lb$_f$/ft.$^3$.

**FIGURE P1.114**

*1.115  If the liquid in pipe $B$ in Problem 1.114 is carbon tetrachloride, whose specific weight is 99 lb$_f$/ft.$^3$, determine the pressure difference between $A$ and $B$.

*1.116  In a U-tube manometer, one end is closed, trapping atmospheric air in the column. The other end is connected to a pressure supply of 5 psig. If the level of mercury in the closed end is 2 in. higher than that in the open end, what is the pressure of the trapped air?

*1.117  For the arrangement shown in Figure P1.117, determine $p_A - p_B$.

**FIGURE P1.117**

*1.118  A bathyscape dives 8000 m below the surface in the Pacific. Atmospheric pressure is 101 kPa, and seawater density is 1.03 g/cm$^3$. Working in SI units, determine the following: (a) absolute pressure (kPa) and (b) force on a 250-mm-diameter window.

*1.119  A fuel tank on the moon ($g$=5.32 ft./s$^2$) holds 2 m$^3$ of jet fuel (specific gravity = 0.70). Water density is 1 g/cm$^3$. Determine the fuel weight, in Newtons (N).

# 2

## Work, Energy, and Heat

### Learning Goals

After reading and studying the material in this chapter, you should be able to

1. State the definitions of work, energy, and heat.
2. Use the fact that both work and heat are forms of energy in transition.
3. Apply the convention that heat into a system is to be taken as a positive quantity and that work out of a system is also to be taken as a positive quantity. (This convention is taken from the customary power-producing cycle in which heat into a system is used to generate useful work.)
4. Show that internal energy is a form of energy in which a body is said to possess internal energy by virtue of the motion of the molecules of the body.
5. Differentiate between a nonflow or closed system and a flow or open system.
6. Understand the origin of the term *flow work* and apply it to a flow system.
7. Use the fact that the work of a quasi-static, nonflow system is the area under the pressure–volume curve.
8. Understand the difference between sensible heat and latent heat.
9. Understand the key aspects of cyclic processes.

## 2.1 Introduction

In Chapter 1, certain concepts were explained such as the motion of gas particles in an enclosure or closed system. Specifically, pressure was found to involve the principle of momentum interchange with the container walls, temperature was associated with the motion of the particles, and density was taken to be a measure of the number of particles per unit volume. This simple analysis followed the history of a single particle, and it was subsequently generalized to all the particles in the enclosure. This type of analysis is representative of a microscopic description of the processes occurring within the boundaries of the defined system, because the history of a single particle was followed in detail. Rather than pursue further the microscopic concept of matter, we shall be concerned with the macroscopic, or average, behavior of the particles composing a system. The macroscopic viewpoint essentially assumes that it is possible to describe the average behavior of these particles at a given time and at some subsequent time after changes have occurred to the system. The system changes that are of concern to us in this study are temperature, pressure, density, work, energy, velocity, and position. The power of the macroscopic approach lies in its ability to describe the changes that have occurred to the system without having to detail all the events of the processes involved.

## 2.2 Work

In thermodynamics, one of the forms of energy is *work*. *Work* by definition is the mathematical product of the displacement of a body multiplied by the component of the force acting on it, in the direction of the displacement, that is, Work = Force · Displacement. Thus, in Figure 2.1, the displacement of the body along the horizontal plane is $x$, and the component of the force in the direction of the displacement is ($F \cos \theta$). Therefore, the work done is ($F \cos \theta$)$x$. For the example shown in Figure 2.1b, the constant force ($F \cos \theta$) is plotted as a function of $x$, and it should be noted that the resulting figure is a rectangle. The area of this rectangle (shaded) is equal to the work done, because it is ($F \cos \theta$)($x$). If the force varies

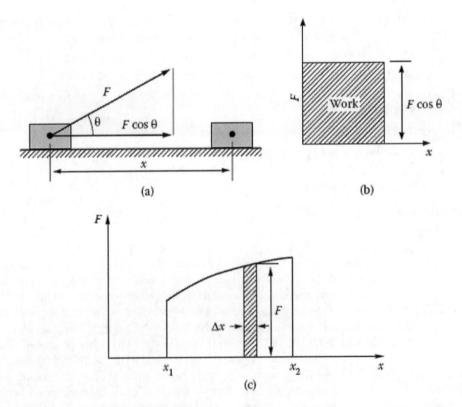

(a)                                                                          (b)

(c)

**FIGURE 2.1**
Work. (a) Force component in $x$ direction. (b) Plot of force vs. displacement. (c) Force a function of displacement.

so that it is a function of the displacement, it is necessary to consider the variation of force with displacement in order to find the overall work done.

Figure 2.1c shows a general plot of force as a function of displacement. If the displacement is subdivided into many small parts, $\Delta x$, and for each of these small parts $F$ is assumed to be very nearly constant, it is apparent that the sum of the small areas $(F)\,(\Delta x)$ will represent the total work done when the body is displaced from $x_1$ to $x_2$. Thus, the area under the curve of $F$ (as a function of $x$, i.e., $F=f(x)$) is the total work done if $F$ is the force component in the direction of the displacement $x$.

---

### ILLUSTRATIVE PROBLEM 2.1

A spring is slowly compressed by varying force $F$ until it reaches an equilibrium position. Assuming that the force on the spring is proportional to the spring displacement, $F=k\Delta x$, determine the work done on the spring. Assume that the constant of proportionality $k$ is constant and expressed in pounds force per foot ($lb_f$/ft.) of spring deflection or Newtons per meter (N/m) spring deflection.

### SOLUTION

As shown in Figure 2.2a, the system consists of a spring and a force $F$ directed along the axis of the spring. A plot of $F$ as a function of $x$ is shown in Figure 2.2b. It will be noted that the force-displacement relation is a linear one, in which the force equals $kx$ at all times. Thus, the work done is represented by the shaded triangular area of Figure 2.2b. Because the base of the triangle is $l$ (i.e., total displacement of the spring) and the height is $kl$, the work done is $\frac{1}{2}(l)(kl)$ or $\frac{1}{2}(k)(l^2)$. A formal derivation of the preceding equation can be obtained by integration (i.e., by finding the antiderivative of the force function, as shown in the proceeding calculus enrichment note).

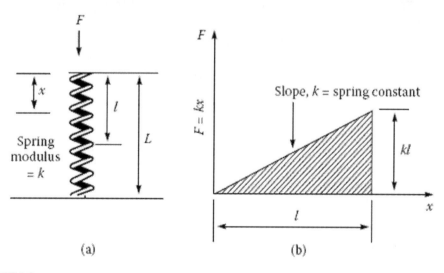

(a)           (b)

**FIGURE 2.2**
Illustrative Problems 2.1 and 2.2. (a) Spring-force system. (b) Spring-force work.

## CALCULUS ENRICHMENT

The work done on a spring by a force can be found with integration of the expression for work, namely,

$$W = \int F \cdot dx \qquad\qquad (a)$$

where the force and displacement are related by the spring constant $k$ as

$$F = kx \qquad\qquad (b)$$

Using the spring equation in conjunction with Equation (a) between the limits $l_1$ and $l_2$,

$$W = \int_{l_1}^{l_2} kx \, dx \qquad\qquad (c)$$

Integrating,

$$W = k\left(l_2^2 - l_1^2\right)/2 \qquad\qquad (d)$$

If the initial length is zero, we obtain the same result as before, namely, $(1/2)kl^2$.

## ILLUSTRATIVE PROBLEM 2.2

A spring has a spring constant $k$ of 100 $lb_f$/in. deflection. How much work is done when the spring is compressed 2 in.?

### SOLUTION
Referring to Figure 2.2 and the results of Illustrative Problem 2.1, we have

$$\text{work} = \frac{1}{2}kl^2$$

For this problem,

$$\text{work} = \frac{1}{2}\left((100)\,lb_f/in.\right)(2)^2 \text{ in.}^2 = 200 \text{ in.} \cdot lb_f$$

---

**ILLUSTRATIVE PROBLEM 2.3**

If the spring constant in Illustrative Problem 2.2 is 20 kN/m, how much work is done when the spring is compressed 75 mm?

**SOLUTION**

As before,

$$\text{work} = \frac{1}{2}kl^2 = \frac{1}{2}\left(20 \times 10^3\right)\text{N/m} \times (0.075)^2 \, (\text{m})^2$$

$$= 56.25 \, \text{N} \cdot \text{m} = 56.25 \, \text{J}$$

As noted before in Chapter 1, 1 J = 1 N·m

---

## 2.3 Energy

At this point, let us define *energy* in terms of work. Energy can be described as the capacity to do work. At first, it may appear that this definition is too restrictive when applied to electrical and magnetic systems. Yet in all instances, the observed effects on a system can (in principle and ideally) be converted to mechanical work. Because work has been defined as the product of a force multiplied by a displacement, it is not stored in a system. It represents a form of energy that must be crossing the boundaries (real or imaginary) of the system and can properly be placed in the category of energy in transition or manifesting itself at the boundaries of the system.

To distinguish between the transfer of energy as work to or from a system, we shall adopt the convention that *the work done by a system on its surroundings is positive, and the work done by the surroundings on the system is negative.* For students, it is best to think of this convention regarding the *useful work out of a system as a conventional, desirable quantity and, therefore, positive.* Thus, in Illustrative Problem 2.1, the spring has work done on it by a force. If we consider the spring as our system, the work is negative; if we consider the system as the external variable force that is compressing the spring, the work is positive. Students will note that the decision whether the work term is positive or negative requires that the system be carefully defined and examined.

The work that a system can perform on its surroundings is not an intrinsic property of the system. The manner in which the process is carried out will determine the effect on the surroundings. As stated earlier, work is a transitory effect and is neither a property of a system nor stored in a system. There is one process, however, that does permit the evaluation of the work done, because the work path is uniquely defined. This is frictionless and quasi-static process, and we shall find it useful in subsequent discussions. To describe this process, we first define *equilibrium state* in the manner given by Hatsopoulos and Keenan (1961): *A state is an equilibrium state if no finite rate of change can occur without a finite change, temporary or permanent, in the state of the environment.* The term *permanent change of state* refers to one that is not canceled out before completion of the process. Therefore, the frictionless *quasi-static process* can be identified as a succession of equilibrium states. Involved in this definition is the concept of a process carried out infinitely slowly, so that it is practically in equilibrium at all times. The utility of the frictionless, quasi-static process lies in our ability to evaluate the work terms involved in it, because its path is uniquely defined in classical terms.

Energy or work done per unit time is called power or the rate of energy change. Energy, work, and power units as well as their conversions are detailed in Table 1.8.

## 2.4 Internal Energy

To this point, we have considered the energy in a system that arises from the work done on the system. However, it was noted in Chapter 1 and earlier in this chapter that a body possesses energy by virtue of the motion of the molecules of the body. In addition, it possesses energy due to the internal attractive and repulsive forces between particles at the molecular level. These forces become the mechanism for energy storage whenever particles become separated, such as when a liquid evaporates or the body is subjected to a deformation by an external energy source. Also, energy may be stored in the rotation and vibration of the molecules. Additional amounts of energy are involved with the electron configuration within the atoms and with the nuclear particles. The energy from all such sources is called the *internal energy* of the body and is designated by the symbol $U$. Per unit mass ($m$), the specific internal energy is denoted by the symbol $u$, where $mu = U$. Thus,

$$mu = U \qquad (2.1)$$

or

$$u = \frac{U}{m} \qquad (2.2)$$

From a practical standpoint, the measurement of the absolute internal energy of a system in a given state presents an insurmountable problem and is not essential to our study of thermodynamics. We are concerned with changes or differences in internal energy, and the arbitrary datum for the zero of internal energy will not enter into these problems.

Just as it is possible to distinguish the various forms of energy, such as work and heat, in a mechanical system, it is equally possible to distinguish the various forms of energy associated with electrical, chemical, and other systems. For the purpose of this book, these forms of energy are not considered. However, students are cautioned that if a system includes any forms of energy other than mechanical, these items must be included. For example, the energy that is dissipated in a resistor as heat when a current flows through it must be taken into account when all the energies of an electrical system are being considered.

## 2.5 Potential Energy

Let us consider the following problem, illustrated in Figure 2.3, where a body of mass $m$ is in a locality in which the local gravitational field is constant and equal to $g$. A force is applied to the body, and the body is raised a distance $Z$ from its initial position. The force is assumed to be only infinitesimally greater than the mass, so the process is carried out on a frictionless, quasi-static basis. In the absence of electrical, magnetic, and other extraneous effects, determine the work done on the body. The solution to this problem is obtained by noting that the equilibrium of the body requires that a force, $F$, be applied to it equal to its weight, $W = m \cdot g$. The weight of the body is given from Chapter 1 as $mg/g_c$ in English units. In moving through a distance $Z$, the work done by this force will then equal

$$\text{Work} = \text{Weight} \cdot \text{Displacement} \qquad (2.3a)$$

**FIGURE 2.3**
Potential energy.

$$\text{Work} = W \cdot Z \tag{2.3b}$$

$$\text{Work} = \frac{mg}{g_c} Z \tag{2.3c}$$

The work done on the body can be returned to the external environment by simply reversing this frictionless, quasi-static process, a feature that is discussed in detail in Chapter 3. Returning to Equation 2.3, we conclude that this system has had work done on it equal to $(mg/g_c)$ (Z) and that, in turn, the system has stored in it an amount of energy in excess of the amount it had in its initial position. The energy added to the system in this case is called *potential energy*. Thus,

$$\text{potential energy} \ (\text{P.E.}) = \frac{mg}{g_c} Z \left(\text{ft.} \cdot \text{lb}_f\right) \tag{2.4}$$

In terms of SI units, Equations 2.3 and 2.4 become

$$\text{work} = mgZ \tag{2.5}$$

and

$$\text{potential energy} \ (\text{P.E}) = mgZ \ (\text{joules}) \tag{2.6}$$

A feature of importance of potential energy is that a system can be said to possess it only with respect to an arbitrary initial or datum plane.

An interesting application of the concept of potential energy storage is the pumped storage in a hydroelectric power plant. The principle of operation can be illustrated using Figure 2.4. In this system, reversible turbine-generators are used as follows: At those times when excess generating capacity is available from other generating stations, the water can be pumped from the lower to the upper reservoir. When additional generating capacity is required, the water is allowed to flow downhill, passing through the generating station where it operates the reversible turbine-generator as a turbine that rotates and power the generator to produce electricity. Thus, the pumped storage station helps to smooth out fluctuations in the electric load demand, resulting in steady, more efficient operation of

**FIGURE 2.4**
Pumped storage concept. (a) Generating mode. (b) Pumping mode.

other stations. In this application, the potential energy of the water stored in the upper reservoir provides the energy to the system as needed. Dams also use the potential energy of water to generate electrical power.

---

### ILLUSTRATIVE PROBLEM 2.4

A pumped storage plant uses water pumped to an elevation of 600 ft. above the turbogenerators. How much work is done lifting the water to this elevation? Assume that local gravity is $g$, which is numerically equal to $g_c$.

**SOLUTION**

Because local gravity is $g$, which is equal to $g_c$, for each pound mass of water,

$$\text{P.E.} = \frac{mg}{g_c}(Z) = 1\text{lb}_m \times \frac{32.174 \text{ft.} / \text{s}^2}{32.174 \text{ lb}_m \cdot \text{ft.} / \text{lb}_f \cdot \text{s}^2} \times 600 \text{ ft.} = 600 \text{ ft.} \cdot \text{lb}_f$$

Note that the use of the term $g/g_c$ is equivalent to a conversion factor $\text{lb}_f/\text{lb}_m$.

---

## ILLUSTRATIVE PROBLEM 2.5

A pump delivers water from a well that is 50 m deep. Determine the change in potential energy per kg of water. Use $g = 9.81$ m/s$^2$.

### SOLUTION

The change in potential energy is $mgZ$, where $Z$ is 50 m, using the bottom of the well as the datum. Therefore,

$$\text{P.E.} = mgZ$$

$$= 1 \text{ kg} \times 9.81 \frac{\text{m}}{\text{s}^2} \times 50 \text{ m}$$

$$= 490.5 \frac{\text{kg} \cdot \text{m}^2}{\text{s}^2}$$

$$= 490.5 \text{ N} \cdot \text{m} \quad \text{because N} = \frac{\text{kg} \cdot \text{m}}{\text{s}^2}$$

$$= 490.5 \text{ J}$$

## ILLUSTRATIVE PROBLEM 2.6

If the pumped storage plant of Illustrative Problem 2.4 has a flow of 10,000 gal./min, determine the power generated. Assume that the local gravity is $g = g_c$ and that the density of water is 62.4 lb$_m$/ft.$^3$ and that fluid levels or elevations both downstream and upstream remain constant.

### SOLUTION

In Illustrative Problem 2.4, we determined the work done to be 600 ft.·lb$_f$/lb$_m$. The power generated is the energy stored per pound mass (also equal to the potential energy) multiplied by the pounds mass per minute that flows through the system $\left( \text{Power} = \dfrac{\text{P.E.}}{\text{time}} \right)$.

One gallon is a volumetric measure equal to 231 in.$^3$. Thus,

$$10\,000 \text{ gal./min} \times \frac{231 \text{ in.}^3}{1728 \text{ in.}^3/\text{ft.}^3} = 1337 \text{ ft.}^3/\text{min}$$

(because the flow is given as 10,000 gal./min).

The mass flow is $m = \rho A V = 62.4$ lb$_m$/ft$^3 \times 1337$ ft.$^3$/min $= 83,429$ lb$_m$/min. The power generated $= 83,429$ lb$_m$/min $\times 600$ ft.·lb$_f$/lb$_m = 50,057,000$ ft.lb$_f$/min. Finally, $50,057,000/33,000 = 1517$ hp, since 1 hp $= 33, 000$ ft.·lb$_f$/min. Again, in this case the potential energy (P.E.) or difference in fluid elevation is assumed to be constant during the operation of the system.

## ILLUSTRATIVE PROBLEM 2.7

Determine the horsepower required by the pump in Illustrative Problem 2.5 if the water has a density of 1000 kg/m³ and the pump delivers 1000 kg/min from the well.

### SOLUTION

From Illustrative Problem 2.5, we found the potential energy to be 490.5 N·m above the bottom of the well. Therefore,

$$\text{power} = (\text{P.E.})\,\dot{m}$$

$$= 490.5\,\frac{\text{N}\cdot\text{m}}{\text{kg}} \times 1000\,\text{kg/min} \times \frac{1}{60\,\text{s/min}}$$

$$= 8175\,\frac{\text{N}\cdot\text{m}}{\text{s}} = 8175\,\frac{\text{J}}{\text{s}} = 8175\,\text{W}$$

Because 1 hp = 746 W,

$$\frac{8175}{746}\,\frac{\text{W}}{\text{W/hp}} = 10.96\,\text{hp}$$

Again, in this case the potential energy or difference in fluid elevation is assumed to be constant during the operation of the system.

---

## 2.6 Kinetic Energy

Let us consider another situation in which a body of mass $m$ is at rest on a frictionless plane (Figure 2.5). If the force $F$ is applied to the mass, it will be accelerated in the direction of the force. After moving through a distance $x$, the velocity of the body will have increased from 0 to $V_2$. The only effect of the work done on the body will be to increase its velocity due to the lack of friction.

The velocity of the mass is simply the time rate of change of displacement,

$$V = \frac{x}{\Delta t} \tag{2.7}$$

And acceleration is the time rate of change of velocity,

$$a = \frac{\Delta V}{\Delta t} \tag{2.8}$$

**FIGURE 2.5**
Kinetic energy.

Combining Equation 2.7 with Equation 2.8 by eliminating the $\Delta t$ terms yields

$$V \cdot \Delta V = a \cdot x \qquad (2.9)$$

If we now multiply both sides of Equation 2.9 by the mass, $m$, we obtain

$$m \cdot V \cdot \Delta V = m \cdot a \cdot x \qquad (2.10)$$

The left side of Equation 2.10 can be evaluated by referring to Figure 2.6, where $V$ is plotted against $V$. It will be noted that $V \cdot \Delta V$ is the area of the shaded rectangle. Summing these small rectangles up to $V_2$ represents the area of the triangle whose base is $V_2$ and whose altitude is $V_2$. This area is $\dfrac{1}{2} V_2^2$ making the left side of Equation 2.10: $m \cdot V_2^2 / 2$.

The right side of Equation 2.10 is interpreted by noting that $ma = F$ (in SI units). Therefore, $m \cdot a \cdot x$ is the work done by the force $F$ to move the body a distance $x$. Thus, we can write

$$\frac{mV_2^2}{2} = Fx \qquad (2.11)$$

The term $m \cdot V^2 / 2$ is called the *kinetic energy* of the body. At the beginning of a process, it is $m \cdot V_1^2 / 2$, and at the end, it is $m \cdot V_1^2 / 2$. Equation 2.11 therefore permits us to evaluate the energy possessed by a body of mass $m$ having a velocity $V$ relative to a stationary reference. It is usual to use the Earth as the reference.

In English units, Equation 2.11 becomes

$$\text{kinetic energy (K.E.)} = \frac{m \cdot V_2^2}{2g_c} \text{ or, in general, } \frac{m \cdot V^2}{2g_c} \qquad (2.12)$$

In terms of SI units,

$$\text{kinetic energy (K.E.)} = \frac{m \cdot V^2}{2} \qquad (2.12a)$$

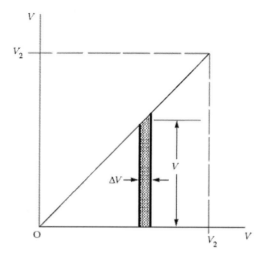

**FIGURE 2.6**
Evaluation of $V \Delta V$.

## ILLUSTRATIVE PROBLEM 2.8

A mass of 10 lb. is slowed from a velocity of 88 ft./s to 10 ft./s. What is the change in the kinetic energy of the system if the body is considered to be the system?

### SOLUTION

The kinetic energy of the body before it is slowed down is

$$\text{K.E.} = \frac{m \cdot V^2}{2g_c} = \frac{10 \text{ lb}_m \times 88 \dfrac{\text{ft.}^2}{\text{s}}}{2 \times 32.174 \dfrac{\text{ft.} \cdot \text{lb}_m}{\text{s}^2 \cdot \text{lb}_f}} = 1203.5 \text{ ft.} \cdot \text{lb}_f$$

After slowing down,

$$\text{K.E.} = \frac{m \cdot V^2}{2g_c} = \frac{10 \text{ lb}_m \times 10 \dfrac{\text{ft.}^2}{\text{s}}}{2 \times 32.174 \dfrac{\text{ft.} \cdot \text{lb}_m}{\text{s}^2 \cdot \text{lb}_f}} = 15.5 \text{ ft.} \cdot \text{lb}_f$$

The change in kinetic energy Δ(K.E.) therefore is

$$\Delta \text{K.E.} = 1203.5 - 15.5 = 1188 \text{ ft.} \cdot \text{lb}_f$$

## ILLUSTRATIVE PROBLEM 2.9

A car with a mass of 1500 kg is slows from 50 km/h to 30 km/h (Figure 2.7). What is the change in its kinetic energy if $g = 9.81$ m/s²?

### SOLUTION

The car's initial kinetic energy is $m \cdot V^2/2$. Therefore,

$$\text{K.E.} = \frac{1500 \text{ kg}}{2} \times \frac{50 \times 1000 \text{ m/h}}{3600 \text{ s/h}}^2$$

$$= 144\,676 \text{ J} = 144.68 \text{ kJ}$$

**FIGURE 2.7**
Illustrative Problem 2.9.

After slowing down, we have

$$\text{K.E.} = \frac{mV^2}{2} = \frac{1500 \text{ kg}}{2} \times \frac{30 \times 1000 \text{ m/h}}{3600 \text{ s/h}}^2$$

$$= 52\,083 \text{ J} = 52.08 \text{ KJ}$$

The change in kinetic energy therefore is

$$\Delta\text{K.E.} = 144.68 - 52.08 = 92.6 \text{ kJ}$$

Note that $g$ did not enter the problem in this system of units.

---

### ILLUSTRATIVE PROBLEM 2.10

A body has a mass of 10 kg and falls freely from rest (Figure 2.8). After falling 10 m, what will its kinetic energy be? Also, what will its velocity be just before impact? Neglect air friction.

### SOLUTION

Because there are no losses in the system, we conclude that the initial potential energy plus the initial kinetic energy must equal the sum of the final potential energy plus the final kinetic energy. Therefore,

$$\text{P.E.}_1 + \text{K.E.}_1 = \text{P.E.}_2 + \text{K.E.}_2 \qquad \text{(a)}$$

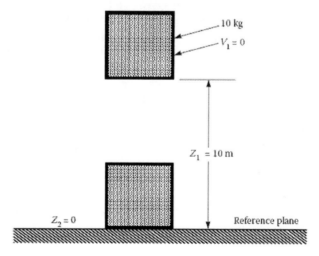

**FIGURE 2.8**
Illustrative Problem 2.10.

By selecting our reference plane as shown in Figure 2.8, K.E.$_1$=0 and P.E.$_2$=0. Equation (a) reduces to

$$P.E._1 = K.E._2 \tag{b}$$

Substituting the data of the problem into Equation (b) gives

$$10 \text{ kg} \times 10 \text{ m} \times 9.81 \text{ m/s}^2 = \frac{10 \text{ kg} \times V^2 \left(\text{m/s}\right)^2}{2}$$

Solving yields

$$V^2 = 2 \times 9.81 \times 10$$

and

$$V = 14.0 \text{ m/s}$$

Note that this result is independent of the mass of the body. The kinetic energy is readily found from Equation (b) to be equal to 10 kg $\times$ 10 m $\times$ 9.81 m/s$^2$ = 981 N·m.

## 2.7 Heat

*Heat* is defined as the energy in transition across the boundaries of a system due to a temperature difference between the system and its surroundings. When a *heat* interaction occurs in a system, two distinct events are observed. The first is an interchange of energy, and the second is that this interchange would not have taken place if there were no temperature difference between the system and its surroundings. That is, *heat* in essence represents energy in transition. In this definition of heat, the transfer of mass across the boundaries of the system is excluded. It should be noted that this indicates a similarity between heat and work because both are energies in transition, and neither is a property of the system in question. Just as in work, heat can transfer quasi-statically to or from a system. The difference in temperature between the system and its surroundings for quasi-static heat transfer can be only an infinitesimal amount at any time. Once again, it is necessary to adopt a convention for the energy interchanged by a system with its surroundings. *We shall use the convention that heat to a system from its surroundings is positive and that heat out of a system is negative.* To learn these conventions, it is convenient to consider the typical situation in which heat is transferred to a system to obtain useful work from the system. This sets the convention that heat into a system is positive and work out of the system is also positive. Positive in this sense means either desirable or conventional from the viewpoint of conventional power cycles. For refrigeration cycles, the opposite of this convention will be more useful.

It is important to recognize the difference between heat and temperature. Temperature is a measure of the energy contained in the molecules of a system due to their motion. When the temperature of a system is greater than that of its surroundings, some of that molecular energy is transferred to the surroundings in what we call heat. Thus, tempera-ture is a property of a system in a given state, whereas heat is associated with a change in the state of a system.

Because work and heat are both forms of energy in transition, it follows that the units of work should be capable of being expressed as heat units, and vice versa. In the English system of units the conversion factor between work and heat, sometimes called *mechanical equivalent of heat*, is 778.169ft.·lb$_f$/Btu and is conventionally given the symbol *J*. We shall use this symbol to designate 778ft.·lb$_f$/Btu, because this is sufficiently accurate for engineering applications of thermodynamics. In the SI system, this conversion factor is not necessary, because the joule (N·m) is the basic energy unit.

There are two forms of heat transfer: *sensible* heat and *latent* heat. Sensible heat transfer occurs when there is a temperature difference between bodies or systems and the amount of heat transferred is related to the magnitude of that difference. This is the most common form. Latent heat transfer occurs when a body is changing state, such as evaporating from liquid to gas or condensing from gas to liquid. The change of phase occurs at constant temperature. This form of heat transfer is utilized in refrigeration and air conditioning systems (see Chapters 7 and 10).

## 2.8 Flow Work

At this time, let us look at two systems, namely, the *nonflow* or *closed system* and the *steady-flow* or *open system*. The nonflow system has boundaries that both heat and work can pen-etrate, but no mass can cross these boundaries. In the steady-flow system, both mass and energy can cross the system boundaries. The term *steady* denotes a process or system that is not time dependent. When a fluid is caused or forced to flow in a system, it is necessary that somewhere in the system work must have been supplied. At this time, let us evaluate the net amount of work required to push the fluid into and out of the system. Consider the system shown in Figure 2.9, where a fluid is flowing steadily across the system boundaries as shown. At the inlet section ①, the pressure is $p_1$, the area is $A_1$, the mass flow rate is $\dot{m}$, and the fluid density is $\rho_1$ (or its reciprocal specific volume, $1/v_1$); at the outlet section ②, the pressure is $\rho_2$, the area is $A_2$, the mass flow rate is still $\dot{m}$, and the fluid density is $\rho_2$ (or its reciprocal specific volume, $1/v_2$). Let us now consider a plug of fluid of length $l_1$ entering

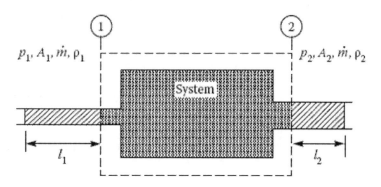

**FIGURE 2.9**
Steady-flow system: flow work.

the system such that the amount of fluid contained in the plug is numerically $\dot{m}$. The force acting on the inlet cross-sectional area $A_1$ is $p_1 A_1$. To push the plug into the system, it is necessary for this force to move the plug a distance equal to $l_1$. In so doing, the work done will be $p_1 A_1 l_1$ (i.e., work = force·distance). However, $A_1 l_1$ is the volume of the plug containing a mass $m$. Using this, we find the work, $W$, to be

$$W = p_1 A_1 l_1 = m\left(p_1 v_1\right) \quad \text{(ft.} \cdot \text{lb}_f \text{ or N} \cdot \text{m)} \tag{2.13}$$

The work per unit mass is $W/m = \mathbf{w}$

$$\mathbf{w} = p_1 v_1 \quad \text{(ft.} \cdot \text{lb}_f / \text{lb}_m \text{ or N} \cdot \text{m/kg)} \tag{2.14}$$

If we now consider the outlet section using the same reasoning, we have

$$W = m\left(p_2 v_2\right) \quad \text{(ft.} \cdot \text{lb}_f \text{ or N} \cdot \text{m)} \tag{2.15}$$

or

$$\mathbf{w} = p_2 v_2 \quad \text{(ft.} \cdot \text{lb}_f / \text{lb}_m \text{ or N} \cdot \text{m/kg)} \tag{2.16}$$

Each of the $pv$ terms is known as the flow work. The net flow work becomes

$$\textbf{net flow work} = p_2 v_2 - p_1 v_1 \quad \text{(ft.} \cdot \text{lb}_f / \text{lb}_m \text{ or N} \cdot \text{m/kg)} \tag{2.17}$$

or in thermal units,

$$\text{net flow work} = \frac{p_2 v_2}{J} - \frac{p_1 v_1}{J} \quad \text{(Btu/lb}_m) \tag{2.18}$$

or

$$\text{net flow work} = p_2 v_2 - p_1 v_1 \quad \text{(N} \cdot \text{m/kg)} \tag{2.19}$$

Equations 2.18 and 2.19 are interpreted to mean that the difference in the $pv$ terms represents the amount of work that is done on a system to introduce a unit mass into it minus the work done on its environment as it leaves the system. However, a word of caution is necessary at this time. Any fluid in any system has both properties, pressure and specific volume, and therefore the product $pv$ can always be evaluated. *The product $pv$ represents flow work only in the steady-flow system.* Thus, even though the term $pv/J$ appears in the nonflow process, it cannot and does not represent flow work, because the system is by definition stationary. Flow work exists only to cause or force fluid to cross the boundaries of a flow system.

In the foregoing derivations, certain assumptions were made, and these are repeated here for emphasis. When applied to a flow situation, the term *steady* means that the conditions at any section of the system is independent of time, that is, it does not vary with time. Even though the velocity, specific volume, and temperature of the fluid can vary in any manner across the stream, they are not permitted to vary with time. The mass entering the system per unit time must equal the mass leaving the system in the same period of time $\left(\dot{m}_{\text{in}} = \dot{m}_{\text{out}}\right)$, otherwise, the system would either store or be depleted of fluid.

---

### ILLUSTRATIVE PROBLEM 2.11

At the entrance to a steady-flow device, it is found that the pressure is 100 psia and that the density of the fluid is $62.4 \text{lb}_m/\text{ft.}^3$. At the exit, the pressure is 50 psia and the corresponding density 30 lb./ft.$^3$. Determine the flow work term at the entrance and exit of the device.

**SOLUTION**

Refer to Figure 2.9. At the entrance,

$$\frac{pv}{J} = \frac{100\,\dfrac{\text{lb}_f}{\text{in.}^2} \times 144\,\dfrac{\text{in.}^2}{\text{ft.}^2}\,\dfrac{1}{62.4\text{lb}_m/\text{ft.}^3}}{778\,\dfrac{\text{ft.}\cdot\text{lb}_f}{\text{Btu}}} = 0.297\ \text{Btu/lb}_m$$

At the exit,

$$\frac{pv}{J} = \frac{50\,\dfrac{\text{lb}_f}{\text{in.}^2} \times 144\,\dfrac{\text{in.}^2}{\text{ft.}^2}\,\dfrac{1}{30\,\text{lb}_m/\text{ft.}^3}}{778\,\dfrac{\text{ft.}\cdot\text{lb}_f}{\text{Btu}}} = 0.308\ \text{Btu/lb}_m$$

---

### ILLUSTRATIVE PROBLEM 2.12

Determine the flow work at the entrance and exit of a steady-flow device in which the entrance pressure is 200 kPa and the density of the fluid is 1000 kg/m$^3$. At the exit, the pressure is 100 kPa and the density 250 kg/m$^3$.

**SOLUTION**

At the entrance,

$$pv = 200\ \text{kPa} \times 1000\,\frac{\text{Pa}}{\text{kPa}}\,\frac{\text{N}}{\text{m}^2} \times \frac{1}{1000\ \text{kg/m}^3} = 200\,\frac{\text{N}\cdot\text{m}}{\text{kg}}$$

At the exit,

$$pv = 100\ \text{kPa} \times 1000\,\frac{\text{Pa}}{\text{kPa}}\,\frac{\text{N}}{\text{m}^2} \times \frac{1}{250\ \text{kg/m}^3} = 400\,\frac{\text{N}\cdot\text{m}}{\text{kg}}$$

Because 1 N·m/kg is 1 J/kg, the answers are in J/kg.

## 2.9  Nonflow Work (Moving Boundary Work)

Let us now consider the case of the piston and cylinder arrangement (i.e., piston-cylinder device) shown in Figure 2.10. We assume that the piston is in equilibrium with the contents initially, located a distance $l$ above the end or bottom of the cylinder. The piston will now be permitted to compress the contents of the cylinder frictionlessly and quasi-statically. After the piston has moved a distance $\Delta l$, a very small change in distance, the pressure in the cylinder will have increased from its initial value of $p$ to a value of $p + \Delta p$ due to the compression process itself, where $\Delta p$ indicates a very small change in pressure. The increase in fluid pressure inside the system is due to the exertion of an external net force, which include the pressure-force due to atmospheric pressure and the weight of the piston itself. Because this process was specified to be quasi-static, it is possible to evaluate the work as the product of the average force multiplied by the displacement. The average forces is $\frac{1}{2}\left[pA + (p + \Delta p)A\right]$, and the displacement is $\Delta l$. Thus,

$$\mathbf{w}\left(\text{per unit mass}\right) = \frac{p + (p + \Delta p)}{2}\left(A\Delta l\right) \tag{2.20a}$$

$$\mathbf{w}\left(\text{per unit mass}\right) = p\left(A\Delta l\right) + \frac{\Delta p}{2}\left(A\Delta l\right) \tag{2.20b}$$

However, the product of $A\Delta l$ is the small change in volume that the piston has swept out. Replacing $A\Delta l$ by $\Delta v$ and saying that the product of two very small numbers ($\Delta v\,\Delta p$) is indeed so small as to be negligible, we have

$$\mathbf{w}(\text{per unit mass}) = p(\Delta v) + \overset{0}{\frac{\Delta p}{2}(\Delta v)} \tag{2.21a}$$

$$\mathbf{w} = p\Delta v \quad \text{per unit mass of working fluid} \tag{2.21b}$$

**FIGURE 2.10**
Quasi-static nonflow compression process. (a) Piston-cylinder system. (b) Differential work.

Equation 2.21b can be interpreted by referring to Figure 2.10b. It will be noted that the term $p\Delta v$ represents a small element of area, and that the total work can be evaluated by summing all the $p\Delta v$ terms under the $p$–$v$ curve. However, instead of summing all the $p\Delta v$ terms under the $p$–$v$ curve in a discrete manner through summation, the total area under the curve is found by integration, as follows:

$$W_{\text{non-flow or boundary work}} = \int_{v_1}^{v_2} p\,dv \tag{2.21c}$$

Therefore, we can conclude that the work done in a quasi-static, frictionless, nonflow, or moving boundary process is the area under the $p$-$v$ curve.

---

### ILLUSTRATIVE PROBLEM 2.13

A process is carried out in a nonflow, quasi-static manner so that the pressure–volume relationship of the fluid is given by

$$pv = \text{constant}$$

where $p$ is the pressure and $v$ is the specific volume of the fluid. Determine the work done on the fluid if it undergoes a process in which its specific volume goes from $v_1$ to $v_2$.

### SOLUTION

To solve this problem, it is necessary to sum all values of $p\Delta v$ over the entire range of the problem. As shown in Figure 2.11b, this corresponds to obtaining the area under the $p$–$v$ curve between the limits of $v_2$ and $v_1$. Therefore, per unit mass,

$$\mathbf{w} = \Sigma p\Delta v$$

where the symbol $\Sigma$ means the sum of all such values. To carry out this summation, we must first express $p$ as a function of $v$. From the given $p,v$ relation, $p=\text{constant}/v$, where the constant is $pv=p_1v_1=p_2v_2$. When this is substituted in the $p\Delta v$ expression, we obtain

$$w = \text{constant}\,\Sigma \frac{\Delta v}{v}$$

where the total work is the sum of these terms. The summation referred to can best be illustrated by plotting $1/v$ as a function of $v$. The shaded area of Figure 2.12 is simply $\Delta v/v$. Thus, the total work per pound of fluid is the area under this curve between $v_1$ and $v_2$ multiplied by a constant. Using the methods of calculus as shown below, we find the summation to be numerically equal to $\ln(v_2/v_1)$. Note that $\ln x=\log_e x= 2.3026 \log_{10}x$. Because the constant was either $p_1v_1$ or $p_2v_2$,

$$w = p_1v_1 \ln \frac{v_2}{v_1}\left(\frac{\text{ft.}\cdot\text{lb}_f}{\text{lb}_m} \text{ or } \frac{\text{N}\cdot\text{m}}{\text{kg}}\right)$$

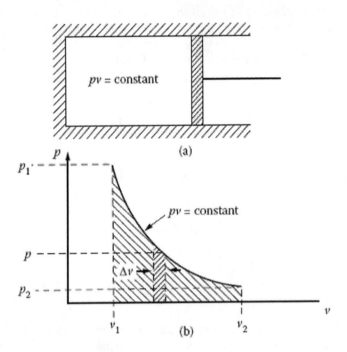

**FIGURE 2.11**
Illustrative Problem 2.13. (a) Piston-cylinder system. (b) Summation of work.

**FIGURE 2.12**
Evaluation of $\Sigma(\Delta v/v)$.

or

$$w = p_2 v_2 \ln \frac{v_2}{v_1} \left( \frac{\text{ft.} \cdot \text{lb}_f}{\text{lb}_m} \text{ or } \frac{\text{N} \cdot \text{m}}{\text{kg}} \right)$$

We shall have occasion to perform this type of summation in connection with other situations, and the procedure should be understood at this time for ease of application in the future.

## CALCULUS ENRICHMENT

The work of the nonflow or moving boundary, quasi-static process, where the equation of or relationship between $p$ and $v$ is known, is

$$w = \int_{v_1}^{v_2} p \, dv \qquad \text{(a)}$$

a. If the given relationship between $p$ and $v$ is $pv =$ constant, then $p =$ constant$/v$. Using this with Equation (a) gives us

$$w = \text{constant} \int_{v_1}^{v_2} dv/v \qquad \text{(b)}$$

The integral or antiderivative of $dv/v$ between the limits of $v_2$ and $v_x$ is $\ln(v_2/v_1)$, or

$$w = (\text{constant}) \ln (v_2/v_1) \qquad \text{(c)}$$

but since $pv =$ constant, then

$$w = (pv) \ln (v_2/v_1) \qquad \text{(d)}$$

b. If the given relationship between $p$ and $v$ is given by $pv^n =$ constant, then $p =$ constant$/v^n$. Again, using this with Equation (a),

$$w = \int_{v_1}^{v_2} p \, dv = \int_{v_1}^{v_2} (\text{constant}/v^n) \, dv \qquad \text{(e)}$$

Integrating, $w = \text{constant} \dfrac{v_2^{1-n} - v_1^{1-n}}{1-n}$

The constant can be determined from either $p_1 v_1^n$ or $p_2 v_2^n$. Using both of these, the work is found to be

$$w = \frac{p_2 v_2 - p_1 v_1}{1-n} \qquad \text{(f)}$$

Note that Equation (f) is valid for every value of $n$ except 1.0. For $n = 1.0$, the results of part (a) must be used.

The previous analyses, both (a) and (b), apply in the same way to problems using total volume $V$ instead of specific volume. Thus for $pV^n =$ constant,

$$w = \frac{p_2 V_2 - p_1 V_1}{1-n}$$

**TABLE 2.1**

Moving Boundary Work Equations for Different $p–v$ Relationships

| $p$-$v$ relationship | Process name | Work equation |
|---|---|---|
| $pv$ = constant | For ideal gases (ideal gas law: $pv$ = RT), constant temperature | $w = (pv)\ln(v_2/v_1)$ |
| $pv^n$ = constant | Polytropic | $w = \dfrac{p_2 v_2 - p_1 v_1}{1 - n}$ |
| $p$ = constant | Isobaric or constant pressure | $w = p(v_2 - v_1)$ |
| $v$ = constant | Isometric or constant volume | $w = 0$ |

Note that pressure must be in lb./ft² and volume in ft.³ to obtain work in foot-pounds. Table 2.1 below shows the corresponding work equations for different processes obtained by solving equation (a).

## ILLUSTRATIVE PROBLEM 2.14

For the process described in Illustrative Problem 2.13, the initial pressure is found to be 100 psia and the initial specific volume 2 ft.³/lb. If the final specific volume is 1 ft.³/lb., how much work was done on the fluid per pound of fluid?

### SOLUTION

From Illustrative Problem 2.13, $w = p_1 v_1 \ln(v_2/v_1)$. For the proper units, it is necessary that pressure be expressed as psfa when the volume is in cubic feet. Thus, 100 psia = (100) (144) psfa, and $v_1 = 2$ ft.³/lb.

$$w = (100 \times 144 \times 2) \ \ln\left(\frac{1}{2}\right)$$

This is evaluated by noting that $\ln \dfrac{1}{2} = \ln 1 - \ln 2$. The ln(2) is 0.693, and the ln 1 is zero. Therefore,

$$w = -100 \times 144 \times 2 \times 0.693 = -19958 \ \text{ft.} \cdot \text{lb}_f / \text{lb}_m$$

This computation is also readily done using the natural log (ln) function of any scientific calculator. In thermal units,

$$w = \frac{-19958 \ \text{ft.} \cdot \text{lb}_f / \text{lb.mass}}{778 \text{ft.} \cdot \text{lb}_f / \text{Btu}} = -25.7 \ \text{Btu/lb}_m$$

The minus sign in the answer indicates work done on the system. As an exercise, it is left to students to solve this problem by graphically evaluating the area under the pressure–volume curve.

### ILLUSTRATIVE PROBLEM 2.15

For the process $pv =$ constant, a gas compression is carried out from an initial pressure of 200 kPa to a final pressure of 800 kPa. If the initial specific volume is 0.1 m³/kg, determine the work done per kilogram of gas.

**SOLUTION**

Because $p_1v_1 = p_2v_2$,

$$v_2 = \frac{p_1v_1}{p_2} = \frac{200}{800} \times 0.1 \, \text{m}^3/\text{kg} = 0.0255 \, \text{m}^3/\text{kg}$$

and

$$w = p_1v_1 \ln\left(\frac{v_2}{v_1}\right)$$

$$= 200 \, \text{kPa} \times 0.1 \, \text{m}^3/\text{kg} \times \ln\left(\frac{0.025}{0.100}\right)$$

$$= -27.7 \, \text{kJ}/\text{kg} \, (\text{into the system})$$

### ILLUSTRATIVE PROBLEM 2.16

A gas with an initial volume of 0.3 m³ is compressed slowly from 500 kPa to 1000 kPa according to $pV^{1.25} =$ constant. Find the work done.

**SOLUTION**

Here $W = (p_2V_2 - p_1V_1)/(1 - n)$, so we must first find $V_2$. Using the $pV$ relationship, $(V_2)^n = p_1(V_1)^n/P_2$. Therefore,

$$(V_2)^{1.25} = \frac{500(0.3)^{1.25}}{1000} = 0.111$$

Taking the 1.25 root of 0.111, we obtain

$$V_2 = 0.1723 \, \text{m}^3$$

Thus,

$$W = \frac{(1000 \times 0.1723) - (500 \times 0.30)}{(1 - 1.25)} = -89.2 \, \text{kJ}$$

The minus sign indicates work into the system as expected for compression of a gas.

**FIGURE 2.13**
Example of a thermodynamic cycle consisting of several processes.

## 2.10 Cyclic Processes

A cycle consists of a series of thermodynamic processes during which the working fluid can be made to undergo changes involving energy transitions and is subsequently returned to its original state. Figure 2.13 shows a thermodynamic cycle (Carnot cycle, discussed in Chapter 8) consisting of several processes connected in sequence (i.e., A to B, B to C, C to D, D back to A) on a $p$–$v$ diagram. As discussed in subsequent chapters, many energy systems follow a particular thermodynamic cycle where energy in converted and transferred to fulfill a specific goal (i.e., cooling effect, power generation, and others).

## 2.11 Review

This chapter has been devoted to developing qualitatively and quantitatively the concepts of work, energy, and heat. A careful study of the contents of this chapter will be invaluable later on in our study of thermodynamics. In addition to defining work, energy, and heat, we also showed that the work of a nonflow or moving boundary, quasi-static process can be taken to be the area under the $p$–$v$ curve. The term *steady* implies a system that does not vary with time. In the steady-flow system, both mass and energy are permitted to cross the boundaries of the system, but such a system is not time dependent. We also noted that when a fluid is caused to flow in a system, somewhere in the system, work must have been supplied to sustain this flow. Using this concept, we demonstrated that the difference in the $pv$ terms equals the amount of work that is done on a system to introduce a unit mass into it minus the work done on its environment as it leaves the system. Even though all systems have pressure and specific volume, the $pv$ term represents flow work only in a flow system. For a nonflow, quasi-static, frictionless process, the work done is in the area under the $pv$ curve.

## Key Terms

Terms used for the first time in this chapter are as follows:

**Closed system:** See nonflow system.

**Cyclic process:** It consists of a series of thermodynamic processes during which the working fluid can be made to undergo changes involving energy transitions and is subsequently returned to its original state.

**Energy:** The capacity to do work.

**Equilibrium state:** A state in which no finite rate of change can occur without a finite change, temporary or permanent, in the state of the environment.

**Flow work:** The product of the pressure and specific volume of a fluid in a given state in a flow process.

**Heat:** A form of energy in transition to or from a system due to the fact that temperature differences exist.

**Internal energy:** The energy possessed by a body by virtue of the motion of the molecules of the body and the internal attractive and repulsive forces between molecules.

**Kinetic energy:** The energy possessed by a body due to its motion.

**Mechanical equivalent of heat:** In the English system of units, 1 Btu = 778 ft.·lb$_f$. This is termed the *mechanical equivalent of heat*.

**Nonflow or closed system:** A system in which there is no mass crossing the boundaries.

**Potential energy:** The energy possessed by a body due to its location with respect to an arbitrary reference plane.

**Quasi-static process:** A frictionless process carried out infinitely slowly so that it is in equilibrium at all times.

**Work:** The product of force and distance in which the distance is measured in the direction of the force; a form of energy in transition that is not stored in a system.

## Equations Developed in This Chapter

| | | |
|---|---|---|
| Internal energy | $mu = U$ | (2.1) |
| Internal energy | $u = \dfrac{U}{m}$ | (2.2) |
| Potential energy | $\text{P.E.} = \dfrac{mg}{g_c} Z \; \left( \text{ft.} \cdot \text{lb}_f \right)$ | (2.4) |
| Potential energy (SI) | $\text{P.E.} = mgZ \; \left( \text{joules} \right)$ | (2.6) |
| Kinetic energy | $\text{K.E.} = \dfrac{mV^2}{2g_c} \left( \text{ft.} \cdot \dfrac{\text{lb}_f}{\text{lb}_m} \right)$ | (2.12) |

| Kinetic energy (SI) | $K.E. = \dfrac{mV^2}{2}\ \left(\text{joules}\right)$ | (2.12a) |
|---|---|---|
| Net flow work | Net flow work $= \dfrac{p_2 v_2}{J} - \dfrac{p_1 v_1}{J}\ (\text{Btu}/\text{lb}_m)$ | (2.18) |
| Net flow work | Net flow work $= p_2 v_2 - p_1 v_1\ \left(\text{N}\cdot\text{m}/\text{kg}\right)$ | (2.19) |
| Work of a nonflow, quasi-static process | $\mathbf{w} = p\Delta v$ | (2.21) |

## QUESTIONS

2.1   A person picks up a package, walks with it for a distance, and then places it on a table. Discuss our concept of work for each of these actions.

2.2   A weight is held stationary at arm's length by a student. Is energy being expended, and is work being done?

2.3   A spring is compressed and placed into an acid solution, where it dissolves. What has happened to the work done on the spring?

2.4   Both heat and work represent forms of energy. Discuss the differences between them.

2.5   What is the value of the mechanical equivalent of heat in the SI system?

2.6   What distinguishes a nonflow system from a flow system?

2.7   Can you describe any quasi-static processes?

2.8   Of what utility is the quasi-static process concept?

2.9   The term $p\Delta v$ is the area under the $p$–$v$ curve. How do you interpret this area?

2.10  In discussing potential energy, what is a necessary requirement?

2.11  Why do we say that a body that is moving possesses energy?

2.12  All substances are subjected to pressure and have a specific volume. Because they all have the product $pv$, do they all have flow work?

2.13  State the convention that we have adopted for net heat and net work.

## PROBLEMS

Unless indicated otherwise, use $g_c = 32.174$ ft. $\text{lb}_m/\text{lb}_f\ s^2$, and $32.174$ ft./$s^2$ or $9.81$ m/$s^2$ for $g$.

### Problems Involving Potential Energy, Kinetic Energy, and Work

2.1   A 2-kg mass is elevated to a distance of 6 m above a reference plane. Determine its potential energy with respect to the plane.

2.2   A mass of 10 lb. is placed at an elevation of 30 ft. above a reference plane. Another mass of 15 lb. is placed at an elevation of 40 ft. above a second reference plane that is 8 ft. below the first plane.

   a.  Compare the potential energy of each mass with respect to its own reference plane.

b. Compare the potential energy of each mass with respect to the second reference plane.

2.3    A body of mass 10 lb. is placed 10 ft. above an arbitrary plane. If the local gravitational field is equivalent to an acceleration of 16.1 ft./s$^2$, how much work (ft.lb.) was done lifting the body above the plane?

2.4    A body of mass of 10 kg is placed 3 m above a plane. How much work was done to lift the body above the plane?

2.5    A body having a mass of 3 lb. is moving with a velocity of 30 ft./s relative to the Earth.

    a. Calculate its kinetic energy.

    b. If the body is resisted by a constant force of 8 lb., how far will it move before coming to rest?

2.6    A body has a mass of 5 kg. If its velocity is 10 m/s, what is its kinetic energy?

2.7    From what height would the body in Problem 2.6 have to fall to attain its velocity?

2.8    A body weighs 25 lb. If it is moving at 10 ft./s, evaluate its kinetic energy.

2.9    From what height would the body in Problem 2.8 have to fall to achieve its velocity?

2.10   A 500-kg mass is lifted 2 m above a datum plane.

    a. Determine its potential energy.

    b. If it is dropped, what will its velocity be at the instant just before it strikes the Earth?

2.11   A jet of air is discharged from a nozzle. If the jet has a velocity of 100 m/s, determine its kinetic energy in joules if 1 kg leaves the nozzle.

2.12   A body weighing 10 lb. is lifted 100 ft. What is its change in potential energy? What velocity will it possess after falling 100 ft?

2.13   A body weighing 10 lb. ($g = g_c$) is moving with a velocity of 50 ft./s. From what height would it have to fall to achieve this velocity? What is its kinetic energy?

2.14   A body weighing 100 N is lifted 3 m. What will its velocity be after a free fall of 3 m?

2.15   Water flows over the top of a dam and falls freely until it reaches the bottom some 600 ft. below. What is the velocity of the water just before it hits the bottom? What is its kinetic energy per pound mass at this point?

2.16   An object is dragged from rest a distance of 40 ft. up an inclined plane that is smooth. During this time, it is elevated 10 ft. and acquires a velocity of 10 ft./s. Given that its mass is 25 lb., perform the following:

    a. Determine the work done in the process.

    b. Determine the change in potential energy.

    c. Determine the increase in kinetic energy.

2.17   A truck having a weight of 50 kN is moving with a velocity of 25 m/s. Determine its kinetic energy.

2.18   If an object has a kinetic energy of 1 kJ when moving at a velocity of 2 m/s, determine its kinetic energy when its velocity is (a) 4 m/s and when it is (b) 6 m/s.

2.19   A force of 100 lb. deflects a spring 5 in. Calculate the spring constant and work done.

2.20   A spring is deflected 250 mm from its free length by a force of 500 N. Calculate the spring constant and the work done.

2.21   A spring is used in a shock absorber. It is found to absorb 2000 J of energy after deflecting 0.35 m. Determine the spring constant.

2.22   A spring is compressed 5 in. from its equilibrium position. If the spring modulus $k$ is 10 $lb_f$/in. of deflection, how much work was done in foot-pounds to deflect the spring?

2.23   A spring is deflected by a weight a distance of 100 mm. If the modulus of the spring is 100 N/m, what is the mass that was placed on the spring?

*2.24  A 100-N weight causes a spring to stretch an unknown distance, $h$. The addition of a weight of 250 N causes the spring to deflect an additional distance 100 mm. Determine the spring constant.

2.25   A spring is initially compressed 75 mm by a 400-N force. If it is compressed an additional distance of 50 mm, determine the work required for this additional compression.

2.26   A 5-lb. mass ($g = g_c$) falls 25 ft. until it strikes a spring. If the spring modulus $k$ is 25 $lb_f$/ft. of deflection, how far will the spring be deflected? Assume that all the energy of the falling body just goes to compress the spring.

2.27   A 10-kg mass falls 3 m until it strikes a spring. If the spring modulus $k$ is 1000 N/m, how far will the spring be deflected if all the energy of the falling body goes into compressing the spring?

2.28   In Figure P2.28, a force of 1000 N moves the block 5 m along the plane. How much work was done by the force on the block?

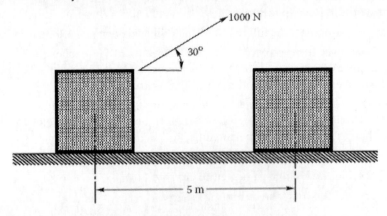

**FIGURE P2.28**

2.29   What is the least amount of work required to move the block 3 m up in the plane shown in Figure P2.29? Assume the plane to be frictionless.

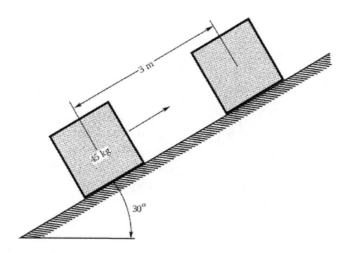

**FIGURE P2.29**

2.30  How much work must be done on a car to decrease its velocity from 90 km/h to 45 km/h? The car has a mass of 1000 kg.

2.31  An automobile has a weight of 3000 lb. and a velocity of 30 mi./h. What constant braking force is required to bring it to a stop in 190 ft.?

*2.32  Show that the kinetic energy of a fluid flowing in a pipe varies inversely as the fourth power of the pipe diameter.

*2.33  If $50 \times 10^6$ $lb_m$/h of water flows over the dam of Problem 2.15, what ideal power can be generated by a power plant located at the base of the fall? Express your answer in kilowatts and horsepower.

*2.34  If $25 \times 10^6$ kg/h flows over a waterfall that is 200 m high, what ideal power in kilowatts can be generated by a power plant located at the base of the fall?

2.35  A bicycle and rider with a combined weight of 800 N travels 40 km/h on a level carriage road. Determine (a) the kinetic energy and (b) the stopping distance if a 200-N braking force is applied to the bicycle (tires do not skid).

## Problems Involving the $p$–$v$ Concept

2.36  A pressure of 500 kPa is presented to the piston of a pump and causes it to travel 100 mm. The cross-sectional area of the piston is 1000 mm². How much work is done by the steam on the piston if the steam pressure is constant?

2.37  A gas expands according to the equation $pv = 100$, where $p$ is the pressure in pounds per square foot absolute and $v$ is the specific volume in cubic feet per pound. If the pressure of the gas drops from 100 to 50 psfa, how much work was done by the gas?

2.38  A gas expands according to the equation $pv = 1000$, where $p$ is the pressure in kPa and $v$ is the specific volume in m³/kg. If the gas pressure drops from 1000 to 500 kPa, how much work was done by the gas?

*2.39  A gas undergoes a process that starts at a pressure of 25 psia and an initial volume of 4 ft.³. Work is done on the gas in the amount of 12 Btu. If the $p$–$v$ curve is given by $pV$ = constant, determine the final pressure of the gas.

*2.40  A gas that is contained in a cylinder undergoes two processes that are in series. The first process, from 1 to 2, consists of a constant-pressure compression at 50 psia, and the volume goes from 0.5 to 0.25 ft.$^3$. The second process consists of a constant-volume heating in which the pressure doubles to 100 psia. Sketch the processes on a $p$–$v$ diagram, and determine the total work of the combined process.

*2.41  A gas undergoes two processes that are in series. The first process is an expansion that is carried out according to the relation $pv$=constant, and the second process is a constant-pressure process that returns the gas to the initial volume of the first process. The start of the first process is at 400 kPa and 0.025 m$^3$ with the expansion to 200 kPa. Sketch the processes on a $p$–$v$ diagram, and determine the work of the combined process.

*2.42  A gas undergoes two processes in series. The first is a constant-pressure expansion that takes the gas from 100 psia and 1 to 2 ft.$^3$. The second process is a constant-volume process during which the pressure decreases to 30 psia. Sketch the processes on a $p$–$v$ diagram, and determine the work of the combined process.

*2.43  A gas expands according to the equation, $pv$=100, where $p$ is the pressure in pounds force per square foot and $v$ is the specific volume in cubic feet per pound mass. The initial pressure of the gas is 100 psfa, and the final pressure is 50 psfa. The gas is then heated at constant volume back to its original pressure of 100 psfa. Determine the work of the combined process.

*2.44  A gas expands according to the equation $p=-250\,V+600$, where $V$ is in m$^3$ and $p$ is in kPa. If the initial pressure is 575 kPa, the initial volume is 0.1 m$^3$, and the final volume is 0.4 m$^3$, determine the work done. (Hint: Plot a $p$–$V$ diagram.)

*2.45  If the gas in Problem 2.44 expands at constant pressure from its initial pressure, determine the work done.

## Problems Involving the Nonflow Work Integral

*2.46  A gas with a specific volume of 2.2 ft.$^3$/lb. and a pressure of 10 psia is compressed slowly to 1.0 ft.$^3$/lb. according to $pv^{1.3}$=constant. Find the work done per pound.

*2.47  A gas with a specific volume of 1.5 ft.$^3$/lb. and a pressure of 50 psia expands slowly to 2.5 ft.$^3$/lb. according to $pv^{1.4}$=constant. Find the work done per pound.

*2.48  A gas initially at 500 kPa and 0.20 m$^3$ expands slowly according to $pV^{1.4}$=constant to 0.40 m$^3$. Find the work done.

*2.49  A gas expands slowly according to $pV^{1.4}$=constant, starting at 1000 kPa and 0.2 m$^3$. If the final volume is 0.5 m$^3$, find the work done.

*2.50  A gas initially at a pressure of 50 psia and a volume of 0.5 ft. is compressed slowly to 100 psia according to $pV^{1.3}$=constant. Find the work done.

*2.51  A gas in a piston-cylinder expands slowly from 500 kPa and 0.15 m$^3$ to a final volume of 0.60 m$^3$. Find the work done if the pressure varies according to (a) $p$=constant, (b) $pV$=constant, and (c) $pV^{1.3}$=constant.

## Cyclic Processes

A cycle consists of a series of thermodynamic processes during which the working fluid can be made to undergo changes involving energy transitions and is subsequently returned to its original state. On a $p-v$ diagram, it is a closed figure.

*2.52 A cycle consists of three processes. The first is a constant-pressure compression at 200 kPa from an initial volume of 0.70 m³ to a final volume of 0.2 m³. The second process takes place at constant volume with the pressure increasing to 600 kPa. The third process completes the cycle and consists of a straight line from the end of the second process to the beginning of the first process. Sketch the cycle on $p-v$ coordinates, and calculate the net work of the cycle.

*2.53 A cycle consists of three processes. The first process consists of a constant-pressure compression from 14.7 psia and 26 to 10 ft.³. The second process is carried out at constant volume to a final pressure of 35 psia. The cycle is closed with a straight line from the end of the second process to the start of the first process. Sketch the cycle on $p-v$ coordinates, and calculate the net work of the cycle.

*2.54 A cycle consists of three processes. The first is an expansion that is carried out according to the relation $pv=$ constant from an initial condition of 500 kPa and 0.2 m³ to a final pressure of 250 kPa. The second process is a constant-pressure compression that returns the gas to the volume that it had at the beginning of the first process. The cycle is closed with a constant-volume process that returns the gas to its state at the beginning of the first process. Sketch the cycle on $p-v$ coordinates, and calculate the net work of the cycle.

# 3

# Properties of Liquids and Vapors

## Learning Goals

After reading and studying the material in this chapter, you should be able to

1. Define the word *phase* and distinguish among the three phases of matter.
2. Understand and use the fact that the vaporization process is carried out at a constant temperature and that less heat is required to vaporize a unit mass of water as the pressure is raised.
3. Define the state that is known as the critical state.
4. Understand and use the nomenclature used in the *Steam Tables*.
5. Understand the use of the triple point as the reference state for zero internal energy and zero entropy.
6. Define the term *quality* and use it to determine the properties in the wet region.
7. Obtain the properties of steam in the subcooled, saturated, and superheated states.
8. Describe the various graphical representations of the properties of steam, especially the temperature–entropy and the enthalpy–entropy (Mollier) charts.
9. Be aware that industry uses computer systems to determine the properties of steam either by use of equations or by use of readily available commercial programs, and be able to use the CRC Press website to obtain the properties of steam in both SI and USCS units.

## 3.1 Introduction

In Chapter 1, a brief, introductory study of the properties of fluids was presented. In this chapter, the properties of liquids and vapors (gases associated with the liquid phase of the same compound such as water) are presented in more depth, because the state of a system can be described in terms of its properties. Based on the observable properties of pressure, temperature, and volume, it is possible to derive other properties that can also suffice to describe the state of a system. The relationship among the pressure, volume, and temperature of a system, when expressed mathematically, is called the *equation of state* for the substance in question. A state is an equilibrium state if no finite rate of change of state can occur without a finite change, temporary or permanent, in the state of the environment.

The ideal gas relation derived in Chapter 1 and used extensively throughout Chapter 6 is an example of such an equation of state (i.e., ideal gas law, see Equation 6.8).

A *phase* of a substance can be defined as that part of a pure substance that consists of a single, homogeneous aggregate of matter. The three common phases that are usually spoken of are *solid, liquid,* and *gaseous.* When dealing with the gaseous phase, a distinction is made between a gas and a vapor that is somewhat artificial but in common usage when using the ideal gas law to estimate physical properties (i.e., pressure, temperature, or volume). The term *vapor* is applied to the gaseous phase that is in contact with saturated liquid and that is not far removed from the saturated state or does not follow the ideal gas law accurately, and the term *gas* is used for the vapor that is either at very low pressure, it t is far removed from the saturated state or follows the ideal gas law accurately.

## 3.2 Liquids and Vapors

The distinction between vapor and liquid is usually made (in an elementary manner) by stating that both will take up the shape of their containers, but the liquid will present a free surface if it does not completely fill its container. The vapor (or gas) will always fill its container entirely.

With the foregoing in mind, let us consider the following system: a container is filled with water, and a movable, frictionless piston is placed on the container, as shown in Figure 3.1. As heat is added to the system, the temperature of the system will increase. Note that the pressure on the system is being kept constant by the fixed weight of the piston and the uniform and constant atmospheric pressure outside the system. The continued addition of heat will cause the temperature of the system to increase until the pressure of the vapor generated exactly balances the pressure of the atmosphere plus the pressure due to the weight of the piston.

At this point, the vapor and liquid are said to be *saturated.* (that is, saturated liquid (SL) and saturated vapor (SV)). As more heat is added, the liquid that was at saturation will start to vaporize. The two-phase mixture (MX) of vapor and liquid has only one degree of freedom since both have the same saturation temperature $(T_{sat})$ and saturation pressure $(p_{sat})$ as long as liquid still is present. Vaporization will continue at a constant temperature $(T_{sat})$ until all the liquid vaporizes. Also, the mixture is said to be *wet,* if liquid still is

**FIGURE 3.1**
Heating water and steam at constant pressure. (a) Liquid water only. (b) Liquid water and steam. (c) Steam only. (d) Superheated steam.

present and both the liquid and vapor are saturated. After all the liquid is vaporized, only the vapor is present, and the further addition of heat will cause the temperature of the vapor to increase at constant system pressure. This state is called the *superheat state*, and the vapor is said to be *superheated* (i.e., superheated vapor (SHV)). If this process is carried out at various pressures, a singular curve of temperature of saturation as a function of pressure will be generated. Such a curve is called a *vapor pressure* or *saturation curve* for the substance, as shown in Figure 3.2. The overall process can be reversed if heat is removed continuously at constant pressure from the system. Initially, the superheated vapor (SHV) will become saturated (SV), and then it will start condensing, forming a liquid–vapor mixture (MX) until all the saturated vapor has become saturated liquid. Upon further cooling, saturated vapor will become *compressed liquid* (CL) or a state in which the fluid temperature is considerably below the saturation temperature for the given pressure.

For a constant rate of heat input, it is possible to plot a curve of temperature as a function of time or heat added at a given system pressure, say, 100 psia. This curve is shown as a solid line in Figure 3.2. If the rate of heating is kept constant and the identical system is made to undergo this process again, but with a system pressure of 1000 psia the dashed curve in Figure 3.2 will represent the heat-added temperature history of the system. The decreased slope at 1000 psia during the heating portion indicates an increased specific heat of the liquid (water). At the end of the heating portion, the liquid is saturated at a higher temperature. The water is subsequently vaporized at a constant temperature, and the length of the horizontal portions of the curves is proportional to the heat necessary to vaporize the fluid at constant pressure (the latent heat of vaporization or LHV). Recall that heat transfer during a change of phase at constant temperature is called *latent heat*, while heat transfer for a single phase with changing temperatures is called *sensible heat*. At 1000 psia, the figure shows that less heat is required to vaporize a unit mass of fluid than at 100 psia. Table 3.1 summarizes and describes each state described above.

By increasing or decreasing the pressure exerted on the system (i.e., changing the weight of the piston) several times and measuring pressure, temperature, and specific volume

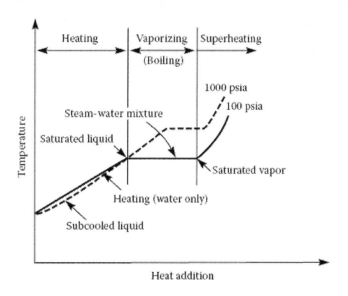

**FIGURE 3.2**
Heating of water.

**TABLE 3.1**

Summary of Liquid and Vapor States

| State | Abbreviation | Conditions | Description |
|---|---|---|---|
| Compressed liquid | CL | $p_{CL} > p_{sat}$<br>$T_{CL} < T_{sat}$ | Liquid not ready to vaporize due to its low temperature or high pressure |
| Saturated liquid | SL | $p_{SL} = p_{sat}$<br>$T_{SL} = T_{sat}$ | Liquid ready to vaporize due to its high temperature or low pressure |
| Liquid–vapor mixture | MX | $p_{SL} = p_{SV} = p_{sat}$<br>$T_{SL} = T_{SV} = T_{sat}$ | Contains both saturated liquid and saturated vapor and both phases have the same saturation pressure and saturation temperature |
| Saturated vapor | SV | $p_{SV} = p_{sat}$<br>$T_{SV} = T_{sat}$ | Vapor ready to condense into liquid if heat is removed at saturation temperature or pressure |
| Superheated vapor | SHV | $p_{SHV} < p_{sat}$<br>$T_{SHV} > T_{sat}$ | Vapor not ready to condense due to its high temperature or low pressure |

*Note:* $p_{sat}$ and $T_{sat}$ stand for saturation pressure and saturation temperature, respectively.

enables us to plot curves of $T\text{–}v$, $p\text{–}T$, and $p\text{–}v$. Figure 3.3a shows a typical $T\text{–}v$ diagram (not to scale) for a substance that expands on freezing, that is, when it goes from liquid to solid. From this diagram, we can see that the water can exist in several states:

1. A pure solid state, usually called ice
2. A pure liquid state
3. A pure gaseous state (vapor), usually called steam
4. Equilibrium mixtures of liquid and vapor states, liquid and solid states, and solid and vapor states

Figures 3.3b and 3.3c show typical $p\text{–}T$ and $p\text{–}v$ diagrams (not to scale) for the vapor–liquid states of a typical substance such as water. The *triple point* is the state in which all three phases are in equilibrium. The *critical point* is the state in which it is impossible to distinguish between the liquid and vapor phases. At the critical point, the properties of the liquid and vapor phases are identical. The temperature and pressure at this point are known, respectively, as the critical temperature and the critical pressure.

All the data can be presented as a single, three-dimensional figure, as shown in Figure 3.4 (not to scale). Figure 3.3a–c can be obtained from Figure 3.4a by projection onto the $T\text{–}v$, $p\text{–}T$, and $p\text{–}v$ planes. Note that the triple line shown in Figure 3.4a projects as a point when it is viewed parallel to the $v$-axis. Because of this, the projection of the triple line is called the triple point. Water is a substance that expands upon freezing, as shown in Figure 3.4a. Other substances that contract on freezing would have a three-dimensional figure such as that shown in Figure 3.4b.

From Figures 3.2 to 3.4, the relationships among the physical variables (i.e., $p$, $v$, and $T$) are shown in Table 3.2.

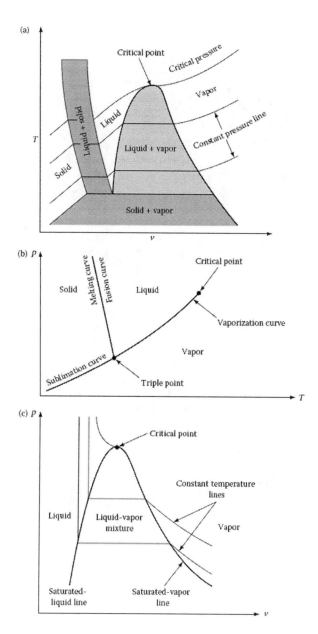

**FIGURE 3.3**
(a) A typical temperature-specific volume (*T–v*) diagram. (Not to scale.) (b) A typical pressure–temperature (*p–T*) diagram. (Not to scale.) (c) A typical pressure–volume (*p–v*) diagram. (Not to scale.).

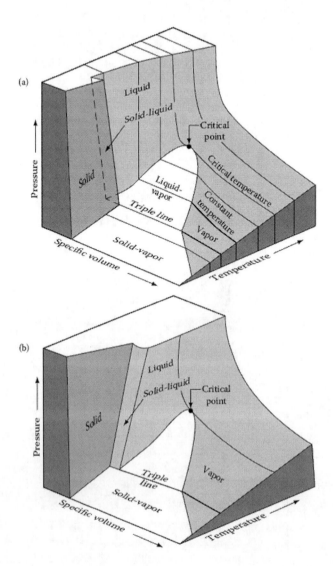

**FIGURE 3.4**
(a) A typical $p$–$v$–$T$ figure for a substance that expands on freezing (not to scale). (b) A typical $p$–$v$–$T$ figure for a substance that contracts on freezing (not to scale).

**TABLE 3.2**

Summary of Relationships among Key Physical Variables Shown in Figure 3.2, 3.3, and 3.4

| Relationships for Saturated Liquid (SL), Liquid-Vapor Mixture (MX), and Saturated Vapor (SV) States | Mathematical Relationships |
|---|---|
| $p_{sat}$ is proportional to $T_{sat}$ and vice versa | $p_{sat} \propto T_{sat}$<br>$T_{sat} \propto p_{sat}$ |
| The difference in specific volumes between the saturated vapor and saturated liquid states is inversely proportional to $p_{sat}$ or $T_{sat}$ | $\left( v_{sv@p_{sat}\,or\,T_{sat}} - v_{sl@p_{sat}\,or\,T_{sat}} \right) \propto \dfrac{1}{p_{sat}}\; or\; \dfrac{1}{T_{sat}}$ |
| LHV is inversely proportional to $p_{sat}$, $T_{sat}$ or $\left( v_{sv@p_{sat}\,or\,T_{sat}} - v_{sl@p_{sat}\,or\,T_{sat}} \right)$ | $LHV \propto \dfrac{1}{p_{sat}}$<br><br>$LHV \propto \dfrac{1}{T_{sat}}$<br><br>$LHV \propto \dfrac{1}{\left( v_{sv@p_{sat}\,or\,T_{sat}} - v_{sl@p_{sat}\,or\,T_{sat}} \right)}$ |

*Note:* $p_{sat}$ and $T_{sat}$ stand for saturation pressure and saturation temperature, respectively. $v_{sl}$ and $v_{sv}$ stand for specific volume of saturated liquid and saturated vapor, respectively. $v_{sv}$ - $v_{sl}$ is the difference in specific volumes between the saturated vapor and saturated liquid states at a given saturation temperature or pressure. LHV stands for latent heat of vaporization, which is the amount of energy per unit mass required to convert saturated liquid (SL) to saturated vapor (SV) at constant pressure or temperature. In a condensation process, the same amount of energy should be removed to reverse the process (from SV to SL).

## 3.3 Thermodynamic Properties of Steam

Water has been used as a thermodynamic working fluid for many centuries, and it has been the subject of extensive research to establish its thermodynamic properties in the past two centuries.

The need to know the thermodynamic properties of water from the subcooled liquid (compressed liquid) state to states above the critical point is obvious if systems such as a steam generator, shown in Figure 3.5, are to be properly designed. Research on the thermodynamic properties of water has been carried out throughout the world, and as a result of this research, extensive tables of the properties of water have been published.* There are nine tables of thermodynamic properties in the *Steam Tables* (as this work is commonly called) that cover the properties of ice, water, and steam. These tables include the transport properties, viscosity, and thermal conductivity. The triple point, that is, the state in which the solid, liquid, and vapor are in equilibrium, is used as the reference state for zero internal energy and zero entropy (discussed later in the book). The use of this reference state gives rise to negative values of internal energy, enthalpy, and entropy of the liquid state at 32°F. The fact that these properties are negative should not cause any difficulty, because we are invariably interested in differences in the values of properties.

---

* See Keenan et al. (1969). These tables are in English units. A separate volume bearing the same title with the additional parenthetical phrase (SI Units) has also been published by the same authors (Keenan et al. 1978).

**FIGURE 3.5**
Water tube boiler.

Tables A.1 and A.2 of the *Steam Tables* (see Appendix A-2) are for the saturated state. Extract of these tables are given in Figures 3.6, 3.7, 3.10, and 3.11. Complete tables are given in Appendix 2. Table A.1 presents the data with temperature as the independent variable; Table A.2 presents essentially the same data with pressure as the independent variable. All the tabulated data are arranged to facilitate linear interpolation between values. The nomenclature used in the *Steam Tables* is shown in Table 3.3.

It should be noted that all the tabulated properties are specific properties and are tabulated per unit mass. In Appendix 2, Table A.1, the temperature table, goes to the critical temperature of 705.44°F, and Table A.2 tabulates data up to the corresponding critical pressure of 3203.6 psia. Portions of these tables are shown in Figure 3.6a (Figure 3.6b for SI units) and Figure 3.7a (Figure 3.7b for SI units). The SI table goes to 374.136°C and 22.09 MPa.

Note that the change in property going from saturated liquid to saturated vapor (the *fg* subscript) is given for internal energy, enthalpy, and entropy, but not for a specific volume; however, it can easily be calculated from the equation for a specific volume. The relation for the saturation state properties, that is, *f* (SL), *g* (SV), and *fg*, is given by

$$h_g = h_f + h_{fg} \tag{3.1}$$

$$v_g = v_f + v_{fg}$$

$$s_g = s_f + s_{fg}$$

$$u_g = u_f + u_{fg}$$

**TABLE 3.3**

Nomenclature Used in the Steam Tables

| | English Units | SI Units |
|---|---|---|
| T = temperature | °F | °C |
| p = absolute pressure | psia or in. Hg | MPa |
| v = specific volume | ft.$^3$/lb$_m$ | m$^3$/kg |
| h = enthalpy | Btu/lb$_m$ | kJ/kg |
| s = entropy | Btu/lb$_m$·R | kJ/kg K |
| u = internal energy | Btu/lb$_m$ | kJ/kg |
| Subscripts | | |
| f = property of the saturated liquid (SL) | | |
| g = property of the saturated vapor (SV) | | |
| fg = property change due to evaporation or condensation | condensation | |

*Note:* It is worth noting, that in this book as in other thermodynamics books, the subscripts $f$ and $g$ are used to denote saturated liquid (SL) and saturated vapor (SV), respectively. The use of $f$ and $g$ was established by German scientists during the development of thermodynamics as a scientific discipline. Also, $fg$ is used to denote the difference between the corresponding $f$ and $g$ values for the same property (i.e., $v_{fg} = v_g - v_f$). Furthermore, Table 3.3 includes enthalpy, h; entropy, s; and internal energy, u; which are properties defined in detail in Chapter 4.

(a)

**Saturation Temperatures**

| | | Specific Volume | | Internal Energy | | | Enthalpy | | | Entropy | | |
|---|---|---|---|---|---|---|---|---|---|---|---|---|
| Temp. (°F) | Press. (lb$_f$/sq. in.) | Sat. Liquid | Sat. Vapor | Sat. Liquid | Evap. | Sat. Vapor | Sat. Liquid | Evap. | Sat. Vapor | Sat. Liquid | Evap. | Sat. Vapor |
| $t$ | $p$ | $v_f$ | $v_g$ | $u_f$ | $u_{fg}$ | $u_g$ | $h_f$ | $h_{fg}$ | $h_g$ | $s_f$ | $s_{fg}$ | $s_g$ |
| 90 | 0.6988 | 0.016099 | 467.7 | 58.07 | 982.2 | 1040.2 | 58.07 | 1042.7 | 1100.7 | 0.11165 | 1.8966 | 2.0083 |
| 91 | 0.7211 | 0.016102 | 454.0 | 59.06 | 981.5 | 1040.6 | 59.07 | 1042.1 | 1101.2 | 0.11346 | 1.8922 | 2.0056 |
| 92 | 0.7439 | 0.016105 | 440.9 | 60.06 | 980.8 | 1040.9 | 60.06 | 1041.5 | 1101.6 | 0.11527 | 1.8877 | 2.0030 |
| 93 | 0.7674 | 0.016108 | 428.2 | 61.06 | 980.2 | 1041.2 | 61.06 | 1041.0 | 1102.0 | 0.11708 | 1.8833 | 2.0003 |
| 94 | 0.8914 | 0.016111 | 415.9 | 62.06 | 979.5 | 1041.5 | 62.06 | 1040.4 | 1102.4 | 0.11888 | 1.8788 | 1.9977 |
| 95 | 0.8162 | 0.016114 | 404.0 | 63.06 | 978.8 | 1041.9 | 63.06 | 1039.8 | 1102.9 | 0.12068 | 1.8744 | 1.9951 |
| 96 | 0.8416 | 0.016117 | 392.4 | 64.05 | 978.1 | 1042.2 | 64.06 | 1039.2 | 1103.3 | 0.12248 | 1.8700 | 1.9925 |
| 97 | 0.8677 | 0.016121 | 381.3 | 65.05 | 977.5 | 1042.5 | 65.05 | 1038.7 | 1103.7 | 0.12427 | 1.8657 | 1.9899 |
| 98 | 0.8945 | 0.016124 | 370.5 | 66.05 | 976.8 | 1042.8 | 66.05 | 1038.1 | 1104.2 | 0.12606 | 1.8613 | 1.9874 |
| 99 | 0.9220 | 0.016127 | 360.1 | 67.05 | 976.1 | 1043.2 | 67.05 | 1037.5 | 1104.6 | 0.12785 | 1.8569 | 1.9848 |

(b)

**Saturation Temperatures**

| | | Specific Volume | | Internal Energy | | | Enthalpy | | | Entropy | | |
|---|---|---|---|---|---|---|---|---|---|---|---|---|
| Temp. (°C) | Press. (MPa) | Sat. Liquid | Sat. Vapor | Sat. Liquid | Evap. | Sat. Vapor | Sat. Liquid | Evap. | Sat. Vapor | Sat. Liquid | Evap. | Sat. Vapor |
| $t$ | $p$ | $10^3 v_f$ | $10^3 v_g$ | $u_f$ | $u_{fg}$ | $u_g$ | $h_f$ | $h_{fg}$ | $h_g$ | $s_f$ | $s_{fg}$ | $s_g$ |
| 30 | 0.004246 | 1.0043 | 32 894 | 125.78 | 2290.8 | 2416.6 | 125.79 | 2430.5 | 2556.3 | 0.4369 | 8.0164 | 8.4533 |
| 31 | 0.004496 | 1.0046 | 31 165 | 129.96 | 2288.0 | 2418.0 | 129.97 | 2428.1 | 2558.1 | 0.4507 | 7.9822 | 8.4329 |
| 32 | 0.004759 | 1.0050 | 29 540 | 134.14 | 2285.2 | 2419.3 | 134.15 | 2425.7 | 2559.9 | 0.4644 | 7.9483 | 8.4127 |
| 33 | 0.005034 | 1.0053 | 28 011 | 138.32 | 2282.4 | 2420.7 | 138.33 | 2423.4 | 2561.7 | 0.4781 | 7.9146 | 8.3927 |
| 34 | 0.005324 | 1.0056 | 26 571 | 142.50 | 2279.5 | 2422.0 | 142.50 | 2421.0 | 2563.5 | 0.4917 | 7.8811 | 8.3728 |
| 35 | 0.005628 | 1.0060 | 25 216 | 146.67 | 2276.7 | 2423.4 | 146.68 | 2418.6 | 2565.3 | 0.5053 | 7.8478 | 8.3531 |
| 36 | 0.005947 | 1.0063 | 23 940 | 150.85 | 2273.9 | 2424.7 | 150.86 | 2416.2 | 2567.1 | 0.5188 | 7.8147 | 8.3336 |
| 37 | 0.006281 | 1.0067 | 22 737 | 155.03 | 2271.1 | 2426.1 | 155.03 | 2413.9 | 2568.9 | 0.5323 | 7.7819 | 8.3142 |
| 38 | 0.006632 | 1.0071 | 21 602 | 159.20 | 2268.2 | 2427.4 | 159.21 | 2411.5 | 2570.7 | 0.5458 | 7.7492 | 8.2950 |
| 39 | 0.006999 | 1.0074 | 20 533 | 163.38 | 2265.4 | 2428.8 | 163.39 | 2409.1 | 2572.5 | 0.5592 | 7.7167 | 8.2759 |

**FIGURE 3.6**

(a) Detailed extract from the saturation table. Greater range of values can be seen in Table A.1 in Appendix 2. (J. H. Keenan, F. G. Keyes, P. G. Hill, and J. G. Moore: *Steam Tables*. 1969. Copyright Wiley-VCH Verlag GmbH & Co. KGaA. Reproduced with permission.). (b) Detailed extract from the saturation table (SI units). Greater range of values can be seen in Table A.1 (SI Units) in Appendix 2. (J. H. Keenan, F. G. Keyes, P. G. Hill, and J. G. Moore: *Steam Tables. Sf Units Edition*. 1978. Copyright Wiley-VCH Verlag GmbH & Co. KGaA. Reproduced with permission.).

(a)

| Press. (lb$_f$/sq. in.) | Temp. (°F) | Specific Volume | | Internal Energy | | | Enthalpy | | | Entropy | | |
|---|---|---|---|---|---|---|---|---|---|---|---|---|
| | | Sat. Liquid | Sat. Vapor | Sat. Liquid | Evap. | Sat. Vapor | Sat. Liquid | Evap. | Sat. Vapor | Sat. Liquid | Evap. | Sat. Vapor |
| $p$ | $t$ | $v_f$ | $v_g$ | $u_f$ | $u_{fg}$ | $u_g$ | $h_f$ | $h_{fg}$ | $h_g$ | $s_f$ | $s_{fg}$ | $s_g$ |
| 115 | 338.12 | 0.017850 | 3.884 | 308.95 | 798.8 | 1107.7 | 309.33 | 881.0 | 1190.4 | 0.48786 | 1.1042 | 1.5921 |
| 116 | 338.77 | 0.017858 | 3.852 | 309.62 | 798.2 | 1107.8 | 310.01 | 880.5 | 1190.5 | 0.48870 | 1.1027 | 1.5914 |
| 117 | 339.41 | 0.017865 | 3.821 | 310.29 | 797.6 | 1107.9 | 310.68 | 880.0 | 1190.7 | 0.48954 | 1.1012 | 1.5907 |
| 118 | 340.04 | 0.017872 | 3.790 | 310.96 | 797.1 | 1108.1 | 311.34 | 879.5 | 1190.8 | 0.49037 | 1.0996 | 1.5900 |
| 119 | 340.68 | 0.017879 | 3.760 | 311.62 | 796.6 | 1108.2 | 312.01 | 879.0 | 1191.0 | 0.49119 | 1.0981 | 1.5893 |
| 120 | 341.30 | 0.017886 | 3.730 | 312.27 | 796.0 | 1108.3 | 312.67 | 878.5 | 1191.1 | 0.49201 | 1.0966 | 1.5886 |
| 121 | 341.93 | 0.017894 | 3.701 | 312.92 | 795.5 | 1108.4 | 313.32 | 877.9 | 1191.3 | 0.49282 | 1.0951 | 1.5880 |
| 122 | 342.55 | 0.017901 | 3.672 | 313.57 | 794.9 | 1108.5 | 313.97 | 877.4 | 1191.4 | 0.49363 | 1.0937 | 1.5873 |
| 123 | 343.17 | 0.017908 | 3.644 | 314.21 | 794.4 | 1108.6 | 314.62 | 876.9 | 1191.6 | 0.49443 | 1.0922 | 1.5866 |
| 124 | 343.78 | 0.017915 | 3.616 | 314.85 | 793.9 | 1108.7 | 315.26 | 876.4 | 1191.7 | 0.49523 | 1.0907 | 1.5860 |

(b)

| Press. (MPa) | Temp. (°C) | Specific Volume | | Internal Energy | | | Enthalpy | | | Entropy | | |
|---|---|---|---|---|---|---|---|---|---|---|---|---|
| | | Sat. Liquid | Sat. Vapor | Sat. Liquid | Evap. | Sat. Vapor | Sat. Liquid | Evap. | Sat. Vapor | Sat. Liquid | Evap. | Sat. Vapor |
| $p$ | $t$ | $10^3 v_f$ | $10^3 v_g$ | $u_f$ | $u_{fg}$ | $u_g$ | $h_f$ | $h_{fg}$ | $h_g$ | $s_f$ | $s_{fg}$ | $s_g$ |
| 1.00 | 179.91 | 1.1273 | 194.44 | 761.68 | 1822.0 | 2583.6 | 762.81 | 2015.3 | 2778.1 | 2.1387 | 4.4478 | 6.5865 |
| 1.02 | 180.77 | 1.1284 | 190.80 | 765.47 | 1818.8 | 2584.2 | 766.63 | 2012.2 | 2778.9 | 2.1471 | 4.4326 | 6.5796 |
| 1.04 | 181.62 | 1.1296 | 187.30 | 769.21 | 1815.6 | 2584.8 | 770.38 | 2009.2 | 2779.6 | 2.1553 | 4.4177 | 6.5729 |
| 1.06 | 182.46 | 1.1308 | 183.92 | 772.89 | 1812.5 | 2585.4 | 774.08 | 2006.2 | 2780.3 | 2.1634 | 4.4030 | 6.5664 |
| 1.08 | 183.28 | 1.1319 | 180.67 | 776.52 | 1809.4 | 2585.9 | 777.74 | 2003.3 | 2781.0 | 2.1713 | 4.3886 | 6.5599 |
| 1.10 | 184.09 | 1.1330 | 177.53 | 780.09 | 1806.3 | 2586.4 | 781.34 | 2000.4 | 2781.7 | 2.1792 | 4.3744 | 6.5536 |
| 1.12 | 184.89 | 1.1342 | 174.49 | 783.62 | 1803.3 | 2586.9 | 784.89 | 1997.5 | 2782.4 | 2.1869 | 4.3605 | 6.5473 |
| 1.14 | 185.68 | 1.1353 | 171.56 | 787.11 | 1800.3 | 2587.4 | 788.40 | 1994.6 | 2783.0 | 2.1945 | 4.3467 | 6.5412 |
| 1.16 | 186.46 | 1.1364 | 168.73 | 790.56 | 1797.4 | 2587.9 | 791.86 | 1991.8 | 2783.6 | 2.2020 | 4.3332 | 6.5351 |
| 1.18 | 187.23 | 1.1375 | 165.99 | 793.94 | 1794.4 | 2588.4 | 795.28 | 1989.0 | 2784.2 | 2.2093 | 4.3199 | 6.5292 |

**FIGURE 3.7**

(a) Detailed extract from the saturation table. Greater range of values can be seen in Table A.2 in Appendix 2. (J. H. Keenan, F. G. Keyes, P. G. Hill, and J. G. Moore: *Steam Tables*. 1969. Copyright Wiley-VCH Verlag GmbH & Co. KGaA. Reproduced with permission.). (b) Detailed extract from saturation table A.2 (SI units). Greater range of values can be seen in Table A.2 (SI Units) in Appendix 2 (J. H. Keenan, F. G. Keyes, P. G. Hill, and J. G. Moore: *Steam Tables. Sf Units Edition*. 1978. Copyright Wiley-VCH Verlag GmbH & Co. KGaA. Reproduced with permission.).

and, in addition, the definition of enthalpy in terms of internal energy, pressure, and specific volume must also be satisfied. Thus,

$$h = u + \frac{pv}{J} \tag{3.2}$$

or in SI units,

$$h = u + pv \tag{3.2a}$$

The following examples will serve to illustrate the use of Tables A.1 and A.2.

## ILLUSTRATIVE PROBLEM 3.1

Determine the enthalpy, $h$, of saturated steam at 90°F using the tabulated properties of pressure, specific volume, and internal energy in Table A.1. Compare the result with the tabulated value of $h$.

### SOLUTION

From Table A.1 (Figure 3.6a),

$$p = 0.6988 \text{ psia}$$

$$v_g = 467.7 \text{ ft.}^3/\text{lb}_m$$

$$u_g = 1040.2 \text{ Btu}/\text{lb}_m$$

Because

$$h = u + \frac{pv}{J} \text{ and } h_g = u_g + \frac{pv_g}{J}$$

therefore,

$$h_g = 1040.2 \text{ Btu}/\text{lb}_m + \frac{0.6988 \text{ lb}_f/\text{in.}^2 \times 144 \text{ in.}^2/\text{ft.}^2 \times 467.7 \text{ ft.}^3/\text{lb}_m}{778 \text{ ft.} \cdot \text{lb}_f/\text{Btu}}$$

$$= 1100.7 \text{ Btu}/\text{lb}_m$$

This value is in agreement with 1100.7 Btu/lb$_m$ for $h_g$ from Table A.1.

## ILLUSTRATIVE PROBLEM 3.2

Determine the enthalpy of saturated steam at 30°C using the tabulated properties of pressure, specific volume, and internal energy in Table A.1. Compare the result with the tabulated value of $h_g$.

### SOLUTION

From Table A.1 (Figure 3.6a),

$$p = 0.004246 \text{ MPa} = 4.246 \text{ kPa}$$

$$v_g = 32894/10^3 \text{ m}^3/\text{kg} = 32.894 \text{ m}^3/\text{kg}$$

$$u_g = 2416.6 \text{ kJ/kg}$$

Because

$$h = u + pv \quad \text{and} \quad h_g = u_g + pv_g$$

therefore,

$$h_g = 2416.6 \text{ kJ/kg} + 4.246 \text{ kN/m}^2 \times 32.894 \text{ m}^3/\text{kg}$$

$$= 2556.27 \text{ kJ/kg}$$

The tabulated value of $h_g$ from Table A.1 is 2556.3 kJ/kg. (Students should carefully note the units used.)

When using thermodynamic tables, it is often necessary to interpolate between adjoining values to find the corresponding value for a property of interest. Let's consider two properties (Prop 1 and Prop 2) with values listed as follows:

| Prop 1 | Prop 2 |
|--------|--------|
| A | X |
| B | Y |
| C | Z |

For Prop 1, all the values are known ($A$, $B$, and $C$), and for Prop 2, the only values known are $X$ and $Z$. That is, $Y$ is unknown, but we suspect its value should be between $X$ and $Z$, since $Y$ falls between them for the corresponding value of Prop 1 ($B$). To find $Y$, the simple interpolation rule based on common ratios should be used, as follows:

$$\frac{(B-A)}{(C-A)} = \frac{(Y-X)}{(Z-X)}$$

Then, solving for $Y$, we obtain:

$$Y = X + \frac{(B-A)(Z-X)}{(C-A)}$$

By using the above equation, we should be able to find values between corresponding adjoining values when using thermodynamic tables. The following example illustrates how interpolation is done.

| Property 1 | Property 2 |
|------------|------------|
| 10 | 50 |
| 15 | Y |
| 20 | 80 |

Using the same variables shown above, let's find the value of $Y$ is the numerical values of $A$, $B$, $C$, $X$, and $Z$ are given, as follows:

$$Y = 50 + \frac{(15-10)(80-50)}{(20-10)}$$

$$Y = 50 + \frac{(5)(30)}{(10)}$$

$$Y = 50 + 15$$

$$Y = 65$$

As the result shows, the numerical value of $Y$ falls in between 50 and 80, which corresponds to the numerical value of Prop 1, 15. That is, if Prop 1 is 15, then Prop 2 should be 65.

---

### ILLUSTRATIVE PROBLEM 3.3

Determine the enthalpy, entropy (property denoted with $s$), specific volume, and internal energy of saturated steam at 118 psia. Assume that the values for 115 and 120 psia are available, and perform the necessary interpolations. Compare the results with the tabulated values for 118 psia.

### SOLUTION

The necessary interpolations are best done in tabular form, as shown in the following:

| $p$ | $h_g$ | |
|-----|-------|---|
| 115 | 1190.4 | Table A.2 |
| 118 | 1190.8 | $(h_g)_{118} = 1190.8$ |
| 120 | 1191.1 | |

$(3/5)(1191.1 - 1190.4) = 0.42$
$(h_g)_{118} = 1190.4 + 0.42$
$= 1190.8$ Btu/lb$_m$

| $p$ | $v_g$ | |
|-----|-------|---|
| 115 | 3.884 | Table A.2 |
| 118 | 3.792 | $(v_g)_{118} = 3.792$ |
| 120 | 3.730 | |

$(3/5)(3.884 - 3.730) = 0.092$
$(v_g)_{118} = 3.884 - 0.092$
$= 3.792$ ft.$^3$/lb$_m$
$= 3.792$ ft.$^3$/lb$_m$

| $p$ | $s_g$ | |
|-----|-------|---|
| 115 | 1.5921 | Table A.2 |
| 118 | 1.5900 | $(s_g)_{118} = 1.5900$ |
| 120 | 1.5886 | |

$(3/5)(1.5921 - 1.5886) = 0.0021$
$(s_g)_{118} = 1.5921 - 0.0021$
$= 1.5900$

| $p$ | $u_g$ | |
|---|---|---|
| 115 | 1107.7 | Table A.2 |
| 118 | 1108.06 | $(u_g)_{118} = 1108.1$ |
| 120 | 1108.3 | |

$(3/5)(1108.3 - 1107.7) = 0.36$

$(u_g)_{118} = 1107.7 - 0.36 = 1108.06$

The interpolation process that was done in tabular form for this Illustrative Problem can also be demonstrated by referring to Figure 3.8 for the specific volume. It will be seen that the results of this problem and the tabulated values are essentially in exact agreement and that linear interpolation is satisfactory in these tables.

**FIGURE 3.8**
Illustrative Problem 3.3.

## ILLUSTRATIVE PROBLEM 3.4

Determine $h_{fg}$ for saturated steam at 115 psia using the pressure, volumes, and internal data given in Table A.2 (Figure 3.7).

**SOLUTION**
By definition,

$$h_g = u_g + \frac{pv_g}{J}$$

$$h_f = u_f + \frac{pv_f}{J}$$

and

$$h_{fg} = h_g - h_f = (u_g - u_f) + \frac{p(v_g - v_f)}{J} = u_{fg} + \frac{p(v_g - v_f)}{J}$$

From Table A.2 at 115 psia,

$$u_{fg} = 798.8 \text{ Btu/lb}_\text{m}$$

$$v_g = 3.884 \text{ ft.}^3 / \text{lb}_\text{m}$$

$$v_f = 0.017850 \text{ ft.}^3 / \text{lb}_\text{m}$$

Note that if the table had $v_{fg}$, we could have read it directly because $v_g - v_f = v_{fg}$. Proceeding gives

$$h_{fg} = 798.8 \text{ Btu/lb}_\text{m} + \frac{115 \text{ lb}_\text{f}/\text{in.}^2 \times 144 \text{ in.}^2/\text{ft.}^2 \, (3.884 - 0.017850) \text{ ft.}^3/\text{lb}_\text{m}}{778 \text{ ft.} \cdot \text{lb}_\text{f}/\text{Btu}}$$

$$= 798.8 + 82.3 = 881.1 \text{ Btu/lb}_\text{m}$$

The tabulated value is 881.0 Btu/lb$_\text{m}$, and the agreement is satisfactory.

---

### ILLUSTRATIVE PROBLEM 3.5

Determine $h_{fg}$ for saturated steam at 1.0 MPa using the pressure, volumes, and internal data given in Table A.2 (Figure 3.7a).

#### SOLUTION

Refer to Illustrative Problem 3.4. From Table A.2 at 1.0 MPa,

$$u_{fg} = 1822.0 \text{ kJ/kg}$$

$$v_g = 194.44/10^3 = 0.19444 \text{ m}^3/\text{kg}$$

$$v_f = 1.1273/10^3 = 0.0011273 \text{ m}^3/\text{kg}$$

$$p = 1.0 \text{ MPa}$$

Because $v_{fg} = v_g - v_f$,

$$v_{fg} = 0.19444 - 0.0011273 = 0.1933127 \text{ m}^3/\text{kg}$$

$$h_{fg} = u_{fg} + p\left(v_{fg}\right)$$

$$= 1822.0 \text{ kJ/kg} + \left(1000 \text{ kN/m}^2\right)\left(0.1933127 \text{ m}^3/\text{kg}\right)$$

$$= 2015.3 \text{ kJ/kg}$$

The tabulated value is 2015.3 kJ/kg, which is in exact agreement with the value calculated.

Between the saturated liquid and the saturated vapor, there exists a mixture (MX) of vapor plus liquid (the *wet* region), as explained earlier. To denote the state of a liquid–vapor mixture, it is necessary to introduce a term describing the relative quantities of liquid and vapor in the mixture since both phases (i.e., liquid and vapor) have the same saturation pressure and saturation temperature. The *quality* of a mixture ($x$) is defined as the ratio of the mass of vapor to the mass of the mixture $\left( x = \dfrac{m_{vapor}}{m_{vapor}+m_{liquid}} \right)$. Thus, in 1 lb. of mixture, there must be $(1 - x)$ lb. of liquid. Another term is used to describe the wet region: *percent moisture*. A mixture whose quality is 80% has 20% moisture by weight and is simply said to have 20% moisture (i.e., saturated liquid).

Consider a mixture weighing 1 lb. and having a quality $x$. In this mixture, there is $x$ lb. of vapor and $(1 - x)$ lb. of liquid. The enthalpy of this mixture per pound of mixture is the sum of the enthalpies of the components. Therefore, the enthalpy of the liquid portion of the mixture is

$$h_l = (1-x)h_f$$

and the enthalpy of the vapor portion is

$$h_v = xh_g$$

The sum of these terms is the enthalpy of the mixture $h_x$:

$$h_x = (1-x)h_f + xh_g \tag{3.3a}$$

By the same reasoning, the entropy, internal energy, and specific volume of the wet mixture are given as

$$s_x = (1-x)s_f + xs_g \tag{3.3b}$$

$$u_x = (1-x)u_f + xu_g \tag{3.3c}$$

$$v_x = (1-x)v_f + xv_g \tag{3.3d}$$

It is sometimes more convenient to express Equations 3.3a–3.3d in terms of the property change during vaporization. By noting that the *fg* property is equal to the *g* value minus the *f* value, it is possible to rewrite Equations 3.3a–3.3d as follows:

$$h_x = h_f + xh_{fg} \tag{3.4a}$$

$$s_x = s_f + xs_{fg} \tag{3.4b}$$

$$u_x = u_f + xu_{fg} \tag{3.4c}$$

$$v_x = v_f + xv_{fg} \tag{3.4d}$$

The relationship among these quantities can be demonstrated by referring to Figure 3.9, which is a *T–s* (temperature vs. entropy) diagram for water. The saturation curve is indicated,

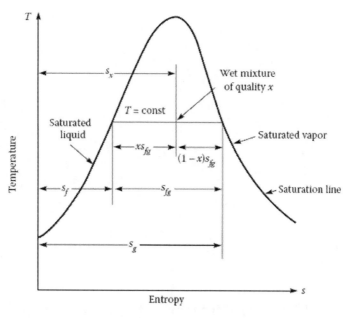

**FIGURE 3.9**
*T–s* diagram showing the relation of properties and quality. (From F. P. Durham, *Thermodynamics*, 2nd ed., Upper Saddle River, N.J.: Prentice Hall, Inc., 1959, p. 43. With permission.).

and a point below the saturation curve is in the wet region. The *T–s* diagram is discussed further later in the book (Chapter 5), but for now, it should be noted that the expressions for entropy in Equations 3.3 and 3.4 can be directly determined from Figure 3.9.

## ILLUSTRATIVE PROBLEM 3.6

A wet steam mixture at 120 psia is found to have a quality of 80%. Determine its entropy (*s*), enthalpy, internal energy, and specific volume.

### SOLUTION

Using Table A.2 and a quality of 80% (*x* = 0.8), we have

$$s_x = s_f + xs_{fg} = 0.49201 + 0.8(1.0966) = 1.3693 \text{ Btu/lb}_m \cdot \text{R}$$

$$h_x = h_f + xh_{fg} = 312.67 + 0.8(878.5) = 1015.47 \text{ Btu/lb}_m$$

$$u_x = u_f + xu_{fg} = 312.27 + 0.8(796.0) = 949.07 \text{ Btu/lb}_m$$

$$v_x = v_f + xv_{fg} = 0.017886 + 0.8(3.730 - 0.017886)$$

$$= 2.988 \text{ ft.}^3/\text{lb}_m$$

As a check on the solution obtained, we calculate

$$u_x = h_x - \frac{p_x v_x}{J}$$

$$= 1015.47 \text{ Btu}/\text{lb}_\text{m} - \frac{120\,\dfrac{\text{lb}_\text{f}}{\text{in.}^2} \times 144\,\dfrac{\text{in.}^2}{\text{ft.}^2} \times 2.988\,\dfrac{\text{ft.}^3}{\text{lb}_\text{m}}}{778\ \text{ft.}\cdot\text{lb}_\text{f}/\text{Btu}}$$

$$= 949.10 \text{ Btu}/\text{lb}_\text{m}$$

which agrees with the value 949.07 Btu/lb$_\text{m}$.

---

## ILLUSTRATIVE PROBLEM 3.7

A wet mixture at 1.0 MPa is found to have a quality of 85%. Determine its entropy ($s$), enthalpy, internal energy, and specific volume.

**SOLUTION**
Using $x = 0.85$ yields

$$s_x = s_f + x s_{fg} = 2.1387 \text{ kJ}/\text{kg}\cdot\text{K} + (0.85)(4.4487 \text{ kJ}/\text{kg}\cdot\text{K})$$

$$= 5.9201 \text{ kJ}/\text{kg}\cdot\text{K}$$

$$h_x = h_f + x h_{fg} = 762.81 \text{ kJ}/\text{kg} + (0.85)(2015.3 \text{ kJ}/\text{kg})$$

$$= 2475.82 \text{ kJ}/\text{kg}$$

$$u_x = u_f + x u_{fg} = 761.68 \text{ kJ}/\text{kg} + (0.85)(1822.0 \text{ kJ}/\text{kg})$$

$$= 2310.38 \text{ kJ}/\text{kg}$$

$$v_x = v_f + x v_{fg} = \left[ 1.1273 \text{ m}^3/\text{kg} + (0.85)(194.44 \text{ m}^3/\text{kg} - 1.1273 \text{ m}^3/\text{kg}) \right] \times 10^{-3}$$

$$= 165.44/10^3 = 0.16544 \text{ m}^3/\text{kg}$$

As a check,

$$u_x = h_x - p_x v_x = 2475.82 \text{ kJ}/\text{kg} - \frac{1.0 \times 10^6 \left(0.16544 \text{ m}^3/\text{kg}\right)}{10^3 \text{ kPa}/\text{MPa}}$$

$$= 2310.38 \text{ kJ}/\text{kg}$$

Again, the agreement is exact. Note the units used for consistency.

---

### ILLUSTRATIVE PROBLEM 3.8

A mixture of wet steam at 90°F is found to have an enthalpy of 900 Btu/lb$_m$. What is its quality?

**SOLUTION**

For the wet mixture, $h_x = h_f + xh_g$. Solving for $x$ gives us

$$x = \frac{h_x - h_f}{h_{fg}}$$

Using the data from Table A.1 (Figure 3.6), we have

$$x = \frac{900 \text{ Btu/lb}_m - 58.07 \text{ Btu/lb}_m}{1042.7 \text{ Btu/lb}_m} = 0.807 = 80.7\%$$

---

### ILLUSTRATIVE PROBLEM 3.9

A mixture of wet steam at 30°C is found to have an enthalpy of 2000.0 kJ/kg. Determine its quality.

**SOLUTION**

Because $h_x = h_f + xh_{fg}$,

$$x = \frac{h_x - h_f}{h_{fg}} = \frac{2000.0 \text{ kJ/kg} - 125.79 \text{ kJ/kg}}{2430.5 \text{ kJ/kg}} = 0.771 \text{ or } 77.1\%$$

---

Steam quality is important in the usage of steam equipment such as a heat exchanger or a steam turbine. These devices require steam as free from moisture as possible to avoid problems such as fouling or metal deterioration.

Table A.3 of the *Steam Tables* gives the properties of the superheated vapor, and it occupies 39 pages in the main table. The beginning of this table is for water vapor at low pressure, which can be treated as an ideal gas. For now, we will not be concerned with this section of Table A.3. The main portion of the table extends to pressures of 1000 psia and 2400°F. In the SI tables, the values go to 60 MPa and 1300°C. For the superheated region, it is necessary to specify two variables, such as pressure (*p*) and temperature (*T*) or enthalpy (*h*) and pressure (*p*), in order to specify the state of the vapor. Table A.3 is organized using pressure and temperature as independent variables, and it lists values of specific volume, internal energy, enthalpy, and entropy corresponding to specified values of temperature and pressure.

Referring to Figure 3.10a (Figure 3.10b gives SI units), which is an extract from Table A.3, it will be noted that the topmost line lists the different pressures in psia and in parentheses displays the saturation temperature corresponding to the pressure. The second horizontal line down lists the column headings. All values are specific values, and the temperature is listed in °F. The third horizontal line shown lists saturation (Sat.) values corresponding to

(a)

| p (t Sat.) | 320 (423.39) | | | | 325 (424.84) | | | | 330 (426.27) | | | |
|---|---|---|---|---|---|---|---|---|---|---|---|---|
| t | v | u | h | s | v | u | h | s | v | u | h | s |
| Sat. | 1.4493 | 1118.6 | 1204.4 | 1.5058 | 1.4274 | 1118.6 | 1204.5 | 1.5044 | 1.4061 | 1118.7 | 2204.6 | 1.5031 |
| 410 | 1.4110 | 1110.8 | 1194.4 | 1.4944 | 1.3852 | 1110.0 | 1193.3 | 1.4917 | 1.3602 | 1109.2 | 1192.2 | 1.4890 |
| 420 | 1.4398 | 1116.6 | 1201.9 | 1.5030 | 1.4139 | 1115.9 | 1200.9 | 1.5004 | 1.3887 | 1118.1 | 1199.9 | 1.4978 |
| 430 | 1.4677 | 1122.2 | 1209.2 | 1.5112 | 1.4416 | 1121.5 | 1208.2 | 1.5087 | 1.4163 | 1120.8 | 1207.3 | 1.5061 |
| 440 | 1.4949 | 1127.7 | 1216.2 | 1.5191 | 1.4686 | 1127.0 | 1215.4 | 1.5166 | 1.4430 | 1126.4 | 1214.5 | 1.5142 |
| 450 | 1.5213 | 1133.0 | 1223.1 | 1.5267 | 1.4948 | 1132.4 | 1222.3 | 1.5243 | 1.4691 | 1131.8 | 1221.5 | 1.5219 |
| 460 | 1.5472 | 1138.1 | 1229.8 | 1.5340 | 1.5205 | 1137.6 | 1229.0 | 1.5316 | 1.4945 | 1137.0 | 1228.2 | 1.5293 |
| 470 | 1.5726 | 1143.2 | 1236.3 | 1.5411 | 1.5456 | 1142.6 | 1235.6 | 1.5387 | 1.5194 | 1142.1 | 1234.9 | 1.5364 |
| 480 | 1.5975 | 1148.1 | 1242.7 | 1.5479 | 1.5702 | 1147.6 | 1242.0 | 1.5456 | 1.5438 | 1147.1 | 1241.4 | 1.5434 |
| 490 | 1.6219 | 1152.9 | 1249.0 | 1.5545 | 1.5944 | 1152.4 | 1248.3 | 1.5523 | 1.5678 | 1152.0 | 1247.7 | 1.5501 |

| p (t Sat.) | 335 (427.68) | | | | 340 (429.07) | | | | 345 (430.45) | | | |
|---|---|---|---|---|---|---|---|---|---|---|---|---|
| t | v | u | h | s | v | u | h | s | v | u | h | s |
| Sat. | 1.3854 | 1118.8 | 1204.7 | 1.5017 | 1.3653 | 1118.9 | 1204.8 | 1.5004 | 1.3457 | 1118.9 | 1204.9 | 1.4991 |
| 410 | 1.3360 | 1108.3 | 1191.2 | 1.4863 | 1.3124 | 1107.5 | 1190.1 | 1.4837 | 1.2894 | 1106.6 | 1189.0 | 1.4810 |
| 420 | 1.3643 | 1114.3 | 1198.9 | 1.4952 | 1.3406 | 1113.6 | 1197.9 | 1.4926 | 1.3175 | 1112.8 | 1196.9 | 1.4901 |
| 430 | 1.3917 | 1120.1 | 1206.4 | 1.5036 | 1.3678 | 1119.4 | 1205.5 | 1.5012 | 1.3445 | 1118.7 | 1204.5 | 1.4987 |
| 440 | 1.4182 | 1125.7 | 1213.6 | 1.5117 | 1.3941 | 1125.0 | 1212.8 | 1.5093 | 1.3707 | 1124.4 | 1211.9 | 1.5069 |
| 450 | 1.4441 | 1131.1 | 1220.7 | 1.5195 | 1.4198 | 1130.5 | 1219.8 | 1.5171 | 1.3962 | 1129.9 | 1219.0 | 1.5148 |
| 460 | 1.4693 | 1136.4 | 1227.5 | 1.5270 | 1.4448 | 1135.8 | 1226.7 | 1.5247 | 1.4210 | 1135.2 | 1226.0 | 1.5224 |
| 470 | 1.4940 | 1141.5 | 1234.2 | 1.5342 | 1.4693 | 1141.0 | 1233.4 | 1.5319 | 1.4453 | 1140.4 | 1232.7 | 1.5297 |
| 480 | 1.5182 | 1146.6 | 1240.7 | 1.5412 | 1.4933 | 1146.0 | 1240.0 | 1.5389 | 1.4691 | 1145.5 | 1239.3 | 1.5368 |
| 490 | 1.5419 | 1151.5 | 1247.1 | 1.5479 | 1.5168 | 1151.0 | 1246.4 | 1.5458 | 1.4924 | 1150.5 | 1245.8 | 1.5436 |

(b)

Vapor

| p (t Sat.) | 2.00 (212.42) | | | | 2.05 (213.67) | | | | 2.10 (214.90) | | | |
|---|---|---|---|---|---|---|---|---|---|---|---|---|
| t | $10^3 v$ | u | h | s | $10^3 v$ | u | h | s | $10^3 v$ | u | h | s |
| Sat. | 99.63 | 2600.3 | 2799.5 | 6.3409 | 97.25 | 2600.7 | 28000.0 | 6.3318 | 94.98 | 2601.0 | 2800.5 | 6.3229 |
| 200 | 95.27 | 2570.6 | 2761.1 | 6.2608 | 92.52 | 2567.6 | 2757.3 | 6.2427 | | | | |
| 205 | 97.06 | 2582.8 | 2777.0 | 6.2941 | 94.29 | 2580.1 | 2773.4 | 6.2765 | 91.65 | 2577.2 | 2769.7 | 6.2591 |
| 210 | 98.80 | 2594.7 | 2792.3 | 6.3259 | 96.01 | 2592.1 | 2788.9 | 6.3089 | 93.35 | 2589.5 | 2785.5 | 6.2920 |
| 215 | 100.50 | 2606.2 | 2807.2 | 6.3566 | 97.69 | 2603.7 | 2804.0 | 6.3399 | 95.01 | 2601.3 | 2800.8 | 6.3235 |
| 220 | 102.15 | 2617.4 | 2821.7 | 6.3861 | 99.32 | 2615.1 | 2818.7 | 6.3699 | 96.62 | 2612.8 | 2815.7 | 6.3538 |
| 225 | 103.77 | 2628.3 | 2835.8 | 6.4147 | 100.92 | 2626.1 | 2833.0 | 6.3987 | 98.20 | 2623.9 | 2830.2 | 6.3831 |
| 230 | 105.36 | 2638.9 | 2849.6 | 6.4423 | 102.48 | 2636.9 | 2847.0 | 6.4267 | 99.74 | 2634.9 | 2844.3 | 6.4113 |
| 235 | 106.91 | 2649.4 | 2863.2 | 6.4691 | 104.02 | 2647.5 | 2860.7 | 6.4538 | 101.25 | 2645.5 | 2858.2 | 6.4387 |
| 240 | 108.45 | 2659.6 | 2876.5 | 6.4952 | 105.52 | 2657.8 | 2874.1 | 6.4801 | 102.74 | 2656.0 | 2871.7 | 6.4653 |
| 245 | 109.95 | 2669.7 | 2889.6 | 6.5205 | 107.01 | 2668.0 | 2887.3 | 6.5057 | 104.20 | 2666.2 | 2885.0 | 6.4911 |

| p (t Sat.) | 2.15 (216.10) | | | | 2.20 (217.29) | | | | 2.25 (218.45) | | | |
|---|---|---|---|---|---|---|---|---|---|---|---|---|
| t | $10^3 v$ | u | h | s | $10^3 v$ | u | k | s | $10^3 v$ | u | h | s |
| Sat. | 92.81 | 2601.4 | 2800.9 | 6.3141 | 90.73 | 2601.7 | 2801.3 | 6.3056 | 88.75 | 2602.0 | 2801.7 | 6.2972 |
| 200 | | | | | | | | | | | | |
| 205 | 89.12 | 2574.4 | 2766.0 | 6.2419 | 86.71 | 2571.5 | 2762.2 | 6.2248 | | | | |
| 210 | 90.81 | 2586.8 | 2782.0 | 6.2753 | 88.38 | 2584.1 | 2778.5 | 6.2587 | 86.06 | 2581.3 | 2774.9 | 6.2423 |
| 215 | 92.45 | 2598.8 | 2797.6 | 6.3073 | 90.01 | 2596.3 | 2794.3 | 6.2912 | 87.66 | 2593.7 | 2790.9 | 6.2753 |
| 220 | 94.05 | 2610.4 | 2812.6 | 6.3380 | 91.58 | 2608.1 | 2809.5 | 6.3223 | 89.23 | 2605.7 | 2806.4 | 6.3068 |
| 225 | 95.60 | 2621.7 | 2827.3 | 6.3676 | 93.12 | 2619.5 | 2824.4 | 6.3523 | 90.75 | 2617.3 | 2821.5 | 6.3372 |
| 230 | 97.13 | 2632.8 | 2841.6 | 6.3962 | 94.63 | 2630.7 | 2838.9 | 6.3812 | 92.23 | 2628.6 | 2836.1 | 6.3664 |
| 235 | 98.62 | 2643.6 | 2855.6 | 6.4238 | 96.10 | 2641.6 | 2853.0 | 6.4092 | 93.69 | 2639.6 | 2850.4 | 6.3947 |
| 240 | 100.08 | 2654.1 | 2869.3 | 6.4506 | 97.54 | 2652.2 | 2866.8 | 6.4362 | 95.11 | 2650.4 | 2864.4 | 2.4220 |
| 245 | 101.52 | 2664.5 | 2882.7 | 6.4767 | 98.96 | 2662.7 | 2880.4 | 6.4626 | 96.51 | 2660.9 | 2878.1 | 6.4486 |

**FIGURE 3.10**

(a) Detailed extract from the superheat table. Greater range of values can be seen in Table A.3 in Appendix 2. (J. H. Keenan, F. G. Keyes, P. G. Hill, and J. G. Moore: *Steam Tables*. 1969. Copyright Wiley-VCH Verlag GmbH & Co. KGaA. Reproduced with permission.). (b) Detailed extract from the superheat table (SI units). Greater range of values can be seen in Table A.3 (SI Units) in Appendix 2. (J. H. Keenan, F. G. Keyes, P. G. Hill, and J. G. Moore: *Steam Tables. Sf Units Edition*. 1978. Copyright Wiley-VCH Verlag GmbH & Co. KGaA. Reproduced with permission.).

the pressure listed on the first line of the table. These values are for convenience so that it is not necessary to turn back to Tables A.1 and A.2 for saturation properties. The desired thermodynamic properties at a given temperature are read horizontally underneath the appropriate pressure heading. It will be noted that at the upper part of the table, values are given in italics for certain temperatures with a horizontal line separating the main portion of the table from the italicized values. The values in italics (above the horizontal dividing line) are for the vapor temperature *below* the saturation temperature corresponding to the pressure listed. These states are metastable states (they are not equilibrium states), and we will not deal with them at all in our study. They should be ignored at this time by the student.

The SI tables are arranged in the same manner as the English tables. Pressures are in megapascals (MPa), temperatures are in degrees Celsius (°C), and the unit mass is the kilogram (kg). The energy unit is the kilojoule (kJ), and the volume unit is $10^3 \times$ cubic meters/kilogram ($10^3 \times m^3/kg$). Figure 3.10b is an extract from the superheat table.

---

### ILLUSTRATIVE PROBLEM 3.10

Determine the specific volume, internal energy, enthalpy, and entropy (s) of superheated steam at 330 psia and 450°F.

#### SOLUTION

The values of temperature and pressure are listed in Table A.3 (Figure 3.10a) and can be read directly.

$$v = 1.4691 \text{ ft.}^3/\text{lb}_m$$

$$u = 1131.8 \text{ Btu}/\text{lb}_m$$

$$h = 1221.5 \text{ Btu}/\text{lb}_m$$

$$s = 1.5219 \text{ Btu}/\text{lb}_m \cdot R$$

---

### ILLUSTRATIVE PROBLEM 3.11

Determine the specific volume, internal energy, enthalpy, and entropy (s) of superheated steam at 2.0 MPa and 240°C.

#### SOLUTION

Reading directly from Figure 3.10b, we have

$$v = 108.45/10^3 = 0.10845 \text{ m}^3/\text{kg}$$

$$u = 2659.6 \text{ kJ/kg}$$

$$h = 2876.5 \text{ kJ/kg}$$

$$s = 6.4952 \text{ kJ/kg} \cdot K$$

## ILLUSTRATIVE PROBLEM 3.12

Determine the specific volume, internal energy, enthalpy, and entropy of superheated steam at 330 psia and 455°F.

### SOLUTION

Because the data in Figure 3.10a do not give the properties at this temperature, it is necessary to interpolate between 450°F and 460°F. Thus, at 330 psia,

| $T$ | $v$ |
|---|---|
| 460 | 1.4945 |
| 455 | 1.4818 |
| 450 | 1.4691 |

$v_{455} = 1.4691 + 1/2$
$(1.4945 - 1.4691)$

| $T$ | $u$ |
|---|---|
| 460 | 1137.0 |
| 455 | 1134.4 |
| 450 | 1131.8 |

$u_{455} = 1131.8 + 1/2$
$(1137.0 - 1131.8)$

| $T$ | $h$ |
|---|---|
| 460 | 1228.2 |
| 455 | 1224.9 |
| 450 | 1221.5 |

$h_{455} = 1221.5 + 1/2$
$(1228.2 - 1221.5)$

| $T$ | $s$ |
|---|---|
| 460 | 1.5293 |
| 455 | 1.5256 |
| 450 | 1.5219 |

$s_{455} = 1.5219 + 1/2$
$(1.5293 - 1.5219)$

Thus,

$$v = 1.4818 \text{ ft.}^3/\text{lb}_m$$

$$u = 1134.4 \text{ Btu}/\text{lb}_m$$

$$h = 1224.9 \text{ Btu}/\text{lb}_m$$

$$s = 1.5256 \text{ Btu}/\text{lb.}\cdot\text{R}$$

## ILLUSTRATIVE PROBLEM 3.13

Determine the specific volume and enthalpy of superheated steam at 465°F and 337 psia.

## SOLUTION

It can be noted from Figure 3.10a that neither the temperature nor the pressure values of the problem are tabulated in Table A.3. Thus, it becomes necessary to interpolate around both the listed temperature and pressure for the desired properties. We will first obtain the properties at 337 psia and 460°F and then at 337 psia and 470°F. Proceeding with the calculation, at 460°F,

| $p$ | $v$ |
|---|---|
| 340 | 1.4448 |
| 337 | 1.4595 |
| 335 | 1.4693 |

$v_{337} = 1.4693 - 2/5\,(1.4693 - 1.4448)$

| $p$ | $h$ |
|---|---|
| 340 | 1226.7 |
| 337 | 1227.2 |
| 335 | 1227.5 |

$h_{337} = 1227.5 - 2/5\,(1227.5 - 1226.7)$

and at 470°F,

| $p$ | $v$ |
|---|---|
| 340 | 1.4693 |
| 337 | 1.4841 |
| 335 | 1.4940 |

$v_{337} = 1.4940 - 2/5\,(1.4940 - 1.4693)$

| $p$ | $h$ |
|---|---|
| 340 | 1233.4 |
| 337 | 1233.9 |
| 335 | 1234.2 |

$h_{337} = 1234.2 - 2/5\,(1234.2 - 1233.4)$

Therefore, at 337 psia and 465°F,

| $T$ | $v$ |
|---|---|
| 470 | 1.4841 |
| 465 | 1.4718 |
| 460 | 1.4595 |

$v_{465} = 1.4595 + 1/2\,(1.4841 - 1.4595)$

| T | h |
|---|---|
| 470 | 1233.9 |
| 465 | 1230.7 |
| 460 | 1227.5 |

$$h_{465} = 1227.5 + 1/2\,(1233.9 - 1227.5)$$

The desired values of specific volume and enthalpy at 465°F and 337 psia are 1.4718 ft.³/lb$_m$ and 1230.7 Btu/lb$_m$.

An extract from Table A.4 of the *Steam Tables* entitled *Liquid* is shown in Figure 3.11a (Figure 3.11b gives SI units). The state of the liquid is the *subcooled* state (i.e., compressed liquid), in which the pressure on the liquid exceeds the saturation pressure corresponding to the temperature of the liquid. In this region, the tabulated properties of specific volume, internal energy, enthalpy, and entropy are exhibited as functions of pressure and temperature similar to the format used for the superheated vapor. It will be noted from Figure 3.11a that italicized values are shown for certain values of pressure and temperature. Once again, these are metastable states, because they correspond to temperatures of the liquid above the saturation temperature. They are not to be used within the scope of this text.

A useful first approximation is often made for the enthalpy of the subcooled (or compressed) liquid by assuming it to be essentially incompressible (i.e., density is independent of pressure). Then

$$h - h_f \approx \frac{(p - p_f)v_f}{J} \tag{3.5}$$

where $p_f$ and $v_f$ are the values of pressure and specific volume corresponding to the saturation condition at the temperature of the fluid. In other words, the change in enthalpy is approximately equal to $v\Delta p$, the work done on the fluid during a process carried out at nearly constant volume. In SI units,

$$h - h_f \approx (p - p_f)v_f \tag{3.5a}$$

---

### ILLUSTRATIVE PROBLEM 3.14

Determine the enthalpy specific volume, internal energy, and entropy of subcooled water at 300°F and 1000 psia.

#### SOLUTION

From Table A.4 (Figure 3.11), we find that the values are directly tabulated. Therefore,

$$v = 0.017379 \text{ ft.}^3/\text{lb}_m$$

$$u = 268.24 \text{ Btu/lb}_m$$

$$h = 271.46 \text{ Btu/lb}_m$$

$$s = 0.43552 \text{ Btu/lb}_m \cdot \text{R}$$

(a)

| | Liquid | | | | | | | | | | | |
|---|---|---|---|---|---|---|---|---|---|---|---|---|
| p (t Sat.) | 0 | | | | 500 (467.13) | | | | 1000 (544.75) | | | |
| t | v | u | h | s | v | u | h | s | v | u | h | s |
| Sat. | | | | | 0.019748 | 447.70 | 449.53 | 0.64904 | 0.021591 | 538.39 | 542.38 | 0.74320 |
| 32 | 0.016022 | -0.01 | -0.01 | 0.00003 | 0.015994 | 0.00 | 1.49 | 0.00000 | 0.015967 | 0.03 | 2.99 | 0.00005 |
| 50 | 0.016024 | 18.06 | 18.06 | 0.0360 | 0.015998 | 18.02 | 19.50 | 0.03599 | 0.015972 | 17.99 | 20.94 | 0.03592 |
| 100 | 0.016130 | 68.05 | 68.05 | 0.12963 | 0.016106 | 67.87 | 69.36 | 0.12932 | 0.016082 | 67.70 | 70.68 | 0.12901 |
| 150 | 0.016343 | 117.95 | 117.95 | 0.21504 | 0.016318 | 117.66 | 119.17 | 0.21457 | 0.016293 | 117.38 | 120.40 | 0.21410 |
| 200 | 0.016635 | 168.05 | 168.05 | 0.29402 | 0.016608 | 167.65 | 169.19 | 0.29341 | 0.016580 | 167.26 | 170.32 | 0.29281 |
| 250 | 0.017003 | 218.52 | 218.52 | 0.36777 | 0.016972 | 217.99 | 219.56 | 0.36702 | 0.016941 | 217.47 | 220.61 | 0.36628 |
| 300 | 0.017453 | 269.61 | 269.61 | 0.43732 | 0.017416 | 268.92 | 270.53 | 0.43641 | 0.017379 | 268.24 | 271.46 | 0.43552 |
| 350 | 0.018000 | 321.59 | 321.59 | 0.50359 | 0.017954 | 320.71 | 322.37 | 0.50249 | 0.017909 | 319.83 | 323.15 | 0.50140 |
| 400 | 0.018668 | 374.85 | 374.85 | 0.56740 | 0.018608 | 373.68 | 375.40 | 0.56604 | 0.018550 | 372.55 | 375.98 | 0.56472 |
| 450 | 0.019503 | 429.96 | 429.96 | 0.62970 | 0.019420 | 428.40 | 430.19 | 0.62798 | 0.019340 | 426.89 | 430.47 | 0.62632 |
| 500 | 0.02060 | 488.1 | 488.1 | 0.6919 | 0.02048 | 485.9 | 487.8 | 0.6896 | 0.02036 | 483.8 | 487.5 | 0.6874 |
| 510 | 0.02087 | 500.3 | 500.3 | 0.7046 | 0.02073 | 497.9 | 499.8 | 0.7021 | 0.02060 | 495.6 | 499.4 | 0.6997 |
| 520 | 0.02116 | 512.7 | 512.7 | 0.7173 | 0.02100 | 510.1 | 512.0 | 0.7146 | 0.02086 | 507.6 | 511.5 | 0.7121 |
| 530 | 0.02148 | 525.5 | 525.5 | 0.7303 | 0.02130 | 522.6 | 524.5 | 0.7273 | 0.02114 | 519.9 | 523.8 | 0.7245 |
| 540 | 0.02182 | 538.6 | 538.6 | 0.7434 | 0.02162 | 535.3 | 357.3 | 0.7402 | 0.02144 | 532.4 | 536.3 | 0.7372 |
| 550 | 0.02221 | 552.1 | 552.1 | 0.7569 | 0.02198 | 548.4 | 550.5 | 0.7532 | 0.02177 | 545.1 | 549.2 | 0.7499 |
| 560 | 0.02265 | 566.1 | 566.1 | 0.7707 | 0.02237 | 562.0 | 564.0 | 0.7666 | 0.02213 | 558.3 | 562.4 | 0.7630 |
| 570 | 0.02315 | 580.8 | 580.8 | 0.7851 | 0.02281 | 576.0 | 578.1 | 0.7804 | 0.02253 | 571.8 | 576.0 | 0.7763 |
| 580 | | | | | 0.02332 | 590.8 | 592.9 | 0.7946 | 0.02298 | 585.9 | 590.1 | 0.7899 |
| 590 | | | | | 0.02392 | 606.4 | 608.6 | 0.8096 | 0.02349 | 600.6 | 604.9 | 0.8041 |
| 600 | | | | | | | | | 0.02409 | 616.2 | 620.6 | 0.8189 |
| 610 | | | | | | | | | 0.02482 | 632.9 | 637.5 | 0.8345 |

(b)

| | Liquid | | | | | | | | | | | |
|---|---|---|---|---|---|---|---|---|---|---|---|---|
| p (t Sat.) MPa | 0 | | | | 2.5 (223.99) | | | | 5.0 (263.99) | | | |
| t | $10^3 v$ | u | h | s | $10^3 v$ | u | h | s | $10^3 v$ | u | h | s |
| Sat. | | | | | 1.1973 | 959.1 | 962.1 | 2.5546 | 1.2859 | 1147.8 | 1154.2 | 2.9202 |
| 0 | 1.0002 | -0.03 | -0.03 | -0.0001 | 0.9990 | -0.00 | 2.50 | -0.0000 | 0.9977 | 0.04 | 5.04 | 0.0001 |
| 20 | 1.0018 | 83.95 | 83.95 | 0.2966 | 1.0006 | 83.80 | 86.30 | 0.2961 | 0.9995 | 83.65 | 88.65 | 0.2956 |
| 40 | 1.0078 | 167.56 | 167.56 | 0.5725 | 1.0067 | 167.25 | 169.77 | 0.5715 | 1.0056 | 166.95 | 171.97 | 0.5705 |
| 60 | 1.0172 | 251.12 | 251.12 | 0.8312 | 1.0160 | 250.67 | 253.21 | 0.8298 | 1.0149 | 250.23 | 255.30 | 0.8285 |
| 80 | 1.1291 | 334.87 | 334.87 | 1.0753 | 1.0280 | 334.29 | 336.86 | 1.0737 | 1.0268 | 333.72 | 338.85 | 1.0720 |
| 100 | 1.0436 | 418.96 | 418.96 | 1.3069 | 1.0423 | 418.24 | 420.85 | 1.3050 | 1.0410 | 417.52 | 422.72 | 1.3030 |
| 120 | 1.0604 | 503.57 | 503.57 | 1.5278 | 1.0590 | 502.68 | 505.33 | 1.5255 | 1.0576 | 501.80 | 507.09 | 1.5233 |
| 140 | 1.0800 | 588.89 | 588.89 | 1.7395 | 1.0784 | 587.82 | 590.52 | 1.7369 | 1.0768 | 586.76 | 592.15 | 1.7343 |
| 160 | 1.1024 | 675.19 | 675.19 | 1.9434 | 1.1006 | 673.90 | 676.65 | 1.9404 | 1.0988 | 672.62 | 678.12 | 1.9375 |
| 180 | 1.1283 | 762.72 | 762.72 | 2.1410 | 1.1261 | 761.16 | 763.97 | 2.1375 | 1.1240 | 759.63 | 765.25 | 2.1341 |
| 200 | 1.1581 | 851.8 | 851.8 | 2.3334 | 1.1555 | 849.9 | 852.8 | 2.3294 | 1.1530 | 848.1 | 853.9 | 2.3255 |
| 210 | 1.1749 | 897.1 | 897.1 | 2.4281 | 1.1720 | 895.0 | 898.0 | 2.4238 | 1.1691 | 893.0 | 898.8 | 2.4195 |
| 220 | 1.1930 | 943.0 | 943.0 | 2.5221 | 1.1898 | 940.7 | 943.7 | 2.5174 | 1.1866 | 938.4 | 944.4 | 2.5128 |
| 230 | 1.2129 | 989.6 | 989.6 | 2.6157 | 1.2092 | 987.0 | 990.1 | 2.6105 | 1.2056 | 984.5 | 990.6 | 2.6055 |
| 240 | 1.2347 | 1037.1 | 1037.1 | 2.7091 | 1.2305 | 1034.2 | 1037.2 | 2.7034 | 1.2264 | 1031.4 | 1037.5 | 2.6979 |
| 250 | 1.2590 | 1085.6 | 1085.6 | 2.8027 | 1.2540 | 1082.3 | 1085.4 | 2.7964 | 1.2493 | 1079.1 | 1085.3 | 2.7902 |
| 260 | 1.2862 | 1135.4 | 1135.4 | 2.8970 | 1.2804 | 1131.6 | 1134.8 | 2.8898 | 1.2749 | 1127.9 | 1134.3 | 2.8830 |
| 270 | 1.3173 | 1186.8 | 1186.8 | 2.9926 | 1.3102 | 1182.4 | 1185.7 | 2.9844 | 1.3036 | 1178.2 | 1184.3 | 2.9766 |
| 280 | 1.3535 | 1240.4 | 1240.4 | 3.0904 | 1.3447 | 1235.1 | 1238.5 | 3.0808 | 1.3365 | 1230.2 | 1236.8 | 3.0717 |
| 290 | 1.3971 | 1297.0 | 1297.0 | 3.1918 | 1.3855 | 1290.5 | 1294.0 | 3.1801 | 1.3750 | 1284.4 | 1291.3 | 3.1693 |
| 300 | 1.4520 | 1358.1 | 1358.1 | 3.2992 | 1.4357 | 1349.6 | 1353.2 | 3.2843 | 1.4214 | 1341.9 | 1349.0 | 3.2708 |
| 310 | | | | | | | | | 1.4803 | 1404.1 | 1411.5 | 3.3789 |

**FIGURE 3.11**

(a) Detailed extract from the subcooled table. Greater range of values can be seen in Table A.4 in Appendix 2. (J. H. Keenan, F. G. Keyes, P. G. Hill, and J. G. Moore: *Steam Tables.* 1969. Copyright Wiley-VCH Verlag GmbH & Co. KGaA. Reproduced with permission.). (b) Extract from the subcooled table (SI units). Greater range of values can be seen in Table A.4 (SI Units) in Appendix 2. (J. H. Keenan, F. G. Keyes, P. G. Hill, and J. G. Moore: *Steam Tables. Sf Units Edition.* 1978. Copyright Wiley-VCH Verlag GmbH & Co. KGaA. Reproduced with permission.).

### ILLUSTRATIVE PROBLEM 3.15

Determine the enthalpy of the subcooled water in Illustrative Problem 3.14 using the approximation of Equation 3.5, and compare the result obtained with the tabulated value.

### SOLUTION

It is necessary to obtain the saturation values corresponding to 300°F. This is done by reading Table A.1 in Appendix 2, which gives us $p_f = 66.98$ psia, $v_f = 0.017448$ ft.$^3$/lb$_m$, and $h_f = 269.73$ Btu/lb$_m$. From Equation 3.5,

$$h = h_f + \frac{(p - p_f)v_f}{J}$$

$$= 269.73 \text{ Btu/lb}_m + \frac{(1000 - 66.98)\text{lb}_f/\text{in.}^2 (0.017448)\text{ft.}^3/\text{lb}_m \times 144 \text{ in.}^2/\text{ft.}^2}{778 \text{ ft.}\cdot\text{lb}_f/\text{Btu}}$$

$$= 272.74 \text{ Btu/lb}_m$$

The difference between this value and the value found in Illustrative Problem 3.15 expressed as a percentage is

$$\text{percent of error} = \frac{272.74 - 271.46}{271.46} \times 100 = 0.47\%$$

For this pressure and temperature, the approximation expressed by Equation 3.5 is obviously within the accuracy required in most engineering calculations.

## 3.4 Computerized Properties

It is obvious from Section 3.3 that a great deal of effort is involved in the interpolation in the *Steam Tables* and that the potential for errors is ever-present. Values for the thermodynamic properties of steam are available in the form of computer programs that can provide the desired properties. Any combination of two independent variables that define a state of steam may be entered, and the program will provide all of the other properties of that state. The user may select from temperature, pressure, specific volume, specific internal energy, specific enthalpy, specific entropy, and, if in saturation, quality. Note that temperature and pressure are not independent in the saturation range. Click on AUTO and follow the instructions. Select the desired fluid, steam, and press "Enter." Use the cursor arrow up/down keys to select the system of units. Use "Page Down" to begin entering values. To quit the program, press "End." Note that this program provides properties of fluids other than steam.

Because the properties are computed from equations, they do not contain any interpolation errors.

---

### ILLUSTRATIVE PROBLEM 3.16

Use the computer program to solve Illustrative Problem 3.1.

### SOLUTION

The problem is to determine the enthalpy of saturated steam at 90°F. We select USCS units and enter 90°F and the quality, $x$, of 1.0. The program gives us the properties of the saturated liquid, the saturated vapor, and $T, p, v, h, s, u$, and $x$ for the desired input. The agreement with Table A.1 is very good.

| Thermodynamic Properties: | |
|---|---|
| $T = 90.00$ | deg F |
| $P = 0.6987$ | psia |
| $v = 467.65$ | ft.$^3$/lb$_m$ |
| $h = 1039.9$ | BTU/lb$_m$ |
| $s = 1100.3$ | BTU/lb$_m$°F |
| $u = 2.0078$ | BTU/lb$_m$ |
| $x = 1.0000$ | |
| Region: saturated | |

---

### ILLUSTRATIVE PROBLEM 3.17

Use the computer program to solve Illustrative Problem 3.3.

### SOLUTION

This problem is similar to Illustrative Problem 3.16. The entries to the program are pressure (118 psia) and quality ($x = 1.000$). The results from the computer program are in excellent agreement with the interpolated values.

| Thermodynamic Properties: | |
|---|---|
| $T = 340.06$ | deg F |
| $P = 118.00$ | psia |
| $v = 3.789$ | ft.$^3$/lb$_m$ |
| $h = 1108.0$ | BTU/lb$_m$ |
| $s = 1190.7$ | BTU/lb$_m$F |
| $u = 1.5899$ | BTU/lb$_m$ |
| $x = 1.0000$ | |
| Region: saturated | |

### ILLUSTRATIVE PROBLEM 3.18

Solve Illustrative Problem 3.6 using the computer program.

### SOLUTION

The input to the program is USCS units; $p = 120$ psia, and $x = 0.8$. The results are in essentially exact agreement with the results of Illustrative Problem 3.6.

---

**Thermodynamic Properties:**

| | |
|---|---|
| $T = 341.32$ | deg F |
| $P = 120.00$ | psia |
| $v = 2.987$ | ft.$^3$/lb$_m$ |
| $h = 949.0$ | BTU/lb$_m$ |
| $s = 1015.3$ | BTU/lb$_m$F |
| $u = 1.3693$ | BTU/lb$_m$ |
| $x = 0.8000$ | |
| Region: saturated | |

---

### ILLUSTRATIVE PROBLEM 3.19

Use the computer program for steam to solve Illustrative Problem 3.7.

### SOLUTION

The input to this problem is 1000 kPa (1 MPa) and $x = 0.85$. The printout agrees with the calculated values in Illustrative Problem 3.7.

---

**Thermodynamic Properties:**

| | |
|---|---|
| $T = 179.92$ | deg C |
| $P = 1000.0$ | kPa |
| $v = 0.16539$ | m$^3$/kg |
| $h = 2310.1$ | kJ/kg |
| $s = 2.475.5$ | kJ/kgK |
| $u = 5.9189$ | kJ/kg |
| $x = 0.8500$ | |
| Region: saturated | |

---

## ILLUSTRATIVE PROBLEM 3.20

Solve Illustrative Problem 3.8 using the computerized properties of steam.

### SOLUTION

This problem has USCS inputs of 90°F and $h = 900$ Btu/lb$_m$. The solution yields the quality and all of the other properties at this state.

**Thermodynamic Properties:**

| | |
|---|---|
| $T = 90.000$ | deg F |
| $P = 0.69878$ | psia |
| $V = 377.77$ | ft.$^3$/lb$_m$ |
| $h = 851.15$ | BTU/lb$_m$ |
| $s = 900.00$ | BTU/lb$_m$F |
| $u = 1.6433$ | BTU/lb$_m$ |
| $x = 0.8078$ | |
| Region: saturated | |

## ILLUSTRATIVE PROBLEM 3.21

Use the computer program to solve Illustrative Problem 3.11.

### SOLUTION

The input for this problem is SI units of 240°C and 2000 kPa (2.0 MPa). The output of the program yields the desired properties.

**Thermodynamic Properties:**

| | |
|---|---|
| $T = 240.00$ | deg C |
| $P = 2.0000$ | kPa |
| $v = 0.10841$ | m$^3$/kg |
| $h = 2658.8$ | kJ/kg |
| $s = 2875.6$ | kJ/kgK |
| $u = 6.4937$ | kJ/kg |
| Region: superheated | |

## ILLUSTRATIVE PROBLEM 3.22

Solve Illustrative Problem 3.13 using the computerized properties.

### SOLUTION

The inputs in USCS units are 465°F and 337 psia. The computer printout yields $v$, $h$, $s$, and $u$ in an instant, while the calculated value in Illustrative Problem 3.13 requires six interpolations. Not only is the work less, but the potential for error in the interpolation is eliminated.

---

**Thermodynamic Properties:**

| | |
|---|---|
| $T = 465.00$ | deg F |
| $P = 337.00$ | psia |
| $v = 1.4713$ | ft.$^3$/lb$_m$ |
| $h = 1138.4$ | BTU/lb$_m$ |
| $s = 1230.2$ | BTU/lb$_m$F |
| $u = 1.5293$ | BTU/lb$_m$ |

Region: superheated

---

### ILLUSTRATIVE PROBLEM 3.23

Use the computer program to solve Illustrative Problem 3.14.

### SOLUTION

With USCS inputs of 300°F and 1000 psia, the program yields the following results.

---

**Thermodynamic Properties:**

| | |
|---|---|
| $T = 300.00$ | deg F |
| $P = 1000.0$ | psia |
| $v = 0.017379$ | ft.$^3$/lb$_m$ |
| $h = 268.30$ | BTU/lb$_m$ |
| $s = 271.51$ | BTU/lb$_m$F |
| $u = 0.4356$ | BTU/lb$_m$ |

Region: subcooled

---

The computer program for steam gives us the accuracy that is equivalent to that of the tables in Appendix 2 and should be used to avoid the tedious calculations entailed when using the tables. However, it is important to obtain an understanding and mastery of basics first and then to use the computerized tables as an invaluable tool.

## 3.5 Thermodynamic Diagrams

Tables of thermodynamic properties provide accurate data for various substances. However, diagrams and charts based on the data of these tables are both useful and desirable. One most important fact must be borne in mind. Thermodynamic data such as those given by the *Steam Tables* are equilibrium data, and charts plotted from these data can represent only the equilibrium states. The path of a process that is not an equilibrium path cannot be drawn on these charts.

Figure 3.12 is a *T–s* diagram (*s* is entropy) for steam showing the liquid and vapor phases. In the wet region (below the saturation curve), lines of constant temperature and lines of constant pressure are horizontal lines. For convenience, lines of constant moisture (constant *x*) and lines of constant volume are also shown in the wet region. In the superheat region above the saturation curve, lines of constant pressure start at the saturation curve, rise steeply, and are almost vertical. Lines of constant enthalpy are nearly horizontal away from the saturation curve. Near the saturation curve, and especially near the critical point, there is a marked change in the curvature of the constant-enthalpy lines, which approach the vertical as the critical pressure is approached. On the large *T–s* diagram that is found in the *Steam Tables*, curves of constant superheat are also shown in the superheat region. A

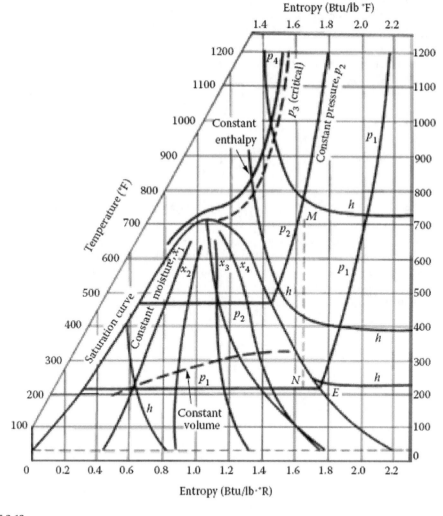

**FIGURE 3.12**
Outline of a temperature–entropy diagram for steam. (From B. F. Dodge, *Chemical Engineering Thermodynamics*, New York: McGraw-Hill Book Co., 1944.).

curve of constant superheat corresponds to a curve plotted parallel to the saturation curve and a fixed number of degrees above the saturation curve. Thus, the term 200 *degrees of superheat* means superheated steam whose temperature corresponds to saturation temperature plus 200°F. While the *T–s* chart is useful in portraying processes, it is not as useful nor as widely used as the *h–s* diagram (or the *Mollier chart* as the *h–s* diagram is usually called).

The Mollier chart is a plot on *h–s* coordinates of the thermodynamic properties of a substance. Figure 3.13 shows the outline of a portion of the Mollier chart for steam. Below the saturation line, lines of constant pressure (which are also lines of constant temperature) and lines of constant moisture are shown. Above the saturation line, curves of constant temperature, constant pressure, and constant superheat are shown. As noted for the *T–s* diagram, the curves of constant superheat correspond to curves plotted parallel to the saturation curve and a fixed number of degrees above the saturation curve. As we shall see, the Mollier chart is particularly suited to obtaining properties, describing flow, or describing constant-pressure processes (see Chapter 4). If a process is not reversible (see Chapter 5), then only its end states can be shown on a thermodynamic diagram such as the *T–s* diagram or the *Mollier chart*. A Mollier chart similar to the one in the *Steam Tables* is found in Appendix 2. Note that the specific volume property is not included in the Mollier chart.

In addition to the *T–s* and *h–s* charts, other diagrams of the thermodynamic properties also have utility. Figure 3.14 shows the specific heat of steam at constant pressure as a function of temperature and pressure. It can be proved by theoretical reasoning that the specific heats at the critical point are infinite, and it will be seen from Figure 3.14 that in the regions of the critical temperature, there is a sharp rise in specific-heat values.

The wide fluctuations in the region of the critical point are more evident from Figure 3.15, which also shows the specific heat at constant pressure for a saturated liquid. At the critical point, the saturated vapor and saturated liquid are indistinguishable, and both curves merge to infinity.

The effect of pressure on the specific heat at constant pressure for a liquid (water) is shown in Figure 3.16. It should be noted that the specific heat at constant pressure and at a given temperature decreases with increasing pressure. At a given pressure, it increases with temperature. It should be noted that Figure 3.12 through exist in the (*SI Units*) *Edition* of the *Steam Tables*.

---

### ILLUSTRATIVE PROBLEM 3.24

Determine the enthalpy of saturated steam at 90°F using the Mollier chart. Compare the result with Illustrative Problem 3.1.

### SOLUTION

On the chart in Appendix 2, it is necessary to estimate the 90°F point on the saturation line. From the chart or the table in the upper left of the chart, we note that 90°F is between 1.4 and 1.5 in. of mercury. Estimating the intersection of this value with the saturation curve yields $h_g = 1100$ Btu/lb$_m$. This is in good agreement with Illustrative Problem 3.1.

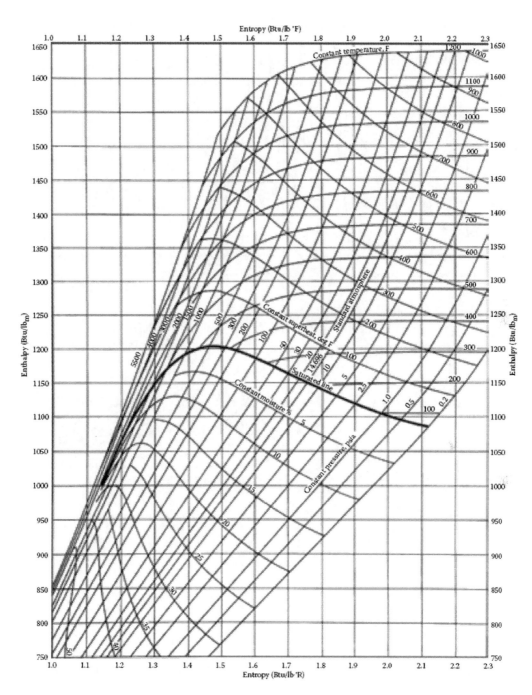

**FIGURE 3.13**
Outline of *h–s* (Mollier) diagram for steam. (Courtesy of Babcock and Wilcox Corp.).

**FIGURE 3.14**
Specific heat of steam at constant pressure. (J. H. Keenan, F. G. Keyes, P. G. Hill, and J. G. Moore: *Steam Tables*. 1969. Copyright Wiley-VCH Verlag GmbH & Co. KGaA. Reproduced with permission.).

---

### ILLUSTRATIVE PROBLEM 3.25

Determine the enthalpy of a wet steam mixture at 120 psia with a quality of 80% by using the Mollier chart. Compare the results with Illustrative Problem 3.6.

### SOLUTION

The Mollier chart has lines of constant moisture in the wet region that correspond to $(1 - x)$. Therefore, we read at 20% moisture (80% quality) and 120 psia an enthalpy of 1015 Btu/$\text{lb}_\text{m}$, which also agrees well with the calculated value in Illustrative Problem 3.6.

**FIGURE 3.15**
Specific heat at constant pressure near the critical point. (J. H. Keenan, F. G. Keyes, P. G. Hill, and J. G. Moore: *Steam Tables.* 1969. Copyright Wiley-VCH Verlag GmbH & Co. KGaA. Reproduced with permission.).

---

### ILLUSTRATIVE PROBLEM 3.26

Using the Mollier chart, determine the quality of a wet steam mixture having an enthalpy of 900 Btu/$lb_m$ and a temperature of 90°F. Compare the result with Illustrative Problem 3.8.

### SOLUTION

Entering the Mollier chart at 900 Btu/$lb_m$ and estimating 90°F (near the 1.5-in. Hg dashed line) yields a constant moisture percentage of 19.2%. The quality is therefore $(1 - 0.192)(100) = 80.8\%$. We again show good agreement with the calculated value.

**FIGURE 3.16**
Specific heat of water at constant pressure. (J. H. Keenan, F. G. Keyes, P. G. Hill, and J. G. Moore: *Steam Tables.* 1969. Copyright Wiley-VCH Verlag GmbH & Co. KGaA. Reproduced with permission.).

---

### ILLUSTRATIVE PROBLEM 3.27

Determine the enthalpy of steam at 330 psia and 450°F using the Mollier chart. Compare the results with Illustrative Problem 3.10.

#### SOLUTION

From the chart, $h = 1220$ Btu/lb$_m$, compared to 1221.5 Btu/lb$_m$ found in Illustrative Problem 3.10.

---

### ILLUSTRATIVE PROBLEM 3.28

Determine the enthalpy and entropy of steam at 2.0 MPa and 240°C using the Mollier chart.

#### SOLUTION

We note that the steam is superheated. From the Mollier chart in SI units, $h = 2880$ kJ/kg and $s = 6.5$ kJ/kg K. These values compare with $h = 2876.5$ kJ/kg and $s = 6.4952$ kJ/kg K found in Illustrative Problem 3.11. For most purposes, this accuracy would be acceptable.

---

### ILLUSTRATIVE PROBLEM 3.29

Determine the enthalpy of steam at 465°F and 337 psia using the Mollier chart. Compare the results with Illustrative Problem 3.13.

#### SOLUTION

Because neither pressure nor temperature is shown directly, it is necessary to estimate to obtain the desired value. Reading the chart in this manner, we obtain $h = 1231$ Btu/$lb_m$. In Illustrative Problem 3.13, we obtained $h = 1230.7$ Btu/$lb_m$. The agreement is good, and the savings in laborious interpolations is considerable.

---

### ILLUSTRATIVE PROBLEM 3.30

Use the Mollier chart to solve Illustrative Problem 3.2.

#### SOLUTION

Reading the chart at 30°C and saturation gives us $h_g = 2556$ kJ/kg, which is in excellent agreement with the tables.

---

### ILLUSTRATIVE PROBLEM 3.31

Use the Mollier chart to solve Illustrative Problem 3.7.

#### SOLUTION

Reading the chart in the wet region at 1.0 MPa and $x = 0.85$ (moisture content of 15%) gives us

$$h_x = 2476 \text{ kJ/kg}$$

$$s_x = 5.92 \text{ kJ/kg} \cdot \text{K}$$

The chart does not give us $u_x$ or $v_x$ directly.

---

### ILLUSTRATIVE PROBLEM 3.32

Solve Illustrative Problem 3.9 using the Mollier chart.

#### SOLUTION

Locate 30°C on the saturation line. Now follow a line of constant pressure, which is also a line of constant temperature in the wet region, until an enthalpy of 2000 kJ/kg is reached. At this point, we find the moisture content to be 23%, or $x = 77\%$.

---

**ILLUSTRATIVE PROBLEM 3.33**

Solve Illustrative Problem 3.11 using the Mollier chart.

**SOLUTION**

We enter the chart in the superheat region at 2.0 MPa and 240°C to read the enthalpy and entropy. This procedure gives

$$h = 2877 \text{ kJ/kg}$$

$$s = 6.495 \text{ kJ/kg} \cdot \text{K}$$

The other properties cannot be obtained directly from the chart.

---

Use of the Mollier chart does not permit us to obtain the specific volume or internal energy directly. Special charts are available that plot enthalpy as ordinate and specific volume as abscissa. These charts are particularly useful in steam turbine computations (see Chapter 4).

In Chapter 4, liquid and vapor properties will be used in different processes by also taking into account the first law of thermodynamics.

## 3.6 Review

Although this chapter is entitled "Properties of Liquids and Gases," it has been devoted primarily to the properties of water because it is a primary working fluid for a range of thermodynamic processes. However, similar tables and diagrams exist for other common fluids, including refrigerants, and other important industrial fluids. Information for water exists in the *Steam Tables* in extensive tabulations and charts. A great deal of time was spent on exercises to become familiar with the nomenclature and the format of the tables and charts. Once familiar with this resource, we were able to apply the tables and charts to solve various problems in subsequent chapters. Unfortunately, there are great amounts of arithmetic computation associated with the application of the data in the *Steam Tables*. The equations for the properties of steam have been programmed and are available in the disk included with the text. In addition, it was shown that these computer programs will yield the desired properties with relative ease.

Whether the *Steam Tables* or a computer program is used, it is important to write the energy equation for the process and to sketch out the process on an appropriate *T–s* or *h–s* diagram as it is explained in the next chapter. This procedure will lead to a better understanding of the problem and eliminate a great deal of unnecessary numerical calculations.

## Key Terms

Terms used for the first time in this chapter are as follows:

**Critical point:** A limiting state of a fluid in which the density of saturated liquid equals the density of the saturated vapor. Also, the latent heat of vaporization is zero at this state. At the critical point, the saturated liquid and saturated vapor states are identical.

**Critical pressure:** The pressure at the critical point.

**Critical temperature:** The temperature at the critical point.

**Gas:** A vapor that is either at very low pressure or far removed from the saturated state.

**Mollier chart:** A chart having enthalpy–entropy coordinates.

**Percent moisture:** The percent by weight of moisture in a wet mixture; equal to where quality is expressed as a percentage.

**Phase:** That part of a pure substance that consists of a single-homogeneous aggregate of matter; for our purposes, these may be characterized as solid, liquid, and gas.

**Quality:** ($x$) fraction by weight of vapor in a mixture.

**Saturation:** The state of a liquid or vapor in which the vapor and liquid coexist in equilibrium in any proportion. Saturation temperature and saturation pressure refer, respectively, to the properties in the saturation state.

**Steam Tables:** J. H. Keenan, F. G. Keyes, P. G. Hill, and J. G. Moore, *Steam Tables— Thermodynamic Properties of Water Including Vapor, Liquid, and Solid Phases* (New York: John Wiley & Sons, Inc., 1969, 1978).

**Subcooled:** That liquid state in which the pressure exceeds the saturation pressure corresponding to the temperature of the liquid.

**Subscripts:** f for saturated liquid, g for saturated vapor, and fg for a property change due to vaporization.

**Superheated vapor:** A vapor whose temperature is greater than the saturation temperature corresponding to the pressure.

**USCS:** United States conventional system of units.

**Vapor:** The gaseous phase that is in contact with saturated liquid or is not far from the saturated state.

**Wet:** A mixture of vapor plus liquid.

## Equations Developed in This Chapter

| Saturation properties | $h_g = h_f + h_{fg}$ | (3.1) |
|---|---|---|
| | $v_g = v_f + v_{fg}$ | |
| | $s_g = s_f + s_{fg}$ | |
| | $u_g = u_f + u_{fg}$ | |

| Definition of enthalpy | $h = u + \dfrac{pv}{J}$ | (3.2) |
| Definition of enthalpy (SI) | $h = u + pv$ | (3.2a) |
| Wet mixture property | $h_x = h_f + xh_{fg}$ | (3.4a) |
| Wet mixture property | $s_x = s_f + xs_{fg}$ | (3.4b) |
| Wet mixture property | $u_x = u_f + xv_{fg}$ | (3.4c) |
| Wet mixture property | $v_x = v_f + xv_{fg}$ | (3.4d) |
| Approximation for the enthalpy of subcooled liquid | $h - h_f \simeq \dfrac{(p - p_f)v_f}{J}$ | (3.5) |

## QUESTIONS

3.1 Define the word *phase*, and list the three common phases of matter.

3.2 State the difference between a gas and a vapor.

3.3 As the saturation pressure is increased, does the saturation temperature (a) increase, (b) decrease, or (c) remain the same?

3.4 What is meant by the *critical point?*

3.5 Define the subscripts f, g, and fg. To what states do these subscripts refer?

3.6 Define the term *wet*. How is this term related to percent moisture?

3.7 How many properties are required to determine (a) the saturation state, (b) the superheat state, and (c) the wet state?

3.8 What are the principal coordinates of the Mollier diagram?

3.9 How is an isentropic process shown on a Mollier diagram?

3.10 How do you show a constant-temperature line in the wet region on the Mollier diagram?

3.11 Can the path of an irreversible process be shown on a *T–s* or a Mollier diagram?

3.12 What is known about the path of a throttling process on the Mollier diagram?

3.13 What is the ideal way that a compressor or an expander should be operated?

3.14 Using a sketch of the Mollier diagram show constant pressure, constant temperature, isentropic expansion, irreversible expansion, and throttling processes.

## PROBLEMS

Wherever appropriate, sketch processes on a *T–s* or *h–s* diagram. Also, use the tables and Mollier charts in Appendix 2 for all problems.

### Problems Involving the Use of the Tables and Charts

3.1 Determine the enthalpy, entropy, specific volume, and internal energy of saturated steam at 100 and 1000 psia.

3.2 Determine the enthalpy, entropy, specific volume, and internal energy of saturated steam at 1.0 and 1.1 MPa.

3.3 Determine the enthalpy, entropy, specific volume, and internal energy of saturated water at 1.0 and 1.1 MPa.

3.4    Determine the pressure, specific volume, and enthalpy of saturated water at 350°F and 500°F.

3.5    Determine the internal energy of steam at 10 MPa and 500°C.

3.6    Determine $h_{fg}$ at 400°F.

3.7    Determine $s_{fg}$ at 150°F.

3.8    Determine $u_{fg}$ at 212°F, knowing $h_{fg}$ and $v_{fg}$. Check your answer against the tabulated value.

3.9    Determine the pressure, specific volume, and entropy of saturated water at 35°C.

3.10   Determine the enthalpy, entropy, specific volume, and internal energy of wet steam at 100 psia if it has 5% moisture.

3.11   Determine the enthalpy, specific volume, entropy, and internal energy of wet steam at 1.12 MPa if the quality is 90%.

3.12   Steam at 200 psia has an enthalpy of 1050 Btu/lb$_m$. Determine its quality.

3.13   If the enthalpy of steam at 1.0 MPa is 1400 kJ/kg, what is its quality?

3.14   Steam at 150 psia has a quality of 97%. What is the specific volume of this wet mixture?

3.15   The internal energy of wet steam is 2000 kJ/kg. If the pressure is 1.18 MPa, what is the quality of the mixture?

3.16   Determine the internal energy for a steam–water mixture for $t = 400°F$ and $x = 0.6$ in Btu/lb$_m$.

3.17   Determine the specific volume of wet steam at 200°C if the quality is 60%.

3.18   Calculate the specific volume of wet steam at 250°F if the mixture has a quality of 80%.

3.19   A steam–water mixture has a specific volume of 0.19 m³/kg. If the pressure of the mixture is 44 kPa, determine the quality and temperature of the mixture.

3.20   The temperature of a steam–water mixture is 270°F. If the specific volume of the mixture is 4.0 ft.³/lb$_m$, determine the quality and pressure of the mixture.

3.21   A tank contains 2 kg of water and 0.2 kg of vapor at 220°C. Determine the quality, the volume of the tank, the volume of the water, and the pressure in the tank.

3.22   Saturated steam has an enthalpy of 2782.0 kJ/kg. What is its pressure?

3.23   If the enthalpy of saturated water is 100 Btu/lb$_m$, determine its pressure, temperature, and specific volume.

3.24   Steam at 500 psia has a specific volume of 0.800 ft.³/lb$_m$. Determine its enthalpy and temperature.

3.25   Steam at 2.0 MPa has a specific volume of 0.08 m³/kg. Determine its temperature and enthalpy.

3.26   Steam at 300 psia is at 600°F. Determine its specific volume.

3.27   Steam at 600 psia has a specific volume of 1.500 ft.³/lb$_m$. Determine its enthalpy.

3.28   Steam at 900°F has an enthalpy of 1465 Btu/lb$_m$. Determine its pressure.

3.29   Steam at 1000 psia has an enthalpy of 1250 Btu/lb$_m$. Determine its specific volume.

3.30   Steam at 1.00 MPa is at 400°C. Determine its specific volume.

3.31    Steam is at 2.0 MPa and 225°C. Determine its enthalpy.

3.32    Steam at 200 psia is at 610°F. Determine its enthalpy.

3.33    Steam is at 2.5 MPa and 227°C. Determine its enthalpy.

3.34    Determine the internal energy of steam at 190 psia and 910°F.

3.35    Calculate the internal energy of a mixture of steam and water for $t = 200°C$ and $x = 0.8$, in kJ/kg.

3.36    Determine the specific volume, enthalpy, and internal energy of steam having a pressure of 5 psia and a quality of 90%.

3.37    What is the internal energy of steam at 2.17 MPa and 225°C?

3.38    Determine the enthalpy, specific volume, internal energy, and entropy of sub-cooled water at 500°F and 4000 psia.

3.39    Determine the enthalpy of subcooled water at 500°F and 4000 psia using Equation 3.5. Compare your result with Problem 3.38.

3.40    Steam at 245°C has a specific volume of 0.10 m$^3$/kg. Determine its pressure.

*3.41   Steam at 500 psia has a specific volume of 1.7500 ft.$^3$/lb$_m$. Determine its temperature and enthalpy.

## Problems Involving the Use of the Computerized Tables

3.42–3.82   Solve assigned problems 3.1–3.41 using the computerized tables on the CRC Press website.

# 4

## First Law of Thermodynamics

---

### Learning Goals

After reading and studying the material in this chapter, you should be able to

1. State the first law of thermodynamics, or energy conservation, for both nonflow and flow systems.
2. Define enthalpy and show that it is a property and therefore does not depend on whether a system is a nonflow or flow system.
3. Apply the first law of thermodynamics to several nonflow systems, such as the adiabatic system, the constant-volume system, and the constant-pressure system.
4. Derive and apply the continuity equation in its several forms to express the conservation of mass in a steady-flow system.
5. Derive the energy equation from the first law of thermodynamics for a steady-flow system
6. Think of the area under the pressure–volume diagram as the work of compression or expansion and the area behind the pressure–volume diagram as the work of the compressor or expander.
7. Understand the shortcomings of the Bernoulli equation.
8. Define the term *specific heat* and show that the specific heats at constant volume and constant pressure are properties.
9. Apply the first law of thermodynamics to the analysis of the steam or gas turbine, pipe flow, boilers, nozzles, throttling, heat exchangers, and the filling of a tank.

---

## 4.1 Introduction

We have discussed properties, state functions, systems, work, energy, and heat in the previous chapters. In this chapter, we will consider the conservation of energy as the basis for the first law of thermodynamics. Although Newton first proposed the equivalence of heat and work, it is only in relatively recent times, as a result of both observation and analysis, that the principle of conservation has become one of the cornerstones of thermodynamics and in science in general. Most of this chapter will be concerned with the application of the first law of thermodynamics to both nonflow (closed) and flow (open) systems and to develop energy equations applicable to these systems. The first law of thermodynamics will enable us to determine the energy quantities passing to or from a system as well as work and/or heat as forms of energy interactions. The first law can also be used to determine the changes in the energy stored in the system.

## 4.2  First Law of Thermodynamics

The first law of thermodynamics can be expressed in the following equivalent statements:

1.  The first law of thermodynamics is essentially the statement of the principle of the conservation of energy for thermodynamical systems. As such, it may be expressed by stating that the variation of energy of a system during any transformation is equal to the amount of energy that the system receives from its environment (Fermi, 1956).

2.  Energy can be neither created nor destroyed but only converted from one form to another (Obert, 1960).

3.  If a system is caused to change from an initial state to a final state by adiabatic means (no heat transfer) only, the work done is the same for all adiabatic paths connecting the two states (Zemansky, 1957).

In statement 3, by Zemansky, the term adiabatic is used. In general, we define an *adiabatic* transformation of a system as a process in which no heat transfer takes place during the process.

For the purpose of this book, the concept of the conservation of energy, explicitly stated in the Fermi and Obert definitions, is essentially the first law of thermodynamics. By combining them, we have for the statement of the first law of thermodynamics that *energy can be neither created nor destroyed but only converted from one form to another*. Because we shall not concern ourselves with nuclear reactions at this time, it will not be necessary to invoke the interconvertibility of energy and mass in our present study.

A system has already been defined as a grouping of matter taken in any convenient or arbitrary manner. However, when dealing with fluids in motion, it is more convenient to utilize the concept of an arbitrary volume in space, known as a *control volume*, that can be bounded by either a real or imaginary surface, known as a *control surface*. By correctly noting all the forces acting on the fluid within the control volume, all the energy forms crossing the control surface, and the mass crossing the control surface, it is possible to derive mathematical expressions capable of predicting the flow of the fluid relative to the control volume. For a system in which fluids are flowing steadily, a monitoring device placed anywhere within the control volume will indicate no change in the fluid properties or energy quantities crossing the control surface with time, even though these quantities can and will vary from position to position within the control volume. As noted by Keenan (1941), *"the first step in the solution of a problem in thermodynamics is the description of a system and its boundaries."*

In general terms, the first law of thermodynamics can be expressed as follows:

$$E_{in} - E_{out} = \Delta E_{system} \tag{4.1}$$

Where $E_{in}$, $E_{out}$, and $\Delta E_{system}$ are the total energy entering the system, total energy leaving the system, and the net change or difference in energy during the energy transfer process, respectively. As mentioned in Chapter 1, energy has different forms, and energy in transition (when crossing a boundary) can be classified either as work ($W$) or heat ($Q$) depending on the nature of the corresponding phenomena. Equation 4.1 on a per mass basis, is:

$$e_{in} - e_{out} = \Delta e_{system} \tag{4.2}$$

where $e_{in}$ and $e_{out}$ can take the form of $W$ or $Q$ depending on the nature of the energy interaction.

## 4.3 Nonflow System (Closed System)

The statements made in Section 4.2 about the first law of thermodynamics are, in essence, equivalent to each other in that they express the concept of energy conservation. To explore the implications and applications of the first law, let us examine the *nonflow* or *closed system* first. This system will have boundaries across which both heat and work can penetrate, but no mass will be permitted to cross them, as previously mentioned in Chapter 1. Note that the boundaries can move with the system as the process proceeds along even when no mass flow crosses the boundaries. An example is a piston-and-cylinder arrangement (piston–cylinder device) in which the piston compresses the working fluid in the cylinder and heat may cross the boundary (e.g., by the cooling of the cylinder) at the same time. This system is shown schematically in Figure 4.1. For convenience, it is assumed that there is a unit mass in the cylinder. Assuming also that there are no (or negligible) changes in elevation during the process ($\Delta Z=0$, therefore, $\Delta P.E.=0$) and no directed (*net*) velocity of the working fluid ($V_{net}=0$, therefore, $\Delta K.E.=0$), we may properly neglect energy terms relating to potential and kinetic energy. Moreover, if the only forms of energy interactions are $W$ and $Q$, Equation 4.1 can be written as follows:

$$Q_{net} - W_{net} = \Delta E_{system} \tag{4.3}$$

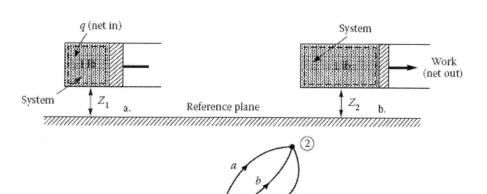

**FIGURE 4.1**

Nonflow system. (a) Initial piston–cylinder system. (b) Final piston–cylinder system. (c) Work as a path function.

Since $\Delta K.E.$ and $\Delta P.E$ .are both zero, then $\Delta E_{\text{system}}$ ($\Delta E_{\text{system}} = \Delta P.E. + \Delta K.E. + \Delta U$) is just equal to $\Delta U$ ($\Delta E_{\text{system}} = \Delta U_{\text{system}}$) or the net difference in internal energy, then the first law for stationary nonflow or closed system becomes:

$$Q_{\text{net}} - W_{\text{net}} = \Delta U_{\text{system}} \tag{4.4}$$

On a per mass basis, Equation 4.4 can be written as follows:

$$q_{\text{net}} - w_{\text{net}} = \Delta u_{\text{system}} \tag{4.5}$$

Where net heat, $q_{\text{net}}$ is equal to ($q_{\text{in}} - q_{\text{out}}$) and net work, $w_{net}$ is equal to ($w_{\text{out}} - w_{\text{in}}$), and if $\Delta u$ represents the difference in internal energy between the final and initial states ($\Delta u_{\text{system}} = u_2 - u_1$), then Equation 4.5 becomes:

$$q_{\text{net}} - w_{\text{net}} = u_2 - u_1 \tag{4.6}$$

Note that net heat added is always positive and will increase the energy in the system. If work is added to the system from the surroundings, this will also increase the energy in the system. For consistency, because from Section 2.3 work added is considered negative, the work term in Equation 4.6 must have a negative sign. In other words, even with the sign convention for heat and work, mathematically speaking, all forms of energy entering the system are considered positive (i.e., gains), while those exiting are considered negative (i.e., losses).

It is interesting and instructive to note that the piston–cylinder arrangement is treated as a nonflow process by the selection of the boundaries of the system, as shown by the dashed lines in Figures 4.1a and 4.1b. However, when the piston and cylinder are part of an internal combustion engine, the selection of the system boundary at the exterior of the engine leads us to a steady-flow system where fuel and air enter while heat and work cross the boundary. This concept will be treated in further detail when we concern ourselves with power cycles in Chapters 8 and 9.

Equation 4.6 is useful in establishing that internal energy is a property of a system and is not dependent on the path taken to place the system in a given configuration. This is true even though both $q$ and $w$ are path functions; that is, their values depend on the method of placing a system in a given configuration. For proof of the fact that internal energy is a property, the following reasoning is used.* Consider a system initially at state 1 that is caused to undergo a change to a second state 2 via path $a$. From Equation 4.6 and Figure 4.1c, we have

$$u_2 - u_1 = q_a - \mathbf{w}_a \tag{4.6a}$$

where the subscript $a$ is used to denote path $a$. If path $b$ is now followed between the same points 1 and 2, Equation 4.6 becomes

$$u_2 - u_1 = q_b - \mathbf{w}_b \tag{4.6b}$$

This procedure can be carried out for any path ($x$), and it must follow that when Equation 4.6 is applied,

$$u_2 - u_1 = q_x - \mathbf{w}_x \tag{4.6c}$$

---

* This material is a modification of the proof in Keenan (1941, p. 12 and following).

Because the left sides of Equations 4.6a, 4.6b, and 4.6c are equal, it follows that the right sides of these equations

$$q_x - \mathbf{w}_x = q_a - \mathbf{w}_a = q_b - \mathbf{w}_b \tag{4.7}$$

are also equal, and that $u_2 - u_1$ is fixed only by the end states of the system and is independent of the process. It can therefore be concluded that internal energy is a property because its values only depend on the conditions at the initial and final states. In other words, it is a state function and independent of the path of any process. There is one point about which students should take careful note. The proof of the fact that internal energy is a property is extremely important. In the calculus, an area can be obtained by integration. This process can be performed for those functions (paths) that can be evaluated as being continuous and dependent only on the end states. Such a function is said to be mathematically exact, and it is possible to perform all operations of the calculus on it. Opposed to this concept is the function whose value between two end states is determined not by the end states but by the path taken to achieve the end state. For example, the work done in moving a given block from one position on a plane to another position on the plane will depend on the amount of work done against friction. If the table is rough, more work is required to go from the initial to the final position than if the table were smooth. Mathematically, such a function is said to be inexact, and in general, it is not possible to evaluate this function directly by the methods of the calculus unless the path is defined. *Work and heat are inexact (path functions), whereas internal energy is exact (a function of the end states only).* The following Calculus Enrichment note explains the concept in a mathematical way.

---

### CALCULUS ENRICHMENT

As we have stated, a property has a unique and singular value when the system in question is in a given state. This value does not depend on the intermediate states that the system has experienced. Thus, a property is not a function of the system's path. The change of a property is a function only of the initial and final states of the system. Mathematically, this can be written as

$$\int_{\phi_1}^{\phi_2} d\phi = \phi_2 - \phi_1 \tag{a}$$

where $\phi$ denotes the property. Notice that the value of the property change is a function only of the limits of the integral. The differential, $d\phi$, is known as an *exact differential*. There are quantities such as heat and work that are not exact differentials, and it becomes necessary to specify the path mathematically in order to evaluate them. These quantities are known as *path functions*.

---

### ILLUSTRATIVE PROBLEM 4.1

If a nonflow, constant-volume process has 10 Btu/lb$_m$ as heat added to the system, what is the change in internal energy per pound of working fluid?

**SOLUTION**

For a constant-volume process, we can consider that the piston in Figure 4.1 has not moved, that is, volume remains constant. Alternatively, we can consider that a tank with a fixed volume has heat added to it. Under these conditions, the mechanical or boundary work done on or by the system must be zero since there is no net change in volume or displacement. The application of Equation 4.6 to this system yields

$$u_2 - u_1 = q$$

and it must be concluded that all the energy crossing the boundary as heat has been converted to internal energy of the working fluid. Therefore,

$$u_2 - u_1 = 10 \text{ Btu/lb}_m$$

---

**ILLUSTRATIVE PROBLEM 4.2**

The working fluid in a nonflow system undergoes an adiabatic change. Determine the work done in this process.

**SOLUTION**

An adiabatic process is defined to be one during which no energy interchange *as heat* crosses the boundary of a system; that is, there is no heat transfer in or out. As shown in Figure 4.2, this is the same as considering the system boundaries to be perfectly insulated. Note, however, that the piston can and does move. Application of Equation 4.6 to this situation yields

$$u_2 - u_1 = -\mathbf{w}$$

Thus, the energy interchange as work to or from the system per unit mass of working substance equals the change in internal energy of the working fluid per unit mass of fluid. The negative sign is taken to mean that work into the system (negative

**FIGURE 4.2**
Illustrative Problem 4.2.

work by convention) will cause an increase in the internal energy of the working fluid and that work out of the system (positive work) will cause a decrease in the internal energy of the working fluid.

In Chapter 2, we concluded that the work done in a quasi-static frictionless, nonflow process is the area under the *p–v* curve. Using this concept and Equation 2.21 for the work in Equation 4.6 gives us the following useful relation:

$$u_2 - u_1 = q - w_b \tag{4.8}$$

where $w_b = \int P dv$ (boundary work). Equation 4.8 is written per unit mass of working fluid, and students are cautioned to understand the reasoning behind this equation fully so that it will not be misapplied. Moreover, boundary work, $w_b$ or term $\int P dv$ implies that calculus may be needed for evaluating Equation 4.8, depending on how pressure, *p*, varies with volume between states 1 and 2, see Table 2.2 in Chapter 2.

Either Equation 4.6 or Equation 4.8 is called the *nonflow energy equation* and expresses the first law as applied to a nonflow process in a closed system. Equation 4.8 is more restrictive and, in the strictest sense, is applicable only to frictionless, quasi-static, nonflow processes.

---

**ILLUSTRATIVE PROBLEM 4.3**

Solve Illustrative Problem 4.1 by the direct application of Equation 4.8.

**SOLUTION**

$$u_2 - u_1 = q - \int p dv$$

But $v_2 - v_x$ (*or* $\Delta v$) is zero; see Table 2.1 in Chapter 2 for boundary work at constant volume. Therefore,

$$u_2 - u_1 = q \quad \text{(as before)}$$

---

There is one nonflow process that will be found to be quite important in our subsequent discussion: the quasi-static, *nonflow, constant-pressure* process. Using Equation 4.8, we will now derive an expression relating heat, work, and internal energy for this process. Refer to the situation illustrated in Figures 4.1 and 4.3 where heat and work can cross the system boundary. Transposing terms in Equation 4.8 yields

$$u_2 - u_1 + w_b = q \tag{4.8a}$$

However, boundary work, $w_b$ or $\int P dvp$ becomes $p(v_2 - v_1)$ for a *constant-pressure* process, as shown in Table 2.2. Therefore,

$$u_2 - u_1 + p(v_2 - v_1) = q \tag{4.8b}$$

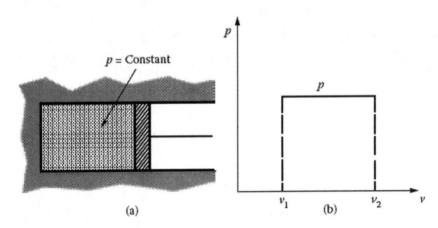

**FIGURE 4.3**
Quasi-static, nonflow, constant-pressure process. (a) Closed piston–cylinder. (b) Constant-pressure process.

The condition of constant pressure permits us to write $p = p_1 = p_2$. Thus,

$$u_2 - u_1 + p_2 v_2 - p_1 v_1 = q \tag{4.8c}$$

Equation 4.8c is consistent in terms of SI units. If the internal energy is in Btu per pound, pressure is in pounds per square foot, and specific volume is in cubic feet per pound, we can regroup terms and place them into consistent thermal units of Btu per pound as follows:

$$u_2 + \frac{p_2 v_2}{J} - u_1 + \frac{p_1 v_1}{J} = q \tag{4.8d}$$

Equation 4.8d in SI units is:

$$u_2 + p_2 v_2 - u_1 + p_1 v_1 = q \tag{4.8e}$$

The composite term $u + pv$ is a property that we briefly introduced in Chapters 2 and 3, and the student will find this to have great utility when flow processes are considered. At present, we will simply define $h$ to be enthalpy expressed as:

$$h = u + pv \tag{4.9}$$

Returning to Equation 4.8e, we now have the following energy equation for the nonflow, quasi-static, constant-pressure process:

$$q = h_2 - h_1 = \Delta h \tag{4.8f}$$

### 4.3.1 Processes of Nonflow System (Closed System)

So far, three nonflow, quasi-static processes have been considered in this chapter. To summarize the energy equations for these processes are

1. Constant-volume process ($\Delta v = 0$)

$$q = u_2 - u_1 = \Delta u \tag{4.10a}$$

2. Adiabatic process ($q = 0$)

$$-\mathbf{w} = u_2 - u_1 = \Delta u \qquad (4.10b)$$

3. Constant-pressure process ($\Delta p = 0$)

$$q = h_2 - h_1 = \Delta h \qquad (4.8f)$$

Further, the work of a quasi-static, nonflow process is

$$\mathbf{w} = \int P dv \qquad (2.21c)$$

*Enthalpy* is defined as

$$h = u + \frac{pv}{J} \qquad (4.9)$$

or

$$h_2 - h_1 = u_2 + \frac{p_2 v_2}{J} - u_1 + \frac{p_1 v_1}{J} \qquad (4.9a)$$

In all the preceding equations, it has been assumed that the mass of fluid was unity. However, the equations derived can be used for any mass simply by multiplying each term in the energy equation by $m$, the mass involved. The enthalpy definition used in this book will include the conversion factor $J$ for convenience and consistency, especially when using English units. It is also important to understand how to use fluid properties and diagrams are used when solving nonflow problems.

**Constant-Volume Process (Isometric Process) in a Nonflow (Closed) System**
The constant-volume process in a nonflow system occurs when a fluid is heated or cooled isometrically (i.e., at constant volume). Figure 4.4 shows a nonflow or closed system that contains a fluid being heated, while the volume inside is held constant by the rigid walls of the container. Also, the mass within the system remains constant since mass has no way to

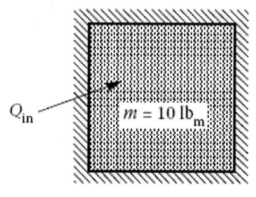

**FIGURE 4.4**
Constant-volume process in a rigid tank.

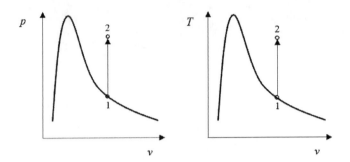

**FIGURE 4.5**
Example of constant-volume process on $p$–$v$ and $T$–$v$ diagrams.

get in or out. The constant-volume process can be seen in the $p$–$v$ and $T$–$v$ diagrams introduced in Chapter 3, as shown in Figure 4.5. Assuming state ① to be a saturated vapor (SV or g) and heated at constant volume ($v$ = constant) in a closed tank, the final state ② will be at the superheated (SHV) condition. The energy equation for this process has already been derived as $q = u_2 - u_1$ (Equation 4.10a).

---

### ILLUSTRATIVE PROBLEM 4.4

A closed tank contains 1 $lb_m$ of saturated vapor at 300°F. If the tank is heated ($q = 50.88$ Btu/$lb_m$) until it is filled with superheated steam, determine the volume of the tank occupied by the fluid, and the initial and final pressure of the saturated vapor and superheated steam, respectively, and final temperature of superheated steam.

### SOLUTION

Because the tank contains a fixed mass of 1 $lb_m$ of fluid, the initial and final specific volumes are the same at 6.472 ft.³/$lb_m$ (see $v_{sv}$ or $_g$ in Appendix A.1 at 300°F). Since the mass and volume are constant during the process, volume can be found using the definition of specific volume ($v = V/m$), as follows:

$$V = v \cdot m = (6.472 \cdot 1) = 6.472 \text{ ft.}^3$$

To find the initial pressure, it should be noticed that for a saturated vapor (SV), there is only one saturation pressure, which at 300°F is 66.98 psi (see $p_{sat}$ in Appendix A.1 at 300°F). To find the final pressure, $p_2$, internal energy, $u_1$, should be found in Table A.2 as well ($u_1 = 1100.0$ Btu/$lb_m$) and then use Equation 4.10a to find $u_2$ first, as follows:

$$q = u_2 - u_1 \rightarrow u_2 = q + u_1$$

$$u_2 = 50.88 + 1059.3 = 1150.88 \text{ Btu/lbm}$$

By taking into account the fact that the initial and final specific volumes are the same (i.e., 6.472 ft.³/$lb_m$ and that $u_2 = 1150.88$ Btu/$lb_m$ from Appendix A.3 we obtain through interpolation that $p_2 = 80$ psi.

To obtain the final temperature, it should be noticed that we already have enough properties to identify the conditions for the final state (i.e., $v_2$, $u_2$, and $p_2$), so through interpolation, we obtain that $T_2 = 431.48°F$. It is interesting to know that since the volume of the closed tank is constant, both temperature and pressure increase as heat goes into the system, as seen in Figure 4.5.

## ILLUSTRATIVE PROBLEM 4.5

A closed tank has a volume of 60 ft.$^3$. It contains 15 ft.$^3$ of steam (saturated vapor) and 45 ft.$^3$ of water (saturated liquid) at 14.7 psia. Heat is added until the pressure increases to 800 psia. Determine the heat added.

## SOLUTION

The mass in the tank is constant, and the heat added will be the change in internal energy of the contents of the tank between the two states (i.e., $q = u_2 - u_1$). The initial mass in the tank is found as follows:

$$m_f = \frac{V_f}{v_f} = \frac{45}{0.016715} = 2692.19 \text{ lb}_m$$

$$m_g = \frac{V_g}{v_g} = \frac{15}{26.80} = 0.56 \text{ lb}_m$$

$$\text{Total mass} = 2692.75 \text{ lb}_m$$

The initial internal energy is the sum of the internal energy of the liquid plus the vapor:

$$U_g = m_g u_g = 0.56 \times 1077.6 = 603.5 \text{ Btu}$$

$$U_f = m_f u_f = 2692.19 \times 180.1 = 484863.4 \text{ Btu}$$

$$\text{Total initial internal energy, } U_{initial} = 485466.9 \text{ Btu}$$

The values of $v_f$, $v_g$, $u_f$, and $u_g$ can be found in Appendix A.2.

Because the mass in the tank is constant, the final specific volume must equal the initial specific volume, or

$$v_{mx} = \frac{V_{mx}}{m} = \frac{60}{2692.75} = 0.022282 \text{ ft}^3/\text{lb}_m$$

But $v_{mx} = v_f + x v_{fg}$. Therefore, using Table A.2 at 800 psia to find $v_{fg}$ ($v_{fg} = v_g - v_f$)

$$x = \frac{v_x - v_f}{v_{fg}} = \frac{0.022282 - 0.02087}{0.5691 - 0.02087} = 0.0025756 = 0.2576\%$$

The final amount of vapor is

$$m_{g,\text{final}} = x \cdot m_{\text{total}} = 0.0025756 \cdot 2692.75 = 6.935 \text{ lb}_{\text{m}}$$

The final amount of liquid is

$$m_{f,\text{final}} = m_{\text{total}} - m_{g,\text{final}} = 2692.75 - 6.935 = 2685.815 \text{ lb}_{\text{m}}$$

The final internal energy is found as before:

$$U_{g,\text{final}} = m_g u_g = 6.935 \times 1115.0 = 7732.5$$

$$U_{f,\text{final}} = m_f u_f = 2685.15 \times 506.6 = 1360297$$

$$U_{\text{final}} = U_{g,\text{final}} + U_{f,\text{final}} = 7732.5 + 1360297 = 1368029.5 \text{ Btu}$$

The difference is

$$Q = U_{\text{final}} - U_{\text{initial}} = 1368029.5 - 485466.9 = 882562.6 \text{ Btu}$$

Per unit mass, the heat added is

$$q = \frac{Q}{m_{\text{total}}} = \frac{882562.6}{2692.75} = 327.76 \text{ Btu/lb}_{\text{m}}$$

The small quantity of vapor mass necessitates the unusual accuracy needed to solve this problem.

---

**ILLUSTRATIVE PROBLEM 4.6**

A rigid container contains 10 lb. of water. (a) If 100 Btu as heat is added to the water, what is its change in internal energy per pound of water? (b) If the 100 Btu is added by the mechanical friction of a paddle wheel stirring the water, what is the change in internal energy of the water? Discuss both processes. Refer to Figure 4.6.

**SOLUTION**

a.  The nonflow energy equation applied to this process yields $q = u_2 - u_1$. Per pound mass of working substance, $q = 100/10$, or 10 Btu/lbm. Therefore, $u_2 - u_1 = 10$ Btu/lbm.

b.  In this process, energy crosses the boundary of the system by means of frictional work. As far as the system is concerned, we note only that energy has crossed its boundaries. The similarity of the terms *work* and *heat* lies in the fact that both are energy in transition. Thus, for the present problem, the contents of the tank will not distinguish between the energy if it is added as heat or the energy added as frictional work. As for part (a), $u_2 - u_1 = 10$ Btu/lbm.

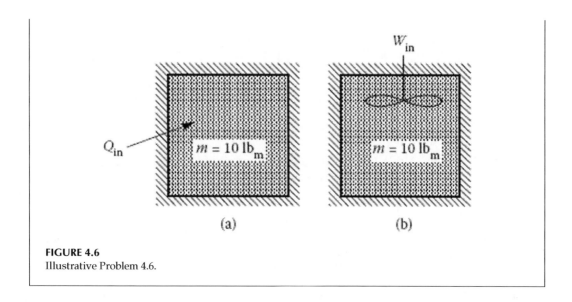

**FIGURE 4.6**
Illustrative Problem 4.6.

## Constant-Pressure Process (Isobaric Process) in a Nonflow (Closed) System

The constant-pressure process in a nonflow system occurs when a fluid is heated or cooled isobarically (i.e., at constant pressure). Figure 4.7 shows a nonflow or closed system that contains a fluid being heated, while the pressure inside is held constant by the piston that is free to move or is completely unconstrained. Also, the mass within the system remains constant since mass has no way to get in or out. The constant-pressure process can be seen in the $p$–$v$ and $T$–$v$ diagrams introduced in Chapter 3, as shown in Figure 4.8. Assuming state ① to be a saturated vapor (SV or g) and heated at constant pressure ($p$ = constant) until it reaches the final state ②, which is at the superheated (SHV) condition. The energy equation for this process has already been derived as $q = h_2 - h_1$ (Equation 4.8f).

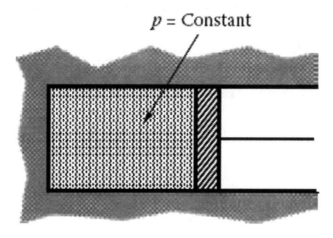

**FIGURE 4.7**
Constant-pressure process in a piston–cylinder device.

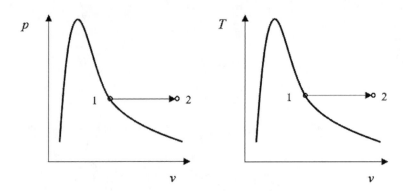

**FIGURE 4.8**
Example of constant-pressure process on $p$–$v$ and $T$–$v$ diagrams.

---

### ILLUSTRATIVE PROBLEM 4.7

A piston–cylinder device, like the one shown in Figure 4.7, contains 1 kg of saturated water vapor (SV) at 100 kPa. The system is heated until the volume inside doubles. The piston is unconstrained (i.e., free to move). Find the initial and final volume, the boundary work, $w_b$, and the total amount of heat transfer, $Q$.

### SOLUTION

The first step is to calculate the initial volume based on the conditions of the initial state (1 kg of saturated water vapor at 100 kPa). From Appendix A.2 (SI), we obtain that the specific volume for state 1 is 1.6940 m³/kg. To find $V_1$, let's use the definition of specific volume ($v = V/m$), as follows:

$$V_1 = v \cdot m = (1.6940 \cdot 1) = 1.694 \text{ m}^3$$

Since the final volume is twice the initial volume, then $V_2 = 2 \cdot 1.694 = 3.39$ m³. To find the amount of boundary work, $w_b$, it is important to notice that pressure remains constant during the whole process, therefore $w_b = p(V_2 - V_1)$. So, the amount of boundary work for the process is:

$$W_b = p(V_2 - V_1) = 100 \text{ kPa} \left( 3.39 \text{ m}^3 - 1.694 \text{ m}^3 \right) = 169.4 \text{ kJ}$$

Lastly, the total amount of heat transfer can be found by using the first law for a constant-pressure process, as follows:

$$Q = m(h_2 - h_1) = 1 \text{ kg} \left( 3408.59 \, \frac{\text{kJ}}{\text{kg}} - 2675.5 \, \frac{\text{kJ}}{\text{kg}} \right) = 733.09 \text{ kJ}$$

The values of $h_2$ and $h_1$ can be found in Table A.2 (SI) at p = 100 kPa at saturated vapor condition and superheated vapor condition ($v_2 = 3.39$ m³/kg), respectively. It is worth noting that $w_b$ can be found using Equation 4.8, as follows

$$u_2 - u_1 = q - w_b$$

or

$$U_2 - U_1 = Q - W_b$$

Solving for Q, we obtain:

$$Q = U_2 - U_1 + W_b = m(u_2 - u_1) + W_b$$

$$Q = 1\,kg\left(3069.59\,\frac{kJ}{kg} - 2506.1\,\frac{kJ}{kg}\right) + 169.4\,kJ = 732.09\,kJ$$

The values of $u_2$ and $u_1$ can be found in Table A.2 (SI) at p = 100 kPa at saturated vapor condition and superheated vapor condition ($v_2 = 3.39$ m³/kg), respectively.

## 4.4 Steady-Flow (Open) System

### 4.4.1 Conservation of Mass—Continuity Equation

In steady-flow systems, both mass and energy are permitted to cross the boundaries of the system, but when the process is steady, we limit ourselves to those systems that are not time-dependent. For steady-flow systems, we can express the principle of conservation of mass for these systems as requiring the total mass of fluid in the control volume at any time be constant. In turn, this requires that the net mass flowing into the control volume must equal the net mass flowing out of the control volume at any instant of time. To express these concepts in terms of a given system, let us consider the system schematically, as shown in Figure 4.9.

Let us assume that at a certain time, fluid starts to enter the control volume by crossing the control surface at ① and that, after a small interval of time, the flowing fluid fills the pipe for a short distance $\Delta x$. If it is further assumed that, in this short section of uniform

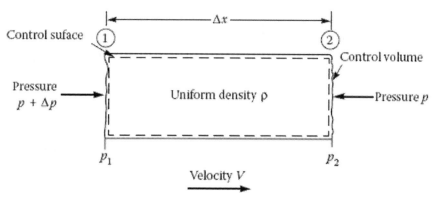

**FIGURE 4.9**
Elementary flow system.

pipe, no heat is added or work is exchanged and that the fluid density stays constant, then we can evaluate the amount of fluid that flowed between ① and ②. The mass contained between these sections is equal to the volume contained between the sections multiplied by the density of the fluid. The volume is $A\Delta x$, and the density is $\rho$. Therefore, the contained mass is $\rho A\Delta x$. The distance between stations, $\Delta x$, is simply $V\Delta t$, where $V$ is the velocity of the fluid and $\Delta t$ is the flow time required to fill the pipe between ① and ②. Substituting this for $\Delta x$ yields

$$m = \rho A V \Delta t \tag{4.9a}$$

or

$$\dot{m} = \rho A V = \rho_1 A_1 V_1 = \rho_2 A_2 V_2 \tag{4.9b}$$

where $\dot{m}$ is the mass rate of flow per unit time, $m/\Delta t$. It can also be expressed as

$$\dot{m} = \frac{AV}{v} \tag{4.9c}$$

where the specific volume replaces the density, $v = 1/\rho$.

Equation 4.9b is known as the *continuity or mass conservation equation,* and as written for a pipe or duct, we have made the assumption that the flow is normal to the pipe cross-section and that the velocity $V$ either is constant across the section or is the average value over the cross-section of the pipe. We shall use the continuity equation quite frequently in the study of open systems, and it is important to note that it expresses the fact that the mass flow into the control volume must equal the mass flow out of the control volume in steady-flow cases.

---

### ILLUSTRATIVE PROBLEM 4.8

At the entrance to a steady-flow device, it is found that the pressure is 100 psia and that the density of the fluid is constant at 62.4 $lb_m/ft.^3$. If 10,000 $ft.^3/min$ of this fluid enters the system and the exit area is 2 $ft.^2$, determine the mass flow rate and the exit velocity (see Figure 4.10).

**FIGURE 4.10**
Illustrative Problem 4.8.

## SOLUTION

Let us first calculate the mass flow into the system. From Equation 4.9b, $\dot{m} = \rho_1 A_1 V_1$. We are told that 10,000 ft.$^3$/min of this fluid enters the system, which is the same as saying $A_1 V_1 = 10\,000$ ft.$^3$/min. Therefore, $\dot{m} = \rho_1 A_1 V_1 = 62.4 \times 10{,}000 = 624{,}000 \text{lb}_m/\text{min}$. Because this mass flow must also leave the system,

$$\dot{m} = \rho_2 A_2 V_2$$

Thus,

$$624\,000\,\frac{\text{lb}_m}{\text{min}} = 62.4\,\frac{\text{lb}_m}{\text{ft.}^3} \times 2\,\text{ft.}^2 \times V_2\,\frac{\text{ft.}}{\text{min}}$$

$$V_2 = 5000\,\text{ft./min}$$

---

### ILLUSTRATIVE PROBLEM 4.9

If the fluid entering the system shown in Figure 4.10 has a density of 1000 kg/m$^3$ and 2000 m$^3$/min enters the system, determine the mass flow rate and the exit velocity if the exit area is 0.5 m$^2$. Assume that the density is constant.

## SOLUTION

Proceeding as in Illustrative Problem 4.8, $m = \rho A V$; $\dot{m} = \rho_1 A_1 V_1$. Therefore,

$$\dot{m} = 1000\,\frac{\text{kg}}{\text{m}^3} \times 2000\,\frac{\text{m}^3}{\text{min}}$$

$$= 2 \times 10^6\,\frac{\text{kg}}{\text{min}}$$

Because $\dot{m} = \rho_2 A_2 V_2$,

$$2 \times 10^6\,\frac{\text{kg}}{\text{min}} = 1000\,\frac{\text{kg}}{\text{m}^3} \times 0.5\text{m}^2 \times V_2$$

and

$$V_2 = 4000\,\text{m/min}$$

---

### ILLUSTRATIVE PROBLEM 4.10

A hose is 1 in. in diameter and has water the density of which is 62.4 lb$_m$/ft.$^3$ flowing steadily in it at a velocity of 100 ft./s. Determine the mass flow of water in the hose (see Figure 4.11).

**FIGURE 4.11**
Illustrative Problem 4.10.

## SOLUTION

From the continuity equation,

$$\dot{m} = \rho A V = 62.4 \, \frac{\mathrm{lb_m}}{\mathrm{ft.}^3} \times \frac{\pi(1 \, \mathrm{in.})^2}{4 \times 144 \, \dfrac{\mathrm{in.}^2}{\mathrm{ft.}^2}} \times 100 \, \frac{\mathrm{ft.}}{\mathrm{s}} = 34.0 \, \mathrm{lb_m}/\mathrm{s}$$

Note that the dimension of the area has been used in square feet for dimensional consistency.

Let's consider a steam turbine (one type of open system) as a flow system. Our present interest is in the flow of steam into and out of the noted boundaries of the system. The actions and interactions within the boundaries of this system will be studied later, but for us at this time, the boundaries can be thought of as defining a "black box." Only the fluid (steam) interfaces are considered. The continuity equation is applied to the inlet and outlet steam flow, as in the following example.

## ILLUSTRATIVE PROBLEM 4.11

A steam turbine has an inlet steam flow of 50,000 $\mathrm{lb_m}$/h of steam whose specific volume is 0.831 $\mathrm{ft.}^3/\mathrm{lb_m}$ as shown in Figure 4.12. The inlet diameter is 6 in. At the outlet, the diameter is 8 in., and the specific volume of the steam is 1.825 $\mathrm{ft.}^3/\mathrm{lb_m}$. Determine the velocity at the inlet and outlet of the turbine in ft./s.

## SOLUTION

We have at the inlet $\dot{m} = 50{,}000 \, \mathrm{lb_m}$/h and $v_1 = 0.831 \, \mathrm{ft.}^3/\mathrm{lbm}$. Therefore,

$$\dot{m}_1 = \frac{A_1 V_1}{V_1}$$

$$V_1 = \frac{\dot{m}_1 v_1}{A_1} = \frac{50000 \, \dfrac{\mathrm{lb_m}}{\mathrm{h}}}{60 \, \dfrac{\mathrm{min}}{\mathrm{h}} \times 60 \, \dfrac{\mathrm{s}}{\mathrm{min}}} \times \frac{0.831 \, \dfrac{\mathrm{ft.}^3}{\mathrm{lb_m}}}{\dfrac{\pi(6 \, \mathrm{in.})^2}{4(144 \, \mathrm{in.}^2/\mathrm{ft.}^2)}} = 58.8 \, \mathrm{ft.}/\mathrm{s}$$

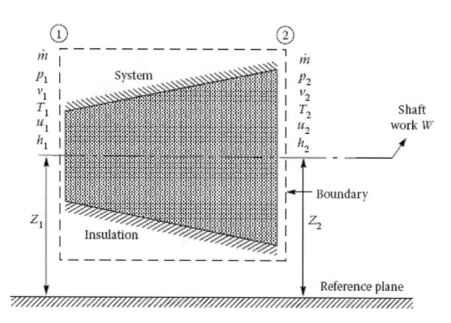

**FIGURE 4.12**
Illustrative Problem 4.11.

At the outlet, the value of $\dot{m}_2$ is the same, that is, 50,000 $lb_m$/h. Therefore,

$$V_2 = \frac{\dot{m}_2 v_2}{A_2} = \frac{50000 \dfrac{lb_m}{h}}{60 \dfrac{min}{h} \times 60 \dfrac{s}{min}} \times \frac{1.825 \ ft.^3/lb_m}{\dfrac{\pi}{4}\left(\dfrac{8 \ in.}{12 \ in./ft.}\right)^2}$$

$$V_2 = 72.6 \ ft./s$$

Again note the conversion factors of $(12 \ in.)^2/ft.^2$ and $60 \ min/h \times 60 \ s/min$ to obtain the desired units of velocity.

---

## ILLUSTRATIVE PROBLEM 4.12

A steam turbine, as the one shown in Figure 4.12, has an inlet steam flow of $10^4$ kg/h whose specific volume is 0.05 $m^3$/kg. The inlet diameter is 100 mm, and the outlet diameter is 200 mm. If the outlet specific volume is 0.10 $m^3$/kg, determine the inlet and outlet velocities.

**SOLUTION**

$$\dot{m}_1 = \dot{m}_2 = 10^4 \ kg/h$$

Because $\dot{m}_1 = AV/v$, $V = \dot{m}v/A$. Therefore,

$$V_1 = \frac{\dot{m}_1 v_1}{A_1} = \frac{10^4 \text{ kg/h}}{(60 \times 60) \text{ s/h}} \times \frac{0.05 \dfrac{\text{m}^3}{\text{kg}}}{\dfrac{\pi}{4}(0.1)^2 \text{ m}^2} = 17.68 \text{ m/s}$$

At the outlet,

$$V_1 = \frac{\dot{m}_2 v_2}{A_2} = \frac{10^4 \text{ kg/h}}{(60 \times 60) \text{ s/h}} \times \frac{0.10 \dfrac{\text{m}^3}{\text{kg}}}{\dfrac{\pi}{4}(0.2)^2 \text{ m}^2} = 8.84 \text{ m/s}$$

### 4.4.2 Steady-Flow Energy Equation

Figure 4.13 shows a steady-flow system in which it is assumed that each form of energy can enter and leave the system. At the entrance, $\dot{m}$ lb. of fluid per second enters, and the same amount leaves at the exit. At the entrance, the fluid has a pressure of $p_1$, a specific volume of $v_1$, an internal energy of $u_1$, and a velocity of $V_1$. At the exit, we have similar quantities expressed as $p_2$, $v_2$, and $V_2$. The fluid enters and leaves at different elevations, and work and heat cross the boundary in both directions.

In applying the first law to a system in which both mass and energy can cross the boundaries, it is necessary to adhere to the mathematical conventions chosen for positive and negative quantities of energy. It is also imperative that all pertinent energy terms be included in any analysis. To summarize, Table 4.1 identifies six energy terms that apply to various situations for steady-flow or open systems.

We express the first law for the system shown in Figure 4.13 by stating that all the energy entering the system must equal all the energy leaving the system under steady-flow conditions ($\Delta e_{\text{system}} = 0$).

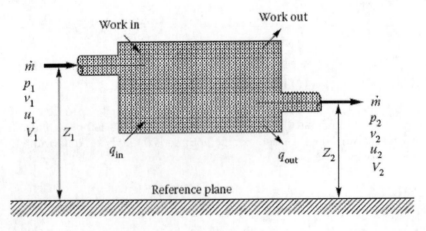

**FIGURE 4.13**
Steady-flow system.

**TABLE 4.1**

Energy Terms

| Item | Value | |
|------|-------|--|
| | **Btu/Unit Mass** | **kJ/Unit Mass** |
| Potential energy | $\dfrac{Zg}{Jg_c}$ | $Zg$ |
| Kinetic energy | $\dfrac{V^2}{2g_cJ}$ | $\dfrac{V^2}{2}$ |
| Internal energy | $u$ | $u$ |
| Flow work | $\dfrac{pv}{J}$ | $pv$ |
| Work | $\dfrac{\mathbf{w}}{J}$ | $w$ |
| Heat | $q$ | $q$ |

$$e_{in} - e_{out} = \Delta e_{system} = 0 \qquad (4.10)$$

$$e_{in} = e_{out}$$

The components of the energy balance above, are shown in Table 4.2. By equating all the terms in Table 4.2 in English units, we obtain

$$\frac{Z_1}{J}\frac{g}{g_c} + \frac{V_1^2}{2g_cJ} + u_1 + \frac{p_1v_1}{J} + \frac{w_{in}}{J} + q_{in}$$

$$= \frac{Z_2}{J}\frac{g}{g_c} + \frac{V_2^2}{2g_cJ} + u_2 + \frac{p_2v_2}{J} + \frac{w_{out}}{J} + q_{out} \qquad (4.11)$$

**TABLE 4.2**

Energy Balance

| | Energy in | | Energy out | |
|---|---|---|---|---|
| | **Btu/lb$_m$** | **kJ/kg** | **Btu/lb$_m$** | **kJ/kg** |
| Potential energy | $\dfrac{Z_1g}{Jg_c}$ | $Z_1g$ | $\dfrac{Z_2g}{Jg_c}$ | $Z_2g$ |
| Kinetic energy | $\dfrac{V_1^2}{2g_cJ}$ | $\dfrac{V_1^2}{2}$ | $\dfrac{V_2^2}{2g_cJ}$ | $\dfrac{V_2^2}{2}$ |
| Internal energy | $u_1$ | $u_1$ | $u_2$ | $u_2$ |
| Flow work | $\dfrac{p_1v_1}{J}$ | $p_1v_1$ | $\dfrac{p_2v_2}{J}$ | $P_2v_2$ |
| Work | $\dfrac{\mathbf{w}_{in}}{J}$ | $\mathbf{w}_{in}$ | $\dfrac{\mathbf{w}_{out}}{J}$ | $w_{out}$ |
| Heat | $q_{in}$ | $q_{in}$ | $q_{out}$ | $q_{out}$ |

In SI units,

$$Z_1 g + \frac{V_1^2}{2} + u_1 + p_1 v_1 + w_{\text{in}} + q_{\text{in}}$$

$$= Z_2 g + \frac{V_2^2}{2} + u_2 + p_2 v_2 + w_{\text{out}} + q_{\text{out}} \tag{4.11a}$$

Equation 4.10 is quite general and expresses the first law for this system. It is sometimes called the *steady-flow energy* or the *general energy equation*. We note that it is also quite proper to call it the *energy equation* applied to a steady-flow system in which it does not gain or lose energy as time goes by.

If at this point we note that both the heat and the work terms can be combined to form individual terms of net heat ($q_{\text{net}} = q_{\text{in}} - q_{\text{out}}$, or namely $q$) and net work ($w_{\text{net}} = w_{\text{out}} - w_{\text{in}}$, or namely $w$), and being careful of the mathematical signs of these net terms, we can write for English units,

$$\frac{Z_1}{J}\frac{g}{g_c} + \frac{V_1^2}{2g_c J} + u_1 + \frac{p_1 v_1}{J} + q = \frac{Z_2}{J}\frac{g}{g_c} + \frac{V_2^2}{2g_c J} + u_2 + \frac{p_2 v_2}{J} + \frac{\mathbf{w}}{J} \tag{4.12}$$

In SI units,

$$Z_1 g + \frac{V_1^2}{2} + u_1 + p_1 v_1 + q = Z_2 g + \frac{V_2^2}{2} + u_2 + p_2 v_2 + w. \tag{4.12a}$$

In Equation 4.11, all terms are written as Btu/lb$_\text{m}$, and $q$ and $\mathbf{w}$ represent *net values* per unit mass of working fluid. If we now note further that the grouping of terms $u + pv/J$ appears on both sides of Equation 4.11, and that this combined term is a property that we have already called enthalpy, then from the definition of the term *enthalpy* we have

$$h = u + \frac{pv}{J} \tag{4.9}$$

(or $h = u + pv$ in SI units). Using Equations 4.5 and 4.11 yields, for English units,

$$\frac{Z_1}{J}\frac{g}{g_c} + \frac{V_1^2}{2g_c J} + h_1 + q = \frac{Z_2}{J}\frac{g}{g_c} + \frac{V_2^2}{2g_c J} + h_2 + \frac{\mathbf{w}}{J} \tag{4.13}$$

By regrouping the terms in Equation 4.12, we obtain

$$q - \frac{w}{J} = \left(h_2 - h_1\right) + \frac{Z_2 - Z_1}{J}\frac{g}{g_c} + \frac{V_2^2 - V_1^2}{2g_c J} \tag{4.14}$$

where each quantity is in Btu/lb$_\text{m}$.

In terms of SI units,

$$q - w = h_2 - h_1 + g\left(Z_2 - Z_1\right) + \frac{V_2^2 - V_1^2}{2} \tag{4.14a}$$

where each quantity is in terms of J/kg or kJ/kg.

Note that when $Z$ is in meters and $V$ is in meters per second, the potential energy and kinetic energy terms will be in J/kg, the equivalent of m²/s². Heat, work, and enthalpy per unit mass are usually given in kJ/kg. Thus, a factor of 1000 is often needed to make all the terms in Equation 4.14a compatible.

Students should note that Equations 4.11 through 4.14 are essentially the same. To apply these equations correctly, it is important that each of the terms be completely understood.

Before illustrating the use of these energy equations, let us consider the following situation. A fluid is flowing steadily in a device in which it undergoes compression. Let us further assume that this process is frictionless and quasi-static. Its work term can be written as the sum of the work done on the fluid (boundary work) plus the flow work. Kinetic and potential energy terms are assumed to be negligible. Thus, the total net work for this process, becomes

$$w = p_1v_1 - p_2v_2 - \int_{v_1}^{v_2} pdv \tag{4.15}$$

Let us now plot a $pv$ diagram for the fluid, as shown in Figure 4.14. As shown in Figure 4.14a, the term $p\Delta v$ is the area under the curve between the limits 1 and 2 ($\text{Area}_1 = \int_{v_1}^{v_2} pdv$), which represents boundary or work done on the fluid. Subtracting $p_1v_1$ and adding $p_2v_2$ (graphically) results in $v\Delta p$ or the area under the $vp$ diagram ($\text{Area}_2 = \int_{p_1}^{p_2} vdp$) as shown in Figure 4.14b. Mathematically,

$$w = -\int_{p_1}^{p_2} vdp = -\left( p_2v_2 - p_1v_1 + \int_{v_1}^{v_2} pdv \right) \tag{4.16}$$

It will be recalled that for the quasi-static, frictionless, nonflow system, $p\Delta v$ evaluated the work done on the working fluid. For the flow system, flow work of an amount of $p_1v_1$ enters the system, and flow work of an amount $p_2v_2$ leaves the system. We can therefore interpret

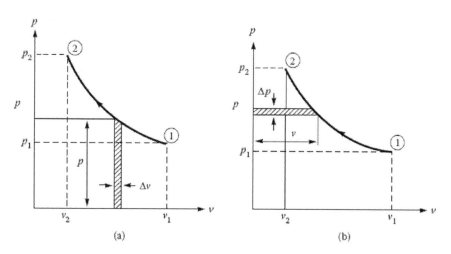

**FIGURE 4.14**
A $p$–$v$ diagram. (a) Work as $p\Delta v$. (b) Work as $v\Delta p$.

the meaning of the areas on Figure 4.14 or the terms of Equation 4.16 as follows: The work of the quasi-static, frictionless flow system (neglecting the kinetic and potential energy terms) is the algebraic sum of the work to induct the fluid into the system plus the work of compression (or expansion) less the flow work to deliver the fluid to the downstream exit. It is sometimes convenient to think of $p\Delta v$ as the work of *compression* (or expansion) and $v\Delta p$ as the work of the *compressor* (or expander) that delivers (or extracts) flow work to (from) the fluid. It is extremely important to note that both $p\Delta v$ ($pdv$) and $v\Delta p$ ($vdp$) evaluate work only for a quasi-static process or system. For systems that are real (or do not follow quasi-static conditions), they cannot be used to evaluate work accurately. However, for a large class of systems, they approximate the work terms, and their use is justified both from this approximation and the relative ease with which they can be evaluated.

The form of Equations 4.11 through 4.13a, where work is expressed per unit mass, is helpful when determining the power associated with a given process. The work per unit mass times the mass flow rate is the power. Work out is associated with power out and likewise for work in. Expressed as an equation,

$$\dot{W} = \dot{m}w \tag{4.17}$$

Typical units are, in the English system, $\dot{m}$ in $lb_m/s$, $w$ in $Btu/lb_m$, and $\dot{W}$ in $Btu/s$. Power may be expressed in horsepower using the conversion factor in Table 1.8. In the SI system, $\dot{m}$ may be in $kg/s$, $w$ in $kJ/kg$, and $\dot{W}$ in $kJ/s$, or kilowatts.

There is a similar relationship between heat transfer per unit mass, q, the mass flow rate, $\dot{m}$, and the rate of heat transfer per unit time. This is given by

$$\dot{Q} = \dot{m}q \tag{4.18}$$

---

## CALCULUS ENRICHMENT

From the general energy equation (Equations 4.14 and 4.14a), we can write in differential form:

$$\delta q - \delta w = dh + d\left(\text{K.E.}\right) + d\left(\text{P.E.}\right) \tag{a}$$

where $\delta$ (delta) denotes an inexact quantity, that is, one that is a function of the path.

Because $h$, enthalpy, is defined to equal $u + pv$,

$$h = u + pv; \quad dh = du + p\,dv + v\,dp \tag{b}$$

Substitute Equation (b) into Equation (a), and

$$\delta q - \delta w = du + p\,dv + v\,dp + d\left(\text{K.E.}\right) + d\left(\text{P.E.}\right) \tag{c}$$

From the energy equation for the nonflow system,

$$\delta q - \delta u = p\,dv \tag{d}$$

Using Equations (d) and (c),

$$\delta q - \delta w = du + \delta q - \delta u + v\,dp + d\left(\text{K.E.}\right) + d\left(\text{P.E.}\right) \tag{e}$$

and

$$\delta w = -v\,dp \tag{f}$$

if the K.E. and P.E. terms are negligible.

From Equation (f), we conclude that the work of a quasi-static flow process is the area *behind* the $p$–$v$ curve, as shown in Figure 4.14.

### 4.4.3 Bernoulli Equation

In fluid mechanics, the *Bernoulli equation* is used since it is a particular form of the energy equation for fluids. Because of the frequent misapplication of this equation, a brief discussion of it will now be undertaken. Let us consider a system in which the flow is steady, there is no change in internal energy, no work is done on or by the system, no energy as heat crosses the boundaries of the system, and the fluid is incompressible (i.e., density remains constant). We further assume that all processes are ideal in the sense that they are frictionless. For this system, the energy equation reduces to the following in mechanical units of foot-pounds per pound mass:

$$Z_1\frac{g}{g_c} + \frac{V_1^2}{2g_c} + p_1v_1 = Z_2\frac{g}{g_c} + \frac{V_2^2}{2g_c} + p_2v_2 \tag{4.19}$$

or in SI units,

$$Z_1 g + \frac{V_1^2}{2} + p_1v_1 = Z_2 g + \frac{V_2^2}{2} + p_2v_2 \tag{4.19a}$$

Note that for an incompressible fluid, $v_1 = v_2$.

Equation 4.1 is the usual form of the Bernoulli equation found in elementary physics and fluid mechanics texts. Each term can represent a height, because the dimension of foot-pounds per pound corresponds numerically (and dimensionally) to a height. Hence, the term *head* is frequently used to denote each of the terms in Equation 4.18. From the derivation of this equation, we note that its use is restricted to situations in which the flow is steady, there is no friction, no shaft work is done on or by the fluid, the flow is incompressible, and there is no change in internal energy during the process. These restrictions are severe, and only under the simplest situations can we hope to apply the Bernoulli equation successfully. It becomes even more difficult to justify the procedure of adding heat and work terms to this equation while maintaining all the other restrictions. Students are cautioned against the use of this equation without a thorough understanding of its restrictions. In every case, it is preferable to write the complete and correct form of the energy equation first and then to make those assumptions that can be justified for each problem.

### 4.4.4 Specific Heat

The term *specific heat* is defined as the ratio of energy as *heat* transferred during a particular process per unit mass of fluid involved divided by the corresponding change of temperature of the fluid that occurs during this process. Because heat can be transferred to or from a fluid and algebraic signs have been adopted for the direction of heat transfer, it is entirely possible for a process to have a negative specific heat. However, students should not confuse the specific heat of a process with the specific heat property, which denotes the amount of energy a substance can contain per unit mass and per unit temperature (i.e., kJ/kg-K or Btu/lb$_m$·R)

This definition of specific heat is important for two processes, because they serve to define a new property of a fluid. These processes are the constant-pressure process (flow or nonflow) and the constant-volume process. Recall from earlier work that the constant-pressure, nonflow process is characterized by the energy equation that $q = h_2 - h_1 = \Delta h$. For the steady-flow process without a change in elevation, in the absence of external work, and with negligible changes in kinetic energy, the steady-flow energy equation leads to the same result. Thus, for *any constant-pressure process* (flow or nonflow with the conditions noted), the specific heat is defined as

$$c_p = \frac{q}{\Delta T_p} = \frac{\Delta h}{\Delta T_p} \qquad (4.20)$$

or

$$\left( c_p \Delta T = \Delta h \right)_p \qquad (4.20a)$$

where the subscript $p$ indicates a constant-pressure process.

For the constant-volume process (which can only be a nonflow process), $q = u_{2v} - u_1 = \Delta u$. Therefore,

$$c_v = \frac{q}{\Delta T_v} = \frac{\Delta u}{\Delta T_v} \qquad (4.21)$$

or

$$\left( c_p \Delta T = \Delta u \right)_v \qquad (4.21a)$$

where the subscript $v$ indicates a constant-volume process.*

These specific heats are the properties of substances (i.e., fluid including liquids and gases) and depend only on the state of the substance. Typical units of specific heat are Btu/lb$_m$·R in English units and kJ/kg·K in SI units. A further discussion of specific heats is given in Chapter 6, where the internal energy and enthalpy of the ideal gas are functions of temperature only, making the definition of Equations 4.20 and 4.21 general for the ideal gas and not restricted to only constant-pressure or constant-volume processes.

---

* Equations 4.20 and 4.21 are more correctly written mathematically as

$$c_p \equiv \frac{\partial h}{\partial t_p} \quad \text{and} \quad c_v \equiv \frac{\partial u}{\partial t_v}$$

## ILLUSTRATIVE PROBLEM 4.13

A gas initially of 100°F and with $c_p = 0.22$ Btu/lb$_m$·R and $c_v = 0.17$ Btu/lb$_m$·R is placed within a piston–cylinder device. If 800 Btu as heat is added to 10 lb$_m$ of the gas in a nonflow, constant-pressure process, determine the final gas temperature. Also, determine the work done on or by the gas. Refer to Figure 4.15.

### SOLUTION

For the nonflow, constant-pressure process, $q = \Delta h = h_2 - h_1$. However, Equation 4.20 allows us to evaluate $\Delta h$ in terms of temperature and the specific heat at constant pressure. Thus,

$$q = h_2 - h_1 = c_p\left(T_2 - T_1\right)$$

Using the data given and $q$ on a unit mass basis,

$$q = \frac{800 \text{ Btu}}{10 \text{ lb}_m} = 0.22\left(T_2 - T_1\right)$$

so $T_2 - T_1 = 363.6$°F. Because the initial temperature is 100°F, the final temperature is 363.6°F + 100 = 463.6°F.

To obtain the work, we apply Equation 4.2 to the nonflow process:

$$q - \mathbf{w} = u_2 - u_1$$

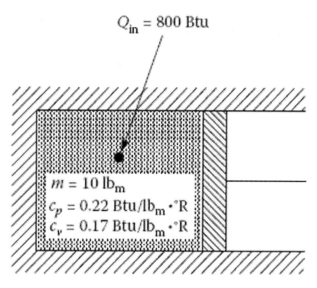

**FIGURE 4.15**
*Illustrative Problem 4.13.*

The term $u_2 - u_1$ can be evaluated in terms of temperature and the specific heat at constant volume from Equation 4.21. Using this, we obtain

$$-\mathbf{w} = (u_2 - u_1) - q = c_v(T_2 - T_1) - q$$

Using $q$ on a unit mass basis and noting that $T_2 - T_1 = 363.6°F$,

$$-\mathbf{w} = 0.17\ \text{Btu/lb}_m \cdot R(363.6\ R) - \frac{800\ \text{Btu}}{10\ \text{lb}_m}$$

$$-\mathbf{w} = 61.8 - 80 = -18.2\ \text{Btu/lb}_m$$

or

$$\mathbf{w} = 18.2\ \text{Btu/lb}_m$$

Thus, 18.2 Btu/lb$_m$ or 182 Btu as work is taken out of the system due to work done by the gas, because there is 10 lb$_m$ in the system.

## 4.5 Applications of First Law of Thermodynamics in Steady-Flow Systems

At this point in our study, we will apply the first law to several steady-flow situations that we will have applications for in later sections of this book. For the present, we restrict ourselves to seven steady-flow processes:

1. The steam or gas turbine
2. Pipe flow
3. The boiler
4. The flow in nozzles
5. The throttling process
6. The heat exchanger
7. Filling a tank

### 4.5.1 Turbine

As our first illustration of the applicability of the first law to steady-flow processes, let us consider the steam turbine shown in Figure 4.16. In this device, steam enters and expands in fixed nozzles to a high velocity. The high-velocity steam is then directed over the turbine blades, where it does work on the turbine wheel. The steam then exhausts from the turbine. The purpose of this machine is to obtain shaft work (mechanical work), and certain features about it should be noted. First, the shaft of the turbine is horizontal. Second,

**FIGURE 4.16**
Single-stage turbine. (Courtesy of Worthington Corp., Columbus, Ohio.)

as the steam expands, its specific volume increases (see Chapter 5), and to keep the exit velocity nearly equal to the entering velocity, the exit pipe area is proportionally greater than the inlet pipe area. The turbine is suitably insulated to minimize heat losses to the surroundings and also to eliminate the possibility of injuring operating personnel working in the vicinity of the hot turbine casing.

With the foregoing in mind, let us now apply the first law to the turbine (either steam or gas) shown schematically in Figure 4.16. The first law is given by

$$q - \frac{\mathbf{w}}{J} = h_2 - h_1 + \frac{Z_2 - Z_1}{J}\frac{g}{g_c} + \frac{V_2^2 - V_1^2}{2g_cJ} \qquad (4.22)$$

or

$$q - \mathbf{w} = h_2 - h_1 + g\left(Z_2 - Z_1\right) + \frac{V_2^2 - V_1^2}{2} \qquad (4.22\text{a})$$

Because the shaft of the machine is horizontal, $[(Z_2 - Z_1)/J]\ (g/g_c)$ can be taken to be zero; that is,

$$\left(\cancel{\frac{Z_2 - Z_1}{J}}\right)^{0}\left(\frac{g}{g_c}\right)$$

where $\nearrow^0$ is taken to mean "goes to zero." Also, as noted, the inlet and outlet velocities are kept nearly equal, leading us to conclude that the kinetic energy difference term goes to zero:

$$\left(\cancel{\frac{V_2^2 - V_1^2}{2g_cJ}}\right)^{0}$$

Finally, the insulation of the turbine would effectively prevent heat losses to the surroundings:

Equation 4.22, as applied to this device, becomes

$$\frac{\mathbf{w}}{J} = h_1 - h_2$$

or from Equation 4.22a,

$$\mathbf{w} = h_1 - h_2$$

All the assumptions made for the steam turbine are equally applicable to the gas turbine. At this point, it is important to note that water and other fluid vapors do not behave exactly as ideal gases (see Chapter 6). In an industrial gas turbine unit, the turbine also drives the compressor as part of a power cycle. Exhaust gases leave at the right of the figure via an exhaust elbow. This unit will be discussed in detail later on, as a prime mover in a power-producing cycle.

---

### ILLUSTRATIVE PROBLEM 4.14

A large steam turbine operates with a steam flow rate of 150,000 $lb_m$/h. The conditions at inlet and outlet are tabulated below. Determine the power produced (a) if heat losses are negligible and (b) if heat losses equal 50,000 Btu/h.

|  | Inlet | Outlet |
|---|---|---|
| Pressure (psia) | 1000 | 1 |
| Temperature (°F) | 1000 | 101.74 |
| Velocity (ft./s) | 125 | 430 |
| Inlet position (ft.) | +10 | 0 |
| Enthalpy (Btu/$lb_m$) | 1505.4 | 940.0 |

### SOLUTION

Let us start this problem by referring to Figure 4.17 and using the form of energy equation given by Equation 4.13,

$$\frac{Z_1}{J}\frac{g}{g_c} + \frac{V_1^2}{2g_cJ} + h_1 + q = \frac{Z_2}{J}\frac{g}{g_c} + \frac{V_2^2}{2g_cJ} + h_2 + \frac{\mathbf{w}}{J}$$

where all terms are per unit mass,

   a.  $q=0$, and assuming that $g=g_c$,

$$\frac{10}{778} + \frac{(125)^2}{2\times32.17\times778} + 1505.4 + 0$$

$$= 0 + \frac{(430)^2}{2\times32.17\times778} + 940.0 + \frac{\mathbf{w}}{J}$$

(a)

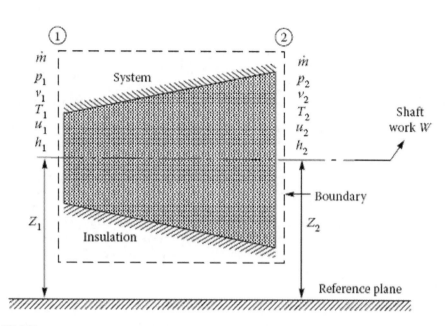

**FIGURE 4.17**
Schematic of a turbine.

$$0.013 + 0.312 + 1505.4 + 0 = 0 + 3.694 + 940.0 + \frac{w}{J} \qquad \text{(b)}$$

or

$$\frac{w}{J} = 562.03 \text{ Btu/lb}_m$$

(Had we neglected potential and kinetic terms, $w/J = 565.4$ Btu/lb$_m$, a difference of 0.6%, which is quite small.)

The total work of the turbine is 150,000 lb$_m$/h × 562.03 Btu/lb$_m$ = 84,304,500 Btu/h. In terms of horsepower,

$$\frac{84\,304\,500 \text{ Btu/h} \times 778 \text{ ft} \cdot \text{lb}_f/\text{Btu}}{60 \text{ min/h} \times 33\,000 \text{ ft} \cdot \text{lb}_f/\text{hp}} = 33\,125.7 \text{ hp}$$

In terms of kilowatts,

$$33\,125.7 \text{ hp} \times 0.746 \text{ kW/hp} = 24\,711.8 \text{ kW}$$

b.  $q_{loss} = \dfrac{500\,000 \text{ Btu/h}}{150\,000 \text{ lb}_m/\text{h}} = 0.333 \text{ Btu/lb}_m$ (heat loss)

Using this with Equation (b), and noting that it is negative by convention, yields

$$0.013 + 0.312 + 15.5.4 - 0.333 = 0 + 3.694 + 940.0 + \frac{w}{J}$$

or

$$\frac{\mathbf{w}}{J} = 561.70 \ \mathrm{Btu/lb_m}$$

which differs from part (a) by 0.06%. For all practical purposes, we could have neglected the potential, kinetic, and heat loss terms.

---

## ILLUSTRATIVE PROBLEM 4.15

A steam turbine operates with an inlet velocity of 40 m/s and an inlet enthalpy of 3433.8 kJ/kg. At the outlet, which is 2 m lower than the inlet, the enthalpy is 2675.5 kJ/kg and the velocity is 162 m/s. If the heat loss from the turbine is 1 kJ/kg, determine the work output per kilogram.

### SOLUTION

If we refer to Illustrative Problem 4.14, it will be noted that these problems are similar except for units. Using Equation 4.12a yields

$$Z_1 g + \frac{V_1^2}{2} + h_1 + q = Z_2 g + \frac{V_2^2}{2} + h_2 + \mathbf{w}$$

For the units of this problem,

$$\frac{2\,\mathrm{m} \times 9.81 \mathrm{m/s^2}}{1000\,\mathrm{J/kJ}} + \frac{\left(40\,\mathrm{m/s}\right)^2}{2 \times 1000\,\mathrm{J/kJ}} + 3433.8\,\frac{\mathrm{kJ}}{\mathrm{kg}} - 1\,\frac{\mathrm{kJ}}{\mathrm{kg}}$$

$$= 0 + \frac{\left(162\,\mathrm{m/s}\right)^2}{2 \times 1000\,\mathrm{J/kJ}} + 2675.5\,\frac{\mathrm{kJ}}{\mathrm{kg}} + \mathbf{w}$$

$$0.0196\,\frac{\mathrm{kJ}}{\mathrm{kg}} + 0.8\,\frac{\mathrm{kJ}}{\mathrm{kg}} + 3433.8\,\frac{\mathrm{kJ}}{\mathrm{kg}} - 1\,\frac{\mathrm{kJ}}{\mathrm{kg}} = 0 + 13.12\,\frac{\mathrm{kJ}}{\mathrm{kg}} + 2675.5\,\frac{\mathrm{kJ}}{\mathrm{kg}} + \mathbf{w}$$

or

$$\mathbf{w} = 744.9976\,\frac{\mathrm{kJ}}{\mathrm{kg}}$$

---

## ILLUSTRATIVE PROBLEM 4.16

A turbine (gas) receives air at 150 psia and 1000 R and discharges to a pressure of 15 psia. The actual temperature at discharge is 600 R. If $c_p$ of the gas can be taken to be constant over this temperature range and equal to 0.24 Btu/lb$_m$·R, determine the work output of the turbine per pound of working fluid. At inlet conditions, the specific volume is 2.47 ft.$^3$/lb$_m$, and at outlet conditions it is 14.8 ft.$^3$/lb$_m$. The data are also shown in Figure 4.18.

## SOLUTION

We have already discussed the turbine in detail and concluded that the first law yields

$$\frac{w}{J} = h_1 - h_2$$

Equation 4.20 enables us to obtain $h_1 - h_2$, because for the gas,

$$h_1 - h_2 = c_p(T_1 - T_2)$$

Therefore,

$$\frac{w}{J} = h_1 - h_2 = c_p(T_1 - T_2) = (0.24)(1000 - 600)$$

and

$$\frac{w}{J} = 96 \text{ Btu/lb}_m$$

Note that the specific volume and pressure given do not enter the solution of the problem. However, a pressure differential (150 − 15 = 135 psia) is required to cause the gas to flow.

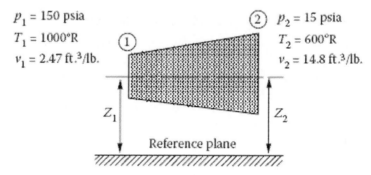

**FIGURE 4.18**
Illustrative Problem 4.16.

---

## ILLUSTRATIVE PROBLEM 4.17

Even though the turbine of Illustrative Problem 4.16 may be well insulated, there will be some heat loss. If the heat loss is found by experiment to be equal to 1.1 Btu/lb$_m$ of gas, determine the work output of the turbine per pound mass of gas.

**SOLUTION**

We must again return to the point in our discussion where we made the assumption that $q$ was equal to zero; however, for this problem the value of q is given and should be taken into account as follows:

$$q - \frac{\mathbf{w}}{J} = h_2 - h_1$$

or

$$\frac{\mathbf{w}}{J} = q + \left(h_1 - h_2\right)$$

Because $q$ is out of the system, it is a negative quantity, and $h_1 - h_2$ is the same as for Illustrative Problem 4.16. Thus,

$$\frac{\mathbf{w}}{J} = -1.1 + 96.0 = 94.9 \text{ Btu/lb}_\text{m}$$

In other words, a heat loss decreases the work output of the turbine, as we have seen earlier.

### 4.5.2 Pipe Flow

As our next illustration of the first law, we will apply the law to the flow of fluids in pipes.

**ILLUSTRATIVE PROBLEM 4.18**

A gas flows in a pipe whose pressure and temperature at one section are 100 psia and 950°F. At a second section of the pipe, the pressure is 76 psia and the temperature is 580°F. The specific volume of the inlet gas is 4.0 ft.³/lb$_\text{m}$, and at the second section it is 3.86 ft.³/lb$_\text{m}$. Assume that the specific heat at constant volume is 0.32 Btu/lb·R. If no shaft work is done and the velocities are small, determine the magnitude and direction of the heat transfer. Assume the pipe to be horizontal, and neglect velocity terms. The data are also shown in Figure 4.19.

**SOLUTION**

We first write out the equation of the first law and then apply it to this process:

$$\frac{Z_1}{J}\frac{g}{g_c} + \frac{V_1^2}{2g_cJ} + u_1 + \frac{p_1v_1}{J} + q = \frac{Z_2}{J}\frac{g}{g_c} + \frac{V_2^2}{2g_cJ} + u_2 + \frac{p_2v_2}{J} + \frac{\mathbf{w}}{J}$$

Because the pipe is horizontal and velocity terms are to be neglected,

$$\left(\frac{Z_1 - Z_2}{J}\right)\left(\frac{g}{g_c}\right)^{\!\!\!\diagup 0}, \quad \left(\frac{V_2^2 - V_1^2}{2g_cJ}\right)^{\!\!\!\diagup 0}$$

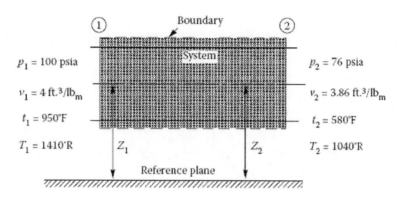

**FIGURE 4.19**
Illustrative Problem 4.18.

Also, $w\!\!\!/\,J^0$, no work crosses the boundaries of the system. The energy equation is reduced to

$$u_1 + \frac{p_1 v_1}{J} + q = u_2 + \frac{p_2 v_2}{J}$$

However, Equation 4.21 permits us to express the internal energy change for the gas in terms of the temperature change as

$$u_2 - u_1 = c_v \left( T_2 - T_1 \right) \quad \text{for constant } c_v$$

Therefore,

$$q = c_v \left( T_2 - T_1 \right) + \frac{p_2 v_2}{J} - \frac{p_1 v_1}{J}$$

By inserting numerical quantities, we obtain

$$q = 0.32 \left( 1040 - 1410 \right) + \frac{76 \times 144 \times 3.86}{778} - \frac{100\left(144\right)\left(4\right)}{778}$$

$$= -118.4 + 54.3 - 74.0 = -138.1 \, \text{Btu/lb}_m$$

Thus, 138.1 Btu/lb$_m$ is transferred *from* the gas.

---

## ILLUSTRATIVE PROBLEM 4.19

If the pipe referred to in Illustrative Problem 4.18 was a vertical run of pipe such that section ② was 100 ft. above section ①, determine the direction and magnitude of the heat transfer.

**SOLUTION**

Using the reference plane of Figure 4.19 to coincide with the elevation of the pipe at section ① makes $Z_1 = 0$. The energy equation becomes

$$u_1 + \frac{p_1 v_1}{J} + q = u_2 + \frac{p_2 v_2}{J} + \frac{Z_2}{J}\frac{g}{g_c}$$

and

$$q = u_2 - u_1 + \frac{p_2 v_2}{J} - \frac{p_1 v_1}{J} + \frac{Z_2}{J}\frac{g}{g_c}$$

Using $u_2 - u_1 = c_v(T_2 - T_1)$ and $Z_2 = 100$ ft. gives us

$$q = 0.32(1040 - 1410) + \frac{76 \times 144 \times 3.86}{778} - \frac{100(144)(4)}{778} + \frac{100}{778}\frac{g}{g_c}$$

Letting $g/g_c = 1$ yields

$$q = -118.4 + 54.3 - 74.0 + 0.13 = -138.0 \text{ Btu/lb}_m$$

For this problem, neglecting the elevation term leads to an insignificant error.

### 4.5.3 Boiler

The next application of the first law will be to the boiler or steam generator. The basic purpose of the steam generator is the turning of water into steam by the application of heat. It consists of a combination of many elements. In this unit, solid (coal or biomass), liquid (oil), or gas (methane or natural gas) is burned in the furnace with air that has been preheated in the air heater. In addition to generating superheated steam, this unit reheats steam from the high-pressure turbine exhaust and may return it to the low-pressure turbine for reheating purposes if the system was designed that way. The energy input to this system comes from the fuel, air, and feed water, and the useful output is steam. Because the purpose of the unit is to generate steam, the energy in the stack gas, unburned fuel, and heat transfer to the surroundings all represent losses that decrease the useful steam output. Again, we shall return to the steam generator in Chapter 8, where it is studied in some detail. For the present, the following problem will serve to illustrate the application of the first law to this unit.

**ILLUSTRATIVE PROBLEM 4.20**

A steam boiler is required to produce 10,000 lb$_m$/h of superheated steam at 1000°F and 1000 psia ($h = 1505.9$ Btu/lb$_m$) from feed water supplied at 1000 psia and 100°F ($h = 70.68$ Btu/lb$_m$). How much energy has been added to the water to convert it to steam at these conditions? The data are shown in Figure 4.20.

**FIGURE 4.20**
Illustrative Problem 4.20.

## SOLUTION

As indicated in Figure 4.20, we can consider this system as a single unit with feed water entering and steam leaving. If well designed, this unit will be thoroughly insulated, and heat losses will be reduced to a negligible amount. Also, no work will be added to the fluid during the time it is passing through the unit, and kinetic energy differences will be assumed to be negligibly small. In large units, the inlet and outlet may be as much as 100 ft. apart. However, 100/778 is 0.129 Btu/lb$_m$, which is quite small compared to the 1000 Btu/lb$_m$ or more involved in this problem. On this basis, differences in elevation will also be considered negligible. Once again, the energy equation is

$$\frac{Z_1}{J}\frac{g}{g_c} + \frac{V_1^2}{2g_cJ} + u_1 + \frac{p_1v_1}{J} + q = \frac{Z_2}{J}\frac{g}{g_c} + \frac{V_2^2}{2g_cJ} + u_2 + \frac{p_2v_2}{J} + \frac{w}{J}$$

and for this problem

$$\left(\frac{Z_1 - Z_2}{J}\right)\left(\frac{g}{g_c}\right)^0, \quad \left(\frac{V_2^2 - V_1^2}{2g_cJ}\right)^0$$

Therefore,

$$u_1 + \frac{p_1v_1}{J} + q = u_2 + \frac{p_2v_2}{J}$$

or because

$$h = u + \frac{pv}{J}$$

$q = h_2 - h_1$ (this is the net value, because we assumed no heat losses). Using the data given, we have

$$q = (1505.9 - 70.68) = 1435.2 \text{ Btu/lb}_m$$

For 10,000 $lb_m/h$

$$10000 \times 1435.2 = 14.35 \times 10^6 \text{ Btu/h}$$

are required.

Note that the working fluid (water) is our system and that the solution gives us the energy added to the water. This is not the energy released by the combustion of fuel.

---

### ILLUSTRATIVE PROBLEM 4.21

Steam that is initially saturated, is superheated in a boiler at constant pressure. Determine the final state if the initial pressure, $p_1$ is 500 psia and the final temperature, $T_2$ is 800°F. How much heat per pound of steam, $q_{in} = h_2 - h_1$ was added?

### SOLUTION

From the saturation table, Appendix 2, 500 psia corresponds to a saturation temperature, $T_{sat}$ of 467.13°F, with a saturated vapor enthalpy, $h_g$ of 1205.3 Btu/$lb_m$. At 500 psia and 800°F, the superheated vapor has an enthalpy of 1412.1 Btu/$lb_m$. Because this process is a steady-flow process at constant pressure, the energy equation becomes $q = h_2 - h_1$ assuming that differences in the kinetic energy and potential energy terms are negligible. Therefore, $q = 1412.1 - 1205.3 = 206.8$ Btu/$lb_m$ added.

---

### 4.5.4 Nozzle

A nozzle is a static device that is used to convert the energy of a fluid into kinetic energy. Basically, the fluid enters the nozzle at a high pressure and leaves at a lower pressure. In the process of expanding, velocity is gained as the fluid progresses through the nozzle. No work is done on or by the fluid in its passage through the nozzle.

---

### ILLUSTRATIVE PROBLEM 4.22

Steam is expanded in a nozzle from an initial enthalpy of 1220 Btu/$lb_m$, to a final enthalpy of 1100 Btu/$lb_m$. (a) If the initial velocity of the steam is negligible, what is the final velocity? (b) If the initial velocity is 1000 ft./s, what is the final velocity? See Figure 4.21.

### SOLUTION

The energy equation for a steady-flow device is

$$\frac{Z_1}{J}\frac{g}{g_c} + \frac{V_1^2}{2g_cJ} + u_1 + \frac{p_1v_1}{J} + q = \frac{Z_2}{J}\frac{g}{g_c} + \frac{V_2^2}{2g_cJ} + u_2 + \frac{p_2v_2}{J} + \frac{w}{J}$$

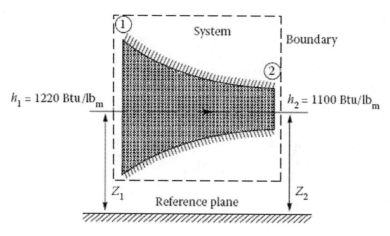

**FIGURE 4.21**
Illustrative Problem 4.21.

For this device, differences in elevation are negligible. No work is done on or by the fluid, friction is negligible, and due to the speed of the fluid flowing and the short length of the nozzle, heat transfer to or from the surroundings is also negligible. Under these circumstances,

$$\left(\frac{Z_1 - Z_2}{J}\right)\!\!\left(\frac{g}{g_c}\right)^{\!\!\!\nearrow}, \quad q^{\nearrow}, \quad \left(\frac{w}{J}\right)^{\!\!\!\nearrow}$$

Therefore,

$$u_1 + \frac{p_1 v_1}{J} + \frac{V_1^2}{2g_c J} = u_2 + \frac{p_2 v_2}{J} + \frac{V_2^2}{2g_c J}$$

or

$$h_1 - h_2 = \frac{V_2^2 - V_1^2}{2g_c J}$$

a.  For negligible entering velocity and

$$h_1 - h_2 = \frac{V_2^2}{2g_c J}$$

and

$$V_2 = \sqrt{2g_c J (h_1 - h_2)}$$

Substituting the data of the problem gives us

$$V_2 = \sqrt{2 \times 32.17\ \text{lb}_m \cdot \text{ft./lb}_f \cdot \text{s}^2 \times 778\ \text{ft.} \cdot \text{lb}_f / \text{Btu} \times (1220 - 1100)\frac{\text{Btu}}{\text{lb}_m}}$$

$$= 2451\ \text{ft./s}$$

b. If the initial velocity is appreciable,

$$h_1 - h_2 + \frac{V_1^2}{2g_cJ} = \frac{V_2^2}{2g_cJ}$$

Again, inserting numerical values yields

$$(1220-1100)\frac{\text{Btu}}{\text{lb}_\text{m}} + \frac{(1000)^2 \frac{\text{ft.}^2}{\text{s}}}{2\times 32.17\ \text{lb}_\text{m}\cdot\text{s}^2 \times 778\ \frac{\text{ft.lb}_\text{f}}{\text{Btu}}}$$

$$= \frac{V_2^2 \frac{\text{ft}^2}{\text{s}}}{2\times 32.17\ \text{lb}_\text{m}\cdot\text{ft./lb}_\text{f}\cdot\text{s}^2 \times 778\ \frac{\text{ft.}\cdot\text{lb}_\text{f}}{\text{Btu}}}$$

$$120 + 19.98 = \frac{V_2^2}{2\times 32.17\times 778}$$

$$V_2 = 2647\ \text{ft./s}$$

Note that in this part of the problem, the entering velocity was nearly 40% of the final velocity, yet neglecting the entering velocity makes approximately a 7.4% error in the answer. It is quite common to neglect the entering velocity in many of these problems.

---

## ILLUSTRATIVE PROBLEM 4.23

Assume steam enters a nozzle with an enthalpy of 3450 kJ/kg and leaves with an enthalpy of 2800 kJ/kg. If the initial velocity of the steam is negligible, what is the final velocity?

**SOLUTION**

Refer to Illustrative Problem 4.21, and note that in SI units,

$$\frac{V_2^2}{2} = h_1 - h_2$$

and

$$V_2 = \sqrt{2h_1 - h_2}$$

Substitution yields

$$V_2 = \sqrt{2\times 1000\ \text{J/kJ}\,(3450-2800)} = 1140.2\ \text{m/s}$$

(Note the need for 1000 J/kJ to obtain m/s.)

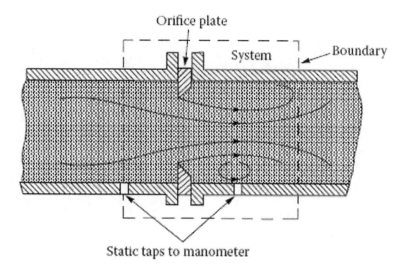

Orifice plate

System

Boundary

Static taps to manometer

**FIGURE 4.22**
Orifice as a meter.

### 4.5.5 Throttling Process

The next device that we consider is the case of an obstruction placed in a pipe (deliberately or as a result of the presence of a valve). One deliberate local obstruction that is often placed in a pipe is an orifice that is used to meter the quantity of fluid flowing through a system. When used in this manner, the orifice usually consists of a thin plate inserted into a pipe and clamped between flanges. The hole in the orifice plate is concentric with the pipe, and static pressure taps are provided upstream and downstream of the orifice. Because of the presence of the orifice, the flow is locally constricted, and a measurable drop in static pressure occurs across the orifice. Static taps placed, as shown in Figure 4.22 are usually used to measure the flow.

The advantages of this device are its relatively small size, the ease of installation in a pipe, and the fact that standard installations can be used without the need for calibration. However, the orifice meter behaves in the same manner as a partly open valve and causes a relatively high-pressure drop. This effect leads to the descriptive term of *throttling* for partly open valves, orifices, or other obstructions in pipes. In effect, the full flow is throttled back to some lesser flow by the obstruction. The throttling process is revisited in Chapter 5 where the concept of irreversibility is considered. Moreover, in Chapter 5 the process is shown on the Mollier chart.

---

### ILLUSTRATIVE PROBLEM 4.24

A fluid is flowing in a pipe. At some section of the pipe, there is an obstruction that causes an appreciable local pressure loss. Derive the energy equation for this process after the flow has become uniform in the downstream section of the pipe. As noted earlier, this process is known as a throttling process and is characteristic of valves and orifices in pipelines. Refer to Figure 4.23.

### SOLUTION

For the conditions of this problem, we can take differences in elevation to be negligible, differences in kinetic energy terms to be relatively small, and no work or heat to cross the system boundary. The complete energy equation is

**FIGURE 4.23**
Illustrative Problem 4.24

$$\frac{Z_1}{J}\frac{g}{g_c}+\frac{V_1^2}{2g_cJ}+u_1+\frac{p_1v_1}{J}+q=\frac{Z_2}{J}\frac{g}{g_c}+\frac{V_2^2}{2g_cJ}+u_2+\frac{p_2v_2}{J}+\frac{\mathbf{w}}{J}$$

and, with the assumption made for this process,

$$u_1+\frac{p_1v_1}{J}=u_2+\frac{p_2v_2}{J}$$

or

$$h_1=h_2$$

In words, a throttling process is carried out at constant enthalpy. One assumption should be verified; that is, the kinetic energy differences at inlet and outlet are indeed negligible. This can easily be done by using the constant-enthalpy condition and using the final properties of the fluid and the cross-sectional area of the duct to determine an approximate final velocity.

## 4.5.6 Heat Exchanger

The heat exchanger, as its name implies, is used to transfer heat from one fluid to another when the fluids are not mixed. In Chapter 11, the heat exchanger is considered in detail, when the problem of calculating the heat transfer is studied. Figure 4.24 shows a typical shell-and-tube heat exchanger. The "tube-side" fluid flows inside the tubes, and the "shell-side" fluid flows outside the tubes but is contained by the shell of the heat exchanger.

Let us now analyze the heat exchanger by referring to the sketch of Figure 4.25. The two fluids flow at rates $\dot{m}_1$ and $\dot{m}_2$, respectively. The temperatures at inlet and outlet are as shown. It is important to know that both fluids enter the heat exchanger but do not mix since they are separated by a solid (usually metal) material. For the purposes of our present study, we will assume that potential and kinetic terms in the energy equation are negligible. Also, we will assume that there are no heat losses from the unit to the surroundings. This requires all of the heat transferred from one fluid to be received by the other. That is,

$$\dot{Q}_1=\dot{Q}_2 \tag{4.23}$$

**FIGURE 4.24**
Shell-and-tube heat exchanger. (Courtesy of Patterson-Kelly Co., Division of Harsco Corp., East Stroudsburg, Pennsylvania, and Wikipedia)

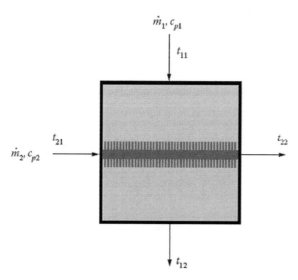

**FIGURE 4.25**
Analysis of heat exchanger.

But

$$\dot{Q}_1 = \dot{m}_1\left(h_{12} - h_{11}\right) = \dot{m}_1 c_{p1}\left(t_{12} - t_{11}\right) \tag{4.23a}$$

and

$$\dot{Q}_2 = \dot{m}_2\left(h_{22} - h_{21}\right) = \dot{m}_2 c_{p2}\left(t_{22} - t_{21}\right) \tag{4.23b}$$

Setting Equations 4.23a and 4.23b equal to each other, the resulting heat balance equation for single-phase fluids (all gas or all liquid) in the exchanger is as follows:

$$\dot{m}_1 c_{p1}\left(t_{12} - t_{11}\right) = \dot{m}_2 c_{p2}\left(t_{22} - t_{21}\right) \tag{4.24a}$$

For fluids undergoing phase change (from liquid to vapor, or from vapor to liquid) in heat exchangers such as in boilers and condensers, enthalpy values of the fluids should be used as follows:

$$\dot{m}_1\left(h_{12} - h_{11}\right) = \dot{m}_2\left(h_{22} - h_{21}\right) \tag{4.24b}$$

The enthalpy values are found by using the property tables in Appendix 2.

---

**ILLUSTRATIVE PROBLEM 4.25**

A heat exchanger is used to cool 400 $lb_m$/mm of lubricating oil. Hot oil enters at 215°F and leaves at 125°F. The specific heat of the oil is 0.85 Btu/$lb_m$·R. Cooling water enters the unit at 60°F and leaves at 90°F. If the specific heat of the water is 1.0 Btu/$lb_m$·R, determine the required water flow rate if heat losses are negligible.

**SOLUTION**

Refer to Figure 4.25:

$$\dot{Q}_{oil} = \dot{m}c_p\Delta t$$

$$= 400\ lb_m/min \times 0.85\ Btu/lb_m \cdot R \times \left(125 - 215\right) R$$

$$= -30\,600\ Btu/min\ \left(\text{out of oil}\right)$$

Because the heat out of the oil is the heat into the water,

$$\dot{m}_w c_{pw}\left(\Delta t\right)_w = 30\,600\ Btu/min$$

$$\dot{m}_w \times 1 \times \left(90 - 60\right) = 30\,600$$

$$\dot{m}_w = 1020\ lb_m/min$$

---

One type of heat exchanger that is used in applications such as power plants and refrigeration systems is the condenser. As the name implies, a condenser is used to condense a flowing vapor essentially at constant temperature (isothermally) during the phase change or transformation process from vapor to liquid. This process is also one of constant pressure

**FIGURE 4.26**
Shell-and-tube condenser.

in the wet or saturated region since the corresponding saturation pressure is a direct function (directly proportional) to the saturation or condensation temperature. The amount of energy transfer from the condensing fluid (fluid 1) to the flowing cooling fluid (fluid 2) can be calculated using Equation 4.23 by knowing the inlet and outlet enthalpies of the fluids.

Figure 4.26 shows a schematic of a shell-and-tube condenser, indicating the steam or vapor flow and cooling water paths. A lengthier discussion about heat exchangers, including condensers, can be found in Chapter 11.

---

### ILLUSTRATIVE PROBLEM 4.26

Steam is initially wet, having a moisture content of 3% (quality, $x$ of 97%) at 1 psia. If it is condensed to saturated liquid, how much heat is removed?

### SOLUTION

From the saturation table, Appendix 2 at 1 psia, $h_f = 69.74$ Btu/lb$_m$, $h_{fg} = 1036.0$ Btu/lb$_m$, and $h_g = 1105.8$ Btu/lb$_m$. Because the condensation process is carried out at constant pressure, the energy equation is $q = \Delta h = h_1 - h_2$. The initial enthalpy for the mixture is

$$h_1 = h_f + x_1 h_{fg} = 69.74 + (0.97)(1036.0) = 1074.66 \text{ Btu/lb}_m$$

The final enthalpy for the saturated (condensed) liquid is $h_2 = h_f = 69.74$. The enthalpy difference ($\Delta h = h_1 - h_2$) is $1074.66 - 69.74 = 1004.92$ Btu/lb$_m$ removed during the condensation process.

The computer solution is given below where $\Delta h = 1074.3 - 69.72 = 1004.6$ Btu/lb$_m$.

| Saturation Properties | | |
|---|---|---|
| $T = 101.71$ | deg F | |
| $P = 1.0000$ | psia | |
| | f | g |
| $v$ (ft$^3$/lb$_m$) | 0.016137 | 333.56 |
| $h$ (BTU/lb$_m$) | 69.72 | 1105.4 |
| $s$ (BTU/lb$_m$F) | 0.1326 | 1.9774 |
| $u$ (BTU/lb$_m$) | 69.72 | 1043.6 |

**Thermo Properties**

| | |
|---|---|
| $T = 101.71$ | deg F |
| $P = 1.0000$ | psia |
| $v = 323.55$ | ft.$^3$/lb$_m$ |
| $h = 1074.3$ | BTU/lb$_m$ |
| $s = 1.9221$ | BTU/lb$_m$F |
| $u = 1014.4$ | BTU/lb$_m$ |
| $x = 0.9700$ | |
| Region: saturated | |

### 4.5.7 Filling a Tank

Consider the following situation, namely, that an evacuated (completely empty with no pressure inside) tank is connected to a large steam source, as shown in Figure 4.27. The valve is opened, and steam is allowed to flow into the tank until the pressure in the tank equals the pressure in the line. Assume that the tank is perfectly insulated and that no shaft work occurs during the process. Also, assume that the kinetic and potential energy terms are negligible. In order to write the energy equation for this process, we will first define our system as consisting of the tank and the valve up to the supply side of the valve. The process of charging the closed tank is an example of an open or flow system in transient flow that can be solved using the information that we have already developed. The supply steam is throttled through the valve, and its enthalpy remains constant during the process. This assumes that the mass of steam withdrawn from the supply line is small enough to leave the conditions in the supply line constant. The change in energy of the mass in the tank (closed system with no boundary work) will be $m_2 u_2 - m_1 u_1$, where $m_1$ and $m_2$ are the initial and final masses in the tank and $u_1$ and $u_2$ are the initial and final internal energy, respectively. The initial mass in the tank, $m_1$, is taken as zero, because the tank was specified as being evacuated (completely empty) at the start of the process. The total energy into the tank will equal $m_2 h_i$, where $h_i$ is the enthalpy of the source. Equating these energy terms gives us

$$m_2 u_2 = m_2 h_i \tag{4.25}$$

or

$$u_2 = h_i \tag{4.26}$$

**FIGURE 4.27**
Filling a tank.

From Equation 4.25, we see that the final internal energy of the mass in the tank equals the enthalpy of the steam in the supply line. As we will see in Chapters 5 and 6, the final temperature in the tank will be greater than the temperature in the supply line.

## 4.6 Review

Using the material developed in Chapter 2, namely, the concepts of energy, work, heat, and the statement of the first law of thermodynamics as being the conservation of energy, we analyzed both nonflow and steady-flow mechanical systems. For the nonflow or closed system, one in which no mass crosses the boundary, we were able to write an energy equation that related the internal energy and heat entering a system to the internal energy and work leaving a system. This equation then permitted us to analyze several nonflow situations, including the adiabatic, constant-volume, and constant-pressure systems. As a consequence of our analysis of the constant-pressure system, we defined a new property, which we called enthalpy, to be the sum of the internal energy plus the product of the pressure and the specific volume of a substance (i.e., flow work). We also showed that the work of a nonflow, quasi-static process can be taken to be the area under the $p$–$v$ curve.

The term *steady* implies a system that does not vary with time. For the steady-flow system, this requires that the rate at which mass enters must equal the rate at which mass leaves the system. We then developed the continuity equation, which expresses this fact mathematically. Then we considered the energy terms required to write the first law of thermodynamics for the steady-flow system. These terms were potential energy, kinetic energy, and flow work, which we considered in Chapter 2. Using these terms with the already developed concepts of internal energy, heat, and work enabled us to write an energy equation that we called the steady-flow energy equation or, more simply, the general energy equation. Before we applied this equation to several industrial situations or systems, we developed two new properties from the definition of the term *specific heat*. These properties were the specific heat at constant pressure ($c_p$) and the specific heat at constant volume ($c_v$), and we were able to show their relation to the properties enthalpy and internal energy, respectively.

Our final part of this chapter was the application of the general energy equation to several common industrial situations: the boiler, the turbine, the nozzle, throttling, pipe flow, the heat exchanger, and the filling of a tank. By making certain simplifications, it was possible to analyze each system quantitatively.

In addition to the foregoing, it should be apparent that the successful application of the material of this chapter requires a systematic approach to problem solving. It is strongly suggested that the following procedure be followed to analyze all problems:

1. Read, visualize, and understand the problem.

2. Draw a schematic sketch of the problem, and indicate all known quantities.

3. Write out the full corresponding (continuity or energy) equation.

4. On the basis of the statement of the problem or knowledge of the device in question, determine those terms that can be omitted or neglected.

5. Solve the problem by performing the necessary algebraic and arithmetic steps. Check carefully for dimensional consistency.

6. Using the solution as a first approximation, check the validity of the assumptions made in step 4.

There are no shortcuts to an understanding of thermodynamics, and it will be found that the procedure outlined here will prove to be invaluable.

## Key Terms

Terms used for the first time in this chapter are as follows:

**Adiabatic process:** A process in which no energy as heat enters or leaves during the specified state change.

**Bernoulli equation:** A very restrictive form of the energy equation that is frequently used in fluid mechanics when dealing with incompressible fluids.

**Boiler:** A device used to convert the chemical energy in a fuel into an increase in the thermal energy of a working fluid through the combustion of the fuel.

**Continuity equation:** An equation expressing the conservation of mass in a steady-flow system.

**Energy equation:** See *general energy equation.*

**Enthalpy:** The sum of the internal energy and the $pv$ (flow work) term.

**First law of thermodynamics:** An expression of the conservation of energy in a system.

**General energy equation:** The quantitative statement of the first law of thermodynamics applied to a steady-flow system.

**Nonflow energy equation:** The quantitative expression of the first law of thermodynamics applied to a nonflow system.

**Nozzle:** A static device used to increase the velocity of a fluid.

**Specific heat:** The ratio of the energy transferred as heat to the corresponding temperature change of a substance. The specific heat at constant volume and the specific heat at constant pressure are both physical properties of any substance.

**Steady-flow energy equation:** See the *general energy equation.*

**Throttling:** An irreversible, adiabatic, steady-flow process in which the fluid is caused to flow through an obstruction in a pipe with a resulting drop in pressure.

**Turbine:** A machine consisting of a rotor and stator whose purpose is to convert the energy in a working fluid to useful work.

## Equations Developed in This Chapter

| | | |
|---|---|---|
| Nonflow energy equation | $q_{net} - w_{net} = u_2 - u_1$ | (4.6) |
| Nonflow energy equation | $u_2 - u_1 = q - w_b$ | (4.8) |
| Enthalpy | $h = u + pv$ | (4.9) |

| | | |
|---|---|---|
| Continuity equation | $\dot{m} = pAV = p_1 A_1 V_1 = p_2 A_2 V_2$ | (4.9b) |
| Continuity equation | $\dot{m} = \dfrac{AV}{v}$ | (4.9c) |
| General energy equation | $\dfrac{Z_1}{J}\dfrac{g}{g_c} + \dfrac{V_1^2}{2g_c J} + u_1 + \dfrac{p_1 v_1}{J} + \dfrac{\mathbf{w}_{in}}{J} + q_{in}$ $= \dfrac{Z_2}{J}\dfrac{g}{g_c} + \dfrac{V_2^2}{2g_c J} + u_2 + \dfrac{p_2 v_2}{J} + \dfrac{\mathbf{w}_{out}}{J} + q_{out}$ | (4.11) |
| General energy equation | $Z_1 g + \dfrac{V_1^2}{2} + u_1 + p_1 v_1 + \mathbf{w}_{in} + q_{in}$ $= Z_2 g + \dfrac{V_2^2}{2} + u_2 + p_2 v_2 + \mathbf{w}_{out} + q_{out}$ | (4.11a) |
| Power | $\dot{W} = \dot{m}\mathbf{w}$ | (4.17) |
| Heat transfer rate | $\dot{Q} = \dot{m}q$ | (4.18) |
| Specific heat at constant pressure | $c_p = \dfrac{q}{\Delta T_p} = \dfrac{\Delta h}{\Delta T_p}$ | (4.20) |
| Specific heat at constant volume | $c_v = \dfrac{q}{\Delta T_v} = \dfrac{\Delta u}{\Delta T_v}$ | (4.21) |
| Heat exchanger balance equation for single-phase fluids | $\dot{m}_1 c_{p1}(t_{12} - t_{11}) = \dot{m}_2 c_{p2}(t_{22} - t_{21})$ | (4.24a) |
| Heat exchanger balance equation for two-phase fluids | $\dot{m}_1(h_{12} - h_{11}) = \dot{m}_2(h_{22} - h_{21})$ | (4.24b) |

## QUESTIONS

4.1 Does the first law of thermodynamics apply to each of the following situations? (a) An atom bomb; (b) a system that is time-dependent; (c) a chemical system that is exothermic, that is, giving off heat as the reaction proceeds; (d) the tides on the Earth; (e) the sun and the reactions that occur on the sun; (f) a permanent magnet.

4.2 An automobile engine consists of a number of pistons and cylinders. If a complete cycle of the events that occur in each cylinder can be considered to consist of a number of nonflow events, can the engine be considered a nonflow device?

4.3 Can you name or describe some adiabatic processes?

4.4 What is unique about the nonflow, constant-pressure process?

4.5 Is enthalpy the sum of internal energy plus flow work?

4.6 In words, what does the continuity equation express?

4.7 What are the limitations of the continuity equation?

4.8 All substances are subjected to pressure and have a specific volume. Because this means that they all have the product $pv$, do they all have flow work?

4.9 Do all substances have enthalpy?

4.10 The term $p\Delta v$ is the area under a $p$–$v$ curve, and the term $-v\Delta p$ is the area behind the $p$–$v$ curve. How do you interpret each of these areas?

4.11  Discuss the problems associated with the Bernoulli equation.

4.12  With all of the problems associated with the Bernoulli equation, why is it still used?

4.13  When a fluid is vaporized, the temperature does not change during the process as heat is added. What is the specific heat for this process?

4.14  Can you name another common process that is carried out at a constant temperature and therefore has a specific heat that is similar to the one in Question 4.13?

4.15  What is the importance of the fact that $c_p$ and $c_v$ are properties?

4.16  A number of assumptions were made when the seven applications of the energy equation were discussed. Is it always necessary to make these assumptions when any of these situations are analyzed?

4.17  In all seven applications, we took the difference in elevations to be negligible. Would this also be true if you were analyzing the hydroelectric power plants at Niagara Falls?

## PROBLEMS

Unless indicated otherwise, use $g_c = 32.174 \ \text{lb}_m \cdot \text{ft.}/\text{lb}_f \text{s}^2$ and $g = 32.174 \ \text{ft.}/\text{s}^2$ or $9.81 \ \text{m/s}^2$.

### Problems Involving Nonflow Systems

4.1  The internal energy of a nonflow system increases by 70 Btu when 82 Btu of work is done by the system on its surroundings. How much heat has been transferred to or from the system?

4.2  If the internal energy of a nonflow system increases by 90 kJ while the system does 125 kJ of work on the surroundings, determine the heat transfer to or from the system.

4.3  A nonflow process is carried out when 1000 Btu as heat is added to 10 $\text{lb}_m$ of hydrogen at constant volume. What is the change in specific internal energy? How much work is done by the gas?

4.4  If a nonflow process is carried out so that 30 Btu/$\text{lb}_m$ of work is removed from the process while 100 Btu/$\text{lb}_m$ is added as heat, determine the change (increase or decrease) in the specific internal energy of the fluid.

4.5  If 10 kg of a gas is heated by the addition of >5 kJ in a nonflow process, what is the change in its specific internal energy if the process is carried out at constant volume?

4.6  If a nonflow process is carried out so that 30 kJ/kg of work is removed while 75 kJ/kg is added as heat, determine the change in internal energy of the fluid.

4.7  A constant-pressure, nonflow process is carried out at 125 kPa. During the process, the volume changes from 0.15 to 0.05 $\text{m}^3$, and 25 kJ of heat is rejected. Evaluate (a) the work done and (b) the change in internal energy of the system.

4.8  A closed gaseous system undergoes a reversible process during which 20 Btu/$\text{lb}_m$ is rejected as heat and the specific volume changes from 0.5 to 0.2 $\text{ft.}^3/\text{lb}_m$. If the pressure is constant at a value of 40 psia, determine the change in specific internal energy.

4.9     In a constant-pressure, nonflow process, 10 $lb_m$ of gas has 500 Btu added to it. During the process, the internal energy decreases by 25 $Btu/lb_m$. How much work was done by the gas per pound of gas?

4.10    A constant-pressure, nonflow system receives heat at a constant pressure of 350 kPa. The internal energy of the system increases by 180 kJ while the temperature increases by 170°C, and the work done is 75 kJ. Determine $c_p$ and the change in volume if there is 1.5 kg in the cylinder.

4.11    A constant-pressure, nonflow process is carried out at a pressure of 200 kPa. If 50 kJ of heat is removed while the volume changes from 0.2 to 0.1 $m^3$, what is the change in the specific internal energy of the working fluid? Assume 0.5 kg of fluid.

4.12    A gas that is contained in a cylinder is made to undergo a constant-pressure expansion from a volume of 0.1 $m^3$ to a volume of 0.4 $m^3$. If the piston moves to keep the pressure constant at 400 kPa, determine the work in kilojoules.

4.13    A nonflow system has a mass of working fluid of 5 kg. The system undergoes a process in which 140 kJ as heat is rejected to the surroundings and the system does 80 kJ of work. Assuming that the initial specific internal energy of the system is 500 kJ/kg, determine the final specific internal energy in kJ/kg.

4.14    A nonflow system has a mass of 5 lb. in it. The system undergoes a process during which 200 $ft.lb_f$ of heat is transferred to the surroundings. If the system does 1000 $ft.lb_f$ of work on the surroundings, determine the change in specific internal energy of the fluid.

4.15    A nonflow process is carried out adiabatically. If 55 kJ of work is removed from 4 kg of fluid, what is the change in the total internal energy and the specific internal energy of the fluid?

4.16    A nonflow process is carried out adiabatically. What is the change in specific internal energy of the fluid if 55,000 ft. $lb_f$ of work is removed from 78 $lb_m$ of fluid in this process?

4.17    During a certain nonflow process, 100 $Btu/lb_m$ is added as heat to the working fluid while 25,000 ft. $lb_f/lb_m$ is extracted as work. Determine the change in the internal specific energy of the fluid.

4.18    If the enthalpy of a gas is 240 $Btu/lb_m$ when the pressure is 100 psia and the specific volume is 3.70 $ft.^3/lb_m$, determine the specific internal energy of the gas.

4.19    What is the specific internal energy of a gas whose enthalpy is 500 kJ/kg? At this condition, the pressure is 100 kPa and the specific volume 0.1 $m^3/kg$.

4.20    A closed system receives 1 kJ as heat, and its temperature rise is 10 K. If 20 kg is in the system, what is the specific heat of the process?

4.21    Determine the specific heat of an adiabatic process. Is this the specific heat at a constant volume or the specific heat at constant pressure?

4.22    Two kilograms of a gas receive 200 kJ as heat at constant volume. If the temperature of the gas increases by 100°C, determine the $c_v$ of the process.

4.23    Heat is supplied to a gas in a rigid container. If the container has 0.6 $lb_m$ of gas in it and 100 Btu is added as heat, determine the change in temperature of the gas and the change in its specific internal energy. For this gas, $c_v$ is 0.35 $Btu/lb_m$· R.

4.24    If 100 kJ/kg of heat is added to 10 kg of a fluid while 25 kJ/kg is extracted as work, determine the change in the specific internal energy of the fluid for a nonflow process.

4.25 Heat is supplied to a gas that is contained in a rigid container. If 0.2 kg of gas has 100 kJ as heat added to it, determine the change in temperature of the gas and the change in its specific internal energy. Use $c_v = 0.7186$ kJ/kg·K and $c_p = 1.0062$ kJ/kg K.

4.26 If $c_p$ and $c_v$ of air are, respectively, 1.0062 and 0.7186 kJ/kg·°K and 1 MJ as heat is added to 10 kg in a nonflow, constant-pressure process, what is the final temperature of the gas, and how much work is done by the gas? Assume that the initial temperature of the gas is 50°C.

4.27 The $c_p$ and $c_v$ of air are 0.24 and 0.17 Btu/lb$_m$· R, respectively. If 1000 Btu is added as heat to 20 lb$_m$ of air in a nonflow, constant-pressure process, what is the final temperature? How much work is done by the gas? The initial temperature is 100°F.

4.28 In a certain process, the working fluid is cooled until 100 Btu/lb$_m$ has been extracted as heat. During this process, 100 ft·lb$_f$/lb$_m$ of fluid is added as work. If the working substance undergoes a change in temperature from 100°F to 50°F, what is the specific heat of the process?

4.29 Air is adiabatically compressed in a nonflow process from a pressure and temperature of 14.7 psia and 70°F to 200 psia and 350°F. If $c_v$ of the air is 0.171 Btu/lb$_m$·R, determine the change in specific internal energy of the air and the work done.

4.30 If a nonflow system has 6 kg in it and undergoes a process in which 175 kJ as heat is transferred to its surroundings, determine the final specific internal energy in kJ/kg if the work done by the system is 100 kJ. Assume that the initial specific internal energy is 400 kJ/kg.

4.31 A nonflow, cylinder-piston apparatus contains 0.5 lb$_m$ of a gas. If 10 Btu is supplied as work to compress the gas as its temperature increases from 70°F to 150°F, determine the energy interchange as heat. For the gas, $c_v$ can be taken as 0.22 Btu/lb$_m$· R.

4.32 A nonflow system undergoes a process in which 42 kJ of heat is rejected. If the pressure is kept constant at 125 kPa while the volume changes from 0.20 to 0.06 m³, determine the work done and the change in internal energy.

*4.33 Four individual nonflow processes numbered as 1, 2, 3, and 4 are carried out. For each of the processes, fill in the following table. Assume all values are in kJ.

| Process | $q$ | $w$ | $u_2$ | $u_1$ | $\Delta u$ |
|---|---|---|---|---|---|
| 1 | 15 | 8 | | 6 | |
| 2 | 22 | | 16 | 4 | |
| 3 | −20 | | 20 | | 15 |
| 4 | | 18 | 14 | 4 | |

*4.34 A closed cycle consists of the three processes listed in the table below. Fill in the missing values if all the quantities are in Btu/lb$_m$. (*Hint:* A closed cycle returns to its starting point.)

| Process | $q$ | $w$ | $\Delta u$ |
|---|---|---|---|
| 1 → 2 | 200 | $x$ | 100 |
| 2 → 3 | $z$ | 100 | $t$ |
| 3 → 1 | $y$ | −150 | 300 |

4.35 A tank contains 200 lbm of water and 20 lbm of steam at 400 psia. Determine the volume of the tank and the quality of the mixture.

4.36   A tank contains a steam–water mixture. The tank volume is 1.5 m³ and the pressure is 700 kPa. Calculate the mass of the mixture in the tank if the quality is 0.85.

4.37   A 5 m³ drum contains saturated steam at 300°C. Determine the pressure in the drum and the mass of steam in the drum.

4.38   A 3 m³ drum contains saturated water at 150°C. Determine the pressure in the drum and the mass of water in the drum.

4.39   90 A steam drum has a volume of 70 ft.³. If 70% of the volume is occupied by the vapor and the contents of the drum are at 500 psia, determine the weight of liquid and vapor in the drum.

4.40   Ten pounds of steam–water mixture occupies a steam drum. If the quality of the mixture is 65%, what is the volume of the drum? The pressure is 100 psia.

4.41   A steam drum has a volume of 7 m³. If the quality is 80% and the drum is at MPa, determine the weight of liquid and vapor in the drum.

4.42   A rigid, closed vessel contains a mixture of liquid and vapor at 400°F. If there is 45% liquid and 55% vapor by volume in the tank, determine the quality and pressure in the vessel.

4.43   A mixture of steam and water at 230°F has a volume of 150 ft.³. If there is 12 $lb_m$ in the mixture, determine its quality.

4.44   Dry saturated steam at 350°C is cooled to 300°C in a tank. Determine the final quality.

4.45   Dry saturated steam at 350°F is cooled in a tank until the temperature is 300°F. Determine the final quality.

4.46   Dry saturated steam at 200°C expands to a specific volume of 1.1 m³/kg at a constant temperature. Determine the final pressure.

4.47   How much heat must be added to 1 $lb_m$ of dry saturated steam at 500 psia in a closed tank to convert it to superheated steam at 900 psia and 1000°F?

4.48   Dry saturated steam fills a 100 ft.³ container at 100 psia. If the container is cooled until the contents are at 50 psia, what is the quality of the final mixture?

4.49   A closed, rigid vessel contains vapor and liquid at 400°F. If there is initially 45% liquid and 55% vapor by volume in the tank, determine the final percent by volume if the tank is cooled to 300°F.

4.50   One pound of superheated steam is initially at 100 psia and 400°F. If it is cooled at constant volume, determine the pressure at which it becomes saturated.

4.51   A rigid (constant volume) tank has a volume of 2.0 ft.³ and contains saturated water at 50 psia. Heat is transferred from the tank until the pressure is 20 psia. Calculate the heat transfer from the tank.

4.52   A closed container has 20 $lb_m$ of liquid and 20 lbm of vapor in it at 400°F. Determine the heat that must be added to raise the temperature to 425°F.

4.53   Heat is added to steam in a closed cylinder. If a movable piston is placed on one end of the cylinder, it is possible (in principle) to carry out the process at constant pressure. If the initial steam is at 250°F and is saturated, how much heat is added if the final temperature is 500°F?

4.54   Steam having a quality of 0.80 is contained in a rigid vessel at a pressure of 300 psia. Heat is added until the temperature reaches 500°F. Determine the final pressure.

4.55   In a constant-pressure piston–cylinder process, steam is compressed from 1 MPa and 500°C to a saturated vapor. Calculate the work done to compress the steam and the heat transfer per kilogram.

## Problems Involving the Continuity Equation

4.56   Ten cubic feet of fluid per second flows in a pipe 1 in. in diameter. If the fluid is water with a density of 62.4 $lb_m/ft.^3$, determine the weight rate of flow in pounds per hour.

4.57   A gallon is a volume measure of 231 $in^3$. Determine the velocity in a pipe 2 in. in diameter when water flows at the rate of 20 gal./min in the pipe.

4.58   When a fluid of constant density flows in a pipe, show that for a given mass flow the velocity is inversely proportional to the square of the pipe diameter.

4.59   Water having a density of 1000 $kg/m^3$ flows in a pipe that has an internal diameter of 50 mm. If 0.5 $m^3/s$ flows in the pipe, determine the mass flow rate in kilograms per hour.

4.60   What is the velocity of the water flowing in Problem 4.38?

4.61   Air flows through a 12 in. × 12 in. duct at a rate of 800 cfm. Determine the mean flow velocity in the duct.

4.62   Air flows through a rectangular duct 0.5 m × 0.75 m at the rate of 200 $m^3/min$. Determine the average velocity of the air.

4.63   Water flows through the pipe shown at the rate of 30 $ft.^3/s$. Determine the velocity at sections ① and ②. Also calculate the mass flow rate per second in the pipe at these sections. The diameters shown in Figure P4.63 are internal diameters.

**FIGURE P4.63**
Sudden expansion pipe problem.

4.64   Air, having a specific weight of 11.8 $N/m^3$, flows in the pipe shown in Figure P4.64 with a velocity of 10 m/s at section ①. Calculate the weight flow rate per second of the air and its specific weight at section ② if the air has a velocity of 3 m/s at section ②.

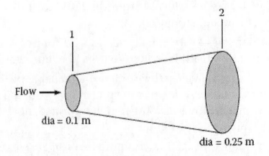

**FIGURE P4.64**
Gradual expansion pipe problem.

4.65 If the water level in the tank remains constant, determine the velocity at section ②, as shown in Figure P4.65. Assume the water to be incompressible.

**FIGURE P4.65**
Water tank flow problem.

4.66 Water flows through a pipe at the rate of 1500 L/min. At one section, the pipe is 250 mm in diameter, and later on, it decreases to a diameter of 100 mm. What is the average velocity in each section of the pipe?

4.67 A pump can fill a tank in 1 min. If the tank has a capacity of 18 gal., determine the average velocity in the 1.25-in. diameter pipe that is attached to the pump.

4.68 For the liquid rocket shown in Figure P4.68, calculate $V_2$ for steady operation.

**FIGURE P4.68**
Nozzle problem.

4.69 Water flows through the nozzle shown in Figure P4.69 at the rate of 100 $lb_m$/s. Calculate the velocity at sections ① and ②.

**FIGURE P4.69**
Nozzle problem.

## Problems Involving the Steady-Flow Energy Equation

4.70 A water turbine operates from a water supply that is 100 ft. above the turbine inlet. It discharges to the atmosphere through a 6-in.-diameter pipe with a velocity of 25 ft./s. If the reservoir is infinite in size, determine the power out of the turbine if the density of water is 62.4 $lb_m$/ft.$^3$.

4.71    A water pump is placed at the bottom of a well. If it is to pump 1.0 ft.$^3$/s to the surface, 100 ft. away, through a 4-in.-inside-diameter pipe, determine the power required in foot-pounds per second and horsepower if the density of water is 62.4 lb$_m$/ft.$^3$.

4.72    Air is compressed until its final volume is half its initial volume. The initial pressure is 100 psia, and the final pressure is 35 psia. If the initial specific volume is 1 ft.$^3$/lbm, determine the difference in the $pv/J$ term in Btu/lb$_m$. Is this difference "flow work" for this process?

4.73    Determine the power of a pump if 2000 kg/min of water is compressed from 100 kPa to 1 MPa. The water density can be taken to be 1000 kg/m$^3$, and its temperature does not change. The inlet to the pump is 100 mm in diameter, and the outlet is 150 mm in diameter. The inlet is 50 m below the outlet.

4.74    A steady-flow device is operated with an entering pressure of 50 psia and a specific volume of 0.8 ft.$^3$/lb$_m$. At the exit of the device, the pressure is 15 psia and the specific volume 4.2 ft.$^3$/lb$_m$. Determine the change in flow work in this device.

4.75    If a fluid flows past a section of pipe with a pressure of 100 kPa and having a specific volume of $10^{-3}$ m$^3$/kg, determine its flow work at this section.

4.76    At the entrance to a steady-flow device, the pressure is 350 kPa and the specific volume 0.04 m$^3$/kg. At the outlet, the pressure is 1 MPa and the specific volume 0.02 m$^3$/kg. Determine the flow work change in this device.

4.77    The enthalpy of a substance is found to be 1000 Btu/lb$_m$ at 1000°F above an arbitrary datum of 0°F. If $c_v$ is a constant and equal to 0.5 Btu/lb$_m$· R, determine the difference in the $pv/J$ terms at zero and 1000°F. Does this difference represent flow work?

*4.78   A steam turbine is supplied with steam at 700°F and 500 psia. The steam is expanded adiabatically until the final condition is saturated vapor at 300°F. If the properties of steam at the initial and final states are as given here, determine the work output of the turbine per pound of steam.

| 500 psia and 700°F | 300°F (saturated vapor) |
|---|---|
| $h = 1356.7$ Btu/lb$_m$ | $h = 1180.2$ Btu/lb$_m$ |
| $v = 1.3040$ ft.$^3$/lb$_m$ | $v = 6.472$ ft.$^3$/lb$_m$ |
| $u = 1236.0$ Btu/lb$_m$ | $u = 1100.0$ Btu/lb$_m$ |

*4.79   A fluid at 1 MPa has a specific volume of 0.2 m$^3$/kg and an entering velocity of 200 m/s. The heat loss is 10 kJ/kg, and the fluid does 180 kJ/kg of work. The fluid leaves the device with a pressure of 200 kPa, a specific volume of 1 m$^3$/kg, and a velocity of 600 m/s. What is the change in the specific internal energy of the fluid?

*4.80   A steam turbine operates adiabatically with inlet and other conditions as given here. Determine the work out of the turbine. State all assumptions.

| At 0.5 MPa and 500°C | At 0.1 MPa (saturated vapor) |
|---|---|
| $h = 3483.9$ kJ/kg | $h = 2675.5$ kJ/kg |
| $v = 0.7109$ m$^3$/kg | $v = 1.6940$ m$^3$/kg |
| $u = 3128.4$ kJ/kg | $u = 2506.1$ kJ/kg |

4.81 A steam turbine receives steam at 110 ft./s velocity and 1525 Btu/lb$_m$ enthalpy. The steam leaves at 810 ft./s and 1300 Btu/lb$_m$ enthalpy. What is the work out in Btu per pound mass flowing?

4.82 A turbine is operated with an enthalpy at the entrance of 1340 Btu/lb$_m$ and an exit enthalpy of 1285 Btu/lb$_m$. If the entrance velocity of the turbine is 150 ft./s and the exit velocity is 500 ft./s, determine the work out of the turbine.

4.83 Solve Problem 4.82 if there is a heat loss of 5.5 Btu/lb$_m$ of fluid by heat transfer from the turbine casing.

4.84 A gas turbine receives an air–fuel mixture having an enthalpy of 550 Btu/lb$_m$ of gas. If 10 Btu/lb$_m$ is lost by heat transfer from the turbine and the enthalpy of the leaving gas is 50 Btu/lb$_m$, how much work can be obtained from the turbine?

4.85 An air compressor takes in air with an enthalpy of 100 kJ/kg. The air is compressed to a final condition in which the enthalpy is 180 kJ/kg. As the air travels through the compressor, it loses 20 kJ/kg as heat. Determine the work per kilogram of the compressor.

4.86 Air having the following properties passes through a steady-flow device. Determine the work in kJ/kg if the heat loss is 20 kJ/kg.

|  | Inlet | Outlet |
|---|---|---|
| Pressure | 700 kPa | 170 kPa |
| Density | 3.0 kg/m³ | 0.60 kg/m³ |
| Velocity | 70 m/s | 200 m/s |
| Internal energy | 2100 kJ/kg | 2025 kJ/kg |

4.87 A gas turbine has gas entering it at 1100°F and leaving at 800°F. If the specific heat $c_p = 0.265$ Btu/lb$_m$· R, determine the work out of the turbine. Neglect losses and velocity effects.

4.88 Solve Problem 4.82 if the entering velocity is 100 ft./s, the leaving velocity is 300 ft./s, and there is a heat loss of 4.8 Btu/lb$_m$ from the turbine.

4.89 Assume 4.8 kg/s of steam enters a turbine. The inlet of the turbine is 2.5 m higher than the outlet. The inlet velocity is 132 m/s, and the outlet velocity is 327 m/s. Heat is lost from the casing of the turbine at the rate of 9.2 kJ/s. At the inlet of the unit, the enthalpy is 3127.4 kJ/kg, and at the outlet, the enthalpy is 2512 kJ/kg. Determine the power out of the turbine in kilowatts.

4.90 Gas enters a turbine at 600°C and leaves at 350°C. If the specific heats of the gas are $c_p = 0.8452$ kJ/kg·K and $c_v = 0.6561$ kJ/kg·K, determine the work out of the turbine. State all assumptions.

4.91 Assume the entering velocity in Problem 4.90 is 86 m/s and the leaving velocity is 232 m/s. If the outlet of the turbine is 3.2 m higher than the inlet, determine the work out of the turbine if the process is adiabatic.

4.92 Air flows in a compressor at the rate of 1.1 kg/s. The air enters the compressor at a pressure of 100 kPa, a temperature of 25°C, and a velocity of 72 m/s. It leaves with a pressure of 225 kPa, a temperature of 84°C, and a velocity of 125 m/s. A cooling jacket surrounds the compressor to cool the air, and it removes 15 kJ/kg as heat from the air. If $c_p = 1.0041$ kJ/kg·K and $c_v = 0.7172$ kJ/kg·K, calculate the power to operate the compressor.

4.93    Air expands through a nozzle from 1000 psia and 500°F to 600 psia and 0°F. If $c_p$ of air can be taken to be constant and equal to 0.24 Btu/lb$_m$· R, what is the final velocity? Assume that the initial velocity is zero.

4.94    If the initial velocity in Problem 4.93 is 9.25 ft./s, what is the final velocity?

4.95    Steam expands in a nozzle from an initial enthalpy of 1300 Btu/lb$_m$ to a final enthalpy of 980 Btu/lb$_m$. Determine the final velocity if the entering velocity and heat losses are negligible.

4.96    Solve Problem 4.95 if the initial velocity is 1100 ft./s.

4.97    Air expands in a nozzle from 7 MPa and 250°C to 3.5 MPa and 50°C. If $c_p = 1.0062$ kJ/kg·K and $c_v = 0.7186$ kJ/kg·K, determine the velocity out of the nozzle. State all assumptions.

4.98    Solve Problem 4.97 if the initial velocity is 325 m/s.

4.99    A nozzle has steam flowing through it. The steam enters the nozzle at 250 psia with a velocity of 100 ft./s. The initial specific volume is 2.31 ft.³/lb$_m$ and the initial internal energy is 1114.1 Btu/lb$_m$. The steam expands to 15.0 psia with a specific volume of 27.34 ft.³/lb$_m$ and a final internal energy of 1080.1 Btu/lb$_m$. Calculate the velocity at the exit of the nozzle.

4.100   Steam flows through a nozzle. If the properties are as given here, determine the exit velocity if the nozzle is well insulated. Neglect the inlet velocity.

|                  | Inlet              | Outlet             |
| ---------------- | ------------------ | ------------------ |
| Pressure         | 300 psia           | 100 psia           |
| Temperature      | 500°F              | 350°F              |
| Internal energy  | 1159.5 Btu/lb$_m$  | 1115.4 Btu/lb$_m$  |
| Enthalpy         | 1257.5 Btu/lb$_m$  | 1200.4 Btu/lb$_m$  |

4.101   Solve Problem 4.100 for an initial velocity of 600 ft./s.

4.102   Steam is generated in a boiler. If the initial water enters the boiler as saturated water at 1000 psia ($h = 542.4$ Btu/lb$_m$) and leaves as superheated steam at 950 psia and 1000°F ($h = 1507.4$ Btu/lb$_m$), how much heat is added to the steam to generate 1 lb. of steam at these conditions?

4.103   Steam is generated in a boiler. Water enters the boiler as a saturated liquid at 7 MPa ($h = 1267.00$ kJ/kg) and leaves as superheated steam at 6 MPa and 500°C ($h = 3422.2$ kJ/kg). Neglecting potential and kinetic terms, determine the amount of heat that is added to generate 1 kg of steam under these conditions. Assume that there are no heat losses.

4.104   Water flows in a heat exchanger, and heat is transferred from the water to heat air. At the inlet, the pressure of the water is 100 psia, and at the outlet, it is 80 psia. The initial specific volume is 0.017 736 ft.³/lb$_m$, and the final specific volume is 0.017 57 ft.³/lb$_m$. The initial enthalpy is 298.61 Btu/lb$_m$, the initial internal energy is 298.28 Btu/lb$_m$, the final enthalpy is 282.21 Btu/lb$_m$, and the final internal energy is 281.95 Btu/lb$_m$. Determine the heat transferred from the water. State all assumptions.

4.105   Air flows in steady flow in a pipeline whose internal diameter is constant. At a particular section of the pipe, the pressure, temperature, and velocity are

150 psia, 500 R, and 30 ft./s. The specific volume corresponding to these conditions is 1.24 ft.$^3$/lb$_m$. At a section farther downstream, the pressure is 300 psia, the velocity is 45 ft./s, and the temperature is 1500 R. Determine the specific volume of the air at the downstream section and also the magnitude and direction of the heat transfer in Btu per pound mass. Assume that $c_p$ is 0.24 Btu/lb$_m \cdot$ R and that it is constant.

4.106   Dry saturated steam at 200 psia flows adiabatically in a pipe. At the outlet of the pipe, the pressure is 100 psia and the temperature 400°F. The initial specific volume of the steam is 2.289 ft.$^3$/lb$_m$, and its internal energy is 1114.6 Btu/lb$_m$. The final enthalpy is 1227.5 Btu/lb$_m$, and the final specific volume is 4.934 ft.$^3$/lb$_m$. How much energy per pound mass was transferred? Was this into or out of the system? Was this heat or work?

4.107   A heat exchanger is used to heat petroleum distillates for use in a chemical process. Hot water enters the unit at 180°F and leaves at 105°F. The distillate enters at 85°F and leaves at 148°F. If the specific heat of water is taken to be 1.0 Btu/lb$_m \cdot$ R and the distillate has a specific heat of 0.82 Btu/lb$_m \cdot$R, determine the pounds of water required per pound of distillate flowing.

4.108   A hydroelectric turbine at the base of a 6 m tall dam receives 130 m$^3$/s of water (density = 1000 kg/m$^3$) from a large reservoir. Turbine discharge velocity is 5 m/s. Neglecting friction, determine the following: (a) turbine specific work and direction, in kJ/kg (b) turbine power, in kW and (c) discharge pipe diameter in m.

4.109   Steam flowing in a pipe is throttled by a partly open valve. The initial steam conditions are 600 psia and 800°F. Determine the final temperature if the final pressure of the steam is 100 psia.

4.110   Wet steam at a pressure of 200 psia and a quality of 90% is throttled to a final pressure of 80 psia. What is the temperature and quality of the final condition? Use the Steam Tables to obtain your solution.

4.111   Solve Problem 4.110 using the Mollier chart.

4.112   Wet steam leaves the exhaust of a turbine and is subsequently condensed. Assuming the wet steam to be at 0.5 psia with 15% moisture, determine the heat extracted in the condenser if the condensation process takes the mixture to saturated liquid.

4.113   Two kilograms of saturated water at 1.0 MPa are heated to saturated vapor at constant pressure. Determine the heat transfer for this process.

4.114   Two lb$_m$ of saturated water at 200 psia is vaporized to saturated steam at constant pressure. Determine the change in volume for this change.

4.115   Steam at 340 kPa and 600°C undergoes a constant-pressure process to 900°C. Calculate the work done per kilogram of steam.

4.116   Determine the change in internal energy of water vapor per lbm as it undergoes a constant-pressure process from 30 psia and 500°F to 700°F.

4.117   Dry saturated steam at 350°F is heated at constant pressure until it reaches a temperature of 600°F. Determine the work for this process in Btu/lbm.

# 5

## Second Law of Thermodynamics

### Learning Goals

After reading and studying the material in this chapter, you should be able to

1. Understand that work can be converted into heat, but that the conversion of heat into useful work may not always be possible.

2. Define a heat engine as a continuously operating system across whose boundaries flow only heat and work.

3. Define thermal efficiency as the ratio of the useful work delivered by a heat engine or cycle to the heat input to the engine or cycle.

4. Understand what is meant by the statement that a reversible process is any process performed so that the system and all its surroundings can be restored to their initial states by performing the process in reverse.

5. State the second law of thermodynamics as "Heat cannot of itself pass from a lower temperature to a higher temperature."

6. Understand that all natural processes are irreversible, and cite some of the effects that cause irreversibility.

7. Explain the four processes that constitute the Carnot cycle.

8. Deduce from the Carnot cycle three important general conclusions concerning the limits of the efficiency of a heat engine.

9. Define the new property that is introduced in this chapter that we have called entropy.

10. Understand that entropy is also a measure of the unavailability of energy that occurs in an irreversible process.

11. Calculate the change in entropy for a process in which there is a temperature change, such as the mixing of two fluids.

12. Understand that the entropy of an isolated system increases or, in the limit, remains the same, which, for a given internal energy, we interpret to mean that the state having the greatest entropy will be the most probable state that the system will assume. This state is called stable equilibrium.

13. Use entropy as a property in the analysis of thermodynamic systems.

## 5.1 Introduction

Thus far, we have considered various forms of energy (including energy in transition as both work and heat) without regard to any limitations on these quantities. It has been assumed that work and heat are mutually interchangeable forms of energy, and it may have appeared to students that the distinction made between these quantities was arbitrary and possibly not necessary. In this chapter, the interconvertibility of these quantities is explored with the object of determining any possible limitations and to express these limitations quantitatively, if they exist. As an example of the point in question, consider the motion of a block sliding along a rough horizontal plane. For motion to proceed along the plane, it is necessary that work is done on the body. All this work subsequently appears as heat at the interface between the block and the plane. There is no question that work has been converted into heat, but can the heat generated in this process be converted into an equivalent amount of work? Let us assume (incorrectly) that this heat can be converted into work without any losses in the process. We know that the energy (as heat) increases the motion of the individual molecules of the body. By increasing the molecular motion within the body, we have, in a general sense, done work, but it is also possible to distinguish that this form of work is not the same as the external work put into the process. The original transitional energy as work has been converted into heat, and this heat can be expressed as molecular work. However, this form of energy will not be available to return the body to its original state. From this simple example, we note that work can be converted into heat, but that the conversion of heat into useful work may not always be possible. Even though the first law states that energy is conserved, it does not furnish the necessary information to enable us to determine whether energy has become unavailable.

It is now necessary to define certain terms. The first of these is the concept of a *heat engine*. As defined by Keenan (1941, p. 58):

> A heat engine may be defined as a continuously operating system across whose boundaries flow only heat and work. It may be used to deliver work to external devices, or it may receive work from external devices and cause heat to flow from a low level of temperature to a high level of temperature. This latter type of heat engine is known as a refrigerator.

In essence, this definition of a heat engine can be taken to be the definition of a thermodynamic *cycle*, which we shall understand to be a series of thermodynamic processes during which the working fluid can be made to undergo changes involving only heat and work interchanges and is then returned to its original state.

The purpose of the conventional engineering thermodynamic cycle is, of course, to convert heat into work. In an air-conditioning or refrigeration cycle, work is used to remove heat from an area in which it is undesirable. Other special cycles exist but are not treated in this text. Associated with the concept of a cycle is the term *efficiency*. Because the usual purpose of a cycle is to produce useful work, the *thermal efficiency* of a cycle is defined as the ratio of the *net work* of the cycle to the *heat added* to the cycle, that is,

$$\eta = \frac{\text{network output}}{\text{heat added}} \times 100 \tag{5.1}$$

Note that the heat term is the heat added and not the net heat of the cycle. For power-producing cycles, the heat is usually added from some high-temperature source. Using the notation that $Q_{in}$ is the heat added to the cycle and that $Q_r$ is the heat rejected by the cycle, the first law applied to the cycle will yield $W / J = Q_{in} - Q_r$. Therefore,

$$\eta = \frac{Q_{in} - Q_r}{Q_{in}} \times 100 = 1 - \frac{Q_r}{Q_{in}} \times 100 \tag{5.2}$$

For cycles whose purpose is not the production of useful work, other standards of comparison have been devised and are in use.

If we consider the large central power station plant, the importance of obtaining high thermal efficiencies is at once obvious. An examination of Equation 5.2 leads us to the conclusion that minimizing the heat rejection of a cycle leads to the maximum conversion of heat to work. This leads us to two questions: (1) Must there be a rejection of heat from a cycle, and if so, (2) what is the best mode of cycle operation to minimize the heat rejected in order to obtain maximum thermal efficiency? These questions will be partially answered in this chapter, and we shall return to them when we study practical engine cycles.

## 5.2 Reversibility—Second Law of Thermodynamics

In Section 5.1, the illustration of a block sliding along a horizontal plane was used to introduce the concept that heat and work are not always mutually convertible without losses. This same body moving along the plane will also serve to answer the following question: By reversing each step of the process that caused the body to move along the plane, is it possible to restore the body to its original state, and at the same time will the surroundings also be restored to the condition that existed before the start of the original process? To answer this question, let us once again consider the forward motion of the body along the plane. We have stated that as the block moves, heat is generated at the interface between the block and the plane due to friction. This energy is transferred to the body and the plane, and will tend to raise the temperature of the body and its surroundings. When the block reaches the end of the plane, let us reverse the force system acting on the body and attempt to restore it to its original position. As the body moves back along the path, heat will again be generated at the interface between the body and the plane due to friction. Obviously, the heat generated on the return path is *in addition* to the heat generated at the interface during forward motion of the body. To a casual observer who viewed the block before the beginning of motion and then viewed it a small amount of time after motion had ceased, it would appear that the body had not moved and that it, as well as the surroundings, had been restored to its original state; however, this is not true. A net transfer of energy has taken place to the body and its surroundings during the downward and upward motions (or complete cycle), and the body or surroundings are not in their original state. Even though the net effect has been an infinitesimal change in the temperature of the body and its surroundings, it is a real effect that precludes us from saying that the system *and its surroundings* have been restored to their original states since a net transfer of energy has taken place. We also notice that each step of the forward motion of the body was not identically reversed because of this effect on the surroundings. The heat generated during the forward motion of the block was not returned to the system as work during the return motion. On the contrary, additional heat was generated during the return motion due to friction, and even if both the plane and the body were perfectly insulated from their surroundings, none of the heat generated would have been returned to the system as work.

The process we have considered is illustrative of an irreversible process. To formalize the concepts of reversibility and irreversibility, the following definition of *reversible process* is used:

A reversible process is any process performed so that the system and all its surroundings can be restored to their initial states by performing the process in reverse.

All processes of a reversible cycle must, therefore, also be reversible. Students should note that the concept of the frictionless, quasi-static process introduced earlier basically implies that such a process is reversible.

The *second law of thermodynamics* is an expression of empirical fact that all forms of energy are not necessarily equivalent in their ability to perform useful work. There are many statements and corollaries of the second law that can be found in the literature on thermodynamics. For the present, the statements of Clausius and Kelvin–Planck will serve to express the second law fully (Howell and Buckius 1992).

It is impossible to construct a device that operates in a cycle and whose sole effect is to transfer heat from a cooler body to a hotter body. (Heat cannot, of itself, pass from a lower temperature to a higher temperature.)

## The Clausius Statement

It is impossible to construct a device that operates in a cycle and produces no other effect than the production of work and exchange of heat with a single reservoir.

## The Kelvin–Planck Statement

One of the many consequences of the second law of thermodynamics is the conclusion that all natural processes are irreversible. It has already been shown that the presence of friction will cause a process to be irreversible. Some examples of irreversible processes are the following:

1. *Any process* in which work is transformed into internal energy via the agency of friction or inelastic action.
2. *Any process* in which inelastic molecular action occurs.
3. *Any process* that transfers heat from one portion of a system to another by virtue of a finite temperature difference (i.e., heat transfer).
4. *Any process* that causes temperature differences between parts of the same system.
5. *Any process* involving combustion or chemical reactions.
6. *Any process* that is not performed quasi-statically; thus, to be reversible, a process must proceed at an infinitesimally slow rate.

It is important for students to fully understand where the irreversibilities occur in the listed processes. Also, by observing the effects on the environment, other irreversible processes will become apparent.

The next question we ask is, under what conditions will a process be reversible? The answer is that, in reality, no process is reversible. However, as an abstract ideal, the reversible process is extremely useful, and this ideal can be achieved only if the process is frictionless and quasi-static—and then only for an isothermal or adiabatic process. The quasi-static process is always in thermodynamic equilibrium and is carried out with infinite slowness so that at any step in the process, it can be reversed and all steps retraced. Also, when such a process is specified to be either isothermal or adiabatic, temperature differences within the system or in parts of the system are precluded. To be general, we must exclude other irreversible effects such as magnetic hysteresis and electrical currents.

## 5.3 The Carnot Cycle

The material discussed so far in this chapter has served to define a cycle, its efficiency, and the concept of a reversible process. It would appear quite natural at this point to combine all these concepts and to discuss reversible cycles and their efficiency. Historically, these concepts were first enunciated by Nicolas Leonard Sadi Carnot in 1824, and the reversible thermodynamic cycle that he proposed now bears his name. It is interesting to note that Carnot did his work over 195 years ago. In this short span of human history, scientific thermodynamics has become a reality. Figure 5.1 shows some important names and dates in the development of the science of thermodynamics.

As noted in Section 5.2, the two frictionless, quasi-static processes that are reversible are the isothermal (constant temperature) and adiabatic (no energy as heat crosses the boundary). Carnot proposed a reversible cycle composed of two reversible isothermal processes and two reversible adiabatic processes, and on the basis of this cycle, he was able to reach certain general conclusions. Let us consider the cycle that has been named for him by describing each step of the cycle. Figure 5.2 (solid line) is a schematic of a direct engine cycle. For the Carnot cycle, we define the following sequence of events:

1. Heat is taken from an infinite reservoir (source) at $T_1$ isothermally and reversibly. Basically, this is equivalent to a quasi-static reception of heat into the cycle without temperature differences.

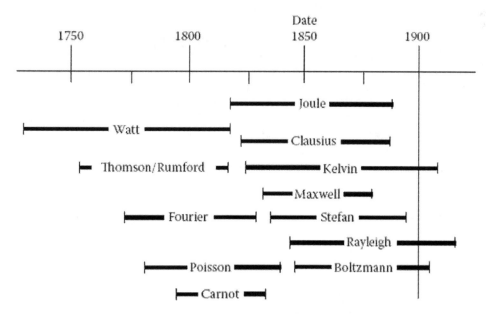

**FIGURE 5.1**
Some important dates and persons in the development of thermodynamics.

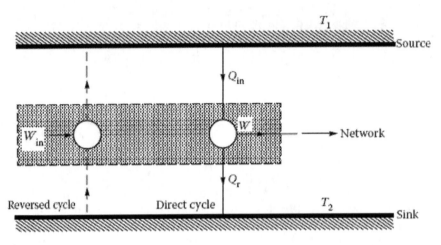

**FIGURE 5.2**
Elements of Carnot cycle.

2. The energy received from step 1 is permitted to produce work by expanding reversibly and adiabatically in an ideal frictionless engine. During this step, net work is produced, but no energy as heat is permitted to cross the boundaries of the system even though the pressure and temperature of the working fluid may have changed.

3. At this point in the cycle, the working fluid is at temperature $T_2$, and we shall want to return it to its starting point. To do this, we first reject heat at constant temperature ($T_2$) reversibly and isothermally to an infinite sink.

4. The final step in the cycle is to cause the working fluid to be adiabatically and reversibly compressed to its initial state.

Note that in every analysis of the Carnot cycle, $T_1$ always represents the higher temperature and $T_2$ always represents the lower temperature.

For a noncondensing gas, the steps of the cycle are portrayed on pressure–volume coordinates, as seen in Figure 5.3.

To visualize what a real engine working on the Carnot cycle could look like, consider a gas contained within a piston cylinder, as shown in Figure 5.4. The numbers refer to the steps in Figure 5.3. Initially, the gas is heated while the piston rises, with just enough heat added to maintain the gas at a constant temperature $T_H$. Normally, this expanding gas would tend to lose temperature as its pressure drops. This is an isothermal expansion process. Next, the piston continues to rise but no heat is transferred, causing the temperature and pressure to fall. This is an adiabatic expansion process. Next, the gas is cooled as heat is removed and the piston is allowed to fall, increasing the gas pressure sufficiently to keep its temperature constant at $T_C$. This is an isothermal compression process. Finally, the cold sink is removed and the piston continues to fall while the temperature and pressure increase back to the condition at the start of the cycle. This is an adiabatic compression process.

The Carnot cycle just described is a reversible cycle, and it is therefore possible to reverse each step in turn and thus reverse the cycle. Such a reversed cycle would effectively take work as an input and pump heat from $T_2$ to $T_1$. The reversed cycle is known as a *heat pump* and is discussed further in Chapter 10. It should be noted that the Carnot cycle is not

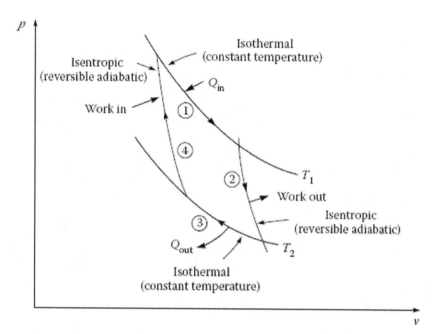

**FIGURE 5.3**
Carnot cycle on $p$–$v$ coordinates; noncondensable gas.

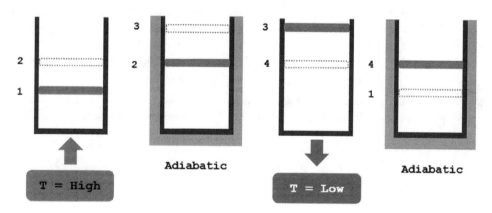

**FIGURE 5.4**
Visualized Carnot engine.

unique, and it is not the only reversible cycle that can be devised. Actually, many reversible cycles have been proposed as prototypes of real cycles. The power of the Carnot cycle is that the following general conclusions can be deduced from it:

1. *No engine operating between fixed source ($T_1$) and sink ($T_2$) temperatures and continuously delivering work can be more efficient than a reversible Carnot engine operating between these same temperature limits.*

   To prove this proposition, let us take two engines, engine 1 with an efficiency *greater* than the efficiency of a reversible Carnot engine, and engine 2, which is a reversible Carnot engine, and operate them between a hot reservoir at $T_1$ and a cold reservoir at $T_2$. If these engines are arranged as shown in Figure 5.5, we have engine 1 operating directly to produce 100 units of work. The output of engine 1 is now used to drive engine 2, the reversible engine, in the reversed direction. This reversed engine will take the same energy values used in the heat engine (*heat*$_{in}$, *work*$_{out}$, *heat*$_{out}$) and reverse their direction (*heat*$_{out}$, *work*$_{in}$, *heat*$_{in}$). Thus, the efficiency of the heat engine will equal the *work*$_{in}$ divided by the *heat*$_{out}$ ($\eta =$ *work*$_{in}$/*heat*$_{out}$) when this engine is operated in reverse. Because the work of engine 1 is used as the input to engine 2, there is no net *work*$_{out}$ of the combined system. Assuming that engine 1 has an efficiency of 50% and engine 2 has an efficiency of 40%, we find that the input to engine 1 is 200 units and that it rejects 100 units of heat. Engine 2, the reversible engine with an assumed efficiency of 40%, has 150 units of heat entering from the cold reservoir at $T_2$ and rejects 250 units to the hot reservoir at $T_1$. Therefore, we have 50 units of heat being delivered from $T_2$, the lower temperature, to $T_1$, the higher temperature, *with no net work being put into the combined cycle*. This directly violates the Clausius statement of the second law of thermodynamics, which states that heat cannot *of itself pass* from a lower to a higher temperature. Therefore, we may reject the assumption that an engine can be more efficient than a reversible engine operating between the same temperature limits. Thus, Carnot principle 1 is proved.

2. *The efficiency of all reversible cycles operating between the same temperature limits is the same.*

   The proof of Carnot principle 2 is essentially the same as that used for Carnot principle 1 and is not given in detail. Students should note that this principle, combined with the first, proves that the reversible cycle and its associated processes indeed serve to establish the index of performance for heat engine cycles.

3. *The thermal efficiency of a reversible engine is a function solely of the upper and lower temperatures of the cycle and is not a function of the working substances used in the cycle.*

   This third principle is somewhat different in its viewpoint, and part of the mathematical reasoning is quite abstract. We can argue this point qualitatively in the following manner. Let us assume that the efficiency of a reversible engine

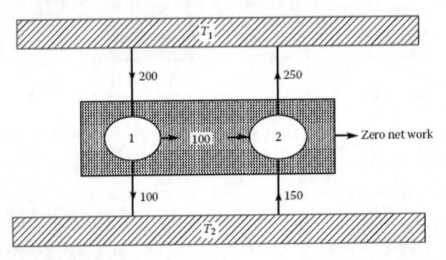

**FIGURE 5.5**
Proof of Carnot principle 1.

is a function of the working substance used in the cycle. By using two reversible cycles, as for principle 1, we can place a different working fluid in each of the cycles. One reversible cycle would be more efficient than the other, and by the identical reasoning used in principle 1, we would arrive at a violation of the Clausius statement. Thus, the efficiency of a reversible-engine cycle cannot be a function of the working substance used in the cycle. By continuing this line of reasoning, we are also directed to the conclusion that the efficiency of a reversible engine is a function only of the upper and lower temperatures used in the cycle.

To establish the temperature function, we can resort to reasoning similar to that used by Fermi (1956) and Dodge (1944). Consider the three heat reservoirs shown in Figure 5.6 and maintained at temperatures $t_1$, $t_2$, and $t_3$, respectively, on some arbitrary absolute temperature scale. Assume that three Carnot heat engines operate between these temperatures. Because the efficiency of the cycle is assumed to be some function of the temperature limits of the cycle, it follows from Equation 5.2 that for each engine, $Q_r/Q_{in}$ is also a function of $t_1$ and $t_2$ for the respective temperatures and heat quantities associated with each engine. Therefore,

$$\frac{Q_1}{Q_2} = \phi_1\left(t_1, t_2\right) \tag{5.3a}$$

$$\frac{Q_2}{Q_3} = \phi_2\left(t_2, t_3\right) \tag{5.3b}$$

$$\frac{Q_1}{Q_3} = \phi_3\left(t_1, t_3\right) \tag{5.3c}$$

The symbol $\phi$ is interpreted to mean "a function of"; thus, from Equation 5.3a, $Q_1/Q_2$ is a function of $t_1$ and $t_2$. Dividing Equation 5.3c by Equation 5.3b and comparing with Equation 5.3a yields

$$\frac{Q_1}{Q_2} = \phi_1\left(t_1, t_2\right) = \frac{\phi_3\left(t_1, t_3\right)}{\phi_2\left(t_2, t_3\right)} \tag{5.4}$$

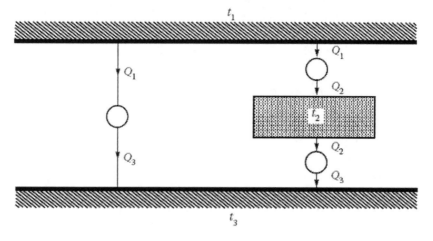

**FIGURE 5.6**
Derivation of the absolute temperature.

The left side of Equation 5.4 indicates that $\phi_3 / \phi_2$ is a function only of $t_1$ and $t_2$. Thus, the function of $t_3$ must cancel out of Equation 5.4, and we obtain

$$\frac{Q_1}{Q_2} = \phi_1(t_1, t_2) = \frac{\phi_3(t_1)}{\phi_2(t_2)} \tag{5.5}$$

At this point, it becomes impossible to determine the function in Equation 5.5 analytically, because it is entirely arbitrary and many temperature functions can satisfy it. Kelvin proposed that the temperature function in Equation 5.5 be taken as

$$\frac{Q_1}{Q_2} = \frac{T_1}{T_2} \tag{5.6}$$

Equation 5.6 is taken to be the definition of the *absolute thermodynamic temperature* scale. This temperature scale is the same as the absolute temperature scale that was defined by the ideal gas (see Chapters 1 and 6). Thus, the temperature functions given by Equations 5.5 and 5.6 are simply the absolute temperatures.

If Equation 5.6 is inverted,

$$\frac{Q_2}{Q_1} = \frac{T_2}{T_1} \text{ or } \frac{Q_L}{Q_H} = \frac{T_L}{T_H} \tag{5.6a}$$

where subscripts $L$ and $H$ correspond to the properties ($Q$ or $T$) of the energy sink and energy source, respectively, when analyzing a reversible heat engine. Subtracting unity from each side of Equation 5.6a gives us

$$\frac{Q_1}{Q_2} - 1 = \frac{T_2}{T_1} - 1 \tag{5.7}$$

Simplifying Equation 5.7 yields

$$\frac{Q_1 - Q_2}{Q_1} = \frac{T_1 - T_2}{T_1} = 1 - \frac{T_2}{T_1} \tag{5.7a}$$

In terms of the previous notation,

$$\eta = \frac{Q_{in} - Q_r}{Q_{in}} \times 100 = \frac{T_1 - T_2}{T_1} \times 100 = 1 - \frac{T_2}{T_1} \times 100 \text{ or } \eta = 1 - \frac{T_L}{T_H} \times 100 \tag{5.7b}$$

From Equation 5.7b, we conclude the following:

1. The efficiency of a reversible-engine cycle is a function only of the upper and lower temperatures of the cycle.
2. Increasing the upper temperature while the lower temperature is kept constant increases the efficiency of the cycle.
3. Decreasing the temperature at which heat is rejected while keeping the upper temperature of the cycle constant increases the efficiency of the cycle.

As noted before, the temperature scale that is defined by Equation 5.6 is called the *absolute thermodynamic scale*, because it does not depend on the working substance.

The Carnot cycle would make for a most efficient engine if it could be utilized as such. The problem is the difficulty of devising heat transfer processes that are constant temperature. The only known system that may be considered equivalent to a Carnot engine is in nature: the hurricane. Both the heat addition to the storm with the evaporation of ocean water and the cooling of the storm in the upper atmosphere occur at approximately constant temperatures. See the list of references (Bluestein, 2006) for an understanding of this phenomenon.

---

## ILLUSTRATIVE PROBLEM 5.1

A reversible engine operates between 1000°F and 80°F. (a) What is the efficiency of the engine? (b) If the upper temperature is increased to 2000°F while the lower temperature is kept constant, what is the efficiency of the engine? (c) If the lower temperature of the cycle is increased to 160°F while the upper temperature is kept at 1000°F, what is the efficiency of the cycle?

### SOLUTION

Referring to Figure 5.7 and converting all temperatures to absolute temperatures, we obtain the following:

a. $\dfrac{T_1 - T_2}{T_1} = \dfrac{1460 - 540}{1460} = 0.63 = 63\%$

b. $\dfrac{T_1 - T_2}{T_1} = \dfrac{2460 - 540}{2460} = 0.78 = 78\%$

c. $\dfrac{T_1 - T_2}{T_1} = \dfrac{1460 - 620}{1460} = 0.575 = 57.5\%$

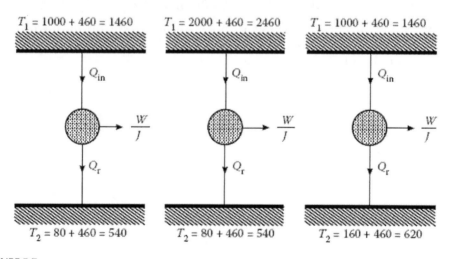

**FIGURE 5.7**
Illustrative Problem 5.1.

---

**ILLUSTRATIVE PROBLEM 5.2**

Assume that 100 units of heat enter the reversible engine in Illustrative Problem 5.1. If the cycle is reversed, determine the amount of work into the cycle and the heat removed from the reservoir at $T_2$.

**SOLUTION**

The quantities of energy in the reversed cycle must equal those of the direct cycle. Several approaches are possible in solving this problem. We can use the efficiency obtained in Illustrative Problem 5.1. Therefore,

$$\frac{W/J}{Q_{in}} = 0.63 \quad \text{and} \quad \frac{W}{J} = 100(0.63) = 63 \text{ units out}$$

$$Q_{in} - Q_r = \frac{W}{J}$$

Therefore,

$$100 - 63 = 37 = Q_r$$

For the reversed cycle, we have only to note that $Q_r$ (or 37 units) is taken into the system from the low-temperature reservoir, 63 units of work enter the system, and 100 units are returned to the high-temperature reservoir.

---

**ILLUSTRATIVE PROBLEM 5.3**

It is desired to heat a house in the winter with a heat pump when the outside air is at 15°F. If the inside of the house is maintained at 70°F while the house loses 125,000 Btu/h, what is the minimum horsepower input required?

**SOLUTION**

Refer to Figure 5.8 and note that the reversed cycle is evaluated by considering the direct cycle first. From Equation 5.6,

$$\frac{Q_{in}}{T_1} = \frac{Q_r}{T_2}$$

where $Q_{in} = 125,000$ Btu/h, $T_1 = 70 + 460 = 530$ R, and $T_2 = 15 + 460 = 475$ R:

$$Q_r = Q_{in} \times \frac{T_2}{T_1} = 125000 \times \frac{475}{530}$$

$$= 112028 \text{ Btu / h}$$

**FIGURE 5.8**
Illustrative Problem 5.3.

$$\text{work} = Q_{in} - Q_r = 125000 - 112028$$

$$= 12972\,\text{Btu}\,/\,\text{h}$$

Therefore, the reversed cycle (heat pump) requires at least 12,972 Btu/h of work input. This is

$$\frac{12972\,\dfrac{\text{Btu}}{\text{h}} \times 778\,\dfrac{\text{ft}\cdot\text{lb}_f}{\text{Btu}}}{60\,\dfrac{\text{min}}{\text{h}} \times 33000\,\dfrac{\text{ft}\cdot\text{lb}_f}{\text{min}\cdot\text{hp}}} = 5.1\,\text{hp}$$

---

## ILLUSTRATIVE PROBLEM 5.4

A Carnot engine operates between a source temperature of 1000°F and a sink temperature of 100°F. If the engine is to have an output of 50 hp, determine the heat supplied, the efficiency of the engine, and the heat rejected. Refer to Figure 5.9.

## SOLUTION

$$50\,\text{hp} = \frac{50 \times 33000}{778} = 2120.8\,\text{Btu/min as output}$$

$$\left(\text{because 1 hp is equivalent to } 33000\,\text{ft} \cdot \text{lbf/min}\right)$$

$$\eta = 1 - \frac{T_2}{T_1} \times 100 = 1 - \frac{560}{1460} \times 100 = 61.6\%$$

$$= \frac{W/J}{Q_{in}} = \frac{2120.8}{Q_{in}}$$

**FIGURE 5.9**
Illustrative Problem 5.4.

$$Q_{in} = \frac{2120.8}{0.616} = 3443 \text{ Btu/min}$$

$$Q_r = 3443(1-0.616) = 1322 \text{ Btu/min}$$

---

## ILLUSTRATIVE PROBLEM 5.5

A Carnot engine operates between a source temperature of 700°C and a sink temperature of 20°C. Assuming that the engine will have an output of 65 hp, determine the heat supplied, the efficiency of the engine, and the heat rejected.

### SOLUTION

We can also refer to Figure 5.9, noting that $T_1 = 700+273 = 973$ K and $T_2 = 20+273 = 293$ K. The efficiency of the Carnot engine is

$$\eta = \frac{T_1 - T_2}{T_1} \times 100 = \frac{973 - 293}{973} \times 100 = 69.9\%$$

The work output of the engine is $65 \times 0.746 = 48.49$ kJ/s. Because efficiency is work out/heat in,

$$\text{heat in} = Q_{in} = \frac{48.49}{0.699} = 69.37 \text{ kJ/s}$$

$$\text{heat rejected} = Q_{out} - \text{work out}$$

$$= 69.37 - 48.49 = 20.88 \text{ kJ/s}$$

As a check,

$$Q_r = Q_{in}(1-\eta) = 69.37(1-0.699) = 20.88 \text{ kJ/s}$$

## ILLUSTRATIVE PROBLEM 5.6 USING CALCULUS ENRICHMENT

If you have the choice of lowering the lowest temperature in a Carnot cycle by 1° or of raising the upper temperature by 1°, which would you do? Base your answer on the one that will yield the higher efficiency.

### SOLUTION

We start the solution to this problem by writing the efficiency of the Carnot cycle as

$$\eta = \left(1 - T_2/T_1\right)$$

Differentiating the efficiency, first with respect to $T_1$ while holding $T_2$ constant and then with respect to $T_2$ while holding $T_1$ constant, yields

$$d\eta = -\left(T_2\right) - dT_1 / \left(T_1\right)^2 = \left(dT_1 / T_1\right)\left(T_2 / T_1\right) \tag{a}$$

and

$$d\eta = -dT_2 / T_1 \tag{b}$$

If we now refer to Equation (b), we see that a 1° decrease in $T_2$ gives us an efficiency increase of $1/T_1$. From Equation (a), we see that a 1° increase in $T_1$ gives us an increase in efficiency of $(1/T_1)$ $(T_2/T_1)$. This is always smaller than the result that we obtained for the 1° decrease in $T_2$ (i.e., $1/T_1$), because the ratio of $(T_2/T_1)$ is always less than unity. Thus, we conclude that if it is feasible, we should strive to decrease the lower temperature of the cycle as much as possible to increase the cycle efficiency. In most cases, the lower temperature is determined by ambient conditions, and it may not be possible to lower it to any extent.

## ILLUSTRATIVE PROBLEM 5.7

Two Carnot engines are operated in series with the exhaust of the first engine being the input to the second engine. The upper temperature of this combination is 700°F, and the lower temperature is 200°F. If each engine has the same thermal efficiency, determine the exhaust temperature of the first engine (the inlet temperature to the second engine). Refer to Figure 5.10.

### SOLUTION

The statement of the problem has the efficiency of both engines being equal. Denoting the efficacy of each engine as $\eta_1$ and $\eta_2$, respectively,

$$\eta_1 = \left(T_1 - T_i\right)T_1,$$

and

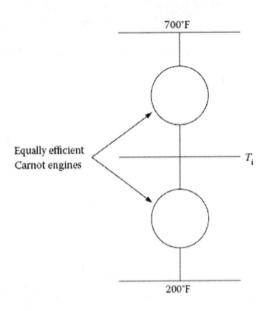

**FIGURE 5.10**
Illustrative Problem 5.48.

$$\eta_2 = \left(T_i - T_2\right) / T_i$$

Equating these efficiencies and using the given data in Rankine ($T[R] = T[F] + 460$),

$$\left(1160 - T_i\right) / 1160 = \left(T_i - 660\right) / T_i$$

Solving,

$$1160 T_i - T_i^2 = 1160 T_i - \left(660\right)\left(1160\right)$$

or

$$T_i^2 = \left(660\right)\left(1160\right)$$

We can generalize this result as

$$T_i = \sqrt{T_1 T_2}$$

Finally, for this problem,

$$T_i = \sqrt{\left(660\right)\left(1160\right)}$$

and

$$T_i = 875 \text{ R} = 415°\text{F}$$

## 5.4 Entropy

If we refer to the working fluid and the changes that occur to it (for any reversible cycle operating between the same temperature limits), as described above and based on Equation 5.6, we have established that

$$\frac{Q_{in}}{T_1} = \frac{Q_r}{T_2} \text{ or } \frac{Q_L}{T_L} = \frac{Q_H}{T_H}$$

In other words, the heat reception ($Q_{in}$ or $Q_H$) or rejection ($Q_r$ or $Q_L$) for the fluid in *any* reversible cycle divided by the temperature (i.e., $T_1$ or $T_H$ and $T_2$ or $T_L$, respectively) at which the heat is interchanged is a constant. The specific reversible paths that constitute the cycle do not change the value of these quantities. If the value of these quantities was a function of the specific reversible paths chosen, we could readily show that a violation of the second law would result. In summary, the equality of the ratios ($Q_H/T_H = Q_L/T_L$) holds only for reversible heat engines.

The uniqueness of these ratios leads us to the conclusion that they may represent state or point functions of the fluid, and as such, we may define them for a reversible process as properties. In fact, it is correct to make this assumption. To generalize the foregoing conclusion, we define the quantity $S$ as referring to this new property and call it *entropy*. On a unit mass basis, the specific entropy is $s$. The defining equation for entropy is given by

$$\Delta S = \frac{Q}{T} \text{ reversible process } \Delta S \text{ in Btu/R or kJ/K} \tag{5.8}$$

$$\Delta s = \frac{q}{T} \text{ reversible process } \Delta s \text{ in Btu/lb} \cdot \text{R or kJ/kg} \cdot \text{K}$$

Entropy is defined as a differential because it is associated with a transfer of heat, which changes the state of the substance. Thus, the absolute value of the entropy of a substance may be assigned a zero value at any arbitrary state, as has been done in the development of the tables in Appendix 2.

To show that entropy is a property of the state (i.e., point function) and not a function of the path chosen, let us consider the following situation: A reversible cycle operates between states *a* and *b* as indicated by the path *a*, 1, *b*, 2, *a* in Figure 5.11. Let us also indicate a second possible return path, *b*, 3, *a*. For the first path (*a*, 1, *b*, 2, *a*),

$$\Delta S_{1,2} = \frac{Q^{\text{path 1}}}{T_{a,b}} + \frac{Q^{\text{path 2}}}{T_{b,a}} \tag{5.9}$$

where the symbol $\dfrac{Q^{\text{path 1}}}{T_{a,b}}$ indicates the sum of the $Q/T$ items from *a* to *b* via path 1, and, $\dfrac{Q^{\text{path 2}}}{T_{b,a}}$ denotes the same from *b* to *a* via path 2. For path *b*, 3, *a*, as the return path,

$$\Delta S_{1,3} = \frac{Q^{\text{path 1}}}{T_{a,b}} + \frac{Q^{\text{path 3}}}{T_{b,a}} \tag{5.10}$$

All the processes in these two cycles are reversible, which permits us to write the foregoing relations for known paths. Let us once again repeat that the two frictionless, quasi-static

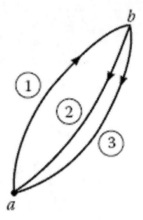

**FIGURE 5.11**
Proof that entropy is a property.

processes that are reversible are the adiabatic and isothermal. For the reversible isothermal process, $Q$ is not a function of $T$, and the reversible adiabatic process requires $Q$ to be zero. Thus, if a cycle is composed of these reversible processes, $\Delta S$ of the cycle is zero. We may generalize this statement for all reversible cycles and say that the summation of $\Delta S$ around the reversible cycle must equal zero ($\Delta S_{\text{reversible}} = 0$). Using this fact and equating Equations 5.9 and 5.10 yields

$$\frac{Q^{\text{path 2}}}{T_{b,a}} = \frac{Q^{\text{path 3}}}{T_{b,a}} \tag{5.11}$$

Thus, we can conclude that the function $(Q/T)_{\text{reversible}}$ represents a property that is a function only of the state of the fluid and is independent of the reversible path taken to reach the particular state. The importance of this statement is not just in the proof of the fact that entropy is a state or point function; it is also important in that it provides a means of calculating the change in entropy for any process. All that is necessary is a knowledge of the initial and final states of the process, because we can always (at least in principle) consider a reversible process between the same initial and final states. It must be emphasized that Q/T can be used to evaluate only the entropy change for a reversible process.

---

### CALCULUS ENRICHMENT

In order to establish entropy as a property, let us first consider a mathematical relation known as *the inequality of Clausius*. Consider the figure, which shows a reversible engine operating in a cycle with a constant upper-temperature reservoir of $T_0$ and rejecting heat as a temperature of $T$ to a second engine operating in a cycle. The second engine converts all of the rejected heat of the first engine into useful work. If we take both engines as our system, the total work output will equal the sum of the work outputs of the individual engines. Thus,

$$\delta W = \delta W_{\text{R}} + \delta W_{\text{E}} \tag{a}$$

For the work of the first engine, we have

$$\delta Q_0 = \delta W_R + \delta Q_R \rightarrow \delta W_R = \delta Q_0 \left(1 - T / T_0\right) \tag{b}$$

That's it, $\delta Q_o/T_o = \delta Q_R/T$ for a reversible engine. For the second engine,

$$\delta W_E = \delta Q_R \tag{c}$$

Adding Equations (b) and (c) and simplifying yields the total work,

$$\delta W = \left(T_0 / T\right) \delta Q_R \tag{d}$$

That's it, $\delta W = \delta W_R + \delta W_E$, we can write Equation (d) for the complete cycle using the notation that $\oint$ indicates an integration around the complete cycle as

$$\oint \delta W = T_0 \oint \left(\delta Q_R / T\right) \tag{e}$$

If the figure is studied, it will be seen that this arrangement cannot produce the net work output, because it would violate the Kelvin–Planck statement of the second law. In other words, no device or heat engine can exchange energy with only one reservoir and produce work, or convert all the incoming energy into useful work. We therefore conclude that the only way this arrangement can operate is with a *net work output* equal to or less than zero. Equation (e), with $T_0$ constant, leads us to conclude

$$\oint \delta W \leq 0 \tag{f}$$

Combining this conclusion with Equation (e) yields

$$\oint \left(\delta Q / T\right) \leq 0 \tag{g}$$

This relation, Equation (g), is known as *the inequality of Clausius.*
For a reversible cycle, it is readily shown that the equality in Equation (g) holds. Thus,

$$\oint \left(\delta Q / T\right)_{rev} = 0 \tag{h}$$

In Chapter 1, we showed that a quantity that is a property is a function only of the state of a system. Such a quantity is also an exact differential and can be represented mathematically as

$$\oint dx = 0 \tag{i}$$

We therefore conclude that Equation (h) defines a property, $S$, that we have called entropy, where the property entropy is defined as

$$dS = (\delta Q / T)_{\text{rev}} \qquad \text{(j)}$$

Therefore, for a reversible process, we have

$$\Delta S = S_2 - S_1 = \int_1^2 (dQ / T)_{\text{rev}} \qquad \text{(k)}$$

Because entropy is a state function, we may use it to graphically portray any equilibrium state of a fluid. Let us plot on temperature–entropy coordinates the reversible processes constituting the Carnot cycle. For convenience, we use a unit mass of working fluid, and the entropy coordinate becomes the specific entropy $s$. Let us consider each step of the cycle and interpret each step with the aid of Figure 5.12.

As Figure 5.12 shows, energy as heat enters the cycle at the constant upper-temperature $T_1$ with an increase in entropy from $B$ to $C$. This is represented by the horizontal line $BC$. Because $\Delta s = q_{\text{in}} / T_1$, $q_{\text{in}} = T_1 \Delta s$, which is equivalent to the area under the line $BC$ or area $A'BCD'$. The fluid is then expanded via a reversible adiabatic process, which cannot give rise to an increase in entropy because no heat enters or leaves the system during its course. The reversible adiabatic process is therefore carried out at constant entropy ($s_C = s_D$). Because of this fact, it is called an *isentropic* process because entropy remains constant. That is that a reversible adiabatic process is isentropic. Therefore, the expansion of this step is the vertical line $CD$. The next step in the cycle is the isothermal rejection of heat represented by the horizontal line $DA$. Because $\Delta s = q_r / T_2$, $q_r = T_2 \Delta s$ and the heat rejected by the cycle is represented by area $A'ADD'$, which is the area under the line $DA$. The final step in the cycle is the isentropic (reversible adiabatic) compression along path $AB$ to return the cycle to its starting point. Because the area $A'BCD'$ is the heat into the cycle ($q_{\text{in}}$) and area $A'ADD'$ is the heat rejected ($q_r$), the net work ($w_{\text{net}} = q_{\text{in}} - q_r$) of the cycle is the area $ABCD$, or

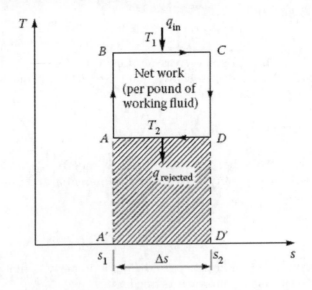

**FIGURE 5.12**
Direct Carnot cycle.

the area enclosed by the paths of the cycle on *T–s* coordinates. It should be noted that we have considered 1 lb. of working fluid because *s* is the specific entropy.

---

### ILLUSTRATIVE PROBLEM 5.8

To vaporize 1 $lb_m$ of saturated water into saturated steam at 200 psia, 843.7 Btu is required. If the temperature of this process is constant and equal to 381.86°F, what is the change in entropy for the process?

### SOLUTION

For the reversible isothermal process, we can write $\Delta s = q/T$. Thus,

$$\Delta s = \frac{843.7}{381.86 + 460} \quad \text{or} \quad \Delta s = 1.002 \frac{\text{Btu}}{lb_m \cdot R}$$

---

### ILLUSTRATIVE PROBLEM 5.9

If the process described in Illustrative Problem 5.8 is the heat reception portion of a Carnot cycle, what is the efficiency of the cycle if the lowest temperature of the cycle is 50°F? How much work is done per pound of fluid? How much energy is rejected?

### SOLUTION

$$\eta = \frac{T_1 - T_2}{T_1} \times 100 = 1 - \frac{T_2}{T_1} \times 100$$

$$= 1 - \frac{460 + 50}{460 + 381.86} \times 100$$

$$= (1 - 0.606) \times 39.4\%$$

$$\frac{w}{J} = \eta\, q_{in} = (0.394)(843.7) = 332.4 \text{ Btu}/lb_m$$

$$q = q_{in} - \frac{w}{J} = 843.7 - 332.4 = 511.3 \text{ Btu}/lb_m$$

As an alternative solution and referring to Figure 5.12, we can write the following:

$$q_{in} = T_1 \Delta s \tag{5.12a}$$

$$q_r = T_2 \Delta s \tag{5.12b}$$

$$\frac{w}{J} = q_{in} - q_r = (T_1 - T_2)\Delta s$$

$$q_r = T_2 \Delta s = (460 + 50)(1.002) = 511 \text{ Btu/lb}_m$$

$$\frac{w}{J} = q_{in} - q_r = 843.7 - 511 = 332.7 \text{ Btu/lb}_m \qquad\qquad (5.12c)$$

$$\eta = \frac{w/J}{q_{in}} \times 100 = \frac{332.7}{843.7} \times 100 = 39.4\%$$

---

### ILLUSTRATIVE PROBLEM 5.10

Determine the change in entropy at 1.4 MPa for the vaporization of 1 kg of saturated water to saturated steam. Compare your answers to the *Steam Tables* in Appendix 2 $\left( h_{fg} = 1959.7 \text{kJ/kg } s_{fg} = 4.1850 \text{kJ/kg} \cdot \text{K}, t = 195.07^\circ\text{C} \right)$.

### SOLUTION

As in Illustrative Problem 5.9, we can consider the vaporization process to be isothermal. Therefore,

$$\Delta s = \frac{q_{in}}{T} = \frac{h_{fg}}{T} = \frac{1959.7}{195.07 + 273}$$

$$= 4.1867 \text{ kJ/kJ} \cdot \text{K}$$

where $q_{in} = h_g - h_f = h_{fg}$. If we use 273.16 to obtain temperature in degrees Kelvin,

$$\Delta s = 4.1853 \text{ kJ/kg} \cdot \text{K}$$

This compares very closely to the *Steam Tables* value.

---

It has already been noted that cycle efficiency can be improved by raising the upper temperature of the cycle or lowering its lowest temperature. From Equation 5.12b, it is noted that the heat rejected during this cycle is equal to the change in entropy during the heat-rejection process multiplied by the temperature at which the heat is rejected ($q_r = T_r \Delta s$). In a general sense, the entropy change of the process becomes a measure of the amount of heat that becomes unavailable (rejected) during the cycle. For an irreversible process, we have already indicated that some energy becomes unavailable because of friction or other irreversible actions that take place during the process. If it is assumed that the irreversible process can be restored to its original state by the input of additional work to the system via a reversible path, it can be shown that the increase in entropy during the irreversible process can be used to evaluate the least amount of work necessary to restore the system to its original state. Thus, a greater entropy change requires more work to restore the system ($\Delta s_{irreversible} > \Delta s_{reversible}$). In this sense, entropy is also a measure of the unavailability of energy that occurs in an irreversible process.

## ILLUSTRATIVE PROBLEM 5.11

One hundred Btu enters a system as heat ($Q_{in}$ = 100 Btu) at 1000°F (1460 R). How much of this energy is unavailable with respect to a receiver at 50°F (510 R) assuming a Carnot (reversible) heat engine? Also, how much of this energy is unavailable with respect to a receiver at 0°F (460 R)?

## SOLUTION

Let us assume that a Carnot engine cycle operates between the two temperatures in each case. Figure 5.13 shows this problem on *T–S* coordinates. The shaded area under the 1460 R (1000°F) line represents the input of 100 Btu. Thus,

$$T_1 \Delta S = Q_{in}$$

$$\Delta S = \frac{Q_{in}}{T_1} = \frac{100}{1460} = 0.0685 \frac{Btu}{R}$$

The crosshatched area represents heat rejected ($Q_r$)

$$Q_r = T_2 \Delta S$$

$$= (510)(0.0685) = 34.9 \text{ Btu}$$

The entropy change ($\Delta s_{in}$) during the heat-addition process is equal to the entropy change ($\Delta s_{out}$) during the heat-rejection process for a Carnot (reversible) heat engine. The crosshatched area ($Q_r$) for a receiver at 0°F (460 R) is

$$Q_r = (460)(0.0685) = 31.5 \text{ Btu}$$

**FIGURE 5.13**
Illustrative Problem 5.11.

Note from this problem that the maximum work output is

$$\text{work}(\text{max.}) = Q_{\text{in}} - Q_{\text{r}} = Q_{\text{in}} - T_2 \Delta S$$

and the *unavailable energy* is the difference between the heat into the cycle and the maximum work output, or $Q_{\text{r}}$. Also, by lowering $T_2$, we decrease the energy rejected and increase the energy available.

---

### ILLUSTRATIVE PROBLEM 5.12

If 1 kJ enters a system as heat at 500°C, how much of this energy is unavailable with respect to (a) a receiver at 20°C and (b) a receiver at 0°C?

**SOLUTION**

Refer to Illustrative Problem 5.11.

$$\Delta S = \frac{Q_{\text{in}}}{T_1} = \frac{1000 \text{J}}{500 + 273} = 1.2937 \frac{\text{J}}{\text{K}}$$

$$Q_{\text{r}} = T_2 \Delta S$$

Therefore,

a.
$$Q_{\text{r}} = (273 + 20)(1.2937)$$
$$= 379.05 \text{ J for the receiver at 20°C}$$

b.
$$Q_{\text{r}} = (273 + 0)(1.2937)$$
$$= 353.18 \text{ J for the receiver at 0°C}$$

*Note that $\Delta S$ for the process is independent of the receiver temperature.*

---

The available energy for a process that receives heat is the amount of net work that would be obtained from the most efficient engine cycle. Similarly, the unavailable energy for a process is the amount of energy that would be rejected from the most efficient engine cycle. Because the atmosphere is essentially the lowest temperature reservoir to which heat is rejected, 77°F (25°C) and 1 atm is often taken as the heat-rejection condition. This has been called the *dead state*. If a state is in equilibrium with the dead state, it can no longer be used to obtain useful work.

Any process can be considered to have a characteristic specific heat associated with it. Therefore, the heat transfer during such a process can be written as

$$q = c \Delta T \tag{5.13}$$

where $c$ is the specific heat of the substance and the process is either reversible or irreversible.

Using Equation 5.13 and the definition of entropy given in Equation 5.8, $s_2 - s_1 =$ summation of $q/T$ or $c\Delta T/T$ for a reversible process between its temperature limits. This relation can be used to evaluate the change in entropy for a process taking place between temperatures $T_1$ and $T_2$, and it will correctly evaluate the change in entropy for any process between the prescribed limits as long as we restrict it to apply to the system and not to the surroundings. Thus, for a constant $c$ value, it can be shown that the summation of $c\Delta T/T$ between the limits of $T_1$ and $T_2$ yields

$$s_2 - s_1 = c \cdot \ln\left(\frac{T_2}{T_1}\right) \tag{5.14}$$

where the symbol ln is the natural logarithm to the base $e$.

---

### ILLUSTRATIVE PROBLEM 5.13

If 6 lb. of a gas undergoes a constant pressure change from 1440°F (1900 R) to some second temperature, determine the final temperature if the entropy change is −0.7062 Btu/R. Assume the specific heat, $c_p = 0.361$ Btu/lb$_m$.R for this gas.

#### SOLUTION

If we multiply both sides of Equation 5.14 by the mass $m$,

$$m\Delta s = \Delta S = mc_p \ln\left(\frac{T_2}{T_1}\right)$$

For this problem,

$$-0.70662 \, \frac{\text{Btu}}{\text{R}} = 6\text{lb}_m \times 0.361 \frac{\text{Btu}}{\text{lb}_m \text{R}} \times \ln\left(\frac{T_2}{1440 + 460}\right)$$

or

$$\ln\left(\frac{T_2}{1440 + 460}\right) = \frac{-0.7062}{6 \times 0.361} = -0.3260$$

From the definition of ln,

$$\left(\frac{T_2}{1440 + 460}\right) = e^{-0.3260}$$

$$T_2 = (1440 + 460)(0.7218)$$

and

$$T_2 = 1371.4 \, \text{R} \text{ or } 911.4°\text{F}$$

### ILLUSTRATIVE PROBLEM 5.14

If 1 lb$_m$ of water at 500°F is adiabatically mixed with 1 lb$_m$ of water at 100°F, determine the change in entropy. Assume that the specific heat of the hot and cold streams can be considered constant and equal to unity. Also, the specific heat of the mixture can be taken to be unity. Refer to Figure 5.14.

### SOLUTION

To solve this problem, it is necessary to first establish the equilibrium temperature of the final mixture. Applying the first law to the closed system of 2 lb. of water, there is no work done and no heat transfer into or out of the system. Thus, total $\Delta U = 0$, and so $m_1 c_1 (t - t_1) + m_2 c_2 (t - t_2) = 0$. Solving for $t$,

$$(1)(1)(t - 500) + (1)(1)(t - 100) = 0 \text{ yielding } t = 300°F$$

For this problem, the hot stream is cooled from 500°F to 300°F. Because we may state that heat $Q$ was removed during this process, $\Delta s$ would be expected to be negative. By applying Equation 5.14,

$$\Delta s = 1 \ln \left( \frac{300 + 460}{500 + 460} \right) = \ln \left( \frac{760}{960} \right) = -\ln \left( \frac{960}{760} \right) = -\ln (1.263)$$

$$= -0.233 \text{ Btu/lb}_m \cdot R$$

The cold stream is heated from 100°F to 300°F; therefore,

$$\Delta s = 1 \ln \left( \frac{760}{560} \right) = \ln (1.357) = 0.305 \text{ Btu/lb}_m \cdot R$$

The net change is $0.305 - 0.233 = +0.072$ Btu/lb$_m$·R. An alternative solution to this problem is found by assuming an arbitrary temperature lower than any other temperature in the system. The change in entropy of each fluid is determined with respect to this arbitrary temperature, and the net change in entropy is the algebraic sum of the values found in this manner. An illustration of the procedure is found in Chapter 7, as well as in Illustrative Problem 5.15. If the specific heat of the substances being mixed varies with temperature, the appropriate average specific heats for the temperature intervals should be used.

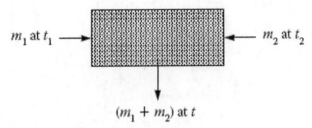

**FIGURE 5.14**
Illustrative Problem 5.14.

### ILLUSTRATIVE PROBLEM 5.15

Solve Illustrative Problem 5.14 using 0°F as a reference temperature for entropy.

### SOLUTION

As a first step, we calculate the final mixture temperature as was done in Illustrative Problem 5.14 and find it to be 300°F. Next, let us calculate the initial entropy of each fluid above the 0°F base. For the "hot" fluid,

$$\Delta s = c \ln\left(\frac{T_2}{T_1}\right) = 1\ln\left(\frac{500 + 460}{0 + 460}\right)$$

$$= 1\ln(2.087) = 0.736 \text{ Btu} / \text{lb}_m \cdot R$$

For the "cold" fluid,

$$s = c \ln\left(\frac{T_2}{T_1}\right) = 1\ln\left(\frac{100 + 460}{0 + 460}\right)$$

$$= 1\ln(1.217) = 0.196 \text{ Btu/lb}_m \cdot R$$

At the final mixture temperature of 300°F, the entropy of each stream above 0°F is, for the "hot" fluid,

$$s = c \ln\left(\frac{T_2}{T_1}\right) = 1\ln\left(\frac{300 + 460}{0 + 460}\right)$$

$$= 1\ln(1.652) = 0.502 \text{ Btu/lb}_m \cdot R$$

and for the "cold" fluid,

$$s = c \ln\left(\frac{T_2}{T_1}\right) = 1\ln\left(\frac{300 + 460}{0 + 460}\right)$$

$$= 1\ln(1.652) = 0.502 \text{ Btu/lb}_m \cdot R$$

The change in entropy of the "hot" fluid is $0.502 - 0.736 = -0.234$ Btu/lb$_m$·R. The change in entropy of the "cold" fluid is $0.502 - 0.196 = 0.306$ Btu/lb$_m$·R.

$$\text{total change} = 0.306 - 0.234 = 0.072 \text{ Btu/ln}_m \cdot R$$

The advantage of this alternative procedure is that by using a convenient, arbitrary datum below the lowest temperature in the system, we avoid negative logarithms. Either method is correct, and the choice of one over the other is purely personal preference.

The fact that entropy is a property leads us to inquire whether there are possible relations between entropy and other properties of a fluid. If they exist, they should prove to be very valuable, because it would then be possible to compute one from the other without resorting to experiment. Let us recall the energy equation applied to the reversible non-flow process:

$$q = \Delta u + \frac{p\Delta v}{J} \tag{5.15}$$

Because the process is assumed to be reversible, we may replace $q$ by $T\Delta s$. Thus,

$$T\Delta s = \Delta u + \frac{p\Delta v}{J} \tag{5.16}$$

In Equation 5.16, $T\Delta s$ evaluates the energy as heat only if the process is reversible, and similarly $p\Delta v/J$ evaluates only the work of a reversible process. However, each term of this equation consists of properties of the fluid that are not functions of the path. Thus, by applying the energy equation and the second law to a reversible, nonflow process, we have been able to arrive at a general equation involving only property terms. The only restriction on this equation occurs when each of the various terms is interpreted as either heat or work.

By referring to the reversible steady-flow process, we can arrive at another relationship between the properties. Alternatively, the same result can be obtained as follows. By definition,

$$h_2 - h_1 = (u_2 - u_1) + \frac{p_2 v_2 - p_1 v_1}{J} \tag{5.17}$$

or

$$\Delta h = \Delta u + \frac{\Delta(pv)}{J} \tag{5.18}$$

The change in the product $pv/J$ equals

$$\frac{(p + \Delta p)(v + \Delta v) - pv}{J} \tag{5.19}$$

Carrying out the multiplication of the terms of Equation 5.19 yields

$$\frac{\Delta(pv)}{J} = \Delta pv + \Delta vp + \Delta p\Delta v \tag{5.20}$$

The product ($\Delta p\Delta v$) is the product of two small terms and can be considered negligible. Thus,

$$\frac{\Delta(pv)}{J} = \frac{\Delta pv + p\Delta v}{J} \tag{5.21}$$

Replacing $\Delta(pv)/J$ in Equation 5.18 with its equivalent from Equation 5.21 yields

$$\Delta h = \Delta u + \frac{\Delta pv + \Delta vp}{J} \tag{5.22}$$

By the substitution of Equation 5.22 into Equation 5.16 and rearranging, we have the desired result:

$$T\Delta s = \Delta h - \frac{v\Delta p}{J} \tag{5.23}$$

We can thus relate entropy to enthalpy by means of Equation 5.23. Once again, each term in this equation is a property, but the interpretation of such terms as $T\Delta s$ as a heat quantity or $v\Delta p$ as a work quantity can be valid only for a reversible process.

The student should note that Equations 5.15 through 5.23 can be written in terms of SI units simply by omitting the conversion factor $J$ wherever it appears. The final result would be

$$T\Delta s = \Delta h - v\Delta p \tag{5.23a}$$

---

## CALCULUS ENRICHMENT

Applying the first and second laws to a nonflow (closed) system undergoing an internally reversible process, we obtain the following relation:

$$\delta q = \delta w + du \tag{a}$$

For this process,

$$\delta w = p\,dv \tag{b}$$

For the heat transfer, d$q$, we can write

$$dq = T\,ds \tag{c}$$

Substitution of Equations b and c into Equation a yields

$$T\,ds = du + p\,dv \tag{d}$$

Equation d is known as the first of the Gibbs or $T\,ds$ equations.

Starting with the definition of enthalpy ($h = u + pv \rightarrow u = h - pv$), we can obtain the second $T\,ds$ equation:

$$du = d(h - pv) = dh - p\,dv - v\,dp \tag{e}$$

Equating d$u$ in Equation (d) with d$u$ in Equation e gives us the desired second $T\,ds$ relation, namely,

$$T\,ds = dh - v\,dp \tag{f}$$

The $T \, ds$ equations involve only properties, and even though they were obtained by considering a nonflow, internally reversible system, they are valid for all systems and processes. Solving these equations for $ds$ gives us

$$ds = du \, / \, T + p \, dv \, / \, T \qquad \qquad \text{(g)}$$

and

$$ds = dh \, / \, T - v \, dp \, / \, T \qquad \qquad \text{(h)}$$

By integration of either $ds$ equation, we can obtain the change in entropy for a process knowing the $p, v, T$ relation and the relation of temperature to either $h$ or $u$.

It has been established that entropy is a property and can be evaluated by considering reversible paths connecting the given end states of a process. Let us consider the situation in which two processes start from a given state. The first one is carried out reversibly until a second state is reached. The second process is carried out irreversibly until the same pressure as the first process is reached. These processes operate between a fixed source at $T_1$ and a fixed sink at $T_2$ as part of a Carnot cycle. For the reversible process and based on Equation 5.6,

$$\frac{Q_{\text{in}}}{T_1} = \frac{Q_r}{T_2}$$

For the irreversible process, less net work is produced, with the consequence that more energy must be rejected as heat. Thus, $Q_r \, / \, T_2$ for the irreversible process can at best equal the equivalent ratio of the reversible process, or as is the case in all instances, it must be greater, $(Q_r \, / \, T_2)_{\text{irreversible}} > (Q_r \, / \, T_2)_{\text{reversible}}$. Using this qualitative argument, we can arrive at another important consequence of the second law:

> The entropy of an isolated system increases or in the limit remains the same.
>
> **Keenan**
> *1941*

This *principle of the increase of entropy* also serves as a criterion of irreversibility. Thus, if we find during a process that the entropy of an isolated system increases, we must conclude that the process is irreversible. Another consequence of the principle of the increase of entropy is that at a given internal energy, that state having the greatest entropy will be the most probable state that the system will assume. At this most probable state, the system is said to be in stable equilibrium.

The foregoing can be summarized by the following simple equation:

$$\Delta S \geq 0 \text{ for isolated systems} \qquad \qquad (5.24)$$

which is interpreted to state that for any reversible change in an isolated system, the total entropy remains unchanged, and for any irreversible change, the total entropy increases.

## CALCULUS ENRICHMENT

From the inequality of Clausius, we can now develop the *principle of the increase of entropy.* Consider points $A$ and $B$ in the figure that are connected by two processes, one reversible and the other irreversible. From the inequality of Clausius, we can write that

$$\oint (\delta Q / T) = \int_A^B (\delta Q / T) + \int_B^A (\delta Q / T) < 0 \qquad \text{(a)}$$

because the cycle is irreversible. For the reversible path, we can use the definition of entropy to give us

$$\int_A^B (\delta Q / T) + S_A - S_B < 0 \qquad \text{(b)}$$

Transposing terms,

$$S_B - S_A \geq \int_A^B (\delta Q / T) \qquad \text{(c)}$$

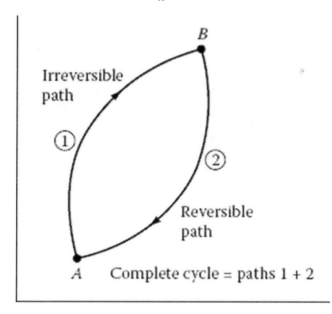

Complete cycle = paths 1 + 2

Note that the equality holds for a reversible process and that the inequality holds for an irreversible process. Equation (c) is the general statement of the *principle of the increase of entropy.* For a reversible adiabatic process, the change in entropy will be zero, while for an irreversible adiabatic process, the entropy will increase. For a system that is isolated,

$$S_B - S_A \geq 0 \qquad \text{(d)}$$

> and we conclude that *the entropy of an isolated system increases or in the limit remains the same.* Thus, if the changes in an isolated system are entirely reversible, the entropy will remain constant.

Many researchers have also referred to entropy as an indication of the arrow of time. Since entropy is never decreasing in a system, it relates to the progress of time. Thus, the universe is considered to be a very low entropy state in its formation and the increase in entropy follows our progress in time.

Another aspect of entropy is its relationship to the amount of disorder or randomness in a system. As entropy must always increase in real systems, so do systems favor disorder. Thus, solids have lower entropy than liquids and liquids lower entropy than gases. As an example, if ice is left at its freezing point, 32°F (0°C) in 14.7 psia (101 kPa) air, it will melt into liquid water with the more disorderly molecules. Likewise, water left to stand in street puddles after rain will eventually evaporate into water vapor.

In general, processes occur that result in increased entropy and randomness. As shown in Illustrated Problem 5.14, when heat is transferred from a hot body to a cold body until an equilibrium temperature is reached, the increase in entropy of the cold body exceeds the decrease in entropy of the hot body so the net entropy has increased.

## 5.5 Entropy as Property in the Analysis of Thermodynamic Systems

Thus far, we have mainly concerned ourselves with how changes in a given system affect the value of entropy as a thermodynamic property. However, in many systems it is important to understand how entropy varies during an actual processes. Some of the processes of interest are the throttling process, which is found in pipe flow and flow measurement; the constant-volume process (accumulator); reversible and irreversible compressions (in steady-flow compressors) and expansions (in pumps and turbines), and the constant-pressure process (in heaters and boilers). For each of these processes, it is necessary to know the energy equation for the path and one of the end states. Furthermore, knowing the value of entropy at the initial state and final state is useful in the analysis of thermodynamic systems.

### 5.5.1 Throttling in Open/Steady-Flow Systems

The throttling process, as already noted in Chapters 2 and 4, is found to occur when an obstruction occurs in a pipe that locally disturbs the flow. This process is an irreversible adiabatic process whose path equation we have seen to be a constant-enthalpy path. Figure 5.15 shows this process on both $T$–$s$ and $h$–$s$ coordinates. The end states, 1 and 2, are known. The path is shown as a dashed line on both diagrams, because it is an irreversible path, in which $s_2$ is greater $s_1$. As will be noted from the diagram, steam that is initially wet will become drier, and depending on the initial and final states, it will become superheated after throttling. The final pressure is always less than the initial pressure due to pressure losses taking place inside the device. Moreover, a throttling device is a steady-flow or open system type with no heat or work input or output, and based on the first law

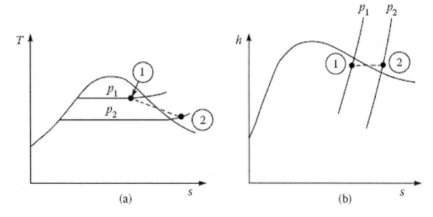

**FIGURE 5.15**
Throttling process. (a) *T–s* diagram. (b) *h–s* diagram.

of thermodynamics, we obtain: $h_{\text{inlet}} = h_{\text{outlet}}$. Simply, the total enthalpy of the fluid remains constant during the process ($h = u + pv$); however, the pressure decreases at the expense of internal energy (i.e., $h_1 = h_2 \rightarrow u_1 + p_1v_1 = u_2 + p_2v_2$).

The superheating of initially wet steam in a throttling process provides the basis for a device known as a *throttling calorimeter* that is used to determine the quality of wet steam flowing in a pipe. Figure 5.16 is a diagram of the throttling calorimeter showing its installation in a vertical run of the pipe. The steam enters the sampling tube and is expanded in the orifice to the main body of the calorimeter. The initial pressure in the pipe is monitored by a pressure gauge, and the final temperature and pressure after expansion are monitored by a thermometer or thermocouple and manometer or pressure transducer. To eliminate the velocity terms in the energy equation, the unit is sized so that the entry flow area and the flow area at the point of temperature measurement are made approximately equal (i.e., $V_1 = V_2$). For reliable operation of the calorimeter, the final state should have a superheat of at least 10°F ($T_{\text{final}} > T_{\text{sat}} + 10°F$). The calculations for the throttling process are greatly simplified by use of the Mollier chart, as will be seen from Illustrative Problem 5.16.

---

### ILLUSTRATIVE PROBLEM 5.16

Steam flows in a pipe is at 150 psia ($p_1 = 150$ psia). If a throttling calorimeter installed in the pipe has a thermometer reading ($T_2$) of 250°F and the calorimeter pressure, $p_2$ is essentially at 14.7 psia, determine the quality (moisture) in the steam flowing in the pipe.

### SOLUTION

As already noted in Chapter 4, $h_1 = h_2$ for this process. On the Mollier chart, $h_2$ is found to be 1170 Btu/lb$_m$ at 14.7 psia and 250°F, which corresponds to a superheated state. Proceeding to the left on the chart, the constant-enthalpy value of 1170 Btu/lb$_m$ to 150 psia yields a moisture of 3% or a quality, $x$ of 97%.

If we use the tables to obtain the solution to this problem, we would first obtain $h_2$ from the superheated vapor tables as 1168.8 Btu/lb$_m$ for $T_2$ of 250°F and $p_2$ of 14.7 psia. At $p_1$ of 150 psia, the enthalpy of saturated vapor, $h_g$ is 1194.9 Btu/lb$_m$; therefore, the fluid in the pipe is a mixture (i.e., wet steam) since $h_1 = 1170$ Btu/lb$_m$., which is less than 1194.9 Btu/lb$_m$. Therefore, the quality of the mixture, $x_1$, is found by using the enthalpy equation for mixture ($h_1 = h_f + x h_{fg}$), to obtain $x_1$ as follows,

$$x_1 = \frac{h_x - h_f}{h_{fg}} \quad \text{where } h_f \text{ and } h_{fg} \text{ are values of 150 psia}$$

$$= \frac{1168.8 - 330.75}{864.2} = 0.97$$

Very often, it is necessary to perform multiple interpolations if the tables are used, and the Mollier chart yields results within the required accuracy for most engineering problems and saves considerable time.

We can also use the computerized properties to solve this problem. We first enter the 250°F and 14.7 psia to obtain $h$ of 1168.7 Btu/lb$_m$. We then continue by entering $h$ of 1168.7 Btu/lb$_m$ and $p$ of 150 psia. The printout gives us $x$ of 0.9699 or 97%. Although the computer solution is quick and easy to use, you should still sketch out the problem on an $h$–$s$ or $T$–$s$ diagram to show the path of the process.

| **Thermo Properties:** | |
|---|---|
| $T = 250.00$ | deg F |
| $P = 14.700$ | psia |
| $v = 28.418$ | ft$^3$/lb$_m$ |
| $h = 1168.7$ | BTU/lb$_m$ |
| $s = 1.7831$ | BTU/lb$_m$F |
| $u = 1091.4$ | BTU/lb$_m$ |
| Region: superheated | |
| **Thermo Properties:** | |
| $T = 358.49$ | deg F |
| $P = 150.00$ | psia |
| $v = 2.925$ | ft$^3$/lb$_m$ |
| $h = 1168.7$ | BTU/lb$_m$ |
| $s = 1.5384$ | BTU/lb$_m$F |
| $u = 1087.5$ | BTU/lb$_m$ |
| $x = 0.9699$ | |
| Region: saturated | |

### 5.5.2 Constant-Volume Process (Isometric Process) in Closed Systems

The constant-volume process in a nonflow or closed system occurs when a fluid is heated in a closed tank. Figure 5.17 shows a constant-volume process in which wet steam ① is heated at constant volume in a closed tank and goes to the superheated state at ②. The energy equation for this process has already been derived as $q = u_2 - u_1$ (there is no boundary work in this case since the volume of the system or fluid remains constant). Because the Mollier chart does not have lines of constant internal energy, it is not suited for calculations of this process. The $T$–$s$ chart does not have lines of constant internal energy, and it is not too useful in calculations involving the constant-volume process. However, the internal energy values can be obtained from the appropriate tables.

**FIGURE 5.16**
Schematic of a throttling calorimeter.

---

**ILLUSTRATIVE PROBLEM 5.17**

A closed tank contains 1 $lb_m$ of saturated liquid at 150°F. If the tank is heated until it is filled with saturated steam, determine the final pressure of the steam and the heat added. The tank has a volume of 10 ft.$^3$. Also, determine the change in entropy for the fluid inside the system.

## SOLUTION

Because the tank volume is 10 ft.$^3$ and contains 1 lb$_m$ of saturated liquid, the final specific volume of the steam is 10 ft.$^3$/lb$_m$ ($v = V/m = v_1 = v_2$). Interpolations in Table A.2 using specific volume ($v_2 = v_g = 10$ ft.$^3$/lb$_m$) yield a final pressure of 42 psia and a final internal energy value, $u_2$ of 1093.0 Btu/lb$_m$ for the saturated steam or vapor. The initial internal energy value, $u_1$ is 117.95 Btu/lb$_m$, which corresponds to $u_f$ at 150°F. The heat added is simply the difference in internal energy between the two states.

$$q = u_2 - u_1 = (1093.0 - 117.95) = 975.05 \text{ Btu/lb}_m \text{ added}$$

To find the difference of entropy between the initial and final states, the entropy value at each state should be found using Table A.1. For the initial state (saturated liquid) which is at a saturation temperature of 150°F, the initial entropy is 0.21503 Btu/lbm·R. For the final state, the entropy value is 1.6724, which can be obtained through interpolation using specific volume $v_g = 10$ ft.$^3$/lb$_m$ for the saturated steam or vapor. Therefore, the difference in entropy is:

$$\Delta S = m(s_2 - s_1) = 10(1.6724 - 0.21503) = 14.574 \text{ Btu/R}$$

As the result shows, entropy increased considerably during the liquid-to-vapor transformation process.

---

### ILLUSTRATIVE PROBLEM 5.18

A closed tank has a volume of 60 ft.$^3$. It contains 15 ft.$^3$ of steam and 45 ft.$^3$ of water at 14.7 psia. Heat is added until the pressure increases to 800 psia. Determine the heat added. Also, determine the change in entropy for the fluid inside the system.

## SOLUTION

The mass in the tank is constant, and the heat added will be the change in internal energy of the contents of the tank between the two states ($q = u_2 - u_1$). The mass in the tank is found as follows:

$$m_f = \frac{V_f}{v_f} = \frac{45}{0.016715} = 2692.19 \text{ lb}_m$$

$$m_g = \frac{V_g}{v_g} = \frac{15}{26.80} = 0.56 \text{ lb}_m$$

$$\text{Total mass} = m_t = 2692.75 \text{ lb}_m$$

By knowing the mass of saturated liquid and mass of saturated vapor, the quality of steam at the initial state can be found as follows:

$$x_1 = \frac{m_g}{m_t} = \frac{0.56}{2692.75} = 0.000208$$

The initial internal energy is the sum of the internal energy of the liquid plus the vapor:

$$U_g = m_g u_g = 0.56 \times 1077.6 = 603.5$$

$$U_f = m_f u_f = 2692.19 \times 180.1 = 484863.4$$

$$\text{Total internal energy} = U_1 = U_f + U_g = 485466.9 \text{ Btu}$$

Because the mass in the tank is constant, the final specific volume, $v_2$ must equal the initial specific volume, $v_1$ or

$$v_1 = v_2 = \frac{60}{2692.75} = 0.022282 \text{ ft}^3/\text{lb}_m$$

At 800 psia, the values of the specific volume of saturated vapor, $v_g$ and saturated liquid, $v_f$ are 0.5691 and 0.02087 ft.3/lb$_m$, respectively. Therefore, state 2 consists of a mixture of vapor and liquid. To find the final internal energy, the final quality, $x_2$, should be found using $v_2 = v_f + x v_{fg}$ as follows,

$$x_2 = \frac{v_2 - v_f}{v_{fg}} = \frac{0.022282 - 0.02087}{0.5691 - 0.02087} = 0.0025756$$

The final amount of vapor is

$$m_g = x m_t = (0.0025756)(2692.75) = 6.935 \text{ lb}_m$$

The final amount of liquid is

$$m_f = m_t - m_g = 2692.75 - 6.935 = 2685.815 \text{ lb}_m$$

The final internal energy is found as before:

$$U_g = m_g u_g = 6.935 \times 1115.0 = 7732.5$$

$$U_f = m_f u_f = 2685.15 \times 506.6 = 1360297$$

$$\text{Total internal energy} = U_1 = U_f + U_g = 1368029.5 \text{ Btu}$$

The difference is

$$\Delta U = U_2 - U_1 = 1368029.5 - 485466.9 = 882562.6 \text{ Btu}$$

Per unit mass, the heat added is

$$q = \frac{\Delta U}{m} = \frac{882899.5}{2692.75} = 327.76 \text{ Btu}/\text{lb}_m$$

The small quantity of vapor mass necessitates the unusual accuracy needed to solve this problem.

To find the difference of entropy between the initial and final states, the entropy value at each state should be found using Table A.2. For the initial state (mixture), which is at a saturation pressure of 14.7 psia, the initial entropy value is found as follows:

$$s_1 = s_f + x_1 s_{fg} = 0.31212 + (0.000208)(1.4446) = 0.3124 \text{ Btu/lbm} \cdot \text{R}$$

For the final state (mixture), which is at a saturation pressure of 800 psia, the final entropy value is found as follows:

$$s_2 = s_f + x_2 s_{fg} = 0.7110 + (0.0025756)(0.7050) = 0.7128 \text{ Btu/lbm} \cdot \text{R}$$

Therefore, the difference in entropy is,

$$\Delta S = m(s_2 - s_1) = 2692.75(0.7128 - 0.3124) = 1078.165 \text{ Btu/R}$$

As the result shows, entropy increased considerably even though most the final state consists of saturated liquid.

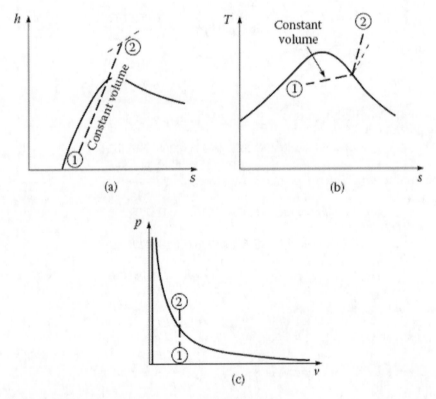

**FIGURE 5.17**
Constant-volume process. (a) *h–s* diagram. (b) *T–s* diagram. (c) *p–v* diagram.

### 5.5.3 Adiabatic Processes in Steady-Flow Open Systems

The adiabatic isentropic (constant entropy) process is one of the most important processes that we shall consider, because most compression and expansion processes can be idealized as adiabatic isentropic processes. Ideally, these processes would be carried out isentropically ($s_{final} = s_{initial}$). Figure 5.18 shows the $h$–$s$ diagram for two differing types of expansion and two different types of compression processes. The solid lines on this figure represent isentropic (reversible adiabatic) paths with $s_{final} = s_{initial}$, while the dashed lines are used to represent irreversible processes ($s_{final} > s_{initial}$). All the processes shown in Figure 5.18 are steady-flow processes.

For the compression processes shown in Figure 5.18(a), in which the fluid pressure increases from $p_1$ to $p_2$, it is evident that the difference in enthalpy for the isentropic process is smaller than for the irreversible process given the shapes of the constant-pressure lines. When analyzing steady-flow compressors, it is convenient to determine the isentropic efficiency of the compression process, which is defined as follows,

$$\eta_{compression} = \frac{w_{c,s}}{w_{c,a}} = \frac{h_{2,s} - h_1}{h_{2,a} - h_1} = \frac{\Delta h_s}{\Delta h_a} \tag{5.25}$$

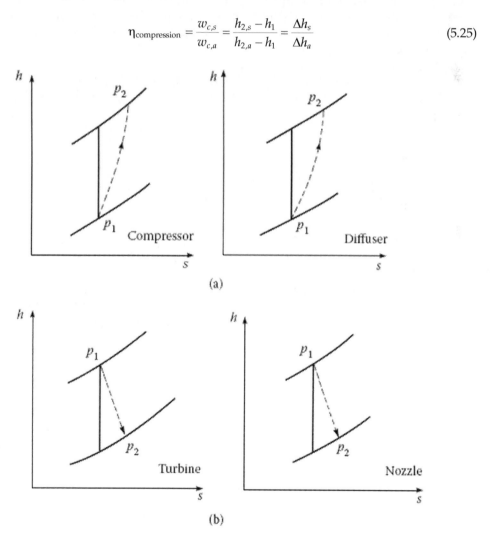

**FIGURE 5.18**
Adiabatic processes. (a) Compression. (b) Expansion.

**TABLE 5.1**

How to Find $h_{2s}$ and $h_{2a}$ Values Using Fluid Properties

| Fluid | Process | State 1 | State 2s | State 2a |
|---|---|---|---|---|
| Water or Refrigerants | Compression or expansion | $\left.\begin{array}{l}p_1\\T_1\end{array}\right\} \to h_1, s_1$ | $\left.\begin{array}{l}p_2\\s_2=s_1\end{array}\right\} \to h_{2,s}$ | $\left.\begin{array}{l}p_2\\T_2\end{array}\right\} \to h_{2,a}$ |
| Gases (i.e., air) | Compression or expansion | $T_1\} \to h_1, p_{r,1}$ | $\left\{(\eta_{c,s}\ \text{or}\ \eta_{t,s})\ \text{and}\ p_{r,1}\right\} \to p_{r,2}, h_{2,s}$ | $T_2\} \to h_{2a}$ |

*Note:* $p_r$ represents relative pressure (fluid property) as defined in Chapter 6.

Where $h_1$, $h_{2s}$ and $h_{2a}$ correspond to the inlet enthalpy, outlet enthalpy under isentropic conditions, and outlet enthalpy under actual conditions, respectively. In the case of water and refrigerants as working fluids, the enthalpy values are found using tables, as indicated in Table 5.1. For gases including air, pressure ratio and relative pressure (described and explained in Chapter 6) are used to find $h_{2s}$, as indicated in Table 5.1.

For the expansion processes shown in Figure 5.18(b), in which the fluid pressure decreases from $p_1$ to $p_2$, it is evident that the difference in enthalpy for the isentropic process is greater than for the irreversible process given the shapes of the constant-pressure lines. When analyzing steady-flow expansion devices such as turbines, it is convenient to determine the isentropic efficiency of the expansion process, which is defined as follows,

$$\eta_{\text{expansion (turbine)}} = \frac{w_{t,a}}{w_{t,s}} = \frac{h_1 - h_{2a}}{h_1 - h_{2s}} = \frac{\Delta h_a}{\Delta h_s} \tag{5.26}$$

If the isentropic efficiency value, $\eta_{c,s}$ or $\eta_{t,s}$ is known *a priori*, $h_{2a}$ and subsequently $T_{2a}$ can be found using Equation 5.25 or 5.26.

---

**ILLUSTRATIVE PROBLEM 5.19**

Steam is expanded isentropically without a change in elevation and with negligible kinetic energy differences between the inlet and outlet sections of a turbine (steady-flow device). If the initial pressure is 800 psia and the initial temperature is 600°F, determine the change in enthalpy if the end state or outlet pressure is 200 psia.

**SOLUTION**

As shown in Figure 5.18b, the process described in this problem is a vertical line on the Mollier chart. For 800 psia and 600°F, the Mollier chart yields $h_1 = 1270$ Btu/lb$_m$ and $s_1 = 1.485$. Proceeding vertically down the chart at constant $s$ to 200 psia yields a final enthalpy $h_2 = 1148$ Btu/lb$_m$. The change in enthalpy for the process, $\Delta h$ is $1270 - 1148 = 122$ Btu/lb$_m$.

We may also solve this problem using the *Steam Tables* in Appendix 2. Thus, the enthalpy at 800 psia and 600°F is 1270.4 Btu/lb$_m$, and its entropy is 1.4861 Btu/lb$_m$·R. Because the process is isentropic ($s_1 = s_2$), the final entropy at 200 psia must be 1.4861. From the saturation table, the entropy of saturated steam at 200 psia, $s_g$ is 1.5464, which indicates that the final steam condition must be wet (mixture) because the final entropy, $s_2 = 1.4861$ is less than the entropy of saturation, $s_g$. Using the wet steam or mixture relation yields

$$s_2 = s_f + x s_{fg} \rightarrow 1.4861 = 0.5440 + x(1.0025)$$

$$x_2 = \frac{1.4861 - 0.544}{1.0025} = 0.95$$

Therefore, the final enthalpy is

$$h_2 = h_f + x_2 h_{fg} = 355.6 + 0.94(843.7) = 1148.7 \text{ Btu/lb}_m$$

The change in enthalpy, $\Delta h = h_2 - h_1$, is $1270.4 - 1148.7 = 121.7$ Btu/lb$_m$. Note the agreement with the Mollier chart solution.

We can also use a computer program to solve this problem. For 600°F and 800 psia, $h_1 = 1270.0$ Btu/lb$_m$ and $s_1 = 1.4857$ Btu/lb$_m$·R. Now, using $p_2 = 200$ psia and $s_2 = 1.4857$, we obtain $h_2 = 1148.1$ Btu/lb$_m$. The change in enthalpy, $\Delta h$ is $1270.0 - 1148.1 = 121.9$ Btu/lb$_m$. Note the effort saved using either the Mollier chart or the computer program.

| Thermo Properties: | |
| --- | --- |
| $T = 600.00$ | deg F |
| $P = 800.00$ | psia |
| $v = 0.6774$ | ft³/lb$_m$ |
| $h = 1270.0$ | BTU/lb$_m$ |
| $s = 1.4857$ | BTU/lb$_m$F |
| $u = 1169.7$ | BTU/lb$_m$ |
| Region: superheated | |
| $T = 381.87$ | deg F |
| $P = 200.00$ | psia |
| $v = 2.151$ | ft³/lb$_m$ |
| $h = 1148.1$ | BTU/lb$_m$ |
| $s = 1.4857$ | BTU/lb$_m$F |
| $u = 1068.5$ | BTU/lb$_m$ |
| $x = 0.9396$ | |
| Region: saturated | |

## ILLUSTRATIVE PROBLEM 5.20

If the process described in Illustrative Problem 5.19 is carried out adiabatically but not reversibly between the same initial conditions ($p_1 = 800$ psia, $T_1 = 600$°F) and the same final pressure ($p_2 = 200$ psia), determine the final state of the steam if the isentropic efficiency of the process, $\eta_{turbine}$ is 80%.

## SOLUTION

Again referring to Figure 5.18(b), it can be seen that the final temperature and enthalpy will both be higher than for the isentropic case. For a turbine, the isentropic efficiency is as follows,

$$\eta_{\text{turbine}} = \frac{w_{t,a}}{w_{t,s}} = \frac{h_1 - h_{2a}}{h_1 - h_{2s}} = \frac{\Delta h_a}{\Delta h_s}$$

where $h_{2s}$ is found at $p_2$ and $s_2 = s_1$; and $h_{2a}$ is found at $T_2$ and $p_2$. However, for this Illustrative Problem, the change in actual enthalpy is simply $\Delta h_a = \eta_{\text{turbine}} \times \Delta h_s = 0.8 \times 121.9 = 97.5$ Btu/lb$_\text{m}$. Therefore, the final enthalpy is $h_{2a} = h_1 - \Delta h_a = 1270 - 97.5 = 1172.4$ Btu/lb$_\text{m}$, with a final pressure of 200 psia. The Mollier chart indicates the final state to be in the wet region (mixture), with a 3.1% moisture content (quality of 96.9%) and an entropy of 1.514 Btu/lb$_\text{m}$·R. Moreover, by looking at the tables (Appendix 2), at 200 psia, the enthalpy of saturated steam, $h_g$ is 1199.3 Btu/lb$_\text{m}$, which is greater than 1172.4 Btu/lb$_\text{m}$; therefore, the final state is a mixture (wet steam). To find quality, $x_2$, we find it using property values from Appendix 2 as follows,

$$x_2 = \frac{h_2 - h_f}{h_{fg}} = \frac{1172.4 - 355.6}{843.7} = 0.968 = 96.8\%$$

As can be seen, the tabulated result compares well with the result obtained using the Mollier chart.

---

### ILLUSTRATIVE PROBLEM 5.21

A turbine expands steam isentropically from an initial pressure of 1 MPa and an initial temperature of 250°C to a final pressure of 0.1 MPa. Neglecting kinetic and potential energy terms, determine the change in enthalpy of the steam.

### SOLUTION

Using the Mollier chart, $h_1 = 2942$ kJ/kg. Proceeding, as shown in Figure 5.18b, that is, vertically at constant entropy to a pressure of 0.1 MPa, gives us a final enthalpy of 2512 kJ/kg. The change in enthalpy is $\Delta h_s = h_1 - h_{2s} = 2942 - 2512 = 430$ kJ/kg. A similar result can be obtained using the steam tables.

---

### ILLUSTRATIVE PROBLEM 5.22

Steam is expanded isentropically in a nozzle from an initial condition of 800 psia and 600°F to a final pressure of 200 psia. Determine the final velocity of the steam as it leaves the nozzle.

## SOLUTION

The conditions given correspond to those in Illustrative Problem 5.19. Therefore, the isentropic change in enthalpy for this process is $\Delta h_s = h_1 - h_{2s} = 122$ Btu/lb$_m$. We now recall from Chapter 4 that the change in velocity in a nozzle with negligible entering velocity, $V_1 = 0$, as given from Illustrative Problem 4.22 is

$$\frac{V_1^2}{2g_cJ} + h_1 = \frac{V_2^2}{2g_cJ} + h_{2s} \rightarrow V_2 = \sqrt{2g_cJ\left(h_1 - h_{2s}\right)}$$

Substituting the value of 122 Btu/lb$_m$ for $h_1 - h_{2s}$ yields

$$V_2 = \sqrt{2 \times 32.17 \text{ lb}_m \cdot \text{ft./lb}_f \cdot \text{s}^2 \times 778\text{ft.} \cdot \text{lb}_f/\text{Btu} \times \left(122 \text{ Btu/lb}_m\right)}$$

$$= 2471.2 \text{ ft./s}$$

---

### ILLUSTRATIVE PROBLEM 5.23

If the entering steam in Illustrative Problem 5.22 is expanded irreversibly in the nozzle to the same final pressure and to saturated vapor (sv or g), determine the final velocity, $V_2$ as the steam leaves the nozzle.

## SOLUTION

Because the process is irreversible, we cannot show it on the Mollier diagram. However, the analysis of Illustrative Problem 4.22 for the nozzle is still valid, and all that is needed is the enthalpy at the beginning and the end of the expansion. From Illustrative Problem 5.38, $h_1$ is 1270 Btu/lb$_m$. For $h_2$, we locate the state point on the Mollier diagram as being saturated vapor at 200 psia. This gives us 1199 Btu/lb$_m$. Alternatively, using Appendix 2, $h_{2a}$ can be found using $h_g$ at 200 psia ($h_{2a} = h_{g@200 \text{ psia}} = 1199.3$ Btu/lb$_m$.). Using the results of Illustrative Problem 4.22 yields

$$V_2 = \sqrt{2g_cJ\left(h_1 - h_{2a}\right)}$$

$$= \sqrt{2 \times 32.17 \text{ lb}_m \cdot \text{ft./lb}_f \cdot \text{s}^2 \times 778\text{ft.} \cdot \text{lb}_f/\text{Btu} \times \left(1270 - 1199\right)\text{Btu/lb}_m}$$

$$= 1885.2 \text{ ft./s}$$

As would be expected, the final velocity in the irreversible process is less than the final velocity that we obtained for the isentropic expansion. It is left as an exercise for the student to use the computer program to solve this problem.

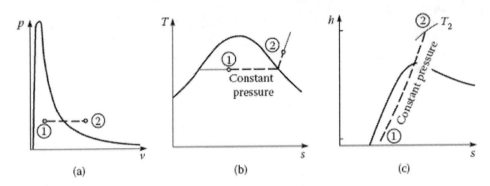

**FIGURE 5.19**
Constant-pressure process. (a) $p-v$ diagram. (b) $T-s$ diagram. (c) $h-s$ diagram.

### 5.5.4 Constant-Pressure Process (Isobaric Process) in Steady-Flow Open Systems

The constant-pressure process is an idealization that can be used to describe the addition of heat to the working fluid in a steady-flow system such as a boiler or the combustion process in a gas turbine. Figure 5.19 shows the $p-v$, $T-s$, and $h-s$ diagrams for a vapor undergoing an irreversible, constant-pressure process. The paths are shown as dashed lines to denote that the process is irreversible. Both the heat-addition portion and the heat-rejection portion of the gas-turbine cycle are considered as constant-pressure flow processes, as discussed in detail in Chapter 9.

Because the exhaust temperature of a gas turbine is relatively high, the exhaust gases can also be used as a heat source and generate steam for a steam turbine generator. Supplemental oil or gas firing keeps steam pressure and temperature at the required level. Because a substantial number of Btu's (i.e., high enthalpy) are recovered from the gas-turbine exhaust as discussed in Chapter 9, the typical unit operates at relatively high efficiency. This combination of gas-turbine and heat-recovery steam generator is responsible for much of the current interest in on-site generating systems for commercial buildings and small industrial plants. In fact, such an arrangement where a gas-turbine cycle is coupled with a steam or Rankine cycle is called a combined cycle, which is discussed in greater detail in Chapter 9.

---

**ILLUSTRATIVE PROBLEM 5.24**

For Illustrative Problem 4.21, in which steam that is initially saturated, is superheated in a boiler at constant pressure. Determine the change in entropy if the initial pressure, $p_1$ is 500 psia and the final temperature, $T_2$ is 800°F.

**SOLUTION**

From the saturation table, Appendix 2, the initial entropy for saturated vapor, $s_g$ is 1.4645 Btu/lb$_m$·R. At 500 psia and 800°F, the final entropy is 1.6571 Btu/lb$_m$·R. Because this process is a steady-flow process at constant pressure, the energy equation becomes $q = h_2 - h_1$ assuming that differences in the kinetic energy and potential energy terms are negligible. Therefore, $q = 1412.1 - 1205.3 = 206.8$ Btu/lb$_m$ added.

$$\Delta s = (s_2 - s_1) = (1.6571 - 1.4645) = 0.1926 \text{ Btu/lbm} \cdot \text{R}$$

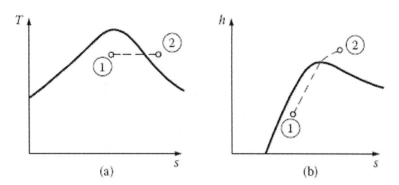

**FIGURE 5.20**
Isothermal process. (a) $T$–$s$ diagram. (b) $h$–$s$ diagram.

The difference in entropy is positive even though the pressure remains constant during the process. As Figure 5.20c shows, entropy increases when superheating steam.

### 5.5.5 Constant-Temperature Process (Isothermal Process) in Steady-Flow Open Systems

At the exhaust of a steam turbine, the steam is usually wet. This steam is subsequently condensed in a unit appropriately known as a condenser, which is basically a heat exchanger as described in Chapter 4. Because the steam is initially wet, this process is carried out essentially at constant temperature (isothermally) during the phase change or transformation process from vapor to liquid. This process is also one of constant pressure in the wet region since the corresponding saturation pressure is a direct function (directly proportional) to the saturation or condensation temperature. Figure 5.20 shows the isothermal process on various diagrams, and Figure 5.21 shows a schematic of a shell-and-tube condenser, indicating the steam flow and cooling water paths. A lengthier discussion about heat exchangers, including condensers, can be found in Chapter 11.

---

### ILLUSTRATIVE PROBLEM 5.25

From Illustrative Problem 4.25, in which steam is initially wet, having a moisture content of 3% (quality, $x$ of 97%) at 1 psia, determine the change of entropy during the complete condensation process.

### SOLUTION

From the saturation table, Appendix 2 at 1 psia, $s_f = 0.13266$ Btu/lb$_m$·R and $s_{fg} = 1.8453$ Btu/lb$_m$·R, so the initial entropy value for the mixture is

$$s_1 = s_f + x_1 s_{fg} = 0.13266 + (0.97)(1.8453) = 1.9226 \text{ Btu/lb}_m \cdot R$$

Since the mixture is fully condensed (i.e., final state corresponds to a saturated liquid), the final entropy value is $s_2 = s_f = 0.13266$ Btu/lb$_m$·R. The entropy difference ($\Delta s = s_2 - s_1$) is $0.13266 - 1.9226 = -1.7899$ Btu/lb$_m$·R during the condensation process.

**FIGURE 5.21**
Shell-and-tube condenser.

The computer solution is given below where $\Delta s = 0.1326 - 1.9221 = -1.7895$ Btu/lb$_m$·R. As the results show, the entropy of the fluid is lower at the outlet of the condenser because heat was removed from it ($\Delta h = 1004.92$ removed from the fluid during the condensation process, see Illustrative Problem 4.25).

**Saturation Properties**

| | | f | g |
|---|---|---|---|
| $T = 101.71$ | deg F | | |
| $P = 1.0000$ | psia | | |
| $v$ (ft$^3$/lb$_m$) | | 0.016137 | 333.56 |
| $h$ (BTU/lb$_m$) | | 69.72 | 1105.4 |
| $s$ (BTU/lb$_m$F) | | 0.1326 | 1.9774 |
| $u$ (BTU/lb$_m$) | | 69.72 | 1043.6 |

**Thermo Properties**

| | |
|---|---|
| $T = 101.71$ | deg F |
| $P = 1.0000$ | psia |
| $v = 323.55$ | Ft.$^3$/lb$_m$ |
| $h = 1074.3$ | BTU/lb$_m$ |
| $s = 1.9221$ | BTU/lb$_m$F |
| $u = 1014.4$ | BTU/lb$_m$ |
| $x = 0.9700$ | |
| Region: saturated | |

A summary of all the foregoing processes on the Mollier diagram is shown in Figure 5.22. The dashed paths are used to indicate that these are not equilibrium paths, and as such, they cannot really be drawn on an equilibrium diagram.

For the irreversible expansion process between the same pressure limits, the final enthalpy (and temperature) is higher than for the same process performed reversibly. For the irreversible compression process between the same pressure limits, the final enthalpy (and temperature) is higher as well than for the same process performed reversibly as shown in Figure 5.18. The constant-pressure process is also one of constant temperature in the wet region during phase transformation. Finally, throttling is an irreversible, constant-enthalpy process, as shown in Figure 5.22.

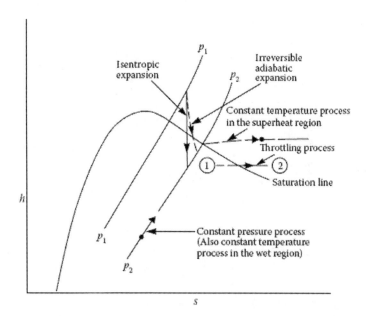

**FIGURE 5.22**
Processes on a Mollier diagram.

## 5.6 Review

The reasoning in this chapter is more abstract than in previous chapters and therefore may present more difficulty for students. Starting with the definition of thermal efficiency, we then continued with the statement of the second law of thermodynamics. For our purposes, the second law of thermodynamics can be stated as "Heat cannot, of itself, flow from a lower temperature to a higher temperature." Based on the second law and the concept of reversibility, we were able to reason out the three general conclusions that Carnot arrived at:

1. No engine operating between fixed source ($T_1$) and sink ($T_2$) temperatures and continuously delivering work can be more efficient than a reversible engine operating between these same temperature limits.
2. The efficiency of all reversible cycles operating between the same temperature limits is the same.
3. The thermal efficiency of a reversible engine is a function solely of the upper and lower temperatures of the cycle and is not a function of the working substance used in the cycle.

In 1824, Sadi Carnot proposed a reversible cycle consisting of two constant-temperature and two adiabatic processes. The utility of this cycle lies in the ease with which we were able to calculate its thermal efficiency using the temperature function proposed by Kelvin. This is the absolute thermodynamic temperature, which coincides with our earlier definition of absolute temperature. We thus became able to determine the limiting efficiency of any cycle operating between upper and lower temperature limits by calculating the thermal efficiency of the Carnot cycle between the same temperature limits. This definition of the absolute thermodynamic temperature combined with the Carnot cycle also enabled us to define a new property called entropy. The property entropy was also shown to be a measure of the unavailability of energy that occurs in an irreversible process. Using the energy equation for a reversible, nonflow process enabled us to derive some general relations among entropy, enthalpy, and internal energy.

In the last part of the chapter, we saw how entropy can be used as a property in the analysis of open and closed systems.

## Key Terms

Terms used for the first time in this chapter are as follows:

**Absolute thermodynamic temperature:** An arbitrary temperature function used with the Carnot cycle. For practical purposes, it is identical to the absolute temperature scales of Kelvin or Rankine.

**Carnot cycle:** A reversible cycle proposed by Sadi Carnot in 1824 that consists of two constant-temperature and two adiabatic processes.

**Cycle:** A series of thermodynamic processes during which the working fluid can be made to undergo changes involving energy transitions and is subsequently returned to its original state.

**Entropy:** A property of a substance; also a measure of the unavailability that occurs in an irreversible process.

**Heat engine:** A continuously operating system across whose boundaries flow only heat and work.

**Isentropic:** A process carried out at constant entropy. A reversible adiabatic process is isentropic.

**Isothermal:** A process carried out at constant temperature.

**Principle of the increase of entropy:** The entropy of an isolated system increases or in the limit remains the same.

**Reversible process:** Any process performed so that the system and all its surroundings can be restored to their initial states by performing the process in reverse.

**Second law of thermodynamics:** Heat cannot, of itself, pass from a lower temperature to a higher temperature.

**Thermal efficiency:** The ratio of the net work of a cycle to the heat added to the cycle.

## Equations Developed in This Chapter

| Net work output | $\eta = \dfrac{\text{net work output}}{\text{heat added}} \times 100$ | (5.1) |
|---|---|---|
| Thermal efficiency | $\eta = \dfrac{Q_{in} - Q_r}{Q_{in}} \times 100 = \left[1 - \dfrac{Q_r}{Q_{in}}\right] \times 00$ | (5.2) |

| | | |
|---|---|---|
| Kelvin temperature function | $\dfrac{0_1}{Q_2} = \dfrac{T_1}{T_2}$ | (5.6) |
| Efficiency of a reversible cycle | $\eta = \dfrac{T_1 - T_2}{T_1} \times 100 \left[ 1 - \dfrac{T_2}{T_1} \right] \times 100$ | (5.7a) |
| Entropy | $\Delta S = \dfrac{Q}{T}$ or $\Delta s = \dfrac{q}{T}$ reversible process | (5.8) |
| Entropy change | $s_2 - s_1 = c \cdot \ln\left( \dfrac{T_2}{T_1} \right)$ | (5.14) |
| General property relation | $T\Delta s = \Delta u + \dfrac{p\Delta v}{J}$ | (5.16) |
| General property relation | $T\Delta s = \Delta h + \dfrac{v\Delta p}{J}$ | (5.23) |
| General property relation (SI) | $T\Delta s = \Delta h - v\Delta p$ | (5.23a) |
| Entropy increase principle | $\Delta S \geq 0$ for all isolated systems | (5.24) |
| Isentropic efficiency during compression | $\eta_{compression} = \dfrac{w_{c,s}}{w_{c,a}} = \dfrac{h_{2,s} - h_1}{h_{2,a} - h_1} = \dfrac{\Delta h_s}{\Delta h_a}$ | (5.25) |
| Isentropic efficiency during expansion | $\eta_{expansion\,(turbine)} = \dfrac{w_{t,a}}{w_{t,s}} = \dfrac{h_1 - h_{2a}}{h_1 - h_{2s}} = \dfrac{\Delta h_a}{\Delta h_s}$ | (5.26) |

## QUESTIONS

5.1   Can a heat engine do anything other than deliver work?

5.2   Define a cycle.

5.3   Thermal efficiency is defined to be the ratio of net work to heat added in a cycle. Would you think that this is an appropriate definition to be used when a refrigerator is being discussed?

5.4   Do you know of any process in nature that is reversible?

5.5   There are four distinct events that occur in the Carnot cycle. Starting with the heat reception event, name and describe each one.

5.6   There are three conclusions reached from the Carnot cycle regarding reversible cycles. What are they?

5.7   What was the contribution by Kelvin?

5.8   What two factors determine the limiting efficiency of any cycle?

5.9   How does the combination of the work of Kelvin and Carnot help in the design of power cycles?

5.10   The statement has been made that entropy is a property. What does this mean to you?

5.11   Other than being a property, what else does entropy represent?

5.12   It is possible to derive a relation among entropy, enthalpy, and internal energy that is 5.12 perfectly general. Under what conditions can you associate heat and work to these terms?

5.13   What is the principle of the increase of entropy?

5.14   Why is the principle of the increase of entropy important?

## PROBLEMS

### Problems Involving Thermal Efficiency

5.1    An engine cycle is operated to produce 12.5 Btu/min as work. If 100 Btu/min as heat enters the cycle, determine the heat rejected and the efficiency of the cycle.

5.2    A reversible engine produces 10 hp as work while 1270 Btu/min enter the engine as heat. Determine the energy rejected and the thermal efficiency of the cycle.

5.3    A heat engine produces 75 hp for a heat addition of 9000 Btu/min. Determine the thermal efficiency and heat rejected by this engine.

5.4    A Carnot power plant operates between a high-temperature reservoir at 1500°F and a low-temperature reservoir at 70°F. Determine the heat addition if the plant produces 1000 MW of power.

5.5    An engine produces 10 kJ of work while 80 kJ enters the engine cycle as heat. Determine the energy rejected and the thermal efficiency of the cycle if it is reversible.

5.6    An inventor claims to have an engine that has a thermal efficiency of 90%. Comment on the claim.

5.7    A diesel engine uses 450 $lb_m$ of fuel per hour. If the burning of the fuel releases 18,000 Btu/$lb_m$ and the engine produces 840 hp, determine its thermal efficiency.

5.8    An internal combustion engine uses 0.38 $lb_m$/h of gasoline for each horsepower that it produces. If the gasoline has an energy content of 19,500 Btu/$lb_m$, determine the thermal efficiency of the engine.

*5.9   An automobile engine has an efficiency of 23%. If the engine produces 100 hp as its output, determine the heat input to the engine. If gasoline has a heat content of 20,750 Btu/$lb_m$ and a specific gravity of 0.74, determine the gallons of gasoline used per hour at this rating.

*5.10  An automobile has an efficiency of 22%. It is rated to give an output of 100 kW. Determine the heat input to the engine and the fuel consumption in liters per hour if the gasoline has a heat content of $4.9 \times 10^4$ kJ/kg and a specific gravity of 0.82.

5.11   A gas turbine has a thermal efficiency of 25% and develops 10,000 hp. If the fuel releases 18,500 Btu/$lb_m$, determine the rate of fuel usage.

5.12   A power plant burns 900 kg of coal per hour and produces 480 kW of power. If each kilogram of coal releases 5.9 MJ, determine the efficiency of the plant.

5.13   A gas expands isothermally in a cylinder, and it is found to deliver 1 kJ of work. During the process, it is found that 1 kJ of heat is added to the gas to keep it isothermal. It would appear that all the heat has been converted to useful work in this process. Is this possible? Base your answer on both the first and second laws of thermodynamics.

## Problems Involving Reversible Heat Engines and the Carnot Cycle

5.14 A Carnot cycle is operated between 1000°F and 500°F. If the upper temperature is increased to 1100°F, what is the efficiency of the cycle? If the lower temperature is decreased to 400°F, what is the efficiency of the cycle?

5.15 A Carnot cycle operates between 900°C and 100°C. Determine the efficiency of the cycle, the heat rejected, and the useful work output if 100 kJ enters the cycle as heat.

5.16 A Carnot engine operates between 940°F and 60°F and produces 80 hp. How much heat is supplied to the engine, how much heat is rejected by the engine, and what is the thermal efficiency of the engine?

5.17 A reversible power cycle has a lower reservoir temperature of 300 K. The cycle produces 42 kW as power and rejects 180 kW to the cold reservoir. Determine the temperature of the high-temperature reservoir.

5.18 A Carnot engine operates between 1350°F and 125°F. If it rejects 55 Btu as heat, determine the work output.

5.19 A Carnot engine produces 40 hp when operated between temperatures of 1500 R and 500 R. Calculate the heat supplied per hour to the engine.

5.20 A Carnot cycle operates between 1800°F and 200°F. Determine the efficiency of the cycle, the heat rejected, and the useful work output if 500 Btu/min enters the cycle as heat.

5.21 Is it possible for a reversible-engine cycle to produce 100 hp if it receives 300 Btu/s at 1000°F and rejects heat at 500°F?

5.22 An inventor tests an engine cycle that she has built and finds that its thermal efficiency is 58%. If the engine operates between 540°F and 60°F, should she release her findings to the press or retest the engine?

5.23 An inventor claims that he can obtain 155 hp as work from an engine with a heat input equivalent to 500 hp. Comment on the validity of his claim if the engine is to operate between 250°F and 10°F. Also comment on the probability of achieving these conditions.

*5.24 Two engines are operated between the same temperature limits, and each engine rejects 1000 kJ as heat to the sink. If engine A is irreversible and has a claimed efficiency of 50% and engine B is reversible and has a claimed efficiency of 40%, prove numerically that a violation of the second law will occur under these conditions.

*5.25 Two reversible engines operate in series between temperatures of 1000°F and 100°F. If the first engine receives 1000 Btu, determine the total work output of this arrangement and the temperature at the inlet to the second engine. Assume that both engines have the same thermal efficiency.

*5.26 A Carnot cycle has an efficiency of 32%. Assuming that the lower temperature is kept constant, determine the percent increase of the upper temperature of the cycle if the cycle efficiency is raised to 48%.

*5.27 A reversible cycle is shown on a $T$–$S$ diagram in Figure P5.27. Calculate the efficiency of this cycle.

**FIGURE P5.27**

*5.28 A house is cooled by the removal of 68,000 Btu/h using a reversed Carnot cycle. Assume that the house is kept at 72°F and that the outside temperature is 90°F. If the motor used is rated at 6.5 hp and electricity costs 14 cents/kWh, determine the cost per day to operate this cycle. Repeat this problem for a 4-hp motor, and compare your results. Note that this is an ideal cycle and that all processes are ideal.

*5.29 A reversible engine has an efficiency of 40%. If it is operated in such a manner that 15% of the work is lost as friction in its moving parts, what would the efficiency be?

5.30 A reversible engine operates between 900°F and 200°F. What is the maximum efficiency of this cycle? If the rating of this engine is 1 hp, determine the heat rejected per minute.

5.31 If the cycle in Problem 5.30 is reversed, how much heat is rejected to the upper-temperature reservoir?

5.32 A reversible engine operates between 750°C and 80°C. Determine the maximum efficiency of the cycle and the heat rejected if the engine is rated at 1 hp.

5.33 If the cycle in Problem 5.32 is reversed, how much heat is rejected to the upper-temperature reservoir?

5.34 In order to keep a freezer at 20°F while the room temperature is 72°F, 13,500 Btu/h must be removed. If the freezer operates as a reversible cycle, determine the minimum work input required.

5.35 A Carnot engine operates between a source temperature of 900°F and a sink temperature of 100°F. If the engine produces 100,000 ft.-lb. of work, determine its thermal efficiency, the heat supplied, and the heat rejected.

5.36 The ocean is an almost infinite energy source. Off the coast of Hawaii, the surface temperature is 86°F, while at a depth of 2200 ft., the temperature is 43°F. Determine the maximum efficiency of a reversible power plant that is operated between these temperature limits. This cycle is described in Chapter 9 as the ocean thermal energy conversion (OTEC) plant.

5.37    A Carnot engine receives 1000 Btu as heat at 1000°F and operates with a sink temperature of 150°F. Determine the efficiency, work out, and heat rejected by this engine.

5.38    A Carnot engine operates between an upper temperature of 700°C and a lower temperature of 30°C. If the engine can produce 150 kJ of work, determine its thermal efficiency, the heat supplied, and the heat rejected.

5.39    A Carnot engine develops 30 hp while rejecting 100 MJ/h as heat to a receiver at 20°C. Determine the upper temperature of the cycle and its efficiency.

5.40    A Carnot engine develops 20 hp while rejecting 70,000 Btu/h as heat to a receiver at 80°F. Determine the efficiency of the engine and the source temperature.

5.41    A reversed Carnot engine operates between 150°F and 60°F; 100 Btu/min is extracted from the cold body. Determine the horsepower required to operate the engine.

5.42    A Carnot engine operates with an efficiency of 30% and, when reversed, removes 1000 Btu/h as heat from the cold sink. Compute the heat rejected to the hotter source.

5.43    A reversed Carnot engine operates between 80°C and 20°C; 100 kJ/min as heat is extracted from the cold body. Determine the horsepower required to operate the engine.

5.44    A reversed Carnot engine operates with an efficiency of 30%, removing 1 MJ as heat from the cold sink. Calculate the heat rejected to the hotter source.

5.45    A solar-powered 10 kW Carnot power cycle is proposed drawing heat from 50°C solar collectors. Determine the required rate of heat rejection to the 20°C surrounding air.

5.46    A reversed Carnot cycle is used to heat a room. If 9000 Btu/h is required, how much energy must be put in electrically if the room is to be at 70°F and a deep well at 40°F is used as the source of working fluid for the cycle?

5.47    A reversed Carnot engine operates between 40°C and 5°C. If the motor input is 4 hp, determine the heat removed from the cold sink.

5.48    A reversed Carnot engine is used to heat a room. If 10 kW is required to replace heat losses from the room, how much energy must be put into the motor if the room is to be kept at 22°C and a deep well at 4°C is used as the source of the working fluid of the cycle?

5.49    A solar-powered 10 kW Carnot power cycle is proposed, drawing heat from 50°C solar collectors. Determine the required rate of heat rejection to the 20°C surrounding air, in kW.

5.50    A Carnot engine operates with an efficiency of 40%. If the upper temperature is 500°C, determine the lower temperature of the cycle.

5.51    A Carnot engine develops 25 hp while operating with an efficiency of 32%. If the engine rejects heat to the atmosphere at 70°F, determine the upper temperature of operation and the heat rejected.

5.52    A reversible-engine cycle operates with an efficiency of 35%. If the engine develops 19 kW while rejecting heat to a reservoir at 5°C, determine the heat input, the heat rejected, and the upper temperature of the cycle.

5.53    A reversed Carnot engine is used to cool a large room in the summer. The room is to be kept at 24°C when the outside temperature is 35°C. It is estimated that 150 kJ/min is the heat transfer from the outside to the room. Calculate the power required to operate the engine.

**Problems Involving Entropy**

5.54    A Carnot engine receives 5800 Btu/min as heat at a temperature of 400°F. If the engine develops 50 hp, what is the receiver temperature? What is the change in entropy during the heat input portion of the cycle?

5.55    What is the change in entropy during the heat-rejection portion of the engine cycle described in Problem 5.54?

*5.56    The *Steam Tables* show that at 100°F, the heat of vaporization, hfg, is 1037.0 Btu/lbm. Determine the change in entropy, sfg, and compare the calculated value to the value given for *sfg* in the Steam Tables. (Note that the Steam Tables use −459.67°F as absolute zero.)

5.57    To change a pound of saturated water to saturated steam at 32°F, 1075.4 Btu/lbm is required. Determine the change in entropy for this process.

5.58    Water is heated from 32°F to 212°F. If the specific heat of this process is taken to be constant and equal to unity, what is the change in entropy for this process per pound of water?

5.59    If the process described in Problem 5.58 is reversed, what is the change in entropy per pound of water?

5.60    For the engine of Problem 5.35, determine the change in entropy for the heat reception and heat-rejection portions of the cycle.

*5.61    A Carnot engine cycle is operated as described in Problem 5.35. If there are irreversibilities in the cycle that cause an entropy change in the heat-rejection portion of the cycle to be 10% greater than in the heat reception portion, determine the thermal efficiency of the cycle.

5.62    One kilogram of air is heated at constant volume from 100°C to 400°C. If $c_v$ is 0.7186 kJ/kgK, determine the change in entropy of the air.

5.63    If 1 $lb_m$ of air is heated at constant volume from 70°F by the addition of 200 Btu as heat, calculate the change in entropy for the process. Assume that $c_v = 0.171$ Btu/$lb_m$·R and remains constant for the entire process.

5.64    If 1 $lb_m$ of air is heated at constant pressure from 70°F by the addition of 200 Btu as heat, calculate the change in entropy for the process. Assume that $c_p = 0.24$ Btu/$lb_m$·R and remains constant for the entire process.

5.65    If the process described in Problem 5.62 is carried out at constant pressure with $c_p = 1.0062$ kJ/kgK, determine the change in entropy of the process.

*5.66    A Carnot engine has an entropy change during the isothermal heat-addition portion of the cycle of 0.15 Btu/$lb_m$·R when it operates between the temperatures of 1500°F and 500°F. How much work does this engine deliver per pound of working fluid?

5.67    One pound of air is heated at constant volume. If the value of $c_v = 0.171$ Btu/$lb_m$·R and is constant, determine the entropy change when the air is heated from 200°F to 800°F.

5.68   If the process in Problem 5.67 is carried out at constant pressure and $c_p = 0.24$ Btu/$lb_m \cdot R$, determine the entropy change.

5.69   A Carnot engine operates between 200°F and 1800°F. If the entropy change during the heat-addition portion of the cycle equals 0.25 Btu/$lb_m \cdot R$, calculate the work per $lb_m$ of working fluid.

5.70   A Carnot power cycle operates between 1200 K and 300 K. If 4.8 MJ of heat is supplied to the cycle, determine the heat rejected and the change in entropy during the heat reception portion of the cycle.

5.71   One $lb_m$ of air is heated at a constant temperature of 100°F by the addition of 250 Btu. Calculate the change in entropy for this process.

5.72   One kilogram of air is heated at a constant temperature of 40°C by the addition of 200 kJ. Determine the change in entropy for this process.

*5.73   Five pounds of water at 200°F is mixed with 2 lb. at 100°F. What is the total change in entropy for this process? Assume that all specific heats are constant and equal to unity. Work this problem by considering each stream separately.

5.74   A process is carried out at constant volume. It is found that the change in entropy for the process is 0.05 Btu/$lb_m R$. If $c_v$ is constant and equal to 0.171 Btu/$lb_m \cdot R$ and the upper temperature of the process is 250°F, determine the initial temperature of the process.

5.75   A gas is cooled from 250°C, and it is found that the entropy change is −0.1427 kJ/kgK. Assuming that the appropriate specific heat for the process is 1.4876 kJ/kgK, determine the temperature after cooling.

5.76   A gas is heated at constant pressure from 100°F. If $c_p = 0.31$ Btu/$lb_m \cdot R$ and the entropy change is 0.2107 Btu/$lb_m \cdot R$, determine the upper temperature.

5.77   A process is carried out at constant volume. It is found that the entropy change is 1 kJ/kg°K. If $c_v$ is constant and equal to 0.7186 kJ/kg·K and the lower temperature of the process is 20°C, determine the upper temperature.

*5.78   An *isentropic* process is carried out with the working fluid expanding from 700°F to 100°F. The specific heat at constant pressure is 0.24 Btu/$lb_m \cdot R$. During the process, the internal energy changes by 102.6 Btu/$lb_m$. Determine the change in entropy for this process.

5.79   A constant-temperature process is carried out with a concurrent change in entropy of 0.1 Btu/$lb_m \cdot R$ at 600°F. Determine the unavailable portion of the energy received with respect to a receiver at 50°F and at 100°F.

5.80   A constant-temperature process is carried out with a concurrent change in entropy of 0.2 kJ/kgK at 400°C. What is the unavailable portion of the energy received with respect to a receiver at 20°C and at 45°C?

*5.81   During a constant-pressure process, it is found that the change in entropy is equal to $\frac{1}{2} c_p$. If $c_p$ is constant and the initial temperature is 450°F, determine the final temperature for the process.

*5.82   One pound of air undergoes two processes that are in series. The first process consists of an isothermal expansion during which the air is supplied with 22 Btu at a temperature of 386°F. This process is then followed by a reversible adiabatic process that takes the gas to a temperature of 134°F. Determine the change in entropy for these combined processes.

## Problems Involving Processes

Use the *Steam Tables* and Mollier chart as applicable to solve the following problems.

5.83   A tank having a volume of 30 ft.$^3$ is half-filled with water, and the remainder is filled with water vapor at 100 psia. Calculate the quality of the mixture and the enthalpy and entropy of the mixture in Btu/lbm and Btu/lbm-R, respectively.

5.84   A 20 ft.$^3$ drum contains saturated steam at 400°F. What is the pressure in the drum, and what is the mass of vapor in the drum?

5.85   Solve Problem 5.84 if the drum contains saturated water.

5.86   A boiler drum contains a steam–water mixture at 100 psia. At this time, there is 10,000 lb$_m$ of water in the drum and 5 lb$_m$ of vapor. If the contents of the drum are now heated until the pressure is 200 psia, how much heat was added?

5.87   Steam at 50 psia undergoes a process in which its temperature goes from 500°F to 700°F. What is the change in enthalpy for this process?

5.88   One pound of saturated water at 200°F is converted to saturated steam at 100 psia. Determine the enthalpy difference between these two states.

5.89   If saturated steam at 500 psia is superheated to a temperature of 1000°F, determine the change in enthalpy and entropy between these two states. Assume the final pressure of the steam to be 500 psia.

5.90   If the final pressure of the steam in Problem 5.89 is 450 psia, determine the difference in enthalpy and entropy between the two states given.

5.91   A steam line contains wet steam at 200 psia. A sample of this steam leaves a throttling calorimeter at 4 psia and 200°F. Determine the percent moisture in the steam line.

5.92   After throttling, it is found that the pressure and temperature of steam are 30 psia and 300°F, respectively. If the pressure before throttling is 400 psia, what is the quality of the initial mixture? Use the *Steam Tables*.

5.93   Solve Problem 5.92 using the Mollier chart.

5.94   Superheated steam is expanded isentropically from 900 psia and 700°F to saturated vapor. Determine the difference in specific enthalpy for this expansion. Use the Steam Tables.

5.95   Solve Problem 5.94 using the Mollier chart.

5.96   One pound mass of steam is expanded isentropically from 500 psia and 800°F to 10 psia. Use the Mollier chart to determine the final enthalpy.

5.97   Solve Problem 5.96 using the Steam Tables.

5.98   Steam expands isentropically from 300 psia and 620°F to saturation. Determine the final enthalpy using the Mollier chart.

5.99   Solve Problem 5.98 using the *Steam Tables*.

5.100  Steam is used for heating a room. Assuming that the required heating load is 10,000 Btu/h and the steam enters at 20 psia and 250°F and is condensed, how many pounds of steam per hour are required? Assume that pressure losses are negligible and that the liquid is just saturated.

5.101  Calculate the "average" specific heat at constant pressure for superheated steam between 400°F and 1000°F if the pressure is 200 psia. The definition of average for this case is $\Delta h/\Delta T$.

5.102  Steam expands isentropically in a turbine from 500 psia and 1000°F to 14.7 psia. Determine the difference in enthalpy between the initial and final conditions.

5.103  If the expansion in Problem 5.102 is not carried out isentropically but is expanded irreversibly to the same final pressure and has 40°F of superheat, determine the difference in enthalpy between the initial and final conditions. Use the Mollier chart.

5.104  One pound of superheated steam at 200 psia and 800°F expands irreversibly and adiabatically to 14.7 psia and 250°F. Determine the change in enthalpy between the initial and final states.

5.105  A kilogram of saturated water at 200°C expands isothermally until it changes into a saturated vapor. Determine the work done.

5.106  Saturated steam at 500°F expands isothermally to 100 psia. Determine the change in enthalpy between the initial and final states.

*5.107  Saturated water enters a boiler at 500 psia. It is vaporized and superheated to a final condition of 500 psia and 1000°F. From this state, it enters a turbine where it is expanded isentropically to 1 in. Hg. What fraction of the energy required to produce the steam is obtained from the turbine if the turbine is 100% efficient?

5.108  Steam is expanded isentropically in an ideal nozzle from a state of 1000 psia and 1000°F to a final state of 20 psia. Determine the final velocity of the steam from the nozzle if the initial velocity is negligible. Use the Mollier chart.

5.109  Solve Problem 5.108 if the initial velocity in the nozzle is 950 ft./s.

5.110  Steam is expanded isentropically in a nozzle from 1 MPa and 500°C to saturated vapor. If the initial velocity is negligible, determine the final velocity from the nozzle. Use the Mollier chart.

5.111  Steam expands irreversibly in a nozzle. If the initial state is 900 psia and 600°F and the steam is expanded to 50 psia and a moisture content of 12%, determine the final velocity. Assume that the initial velocity is negligible. Use the Mollier chart.

5.112  Steam expands irreversibly in a nozzle. This initial state is 1 MPa and 500°C. The final state is 20 kPa and a moisture content of 2%. If the initial velocity can be neglected, determine the final velocity from the nozzle. Use the Mollier chart.

## Use the Computerized Properties to Solve Any of the Following Problems

5.113–5.142  Solve assigned problems 5.83–5.112 using the computerized properties.

# 6

## The Ideal Gas

### Learning Goals

After reading and studying the material in this chapter, you should be able to

1. Understand the terms *ideal* and *perfect* as applied to gases.
2. Express the units of the gas constant in both SI and English systems, and use the molecular weight (MW) to determine the gas constant for a given gas.
3. Calculate the average specific heat of a gas for a gas over a temperature range.
4. Use the *Gas Tables* to solve gas processes.
5. Write the general expression for the entropy change of an ideal gas from which the equation of path for an isentropic process is obtained.
6. Derive relations for constant-volume, constant-pressure, constant-temperature, isentropic, and polytropic processes, for both open and closed systems.
7. Understand the difference between real and ideal gases.
8. Apply the equation of state for real gases.
9. Understand how the paddle-work problem relates to frictional effects.

### 6.1 Introduction

In Chapter 1, we derived a simple equation for the pressure–volume relation of a gas based on elementary considerations. Because many oversimplifications were involved in this derivation, it could not be reasonably expected that the result obtained in this manner would even closely approach the behavior of a real gas. Surprisingly, it has been found that the equation can be used to represent the behavior of a large class of actual gases with accuracy usually sufficient for engineering applications. In any case, this equation can be used to predict qualitatively the behavior of most gases, and the results so obtained can be employed as a guide for design or performance purposes.

The expressions *ideal gas* and *perfect gas* appear in many textbooks on thermodynamics, and unfortunately, some confusion has developed regarding the exact definition of these terms. For clarity and consistency, these terms will be given identical meanings and defined as gases having equations of state that correspond to Equation 6.7 or 6.8.

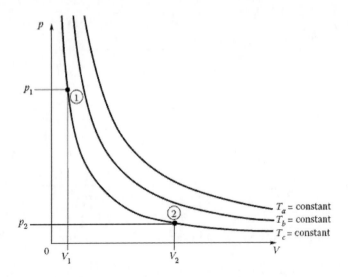

**FIGURE 6.1**
Boyle's law.

## 6.2 Basic Considerations

The first observations concerning the equation of state of a gas were made by Robert Boyle (1629–1691), an English chemist, in the year 1662. He observed experimentally that *the volume of a given quantity of gas varies inversely with absolute pressure if the temperature of the gas is held constant.*

We can write *Boyle's law* as

$$\frac{p_1}{p_2} = \frac{V_2}{V_1} \quad \text{or} \quad p_1 V_1 = p_2 V_2 = \text{constant} \tag{6.1}$$

Because $T = \text{constant}$, we can also write Equation 6.1 as

$$pV = C \tag{6.1a}$$

where $C$ represents a constant for a given temperature. A family of curves for different temperatures is shown in Figure 6.1. Each of these curves is an equilateral hyperbola, because they are of the form $xy = C$. It is important to keep in mind that pressure in Equation 6.1 must be expressed in absolute units (kPa, absolute or psia).

---

### ILLUSTRATIVE PROBLEM 6.1

A gas occupies a volume of 100 ft.³ and is at a pressure of 100 psia. If the pressure is reduced to 30 psia, what volume will the gas occupy? Assume that the gas temperature is kept constant.

**SOLUTION**

For a constant-temperature process, Boyle's law is

$$p_1 V_1 = p_2 V_2$$

Therefore,

$$V_2 = \frac{p_1}{p_2} V_1$$

and

$$V_2 = \frac{100}{30} \left( 100 \text{ ft.}^3 \right) = 333 \text{ ft.}^3$$

---

**ILLUSTRATIVE PROBLEM 6.2**

A gas occupies a volume of 2 m³ at a pressure of 1 MPa. If the pressure is increased to 8 MPa, what volume will the gas occupy if the gas temperature is kept constant?

**SOLUTION**

For the constant-temperature process, we again have from Boyle's law

$$p_1 V_1 = p_2 V_2$$

and

$$V_2 = \frac{p_1}{p_2} V_1 = \frac{10^6}{8 \times 10^6} \left( 2 \, \text{m}^3 \right) = 0.25 \, \text{m}^3$$

---

In 1787, approximately 100 years after Boyle's law was formulated, the French physicist Jacques Charles (1747–1832) investigated and published his experiments that showed that *the volume of a gas was directly proportional to its absolute temperature if the pressure was held constant.* He also demonstrated that *the pressure of a gas was directly proportional to the absolute temperature if the volume was held constant.* Mathematically, Charles' laws can be written as

$$\frac{V}{T} = \text{constant (pressure constant)} \tag{6.2}$$

$$\frac{P}{T} = \text{constant (volume constant)} \tag{6.3}$$

---

### ILLUSTRATIVE PROBLEM 6.3

A given mass of gas occupies 150 ft.$^3$ at 32°F. If heat is added while the gas pressure is kept constant, determine the volume if its temperature is 100°F.

#### SOLUTION

For the conditions given, $T_1 = 32 + 460 = 492$R and $T_2 = 100 + 460 = 560$R. Thus, for a constant-pressure process,

$$\frac{V_1}{V_2} = \frac{T_1}{T_2} \text{ or } V_2 = V_1\left(\frac{T_2}{T_1}\right) = \frac{\left(150\,\text{ft.}^3\right)\left(560\right)}{492} = 170.7 \text{ ft.}^3$$

---

### ILLUSTRATIVE PROBLEM 6.4

If after the process performed in Illustrative Problem 6.3 the gas is contained at constant volume and its absolute temperature is increased by 25%, what percent increase in its solute pressure will occur?

#### SOLUTION

If for this process $T_2 = 1.25T_1$,

$$\frac{T_2}{T_1} = 1.25$$

Therefore,

$$\frac{p_1}{p_2} = \frac{T_1}{T_2} \text{ or } \frac{p_2}{p_1} = \frac{T_2}{T_1}$$

Thus, $p_2/p_1 = T_2/T_1 = 1.25$, and the absolute gas pressure increases by 25%.

---

### ILLUSTRATIVE PROBLEM 6.5

A gas is cooled at constant pressure from 100°C to 0°C. If the initial volume is 4 m$^3$, what will its final volume be?

#### SOLUTION

$$V_2 = V_1\left(\frac{T_2}{T_1}\right) = \left(4 \text{ m}^3\right)\left(\frac{0 + 273}{100 + 273}\right) = 2.93 \text{ m}^3$$

---

By combining Boyle's law and Charles' laws into a single relation, we can obtain a general gas law. We can do this by considering a gas that goes from one state, state ①, to a second state, state ②. To go from state ① to state ②, we shall first assume that the pressure is increased from state ① to the pressure of state ② by heating at constant volume. The gas is then cooled at constant pressure (the pressure at ②) until its volume reaches the volume of state ②. These events are portrayed in Figure 6.2. From this figure, we can write for the constant-volume path $1 - a$,

$$\frac{p_a}{p_1} = \frac{T_a}{T_1} \tag{6.4}$$

and for path $a - 2$, the constant-pressure path,

$$\frac{T_a}{T_1} = \frac{V_a}{V_2} \tag{6.5}$$

Solving for $T_a$ from Equation 6.4 and substituting into Equation 6.5 yields

$$\frac{p_a T_1}{p_1} = \frac{T_2 V_a}{V_2} \tag{6.6}$$

but $p_a = p_2$ and $V_a = V_1$. Using this and rearranging gives us

$$\frac{p_1 V_1}{T_1} = \frac{p_2 V_2}{T_2} \tag{6.7}$$

Because we selected points ① and ② arbitrarily, it follows that the term $pV/T$ must be a constant for a given gas. For convenience, it is easier to express this constant on the basis of a unit mass of gas. Denoting the gas constant per unit mass by the symbol $R$, we have

$$pV = pmv = mRT \tag{6.8}$$

It will be recalled from Chapter 1 that we identified the velocity term in the equation derived on the basis of elementary kinetic theory as being proportional to the temperature of the gas (the random motion of the gas molecules). Using this, we can also obtain the equation of state of a gas as Equation 6.8.

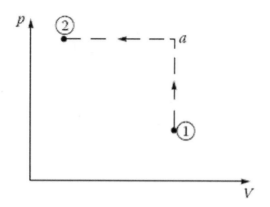

**FIGURE 6.2**
Ideal gas law derivation.

In English engineering units, $p$ is the pressure in pounds force per square foot absolute, $V$ is the total volume of the gas in cubic feet, $T$ is the absolute temperature in Rankine, $m$ is the mass of gas in pounds mass, and $R$ is a constant of proportionality in ft. $lb_f/lb_m \cdot R$. If $m$ is eliminated in Equation 6.8, it becomes

$$pv = RT \tag{6.9}$$

in which $p$, $R$, and $T$ correspond to the terms in Equation 6.8, and $v$ represents the specific volume. In SI units, $p$ is in kilopascals, $T$ is in degrees Kelvin, $m$ is in kilograms when $R$ is in kJ/kg·K, and $v$ is in cubic meters per kilogram.

$R$ for actual gases is not a constant and varies from gas to gas, as shown in Table 6.1. However, it has been found experimentally that most gases at very low pressure or with high degrees of superheat exhibit nearly constant values of $R$. As further indicated in Table 6.1, the product of molecular weight and $R$ for most gases is nearly constant and, in usual English engineering units, equals 1545.3 ft. $lb_f/lb_m$ mol·R. For any other combination of pressure, volume, and temperature units, it is a relatively straightforward calculation to derive $R$. Some values of this constant in other systems are listed in Table 6.2. For the purposes of this book, a value of $MW \times R$ of 1545 ft. $lb_f/lb_m$ mol·R is used. In SI units, the value of this product is 8.314 kJ/kg mol·K. These values are often called the universal gas constant in their respective unit systems with the notation $R_o$.

The pound mole, or more briefly, the *mole*, is the amount of a substance whose mass is equal to its MW in pounds mass. Similarly, the kilogram mole is the amount of a substance

**TABLE 6.1**

Gas Constant Data

| Gas | Gas MW | Product MW × R (rounded) | R, psia.ft.³/lbm·R | R, kJ/kg·K |
|---|---|---|---|---|
| Air | 28.97 | 1545 ft. $lb_f/lb_m$·mol·R | 0.3704 | 0.2870 |
| Ammonia ($NH_3$) | 17.02 | 1520 | 0.6301 | 0.4882 |
| Carbon dioxide ($CO_2$) | 44.01 | 1532 | 0.2438 | 0.1889 |
| Carbon monoxide (CO) | 28.01 | 1545 | 0.3831 | 0.2968 |
| Hydrogen ($H_2$) | 2.02 | 1535 | 5.3224 | 4.1240 |
| Nitrogen ($N_2$) | 28.02 | 1537 | 0.3830 | 0.2968 |
| Oxygen ($O_2$) | 32.00 | 1543 | 0.3353 | 0.2598 |

**TABLE 6.2**

Universal Gas Constant Values

| Molecular Weight × R |
|---|
| 1545.3 ft. $lb_f/lb_m$ mol·R |
| 0.082 L·atm/g mol·K |
| $8.3143 \times 10^7$ erg/g mol·K |
| 8.314 J/g mol·K |
| 1.986 Btu/$lb_m$ mol·R |
| 1.986 cal/g mol·K |

whose mass is its MW in kilograms. Thus, the mole is essentially a unit of volume, but the pound mole and the kilogram mole occupy different volumes. The number of moles of a gas ($n$) equals the mass of the gas divided by its MW. Thus, 28 $lb_m$ of nitrogen is 1 lb.-mole of nitrogen. Two moles of nitrogen would have a mass of $2 \times 28$, or 56 $lb_m$. MWs and gas constant values for many gases can be found in Appendix 2, Table A.13.

The ideal gas equation can now be written in two ways:

$$pV = mRT \tag{6.9a}$$

or

$$pV = nR_oT \tag{6.9b}$$

where $n$ equals the number of moles.

---

**ILLUSTRATIVE PROBLEM 6.6**

Nitrogen at 200 psig is used to fill a container of 120 in.³. The filling process is very slow, and the contents of the tank attain the room temperature of 73°F. How much gas is there in the container? Assume atmospheric pressure to be 14.7 psia.

**SOLUTION**

Let us first put each of the given variables into a consistent set of units:

$$p = (200 + 14.7)(144)\, psfa\left(lb_f / ft.^2\right)$$

$$T = (460 + 73)R$$

$$V = \frac{120}{1728}\, ft.^3$$

$$R = \frac{1545}{28}\, ft.lb_f / lb_m \cdot R\left(\text{because the molecular weight of nitrogen is } 28\right)$$

Applying Equation 6.9, $pv = RT$, $v = RT/p$, and

$$v = \frac{(1545 / 28)(460 + 73)}{(200 + 14.7)(144)}$$

$$= 0.951\ ft.^3 / lb_m$$

The mass of gas is the total volume divided by the specific volume. Thus,

$$\frac{(120 / 1728)ft.^3}{0.951 ft.^3 / lb_m} = 0.073\ lb_m$$

The same result is obtained by direct use of Equation 6.8: $pV = mRT$, $m = pV/RT$, and

$$m = \frac{(200 + 14.7)(144)(120/1728)}{(1545/28)(460 + 73)} = 0.073 \text{ lb}_m$$

---

### ILLUSTRATIVE PROBLEM 6.7

If the gas in Illustrative Problem 6.6 is now heated until the temperature is 200°F, what is the pressure?

#### SOLUTION

Apply Equation 6.8 and note that the volume is constant.

$$\frac{p_1 V_1}{T_1} = \frac{p_2 V_2}{T_1} \quad \text{and} \quad p_2 = p_1 \left( \frac{T_2}{T_1} \right) \quad \text{because} \quad V_1 = V_2$$

Therefore,

$$p_2 = (200 + 14.7) \left( \frac{460 + 200}{460 + 73} \right) = 266 \text{ psia}$$

---

### ILLUSTRATIVE PROBLEM 6.8

Carbon dioxide (MW = 44) occupies a tank at 100°C. If the volume of the tank of 0.5 m³ and the pressure is 500 kPa, determine the mass of gas in the tank.

#### SOLUTION

For $CO_2$,

$$R = \frac{8.314 \text{ kJ/kg} \cdot \text{mol} \cdot \text{K}}{44 \text{kg/kg} \cdot \text{mol}} = 0.1890 \text{ kJ/kg} \cdot \text{K}$$

Applying Equation 6.8, $pV = mRT$, and

$$m = \frac{pV}{RT} = \frac{(500)(0.5)}{0.1890(273 + 100)} = 3.546 \text{ kg}$$

---

It is sometimes convenient to express the volume of a gas in terms of the volume that 1 mol of gas will occupy at a given temperature and pressure. We can readily obtain this value by considering Equation 6.9. Let us multiply both sides of this equation by MW, the molecular weight, of a given gas. Thus,

$$(MW)pv = (MW)RT \tag{6.10}$$

Because $R = 1545/MW$,

$$pv(MW) = 1545T \tag{6.11}$$

Because $v(MW)$ equals $V(MW)/m$ and $m/(MW)$ is the number of moles, $n$, 1 mol of a gas would have a volume, called $V_{molar}$, of

$$V_{molar} = \frac{1545T}{p} \quad \text{or} \quad \frac{8.314T}{p} \text{ in SI units} \tag{6.12}$$

It is apparent from Equation 6.12 that for a given pressure and temperature, a mole of *any* gas will occupy the *same volume*. Quite often, the term *standard state* appears in the engineering and scientific literature, and some confusion exists as to its specific meaning. This is due to the lack of general agreement on the definition of temperature and pressure at this state. The most common standard state in use is 32°F and 14.7 psia. For this state, it will be found that 1 $lb_m$ mol occupies 358 ft.$^3$ or 22.4 L/g mol. The use of the term *standard state* should be avoided unless the conditions of this state are specified.

## 6.3 Specific Heat

In the general case of a fluid involved in a thermodynamic process, both heat and work energy interchanges are involved, and concurrently, there is a temperature change of the working fluid. The term *specific heat* has been defined in Chapter 4 as the ratio of the heat transferred per unit mass by the working fluid in a process to the corresponding change in temperature of the fluid. Mathematically,

$$c = \frac{q}{\Delta T} \tag{6.13}$$

Several features of Equation 6.13 have already been noted in Chapter 4 and are repeated here for emphasis. Heat is transferred energy and, by convention, is positive if added to a system and negative if extracted from a system. Also, the addition or removal of energy as work does not enter into the definition of specific heat. These concepts lead us to the conclusion that the specific heat of a process can be zero, positive, negative, or even infinite. Two processes proved themselves to be of particular interest: the constant-pressure process and the constant-volume process. For these processes, the respective specific heats are

$$c_p = \frac{q}{\Delta T}\bigg|_{p=\text{constant}} \quad \text{(flow or nonflow)} \tag{6.14a}$$

$$c_v = \frac{q}{\Delta T}\bigg|_{v=\text{constant}} \tag{6.14b}$$

Throughout this chapter, it will be necessary to invoke both the first and second laws to specify the path and state conditions as a gas undergoes a change of state. In Chapter 4, the first law was applied to the steady-flow (open system) and nonflow (closed system), constant-pressure processes, and the energy relation $q = \Delta h$ resulted in both processes. Although a complete discussion of the temperature and pressure dependence of the specific heat of a real gas is beyond the scope of this book, it may be noted that a gas whose equation of state is given by Equation 6.9 has both its internal energy and enthalpy independent of pressure and that these properties depend solely on temperature. Therefore, for the ideal gas for any process, flow or nonflow,

$$c_p = \frac{\Delta h}{\Delta T} \tag{6.15a}$$

and

$$c_v = \frac{\Delta u}{\Delta T} \tag{6.15b}$$

Note that these equations are valid for any ideal gas in any process. Thus, $\Delta h$ always equals $c_p \Delta T$ even for a constant-volume process and $\Delta u$ always equals $c_v \Delta T$ even for a constant-pressure process.

---

### CALCULUS ENRICHMENT

Because temperature and specific volume are independent, measurable properties, they can be used to establish the thermodynamic state of a substance. Therefore, we can state that the internal energy, $u$, is given as a function of $T$ and $v$. Mathematically,

$$u = u(T, v) \tag{a}$$

The total differential of $u$ is given in terms of the partial derivatives as

$$du = \frac{\delta u}{\delta T}_v \, dT + \frac{\delta u}{\delta T}_T \, dv \tag{b}$$

The first partial derivative in Equation a is the change in internal energy per degree of temperature change for a constant-volume process, and it is called the *specific heat at constant volume*. Thus,

$$c_v \equiv \frac{\delta u}{\delta T}_v \tag{c}$$

In a series of famous experiments, Sir James Prescott Joule (1818–1889), an English physicist, determined that the internal energy of a gas depended only on its temperature. Joule performed his experiments on an apparatus such as the one shown schematically in the figure below. Basically, it consisted of two tanks connected by a valve that were placed in a tank of water. The temperature of the water was monitored with a thermometer. One tank was evacuated while a gas was sealed in the other tank. The valve was slowly opened, and after equilibrium was attained, the temperature of the water was found to be unchanged. There was no heat transfer to or from the system, and no work was performed on or by the system. From the first law for a closed system, we conclude that there was no change in the internal energy of the system, even though both the pressure and specific volume of the gas changed. Joule concluded that the internal energy of the gas was a function only of temperature, making Equation (a)

$$u = u(T) \tag{d}$$

Equation (d) can also be derived mathematically for a gas whose equation of state corresponds to Equation 6.9.

Equation (c) can be changed as it is applied to an ideal gas as

$$c_v = \frac{du}{dT} \text{ (ideal gas)} \tag{e}$$

We can consider the change in enthalpy of an ideal gas in the same manner as we considered the internal energy. Thus,

$$h = h(T,p) \tag{f}$$

where the independent variables chosen are pressure and temperature. The change in enthalpy is

$$dh = \frac{\delta h}{\delta T_p} dT + \frac{\delta h}{\delta p_T} dv \tag{g}$$

The first partial derivative in Equation (g) is the change in enthalpy per degree of temperature change for a constant-pressure process, and it is called the specific heat at constant-pressure. Thus,

$$c_p \equiv \frac{\delta h}{\delta T}_p \tag{h}$$

Because the definition of enthalpy is

$$h \equiv u + pv \tag{i}$$

we can write for the ideal gas that

$$h = u + RT \tag{j}$$

Therefore, because $R$ is a constant and $u$ is only a function of temperature, it follows that for the ideal gas, the enthalpy is a function only of temperature. The specific heat at constant pressure for the ideal gas is

$$c_p = \frac{dh}{dT} \tag{k}$$

When internal energy or enthalpy data are available (e.g., in the work of Keenan and Kaye [1948]), an "average" value of $c_p$ and $c_v$ can readily be obtained.

$$\bar{c}_p = \frac{h_2 - h_1}{T_2 - T_1} \tag{6.16}$$

$$\bar{c}_v = \frac{u_2 - u_1}{T_2 - T_1} \tag{6.17}$$

For cases in which internal energy or enthalpy data are not available, equations have been developed from either empirical or spectroscopic data expressing the dependence of the specific heat on temperature. These equations are common in the following forms, with $A$, $B$, and $D$ as constants:

$$c = A + BT + DT^2 \tag{6.18a}$$

$$c = A + \frac{B}{T} + \frac{D}{T^2} \tag{6.18b}$$

The mean or average specific heat of a substance can be defined as that value, when multiplied by the temperature interval, that will give the energy as heat interchanged during a process. Thus, if the specific heat varies, as shown in Figure 6.3, the mean value occurs when areas *abcd* and *aefd* are equal. For gases whose specific heats are of the form given by Equation 6.18, mean specific heats are, respectively,

$$\bar{c} = A + \frac{B}{2}(T_2 + T_1) + \frac{D}{3}\left(T_2^2 + T_2 T_1 + T_1^2\right) \tag{6.19a}$$

$$\bar{c} = A + \frac{B \ln(T_2 / T_1)}{T_2 - T_1} + \frac{D}{T_2 T_1} \tag{6.19b}$$

Area $A$ = area $B$
Area $abcd$ = area $aefd$

**FIGURE 6.3**
Mean specific heat.

An alternative interpretation of the average specific heat can be illustrated graphically if reference is again made to Figure 6.3. The $c$ and $T$ scales are plotted linearly, and $\bar{c}$ expresses the condition that the equal positive and negative areas cancel each other. In other words, area $A$ equals area $B$. Note that the mean specific heat does not necessarily occur at the arithmetic mean temperature unless $c$ is constant or a linear function of $T$.

---

### CALCULUS ENRICHMENT

The mathematical definition of the average value of a function can be written as

$$\bar{y} = \frac{\displaystyle\int_{x_1}^{x_2} y\,dx}{x_2 - x_1} \tag{a}$$

Using Equation 6.18a,

$$\bar{c} = \frac{\displaystyle\int_{T_1}^{T_2} \left(A + BT + DT^2\right)dT}{\left(T_2 - T_1\right)} \tag{b}$$

Integrating,

$$\bar{c} = \frac{A\left(T_2 - T_1\right) + \dfrac{B\left(T_2^2 - T_1^2\right)}{2} + \dfrac{D\left(T_2^3 - T_1^3\right)}{3}}{\left(T_2 - T_1\right)} \tag{c}$$

Factoring and dividing by $T_2 - T_1$,

$$\bar{c} = A + \frac{B}{2}\left(T_2 + T_1\right) + \frac{D}{3}\left(T_2^2 + T_2 T_1 + T_1^2\right) \tag{6.19a}$$

---

Using Equation 6.18b,

$$\bar{c} = \frac{\int_{T_1}^{T_2} A + \dfrac{B}{T} + \dfrac{D}{T^2}\, dT}{\left(T_2 - T_1\right)} \tag{d}$$

Integrating,

$$\bar{c} = \frac{A\left(T_2 - T_1\right) + B\ln\left(\dfrac{T_2}{T_1}\right) - \left(T_2^{-1} - T_1^{-1}\right)}{T_2 - T_1} \tag{e}$$

and

$$\bar{c} = A + \frac{B\ln\left(T_2 / T_1\right)}{T_2 - T_1} - \frac{\dfrac{1}{T_2} - \dfrac{1}{T_1}}{T_2 - T_1} \tag{f}$$

Rearranging the last term in Equation f,

$$\bar{c} = A + \frac{B\ln\left(T_2 / T_1\right)}{T_2 - T_1} - \frac{1}{T_2 T_1} \tag{6.19b}$$

---

## ILLUSTRATIVE PROBLEM 6.9

For nitrogen between 540R and 9000R, an equation for the instantaneous specific heat at constant pressure is

$$\bar{c}_p = 0.338 - \frac{1.24 \times 10^2}{T} + \frac{4.15 \times 10^4}{T^2}\ \text{Btu/lb}_\text{m} \cdot \text{R}$$

Determine the mean specific heat at a constant pressure between 80°F and 500°F.

### SOLUTION

This equation has the form of Equation 6.18b, with $A' = 0.338$, $B' = -1.24 \times 10^2$, and $D' = 4.15 \times 10^4$. Therefore, from Equation 6.19b, with temperatures in degrees Rankine,

$$\bar{c}_p = 0.338 - \frac{1.24 \times 10^2\ \ln\left(960 / 540\right)}{960 - 540} + \frac{4.15 \times 10^4}{\left(540\right)\left(960\right)}$$

$$= 0.248\ \text{Btu/lb}_\text{m} \cdot \text{R}$$

## ILLUSTRATIVE PROBLEM 6.10

Using the data from the properties of some gases at low pressure in Appendix 2, solve Illustrative Problem 6.9.

## SOLUTION

The table in Appendix 2 does not give us the enthalpy data at 960R and 540R that we need. Interpolating yields

| $T$ | $\bar{h}$ | $T$ | $\bar{h}$ |
|---|---|---|---|
| $3\begin{pmatrix}537\\540\\600\end{pmatrix}63$ | 3729.5<br>3750.4<br>4167.9 | $60\begin{pmatrix}900\\960\\1000\end{pmatrix}100$ | 6268.1<br>6694.0<br>6977.9 |

$$\bar{h}_{540} = 3729.5 + \frac{3}{36}(4167.9 - 3729.5)$$

$$\bar{h}_{960} = 6268.1 + \frac{60}{100}(6977.9 - 6268.1)$$

Note that $\bar{h}$ is given for a mass of 1 lb mol. To obtain the enthalpy per pound, it is necessary to divide the values of $\bar{h}$ by the MW, 28.

$$\bar{c} = \frac{h_2 - h_1}{T_2 - T_1} = \frac{6694.0 - 3750.4}{28(960 - 540)} = 0.250 \text{ Btu/lb}_m \cdot \text{R}$$

With the more extensive *Gas Tables*, these interpolations are avoided. The *Gas Tables* provide a relatively easy and accurate method of obtaining average specific heats. Also, these tables have been computerized for ease of application.

## ILLUSTRATIVE PROBLEM 6.11

The specific heat of air between 500R and 2700R is given by

$$c_p = 0.219 + 3.42 \times 10^{-5}T - 2.93 \times 10^{-9}T^2 \text{ Btu/lb}_m \cdot \text{R}$$

Determine the mean specific heat at constant pressure for air between 80°F and 500°F.

## SOLUTION

The equation for the instantaneous specific heat is of the form of Equation 6.18a, with $A = 0.219$, $B = 3.42 \times 10^{-5}$, and $D = -2.93 \times 10^{-9}$. Using these values and Equation 6.19a yields

$$\overline{c}_p = 0.219 + \frac{3.42 \times 10^{-5}}{2}(960 + 540)$$

$$- \frac{2.93 \times 10^{-9}}{3}\left(960^2 + 960 \times 540 + 540^2\right)$$

$$= 0.243 \text{ Btu/lb}_m \cdot \text{R}$$

For the ideal gas, certain unique and simple relations relating the specific heat at constant pressure to the specific heat at constant volume can be derived. Let us first recall the general definition of the term enthalpy,

$$h = u + \frac{pv}{J} \tag{6.20}$$

The change in enthalpy between any two states therefore is

$$h_2 - h_1 = (u_2 - u_1) + \frac{p_2 v_2}{J} - \frac{p_1 v_1}{J} \tag{6.21}$$

But

$$h_2 - h_1 = c_p(T_2 - T_1); \quad u_2 - u_1 = c_v(T_2 - T_1)$$

and

$$pv = RT \tag{6.22}$$

Substituting Equation 6.22 into Equation 6.21 gives

$$c_p(T_2 - T_1) = c_v(T_2 - T_1) + \frac{R}{J}(T_2 - T_1) \tag{6.23}$$

Rearranging and simplifying Equation 6.23 yields the desired result:

$$c_p - c_v = \frac{R}{J} \tag{6.24}$$

In SI units, the conversion factor $J$ is not included. Therefore,

$$c_p - c_v = R \text{ SI units} \tag{6.24a}$$

At this point, it is often found convenient to express the results of Equation 6.24 in terms of the ratio of the specific heats. By defining

$$k = \frac{c_p}{c_v} \tag{6.25}$$

and dividing Equation 6.24 by $c_v$, we obtain

$$\frac{c_p}{c_v} - \frac{c_v}{c_v} = \frac{R}{Jc_v} \tag{6.26}$$

Using the definition of $k$ (specific heat ratio) given by Equation 6.25, we have

$$k - 1 = \frac{R}{Jc_v} \tag{6.27}$$

Rearranging yields

$$c_v = \frac{R}{J(k-1)} \tag{6.28}$$

In SI units,

$$c_v = \frac{R}{k-1} \tag{6.28a}$$

Because $c_p = kc_v$,

$$c_p = \frac{R}{J}\frac{k}{k-1} \tag{6.29}$$

In SI units,

$$c_p = R\left(\frac{k}{k-1}\right) \tag{6.29a}$$

---

### ILLUSTRATIVE PROBLEM 6.12

Oxygen has a $c_p$ of 0.24 Btu/lb$_m$·R at a given temperature. Determine $c_v$ if its MW is 32.

**SOLUTION**

The MW of oxygen is 32. Therefore, $R$ is 1545/32 = 48.28. Because $c_p - c_v = R/J$,

$$c_v = c_p - \frac{R}{J}$$

$$= 0.24 - \frac{48.28}{778} = 0.178 \text{ Btu/lb}_m \cdot R$$

---

### ILLUSTRATIVE PROBLEM 6.13

If $k$ of oxygen is 1.4, determine $c_p$ and $c_v$ in SI units.

**SOLUTION**

From Equation 6.28a,

$$c_v = \frac{R}{k-1} = \frac{8.314/32}{1.4-1} = 0.6495 \frac{\text{kJ}}{\text{kg} \cdot \text{K}}$$

Because $c_p = kc_v$,

$$c_p = (1.4)(0.6495) = 0.9093 \frac{kJ}{kg \cdot K}$$

---

### ILLUSTRATIVE PROBLEM 6.14

A gas with an $R = 60$ ft. $lb_f/lb_m$·R undergoes a process in which $\Delta h = 500$ Btu/$lb_m$ and $\Delta u = 350$ Btu/$lb_m$. Determine $k$, $c_p$, and $c_v$, for this gas.

**SOLUTION**
Because $\Delta h = c_p \Delta T$ and $\Delta u = c_v \Delta T$,

$$\frac{\Delta h}{\Delta u} = \frac{c_p \Delta T}{c_v \Delta T} = \frac{c_p}{c_v} = k$$

Therefore,

$$k = \frac{500}{300} = 1.429$$

From Equation 6.28,

$$c_v = \frac{R}{J(k-1)} = \frac{60}{(778)(1.429-1)}$$

$$= 0.180 \text{ Btu/lb}_m \cdot R$$

and

$$c_p = kc_v$$

yielding

$$c_p = (1.429)(0.180) = 0.257 \text{ Btu/lb}_m \cdot R$$

---

### ILLUSTRATIVE PROBLEM 6.15

A closed, rigid container with a volume of 60 in.$^3$ contains 0.0116 $lb_m$ of a certain gas of 90 psia and 40°F (Figure 6.4). The gas is heated to 140°F and 108 psia, and the heat input is found to be 0.33 Btu. Assuming this to be a perfect gas, find the specific heat at constant pressure $c_p$.

Initial conditions

$$V = 60 \text{ in}^3$$
$$m = 0.0116 \text{ lbs}$$
$$p = 90 \text{ psia}$$
$$T = (460 + 40)$$

+ Heat

Final conditions

$$V = 60 \text{ in}^3$$
$$m = 0.0116 \text{ lbs}$$
$$p = 108 \text{ psia}$$
$$T = (460 + 140)$$

**FIGURE 6.4**
Illustrative Problem 6.15.

## SOLUTION

When solving this type of problem, it is necessary to note carefully the information given and to write the correct energy equation for the process. Because the process is carried out at constant volume, the heat added equals the change in internal energy. Because the change in internal energy per pound for the ideal gas is $c_v(T_2 - T_1)$, the total change in internal energy for $m$ pounds must equal the heat added. Thus,

$$Q = m(u_2 - u_1) = mc_v(T_2 - T_1)$$

Using the data of the problem, $0.33 = 0.0116c_v(600 - 500)$ and

$$c_v = 0.284 \text{ Btu/lb}_m \cdot \text{R}$$

To obtain $c_p$, it is first necessary to obtain $R$. Enough information was given in the initial conditions of the problem to apply Equation 6.8 for $R$, $pV = mRT$,

$$(90)(144)\frac{60}{1728} = 0.0116(R)(460 + 40)$$

and

$$R = 77.6 \text{ ft.lb}_f/\text{lb}_m \cdot \text{R}$$

Because

$$c_p - c_v = \frac{R}{J}$$

$$c_p = c_v + \frac{R}{J}$$

$$c_p = 0.284 + \frac{77.6}{778} = 0.384 \text{ Btu/lb}_m \cdot \text{R}$$

## 6.4 Entropy Changes of Ideal Gas

In the rest of this chapter, various nonflow (closed system) processes of the ideal gas will be studied. It must be remembered that the energy quantities called work and heat depend on the path that the gas undergoes, whereas enthalpy, entropy, and internal energy are determined by the final and initial states of the fluid because they are properties.

Consider a gas that undergoes a change in state from A to B, as shown in Figure 6.5a, on pressure–volume coordinates. In this process (path A, B), the pressure, volume, and temperature change to the final values so that $p_2$, $v_2$, and $T_2$ are greater than $p_1$, $v_1$, and $T_1$.

To go from state A to state B, let us select the two separate reversible paths A, C and C, B. Path A, C is carried out at constant volume, and path C, B is carried out at constant pressure. These paths are shown in Figure 6.5a as dashed lines and are similarly shown on $T$–$s$ coordinates in Figure 6.5b. It is important to note that the properties at the initial and final states will determine the changes in enthalpy, entropy, and internal energy, but that the path selected between these states will determine the energy interchange as both heat and work.

The change in entropy for a gas undergoing the reversible paths A, C and C, B can readily be determined. For the constant-volume path, $q = \Delta u = c_v \Delta T$, and for the constant-pressure path, $q = \Delta h = c_p \Delta T$. Therefore, from the definition of the property entropy,

$$\Delta s = \frac{q}{T} = \left( \frac{c_v \Delta T}{T} \right)_v + \left( \frac{c_p \Delta T}{T} \right)_p \tag{6.30}$$

(a)

(b)

**FIGURE 6.5**
General gas process. (a) $p$–$v$ diagram. (b) $T$–$s$ diagram.

where the subscripts $v$ and $p$ refer to the constant-volume and constant-pressure processes, respectively.

To obtain the total change in entropy for a finite process between the temperature limits of $T_2$ and $T_1$, it is necessary to sum all the terms of Equation 6.30 over the entire temperature range. Let us first assume that the specific heats are independent of temperature. It is necessary only to sum the $\Delta T/T$ values. This process has already been illustrated, but it is worth repeating at this point. Figure 6.6 shows a plot of $1/T$ as a function of $T$. By selecting a value of $\Delta T$, as shown, the shaded area is $\Delta T/T$. Thus, the sum of the $\Delta T/T$ values is the area under the curve between the desired temperature limits. By the methods of the calculus, it can be shown that this area is $\ln T_2/T_1$.

By performing the summation required by Equation 6.30 and again noting that the specific heats are constant,

$$\Delta s = c_v \ln\left(\frac{T_c}{T_1}\right)_v + c_p \ln\left(\frac{T_2}{T_c}\right)_p \tag{6.31}$$

Equation 6.31 can be put into a more useful form by noting that for the constant-volume process, Equation 6.9 yields $T_c/T_1 = p_2/p_1$, and for the constant-pressure process, $T_2/T_c = v_2/v_1$. Thus,

$$\Delta s = c_v \ln\left(\frac{p_2}{p_1}\right) + c_p \ln\left(\frac{v_2}{v_1}\right) \tag{6.32}$$

By dividing both sides of Equation 6.32 by $c_v$ and noting that $k = c_p/c_v$, we obtain

$$\frac{\Delta s}{c_v} = k \ln\left(\frac{v_2}{v_1}\right) + \ln\left(\frac{p_2}{p_1}\right) \tag{6.33}$$

Another form of Equation 6.32 can be derived by using Equations 6.24 and 6.33:

$$\Delta s = c_p \ln\left(\frac{T_2}{T_1}\right) - \frac{R}{J} \ln\left(\frac{p_2}{p_1}\right) \tag{6.34}$$

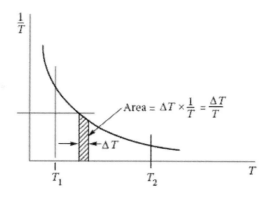

**FIGURE 6.6**
Evaluation of $\Delta T/T$.

Equation 6.34 can also be derived in the following alternative way, from Chapter 5:

$$T\Delta s = \Delta h - \frac{v\Delta p}{J}$$

Dividing through by $T$ and noting that $\Delta h = c_p \Delta T$ and $v/T = R/p$,

$$\Delta s = c_p \frac{\Delta T}{T} - \frac{R}{J}\frac{\Delta p}{p}$$

Summing each term on the right-hand side of this equation gives us Equation 6.34.

The utility of these considerations and equations will become apparent as the individual gas processes are studied. Students should note that Equations 6.31 through 6.34 are valid for *any process* of an *ideal gas* when the specific heats are *constant*. Note that in SI units, the $J$ in the preceding equations is not necessary.

---

### CALCULUS ENRICHMENT

In Chapter 5, we wrote the first of the Gibbs or $T\,ds$ equations as

$$T ds = du + p\,dv \qquad\qquad\qquad\qquad (a)$$

Assuming that we are dealing with an ideal gas,

$$du = c_v dT \qquad\qquad\qquad\qquad (b)$$

and

$$pv = RT \qquad\qquad\qquad\qquad (c)$$

Inserting Equations (b) and (c) into Equation a and solving for $ds$,

$$\int_{s_1}^{s_2} ds = \int \frac{du}{T} + \int \frac{p\,dv}{T} = \int_{T_1}^{T_2} c_v \frac{dT}{T} + \int_{v_1}^{v_2} R\frac{dv}{v} \qquad\qquad (d)$$

For constant-specific heats, Equation (d) is integrated to yield

$$s_2 - s_1 = \Delta s = c_v \ln\left(\frac{T_2}{T_1}\right) + R\ln\left(\frac{v_2}{v_1}\right) \qquad\qquad (e)$$

Equation (e) can be rearranged to yield Equation 6.32 using $c_p - c_v = R$. Thus,

$$\Delta s = c_v \ln\left(\frac{T_2}{T_1}\right) + \left(c_p - c_v\right)\ln\left(\frac{v_2}{v_1}\right) \qquad\qquad (f)$$

and

$$\Delta s = c_v \ln\left(\frac{T_2}{T_1}\right) + c_p \ln\left(\frac{v_2}{v_1}\right) - c_v \ln\left(\frac{v_2}{v_1}\right)$$ (g)

Combining the $c_v$ terms and using $\dfrac{p_1 v_1}{T_1} = \dfrac{p_2 v_2}{T_2}$ yields

$$\Delta s = c_v \ln\left(\frac{p_2}{p_1}\right) + c_p \ln\left(\frac{v_2}{v_1}\right)$$ (h)

which is Equation 6.32.

---

## ILLUSTRATIVE PROBLEM 6.16

One pound of air (MW = 29) expands from 100 psia and 100°F to 15 psia and 0°F. Determine the change in entropy. Assume that $c_p$ of air is 0.24 Btu/lb$_m$·R and is constant.

### SOLUTION

On the basis of the data given, we can use Equation 6.34 to solve this problem. Thus,

$$\Delta s = 0.24 \ln\left(\frac{460}{460+100}\right) - \frac{R}{J} \ln\left(\frac{15}{100}\right)$$

We can make these logarithms positive quantities by noting that $\log x = -\log 1/x$. Therefore,

$$-\Delta s = 0.24 \ln\left(\frac{460+100}{460}\right) - \frac{R}{J} \ln\left(\frac{100}{15}\right)$$

$$= 0.24 \ln\left(\frac{560}{460}\right) - \frac{1545/29}{778} \ln\left(\frac{100}{15}\right)$$

and

$$-\Delta s = 0.04721 - 0.1299$$

$$= 0.0827 \quad \text{or} \quad \Delta s = 0.0827 \text{ Btu/lb}_m \cdot \text{R}$$

---

## ILLUSTRATIVE PROBLEM 6.17

Solve Illustrative Problem 6.16 using Equation 6.32.

### SOLUTION

Because $c_p$ and $R$ are given, let us first solve for $c_v$.

$$c_p = \frac{R \cdot k}{J(k-1)}$$

$$0.24 = \frac{(1545/29)(k)}{778(k-1)}$$

Solving for $k$ yields

$$k = 1.399 \text{ and } c_v = \frac{c_p}{k} = \frac{0.24}{1.399} = 0.1716 \text{ Btu/lb}_m \cdot \text{R}$$

Using Equation 6.32 gives us

$$\Delta s = c_v \ln\left(\frac{p_2}{p_1}\right) + c_p \ln\left(\frac{v_2}{v_1}\right)$$

But

$$\frac{v_2}{v_1} = \frac{T_2}{T_1}\frac{p_2}{p_1} = \frac{460+0}{460+100}\frac{100}{15} = 5.476$$

Thus,

$$\Delta s = 0.1716 \ln\left(\frac{15}{100}\right) + 0.24 \ln(5.476)$$

$$= -0.3255 + 0.4081 = 0.0826 \text{ Btu/lb}_m \cdot \text{R}$$

The agreement is very good.

---

### ILLUSTRATIVE PROBLEM 6.18

Two kilograms of oxygen expand from 500 kPa and 100°C to 150 kPa and 0°C. Determine the change in entropy if $c_p = 0.9093$ kJ/kg·K and remains constant.

### SOLUTION

Using Equation 6.34 and dropping $J$ gives us

$$\Delta s = 0.9093 \ln\left(\frac{273}{273+100}\right) - \frac{8.314}{32} \ln\left(\frac{150}{500}\right) = 0.029 \text{ kJ/kgK}$$

For 2 kg,

$$\Delta S = 2(0.02901) = 0.05802 \text{ kJ/K}$$

---

**ILLUSTRATIVE PROBLEM 6.19**

An ideal gas with constant-specific heats undergoes a change during which its specific volume is halved and its entropy increases by an amount equal to one-quarter of its specific heat at constant volume. Assuming that $k$ is 1.4, what was the increase in pressure for this process?

**SOLUTION**

From Equation 6.33 and the data given,

$$\frac{\Delta s}{c_v} = k \ln\left(\frac{v_2}{v_1}\right) + \ln\left(\frac{p_2}{p_1}\right)$$

$$\frac{1/4c_v}{c_v} = 1.4 \ln\left(\frac{1/2}{1}\right) + \ln\left(\frac{p_2}{p_1}\right)$$

or

$$\frac{1}{4} - 1.4 \ln\left(\frac{1}{2}\right) = \ln\left(\frac{p_2}{p_1}\right)$$

But $-1.4 \ln\frac{1}{2} = -1.4(\ln 1 - \ln 2) = -1.4(0 - 0.693) = 1.4(0.693)$. Thus,

$$\frac{1}{4} + 1.4(0.693) = \ln\left(\frac{p_2}{p_1}\right)$$

$$\ln\left(\frac{p_2}{p_1}\right) = 1.22$$

and by taking antilogarithms,

$$\frac{p_2}{p_1} = 3.4 \text{ or } p_2 = 3.4\, p_1$$

---

## 6.5 Nonflow (Closed System) Gas Processes

In the following portions of this section, five nonflow (closed system) gas processes are analyzed in detail. These processes are constant-volume, constant-pressure, isothermal, isentropic, and polytropic. The derivations in each case proceed from a consideration of the equation of state and the equation of the path. Paths shown as dashed lines indicate nonequilibrium paths.

### 6.5.1 Constant-Volume Process (Isometric Process)

The constant-volume process is best exemplified by a closed tank from which heat is either added or removed. In Figure 6.7, the indicated path is one in which heat is being added.

From Equation 6.9 (ideal gas law), the equation of the path can readily be written as

$$\frac{p_1}{T_1} = \frac{p_2}{T_2} \tag{6.35}$$

The applications of the first law yield the relationship between the heat added and the final state condition. Because there is no volume change, no boundary work is done. Thus,

$$q = u_2 - u_1 = c_v \left( T_2 - T_1 \right) \tag{6.36}$$

Also,

$$\mathbf{w} = 0 \tag{6.37}$$

because there is no change in volume.

The change in entropy for this process is given directly by the defining equation for entropy or as the volume part of Equation 6.31. In either case,

$$\Delta s = c_v \ln\left(\frac{T_2}{T_1}\right) \tag{6.38}$$

Two further considerations complete all the necessary relations for the constant-volume process. The first of these states that the specific heat for this process is by definition $c_v$ and that we have assumed it to be constant. The second consideration (whose utility will appear later) determines the exponent in the equation $pv^n = $ constant, which will make this equation fit the constant-volume process. Rearranging this equation gives us

$$\left(\frac{p_1}{p_2}\right)\left(\frac{v_1}{v_2}\right)^n = 1 \ \text{ or } \ \frac{v_1}{v_2} = \left(\frac{p_1}{p_2}\right)^{1/n} \tag{6.39}$$

Because $v_1/v_2 = 1$, it becomes necessary for $1/n$ to be zero or

$$n \rightarrow \infty \tag{6.40}$$

**FIGURE 6.7**
Constant-volume process. (a) $p$–$v$ diagram. (b) $T$–$s$ diagram.

---

### ILLUSTRATIVE PROBLEM 6.20

If 1/2 lb of gas is heated at constant volume from 70°F to 270°F, determine the change in entropy for this process. Assume that $c_v = 0.17$ Btu/lb$_m$·R.

**SOLUTION**

For this process, $T_1 = 530$R and $T_2 = 730$R. Therefore,

$$\Delta s = c_v \ln\left(\frac{T_2}{T_1}\right) = 0.17 \ln\left(\frac{730}{530}\right)$$

$$\Delta s = 0.0544 \text{ Btu/lb}_m \cdot \text{R}$$

$$\Delta S = \left(\frac{1}{2}\right)(0.0544) = 0.0270 \text{ Btu/R}$$

---

### ILLUSTRATIVE PROBLEM 6.21

If 0.2 kg of air is heated at constant volume from 20°C to 100°C, determine the change in entropy for the process. Assume that $c_v$ is constant and equal to 0.7186 kJ/kg·K.

**SOLUTION**

$$\Delta s = c_v \ln\left(\frac{T_2}{T_1}\right) = 0.7186 \ln\left(\frac{100 + 273}{20 + 273}\right)$$

$$= 0.1735 \frac{\text{kJ}}{\text{kg} \cdot \text{K}}$$

For 0.2 kg,

$$\Delta S = (0.2)(0.1735) = 0.03470 \frac{\text{kJ}}{\text{K}}$$

---

### ILLUSTRATIVE PROBLEM 6.22

The entropy of 1 lb. of gas changes by 0.0743 Btu/lb$_m$·R when heated from 100°F to some higher temperature at constant volume. If $c_v = 0.219$ Btu/lb$_m$·R, determine the higher temperature for this process.

**SOLUTION**

$$\Delta s = c_v \ln\left(\frac{T_2}{T_1}\right)$$

$$0.0743 = 0.219 \ln\left(\frac{T_2}{100 + 460}\right)$$

or

$$\ln\left(\frac{T_2}{560}\right) = 0.3393$$

Taking antilogarithms gives us

$$\frac{T_2}{560} = e^{0.3393}$$

$$= 1.404$$

$$T_2 = 786.2\text{R}$$

## ILLUSTRATIVE PROBLEM 6.23

If 1.5 kg of a gas is heated at constant volume to a final temperature of 425°C and the entropy increase is found to be 0.4386 kJ/K, determine the initial temperature of the process. Use $c_v = 0.8216$ kJ/kg·K.

**SOLUTION**

As before,

$$\Delta S = mc_v \ln\left(\frac{T_2}{T_1}\right)$$

$$0.4386 = 1.5 \times 0.8216 \ln\left(\frac{273 + 425}{T_1}\right)$$

or

$$\ln\left(\frac{698}{T_1}\right) = 0.35589$$

$$\frac{698}{T_1} = e^{0.35589} = 1.4275$$

and

$$T_1 = \frac{698}{1.4275} = 489\text{K} = 216°\text{C}$$

### 6.5.2 Constant-Pressure Process (Isobaric Process)

The constant-pressure process is a good approximation to many of the common physical processes with which we are familiar. The combustion of fuel in a boiler or a gas turbine engine, the flow of fluids, the flow of air in ducts, and other processes can be used to illustrate constant pressure. Both the nonflow (closed system) and flow processes (open system) yield to the following analysis: For an ideal gas, Equation 6.9 once again gives us the equation of the path for this process.

$$\frac{v_1}{v_2} = \frac{T_1}{T_2} \tag{6.41}$$

From the first law, the constant-pressure process has as its energy equation (neglecting kinetic and potential energy terms)

$$q = h_2 - h_1 \tag{6.42}$$

and consequently,

$$q = c_p \left(T_2 - T_1\right) \tag{6.43}$$

where $c_p$ is the specific heat of the process.

During the constant-pressure process shown in Figure 6.8, work is removed as heat is added. The amount of work is the area under the $pv$ curve, which is

$$p\left(v_2 - v_2\right) = p_2 v_2 - p_1 v_1 \tag{6.44}$$

Application of Equation 6.9 and reduction of the mechanical units of Equation 6.44 to thermal units yields

$$p_2 v_2 - p_1 v_1 = \frac{R}{J}\left(T_2 - T_1\right) \tag{6.45}$$

The exponent of $pv^n$ that is applicable to this process is found as follows:

$$\frac{p_2}{p_1}\left(\frac{v_2}{v_1}\right)^n = 1 \text{ but } \frac{p_2}{p_1} = 1$$

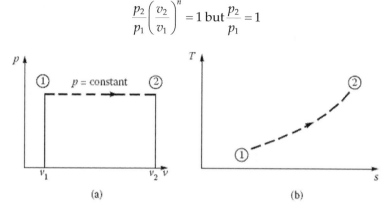

(a)

(b)

**FIGURE 6.8**
Constant-pressure process. (a) $p$–$v$ diagram. (b) $T$–$s$ diagram.

Therefore, $n=0$ for a constant-pressure process.

Finally, the change in entropy for this process is

$$\Delta s = c_p \ \ln\left(\frac{T_2}{T_1}\right) \tag{6.46}$$

---

### ILLUSTRATIVE PROBLEM 6.24

In a gas turbine cycle, heat is added to the working fluid (air) at constant pressure. Determine the heat transferred, the increase in entropy, and the flow work change per pound of air if the initial pressure is 100 psia and the temperature goes from 70°F to 400°F. The specific heat $c_p$ for this process is constant and equals 0.24 Btu/lb$_m$·°F.

### SOLUTION

This process is shown graphically in Figure 6.11. From the energy equation for the constant-pressure process, the heat transferred is $\Delta h$. Therefore,

$$q = \Delta h = c_p \left(T_2 - T_1\right) = (0.24)(860 - 530)$$

$$= 79.2 \ \text{Btu/lb} \left(\text{into the system}\right)$$

$$\Delta s = c_p \ \ln\left(\frac{T_2}{T_1}\right) = 0.24\ln\left(\frac{860}{530}\right)$$

$$= 0.116 \ \text{Btu/lb}_m \cdot \text{R}$$

The flow work change is

$$\frac{p_2 v_2}{J} - \frac{p_1 v_1}{J} = \frac{R}{J}\left(T_2 - T_1\right)$$

$$= \frac{(1545 / 29)(860 - 530)}{778}$$

$$= 22.6 \ \text{Btu/lb}_m$$

In addition to each of the assumptions made in all the processes being considered, it has further been tacitly assumed that these processes are carried out quasi-statically and without friction.

---

### ILLUSTRATIVE PROBLEM 6.25

Heat is added to a turbofan engine at constant pressure. If the air temperature is raised from 20°C to 500°C in the combustion chamber, which is operated at 1 MPa, determine the heat transferred and the increase in entropy per kilogram of air. Use $c_p = 1.0062$ kJ/kg·K.

**SOLUTION**

We proceed as in Illustrative Problem 6.24.

$$q = \Delta h = c_p \left( T_2 - T_1 \right) = \left( 1.0062 \right)\left( 500 - 20 \right) = 483.0 \text{ kJ/kg}$$

and

$$\Delta s = c_p \, \ln\left( \frac{T_2}{T_1} \right) = 1.0062 \ln\left( \frac{500 + 273}{20 + 273} \right)$$

$$= 0.9761 \text{ kJ/kg} \cdot \text{K}$$

### 6.5.3 Constant-Temperature Process (Isothermal Process)

The expansion process shown in Figure 6.9 could conceivably be one in which a gas expands in a hot cylinder. Heat would be transferred from the hot cylinder walls to maintain the gas temperature constant. Since temperature is constant, from the ideal gas law ($Pv = RT = $ constant), the path equation is

$$p_1 v_1 = P_2 v_2 = \text{constant} \tag{6.47}$$

and from the first law, it can be deduced that both the changes in enthalpy and internal energy for this process are zero ($q - w = 0$) because the temperature is constant. Therefore, the work of the isothermal process must exactly equal the heat transferred. Thus,

$$q = \frac{w}{J} \tag{6.48}$$

or

$$q = w \quad \text{in SI units} \tag{6.49}$$

Let us evaluate the change in entropy for the isothermal nonflow (closed system) process from Equation 6.32 as follows:

$$\Delta s = c_p \, \ln\left( \frac{v_2}{v_1} \right) + c_v \, \ln\left( \frac{p_2}{p_1} \right) \tag{6.50}$$

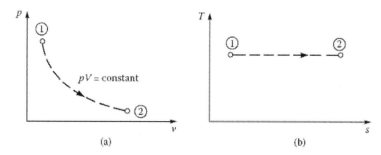

(a)            (b)

**FIGURE 6.9**
Isothermal process. (a) *p–v* diagram. (b) *T–s* diagram.

Because $p_1 v_1 = p_2 v_2$,

$$\frac{v_1}{v_2} = \frac{p_2}{p_1} \tag{6.51}$$

$$\Delta s = c_p \, \ln\left(\frac{v_2}{v_1}\right) + c_v \ln\left(\frac{v_1}{v_2}\right) \tag{6.52}$$

However,

$$\ln\left(\frac{v_1}{v_2}\right) = -\ln\left(\frac{v_2}{v_1}\right) \tag{6.53}$$

Therefore,

$$\Delta s = c_p \, \ln\left(\frac{v_1}{v_2}\right) - c_v \ln\left(\frac{v_2}{v_1}\right) \tag{6.54}$$

Factoring yields

$$\Delta s = \ln\left(\frac{v_2}{v_1}\right)\left(c_p - c_v\right) \tag{6.55}$$

But $c_p - c_v = R/J$. Therefore,

$$\Delta s = \frac{R}{J} \ln\left(\frac{v_2}{v_1}\right) \tag{6.56}$$

Equation 6.56 can also be written as

$$\Delta s = \frac{pv}{TJ} \ln\left(\frac{v_2}{v_1}\right) = \frac{R}{J} \ln\left(\frac{v_2}{v_1}\right) \tag{6.57}$$

In SI units, Equation 6.57 is

$$\Delta s = \frac{pv}{T} \ln\left(\frac{v_2}{v_1}\right) = R \ln\left(\frac{v_2}{v_1}\right) \tag{6.57a}$$

We can now determine the energy interchange as heat and also the work by noting that

$$\Delta s = \frac{q}{T} \quad \text{or} \quad q = T\Delta s \tag{6.58}$$

But from Equation 6.48, $q = w/J$. This gives us

$$\frac{w}{J} = q = T\Delta s = \frac{RT}{J} \ln\left(\frac{v_2}{v_1}\right) = \frac{p_1 v_1}{J} \ln\left(\frac{v_2}{v_1}\right) = \frac{p_2 v_2}{J} \ln\left(\frac{v_2}{v_1}\right) \tag{6.59}$$

In SI units,

$$w = q = p_2 v_2 \, \ln\left(\frac{v_2}{v_1}\right) \tag{6.59a}$$

Because there is no change in temperature for this process, the process follows $pv^n$ as the equation of the path, where $n = 1.0$.

---

**ILLUSTRATIVE PROBLEM 6.26**

If 0.1 lb of nitrogen is kept at a constant temperature of 200°F while its volume is increased to twice its initial volume, determine the heat added and work out of the system.

**SOLUTION**

From Equation 6.59, we have

$$q = \frac{RT}{J} \ln\left(\frac{v_2}{v_1}\right)$$

$$= \frac{(1545/28)(460 + 200)}{778} \ln\left(\frac{2}{1}\right)$$

$$= 32.4 \text{ Btu/lb}_m$$

For 0.1 lb., $Q = 0.1 \times 32.4 = 3.24$ Btu (added to the system). The work out of the system is equal to the heat added; thus,

$$\frac{W}{J} = 3.24 \text{ Btu (out of system)}$$

---

**ILLUSTRATIVE PROBLEM 6.27**

If 1 kg of oxygen has its volume halved at a constant temperature of 50°C, determine the heat added and the work out of the system.

**SOLUTION**

From Equation 6.59a, we have

$$q = RT \ln\left(\frac{v_2}{v_1}\right) = \frac{8.314}{32}(273 + 50)\ln\left(\frac{1}{2}\right)$$

$$= -58.17 \frac{\text{kJ}}{\text{kg}} \text{ (heat out of system)}$$

Therefore,

$$w = -58.17 \frac{\text{kJ}}{\text{kg}} \text{ (work into system)}$$

---

### ILLUSTRATIVE PROBLEM 6.28

Determine the change in entropy in Illustrative Problem 6.27.

### SOLUTION

For constant temperature,

$$\Delta s = \frac{q}{T}$$

Therefore,

$$\Delta s = \frac{-58.17 \text{ kJ/kg}}{(50+273)\text{K}} = -0.1801 \text{ kJ/kg} \cdot \text{K}$$

---

#### 6.5.4  Constant-Entropy Process (Isentropic Process)

The turbofan engine and the gas turbine engine produce useful work by the expansion of hot combustion gases in turbines. These processes are ideally isentropic. It will be recalled that the isentropic process is a reversible, adiabatic process. An alternative definition is a process carried out with no change in entropy. Equation 6.33 serves to define this path:

$$\frac{\Delta s}{c_v} = k \ln\left(\frac{v_2}{v_1}\right) + \ln\left(\frac{p_2}{p_1}\right) = 0 \tag{6.33}$$

or

$$\ln\left(\frac{v_2}{v_1}\right)^k = \ln\left(\frac{p_2}{p_1}\right) \tag{6.60}$$

Taking antilogarithms and rearranging yields

$$p_1 v_1^k = p_2 v_2^k \tag{6.61}$$

Equation 6.61 is the path equation for this process, and by using the equation of state (Equation 6.9), it is possible to rewrite Equation 6.61 in terms of pressures and temperatures as

$$\frac{v_2}{v_1} = \left(\frac{p_1}{p_2}\right)^{1/k} \tag{6.62}$$

and $p_1 v_1 / T_1 = p_2 v_2 / T_2$ from Equation 6.9. Therefore,

$$\frac{p_1}{p_2}\frac{T_2}{T_1} = \left(\frac{p_1}{p_2}\right)^{1/k}$$

Rearranging gives us

$$\frac{T_1}{T_2} = \left(\frac{p_1}{p_2}\right)^{(k-1)/k} \tag{6.63}$$

Similarly, we obtain the $p, v$ relation from Equation 6.61 as

$$\frac{p_1}{p_2} = \left(\frac{v_2}{v_1}\right)^k \quad \text{or} \quad p_1 v_1^k = p_2 v_2^k \tag{6.64}$$

From Equation 6.9, we have $p_1/p_2 = (v_2/v_1)(T_1/T_2)$. Substitution into Equation 6.64 yields

$$\frac{T_1}{T_2} = \left(\frac{v_2}{v_1}\right)^{k-1} \tag{6.65}$$

Equations 6.61, 6.63, and 6.65 define both the state and path of an ideal gas undergoing an isentropic change and are called *isentropic relationships*.

It can be shown that the process follows $pv^n$ with $n = k$, where $k$ (specific heat ratio) is the *isentropic exponent* for the isentropic process.

The work of the nonflow process shown in Figure 6.10 can be evaluated by noticing that the work done in a nonflow process in the absence of heat transfer ($q = 0$) equals the change in internal energy $\left(\text{i.e. } q - w = \Delta u \rightarrow -w = \Delta u\right)$. Therefore,

$$\text{work} = u_1 - u_2$$

$$\text{work} = c_v\left(T_1 - T_2\right)$$

$$c_v = \frac{R}{J(k-1)}$$

Therefore,

$$\frac{w}{J} \frac{R\left(T_1 - T_2\right)}{J(k-1)} \quad \text{or} \quad \frac{R\left(T_2 - T_1\right)}{J(1-k)} \text{ Btu/lb}_m \tag{6.66}$$

Again, note that in SI units, the conversion factor $J$ is not included.

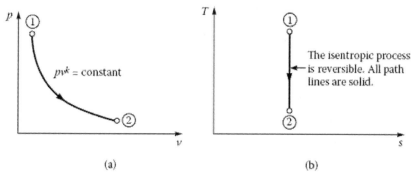

FIGURE 6.10
Isentropic process. (a) $p-v$ diagram. (b) $T-s$ diagram.

## CALCULUS ENRICHMENT

The boundary work of the nonflow, isentropic process can be obtained by integration of the area under the $p$–$v$ curve, as shown in the figure below. Therefore,

$$w = \int_1^2 p\,dv \qquad\qquad (a)$$

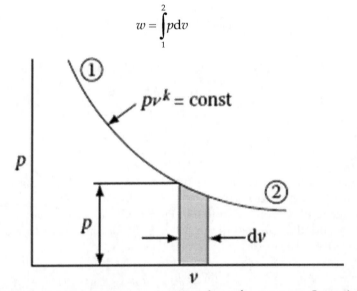

The path of the isentropic process is given by $pv^k$ = constant. Inserting this into Equation (a),

$$w = \int_1^2 \frac{c}{v^k}\,dv \qquad\qquad (b)$$

Integrating,

$$w = c\left.\frac{v^{-k+1}}{-k+1}\right|_1^2 = \frac{c}{-k+1}\left(v_2^{-k+1} - v_1^{-k+1}\right) \qquad\qquad (c)$$

Because $c = pv^k = p_1 v_1^k = p_2 v_2^k$, Equation (c) becomes

$$w = \frac{p_2 v_2 - p_1 v_1}{-k+1} \qquad\qquad (d)$$

but $pv = RT$;

$$w = \frac{R\left(T_2 - T_1\right)}{1-k} \qquad\qquad (e)$$

Equation (e) is the same as Equation 6.66 in appropriate units.

---

## ILLUSTRATIVE PROBLEM 6.29

One pound of air expands isentropically from 5 atm absolute to 1 atm absolute. If the initial temperature is 1000R, determine the final state and the work done by the air if $k$ is 1.4 over this range of temperature.

**SOLUTION**

From Equation 6.63,

$$T_2 = T_1 \left( \frac{p_2}{p_1} \right)^{(k-1)/k}$$

$$= 1000 \left( \frac{1}{5} \right)^{(1.4-1)/1.4} = 631\,\text{R}$$

From Equation 6.66,

$$\text{work} = \frac{R(T_2 - T_1)}{J(1-k)} = \frac{(1545)(631-1000)}{(29)(778)(1-1.4)}$$

$$= 63\,\text{Btu/lb}_m\,(\text{out})$$

---

## ILLUSTRATIVE PROBLEM 6.30

If the initial temperature in Illustrative Problem 6.29 is 500°C, determine the final state and the work done. Assume a mass of 1 kg.

**SOLUTION**

$$T_2 = T_1 \left( \frac{p_2}{p_1} \right)^{(k-1)/k}$$

$$= (500 + 273)\frac{1}{5}^{(1.4-1)/1.4}$$

$$= 488.60\,\text{K} = 215.06°\text{C}$$

$$\text{work} = \frac{R(T_2 - T_1)}{1-k} = \frac{8.314}{29}\frac{215.06 - 500}{1-1.4}$$

$$= 204.2\,\frac{\text{kJ}}{\text{kg}}\,(\text{out})$$

## ILLUSTRATIVE PROBLEM 6.31

A gas expands isentropically from 5 to 1 MPa. If the initial temperature is 800°C and the final temperature is 500°C, determine $k$ for this gas.

### SOLUTION
Because

$$\frac{T_2}{T_1} = \left(\frac{p_2}{p_1}\right)^{(k-1)/k}$$

$$\frac{273+500}{273+800} = \left(\frac{1}{5}\right)^{(k-1)/k}$$

Taking the logarithms of both sides of this equation yields

$$\ln\left(\frac{273+500}{273+800}\right) = \frac{k-1}{k}\ln\left(\frac{1}{5}\right)$$

or

$$\frac{k-1}{k} = \frac{\ln\left(\dfrac{273+500}{273+800}\right)}{\ln\left(\dfrac{1}{5}\right)} = 0.2038$$

Solving,

$$k = 1.256$$

### 6.5.5 Polytropic Process

Each of the processes discussed in this section can have its path equation written in terms of

$$pv^n = \text{constant} \tag{6.67}$$

which is known as the polytropic equation and is used, in general, to characterize any mechanically reversible process. It is also used as an approximation for real processes. A diagram is shown in Figure 6.11. To recapitulate those processes already considered, see Table 6.3. If each of these processes is plotted on $pv$ and $TS$ coordinates (basically, the superposition of Figures 6.7 through 6.10), it is possible to show the entire spectrum of $n$ values.

Observe in Figure 6.12 that going from an initial state of pressure and volume, represented by the point where $n = \infty$ and $n = 0$ axes meet, to a second state, the exponent $n$

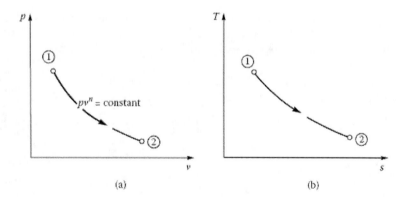

**FIGURE 6.11**
Polytropic process. (a) $p$–$v$ diagram. (b) $T$–$s$ diagram.

**TABLE 6.3**

Values of $n$ for Polytropic Processes

| Process | $n$ |
|---|---|
| Constant volume | $\infty$ |
| Constant pressure | 0 |
| Isothermal | 1 |
| Isentropic | $k$ |
| Polytropic | $-\infty$ to $+\infty$ |

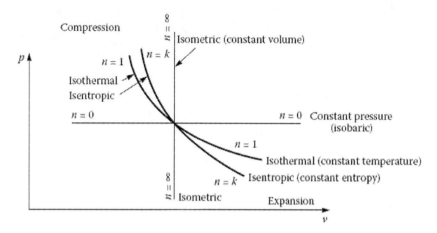

**FIGURE 6.12**
Polytropic processes.

determines the slope of the $p$–$v$ path. At greater values of $n$, the slope is steeper. Since work is the area under the $p$–$v$ curve, one should strive for the isothermal process in compression (to minimize work) or in expansion (to maximize work). Figure 6.13 shows these relationships on a $T$–$s$ diagram.

The pressure, temperature, and volume relations for the polytropic process corresponding to Equations 6.62, 6.63, and 6.65 can be written as follows:

$$\frac{v_2}{v_1} = \left(\frac{p_1}{p_2}\right)^{1/n} \tag{6.68}$$

$$\frac{p_1}{p_2} = \left(\frac{v_2}{v_1}\right)^{n} \text{ or } p_1 v_1^n = p_2 v_2^n \tag{6.69}$$

$$\frac{T_1}{T_2} = \left(\frac{p_1}{p_2}\right)^{(n-1)/n} \tag{6.70}$$

$$\frac{T_1}{T_2} = \left(\frac{v_2}{v_1}\right)^{n-1} \tag{6.71}$$

The work of the polytropic process can be determined by the same method used to obtain the work of the isentropic process. The result is

$$\text{work} = \frac{p_2 v_2 - p_1 v_1}{1 - n} \tag{6.72}$$

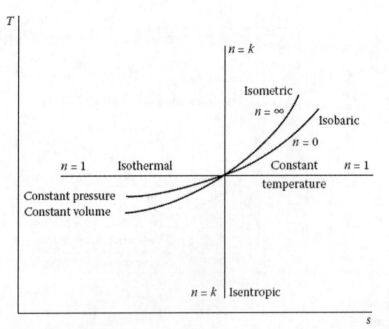

**FIGURE 6.13**
Polytropic processes.

and in terms of thermal units

$$\text{work} = \frac{R(T_2 - T_1)}{J(1-n)} \tag{6.73}$$

To determine the effective specific heat of the polytropic process, we shall use the energy equation for the nonflow process:

$$q = (u_2 - u_1) + \frac{w}{J} \tag{6.74}$$

It is convenient to define the heat transferred by

$$c_n(T_2 - T_1) = q \tag{6.75}$$

where $c_n$ is the polytropic specific heat. Substitution of Equations 6.73 and 6.75 into Equation 6.74 yields

$$c_n(T_2 - T_1) = c_v(T_2 - T_1) + \frac{R(T_2 - T_1)}{J(1-n)} \tag{6.76}$$

$$c_n = c_v + \frac{R}{J(1-n)} \tag{6.77}$$

and

$$c_n = \frac{Jc_c - nJc_v + R}{J(1-n)} \tag{6.78}$$

Using Equation 6.29, we have

$$c_n = \frac{c_p - nc_v}{(1-n)} \tag{6.79}$$

or

$$c_n = c_v \frac{k-n}{1-n} \tag{6.80}$$

Note that $c_n$ can be positive or negative depending on the values of $n$ and $k$.

Substitution of $c_n$ into Equation 6.75 gives the heat transferred in the polytropic process:

$$q = c_v \frac{k-n}{1-n}(T_2 - T_1) \tag{6.81}$$

The change in entropy for the polytropic process can be arrived at by using Equation 6.30 and assuming that $c_n$ is independent of temperature:

$$\Delta s = \frac{q}{T} = \frac{c_n \Delta T}{T} \tag{6.82}$$

Evaluating Equation 6.82 as done previously by summing the $\Delta T/T$ terms yields

$$\Delta s = c_n \ln\left(\frac{T_2}{T_1}\right) \tag{6.83}$$

---

### ILLUSTRATIVE PROBLEM 6.32

For an internal combustion engine, the expansion process can be characterized by $pv^{1.3} = $ constant. If the ratio of specific heats for this gas is 1.4 and the specific heat at constant pressure is 0.24 Btu/lb$_m$·R, determine the heat transferred as the gas expands from 1500R. Also evaluate the work done and the change in entropy for the process. $R$ equals 53.3 for this gas.

**SOLUTION**

From

$$\frac{c_p}{c_v} = k \qquad c_v = \frac{0.24}{1.4} = 0.171$$

Therefore

$$c_n = c_v \frac{k-1}{1-n} = 0.171 \; \frac{1.4-1.3}{1-1.3} = -0.0570 \; \text{Btu/lb}_m \cdot R$$

The negative sign of $c_n$ indicates that either the heat transfer for the process comes from the system or there is a negative temperature change while heat is transferred to the system.

The heat transferred is $c_n(T_2 - T_1)$. Therefore,

$$q = -0.0570(600 - 500)$$

$$= 51.3 \; \text{Btu/lb}_m \; (\text{to the system})$$

The work done can be found using Equation 6.73, giving us

$$\text{work} = \frac{R(T_2 - T_1)}{J(1-n)} = \frac{53.3}{778} \; \frac{600 - 1500}{1-1.3}$$

$$= 205.5 \; \text{Btu/lb}_m \; (\text{from the system})$$

The change in entropy $\Delta s$ is given by

$$\Delta s = -0.0570 \ln\left(\frac{600}{1500}\right)$$

$$= -0.0570 - \ln\left(\frac{1500}{600}\right) = 0.0522 \; \text{Btu/lb}_m \cdot R$$

---

### ILLUSTRATIVE PROBLEM 6.33

Determine the ratio of the inlet to the outlet pressure in Illustrative Problem 6.32.

### SOLUTION

From Equation 6.70,

$$\frac{T_1}{T_2} = \left(\frac{p_1}{p_2}\right)^{(n-1)/n}$$

For this problem,

$$\frac{1500}{600} = \left(\frac{p_1}{p_2}\right)^{(1.3-1)/1.3}$$

$$2.5 = \left(\frac{p_1}{p_2}\right)^{0.3/1.3} = \left(\frac{p_1}{p_2}\right)^{0.2308}$$

Taking the logarithms of both sides of this equation yields

$$\ln 2.5 = 0.2308 \ln\left(\frac{p_1}{p_2}\right)$$

$$\frac{0.9162}{0.2308} = \ln\left(\frac{p_1}{p_2}\right) = 3.97$$

and

$$\frac{p_1}{p_2} = 52.99$$

---

Keep in mind that when deriving the equations in this chapter, the specific heats were assumed to be constant and independent of temperature. The same assumption was made for $k$. When the variation of these quantities does not permit us to assume an essentially constant average value, the methods of the calculus must be employed. For these techniques, students are referred to the references.

Table 6.4 summarizes the results of the derivations made for nonflow processes. In this table, the enthalpy and internal energies have also been tabulated, and it will be noted that these quantities depend only on the end states of the specified process and are independent of the path. This table has been included as a convenient reference. Students are cautioned against memorizing the table or using it without fully understanding the restrictions involved in each of the processes discussed.

**TABLE 6.4**

Ideal Gas Relations (per Unit Mass of Gas, Closed System)[a]

| Process → | Constant Volume $V$=Constant (Isometric) | Constant Pressure $p$=Constant (Isobaric or Isopiestic) | Isothermal $T$=Constant | Isentropic $s$=Constant | Polytropic $pv^n$=Constant |
|---|---|---|---|---|---|
| → $p,v,T$ | $\dfrac{T_1}{T_2}=\dfrac{p_1}{p_2}$ | $\dfrac{T_1}{T_2}=\dfrac{v_1}{v_2}$ | $p_1v_1=p_2v_2$ | $p_1v_1^k=p_2v_2^k$ | $p_1v_1^n=p_2v_2^n$ |
| | | | | $\dfrac{T_1}{T_2}=\left(\dfrac{v_1}{v_2}\right)^{k-1}$ | $\dfrac{T_1}{T_2}=\left(\dfrac{v_1}{v_2}\right)^{n-1}$ |
| | | | | $\dfrac{T_1}{T_2}=\left(\dfrac{p_1}{p_2}\right)^{(k-1)/k}$ | $\dfrac{T_1}{T_2}=\left(\dfrac{p_1}{p_2}\right)^{(n-1)/n}$ |
| $w$ | $0$ | $p(v_2-v_1)$ | $p_1v_1\ln\left(\dfrac{v_2}{v_1}\right)$ | $\dfrac{p_2v_2-p_1v_1}{1-k}$ | $\dfrac{p_2v_2-p_1v_1}{1-n}$ |
| $u_2-u_1$ | $c_v(T_2-T_1)$ | $c_v(T_2-T_1)$ | $0$ | $c_v(T_2-T_1)$ | $c_v(T_2-T_1)$ |
| $q$ | $c_v(T_2-T_1)$ | $c_p(T_2-T_1)$ | $p_1v_1\ln\left(\dfrac{v_2}{v_1}\right)$ | $0$ | $c_n(T_2-T_1)$ |
| $n$ | $\infty$ | $0$ | $1$ | $k$ | $-\infty$ to $+\infty$ |
| $c$ | $c_v$ | $c_p$ | $\infty$ | $0$ | $c_n=c_v\dfrac{k-n}{1-n}$ |
| $h_2\,h_1$ | $c_p(T_2-T_1)$ | $c_p(T_2-T_1)$ | $0$ | $c_p(T_2-T_1)$ | $c_p(T_2-T_1)$ |
| $s_2-s_1$ | $c_v\ln\left(\dfrac{T_2}{T_1}\right)$ | $c_p\ln\left(\dfrac{T_2}{T_1}\right)$ | $R\ln\left(\dfrac{v_2}{v_1}\right)$ | $0$ | $c_n\ln\left(\dfrac{T_2}{T_1}\right)$ |

[a] The conversion factor $J$ has been omitted from all equations. Care should be taken when using this table with English units.

The change in enthalpy and the change in internal energy for the ideal gas are useful quantities, and we can obtain these readily from the previous considerations in this chapter. Thus, for enthalpy, we can write

$$h_2-h_1=c_p\left(T_2-T_1\right) \tag{6.84}$$

From Equation 6.9, we have $T=pv/R$, and from Equation 6.29, we have $c_p=R/J(k-1)$. Substitution of these quantities in Equation 6.84 gives us

$$h_2-h_1=\frac{R}{J}\frac{k}{k-1}\left(\frac{p_2v_2}{R}-\frac{p_1v_1}{R}\right) \tag{6.85}$$

Simplifying yields

$$h_2-h_1=\frac{k}{J(k-1)}\left(p_2v_2-p_1v_1\right) \tag{6.86}$$

or

$$h_2 - h_1 = \frac{k}{k-1}\left(p_2 v_2 - p_1 v_1\right) \text{SI units} \tag{6.86a}$$

Similarly, for internal energy,

$$u_2 - u_1 = c_v\left(T_2 - T_1\right)$$

and

$$u_2 - u_1 = \frac{k}{J(k-1)}\left(p_2 v_2 - p_1 v_1\right) \tag{6.87}$$

or

$$u_2 - u_1 = \frac{k}{k-1}\left(p_2 v_2 - p_1 v_1\right) \text{SI units} \tag{6.87a}$$

As a check on Equations 6.86 and 6.87, note that the ratio of $h_2 = -h_1$ to $u_2 - u_1$ is the ratio of $c_p$ to $c_v$, which it should be.

## 6.6 The *Gas Tables*

The *Gas Tables* provide a convenient and most useful set of data in tabular form for the computation of gas processes. By designating Equation 6.9 as the equation of state, it has been possible to express the various thermodynamic properties of these gases in terms of their temperatures. For air at 32°F, the error in using this equation of state is of the order of 1% at 300 psia and only 0.1% at atmospheric pressure. In addition to using the equation of state of an ideal gas, the specific heats listed in these tables were based on spectroscopic data at zero pressure. The zero of entropy and enthalpy were chosen to be zero at zero degrees absolute. Use of the specific heats at zero pressure is equivalent to the assumption that the specific heat is a function of temperature only.

The outline of the *Gas Tables* can be easily developed by considering that the specific heats are temperature functions. The enthalpy values are listed as functions of the absolute temperature, and the change in enthalpy for a process is just the difference for the values tabulated at the final and initial temperatures. Because ideal gas relations are used,

$$u_2 - u_1 = \left(h_2 - h_1\right) - \frac{R}{J}\left(T_2 - T_1\right) \tag{6.88}$$

The general expression for the change in entropy of an ideal gas with a variable specific heat can be obtained, and the change in entropy per unit mass between states 1 and 2 is then

$$s_2 - s_1 = \phi_2\left(T_2\right) - \phi_1\left(T_1\right) - \frac{R}{J} \ln\left(\frac{p_2}{p_1}\right) \tag{6.89}$$

where $\phi$ is a function of the temperature (i.e., $\phi = f(t)$ or simply $\phi(T)$) given in the tables for each gas at each temperature. For an isentropic change,

$$\ln\left(\frac{p_2}{p_1}\right) = \frac{J}{R}\left(\phi_2\left(T_2\right) - \phi_1\left(T_1\right)\right) \tag{6.90}$$

---

### CALCULUS ENRICHMENT

When the specific heat of an ideal gas is not constant, we need to return to the second Gibbs or $T\,ds$ equation, namely,

$$T ds = dh - v\,dp \tag{a}$$

Using $dh = c_p\,dT$ and $pv = RT$ for the ideal gas yields

$$ds = c_p\frac{dT}{T} - \frac{R}{p}dp \tag{b}$$

It is possible to remove the constant $R$ from the integral of Equation (b). However, with $c_p = c_p(T)$, that is, a function of temperature, $c_p$ cannot be removed from the integral. Therefore, we write

$$\Delta s = \int_{s_1}^{s_2} ds = \int_{T_1}^{T_2}\frac{c_p dT}{T} - R\ln\left(\frac{p_2}{p_1}\right) \tag{c}$$

The integral in Equation c is a function only of temperature. In the *Gas Tables*, the tabulated function is

$$\int_{T_1}^{T_2}\frac{c_p dT}{T} = \phi_2\left(T_2\right) - \phi_1\left(T_1\right) \tag{d}$$

The change in entropy for an ideal gas with variable specific heats therefore is

$$\Delta s = s_2 - s_1 = \phi_2\left(T_2\right) - \phi_1\left(T_1\right) - R\ln\left(\frac{p_2}{p_1}\right) \tag{e}$$

In English units,

$$s_2 - s_1 = \phi_2\left(T_2\right) - \phi_1\left(T_1\right) - \frac{R}{J}\ln\left(\frac{p_2}{p_1}\right) \tag{f}$$

where Equation f is the same as Equation 6.89.
If we now use the first Gibbs or $T ds$ equation

$$T ds = du + p\,dv \tag{g}$$

Dividing by $T$ and using $du = c_v dT$,

$$\int_{s_1}^{s_2} ds = \int_{T_1}^{T_2} \frac{c_v dT}{T} + R\ln\left(\frac{v_2}{v_1}\right) \qquad \text{(h)}$$

or

$$s_2 - s_1 = \int_{T_1}^{T_2} \frac{c_v dT}{T} + R\ln\left(\frac{v_2}{v_1}\right) \qquad \text{(i)}$$

For an isentropic process,

$$-R\ln\frac{v_2}{v_1} + \int_{T_1}^{T_2} \frac{c_v dT}{T} \qquad \text{(j)}$$

If we now define the relative specific volume, $v_r = \ln\left(\frac{v_2}{v_1}\right)$,

$$v_r = \frac{1}{R}\int_{T_1}^{T_2} \frac{c_v dT}{T} \qquad \text{(k)}$$

The *Gas Tables* tabulate two other useful properties: the relative pressure $p_r$ and the relative specific volume $v_r$. If we define the relative pressure $p_r$ as

$$\ln p_r = \frac{\phi(T)}{R} \qquad \text{(6.90a)}$$

then Equation 6.90 becomes

$$\ln\left(\frac{p_{r,2}}{p_{r,1}}\right) = \left(\phi_2(T_2) - \phi_1(T_1)\right) \qquad \text{(6.90b)}$$

Equating Equations 6.90 and 6.90b,

$$\ln\left(\frac{p_{r,2}}{p_{r,1}}\right) = \ln\left(\frac{p_2}{p_1}\right) \qquad \text{(6.90c)}$$

or

$$r_p = \frac{p_2}{p_1} = \frac{p_{r,2}}{p_{r,1}} \text{ for an isentropic process} \qquad \text{(6.90d)}$$

where $r_p$ is called the pressure ratio, which is used in the analysis of turbines and compressors. If we now use the definition of the relative specific volume $v_r$ as

$$v_r = \ln\left(\frac{v_2}{v_1}\right) \qquad \text{(6.90e)}$$

we can show that

$$r_v = \frac{v_{r,2}}{v_{r,1}} = \frac{v_2}{v_1} \text{ for an isentropic process} \qquad (6.90\text{f})$$

where $r_v$ is called the compression ratio, which is used in the analysis of gasoline and diesel engines. The *Gas Tables* also have adopted the definition that

$$v_r = \frac{RT}{p_r} \qquad (6.91)$$

with the units of $R$ selected so that $v_r$ is the specific volume in cubic feet per pound when the pressure is in psia for those tables based on a unit mass. When the mass is the pound-mole, $v_r$ is the molal specific volume in cubic feet per pound-mole when the pressure is psia.

The following problems will serve to illustrate the use of the abridged air table, which is given in Appendix 2, Table A.20.

---

### ILLUSTRATIVE PROBLEM 6.34

Determine the change in enthalpy, internal energy, and entropy when air is heated at constant pressure from 500R to 1000R.

### SOLUTION

| From the Table at 1000R | From the Table at 500 R |
|---|---|
| $h = 240.98$ Btu/lb$_m$ | $h = 119.48$ Btu/lb$_m$ |
| $u = 172.43$ Btu/lb$_m$ | $u = 85.20$ Btu/lb$_m$ |
| $\phi = 0.75042$ Btu/lb$_m$·R | $\phi = 0.58233$ Btu/lb$_m$·R |

The change in enthalpy is $h_2 - h_1 = 240.98 - 119.48 = 121.5$ Btu/lb$_m$. The change in internal energy is $u_2 - u_1 = 172.43 - 85.20 = 87.23$ Btu/lb$_m$. Because in the constant-pressure process $- R \ln(p_2/p_1)$ is zero,

$$\Delta s = \phi_2 - \phi_1 = 0.75042 - 0.58233 = 0.16809 \text{ Btu/lb}_m \cdot R$$

---

### ILLUSTRATIVE PROBLEM 6.35

Solve Illustrative Problem 6.29 using the *Gas Tables*.

### SOLUTION

In this problem, air expands from 5 atm absolute to 1 atm absolute from an initial temperature of 1000R. At 1000R, $p_{r,1} = 12.298$ and $h_1 = 240.98$ Btu/lb$_m$. Since the pressure ratio $r_p = \frac{1}{5} = \frac{p_{r,2}}{p_{r,1}}$, the value of the final $p_{r,2} = 12.298/5 = 2.4596$. Interpolation in the air table yields the following:

| $T$ | $p_r$ |
|-----|-------|
| 620 | 2.249 |
|     | 2.4596 |
| 640 | 2.514 |

$$620 + \frac{2.4596 - 2.249}{2.514 - 2.249} \times 20 = 635.9 \text{ or } 636 \text{ R}$$

The work done in an isentropic nonflow expansion ($q - w = \Delta u, q = 0 \rightarrow -w = \Delta u$) is

$$\text{work} = u_1 - u_2 = 172.43 - 108.51 = 63.92 \text{ Btu/lb}_\text{m}$$

where the value of $u_2$ is obtained by interpolation at 636R, and the value of $u_1$ is read from the air table at 1000R.

The fact that the results found by using the *Gas Tables* agree well with those found from the ideal gas relations is not a coincidence. In all these problems, the pressures were low and the temperatures relatively high. Under these conditions, it would be expected that these gases would behave much like ideal gases. The principal differences are caused by the fact that the specific heats vary with temperature. The *Gas Tables* take this variation into account and reduce the work of computations considerably.

A computerized table of properties of air is available at the CRC Press website. Use the Downloads/Updates tab at http://www.crcpress.com/product/isbn/9781482238556. Properties of other common gases (carbon dioxide, carbon monoxide, hydrogen, nitrogen, oxygen, and water vapor) are available as well.

## 6.7 Steady-Flow Gas Processes

In a manner similar to that for nonflow (closed system) gas processes, the steady flow (open system) of an ideal gas can be analyzed. The steady-flow energy equation, the ideal gas law, and the definitions of specific heat and entropy are utilized. These are applied to a control volume such as that shown in Figure 4.13. In most ideal gas flow processes, the kinetic energy, K.E. and potential energy, P.E. terms are negligible. Thus, as shown in the calculus enrichment of Section 4.4.2, the differential reversible work per unit mass flowing is

$$dw = -v\,dp \tag{6.92}$$

From this equation, we see that reversible work is a function of specific volume, the inverse of fluid density. Thus, for turbomachinery such as a compressor, which consumes work, the greater the density of the fluid, the less the work required for a given pressure differential. Conversely, for devices that produce work such as a turbine, the least density fluid is desired. Thus, the steam turbine cycle described in Chapter 8 expands a gas (steam) to produce work and compresses a liquid (water) when input work is required.

Equation 6.92 may be integrated to obtain relationships for flow work for the same various processes that were analyzed in Section 6.5. The ideal gas relationships for pressure, temperature, specific volume, specific heats, and entropy apply to flow as well as nonflow processes.

To solve steady-flow problems, note that the mass flow rate is assumed to be constant; the volume flow rates may vary from the inlet to the outlet of the device due to density differences. The power developed or consumed by the device is found by multiplying the specific work $w$ by the mass flow rate. For conservation, all the energy entering the system must equal that leaving the system $\left( \dot{E}_{in} - \dot{E}_{out} = \Delta \dot{E}_{system}; \text{ For } \Delta \dot{E}_{system} = 0 \rightarrow \dot{E}_{in} = \dot{E}_{out} \right)$. For most steady-flow gas problems, the potential and kinetic energy changes are insignificant relative to the enthalpy changes. Thus, the energy equation becomes

$$dq = dh - dw = dh - vdp \tag{6.93}$$

This equation is solved by integration. For constant-specific heat, the integral of $dh$ is

$$h_2 - h_1 = c_p \left( T_2 - T_1 \right)$$

The integral of the work term depends on the process, which will now be discussed. Note that after finding the work per unit mass, $w$, the power is found by multiplying $w$ by the appropriate mass flow rate, $\dot{m}$. Heat transfer can then be found from Equation 6.93.

### 6.7.1 Constant-Specific Volume Process

When the specific volume of the fluid remains constant, this is equivalent to an incompressible flow problem, as seen in liquid pumps. Even when a gas is compressed just slightly, the assumption of incompressibility still is valid. Integration of Equation 6.92 shows the work to be

$$w = -v \left( p_2 - p_1 \right) \tag{6.94}$$

Here, subscripts 1 and 2 refer to the entrance and exit conditions, respectively. Utilizing the ideal gas law $\left( pv = RT \right)$, this is equivalent to a process that depends only on the temperature difference between the inlet and the outlet, as, in SI units:

$$w = -R \left( T_2 - T_1 \right) \tag{6.95}$$

In solving problems involving a constant-specific volume $\left( v_1 = v_2 = v \right)$ of an ideal gas $\left( \dfrac{p_1 v}{T_1} = \dfrac{p_2 v}{T_2} \right)$, the ideal gas law yields the relationship

$$\frac{p_1}{p_2} = \frac{T_1}{T_2}. \tag{6.96}$$

---

**ILLUSTRATED PROBLEM 6.36**

A small compressor receives 0.70 m³/s $\left(\dot{V} = 0.7\,{}^{m^3}\!\!/_s\right)$ of air at 70°C and 101 kPa. It discharges the air at 100°C and 110 kPa. If the air is considered incompressible, find the compressor power required.

**SOLUTION**

Assuming air as an ideal gas, the first law for a steady-flow system can be written as

$$q = c_p\left(T_2 - T_1\right) + w$$

To find power, we must find $\dot{m}$ since $\dot{W} = \dot{m}w$. We find $m$ from the equation of state (ideal gas law) $(pv = RT)$ as follows

$$\dot{m} = \frac{\dot{V}}{v} = \frac{p\dot{V}}{RT}$$

Thus,

$$\dot{m} = \left(101\ \text{kPa}\right)\left(0.70\ \text{m}^3/\text{s}\right)/\left(0.287\ \text{kJ}/343\ \text{K}\right) = 0.718\ \text{kg/s}$$

Solving for $w$ yields

$$w = -R\left(T_2 - T_1\right) = -0.287\ \text{kJ/kg K} \times \left(373\ \text{K} - 343\ \text{K} = -8.16\ \text{kJ/kg}\right)$$

$$\dot{W} = \dot{m}w = 0.718\ \text{kg/s} \times \left(-8.61\ \text{kJ/kg}\right) = -6.18\ \text{kW}$$

A negative answer is expected since a compressor takes in power.

---

## 6.7.2 Constant-Pressure Process

It can be seen from Equation 6.93 that in the absence of kinetic and potential energy effects, with no elements to add or extract work from a constant-pressure open system, the first law then yields

$$Q = \dot{m}\left(h_2 - h_1\right) = \dot{m}c_p\left(T_2 - T_1\right) \tag{6.97}$$

One must bear in mind that flow work $(pv \rightarrow h = u + pv)$ still is present in an open systems under steady-flow conditions. Open systems that typify the constant pressure types are heat exchangers where a cold fluid is used to cool a hotter fluid flowing through it. Both fluids flow at constant pressure, with heat being exchanged as the fluids go from the corresponding entrance points to their exit points. Most heat exchangers are well insulated and may be considered adiabatic (i.e., no heat being exchanged with the environment or outside). Thus,

$$\dot{m}c_{ph}\left(T_{h,2} - T_{h,1}\right) = -\dot{m}c_{pc}\left(T_{c,2} - T_{c,1}\right)$$

where $c_{ph}$ is the specific heat of the hot fluid and $c_{pc}$ is the specific heat of the cold fluid. Note that the negative sign is needed because the heat transfers are going in opposite directions (one fluid gains energy while the other loses energy). Heat exchangers are discussed in more detail in Chapter 11.

### 6.7.3 Constant-Temperature Process

Next, the isothermal steady-flow process for the ideal gas is considered. Here, the relationship $pv =$ constant is substituted into Equation 6.92, which is then integrated to yield

$$w = p_1 v_1 \ln\left(\frac{p_1}{p_2}\right) \qquad (6.98)$$

Using the ideal gas law $\left(pv = RT\right)$, this relationship can also be written as

$$w = RT \ln\left(\frac{p_1}{p_2}\right) \qquad (6.99)$$

Since the temperature is constant, there is no change in enthalpy for such processes since internal energy for an ideal gas is only a function of temperature $\left(h = u + pv = u + RT\right)$. In practice, reversible constant-temperature processes are never realized but are a goal to shoot for as they represent the minimum work input for compression and the maximum work output for expansion (see Section 6.5.5). This is also what makes the Carnot cycle with its two constant-temperature processes exist only in the ideal state. Note that even though the temperature may be constant, there can still be heat transfer equal to the work being done.

### 6.7.4 Isentropic Process

Next, the isentropic steady-flow ideal gas process is considered. The general path equation is given in Equation 6.67, which may be substituted into Equation 6.92 with $n = k$. The integration of this equation yields

$$w = \frac{k\left(p_2 v_2 - p_1 v_1\right)}{1 - k} \qquad (6.100)$$

The ideal gas law may be used to find alternate expressions for Equation 6.98 that are more useful for the values given in a particular problem. The following is a useful relationship between pressures and temperatures for a process from state 1 to state 2:

$$\frac{T_2}{T_1} = \left(\frac{p_1}{p_2}\right)^{\frac{k-1}{k}} \qquad (6.101)$$

If only temperatures are known, the following expression may be used to find work per unit mass flowing:

$$w = \frac{kR\left(T_2 - T_1\right)}{1 - k} \qquad (6.102)$$

In many cases, only the initial temperature and pressure ratio of a process are known. In such cases, the manipulation of the work expression yields

$$w_{12} = \frac{kRT_1}{1-k}\left[\left(\frac{P_2}{P_1}\right)^{\frac{k-1}{k}} - 1\right] \tag{6.103}$$

These equations are used to solve reversible adiabatic (isentropic) processes, which frequently apply to compressor and turbine problems, which often provide only initial conditions and pressure ($r_p$) or compression ratios ($r_v$).

---

### ILLUSTRATIVE PROBLEM 6.37

The input to a compressor is air at 101 kPa and 30°C. The pressure ratio, $r_p$ is 5. If the flow rate of the air is 100 kg/min, find (a) exit air temperature and (b) power consumed.

**SOLUTION**

Using Equation 6.101, the exit temperature is

$$T_2 = T_1\left(\frac{p_2}{p_1}\right)^{\frac{k-1}{k}} = 303(5)^{0.2857} = 479.9 \text{ K} \quad \text{or} \quad 206.9°C$$

The work is found using Equation 6.103:

$$w_{12} = \frac{kRT_1}{1-k}\left(\frac{P_2}{P_1}\right)^{\frac{k-1}{k}} - 1 = \frac{1.4 \times 0.287 \times 303}{-0.4}\left[(5)^{0.2857} - 1\right] = -177.7 \text{ kJ/kg}$$

The power consumed is

$$\dot{W} = \dot{m}w = 100 \times (-177.7)/60 = -296.16 \text{ kW}$$

---

## 6.7.5 Polytropic Process

Finally, the polytropic process may be analyzed as Section 6.7.4 with the exponent $n$ instead of $k$. The integration of Equation 6.92 yields, for this case,

$$w = \frac{n(p_2v_2 - p_1v_1)}{1-n} \tag{6.104}$$

Note that in the equations for polytropic processes, they may be stated in terms of temperatures only by invoking the ideal gas law $pv = RT$ as in Equation 6.100 with $n$ instead of $k$. For problems involving turbomachinery, the most commonly used equation for work is Equation 6.103 with $n$ instead of $k$:

$$w = \frac{nRT_1}{1-n}\left[\left(\frac{P_2}{P_1}\right)^{\frac{n-1}{n}} - 1\right] \tag{6.105}$$

Also note that in polytropic processes, there is usually heat transferred into or out of the control system.

For steady-flow (open system) processes, the relationships for specific heat, internal energy change, enthalpy change, and entropy change are the same as for the nonflow (closed system) processes. It is suggested that students utilize these results to construct a chart similar to that of Table 6.4 for steady-flow processes. It should be remembered that these results are for work per unit mass and are given in SI units. To find the power or the heat transfer rate, multiply the equations given above by the mass flow rate $\dot{m}$.

Heat transfer in a polytropic process is found from the first law, Equation 6.93, finding work and the change in enthalpy as a function of the temperature change. There is a short-cut found by creating an artificial specific heat, $c_n$ defined as follows:

$$c_n = c_v \frac{k-n}{1-n} \tag{6.106}$$

which allows heat transfer to be calculated directly as

$$q = c_n (T_2 - T_1) \tag{6.107}$$

---

### ILLUSTRATIVE PROBLEM 6.38

A turbine receives 120 $lb_m$/min of air at 60 psia and 2000R and expands it polytropically ($n=1.5$) to 14.7 psia. Find the power generated and the heat transfer rate.

#### SOLUTION

The exit temperature is found from Equation 6.101:

$$T_2 = T_1 \left( \frac{p_2}{p_1} \right)^{\frac{k-1}{k}} = 2000 \left( \frac{14.7}{60} \right)^{\frac{0.5}{1.5}} = 1251.47 \text{ R}$$

The work is found from Equation 6.105:

$$w = \frac{nRT_1}{1-n} \left( \frac{P_2}{P_1} \right)^{\frac{n-1}{n}} - 1 \frac{1.5 \times 53.34 \times 2000}{0.5 \times 778} \left( \frac{14.7}{60} \right)^{\frac{0.5}{1.5}} - 1 = 153.97 \text{ Btu/min}$$

Power is found by multiplying work by the mass flow rate:

$$\dot{W} = \dot{m}w = 120 \times 153.97 = 18,476.76 \text{ Btu/min or } 453.3 \text{ hp}$$

The heat transfer rate is found from the integration of Equation 6.93, with each term multiplied by the mass flow rate:

$$\dot{Q} = \dot{m}c_p (T_2 - T_1) - \dot{W} = 120 \times 0.24 = (1251.47 - 2000) - 18,476.76$$

$$= -3080 \text{ Btu/min}$$

The negative sign for the heat rate means that the turbine loses heat to its surroundings. Equation 6.107 may be used to verify this result.

## 6.8 Real Gases

Many equations have been proposed to describe the pressure, volume, and temperature behavior of real gases where the simple equation of state for the ideal gas is inadequate. The earliest *equation of state*, based on elementary kinetic theory, was proposed by van der Waals. This equation is

$$p + \frac{a}{v^2}(v-b) = RT \tag{6.108}$$

where $v$ is the volume occupied by 1 lb mol of gas, $R$ is the universal gas constant (1545), and $a$ and $b$ are constants for each gas. In the limit, as $a$ and $b$ go to zero, the van der Waals equation reduces to the ideal gas relation. This equation of state is most accurate when applied to gases that are removed from the saturated vapor state. It is possible to evaluate $a$ and $b$ in terms of the conditions at the critical state (the $c$ state). When this is done, it will be found that the van der Waals equation of state can be rewritten in terms of the *reduced properties* $p/p_c$, $T/T_c$, and $v/v_c$. These ratios are known, respectively, as the *reduced pressure*, the *reduced temperature*, and the *reduced specific volume* and are indicated by the symbols $p_r$, $T_r$, and $v_r$. The values with subscript c correspond to properties at the critical state of the gas, when its velocity is the speed of sound. Using this notation, we can show from the van der Waals equation that all gases have the same $p, v, T$ relation when their reduced properties are the same. This is essentially the *law of corresponding states*. This law is not universally correct, and it has been modified to make it more generally applicable.

The utility of the law of corresponding states lies in the fact that a single curve (called a generalized compressibility chart) can be used to evaluate the $p, v, T$ data for all gases. Most charts (made for various classes of gases) based on this principle use $Z$ as the ordinate, where $Z = pv/RT$. A generalized compressibility chart is shown in Figure 6.14.

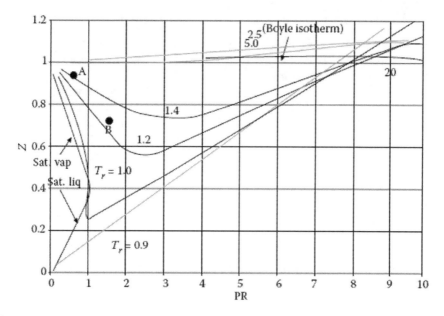

**FIGURE 6.14**

Compressibility factor Z versus reduced pressure PR. (From Annamalai, K. et al., *Advanced Thermodynamics Engineering*, 2nd edition, CRC Press, Boca Raton, Florida, 2011.)

**ILLUSTRATIVE PROBLEM 6.39**

Methane ($CH_4$, MW = 16) has a critical temperature of 343R and a critical pressure of 674 psia. Determine the specific volume of methane at 50°F and 500 psia using both the ideal gas relation and Figure 6.14.

**SOLUTION**

$$p_r = \frac{p}{p_c} = \frac{500}{674} = 0.742$$

$$T_r = \frac{T}{T_c} = \frac{460 + 50}{343} = 1.487$$

Reading Figure 6.14 at these values gives Z equal to 0.93. Therefore,

$$Z = 0.93 = \frac{pv}{RT} \qquad v = 0.93 \frac{RT}{P}$$

and

$$v = \frac{0.93(1545/16)(510)}{500 \times 144} = 0.635 \text{ ft}^3/\text{lb}_m$$

For the ideal gas,

$$v = \frac{RT}{p} = \frac{1545 \times 510}{16 \times 500 \times 144} = 0.683 \text{ ft}^3/\text{lb}_m$$

To obtain the other thermodynamic properties of the real gas, similar procedures based on the law of corresponding states have been developed. The interested reader is referred to the references for this information.

In general, the ideal gas relations yield more nearly correct results when the gas under consideration is at a pressure that is small compared to the critical pressure and when its temperature is much greater than the critical temperature. In other words, when $T_r \gg 1$ and $p_r \ll 1$, the ideal gas relations are usually applicable with negligible error.

## 6.9 Frictional Effects

Up to now, we have concentrated on reversible systems that exclude friction. We can, however, account for irreversibility due to friction using the concept of *paddle work*. Considering this for a closed system, imagine a closed container into which a paddle is inserted, as in Figure 4.6. The spinning paddle wheel does frictional work, and this is always negative, going into the system when following the sign convention. It is analogous to what friction always does in a system, and in all cases, it effectively causes the system to heat up.

This can be seen by writing the first law for a closed system with this additional frictional paddle-work term $W_p$:

$$Q - W - W_p = \Delta U \qquad (6.109)$$

Paddle work is always negative, and with the minus sign before it, it will always tend to add as a heat input term. $W$ is the useful reversible work (i.e., boundary work) of the system, if any.

---

### ILLUSTRATED PROBLEM 6.40

A closed container with air initially at 100 kPa pressure, 1 m³ fixed volume, and 50°C receives 100 kJ of heat and 50 kJ of paddle work. Find the final temperature of the air.

#### SOLUTION

Note that, in this case, $Q$ is not equal to $\Delta U$ (which is always $mc_v\Delta T$) because of the paddle-work term. The first law is written as

$$Q - W - W_p = \Delta U = mc_v\Delta T$$

The work $W$ (boundary work) is zero since the volume of this closed system is fixed ($p\Delta V$ is zero). The mass is found from the ideal gas equation.

$$m = \frac{pV}{RT} = \frac{100 \times 1.0}{0.287 \times 323} = 1.079 \text{ kg}$$

Solving the first law equation for $\Delta T$:

$$\Delta T = \frac{Q - W_p}{mC_v} = \frac{100 - (-50)}{1.079 \times 0.718} = 193.62$$

Thus, $T_2 = 50 + 193.62 = 243.62°C$.

---

### ILLUSTRATED PROBLEM 6.41

If the closed container of air in Illustrated Problem 6.40 is allowed to expand at constant pressure after receiving 100 kJ of heat and 50 kJ of paddle work, find the final temperature and final volume of the container.

#### SOLUTION

Here $\Delta V$ is not zero, therefore, there is boundary work at constant pressure. So the first law becomes

$$Q - p\Delta V - W_p = mC_v\Delta T$$

For ideal gas at constant pressure, $p\Delta V$ equals $mR\Delta T$ (i.e. $pV_2 - pV_1 = mRT_2 - mRT_1$), so the first law equation may be solved for one variable, $\Delta T$.

The equation becomes

$$100 - 1.079(0.287)(\Delta T) - (-50) = 1.079(0.718)(\Delta T)$$

Solving, $\Delta T = 138.33$, so $T_2 = 188.33°C$. Note that the final temperature in this case is lower than that for the previous problem because the work done by the expansion uses up energy.

The final volume is found from

$$\Delta V = V_2 - V_1 = \frac{mR}{p}(T_2 - T_1) = \frac{mR\Delta T}{p} = \frac{1.079 \times 0.287 \times 138.33}{100} = 0.428 \text{ m}^3$$

Thus, $V_2 = 1.428 \text{ m}^3$.

## 6.10 Review

Gases have been empirically found to have a simple equation that relates the properties, pressure, temperature, and specific volume. We called this relation the equation of state for an ideal gas and were able to express it mathematically in several equivalent forms. We also determined that the constant in the equation of state could be expressed as a constant divided by the MW of the gas. For the ideal gas, the specific heats at constant pressure and constant temperature are functions only of the temperature. We then discussed the concept of average specific heat. Because the specific heats are functions only of temperature, a simple relation was developed relating the specific heats and the universal gas constant. Several other useful equations were subsequently developed using the ratio of the two specific heats, $k$, as a useful parameter. A general relation for the change in entropy of the ideal gas was evaluated based on the general concepts that were discussed in Chapter 5. From this general relation, we then were able to write the equation of path for an isentropic process.

Once the foregoing was established, five nonflow processes were studied in detail: the constant-volume, constant-pressure, isothermal, isentropic, and polytropic processes. The relations developed were summarized in a convenient table. Much of the tedious work required to use the many equations developed can be minimized using the *Gas Tables*. The use of these tables was discussed and applied to several situations and compared to the solutions obtained using the equations developed earlier.

Our next discussion in this chapter involved the flow of gases in open systems for the same five processes noted above. This was followed by a discussion of real gases and their equations of state. Finally, we used the concept of paddle work as a way of integrating friction into the first law of thermodynamics.

The material covered in this chapter will probably be used either directly or as a basis for further study as frequently as the material in any chapter in this book. At first glance, it may appear to be formidable, because a large number of equations are developed. In actuality, for each of these situations, we have always invoked three basic concepts: an equation of state, an equation of path, and an energy equation. Rather than attempting to memorize formulas, each situation should be studied and the three basic concepts then applied. The definitions given throughout the chapter should be thoroughly understood as one proceeds in the chapter.

The material on gas flow and real gases is intended as a brief introduction to topics about which volumes have been written. This material will, however, provide the background for further studies in these fields.

## Key Terms

Terms used for the first time in this chapter are as follows:

**Boyle's law:** The path of a constant-temperature ideal gas process; $pv =$ constant.

**Charles' laws:** $p_1/p_2 = T_1/T_2$ for a constant-volume process; $v_1/v_2 = T_1/T_2$ for a constant-pressure process.

**Equation of state:** An equation relating $p$, $v$, and $T$ for a gas.

**Gas constant (R):** Either 1545/MW ft. $lb_f/lb_m \cdot R$ or 8.314/MW kJ/kg·K.

**Gas Tables:** J. H. Keenan and Joseph Kaye, *Gas Tables* (New York: John Wiley & Sons, Inc., 1948).

**Ideal gas:** A gas having the equation of state $pv = RT$.

**Isentropic exponent (k), specific heat ratio:** $k = c_p/c_v$.

**Isobaric process:** A constant-pressure process.

**Isometric process:** A constant-volume process.

**Law of corresponding states:** All gases have the same $p$, $v$, $T$ relation when their reduced properties are the same.

**Mole:** The amount of a substance whose mass is equal to its molecular weight in pounds mass or in kilograms.

**Polytropic process:** A mechanically reversible, nonadiabatic process characterized by the equation of state, $pv^n =$ constant.

**Reduced properties:** The ratios of $p/p_c$, $v/v_c$, or $T/T_c$, where the subscript c refers to the critical state.

## Equations Developed in This Chapter

| | | |
|---|---|---|
| Ideal gas law | $pV = pmv = mRT$ | (6.8) |
| Ideal gas law | $pv = RT$ | (6.9) |
| Molar volume | $V_{molar} = \dfrac{1545T}{p}$ or $\dfrac{8.314T}{p}$ (in SI units) | (6.12) |

| Specific heat of the ideal gas | $c_p = \dfrac{\Delta h}{\Delta T}$ | (6.15a) |

| Specific heat of the ideal gas | $c_v = \dfrac{\Delta u}{\Delta T}$ | (6.15b) |

| Property relation | $c_p - c_v = \dfrac{R}{J}$ | (6.24) |

| Property relation (SI) | $c_p - c_v = R$ | (6.24a) |

| Adiabatic exponent | $k = \dfrac{c_p}{c_v}$ | (6.25) |

| Property relation | $c_v = \dfrac{R}{J(k-1)}$ | (6.28) |

| Property relation (SI) | $c_v = \dfrac{R}{k-1}$ | (6.28a) |

| Property relation | $c_p = \dfrac{R}{J}\dfrac{k}{k-1}$ | (6.29) |

| Property relation (SI) | $c_p = R\dfrac{k}{k-1}$ | (6.29a) |

| Entropy change | $\Delta s = c_v \ln\left(\dfrac{p_2}{p_1}\right) + c_p \ln\left(\dfrac{v_2}{v_1}\right)$ | (6.32) |

| Entropy change | $\dfrac{\Delta s}{c_v} = k\ln\left(\dfrac{v_2}{v_1}\right) + \ln\left(\dfrac{p_2}{p_1}\right)$ | (6.33) |

| Isentropic path | $p_1 v_1^k = p_2 v_2^k$ | (6.61) |

| Isentropic path | $\dfrac{v_2}{v_1} = \left(\dfrac{p_1}{p_2}\right)^{1/k}$ | (6.62) |

| Isentropic path | $\dfrac{T_1}{T_2} = \left(\dfrac{p_1}{p_2}\right)^{(k-1)/k}$ | (6.63) |

| Isentropic path | $\dfrac{T_1}{T_2} = \left(\dfrac{v_2}{v_1}\right)^{k-1}$ | (6.65) |

| Polytropic process | $pv^n = \text{constant}$ | (6.67) |

| Polytropic process | $\dfrac{v_2}{v_1} = \left(\dfrac{p_1}{p_2}\right)^{1/n}$ | (6.68) |

| Polytropic process | $\dfrac{T_1}{T_2} = \left(\dfrac{p_1}{p_2}\right)^{(n-1)/n}$ | (6.70) |

| Polytropic process | $\dfrac{T_1}{T_2} = \left(\dfrac{v_2}{v_1}\right)^{(n-1)}$ | (6.71) |

| Polytropic process | $c_n = c_v = \dfrac{k-n}{1-n}$ | (6.80) |

| Property relation | $u_2 - u_1 = (h_2 - h_1) - \dfrac{R}{J}(T_2 - T_1)$ | (6.88) |

| Gas Tables | $s_2 - s_1 = \phi_2\left(T_2\right) - \phi_1\left(T_1\right) - \dfrac{R}{J}\ln\left(\dfrac{p_2}{p_2}\right)$ | (6.89) |
|---|---|---|
| Gas Tables | $\ln\dfrac{p_2}{p_1} = \dfrac{J}{R}\left(\phi_2\left(T_2\right) - \phi_1\left(T_1\right)\right)$ | (6.90) |
| Gas Tables | $v_r = \dfrac{RT}{p_r}$ | (6.91) |
| Constant-volume process | $w = -v\left(p_2 - p_1\right)$ | (6.94) |
| Constant-volume process | $w = -R\left(T_2 - T_1\right)$ | (6.95) |
| Constant-pressure process | $Q = \dot{m}c_p\left(T_2 - T_1\right)$ | (6.97) |
| Isothermal process | $w = p_1 v_1 \ln\left(\dfrac{p_1}{p_2}\right)$ | (6.98) |
| Isentropic process | $w = \dfrac{k\left(p_2 v_2 - p_1 v_1\right)}{1-k}$ | (6.100) |
| Isentropic process | $w = \dfrac{kR\left(T_2 - T_1\right)}{1-k}$ | (6.102) |
| Isentropic process | $w_{12} = \dfrac{kRT_1}{1-k}\left[\left(\dfrac{p_2}{p_1}\right)^{\frac{k-1}{k}} - 1\right]$ | (6.103) |
| Polytropic process | $w = \dfrac{n\left(p_2 v_2 - p_1 v_1\right)}{1-n}$ | (6.104) |
| Polytropic process | $w = \dfrac{nRT_1}{1-n}\left[\left(\dfrac{p_2}{p_1}\right)^{\frac{n-1}{n}} - 1\right]$ | (6.105) |
| Specific heat | $c_n = c_v\dfrac{k-n}{1-n}$ | (6.106) |
| Real gas | $p + \dfrac{a}{v^2}\left(v - b\right) = RT$ | (6.108) |
| Paddle-work first law | $Q - W - W_p = \Delta U$ | (6.109) |

## QUESTIONS

6.1 Define the term *ideal gas*.

6.2 How does an ideal gas differ from a perfect gas?

6.3 Define the term *mole*.

6.4 Write an expression for $R$ in both systems of units. Clearly indicate the units in each case.

6.5 Air is a mechanical mixture of several gases. Which two gases make up the major part of air? What is the approximate MW of air?

6.6 *True or false?* One mole of air at room pressure and temperature will occupy more volume than 1 mole of hydrogen at the same conditions.

6.7     The MW of a gas in kilograms is that mass that contains $6.02 \times 10^{26}$ molecules, where $6.02 \times 10^{26}$ is called Avogadro's number. From this, estimate the mass of a nitrogen atom.

6.8     Define the specific heat at constant volume and the specific heat at constant pressure. For the ideal gas, what can be stated about the effects of temperature and pressure on these properties?

6.9     Illustrate what is meant by the term *mean specific heat*.

6.10    What is the lower limit of $k$?

6.11    Write the equation of the path for a constant-volume process.

6.12    Using the $p$–$v$ diagram, show that the work of a constant-volume process is zero.

6.13    Define a *polytropic process*.

6.14    Write the equation of the path for a constant-pressure process.

6.15    Using the $p$–$v$ diagram, show how the work of a constant-pressure process can be obtained.

6.16    Write the equation of the path for an isothermal process.

6.17    Write the equation of the path for an isentropic process.

6.18    How does a polytropic process differ from an isentropic process?

6.19    What is the equation of the path for a polytropic process?

## PROBLEMS

Unless otherwise noted, use 29 for the MW of air. Also, use $c_p = 0.24$ Btu/lbm·R (1.0061 kJ/Kg·K) and $c_v = 0.171$ Btu/lb$_m$·R (0.7186 kJ/kg·K).

### Problems Involving the Ideal Gas Law

6.1     One pound of nitrogen initially at 14.7 psia and 32°F is compressed until its volume is halved and its pressure is increased by 50%. What is its final temperature?

6.2     One kilogram of nitrogen initially at 100 kPa and 0°C is expanded so that its volume is doubled while its pressure decreases to 0.3 times its initial value. What is the final temperature?

6.3     Basing your argument on the MWs of dry air and of water vapor, which would you expect to be heavier, dry air or moist air?

6.4     Two pounds of air is in a tank having a volume of 5 ft.³. What is the density of air in the tank?

6.5     Air is at 101.3 kPa and 20°C. Determine its specific volume.

6.6     Calculate the specific volume of air at 14.7 psia and 68°F.

6.7     Air at 40°C is in a tank at 1 MPa. If there is 1.5 kg of air in the tank, what is the volume of the tank?

6.8     What is the density of air at 72°F and 160 psia (MW = 29)?

6.9     Determine the density of air at 20°C and 100 kPa (MW = 29).

6.10    A closed tank having a volume of 5 ft.³ is filled with methane ($CH_4$, MW = 16) at 68°F. If the tank contains 6 lb$_m$, determine the pressure in the tank.

6.11 A closed cylinder having a volume of 0.14 m³ is filled with a gas at 20°C. If the cylinder contains 2.75 kg of methane (MW = 16), determine the pressure in the cylinder.

6.12 A gas mixture occupies a volume of 1.4 m³ at a temperature of 65°C and a pressure of 350 kPa. Determine $R$ for the gas if the mass of gas in the container is 1.13 kg. Also, determine the MW of the gas. Assume that the mixture behaves as a single ideal gas.

6.13 If the MW of refrigerant HFC-134a is 102, determine its specific volume at 20 psia and 60°F.

6.14 If the MW of refrigerant HFC-134a is 102, determine its specific volume at 140 kPa and 20°C.

6.15 A gas occupies 50 ft.³ at a temperature of 150°F and a pressure of 50 psia. Determine $R$ for the gas if there is a mass of 2.5 lb. in the tank.

6.16 A gas at 450 kPa occupies 1.5 m³ at a temperature of 80°C. If there is a mass of 1.8 kg in the tank, determine $R$ for this gas.

6.17 Three kilograms of $CO_2$ fills a tank having a volume of 0.2 m³ at 20°C. What is the pressure in the tank?

6.18 A tank of $CO_2$ is filled until the pressure is 5 MPa and the temperature is 100°C. If the volume of the tank is 1 m³, how much mass is there in the tank?

6.19 A tank contains nitrogen at a pressure of 30 MPa and 250 K. If the tank has a volume of 100 L, determine the mass in the tank.

6.20 Determine the specific volume of $CO_2$ at a pressure of 190 psia and a temperature of 550R.

6.21 Calculate the density and volume of 29 lb. of air (1 lb. mol) at 620R and 14.7 psia.

6.22 A tank contains 1 kg of oxygen at 600 psia and 32°F. Determine the volume of the tank.

6.23 A tire is inflated to 30 psig when the temperature is 90°F. Determine (a) the tire pressure if the temperature drops to −20°F and (b) the initial air density in the tire if its volume is 1 ft.³

6.24 Determine the temperature of 2 lb$_m$ of air that is in a tank whose volume is 5 ft.³ at a pressure of 50 psia.

*6.25 A "rule of thumb" has it that the pressure in a tire will change approximately 1 psi for every 10°F temperature change. Starting with the assumption that the tire is rigid and that the ambient temperature is 60°F, show how close this rule of thumb is for a tire that is initially at 30 psig. Also, show that the % change in pressure is $(\Delta/T_1)$ 100 where $\Delta$ is the temperature rise and is the initial absolute temperature.

6.26 An automobile tire contains air at 180 kPa and 0°C. After prolonged high-speed driving, the tire temperature is found to be 80°C. Assuming the tire to be rigid, determine its pressure when hot.

6.27 An automobile tire is inflated to 26 psig on a cold morning when the temperature is 32°F. After prolonged high-speed driving, the tire temperature is 140°F. If the tire is considered to be rigid, what is its hot pressure? Use Problem 6.25 as a check on your answer.

*6.28    A tank with a fixed volume contains $O_2$ at 30 psia and 60°F. Two pounds of $O_2$ is added to the contents so that the final pressure is 40 psia and the final temperature is 80°F. What is the volume of the tank?

## Problems Involving Property Relations

6.29    An ideal gas has $R = 80$ ft. lbf/$lb_m$·R and $k = 1.4$. Determine $c_p$ and $c_v$ for this gas.

6.30    An ideal gas has $k = 1.4$ and $R = 0.1890$ kJ/kg·K. Determine $c_p$ and $c_v$ for this gas.

6.31    An ideal gas has MW equal to 29 and $k$ equal to 1.3. Determine $c_p$ and $c_v$ of this gas in English units.

6.32    Solve Problem 6.31 in SI units.

6.33    An ideal gas has $c_p = 0.312$ Btu/$lb_m$·R and $c_v = 0.240$ Btu/$lb_m$·R. Determine $R$ and the MW of the gas.

6.34    An ideal gas has $c_p = 1.302$ kJ/kg·K and $c_v = 0.986$ kJ/kg·K. Determine $R$ and the MW of the gas.

6.35    If the specific heat at constant pressure of a gas having a MW of 18 is 0.45 Btu/$lb_m$·R, what is its specific heat at constant volume?

6.36    A gas has a specific heat at constant pressure of 0.34 Btu/$lb_m$·R, and a specific heat at constant volume of 0.272 Btu/$lb_m$·R. Determine the MW of this gas.

6.37    If $c_p$ of a gas is 5.2028 kJ/kg·K and its MW is 4, determine $c_v$.

6.38    If $c_p$ of a gas is 1.0426 kJ/kg·K and $c_v$ is 0.7454 kJ/kg·K, determine the MW and $k$ of the gas.

6.39    If $c_p$ of a gas is 0.3 Btu/$lb_m$·R, what is $c_v$ if the gas has a MW of 28?

6.40    An ideal gas has $c_p = 1.2$ kJ/kg·K and $k = 1.32$. Determine the MW, $R$, and $c_v$ for the gas.

6.41    An ideal gas has $c_p = 0.286$ Btu/$lb_m$·R and $k = 1.29$. Determine the MW, $R$, and $c_v$ for this gas.

## Problems Involving Isothermal Processes

6.42    Nitrogen initially has a pressure of 55 psia and a temperature of 80°F. If its final pressure after an isothermal expansion is 18 psia, what is its final specific volume?

6.43    Oxygen is initially at 30°C and 350 kPa. What is its final specific volume after it undergoes an isothermal expansion to 120 kPa?

6.44    How much work is required to isothermally compress 1 kg of air from 200 kPa to 2 MPa? Assume that $c_p$ is 1.0061 kJ/kg·K, $c_v$ is 0.7186 kJ/kg·K, and the air is at 20°C (MW = 29).

6.45    Determine the change in entropy for Problem 6.43.

6.46    Determine the change in entropy for Problem 6.44.

6.47    Air is compressed from 20 to 120 psia at a constant temperature of 75°F. Calculate the work required for compression per $lb_m$.

6.48    How much work is required to isothermally compress 1 $lb_m$ of air from 25 psia and 125°F to 250 psia?

6.49 Air initially at 100°F is isothermally compressed from a pressure of 1 atm to a pressure of 5 atm. Determine the work done on 1 lb$_m$ of air and the heat removed during the process.

6.50 Air initially at 200°C is isothermally compressed from a pressure of 1 atm to a pressure of 5 atm. Determine the work done on 1 kg of air and the heat removed during the process.

6.51 Nitrogen (MW = 28) is isothermally compressed from 200°F and 40 psia to 1000 psia. Calculate the final specific volume, the change in entropy, and the heat transferred in this process.

6.52 One pound mass of air is expanded isothermally until its final volume is 50% greater than its initial volume. If the temperature of the air is 150°F, how much heat was supplied and what is the change in entropy of this process?

6.53 One kilogram of nitrogen (MW = 28) is expanded isothermally until its final volume is 100% greater than its initial volume. If the temperature of the nitrogen is 80°C, how much heat was supplied and what is the change in entropy for this process?

6.54 One pound of argon (MW = 39.90) is isothermally compressed from 100°F and 20 psia to 600 psia. Calculate the final specific volume, the change in entropy, and the heat transferred.

6.55 Heat is added to air at constant temperature from initial conditions of 80 psia and 250°F to a final pressure of 25 psia. Determine the heat added and the change in entropy per pound mass of air for this process (MW = 29).

6.56 Air is expanded at constant temperature from 200 psia and 2000R to a final volume of 10 ft.³. If the initial volume was 4 ft.³, calculate the final pressure, the total work done, the total change in internal energy, and the total heat transfer for the process.

*6.57 A cylinder of nitrogen is 2 m high and has a diameter of 0.5 m. It is initially filled with gas until the pressure is 200 psia and the temperature is 70°F. Some of the gas is used in a welding process until the pressure is 50 psia while the temperature remains constant at 70°F. Calculate the mass of nitrogen that has been used.

*6.58 A cylinder is filled with air at 200°F and 150 psia. If 5 lb$_m$ is removed from the cylinder while the temperature is kept constant at 200°F and the pressure falls to 100 psia, determine the volume of the cylinder and the original mass in it.

6.59 Five pounds of air at 120°F and 100 psia undergo an isothermal process until the final pressure is 40 psia. Determine the work done, the change in internal energy, the heat removed, and the change in entropy for this process.

6.60 Two kilograms of a gas having $c_p = 1.205$ kJ/kg·K and $c_v = 0.8607$ kJ/kg·K undergo an isothermal process from initial conditions of 500 kPa and 200°C to a final pressure of 100 kPa. Determine the work done, the change in internal energy, the heat removed, and the change in entropy for the process.

6.61 A gas has $c_p = 0.3215$ Btu/lb$_m$·R and $c_v = 0.2296$ Btu/lb$_m$·R. Determine the work done, the heat removed, the change in internal energy, and the change in entropy for a pound mass that is initially at 85 psia and 100°F and is compressed to isothermally 255 psia.

*6.62  Ten pounds of carbon monoxide at 86°F occupies 100 ft.$^3$ in a tank. An additional amount of carbon monoxide is added slowly to the tank, raising the pressure to 150 psia. Assuming the process is carried out isothermally, determine the mass of gas added.

*6.63  A gas has a MW of 30 and is contained in a 1 m$^3$ tank at 350 kPa and 25°C. A second evacuated tank having a volume of 0.15 m$^3$ is attached to the first tank. A connecting valve is opened and the gas flows into the second tank until the pressures in both tanks are equal. If the second tank was initially at 25°C and the temperature remains constant, determine the final pressure in the tanks.

## Problems Involving Constant-Volume Processes

6.64  A closed tank contains 8 ft.$^3$ of air at 17 psia and 125°F. If 60 Btu is added to the air, determine the final pressure. Assume that $c_v = 0.17$ Btu/lb$_m$·R.

6.65  What is the final pressure in a closed tank after 60 kJ is added to 0.3 m$^3$ of air that is at 150 kPa and 40°C? Assume that $c_p = 1.0061$ kJ/kg·K and $c_v = 0.7186$ kJ/kg·K.

6.66  If 0.1 lb$_m$ of a gas is heated at constant volume from 90°F to 350°F, determine the increase in entropy for the process. Assume that $c_v = 0.22$ Btu/lb$_m$·R. Also, determine the ratio of the final pressure to the initial pressure for this process. Assume that $c_p = 0.304$ Btu/lb$_m$·R.

6.67  Determine the change in internal energy and enthalpy for the gas in Problem 6.66.

6.68  Air is heated at constant volume from 40°F to 1000R. Calculate the change in entropy for this process.

6.69  Solve Problem 6.68 using the *Gas Tables*.

6.70  Calculate the change in internal energy for the gas in Problem 6.68.

6.71  Solve Problem 6.68 for the average specific heat at constant pressure for the process using the *Gas Tables*.

6.72  Solve Problem 6.68 for the average specific heat at constant volume for the process using the *Gas Tables*.

6.73  An air compressor is used to fill a rigid tank to 80 psia at a temperature of 70°F. Determine the temperature of the gas that will cause the relief valve to open if the valve is set at 150 psia.

6.74  If 0.05 kg of a gas is heated from 30°C to 175°C, what is the entropy change if the process is carried out at constant volume and $c_v = 0.90$ kJ/kg·K? $\kappa$ for this gas is 1.3.

6.75  Determine the change in internal energy and enthalpy for the gas in Problem 6.74.

6.76  One hundred Btu is added to 1 lb$_m$ of air at constant volume whose initial conditions are 85°F and 35 psia. If $c_v = 0.19$ Btu/lb$_m$·R, determine the final pressure and final temperature and the change in entropy for the process (MW = 29).

6.77  If 100 kJ is added to 0.5 kg of air at constant volume and the initial conditions are 30°C and 200 kPa, and $c_v = 0.7$ kJ/kg·K, determine the final pressure, the final temperature, and the change in entropy for the process (MW = 29).

6.78  One pound of a gas at 80°F is heated until its entropy increase is 0.15 Btu/lb$_m$·R. If $c_v$ is 0.21 Btu/lb$_m$·R and the process is carried out at constant volume, determine the final gas temperature.

*6.79  A closed, rigid, well-insulated tank has a volume of air of 5 ft.$^3$ that is initially at 150 psia and 50°F. The air is stirred by a paddle wheel until the pressure increases to 400 psia. Determine the final temperature and the paddle-work input in Btu.

*6.80  A closed, rigid, well-insulated tank has a volume of air of 0.15 m$^3$ that is initially at 1 MPa and 10°C. The gas is stirred by a paddle wheel until the pressure increases to 2.8 MPa. Determine the final temperature and the paddle-work input in kJ.

## Processes Involving Constant Pressure

6.81  A process is carried out at a constant pressure of 350 kPa. If the volume changes from an initial value of 2.5 to 7.5 m$^3$, determine the work of the process.

6.82  Air is expanded from 1 to 8 ft.$^3$ in a piston cylinder apparatus at a constant pressure of 100 psia from an initial temperature of 100°F. Calculate the final air temperature and the work per lb$_m$ in the cylinder.

6.83  Air at 700 kPa is heated from 20°C to 120°C at constant pressure. Determine the heat required per kilogram.

6.84  Air at 100 psia is heated from 80°F to 200°F at constant pressure. Determine the heat required per pound mass.

6.85  Solve Problem 6.84 using the *Gas Tables*.

6.86  Solve Problem 6.84 for the average specific heat for the process using the *Gas Tables*.

6.87  Use the *Gas Tables* to determine the change in enthalpy of air when the temperature changes from 540R to 1300R.

6.88  Solve Problem 6.87 for the average specific heat for the process.

6.89  Air is heated at constant pressure from 40°F to 1240°F. Determine the change in entropy for this process.

6.90  Solve Problem 6.89 using the *Gas Tables*.

*6.91  If 4.5 lb$_m$ of oxygen is at 100°F, determine the change in density expressed as a percentage of the initial density when the temperature is changed to 250°F at a constant pressure of 100 psia.

*6.92  One kilogram of a gas is heated at a constant pressure of 425 kPa until the internal energy increases by 250 kJ/kg and the temperature increases by 100°C. If there is work done by the gas that equals 150 kJ/kg, determine the change in volume and $c_p$.

*6.93  If 2 kg of oxygen occupies 0.25 m$^3$ at 40°C, determine the change in density expressed as a percentage of the initial density when the temperature is changed to 125°C at a constant pressure of 0.7 MPa.

6.94  Compute the new volume in a cylinder if air is cooled at constant pressure until its initial temperature of 500°F is reduced to 100°F. The initial volume of the cylinder is 10 ft.$^3$.

6.95  An ideal gas has $R = 80$ ft. lb$_f$/lb$_m$·R and $k = 1.4$. If the gas undergoes a nonflow, constant-pressure process from 1000°F to 500°F, determine the work done, the change in internal energy, and the change in enthalpy for the process. All answers should be given in Btu/lb$_m$.

6.96     Determine the heat removed from the cylinder of Problem 6.94. Assume that $c_p = 0.24$ Btu/lb$_m$·R and $c_v = 0.171$ Btu/lb$_m$·R.

6.97     An ideal gas (MW = 20) undergoes a nonflow, constant-pressure process from 500°C to 250°C. If $k$ is 1.4, determine the work done, the change in internal energy, and the change in enthalpy for the process.

6.98     Determine the work done, the change in internal energy, and the heat that is supplied when air initially at 200 psia and 200°F expands in a cylinder at constant pressure from an initial volume of 20 ft.$^3$ to triple its initial volume. Assume that $c_p = 0.24$ Btu/lb$_m$·R and $c_v = 0.171$ Btu/lb$_m$·R.

6.99     A piston cylinder contains 2 lb$_m$ of air and operates at a constant pressure of 400 psia. The initial volume is 1 ft.$^3$ and the initial temperature is 100°F. The final volume is 4 ft.$^3$ and the final temperature is 1780°F. Determine the work and its direction (in or out) and the heat transfer and its direction, in Btu/lb$_m$.

6.100    Carbon dioxide (MW = 44) is heated at constant pressure from 14.7 psia and 90°F to 500°F. If $k = 1.4$, determine the heat transferred and the change in entropy for the process. Assume that $c_p$ is constant.

6.101    Carbon dioxide is heated from 100 kPa and 35°C to 250°C at constant pressure. If $k$ equals 1.3, determine the heat transferred and the change in entropy for the process if $c_p$ is constant (MW = 44).

6.102    One pound of nitrogen occupies a volume of 5 ft.$^3$ at a temperature of 200°F. If the temperature is increased to 300°F at constant pressure, determine the work and the final volume.

*6.103   Ten cubic feet of air is compressed until the volume is 5 ft.$^3$. If the pressure is kept constant at 190 psia, determine the work, the change in internal energy, and the heat transfer for this process.

## Problems Involving Isentropic and Polytropic Processes

Use Table A.13 for physical data (except for air) as necessary.

6.104    Argon is compressed isentropically from 80°F and 20 psia to 1200°F. Determine the final pressure.

6.105    Helium is compressed isentropically from 100 kPa and 0°C to 2 MPa. Calculate the final temperature.

6.106    Nitrogen is compressed isentropically from 14.7 psia and 70°F to 1400 psia. What is the final temperature of the nitrogen?

6.107    Air at 110 kPa and 20°C is isentropically compressed to 10 MPa. Determine the final temperature.

6.108    Air is expanded isentropically from a volume of 10 ft.$^3$ to a volume of 30 ft.$^3$. If the initial temperature is 200°F, determine the final temperature.

6.109    An ideal gas with $k = 1.4$ undergoes a reversible adiabatic expansion from 100 psia and 700°F to 350°F. Determine the final pressure of the gas.

6.110    An ideal gas expands isentropically from 120 psia and 200°F to 20 psia. Determine the final temperature for this nonflow process if $k = 1.4$.

6.111    Air is compressed isentropically from 1 atm absolute to 10 atm absolute. If the initial temperature is 70°F, determine the final temperature and the work required for this nonflow process if $k = 1.4$.

6.112 Solve Problem 6.110 using the *Gas Tables*.

6.113 Solve Problem 6.111 using the *Gas Tables*.

*6.114 An ideal gas expands isentropically from 150 psia and 450°F to 25 psia and 100°F. Determine $k$ for this process.

*6.115 An ideal gas expands isentropically from 600 kPa and 195°C to 110 kPa and 40°C. What is the $k$ of this gas process?

*6.116 A gas expands isentropically from 90 psia and 380°F to 15 psia and 100°F. Determine $k$ for this gas.

*6.117 Determine the final state and change in enthalpy and internal energy when air is expanded isentropically from 100 psia and 1000R to 75 psia. Assume that $k = 1.4$. If all the work of this nonflow process could be recovered, how much work would be available?

6.118 Air undergoes a polytropic expansion from 2500R to 1500R. The final pressure of the air is 20 psia, and $n$ for the process is 1.32. Determine the work output per pound mass of air.

*6.119 One pound mass of air is compressed from 20 to 60 psia polytropically from an initial temperature of 100°F. If $n = 1.26$, determine the final temperature and the heat transfer for the process.

*6.120 A gas initially at 170 psia and 270°F is expanded to 40 psia and 90°F. Determine $n$ for this process.

*6.121 A process is carried out in which a gas is compressed from 14 to 28 psia, and its final volume is decreased to 60% of its initial volume. Determine $n$ for this process.

*6.122 One pound mass of nitrogen undergoes a process during which it changes from an initial state of 100 psia and 400°F to a final state of 300 psia and 650°F. Determine $n$ for this process.

*6.123 A gas that is initially at 1.16 MPa and 133°C is expanded to 275 kPa and 32°C. What is $n$ for this process?

6.124 Air is compressed polytropically from 1 atm to 10 atm absolute. The initial temperature is 70°F and $n = 1.2$. Determine the work per pound required for the compression if $c_v = 0.171$ Btu/lb$_m$·R.

6.125 Determine the entropy change and heat transferred for the polytropic process described in Problem 6.124.

6.126 If $n$ of a nonflow process is 1.2, how much work is required to compress 1 kg of air from 120 to 600 kPa if the air is initially at 30°C? Assume that $k$ is 1.4 and MW is 29.

6.127 If $n$ of a nonflow polytropic process is 1.18, determine the work required to compress 1 lb$_m$ of air from 20 to 100 psia if the initial air temperature is 80°F.

6.128 One pound mass of air undergoes a polytropic process with $n = 1.33$. The air is initially at 14.7 psia and 68°F. During the process, 12 Btu/lb$_m$ is transferred from the air as heat. Determine the final temperature and pressure for the process.

6.129 A polytropic process for air has $n = 1.5$. If the initial state of the air is 80 psia and 900°F and the final state is at 20 psia, determine the final temperature and the work done per unit mass.

## Problems Involving Average Specific Heat and Real Gases

6.130   An equation for the instantaneous specific heat of air is given as

$$c_p = 0.219 - \frac{0.342T}{10^4} - \frac{0.293T}{10^8} \quad \text{where } T \text{ is in } °$$

Determine the average specific heat of air between 340°F and 840°F.

6.131   Using the *Gas Tables*, solve Problem 6.130.

*6.132   The instantaneous specific heat of carbon dioxide is given by

$$c_p = 0.368 - \frac{148.4}{T} + \frac{3.2 \times 10^4}{T^2} \quad \text{where } T \text{ is in R}$$

Determine the average specific heat of carbon dioxide between 500R and 1000R.

# 7

## Mixtures of Ideal Gases

### Learning Goals

After reading and studying the material in this chapter, you should be able to

1. Apply the ideal gas relation to a gas mixture.
2. Understand and use the term *partial pressure*.
3. Apply Dalton's law to a gas mixture.
4. Define the *mole fraction* of a gas.
5. Arrive at the simple conclusion that the total number of moles of gas in a mixture is the sum of the moles of each of its constituents.
6. Use the fact that the mole fraction of a component of a gas mixture equals the partial pressure of that component.
7. Apply the ideal gas equation to both the components and the entire mixture.
8. Utilize the conclusion that the mole fraction is also equal to the volume fraction of a component of the mixture.
9. Convert from volume fraction to weight fraction and from weight fraction to volume fraction.
10. Recall that the thermodynamic properties of a mixture are each equal to the sum of the mass average of the individual properties of the components of the mixture.
11. Understand and use the terms that are used in conjunction with air–water vapor mixtures.
12. Determine the properties of air–water vapor mixtures.
13. Understand and use the psychrometric chart.

## 7.1 Introduction

Many mixtures of gases are of thermodynamic importance. For example, atmospheric air is principally a mixture of oxygen and nitrogen with some water vapor and other gases in minute quantities. The products of combustion from power plants and internal combustion engines are also mixtures of gases that are of interest. From the simple kinetic considerations of Chapter 1, it was possible to deduce an equation of state to relate the pressure, temperature, and volume of a gas. This particular equation, $pv = RT$, formed the basis of most of the work in Chapter 6. As we have remarked several times, it is surprising that so simple an equation can represent real gases with any degree of accuracy. The success of this equation for a single gas leads us to hope that mixtures of gases can be represented equally well by an equally simple equation of state. To some extent, this hope is realized for some real gaseous mixtures, but for others, it has not been possible to write a simple $p$, $v$, $T$ relation in terms of the properties of the individual components of the mixture. With the use of semiempirical methods, it has been found possible to correlate experimental $p$, $v$, $T$ data for some gases, but no single method of correlation has yet been devised to correlate all gaseous mixtures satisfactorily.

In this chapter, we will treat gas mixtures as if they were pure substances to which the ideal gas relation can be applied. This will be taken to be true of the components of the mixture as well as of the mixture as a whole.

## 7.2  Pressure of a Mixture

Consider that a mixture of several gases is contained in volume $V$ at temperature $T$ and that an absolute pressure gauge placed in the volume would read a total mixture pressure $p_m$, as shown in Figure 7.1. In this gas mixture, the various component gas molecules are in constant random motion, and the gas mixture is assumed to be homogeneous. For such a mixture, *Dalton's law* (or Dalton's rule of additive pressures) is found to apply exactly if the ideal gas relation holds for the components and for the mixture. Dalton's law can be stated as follows:

1. Each gas behaves as if it alone occupied the volume.
2. Each gas behaves as if it exists at the temperature of the mixture and filled the entire volume by itself.
3. The pressure of the gas mixture is the sum of the pressures of each of its components when each behaves as described in statements 1 and 2.

Applying Dalton's law to the situation shown in Figure 7.1, where the mixture $m$ and each of the components are all shown to occupy equal volumes, yields

$$p_m = p_a + p_b + p_c \tag{7.1}$$

where $p_m$ is the total pressure of the mixture, and $p_a$, $p_b$, and $p_c$ represent the pressures that each of the constituent gases would exert if they alone occupied the total volume at the temperature of the mixture. These pressures are called the *partial pressures* of each gas. By assuming that the pressure–volume relationship for an ideal gas is applicable and denoting the molecular weight of the mixture as $MW_m$, we have, in English units, per Section 6.2,

$$p_m V = m_m RT = m_m \frac{1545T}{MW_m} \tag{7.2}$$

| mass *m* | mass *a* | mass *b* | mass *c* |
|---|---|---|---|
| $p_m$ = Pressure of mixture <br> $V$ = Volume <br> $T$ = Temperature | $p_a$ <br> $V$ <br> $T$ | $p_b$ <br> $V$ <br> $T$ | $p_c$ <br> $V$ <br> $T$ |

**FIGURE 7.1**
Dalton's law.

where $m_m$ is the mass of the mixture equal to the sum of the masses of the individual gases. The partial pressure of each gas can be substituted for $p_m$, the mass of each gas for $m_m$, and the molecular weight of each gas for $MW_m$ in Equation 7.2. This yields a set of ideal gas equations for each component gas. Using each of these equations in Equation 7.1 yields

$$\frac{m_m}{MW_m} = \frac{m_a}{MW_a} + \frac{m_b}{MW_b} + \frac{m_c}{MW_c} \tag{7.2a}$$

Before continuing, it will be convenient to recall certain terms that were discussed briefly in Chapters 1 and 6. If the molecular weight of a substance is expressed in pounds, the resulting quantity is known as the pound molecular weight, a pound mole, or simply a mole. Thus, 32 $lb_m$ of oxygen constitutes 1 lb. mol of oxygen, and 28.02 $lb_m$ of nitrogen is 1 lb. mol of nitrogen. It has also been found that 1 mol of a substance, whether it be solid, liquid, or gas, contains the same number of molecules as 1 mol of any other substance. The number of molecules in 1 mol is known as the Avogadro number or Avogadro's constant ($6.02 \times 10^{23}$). Thus, by application of the ideal gas equation of state, it is found that at a given pressure and temperature, 1 mol of gas will occupy a fixed volume, regardless of the gas. This volume is known as the molar volume, and at 14.7 psia and 32°F, it is 358 ft.$^3$. Table 7.1 is a convenient listing of the molecular weights of various gases.

Recall from Section 6.2 that a kilogram mole is that quantity of gas whose mass equals its molecular weight in kilograms. Thus, 1 kg mol of oxygen has a mass of 32 kg. The mass of a kilogram mole is greater than the mass of a pound mole.

It will now be noted that the ratio of $m/MW$ in Equation 7.2a is simply the number of moles of each component of the mixture. Denoting the moles of each constituent as $n$, with the appropriate subscripts,

$$n_m = n_a + n_b + n_c \tag{7.2b}$$

In words, Equation 7.2b states that the number of moles of gas in a mixture is equal to the sum of the moles of its constituent gases. It is also evident that the total mass of a gas mixture must equal the sum of the component masses. Referring to Figure 7.1, we have

$$m_m = m_a + m_b + m_c \tag{7.3}$$

The ratio $n_a/n_m$ is the *mole fraction* $x_a$ of gas $a$, and the ratio $m_a/m_m$ is the mass fraction of gas $a$. From Equations 7.1 and 7.2,

$$\frac{p_a}{p_m} = \frac{n_a}{n_m} = x_a$$

**TABLE 7.1**

Molecular Weights of Gases

| Gas | Chemical Formula | Molecular Weight |
|---|---|---|
| Acetylene | $C_2H_2$ | 26.02 |
| Air | | 28.96 |
| Ammonia | $NH_3$ | 17.024 |
| Argon | Ar | 39.90 |
| Butane | $C_4H_{10}$ | 58.08 |
| Carbon dioxide | $CO_2$ | 44.00 |
| Carbon monoxide | CO | 28.00 |
| Dodecane | $C_{12}H_{26}$ | 170.3 |
| Ethane | $C_2H_6$ | 30.05 |
| Ethylene | $C_2H_4$ | 28.03 |
| Helium | He | 4.00 |
| Hydrogen | $H_2$ | 2.016 |
| Methane | $CH_4$ | 16.03 |
| Nitrogen | $N_2$ | 28.02 |
| Octane | $C_8H_{18}$ | 114.14 |
| Oxygen | $O_2$ | 32.00 |
| Propane | $C_3H_8$ | 44.06 |
| Sulfur dioxide | $SO_2$ | 64.07 |
| Water vapor | $H_2O$ | 18.01 |

and

$$p_a = x_a p_m \tag{7.4}$$

and from Equations 7.2a and 7.3,

$$n_m \left( MW_m \right) = n_a \left( MW_a \right) + n_b \left( MW_b \right) + n_c \left( MW_c \right)$$

or

$$MW_m = x_a MW_a + x_b MW_b + x_c MW_c \tag{7.5}$$

Because Equation 7.5 is based on the fact that the mixture and its components are ideal gases and that the mixture can also be treated as a single ideal gas, the gas constant for the mixture can be written as

$$R_m = \frac{1545}{MW_m} \tag{7.6}$$

To sum up the foregoing:

1. The total pressure of a mixture of ideal gases is the sum of its partial pressures.
2. The mass of the mixture is the sum of the masses of the components of the mixture.
3. The total number of moles in the mixture equals the sum of the number of moles of the component gases of the mixture.

4. The partial pressure of a component equals the mole fraction of the component multiplied by the total pressure of the mixture.

5. The molecular weight of the mixture equals the sum of the products of mole fraction of each gas component multiplied by the molecular weight of each component gas.

6. The ideal gas equation of state is applicable to the components as well as to the entire mixture.

---

### ILLUSTRATIVE PROBLEM 7.1

Dry air is a mixture of oxygen and nitrogen as the principal gases. Per pound of mixture, there is 0.2315 lb. of oxygen and 0.7685 lb. of nitrogen. Determine (a) the number of moles of each gas, the total number of moles, and the mole fraction of each component; (b) the partial pressure of each component if the air is at 14.7 psia; (c) the molecular weight of the air; and (d) the gas constant of the air.

### SOLUTION

As the basis of this calculation, assume that we have 1 $lb_m$ of mixture. Also, take the molecular weight of oxygen to be 32.00 and nitrogen to be 28.02 (from Table 7.1).

a.  $n_{O_2} = \dfrac{0.2315}{32} = 0.00723$ mol of oxygen/$lb_m$ of mixture

$n_{N_2} = \dfrac{0.7685}{28.02} = 0.2743$ mol of nitrogen/$lb_m$ of mixture

$n_m = n_{O_2} + n_{N_2} = 0.00723 + 0.02743$

$n_m = 0.03466$ mol/$lb_m$ of mixture

$x_{O_2} = \dfrac{n_{O_2}}{n_m} = \dfrac{0.00723}{0.03466} = 0.209$

$x_{N_2} = \dfrac{n_{N_2}}{n_m} = \dfrac{0.02743}{0.03466} = 0.791$

Check : $\left( x_{O_2} + x_{N_2} = 0.209 + 0.791 = 1.000 \right)$

b.  $p_{O_2} = x_{O_2}\left(14.7\right) = 0.209 \times 14.7 = 3.07$ psia

$p_{N_2} = x_{N_2}\left(14.7\right) = 0.791 \times 14.7 = 11.63$ psia

334 Thermodynamics and Heat Power

c. $MW_m = x_{O_2}(32) + x_{N_2}(28.02)$

$$= 0.209 \times 32 + 0.791 \times 28.02 = 28.85$$

d. $R_m = \dfrac{1545}{MW_m} = \dfrac{1545}{28.85} = 53.55 \text{ or } 53.6$

---

### ILLUSTRATIVE PROBLEM 7.2

Gaseous Freon-12 ($CCl_2F_2$) is used to slowly fill a tank that initially contained air at atmospheric pressure and temperature to a final pressure of 1000 psia and room temperature. What is the mole fraction of air in the final mixture? Also, find the molecular weight of the mixture and the partial pressure of the gases. Assume that the molecular weight of air is 29 and the molecular weight of Freon-12 is 120.9.

**SOLUTION**

The partial pressure of the air is, by definition, 14.7 psia. Therefore, the partial pressure of the Freon-12 is $1000 - 14.7 = 985.3$ psia. The mole fraction of air is $14.7/1000 = 0.0147$, and the mole fraction of Freon-12 is $985.3/1000 = 0.9853$.

The molecular weight of the mixture is

$$MW_m = 0.0147 \times 29 + 0.9853 \times 120.9 = 119.5$$

---

### ILLUSTRATIVE PROBLEM 7.3

Ten pounds of air, 1 lb. of carbon dioxide, and 5 lb. of nitrogen are mixed at constant temperature until the mixture pressure is constant. The final mixture is at 100 psia. Determine the moles of each gas, the mole fraction of each gas, the partial pressure of each gas, the molecular weight of the mixture, and the gas constant of the mixture. Assume that the molecular weight of air is 29; of carbon dioxide, 44; and of nitrogen, 28.

**SOLUTION**

Let us consider as a basis for the calculation that we have 16 $lb_m$ $(10+1+5=16)$ of mixture. Therefore,

$$n_{air} = \frac{10}{29} = 0.345 \quad n_{CO_2} = \frac{1}{44} = 0.023 \quad n_{N_2} = \frac{5}{28} = 0.179$$

The total number of moles is $0.345 + 0.023 + 0.179 = 0.547$. The mole fractions are

$$x_{air} = \frac{0.345}{0.547} = 0.631 \quad x_{CO_2} = \frac{0.023}{0.547} = 0.042$$

$$x_{N_2} = \frac{0.179}{0.547} = 0.327$$

$$p_{air} = (0.631)(100) = 63.1 \text{ psia}$$

$$p_{CO_2} = (0.042)(100) = 4.1 \text{ psia}$$

$$p_{N_2} = (0.327)(100) = 32.7 \text{ psia}$$

The molecular weight of the mixture is found from

$$MW_m = (0.631)(29) + (0.042)(44) + (0.327)(28) = 29.3$$

and

$$R_m = \frac{1545}{29.3} = 52.7$$

## 7.3 Volume of a Mixture

In arriving at Dalton's law in Section 7.2, it was assumed that each gas in a mixture expands to fill the entire volume of the container at the temperature of the mixture. Let us assume that three gases, *a*, *b*, and *c*, are placed in separate containers, each having different volumes but each at the same pressure and temperature. This condition is shown in Figure 7.2. If the three containers are brought together and the partitions are removed, the pressure and temperature of the mixture will remain constant, if these gases are ideal, and Dalton's law can be applied. The partial pressures will equal the initial pressure multiplied by the ratio of the initial to final volume. Therefore,

$$p_m = p_{oa}\frac{V_a}{V_m} + p_{ob}\frac{V_b}{V_m} + p_{oc}\frac{V_c}{V_m} \tag{7.7}$$

where $p_{oa}$ is the initial pressure of *a* in its container, $V_a$ is the *volume of a before mixing*, and $V_m$ is the *volume of the mixture after mixing*. Because $p_o$ of each gas is the same as the final pressure $p_m$,

$$1 = \frac{V_a}{V_m} + \frac{V_b}{V_m} + \frac{V_c}{V_m} \quad \text{or} \quad V_m = V_a + V_b + V_c \tag{7.8}$$

Equation 7.8 is an expression of *Amagat's law,* which states that the volume of a mixture is the sum of the volumes that each constituent gas would occupy if each were at the pressure and temperature of the mixture.

**FIGURE 7.2**
Law of additive volumes.

By applying the equation of state of an ideal gas to Equation 7.8, we obtain

$$\frac{V_a}{V_m} = \frac{n_a}{n_m} = x_a \tag{7.9}$$

That *is, the mole fraction of a component also equals the volume fraction of that component in the mixture.* The volume fraction is defined as the ratio of the partial volume of a constituent to the total volume of the mixture. Amagat's law and Dalton's law are equivalent to each other if the gases and the mixture are ideal gases. The following two examples will serve to illustrate the concept of mixture volume.

---

### ILLUSTRATIVE PROBLEM 7.4

Five (5) moles of oxygen at 14.7 psia and 70°F are adiabatically mixed with ten (10) moles of hydrogen at 14.7 psia and 70°F. Determine the mixture volume, the partial volume of the constituents, the mole fraction of each constituent, and the molecular weight of the resulting mixture. The molecular weight of oxygen is 32.6 and of hydrogen, 2.016 (see Table 7.1). Conversion factor and equations used: 1 ft$^2$ = 144 in$^2$, R = °F + 460, $V = v \cdot m$.

### SOLUTION

The total number of moles is $10 + 5 = 15$. Therefore, the mole fraction of each constituent is

$$x_{O_2} = \frac{5}{15} = \frac{1}{3} \quad x_{H_2} = \frac{10}{15} = \frac{2}{3}$$

The molecular weight of the final mixture is

$$\frac{1}{3}(32) + \frac{2}{3}(2.016) = 12.01$$

The partial volume of the oxygen can be found as follows: per pound of oxygen,

$$(pv)_{O_2} = RT$$

$$14.7 \times 144 \times v_{O_2} = \frac{(1545)(460+70)}{32}$$

$$v_{O_2} = 12.09 \text{ ft.}^3/\text{lb}_m$$

and because there are 5 mol of oxygen, each containing 32 $\text{lb}_m$,

$$V_{O_2} = 12.09 \times 5 \times 32 = 1934 \text{ ft.}^3$$

For the hydrogen, we can simplify the procedure by noting that the fraction of the total volume occupied by the oxygen is the same as its mole fraction. Therefore, the total volume is three times the volume of oxygen $\left( \text{or } V_m = \dfrac{V_{O_2}}{x_{O_2}} \right)$, then

$$V_m = (3)(1934) = 5802 \text{ ft.}^3$$

and the hydrogen volume $\left( V_{H_2} = V_m - V_{O_2} \right)$, is

$$V_{H_2} = 5802 - 1934 = 3868 \text{ ft.}^3$$

We could obtain the partial volume of hydrogen by proceeding as we did for the oxygen. Thus,

$$(pv)_{H_2} = RT$$

$$14.7 \times 144 \times v_{H_2} = \frac{(1545)(460+70)}{2.016}$$

and

$$v_{H_2} = 191.88 \text{ ft.}^3/\text{lb}_m$$

For 10 mol of hydrogen, each containing 2.016 $\text{lb}_m$, we have a total hydrogen volume of

$$V_{H_2} = 10 \times 2.016 \times 191.88 = 3868 \text{ ft.}^3$$

which checks with the previous value. As an alternative to the foregoing, we could also use the fact that at 14.7 psia and 32°F, a mole of *any* gas occupies a volume of 358 ft.³. At 70°F and 14.7 psia, a mole occupies $358 \times [(460+70)/(460+32)] = 385.7$ ft.³. Therefore, 5 mol of oxygen occupy

$$V_{O_2} = 5 \times 385.7 = 1928 \text{ ft.}^3$$

and 10 mol of hydrogen occupy

$$V_{H_2} = 10 \times 385.7 = 3857 \text{ ft.}^3$$

Both values are in good agreement with the previous calculations.

## ILLUSTRATIVE PROBLEM 7.5

A mixture contains 10 lb. of carbon dioxide and 5 lb. of nitrogen. If the mixture is at 70°F and 100 psia, determine the mixture volume, the partial volume of each constituent, the partial pressure of each constituent, the mole fraction of each constituent, and the molecular weight of the mixture (MW of $CO_2 = 44.0$; MW of $N_2 = 28.02$).

**SOLUTION**

Referring to Figure 7.3, we have for the $CO_2$

$$n_{CO_2} = \frac{10}{44} = 0.227 \text{ mole}$$

and for $N_2$

$$n_{N_2} = \frac{5}{28.02} = 0.178 \text{ mole}$$

The total moles in the mixture is therefore $0.227 + 0.178 = 0.405$ mole. Therefore,

$$x_{CO_2} = \frac{0.227}{0.405} = 0.560$$

and

$$x_{N_2} = \frac{0.178}{0.405} = 0.440$$

The molecular weight of the mixture is found from Equation 7.5 as

| Mixture | | CO$_2$ | | N$_2$ | |
|---|---|---|---|---|---|
| Pressure | $= p_m$ | M | $= 10 \text{ lb}_m$ | M | $= 5 \text{ lb}_m$ |
| Temperature | $= T$ | Volume | $= V_{CO_2}$ | Volume | $= V_{N_2}$ |
| Volume | $= V_m$ | Temperature | $= T$ | Temperature | $= T$ |
| | | Pressure | $= p_m$ | Pressure | $= p_m$ |

**FIGURE 7.3**
Illustrative Problem 7.5.

$$MW_m = (0.560)(44) + (0.440)(28.02) = 37.0$$

Because the mixture is 15 $lb_m$ (10 $CO_2$ + 5 $N_2$), the volume of the mixture is found from

$$p_m V_m = m_m R_m T_m \quad V_m = \dfrac{\dfrac{15 \times 1545}{37.0} \times (460 + 70)}{100 \times 144}$$

$$= 23.05 \text{ ft}^3$$

The partial volume of carbon dioxide is the total volume multiplied by the mole fraction. Thus,

$$V_{CO_2} = (23.05)(0.560) = 12.91 \text{ ft.}^3$$

and $\quad V_{N_2} = (23.05)(0.440) = 10.14 \text{ ft.}^3$

The partial pressure of each constituent is proportional to its mole fraction. For these conditions,

$$p_{CO_2} = (100)(0.560) = 56.0 \text{ psia}$$

and $\quad p_{N_2} = (100)(0.440) = 44.0 \text{ psia}$

## 7.4 Mixture Composition

When a gas mixture is analyzed, it is possible to state its composition correctly in several different ways. For instance, the moles of each constituent gas could be given; alternatively, the weight or partial volumes of each gas could be stated. The manner of expressing the analysis is completely arbitrary. Therefore, it is necessary to be able to convert from one form of analysis to another.

If the moles (or mole fractions) are given, the volume fractions are immediately known. Conversely, if the volume fractions are known, the mole fractions are known. Because the mass of a gas is the product of the number of moles multiplied by the molecular weight, the conversion to a weight (gravimetric) analysis is readily achieved. The molecular weight of the mixture is also readily determined from Equation 7.5.

### ILLUSTRATIVE PROBLEM 7.6

A gas mixture consists of the following volume percentages: 40% $CO_2$, 10% $N_2$, 40% $O_2$, and 10% $H_2$. Determine the mass fraction of each gas in the mixture, the molecular weight of the mixture, and the gas constant of the mixture.

### SOLUTION

In solving this type of problem, we shall first assume a convenient volume for the gas mixture. Because we have percentages by volume for each constituent, we will assume that we have 100 volumes of gas mixture and set up Table 7.2. In the first column, we tabulate the gas, and in the second column, we tabulate the given volume fractions. Because the mole fraction equals the volume fraction, the values in column 3 are the same as those in column 2. The molecular weight is obtained from Table 7.1. Because the molecular weight of the mixture is the sum of the individual mole fractions multiplied by the respective molecular weights, the next column tabulates the product of mole fraction multiplied by molecular weight (column 3 multiplied by column 4). The sum of these entries is the molecular weight of the mixture, which for this case is 33.4. If we now consider that we have 33.4 $lb_m$ of mixture and that the mass of each gas in the mixture is given by its product $(x)$ (MW), the mass fraction of each constituent is the entry in column 5 divided by 33.4. The value of $R$ for the mixture is simply $1545/33.4 = 46.3$. The use of a tabular solution helps to simplify the calculations and is also a timesaver when this type of repetitive calculation is involved.

**TABLE 7.2**

Illustrative Problem 7.6

| | Basis: 100 Volumes of Gas Mixture | | | | |
|---|---|---|---|---|---|
| | **(2)** | **(3)** | **(4)** | **(5)** | |
| **Gas** | **Volume Fraction** | **Mole Fraction, $x$** | **Molecular Weight, MW** | **$(3)\times(4)$ $(x)$ (MW)** | **Mass Fraction** |
| $CO_2$ | 0.40 | 0.40 | 44.0 | 17.6 | $17.6/33.4 = 0.527$ |
| $N_2$ | 0.10 | 0.10 | 28.02 | 2.8 | $2.8/33.4 = 0.084$ |
| $H_2$ | 0.10 | 0.10 | 2.016 | 0.2 | $0.2/33.4 = 0.006$ |
| $O_2$ | 0.40 | 0.40 | 32.0 | 12.8 | $12.8/33.4 = 0.383$ |
| | 1.00 | 1.00 | | $33.4 = MW_m$ | $= 1.000$ |

The procedure when the weight (gravimetric) analysis is known and conversion to a volumetric basis is desired is similar to that given in the table. However, rather than assuming 100 volumes of gas, it is more convenient to use 100 $lb_m$ of gas as the basis of the computations.

- ## ILLUSTRATIVE PROBLEM 7.7

The result of Illustrative Problem 7.6 shows that the mass fractions of gases are $CO_2 = 0.527$, $N_2 = 0.084$, $H_2 = 0.006$, and $O_2 = 0.383$. Using these gases and mass fractions, determine the fraction by volume of each of the constituent gases of this mixture.

### SOLUTION

We will again use a tabular solution, as we did for Illustrative Problem 7.6, but for this calculation, we will take as a basis 100 $lb_m$ of mixture. The first three columns of Table 7.3 are similar to the items in Table 7.2 and are self-explanatory.

**TABLE 7.3**

Illustrative Problem 7.7

| | | (3) | (4) | (5) | (6) |
|---|---|---|---|---|---|
| | | | **Basis: 100 Pounds of Mixture** | | |
| Gas | (2) Mass $lb_m$ | Molecular Weight, MW | Mass Molecular Weight = moles | Mole Fraction (moles/mole) | Percent Volume |
| $CO_2$ | 52.7 | 44 | 1.2 | 1.2/3 = 0.4 | 40 |
| $N_2$ | 8.4 | 28.02 | 0.3 | 0.3/3 = 0.1 | 10 |
| $H_2$ | 0.6 | 2.016 | 0.3 | 0.3/3 = 0.1 | 10 |
| $O_2$ | 38.3 | 32 | 1.2 | 1.2/3 = 0.4 | 40 |
| | 100.0 | | Total moles = 3.0 | 1.0 | 100 |
| | | | $MW_m = 100/3 = 33.3$ | | |

Dividing column 2 by column 3 gives us the mass/molecular weight or moles of each constituent. The total number of moles in the mixture is the sum of column 4, and the molecular weight of the mixture is the mass of the mixture (100 $lb_m$) divided by the number of moles, 3 or 33.3. Column 5, the mole fraction, is the moles listed in column 4 divided by the total moles in the mixture, which for this problem is 3.0. Because mole fraction is also volume fraction, the last column, percent volume, is just the fifth column multiplied by 100. It should also be noted that the volume percent is also the pressure percent of the various constituents of the mixture. Because this problem is the inverse of Illustrative Problem 7.6, the results obtained are found to be in good agreement for both problems.

## ILLUSTRATIVE PROBLEM 7.8

Air consists of 23.18% $O_2$, 75.47% $N_2$, 1.30% Ar, and 0.05% $CO_2$ by weight. Determine the volumetric analysis.

### SOLUTION

This problem is a weight-to-volume conversion similar to Illustrative Problem 7.7. We set up Table 7.4 in the same manner as Table 7.3 and use the same procedure to solve the problem.

**TABLE 7.4**

Illustrative Problem 7.8

| | | Basis: 100 Pounds of Mixture | | | |
|---|---|---|---|---|---|
| Gas | Mass ($lb_m$) | Molecular Weight, MW | Mass Molecular Weight = moles | Moles Fraction | Percent Volume |
| $O_2$ | 23.18 | 32.00 | 0.724 | 0.210 | 21 |
| $N_2$ | 75.47 | 28.02 | 2.693 | 0.780 | 78 |
| Ar | 1.30 | 39.90 | 0.033 | 0.010 | 1 |
| $CO_2$ | 0.05 | 44.00 | — | — | |
| | 100.00 | | 3.45 | 1.00 | 100 |

$$MW_m = \frac{100}{3.45} = 28.99$$

## 7.5 Thermodynamic Properties of a Gas Mixture

When several gases are mixed, it is assumed that the gas molecules will disperse until the mixture is completely homogeneous. It is also assumed that the components of the mixture will reach the same average temperature as the mixture. In Chapter 6, it was noted that the internal energy and enthalpy of the ideal gas are solely functions of the gas temperature. With this in mind, it would seem reasonable to state that these thermodynamic properties of a gas mixture (internal energy and enthalpy) are simply the sum of the properties of the gases of which the mixture is composed. This statement, combined with Dalton's law, can be considered as the Gibbs–Dalton law of mixtures. As before, this law is valid for the ideal gas, and deviations will be found when the components of a mixture differ considerably from the ideal gas law.

As a consequence of the Gibbs–Dalton law, we can write the following property relations for a mixture m composed of three gases, *a*, *b*, and *c*, on the basis that the properties of the mixture equal the sum of the properties of each component of the mixture when each of the components is assumed to occupy the total volume at the temperature of the mixture and the gases behave as ideal gases. Thus, for the internal energy and enthalpy of a mixture, we write

$$U_m = m_a u_a + m_b u_b + m_c u_c \tag{7.10}$$

and

$$H_m = m_a h_a + m_b h_b + m_c h_c \tag{7.11}$$

where $U_m$ and $H_m$ are the total internal energy and the total enthalpy of the mixture, respectively. Because the total value of these properties is the product of the mass and the specific value of each property,

$$U_m = m_m u_m = m_a u_a + m_b u_b + m_c u_c \tag{7.12}$$

and

$$H_m = m_m h_m = m_a h_a + m_b h_b + m_c h_c \tag{7.13}$$

From Equations 7.12 and 7.13, the specific internal energy and the specific enthalpy of the mixture are

$$u_m = \frac{m_a}{m_m} u_a + \frac{m_b}{m_m} u_b + \frac{m_c}{m_m} u_c \tag{7.14}$$

and

$$h_m = \frac{m_a'}{m_m} h_a + \frac{m_b}{m_m} h_b + \frac{m_c}{m_m} h_c \tag{7.15}$$

where $m_m = m_a + m_b + m_c$.

From Equations 7.14 and 7.15, we can also obtain the specific heat at constant volume and the specific heat at constant pressure for the mixture, because both internal energy and enthalpy for the ideal gas are functions only of the temperature of the mixture. Therefore,

$$c_{vm} = \frac{m_a}{m_m} c_{va} + \frac{m_b}{m_m} c_{vb} + \frac{m_c}{m_m} c_{vc} \tag{7.16}$$

and

$$c_{pm} = \frac{m_a}{m_m} c_{pa} + \frac{m_b}{m_m} c_{pb} + \frac{m_c}{m_m} c_{pc} \tag{7.17}$$

It should be noted that Equations 7.14 through 7.17 state that the mixture properties in question are each equal to the sum of the mass average of the individual properties of the components of the mixture.

---

### ILLUSTRATIVE PROBLEM 7.9

In a steady-flow (open system) process, 160 $\mathrm{lb_m}$/h of oxygen and 196 $\mathrm{lb_m}$/h of nitrogen are mixed adiabatically. Both gases are at atmospheric pressure, but the oxygen is at 500°F and the nitrogen is at 200°F. Assume $c_p$ of oxygen to be constant and equal to 0.23 Btu/$\mathrm{lb_m}$·R and $c_p$ of nitrogen also to be constant and equal to 0.25 Btu/$\mathrm{lb_m}$·R; determine the final temperature of the mixture (see Figure 7.4).

**FIGURE 7.4**
Illustrative Problem 7.9.

## SOLUTION

The energy equation for the steady-flow, adiabatic mixing process with no work input or output gives us the requirement that the enthalpy of the mixture must equal the enthalpies of the components $\left(q_{in} + w_{in} + h_{N_2} + h_{O_2} = q_{out} + w_{out} + h_m; \text{ with } q_{in} = w_{in} = q_{out} = w_{out} = 0 \rightarrow h_{N_2} + h_{O_2} = h_m\right)$. An alternative statement of this requirement is that the gain in enthalpy of the nitrogen $\left(\Delta h_{N_2} = h_{N_2,2} - h_{N_2,1} = c_{p,N_2}\left(T_{N_2,2} - T_{N_2,1}\right)\right)$ must equal the decrease in enthalpy of the oxygen $\left(\Delta h_{O_2} = h_{O_2,1} - h_{O_2,2} = c_{p,O_2}\left(T_{O_2,1} - T_{O_2,2}\right)\right)$, where $T_{N_2,2} = T_{O_2,2} = T_m$. Using the latter statement, that the change in enthalpy of the oxygen $\left(\Delta \dot{H}_{O_2} = \dot{m}\Delta h_{O_2} = \dot{m}c_{p,O_2}\Delta T_{O_2}\right)$ must equal the change in enthalpy of the nitrogen $\left(\Delta \dot{H}_{N_2} = \dot{m}\Delta h_{N_2} = \dot{m}c_{p,N_2}\Delta T_{N_2}\right)$, yields

$$(160)(0.23)(500 - t_m) = (196)(0.25)(t_m - 200)$$

where $mc_p\Delta t$ has been used for $\Delta h$, as shown above. Solving yields

$$t_m = 328.7°F$$

Using the requirement that the enthalpy of the mixture must equal the sum of the enthalpies of the components yields an alternative solution to this problem. Let us assume that at 0°F, the enthalpy of each gas and of the mixture is zero. The enthalpy of the entering oxygen is $(160)(0.23)(500 - 0)$, and the enthalpy of the entering nitrogen is $(196)(0.25)(200 - 0)$. The enthalpy of the mixture is $(160 + 196)c_{pm}(t_m - 0)$. Therefore, $(160)(0.23)(500 - 0) + (196)(0.25)(200 - 0) = (160 + 196)c_{pm}(t_m - 0)$.

The required $c_{pm}$ is obtained from Equation 7.17:

$$c_{pm} = \frac{160}{160 + 196}(0.23) + \frac{196}{160 + 196}(0.25)$$

$$= 2.41 \text{ Btu/lb}_m \cdot R$$

Therefore,

$$t_m = \frac{(160)(0.23)(500 - 0) + (196)(0.25)(200 - 0)}{(160 + 196)(0.241)}$$

$$= 328.7°F$$

The use of 0°F as a base was arbitrary but convenient. Any base would yield the same results. Also, either method used to solve this problem yields the same mixture temperature. The specific heat data of Table A.13 are not applicable to this problem because Table A.13 gives data at 77°F.

## ILLUSTRATIVE PROBLEM 7.10

Assume that the process described in Illustrative Problem 7.9 is carried out as a non-flow (closed system) mixing process. Also, assume that $c_v$ of the oxygen is 0.164 Btu/$lb_m$·R and the $c_v$ of the nitrogen is 0.178 Btu/$lb_m$·R. Determine the final temperature of the mixture.

### SOLUTION

Because this is a nonflow (closed system) process, the energy equation for this process requires the internal energy of the mixture to equal the sum of the internal energies of its components $\left(Q - W = \Delta U, \text{ with } Q = W = 0 \rightarrow \Delta U_m = 0 \rightarrow \Delta U_{O_2} + \Delta U_{N_2} = 0\right)$. Alternatively, the decrease in internal energy of the oxygen must equal the increase in internal energy of the nitrogen $\left(\Delta U_{O_2} = \Delta U_{N_2} \rightarrow mc_{v,O_2}\left(T_{O_2,1} - T_m\right) = mc_{v,N_2}\left(T_m - T_{N_2,1}\right)\right)$. Using the latter statement gives us

$$(160)(0.164)(500 - t_m) = (196)(0.178)(t_m - 200)$$

$$t_m = 328.8°F$$

We obtained the same temperature for Illustrative Problems 7.9 and 7.10 because the ratio of $c_p/c_v$ for both gases was deliberately taken to be the same. In effect, this constant was canceled from both sides of the equation.

In Chapter 6, we noted that the entropy of an ideal gas is, in general, a function of both pressure and temperature. For the mixture of ideal gases where each component of the mixture is taken to exist at its own partial pressure, we can write the following equation:

$$m_m s_m = m_a s_a + m_b s_b + m_c s_c \tag{7.18}$$

where the individual entropies $s_a$, $s_b$, and $s_c$ are evaluated per pound at the mixture temperature and the component partial pressure.

## ILLUSTRATIVE PROBLEM 7.11

Two ideal gases having constant specific heats each at the same pressure and temperature are adiabatically mixed by removing the partition shown in Figure 7.5. Assume gas a to have a molecular weight of $MW_a$ and a mass of $m_a$ lb. Assume gas b to have a molecular weight of $MW_b$ and a mass of $m_b$ lb. Determine the entropy change for this mixing process.

### SOLUTION

The general equation applicable to the change in entropy of an ideal gas due to both a pressure and temperature change is

$$\Delta s = c_p \ln \frac{T_2}{T_1} - \frac{R}{J} \ln \frac{p_2}{p_1} \text{ Btu/lb}_m \cdot R \tag{a}$$

Because the process is carried out at constant temperature, the first term of Equation (a) applied to this process is zero. Also, the final partial pressure of gas $a$ is $p_a$, and the final partial pressure of gas $b$ is $p_b$. Denoting the total pressure of the mixture as $p_m$, we obtain

(a)

| $MW_a$ | $MW_b$ |
|---|---|
| $m_a$ | $m_b$ |
| $T$ | $T$ |
| $V_a$ | $V_b$ |
| $p_a$ | $p_b$ |

(b)

| |
|---|
| $V = V_a + V_b$ |
| $m = m_a + m_b$ |
| $T = T$ |
| $p_{in} = p_{in}$ |

**FIGURE 7.5**
Illustrative Problem 7.11. (a) Before mixing. (b) After mixing.

$$\Delta S_a = -m_a \frac{R_a}{J} \ln \frac{p_a}{p_m}$$

$$\text{and} \quad \Delta S_b = -m_b \frac{R_b}{J} \ln \frac{p_b}{p_m}$$

The total change in entropy for the mixture for $m_a + m_b$ lb$_m$ of mixture is

$$\Delta S = -m_a \frac{R_a}{J} \ln \frac{p_a}{p_m} - m_b \frac{R_b}{J} \ln \frac{p_b}{p_m}$$

However, $R = 1545/MW$. Therefore,

$$\Delta S = -\frac{m_a(1545)}{(MW_a)J} \ln \frac{p_a}{p_m} - \frac{m_b(1545)}{(MW_b)J} \ln \frac{p_b}{p_m}$$

We now note that $m_a/MW_a = n_a$, the number of moles of $a$; $m_b/MW_b = n_b$, the number of moles of $b$; $p/p_m = x_a$, the mole fraction of a; and $p_b/p_m = x_b$, the mole fraction of b:

$$\Delta S = -\frac{n_a(1545)}{J} \ln x_a - \frac{n_b(1545)}{J} \ln x_b \quad\quad (b)$$

Equation (b) is the desired result for the problem. However, it has an interesting conclusion in that the entropy increase due to the mixing process depends only on the moles of each gas present, not on the gas itself. Thus, the mixing of 1 mole of oxygen with 1 mole of hydrogen will yield the same entropy increase as the mixing of 1 mole of argon with 1 mole of Freon. One point that should be noted at this time is that the adiabatic mixing of *identical* gases initially at the same pressure and temperature does not result in an increase in entropy; Equation (b) is not applicable to this process. This has been called *Gibbs Paradox*, and further discussion on this can be found in the references.

---

### ILLUSTRATIVE PROBLEM 7.12

Determine the change in entropy for the process described in Illustrative Problem 7.9.

### SOLUTION

The change in entropy of the mixture is the sum of the changes in the entropy of each component. For the oxygen, the temperature starts at 500°F (960R) and decreases to 328.7°F. For the nitrogen, the temperature starts at 200°F (660R) and increases to 328.7°F. Applying Equation (a) of Illustrative Problem 7.11 and noting that the process is carried out at constant pressure, which makes the pressure term zero (ln 1 = 0), *for the oxygen,*

$$\Delta S = c_p \ \ln \frac{T_2}{T_1} = 0.23 \ln \frac{328.7 + 460}{500 + 460}$$

$$= 0.23 \left( \ln 788.7 - \ln 960 \right) = -0.045 \ \text{Btu/lb}_m \cdot R$$

The total change in entropy of the oxygen is

$$\Delta S = \left( 160 \right) \left( -0.045 \right) = -7.20 \ \text{Btu/R}$$

*For the nitrogen,*

$$\Delta s = c_p \ln \frac{T_2}{T_1} = 0.25 \ln \frac{328.7 + 460}{200 + 460}$$

$$= -0.045 \ \text{Btu/lb}_m \cdot R$$

The total change in entropy of the nitrogen is

$$\Delta S = \left( 196 \right) \left( 0.045 \right) = 8.82 \ \text{Btu/R}$$

*For the mixture,* the total change in entropy is the sum of the changes of the constituent entropies. Therefore,

$$\Delta S = -7.20 + 8.82 = 1.62 \ \text{Btu/R}$$

Per pound of mixture,

$$\Delta s_m = \frac{1.62}{196 + 460}$$

$$= 0.00455 \ \text{Btu/lb}_m \cdot R \left( \text{increase per pound mass of mixture} \right)$$

As an alternative solution, assume an arbitrary datum of 0°F (460R). The initial entropy of the oxygen above this base is

$$\Delta s = c_p \ \ln \frac{T_2}{T_1} = 0.23 \ln \frac{500 + 460}{460} = 0.169 \ \text{Btu/lb}_m \cdot R$$

The final entropy of the oxygen above this base is

$$\Delta s = c_p \ln \frac{T_2}{T_1} = 0.23 \ln \frac{328.7 + 460}{460} = 0.1240 \ \text{Btu}/\text{lb}_m \cdot R$$

The entropy change of the oxygen therefore is

$$\Delta s = 0.124 - 0.169 = -0.045 \ \text{Btu}/\text{lb}_m \cdot R$$

*For the nitrogen,*

$$\Delta s = c_p \ln \frac{T_2}{T_1} = 0.25 \ln \frac{660}{460}$$

$$= 0.090 \ \text{Btu}/\text{lb}_m \cdot R \ (\text{above a base of } 0°F) \ \text{initially}$$

$$\Delta s = c_p \ln \frac{T_2}{T_1} = 0.25 \ln \frac{460 + 328.7}{460}$$

$$= 0.135 \ \text{Btu}/\text{lb}_m \cdot R \ (\text{above a base of } 0°F)$$

The entropy change of the nitrogen therefore is

$$\Delta s = 0.135 - 0.090 = 0.045 \ \text{Btu}/\text{lb}_m \cdot R$$

The remainder of the problem is as before. The advantage of this alternative method is that negative logarithms are avoided by choosing a reference temperature lower than any other temperature in the system.

## 7.6 Air–Water Vapor Mixtures*

A gas may be considered to be a vapor that is superheated. If the degree of superheat is high and the vapor is not close to critical conditions, it will usually behave as if it were an ideal gas. Although there are many exceptions to this generalization, it is found to be useful when treating air–water vapor mixtures. Air is a mixture of many gases, and dry air is defined as a mixture having the following composition by volume: oxygen, 20.95%; nitrogen, 78.09%; argon, 0.93%; carbon dioxide, 0.03%. In the treatment of air–water vapor mixtures, it is common to treat air as a single gas with a molecular weight of 28.966.

Some of the concepts involved in dealing with air–water vapor mixtures can best be illustrated by referring to Figure 7.6, a $T$–$s$ diagram for water vapor. In general, atmospheric air contains superheated water vapor, which may typically be at state $B$ in Figure 7.6.

The *dry-bulb temperature* ($T_1$) is the temperature of the mixture of air and water vapor at rest, and it is measured by an instrument (thermometer) that is not affected by the amount

---

* The material in this section has been based largely on the *ASHRAE Guide and Data Book* published by the American Society of Heating, Refrigerating, and Air-Conditioning Engineers, with permission.

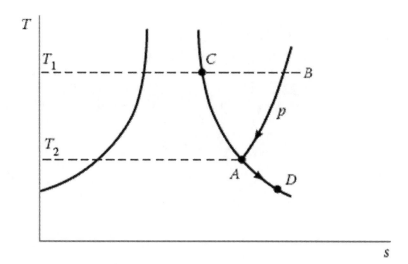

**FIGURE 7.6**
The water vapor portion of an air–water vapor mixture.

of moisture in the air or by thermal radiation. If the mixture of air and water vapor that is originally at $B$ is cooled at constant pressure, it will follow the path from $B$ to $A$ along the constant-pressure line. In practice, the *dew-point temperature* is the temperature ($T_2$) at which condensation will just begin when the moist air mixture under consideration is cooled at constant pressure. If the mixture is cooled further, the water vapor will follow along the saturation line to $D$, and some moisture will condense. If water is added to the mixture at a constant temperature, it will follow the path from $B$ to $C$. Moist air is said to be saturated ($C$) when its condition is such that it can coexist in neutral equilibrium with an associated condensed moisture phase presenting a flat surface to it. The word *neutral* is used to exclude unstable states from the definition, and the flat surface is used to exclude surface-tension effects.

Note that referring to "saturated air" is really a misnomer because it is the water vapor in the air that is saturated. Referring to Chapter 3, the water vapor is then at the "g" point of the saturation curve.

In addition to the quantities already defined, certain other terms describing air–water vapor mixtures require definition.

*Relative humidity* ($\phi$). The relative humidity of any mixture of air and water vapor is defined as the ratio of the mole fraction of water vapor in the mixture to the mole fraction of water vapor in saturated air at the same dry-bulb temperature and barometric pressure. It is also equal to the ratio of the partial pressure of the water vapor to the saturation pressure of water vapor at the temperature of the mixture $\left( \phi = \dfrac{x_{w,m}}{x_{w,\text{sat}}} = \dfrac{p_w}{p_{w,\text{sat}}} \right)$. Relative humidity can be understood as the ratio of the actual water vapor in the air to that amount the air could hold if it were saturated at the actual air temperature.

*Humidity ratio* ($W$). The humidity ratio of any mixture of air and water vapor is defined as the ratio of the mass of water vapor in the mixture to the mass of dry

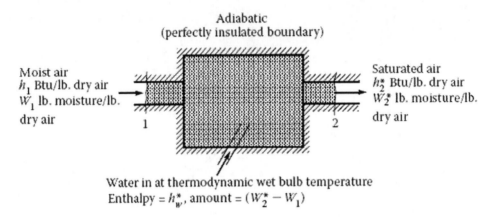

**FIGURE 7.7**
Adiabatic saturation process.

air in the mixture $\left(W = \dfrac{m_w}{m_a}\right)$. The units of humidity ratio are pounds of water vapor per pound of dry air.

*Specific humidity.* The term *specific humidity* has been and still is sometimes used for humidity ratio. They are synonymous.

*Thermodynamic wet-bulb temperature.* The thermodynamic wet-bulb temperature is the temperature at which liquid water, by evaporating into air, can bring the air to saturation adiabatically at the same temperature. To illustrate the definition of the thermodynamic wet-bulb temperature, consider the idealized system shown in Figure 7.7. It is a steady-flow system in which no work is done between sections 1 and 2 $(w_{in} = w_{out} = 0)$, and no energy as heat enters or leaves $(q_{in} = q_{out} = 0)$. Water enters the system at the thermodynamic wet-bulb temperature, and air leaves the system saturated at the thermodynamic wet-bulb temperature. The energy equation for this process is, per pound of dry air,

$$h_1 + h_w^* \left(W_2^* - W_1\right) = h_2^* \qquad\qquad (7.19)$$

The * notation is used to denote properties at the thermodynamic wet-bulb temperature, and the subscript w is used to denote water. The temperature corresponding to $h_2^*$ for given values of $h_1$ and $W_1$ is the defined thermodynamic wet-bulb temperature $\left(T_{h_2^*, p_2}\right)$. The process that we have just considered is also known as the *adiabatic saturation process*, and the thermodynamic wet-bulb temperature is also known as the adiabatic saturation temperature. The usefulness of the foregoing discussion lies in the fact that the temperature of the saturated air–water vapor mixture leaving the duct is a function of the temperature, pressure, relative humidity of the entering mixture, and the exit pressure. Conversely, knowing the entering and exit pressures and temperatures, we can determine the relative humidity and humidity ratio of the entering air–water vapor mixture.

## ILLUSTRATIVE PROBLEM 7.13

If the relative humidity of moist air is 50% when the mixture temperature is 80°F, determine the dew point of the air. Assume that the barometric pressure is 14.7 psia.

### SOLUTION

Referring to Figure 7.6, it will be seen that the cooling of an air–water vapor mixture from $B$ to $A$ proceeds at constant pressure until the saturation curve is reached. The dew-point temperature of the air–water vapor mixture is equal to the saturation temperature, corresponding to the partial pressure of the water vapor $\left(T_{dp,\,\text{air-water}} = T_{\text{sat, water @ partial}}\right)$. The relative humidity has been defined to be the ratio of the mole fraction of water vapor in the mixture to the mole fraction of water vapor in saturated air at the same dry-bulb temperature and barometric pressure $\left(\phi = \dfrac{x_{w,m}}{x_{w,\text{sat}}} = \dfrac{p_w}{p_{w,\text{sat}}}\right)$. It will be recalled from Section 7.2 that the partial pressure of a component equals the mole fraction of the component multiplied by the total pressure of the mixture $\left(p_a = x_a \cdot p_m \rightarrow p_w = x_w \cdot p_m\right)$. Because the relative humidity is a mole fraction ratio, it is also the ratio of the partial pressure of the vapor in the air to the saturation pressure corresponding to the air temperature $\left(\phi = \dfrac{p_w}{p_{w,\text{sat}}}\right)$. At 80°F, the *Steam Tables* give us a saturation pressure, $p_{w,\text{sat}}$ of 0.5073 psia, and because the relative humidity is 50% or 0.5, the vapor pressure of the water, $p_w$ is $0.5 \times (0.5073) = 0.2537$ psia. Using the full *Steam Tables* or interpolating in the table in Appendix 2 yields a saturation temperature corresponding to 0.2537 psia as close to 60°F $\left(T_{w,\text{ sat @ 0.2537 psia}} = 60°F\right)$. This temperature is the desired dew-point temperature.

## ILLUSTRATIVE PROBLEM 7.14

Air at 90°F contains 0.005 $\text{lb}_m$ of water vapor per pound of dry air. Determine the partial pressure of each component, the relative humidity, and the dew-point temperature of the mixture if the mixture is at 14.7 psia.

### SOLUTION

To solve this problem, it is necessary to determine the properties of the saturated mixture at 90°F. If the air is saturated at 90°F, the partial pressure of the water vapor, $p_w$ is found directly from the *Steam Tables* as 0.6988 psia, and the specific volume of the water vapor, $v_w$ is 467.7 ft.$^3$/$\text{lb}_m$ of vapor. At this point, let us assume that we have a container having a volume of 467.7 ft.$^3$ (i.e., 1 $\text{lb}_m$ of water vapor) and that it contains a saturated air–water vapor mixture at 90°F. Because the mixture is assumed to be saturated, there must be 1 $\text{lb}_m$ of water vapor in the container, and the partial pressure of the dry air must be $(p_a = p_m - p_w)$: $14.7 - 0.6988 = 14.0$ psia. By applying the ideal gas equation to the air, we can determine the specific volume and mass of air in the container. Thus,

$$v_{\text{dry air}} = \frac{RT}{p_{\text{dry air}}} = \frac{1545}{28.966} \times \frac{460 + 90}{14.0 \times 144}$$

$$= 14.55 \text{ ft.}^3/\text{lb}_m\text{dry air}$$

The mass of dry air in the 467.7 ft.³container is $(m = V/v)$: 467.7/14.55 = 32.1 lb$_m$. The saturated mixture therefore contains 1 lb$_m$ of water vapor per 32.1 lb$_m$ of dry air or $(W = m_w/m_a)$: 0.312 lb$_m$ water vapor/lb$_m$ dry air.

To obtain the relative humidity ($\phi$), it is necessary to determine the mole fraction of water vapor for both the saturated mixture and the mixture in question $\left( \phi = \dfrac{x_{w,m}}{x_{w,\text{sat}}} \right)$.

The saturated mixture contains 1 lb$_m$ of water vapor or $(m_w/MW_w)$: 1/18.016 moles = 0.0555 mole of water vapor, and $(m_a/MW_a)$: 32.1/28.966 = 1.108 moles of dry air. Therefore, for the saturated mixture, the ratio of moles of water vapor to moles of mixture is $(x_{w,\text{sat}} = \text{moles}_w/\text{moles}_m)$: 0.0555/(1.108 + 0.0555) = 0.0477. For the actual mixture, the moles of water vapor per pound of dry air is $(m_w/MW_w)$: 0.005/18.016 = 0.000278, and 1 lb$_m$ of dry air is $(m_a/MW_a)$: 1/28.966 = 0.0345 mole. The moles of water vapor per mole of mixture at the conditions of the mixture is $(x_{w,m} = \text{moles}_w/\text{moles}_m)$: 0.000278/ (0.0345 + 0.000278) = 0.00799. From the definition of relative humidity, the relative humidity of this mixture is $\left( \phi = \dfrac{x_{w,m}}{x_{w,\text{sat}}} \right)$: 0.00799/0.0477 = 0.168 or 16.8%.

Because the mole ratio is also the ratio of the partial pressures for the ideal gas, $\phi$ can be expressed as the ratio of the partial pressure of the water vapor in the mixture to the partial pressure of the water vapor at saturation $\left( \phi = \dfrac{x_{w,m}}{x_{w,\text{sat}}} = \dfrac{p_w}{p_{w,\text{sat}}} \right)$. Therefore, the partial pressure of the water vapor is $(p_w = \phi \cdot p_{w,\text{sat}})$: 0.168 × 0.6988 = 0.1173 psia, and the partial pressure of the dry air in the mixture is $(p_a = p_m - p_w)$: 14.7 − 0.1173 = 14.58 psia.

The dew-point temperature, $T_{dp}$ is the saturation temperature corresponding to the partial pressure of the water vapor in the mixture $(T_{dp} = T_w @ p_{sat})$. The saturation temperature corresponding to 0.1173 psia is found from the *Steam Tables* to be close to 39°F.

The adiabatic saturation of air is a process that is useful to analytically study for the thermodynamics of air–water vapor mixtures. If this process had to be used each time one wished to obtain the relative humidity of an air–water vapor mixture, it would be impractical. Instead, a relatively simple method is used to establish the partial pressure of the water vapor based upon a temperature known as the *wet-bulb temperature*. Two thermometers are placed in a moving airstream. The bulb of the wet-bulb thermometer is covered with a gauze that is immersed in clean water and acts to wick the water to the bulb. Air is continuously blown over the thermometer by a fan. At the wet-bulb thermometer, three actions occur. The first is water evaporation from the wick due to heat transfer from the air at the dry-bulb temperature; the second action is water evaporation from the wick due to radiant heat transfer from the surroundings; and the last action is a diffusion of water vapor from the wick to the surrounding air–water vapor mixture. As a result of these actions, the temperature of the wet-bulb thermometer will be lower than the dry-bulb temperature.

In principle, there is a difference between the wet-bulb temperature and the temperature of adiabatic saturation (the thermodynamic wet-bulb temperature). The wet-bulb temperature is a function of both heat and mass diffusion rates, while the adiabatic saturation temperature is a function of a thermodynamic equilibrium process. However, in practice, it has been found that for air–water vapor mixtures at atmospheric pressures and temperatures, the wet-bulb and adiabatic saturation temperatures are essentially the same. Thus, the observed readings of the wet-bulb thermometer can be taken to be the same as the thermodynamic wet-bulb temperature defined by the adiabatic saturation process if the air velocity past the wick is approximately 1000 ft./min and if the wick is exposed to radiation exchange with the surroundings at temperatures differing from the dry-bulb temperature by no more than 20°F. It must be noted as a caution that the wet-bulb temperature and adiabatic saturation temperatures deviate considerably from each other for other than dry air–water vapor gas–vapor mixtures and for air–water vapor mixtures at conditions other than usual atmospheric conditions.

Problems concerning air–water vapor mixtures can be considerably simplified by using the ideal gas equations for each of the components. By denoting the subscripts $a$ to be dry air, $w$ to be water vapor, $sat$ to be saturation, and $m$ to mean mixture, the following relations can be derived:

Gas constant of the mixture.

$$R_m = \frac{m_a}{m_a + m_w} R_a + \frac{m_w}{m_a + m_w} R_{w\,(\text{vapor})} \tag{7.20}$$

By dividing the numerator and denominator of each of the terms in parentheses by $m_a$,

$$R_m = \frac{1}{1 + \dfrac{m_w}{m_a}} R_a + \frac{m_w / m_a}{1 + m_w / m_a} R_w$$

But $m_w/m_a$ is the humidity ratio W.

$$R_m = \frac{1}{1 + W} R_a + \frac{W}{1 + W} R_{w\,(\text{vapor})} \tag{7.21}$$

Because W is a small number compared to unity, it is possible at a first approximation to use the gas constant of the mixture as the gas constant of dry air $\left(R_m = R_a\right)$.

*Relative humidity.* By using the ideal gas law at constant temperature $\left(p_w v_w = p_{w,\text{sat}} v_{w,\text{sat}}\right)$, the relative humidity $\phi$ is

$$\phi = \frac{p_w}{p_{w,\,\text{sat}}} = \frac{v_{w,\,\text{sat}}}{v_w} = \frac{\rho_w}{\rho_{w,\,\text{sat}}} \tag{7.22}$$

When the mixture is at the dew-point temperature, $\phi$ is 1.

*Humidity ratio.* For a mixture where both air and water vapor are at the same temperature $\left(T_w = T_a\right)$, and occupy the same volume $\left(V_w = V_a\right)$, using the ideal gas law in the definition of humidity ratio leads to,

$$W = \frac{m_w}{m_a} = \left(\frac{p_w V_w MW_W}{1545 T_w}\right) \cdot \left(\frac{1545 T_a}{p_a V_a MW_a}\right) = \frac{18.016}{28.966} \frac{p_w}{p_a} = 0.622 \frac{p_w}{p_m - p_w} \tag{7.23}$$

In Equation 7.23, where 18.016 is the molecular weight of water vapor and 28.966 is the molecular weight of dry air.

Relative humidity and humidity ratio. From Equations 7.22 and 7.23,

$$\phi \cdot p_{w,\text{sat}} = p_w \qquad (7.24)$$

Recalling $R_a = \dfrac{1545}{MW_a}$ and $R_{w(\text{vapor})} = \dfrac{1545}{MW_{w(\text{vapor})}}$ and rearranging both leads to $\dfrac{R_a}{R_w} = \dfrac{MW_w}{MW_a} = 0.622$, therefore,

$$W = 0.622 \frac{p_w}{p_a} = 0.622\phi \frac{p_{w,\text{sat}}}{p_a} = \phi \frac{p_{w,\text{sat}}}{p_a} \frac{R_a}{R_{w(\text{vapor})}} = \phi \frac{v_a}{v_{w,\text{sat}}}$$

Where $\dfrac{p_{w,\text{sat}} v_{w,\text{sat}}}{R_w} = \dfrac{p_a v_a}{R_a}$ under isothermal conditions using the ideal gas law, which leads to $\dfrac{p_{w,\text{sat}}}{P_a} \dfrac{R_a}{R_{w,\text{vapor}}} = \dfrac{v_a}{v_{w,\text{sat}}}$.

---

### ILLUSTRATIVE PROBLEM 7.15

Solve Illustrative Problem 7.14 using Equations 7.20 through 7.24.

#### SOLUTION

$W$ is given as 0.005. Therefore,

$$W = 0.622 \frac{p_w}{p_m - p_w} \rightarrow p_w = \frac{W \cdot p_m}{(0.622 + W)}$$

with $p_m = 14.7$ psia. Thus, $p_w = (0.005 \times 14.7)/(0.622 + 0.005) = 0.117$ psia, and $p_a = p_m - p_w$: $14.7 - 0.117 = 14.58$ psia. Because $\phi = p_w p_{w,\text{sat}}$, it is necessary to obtain $p_{w,\text{sat}}$ *from the Steam Tables* at 90°F. This is 0.6988 psia. Therefore, $\phi = 0.117/0.6988 = 0.167 = 16.7\%$. The dew-point temperature, $T_{dp}$ is the saturation temperature corresponding to 0.117 psia, which is found from the *Steam Tables* to be 39°F. The results of this problem and Illustrative Problem 7.14 are in good agreement ($\phi$: 16.7% vs. 16.8%; $T_{dp}$: 39°F vs. 39°F).

---

### ILLUSTRATIVE PROBLEM 7.16

Air at 90°F and 70% relative humidity is conditioned so that its final state is 80°F and 40% relative humidity. How much water was removed from the air? Assume that the barometer is at 14.7 psia.

#### SOLUTION

The amount of water vapor removed (per pound of dry air) is the difference between the humidity ratio (specific humidity) at the inlet and the outlet of the conditioning unit. We shall therefore evaluate $W$ for both specified conditions. Because $\phi = p_w/p_{w,\text{sat}}$,

$$\text{at } 90°F : p_{w,sat} = 0.6988 \text{ psia}$$

$$\text{at } 80°F : p_{w,sat} = 0.5073 \text{ psia}$$

$$\text{at } 90°F : p_w = \phi \cdot p_{w,sat} = 0.7 \times 0.6988 = 0.4892 \text{ psia}$$

$$\text{at } 80°F : p_w = \phi \cdot p_{w,sat} = 0.4 \times 0.5073 = 0.2029 \text{ psia}$$

But $p_a = p_m - p_w$. Thus,

$$\text{at } 90°F : p_a = 14.7 - 0.4892 = 14.2108 \text{ psia}$$

$$\text{at } 80°F : p_a = 14.7 - 0.2029 = 14.4971 \text{ psia}$$

Because $W = 0.622\, p_w/(p_m - p_w) = 0.622\, p_w/p_a$,

$$\text{at } 90°F : W = 0.622 \frac{0.4892}{14.2108}$$

$$= 0.0214 \text{ lb.water/lb.dry air}$$

$$\text{At } 80°F : W = 0.622 \frac{0.2029}{14.4971}$$

$$= 0.0087 \text{ lb.water/lb.dry air}$$

The amount of water removed per pound of dry air is

$$0.0214 - 0.0087 = 0.0127 \text{ lb.water removed/lb.dry air}$$

A unit of weight commonly used in air-conditioning work is the *grain*. By definition, there are 7000 grains per pound. Using this definition, the water removal per pound of dry air in Illustrative Problem 7.16 is $7000 \times 0.0127 = 88.9$ grains per pound of dry air.

## 7.7 Thermodynamic Properties of Air–Water Vapor Mixtures

The thermodynamic properties of air–water vapor mixtures can be determined from tabulated data available in the latest edition of the *ASHRAE Guide and Data Book*, published by the American Society of Heating, Refrigerating, and Air-Conditioning Engineers (ASHRAE). These data are tabulated from –160°F to 200°F for moist air at 29.921 in. Hg and are given for either perfectly dry or completely saturated air. By simple interpolation, the specific volume, enthalpy, or entropy of mixtures can be obtained. The interpolation formulas are given in the *Guide* or as shown in Chapter 3.

For those designs or analyses that require accuracies that cannot be achieved by use of the ideal gas relations, these tables are invaluable. However, for those cases that can be dealt with by using ideal gas relations, it is possible to express the thermodynamic properties of air–water vapor mixtures in a relatively simple manner. It is convenient to express these properties per unit mass of dry air in the mixture, for example, Btu per pound of dry air.

*Specific heat of the mixture.* The specific heat of the mixture in Btu per pound mass per degree Fahrenheit (of dry air) is

$$c_{pm} = c_{pa} + W c_{pw} \tag{7.25}$$

where $c_{pa} = 0.24$ Btu/lb$_m$·R and $c_{pw} = 0.44$ Btu/lb$_m$·R in the ordinary temperature range.

*Enthalpy of the mixture.* The enthalpy of the mixture is given by

$$h_m = h_a + W h_w \tag{7.26}$$

It is possible to express $h_m$ accurately to within 0.1% between 32°F and 100°F at 1 atm (29.921 in. Hg) by

$$h_m = (0.2402 + 0.44\,W)t + 1061\,W \tag{7.27}$$

where $t$ is the dry-bulb temperature in degrees Fahrenheit.

*Entropy of the mixture.* The entropy of the mixture can be given as

$$s_m = s_a + W s_w \tag{7.28}$$

where $s_a$ and $s_w$ are evaluated at the respective partial pressures in the mixture.

## 7.8 Psychrometric Chart

Just as the Mollier chart was found to be useful in plotting the thermodynamic properties of a fluid, a *psychrometric chart* is found to be equally useful when dealing with air–water vapor mixtures (see Figure 7.8). To understand the chart and its limitations fully, let us construct one in outline form. From the *Steam Tables*, we have the saturation data given in Table 7.5.

If these data are plotted with the partial pressure of water vapor ($p_w$, $p_v$ in Figure 7.9) as the ordinate and the temperature (dry-bulb) as abscissa, the saturation curve is established as shown in Figure 7.9 (100% relative humidity). From Equation 7.22, the relative humidity is the ratio of $p_w/p$w,sat ($p_v/p_{vs}$ in Figure 7.10). Thus, by plotting relative humidity at each dry-bulb temperature and connecting corresponding points, it is possible to construct lines of constant relative humidity. For instance, the outline of the 50% relative humidity line has been drawn in Figure 7.10.

Let us further assume that the chart will apply for a given barometric pressure, say, 29.921 in. Hg. From Equation 7.23,

$$W = 0.622 \frac{p_w}{p_m - p_w}$$

**FIGURE 7.8**
Psychrometric chart for air–water vapor mixtures.

**TABLE 7.5**

Temperature and Pressure Data

| Temperature (°F) | Pressure (psia) |
| --- | --- |
| 32 | 0.08859 |
| 40 | 0.12166 |
| 50 | 0.17803 |
| 60 | 0.2563 |
| 70 | 0.3632 |
| 80 | 0.5073 |
| 90 | 0.6988 |
| 100 | 0.9503 |
| 110 | 1.2763 |
| 120 | 1.6945 |

Because the mixture pressure has been assumed to be constant, $W$ becomes a simple linear function of $p_w$ ($p_v$ in the figures). Thus, the ordinate also becomes proportional to the weight of water vapor per pound of dry air. Because by definition there are 7000 grains/lb, we have plotted on the right ordinate of Figure 7.9 an auxiliary scale of $W$ in grains of water vapor per pound of dry air.

From Equation 7.24, it is possible to evaluate the specific volume of the mixture per pound of dry air. Because ideal gas relations have been assumed, the specific volume of the air–water vapor mixture is a linear function of the humidity ratio $W$. Lines of constant specific volume are shown in Figure 7.10.

By using the ideal gas relation and Equation 7.27, the enthalpy of the mixture is also found to be linear with dry-bulb temperature and humidity ratio. Typical lines of constant

**FIGURE 7.9**
Skeleton psychrometric chart (showing constant ϕ).

**FIGURE 7.10**
Skeleton psychrometric chart (with lines of constant $v$).

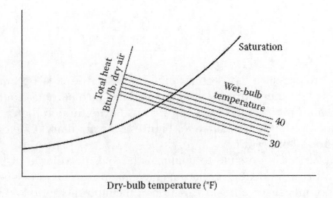

**FIGURE 7.11**
Skeleton psychrometric chart (with lines of constant $h$).

enthalpy (total heat) are indicated in Figure 7.11. The figure also shows the lines of constant "total heat" and wet-bulb temperatures as the same. Actually, wet-bulb temperature lines are straight, but they are not parallel. However, the error is small enough to be negligible. For a further discussion of the conclusions regarding the relation of total heat to enthalpy and so on, the reader is referred to the ASHRAE publications.

---

### ILLUSTRATIVE PROBLEM 7.17

Solve Illustrative Problem 7.13 using the psychrometric chart.

#### SOLUTION

Entering Figure 7.8 at a dry-bulb temperature of 80°F, we proceed vertically until we reach the 50% humidity curve. At this intersection, we proceed horizontally and read the dew-point temperature as approximately 60°F.

---

### ILLUSTRATIVE PROBLEM 7.18

Solve Illustrative Problem 7.14 using the psychrometric chart.

#### SOLUTION

In this problem, we are given the moisture content of the air to be 0.005 lb. per pound of dry air. This corresponds to $0.005 \times 7000 = 35$ grains per pound of dry air. Entering the chart at 90°F and proceeding vertically to 35 grains per pound of dry air, we find the dew point to be 39°F by proceeding horizontally to the intersection with the saturation curve. The relative humidity is approximately 17%. From the leftmost scale, we read the pressure of water vapor to be 0.12 psia. The partial pressure of the air is $p_a = p_m - p_w$: $14.7 - 0.12 = 14.58$ psia. Comparing these results to Illustrative Problem 7.14 indicates good agreement between the results obtained by chart and by calculation.

---

### ILLUSTRATIVE PROBLEM 7.19

Solve Illustrative Problem 7.16 using the psychrometric chart.

#### SOLUTION

The initial conditions are 90°F and 70% relative humidity. Entering the chart at 90°F dry-bulb temperature and proceeding vertically to 70% relative humidity, we find the air to have 150 grains water vapor per pound of dry air. At the final condition of 80°F and 40% relative humidity, we read 61 grains of water/lb of dry air. The water removed is $150 - 61 = 89$ grains per pound of dry air or $89/7000 = 0.0127$ lb. of water per pound of dry air removed.

Having constructed the chart, we can illustrate its usefulness best by showing typical processes on the chart. Remember, the chart in Figure 7.8 can only be used for atmospheric air at approximately 14.7 psia. Note that to find all psychrometric properties, any two of the following will suffice: specific humidity $W$, dry- and wet-bulb temperatures, relative humidity $\phi$, dew point ($T_{dp}$), vapor pressure $p_g$ or $p_w$, and/or enthalpy, $h$. As an example, Figure 7.12 shows the resultant properties for air with a dry-bulb temperature of 80°F and a wet-bulb temperature of 60°F ($W = 46$ gr/lb; $\phi = 30\%$; $h = 26.5$ Btu/lb.; $p_w = 0.15$ psi; $T_{dp} = 46$°F).

*Dew-point determination—heating and cooling without change in moisture content.* From the definition of the dew point, it is apparent that a horizontal line on the chart (cooling at constant pressure of water vapor) extended to the saturation line will yield the dew-point temperature. This process of "sensible" heating or cooling (Figure 7.13) is distinguished by a change in dry-bulb temperature, relative humidity, wet-bulb temperature, total heat, and specific volume, and by no change in moisture content, dew-point temperature, and vapor pressure of the moisture in the air. Sensible heat is heat transfer due to a temperature difference without condensation or evaporation of the moisture in the air. Latent heat is heat transfer resulting from evaporation or condensation of the moisture in the air at saturation temperature.

**FIGURE 7.12**
Psychrometric relationships.

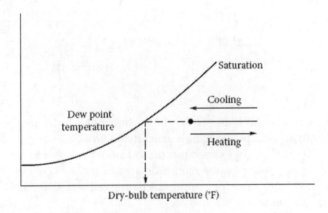

**FIGURE 7.13**
Sensible heating or cooling process.

**FIGURE 7.14**
Air cooling and heating coils.

Figure 7.14 shows an example of chilled-water coils using extended surfaces (fins) to increase the heat transfer of the unit. This type of construction is usually used in larger installations where cooling (or heating) is accomplished by flowing the heat-transfer fluid inside the tubes and passing the air over the external tube surface. When extended surfaces are used, they may be staggered or placed in line with the airflow. In those cases where frost may accumulate on the tube surface, plain tubes rather than extended surface elements are used. Coil performance is dependent on the way air flows over the outside of the tubes in relation to the way the heating or cooling fluid flows on the inside of the tubes. A discussion of the heat transfer in heat exchangers and the use of fins is given in Chapter 11.

---

**ILLUSTRATIVE PROBLEM 7.20**

How much heat is required to sensibly heat air having a dry-bulb temperature of 50°F and a relative humidity of 50% to 80°F?

**SOLUTION**

We first locate 50°F and 50% relative humidity in Figure 7.8. At this state, we read 26 grains of water per pound of dry air and a total heat, $h_1$ of 16.1 Btu per pound of dry air. We now proceed horizontally to 80°F at a constant value of 26 grains of water per pound of dry air and read a total heat $h_2$ of 23.4 Btu per pound of dry air. The heat required therefore is $q = h_2 - h_1$: 23.4 − 16.1 = 7.3 Btu per pound of dry air.

---

*Humidifying of air with no change in dry-bulb temperature.* As shown in Figure 7.15, this process is a vertical line on the psychrometric chart between the desired moisture limits. During this process, there is an increase in relative humidity, wet-bulb temperature, total heat, specific volume, moisture content, dew-point temperature, and vapor pressure of the moisture in the air. The addition of moisture to air is usually necessary for winter operation of air-conditioning systems. The process shown in Figure 7.15 can be accomplished by using heated spray water whose temperature is kept above the dry-bulb temperature of the air. In the limit, the air temperature approaches the final water temperature.

*Evaporative cooling of air.* If air is brought into contact with water that is at the wet-bulb temperature of the air, moisture will be added to the air at constant wet-bulb temperature.

**FIGURE 7.15**
Humidification of air: constant dry-bulb temperature.

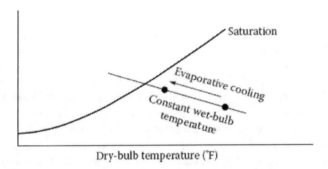

**FIGURE 7.16**
Evaporative cooling of air.

Because a line of constant wet-bulb temperature is also a line of constant total heat, the total heat for this process is constant. The heat "lost" by the incoming air serves to increase the moisture content of the final air, as given by Equation 7.27. Although both the wet-bulb temperature and total heat of the air remain constant, the dry-bulb temperature and the specific volume decrease as the relative humidity, the moisture content, and the vapor pressure of the moisture in the air increase. This process is shown in Figure 7.16. As the process proceeds along the line of the constant wet-bulb temperature, the dry-bulb temperature and the dew point of the leaving air approach the wet-bulb temperature of the air. A spray-type air washer can be used to accomplish this process. Evaporative cooling is one of the oldest methods of cooling known to man and is most successfully used in regions that are dry, such as in deserts. Portable evaporative coolers have been used in automobile and truck cabs in arid and desert regions for comfort control. For this method to be effective, there must be a large difference between the wet- and dry-bulb temperatures of the air, making it ineffective in humid, tropic areas.

---

### ILLUSTRATIVE PROBLEM 7.21

In an evaporative cooling process, the exit air is found to be saturated at 50°F. If the entering air has a dry-bulb temperature of 80°F, determine the relative humidity of the entering air.

**SOLUTION**

Because the exit air is saturated, we find the exit condition on the saturation curve corresponding to a wet-bulb temperature of 50°F. The process is carried out at constant total enthalpy, which is along a line of constant wet-bulb temperature. Proceeding along the 50°F wet-bulb temperature line of Figure 7.8 diagonally to the right until it intersects with the vertical 80°F dry-bulb temperature line yields a relative humidity of approximately 4%.

*Chemical drying of air.* If air is passed over a drying agent that is not dissolved by the moisture extracted from the air and does not retain an appreciable amount of the heat of vaporization liberated when the water is condensed, the process can be said to be carried out along a wet-bulb temperature line. If the adsorber retains an appreciable amount of this heat, the process takes place along a line below the wet-bulb temperature line. When the adsorber is soluble in water (e.g., calcium chloride), the process line is above or below the wet-bulb temperature line depending on whether heat is liberated or adsorbed when the adsorber goes into solution. This process is shown graphically in Figure 7.17. Typical solid adsorbents are activated alumina, silica gel, activated carbon, and bauxite. These remove water from an airstream only if the vapor pressure of the water in the adsorbent is less than the partial pressure of the water vapor in the surrounding airstream. In other words, the process is primarily one of condensation. The latent heat of vaporization and heat of adsorption have to be taken away by the surroundings. Thus, the airstream, the adsorbent itself, and the adsorbent container all increase in temperature. Eventually, the sorbent temperature reaches a state of equilibrium, and the material has to be reactivated—heated to drive off the adsorbed moisture and cooled down for reuse.

*Cooling and dehumidifying of air.* If the cooling surface temperature is below the initial dew-point temperature, this process can be portrayed as a straight line extending from the initial condition to the surface temperature on the saturation curve. The final condition of the air will depend on the total heat extracted from the air. During this process, the dry-bulb temperature, wet-bulb temperature, moisture content, specific volume, vapor pressure of the moisture, and total heat all decrease, as indicated in Figure 7.18. The coils shown in Figure 7.14 can also be used for cooling and dehumidifying air by keeping the coil temperature below the dew point of the entering air.

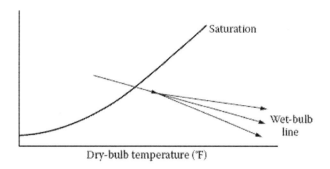

**FIGURE 7.17**
Chemical drying of air.

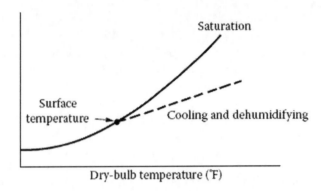

**FIGURE 7.18**
Cooling and dehumidifying process.

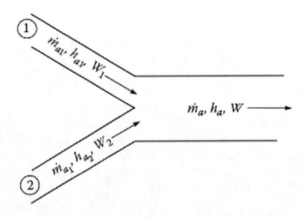

**FIGURE 7.19**
Adiabatic mixing of airstreams.

*Adiabatic mixing of airstreams.* The mixing of two airstreams at different initial conditions is quite commonly used in air-conditioning work. Let us assume that no condensation takes place during this process. Then, from Figure 7.19, a mass balance on the dry air is

$$\dot{m}_{a1} + \dot{m}_{a2} = \dot{m}_a \qquad (7.29)$$

where $\dot{m}_{a1}$ is the mass flow rate of dry air in stream 1, $\dot{m}_{a2}$ is the mass flow rate of dry air in stream 2, and $\dot{m}_a$ is the mass flow rate of the combined streams. A mass balance on the water vapor yields

$$\dot{m}_{a1}W_1 + \dot{m}_{a2}W_2 = \dot{m}_a W \qquad (7.30)$$

For an adiabatic mixing process ($q=0$) in which velocity and elevation terms are negligible and in which no work ($w$) enters or leaves the system, an energy balance yields

$$\dot{m}_{a1}h_{a1} + \dot{m}_{a2}h_{a2} = \dot{m}_a h_a \qquad (7.31)$$

**FIGURE 7.20**
Adiabatic mixing of airstreams.

A combination of these equations yields

$$\dot{m}_{a1}W_1 + \dot{m}_{a2}W_2 = \left(\dot{m}_{a1} + \dot{m}_{a2}\right)W \rightarrow \frac{W_2 - W}{W - W_1} = \frac{\dot{m}_{a1}}{\dot{m}_{a2}}$$

$$\dot{m}_{a1}h_{a1} + \dot{m}_{a2}h_{a2} = \left(\dot{m}_{a1} + \dot{m}_{a2}\right)h_a \rightarrow \frac{h_a - h_{a2}}{h_a - h_{a1}} = \frac{\dot{m}_{a1}}{\dot{m}_{a2}} \qquad (7.32)$$

$$\frac{h_a - h_{a2}}{h_a - h_{a1}} = \frac{W_2 - W}{W - W_1} = \frac{\dot{m}_{a1}}{\dot{m}_{a2}} = \frac{l_1}{l_2}$$

where $l_1$ and $l_2$ are the corresponding line segment lengths shown in Figure 7.20. Equation 7.32 can be interpreted to mean that the mixing of air at one condition with air at another condition can be represented by a straight line connecting the initial and final conditions. The final mixture will lie along this line at a point, determined by the relative quantities of air being mixed, as shown in Figure 7.20.

---

**ILLUSTRATIVE PROBLEM 7.22**

Indoor air at 75°F dry bulb and 50% relative humidity is to be mixed with outdoor air at 90°F dry bulb and 60% relative humidity. If 4 parts of indoor air is mixed with 1 part of outdoor air (by weight), what is the final mixture composition (4 parts + 1 part = 5 parts)?

**SOLUTION**

As noted from Figure 7.21, 1 lb. of mixture, $\frac{4}{5}$ lb. of indoor air, and $\frac{1}{5}$ lb. of outdoor air are mixed per pound of mixture. We now locate the two end states on the psychrometric chart and connect them with a straight line. The line connecting the end states is divided into 5 equal parts. Using the results of Equation 7.32, we now proceed *from* the 75°F indoor air state 1 part *toward* the 90°F outdoor air state. This locates the state of the mixture, which is found to be a dry-bulb temperature of approximately 78°F, a wet-bulb temperature of approximately 66°F, and a relative humidity of approximately 54%.

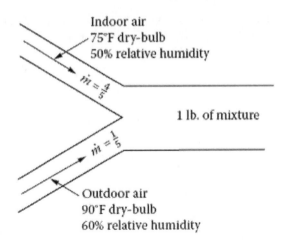

Indoor air
75°F dry-bulb
50% relative humidity

$\dot{m} = \frac{4}{5}$

1 lb. of mixture

$\dot{m} = \frac{1}{5}$

Outdoor air
90°F dry-bulb
60% relative humidity

**FIGURE 7.21**
Illustrative Problem 7.22.

*Cooling towers.* Cooling towers are used to cool the heat-transfer fluid (usually water) used to cool condensers in power plants and refrigeration systems. To minimize the environmental impact (i.e., thermal pollution) on rivers or other water sources when used as heat sinks, cooling towers are used to cool the water discharged from the condensers, and this water is then recirculated or reused on a continuous basis. In effect, instead of using river or local water to carry off the heat rejected from a typical power plant, the atmosphere (surrounding air) is used as a cooling medium. The cooling tower is simply a device in which water is evaporatively cooled by air. In the natural-draft cooling tower, air is circulated through the tower in a horizontal direction while water is sprayed or trickled over wood filling or other materials. In the forced-draft cooling tower, a fan is used to positively circulate air countercurrently to the falling water. The fan can be located at the bottom of the tower, and this arrangement is known as a forced-draft tower. When the fan is located at the top to prevent recirculation of the hot moist air, it is called an induced-draft tower. Figure 7.22 shows a large bank of cooling towers used in a power plant installation.

We can analyze the action in the cooling tower by referring to the schematic diagram in Figure 7.23. In this diagram, air flows from the bottom to the top of the tower, and water droplets flow counter to the airstream and, therefore, are in intimate contact with the counterflowing airstream. Heat is transferred to the flowing air, raising its wet-bulb temperature, its dry-bulb temperature, and its moisture content. The evaporation of a small portion of the water accounts for the water-cooling effect. The water that evaporates into the air stream must be replaced, and this replacement water is known as *makeup or makeup water.* Referring to the symbols in Figure 7.23, we can write a heat balance for the tower by considering the air and water streams separately. *For the air,*

heat pickup per lb. dry air, $\dot{Q}_a / \dot{m}_a = H_2 - H_2$ where $H$ = total heat/lb. dry air

total heat pickup, $\dot{Q}_a = \dot{m}_a \left( H_2 - H_1 \right)$

**FIGURE 7.22**
Large cooling tower installation.

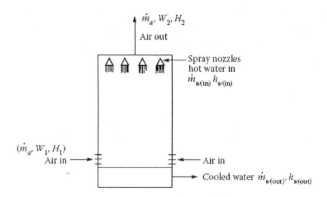

**FIGURE 7.23**
Schematic diagram of a cooling tower.

$$\text{moisture pickup per lb. dry air, } \dot{m}_w / \dot{m}_a = W_2 - W_1$$

$$\text{total moisture pickup, } \dot{m}_w = \dot{m}_a \left( W_2 - W_1 \right)$$

*For the water,*

$$\text{entering total enthalpy, } \dot{H}_{w(\text{in})} = \dot{m}_{w(\text{in})} h_{w(\text{in})}$$

$$\text{leaving total enthalpy, } \dot{H}_{w(\text{out})} = \dot{m}_{w(\text{out})} h_{w(\text{out})}$$

$$\text{total water enthalpy change, } \dot{Q}_{w(\text{net})} = \dot{m}_{w(\text{in})} h_{w(\text{in})} - \dot{m}_{w(\text{out})} h_{w(\text{out})}$$

$$\text{water evaporated, } \dot{m}_{w,\text{evap}} = \dot{m}_{w(\text{in})} - \dot{m}_{w(\text{out})}$$

Because the heat picked up by the air, $\dot{Q}_a$ must equal the heat exchanged by the water $\dot{Q}_{w(\text{net})}$, we have

$$\dot{m}_a \left( H_2 - H_1 \right) = \dot{m}_{w(\text{in})} h_{w(\text{in})} - \dot{m}_{w(\text{out})} h_{w(\text{out})} \tag{7.33}$$

The water picked up by the air, $\dot{m}_w$ must equal the water loss, $\dot{m}_{w,\text{evap}}$ in the tower. Therefore,

$$\dot{m}_a \left( W_2 - W_1 \right) = \dot{m}_{w(\text{in})} - \dot{m}_{w(\text{out})} \tag{7.34}$$

Equations 7.33 and 7.34 provide the basic relations for the solution of cooling tower problems.

---

### ILLUSTRATIVE PROBLEM 7.23

Hot water enters a cooling tower at the rate of 200,000 lb/h at 100°F and the water leaving is at 70°F. Air enters the tower at a dry-bulb temperature of 60°F with a 50% relative humidity and leaves at a dry-bulb temperature of 90°F with a relative humidity of 90%. Assuming that atmospheric pressure is 14.7 psia and the psychrometric chart (Figure 7.8) can be used, determine the amount of water lost per hour due to evaporation and the amount of air required per hour. The situation is summed up in Figure 7.24.

### SOLUTION

Let us first obtain the necessary physical data required from the *Steam Tables* and the psychrometric chart. For the water,

$$h_{100°F} = 68.05 \text{ Btu/lb}_m$$

$$h_{70°F} = 38.09 \text{ Btu/lb}_m$$

**FIGURE 7.24**
Illustrative Problem 7.23.

For the air,

at inlet: $H = 20.4$ Btu/lb dry air

$W = 38.2$ grains/lb dry air (at 60°F dry-bulb and 50% relative humidity)

at outlet: $H = 52.1$ Btu/lb. dry air

$W = 194.0$ grains/lb dry air (at 90°F dry-bulb and 90% relative humidity)

Per pound of dry air, the heat interchange is, $\dot{Q}_a / \dot{m}_a = H_2 - H_2 : 52.1 - 20.4 = 31.7$ Btu per pound of dry air. Per pound of dry air, the moisture increase is $\dot{m}_w / \dot{m}_a = W_2 - W_1$: $(194.0 - 38.2)/7000 = 0.0223$ lb. per pound of dry air. From Equation 7.33,

$$\dot{m}_a (31.7) = (200000)(68.05) - \dot{m}_{w(\text{out})} (38.09)$$

From Equation 7.34,

$$\dot{m}_a (0.0223) = (200000) - \dot{m}_{w(\text{out})}$$

Solving the latter equation for $\dot{m}_{w(\text{out})}$, we have

$$\dot{m}_{w(\text{out})} = 200000 - \dot{m}_a (0.0223)$$

Substituting this into the heat balance yields

$$\dot{m}_a (31.7) = (200000)(68.05) - (200000)(38.09) + \dot{m}_a (0.0223)(38.09)$$

Solving gives us

$$\dot{m}_a (31.7 - 0.85) = (200000)(68.05 - 38.09)$$

$$= 5992000$$

$$\dot{m}_a = 194230 \text{ lb}_\text{m}/\text{h of dry air}$$

$$\text{water evaporated (lost)} = 194230 \times 0.0223$$

$$= 4331 \text{ lb}_\text{m}/\text{h}$$

Note that the water evaporated is slightly over 2% of the incoming water, and this is the makeup that has to be furnished to the tower.

Cooling towers are also utilized in cooling apartments in large high-rise buildings, such as in condominiums. Each apartment has its own air conditioner, which is tied into the building's circulating water system. The condenser coil transfers heat to the water, which is carried to the cooling tower. Replacement or cooled water is returned to each apartment's condenser. Valves are utilized in the condenser's output water line to maintain equal flows to all apartments no matter how high they are located. The system is illustrated in Figure 7.25.

**FIGURE 7.25**
Cooling tower usage in a high-rise apartment building.

## 7.9 Air Conditioning

In Sections 7.6 to 7.8, we have studied the properties of air–water vapor mixtures in some detail, the processes that these mixtures undergo, and the equipment in which the processes are carried out. The purpose of this section is to briefly study the factors that are involved in the design of air-conditioning systems that combine the elements that we have individually considered earlier in this chapter.

The term *air conditioning*, as we will use it, is meant to encompass the control of the properties of air in an enclosure or closed system such as a room or a building. If the objective is to provide for the comfort of humans within an environment, it may be necessary to effectively control the dry-bulb temperature, the wet-bulb temperature, humidity, dust content, odors, bacteria, and toxic gases. The processes by which these objectives are attained are heating, cooling, ventilation, humidification, cleaning, and dehumidification. We have already considered the elements required for the control of dry-bulb temperature, wet-bulb temperature, and humidity. The removal of odors, gases, and dust is usually accomplished by passing the air through filter elements.

Figure 7.26 shows schematically the elements of a year-round air-conditioning unit. This unit controls the basic items for both summer and winter use. In addition to control of temperature and humidity, air motion and distribution are controlled, and air purity is controlled using filters. There are many variations in the placement of the parts of such a system. Under certain conditions, depending on the condition of the entering air, one or more parts may not be used. As an example, in a moist climate, the spray would be eliminated. In general, in moderate climates, all of the elements shown would be used for all-year-round air conditioning.

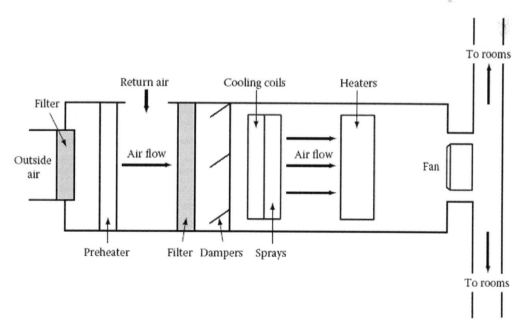

**FIGURE 7.26**
A year-round air-conditioning unit.

The control of the moisture content of air is one of the principal functions of air-conditioning systems. The addition of moisture is known as *humidification*, and the removal of moisture is known as *dehumidification*. The apparatus known as an air washer can be used to perform both processes as well as to clean the air. In essence, an air washer consists of a casing with one or more spray bands through which air flows, and at the outlet, there is an eliminator to remove any entrained moisture. The operation of the spray washer can be summarized by noting that (1) if the final water temperature is held above the entering air dry-bulb temperature, the air is both heated and humidified; (2) if the final water temperature is kept below the entering dry-bulb temperature, the air will be cooled and humidified; and (3) if the final water temperature is held below the air's dew point, the air is both cooled and dehumidified.

Year-round air conditioning typically involves cooling and dehumidifying the incoming air in the summer and heating and humidifying the air in the winter. In the case of summer, the process would appear on the psychrometric chart, as shown in Figure 7.27. Point 1 represents the state of the incoming air, which is cooled to point 2, the dew point, at a constant humidity ratio. The saturated air is further cooled to point 3 as the excess moisture in the air is condensed and removed. At this point, the air is usually too cool for comfort and so is either mixed with recirculated room air or reheated to a comfortable temperature shown as point 4. Note that the relative humidity at point 4 may be the same as that of the incoming air, but it feels more comfortable because, at this lower temperature, there is less moisture in the air.

In the case of winter, the process would appear on the psychrometric chart, as shown in Figure 7.28. The incoming air is heated from point 1 to point 2 at a constant humidity ratio. The addition of a water spray increases the humidity ratio to point 3, which can vary as shown depending on the amount and state of the added water. The result can be room air with a higher temperature and lower relative humidity (3a) or a lower temperature and a higher relative humidity (3b), so long as the human comfort zone requirements are met.

It is necessary to maintain adequate ventilation in indoor spaces for both health and comfort. In recent years, the concern for energy conservation has caused the amount of outside air entering buildings to be reduced. Many of the modern buildings have materials and furnishings that outgas volatile organic compounds (VOCs) that can be harmful to

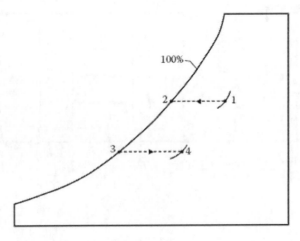

**FIGURE 7.27**
Psychrometric chart for summer.

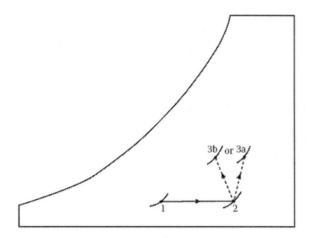

**FIGURE 7.28**
Psychrometric chart for winter.

humans. Terms such as "sick building syndrome" have been used to indicate poor indoor air quality (IAQ).

When we consider human beings and their effect on the environment when they are in a relatively confined space, we must consider the bodily processes that occur. Oxygen is inhaled, and a moist mixture less rich in oxygen and richer in carbon dioxide is exhaled into the enclosure. In addition, heat is transferred from the human, and moisture is evaporated into the environment. Body odors, smoke, and organic material are also added to the air. For a person at rest, approximately 180 Btu/h is given off as latent heat and 220 Btu/h as sensible heat. For a person working, these values increase by approximately 25%. The term *latent heat* as used in air-conditioning work refers to the changes of liquid water to water vapor in humidification or the change of water vapor to liquid water in dehumidification. *Sensible heat* refers to the heating of dry air. One method of removing the heat, moisture, and odors that exist in a room is to ventilate the room. This process consists of supplying air to, or removing air from, the room. In ordinary residences, air infiltration (through leaks, closing and opening doors and windows) usually provides sufficient ventilation.

Current ventilation standards are based upon the amount of air entering a room divided by the number of occupants in that room. The unit of ventilation is cubic feet per minute per person, or cfm/p. The ASHRAE has set standards for new construction. However, many states do not require that ventilation in older buildings be improved to meet current standards. This can present a major problem because some of the current ventilation standards are as much as four times stricter than they were in the years 1975–1989, when they were relaxed to conserve energy. Table 7.6 shows some current ventilation standards for commercial uses.

In hot weather, air-conditioning plants have to remove both sensible and latent heat (moisture). Some sources of the heat are as follows:

1. Heat released by occupants—sensible and latent
2. Infiltration of outside air—sensible and latent
3. Process heat (industrial, cooking, and so on)—sensible and latent
4. Heat infiltration from walls and partitions—sensible and latent
5. Solar heat absorption through walls, roofs, windows, and so on—sensible

**TABLE 7.6**

Some Current Ventilation Standards for Commercial Uses

| | |
|---|---|
| Smoking lounges | 60 cfm/person |
| Public restrooms | 50 cfm/person |
| Hotel rooms | 35 cfm/person |
| Offices and conference rooms | 20 cfm/person |
| Classrooms and libraries | 15 cfm/person |
| Hospital rooms | 25 cfm/person |
| Beauty salons | 25 cfm/person |
| Ballrooms, discos | 25 cfm/person |

*Source:* *ASHRAE Guide and Data Book*, Standard 62–89, 1989.

6. Heat transfer between the exterior and the interior of the building due to temperature differences—sensible

7. Air brought in for ventilation—sensible and latent

The detailed calculation of these heat loads is given in the latest edition of the *ASHRAE Guide* and will not be considered further. The references should be consulted as well as the *ASHRAE Guide* for these details. Also, Chapter 11 on heat transfer should be studied prior to performing such calculations.

Securing proper distribution of air within a large room without creating drafts is difficult. Any air movement at a rate greater than 30 ft./min in a room of seated people may seem uncomfortable to many. If the temperature and relative humidity are within comfort zone limits, somewhat higher velocities can be tolerated. Where people are moving about, velocities as high as 120 ft./min can be used. Air can be withdrawn through grilles at up to 70 ft./min without discomfort.

Over and above the precautions taken to prevent undue drafts, it is desired to get even distribution of air despite the fact that incoming air may be different in temperature from the room air. Furthermore, some means of controlling air quantity at room inlets may be desirable. Thus, a considerable variety of grilles and diffusers is available.

In the design of the air-conditioning system, the design outdoor dry-bulb and wet-bulb temperatures need to be known, and so do the dry- and wet-bulb temperature range in which humans can feel comfortable. For New York City, the design dry-bulb temperature is 95°F and the design wet-bulb temperature is 75°F. In Mobile, Alabama, the corresponding values are 95°F and 80°F. Extensive tabulations of the design dry-bulb and design wet-bulb temperatures for many cities in the United States can be found in the *ASHRAE Guide*. This document includes a recommended human comfort zone for summer and one for winter. These are shown as the shaded areas in Figure 7.29 with a slight overlap. The zones apply to adults who are sedentary or slightly active, wearing appropriate seasonal clothing and exposed to moderate air speeds. It has been assumed that persons are not exposed to direct solar radiation; these zones are used primarily for designing building interiors.

The design of any air-conditioning system is a complex interaction of many considerations, and the *ASHRAE Guide* is an invaluable source of data for any design study. The psychrometric chart may be used to analyze the requirements for an air-conditioning system. Given the design temperatures for a region and the human comfort goals, applying mass and energy balances can provide rates of heat and moisture removal required. One assumes a constant flow of dry air at atmospheric pressure into the unit. Illustrative Problem 7.24 describes the procedure.

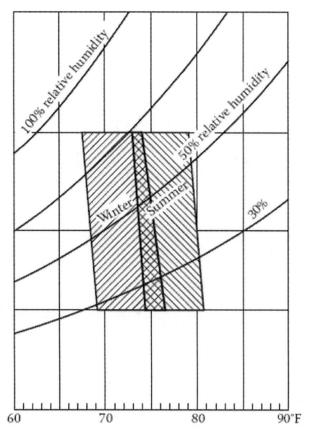

**FIGURE 7.29**
Human comfort zones. (Reprinted from Pita, E.G., *Air Conditioning Principles and Systems*, 3rd ed., © 1998. Pearson Education, Inc., Upper Saddle River, NJ. With permission.)

---

### ILLUSTRATIVE PROBLEM 7.24

An air conditioner with reheat takes in air at 95°F dry bulb and 79°F wet bulb at 350 cfm (inlet flowrate) and provides cooled and dehumidified air at 77°F and 40% relative humidity. Find the rates of heat and water vapor removed.

### SOLUTION

Using the psychrometric chart (see Figure 7.30), the properties of the entering air are $h_1 = 42.5$ Btu/lb., $v_1 = 14.37$ ft.3/lb., and $W_1 = 124$ grains/lb., or 0.0177 lb./lb. To be able to provide air at 77°F and 40% relative humidity, the air needs to be cooled until the desire humidity ratio of 56 gr/lb. is obtained. As Figure 7.30 shows, first the air is cooled sensibly until the air reaches saturation or point 2. The cooling process is continued until the desired humidity ratio is obtained by removing moisture from the air as indicated by point 3. Point 3 is determined from the required end point of 77°F and 40% relative humidity: $h_3 = 20.8$ Btu/lb. and $W_3 = 56$ grains/lb., or 0.008 lb./lb. At point 3, the air is completely saturated and the condensed moisture has been removed. The enthalpy of this moisture is $h_w$ at 51°F, found from the saturation table as 19.06 Btu/lb.

**FIGURE 7.30**
Psychrometric chart for Illustrative Problem 7.24.

The mass flow rate of the air is

$$\dot{m}_a = \frac{\dot{V}_1}{v_1} = \frac{350 \text{ ft.}^3/\text{min}}{14.37 \text{ ft.}^3/\text{lb.}} = 24.356 \text{ lb./min}$$

The rate of moisture removed is found from a mass balance for the water:

$$\dot{m}_w = \dot{m}_a (W_1 - W_3) = 24.356(0.0177 - 0.008) = 0.236 \text{ lb./min}$$

The heat removal rate is found from energy balance:

$$\dot{Q}_{\text{out}} = (\dot{m}_a h_1)_{\text{in}} - (\dot{m}_a h_3 - \dot{m}_w h_w)_{\text{out}} \rightarrow \dot{Q}_{\text{out}} = \dot{m}_a (h_1 - h_3) - \dot{m}_w (h_w)$$

$$= 24.356(42.5 - 20.8) - 0.236(19.06) = 524.02 \text{ But/min}$$

or 31441 Btu/h

Thus, a 3-ton air-conditioning system would be appropriate (see Section 10.3 for the definition of a ton of air conditioning, i.e., 1 ton = 12,000 Btu/h). Note that a practical problem would involve analysis of heat loads, not covered in this text, which would determine the cfm required.

## 7.10 Review

As a logical extension of the material that we studied in Chapter 6, we have considered that a mixture of gases can be treated as a single gas having the equation of state of an ideal gas. With this as our starting point, we derived Dalton's law for the pressure of a gas mixture and Amagat's law for the volume of a gas mixture. From Dalton's law, we obtained the following conclusions:

1. The total pressure of a mixture of ideal gases is the sum of its partial pressures.

2. The mass of the mixture is the sum of the masses of the components of the mixture.

3. The total number of moles in the mixture equals the sum of the number of moles of the component gases of the mixture.

4. The partial pressure of a component equals the mole fraction of the component multiplied by the total pressure of the mixture.

5. The molecular weight of the mixture equals the sum of the products of mole fraction of each gas component multiplied by the molecular weight of each component gas.

6. The ideal gas equation of state is applicable to all components as well as to the entire mixture.

From Amagat's law, we obtained the following:

1. The mole fraction of a component equals the volume fraction of that component in the mixture.

2. The volume of a mixture is the sum of the volumes that each component gas would occupy if each were at the pressure and temperature of the mixture.

We then applied both of these laws to the determination of the composition of a gas. A gas analysis given on a weight basis was converted to a volumetric basis, and a gas analysis on a volumetric basis was converted to a weight basis. Having these analyses, it became a logical extension to ask for the thermodynamic properties of gas mixtures. By a relatively straightforward approach, we demonstrated that the mixture properties are each equal to the sum of the mass average of the individual properties of the components of the mixture.

Air–water vapor mixtures can be treated from fundamental considerations or by use of the psychrometric chart. The psychrometric chart is very useful, because various processes can be portrayed on it. The chart given is based on the ideal gas relations and a total pressure of 29.921 in. Hg and is usable for total pressures varying ±1 in. Hg from 29.921 in. After considering many processes on the psychrometric chart and the typical equipment used, we considered cooling towers. Cooling towers are used to reject heat to the atmosphere while using a small amount of water. Consideration of the airstream and the water stream separately permitted us to write a heat balance for a typical cooling tower and to analyze its performance fully.

Our final brief discussion was on air conditioning. This topic was treated qualitatively with an example of how the psychrometric chart may be used to size an air-conditioning system; quantitative considerations, including the determination of heat loads, were not covered. After studying Chapter 11, the student will be in a better position to delve further into this subject.

---

## Key Terms

Terms used for the first time in this chapter are as follows:

**Air conditioning:** The control of the properties of air in an enclosure such as a room or a building.

**Amagat's law:** The volume of a mixture is the sum of the volumes that each constituent gas would occupy if it were at the pressure and temperature of the mixture.

**Cooling tower:** A device that uses air to cool water evaporatively.

**Dalton's law:** The pressure of a gas mixture is the sum of the partial pressures of each of its components.

**Dew-point temperature:** The temperature that a gas–vapor mixture starts condensing at constant local mixture pressure.

**Dry-bulb temperature:** The temperature of a mixture of air and water vapor as measured by an ordinary thermometer.

**Grain:** 1/7000 of a pound.

**Humidity ratio:** See *specific humidity*.

**Mole fraction:** The ratio of the number of moles of a constituent gas to the total number of moles in the mixture.

**Partial pressure:** The pressure that a constituent gas would exert if it alone occupied the total volume at the temperature of the mixture.

**Psychrometric chart:** A chart giving the thermodynamic properties of air–water vapor mixtures.

**Relative humidity:** The ratio of the partial pressure of vapor to the saturation pressure of the vapor at the same temperature in an air–water vapor mixture.

**Specific humidity:** The ratio of the weight of vapor to the weight of the gas in a unit volume of an air–water vapor mixture.

**Wet-bulb temperature:** The equilibrium temperature registered by a thermometer wetted by a wick in a stream of an air–water vapor mixture. Also the temperature at which water, by evaporating into air, can bring the air to saturation adiabatically at the same temperature.

## Equations Developed in This Chapter

| | | |
|---|---|---|
| Mixture pressure | $p_m = p_a + p_b + p_c$ | (7.1) |
| Moles of gas | $n_m = n_a + n_b + n_c$ | (7.2b) |
| Partial pressure | $\dfrac{p_a}{p_m} = \dfrac{n_a}{n_m} x_a$ | (7.4) |
| Mixture molecular weight | $MW_m = x_a MW_a + x_b MW_b + x_c MW_c$ | (7.5) |
| Mixture gas constant | $R_m = \dfrac{1545}{MW_m}$ | (7.6) |
| Amagat's law | $V_m = V_a + V_b + p_c$ | (7.8) |
| Amagat's law | $\dfrac{V_a}{V_m} = \dfrac{n_a}{n_m} x_a$ | (7.9) |

| Mixture internal energy | $u_m = \dfrac{m_a}{m_m} u_a + \dfrac{m_b}{m_m} u_b + \dfrac{m_c}{m_m} u_c$ | (7.14) |
|---|---|---|
| Mixture enthalpy | $h_m = \dfrac{m_a}{m_m} h_a + \dfrac{m_b}{m_m} h_b + \dfrac{m_c}{m_m} h_c$ | (7.15) |
| Specific heat | $c_{vm} = \dfrac{m_a}{m_m} c_{va} + \dfrac{m_b}{m_m} c_{vb} + \dfrac{m_c}{m_m} c_{vc}$ | (7.16) |
| Specific heat | $c_{pm} = \dfrac{m_a}{m_m} c_{pa} + \dfrac{m_b}{m_m} c_{pb} + \dfrac{m_c}{m_m} c_{pc}$ | (7.17) |
| Mixing entropy increase | $\Delta S = -\dfrac{n_a(1545)}{J}\ln x_a - \dfrac{n_b(1545)}{J}\ln x_b$ | Illustrative Problem 7.11 |
| Mixture gas constant | $R_m = \dfrac{1}{1+W}R_a + \dfrac{W}{1+W}R_v$ | (7.21) |
| Relative humidity | $\phi = \dfrac{p_w}{p_{w,\text{sat}}} = \dfrac{v_{w,\text{sat}}}{v_w} = \dfrac{\rho}{\rho_{w,\text{sat}}}$ | (7.22) |
| Humidity ratio | $W = 0.622\dfrac{p_w}{p_m - p_w}$ | (7.23) |
| Relative humidity | $\phi p_{w,\text{sat}} = p_w$ | (7.24) |
| Mixture specific heat | $c_{pm} = c_{p,a} + Wc_{p,w}$ | (7.25) |
| Mixture enthalpy | $h_m = h_a + Wh_w$ | (7.26) |
| Mixture entropy | $s_m = s_a + Ws_w$ | (7.28) |
| Adiabatic mixing | $\dfrac{h_a - h_{a2}}{h_a - h_{a1}} = \dfrac{W_2 - W}{W - W_1} = \dfrac{\dot{m}_{a1}}{\dot{m}_{a2}} = \dfrac{l_1}{l_2}$ | (7.32) |
| Cooling tower | $\dot{m}_a(H_2 - H_1) = \dot{m}_{w(\text{in})}h_{w(\text{in})} - \dot{m}_{w(\text{out})}h_{w(\text{out})}$ | (7.33) |
| Cooling tower | $\dot{m}_a(W_2 - W_1) = \dot{m}_{w(\text{in})} - \dot{m}_{w(\text{out})}$ | (7.34) |

## QUESTIONS

7.1 State Dalton's law and show how it leads to the statement that the total pressure of an ideal gas mixture is the sum of the partial pressures that each gas exerts.

7.2 The mole fraction of an ideal gas mixture is numerically equal to the fraction of at least three other items in the mixture. What are these items?

7.3 Would you use the mole fractions of the constituents to determine the specific heat of an ideal gas mixture?

7.4 In the determination of the enthalpy of an ideal gas mixture, is it necessary to know that the datum for each of the components is the same?

7.5 State Amagat's law in words.

7.6 The statement is made in the text that Amagat's law and Dalton's law are equivalent to each other if the gases and the mixture are ideal gases. Why is this so?

7.7 Are the gravimetric and volumetric analyses the same?

7.8     Give a reason why the entropy does not change when two identical ideal gases are mixed.

7.9     Define the dew point of an air–water vapor mixture.

7.10    Under what conditions will the wet-bulb temperature of an air–water vapor mixture equal its dry-bulb temperature?

7.11    What is meant by the term adiabatic saturation?

7.12    What is the difference between relative humidity and humidity ratio?

7.13    Is a person's comfort a function of the humidity ratio, the relative humidity, both, or neither?

7.14    Sketch an outline of a psychrometric chart. Label each of the coordinates and scales on the chart with the customary set of units.

7.15    Illustrate the typical processes on the psychrometric chart.

7.16    Using the psychrometric chart, sketch those processes that can be used for dehumidification.

7.17    Why is evaporative cooling ineffective in humid tropical areas?

7.18    When would you use a cooling tower?

7.19    How does the use of a cooling tower lessen the effect on the environment?

7.20    What is the principle on which the cooling tower is based?

7.21    What is the usual percentage of makeup in a cooling tower?

7.22    What is the difference between latent and sensible heat?

7.23    What air ventilation requirement is needed for a store?

7.24    What is the average relative humidity for the summer months where you live?

## PROBLEMS

Use Table 7.1 for the molecular weights of gases.

### Problems Involving Gas Mixtures

7.1     Show that at 14.7 psia and 32°F, a mole of any ideal gas occupies a volume of 359 ft.$^3$.

7.2     A gas mixture consists of 1 lb. of methane ($CH_4$) and 1 lb. of oxygen ($O_2$). Determine the number of moles of each gas, the total number of moles, and the mole fraction of each component. Also, determine the molecular weight of the mixture.

7.3     A gas mixture consists of 2 kg of methane ($CH_4$) and 2 kg of carbon dioxide ($CO_2$). Determine the mole fraction of each component and the molecular weight of the mixture.

7.4     A certain gas mixture consists of 25% carbon dioxide, 55% nitrogen, and 20% helium by volume. Determine the gravimetric analysis of the mixture.

7.5     A gas mixture has a volumetric analysis of 35% $CO_2$, 35% $N_2$, and 30% $O_2$. If the mixture pressure is 14.7 psia and its temperature is 60°F, determine the partial pressure of each gas.

7.6    A mixture contains 1 kg of acetylene ($C_2H_2$) and 1 kg of butane ($C_4H_{10}$). If the total pressure of the mixture is 50 MPa, determine the partial pressure of each component.

7.7    What is the molecular weight of the mixture in Problem 7.6? Also, what is the gas constant of this mixture?

7.8    A mixture consists of 0.4 lb. of $CO_2$, 0.2 lb. of $CH_4$, and 0.4 lb. of $O_2$. Determine the molecular weight and the gas constant of the mixture.

7.9    A mixture contains 1 $lb_m$ of $CO_2$ and 1 $lb_m$ of helium. If the mixture has a volume of 5 ft.$^3$ at 90°F, determine its pressure.

7.10   Determine the molecular weight and volumetric analysis of the mixture in Problem 7.9.

7.11   A mixture of 60 lb. of $O_2$, 20 lb. of $N_2$, and 10 lb. of $H_2$ is at 140°F and 17 psia. Determine the volume of the mixture.

7.12   A mixture of 20 kg of $CO_2$, 10 kg of $O_2$, and 15 kg of He is at 60°C and 0.2 MPa. Determine the volume of the mixture.

*7.13  Three pounds of carbon dioxide ($CO_2$) is mixed with 7 lb. of an unknown gas. If the resulting mixture occupies a volume of 50 ft.$^3$ when the mixture is at 50 psia and 200°F, determine the molecular weight of the unknown gas.

*7.14  Six kilograms of CO is mixed with 8 kg of an unknown gas. If the resulting mixture occupies a volume of 5 m$^3$ when the mixture is at 0.3 MPa and 150°C, determine the molecular weight of the unknown gas.

7.15   If a certain gas is CO (20% by weight), $N_2$ (50%), $O_2$ (10%), and $CH_4$ (20%), compute the partial pressure of each constituent if the total mixture pressure is 100 psia.

*7.16  A mixture contains 20 $lb_m$ of $N_2$ and the rest $CO_2$. If the volume of the mixture is 100 ft.$^3$ at 70 psia and 75°F, determine the average molecular weight of the mixture.

7.17   A gasoline engine's exhaust analysis is found to be $CO_2$ (10% by volume), CO (0.5%), $N_2$ (73%), $H_2O$ (14.5%), and $O_2$ (2%). Determine the gravimetric analysis and the molecular weight of this mixture.

7.18   The volumetric analysis of a gas is $CH_4$ (86%), $CO_2$ (1%), $N_2$ (1%), and $H_2$ (12%). Determine the gravimetric analysis, the molecular weight, and the gas constant for the mixture.

*7.19  The density of a mixture of $CO_2$ and $N_2$ is 0.08 $lb_m$/ft.$^3$ at 14.7 psia and 70°F. Determine the mass of $CO_2$ present in 1 $lb_m$ of the mixture.

7.20   A mixture contains 10 lb. of methane ($CH_4$) and 6 lb. of oxygen ($O_2$). If the mixture is at 140°F and 75 psia, determine the mixture volume, the partial volume of each constituent, the partial pressure of each constituent, the mole fraction of each constituent, and the molecular weight of the mixture.

7.21   A diver breathes a gas mixture containing 0.02 kg $O_2$ and 1 kg He. The mixture has a pressure of 1000 kPa absolute and temperature of 15°C. Determine the mixture volume and the partial pressure of oxygen.

7.22   If a tank has a volume of 2 m$^3$ and contains 1 kg of CO and 2 kg of He, compute the partial pressure of each constituent and the total mixture pressure when the temperature is 30°C.

7.23   It is desired to produce a gas mixture containing 50% by volume of $CO_2$ and $O_2$. If this mixture contains 4 lb. of $CO_2$, how many pounds of $O_2$ are there in this mixture?

7.24   Determine the molecular weight and the gas constant of the mixture in Problem 7.23.

7.25   It is desired to produce a gas mixture containing 40% by volume of CO and 60% by volume of $O_2$. If there are 4 kg of CO in the mixture, how many kilograms of $O_2$ are there in this mixture?

7.26   What is the molecular weight and gas constant of the mixture in Problem 7.25?

*7.27   Two hundred cubic feet of hydrogen at 60 psia and room temperature is adiabatically mixed with 400 ft.$^3$ of nitrogen at 15 psia and room temperature. If the temperature of the room is 70°F, determine the molecular weight, $R$, and the final pressure of the mixture.

*7.28   If 10 m$^3$ of hydrogen at 0.4 MPa and room temperature is adiabatically mixed with 20 m$^3$ of nitrogen at 0.1 MPa and room temperature, determine the molecular weight, $R$, and the final pressure of the mixture. Room temperature is 20°C.

7.29   A mixture consists of seven moles of oxygen ($O_2$), three moles of methane ($CH_4$), and 1 mole of nitrogen ($N_2$). Determine the volumetric analysis, the weight of the mixture, the molecular weight of the mixture, and the gas constant of the mixture.

7.30   A gas mixture consists of the following volume percentages: 40% CO, 20% $CH_4$, 30% $N_2$, and 10% $O_2$. Determine the weight fraction of each gas, the molecular weight of the mixture, and the gas constant of the mixture.

7.31   A gas mixture has a volumetric analysis of 25% $N_2$, 35% $O_2$, and 40% $CH_4$. Determine the gravimetric analysis, the molecular weight of the mixture, and the gas constant of the mixture.

7.32   The volumetric analysis of a gas mixture is 40% $N_2$, 40% $O_2$, and 20% $CO_2$. The mixture is at 100 psia and 100°F. Determine the partial pressure of each gas, the molecular weight of the mixture, the gravimetric (weight) analysis, and the gas constant of the mixture.

7.33   A gas mixture of oxygen, nitrogen, and methane occupies a container, and each constituent has a partial pressure of 50, 20, and 65 psia, respectively. Determine the gravimetric analysis, the volumetric analysis, the molecular weight of the mixture, and the gas constant of the mixture.

*7.34   Determine the final temperature and pressure of a mixture of oxygen and nitrogen if 6 lb. of oxygen at 70°F and 200 psia is mixed with 1 lb. of nitrogen at 200°F and 100 psia. Use $c_v$ of oxygen as 0.164 Btu/lb$_m$ R and $c_v$ of nitrogen as 1.79 Btu/lb$_m$ R. The process is adiabatic.

7.35   Determine $c_v$ of the mixture in Problem 7.34.

*7.36   Determine the final temperature and pressure of a mixture of $CO_2$ and nitrogen if 3 kg of $CO_2$ at 20°C and 1.5 MPa is mixed with 1 kg of nitrogen at 100°C and 0.75 MPa. Use $c_v$ of $CO_2$ as 0.656 kJ/kg K and $c_v$ of nitrogen as 0.752 kJ/kg K. The process can be considered to be adiabatic.

7.37   Compute $c_v$ of the mixture in Problem 7.36.

*7.38 Determine the final temperature and pressure of a mixture if it is made by mixing 10 ft.$^3$ of nitrogen at 120°F and 70 psia and 5 ft.$^3$ of oxygen at 60°F and 120 psia. Use $c_v$ of oxygen as 0.164 Btu/lb$_m$ R and $c_v$ of nitrogen as 0.179 Btu/lb$_m$ R. The process is adiabatic.

7.39 Determine $c_v$ of the mixture in Problem 7.38.

7.40 One pound of nitrogen at 75°F and 14.7 psia is adiabatically mixed with 1 lb. of helium at 80°F and 250 psia. Determine the final mixture pressure and temperature. For nitrogen, $c_v = 0.177$ Btu/lb$_m$ R, and for helium, $c_v = 0.754$ Btu/lb$_m$ R.

7.41 Determine the specific heat at constant volume for the mixture in Problem 7.40.

## Problems Involving Air–Water Vapor Mixtures

7.42 An air–water vapor mixture at 100°F has a relative humidity of 40%. Calculate the partial pressure of the water vapor and its dew-point temperature.

7.43 Solve Problem 7.42 using the psychrometric chart.

7.44 A room contains air at 70°F with a humidity ratio of 0.008 lb$_m$ water vapor/lb$_m$ dry air. Determine the relative humidity and the dew point of the air.

7.45 Solve Problem 7.44 using the psychrometric chart.

7.46 An air–water vapor mixture at a dry-bulb temperature of 90°F and 14.7 psia is found to have a relative humidity of 50%. Calculate the dew point, the partial pressure of the water vapor, and the humidity ratio.

7.47 Solve Problem 7.46 using the psychrometric chart.

7.48 The air in a room is at 70°F. A bare pipe running through the room has a surface temperature of 55°F. At what relative humidity will condensation occur on the surface of the pipe?

7.49 Solve Problem 7.48 using the psychrometric chart.

*7.50 An air–water vapor mixture having a total pressure of 14.7 psia is found to have a humidity ratio of 0.010 lb. of water vapor per pound of dry air. Calculate the relative humidity, the dew point, and the pressure of the water vapor if the dry-bulb temperature is 80°F.

7.51 Solve Problem 7.50 using the psychrometric chart.

7.52 The partial pressure of water vapor in an atmospheric pressure (14.7 psia) mixture is 0.4 psia. Calculate the relative humidity and the dew-point temperature if the dry-bulb temperature is 100°F.

7.53 Solve Problem 7.52 using the psychrometric chart.

7.54 Moist air at 14.7 psia has a dry-bulb temperature of 80°F and a wet-bulb temperature of 60°F. Determine the relative humidity, the enthalpy, and the specific humidity of the air using the psychrometric chart.

7.55 Moist air at 14.7 psia has a wet-bulb temperature of 68°F and a dew point of 50°F. Determine its dry-bulb temperature and relative humidity using the psychrometric chart.

7.56 Moist air at 14.7 psia has a dew-point temperature of 50°F and a dry-bulb temperature of 86°F. Determine the relative humidity and the wet-bulb temperature of the air using the psychrometric chart.

*7.57    A closed tank has a volume of 200 ft.$^3$. The tank contains atmospheric air at 70°F with a humidity ratio of 0.01 lb$_m$ water vapor/lb$_m$ air. Determine the mass of water that must be removed to obtain a relative humidity of 30%.

*7.58    Solve Problem 7.57 for a total pressure of 12 psia.

*7.59    Solve Problem 7.57 for a total pressure of 10 psia.

7.60    Atmospheric air at 14.7 psia and 80°F has a relative humidity of 30%. Calculate its dew point and its specific humidity.

7.61    Solve Problem 7.60 using the psychrometric chart.

*7.62    Atmospheric air is sensibly cooled at constant pressure from an initial dry-bulb temperature of 80°F and 50% relative humidity to a final dry-bulb temperature of 60°F. Calculate the heat removed per pound of dry air. Assume that the mixture is at 14.7 psia. Use Equation 7.27 for the enthalpy of the mixture.

7.63    Solve Problem 7.62 using the psychrometric chart.

*7.64    Air is dehumidified at constant dry-bulb temperature. Calculate the heat removed per pound of dry air if the initial conditions are 90°F dry-bulb temperature and 40% relative humidity, and the final relative humidity is 20%. The mixture is at 14.7 psia.

7.65    Solve Problem 7.64 using the psychrometric chart.

7.66    Air is cooled adiabatically from 70°F dry bulb and 57% relative humidity until 90% relative humidity is reached. Calculate the final dry-bulb temperature and the initial and final dew-point temperatures.

7.67    Solve Problem 7.66 using the psychrometric chart.

7.68    Atmospheric air at 10 psia and 70°F dry bulb has a relative humidity of 50%. If the air is sensibly heated at constant pressure to 80°F dry bulb, what is the final relative humidity?

7.69    Even though the psychrometric chart is not applicable at this pressure, determine the final relative humidity in Problem 7.68 using the chart and compare results.

7.70    Determine the humidity ratio and dew-point temperature for air at a total pressure of 30.921 in. Hg if it is at 80°F dry-bulb temperature and 60% relative humidity.

7.71    Solve Problem 7.70 using the psychrometric chart. Comment.

7.72    To condition it, air is passed over a coil at 50°F. If the initial air is at 85°F dry bulb and 70% relative humidity and a final 75°F dry-bulb air is specified, determine the final relative humidity, the amount of moisture removed per pound of dry air, and the amount of heat extracted per pound of dry air. Use the psychrometric chart.

*7.73    Determine whether it is possible to obtain air at a dry bulb of 60°F and 70% relative humidity by evaporative cooling and then sensible heating if the initial air is at 70°F dry bulb and 10% relative humidity.

7.74    In Problem 7.73, what is the least value of the initial relative humidity that will permit the process to be completed?

7.75    What final conditions will be obtained when 200 lb$_m$/min of air at 85°F and 60% relative humidity is mixed with 300 lb$_m$/min of air at 50°F and 20% relative humidity? (Temperatures are dry-bulb values.) Use the psychrometric chart.

*7.76   If the flow quantities in Problem 7.75 were given in cubic feet per minute, that is, 200 ft.$^3$/min and 300 ft.$^3$/min, determine the final mixture conditions.

*7.77   A cooling tower operates as follows: tower heat-transfer rate = 3,000,000 Btu/h; air in = 70°F, 30% relative humidity; air out = 93°F, 90% relative humidity; water in = 50,000 lb$_m$/h at 100°F. Determine the air flow rate needed and the makeup water needed, both in lb$_m$/h.

*7.78   Water is cooled in a cooling tower from 110°F to 80°F. Air enters the tower at 80°F and a relative humidity of 40%, and it leaves at 95°F with a relative humidity of 90%. Determine the water entering that is cooled per pound of dry air and the makeup water required per pound of dry air.

*7.79   Air flows through a cooling tower to cool 30,000 lb$_m$/h of water from 100°F to 80°F. Air enters the tower at 70°F with a relative humidity of 30% and leaves at 95°F with a relative humidity of 90%. Determine the makeup water required and the amount of air required.

*7.80   A cooling tower operates with 368 lb$_m$/s of 100°F water entering. The water is cooled to 70°F. Air enters at 60°F with 50% relative humidity and leaves at 90°F and 98% relative humidity. Determine the makeup water requirements and the weight of air required.

*7.81   A cooling tower cools 2400 lb$_m$/min of water from 85°F to 65°F. The air used for cooling flows at the rate of 4000 lb$_m$/min. The air enters the tower at a dry-bulb temperature of 65°F and a wet-bulb temperature of 55°F. If the air leaves the tower with 100% relative humidity, determine the temperature of the air leaving and the amount of makeup water.

# 8

## Vapor Power Cycles and Alternative Power Systems

### Learning Goals

After reading and studying the material in this chapter, you should be able to

1. Understand the definition of the term *cycle*, and differentiate between gas and vapor cycles.
2. Recall that our conclusions regarding the Carnot cycle were independent of the working fluid used in the cycle.
3. Sketch and analyze the elements of the simple Rankine cycle.
4. Conclude that for the same pressures, the efficiency of the Rankine cycle is less than that of a Carnot cycle.
5. Define the type efficiency as the ratio of the ideal thermal efficiency of a given cycle divided by the efficiency of a Carnot cycle operating between the same maximum and minimum temperature limits.
6. Sketch the simple reheat cycle elements and $T$–$s$ and $h$–$s$ diagrams for this cycle.
7. Conclude that reheat does not greatly increase the efficiency of the Rankine cycle but does decrease the moisture content of the steam in the later stages of the turbine.
8. Understand that regeneration is a method of heating the feedwater with steam that has already done some work and that in the limit, regenerative cycle approaches the efficiency of the Carnot cycle operating between the same temperature limits.
9. Apply regeneration to the Rankine cycle, sketch the $T$–$s$ diagram, and show how regeneration improves the Rankine cycle efficiency.
10. Qualitatively understand the construction and operation of the major pieces of equipment used in commercial steam-generating plants.
11. Understand other methods that have been proposed and/or constructed for the production of power, including direct energy devices.
12. Understand the use of cogeneration as a combination of power production and process or space use.

## 8.1 Introduction

A cycle has been defined as a series of thermodynamic processes during which the working fluid can be made to undergo changes involving energy transitions and subsequently is returned to its original state. The objective of any practical cycle is to convert energy from one form to another, more useful form. For instance, the energy bound in a fossil fuel is released by the chemical process of combustion and, by undergoing appropriate thermodynamic processes, is made to yield useful work at the shaft of an engine. Similarly, the energy of the nuclear-fission process is made to yield useful work that ultimately appears as electrical energy.

   In the following, several ideal cycles are discussed. They are "ideal" in the sense that they have been proposed as prototypes of practical cycles, and in the limit, their efficiencies approach, but are always less than the Carnot cycle efficiency. These cycles usually bear the name of the person who either proposed them or developed them, and their composite study represents a large portion of applied thermodynamics. In practice, actual cycles deviate from the ideal because of unavoidable irreversibilities and for other practical reasons. The study of the ideal cycle, however, can and does yield invaluable results that are applicable to real cycles. Additionally, we discuss the actual devices (hardware) used in the practical realization of these ideal cycles. Along with these descriptions, we see where the actual cycles deviate from the "ideal" cycles and the attempts made to make actual cycles approach ideal performance.

   Power cycles are often classified by the character of the working fluid in the cycle. The two general classes of cycle are the vapor cycle and the gas cycle. The vapor cycle differs from the gas cycle in two respects. In the gas cycle, there is no change of phase of the working substance, and the compression work of the gas cycles can therefore represent a large percentage of the useful work output of the cycle. In the vapor cycle, the working substance is condensed to a liquid at the lower temperature of the cycle. This liquid is pumped to the desired delivery pressure. Because the liquid is essentially incompressible, it would be expected that the pump work of the vapor cycle would represent a small percentage of the useful work output of the cycle. Also, in the vapor cycle, the working substance may contain moisture when it is expanded in a turbine. Because this is undesirable, modifications are made in the vapor cycle to alleviate this condition.

## 8.2 Carnot Cycle

The conclusions reached in the study of the Carnot cycle were independent of the working medium, and it is pertinent to review this cycle briefly, as applied to both vapors and gases. It will be recalled that a Carnot cycle consists of two reversible isothermal stages and two isentropic stages. Figure 8.1a shows a Carnot cycle in which the working fluid is indicated as being in the wet vapor region. This is because the Carnot cycle requires constant temperature heat addition and rejection, which can be achieved by adding heat during the saturated (wet) mixture phase, causing evaporation, and removing heat during a wet mixture phase, causing condensation. Both of these processes occur at a constant temperature. The line *A, B* represents the isentropic compression of the fluid from the lower temperature (and pressure) to the upper temperature of the cycle. At the upper temperature, there is an isothermal and reversible reception of energy as heat from some reservoir.

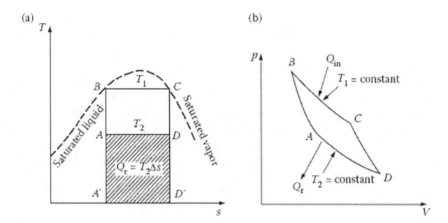

**FIGURE 8.1**
Carnot cycle. (a) Vapor cycle. (b) Gas cycle.

This transfer of the heat proceeds along path $B, C$ with a concurrent increase in the volume of the vapor. The fluid then expands isentropically along path $C, D$ with energy extracted as work. As the fluid expands, the volume of the fluid increases. Finally, the wet fluid is condensed isothermally and reversibly at the lower temperature of the cycle. During the condensation process, the volume of the fluid decreases as heat is rejected to the sink of the system. The energy available is represented by area $A, B, C, D$, and the energy rejected is represented by area $A', A, D, D'$ on the $T$–$s$ diagram in Figure 8.1a. The efficiency of a Carnot cycle is solely a function of the upper and lower temperatures and is given by

$$\eta_c = \frac{T_1 - T_2}{T_1} \times 100 \tag{8.1}$$

Equation 8.1 is also the efficiency of any reversible heat engine operating between these constant temperature limits. All the heat is taken in or rejected at $T_1$ and $T_2$, respectively. Figure 8.1b shows the Carnot gas cycle plotted on $pV$ coordinates. All the conclusions pertaining to the efficiency of the vapor cycle also pertain to the gas cycle, because we have already noted that the efficiency of a Carnot cycle is independent of the working fluid in the cycle.

Unfortunately, there are certain practical limitations on the upper and lower temperatures. The upper temperature is limited by the strength of available materials and the lower by ambient conditions. Recent advances in the field of thermonuclear reactions have indicated that the upper-temperature limitation may be removed by using magnetic fields to contain the working fluid. To date, the limits on the upper temperature of power cycles have depended on advances in the field of metallurgy.

## 8.3 The Rankine Cycle

The Carnot cycle described in Section 8.2 cannot be used in a practical device for many reasons. Fluid expansion ($C$ to $D$) would require the handling of relatively wet steam, which is detrimental to modern equipment such as turbines. Further, fluid compression (*A* to *B*) would also involve wet steam, which requires more work than that needed to compress a liquid. Thus, point *C* should be stretched out into the superheat region so expansion involves only the gas phase, and point *A* should be extended to the saturated liquid point so compression involves only a liquid. Historically, the prototype of actual vapor cycles was the simple Rankine cycle. The elements of this cycle are shown in Figure 8.2; the *T–s* and *h–s* diagrams for the ideal *Rankine cycle* are illustrated in Figure 8.3.

As indicated in the schematic of Figure 8.2, this cycle consists of four distinct processes. Starting with the feed pump, the liquid supplied to the boiler is first brought to the boiler pressure. In the ideal cycle, the liquid supplied to the pump is assumed to be saturated at the lowest pressure of the cycle. In an actual cycle, the liquid is usually slightly subcooled to prevent vapor bubbles from forming in the pump (which causes a process known as *cavitation*, which will subsequently damage the pump). For the ideal cycle, the compression process

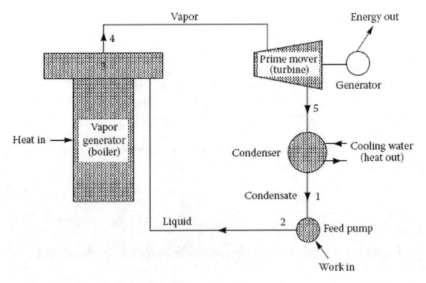

**FIGURE 8.2**
Elements of the simple Rankine cycle.

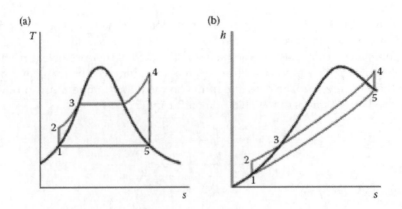

**FIGURE 8.3**
Rankine cycle. (a) *T–s* diagram for the Rankine cycle. (b) *h–s* diagram for the Rankine cycle.

is taken to be isentropic, and the final state of the liquid supplied to the boiler is subcooled (compressed liquid) at the boiler pressure. This subcooled liquid is heated to saturation in the boiler, and it is subsequently vaporized to yield the steam for the prime mover in the cycle (i.e., steam turbine). The energy for the heating and vaporizing of the liquid is provided by the combustion of a fuel sources in the boiler or using an external energy to generate superheated steam. If superheating of the vapor is desired, it is also accomplished in the boiler (also called a steam generator). The vapor leaves the steam generator and is expanded isentropically in a prime mover (e.g., a turbine or steam engine) to provide the work output of the cycle. After the expansion process is completed, the working substance flows to the condenser, where it rejects heat to the cooling water or cooling medium.

Let us study the simple Rankine cycle shown in Figures 8.2 and 8.3 and consider the processes that occur.

### 8.3.1 Process 1–2

For steady-flow isentropic pumping of the feedwater, it is assumed that the inlet condition to the pump corresponds to saturation of the lowest pressure of the cycle. The compression is assumed to be isentropic, and differences in potential and kinetic energies as inlet and outlet to the pump are negligible. For this ideal process, the work in is

$$w = \left(h_2 - h_1\right)_s \text{ Btu/lb}_m \tag{8.2}$$

or

$$w = \left(h_2 - h_1\right)_s \text{ kJ/kg} \tag{8.2a}$$

where the subscript s denotes an isentropic compression (i.e., entropy remains constant: $s_1 = s_2$) Because the water is essentially incompressible, and the specific volume can be assumed to be constant, it is possible to approximate Equation 8.2 as

$$w = \left(p_2 - p_1\right)v_f / J \text{ Btu/lb}_m \tag{8.3}$$

or

$$w = \left(p_2 - p_1\right)v_f \text{ kJ/kg} \tag{8.3a}$$

where $p_2$ is the high pressure of the cycle, $p_1$ is the low pressure of the cycle (both in $\text{lb}_f/\text{ft.}^2$ or kPa), and $v_f$ is the specific volume of saturated liquid in $\text{ft.}^3/\text{lb}_m$ or $\text{m}^3/\text{kg}$ at the inlet pressure of the pump.

### 8.3.2 Process 2–3

This process is the heating of the subcooled water to saturation by steam in the drum. We will consider that this occurs as part of the boiler process (2–4) and not evaluate it separately. However, it is considered as part of the boiler design by the designer.

### 8.3.3 Process 2–4

The liquid leaves the pump at the delivery pressure to the boiler. In the ideal cycle, pressure losses in the pipelines are negligible. Thus, the process of heating in the boiler and subsequent vaporization (and superheat) can be considered to be a steady-flow process carried out at constant pressure. For this process, neglecting differences in potential and kinetic energies at the inlet and the outlet of the boiler,

$$q = (h_4 - h_2) \text{ Btu/lb}_m \text{ or kJ/kg} \tag{8.4}$$

It should be noted that $q$ is the energy required by the working fluid, not the energy released in the steam generator. Also, devices such as attemperators or desuperheaters, which are used to control the final steam temperature leaving the superheater, have to be accounted for in an energy balance of a steam generator.

### 8.3.4 Process 4–5

After leaving the boiler, the high-pressure steam flows to the turbine. Once again, friction losses will be neglected, as will differences in potential and kinetic energies at the inlet and the outlet of the turbine. For the turbine, assuming an isentropic expansion (entropy remains constant: $s_4 = s_5$)

$$w = (h_4 - h_5)_s \text{ Btu/lb}_m \text{ or kJ/kg} \tag{8.5}$$

The presence of moisture in the turbine can lead to mechanical difficulties such as excessive erosion of the turbine blades. This is highly undesirable, and to alleviate this condition, superheating of the steam is often used. This increases the efficiency of the unit (as deduced from the Carnot cycle) and decreases the amount of moisture in the turbine exhaust (see Figure 8.3).

Figure 8.4 illustrates an expansion of steam from 500 psia to a final pressure of 1 1/2 in. Hg. On $h$–$s$ coordinates, this ideal expansion is a vertical line. If the expansion is not isentropic, there is an increase in entropy ($s_5 > s_4$), and the final state point must lie on the 1 1/2 in. Hg line at an increased value of $s$. As shown in Figure 8.4, the final enthalpy is required to be higher for the nonisentropic expansion than for the isentropic case ($h_{5,a} > h_{5,s}$, where $a$ and $s$ stand for actual and isentropic, respectively). In other words, the change in enthalpy for the isentropic expansion ($\Delta h_s = h_4 - h_{5,s}$) is greater than the change in enthalpy for the nonisentropic between ($\Delta h_a = h_4 - h_{5,a}$) the same pressure limits. The nonisentropic expansion has been shown as a dashed line. Actually, this line is not the path of the expansion, because the expansion is irreversible and nonequilibrium states cannot be drawn on an $h$–$s$ diagram. The difference in enthalpy ($h_{5'} - h_5$ or $h_{5,a} - h_{5,s}$) is called *reheat*, because all the mechanical irreversibilities increase the final enthalpy, or "reheat" the fluid. To account for the nonideal effects in the turbine, the internal turbine efficiency or isentropic efficiency is introduced as the ratio of the work actually done ($w_{t,a}$) to the work that could be done ideally ($w_{t,s}$). Students should note that this procedure corresponds to the method used to define nozzle efficiencies in Chapter 6. Therefore, the internal turbine efficiency or isentropic efficiency, $\eta_t$ is

$$\eta_t = \frac{\text{actual change in enthalpy}}{\text{isentropic change in enthalpy}} = \frac{h_4 - h_{5,a}}{h_4 - h_{5,s}} \tag{8.5a}$$

Where $h_{5,s}$ is evaluated at $p_5$ and $s_5 = s_4$, while $h_{5,a}$ comes from the actual ($a$) conditions at the outlet of the turbine. Note that point 5 can be saturated vapor (SV), wet steam (mixture or mx), or superheated steam (SHV), depending on the location of point 4.

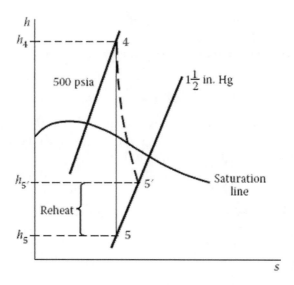

**FIGURE 8.4**
Rankine turbine expansion.

### 8.3.5 Process 5–1

After leaving the turbine, the fluid (which is usually wet steam) is piped to the condenser. Because the specific volume is high and we wish to keep the pressure losses in pipelines and equipment small, the piping and condenser are usually quite large at this part of the cycle. If pressure losses, kinetic terms, and potential terms are negligible and it is assumed that the equipment is well insulated, the heat rejected to the cooling water is

$$q_r = (h_5 - h_1) \text{ Btu/lb}_m \text{ or kJ/kg} \tag{8.6}$$

This energy is rejected to the flowing cooling water and raises the temperature of the cooling water. In the ideal cycle, the condensed steam is assumed to be saturated (i.e., saturated liquid) and is returned to the inlet of the circulating pump with no pressure drop in the piping. Actually, as we have noted, the condensate is usually slightly subcooled before leaving the condenser.

By definition, the thermal efficiency of any cycle is the useful (net) work out of the cycle divided by the thermal energy supplied (input) to the cycle. Therefore, the efficiency of the ideal Rankine cycle is

$$\eta_R = \frac{\text{turbine work out} - \text{pump work}}{\text{heat supplied in boiler}}$$

$$\eta_R = \frac{(w_t)_s - (w_p)_s}{q_{in}} \tag{8.7}$$

where $w_p = (h_{2,s} - h_1) = v_1(p_2 - p_1)$; $w_{t,s} = h_4 - h_{5,s}$; and $q_{in} = h_4 - h_2$. Then, the efficiency of the ideal Rankine cycle becomes

$$\eta_R = \frac{(h_4 - h_{5,s}) - v_1(p_2 - p_1)}{h_4 - h_{2,s}} \tag{8.8}$$

The engine that operates on the expansion of steam between the boiler and the condenser in the Rankine cycle is called the *Rankine engine*. If the expansion is isentropic, the maximum amount of work obtainable from the expansion is $(h_4 - h_{5,s})$.

---

### ILLUSTRATIVE PROBLEM 8.1

A Rankine cycle operates with its upper temperature at 400°C and its upper pressure 3 MPa. If its lower pressure is 50 kPa, determine the thermal efficiency of the cycle (a) neglecting pump work and (b) including pump work.

### SOLUTION

From the *Steam Tables* or Mollier chart in Appendix 2, and assuming saturated liquid water enters the pump, we find that

$$h_1 = h_f \text{ at } 50 \text{ kPa} = 340.49 \text{ kJ/kg}; \quad v_1 = v_f \text{ at } 50 \text{ kPa} = 0.001030 \text{ m}^3/\text{kg}$$

At $p_4 = 3$ MPa and $T_4 = 400°C \rightarrow h_4 = 3230.9$ kJ/kg; $s_4 = 6.9212$ kJ/kg·K

By taking into account the outlet pressure of the turbine, $p_5 = 50$ kPa, and assuming that the expansion process across the steam turbines is isentropic ($s_4 = s_5 = 6.9212$ kJ/kg·K), $h_5$ can be found by using the *Steam Tables* (Appendix 2). First, the phase of water at the turbine outlet should be identified by comparing the value of $s_g$ at 50 kPa with $s_5$, as follows

$$\text{At } p_5 = 50 \text{ kPa} \rightarrow s_g = 7.5939 \text{ kJ/kg·K}$$

Since $s_g > s_5$, then it can be concluded that the steam coming out of the turbine is wet (i.e., it's a mixture); therefore, the value of $x_5$ (quality) should be found to be able to find $h_5$, as follows

$$s_5 = s_f + x_5 s_{fg} \rightarrow x_5 = \frac{s_5 - s_f}{s_{fg}}$$

By obtaining $s_f$ and $s_{fg}$ from Appendix 2, $x_5$ is

$$x_5 = \frac{6.9212 - 1.0910}{6.5029} = 0.8966$$

By obtaining $h_f$ and $h_{fg}$ from Appendix 2, then, $h_5$ can be found as follows

$$h_5 = h_f + x_5 h_{fg} \rightarrow h_5 = 340.49 + (0.8966)(2305.4) = 2407.4 \text{ kJ/kg}$$

Here, point 5 is in the wet steam region (i.e., it is a mixture).

    a.    Neglecting pump work ($h_2 = h_1$) gives

$$\eta_R = \frac{w_t}{q_{in}} = \frac{(h_4 - h_5)}{(h_4 - h_2)} = \frac{3230.9 - 2407.4}{3230.9 - 340.49} \times 100 = 28.49\%$$

b. Taking into account pump work, $w_p = (p_2 - p_1)v_f = (3000 - 50)(0.001030) = 3.04$ kJ/kg, and finding $h_2$, $w_p = h_2 - h_1 \rightarrow h_2 = w_p + h_1 = 3.04 + 340.49 = 343.53$ kJ/kg

The efficiency of the cycle including pump work is

$$\eta_R = \frac{w_t - w_p}{q_{in}} = \frac{(h_4 - h_5) - w_p}{(h_4 - h_2)}$$

$$= \frac{(3230.9 - 2407.4) - 3.04}{3230.9 - 343.43} \times 100 = 28.42\%$$

It is assumed that there is no energy loss between the pump outlet and the boiler inlet.

---

### ILLUSTRATIVE PROBLEM 8.2

Solve Illustrative Problem 8.1 using the computer disk to obtain the necessary properties.

**SOLUTION**

a. For the conditions given in Illustrative Problem 8.1, the properties are found to be

$$h_2 = 340.54 \text{ kJ/kg} \quad h_4 = 3230.7 \text{ kJ/kg} \quad h_5 = 2407.2 \text{ kJ/kg}$$

$$\eta_R = \frac{3230.7 - 2407.2}{3230.7 - 340.54} \times 100 = 28.49\%$$

b. For the pump work, we do not need the approximation, because the computerized tables give us the necessary values directly. Assuming that the condensate leaving the condenser is saturated liquid, this gives us an enthalpy of 340.54 kJ/kg and an entropy of 1.0912 kJ/kgK. For an isentropic compression, the final condition is the boiler pressure of 3 MPa and an entropy of 1.0912 kJ/kgK. For these values, the program yields an enthalpy of 343.59 kJ/kgK. The isentropic pump work is equal to $(343.59 - 340.54) = 3.05$ kJ/kg, which is used to find $h_2$ as follows: $w_p = h_2 - h_1 \rightarrow h_2 = w_p + h_1 = 3.05 + 340.54 = 343.59$ kJ/kg

The efficiency of the cycle including pump work is

$$\eta_R = \frac{w_t - w_p}{q_{in}} = \frac{(h_4 - h_5) - w_p}{(h_4 - h_2)}$$

$$= \frac{3230.7 - 2407.2 - 3.05}{3230.7 - 343.59} \times 100 = 28.42\%$$

Although individual values vary slightly, the final results in this problem agree with the results in Illustrative Problem 8.1.

### ILLUSTRATIVE PROBLEM 8.3

The maximum and minimum pressures in a Rankine cycle are 400 and 14.696 psia, respectively. The steam is exhausted from the engine as saturated vapor. (a) Sketch the cycle on temperature–entropy and enthalpy–entropy diagrams. (b) Find the thermal efficiency of the cycle.

### SOLUTION

a.  Figure 8.3 with the cycle extending into the superheat region and expanding along 4→5 is the appropriate diagram for this process.

b.  This problem can be solved either by use of the Mollier chart or the *Steam Tables*. If the chart is used, 14.696 psia is first located on the saturated vapor line. Because the expansion, 4→5, is isentropic, a vertical line on the chart is the path of the process. The point corresponding to 4 in Figure 8.3 is found where this vertical line intersects 400 psia. At this point, the enthalpy is 1515 Btu/lb$_m$, and the corresponding temperature is approximately 980°F. Saturated vapor at 14.696 psia has an enthalpy of 1150.5 Btu/lb$_m$ (from the Mollier chart). The *Steam Tables* (Appendix 2) show that saturated liquid at 14.696 psia has an enthalpy of 180.15 Btu/lb$_m$. In terms of Figure 8.3, and neglecting the pump work, we have

$$h_1 = h_2 = 180.15 \text{ Btu/lb}_m \quad h_4 = 1515 \text{ Btu/lb}_m \quad h_5 = 1150.5 \text{ Btu/lb}_m$$

Neglecting the pump work yields

$$\eta_R = \frac{w_t}{q_{in}} = \frac{h_4 - h_5}{h_4 - h_2} = \frac{1515 - 1150.5}{1515 - 180.15} \times 100 = 27.3\%$$

The pump work is given approximately by Equation 8.3 as

$$\frac{(p_2 - p_1)v_f}{J} = \frac{(400 - 14.696)144}{778} \times 0.0167$$

$$= 1.19 \text{ Btu/lb}_m$$

After finding $h_2$ ($h_2 = w_p + h_1 = 1.19 + 180.15 = 181.34$), the efficiency of the cycle including pump work is

$$\eta_R = \frac{w_t - w_p}{q_{in}} = \frac{(1515 - 1150.5) - (1.19)}{1515 - 181.34} \times 100 = 27.2\%$$

Neglecting the pump work is obviously justified in this case since the difference in efficiency is quite small. An alternative solution is obtained by using the *Steam Tables*: at 14.696 psia and saturation, $s_g = 1.7567$; therefore, at 400 psia, $s = 1.7567$. From Table A.3 (at 400 psia),

| $s$ | $h$ | $T$ |
|---|---|---|
| 1.7632 | 1523.6 | 1000 |
| 1.75367 | 1514.2 | 982.4 |
| 1.7558 | 1512.9 | 980 |

The remainder of the problem proceeds along similar lines with essentially the same results. The use of the Mollier chart facilitates the solution of this problem.

---

### ILLUSTRATIVE PROBLEM 8.4

Solve Illustrative Problem 8.3 using the computer-generated property values.

### SOLUTION

Refer to Figure 8.3. The desired quantities are obtained as follows:

at 14.696 psia, saturated vapor ($x=1$), $s=1.7566$ Btu/lb$_m$·R, $h=1150.4$ Btu/lb$_m$

at 14.696 psia, saturated liquid ($x=0$), $s=0.3122$ Btu/lb$_m$·R, $h=180.17$ Btu/lb$_m$

at 400 psia, $s=1.7566$ Btu/lb$_m$·R, $h=1514.0$ Btu/lb$_m$, $T=982.07°F$

at 400 psia, $s=0.3122$ Btu/lb$_m$·R, $h=181.39$ Btu/lb$_m$

Note the agreement of these values with the ones obtained for Illustrative Problem 8.3. Also, note the temperature of 982.07°F compared to 982.4°F. Continuing,

$$\eta_R = \frac{w_t}{q_{in}} = \frac{1514.0 - 1150.4}{1514.0 - 180.17} \times 100 = 27.26\%$$

The isentropic pump work: $w_p = (181.39 - 180.17) = 1.22$ Btu/lb$_m$, then $h_2 = w_p + h_1 = 1.22 + 180.17 = 181.39$. Therefore,

$$\eta_R = \frac{w_t - w_p}{h_4 - h_2} = \frac{(1514.0 - 1150.4) - 1.22}{1514.0 - 181.39} \times 100 = 27.19\%$$

The agreement with Illustrative Problem 8.3 is excellent. The effort required to obtain the necessary properties is greatly reduced by the use of the computer-generated values. However, as a word of caution, do not forget to draw a sketch of the cycle, label the points, and then use the computer program as needed.

---

Figure 8.5 shows a schematic of the Rankine cycle with the pertinent energy quantities shown for each of the principal units where SL, MX, SV, and SHV stand for saturate liquid, mixture of saturated liquid and saturated vapor, saturated vapor and superheated vapor. Figure 8.6 shows the ideal Rankine cycle compared to a Carnot cycle on a $T$–$s$ plot. The plot clearly shows that for the same pressures, the efficiency of the Rankine cycle will be less than that of a Carnot cycle since the Rankine cycle would experience an entropy increase due to the slope of the saturation line. If the saturation line were vertical, then the

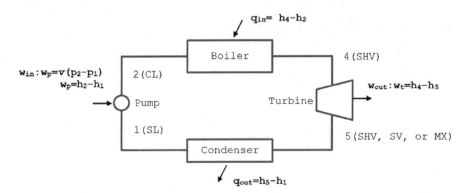

**FIGURE 8.5**
Elementary Rankine cycle.

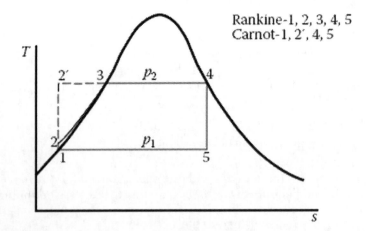

**FIGURE 8.6**
Carnot and Rankine cycle comparison.

thermodynamic efficiency of the Rankine and Carnot cycles would be the same. However, a vertical line on the $T$–$s$ diagram would require the liquid to have a specific heat of zero, which is not possible, but a fluid having a low specific heat compared to the value of $h_{fg}$ would give us a higher Rankine cycle efficiency. We consider this point later when we consider the regenerative cycle.

## 8.4 Rating of Power-Plant Cycles

To evaluate the merit of the power-plant cycles, several yardsticks are employed. One of these, the type efficiency, is quite useful. The *type efficiency* is defined as the ratio of the ideal thermal efficiency of a given cycle divided by the efficiency of a Carnot cycle operating between the same maximum and minimum temperature limits. In essence, it measures the approach of a prototype ideal cycle to the Carnot cycle between the same maximum and minimum temperature limits.

---

### ILLUSTRATIVE PROBLEM 8.5

What is the type efficiency of the ideal Rankine cycle of Illustrative Problem 8.3?

#### SOLUTION

The Carnot cycle would operate between 982.4°F and 212°F. Its efficiency is

$$\eta_c = 1 - \frac{T_2}{T_1} = \frac{T_1 - T_2}{T_1} \times 100$$

$$= \frac{(982.4 + 460) - (212 + 460)}{982.4 + 460} \times 100 = 53.4\%$$

The type efficiency would be

$$\frac{\eta_R}{\eta_c} \times 100 = \frac{27.3}{53.4} \times 100 = 51.1\%$$

In other words, at best, this cycle is only 51.1% as efficient as the Carnot cycle operating between the same temperature limits

---

### ILLUSTRATIVE PROBLEM 8.6

Calculate the type efficiency for the Rankine cycle of Illustrative Problem 8.1.

#### SOLUTION

For the upper temperature of the cycle, we have 400°C, and for 50 kPa, the *Steam Tables* give us a saturation temperature of 81.33°C for the lower or condensation temperature. The efficiency of a Carnot cycle operating between the limits would be

$$\eta_c = 1 - \frac{T_2}{T_1} = \frac{T_1 - T_2}{T_1} \times 100$$

$$= \frac{(400 + 273) - (81.33 + 273)}{400 + 273} \times 100 = 47.4\%$$

The type efficiency is

$$\frac{\eta_R}{\eta_c} \times 100 = \frac{28.5}{47.4} \times 100 = 60.1\%$$

---

For actual cycles, the thermal efficiency is defined as the net work output of the system divided by the energy as heat supplied to the system. In evaluating the thermal efficiency, it is necessary to take care that the proper accounting of energies is carried out. This can best be illustrated by reference to Illustrative Problem 8.7.

## ILLUSTRATIVE PROBLEM 8.7

In the Rankine cycle of Illustrative Problem 8.3, 50 Btu/$lb_m$ of steam is lost by heat transfer from the turbine to the surroundings. What is the thermal efficiency of this cycle?

### SOLUTION

From Illustrative Problem 8.3, it was found that the useful (ideal) work, $w_t$ is 1515 − 1150.5 = 364.5 Btu/$lb_m$ of steam if the pump work is neglected. Because of the heat losses, 50 Btu/$lb_m$ of the 364.5 Btu/$lb_m$ becomes unavailable. Thus, 364.5 − 50 = 314.5 Btu/$lb_m$ is available. The thermal efficiency of the cycle (neglecting the pump work) is

$$\eta = \frac{314.5}{1515 - 180.15} \times 100 = 23.6\%$$

The heat rate and steam rate of a cycle are also figures of merit and have been used to describe cycle performance. The heat rate (actual or ideal) is defined as the amount of energy supplied per horsepower-hour or kilowatt-hour of the net output of the cycle. Therefore, the heat rate equals the energy supplied per horsepower-hour (or kilowatt-hour) divided by $\eta_R$(or $\eta_R$). Thus,

$$\text{heat rate} = \frac{2545}{\eta_R \left( \text{or } \eta_R \right)} \text{Btu/hp} - \text{h} \tag{8.9a}$$

$$= \frac{3413}{\eta_R \left( \text{or } \eta_R \right)} \text{Btu/kWh} \tag{8.9b}$$

The denominator in Equations 8.9a and 8.9b can be the thermal efficiency of any cycle. It is not restricted to the Rankine cycle. It is left as an exercise for the student to verify the constants 2545 and 3413 used in Equations 8.9a and 8.9b.

The steam rate of a cycle is defined as the ratio of the mass of steam supplied per net horsepower-hour or kilowatt-hour. Therefore,

$$\text{steam rate} = \frac{\text{pounds of steam supplied per hour}}{\text{net output hp} - \text{h or kWh}} \tag{8.10}$$

## ILLUSTRATIVE PROBLEM 8.8

Determine the heat rate and the steam rate per kilowatt-hour of the plant output in Illustrative Problem 8.3.

### SOLUTION

Neglecting pump work, we have

$$\text{heat rate} = \frac{3413}{0.273} = 12\,502\ \text{Btu/kWh}$$

Per pound of steam, $1515 - 1150.5 = 364.5$ Btu is delivered. Because $1\ \text{kWh} = 3413$ Btu, the steam rate becomes $3413/364.5 = 9.36\ \text{lb}_m$ of steam per kilowatt-hour.

## 8.5 The Reheat Cycle

The simple Rankine cycle suffers from the fact that heat is added while the temperature varies, and excess moisture during the expansion process is detrimental to the performance and life of the turbine. Superheating tends to help the turbine, but thermodynamically, it would not appear to yield much of an increase in the thermal efficiency of the cycle. To achieve a higher thermal efficiency and to help solve the turbine problem, the *reheat cycle* was proposed. The basis of this cycle is an attempt to approach the Carnot cycle efficiency by adding heat in increments at the highest possible temperature level. The steam is permitted to expand part of the way in the turbine and is then returned to the boiler, where it is heated again (reheated) and subsequently re-expanded through the turbine. Although there is no theoretical limit to the number of stages of reheat that can be employed in a cycle, two, or at most three, are used in practice. A schematic and a *T–s* diagram of this cycle are shown in Figure 8.7a.

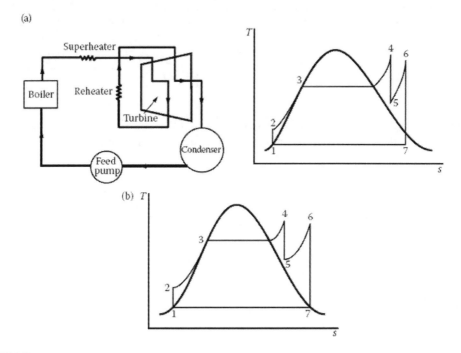

**FIGURE 8.7**
(a) Simple reheat cycle. (b) Reheat cycle with turbine exit at saturation.

It will be noted that this cycle is similar to the Rankine cycle, with the addition of the constant-pressure heating from 5→6 and the second isentropic expansion from 6→7. In an actual cycle, pressure and heat losses will occur in each part of the cycle, but for the present, it will be assumed that all processes are carried out to prevent these losses from occurring. Neglecting pump work, the energy supplied to the cycle per pound of fluid circulated is the sum of the heat supplied during the original heating plus that supplied during reheating. Thus, the heat in is $(h_4 - h_2) + (h_6 - h_5)$, which can also be written as $(h_4 - h_1) + (h_6 - h_5)$ when pump work is neglected ($h_2 = h_1$). The work out of the cycle is due to the two expansions. Thus, the work out is $(h_4 - h_5)_s + (h_6 - h_7)_s$. The efficiency of the cycle is

$$\eta_{reheat} = \frac{w_{t,4\to5} + w_{t,6\to7}}{q_{in,1\to4} + q_{in,5\to6}}$$

$$= \frac{(h_4 - h_5)_s + (h_6 - h_7)_s}{(h_4 - h_1) + (h_6 - h_5)} = \frac{(h_4 - h_7) + (h_6 - h_5)}{(h_4 - h_1) + (h_6 - h_5)}$$

(8.11)

A reheat cycle is sometimes called an "ideal" reheat cycle if, in the first expansion, it is expanded isentropically to saturated vapor and, during the subsequent reheat, it is brought back to the same enthalpy that it had been before expansion, that is, $h_4 = h_6$.

### ILLUSTRATIVE PROBLEM 8.9

Assume that the simple Rankine cycle described in Illustrative Problem 8.3 is made into a reheat cycle by first expanding the steam to 200 psia and then reheating it to the enthalpy that it had prior to the first expansion. If the final expansion is carried out to 14.696 psia, determine the efficiency of the reheat cycle and compare it to the simple cycle.

#### SOLUTION

The Mollier chart provides a convenient way of solving this problem. Expanding from 980°F, 400 psia, $s = 1.7567$ to 200 psia yields a final enthalpy of 1413 Btu/lb$_m$. Expanding from 200 psia and an enthalpy of 1515 Btu/lb$_m$ to 14.696 psia yields a final enthalpy of 1205 Btu/lb$_m$.

$$\eta_{reheat} = \frac{(1515 - 1205) + (1515 - 1413)}{(1515 - 180.15) + (1515 - 1413)} \times 100 = 28.7\%$$

It is apparent that for the conditions of this problem, the increase in efficiency is not very large. The final condition of the fluid after the second expansion is superheated steam at 14.696 psia. By condensing at this relatively high-pressure condition, a large amount of heat is rejected to the condenser cooling water.

---

### ILLUSTRATIVE PROBLEM 8.10

Solve Illustrative Problem 8.9 using the computerized properties.

#### SOLUTION

Some of the property data required were found in Illustrative Problem 8.4. In addition, we need

at 200 psia and $s = 1.7566$ Btu/lb$_m$·R, $h = 1413.6$ Btu/lb$_m$

at 200 psia and $h = 1514.0$ Btu/lb$_m$, $s = 1.8320$ Btu/lb$_m$·R

at 14.696 psia and $s = 1.8320$ Btu/lb$_m$·R, $h = 1205.2$ Btu/lb$_m$

Using these data,

$$\eta_{reheat} = \frac{(1514.0 - 1205.2) + (1514.0 - 1413.6)}{(1514.0 - 180.17) + (1514.0 - 1413.6)} \times 100 = 28.53\%$$

Again, it is cautioned that a sketch of the problem on $h$–$s$ or $T$–$s$ coordinates should precede the use of the computerized *Steam Tables*.

---

There are situations where reheat is accomplished up to the same temperature as the fluid had when it left the boiler (state 4). To get the most out of the turbine, it may be desired to expand the vapor following reheat down to saturated vapor; here, state 7 would be on the saturation line. In such a problem, it is necessary to determine the pressure at which reheat should occur. This cycle is shown in Figure 8.7b.

In the analysis of this cycle, state 7 is found from the saturated vapor ($g$ point) of the condenser pressure. The entropy at state 7 must equal that at state 6. Knowing the temperature and entropy at state 6 permits the determination of its pressure, most easily by using computerized properties. This is the same pressure for state 5, whose enthalpy can now be found since the entropy at state 5 equals that at state 4. Illustrative Problem 8.11 provides an example of such a cycle.

---

### ILLUSTRATIVE PROBLEM 8.11

In a reheat cycle, steam leaves the boiler at 1000°F and 1000 psia and is reheated back to 1000°F. The steam is exhausted from the turbine as saturated vapor at 14.696 psia. Find the reheat pressure and the efficiency of this cycle using computerized properties.

#### SOLUTION

Refer to Figure 8.7b for this cycle. Using the computer, the enthalpies of points 1 and 2 are found (the pump work is neglected, so $h_1 = h_2$):

$$h_1 = h_2 = 180.17 \text{ Btu/lb}_m \quad h_4 = 1505.9 \text{ Btu/lb}_m \quad T_4 = 1000°F$$

By computer,

$$h_7 = h_g@14.696 \text{ psia} = 1150.4 \text{ Btu/lb}_m$$
$$s_7 = s_g@14.696 \text{ psia} = 1.7566 \text{ Btu/lb·R}$$
$$s_6 = s_7 \text{ and } T_6 = T_4 = 1000°F$$

By computer, $h_6 = 1522.9 \text{ Btu/lb}_m$ and the reheat pressure $p_6 = 423.47$ psia.
For point 5, $s_5 = s_4 = 1.6530 \text{ Btu/lb·R}$ and $p_5 = p_6$.
By computer using $s_5$ and $p_5$, $h_5 = 1385.4 \text{ Btu/lb}_m$.
Using Equation 8.11,

$$\eta_{reheat} = \frac{(h_4 - h_7) + (h_6 - h_5)}{(h_4 - h_1) + (h_6 - h_5)} = \frac{(1505.9 - 1150.4) + (1522.9 - 1385.4)}{(1505.4 - 180.17) + (1522.9 - 1385.4)} \times 100$$

$$\eta_{reheat} = 33.7\%$$

## 8.6 The Regenerative Cycle

As indicated in the preceding discussion, reheating is limited in its ability to improve the thermodynamic cycle efficiency and finds greatest usefulness in the reduction of moisture in the turbine. However, because the largest single loss of energy in a power plant occurs at the condenser in which heat is rejected to the coolant, it is pertinent to consider methods of reducing this rejected heat and improving cycle efficiency.

In both the ideal Rankine and reheat cycles, the condensate is returned to the boiler at the lowest temperature of the cycle. The fluid is heated to saturation by direct mixing in the steam drum of the boiler by furnace radiation in the boiler tubes, or by gas convection heating by the flue gases in the economizer section of the unit. These methods depend on large temperature differences and are inherently irreversible. Rather than resorting to such procedures, which involve the internal arrangement of the boiler and its heat-transfer circuit, let us consider a method of feedwater heating that, although idealized, leads to certain interesting and practical conclusions.

The most desirable method of heating the condensate would be by a continuous reversible method. Presuming that this is possible and imagining that the heating can be made to occur reversibly by an interchange within the turbine and in equilibrium with the expanding fluid, the *T–s* diagram will be given as shown in Figure 8.8. This diagram assumes saturated vapor at the initiation of expansion. Curve *AB'* is parallel to *CD*, because it was postulated that the heating is reversible. It will be noted that the increase in entropy during heating equals the decrease during the expansion and cooling of the vapor, and area *A'ABB'* equals *CD'D*. Thus, the cycle is equivalent to a Carnot cycle operating between the maximum and minimum temperatures $T_1$ and $T_2$. The efficiency is therefore given by Equation 8.1.

It is interesting to note that by allowing the working fluid to deliver work and to transfer heat simultaneously in the manner indicated, the ideal-cycle efficiency approaches the Carnot efficiency as a limit.

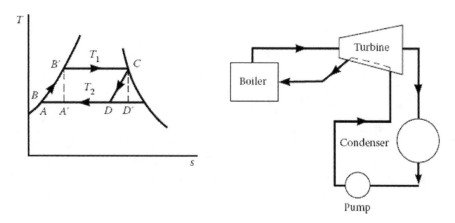

**FIGURE 8.8**
Ideal regenerative cycle.

In practice, this ideal cycle is approached by allowing the condensate from the feed pump to be heated in a separate heater (or heaters) by steam extracted from the turbine after it has partly expanded and done work. The extracted steam may mix directly with the condensate (as in an open heater) or may exchange heat indirectly and condense (as in a closed heater). A schematic of a practical cycle is shown in Figure 8.9a.

The *T–s* diagram for a regenerative Rankine cycle using an open feedwater heater is shown in Figure 8.9b. Here, the feedwater heater takes regenerated steam from the turbine at an intermediate state 5 to be mixed with water pumped up from the condenser at state 8 to become hotter water at state 1. From there, a second feed pump delivers the liquid to the boiler at state 2. Superheated steam exits the boiler at state 4 and expands in the turbine to state 5, where a portion of the superheated steam is extracted and fed to the feedwater heater. The remaining steam completes its expansion in the turbine to state 6, where it enters the condenser to become liquid at state 7. From there, it is pumped up to the feed water heater at state 8.

The *T–s* diagram for a Rankine cycle with two regenerative stages is shown in Figure 8.9c. Multiple regenerative stages are often used to get the maximum efficiency possible.

The cycles shown in Figures 8.9b and 8.9c utilize an open feedwater heater. Each requires two pumps, one into and one out of the heater. A cycle utilizing a closed feedwater heater is shown in Figure 8.10. The arrow indicates that the regenerated steam, once it becomes saturated liquid (state f), is returned to the condenser where it joins the nonregenerated steam as liquid to be pumped up from state 7. The regenerated steam provides heating of the feedwater from state 1 to state 2 where it enters the boiler. Thus, the closed heater cycle requires only one pump. This advantage is offset by the lower efficiency of heating in the closed versus the open feedwater heater.

In large central station installations, 1 to 12 feedwater heaters are often used. These can attain lengths of more than 60 ft., diameters up to 7 ft., and have more than 30,000 ft.$^2$ of surface. Figure 8.11 shows a straight condensing type of feedwater heater with steam entering at the center and flowing longitudinally, in a baffled path, on the outside of the tubes. Vents are provided to prevent the buildup of noncondensable gases in the heater and are especially important where pressures are less than atmospheric pressure. Also, deaerators are often used in conjunction with feedwater heaters or as separate units to reduce the quantity of oxygen and other noncondensable gases in the feedwater to acceptable levels.

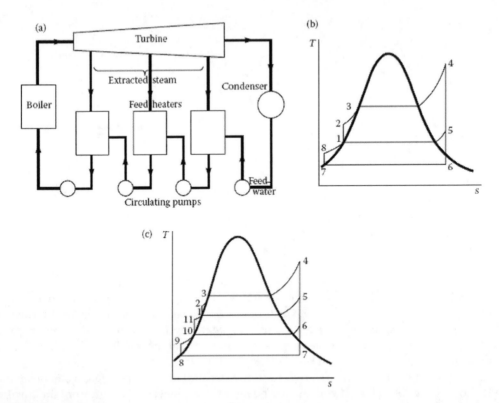

**FIGURE 8.9**
(a) Regenerative cycle. (b) Regenerative cycle with one open feedwater heater. (c) Regenerative cycle with two open feedwater heaters.

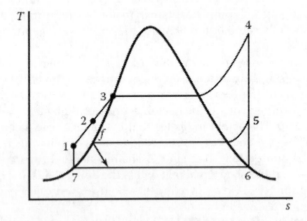

**FIGURE 8.10**
Regenerative cycle with closed feedwater heater.

By combining the regenerative cycle with the reheat cycle, the modern regenerative–reheat cycle is obtained. An analysis of such a cycle involves writing a heat balance around each of the heaters to determine the quantities extracted. A subsequent heat balance around the cycle then establishes the cycle output and efficiency.

**FIGURE 8.11**
Straight condensing feedwater heater.

The heat balance around a heater follows the method described in Section 4.5.6, except that here it is mass rather than mass flow rate that is utilized since only one cycle is analyzed. In a well-insulated feedwater heater, there is no net heat transfer to the outside nor any net-work done since there are no moving parts. The first law for an open system applies here since fluid is moving in and out of the heater. With $Q=0$ and work$=0$, and negligible kinetic energy changes, the first law yields $\Delta H=0$. Thus, the enthalpy entering the heater must equal that leaving, or $\Sigma mh=0$. For an open feedwater heater, using the formulations of Figure 8.9b,

$$Wh_5 +(1-W)h_8 = 1(h_1)$$

(8.12)

This permits solving for $W$, the mass of steam regenerated per pound of steam leaving the generator. This is needed in the efficiency calculation since only $1 - W$ produces work in the low-pressure portion of the turbine.

For the closed feedwater heater, the formulations in Figure 8.10 yield the following heat balance equation from the first law:

$$Wh_5 +1(h_1)=1(h_2)+Wh_f$$

(8.13)

If the state of the liquid at point 2 is not known, its enthalpy is often assumed to be equal to that of the saturated regenerated liquid $h_f$.

---

### ILLUSTRATIVE PROBLEM 8.12

A Rankine cycle operates with an upper pressure of 1000 psia and an upper tempera-ture of 1000°F. (a) If the condenser is at 1 psia, determine the efficiency of this cycle.

(b) Compare it to a regenerative cycle operating under the same conditions with a single heater operating at 50 psia. Neglect the pump work.

### SOLUTION

a. For the Rankine cycle, the Mollier chart gives

$$h_4 = 1505 \text{ Btu/lb}_m \quad h_5 = 922 \text{ Btu/lb}_m$$

and at the condenser,

$$h_1 = 69.14 \text{ Btu/lb}_\text{m}$$

Then

$$\eta_R = \frac{w_t}{q_\text{in}} = \frac{h_4 - h_5}{h_4 - h_1} = \frac{1505 - 922}{1505 - 69.74} \times 100 = 40.6\%$$

b. Figure 8.12 shows the regenerative cycle. After doing work (isentropically), *W* pounds of steam is bled from the turbine at 50 psia for each lb$_\text{m}$ of steam leaving the steam generator, and (1 − *W*) pounds goes through the turbine and is condensed in the condenser to saturated liquid at 1 psia. This condensate is pumped to the heater, where it mixes with the extracted steam and leaves as saturated liquid at 50 psia. The required enthalpies are

Leaving turbine: $h_5 = 1168$ Btu/lb$_\text{m}$ at 50 psia
Leaving turbine: $h_6 = 922$ Btu/lb$_\text{m}$ at 1 psia
Leaving condenser: $h_7 = 69.74$ Btu/lb$_\text{m}$ at 1 psia
$h_7 = h_8$ if pump work is neglected
Leaving heater: $h_1 = 250.24$ Btu/lb$_\text{m}$ at 50 psia
$h_1 = h_2$ if pump work is neglected (saturated liquid)

A heat balance around the heater assuming $h_7 = h_8$ gives

$$Wh_5 + (1-W)h_8 = 1(h_1) \rightarrow W(1168) + (1-W)(69.74) = (1)(250.24)$$

**FIGURE 8.12**
Illustrative Problem 8.12.

$$W = 0.1644 \text{ lb}_m$$

The work output is

$$\text{work} = h_4 - Wh_5 - (1-W)h_6$$

$$\text{work} = 1505 - (0.1644)1168 - (1-0.1644)$$

$$= 542.56 \text{ Btu/lb}_m$$

Heat into the steam generator equals the enthalpy leaving minus the enthalpy of the saturated liquid entering at 50 psia:

$$q_{in} = (h_4 - h_1) = 1505 - 250.24 = 1254.8 \text{ Btu/lb}_m$$

The efficiency is

$$\eta = \frac{\text{work}}{\text{heat in}} = \frac{542.56}{1254.8} \times 100 = 43.2\%$$

When compared to the simple Rankine cycle, it is seen that a single heater raises the efficiency by 2.6% for the same operating limits. This is a considerable efficiency increase.

It can also be shown that the efficiency of a regenerative cycle with one open heater is given by

$$\eta = 1 - \frac{(h_5 - h_1)(h_6 - h_7)}{(h_4 - h_1)(h_5 - h_7)} \times 100 \tag{a}$$

where the subscripts refer to those shown in Figure 8.12 with pump work neglected. Also,

$$W = \frac{(h_1 - h_7)}{h_5 - h_7} \tag{b}$$

For this problem, $W = (250.24 - 69.74)/(1168 - 69.74) = 0.1644 \text{ lb}_m$.

$$\eta = 1 - \frac{(1198 - 250.24)(922 - 69.74)}{(1505 - 250.24)(1168 - 69.74)} \times 100 = 43.2\%$$

---

## ILLUSTRATIVE PROBLEM 8.13

If a second open heater is used in Illustrative Problem 8.12 with extraction from the turbine at 100 psia, determine the efficiency of the cycle. Compare your results to Illustrative Problem 8.12. The *T–s* diagram for this problem is shown in Figure 8.9c.

## SOLUTION

Figure 8.13 shows the cycle. For this cycle, $W_2$ pounds are extracted at 100 psia, and $W_1$ pounds are extracted at 50 psia for each pound produced by the steam generator. The enthalpies that are required are

| | |
|---|---|
| Leaving turbine: | $h_7 = 922$ Btu/lb$_m$ at 1 psia |
| Leaving condenser: | $h_8 = 69.74$ Btu/lb$_m$ at 1 psia (saturated liquid) |
| Leaving low-pressure heater: | $h_{10} = 250.24$ Btu/lb$_m$ at 50 psia (saturated liquid) |
| Leaving high-pressure heater: | $h_1 = 298.61$ Btu/lb$_m$ at 100 psia |
| At low-pressure extraction: | $h_6 = 1168$ Btu/lb$_m$ at 50 psia |
| At high-pressure extraction: | $h_5 = 1228.6$ Btu/lb$_m$ at 100 psia |
| Entering turbine: | $h_4 = 1505$ Btu/lb$_m$ |

A heat balance around the high-pressure heater gives us

$$W_2 h_5 + (1 - W_2) h_{10} = 1(h_1) \rightarrow W_2(1228.6) + (1 - W_2)250.24 = 1(298.61)$$

$$W_2 = 0.049 \text{ lb}_m$$

A heat balance around the low-pressure heater yields

$$W_1 h_6 + (1 - W_1 - W_2) h_8 = (1 - W_2)(h_{10}):$$

$$W_1(1168) + (1 - W_1 - W_2)69.74 = (1 - W_2)250.24$$

$$W_1 = 0.156 \text{ lb}_m$$

**FIGURE 8.13**
Illustrative Problem 8.13

The work output is

$$\text{work} = h_4 - W_2 h_5 - W_1 h_6 - (1 - W_1 - W_2) h_7$$

$$\text{work} = 1505 - (0.049)1228.6 - (0.156)1168 - (1 - 0.156 - 0.049)922$$

$$= 529.6 \text{ Btu/lb}_m$$

Heat into the steam generator equals the enthalpy leaving minus the enthalpy of saturated liquid at 100 psia:

$$q_{in} = (1505 - 298.61) = 1206.4 \text{ Btu/lb}_m$$

The efficiency is

$$\eta = \frac{\text{work}}{\text{heat in}} = \frac{529.6}{1206.4} \times 100 = 43.9\%$$

The addition of a second heater increases the thermal efficiency of the cycle by 0.7% over the cycle with one heater.

It is possible to obtain analytical expressions for $W_1$, $W_2$, and $\eta$. This entails considerable algebraic manipulation. However, the results are useful for computer programming. In terms of Figure 8.13,

$$W_2 = \frac{h_1 - h_{11}}{h_5 - h_{11}} \tag{a}$$

$$W_1 = \frac{h_5 - h_1}{h_6 - h_9} \quad \frac{h_{10} - h_9}{h_5 - h_{10}} \tag{b}$$

Neglecting the pump work,

$$\eta = 1 - \frac{h_7 - h_8}{h_4 - h_1} \quad \frac{h_5 - h_1}{h_5 - h_{10}} \quad \frac{h_6 - h_{10}}{h_6 - h_8} \tag{c}$$

For this problem, $h_8 = h_9$, $h_{10} = h_{11}$, and $h_1 = h_2$. Thus,

$$W_2 = \frac{298.61 - 250.24}{1228.6 - 250.24} = 0.049 \text{ lb}_m$$

$$W_1 = \frac{1228.6 - 298.61}{1168 - 69.74} \quad \frac{250.24 - 69.74}{1228.6 - 250.24} = 0.156 \text{ lb}_m$$

$$\eta = 1 - \frac{922 - 69.74}{1505 - 298.61} \quad \frac{1228.6 - 298.61}{1228.6 - 250.24} \quad \frac{1168 - 250.24}{1168 - 69.74}$$

$$= 43.9\%$$

It is strongly suggested that a sketch similar to Figure 8.13 be used for all regenerative problems to avoid errors in terms.

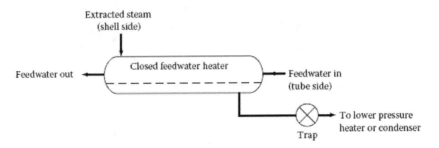

**FIGURE 8.14**
Straight condensing, closed feedwater heater with trap.

As noted earlier, the use of several feedwater heaters in a regenerative cycle increases the efficiency of the cycle. Most power plants incorporate at least one open heater, one in which the fluid streams are mixed, operating at a pressure greater than atmospheric pressure to vent oxygen and other dissolved gases from the cycle in order to minimize corrosion. This process of removing dissolved gases is known as deaeration. Closed feedwater heaters are essentially shell-and-tube heat exchangers (discussed in Chapter 11) with the extracted steam condensing outside the tubes on the shell side of the heat exchanger and feedwater being heated while flowing through the tubes. The two-fluid streams do not mix and are at different pressures. As shown in Figure 8.14, the condensate from a straight condensing, closed feedwater heater passes through a trap to a feedwater heater operating at a lower pressure or into the condenser. The trap is a valve that permits liquid only to pass to a lower pressure region. The liquid undergoes a throttling process in the trap as it flows to the lower pressure region. The operation of an open and closed heater in a regenerative cycle is shown in Illustrative Problem 8.14.

---

### ILLUSTRATIVE PROBLEM 8.14

A regenerative cycle operates with an upper pressure of 1000 psia and an upper temperature of 1000°F. Determine the efficiency of the cycle if it has a single open heater at 50 psia and a single closed heater at 100 psia. The condenser operates at 1 psia. Assume that the condensate from the closed heater is saturated liquid at 100 psia and that it is trapped (throttled) into the open feedwater heater. The enthalpy of the feedwater leaving the closed heater can be taken as that of saturated liquid at 310°F. Pump work can be neglected. Compare your results with Illustrative Problem 8.12.

#### SOLUTION

Figure 8.15 shows the regenerative cycle for this problem. Assume that 1 $lb_m$ of steam leaves the steam generator and that $W_1$ $lb_m$ is bled off to the closed heater at 100 psia and that $W_2$ $lb_m$ is bled off to the open heater at 50 psia. Also, assume that the feed water leaving the closed heater is at 310°F, 18°F less than the saturation temperature corresponding to 100 psia. For calculation purposes, we will use $h_f$ at 310°F for this enthalpy. Using the Mollier diagram and the *Steam Tables*, we find the following values of enthalpy:

$h$ to turbine = 1505 Btu/$lb_m$ (at 1000 psia and 1000°F)

$h$ at first extraction = 1228 Btu/$lb_m$ (isentropically to 100 psia, $s_{1000\ psia,\ 1000°F} = s_{100\ psia}$)

$h$ at second extraction $= 1168$ Btu/lb$_m$ (isentropically to 50 psia, $s_{1000\ psia,\ 1000°F} = s_{50\ psia}$)

$h$ at turbine exit $= 922$ Btu/lb$_m$ (isentropically to 1 psia,

$$s_{1000\ psia,\ 1000°F} = s_{1\ psia} \rightarrow x = \frac{s - s_{f@1\ psia}}{s_{fg@1\ psia}} \rightarrow x = 0.824 \rightarrow h = h_{f@1\ psia} + xh_{fg@1\ psia})$$

$h_f = 298.61$ Btu/lb$_m$ (at 100 psia)

$h_f = 250.24$ Btu/lb$_m$ (at 50 psia)

$h_f = 280.06$ Btu/lb$_m$ (at 310°F)

$h_f = 69.74$ Btu/lb$_m$ (at 1 psia)

A heat balance around the high-pressure heater gives us

$$h_{f@50\ psia} + W_1 h_{1@100\ psia} = W_1 h_{f@100\ psia} + h_{f@310°F}$$

Solving for $W_1$,

$$W_1 = \frac{h_{f@310°F} - h_{f@50\ psia}}{h_{1@100\ psia} - h_{f@100\ psia}}$$

$$W_1 = \frac{280.06 - 250.24}{1228 - 298.61}$$

$$W_1 = 0.032\ lb_m$$

a heat balance around the open heater yields

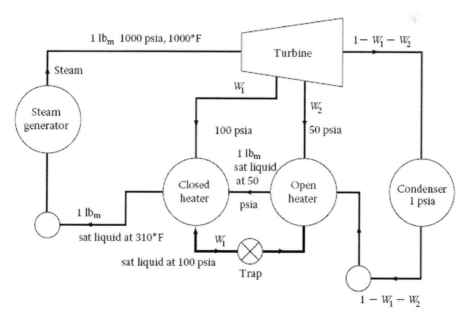

**FIGURE 8.15**
Illustrative Problem 8.14.

$$W_2 h_{2@50\text{ psia}} + \left(1 - W_1 - W_2\right) h_{@1\text{ psia}} + W_1 h_{f@100\text{ psia}} = h_{f@50\text{ psia}}$$

Solving for $W_2$,

$$W_2 = \frac{h_{f@50\text{ psia}} - h_{f@1\text{ psia}} + W_1\left(h_{f@1\text{ psia}} - h_{f@100\text{ psia}}\right)}{h_{2@50\text{ psia}} - h_{f@1\text{ psia}}}$$

$$W_1 = \frac{250.24 - 69.74 + 0.032\left(69.74 - 298.61\right)}{1168 - 69.74}$$

$$W_2 = 0.159\text{ lb}_m.$$

The work output of the cycle consists of the work that 1 $\text{lb}_m$ does in expanding isentropically to 100 psia, plus the work done by $(1 - W_1)$ $\text{lb}_m$ expanding isentropically from 100 to 50 psia, plus the work done by $(1 - W_1 - W_2)$ $\text{lb}_m$ expanding isentropically from 50 to 1 psia. The work is as follows

$$\text{work} = h_{1000\text{ psia},1000°\text{F}} - W_1 h_{100\text{ psia}} - W_2 h_{50\text{ psia}} - \left(1 - W_1 - W_2\right) h_{1\text{ psia}}$$

$$\text{work} = 1505 - \left(0.032\right)1228 - \left(0.159\right)1168 - \left(1 - 0.032 - 0.159\right)922$$

$$\text{work} = 534.34\text{ Btu}/\text{lb}_m.$$

The heat input is

$$q_{in} = h_{1000\text{ psia, }1000°\text{F}} - h_{f,\,310°\text{F}} = 1505 - 280.06 = 1224.94\text{ Btu}/\text{lb}_m$$

The efficiency of the cycle is

$$\eta = \frac{\text{work output}}{\text{heat input}} = \frac{534.34\text{ Btu}/\text{lb}_m}{1224.94\text{ Btu}/\text{lb}_m} \times 100 = 43.6\%$$

When compared to Illustrative Problem 8.12, we see that one open heater at 50 psia raised the efficiency of the simple Rankine cycle by 2.6%. The addition of a closed heater at 100 psia raises the efficiency by an additional 0.4%.

## 8.7 The Steam Generator

At this time, we will consider the major pieces of equipment that are used in steam power plants. The first unit will be the boiler. The term *boiler* is often used to broadly describe the device known more properly as a steam-generating unit. The boiler section of a steam generator refers to the elements in which the change of state from water to steam takes place. The steam generator consists of the boiler section, the superheater, the reheater, the economizer, the air heater, the fuel system, the air system, and the ash removal system. Thus, a complete steam generator is a combination of many elements, all integrated to yield an economical, efficient, reliable unit.

**FIGURE 8.16**
Modern water-tube steam generator.

Let us examine the unit shown in Figure 8.16 and trace out all the elements that constitute this type of a steam generator. The unit shown is known as a water-tube unit, in which steam is generated within the tubes and the combustion gases are outside the tubes. The path of the water can be traced out by noting that after leaving the feed pumps, it is usually pumped into the steam drum. The water is subcooled at the pump exit, and in the steam drum, it mixes with a steam–water mixture. This causes some of the steam to condense in the steam drum in order to heat the feed to saturation temperature. Depending on the particular design, it is possible for as much as 30% of the steam actually generated to be condensed to heat the feedwater in the steam drum. The steam drum acts as a collecting device and releasing point for the steam generated.

For details of the design of the internal components of steam generators, the reader should consult *Steam Power* by the Babcock and Wilcox Company.

## 8.8 The Steam Turbine

The steam leaves the steam generator and goes to the turbine where it expands to give useful work. There are many different types, sizes, and shapes of turbines, depending on the specific application. Because of the large number of combinations possible, let us start our discussion with the blading and the basic principles used to convert the energy in the steam to useful work. The two classifications of blading used are *impulse blading* and *reaction blading*. The basic action that occurs in impulse blading is that the expansion of steam takes place in the nozzles of the turbine and not in the blading. In the simple single impulse stage shown in Figure 8.17a, the nozzles direct the steam into the moving blades (buckets) mounted on a rotor. In the single impulse stage, all the pressure drop is shown to occur in the nozzles. As the steam flows through the passages of the moving vanes, it is

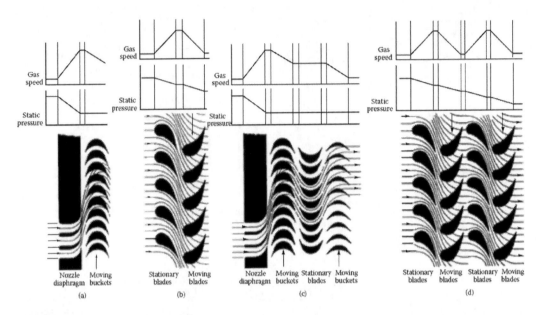

**FIGURE 8.17**
Basic turbine blading configurations, (a) Single impulse stage, (b) Pressure-compounded, (c) Velocity-compounded, (d) Reaction staging.

turned by the symmetrical moving blades in an axial direction, and at the same time, the steam does work on the blades. Thus, the final velocity of the steam is lower than at entry, and also, the enthalpy of the steam is lower. When two or more such stages are placed in series, this is known as *pressure compounding* (also known as Rateau staging). In a pressure-compounded impulse turbine, Figure 8.17b, the pressure drop is taken in equal steps in the alternate fixed nozzles. In Figure 8.17c, we have a velocity-compounded impulse blading situation (also known as Curtis staging). In this arrangement, the complex expansion of the steam takes place in the fixed nozzles before the first stage. The steam gives up part of its kinetic energy in the first row of moving blades and is reversed in direction in the fixed blades. The fixed blading directs the steam into the second row of blades, where the remainder of the kinetic energy is absorbed.

In the reaction blading (also known as Parson's blading), a partial pressure drop occurs in the fixed blades, and the velocity of the steam is increased. The steam then flows between moving blades that are shaped to act as nozzles. A further pressure drop occurs in the moving blades, increasing the velocity of the steam relative to the blades. This expansion in the moving blades causes a reaction against the blades that does work in driving the rotor. Sufficient stages are employed to expand the steam down to the desired exhaust pressure. The right side of Figure 8.17d shows reaction stages being fed from velocity-compounded impulse staging. This is frequently done in multistage turbines where reaction staging is used in the latter stages due to their increased efficiency under these conditions.

There are two basic turbine types that depend on the exhaust-steam conditions. In the condensing type, the exhaust steam leaves at below atmospheric pressure, while in the noncondensing type, the steam is exhausted at pressures above atmospheric. In addition, extraction methods can be further categorized as straightflow, reheat, automatic, or nonautomatic.

## 8.9 Cogeneration

*Cogeneration* is not a new idea; it is a method of energy conservation that has been used for a long time. Basically, cogeneration plants use the same fuel source to produce both electricity and steam. The steam can be used for heating or in a process facility adjacent to the cogeneration facility, with the excess electricity sold to a local utility for its use. The utility benefits from cogeneration by gaining an additional generating source to help it meet its demand and also by purchasing the power at a rate lower than the rate at which it sells power. Whether the system is based on steam turbines, gas turbines, or a combination (see Section 9.6 in Chapter 9) of both, cogeneration can achieve a significant increase in overall fuel efficiency over conventional power plants and steam generators. The advantage of cogeneration lies in the fact that both power and a transfer of heat for an intended use can be accomplished with a total fuel expenditure that is less than would be required to produce them individually.

The economic incentives that have made cogeneration more attractive in recent years include the following:

1. The Public Utility Regulatory Policy Act (PURPA) of 1978, which requires utilities to buy back excess power generated by cogeneration plants.

2. Cogeneration plants are easier and faster to install and do not require as much capital expenditure. Because components approach "off-the-shelf" availability, the lead time from inception to implementation is much shorter than for conventional plants.

3. A company can now be more independent of the local utility company by generating much of its own power at rates lower than the rates charged by the utility. Also, costly cutbacks in power by the company when the utility is having difficulty meeting demand need not occur.

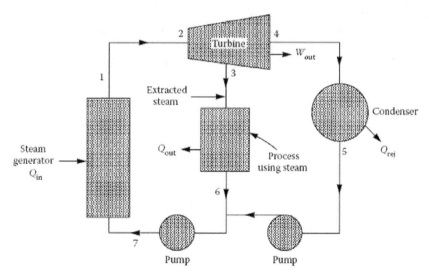

**FIGURE 8.18**
Cogeneration plant using process steam.

4. A company can tailor its cogeneration facility to its own particular needs and project load growth and capital expenditures more accurately.

5. A cogeneration facility can easily add on additional capacity with a minimum of shutdown time and, by using available equipment, with a minimum of additional expenditure.

Let us consider the cogeneration system that is shown schematically in Figure 8.18. In this system, steam is extracted from the turbine and is used for heating or as process steam. The portion of the steam that goes to the condenser eventually combines with the extracted steam and is returned to the steam generator.

---

### ILLUSTRATIVE PROBLEM 8.15

Assume that the regenerative cycle of Illustrative Problem 8.12 is operated with a heat exchanger replacing the heater similar to Figure 8.18. The heat exchanger will use the extracted steam to heat a building, and the condensate from the exchanger will be assumed to be saturated liquid at 50 psia. This liquid will mix with the condensate from the condenser and return as boiler feed. Assuming that 16% of the steam is extracted from the turbine and neglecting pressure drops and pump work, determine the conventional thermal efficiency of the cycle and the efficiency of energy utilization of the cycle. Compare your results with the results found in Illustrative Problem 8.12.

### SOLUTION

Let us first determine the enthalpy values at the various points in the cycle:

1. At 1000 psia and 1000°F, $h_1 = 1505$ Btu/$lb_{ra}$.

2. At turbine inlet, $h_2 = 1505$ Btu/$lb_m$.

3. At turbine extraction of 50 psia, $h_3 = 1168$ Btu/$lb_m$ (after isentropic expansion, $s_{1000\text{ psia, }1000°F} = s_{50\text{ psia}}$).

4. At condenser inlet, $h_4 = 922$ Btu/$lb_m$ (after an isentropic expansion, $s_{1000\text{ psia, }1000°F} = s_{1\text{ psia}} \rightarrow$ mixture with $x_4 = 0.823$)

5. Saturated liquid at condenser pressure, $h_5 = 69.14$ Btu/$lb_m$.

6. Saturated liquid at the exchanger outlet, $h_6 = 250.24$ Btu/$lb_m$ (50 psia).

7. Assuming that the condenser condensate mixes adiabatically (no heat transfer with the surroundings) with the liquid leaving the heat exchanger, $(0.16)h_6 + (0.84)h_5 = h_7 \rightarrow (0.16)(250.24) + (0.84)(69.74) = 98.62$ Btu/$lb_m$ based on 0.16 $lb_m$ extracted and 0.84 $lb_m$ going through the condenser.

Using these data, we have

$$\text{heat in at boiler} = h_1 - h_7 = 1505 \text{ Btu/lb}_m - 98.62 \text{ Btu/lb}_m$$

$$= 1406.38 \text{ Btu/lb}_m$$

$$\text{work out of turbine} = h_2 - (0.84)h_4 - (0.16)h_3$$

$$\text{work out of turbine} = 1505 - (0.84\text{lb}_m)(1168)\text{Btu/lb}_m$$

$$- (0.16)(1168)\text{Btu/lb}_m$$

$$= 543.64 \text{ Btu/lb}_m$$

The "conventional" thermal efficiency is defined as work out/heat in,

$$\eta = \frac{543.64 \text{ Btu/lb}_m}{1406.38 \text{ Btu/lb}_m} \times 100 = 38.66\%$$

If at this time we define the efficiency of energy utilization to be the ratio of the work out plus the useful heat out divided by the heat input to the cycle,

$$\eta_{\text{energy utilization}} = \frac{w + q_{\text{out(useful)}}}{q_{\text{in}}} \times 100$$

and $q_{\text{out(useful)}}$ $(0.16)(h_3 - h_6)$ or $(0.16 \text{ lb.})(1168 - 250.24)\text{Btu/lb}_m = 146.84 \text{ Btu/lb}_m$ of steam entering the turbine. Therefore,

$$\eta_{\text{energy utilization}} = \frac{(543.64 + 146.84) \text{ Btu/lb}_m}{1406.38 \text{ Btu/lb}_m} \times 100$$

$$\eta = 49.1\%$$

From these results, we see that when this cycle is compared with the regenerative cycle of Illustrative Problem 8.12, the "conventional" thermal efficiency is *decreased* by 4.5%. However, the efficiency based on the utilization of energy is *increased* by approximately 6%. Depending on the need, more or less steam can be bypassed to the heat exchanger or allowed to expand through the turbine.

## 8.10 Direct Energy Conversion

In all the devices and cycles that we have discussed thus far, the generation of electrical power from an energy source required the transfer of the energy from the source to a working fluid, which by undergoing circulation and other manipulation in a complex "prime mover" (e.g., turbine, engine, and so on) ultimately yielded the desired electrical output. Recent progress has been made in the development of direct energy conversion systems in which the energy of the fuel is converted directly to electrical energy without the use of a circulating fluid or any moving parts. At present, we will just qualitatively describe some of these systems and their limitations.

In 2007, the United States and most other industrialized nations recognized the need to address global warming. Thus, there has been an increased emphasis on these direct energy conversion systems, which can reduce the usage of fossil fuels, whose combustion puts carbon dioxide, a recognized contributor to global warming, into the atmosphere. As of this writing, the globe emits about 37 gigatons of $CO_2$ into the air each year from the burning of fossil fuels and cement production, of which about 14% of the world's production comes from the United States, second only to China. Of this amount, one-third comes from the electrical industry, which in the United States is based mainly on the combustion of coal, one of the worst polluters of the atmosphere. Some 57% of these emissions are absorbed by the biosphere and the oceans. In 2019, the concentration of carbon dioxide in the atmosphere was over 410 parts per million, an increase of 85% since 1970, according to a United Nations panel. In 2013, this concentration reached 400 parts per million, a critical level according to most researchers. This accelerated growth could be arrested with the increased usage of the systems described in this section.

### 8.10.1 Thermoelectrical Converter

The first of these direct energy conversion devices, the *thermoelectric converter*, is based on the fact that a voltage is generated when two unlike conductors are connected at their ends with the ends kept at different temperatures. If the circuit is closed through a load resistance, current will flow through the load. This phenomenon has already been noted and is used to measure temperature in thermocouples, which we described in Chapter 1. An associated phenomenon, known as the Peltier effect, is discussed in Chapter 10 in conjunction with the thermoelectric refrigerator. Figure 8.19 shows a simple thermoelectric energy converter. The desirable characteristics of the materials used in such a device are as follows:

1. Low internal electrical resistance to reduce internal heat generation resulting from current flow in the materials.
2. Low thermal conductivity (high thermal resistance) to reduce heat conduction from the heat source to the sink.

**FIGURE 8.19**
Thermoelectric converter.

3. High values of open-circuit voltage. Most metals produce open-circuit voltages of the order of microvolts per degree of temperature difference.

The most suitable materials are the semiconductors, such as lead telluride, germanium–silicon alloys, and germanium telluride. The thermoelectric converter is basically a form of heat engine receiving heat from a source, rejecting heat to a sink, and converting heat to electrical work. Thus, it has as its theoretical upper limit the Carnot cycle efficiency. Because of losses, the Carnot cycle limit is not even approached, and although in theory the thermoelectric converter is capable of operating at efficiencies greater than 15%, most of the devices built to date have shown efficiencies much lower than this figure. With development of new materials and using semiconductors, a thermal efficiency approaching 25% is a reasonable objective. Because of their reliability and lack of moving parts, they have been used in such diverse applications as converting waste heat from kerosene lamps to power radio receivers in rural areas and in conjunction with radioisotope heat sources to power long-life, unattended ocean buoys.

## 8.10.2 Fuel Cell

Concern about air pollution, caused in great measure by emissions from automobile engines, has given rise to the importance of the *fuel cell* as an alternative energy source. Fuel cells have been tried in vehicles since 1959 and have been used extensively in the space program.

The fuel cell is an electrochemical device that resembles a battery. However, it is unlike the battery in that it does not feed on itself; a fuel and an oxidant are consumed, but the electrodes remain intact. The fuel cell is a steady-flow, isothermal device, whereas the battery merely stores energy. There is no intermediate working fluid in the fuel cell nor are there energy reservoirs for heat exchange as in an engine. Thus, the Carnot principle does not apply to fuel cells, and there is no theoretical limit to efficiency.

A typical hydrogen–oxygen fuel cell is shown in Figure 8.20. Hydrogen is the fuel and oxygen is the oxidizer. The fuel enters the cell and is ionized at the anode, the negative electrode, releasing positive ions or protons into the electrolyte solution. This leaves excess electrons on the anode, which travel through an external circuit to the cathode, the positive electrode. This circuit provides the direct current electricity to drive a motor or power generator. The positive ions travel through the electrolyte, which permits only protons but not electrons to pass through, to the cathode. There the positive ions and the electrons from the external circuit join with the oxygen to produce water as the by-product. The chemical reaction utilizes two hydrogen molecules and one molecule of oxygen to produce two molecules of water. The chemical equations are as follows:

| | |
|---|---|
| At the anode: | $2H_2 \rightarrow 4H^+ + 4e^-$ |
| At the cathode: | $O_2 + 4H^+ + 4e^- \rightarrow 2H_2O$ |
| Net reaction: | $2H_2 + O_2 \rightarrow 2H_2O$ |

The reaction is accelerated by a catalyst, typically platinum, which encourages each hydrogen molecule to split into two protons and two electrons. A catalyst at the cathode, often nickel, is used to help in the formation of water. The electrolyte is usually acidic, and the voltage produced by a single hydrogen–oxygen cell is of the order of 1 V. Thus, an operational fuel cell system requires a combination of many cells called *a fuel cell stack*. Such a

**FIGURE 8.20**
Basic fuel cell. (Reprinted from Jones, Dugan, *Engineering Thermodynamics*, © 1996. Pearson Education, Inc., Upper Saddle River, N.J. With permission.)

stack would occupy a significant volume if the original fluid electrolyte were employed. Modern fuel cells use a *proton exchange membrane* as the electrolyte, which is more like a thin plastic wrap and so greatly reduces the size of such cells. These cells operate at modest temperatures, in the range of 40–120°C; they do not need extensive containment vessels, and they warm up quickly. A stack the size of a modest suitcase can power a small automobile. The output of the cells at such temperatures is liquid water. Such systems have a high power density and are good for transportation as they have a quick start-up. These systems should not be used in cold climates where the water output can freeze.

Other types of fuel cells have been developed with different electrolytes. If an alkaline solution is used, the ions are hydroxyls and the cell efficiency can be quite high. Such cells have been used for space and defense systems. Other electrolytes that have been used include phosphoric acid, molten carbonates, and even solid chemicals. These systems run at much higher temperatures and have been used for electricity and heat power generation. They produce water in a gaseous form. These systems have higher efficiency and are best for stationary applications.

While hydrogen may not be as dangerous as originally thought since it disperses easily, it is not readily available. The oxygen may be taken from the air, but the hydrogen is typically produced from hydrocarbons that are easily handled, not too expensive, and readily available. This requires a process called *reforming*, which generates heat and other gases. The most promising sources are natural gas, propane, and methanol. Thus, it must be recognized that an energy cost is attached to the production of the hydrogen fuel. In addition, the production of hydrogen typically yields by-product gases that contribute to pollution and global warming, things that the fuel cell concept was designed to reduce.

A practical fuel cell can be as much as 80% efficient. Adding the reforming process reduces the overall efficiency to perhaps 30–40%. The remaining output power equipment to utilize the electricity produced adds another loss factor to the overall efficiency calculation. By comparison, the gasoline-powered automobile engine system is about 20% efficient. A battery-powered electric car has a much higher efficiency, but electricity for it must be produced somehow, so the efficiency of the electric power plant must be added in. For now, fuel cells are considered for automotive applications as powering auxiliaries only. In any case, the usage of fuel cells in automobiles would require an extensive network of fueling stations. As of 2013, there have been 53 hydrogen fueling stations in the United States, 21 of which are in California. There are over 100 overseas.

It should be noted that the usage of the fuel cell is not so much based on efficiency as it is based on the prospect of reducing air pollution.

### 8.10.3 Thermionic Converter

The *thermionic converter* is basically a high-temperature device that converts thermal energy into electricity by literally boiling off electrons from the hot cathode, which then travel to the cold anode. Figure 8.21 shows a schematic of this device in which the electrical circuit is completed by the external load, $R_L$. In principle, this device is a heat engine that uses electrons as the working fluid. Therefore, the upper limit of its efficiency is that of a Carnot engine operating between the temperature limits of the cathode and anode. Efficiencies of 10% have been achieved with cathode temperatures of 3200°F when the Carnot efficiency was 50%. Because electron emission is an exponential function of temperature, high cathode temperatures yield high power densities. Efficiencies of the order of 18% have been achieved in a thermionic converter with a cathode temperature of 4500°F.

There are several problems associated with the operation of the thermionic converter. The first of these arises from the distribution of the electrons in the gap between electrodes. When the gap is maintained as nearly as possible as a "perfect" vacuum, it is necessary to keep the spacing of the electrodes as small as 0.001 in.; otherwise, the retarding potential (space-charge effects) becomes too large for satisfactory operation of the converter. These spacings are extremely difficult to maintain, especially with the large temperature differences between the anode and the cathode. To reduce the space-charge effect and, at the same time, maintain reasonable spacings (of the order of 0.040 in.), most present-day thermionic converters use cesium gas in the interelectrode space. Cesium is relatively

**FIGURE 8.21**
Schematic of a thermionic converter.

easily ionized with positively charged ions, which tend to neutralize and negate the space-charge effect.

In addition to the space-charge problem, the thermionic converter performance is also degraded by heat radiated and conducted from the cathode to the anode, which is ultimately lost from the anode. Thus, to achieve high efficiencies in a thermionic converter, it is necessary to reduce the effects of space charge, to minimize radiation and conduction losses, and to keep the internal resistance of the converter to a minimum. It will be noted from the foregoing that mechanical design plays a vital role in the ultimate performance of a thermionic converter.

## 8.10.4 Magnetohydrodynamic Generator

When a conductor is moved through a magnetic field, an electromotive force results, and if the circuit is closed, a current will flow. This phenomenon is the basis of the operation of all rotating electrical generators in which the rotor turns conductors in a magnetic field to produce electricity. The basic concept of the magnetohydrodynamic (MHD) converter is exactly this principle, with the conductor being a high-speed ionized gas. Figure 8.22 shows the replacement of the conducting wires of a conventional electric generator by a highly ionized gas called a *plasma*. The electrical properties of a plasma are determined by the density of free electrons in the gas. The ionization of the gas (and consequently its electrical conductivity) can be achieved in several ways (high gas temperatures, seeding of

**FIGURE 8.22**
MHD generator principle. (a) Conventional generator. (b) Plasma generator.

a gas with readily ionized atoms such as cesium, nuclear radiation of the gas, and others). When this ionized gas (plasma) flows at right angles to the magnetic field (Figure 8.22), a force is generated at right angles to both the velocity of flow and the magnetic field. This force drives the electrons and ions to the pickup electrodes, giving rise to a current when the electrodes are connected to an external load.

In the MHD cycle, the gas is first compressed, heat is added at constant pressure, and the gas is then accelerated in a nozzle before passing through the magnetic field. If we consider the overall cycle, such as the Rankine cycle or the Brayton cycle, which will be considered in Chapter 9, it will be seen that the MHD converter has basically replaced the turbine and the generator. This device is obviously a heat engine device, and it is subject to the efficiency limitations of the Carnot cycle. Some of the factors that cause losses in the MHD converter are heat-transfer losses through the electrodes, internal electrical resistance losses in the plasma, energy losses in the magnets, and fluid friction losses associated with high-speed gas flow.

Because of the losses noted, it is found that the MHD converter is best suited for large central stations in which the rejected heat from the MHD converter is used as the input to a conventional power cycle. Using this binary type of cycle, it is estimated that practical efficiencies for the combined cycle approaching 50% can be achieved.

### 8.10.5 Solar Energy

The amount of solar energy potentially available on the earth's surface is many times greater than the present energy needs of humankind. Although the input of solar energy to the earth has been estimated at over 1000 W/m², in North America, the average solar radiation (also called *insolation*) reaching the ground over a year is estimated to be between 125 and 375 W/m², the variation due to clouds and night times. In order to use this energy, enormous amounts of money must be expended. Useful amounts of energy have been generated by either *direct energy conversion* using photovoltaic devices (also called solar cells) or large arrays of reflectors used to concentrate the solar energy for use as the heat source in a conventional steam power plant (solar thermal plants).

When solar energy is converted directly into electricity, the photovoltaic effect is utilized. Figure 8.23 shows the principle of the photovoltaic effect. Sunlight falls on the semiconductor cell, and its photons are absorbed in the base material. This releases electrons that migrate to the "n" layer. The resulting imbalance of charge between the two regions creates a voltage potential like that in a battery. Connection to an external load results in the flow of dc electricity. While silicon solar cells have been the dominant type of cell used, materials such as cadmium sulfide/cadmium telluride, germanium, selenium, and gallium arsenide have also been used for photovoltaic cells. Of these materials, cadmium telluride is less efficient but cheaper to manufacture. A typical cell produces 1–2 W, and so arrays or modules of cells are created to produce enough power. Photovoltaic units can work even in partial sun. The best of the current photovoltaic cells has an efficiency of 15%. Excessive pollutants and dust can reduce these amounts further. It should be noted that the electricity produced is dc, so if installed for home use, it needs to be converted to ac, thereby incurring another inefficiency and requiring more equipment. Today, photovoltaics are used to power small systems like water pumps, and in lighting applications where electricity is stored during the day and used at night.

The efficiency of current photovoltaic units in converting sunlight into electricity is about 10–15%. Arizona, with its abundant sunshine, has been a leader in photovoltaic technology applications in the United States. As of 2012, the largest photovoltaic plants produce

**FIGURE 8.23**
Structure of a photovoltaic cell.

200–250 MW. A 500 MW plant is anticipated to open in the near future. Worldwide peak availability of photovoltaic power is currently about 100,000 MW. Of this, the United States accounts for 10%.

The Industrial Revolution saw the development of a device to focus the sun's rays onto a boiler assembly to produce heated water that could power an engine. This concept is utilized today in the other means of solar power, Concentrated Solar Power, or CSP. Today, CSP units use mirrors to track and focus the sun's rays that superheat water in towers, which are used to power turbine generator sets. There are major plants converting sunlight directly into electricity in at least seven Arizona cities, producing a total of about 1 MW of electricity.

During California's energy crisis in the summer of 2001, the fact that this state has the greatest output of electricity from solar energy plants helped avoid a catastrophic situation. These plants utilize the sun to heat water as part of an otherwise conventional steam power plant. The sun typically heats an intermediate substance that can store heat during non-sunlight hours. The prototype plant, comprising five solar generating systems (SEGS), was made up of parabolic mirrors that reflect sunlight onto piping filled with a high-temperature heat-transfer fluid (i.e., oil). The heat-transfer fluid would heat up and turn water into steam that drove a turbine to produce electricity. This plant in the Mojave Desert, called Solar One, produced 10 MW of electricity when operating in the mid-1980s. Unfortunately, the oil used in the system as heat-transfer fluid lost heat too quickly when the sun went down, and so power output was compromised on cloudy days. Today, there are already CSP plants generating 300 MW of electricity. Spain is one of the leaders in the field with large CSP plants and also photovoltaic plants generating 20 MW. California has the Ivanpah solar plant in the Mojave Desert, which uses three towers over 4000 acres and is one of the largest solar thermal power plants in the world. It can produce almost 400 MW of electricity, enough to power 140,000 homes. The expected cost for these systems is estimated at 6 cents/kWh for CSP plants and 8–10 cents/kWh for photovoltaic systems. This is still much more expensive than the costs of gas or oil power plants and more costly than land-based wind power units.

An international consortium has developed a 19 MW plant in Spain now called Gemasolaire, which was originally called Solar Tres. It uses molten salt as the heat-transfer medium in a 140 m tower to produce steam. An array of mirrors called heliostats surround the tower and focus the solar rays. Future systems are expected to utilize parabolic mirrors shaped like a satellite dish. The greatest potential is in dry equatorial climates, with major activity ongoing in Australia, Italy, India and South Africa, as well as in the United States. Other solar energy systems are used for hot water heating.

Another new application for solar energy usage is a solar-powered sterilization system. It uses nanomaterials that create steam from water almost instantaneously. These materials, developed in the United States, heat up quickly when exposed to sunlight. The water is vaporized into steam below its normal boiling point. Initial results show an efficiency of 24%. Sunlight has also been used to convert biomass into a synthetic gas that can be burned more efficiently than the biomass itself.

Other solar energy systems are used for hot water heating. Further information on the status of solar energy systems is available from the American Solar Energy Society and the Electric Power Research Institute.

### 8.10.6 Wind Power

An estimated 1% to 3% of energy from the sun that hits the earth is converted into wind energy. Most of this wind energy can be found at high altitudes where continuous wind speeds of over 160 km/h (100 mph) occur. Wind speeds increase with altitude because of the lower viscosity of air and the reduction of drag from the earth's surface. Eventually, the wind energy is converted through friction into diffuse heat throughout the earth's surface and the atmosphere. The origin of wind is complex. Different portions of the earth are unevenly heated by the sun; for example, dry land heats up (and cools down) more quickly than the seas do. Thus, there is usually a breeze at the beach on even the warmest days. Winds are also created by pressure differentials in the atmosphere.

Since the beginning of recorded history, humans have used the wind to move ships on the seas and windmills to pump water and grind grain. In 1910, wind turbine generators were used to produce usable amounts of electricity in Denmark. In the United States, the first large experimental wind turbine was built on Grandpa's Knob near Rutland, Vermont. Interest in wind power declined until the oil shortages of the 1970s occurred, because the total amount of convertible energy in the wind is tremendous and inexhaustible. However, not all of the kinetic energy of the wind can be captured by the wind turbine. Because the wind exiting the turbine retains some, albeit reduced, velocity, only about 60% of the incoming kinetic energy of the wind can be captured. This assumes no drag, incompressible axial air flow, and no heat transfer. For a constant mass flow rate, the cross-sectional area of the exiting air must be greater than the area of the incoming air. The result is that the maximum uptake of 60% occurs when the exit velocity is one-third of the velocity of the incoming air. The reader is referred to Betz's law published in 1919 for a detailed analysis. The net electrical power available is a function of the density of the air, the air speed to the third power, the rotor area, the generator and mechanical efficiencies, and the 60% factor. When friction is included, the best turbines can do is a 50% efficiency.

The newest turbines typically have three vanes and rise up to 300 ft. above the ground. The higher tower means catching stronger and steadier winds. Better materials provide stability even in high winds, and tiny fins added to the rotor surface induce turbulence to avoid stalling at low speeds. The diameter of the rotors can be as much as 400 ft. They turn at a rate of 20–40 rpm and can generate 2 MW of electricity through a gear box and generator mounted atop the structure. The turbine includes a yaw control to point the blades in the direction of the wind. The blades act as airfoils in transmitting the wind's energy into torque. A typical turbine is shown in Figure 8.24.

Modern turbines are relatively quiet, producing 45 dB at 250 m from a residence, equivalent to the noise from a kitchen refrigerator. Still many environmentalists are concerned about its place in the landscape and its somewhat fatal attraction for birds. Environmental studies have put the annual death rate of birds due to wind turbines at over 500,000. This is far less than bird deaths caused by cats: over 1 billion per year; by building and windows: up to 1 billion per year; by communication towers and power lines: at least 130 million per year; by hunting: 120 million per year. These values for 2012 are the best estimates from various sources for the United States, which has of this date over 45,000 towers in the mainland United States and in Puerto Rico. As of 2012, these towers are producing 60,000 MW, enough to power 15 million homes. The United States has 22% of the world's wind energy production of 282,000 MW, second only to China. Still future developments of wind power must take into consideration that turbines can cause the deaths of protected species of birds such as golden eagles, for which, in the United States, developers can be fined, and that homeowners in areas near wind power farms often complain of noise and the blinking red lights atop the towers at night.

Conventional wind turbines have their rotors facing into the wind with their energy conversion unit mounted behind. This approach captures the full cross-sectional area of the wind but puts a significant load on the support structures. A new company, Wind Turbine

**FIGURE 8.24**
MOD-1, 2 MW wind power generator.

Company of Bellevue, Washington, has reversed the design with a rotor mounted "down-wind." Such a machine can use lighter and more flexible blades and can be made effective with a two-blade design. However, there is usually more turbulence with the down-wind design, creating fatigue in the structure, so most turbines are built with the upwind design.

One of the largest wind turbines currently in use is a 1.5 MW unit produced by the General Electric Company. General Electric is also evaluating a 3.6 MW unit in Spain; it has a rotor with a diameter of 330 ft. mounted on a tower almost 40 stories high. The larger rotors turn more slowly and present less danger to passing birds.

As of October 2002, the total installed wind energy capacity in the United States has been 4329 MW, producing over 11 billion kWh of electricity annually. Most installations are in California (1716 MW) and Texas (1096 MW). Worldwide wind power capacity in 2002 reached 27,000 MW. The units are considered cost-effective if average wind speeds are at least 5–6 m/s, about 11–13 mph. Current state-of-the-art plants are producing electricity for under 5 cents/kWh.

Recent developments in wind energy production have introduced new concepts that can expand its availability. Offshore units can be larger, such as using 500 ft. diameter rotors and over 10 MW output. Such units would be more costly and would need to provide transport of power to the land.

Another development is the wind power kite. Such units would be floated up into the atmosphere with cables to the ground carrying power. The generator could be on the ground with a tether to airborne blades. They would be small, with perhaps 30 ft. diameter rotors generating 30 kW of power. The kites would have much lower building and operating costs. And if the wind dies, the blades could act as propellers to keep the kite afloat.

Some researchers have proposed the use of vertical blades rather than the horizontal rotors. The tower would look like a giant eggbeater. Its advantages would be in not having to face the wind, and its mechanisms would be at the ground level for easy servicing.

Wind power units are considered cost-effective if average wind speeds are at least 56 m/s, about 11–13 mph. Current (2012) state-of-the-art plants are producing electricity for 4–5 cents/kWh, but installation costs have currently been rising to about \$1900/kW. Operational costs are at about \$100/MW-h and about \$200–300/MW-h for offshore units. For further information, consult the American Wind Energy Association.

Despite its environmental attractiveness as a nonpolluting power source, wind power probably could not provide a large fraction of the total U.S. energy requirements because of its variability. However, it can provide a much needed additional capacity for utilities and save large amounts of fuel by becoming a significant source of energy for smaller utilities, rural homes, and agricultural areas.

### 8.10.7 Waste-to-Energy Resource Recovery

As a "rule of thumb," waste managers estimate that each person generates approximately 1 ton of refuse (garbage) a year. This includes household waste and waste produced by businesses. Waste disposal in the past consisted of using dumps to get rid of this material. Because of environmental and financial considerations, dumps are being phased out and waste-to-energy resource recovery plants are being used to deal with the problem of refuse disposal. Waste is fed into a furnace on rolling drums or grates. Computers that monitor the temperature and pollutants adjust the speed of the grates and the air mixture to obtain optimum combustion conditions. Next, "scrubbers" inject lime into the flue gas to neutralize acid rain components. Finally, a baghouse containing more than 1000 vacuum cleaner–like bags captures any ash particles.

From the refuse bunker, the waste is fed directly into the combustion chambers of three independent furnace/boiler units. Hot combustion gases pass through the water wall boiler sections to generate steam. The steam drives a 72 MW turbine generator, which produces electricity for in-plant use and with enough excess to keep the lights burning in 65,000 homes. In addition to solving a difficult environmental problem, this plant is expected to earn more than $500 million over its lifetime from the sale of power.

The discussion of waste-to-energy systems is now included in the term *biomass*. This would include usage of edible plants such as corn and nonedible plants such as algae. Corn is currently used to make ethanol an additive to gasoline and biodiesel for auto fuels. As will be shown in Chapter 9, ethanol produces much less energy than gasoline or diesel fuel.

### 8.10.8 Geothermal Energy

The thermal energy that is stored in the earth's crust is known as *geothermal energy*, and it can be harnessed for different applications. For instance, geothermal energy is being used to heat buildings and to generate electricity. A geothermal reservoir is formed when a layer of rock traps the heated water beneath it. While this water can be tapped with a well, it can also rise through porous rocks above the storage level to emerge from natural surface openings called *fumaroles*. The heated water will rise to the surface in the form of either geysers or hot springs. As the water rises, some of it will turn to steam as a result of the decrease in pressure. This steam is used to generate electricity using turbines. Low-temperature steam and hot water from the wells are used for local heating only.

There are two general types of geothermal energy systems: vapor and hot water. In the vapor system, the earth's heating is sufficient to provide superheated steam at the well head. This steam can be piped directly to a turbine to produce power. A system of this type is in operation at the geysers, in California, just north of San Francisco.

The most common geothermal system is the hot water system in which hot water is obtained from the wells. In one system, the hot water is then used to vaporize a working fluid that is piped to a turbine to produce power. In a second system, the flashed steam system, the water from the well is fed into a separator where it is throttled to a lower pressure to generate steam. The liquid, brine, is separated from the steam and discarded. The steam is then used to operate a turbine.

At present, geothermal energy accounts for a small percentage of the world's energy production. In the future, this percentage of energy production may increase. In order for this to occur, the problems associated with geothermal energy production must be overcome. Some of these problems are:

1. Low pressure and temperature exist at the well head.
2. The water has high concentrations of salt and minerals.
3. The discarded brine solution is harmful to the ecology of the earth's surface.
4. Most regions where geothermal energy is available are geologically unstable.

### 8.10.9 Nuclear Power

An alternative energy source is the use of a nuclear fuel to heat the water in a thermodynamic system. Heat is produced by fission in a nuclear reactor, which provides steam for a turbine that runs a generator to produce electricity. As Figure 8.25 shows, a nuclear

**FIGURE 8.25**
Pressurized Water Nuclear Reactor. (Courtesy of U.S.NRC.)

power plant consists of primary and secondary systems or loops for generating power. In the primary system or loop, radiative material such a uranium decays releasing heat to a flowing fluid such as water, which transfer the heat to a steam generator (boiler). On the secondary system side, the produced steam powers a Rankine cycle. The fission nuclear reaction within the primary system is controlled using control rods that diminish the reaction rate. Both cooling systems and reactor shields are essential for proper and safe operation. Construction costs are high, but lifetime fuel costs are low. A new nuclear plant being built in Finland is expected to have building and operating costs to total $15 billion over 20 years and produce electricity at a cost of 7 cents/kWh.

The United States is the world leader in nuclear power plants, with over 100 in operation. There are currently five under construction. Total capacity is over 100,000 MW, representing about 20% of the nation's electrical output. France is next with about 63,000 MW capacity, but this accounts for about 75% of that nation's electrical output.

While nuclear power represents the most efficient method of electrical production, there are major environmental and safety concerns such as disposal of the spent nuclear fuel for thousands of years, and the history of breakdowns or accidents of nuclear plants in Chernobyl, former USSR, and Fukushima, Japan.

### 8.10.10 Motion-Generated Energy

Systems in motion have a kinetic energy that can potentially be harnessed as an energy source. Most prominent of these are ocean waves; considering the size of the world's oceans, the potential here is enormous. The difficulty is to bring the energy to land.

At much smaller levels, inventors have developed energy-producing devices that utilize a person's walking, biking, or running motion; even a bouncing soccer ball has been invented to provide electricity for lamps in remote areas. The ball is called *SOCCKET* and utilizes a pendulum-type mechanism inside the ball to provide energy to charge LED lights or batteries.

## 8.11 Review

The stated purpose of this chapter is the study of the conversion of energy from one form to a more useful form. We started our study with a review of the Carnot cycle and then went on to the prototype Rankine, reheat, and regenerative steam cycles. These are "ideal" cycles in the sense that we ignored irreversible effects such as friction, pressure drops, heat losses, turbulence, and so on. Having studied these cycles, we then turned our attention to the equipment that is used for their practical realization.

Many other fluids and cycles using them have been proposed and constructed in addition to those cycles that use either water or gas as their working fluid. Binary fluid cycles using mercury or ammonia and water are examples of such alternates.

The direct conversion of the energy of a source to electrical energy without the use of a circulating fluid or moving parts has come under intensive development in recent years. In this chapter, we looked briefly at the thermoelectric converter, fuel cell, thermionic converter, MHD generation, solar energy, wind, and nuclear power. Most of these sources are not suitable at present for the large-scale generation of power. However, they are a welcome supplement to our energy supplies and, with further development, will become increasingly economically useful.

Cogeneration has recently become a feasible method of energy utilization. By the simple expediency of using the low-temperature energy for process or heating requirements, it is possible to combine electrical energy generation with heat requirements to obtain greater energy utilization than when each requirement is performed separately. While cogeneration does not introduce any new concepts, the savings in energy can be considerable. Most of the equipment needed to implement cogeneration can be obtained as standard "off-the-shelf" items.

## Key Terms

Terms used for the first time in this chapter are as follows:

**Cogeneration:** The simultaneous generation of power and heating or process steam to increase the overall utilization of energy.

**Direct energy conversion:** The conversion of a fuel or energy source directly to electrical energy without the use of a circulating fluid or any moving parts.

**Fuel cell:** An electrochemical device in which the fuel and the oxidant are consumed to produce a voltage across an electrical load.

**Geothermal energy:** The energy stored in the earth's crust.

**Rankine cycle:** A vapor cycle consisting of four processes: a constant-pressure vaporization, an isentropic expansion, a constant-pressure condensation, and a return to the initial conditions of the cycle.

**Regenerative cycle:** A cycle in which the condensate from the condenser is ideally continuously heated by an interchange with the steam as it expands in the turbine before it is returned to the boiler.

**Reheat cycle:** A Rankine cycle in which the steam is expanded in a turbine to an intermediate pressure, is returned to the steam generator, and then returned to the turbine to expand to condenser pressure.

**Solar energy conversion:** The use of a photovoltaic device to convert incident solar energy directly to electricity.

**Thermionic converter:** A high-temperature device in which thermal energy is converted to electricity by boiling off electrons from the hot cathode, which then travel to the cold anode.

**Thermoelectric converter:** A device that creates a voltage when there is a different temperature on each side of it. Conversely, when a voltage is applied to the device, heat is transferred from one side to the other, inducing a temperature difference.

**Type efficiency:** The ratio of the ideal efficiency of a given cycle divided by the efficiency of a Carnot cycle operating between the same maximum and minimum temperature limits.

**Wind power:** The use of the energy contained by the prevailing wind at a given location to operate a windmill to generate power.

## Equations Developed in This Chapter

| | | |
|---|---|---|
| Rankine cycle thermal efficiency | $\eta_R = \dfrac{(h_4 - h_5) - (h_2 - h_1)_s}{h_4 - h_2}$ | (8.7) |
| Heat rate | $\text{heat rate} = \dfrac{2545}{\eta_R \,(\text{or } \eta_R)}\,\text{Btu/hp-h}$ | (8.9a) |
| Heat rate | $\text{heat rate} = \dfrac{341.3}{\eta_R \,(\text{or } \eta_R)}\,\text{Btu/kWh}$ | (8.9b) |
| Steam rate | $\text{steam rate} = \dfrac{\text{pounds of steam supplied per hour}}{\text{net output}\,(\text{hp-h or kWh})}$ | (8.10) |

## QUESTIONS

8.1 In what two ways does a gas cycle differ from a vapor cycle?

8.2 How are the temperature limits of the Carnot cycle determined?

8.3 Cite some reasons why the practical realization of the Carnot cycle would be almost impossible.

8.4 How does the Rankine cycle differ from the Carnot cycle?

8.5 What is the basic reason that pump work is usually negligibly small? Would this be true if the steam at the outlet of the turbine was pumped directly back to the boiler without condensing it?

8.6     The reheat cycle causes two things to differ from the Rankine cycle. Indicate what these are.

8.7     What limits the number of reheats that are used in a given cycle?

8.8     The regenerative cycle has the potential of achieving the Carnot cycle efficiency when operated between the same upper and lower temperatures. Basically, describe what is being done in the regenerative cycle that brings this about.

8.9     What factor limits the number of heaters used in the regenerative cycle?

8.10    What has made the concept of cogeneration attractive at this time?

8.11    Does the conventional thermal efficiency of a cogeneration plant increase, decrease, or remain the same as a conventional power plant operating between the same limits?

8.12    What does direct energy conversion attempt to do?

8.13    What makes the reversed thermoelectric converter attractive for use in space vehicles?

8.14    What is the limiting efficiency of a fuel cell?

8.15    Where would you expect to find a MHD power plant?

8.16    Solar energy conversion and wind power plants give us "something for nothing"; that is, with no energy input by humankind, we obtain energy output. Comment on this statement.

## PROBLEMS

Wherever appropriate, sketch the cycle on $h–s$ or $T–s$ coordinates. Also, draw a sketch of the equipment showing the flow of the working fluid.

### Problems Involving the Carnot Cycle

8.1     A Carnot engine operates between 1300 K and 300 K. Determine its thermal efficiency.

8.2     A Carnot engine operates between 60°F and 1200°F and delivers 400 Btu as work. Determine the heat rejected and the thermal efficiency of the engine.

8.3     A Carnot cycle is operated with its maximum temperature equal to 1000 R. If the cycle efficiency is such that 100 Btu/h is rejected, what is the sink temperature if 500 Btu/h enters at the upper temperature?

8.4     A Carnot cycle is operated with its maximum temperature 700 K. If the cycle efficiency is such that 300 kW is rejected, what is the sink temperature if 450 kW enters at the upper temperature?

8.5     A Carnot vapor cycle uses water as the working fluid. If the cycle operates within the wet region and the upper pressure is 500 psia while the lower pressure is 20 psia, determine the cycle efficiency.

*8.6    Two ideal reversible heat engines are operated in series, the first receiving heat at a temperature $T_1$ and the second rejecting heat at a temperature $T_2$. Prove that this arrangement is as efficient as a single reversible engine operating between the same minimum and maximum temperatures.

*8.7    Prove that a single Carnot engine operating between temperatures $T_1$ and $T_2$ has an efficiency of $n_0$, which is related to the efficiencies of two Carnot engines operated in series between the same temperature limits by

$$\eta_0 = \eta_1 + \eta_2 - \eta_1\eta_2$$

where $\eta_1$ is the efficiency of Carnot engine 1 operated between $T_1$ and an intermediate temperature $T$, and $\eta_2$ is the efficiency of Carnot engine 2 that is operated in series with Carnot engine 1 between temperatures $T$ and $T_2$.

8.8    A Carnot engine rejects 50 Btu of heat to a reservoir at 50°F. If this engine produces 80 Btu as work, determine the heat source temperature.

8.9    A Carnot cycle is operated between 2000 R and 400 R. If 500 kW enters as heat, how much heat is rejected, what is the thermal efficiency of the cycle, and how much work is obtained from the cycle?

8.10   A Carnot cycle is operated between 1000 K and 200 K. If 1 MJ/h enters as heat, how much heat is rejected, what is the thermal efficiency of the cycle, and how much work is obtained from the cycle?

8.11   A Carnot cycle rejects 350 Btu/min while producing 5.5 hp of work. Determine the heat added, the lower temperature, and the efficiency of the cycle if the maximum temperature is 625°F.

## Problems Involving the Rankine Cycle

8.12   A turbine manufacturer specifies that the maximum steam temperature in the turbine is to be 1000°F and that the last stage of the turbine is to have a moisture content not exceeding 10% by weight. If the expansion is carried out reversibly and adiabatically from 800 psia, what is the state of the steam as it leaves the turbine?

8.13   Solve Problem 8.12 using the Mollier chart.

8.14   Solve Problem 8.12 using the computerized *Steam Tables*.

8.15   In Problem 8.12, the steam is expanded irreversibly. If the actual $\Delta h = 90\%$ of the $\Delta h$ that would be present if the process were carried out isentropically, what is the final moisture that would leave the turbine if the final pressure was that of Problem 8.12?

8.16   What is the internal efficiency of the turbine in Problem 8.15?

8.17   In a Rankine cycle, steam enters the turbine at 900 psia and 1000°F and exhausts at 1 psia. Determine (a) the turbine work, (b) the heat supplied, and (c) the thermal efficiency of the cycle. Neglect pump work. Use the *Steam Tables* and Mollier chart.

8.18   Solve Problem 8.17 using the computerized Steam Tables.

8.19   Calculate the pump work in Problem 8.17 using the approximate formula and the computerized Steam Tables.

8.20   If the condenser pressure in a Rankine cycle is 1 psia and the maximum pressure in the cycle is 600 psia, calculate the efficiency of the ideal cycle if the steam at 600 psia is saturated vapor. Use the Mollier chart and neglect pump work.

8.21    Solve Problem 8.20 using the computerized *Steam Tables*.

8.22    Use the approximate formula to determine the pump work in Problem 8.20.

8.23    If the vapor in Problem 8.20 has 200°F superheat, calculate the efficiency of the ideal Rankine cycle.

8.24    What is the pump work in Problem 8.20? Use the computerized *Steam Tables*.

8.25    What is the thermal efficiency of the Rankine cycle in Problem 8.20 if pump work is included?

8.26    Repeat Problem 8.20 for an initial pressure of 400 psia.

8.27    Repeat Problem 8.23 for an initial pressure of 400 psia.

8.28    A Rankine cycle operates between pressures of 2 and 1500 psia. If the steam entering the turbine is superheated to 800°F, calculate the efficiency of the cycle and the work produced per pound of steam. Neglect pump work. Use the *Steam Tables*.

8.29    Determine the thermal efficiency of the Rankine cycle in Problem 8.28 if pump work is included. Use the computerized *Steam Tables*.

8.30    A Rankine cycle is operated with a turbine inlet pressure of 5.0 Mpa and 300°C. Determine the efficiency of the cycle if the expansion is to 20 kPa. Neglect pump work.

8.31    If the expansion in Problem 8.30 is carried out to 50 kPa, determine the efficiency of the cycle.

8.32    If pump work is included in Problem 8.30, what is the cycle efficiency? Use the computerized *Steam Tables*.

8.33    If pump work is included in Problem 8.31, what is the efficiency of the cycle? Use the computerized *Steam Tables*.

8.34    A Rankine cycle is operated with a turbine inlet pressure of 600 psia and 600°F. Determine the efficiency if the expansion is to 1 psia. Use the Mollier diagram, and neglect pump work.

8.35    Determine the efficiency of the cycle in Problem 8.34 if the expansion is carried out to 10 psia. Compare results with those obtained in Problem 8.34.

8.36    If the turbine inlet pressure in a Rankine cycle is 5 Mpa and the inlet temperature is 350°C, determine the efficiency of the cycle if pump work is neglected and the engine expansion is to 10 kPa.

8.37    If pump work is included in Problem 8.36, what is the efficiency of the cycle? Use the computerized *Steam Tables* for the pump work.

8.38    A Rankine cycle is operated with a turbine inlet pressure of 500 psia and a temperature of 620°F. If the engine expansion is to 2 psia, what is the efficiency of the cycle? Neglect pump work.

8.39    If pump work is included in Problem 8.38, what is the efficiency of the cycle?

8.40    Steam is expanded to 10% moisture at 2 psia in a Rankine cycle. If the initial pressure is 500 psia, what is the efficiency of the cycle? Neglect pump work.

8.41    If pump work is included in Problem 8.40, what is the efficiency of the cycle?

8.42    Steam is expanded to 12% moisture at 1 psia in a Rankine cycle. If the initial pressure is 400 psia, what is the efficiency of the cycle? Neglect pump work.

8.43 Include pump work in Problem 8.42 and determine the cycle efficiency. Use the computerized *Steam Tables* for the pump work and for the cycle.

8.44 Determine the type efficiency of the cycle of Problem 8.20.

8.45 Determine the type efficiency of the cycle of Problem 8.26.

8.46 Determine the type efficiency of the cycle of Problem 8.30.

8.47 Determine the type efficiency of the cycle of Problem 8.34.

8.48 Determine the type efficiency of the cycle of Problem 8.36.

8.49 Determine the type efficiency of the cycle of Problem 8.38.

8.50 Determine the type efficiency of the cycle of Problem 8.40.

8.51 Determine the heat rate in Btu/kWh for Problem 8.34.

8.52 Determine the heat rate in Btu/kWh for Problem 8.35.

8.53 Determine the ideal steam rate for Problem 8.34.

8.54 Determine the ideal steam rate for Problem 8.35.

8.55 If the specific heat of the condenser cooling water is assumed to be unity and it is desired to limit the condenser cooling water temperature change to 50°F in the condenser, determine the pounds of cooling water required per pound of steam for the ideal Rankine cycle operated from 600 psia and 600°F to 2 psia. The condensate is assumed to be saturated liquid.

8.56 Steam is supplied to the turbine of a Rankine cycle at 500 psia, and the turbine exhausts at 1 psia. Determine the turbine inlet temperature if there is 10% moisture in the exhaust steam. Use the Mollier chart.

8.57 A university steam plant produces 1 000,000 kg/day of steam using natural gas as the fuel. Inlet water is saturated liquid at 1.40 Mpa. Outlet is 10% moisture at 1.40 Mpa. Determine the enthalpies at the inlet and the outlet in kJ/kg and the heat-transfer rate.

## Problems Involving the Reheat Cycle

8.58 A reheat cycle is operated to decrease the amount of moisture in the final steam from a turbine. If the steam, initially at 1000 psia and 800°F, is permitted to expand only to saturation and is reheated to its initial enthalpy, what will the moisture content of the final steam be if it is expanded reversibly to 2 in. Hg? Use the Mollier chart.

8.59 Solve Problem 8.58 using the computerized *Steam Tables*.

8.60 A reheat cycle operates with initial steam conditions of 1500 psia and 1000°F at the inlet to the turbine. The steam expands isentropically to 400 psia and is reheated at constant pressure to 1000°F. If the exhaust pressure of the cycle is 1 psia, determine the cycle efficiency. Use the Mollier chart.

8.61 Solve Problem 8.60 using the computerized *Steam Tables*.

8.62 Determine the heat rate in Problem 8.60.

8.63 If the steam in Problem 8.58 is reheated by the addition of one-half of the initial decrease in enthalpy, what will the moisture content of the final steam be after a reversible expansion to 2 in. Hg?

8.64    Steam in a reheat cycle is initially at 5 Mpa and 500°C. It expands to saturation and is reheated to its initial enthalpy. If the final expansion is to 5 kPa, what is the moisture content of the steam leaving the turbine?

8.65    A reheat cycle is operated from 500 psia and 500°F to a final pressure of 1 in. Hg. If the amount of reheat added is $(1/2)\Delta h$ from the initial conditions to saturated vapor, determine the efficiency of the cycle ($h_f$ at 1 in. Hg is 47.09 Btu/lb; temperature is close to 79°F).

8.66    What is the type efficiency of the cycle in Problem 8.65?

8.67    Steam at 10 Mpa and 500°C enters the turbine of a Rankine cycle that has one stage of reheat. The reheat is from saturated steam to the original temperature. If the final expansion is to 10 kPa, determine the efficiency of the cycle. Use the Mollier chart.

8.68    Solve Problem 8.67 using the computerized *Steam Tables*.

8.69    An ideal Rankine cycle with reheat operates with a turbine inlet pressure of 1300 psia and 1000°F. The condenser pressure is 1 psia. If the steam is first expanded to saturated vapor and then reheated to 900°F, determine the thermal efficiency of the cycle. Use the Mollier chart.

8.70    Solve Problem 8.69 using the computerized *Steam Tables*.

*8.71   Steam at 10 Mpa and 600°C enters the turbine of an ideal Rankine cycle. The steam is reheated to 500°C. If saturated vapor enters the condenser at 5 kPa, determine the cycle efficiency. Use the Mollier chart.

*8.72   Solve Problem 8.71 using the computerized *Steam Tables*.

8.73    In a reheat cycle, the steam is expanded from 700 psia and 600°F to 100 psia. It is subsequently reheated until the steam has the same degree of superheat as it had initially. If the final pressure of the cycle is 1 psia, determine its efficiency.

8.74    Solve Problem 8.73 using the computerized *Steam Tables*.

8.75    Solve Problem 8.73 including the pump work.

8.76    Solve Problem 8.74 including the pump work.

8.77    In an ideal Rankine cycle with reheat, steam enters the turbine at 10 Mpa and 600°C. The steam is reheated to the initial temperature at a pressure of 1 Mpa. If the condenser is operated at 5 kPa, determine the cycle efficiency. Use the Mollier chart. Neglect pump work.

8.78    Use the computerized *Steam Tables* to solve Problem 8.77.

8.79    Solve Problem 8.78 including pump work.

8.80    Starting with 500°F and 500 psia into the turbine and assuming that the steam in a reheat cycle is reheated to an enthalpy equal to its initial enthalpy, determine the efficiency of this cycle if the final pressure is 1 psia and the initial expansion is carried out to 100 psia.

8.81    Assume that the final pressure in a reheat cycle is 1 psia and the first expansion is to saturated vapor. If the steam is reheated to its initial enthalpy, determine the efficiency of the cycle for an initial pressure of 500 psia and an initial temperature of 600°F.

## Problems Involving the Regenerative Cycle

8.82    Determine the efficiency of the cycle in Problem 8.20 if it operates as a regenerative cycle with one open heater at 100 psia. Compare your answers to both problems.

8.83    Solve Problem 8.82 using the computerized *Steam Tables*.

8.84    Use formulas (a) and (b) of Illustrative Problem 8.12 to solve Problem 8.82.

8.85    Determine the efficiency of the cycle in Problem 8.34 if it operates with one open feedwater heater at 50 psia. Compare your answers to both problems.

8.86    Solve Problem 8.85 using the computerized *Steam Tables*.

8.87    A regenerative cycle operates with a turbine inlet temperature of 600°C and a pressure of 10 Mpa. The condenser is operated at 5 kPa. If there is one open heater in the cycle with extraction at 1 Mpa, determine the efficiency of the cycle. Use the Mollier chart.

8.88    Solve Problem 8.87 using the computerized *Steam Tables*.

8.89    Use Equations (a) and (b) of Illustrative Problem 8.12 to solve Problem 8.88.

8.90    Steam enters the turbine of a regenerative cycle at 900 psia and 900°F. The condenser is operated at 1 psia. If extraction to the single open heater is at 200 psia, determine the thermal efficiency of the cycle. Use the Mollier chart.

8.91    Solve Problem 8.90 using the computerized *Steam Tables*.

8.92    Solve Problem 8.91 using Equations (a) and (b) of Illustrative Problem 8.12.

*8.93    Solve Problem 8.90 if a second open heater is used with extraction at 300 psia. Use the Mollier chart.

*8.94    Use the computerized *Steam Tables* to solve Problem 8.93.

8.95    Use Equations (a), (b), and (c) of Illustrative Problem 8.13 to solve Problem 8.94.

*8.96    A regenerative cycle operates with two open heaters, one at 50 psia and the other at 100 psia. The inlet to the turbine is superheated steam at 1000°F and 1000 psia, and the outlet of the turbine is at 1 psia. Using an ideal cycle, determine the efficiency and compare it to Illustrative Problem 8.12.

*8.97    An ideal regenerative cycle operates with reheat. The high-pressure turbine is supplied with steam at 2000 psia and 1000°F. The steam expands to 200 psia and is removed to a reheater where it is reheated back to 1000°F. Part of the steam leaving the high-pressure turbine is used in an open heater for feedwater heating. Steam is extracted from the low-pressure turbine at 50 psia and is used for feedwater heating in a second open feedwater heater. Determine the efficiency of the cycle if the condenser pressure is 1 psia. Use the Mollier chart.

*8.98    Solve Problem 8.97 using the computerized *Steam Tables*.

8.99    A regenerative Rankine cycle uses one closed feedwater heater. Inlet to the turbine is at 10 MPa and 550°C. Steam expands isentropically to 1 Mpa where some is extracted; the remainder expands isentropically to 7.5 kPa where it enters the condenser. The condensed extraction steam is passed through a trap to the condenser. Determine the cycle efficiency and the fraction of steam regenerated.

8.100   A regenerative Rankine cycle uses one closed feedwater heater. Inlet to the turbine is at 800 psia and 1100°F. Steam expands isentropically to 100 psia where some is extracted; the remainder expands isentropically to 1 psia where it enters the condenser. The condensed extraction steam is passed through a trap to the condenser. Determine the cycle efficiency and the fraction of steam regenerated.

# 9

## Gas Power Cycles

### Learning Goals

After reading and studying the material in this chapter, you should be able to

1. Differentiate between internal-combustion and external-combustion heat engines.
2. Describe the sequence of events in the four-stroke and two-stroke cycles.
3. State the assumptions made in air-standard cycle analyses.
4. Sketch the ideal Otto cycle on both $p$–$v$ and $T$–$s$ diagrams.
5. Show how an actual cycle differs from the ideal Otto cycle.
6. Derive the expression for the efficiency of the ideal air-standard Otto cycle and show that it is a function only of the compression ratio.
7. Sketch the ideal Diesel cycle on both $p$–$v$ and $T$–$s$ diagrams.
8. Show that for the same compression ratio, the efficiency of the ideal Otto cycle is greater than the efficiency of the ideal Diesel cycle.
9. Understand why the Diesel cycle can be operated at higher compression ratios than the Otto cycle.
10. Analyze automotive engines for mean effective pressure and efficiencies.
11. Sketch the ideal Brayton cycle on both $p$–$v$ and $T$–$s$ diagrams.
12. Learn to analyze the Brayton cycle and find its ideal and actual efficiencies.
13. Understand how intercooling, reheat, and regeneration affect the efficiency of the Brayton cycle.
14. Understand why the dual cycle was proposed as the prototype for actual internal-combustion engines.
15. Show that the efficiency of the ideal dual-cycle engine lies between the efficiency of the Otto and Diesel cycles.
16. Understand that regeneration can be applied to internal-combustion cycles with the possibility of approaching the Carnot cycle efficiency, but that the practical realization of this goal is difficult because of the requirements and the complexity of the equipment needed.

## 9.1 Introduction

The power-plant cycles studied in Chapter 8 are known as *external-combustion cycles* because heat is supplied from a source that is external to the engine and is rejected to a sink that is also external to the engine. However, the power plants that we use for propelling cars, boats, and so on are commonly known as *internal-combustion engines* because the release of energy is by combustion within the engine. Mobile power plants have been developed to a high state of reliability and are commonplace in everyday life. Reliability is essentially a term that describes the level of development of the mechanical system. Thus, as the usual occurrence, we expect that a car will start easily and that a trip will be completed without mechanical failure. However, the size, weight, and cost of fuel for a given engine are all functions of the thermal efficiency of the unit. It should be noted that external-combustion engines have been studied more extensively in recent times with the aim of reducing air pollution, a problem in modern industrial societies.

In Chapter 8, our attention was focused on systems in which the release of energy by combustion or other means occurred external to the device or engine that converted this energy to useful work. Several limitations occur when this is done. In the combustion device, temperatures are maintained as high as possible. This leads to problems associated with the strength of the heat-transfer surfaces (e.g., tubing) and requires large amounts of expensive, bulky insulation to avoid unnecessary heat losses. The working fluid of the cycle is contained within tubular elements, and to keep thermal and pressure stresses within the strength capabilities of the tube material, the heat-transfer rates must be limited. This leads to the need for large heat-transfer surfaces requiring large units. Many of these objections can be overcome by releasing the energy in the fuel within the engine. In any device, it is usually true that the average temperature rather than the peak temperature of the working fluid will govern the strength limits of the material. In a periodic or cyclic device (e.g., the automotive engine), the average temperature is kept reasonable or acceptable level by maintaining the duration of the peak temperature for a small fraction of the total cycle time.

Before considering the prototype ideal cycles of the internal-combustion engine, let us first describe the engine and then make those idealizations required for the analysis of the cycle. These engines are mechanically complex and have hundreds of moving parts. To understand this engine, we will consider the four-stroke reciprocating automobile engine, which is the most commonly used.

The four events that take place in the cycle are illustrated in Figure 9.1 and are as follows:

1. *Intake.* Consider the piston to be at the uppermost part of its stroke known as top dead center (t.d.c.). As the piston starts downward, it creates a partial vacuum in the cylinder. At this time, the intake valve is open, and a mixture of air and vaporized gasoline is drawn into the cylinder from the intake manifold past the open intake valve.

2. *Compression.* When the piston reaches the end of its downward stroke, the piston is said to be at bottom dead center (b.d.c.). At this point, the intake valve closes and stays closed as the piston moves upward. Because the fuel–air mixture is totally confined within the cylinder, it is compressed as the piston moves upward. Both intake and exhaust valves remain closed.

3. *Power.* Ideally, when the piston again reaches top dead center and the fuel–air mixture is at its maximum compression, an electrical spark ignites the fuel–air mixture, causing combustion to occur quite rapidly. The large pressure force created

**FIGURE 9.1**
Four-stroke internal-combustion engine. (From "How an Internal Engine Works." Courtesy of Renewal Products Co., Fairless Hills, PA.)

by the combustion causes the piston to be driven downward. This driven-downward motion of the piston is the power stroke of the cycle. In actual engines, ignition can occur before, at, or even after top dead center, based on details that we will not consider here. Also, both the intake and exhaust valves remain closed during the power stroke.

4. *Exhaust.* At the end of the power stroke, the gases have fully expanded, and the exhaust valve opens. The upward motion of the piston causes the spent combustion gases to flow out of the cylinder past the open exhaust valve and into the exhaust manifold. At top dead center, the exhaust valve closes, and the cycle starts again.

The four-stroke cycle just described produces one power stroke for each piston for two revolutions of the main shaft (crankshaft). The same sequence of events can also be accomplished in one revolution of the crankshaft to give one power stroke per revolution. This is the two-stroke engine cycle. The sequence of events is shown in Figure 9.2 and consists of the following for a fuel-injected engine:

**FIGURE 9.2**
Two-stroke internal-combustion engine.

1. *Air intake.* With transfer and exhaust ports open, air from the crankcase fills the cylinder.

2. *Compression.* With all ports covered, the rising piston compresses air and creates suction in the crankcase. Injection starts.

3. *Expansion.* The burning mixture expands, forcing the piston down. Air flows into the crankcase to be compressed as the piston descends.

4. *Exhaust.* The descending piston uncovers the exhaust port. Slight pressure builds up in the crankcase, which is enough to force air into the cylinder.

The distinguishing feature of the two-stroke engine is that *every* outward stroke of the piston is a power-producing stroke. It should be noted that for *a given output*, a definite air capacity is required, requiring the two-stroke engine to have an air input per unit time at least equal to the four-stroke equivalent engine.

In a modern engine, the cylinder block is usually an accurately machined housing for the cylinders, pistons, and connecting rods. All the other elements are supported by the cylinder block (or blocks) or bolted to it. At the top of the block is the cylinder head. To prevent melting of the metal parts, cooling must be provided. This can be done by either air cooling or water cooling. Water cooling is the commonest method used, and this water must circulate within the block and the heads. Thus, a series of matching openings is provided in the blocks and heads for water circulation.

The valves in the internal-combustion engine are subjected to the severest conditions of all the elements of the engine. Operating as often as 2000 times per minute, they are required to be leakproof even though they are subjected to the direct pressure and temperature of the combustion occurring within the cylinders. Also, they must operate at the precisely correct moment in the cycle without valve bounce or chatter. The actuation of the valves is performed by one cam for each valve on the camshaft that is driven by the crankshaft. Because the four-cycle engine produces one power stroke per piston for two revolutions of the crankshaft, the camshaft is geared to rotate one revolution for every two revolutions of the crankshaft. The cam shape must be accurately designed to open the intake and exhaust valves precisely, hold them open for the necessary length of time, and close them all as dictated by the cycle.

Many different arrangements of the cylinders and the manner in which they are grouped are possible. We have already seen and described the four-cylinder engine and the horizontally opposed, six-cylinder engine. In addition to the arrangements previously discussed, there are others, as shown in Figure 9.3, that are used in specific applications.

To this point, our discussion of internal-combustion engines has considered only reciprocating engines. We will now consider the rotating-combustion engine, which is also known as the Wankel engine. The rotating-combustion engine was first conceived by a German inventor, Felix Wankel, in 1954. Since the early 1960s, intense development work on this engine has taken place throughout the world. At present, the Wankel has been produced in sizes ranging from 1/2 to 400 hp and powers approximately 500,000 automobiles worldwide. To obtain an idea of the relative size and simplicity of the rotating-combustion engine compared to a conventional reciprocating automotive engine, one only has to note the following: A V-8 engine of 195 hp has 1029 parts, and 388 of these parts move. The engine weighs more than 600 lb. and takes up 15 ft.[3] in the engine compartment. In contrast, a 185 hp Wankel engine has only 633 parts, 154 of which move, and it only weighs 237 lb. and occupies 5 ft.[3]. In addition, there are no valves in the Wankel engine.

**FIGURE 9.3**
Internal-combustion engine cylinder arrangement. (From "How an Internal Engine Works." Courtesy of Renewal Products Co., Fairless Hills, PA.)

The Wankel engine operates on the four-stroke Otto cycle, as indicated in Figure 9.4. As shown in the figure, there are two major moving assemblies in the Wankel engine: a three-sided rotor and a main shaft. The rotor operates inside a chamber that is essentially an ellipse with sides pointed in. The mathematical name of the shape of the chamber is an epitrochoid. This shape is needed due to the motion of the rotor, which rotates on an eccentric axis off the main shaft in a wobbling motion. The cycle shown in Figure 9.4 starts when one of the apex points passes the intake port, causing the fuel–air mixture to flow into the housing in the volume between the rotor and housing. As the rotor turns, the volume between the rotor and chamber decreases, which effectively compresses the fuel–air mixture. The fuel–air mixture is ignited when its volume is at its minimum. The combustion of the fuel–air mixture causes a thrust on the rotor to keep it turning and, at the same time, causes the combustion gases to expand. At the end of expansion, the exhaust port is uncovered, and the combustion gases leave through this port.

Thus far, we have considered only one face of the three-lobed rotor during the course of one revolution. It must be remembered that during this same revolution, the other two faces of the rotor are undergoing the same cycle. Thus, the torque output of the Wankel engine is positive for about two-thirds of one shaft revolution as compared to one-quarter of the two shaft revolutions required to complete a single cycle of a four-cycle reciprocating engine. This is due to the fact that there is a 1:3 rotor-to-crankshaft speed ratio in the Wankel engine, requiring three shaft revolutions for one complete rotor revolution to fire all three rotor flanks. A single-cylinder reciprocating engine fires once in two shaft revolutions, giving a three-cylinder reciprocating engine three power pulses in two shaft revolutions. On the basis of torque output, the Wankel single-rotor engine approaches the

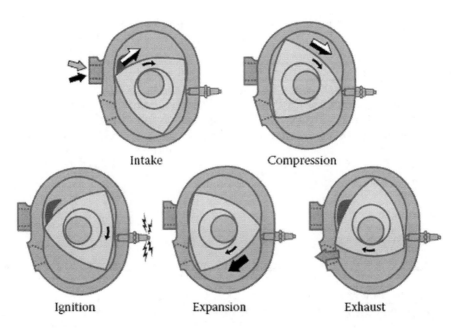

Intake          Compression

Ignition     Expansion     Exhaust

**FIGURE 9.4**
Wankel engine combustion cycle. (Courtesy of Curtiss-Wright Corp.)

three-cylinder reciprocating engine. If we consider power, the single-rotor Wankel engine power output is twice the power output of the same displacement reciprocating the four-cycle engine. This is due to the fact that for every main-shaft revolution, there is one firing, similar to the two-cycle engine. Also, it must be noted that in the piston engine, the piston comes to a complete halt every time it reverses direction and that due to its linear motion, a connecting rod and crank are needed to convert its up-and-down motion into torque. As a result of these factors, the Wankel engine gives smoother torque output at all speeds without dynamic unbalance even when one rotor is used.

The advantages of the Wankel engine can be summarized as follows:

1. There are fewer moving parts: no valves, connecting rods, and so on.
2. There is a high power-to-weight ratio and a small volume.
3. The rotor center of gravity and rotational axis remain at a fixed eccentricity from the shaft centerline so that even a single-rotor engine can attain complete dynamic balance with simple shaft counterweights.
4. Smooth torque and power output exist at all speeds.

One of the Wankel engine's most serious engineering problems was the design of the apex seals that are used to seal off the three compartments from each other. Some of the initial engines built for automotive use seldom ran for more than 15,000 to 20,000 mi. before the seals wore out. At present, it is claimed that seal lifetimes in automotive applications are in the 100,000 mi. range, which is nearing the life of piston rings in conventional engines. Another problem that occurred during the initial development of the Wankel engine was its poor fuel economy, which was about half the miles per gallon of a comparative conventional automotive engine. The principal factors contributing to this poor mileage were the positioning of the spark plugs, insufficient distance between the inlet and exhaust ports, and the already-mentioned seal problem. Using a modified side entry port, a repositioned spark plug, and the latest seals, current versions of the Wankel engine indicate that they are reaching the same fuel consumption levels *as comparably powered, conventional reciprocating engines*. The only commercially produced rotary engine was used in the Mazda RX-7 sports car. This car is no longer available in the United States but continues to be sold in other countries. A similar rotary engine car was introduced in the US market in 2003.

## 9.2 Air-Standard Analysis of the Otto Cycle

The introduction of fuel into an internal-combustion engine gives rise to a variable mass in the cycle. Also, at the end of the cycle, the entire charge is discarded, and a fresh charge is introduced to undergo a new cycle identical to the first. During the actual processes, both heat and work are interchanged at each portion of the cycle. The mass, specific heats, and state of the working fluid are all variable. Under these circumstances, the analysis of a cycle becomes quite difficult. To simplify the analysis, certain idealized cycles have been proposed as prototypes of the actual engine cycles. These models are considered for the rest of this chapter. In the analysis of each of the cycles, the following assumptions are made:

1. Each process is carried out reversibly. Friction, pressure differences, turbulence, and the like are neglected.
2. The working fluid is an ideal gas, and all relations already derived for the ideal gas are applicable.
3. The necessary energy is added or removed externally to achieve the desired state changes.
4. The working fluid is a gas with constant specific heat values.
5. The amount of working fluid in the cycle is constant. In effect, the same air remains in the engine through each cycle.

It is apparent from these assumptions (and their implications) that the analysis of an engine cycle based on them is quite artificial. However, certain generalizations can be arrived at from this analysis, called the *air-standard analysis*, and as such, it can be quite useful.

The *Otto cycle* is the prototype for most internal-combustion engines commonly in use. Regardless of the number of strokes required to complete the cycle, it is conceived of as consisting of four separate and distinct processes. As shown in the *T–s* and *p–v* diagrams in Figure 9.5, the Otto cycle consists of (after induction of the gas) an isentropic compression followed by the reversible constant-volume addition of heat, then an isentropic expansion from which work is extracted, and finally, a reversible constant-volume rejection of heat. Subsequently, the cycle is repeated. It will be noted that each step is an idealization

**FIGURE 9.5**
Air-standard Otto cycle. (a) $p–v$ diagram. (b) $T–s$ diagram. (c) Otto engine cycle

of the events that we have previously described for the internal-combustion engine. Also, note that each portion of the cycle is a nonflow (closed system) process.

Referring to Figure 9.5, we have the nonflow or closed system energy equation for the nonflow reception of heat (step ③ to ④) as

$$q_{in} = c_v \left( T_4 - T_3 \right) \text{Btu/lb}_m, \text{kJ/kg} \tag{9.1}$$

Similarly, the heat rejected during the constant-volume, reversible expansion (step ⑤ to ②) is

$$q_r = c_v \left( T_5 - T_2 \right) \text{Btu/lb}_m, \text{kJ/kg} \tag{9.2}$$

The net work in thermal units available from the cycle is the difference between $q_{in}$ and $q_r$:

$$w = \left( q_{in} - q_r \right) \text{Btu/lb}_m, \text{kJ/kg} \tag{9.3}$$

The efficiency of the cycle is

$$\eta_{Otto} = \frac{w}{q_{in}} = \frac{q_{in} - q_r}{q_{in}} = 1 - \frac{q_r}{q_{in}}$$ (9.4a)

and

$$\eta_{Otto} = 1 - \frac{c_v(T_5 - T_2)}{c_v(T_4 - T_3)} = 1 - \frac{T_5 - T_2}{T_4 - T_3}$$ (9.4b)

From Figure 9.5, it will be noted that the volumetric limits of the expansion and compression portions of the cycle are equal. From equations developed in Chapter 6,

$$\frac{T_4}{T_5} = \frac{T_3}{T_2} \quad \text{or} \quad \frac{T_2}{T_5} = \frac{T_3}{T_4}$$ (9.5)

By adding unity to both sides of Equation 9.5,

$$1 - \frac{T_2}{T_5} = 1 - \frac{T_3}{T_4} \quad \text{or} \quad \frac{T_5 - T_2}{T_5} = \frac{T_4 - T_3}{T_4}$$ (9.6)

By substituting Equation 9.6 into Equation 9.4b, we obtain

$$\eta_{Otto} = 1 - \frac{T_5}{T_4} = 1 - \frac{T_2}{T_3} = 1 - \left(\frac{v_3}{v_2}\right)^{k-1}$$ (9.7)

where we have used the equation of path for the isentropic compression to relate temperature and volumes from Table 6.4.

At this time, we will introduce a term called the *compression ratio*, $r_c$, which is defined as

$$r_c = \frac{v_2}{v_3}$$ (9.8)

Note that the compression ratio defined in this manner is the ratio of two volumes, not two pressures as it is sometimes thought. Using this definition of compression ratio, we have the efficiency of the Otto cycle as

$$\eta_{Otto} = 1 - \left(\frac{1}{r_c}\right)^{k-1} \times 100$$ (9.9)

It is emphasized that the term *compression ratio* is not the ratio of the peak-to-inlet pressure; it is the ratio of the volume before compression (b.d.c) to the volume after compression (t.d.c.).

Figure 9.6 is a plot of Equation 9.9 for $k = 1.4$ (for air as working fluid). It is apparent that an increase in compression ratio yields an increased cycle efficiency. However, it will be noted that the efficiency increases at a decreasing rate, and an increase in compression ratio from 10 to 15 yields an increase in efficiency of only 6%. Until the mid-1960s, the compression ratio of the automotive engine was generally being increased in an effort to get more power out of a given engine. It was not uncommon to have engines with compression ratios in excess of 10, requiring highly leaded fuels to prevent preignition and detonation of the fuel–air mixture in the compression and ignition portions of the cycle. Environmental

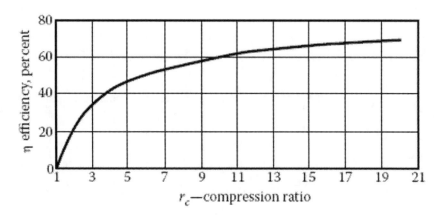

**FIGURE 9.6**
Efficiency of the Otto cycle.

considerations, with the attendant desire to limit the pollutants from these engines, have led to engines having lower compression ratios. The use of pollution controls, fuel injection, multiple valves, and turbo-supercharging has enabled modern engines to achieve high power outputs while still being environmentally acceptable. The use of multipoint fuel injection, electronic ignition, and electronic engine management systems has made it possible to increase the compression ratio on modern engines.

Returning to the air-standard Otto cycle, another point must be emphasized. The Otto air-standard cycle efficiency is determined only by the ratio of $v_2$ to $v_3$. Thus, the cycle operating between the limits of ②, ③, ④, and ⑤ in Figure 9.5 will have the same efficiency as the cycle operated between ②, ③,④, and⑤ The heat in, work out, and heat rejected will be different for each cycle, but the efficiency will be the same. However, a Carnot cycle having as its maximum temperature the upper temperature of the Otto cycle and as its lower temperature the lowest temperature of the Otto cycle will always have a thermal efficiency that is greater than the Otto cycle. This can be determined from Figure 9.5 as follows: As heat is added to the Otto cycle, its maximum temperature $T_4$ is increased, but its efficiency is constant if $v_2/v_3$ is constant. On the other hand, the Carnot cycle efficiency increases as its peak cycle temperature is increased. In the limit, the least value of the efficiency of a Carnot cycle will be $\eta = 1 - T_2/T_3$, which is the maximum efficiency of the Otto cycle.

## ILLUSTRATIVE PROBLEM 9.1

Compute the efficiency of an air-standard Otto cycle if the compression ratio is 7 and $k = 1.4$. Compare this with a Carnot cycle operating between the same temperature limits if the lowest temperature of the cycle is 70°F and the peak temperature is 700°F, 1000°F, and 3000°F. Plot $\eta$ versus temperature.

### SOLUTION

The efficiency of the Otto cycle is a function only of the compression ratio. Therefore, for the entire range of this problem, $\eta$ is constant and equal to

$$\eta_{Otto} = 1 - \frac{1}{7^{0.4}} \times 100 = 54.1\%$$

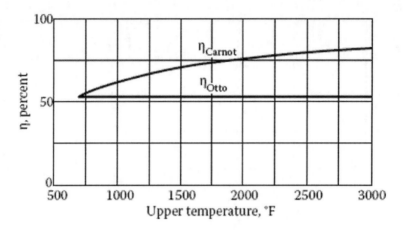

**FIGURE 9.7**
Solution of Illustrative Problem 9.1.

For the Carnot cycle,

$$\eta_c = 1 - \frac{T_2}{T_4}$$

At 700°F (1160 R),

$$\eta_c = 1 - \frac{530}{1160} \times 100 = 54.3\%$$

At 1000°F (1460 R),

$$\eta_c = 1 - \frac{530}{1460} \times 100 = 63.7\%$$

At 3000°F (3460 R),

$$\eta_c = 1 - \frac{530}{3460} \times 100 = 84.7\%$$

These data are plotted in Figure 9.7.

---

## ILLUSTRATIVE PROBLEM 9.2

Assuming $c_v$ to be 0.172 Btu/lb$_\text{m}$·R, determine the heat added ($q_{in}$) and the heat rejected per pound of fluid ($q_r$) during the constant-volume portions of Illustrative Problem 9.1 for the case in which the peak temperature of the cycle, $T_4$ is 1000°F (1460 R). Also, compute the work output per pound of fluid ($w = q_{in} - q_r$). Determine the efficiency of the Otto cycle, $\eta_{Otto}$, based on these values, and compare it with the results of Illustrative Problem 9.1.

## SOLUTION

By referring to Figure 9.5 and using the relations developed in Chapter 6,

$$\frac{T_3}{T_2} = \left(\frac{v_2}{v_3}\right)^{k-1} = (7)^{0.4} = 2.18$$

$$T_3 = (2.18)(530) = 1155.4°R$$

$$q_{in} = c_v\left(T_4 - T_3\right) = (0.172)(1460 - 1155.4)$$
$$= 52.39 \text{ Btu/lb}_m$$

To determine $T_5$, further use is made of Table 6.4:

$$\frac{T_5}{T_4} = \left(\frac{v_4}{v_5}\right)^{k-1} = \left(\frac{v_3}{v_2}\right)^{k-1} = \left(\frac{1}{7}\right)^{1.4-1} \frac{1}{2.18}$$

Therefore,

$$T_5 = \frac{1460}{2.18} = 669.72$$

$$q_r = c_v\left(T_5 - T_2\right) = (0.172)(669.72 - 530) = 24.03 \text{ Btu/lb}_m$$

The net work out is

$$q_{in} - q_r = 52.39 - 24.03 = 28.36 \text{ Btu/lb}_m$$

Therefore,

$$\eta_{Otto} = \frac{28.36}{52.39} \times 100 = 54.1\%$$

This value of efficiency agrees with the results of Illustrative Problem 9.1.

---

### ILLUSTRATIVE PROBLEM 9.3

An Otto cycle operates with a compression ratio, $r_c$ of 8. If $q_{in}$ of 50 kJ is added to the cycle, which has as its lower pressure 150 kPa and as its lower temperature, $T_2 = 20°C$, determine the peak temperature of the cycle, $T_4$, if $c_v = 0.7186$ kJ/kg·K and $k = 1.4$. Assume 1 kg of working fluid.

## SOLUTION

Refer to Figure 9.5 and

$$\frac{T_3}{T_2} = \left(\frac{v_2}{v_3}\right)^{k-1} = 8^{0.4} = 2.297$$

Therefore,

$$T_3 = (2.297)(20 + 273) = 673.14 \text{ K}$$

But

$$q_{in} = c_v (T_4 - T_3)$$

Therefore,

$$50 = 0.7186 (T_4 - T_3)$$

$$69.58 = T_4 - T_3$$

and

$$T_4 = 742.72 \text{ K} = 469.72°C$$

---

### ILLUSTRATIVE PROBLEM 9.4

An Otto cycle operates with a compression ratio, $r_c$ of 7. If $q_{in}$ of 50 Btu/lb$_m$ is added to the cycle, which begins at 60°F and 14.7 psia, determine the pressure, temperature, and specific volume at each point of the cycle. Also, calculate the efficiency of the cycle. Assume that $c_p = 0.24$ Btu/lbm·R, $c_v = 0.171$ Btu/lbm·R, $R = 53.3$ ft.-lb$_m$·R, and $k = 1.4$.

### SOLUTION

We will refer to Figure 9.5 and go through the cycle point to point. At ②, we need $v_2$. This will be found from the equation of state or ideal gas law, $p_2 v_2 = RT_2$:

$$v_2 = \frac{53.3 \times 520}{14.7144} = 13.09 \text{ ft.}^3/\text{lb}_m$$

Because the path from ② to ③ is isentropic,

$$p_3 v_3^k = p_2 v_2^k$$

$$p_3 = p_2 \frac{v_2^k}{v_3} = p_2 r_c^k = (14.7)(7)^{1.4} = 224.1 \text{ psia}$$

Using compression ratio

$$v_3 = \frac{v_2}{r_c} = \frac{13.09}{7} = 1.87 \text{ ft.}^3 / \text{lb}_m$$

Using ideal gas law

$$T_3 = \frac{p_3 v_3}{R} = \frac{224.1 \times 144 \times 1.87}{53.3} = 1132.2 \text{ R}$$

To obtain the values at ④, we note that $v_3 = v_4$ and that $q_{in} = c_v(T_4 - T_3)$. Thus,

$$T_4 = T_3 + \frac{q}{c_v} = 1132.2 + \frac{50}{0.171} = 1424.6 \text{ R}$$

For $p_4$, we use the equation of state or ideal gas law to obtain

$$p_4 = \frac{RT_4}{v_4} = \frac{53.3 \times 1424.5}{144 \times 1.87} = 281.98 \text{ psia}$$

The last point has the same specific volume as ②, giving $v_5 = v_2 = 13.09$ ft.³/lb$_m$. The isentropic path equation gives

$$p_5 v_5^k = p_4 v_4^k \quad p_5 = p_4 \left(\frac{v_4}{v_5}\right)^k = p_4 \left(\frac{1}{r_c}\right)^k$$

$$p_5 = 281.96 \left(\frac{1}{7}\right)^{1.4} = 18.5 \text{ psia}$$

$T_5$ is found from the equation of state (ideal gas law) as

$$T_5 = \frac{p_5 v_5}{R} = \frac{18.49 \times 144 \times 13.09}{53.3} = 654.1 \text{ R}$$

The efficiency can be expressed in terms of the temperature as

$$\eta = \frac{(T_4 - T_3) - (T_5 - T_2)}{T_4 - T_3} \times 100$$

$$= \frac{(1424.6 - 1132.2) - (654.1 - 520)}{1424.6 - 1132.2} \times 100$$

$$= 54.1\%$$

The *mean effective pressure* (mep) of a cycle is defined as the work out of a cycle divided by the volume swept out by the piston. In Figure 9.8, it is the shaded area (work) divided by $(v_2 - v_3)$. It becomes that number (the horizontal dashed line) that, when multiplied by the base, yields the same shaded area. As such, it is the mathematical mean ordinate of Figure 9.8; the area of the rectangle $[w_{out} = (mep) \times (v_2 - v_3)]$ equals the area enclosed by the cycle ②, ③, ④, and ⑤. The value of the mep can be obtained by integrating the area shown in Figure 9.8 and dividing by $(v_2 - v_3)$. However, it can also be obtained as follows:

The work out, $w_{out}$ of the cycle equals $q_{in} - q_r$ in thermal units of Btu per pound:

$$w_{out} = q_{in} - q_r = c_v\left(T_4 - T_3\right) - c_v\left(T_5 - T_2\right)$$

In terms of pressure and volumes, we can show that

$$w_{out} = q_{in} - q_r = \frac{1}{k-1}\left[v_3\left(p_4 - p_3\right) - v_2\left(p_5 - p_2\right)\right]$$

in mechanical units of ft.-lb$_f$/lb$_m$ when pressure is in lb$_f$/ft.$^2$ and specific volume in ft.$^3$/lb$_m$. Dividing by $(v_2 - v_3)$, we obtain

$$mep = \frac{1}{k-1}\frac{v_3\left(p_4 - p_3\right) - v_2\left(p_5 - p_2\right)}{v_2 - v_3} \tag{9.10}$$

or

$$mep = \frac{1}{k-1}\frac{\left(p_4 - r_c^k p_2\right)\left[1 - \left(1/r_c\right)^{k-1}\right]}{r_c - 1} \tag{9.11}$$

The mep is useful in evaluating the ability to produce power. Because the product of piston travel (stroke) and piston area is piston volume (displacement), the relative power-producing capability of two pistons can be established by comparing the product of mep times displacement in each case.

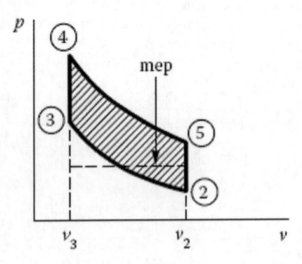

**FIGURE 9.8**
A *p–v* diagram of the Otto cycle.

## ILLUSTRATIVE PROBLEM 9.5

Determine the power output of a cylinder having a cross-sectional area of $A$ square inches, a length of stroke $L$ inches, and a mep of $p_m$ psi, and making $N$ power strokes per minute.

### SOLUTION

The force on the piston is $(F_p = p_m A)$ pounds. The work done is equivalent to a constant force on the piston $(F_p)$ over the length of the piston stroke, the work done per power stroke is $W = F_p L = (P_m A)L$. For $N$ power strokes per minute,

$$\text{hp} = \frac{p_m L A N}{33000}$$

Note that if $p_m$ is in psi, $A$ should be in square inches, while $L$ is in feet. Also, $N$ is *not* the rpm of the engine; it is the number of power strokes per minute, which for a four-cycle engine is rpm/2.

## ILLUSTRATIVE PROBLEM 9.6

It is common practice to denote the total volumetric displacement in liters in internal-combustion engines. Using this notation to mean $(LA)$, determine the power output of an engine having a mep of $p_m$ kPa and making $N$ power strokes per minute.

### SOLUTION

The basic solution is the same as for Illustrative Problem 9.5 with due respect to units. The displacement $(LA)$ in $cm^3/1000$ needs to be converted to $m^3$.

$$cm^3 \times \frac{1}{(cm/m)^3} = cm^3 \times \frac{1}{(100)^3} = cm^3 \times 10^{-6}$$

or

$$\text{liters} \times 10^{-3} = m^3$$

One horsepower is equal to 746 W = 746 N·m/s. per minute,

$$1\text{hp} = 746 \times 60 = 44760 \frac{N \cdot m}{min}$$

Therefore,

$$\text{hp} = \frac{p_m \times 10^3 \times (LA)10^{-3} \times N}{44760} = \frac{p_m L A N}{44760}$$

where $LA$ is expressed in liters, $p_m$ in kilopascals, and $N$ in power strokes per minute.

### ILLUSTRATIVE PROBLEM 9.7

Using the results of Illustrative Problem 9.6, determine the horsepower of a 2.0 L engine that has a mep of 1 MPa operating at 4000 rpm. The engine is a four-cycle engine.

**SOLUTION**

$$hp = \frac{p_m LAN}{44760} = \frac{1000\,\text{kPa} \times 2\,\text{L} \times \dfrac{4000\,\text{rpm}}{2}}{44760\,\dfrac{\text{N}\cdot\text{m}}{\text{min}\cdot\text{hp}}} = 89.4\,\text{hp}$$

### ILLUSTRATIVE PROBLEM 9.8

Refer to Figure 9.9, and note that the difference in volume, $V_2 - V_3$ is commonly called the *displacement volume* and that $V_3$ is the *clearance volume*. Derive an expression for the compression ratio of an Otto engine in terms of the displacement volume and clearance volume. Also, derive an expression for the mep of an Otto engine in terms of the work, the compression ratio, and the initial specific volume.

**SOLUTION**

To simplify the analysis, let us call the clearance volume a fraction ($c$) of the displacement volume $D$. Thus, $V_3 = cD$, and $(V_2 - V_3) = D$. By our definition of compression ratio,

$$r_c = \frac{v_2}{v_3} = \frac{V_2}{V_3} = \frac{D + cD}{cD} = \frac{1+c}{c}$$

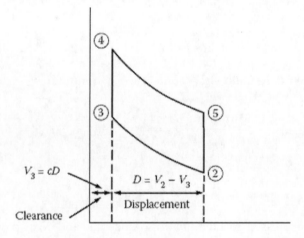

**FIGURE 9.9**
Illustrative Problem 9.8.

The mean effective pressure $(\text{mep}) = \dfrac{\text{work}}{\text{displaced volume}} = \dfrac{w}{v_2 - v_3}$

$$\text{mep} = \frac{w}{v_2 \left(1 - \dfrac{v_3}{v_2}\right)}$$

$$\text{mep} = \frac{w}{v_2 \left(1 - \dfrac{1}{r_c}\right)} = \frac{w}{v_2 - v_3} = \frac{w}{(r_c - 1)v_3}$$

---

### ILLUSTRATIVE PROBLEM 9.9

An Otto engine has a clearance equal to 20% of its displacement. What is its compression ratio?

**SOLUTION**

From Illustrative Problem 9.8,

$$r_c = \frac{1+c}{c} = \frac{1+0.2}{0.2} = \frac{6}{1}$$

---

### ILLUSTRATIVE PROBLEM 9.10

Determine the mep of a four-cycle, six-cylinder engine that has a 100 hp output at 4000 rpm. The cylinder bore is 3 in., and the stroke is 4 in.

**SOLUTION**

Based on the results of Illustrative Problem 9.5, we have

$$\text{hp} = \frac{p_m LAN}{33000}$$

with $L = 4/12$ ft., $A = (\pi/4)(3)^2 \times 6$, and $N = 4000/2$ (because there are six cylinders and a four-cycle engine). Therefore,

$$p_m = \frac{100 \times 33\,000}{\dfrac{4}{12} \times \dfrac{\pi(3)^2 \times 6}{4} \times \dfrac{4000}{2}} = 116.7 \text{ psia}$$

**ILLUSTRATIVE PROBLEM 9.11**

An engine has an output of 230 hp at 5500 rpm. It is a six-cylinder engine with a displacement of 3.3 L. Determine the mep of this engine at this engine speed.

**SOLUTION**

The displacement of 3.3 L can be converted to cubic inches as follows:

$$3.3 \text{L} \times 1000 \, \frac{\text{cm}^3}{\text{L}} \times \frac{\text{in.}^3}{2.54 \text{ cm}} = 201.4 \text{ in.}^3$$

From the results of Illustrative Problem 9.10, we have

$$p_m = \frac{\text{hp} \times 33\,000}{LAN} \text{ where } LA \text{ is the displacement}$$

$$p_m = \frac{230 \times 33\,000 \times 12}{201.4 \times \dfrac{5500}{2}}$$

$$p_m = 164.4 \, \text{psia}$$

## 9.3 Diesel Engine (Compression Ignition Engine)

In the Otto (or spark ignition) cycle, the fuel is mixed with the air prior to the compression stroke of the cycle. Ignition occurs due to an externally timed electrical spark. Because the fuel–air mixture is compressed, it is necessary to use volatile, readily vaporized fuels that can be distributed uniformly into the incoming air. Also, the compression of the fuel–air mixture can cause it to become prematurely ignited during the compression of the cycle, leading to the familiar phenomenon of "knock" in the engine. Recent emphasis on ecological considerations has placed severe limitations on the use of leaded fuels to prevent preignition and nonuniform burning of the fuel.

In the *Diesel engine*, the air is first compressed to a pressure and temperature sufficient to ignite the fuel, which is injected at the end of the compression stroke. Because there is no fuel present during the compression stroke, much higher compression ratios are used in the compression ignition engine than in the spark ignition engine. This engine uses glow plugs (resistance heaters) for starting; the entire spark ignition system is eliminated.

The operation of the Diesel engine can be best illustrated by referring to Figure 9.10. The operation proceeds in the following manner:

1. *Air intake.* Rotation of the crankshaft (bottom of drawing) drives a toothed belt that turns the overhead camshaft that opens the intake valve. As the piston moves downward, fresh air is sucked into the cylinder.
2. *Air compression.* As the piston moves upward, both the intake and exhaust valves are closed, and compression of the trapped air begins. Because it is being compressed into a space 23.5 times smaller than its original volume, the air becomes

1
Air intake

2
Air compression

3
Fuel injection
and combustion

4
Exhaust

**FIGURE 9.10**
Operation of an automotive Diesel engine.

very hot, reaching a maximum temperature of 1650°F (900°C)—far above the flash-point for Diesel fuel.

3. *Fuel injection and power.* As the piston reaches the top of its stroke and the temperature of the compressed air is at its maximum, a mist of fuel is injected into the spherical pre-chamber. The hot air ignites the fuel, and its burning causes the mixture to expand while the flame front spreads quickly from the pre-chamber to the cylinder. As the piston is now past its topmost point, it is driven downward by the force of the expanding gases produced by combustion, driving the crankshaft and the car.

4. *Exhaust.* As the energy of the fuel–air combustion is spent and the piston begins to move upward again, the exhaust valve opens, clearing the cylinder of burned gases, and as the piston starts downward, the air intake stroke (no. 1) begins again.

Before proceeding further, let us compare the features of the spark ignition and compression ignition engines:

## Common Features

1. Both utilize the same mechanical patterns, that is, in-line, V, and so on.
2. Both are similarly cooled and lubricated.
3. The valving and valve actions are similar.
4. Both use external starting systems.
5. Both are made in two-and four-cycle types.

## Differing Features

1. Ignition in the spark ignition engine is initiated by an electrical spark at or near the completion of the compression stroke of the fuel–air mixture. In the compression ignition engine, air alone is compressed, and fuel ignition occurs when the fuel is injected into the cylinder at or near top dead center.
2. The prototype of the spark ignition engine consists of a constant-volume combustion process. In the compression ignition engine, constant-pressure combustion is the prototype.
3. Control of the spark ignition engine is accomplished by varying the quantity of the fuel–air mixture while keeping a constant mixture composition. In the compression ignition engine, control is achieved by varying the fuel input to a constant mass of charging air.
4. Spark ignition engines use an electrical ignition system and a fuel injector. Compression ignition engines have no electrical ignition system, but they do have some form of high-pressure fuel injection system.
5. Spark ignition engines are restricted either to gases or readily vaporized fuels. Compression ignition engines do not have this limitation. The usual fuel is a refined crude petroleum oil. Surprisingly, both compression ignition and spark ignition engines are prone to detonation. The fuel qualities that suppress detonation in the spark ignition engine aggravate this condition in the compression ignition engine.

Although Diesel engines have been built in small sizes, such as for automotive use, most diesels are large, relatively slow machines. To increase the performance of diesels, a turbine-type supercharger is frequently used. This is a self-contained unit composed of a gas turbine and a centrifugal blower mounted on a common shaft. The exhaust gases from the cylinders flow to the gas turbine where they drive the blower, which provides air for combustion and scavenging of exhaust gases from the cylinders. The use of a supercharger can increase the output of an engine as much as 50%. The injection of fuel into the cylinders of the compression ignition engine is one of the most exacting and difficult requirements for this type of engine. As an example, a 100 hp Diesel operating at 750 rpm requires the droplets of oil to be injected with the injection and burning completed in 0.01 s. The fuel system must have the following characteristics:

1. It must accurately measure the fuel.
2. Because the air is compressed to at least 500 psi, the fuel must be delivered under high pressure. The fuel system in some engines compresses the fuel to pressures up to 30,000 psi.
3. The timing of the fuel injection into the cylinder is critical.
4. The rate of fuel injection must be accurately controlled.
5. The fuel must be atomized to ensure uniform combustion.
6. The fuel must be dispersed properly in the chamber.

In general, three types of fuel systems are used for Diesel engines:

1. *Unit system.* In this system, a single injection pump is used per cylinder. This pump provides the fuel pressurization, metering, and timing function. Each unit is usually mechanically operated off a cam shaft.
2. *Common rail system.* In this system, the fuel is kept under pressure at all times in a plenum chamber. Timed valves permit flow during the injection portion of the cycle, and the mechanically operated injection valve provides the timing and metering functions.
3. *Distributor system.* The metering of the fuel for all cylinders is carried out at low pressure by a single injection pump. Fuel is distributed through a distribution valve, which directs the metered fuel to the individual injection valves. The injection valves have mechanically operated plungers to raise the oil pressure to the required injection pressure. In this system, the injection valve pressurizes the fuel and also times the fuel injection. However, it does not meter the fuel.

Because of the exacting requirements for fuel injection in a Diesel engine, these engines run in a narrow range of speeds. Thus, Diesel engines used for transportation require more transmission gearing than comparable Otto cycle engines.

## 9.4 Air-Standard Analysis of the Diesel Cycle

All the assumptions made for the air-standard analysis of the Otto cycle regarding the working fluid and its properties apply to the present analysis of the idealized Diesel cycle.

The idealized air-standard Diesel cycle consists of four processes. The first is an isentropic compression of the air after it has been inducted into the cylinder. At the end of the compression process, fuel is injected, and combustion is assumed to occur at constant pressure. Subsequent to the heat release by combustion, the gas is expanded isentropically to produce work, and finally, heat is rejected at constant volume. The gas is assumed to be recycled rather than rejected. Figure 9.11 shows the ideal Diesel cycle on both $p$–$v$ and $T$–$s$ coordinates.

Heat is received (reversibly) during the nonflow, constant-pressure process, © to ®. The energy equation for a constant-pressure (flow or nonflow) yields

$$q_{in} = c_p \left( T_4 - T_3 \right) \text{Btu/lb}_m, \text{kJ/kg} \tag{9.12}$$

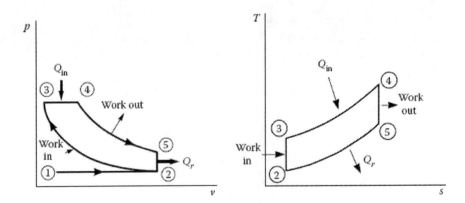

**FIGURE 9.11**
Diesel cycle.

The energy rejected during the constant-volume process is

$$q_r = c_v \left( T_5 - T_2 \right) \text{ Btu/lb}_\text{m}, \text{ kJ/kg} \tag{9.13}$$

The net work available from the cycle is

$$W = q_\text{in} - q_r = c_p \left( T_4 - T_3 \right) - c_v \left( T_5 - T_2 \right) \text{ Btu/lb}_\text{m}, \text{ kJ/kg} \tag{9.14}$$

The efficiency thus becomes

$$\eta_\text{Diesel} = \frac{c_p \left( T_4 - T_3 \right) - c_v \left( T_5 - T_2 \right)}{c_p \left( T_4 - T_3 \right)} = 1 - \frac{1}{k}\frac{T_5 - T_2}{T_4 - T_3} \tag{9.15}$$

Where $k = c_p/c_v$. At this point, it becomes conventional to introduce two terms and to define them as follows:

$$\text{compression ratio } r_c = \frac{v_2}{v_3} \tag{9.16}$$

and

$$\text{expansion ratio } r_e = \frac{v_5}{v_4} \tag{9.17}$$

Based on the nonflow processes discussed in Chapter 6, we can write

$$\frac{T_3}{T_2} = \left( r_c \right)^{k-1} \text{ and } \frac{T_5}{T_4} = \frac{1}{\left( r_e \right)^{k-1}} \tag{9.18}$$

Because heat is received at constant pressure,

$$\frac{T_4}{T_3} = \frac{v_4}{v_3} \tag{9.19}$$

The ratio of

$$\frac{r_c}{r_e} = \frac{v_2 / v_3}{v_5 / v_4} = \frac{v_4 / v_2}{v_3 / v_5} \left( \text{and because } v_2 = v_5 \right) = \frac{v_4}{v_3} \tag{9.20}$$

This ratio $v_4/v_3$ is called the *cutoff ratio*. Therefore,

$$\frac{T_4}{T_3} = \frac{r_c}{r_e} = r_{c.o.} \tag{9.21}$$

By substituting Equations 9.16 through 9.21 into Equation 9.15 and rearranging, we obtain

$$\eta_{\text{Diesel}} = 1 - \frac{1}{k} \frac{\left( r_c / r_e \right)^k - 1}{\left( r_c / r_e - 1 \right)\left( r_c^{k-1} \right)} = 1 - \left( \frac{1}{r_c} \right)^{k-1} \frac{\left( r_{c.o.} \right)^k - 1}{k \left( r_{c.o.} - 1 \right)} \tag{9.22}$$

Equation 9.22 shows that the efficiency of the Diesel cycle is a function only of the two ratios $r_c$ and $r_e$. The term $[(r_c/r_e)^k - 1]/k[(r_c/r_e - 1)]$ requires examination to determine whether the Diesel cycle is more or less efficient than the Otto cycle. Because $v_4/v_3$ is always greater than unity, it follows that this term is greater than unity. Therefore, by comparing Equation 9.22 with Equation 9.9, it can be concluded that the efficiency of the Otto cycle is greater than the efficiency of the Diesel cycle *for the same compression ratio*. We can also arrive at this conclusion by referring to Figure 9.12, which is a plot of the efficiency of the Diesel cycle as a function of the compression and cutoff ratios. When the cutoff ratio is unity, both cycles

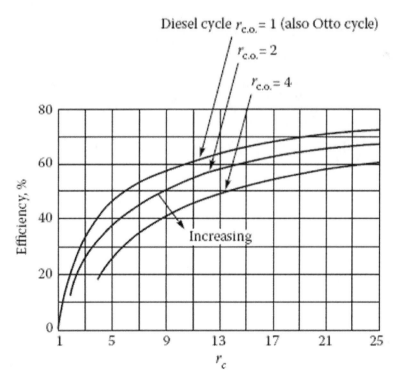

**FIGURE 9.12**
Efficiency of the Diesel cycle.

have equal efficiencies. For any other cutoff ratio, $v_4/v_3$, the efficiency of the Diesel is less than that of the Otto cycle.

However, as noted earlier in this section, the Diesel cycle can be operated at much higher compression ratios than the Otto cycle and therefore can have a higher efficiency than the Otto cycle. It should be further noted that as $v_4/v_3$ is increased, there is also a decrease in the efficiency of the Diesel cycle. Values between 2 and $2\frac{1}{2}$ are generally used as the upper limits of $v_4/v_3$. The mep of the Diesel cycle is obtained in a similar manner to that used to obtain the mep of the Otto cycle. The work of the Diesel cycle is given by Equation 9.14. Dividing it by the volume swept out by the piston per unit mass yields

$$\text{mep}_{\text{Diesel}} = \frac{c_p\left(T_4 - T_3\right) - c_v\left(T_5 - T_2\right)}{v_2 - v_3} J \tag{9.23}$$

By suitably rearranging Equation 9.23, the mep of the Diesel cycle can be expressed in terms of pressures and volumes. Also, in SI units, $J$ is not needed.

---

### ILLUSTRATIVE PROBLEM 9.12

An air-standard Diesel engine has a compression ratio of 16 and a cutoff ratio of 2. Find the efficiency and temperature of the exhaust using $k = 1.4$ with the cycle starting at 14 psia and 100°F.

### SOLUTION

From Figure 9.13, values of the volumes can be assigned as shown. The cutoff ratio $v_4/v_3$ is given as 2. Therefore, $v_4 = 2v_3$. From Equation 9.22,

$$\eta_{\text{Diesel}} = 1 - \left(\frac{1}{r_c}\right)^{k-1} \frac{\left(v_4/v_3\right)^k - 1}{k\left(v_4/v_3 - 1\right)}$$

**FIGURE 9.13**
Illustrative Problem 9.12.

For $k = 1.4$,

$$\eta_{\text{Diesel}} = 1 - \left(\frac{1}{16}\right)^{0.4} \frac{(2)^{1.4} - 1}{(1.4)(2-1)} = 0.614 = 61.4\%$$

From Equation 9.18,

$$\frac{T_3}{T_2} = r_c^{k-1} \quad \text{and} \quad \frac{T_5}{T_4} = \frac{1}{r_e^{k-1}}$$

But

$$\frac{T_4}{T_3} = \frac{v_4}{v_3} = \frac{r_c}{r_e}$$

Substituting for $T_3$ and $T_4$ in terms of $T_2$, $v_4$, and $v_3$ yields

$$T_5 = T_2 \left(\frac{v_4}{v_3}\right)^k$$

Therefore,

$$T_5 = (100 + 460)(2)^{1.4} = 1478 \, R = 1018°F$$

---

## ILLUSTRATIVE PROBLEM 9.13

In Illustrative Problem 9.12, determine the net work per pound of gas and the mep. Assume that $c_v = 0.172$ Btu/lb$_m$·R and $c_p = 0.24$ Btu/lb$_m$·R.

### SOLUTION

From Illustrative Problem 9.12,

$$q^r = c_v (T_5 - T_2) = (0.172)(1018 - 100) = 157.9 \text{ Btu/lb}_m$$

Note that

$$\eta_{\text{Diesel}} = 1 - \frac{q_r}{q_{in}}$$

Therefore,

$$\frac{q_r}{1-\eta} = q_{in}$$

$$q_{in} = \frac{157.9}{1 - 0.614} = 409.1 \text{ Btu/lb}_m$$

net work out $= J\left(q_{in} - q_r\right)$

$$= (778)(409.1 - 157.9)$$

$$= 195,433.6 \text{ ft. lb}_f/\text{lb}_m$$

The mep is net work divided by $(v_2 - v_3)$:

$$v_2 = \frac{RT_2}{p_2} = \frac{0.068\,J(560)}{14(144)} = 14.696 \text{ ft.}^3/\text{lb}_m$$

$$v_3 = \frac{14.696}{16} = 0.918 \text{ ft.}^3/\text{lb}_m$$

$$\text{mep} = \frac{195\,433.6 \text{ ft. lb}_f/\text{lb}_m}{(14.696 - 0.918)\dfrac{\text{ft.}^3}{\text{lb}_m} \times \dfrac{144 \text{ in.}^2}{\text{ft.}^2}} = 98.51 \text{ psia}$$

The Diesel cycle adds heat at higher temperatures, so it tends to be more efficient than the Otto cycle. However, at this higher temperature, the Diesel cycle tends to produce more nitrous oxide, which contributes to smog and more unburned fuel, which contributes to soot. Still many environmentalists favor the Diesel since it produces less carbon dioxide, a suspected "greenhouse gas," per unit of work.

Modern Diesel engines have fuel injectors that can eliminate knock and rattling. They are favored in Europe because they offer more torque for the same size engine and fuel taxes are lower. Diesels are more rugged and so tend to outlast the gasoline engine.

## 9.5 Automotive Engine Analysis

It is instructive at this point to see how an actual cycle compares to the air-standard Otto cycle. Figure 9.14 shows a comparison of an actual and an Otto cycle engine on $p$–$v$ coordinates. In the actual cycle, the air–fuel mixture is drawn into the cylinder from ① to ② with the cylinder pressure below atmospheric pressure. At a point before the piston reaches top dead center, or ③, the spark plug initiates ignition. Combustion is essentially complete at ④, with the piston starting down. Note that the peak pressure of the Otto cycle is not reached in the actual cycle. At ⑤, before bottom dead center, the exhaust valve opens. The pressure in the cylinder is always above atmospheric pressure during the exhaust process to overcome pressure drops in the exhaust system. When ① is reached, the process is repeated. The area noted as B (intake and exhaust) represents negative work, that is, work done by the engine, and the total indicated work in Figure 9.14 is equal to area A minus area B. In order to minimize area B, the intake and exhaust systems should be designed so as to minimize flow resistance.

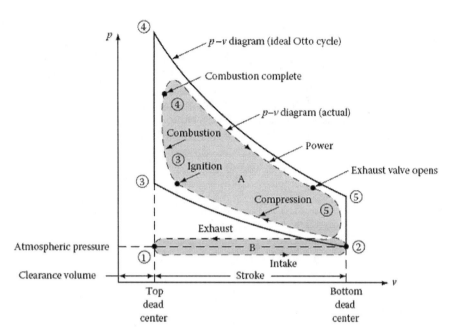

**FIGURE 9.14**
Comparison of Otto and actual engine cycles.

The actual Diesel engine cycle likewise differs from its air-standard version. These variations can be considered as losses in actual cycle efficiency. The air-standard efficiency for an engine cycle is called the "thermodynamic cycle efficiency," designated by the symbol $\eta_c$. The actual engine cycle efficiency is less than this because the segments of the cycle do not reach their ideal "air-standard" limits. The efficiency of the actual engine cycle is called the "indicated cycle efficiency," designated. It is called this because it is determined from a test using an "indicator card" that describes the $p$–$V$ diagram of the actual cycle by tracing out the cylinder pressure as the piston travels. This tracing can be integrated to provide the actual cycle work and mep. This procedure was originally developed by James Watt to analyze steam engines.

Automotive engines, be they Otto cycle or Diesel cycle, are used to provide power to a set of drive wheels. The transmittal of power, the work per cycle times the rate of power cycles, to the drive wheels involves additional losses through the mechanical parts of the system. These mechanical losses result in an overall efficiency called the "brake efficiency," $\eta_b$. If the transmission mechanical efficiency is $\eta_m$, these efficiencies are related as

$$\eta_b = \eta_i \times \eta_m \tag{9.24}$$

The concept of mep can be utilized in the analysis of actual engine power. Analysis of an actual cycle, as shown in Figure 9.14, yields an indicated mean effective pressure $\text{mep}_i$, which is the indicated work divided by the volume displaced. This results in an indicated horsepower for one cylinder, following the equation in Illustrated Problem 9.5, of

$$ihp = \frac{\text{mep}_i \, LAN}{33000} \tag{9.25}$$

Note that $N$ is rpm/2 for Otto or Diesel engines. If the units for mep are psia, area units should be in square inches and stroke length $L$ in feet.

For a real automotive engine with $n$ cylinders, the engine output is the value of Equation 9.25 times $n$. For the actual vehicle's power output, the power to the drive wheels, the "brake horsepower," is the most important factor. This can be measured by a test device such as a dynamometer or a Prony brake. Again, a mep can be determined for this measurement, the brake mep, written as $\text{mep}_b$. Thus, the overall auto engine's power output, the brake horsepower, is expressed as

$$bhp = \frac{\text{mep}_b \, LANn}{33000} \tag{9.26}$$

Remember that for small, two-stroke cycle engines, $N$ equals the rpm.

---

**ILLUSTRATIVE PROBLEM 9.14**

A six-cylinder automobile Diesel engine has an indicated mep of 100 psia and a bore and stroke of 3 and 5 in., respectively, and runs at 3500 rpm. Find the indicated horsepower.

**SOLUTION**

$$ihp = \frac{\text{mep}_i LANn}{33000} = \frac{100 \text{ psi} \times \dfrac{5}{12} \text{ ft.} \times \dfrac{\pi(3)^2}{4} \text{ in.}^2 \times \dfrac{3500}{2} \times 6}{33000} = 93.7 \text{ hp}$$

---

In the foregoing analysis of internal-combustion engines and cycles, a value of heat input has been assumed. With the modern world's concern for automotive efficiency as indicated by miles per gallon of fuel, the heat input should be understood as determined by the process of combustion of engine fuels. Some basic knowledge of chemistry is required. For Otto cycle engines, the fuel is gasoline, a mixture of hydrocarbons that can be considered chemically as octane, a liquid whose chemical formula is $C_8H_{18}$; that is, each molecule contains 8 atoms of carbon and 18 atoms of hydrogen. For Diesel cycles, the fuel is considered as dodecane, or $C_{12}H_{26}$. The fuel burns in the presence of oxygen from the intake air to form carbon dioxide and water. The formula for octane combustion, as an example, is

$$2C_8H_{18} + 25O_2 \rightarrow 16CO_2 + 18H_2O$$

The numbers in front of the chemical symbols indicate the number of molecules of each needed to balance the chemical equation. Thus, the equation may also be read as the number of moles of each for the complete combustion of octane into its products. From this, it is seen that 25 moles of oxygen is required to burn two moles of octane, or 12.5 per mole of fuel. From Chapter 7, a mole of air contains 21% oxygen, so 1 mole of oxygen requires 4.76 moles of air. Thus, 12.5 moles of oxygen requires 59.5 moles of air, yielding an air–fuel (AF) ratio of 59.5 by volume. To convert this ratio into a mass ratio, the molecular weights of the gases, as given in Chapter 7, can be used. The AF ratio is important in evaluating the

**TABLE 9.1**

Heating Values of Common Fuels

| Fuel | HHV (Btu/lb$_m$) | LHV (Btu/lb$_m$) |
|------|------------------|------------------|
| Diesel oil (light) | 19,800 | 18,600 |
| Diesel oil (heavy) | 19,600 | 18,400 |
| Ethanol | 12,760 | 11,530 |
| Gasoline | 20,300 | 18,900 |
| Hydrogen gas | 60,970 | 51,600 |
| Methane gas | 23,880 | 21,520 |
| Methanol | 9740 | 8570 |
| Natural gas | 21,500 | 19,400 |
| Propane liquid | 21,640 | 19,510 |

effectiveness of the combustion process. For the typical Otto engine cycle, the AF ratio by mass is around 15.

The actual determination of the AF ratio is complicated by some incomplete combustion of the fuel, which may be due to variable ignition temperatures, other chemicals in the fuel, or uneven distribution of fuel in the cylinder. Thus excess air, more than that required by the theoretical chemical analysis, may be utilized.

Another variable is the state of the resultant water of the chemical reaction. If it is in the gas state, some energy was utilized to vaporize the water. If it is liquid, no energy was required. This determines the energy available from the combustion of the fuel, called its *heating value*. The fuel's *higher heating value*, or HHV, is used if the water in the products is liquid; its *lower heating value*, or LHV, is used if the water in the products is a gas. Some heating values of common fuels are given in Table 9.1. Since automotive exhaust is usually in the vapor phase, the LHV is used in evaluating vehicles.

Note that ethanol, currently used as an additive to gasoline to reduce carbon emissions, will exhibit reduced heating value of the automotive fuel blends. In selecting a fuel, it is important to use system analysis to make the best choice. The cost of producing the fuel should be considered. As an example, fuels are being extracted from shale and other deep ground structures using a method called *fracking*. This involves the injection of large amounts of high-pressure water with solids to release the fuels. The energy costs of the pumps and effects on the environment should be considered in such processes.

## 9.6 Brayton Cycle

The discussion of gas cycles has so far been limited to intermittent cycles. However, the gas turbine has recently become quite important, both from an aircraft propulsion standpoint and for the production of power in stationary power plants. Because the working processes occur continuously in a gas turbine engine (no "idle" strokes), it is capable of generating a higher power output–weight ratio and so is ideal for use in aircrafts. The emerging prominence of this device is due primarily to the mechanical development of its components and the metallurgical development of alloys that can be used at elevated

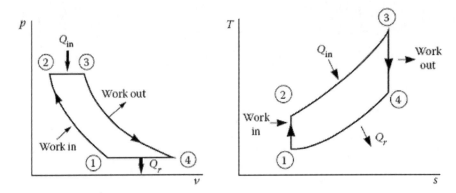

**FIGURE 9.15**
Ideal closed Brayton cycle.

temperatures. The prototype cycle for this device is the Brayton (or Joule) cycle. Its elements are shown in the $p$–$v$ and $T$–$s$ diagrams in Figure 9.15.

The gas is isentropically compressed along path ① to ②, and heat is added at constant pressure along path ② to ③. The gas then undergoes an isentropic expansion to its initial pressure, and heat is rejected at constant pressure. All the processes are reversible in the ideal cycle.

Before we analyze the Brayton cycle, let us look at the hardware used in conjunction with this cycle. There are three basic elements: the compressor, the combustor, and the turbine in the simple cycle. Of these, the compressor and turbine play the key role in the power-generation cycle. The gas turbine cycle displayed in Figure 9.15 shows a working fluid that is returned to pass through the cycle after each sequence of events is completed. Such a cycle is known as a *closed cycle*. The elements of this cycle are shown in Figure 9.16, with a reactor or heat exchanger being the heat source.

The *open-cycle* gas turbine system is a steady-flow system in which the working fluid is continuously discharged and fresh fuel and air are continuously added to maintain a steady flow of gases. The elements of this cycle are indicated in Figure 9.17. The cold air is compressed and flows to the combustion chamber, where fuel is continuously supplied and burned to heat the fluid. The combustion gases are then expanded in the turbine to provide work. The successful operation of the gas turbine requires that the net output of the turbine be greater than the work required by the compressor. In other words, the back

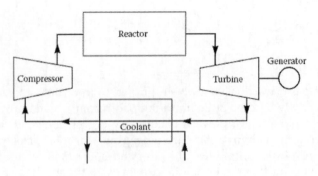

**FIGURE 9.16**
Elements of a closed-cycle gas turbine plant.

**FIGURE 9.17**
Open-cycle gas turbine plant.

work ratio defined as $r_{bw} = w_{compressor}/w_{turbine}$ should always be less than 1.0. It also requires that the mechanical efficiencies of these components be high; otherwise, the losses in these pieces of equipment may decrease useful output of the system to a point at which it might become uneconomical.

The use of regeneration in vapor cycles has been shown to increase the efficiency of these cycles. Similarly, regeneration can be used to increase the efficiency of the Brayton cycle. Such an arrangement is indicated in Figure 9.17 by the dashed lines. The exhaust gas leaving the turbine is used to preheat the air after it leaves the compressor and before entering the combustion chamber in a device called the *regenerator*. If the same quantity of fuel is used, a higher gas temperature is obtained at the inlet to the turbine. If a constant temperature to the turbine is desired, regeneration decreases the amount of fuel required to achieve this temperature.

The compression of the air in a gas turbine cycle is usually accomplished in two types of compressors, the centrifugal compressor and the axial flow compressor. Analysis of compressors and their design is given in Section 10.5.

In the gas turbine engine, the compression is usually done in stages to make it more effective. Each stage imparts approximately the same pressure ratio as the preceding stage. Thus, if each stage produces a pressure ratio of 1.1:1 (that is, a 10% increase), 8 stages will produce a total pressure rise of $(1.1)^8 = 2.14$, and 16 stages will produce an overall pressure rise of $(1.1)^{16} = 4.6$.

The gas turbine is similar to the steam turbine. Gas turbines have lower overall pressure drops than steam turbines, and consequently, they have fewer stages.

The modern jet engine uses the power remaining from the gas turbine after it supplies the compressor to accelerate the air moving through it. This accelerated air exits the engine and, by momentum principles, provides the thrust (force) to propel the aircraft.

## 9.7 Air-Standard Brayton Cycle Analysis

Let us now return to the ideal Brayton cycle shown in Figure 9.15. The heat input in the constant-pressure, steady-flow process per pound is

$$q_{in} = c_p \left( T_3 - T_2 \right) \tag{9.27}$$

and for the constant-pressure, steady-flow heat rejection part of the cycle ④ → ①,

$$q_r = c_p \left(T_4 - T_1\right) \tag{9.28}$$

The net work is

$$q_{\text{in}} - q_r = c_p \left(T_3 - T_2\right) - c_p \left(T_4 - T_1\right) \text{ in thermal units} \tag{9.29}$$

Thus,

$$\eta_{\text{Brayton}} = \frac{c_p \left(T_3 - T_2\right) - c_p \left(T_4 - T_1\right)}{c_p \left(T_3 - T_2\right)} = 1 - \frac{T_4 - T_1}{T_3 - T_2} \tag{9.30}$$

Which can be rewritten as follows,

$$\eta_{\text{Brayton}} = 1 - \frac{T_1 \left(\frac{T_4}{T_1} - 1\right)}{T_2 \left(\frac{T_3}{T_2} - 1\right)} \tag{9.31}$$

Since the Brayton cycle consists of two isentropic processes (1–2, 3–4) as shown in Figure 9.15, the following isentropic relationships apply,

$$\frac{T_2}{T_1} = \left(\frac{p_2}{p_1}\right)^{(k-1)/k} \; ; \; \frac{T_3}{T_4} = \left(\frac{p_3}{p_4}\right)^{(k-1)/k} \tag{6.63}$$

Since $p_2/p_1 = p_3/p_4$, therefore, $T_2/T_1 = T_3/T_4$ or $T_4/T_1 = T_3/T_2$, then Equation 9.31 becomes,

$$\eta_{\text{Brayton}} = 1 - \frac{T_1}{T_2} = 1 - \left(\frac{p_1}{p_2}\right)^{(k-1)/k} \tag{9.32}$$

Noting that $p_2/p_1$ as $r_p$ (pressure ratio), we have

$$\eta_{\text{Brayton}} = 1 - \left(\frac{1}{p_2/p_1}\right)^{(k-1)/k} \times 100 \tag{9.33}$$

With $r_p$ is as follows,

$$\eta_{\text{Brayton}} = 1 - \left(\frac{1}{r_p}\right)^{(k-1)/k} \times 100 \tag{9.34}$$

By looking at Equation 9.34, it is evident that the efficiency of the ideal Brayton cycle is solely a function of the pressure ratio and not of the peak temperature of the cycle. However, when the isentropic efficiency values of the compressor and turbine are taken into account, the overall efficiency of the actual Brayton cycle will be lower than the one predicted by Equation 9.34. The actual overall efficiency of the Brayton cycle can be found as follows,

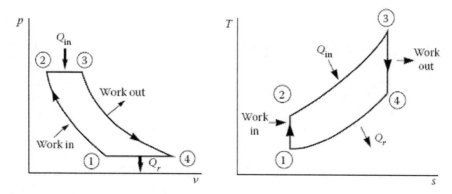

**FIGURE 9.15**
**(Repeated)** Ideal closed Brayton cycle.

$$\eta_{\text{Brayton, actual}} = \frac{w_{\text{turbine, actual}} - w_{\text{compressor, actual}}}{q_{\text{in, actual}}} = \frac{\left(h_3 - h_{4,a}\right) - \left(h_{2,a} - h_1\right)}{\left(h_3 - h_{2,a}\right)} \qquad (9.35)$$

To find $(h_{2,a} - h_1)$ and $(h_3 - h_{4,a})$, we need to use equations 5.25 and 5.26 from Chapter 5 as follows,

$$\eta_{\text{compression}} = \frac{w_{c,s}}{w_{c,a}} = \frac{h_{2,s} - h_1}{h_{2,a} - h_1} \rightarrow \left(h_{2,a} - h_1\right) = \frac{\left(h_{2,s} - h_1\right)}{\eta_{\text{compression}}}$$

and

$$\eta_{\text{turbine}} = \frac{w_{t,a}}{w_{t,s}} = \frac{h_3 - h_{2,a}}{h_3 - h_{2,s}} \rightarrow \left(h_3 - h_{2,a}\right) = \eta_{\text{turbine}} \left(h_3 - h_{2,s}\right)$$

To find $q_{in,a}$ (i.e., $h_3 - h_{2,a}$), $h_{2,a}$ needs to be found as follows,

$$h_{2,a} = h_1 + \frac{\left(h_{2,s} - h_1\right)}{\eta_{\text{compression}}}$$

As can be seen, by using the isentropic efficiency equations, a range of Brayton cycle problems can be solved accurately.

---

### ILLUSTRATIVE PROBLEM 9.15

A Brayton cycle is operated with a pressure ratio of 15.245. Assuming that $k = 1.4$ and $c_p = 0.24$ Btu/lbm·R, determine the heat in, $q_{in}$, the work out, $w_{net}$, the heat rejected, $q_r$, and the efficiency of the cycle, $\eta_{\text{Brayton}}$, if the peak temperature of the cycle is 1500°F (1960 R) and the initial conditions are 14.7 psia and 70°F. $R$ for air is 53.3 ft.-lb$_f$/lb$_m$·R.

## SOLUTION

To determine heat in, first $T_2$ is found using the isentropic relationship (Equation 6.63) as follows,

$$\frac{T_2}{T_1} = \left(\frac{p_2}{p_1}\right)^{(k-1)/k} \rightarrow T_2 = T_1\left(\frac{p_2}{p_1}\right)^{(k-1)/k}$$

$$T_2 = (70 + 460)(15.245)^{(1.4-1)/1.4} = 1154 = 694°F$$

Since $T_3$ is given at 1500°F. Therefore, the heat in is

$$q_{in} = c_p\left(T_3 - T_2\right) = (0.24)(1500 - 694) = 193.4 \text{Btu} / \text{lb}_m$$

We can calculate the efficiency as

$$\eta_{Brayton} = 1 - \left(\frac{1}{r_p}\right)^{k-1/k} = 1 - \left(\frac{1}{15.245}\right)^{(1.4-1)/1.4} = 0.541 \text{ or } 54.1\%$$

Because efficiency can be stated to be work out divided by heat in,

$$0.541 = \frac{w/J}{q_{in}} \frac{w}{J} \rightarrow w = (0.541)(193.4) = 104.6 \text{ Btu/lb}_m$$

The heat rejected is

$$q_r = q_{in} - \frac{w}{J} = 193.4 - 104.6 = 88.8 \text{ Btu/lb}_m$$

This problem can also be solved using the *Gas Tables*, with $c_p$ a function of temperature. In addition, a check is provided by solving for all the temperatures of the cycle and then solving for the efficiency.

---

### ILLUSTRATIVE PROBLEM 9.16

A Brayton cycle is operated with a pressure ratio of 15.245. Find its actual cycle efficiency if the isentropic efficiencies of the compressor and turbine are 80% and 85%, respectively. The temperatures of the working fluid (air) at the inlet of the compressor ($T_1$) and turbine ($T_3$) are 70°F (530 R) and 1500°F (1960 R), respectively. Find the ideal and actual back work ratios.

## SOLUTION

In this Illustrative Problem, it is better to find the enthalpy values and relative pressures of air at different temperatures using Appendix 2 (Table A.20) as follows,

$$T_1 = 530\,R \rightarrow h_1 = 126.67\ \text{Btu/lb}_\text{m}\,; p_{r,1} = 1.3$$

$$T_3 = 1960\,R \rightarrow h_3 = 493.64\ \text{Btu/lb}_\text{m}\,; p_{r,3} = 160.37$$

To find $h_{2,s}$ and $h_{4,s}$, $p_{r,2}$ and $p_{r,4}$ are found using the pressure ratio, $r_p$, and $p_{r,1}$ and $p_{r,3}$ as follows,

$$r_p = \left(\frac{p_{r,2}}{p_{r,1}}\right) \rightarrow p_{r,2} = r_p \cdot p_{r,1} = (15.245)(1.3) = 19.82$$

$$r_p = \frac{p_{r,3}}{p_{r,4}} \rightarrow p_{r,4} = \frac{p_{r,3}}{r_p} = \frac{160.37}{15.245} = 10.52$$

With the values of $p_{r,2}$ and $p_{r,4}$, $h_{2,s}$ and $h_{4,s}$ can be found through interpolation using Table A.20 as follows,

$$p_{r,2} = 19.82 \rightarrow h_{2,s} = 275.93\ \text{Btu/lb}_\text{m}$$

$$p_{r,4} = 10.52 \rightarrow h_{4,s} = 230.49\ \text{Btu/lb}_\text{m}$$

Since all the enthalpy values are known for the isentropic case, $w_{c,s}$ and $w_{t,s}$ can be found as follows,

$$w_{c,s} = h_{2,s} - h_1 = 275.93 - 126.67 = 149.26\ \text{Btu/lb}_\text{m}$$

$$w_{t,s} = h_3 - h_{4,s} = 493.64 - 230.49 = 263.15\ \text{Btu/lb}_\text{m}$$

Then, the ideal back work ratio can be found as follows,

$$r_{bw} = \frac{w_{c,s}}{w_{t,s}} = \frac{149.26}{263.15} = 56.7\%$$

As the back work ratio indicates, a large portion of the power generated by the turbine is used to power the compressor through a shaft connecting both devices. Given the values of $w_{c,s}$ and $w_{t,s}$, $w_{net}$ can also be calculated as follows,

$$w_{net} = w_{t,s} - w_{c,s} = 263.15 - 149.26 = 113.89\ \text{Btu/lb}_\text{m}$$

In this problem, $w_{net}$ was slightly higher (8.9%) that the value shown in illustrative example 9.15 despite having the same pressure ratio and inlet temperatures. This is because temperature-dependent enthalpy values for air at different temperatures were used for finding $w_{net}$. Furthermore, in Illustrative Problem 9.15, it was assumed that both $c_p$ and $k$ used for finding $w_{net}$ were independent of temperature. Nevertheless, either approach is acceptable in most Brayton cycle applications.

To find the overall efficiency using $w_{c,s}$ and $w_{t,s}$, $q_{in}$ needs to be found first, as follows,

$$q_{in} = (h_3 - h_{2,s}) = (493.64 - 275.93) = 217.71 \text{ Btu/lb}_m$$

Then, the overall cycle efficiency is as follows,

$$\eta_{\text{Brayton, isentropic}} = \frac{w_{t,s} - w_{c,s}}{q_{in}} = \frac{w_{\text{net}}}{q_{in}} = \frac{113.89}{217.71} = 52.3\%$$

The isentropic efficiency of the cycle compares well with the value calculated in Illustrative Problem 9.15.

To find the actual efficiency, $w_{c,a}$, $w_{t,a}$, and $q_{in,a}$ need to be found first, as follows,

$$w_{c,a} = \frac{w_{c,s}}{\eta_{\text{compression}}} = \frac{149.26}{0.8} = 186.58 \text{ Btu/lb}_m$$

$$w_{t,a} = \eta_{\text{turbine}} w_{t,s} = 0.85 \cdot 263.15 = 223.68 \text{ Btu/lb}_m$$

$$w_{c,a} = h_{2,a} - h_1 \rightarrow h_{2,a} = h_1 + w_{c,a} = 126.67 + 186.58 = 313.25 \text{ Btu/lb}_m$$

$$q_{in,a} = (h_3 - h_{2,a}) = (493.64 - 313.25) = 180.39 \text{ Btu/lb}_m$$

Then, the actual overall cycle efficiency is as follows,

$$\eta_{\text{Brayton, actual}} = \frac{w_{t,a} - w_{c,a}}{q_{in,a}} = \frac{223.68 - 186.58}{180.39} = 20.56\%$$

As the result shows, taking the isentropic efficiencies of the compressor and turbine into account leads to a significant reduction in the overall efficiency of the cycle. Lastly, the actual back work ratio is as follows,

$$r_{bw,a} = \frac{w_{c,a}}{w_{t,a}} = \frac{186.58}{223.68} = 83.4\%$$

As the actual back work ratio indicates, a larger portion of the power generated by the turbine stage(s) is used to power the compressor stages(s) through a connecting shaft when taking into account isentropic efficiencies.

In addition to regeneration, the performance of the Brayton cycle can be improved with *intercooling* and *reheating*. In intercooling, the compression work is divided into stages, each stage doing a portion of the overall pressure ratio desired. After an intermediate stage, the air is cooled before entering the next stage, reducing the specific volume of the air, so less work is required to compress it further. In reheating, the air is removed from the first turbine stage and passed through a heater before being returned to a second turbine stage. Figure 9.18 shows how the components are arranged in a gas turbine engine with regeneration, two-stage intercooling, and reheating. Figure 9.19 shows a *p–v* diagram, and

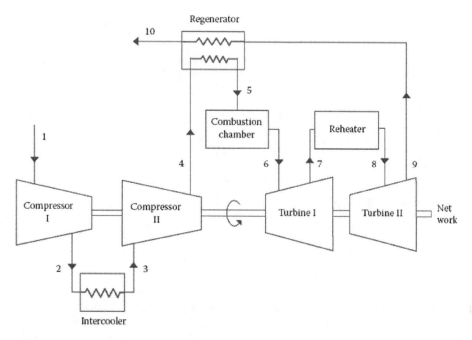

**FIGURE 9.18**

Gas turbine engine with intercooling, reheat, and regeneration.

Figure 9.20 shows a *T–s* diagram for such a cycle. In the ideal cycle, all heating and cooling portions are assumed to be at constant pressure, and all work portions are assumed to be isentropic.

In the ideal regenerator, the air will leave it to enter the combustion chamber at the temperature of the second turbine exhaust. It is also assumed that the air enters each stage of the compressor at the same temperature as a result of intercooling and that air enters each stage of the turbine at the same temperature as a result of reheating. Likewise, the temperatures of the air leaving each stage of compression are the same, as is true of the

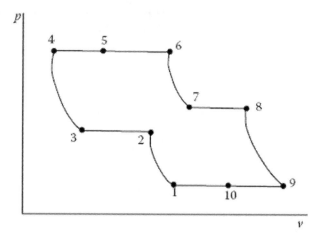

**FIGURE 9.19**

*p–v* diagram for the Brayton cycle with intercooling, reheat, and regeneration.

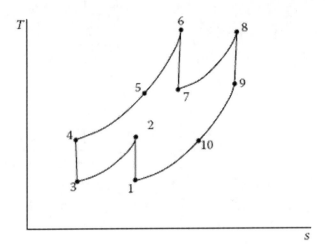

**FIGURE 9.20**
*T–s* diagram for the Brayton cycle with intercooling, reheat, and regeneration.

air leaving each turbine stage. Thus, $p_1 = p_{10} = p_9$, $p_2 = p_3$, $p_4 = p_5 = p_6$, and $p_7 = p_8$. Also, $T_1 = T_3$, $T_4 = T_2 = T_{10}$, $T_5 = T_7 = T_9$, and $T_6 = T_8$.

Note that the addition of intercooling and reheating serves to reduce the ratio of the compressor work to the turbine work. It is, rather, the addition of the regenerator that improves the efficiency of the cycle by reducing the heat input required. This is demonstrated in Illustrated Problem 9.16. The cost of adding a second stage to the compressor and turbine is more than made up for with the significant increase in efficiency. Additional staging improves efficiency further but usually not enough to offset the additional costs. In a two-stage system, the efficiency is maximized when the stages of both the compressor and turbine have the same pressure ratio (see Illustrative Problem 9.16).

---

### ILLUSTRATIVE PROBLEM 9.17

The Brayton cycle of Illustrative Problem 9.15 is modified by using two stages of compression and expansion such that the pressure ratios of each stage are the same. Refer to Figures 9.19 and 9.20 for points in the cycle. Find the efficiency of this modified cycle with and without a regenerator.

### SOLUTION

Assume that $c_p$ is again constant. Temperatures $T_1 = T_3 = 70°F$ and $T_6 = T_8 = 1500°F$. Also $p_1 = p_{10} = p_9 = 14.7$ psia. The overall pressure ratio over the two stages is 15.245.

Since there are two identical compression stages (i.e., $p_2/p_1 = p_4/p_3$ with $p_3 = p_2$), the pressure ratio for each of them is found as follows,

$$\frac{p_4}{p_1} = \left(\frac{p_2}{p_1}\right)\left(\frac{p_4}{p_3}\right) = 15.245 \rightarrow \frac{p_4}{p_1} = \left(\frac{p_2}{p_1}\right)^2 \rightarrow \frac{p_2}{p_1} = \sqrt{15.245} = 3.9045$$

The process from 1 to 2 is isentropic; thus

$$\frac{T_2}{T_1} = \left(\frac{p_2}{p_1}\right)^{\frac{k-1}{k}}$$

Solving for $T_2$ yields

$$T_2 = 530(3.9045)^{0.2857} = 782.143\ R$$

This is the same temperature as $T_4$ and $T_{10}$ due to the same pressure ratios.

We now solve for $T_7$, which is the same as $T_5$ and $T_9$ for the same reason. The process from 6 to 7 is isentropic with $p_6/p_7 = 3.9045$ just like the other compression and expansion stages,

$$\frac{T_6}{T_7} = \left(\frac{p_6}{p_7}\right)^{\frac{k-1}{k}}$$

Solving for $T_7$ yields

$$T_7 = \frac{1960}{(3.9045)^{0.2857}} = 1328.15\ R$$

For an isentropic process from points 1 to 2, when the pressure ratio and the initial temperature are known, the expression for work becomes

$$w_{12} = \frac{kRT_1}{1-k}\left(\frac{p_2}{p_1}\right)^{\frac{k-1}{k}} - 1$$

Solving for the compression work for each stage,

$$w_{12} = \frac{1.4 \times 53.3 \times 530}{-0.4 \times 778}(3.9045)^{0.2857} - 1 = -60.46\ \text{Btu/lb.}$$

The isentropic turbine work is

$$w_{67} = \frac{kRT_6}{1-k}\left(\frac{p_7}{p_6}\right)^{\frac{k-1}{k}} - 1$$

Solving for the turbine work for each stage,

$$w_{67} = \frac{1.4 \times 53.3 \times 1960}{-0.4 \times 778}\left(\frac{1}{3.9045}\right)^{0.2857} - 1 = 151.51\ \text{Btu/lb.}$$

Thus, the net work for the cycle is

$$w = 151.51 \times 2 - 60.46 \times 2 = 182.10\ \text{Btu/lb.}$$

The heat input at constant pressure for the cycle with a regenerator is

$$q_{in} = c_p \left( T_6 - T_5 \right) + c_p \left( T_8 - T_7 \right)$$

since the heat input from 4 to 5 is provided by the heat from 9 to 10. Solving for the total net heat input,

$$q_{in} = 0.24 \left( 1960 - 1328.15 \right) + 0.24 \left( 1960 - 1328.15 \right) = 303.29 \text{ Btu/lb.}$$

The efficiency of the cycle with intercooling, reheat, and regeneration is

$$\eta = \frac{\text{net } w}{q_{in}} = \frac{182.10}{303.29} \times 100 = 60.04\%$$

Without regeneration, the heat input becomes

$$q_{in} = c_p \left( T_6 - T_4 \right) + c_p \left( T_8 - T_7 \right)$$

Solving yields

$$q_{in} = 0.24 \left( 1960 - 782.143 \right) + 0.24 \left( 1960 - 1328.15 \right) = 434.33 \text{ Btu/lb.}$$

The efficiency of this cycle without regeneration is

$$\eta = \frac{\text{net } w}{q_{in}} = \frac{182.10}{434.33} \times 100 = 41.92\%$$

Comparing these values to the basic cycle efficiency from Illustrative Problem 9.15, we see that efficiency can be significantly improved, but only if regeneration is utilized in addition to intercooling and reheat, which, by themselves, actually reduce efficiency.

The peak temperature affects the amount of work produced in a given Brayton cycle. If more heat can be added, with the corresponding increase in the temperature $T_3$ leaving the combustor, more work could be produced. This can be visualized by noting a larger area within the $T$–$s$ diagram. However, industrial turbines have a temperature limit, perhaps 1500 K, based on maintaining the strength of the turbine blades.

For a given maximum temperature $T_3$, fixed by metallurgical considerations, and a minimum $T_1$, typically atmospheric temperature, there is an optimum pressure ratio that results in maximum cycle work. Considering the isentropic compression, for example, one can see that the pressure ratio is related to the temperature ratio $T_2/T_1$. Calculus can be used to determine the optimum value of $T_2$ that maximizes the work (see the following enrichment section). It is found that optimum cycle work occurs when the air temperature leaving the compressor equals that at the turbine exit. This affects the equation for the pressure ratio $r_p$. For maximum work,

$$r_p = \frac{p_2}{p_1} = \left(\frac{T_2}{T_1}\right)^{\frac{k}{k-1}} = \left(\frac{T_3}{T_1}\right)^{\frac{k}{2(k-1)}} \tag{9.35}$$

Note that the relationship in Equation 9.35 is only applicable for the maximum work cycle.

---

**CALCULUS ENRICHMENT**

The net work of the Brayton cycle is given by

$$W = mc_p\left(T_3 - T_2\right) - mc_p\left(T_4 - T_1\right) \tag{a}$$

Using Equation 6.63,

$$\frac{T_4}{T_1} = \frac{T_3}{T_2} \rightarrow T_4 = \frac{T_3 T_1}{T_2} \tag{b}$$

Thus,

$$W = mc_p\left(T_3 - T_2 - \frac{T_3 T_1}{T_2} + T_1\right) \tag{c}$$

For maximum work, the derivative of work with respect to the variable $T_2$ is set equal to 0. With $T_1$ and $T_3$ fixed, we obtain

$$\frac{dW}{dT_2} = mc_p\left(-1 + \frac{T_3 T_1}{T_2^2}\right) = 0 \tag{d}$$

Solving this equation yields $T_3 T_1 = T_2^2$; therefore,

$$T_2 = \sqrt{T_3 T_1} \tag{e}$$

Using Equation (b) yields, by substitution,

$$T_2 = T_4 \tag{f}$$

It can be shown by taking the second derivative of the work with respect to $T_2$ that this relationship results in maximum work; the second derivative is negative.

---

Gas turbine engines are used to power aircraft because the open Brayton cycle engines can use atmospheric air and have a high power-to-weight ratio. The power output is transformed into a thrust force that propels the aircraft. In jet engines, the thrust comes from a high-speed turbine exhaust and its equal and opposite reaction on the aircraft. In propeller jets, called *propjets or turboprop*, the turbine output is used via a shaft to drive the propeller, which moves the air to provide thrust. In this case, the turbine exhaust offers just minimal additional thrust.

Modern jet engines, called *fanjets or turbofans*, which have a fan at the intake to push more air through the engine, bypassing the compressor, combustor, and turbine, and out

**FIGURE 9.21**
A combined gas-steam power plant.

the exhaust to improve thrust. The fanjet moves less mass of air through it than the propjet but at much higher speeds and it can function well at high altitudes. The propjet is efficient at low altitudes where there is more air to work with and where the propeller blades are not subject to the high stress that comes with high speeds.

Note that in the fanjet, the turbine work goes into running the compressor, whereas in the propjet, the turbine drives both the compressor and the propeller.

Modern jet engines add features to increase thrust, including: *afterburning*, whereby combustion of fuel occurs behind the turbine to increase the temperature and velocity of the exhaust gases, and *water injection*, whereby water is sprayed onto the exhaust to increase the mass of the expelled gases. Both of these processes increase the aircraft momentum.

In many power-producing plants, the Rankine cycle is combined with the Brayton cycle with the objective of harnessing the waste heat produced by the Brayton cycle. Figure 9.21 shows how the exhaust gases from the gas turbine are used to make steam in the boiler (steam generator). By combining both cycles, the combined cycle efficiency should exceed 60% in most power plants that use both systems concurrently.

## 9.8 The Dual Combustion Cycle (The Dual Cycle)

A useful method of studying and comparing the performance of an engine experimentally is the indicator diagram. This diagram is a pressure–volume record of the events that are occurring within the cylinder of an operating engine. Indicator diagrams taken from both Otto and Diesel engines show a rounded top, as can be seen from Figure 9.14. For the ideal cycle to more closely represent the actual cycle events, it has been proposed that the ideal cycle have some combustion at constant volume and some at constant pressure. Such a cycle is called the dual combustion cycle, or simply the dual cycle. The $p$–$v$ and $T$–$s$ diagrams for the dual cycle are shown in Figure 9.22. Only the combustion or heat-addition portion of the cycle differs from either the Otto or Diesel cycles.

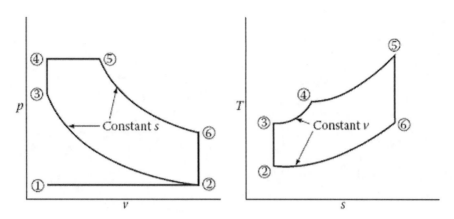

**FIGURE 9.22**
Dual combustion cycle.

Using the points indicated in Figure 9.22, we can write

$$q_{in} = c_v \left(T_4 - T_3\right) + c_p \left(T_5 - T_4\right) \tag{9.36}$$

and

$$q_r = c_v \left(T_6 - T_2\right) \tag{9.37}$$

The efficiency is given in terms of $q_{in}$ and $q_r$ as

$$\eta_{dualcycle} = \left(q_{in} - q_r\right) / q_{in}$$

$$\eta_{dual\ cycle} = \left[c_v \left(T_4 - T_3\right)\right] + c_p \left(T_5 - T_4\right) - c_v \left(T_6 - T_2\right) / \left[c_v \left(T_4 - T_3\right) + c_p \left(T_5 - T_4\right)\right] \tag{9.38}$$

Simplifying,

$$\eta_{dual\ cycle} = 1 - \left(T_6 - T_2\right) / \left[\left(T_4 - T_3\right) + k\left(T_5 - T_4\right)\right] \tag{9.39}$$

If Equation 9.39 is examined, it is found that for a given compression ratio, the efficiency of the dual cycle is less than the efficiency of the Otto cycle but greater than the efficiency of the Diesel cycle.

## 9.9 Stirling Cycle and Ericsson Cycle (Regeneration)

The *Stirling cycle* has, for some time, been only of historical interest, because it has not been used as the prototype for practical power-producing systems. Recently, in this country and abroad, interest has been revived, and engines based on the Stirling cycle have been built

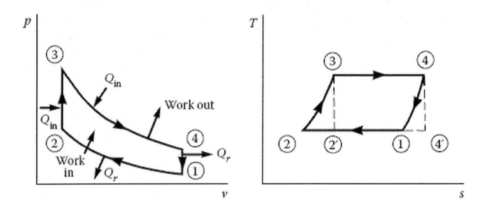

**FIGURE 9.23**
Stirling cycle.

and successfully operated. This development has been partially due to the possibility of using these engines for power production in space.

The elements of this cycle are shown in Figure 9.23. It consists of two isothermal and two constant-volume processes. During the isothermal compression, heat is rejected by cooling the air in the compressor. Heat is added during the constant-volume portion of the cycle (②→③). The gas is then isothermally expanded, and more heat is added. The gas finally rejects heat during the constant-volume position of the cycle (④→①). If all these paths were carried out reversibly, the efficiency of the Stirling cycle would equal that of the Carnot cycle between the same maximum and minimum temperature limits. This can be seen from the $T$–$s$ diagram in Figure 9.22. Both the Stirling and the Ericsson cycles rely on a regenerator that reversibly takes the heat rejected in the constant-volume or constant-pressure process (④→①) and returns it to the engine in the other constant-volume or constant-pressure process (②→③). Paths (②→③) and (④→①) are parallel, and the cycle can be "squared off" to yield the Carnot cycle, shown as ②-③-④-④. The inability of actual devices to achieve this limiting efficiency arises principally from the heat-transfer processes and pressure losses in the engine and its associated piping.

Both the Stirling cycle and Ericsson cycle were originally developed as an effort to find a practical external heat-transfer engine using air as the working fluid. The *Ericsson engine* consists of two reversible constant-pressure processes and two reversible constant temperature energy-addition and energy-rejection processes. The $T$–$s$ diagram is similar to that of the Stirling cycle $T$–$s$ diagram of Figure 9.23. The air-standard efficiency of the ideal Ericsson engine therefore is equal to the Carnot efficiency.

The use of regeneration in the steam cycle or in the Stirling and Ericsson cycles is seen to cause the efficiencies of these cycles to approach the efficiency of the Carnot cycle. In the case of the Stirling and Ericsson cycles, the reduction of theory to actual working engines presents practical difficulties that cause the actual engine efficiencies to be below that of the Carnot cycle.

## 9.10  Review

In Chapter 8, we considered vapor cycles in which the working fluid underwent a change in phase. In this chapter, we have concerned ourselves with cycles in which the working fluid is a gas and remains a gas throughout the cycle. Some of the cycles considered in this chapter have internal combustion; others have external combustion. Some have spark ignition; others have compression ignition. In order to simplify the calculations and to permit us to arrive at general conclusions, we utilized the air-standard cycle analysis. The air-standard cycle analysis assumes that air is the working fluid and that it behaves as an ideal gas. Each process of the cycle is assumed to be internally reversible, and the specific heats of the air are assumed to be constant throughout the cycle. In effect, the same air remains in the engine from cycle to cycle in the air-standard analysis.

Based on the air-standard analysis, the efficiency of the Otto cycle was determined to be a function only of the compression ratio, $r_c$. The efficiency of the Diesel cycle was found to be less than the efficiency of the Otto cycle for the same compression ratio. However, the compression ratio of Diesel engines is much greater than the compression ratio of the Otto cycle because the fuel is injected at the end of the compression stroke rather than compressing the fuel–air mixture. The efficiency of the Brayton cycle was found to be proportional to the pressure ratio, $r_p$. The efficiency of the Brayton cycle can be improved with regeneration. The dual cycle was proposed in an effort to come closer to actual indicator diagrams. In the dual cycle, heat is added at both constant volume and constant pressure. The efficiency of the dual cycle lies between the efficiency of the Otto and Diesel cycles for the same compression ratio.

The Stirling and Ericsson cycles both have thermal efficiencies that equal the thermal efficiency of the Carnot cycle. This performance is achieved through the use of regeneration, during which heat is transferred to an energy storage device during one portion of the cycle and is returned to the working fluid during another portion of the cycle. The hardware required to convert these theoretical cycles to working engines is complex and difficult to build. However, Stirling engine development has recently accelerated as a result of its potential space applications.

## Key Terms

Terms used for the first time in this chapter are as follows:

**Brake horsepower:** The actual power delivered to the drive wheels.

**Brayton cycle:** An ideal four-cycle gas engine consisting of two isentropic and two constant-pressure processes.

**Compression ratio:** The ratio between the volume at the bottom of the stroke and the volume at the top of the stroke.

**Diesel cycle:** An ideal four-cycle gas engine consisting of two isentropic, a constant-pressure, and a constant-volume process.

**Dual cycle:** An ideal gas engine cycle in which heat is added at both constant volume and constant pressure in order to come closer to actual indicator $p$–$v$ diagrams.

**Ericsson engine:** A regenerative internal-combustion engine consisting of two constant temperature and two constant-pressure processes.

**Expansion ratio:** The ratio of the volume prior to expansion divided by the volume after compression in a Diesel cycle.

**Indicated horsepower:** The actual power output of an engine.

**Intercooling:** The cooling of gas between stages in a multistage compressor.

**Internal-combustion engine:** An engine in which the energy release occurs internally.

**Mean effective pressure (mep):** Defined as the work out of a cycle divided by the volume swept out by the piston.

**Otto cycle:** An ideal four-cycle gas engine consisting of two isentropic and two constant-volume processes.

**Pressure ratio:** The ratio between pressure at the outlet of the compressor or inlet of the turbine to the pressure at the inlet of the compressor or outlet of the turbine in a Brayton cycle.

**Stirling cycle:** A regenerative internal-combustion engine cycle consisting of two isothermal and two constant-volume processes.

## Equations Developed in This Chapter

| | | |
|---|---|---|
| Compression ratio | $r_c = \dfrac{v_2}{v_3}$ | (9.8) |
| Efficiency of the Otto cycle | $\eta_{\text{Otto}} = 1 - \left(\dfrac{1}{r_c}\right)^{k-1} \times 100$ | (9.9) |
| Mean effective pressure | $\text{mep} = \dfrac{\text{workout}}{\text{volume swept out by piston}}$ | (Definition) |
| Mean effective pressure | $\text{mep} = \dfrac{1}{k-1} \dfrac{v_3\left(p_4 - p_3\right) - v_2\left(p_5 - p_2\right)}{v_2 - v_3}$ | (9.10) |
| Mean effective pressure | $\text{mep} = \dfrac{1}{k-1} \dfrac{\left(p_4 - r_c^k p_2\right)\left[1 - \left(1/r_c\right)^{k-1}\right]}{r_c - 1}$ | (9.11) |
| Horsepower | $\text{hp} = \dfrac{p_m LAN}{33000}$ | Illustrative Problem 9.5 |
| Horsepower | $\text{hp} = \dfrac{p_m LAN}{44760}$ | Illustrative Problem 9.6 |
| Compression ratio | $r_c = \dfrac{1+c}{c}; \text{mep} = \dfrac{w}{v_2\left(1 - \dfrac{1}{r_c}\right)} = \dfrac{w}{\left(r_c - 1\right)v_3}$ | Illustrative Problem 9.8 |
| Diesel cycle | $r_c = \dfrac{v_2}{v_3}$ | (9.16) |
| Expansion ratio | $r_e = \dfrac{v_5}{v_4}$ | (9.17) |
| Cutoff ratio | $\dfrac{r_c}{r_e} = \dfrac{v_4}{v_3}$ | (9.20) |

| | | |
|---|---|---|
| Cutoff ratio | $$r_{c.o.} = \frac{r_c}{r_e}$$ | (9.21) |
| Diesel efficiency | $$\eta_{Diesel} = 1 - \frac{1}{k} \frac{\left(r_c / r_e\right)^k - 1}{\left(r_c / r_e - 1\right)\left(r_c^{k-1}\right)}$$ | (9.22) |
| | $$= 1 - \left(\frac{1}{r_c}\right)^{k-1} \frac{\left(r_{c.o.}\right)^k - 1}{k\left(r_{c.o.} - 1\right)}$$ | (9.23) |
| Brayton efficiency | $$\eta_{Brayton} = 1 - \left(\frac{1}{p_2 / p_1}\right)^{(k-1)/k} \times 100$$ | (9.33) |
| Brayton efficiency | $$\eta_{Brayton} = 1 - \left(\frac{1}{r_p}\right)^{(k-1)/k} \times 100$$ | (9.34) |
| Pressure ratio | $$r_p = \frac{p_2}{p_1}$$ | (9.34) |
| Brayton actual efficiency | $$\eta_{Brayton,\,actual} = \frac{\left(h_3 - h_{4,a}\right) - \left(h_{2,a} - h_1\right)}{\left(h_3 - h_{2,a}\right)}$$ | (9.35) |

## QUESTIONS

9.1    In what two ways does a gas cycle differ from a vapor cycle?

9.2    List the four events in the four-stroke internal-combustion cycle.

9.3    How does the two-stroke cycle differ from the four-stroke cycle?

9.4    What is the ratio of the number of power strokes in the two-stroke cycle divided by the number of power strokes in the four-stroke cycle at a given value of engine revolutions per minute?

9.5    Describe the operation of the Wankel engine.

9.6    Describe the problems that the practical Wankel engine encountered and that have been overcome.

9.7    Describe the four events in the Otto cycle.

9.8    On what single factor does the efficiency of the Otto cycle depend?

9.9    What limits the practical realization of higher efficiencies in the Otto cycle?

9.10   How does the modern internal-combustion engine achieve higher power outputs without the use of higher compression ratios?

9.11   Of what utility is the concept of mean effective pressure?

9.12   How does the clearance volume affect the efficiency of the Otto cycle?

9.13   In what manner does the Diesel cycle differ from the Otto cycle?

9.14   How is the fuel introduced into the Diesel engine?

9.15   How does the efficiency of the Diesel engine compare to the efficiency of the Otto cycle for the same compression ratio?

9.16   How is it possible for a Diesel engine to operate at efficiencies greater than the efficiency of an Otto cycle?

9.17  How does the Diesel engine avoid the problem of preignition or detonation?

9.18  Describe the Brayton cycle.

9.19  How does the operation of the closed-cycle Brayton plant differ from the open-cycle Brayton plant?

9.20  Show how the dual cycle is a compromise between the Otto and Diesel cycles.

9.21  The Stirling and Ericsson cycles use a common concept in their operation. Describe this concept and its principle.

## PROBLEMS

Wherever appropriate, sketch the cycle on $p$–$v$, $h$–$s$, or $T$–$s$ coordinates. Use $c_p = 0.24$ Btu/$lb_m \cdot °c_v = 0.171$ Btu/$lb_m \cdot$ R, $c_p = 1.0061$ kJ/kg·K, and $c_v = 0.7186$ kJ/kg·K for air unless otherwise noted Assume that these specific heats are constant and that the molecular weight of air is 29.

### Problems Involving the Otto Cycle

9.1  An air-standard Otto cycle is operated with a compression ratio of 6. If $k$ has values of 1.1, 1.2, 1.3, and 1.4, determine the efficiency of the cycle. Plot the efficiency as a function of $k$.

9.2  An air-standard Otto cycle operates with 520 R and 14.7 psia at the beginning of the compression portion of the cycle. If 300 Btu/$lb_m$ is added as heat and the maximum temperature of the cycle is 2900 R, determine the efficiency of the cycle.

9.3  The pressure at the beginning of the compression stroke in an air-standard Otto cycle is 15 psia. After compression, the pressure is 140 psia. Determine the thermal efficiency of the cycle.

9.4  An air-standard Otto cycle has a clearance volume equal to 10% of its displacement volume. Calculate the compression ratio and the efficiency of the engine.

9.5  At the beginning of the compression stroke in an air-standard Otto cycle, the temperature is 68°F, the pressure is 15 psia, and the compression ratio is 7. If the heat rejected by the engine is 250 Btu/$lb_m$, determine the heat input to the cycle and the thermal efficiency of the cycle.

9.6  An air-standard Otto cycle has a compression ratio of 9. At the beginning of the compression process, the temperature is 20°C, and the pressure is 100 kPa. The heat added is 500 kJ/kg. Determine the work output, the heat rejected, and the cycle efficiency.

9.7  At the beginning of the compression stroke of an air-standard Otto cycle, the volume is 0.1 ft.$^3$, the temperature is 90°F, and the pressure is 14 psia. At the end of compression, the pressure is 200 psia. Calculate the efficiency of the engine.

9.8  Determine the heat rejected, the net work, and the compression ratio in Problem 9.7 if 250 Btu/$lb_m$ is added as heat to the cycle.

9.9  An air-standard Otto cycle starts its compression stroke at 75°F. Assuming that the highest temperature that can be reached before the gasoline–air mixture will ignite without an electrical spark is 750°F, determine the maximum compression ratio that can be used in this engine.

9.10 An air-standard Otto cycle operates with 25% clearance. At the beginning of the compression stroke, the pressure is 14 psia, and the temperature is 100°F. The maximum temperature in the cycle is 4040°F. Determine the efficiency of the cycle.

9.11 An air-standard Otto cycle with a compression ratio of 9 rejects 100 kJ/kg as heat. At the end of the compression process, the pressure is 2.2 MPa, and the temperature is 720 K. Determine the heat input to the cycle, the work output, and the efficiency of the cycle.

9.12 Determine the mean effective pressure in the cycle of Problem 9.11.

9.13 An air-standard Otto cycle operates with the compression and expansion strokes being polytropic, with $n = 1.35$ for each stroke. Calculate the thermal efficiency of the cycle for a compression ratio of 9.

9.14 An air-standard Otto cycle operates with a compression ratio of 8. At the start of the compression stroke, the pressure is 15 psia, and the temperature is 80°F. The maximum pressure in the cycle is 550 psia. Calculate the maximum temperature in the cycle and the thermal efficiency.

*9.15 An air-standard Otto cycle operates with a compression ratio of 9.5. At the start of compression, the conditions are 14.7 psia and 80°F. During the combustion portion of the cycle, 250 Btu/lb$_m$ is added as heat. Calculate $p$, $v$, and $t$ at each point of the cycle and the thermal efficiency of the cycle.

*9.16 An air-standard Otto cycle operates with a minimum temperature of 300 K and a maximum temperature of 1700 K. The compression ratio of the cycle is 7. At the beginning of the compression process, the pressure is 105 kPa. Calculate $p$, $v$, and $t$ at each point in the cycle, the mean effective pressure, and the thermal efficiency of the cycle.

*9.17 An air-standard Otto cycle has 2000 kJ/kg added as heat. The initial conditions are 20°C and 150 kPa. If the compression ratio is 7, determine the thermal efficiency of the cycle, the temperature and pressure at each point in the cycle, and its mean effective pressure.

*9.18 An air-standard Otto cycle with a compression ratio of 7 has air at 70°F and 15 psia at the start of the compression portion of the cycle. If 800 Btu/lb$_m$ is added as heat, determine the pressure and temperature in each portion of the cycle, the thermal efficiency of the cycle, and its mean effective pressure.

*9.19 An air-standard Otto cycle has 1400 kJ/kg added as heat. The initial conditions are 20°C and 100 kPa. If the compression ratio is 7, determine the thermal efficiency of the cycle, the temperature and pressure at each point in the cycle, and its mean effective pressure.

*9.20 An air-standard Otto cycle has 500 Btu/lb$_m$ added as heat. If the initial conditions are 70°F and 20 psia and the compression ratio is 7, determine the thermal efficiency of the cycle, the temperature and pressure at each point in the cycle, and its mean effective pressure.

9.21 An air-standard Otto cycle has 300 Btu/lb$_m$ added as heat. If the initial conditions are 50°F and 15 psia and the compression ratio is 8, determine the thermal efficiency of the cycle, the temperature and pressure at each point in the cycle, and its mean effective pressure.

9.22    The mean effective pressure of an Otto cycle is 100 psia. If the displacement of an engine is 144 in.$^3$, what horsepower can the engine deliver? This is a four-cycle engine operating at 3600 rpm.

9.23    The mean effective pressure of an Otto engine is 1 MPa. The displacement of the engine is 3.8 L. If the engine is operated as a four-cycle engine at 4000 rpm, what is its horsepower?

## Problems Involving the Diesel Cycle

9.24    An air-standard Diesel cycle is operated with a compression ratio of 25. The inlet conditions to the cycle are 14.7 psia and 70°F. The gas is to be used after the expansion cycle to heat some process liquid in a heat exchanger. For this purpose, the temperature at the end of the expansion is to be 500°F. Determine the efficiency of the cycle.

9.25    Determine the work out of the engine in Problem 9.24. Is this an efficient way to operate this engine?

9.26    An air-standard Diesel cycle has a compression ratio of 22 and a cutoff ratio of 2.2. Determine the thermal efficiency of the cycle.

9.27    An air-standard Diesel cycle has a compression ratio of 18. At the beginning of compression, the pressure is 100 kPa, and the temperature is 25°C. At the end of the combustion process, the temperature is 1300°C. Determine the thermal efficiency of the cycle and the mean effective pressure.

9.28    An air-standard Diesel cycle is operated with a compression ratio of 20. The ratio of $(v_4 - v_3)/(v_2 - v_3)$ is variable. Determine the efficiency of the cycle for values of the volume ratio of 0.01, 0.04, 0.08, and 0.1. Plot the results. Use $k = 1.4$.

9.29    An air-standard Diesel engine has 1000 kJ/kg added as heat. At the beginning of the compression, the temperature is 20°C, and the pressure is 150 kPa. If the compression ratio is 20, determine the maximum pressure and temperature in the cycle.

9.30    An air-standard Diesel cycle is operated with a compression ratio of 20. The heat input is 300 Btu/lb$_m$. At the beginning of the compression stroke, the conditions are 15 psia and 70°F. Calculate the thermal efficiency of the engine and the mean effective pressure.

9.31    An air-standard Diesel cycle is operated with a compression ratio of 20. At the beginning of the compression stroke, the conditions are 15 psia and 80°F. The maximum temperature in the cycle is 3400 R. Calculate the thermal efficiency of this cycle.

*9.32   An air-standard Diesel engine has a temperature of 70°F at the beginning of compression and a temperature of 1400°F at the end of compression. If the cycle receives 500 Btu/lb$_m$ as heat, determine the compression ratio, the net work of the cycle in Btu/lb$_m$, and the thermal efficiency of the cycle.

*9.33   An air-standard Diesel cycle operates with a compression ratio of 18. The maximum and minimum temperatures in the cycle are 2000 K and 293 K, respectively. The maximum pressure in the cycle is 50 MPa. Calculate the $p$, $v$, and $t$ at each point in the cycle, the thermal efficiency, and the mean effective pressure.

*9.34 Derive the following expression for the mean effective pressure in an air-standard Diesel:

$$\text{mep}_{\text{Diesel}} = p_1 \frac{k_{rc}^k \left(v_3 / v_2 - 1\right) - r_c \left[\left(v_3 / v_2\right)^k - 1\right]}{(k-1)(r_c - 1)}$$

*9.35 The net heat input to a Diesel cycle is 100 Btu/lb$_m$. If the initial conditions to the cycle are 14.7 psia and 60°F and the compression ratio is 20, determine the efficiency of the cycle, the net work out of the cycle, and its mean effective pressure. Assume that 1 lbm of air is used per cycle.

9.36 An air-standard Diesel engine has 500 Btu/lb$_m$ added as heat. At the beginning of the compression, the temperature is 70°F, and the pressure is 20 psia. If the compression ratio is 20, determine the maximum pressure and temperature in the cycle.

9.37 An air-standard Diesel engine has a compression ratio of 20. The cutoff ratio is 2.4. At the beginning of the compression stroke, the pressure is 100 kPa, and the temperature is 293 K. Calculate the thermal efficiency and the heat added per kilogram of air.

## Problems Involving the Brayton Cycle

9.38 A gas turbine cycle is operated with a pressure ratio of 5:1; that is, the pressure after compression is five times the pressure before compression. What is the efficiency of this cycle?

9.39 A Brayton cycle is operated to yield maximum work out of the cycle. If the pressure ratio of the cycle is 4, the initial temperature to the cycle is 70°F, and the temperature at the start of the heat rejection portion of the cycle is 500°F, what is the peak temperature in the cycle?

9.40 An air-standard Brayton cycle has inlet air at 15 psia and 520 R. At the inlet to the turbine, the pressure is 150 psia, and the temperature is 2500 R. Determine the heat input in Btu/lbm, the outlet temperature from the turbine, and the thermal efficiency of the cycle.

9.41 An air-standard Brayton cycle receives air at 15 psia and 80°F. The pressure ratio is 6. If the upper temperature of the cycle is 2300 R, determine the thermal efficiency of the cycle

9.42 An air-standard Brayton cycle is operated with a pressure ratio of 8 and a maximum temperature of 1400 K. Air enters the cycle at 30°C and 100 kPa. Determine the efficiency of the cycle.

9.43 An air-standard Brayton cycle operates with air entering at 14 psia and 80°F. At the turbine inlet, the temperature is 2200°F, and the pressure is 140 psia. Determine the thermal efficiency of the cycle.

9.44 An air-standard Brayton cycle has an inlet temperature of 25°C and an inlet pressure of 110 kPa. The maximum pressure and temperature in the cycle are 1.1 MPa and 1450 K, respectively. Determine the heat input, the work output, and the thermal efficiency of the cycle.

*9.45    In an air-standard Brayton cycle, air enters the compressor at 15 psia and 60°F. The pressure leaving the compressor is 90 psia. The maximum temperature in the cycle is 2000 R. Determine $p$, $v$, and $t$ at each point in the cycle and the thermal efficiency of the cycle.

*9.46    A Brayton cycle is operated to yield maximum work out of the cycle. If the pressure ratio of the cycle is 4, the initial temperature of the cycle is 20°C, and the temperature at the start of the heat rejection portion of the cycle is 240°C, what is the peak temperature in the cycle?

9.47    A Brayton cycle has a pressure ratio of 5. If the initial temperature is 70°F, the initial pressure is 15 psia, and 1000 Btu/lbm is added as heat, what is the thermal efficiency of the cycle?

*9.48    A Brayton cycle has a pressure ratio of 5. If the initial temperature is 20°C, the initial pressure is 100 kPa, and 1000 kJ/kg is added as heat, what is the temperature at each point in the cycle, and what is the thermal efficiency of the cycle?

9.49    A Brayton cycle is operated with pressure ratio of 8. The inlet conditions are 14.7 psia and 90°F. Determine the actual thermal efficiency of this cycle if the air temperature at the inlet of the turbine is 1500°F. The isentropic efficiency of the compressor is 90% and the isentropic turbine efficiency is also 90%. Determine the actual back work ratio of the system.

9.50    A turbine receives air at 1 MPa, 600 K, and discharges to a pressure of 150 kPa. The actual temperature at discharge is 400 K. What is the internal engine efficiency of the turbine?

9.51    A turbine receives air at 150 psia, 1000 R, and discharges to a pressure of 15 psia. The actual temperature at discharge is 600 R. What is the internal engine efficiency of the turbine?

## Problems Involving Other Cycles

9.52    A Stirling cycle operates with a maximum temperature of 900°F and a minimum temperature of 40°F. Determine the efficiency of the cycle.

9.53    If the engine in Problem 9.52 is an Ericsson engine, determine its efficiency.

*9.54    A Stirling engine operates with temperatures of 900°F and 70°F. The maximum pressure in the cycle is 700 psia, and the minimum pressure in the cycle is 14 psia. Determine the efficiency and work output of this engine per cycle.

*9.55    For an air-standard dual cycle, the conditions before compression are 15 psia and 100°F. After compression, the specific volume is 1.04 ft.³/lbm. If 150 Btu/lbm is received as heat during the constant-pressure portion of the cycle, determine the efficiency of the cycle.

*9.56    A cycle operates on air for which the constant-volume and constant-pressure specific heats are 0.17 and 0.24 Btu/lbm· R, respectively. All processes contained within the cycle are reversible and consist of (1) a constant-volume heat addition of 300 Btu/lbm of air, which was initially at 15 psia and 70°F; (2) an adiabatic expansion to a temperature of 800°F; and (3) a constant-pressure heat rejection of a magnitude that will return the air to 15 psia and 70°F. Determine the net work of this cycle in Btu per pound of air. (*Note:* This cycle is not one of the cycles in this chapter. Draw a $p$–$v$ diagram first before trying to solve it.)

# 10

## Refrigeration

### Learning Goals

After reading and studying the material in this chapter, you should be able to

1. Understand and use the fact that it is possible to reverse a power-producing cycle to remove heat from a place where it is undesirable and to reject it by the input of mechanical work.

2. Recall the elements of the Carnot cycle and operate it as a reversed cycle.

3. Define the figure of merit for a refrigeration cycle, known as the coefficient of performance.

4. Conclude that the coefficient of performance of a reversed Carnot cycle is a function of the upper and lower temperatures of the cycle.

5. Use the defined ratings used for tons of refrigeration, horsepower per ton of refrigeration, and kilowatts per ton of refrigeration.

6. Show how the reversed Rankine cycle operates using a sketch of the equipment and the $T$–$s$ diagram.

7. Use the ammonia and hydrofluorocarbon (HFC)-134a tables to calculate the ideal performance of a refrigeration cycle.

8. Be aware of the many refrigerants that have been used and their advantages and disadvantages.

9. Explain the elements of the gas compression cycle and how it works.

10. Analyze the vacuum refrigeration system and calculate the refrigerating effect of this system.

11. Show the thermoelectric effect and how it is used to produce a refrigerating effect.

12. Analyze the cycle of a reciprocating compressor and determine its volumetric efficiency.

13. Calculate the power required to drive a reciprocating compressor.

14. Derive the performance index for a heat pump and relate it to the performance index of a refrigeration cycle.

15. Show how the performance of a heat pump changes with a change in ambient temperatures.

16. Demonstrate that the use of well water or the Earth as a constant-temperature heat source is more desirable than the use of atmospheric air.

17. Explain why the use of a heat pump for a given application depends on economic factors more than it does on thermodynamic considerations alone.

## 10.1 Introduction

In previous chapters, consideration was given to those cycles that (by the proper placing of elements) could yield useful work from a heat source. By the simple expedient of rearranging the sequence of events in these cycles, it is possible (in principle) to remove heat from a region of lower temperature and to deliver it to a region of higher temperature by the input of mechanical work. This removal of heat by the use of mechanical energy has been called the *refrigerating effect*. In more general terms, refrigeration can be defined as the art of maintaining a body at temperatures below its surroundings or, alternatively, as the removal of heat from a place in which it is undesirable to a place in which it is not.

The history of refrigeration can be traced back thousands of years, with natural ice providing the cooling effect desired. The field of refrigeration on a large scale was first developed in the nineteenth century, and in the mid-nineteenth century, the harvesting, storing, and shipping of natural ice became one of the leading industries of the New England states. By the end of the nineteenth century, mechanical refrigeration had become a practical reality, and the refrigeration industry as we know it today was in existence. Along with the use of industrial refrigeration for food preservation, chemical production, metallurgical applications, medicine, and so on, another facet of the refrigeration process appeared: the control of the temperature and humidity of the environment, which is commonly called air conditioning. It is interesting to note that in 1904, a 450-ton air-conditioning system had been installed to air-condition the New York Stock Exchange.

The Carnot cycle has served to establish the performance criteria of power cycles. In this chapter, the study of the reversed Carnot cycle will yield many of the thermodynamic limitations and performance criteria for refrigeration cycles.

## 10.2 Reversed Carnot Cycle

Let's remember that the Carnot cycle consists of four reversible processes: two isothermals and two isentropic. In the direct cycle, in which the production of useful work is the primary objective, the elements are arranged so that the energy-flow diagram is as shown in Figure 10.1. Energy flows into the system from a reservoir at constant temperature $T_1$, work leaves the system through the agency of the prime mover, and heat is rejected to the receiver at constant temperature $T_2$. The $T$–$s$ diagram of the cycle is repeated in Figure 10.2. Because the ideal cycle can be treated independently of the working fluid, property lines are not shown on the diagram.

For the reversed cycle, the energy-flow diagram is as shown in Figure 10.3. Note that in this case, the work of the engine serves to take heat from the reservoir (sink at $T_2$) and rejects it to the source at $T_1$. The $T$–$s$ diagram for the reversed cycle is shown in Figure 10.4, and the areas are interpreted as various energy items corresponding to Figure 10.3. By considering these figures, certain general and important conclusions can be obtained for the reversed cycle. First, however, notice that the area $ABCD$ of Figure 10.4 measures the net work supplied to the cycle, and the area $A'ADD'$ on this same figure represents the heat removed from the colder region (at $T_2$). If the interest is in removing the greatest amount of heat from the sink, it is desirable to do so with the least possible energy input to the system. This mode of operation is equivalent to operating at maximum efficiency.

**FIGURE 10.1**
Direct Carnot cycle.

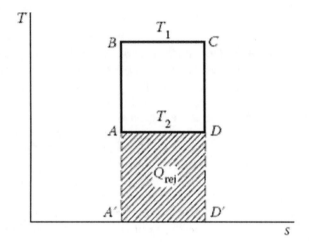

**FIGURE 10.2**
Direct Carnot cycle.

**FIGURE 10.3**
Reversed Carnot cycle.

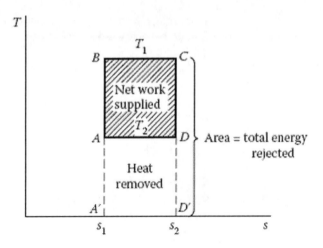

**FIGURE 10.4**
Reversed Carnot cycle.

To evaluate the maximum efficiency system, the same reasoning used for the direct cycle will serve to answer the question of the optimum efficiency conditions for operation of the reversed cycle. Consider a reversible engine operating, as shown schematically in Figure 10.3. The amount of energy rejected at $T_1$ is equal to the work in plus the heat removed from the lower-temperature region at $T_2$ (i.e., $q_r = q_{in} + w$). It has already been demonstrated that a reversible engine operating between two temperatures is the most efficient engine possible. Therefore, the energy rejected by the reversed cycle can be used in a reversible engine operating in a direct manner (Figure 10.1). In essence, the work out of the direct cycle is equal to that required by the reversed cycle, and the energy rejected by the direct engine cycle is equal to that removed by the reversed cycle. However, the inescapable conclusion is that without the use of energy or an external agency, heat cannot be removed continuously from a lower-temperature region and made to flow to a higher-temperature region because it would constitute a violation of the second law. Therefore, it must be concluded that the most efficient reversed engine cycle operating between two temperatures is a reversible engine; therefore, a reversed Carnot cycle represents the most efficient possible refrigerating cycle.

Based on the foregoing, let us now establish a figure of merit for the refrigeration cycle. For the direct power cycle, we were interested in the amount of work that could be obtained from a given heat supply, and this ratio was termed the *efficiency* of the power cycle. For the reversed cycle, we are interested in the amount of work required to remove a given amount of heat from the low-temperature reservoir. This ratio of refrigeration effect to work input is known as the *coefficient of performance* (COP). The COP is a measure of the efficiency of the refrigeration cycle. However, it is not called efficiency, because it is usually greater than one. Thus,

$$\text{COP} = \frac{\text{refrigeration effect}}{\text{work input}} \tag{10.1}$$

Note that Equation 10.1 applies as well to rates of refrigeration effect per unit time and work per unit time (i.e., power) and also to refrigeration effect and work per unit mass of fluid flowing. Thus, the COP always equals the cooling rate divided by the input power.

For the reversed Carnot cycle shown in Figures 10.3 and 10.4, the heat removed from the reservoir at $T_2$ is $T_2(s_2 - s_1)$, and the work supplied is $(T_1 - T_2)(s_2 - s_1)$.

$$\text{COP}_{\text{Carnot refrigeration}} = \frac{T_2(s_2 - s_1)}{(T_1 - T_2)(s_2 - s_1)} = \frac{T_2}{T_1 - T_2} = \frac{1}{T_1/T_2 - 1} = \frac{1}{T_H/T_l - 1} \tag{10.2}$$

Where $T_H$ and $T_L$ represent the temperatures of the high-temperature sink and cold-temperature source, respectively. Note that only absolute temperatures (K or R) are to be used in Equation 10.2.

Note also that the COP for an ideal refrigeration cycle is always greater than unity. To summarize the foregoing,

1. No refrigeration cycle can have a higher COP than a reversible cycle operating between the same temperatures.
2. All reversible refrigeration cycles operating between the same temperature limits have the same COP.
3. The COP of a Carnot cycle, a reversible cycle, is a function only of the upper and lower temperatures of the cycle and increases as the difference between the upper and lower temperatures is decreased. These conclusions are independent of the working fluid of the cycle.
4. Equation 10.2 indicates that, for maximum COP, $T_1$ should be kept to a minimum. In most cases, either the atmosphere or some nearby body of water is the practical heat sink.
5. The value of $T_2$ has a more pronounced effect than $T_1$ on the COP of a Carnot cycle.
6. Any deviations of the actual cycle from the ideal processes predicated for the Carnot cycle lead to values of COP less than the ideal.

### ILLUSTRATIVE PROBLEM 10.1

A Carnot refrigeration cycle is used to keep a freezer at 32°F (492 R). If the room to which the heat is rejected is at 70°F (530 R), calculate for a heat removal of 1000 Btu/min (a) the COP of the cycle, (b) the power required, and (c) the rate of heat rejected to the room. The data are shown in Figure 10.5.

#### SOLUTION

a. The COP is given by Equation 10.2 for the reversed cycle. Thus,

$$\text{COP} = \frac{T_2}{T_1 - T_2} = \frac{492}{530 - 492} = 12.95$$

b. Because the COP is defined as the ratio of the rate of the refrigeration effect to the power input, the power input is the rate of the refrigeration effect divided by the COP. Therefore,

$$\text{COP} = \frac{\dot{Q}_{\text{ref}}}{\dot{W}} \rightarrow \dot{W} = \frac{\dot{Q}_{\text{ref}}}{\text{COP}} \rightarrow \frac{\dot{W}}{J} = \frac{1000}{12.95} = 77.2 \text{ Btu/min}$$

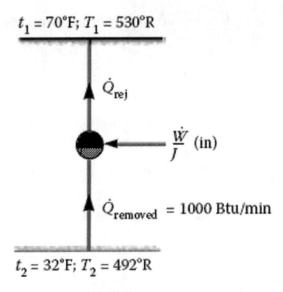

**FIGURE 10.5**
Illustrative Problem 10.1.

   c.   $\dot{Q}_{rej}$ is the sum of the refrigeration effect plus the power into the cycle. Thus,

$$\dot{Q}_{rej} = \dot{Q}_{ref} + \dot{W} = 1000 + 77.2 = 1077.2 \text{ Btu/min}$$

---

**ILLUSTRATIVE PROBLEM 10.2**

A Carnot refrigeration cycle is used to keep a freezer at $-5°C$ ($T_L = -5°C = 268$ K). Heat is rejected to a room at $20°C$ ($T_H = 20°C = 293$ K). If 30 kW is removed as heat, determine (a) the COP of the cycle, (b) the power required, and (c) the rate of heat rejected to the room. Refer to Illustrative Problem 10.1 and Figure 10.5.

**SOLUTION**

Proceeding as in Illustrative Problem 10.1, we obtain the following:

   a.   $COP = \dfrac{T_L}{T_H - T_L} = \dfrac{268}{293 - 268} = 10.72$

   b.   $\dot{W} = \dfrac{\dot{Q}_L}{COP_{ref}} = \dfrac{30 \text{ kW}}{10.72} = 2.8 \text{ kW}$

   c.   $\dot{Q}_{rej} = 30 + 2.8 = 32.8 \text{ kW}$

---

## 10.3 Defined Ratings

The use of mechanical means of producing refrigeration replaced the use of natural ice for this purpose. As a result, the capacity of a machine to extract heat in a refrigeration cycle was expressed in terms of equivalent ice-producing (or melting) capacity. The unit of refrigeration chosen was the amount of heat that is necessary to freeze 1 ton of water (2000 $lb_m$) into ice in one day. Using approximately 144 Btu/$lb_m$ as the latent heat of fusion of water, the unit of refrigeration is 2000 $lb_m$ × 144 Btu/$lb_m$ = 288,000 Btu/day. This rate of steady heat removal is defined to be a *ton* of refrigeration. Thus,

$$1 \text{ ton of refrigeraion} = 288000 \text{ Btu/day}$$

For other times, we have 1 ton = 12,000 Btu/h or 200 Btu/min. This is equivalent to 3.514 kW.

From the definition of COP, it is evident that the product of COP and power input yields the rate at which heat is being removed, or

$$\text{rate of heat removal, } \dot{Q}_{ref} = \text{COP} \times \left( \dot{W}, \text{power input} \right) \left[ \text{hp or kW} \right] \qquad (10.3)$$

Equation 10.3 leads to the following ratings:

$$\frac{\text{horsepower}}{\text{ton of refrigeration}} = \frac{200}{42.4 \times \text{COP}} = \frac{4.717}{\text{COP}} \qquad (10.4)$$

$$\frac{\text{kilowatts}}{\text{ton of refrigeration}} = \frac{200}{56.93 \times \text{COP}} = \frac{3.514}{\text{COP}} \qquad (10.5)$$

It is left as an exercise for students to derive Equations 10.4 and 10.5. The following example will serve to illustrate the foregoing concepts.

---

**ILLUSTRATIVE PROBLEM 10.3**

A refrigeration cycle operates between 20°F (480 R) and 70°F (530 R). (a) Determine the maximum COP for the cycle. (b) It is found that the actual COP is 2. How much more horsepower per ton of refrigeration effect is required by the actual cycle over the minimum possible requirement?

**SOLUTION**
Refer to Figure 10.3.

a. $\text{COP}_{ideal} = \dfrac{T_L}{T_H - T_L} = \dfrac{460 + 20}{(460 + 70) - (460 + 20)} = 9.6$

b. The minimum horsepower per ton of refrigeration (Equation 10.4) is therefore 4.717/9.6 = 0.49 hp/ton. The actual COP is stated to be 2; therefore, the actual horsepower per ton of refrigeration is 4.717/2 = 2.36 hp/ton. The horsepower required by the actual cycle over the minimum is

$$\begin{array}{r} 2.36 \\ -0.49 \\ \hline 1.87 \text{ hp/ton} \end{array}$$

## ILLUSTRATIVE PROBLEM 10.4

If the actual COP for Illustrative Problem 10.1 is 4.5, determine the power required. Compare this to the results found in Illustrative Problem 10.1.

### SOLUTION

From Equation 10.4,

$$\frac{\text{horsepower}}{\text{ton}} = \frac{4.717}{4.5} = 1.048$$

In Illustrative Problem 10.1, 1000 Btu/min is the heat removed. This is (1000 Btu/min)/(200 Btu/min ton), or 5 tons of refrigeration. The horsepower required is $1.048 \times 5 = 5.24$ hp. In Illustrative Problem 10.1, we found that 77.2 Btu/min was required. In terms of horsepower,

$$77.2 \frac{\text{Btu}}{\text{min}} \times 778 \frac{\text{ft.lb}_f}{\text{But}} \times \frac{1}{33000 \dfrac{\text{ft.lb}_f}{\text{min hp}}} = 1.82 \text{ hp}$$

The ratio of the power required in each problem is the same as the inverse ratio of the COP values, that is, $(4.5/12.95)(5.24) = 1.82$, which checks with our results.

## ILLUSTRATIVE PROBLEM 10.5

Using the conclusion of Illustrative Problem 10.4, determine the power required in Illustrative Problem 10.2 if the actual COP is 3.8.

### SOLUTION

In Illustrative Problem 10.2, it was found that the COP was 10.72 and the power was 2.8 kW. Therefore,

$$\text{power} = 2.8 \times \frac{10.72}{3.8} = 7.9 \text{ kW}$$

## 10.4 Refrigeration Cycles

To discuss the various refrigeration cycles that are in common use, we will first look at the "ideal" or prototype of each cycle to understand the basic thermodynamics involved and then consider how the practical realization of these cycles is accomplished.

At present, there are several common refrigeration cycles in use:

1. Vapor-compression cycle
2. Gas-cycle refrigeration
3. Absorption refrigeration cycle
4. Vacuum refrigeration cycle
5. Thermoelectric refrigerator

### 10.4.1 Vapor-Compression Cycle

Just as it was found possible to reverse the Carnot cycle, it is equally possible in principle to reverse the Rankine cycle. An elementary vapor-compression cycle is shown in Figure 10.6, and a corresponding *T–s* diagram is shown in Figure 10.7.

The wet cycle consists of an expansion of the fluid from the saturation point (saturated liquid at high pressure) to the wet region (path ①→②). During this process (throttling), the enthalpy stays essentially constant (see Chapter 4). However, the pressure and temperature of the working fluid decrease, and the fluid becomes a vapor–liquid mixture at state ②. The cooled working fluid (*refrigerant*) then passes to the evaporator, and there (path ②→③) heat enters from the region or fluid to be cooled. This part of the process is carried out at a constant temperature and constant pressure (ideally) because the working fluid is in the wet region. Furthermore, it is assumed that the refrigerant comes out as saturated vapor from the evaporator. The next part of the cycle (path ③→④) is a compression phase. If the compression proceeds from point ③ to point ④ in Figure 10.7, the refrigerant will start from the saturated vapor point and then proceed into the superheated vapor range. This path is called *dry compression*. An alternative path (③a→④a) is shown where the refrigerant is initially "wet" and is just brought to saturation conditions by the compression process.

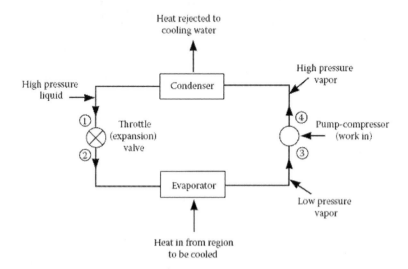

**FIGURE 10.6**
Simple vapor-compression cycle.

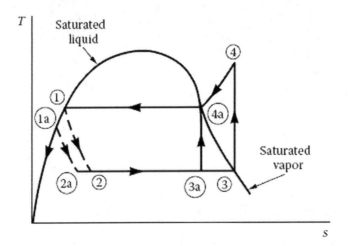

**FIGURE 10.7**
Simple vapor-compression cycle.

This path has been called *wet compression* for obvious reasons. From considerations of the cycle and the *T–s* diagram (Figure 10.7), it can be demonstrated that it is more efficient to operate the cycle with wet compression. However, most vapor-compression cycles operate with dry compression because the loss in efficiency is small while the practical problems of the compressor design and operation are considerably eased. The last path of the cycle (path ④→①) rejects heat at essentially constant pressure until the saturated liquid line is reached in the condenser. Once again, the *T–s* diagram shows an alternate path (①a→②a). In practice, it would be difficult to just achieve point ① for all loads on the system, and some further cooling to point ①a would be expected.

Assuming that all processes are carried out without pressure or heat losses (other than specified) and neglecting kinetic energy and potential energy differences, it is possible to analyze the system as follows:

1. Path ①→②, throttling:

$$h_1 = h_2 \tag{10.6}$$

2. Path ②→③, evaporator (cooling effect):

$$q_{in} = h_3 - h_2 = h_3 - h_1 \tag{10.7}$$

3. Path ③→④, compressor:

$$\text{work} = h_4 - h_3 \tag{10.8}$$

(For the refrigeration cycle, work into the cycle is taken to be positive.)

4. Path ④→①, condenser (heat removal):

$$q_{rej} = h_4 - h_1 \tag{10.9}$$

From these quantities, it will be noted that the energy removed in the condenser ($q_{rej}$) must numerically equal the heat gain in the evaporator ($q_{in}$ or $q_{evap}$) plus the work of the compressor. The refrigerating capability, however, is only the evaporator effect.

The following table should be used as a guide when looking for the corresponding enthalpy values.

| State | Pressure | Description | $h$ value |
|-------|----------|-------------|-----------|
| 1 | High | Saturated liquid | $h_1 = h_{f@p1}$ |
| 2 | Low | Mixture | $h_2 = h_1$ |
| 3 | Low | Saturated Vapor | $h_3 = h_{g@p3}$ |
| 4 | High | Superheated Vapor | $h_{4@p4}$ and $s_4 = s_3$ |

The mass of refrigerant circulated per ton of refrigeration can readily be found from the definition of the tons and the heat absorbed by the evaporator:

$$\frac{\text{mass of refrigerant}}{\text{minute}} = \frac{200 \times \text{tons}}{h_3 - h_2} = \frac{200 \times \text{tons}}{h_3 - h_1} \tag{10.10a}$$

In general terms, the mass flow rate of refrigerant can be found as follows,

$$\dot{m}_{ref} = \frac{\dot{Q}_{evap}}{h_3 - h_2} = \frac{\dot{Q}_{evap}}{h_3 - h_1} \tag{10.10b}$$

The definition of the COP (in terms of this cycle) is the ratio of the heat absorbed in the evaporator to the net work into the cycle. In consistent heat units, this can be written as

$$COP = \frac{h_3 - h_1}{h_4 - h_3} \tag{10.11}$$

For the ideal cycle, the work per ton of refrigeration will be given by

$$\text{horsepower/ton} = 4.717 \frac{h_4 - h_3}{h_3 - h_1} \tag{10.12}$$

Note that this is the definition given in Equation 10.4.

---

**ILLUSTRATIVE PROBLEM 10.6**

An ammonia refrigerator plant is to operate between a saturated liquid at 120 psia at the condenser outlet and a saturated vapor at 15 psia at the evaporator outlet. If a capacity of 30 tons is desired, compute the following on the basis of an ideal cycle only: (a) COP; (b) work of compression, Btu per pound; (c) refrigerating effect, Btu per pound; (d) pounds per minute of ammonia required; and (e) ideal horsepower per ton of refrigeration.

**SOLUTION**

At 120 psia, the corresponding saturation temperature is 66°F. Referring to Figure 10.7, the enthalpies are

$h_1 = 116.0 \text{ Btu/lb}_m$ (values of ammonia properties, from Appendix 2)

$h_2 = 116.0 \text{ Btu/lb}_m$ (throttling gives us $h_1 = h_2$)

$h_3 = 602.4 \text{ Btu/lb}_m$

From the consideration that $s_3 = s_4$, $h_4$ is found: at 15 psia, $s_3 = 1.3938$; therefore, by interpolation in the superheat tables at 120 psia, $t_4 = 237.4°F$ and $h_4 = 733.4 \text{ Btu/lb}_m$.

a.  $\text{COP} = \dfrac{602.4 - 116.0}{733.4 - 602.4} = 3.71$

b.  The work of compression is

$$h_4 - h_3 = 733.4 - 602.4 = 131.0 \text{ Btu/lb}_m$$

c.  The refrigerating effect is

$$h_3 - h_1 = 602.4 - 116.0 = 486.4 \text{ Btu/lb}_m$$

d.  The pounds per minute of ammonia required for circulation equals

$$\frac{200 \times 30}{602.4 - 116.0} = 12.34 \text{ lb}_m/\text{min}$$

e.  The ideal horsepower per ton of refrigeration equals

$$4.717 \frac{733.4 - 602.4}{602.4 - 116.2} = 1.27 \text{ hp/ton}$$

---

## ILLUSTRATIVE PROBLEM 10.7

Solve Illustrative Problem 10.6 if the refrigerant is Freon-12, with the high-pressure side at 110 psig. The Freon-12 enters the expansion valve as a saturated liquid. It leaves the evaporator as a dry saturated vapor at −20°F.

## SOLUTION

From Appendix 2, 110 psig corresponds to 96°F.

$h_1 = 30.14 \text{ Btu/lb}_m$

$h_2 = 30.14 \text{ Btu/lb}_m$

$h_3 = 75.110 \text{ Btu/lb}_m$

To determine $h_4$, we note that $s_3 = s_4$. At −20°F, $s_3 = 0.17102 \text{ Btu/lb}_m°F$. By interpolation in the Freon-12 superheat table at these values, $h_4 = 89.293 \text{ Btu/lb}_m$.

a.  $\text{COP} = \dfrac{75.11 - 30.14}{89.293 - 75.11} = 3.17$

b.  Work of compression $= 89.293 - 75.11 = 14.183\ \text{Btu/lb}_m$

c.  Refrigerating effect $= 75.11 - 30.14 = 44.97\ \text{Btu/lb}_m$

d.  Pounds of Freon-12 required for circulation:

$$\frac{200 \times 30}{44.97} = 133.42$$

e.  Ideal horsepower per ton of refrigeration:

$$4.717\ \frac{89.293 - 75.11}{75.11 - 30.14} = 1.49\ \text{hp/ton}$$

Before proceeding further with our discussion of refrigeration cycles, let us first look at some of the components used and their arrangement in practice. In a typical home refrigerator, the compressor is usually physically in the rear, near the bottom of the unit. The motor and compressor are usually located in a single housing, with the electrical leads for the motor passing through the housing. While this is done to prevent leakage of the refrigerant, it imposes the requirement that the refrigerant should be inert with respect to the insulation. One common refrigerant, Freon, cannot be used with natural rubber, because this class of refrigerant acts as a solvent with certain types of insulating materials and varnishes. Also, the electrical resistance of the refrigerant is of the utmost importance in hermetically sealed units, such as those used in home refrigerators, where the motor windings are exposed to the refrigerant. Figure 10.8 shows a diagrammatic layout of a refrigerating system for a domestic-type refrigerator.

The condenser is physically arranged so that room air flows past the condenser by natural convection. The expansion valve is a long capillary tube, which induces high-pressure drop, and the evaporator is shown around the outside of the freezing compartment inside the refrigerator. The pressures shown in Figure 10.8 are typical for the use of sulfur dioxide as the refrigerant. The compressor shown schematically is a reciprocating unit, but small, efficient, and economical rotary such as scroll compressor units are available for home refrigeration use.

Starting at point ③ of Figure 10.8, the vaporous refrigerant enters the compressor at low pressure and temperature. An isentropic compression (ideally) raises the pressure and temperature to $p_2$ and $t_2$. The saturation temperature corresponding to $p_2$ must be some value above atmospheric temperature or above the temperature of the cooling water that may be used in the condenser. Leaving the compressor in condition ④, the vapor enters the condenser, where it is condensed to a liquid at some temperature $t_1$. After the condenser, the liquid enters an expansion valve that separates the high- and low-pressure regions and passes through the valve in a throttling process with $h_1 = h_2$. The refrigerant then enters the evaporator (or freezing compartment), where it boils because it is receiving heat from the refrigerator and its contents. The vapor from the evaporator enters the compressor, and the cycle starts over.

The heart of any refrigeration system is the compressor. While the types of compressors used are similar to those used for air or other gases, the positive-displacement reciprocating compressor is the unit most widely used in industrial vapor-compression refrigeration installations. Centrifugal and gear-type positive-displacement compressors are also used. The gear-type compressor shows good volumetric efficiencies, but the centrifugal

**FIGURE 10.8**
Diagrammatic layout of a refrigerating system. (From Faires, V.M., *Theory and Practice of Heat Engines*, Macmillan, Inc., New York, 1948. With permission.)

compressors are usually inefficient in smaller sizes and are used when the size of a recip-rocating unit would be excessively large. A discussion of compressors and an analysis of the reciprocating compressor are given in Section 10.5.

The evaporator is another component of the refrigeration system that has to be consid-ered. Many types of evaporators are used depending on the specific application. The evap-orator should be designed so that all the liquid is converted to vapor before the refrigerant reaches the outlet connection of the evaporator.

Many considerations enter into the choice of refrigerant for a given application. Some of the refrigerants that are used commercially are ammonia, butane, carbon dioxide, Carrene, Freon, methyl chloride, sulfur dioxide, and propane. Based on the thermodynamics of the vapor-compression cycle that we have studied thus far, the following properties are desir-able in a refrigerant:

1. The heat of vaporization, $h_{fg}$ of the refrigerant should be high. The higher the $h_{fg}$, the greater the refrigerating effect per pound of fluid circulated.

2. The specific heat of the liquid should be low. The lower the specific heat of the liquid, the less heat it will pick up for a given change in temperature either during throttling or in flow through the piping and, consequently, the greater the refrig-erating effect per pound of refrigerant.

3. The specific volume of the refrigerant should be low to minimize the work required per pound of refrigerant circulated.

4. The critical temperature of the refrigerant should be higher than the condensing temperature to prevent excessive power consumption.

In addition to the foregoing, it is necessary to consider toxicity, corrosiveness, dielectric strength for a hermetically sealed system, viscosity, thermal conductivity, explosiveness, effect on foods, stability, inertness, and cost. No one refrigerant has been found to meet all these requirements. Of the many possible refrigerants, ammonia, Freon, brine, methyl chloride, and sulfur dioxide are the most common that have been used.

Two environmental concerns have recently come to the attention of both industry and government, namely, the potential of a refrigerant to deplete ozone, $O_3$ in the upper atmo-sphere and the potential of a refrigerant to persist in the upper atmosphere, trapping radiation emitted by the Earth (the greenhouse effect). A major effort has been exerted to develop and use refrigerants having low ozone-depletion levels and low greenhouse-warming potential and on developing refrigerant recovery systems and reducing refriger-ant leakage.

In September 1987, an international meeting was held in Montreal, Canada, to discuss the effects of chlorofluorocarbon (CFC) refrigerants on the atmosphere. In June 1990, rep-resentatives of 93 nations met in London and amended the 1987 Montreal Agreements. Basically, the London Amendments called for the complete phasing out of CFC refriger-ants by the year 2000. Hydrochlorofluorocarbon (HCFC) refrigerants are to be phased out by the year 2020, with the absolute deadline being the year 2040. The refrigerant hydro-fluorocarbon (HFC) contains no chlorine and poses no threat to the ozone layer. However, HFC-134a, a replacement for R-11 and R-12 refrigerants and used primarily in autos and refrigerators, has a global warming potential that is 420 times higher than that of carbon dioxide. Further study could lead to regulation on the use of HFC-134a as well as other HFCs. Household units have been using R-22, which is in the process of being replaced by a new refrigerant, R-410A.

Some of the major engineering problems that have arisen and have been overcome in the application of CFC replacement refrigerants are

1. The immiscibility of HFC-134a with mineral oil lubricants previously used in vapor-compression systems
2. The need for a new and different thermal expansion valve
3. The need to use nylon as an interliner in the refrigerant hoses
4. The need for a new desiccant to absorb water vapor
5. The need for a new parallel-flow condenser to reduce system head pressure

Table 10.1 summarizes some of the physical and thermal properties of commonly used refrigerants. Currently, HFC-134a is widely used in commercial systems, and R-22 is used in most residential units. HFC-134a tables are in Appendix 2 along with tables for Freon-12, which is still used in older systems. Ammonia is one of the most effective refrigerants on a per-pound basis and so is also included in the tables in Appendix 2. It is, however, highly toxic. Computerized properties of some refrigerants are available at the CRC Press website. Use the downloads/updates tab at http://www.crcpress.com/product/isbn/9781482238556.

In the actual cycle, there are deviations from the ideal cycle due to compression inefficiencies, pressure drops in piping, temperature differences between the elements interchanging heat, and so on. Because heat is transferred in both the condenser and the evaporator of the vapor-compression system, temperature differences exist between these units and the medium to which they are transferring heat. In the case of the evaporator, the coils must be at a lower temperature than the region being cooled. The temperature difference for the evaporator usually is of the order of 5°F, while in the condenser, the refrigeration fluid is of the order of 10°F higher than the coolant fluid. However, among these effects, those usually considered are the inefficiency in the compression part of the cycle and pressure differences in long piping systems. The other effects are usually considered to be either too difficult to analyze generally or small enough to be safely ignored.

---

### ILLUSTRATIVE PROBLEM 10.8

An ideal refrigeration cycle uses Freon-12 as the working fluid. The temperature of the refrigerant in the condenser is 90°F, and in the evaporator, it is 10°F. Determine the COP of this ideal cycle.

### SOLUTION

Using the Freon-12 tables in Appendix 2, we have

$h_1 = 28.713 \text{ Btu/lb}_m$
$h_2 = h_1 = 28.713 \text{ Btu/lb}_m$
$h_3 = 78.335, s_3 = 0.16798 = s_4$

From the superheat table, at 90°F and $s = 0.16798$, $h_4 = 87.192 \text{ Btu/lb}_m$.

$$\text{The heat extracted} = (78.335 - 28.713) = 49.622 \text{ Btu/lb}_m$$

**TABLE 10.1**

Physical and Thermal Properties of Common Refrigerants

| Refrigerant | Chemical Formula | Critical Points | | | | Specific Gravity of Liquid at Atmos. Pressure | Specific Heat of Liquid, Ave. 5°F to 86°F |
| --- | --- | --- | --- | --- | --- | --- | --- |
| | | Boiling Point (°F) | Freezing Point (°F) | Temperature (°F) | Pressure (psia) | | |
| Ammonia | $NH_3$ | −28.0 | −107.9 | 271.4 | 1657.0 | 0.684 | 1.12 |
| Azeotropic mixture Freon-12 and unsymmetrical difluoroethane (Carrene-7) G-100 | 78.3% F-12, 26.2% G-100 | −28.0 | −254.0 | 221.1 | 631.0 | — | — |
| Bromotrifluoromethane (Kulene-131) | $CF_3Br$ | −73.6 | −226.0 | 153.5 | 587.0 | — | 0.19 |
| Carbon dioxide | $CO_2$ | −109.3 | −69.0 | 87.8 | 1069.9 | 1.560 | 0.77 |
| Dichlorodifluoromethane (Freon-12, Genetron-12) | $CCl_2F_2$ | −21.6 | −252.0 | 233.6 | 596.9 | 1.480 | 0.23 |
| Ethylene | $C_2H_4$ | −155.0 | −272.0 | 48.8 | 731.8 | — | — |
| Isobutane | $C_4H_{10}$ | −10.3 | −229.0 | 272.7 | 537.0 | 0.549 | 0.62 |
| Methyl chloride | $CH_3Cl$ | −10.8 | −144.0 | 289.4 | 968.7 | 1.002 | 0.38 |
| Methylene chloride (Carrene-1) | $CH_3Cl_2$ | 105.2 | −142.0 | 480.0 | 670.0 | 1.291 | 0.33 |
| Monochlorodifluoromethane (Freon-12, Genetron-141) | $CHClF_2$ | −41.4 | −256.0 | 204.8 | 716.0 | 1.411 | 0.30 |
| Sulfur dioxide | $SO_2$ | 14.0 | −103.9 | 314.8 | 1141.5 | 1.357 | 0.34 |
| R-410A | 50% $CH_2F_2$,50% $CHF_2CF_3$ | −60.6 | −247.0 | 162.0 | 715.0 | 1.35 | |
| Trichlorotrifluoroethane (Freon-113) | $CCl_2F-CClF_2$ | 117.6 | −31.0 | 417.4 | 495.0 | 1.559 | 0.21 |
| HF C-134a (1, 1, 1, 2-tetrafluoroethane) | $CH_2FCF_3$ | −15.0 | −14.9 | 213.9 | 588.9 | 1.376 | 0.33 |
| R-22 | $CHClF_2$ | −41.4 | −256 | 205 | 722 | 1.41 | |
| Water | $H_2O$ | 212.0 | 32.0 | 706.1 | 3226.0 | 1.000 | 1.00 |

The work required $= (87.196 - 78.335) = 8.861 \text{ Btu/lb}_\text{m}$

Therefore, $\text{COP} = \dfrac{49.622}{8.861} = 5.60$

---

## ILLUSTRATIVE PROBLEM 10.9

Solve Illustrative Problem 10.8 if the refrigerant is HFC-134a.

**SOLUTION**

Using the HFC-134a tables in Appendix 2,

$h_1 = 41.6 \text{ Btu/lb}_\text{m}$

$h_2 = h_1 = 41.6 \text{ Btu/lb}_\text{m}$

$h_3 = 104.6, s_3 = 0.2244, s_4 = s_3$

$h_4 = 116.0 \text{ Btu/lb}_\text{m}$

The heat extracted $= (104.6 - 41.6) = 63 \text{ Btu/lb}_\text{m}$

The work required $= (116.0 - 104.6) = 11.4 \text{ Btu/lb}_\text{m}$

The work required $= (116.0 - 104.6) = 11.4 \text{ Btu/lb}_\text{m}$

Therefore, $\text{COP} = \dfrac{63}{11.4} = 5.53$

---

## ILLUSTRATIVE PROBLEM 10.10

A Freon-12 vapor-compression refrigeration system has a compressor of 80% adiabatic or isentropic efficiency. At the inlet to the compressor, the refrigerant has a pressure of 5.3 psia and 40°F superheat; at discharge from the compressor, the pressure is 100 psia.

Saturated liquid Freon at 100 psia leaves the condenser and passes through an expansion valve into the evaporative coils. The condenser is supplied with cooling water, which enters at 60°F and leaves at 70°F. For a plant capacity of 5 tons, find (a) the horsepower required to drive the compressor if it has a mechanical efficiency of 100% and (b) the required capacity in gallons per minute of cooling water to the pump.

## SOLUTION

A *T–s* diagram for this problem is shown in Figure 10.9. This problem will require both the tables and the *p–h* chart for Freon-12 given in Appendix 2.

a. By definition, the efficiency of the compressor is the ratio of the ideal compression work (under isentropic conditions) to actual compression work. Based on the points in Figure 10.9,

$$\eta_{\text{isentropic, compressor}} = \frac{w_{c,s}}{w_{c,a}} = \frac{h_{4,s} - h_3}{h_{4',a} - h_3}$$

where $h_{4s}$ is found by knowing $p_4$ and $s_3$ ($s_4 = s_3$), and $h_{4,a}$ is found in tables using $p_4$ and $T_4$. There is close correspondence between 5.3 psia and −60°F for saturated conditions. Therefore, state ③ is a superheated vapor at 5.3 psia and approximately −20°F because the problem states that state ③ has a 40°F superheat. Interpolation in the Freon tables in Appendix 2 yields

$$T = -20°F$$

| $p$ | $h$ | $s$ | |
|---|---|---|---|
| 7.5 | 75.719 | 0.18371 | |
| 5.3 | 75.886 | 0.18985 | $h_3 = 75.886$ Btu/lbm |
| 5.0 | 75.909 | 0.19069 | |

at 100 psia and $s_3 = 0.18985$,

| $t$ | $s$ | $h$ | |
|---|---|---|---|
| 170°F | 0.18996 | 100.571 | |
| 169.6°F | 0.18985 | 100.5 | $h_{4,s} = 100.5$ Btu/lbm |
| 160°F | 0.18726 | 98.884 | |

**FIGURE 10.9**
Illustrative Problem 10.10.

The weight of the refrigerant is given by

$$\frac{200(\text{tons})}{h_3 - h_1} = \frac{(200)(5)}{75.886 - h_1}$$

In the saturated tables, $h_1$ is

| $p$ | $h$ |
|---|---|
| 101.86 | 26.832 |
| 100 psia | 26.542 |
| 98.87 | 26.365 |

and

$$\dot{m} = \frac{\text{mass flow}}{\text{min}} = \frac{(200)(5)}{75.886 - h_1} = \frac{(200)(5)}{75.886 - 26.542} = 20.3 \text{ lb}_\text{m}/\text{min}$$

total work of compression $= 20.3\left(h_{4',a} - h_3\right)$

where

$$w_{c,a} = h_{4',a} - h_3 = \frac{h_{4,s} - h_3}{\eta} = \frac{100.5 - 75.886}{0.8} = 30.8 \text{ Btu}/\text{lb}_\text{m}$$

Therefore,

$$\text{work} = \frac{(20.3)(30.8)(778)}{33000} = 14.7 \text{ hp}$$

b. Assuming the specific heat of the water as unity, we obtain

$$\dot{m} = \frac{20.3\left(h'_{4,a} - h_1\right)}{(70 - 60)}$$

Form part (a),

$$h_{4',a} - h_3 = 30.8$$

$$h_{4',a} = 30.8 + 75.886 = 106.69 \text{ Btu}/\text{lb}_\text{m}$$

Therefore,

$$\dot{m} = \frac{(20.3)(106.59 - 26.542)}{70 - 60}$$

$$= 162.5 \text{ lb}_\text{m}/\text{min of cooling water}$$

$$= 19.5 \text{ gal}/\text{min}$$

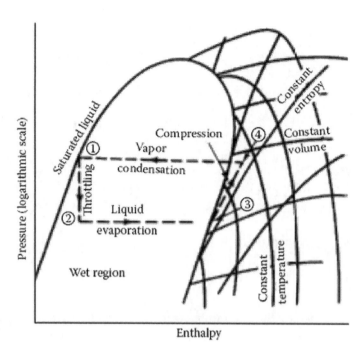

**FIGURE 10.10**
Pressure–enthalpy diagram.

Students will note the effort involved in obtaining this answer when working directly from tables. Just as the Mollier chart simplifies calculations of power cycles, problems in refrigeration are simplified considerably by the use of a modified Mollier chart. Such a chart uses pressure–enthalpy coordinates and usually has other curves of entropy and temperature on it. Figure 10.10 shows a skeleton chart using $p$–$h$ coordinates of a refrigerant. The ordinate, pressure, is usually plotted on a logarithmic scale for convenience. In the superheat region, lines of constant volume, constant entropy, and constant temperature are displayed. Shown by dashed lines is an ideal simple vapor-compression cycle with end states corresponding to the events shown on the simplified schematic diagram of Figure 10.6. The following problem will serve to illustrate the use of the $p$–$h$ diagram as well as the savings obtainable in both time and effort by using this modified Mollier chart.

---

**ILLUSTRATIVE PROBLEM 10.11**

Solve Illustrative Problem 10.10 using the $p$–$h$ chart for Freon-12 in Appendix 2.

**SOLUTION**

a. From Appendix 2, reading the $p$–$h$ diagram directly, we have

$$h_3 = 76.2 \text{ Btu/lb}_m$$

$$h_{4,s} = 100.5 \text{ Btu/lb}_m$$

$$\text{work of compression} = \frac{100.5 - 76.2}{0.8} = 30.4 \text{ Btu/lb}_m$$

The enthalpy of saturated liquid at 100 psia is given at 26.5 Btu/lb$_m$. Proceeding as before yields

$$\frac{\text{mass flow}}{\text{min}} = \frac{(200)(5)}{76.2 - 26.5} = 20 \text{ lb}_m/\text{min}$$

$$\text{total ideal work of compression} = \frac{(20)(30.4)(778)}{33000} = 14.3 \text{ hp}$$

b.  $h_{4',a} = 76.2 + 30.4 = 106.8 \text{ Btu/lb}_m$

$$\dot{m} = \frac{(20)(106.8 - 26.5)}{70 - 60} = 160.6 \text{ lb}_m/\text{min or } 19.3 \text{ gal/min}$$

The accuracy afforded by the diagram is sufficient for engineering applications.

### 10.4.2 Gas-Cycle Refrigeration

Just as it was possible to reverse the Carnot and Rankine cycles to obtain a refrigerating effect, it should be possible to reverse any cycle to obtain a similar effect. Among the many gas cycles that have been proposed as prototypes for practical engines, the reversed Brayton cycle has been used to produce refrigeration. The earliest mechanical refrigeration system used air as the refrigerant because of its availability and safety. However, as a result of its high operating costs and low COP, it has largely become obsolete. Yet the introduction of high-speed aircraft and missiles has required lightweight, high refrigeration-capacity systems to cause a minimum reduction in payload. For example, a jet fighter traveling at 600 mph requires a refrigeration system having a capacity of 10 to 20 tons. In passenger aircraft, solar radiation, the occupants, and electrical and mechanical equipment all cause the heating of cabin air. In addition, the adiabatic stagnation of the air relative to the moving airplane (see Chapter 6) can cause temperature increases of the air of approximately 65°F at a relative air speed of 600 mph. Also, all modern aircraft operate with pressurized cabins, which requires compression of outside air to nearly atmospheric pressure, and in the process, the temperature of the compressed air increases. The gas-cycle refrigeration system using air as the refrigerant has become generally used for this application because of the small size and weight of the components, which is of prime importance. The power to operate the system is of secondary importance in this application. In addition, the power source used for the air-cycle aircraft refrigeration system is the same as is used for pressurization of the cabin. Because pressurization is not needed at sea level but refrigeration is and, at high altitudes, pressurization is required but less refrigeration is needed, it is found that the power requirement for the combined pressurization and cooling is approximately constant.

Let us now consider the reversed Brayton cycle shown schematically in Figure 10.11. This cycle is a closed cycle in that the working fluid (air) is continuously recycled. As shown in Figure 10.11, the cycle consists of an expander (turbine), a compressor (centrifugal fan), and

**FIGURE 10.11**
Schematic of simple, closed-cycle Brayton cycle.

two heat exchangers. Figure 10.12 shows the $p$–$v$ and $T$–$s$ diagrams for the cycle shown in Figure 10.11, with corresponding points ①, ②, ③, and ④ noted on each of these figures. As noted in Chapter 9, the Brayton cycle consists of two isentropic and two constant-pressure (isopiestic or isobaric) portions.

Consider that a single charge of working fluid is in the device and that the cycle starts from the condition denoted as ① in Figures 10.11 and 10.12. At this point, the gas must be colder than the room (or second fluid) that is to be cooled. It flows at constant pressure through the required piping and absorbs heat from the area to be cooled. During this part of the cycle, the temperature of the cooled area will be greater than $T_2$. After being heated, the gas is compressed isentropically along path ②→③. In general, $T_3$ will be greater than room temperature (or local cooling water temperature), and the gas therefore is capable of being cooled in the next part of the path. This is accomplished at constant pressure ③→④, in which $T_4$ is always greater than the temperature of the available coolant. Finally, the gas is cooled to state ① by an isentropic expansion during which work is removed from the system. There is one basic point, however, that causes this cycle to deviate from the previously considered vapor cycle. The vapor cycle employs a throttling process to reduce the temperature of the working fluid. The same effect could be achieved by expanding the liquid in an engine, but this engine is not practical and would be prohibitively expensive. It will be recalled that a throttling process is carried out at constant enthalpy. For an ideal gas, the throttling process is carried out at a constant temperature and therefore is incapable of lowering the temperature for the required refrigerating effect. Even in real gases, the throttling process does not provide much of a refrigerating effect. To obtain this effect when gas is used as the refrigerant, it is necessary to create the required low-temperature level by letting the refrigerant do work by expanding in an engine.

Each of the processes constituting this gas cycle can be evaluated from the considerations given in Chapter 6. Thus, from the definition of COP and assuming an ideal gas with constant specific heat,

$$\text{COP}_{\text{Brayton}} = \frac{\text{heat removed}}{\text{net work required}} = \frac{c_p\left(T_2 - T_1\right)}{w_{3-2} - w_{1-4}} \tag{10.13}$$

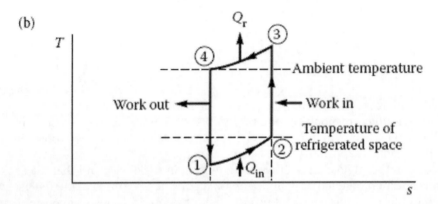

**FIGURE 10.12**
*T–s* and *p–v* diagrams for reversed Brayton cycle. (a) *p–v* diagram. (b) *T–s* diagram.

The denominator of Equation 10.13 represents the net area behind the ③→② and ④→①
curves. Because all elements of the cycle are reversible, it follows that the net work of the
cycle must equal the difference between the heat rejected by the cycle and the heat into the
cycle. Therefore,

$$\text{COP}_{\text{Brayton}} = \frac{c_p(T_2 - T_1)}{c_p(T_3 - T_4) - c_p(T_2 - T_1)} = \frac{T_1\left(\dfrac{T_2}{T_1} - 1\right)}{T_4\left(\dfrac{T_3}{T_4} - 1\right) - T_1\left(\dfrac{T_2}{T_1} - 1\right)} \tag{10.14a}$$

Since $p_4/p_1 = p_3/p_2$, and processes 1–2 and 3–4 are isentropic (see Equation 6.63), Equation
10.14a can be simplified using $T_4/T_1 = T_3/T_2$ (see Section 9.7 in Chapter 9) as follows,

$$\text{COP}_{\text{Brayton}} = \frac{T_1}{T_4 - T_1} = \frac{1}{\dfrac{T_4}{T_1} - 1} = \frac{1}{\dfrac{T_3}{T_2} - 1} = \frac{T_2}{T_3 - T_2} \tag{10.14b}$$

Again, since processes 1–2 and 3–4 are isentropic, and $p_b = p_4 = p_3$, $p_a = p_1 = p_2$, then the COP
for the Brayton refrigerator in terms of pressure based on Equation 10.14b is,

$$\text{COP}_{\text{Brayton}} = \frac{1}{\left(p_4 / p_1\right)^{(k-1)/k} - 1} = \frac{1}{\left(p_b / p_a\right)^{(k-1)/k} - 1} \tag{10.14c}$$

By comparing the foregoing relations with a reversed Carnot cycle operating between the same limits, it will be noted that the index of performance (COP) of the reversed Brayton cycle is less than that of the Carnot cycle. The fundamental difference between the Brayton and Carnot cycles is that the Brayton does not operate at constant heat-rejection and constant heat-absorption temperatures, whereas the Carnot does.

Because it may be of interest to know the work of the compressor and the turbine, it is necessary only to note that each of these processes represents a steady-flow isentropic process; thus,

$$\text{work of compressor per pound}, w_c = c_p \left(T_3 - T_2\right) \tag{10.15a}$$

$$\text{work of expander per pound}, w_t = c_p \left(T_4 - T_1\right) \tag{10.15b}$$

By proper substitution of the path equation, $pv^k$, these equations can be put in terms of pressure.

---

**ILLUSTRATIVE PROBLEM 10.12**

An ideal gas refrigeration cycle is operated so that the upper temperature of the cycle is 150°F (610 R), the lowest temperature of the cycle is −100°F (360 R), and the COP is 2.5. Determine the work of the expander, $w_t$, the work of the compressor, $w_c$, and the net work of the cycle per pound of air. Also, determine the airflow required per ton of refrigeration. Take $c_p = 0.24$ Btu/lb$_m$·R.

**SOLUTION**

From Equation 10.14c,

$$\text{COP}_{\text{Brayton}} = \frac{T_1}{T_4 - T_1} = 2.5 \rightarrow 2.5\left(T_4 - T_1\right) = T_1 \text{ but } T_1 = 360 \text{ R}$$

Therefore,

$$\left(T_4 - 360\right) = \frac{360}{2.5} \rightarrow T_4 = 504 \text{ } R\left(44°F\right)$$

Also,

$$\text{COP}_{\text{Brayton}} = \frac{T_2}{T_3 - T_2} = 2.5 \rightarrow \left(2.5\right)\left(610 - T_2\right) = T_2$$

$$T_2 = 435.7 \text{ } R\left(-24.3°F\right)$$

The work of the expander is

$$c_p \left( T_4 - T_1 \right) = \left( 0.24 \right) \left( 504 - 360 \right) = 34.6 \text{ Btu/lb}_m \text{ of air}$$

The work of the compressor is

$$c_p \left( T_3 - T_2 \right) = \left( 0.24 \right) \left( 610 - 436 \right) = 41.8 \text{ Btu/lb}_m \text{ of air}$$

The net work required by the cycle is

$$41.8 - 34.6 = 7.2 \text{ Btu / lb}_m \left( \text{work in} \right)$$

Per ton of refrigeration, the required airflow is

$$\frac{200}{c_p \left( T_2 - T_1 \right)} = \frac{200}{0.24 \left( 436 - 360 \right)} = 10.96 \text{ lb}_m / \text{min per ton}$$

In an actual cycle, the $\text{COP}_{\text{Brayton}}$ would be much lower because of the irreversible effects in each part of the cycle.

The air cycle described earlier for aircraft use is an example of an open cycle in which the working fluid is not caused to continually recycle. Instead, it is continuously discharged, and a fresh air charge is continuously introduced into the cycle at the entering conditions of the cycle. Figure 10.13 shows a simple, open-air cycle used for refrigeration. Students should compare this cycle to the aircraft refrigeration cycle described earlier.

In addition to the simple Brayton refrigeration cycle, two other versions of this cycle are also used. The first of these is known as the *bootstrap system*. In this system, as shown in Figure 10.14, there are two heat exchangers with cooling air provided by the ram pressure of the aircraft forcing outside air through the exchangers. Air bled from the gas turbine engine compressor is first cooled in the initial heat exchanger, further compressed by the secondary compressor, then cooled further in the secondary heat exchanger, and finally expanded in the cooling turbine and delivered to the cabin of the airplane. The work delivered by the cooling turbine is used to drive the secondary compressor. The bootstrap system has the disadvantage that the airplane must be in flight to obtain the required ram air. To overcome this, a fan is provided to pull air over the secondary heat exchanger, giving a system having the features of both the simple and bootstrap systems.

The second version of the reversed Brayton cycle that is used is the regenerative system. This system is used when the discharge temperature of the simple cycle is too high, as may be the case for very high-speed aircraft. In this cycle, as shown in Figure 10.15, some of the turbine discharge air is diverted back to cool the air entering the turbine. This serves to cool the air entering the turbine to a temperature lower than that which would be obtained by using only ambient air.

### 10.4.3 Absorption Refrigeration Cycle

One of the fundamental concepts of thermodynamics is the equivalence of various forms of energy. A good portion of this study has been devoted to the concept that heat and work

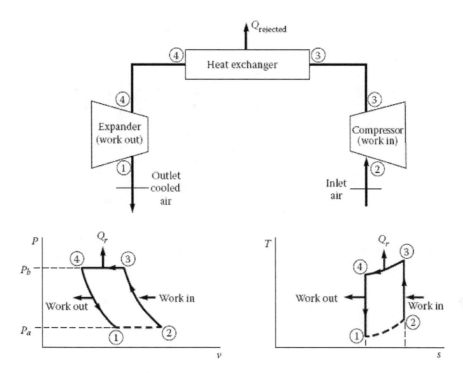

**FIGURE 10.13**
Open Brayton refrigeration cycle.

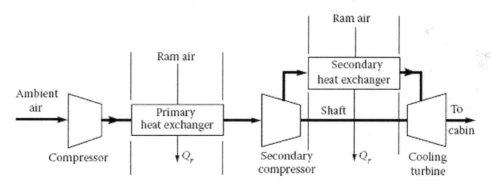

**FIGURE 10.14**
Bootstrap system.

are forms of energy in transition. The *absorption refrigeration* cycle takes advantage of this concept by essentially replacing the vapor-compression pump compressor of the simple vapor-compression cycle with an "equivalent" heat source. In practice, this cycle leads to certain complications in "plumbing," and it is usual to find that the vapor-compression system has a higher index or coefficient of performance than the absorption system.

Figure 10.16 shows a simple absorption system in which the pump compressor of the vapor-compression system (shown dashed for reference in Figure 10.16) has been replaced by an absorber, a pump, and a vapor generator. The rest of the cycle is the same as the simple vapor-compression cycle.

**FIGURE 10.15**
Regenerative system.

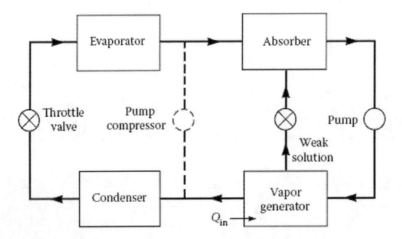

**FIGURE 10.16**
Simple absorption refrigeration cycle.

The most common absorption refrigeration cycle is one using ammonia as the refriger-ant. In terms of the cycle shown in Figure 10.16, the ammonia solution in the vapor genera-tor is heated to create ammonia vapor. The ammonia vapor liberated in the vapor generator proceeds along the indicated path to the condenser and then goes through the conven-tional part of the cycle. After leaving the evaporator, the ammonia enters the absorber. The weak solution in the vapor generator is mixed with the ammonia in the absorber, where the weak solution absorbs the ammonia and the resulting strong solution is pumped to the generator. In this cycle, the work of the circulating pump is very small for a given refrigerating effect because the pump is pumping a liquid that has a small specific volume. To reduce the steam and cooling water requirements of the simple absorption system, a regenerative heat exchanger may be placed between the absorber and generator.

### 10.4.4 Vacuum Refrigeration Cycle

To understand the physical processes involved in this refrigeration cycle, let us start with a simple experiment. Take a flask half-filled with water, say, at 70°F, and place it under a bell jar that can be evacuated. The partial pressure of the water vapor in the air in the flask will correspond to its initial temperature. As the pressure in the bell jar is lowered, no change can be seen in the flask. For all practical purposes, only the air is evacuated, and the water stays essentially isothermal. As the bell jar pressure reaches the saturation pressure (approximately 19 mm Hg absolute), the water will appear to "boil" violently. Actually, the violent agitation of the water is due to the air that was absorbed in the water *outgassing*. Once this violent motion has subsided, the continued pumping will lower the pressure below the initial saturation value. As this occurs, water vapor will be drawn off the surface of the liquid, and the temperature of the remaining fluid will be lowered until it corresponds to the pressure existing in the bell jar. The lowering of the temperature is caused by the decrease in internal energy of the remaining liquid necessitated by the energy required to vaporize some of the liquid at the lower pressure. During the period from the time the outgassing ceases until approximately 4.5 mm Hg absolute is reached, the water will be still, and no visual change will be observed. At approximately 4.5 mm Hg absolute, the surface of the water will quite suddenly become coated with a layer of ice. Thus, by lowering the pressure, it has been possible to lower the temperature of the liquid below its surroundings. In this case, a change of phase from liquid to solid has been achieved.

A schematic of a steam-jet refrigeration system in which water is used as the refrigerant is shown in Figure 10.17. In this system, a steam ejector is used to establish and maintain the required vacuum, which causes a portion of the water to vaporize and cool the rest of the remaining water. The steam-jet ejector is used as the vacuum pump, drawing off both vaporized water and any entrained air and compressing them to a higher pressure. The compressed vapor and steam are then condensed. This condenser requires considerably more cooling water than the condenser for a conventional mechanical compression

**FIGURE 10.17**
Steam-jet refrigeration system.

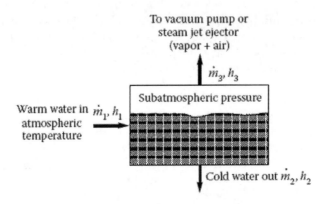

**FIGURE 10.18**
Analysis of vacuum refrigeration system.

system because it must remove the heat of the power steam as well as that liberated from the chilled water.

When the steam-jet ejector system is used, it is common to utilize low-pressure plant steam for reasons of economy. It is necessary to provide a condenser on the steam jet as well as an air ejector to take care of outgassed air and any air that may leak into the subatmospheric parts of the system. The operation of the steam-jet ejector system is simple and does not require moving parts, but the jet is a poor compressor and requires large amounts of steam. However, solar thermal energy can be used to provide the necessary amount of steam at low operating cost.

We can analyze the refrigeration process in the simple *vacuum refrigeration* system by referring to Figure 10.18. Because this is a steady-flow system, water is continuously supplied and withdrawn, with the difference in the mass of water vapor going to the pumping system. Writing the mass balance for this system yields

$$\dot{m}_1 = \dot{m}_2 + \dot{m}_3 \tag{10.16}$$

and

$$\dot{m}_1 h_1 = \dot{m}_2 h_2 + \dot{m}_3 h_3 \tag{10.17}$$

The refrigerating effect (or heat removed from the system) can be obtained from the vapor as $\dot{m}_3(h_3 - h_1)$ or from the liquid as $\dot{m}_2(h_1 - h_2)$.

---

### ILLUSTRATIVE PROBLEM 10.13

A vacuum refrigeration system is used to cool water from 90°F to 45°F. Calculate the mass of water vapor that must be removed per pound of entering water.

### SOLUTION

Refer to Figure 10.19 and Equations 10.16 and 10.17. Assume that 1 lb$_m$ of liquid enters the system and that the vapor leaves as saturated vapor at 45°F.

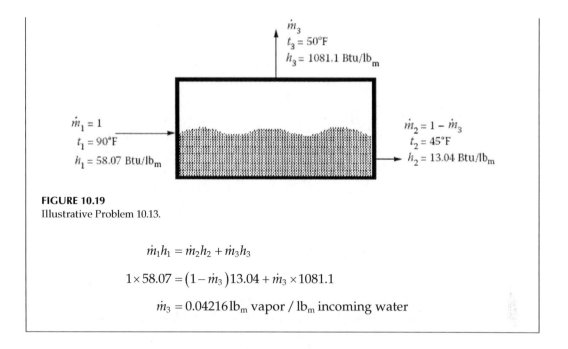

**FIGURE 10.19**
Illustrative Problem 10.13.

$$\dot{m}_1 h_1 = \dot{m}_2 h_2 + \dot{m}_3 h_3$$

$$1 \times 58.07 = (1 - \dot{m}_3) 13.04 + \dot{m}_3 \times 1081.1$$

$$\dot{m}_3 = 0.04216\,\text{lb}_m \text{ vapor / lb}_m \text{ incoming water}$$

---

**ILLUSTRATIVE PROBLEM 10.14**

Determine the refrigeration effect in Illustrative Problem 10.13.

**SOLUTION**

The refrigeration effect can be determined as $\dot{m}_3 (h_3 - h_1)$ or $\dot{m}_2 (h_1 - h_2)$. For this problem,

$$\dot{m}_3 (h_3 - h_1) = 0.04216(1081.1 - 58.07) = 43.13 \text{ Btu/lb}_m \text{ of water}$$

$$\dot{m}_2 (h_1 - h_2) = (1 - 0.04216)(58.07 - 13.04) = 43.13 \text{ Btu/lb}_m \text{ of water}$$

Because of the magnitude of the small numbers involved in this problem and Illustrative Problem 10.13, more significant places have been carried in the calculation than is usually necessary.

---

### 10.4.5 Thermoelectric Refrigerator

In 1822, Thomas J. Seebeck found that an electromotive force is generated between the junctions of dissimilar metals at different temperatures. This is the basis of the thermocouple as a temperature-measuring device (discussed in some detail in Chapter 1). For metals, the electromotive force is on the order of microvolts per degree of temperature difference, but recent advances in semiconductor materials have produced materials having much larger voltage outputs per degree of temperature difference. While such systems are relatively inefficient direct converters of heat to electricity, they are reliable and have no moving parts. Figure 10.20a shows a schematic of a thermoelectric converter.

**FIGURE 10.20**
Thermoelectric converters. (a) Direct and (b) reversed cycle.

   In 1834, Jean Peltier found that when a current flows around a loop of two dissimilar metals, one end warms up, while the other end cools. In effect, the thermoelectric converter has been reversed in a manner similar to the reversal of a power cycle to obtain a refrigeration cycle. Thus, if one junction (the hot one) is kept in the ambient surroundings while the other (the cold one) is kept in the refrigerated space and a potential is applied, as shown schematically in Figure 10.20b, heat will be transferred from the space to be refrigerated to the cold junction. The hot junction will reject heat to the surroundings because its temperature will rise above the ambient temperature.
   At present, the thermoelectric refrigerator cannot compete with the more conventional refrigeration systems. However, in such space-limited applications as the removal of heat from electronic devices and assemblies, the thermoelectric refrigerator is in commercial use. More commercial applications will become economical as this refrigeration effect is developed further.

## 10.5 Compressors

The most important part of a refrigeration or air-conditioning system is the compressor. It represents the addition of mechanical work that allows the transfer of heat from a low-temperature reservoir to a high-temperature one. Compressors are also used in many pneumatic systems to raise air pressure to levels that can drive the air through pneumatic circuits. In refrigeration systems, the compressor increases the temperature and pressure of the gaseous refrigerant, thereby increasing its thermal energy. That energy, when added to the energy extracted from the cold reservoir, is exhausted to a hot reservoir as the refrigerant condenses.
   Compressors are categorized as either positive displacement or centrifugal. The former include reciprocating and rotary devices; the latter include axial units. Centrifugal compressors are utilized where space and weight are limited, as in a jet engine. The reader is referred to the many specialized texts on refrigeration and air conditioning listed in the references for details on the types of compressors. The most common type of compressor used in refrigeration systems is the reciprocating type, which will now be analyzed.

In the reciprocating compressor, a piston moves up and down, alternately sucking in and discharging the gaseous refrigerant. The compressor can be single stage, with one compression per cycle, or dual acting, with two compressions per cycle. To understand the thermodynamics of the compressor, the device will be examined as if it were a self-contained system undergoing a gas cycle. The reciprocating compressor can be thought of as a reversed open heat engine with gas taken in at low pressure and exhausted at high pressure. As in the gas engine, the reciprocating compressor has a clearance volume at the top dead center of the piston travel where the valves are located. The amount of clearance determines the effectiveness of the compressor as to how much gas it can pressurize with each cycle.

To analyze the compressor cycle, it will be assumed that each of the processes are ideal, that is, frictionless. The $p$–$V$ diagram for the cycle is shown in Figure 10.21. The gas is compressed from state 1 to state 2 as a polytropic process. At state 2, the exhaust valve is opened once the desired pressure is reached. The piston's continued movement exhausts the gas up to state 3, the end of the piston's stroke. Here, the exhaust valve closes, leaving a clearance volume of gas, $V_3$. As the piston now moves in the opposite direction, the clearance gas expands, and its pressure drops polytropically until it is low enough to open the intake valve at state 4. Fresh gas enters from state 4 to state 1, where the piston again reverses and the intake valve closes.

The gas intake and exhaust are assumed to be at constant pressure in this idealized cycle. In the actual compressor, there is some drop in pressure as the gas passes through each valve. This pressure drop is similar to that of a throttling process, which thus raises the temperature of the gas as it reaches point 1.

Note that this is a cycle that consumes work; the direction of the processes is counterclockwise on the $p$–$V$ diagram. The compression work is done from state 1 to state 2, and this path is polytropic, $pV^n$ = constant. If compression can be accomplished adiabatically, the path is indicated by $pV^k$ = $C$, where k is the specific heat ratio of the working fluid. Since the valves are closed during compression, this process is a closed system. From Chapter 6, the work for a polytropic process is given by

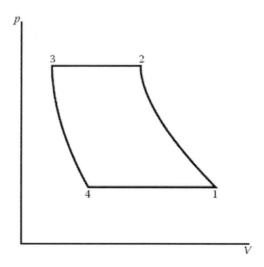

**FIGURE 10.21**
Ideal compressor cycle.

$$W = \frac{p_2 V_2 - p_1 V_1}{1 - n} \tag{10.18}$$

The work of the entire cycle is as follows:

Since process 4→1 is constant pressure,

$$W_{41} = p_1 \left( V_1 - V_4 \right) \tag{10.18a}$$

Since process 1→2 is polytropic,

$$W_{12} = \frac{p_2 V_2 - p_1 V_1}{1 - n} \tag{10.18b}$$

Since process 2→3 is constant pressure,

$$W_{23} = p_2 \left( V_3 - V_2 \right) \tag{10.18c}$$

Process 3→4 is likewise assumed to be polytropic with the same exponent $n$, so

$$W_{34} = \frac{p_4 V_4 - p_3 V_3}{1 - n} \tag{10.18d}$$

Adding the work terms, noting that $p_1 = p_4$ and $p_2 = p_3$,

$$W_{\text{cycle}} = \frac{p_1 n}{n - 1} \left( V_1 - V_4 \right) + \frac{p_2 n}{n - 1} \left( V_3 - V_2 \right) \tag{10.19}$$

Using the polytropic relationships from Chapter 6,

$$\frac{V_2}{V_1} = \left( \frac{p_2}{p_1} \right)^{-\frac{1}{n}} \quad \text{and} \quad \frac{V_3}{V_4} = \left( \frac{p_3}{p_4} \right)^{-\frac{1}{n}} = \left( \frac{p_2}{p_1} \right)^{-\frac{1}{n}}$$

The total cycle work of the compressor then becomes

$$W = \frac{n}{n - 1} p_1 \left( V_1 - V_4 \right) \left( 1 - \left( \frac{p_2}{p_1} \right)^{\frac{n-1}{n}} \right) \tag{10.20}$$

For the compressor power consumed, multiply the work per cycle by the compressor speed, $N$, in rpm. With the pressure in lb./ft.$^2$ and the volume in ft.$^3$, the work is in ft.-lb. Multiplying by rpm gives the power in ft.-lb/min, which, if divided by 33,000, yields horsepower. The resultant equation is

$$\dot{W} = \frac{n}{n - 1} p_1 \left( V_1 - V_4 \right) N \left( 1 - \left( \frac{p_2}{p_1} \right)^{\frac{n-1}{n}} \right) \tag{10.21}$$

Note that $(V_1 - V_4) N$ is the volume flow rate $\dot{V}_1$ entering the compressor. If the compressor is double acting, whereby both sides of the piston are utilized, the work per cycle is doubled. In addition, if the compressor unit includes more than one piston cylinder, the total work is multiplied by the number of cylinders. In any case, since pressure $p_2$ is always greater than $p_1$, the work of the cycle will always be negative.

### 10.5.1 Volumetric Efficiency of Compressor

The compressor's clearance volume allows space around the valves for intake and exhaust. This volume is usually expressed as a percentage clearance $c$ defined as

$$c = \frac{\text{clearance volume}}{\text{volume displaced}} = \frac{V_3}{V_1 - V_3} \tag{10.22}$$

The amount of fresh gas taken in per cycle is $V_1 - V_4$. The effectiveness of the compressor depends on the amount of intake gas that can be compressed relative to the total volume displaced by the piston's stroke. The piston's stroke volume is found by multiplying the piston bore area by its stroke length and equals $V_1 - V_3$. The effectiveness of a compressor is referred to as its *volumetric efficiency* and is defined as

$$\eta_v = \frac{V_1 - V_4}{V_1 - V_3} \tag{10.23}$$

This volumetric efficiency can be expressed in terms of $c$ and the pressure ratio. Since the compression follows a polytropic process, it is assumed that the expansion portion of the cycle is likewise polytropic with the same exponent, that is, $p_3(V_3)^n = p_4(V_4)^n$. If the volume displaced by the piston is called $V_{PD}$ ($V_{PD} = V_1 - V_3$.) the volumetric efficiency can be considered as

$$\eta_v = \frac{(V_{PD} + cV_{PD}) - V_4}{V_{PD}} \tag{10.23a}$$

Dividing through by the displacement volume yields

$$\eta_v = 1 + c - \frac{V_4}{V_{PD}} = 1 + c - \frac{V_4 c}{V_3} = 1 + c - c\frac{V_4}{V_3} = 1 + c - c\left(\frac{p_3}{p_4}\right)^{\frac{1}{n}}$$

Since $p_4 = p_1$,

$$\eta_v = 1 + c - c\left(\frac{p_3}{p_1}\right)^{\frac{1}{n}} \tag{10.24}$$

The mass flow rate of gas through the compressor is assumed to be constant as it is a steady-flow device. The volume flow rate, however, changes with changes in pressure. The input temperature and pressure determine the mass flow rate according to the ideal gas law.

$$\dot{m} = \frac{p_1 \dot{V}_1}{RT_1} \tag{10.25}$$

$\dot{V}_1$ is found by $\dot{V}_1 = (V_1 - V_4)$ times the compressor speed, $N$. At the outflow of the compressor, both pressure and temperature have changed from the inlet conditions. The outlet temperature, $T_3$, can be found by the polytropic relationship, a variant of Equation 6.70,

$$\frac{T_3}{T_1} = \left(\frac{p_3}{p_1}\right)^{\frac{n-1}{n}}$$

(10.26)

Knowing $T_3$, one can now find the output volume flow rate by

$$\dot{V}_3 = \frac{\dot{m}RT_3}{P_3}$$

(10.27)

---

### ILLUSTRATIVE PROBLEM 10.15

A reciprocating compressor with a bore of 15 in. and a stroke of 12 in. has a clearance of 5%. The compressor, running at 500 rpm, takes in air at 15 psia and 70°F and compresses it adiabatically to 90 psia. Assuming air as an ideal gas, find the volumetric efficiency, the power required, the mass flow rate, and the volume flow rate of the air at discharge.

### SOLUTION

The volumetric efficiency is found from the clearance and the pressure ratio,

$$\eta_v = 1 + c - c\left(\frac{p_3}{p_1}\right)^{\frac{1}{n}} = 1 + 0.05 - 0.05(6)^{\frac{1}{1.4}} = 0.87$$

The volume displaced by the piston is

$$V_{PD} = \frac{\pi}{4}D^2L = 1.227 \text{ ft.}^3/\text{cycle}$$

The intake volume is now found from the above values,

$$\eta_v V_{PD} = V_1 - V_4 \rightarrow 0.87\left(1.227 \text{ ft.}^3/\text{cycle}\right) = V_1 - V_4 = 1.0677 \text{ ft.}^3/\text{cycle}$$

The volume flow rate into the compressor is now calculated using the compressor speed,

$$\dot{V}_1 = \frac{N \times (V_1 - V_4)}{60} = \frac{500 \times 1.0677}{60} = 8.897 \text{ ft.}^3/\text{s}$$

The mass flow is now found,

$$\dot{m} = \frac{p_1 \dot{V}_1}{RT_1} = \frac{15 \times 144 \times 8.897}{53.34 \times 530} = 0.68 \text{ lb/s or } 40.79 \text{ lb./min}$$

Now the outlet volume flow rate is found after the outlet temperature is calculated,

$$\frac{T_3}{T_1} = \left(\frac{p_3}{p_1}\right)^{\frac{k-1}{k}} = (6)^{\frac{0.4}{1.4}} = 1.67; \text{thus } T_3 = 884.31R$$

$$\dot{V}_3 = \frac{\dot{m}RT_3}{p_3} = \frac{0.68 \times 53.34 \times 884.31}{90 \times 144} = 2.47 \text{ ft.}^3/\text{s}$$

Finally, the power required is found from Equation 10.21,

$$\dot{W} = \frac{1.4}{1.4-1} 15 \times \left(144 \text{ in}^2/\text{ft}^2\right)(1.0677) \times 500 \left(1 - (6)^{\frac{1.4-1}{1.4}}\right)$$

$$= -2697873 \text{ ft.lb}_f/\text{min} \times \left(\frac{1 \text{ min}}{60 \text{ sec}}\right) \times \left(\frac{1 \text{ hp}}{550 \text{ lb} - \text{ft/sec}}\right)$$

$$= 81.75 \text{ hp}$$

When there is a need for a large increase in pressure, a single-stage compressor will have lowered volumetric efficiency since the greater pressure change will result in increased expansion of the clearance volume. To overcome this, multistage compression is utilized whereby each stage produces a portion of the total pressure increase required. Each stage can thus have a better volumetric efficiency. In multistage compression, the gas is usually cooled between stages, reducing its specific volume and reducing the total work required; this is called *intercooling*.

## 10.6 The Heat Pump

Thus far, we have directed our attention to the heat removed from the region to be cooled and denoted this as the refrigeration effect. If instead we consider the energy rejected as heat as the desired outcome from the refrigeration cycle, we call such a device a *heat pump*. To understand the basic principles of operation of this device, let us once again consider the reversed Carnot cycle shown in Figures 10.3 and 10.4. During the process of cooling a given space, energy (heat) is rejected at some higher temperature. As can be seen from Figure 10.3, the heat rejected is equal to the heat removed from the cold region plus the work into the system. For the heat to be rejected, say, as in a household refrigerator, the temperature of the fluid in the cycle must be above the temperature of the surroundings. As shown in Figure 10.4, the total energy rejected is given by the rectangle $A'BCD'$, which is also equal to $T_1(s_2 - S_1)$. The energy (or work) supplied is equal to $(T_1 - T_2)(s_2 - s_1)$. If we now define the performance index or coefficient of performance for such a system as the ratio of the energy delivered (the heating effect or total energy rejected) to the net work supplied as the *COP*, we have

**FIGURE 10.3**
**(Repeated)** Reversed Carnot cycle.

**FIGURE 10.4**
**(Repeated)** Reversed Carnot cycle.

$$\text{COP}_{\text{Carnot heat pump}} = \frac{Q_{\text{rej}}}{W_{\text{in}}} = \frac{T_1(s_2 - s_1)}{(T_1 - T_2)(s_2 - s_1)} = \frac{T_1}{T_1 - T_2} = \frac{1}{1 - \dfrac{T_2}{T_1}} = \frac{1}{1 - \dfrac{T_L}{T_H}} \quad (10.28)$$

Where $T_H$ and $T_L$ represent the temperatures of the high-temperature sink and cold-temperature source, respectively. Note that only absolute temperatures (K or R) are to be used in Equation 10.28. Also, Equation 10.28 applies *only* to a Carnot device, not to a heat pump operating on another cycle, such as the vapor-compression cycle.

By comparing Equations 10.2 (COP for Carnot refrigerator) and 10.28 (COP for Carnot heat pump), the following relationship between the heat pump and refrigerator in terms of COP can be found as follows,

$$\text{COP}_{\text{heat pump}} = \left(1 + \text{COP}_{\text{refrig}}\right)_{\text{Carnot cycle}} \quad (10.28a)$$

In an actual machine, the necessity for finite temperature differences to transfer the heat, as well as other irreversibilities, will make the actual performance less than the optimum performance of a heat pump.

---

### ILLUSTRATIVE PROBLEM 10.16

Determine the COP as a heat pump for the cycle described in Illustrative Problem 10.1.

### SOLUTION

We can solve this problem by using the cycle temperatures, that is, a rejection temperature of 530 R and a cold temperature of 492 R. Thus,

$$\text{COP}_{\text{Carnot heat pump}} = \frac{T_1}{T_1 - T_2} = \frac{530}{530 - 492} = 13.95$$

The COP can also be obtained from the energy items solved for in Illustrative Problem 10.1. The power was found to be 77.2 Btu/min, and the total rate of heat rejection was 1077.2 Btu/min. Therefore, the COP = 1077.2/77.2 = 13.95, as before for the Carnot heat pump.

---

### ILLUSTRATIVE PROBLEM 10.17

If the cycle described in Illustrative Problems 10.1 and 10.16 has its lower temperature reduced to 0°F (460 R), determine the work in, the heat rejected, and the COP as a heat pump.

### SOLUTION

Let us first consider the cycle as a refrigeration cycle. The COP as a refrigeration cycle is $(0+460)/(530 - 460) = 6.57$. The power input would be $1000/6.57 = 152.2$ Btu/min, and the total rate of heat rejection would be $1000 + 152.2 = 1152.2$ Btu/min. The COP as a heat pump is $1152.2/152.2 = 7.57$. As a check, $6.57 + 1 = 7.57$ (Equation 10.28a).

---

Note, as a result of Illustrative Problems 10.16 and 10.17, that as the outside temperature decreased from 32°F to 0°F, the COP as a heat pump was almost halved, and that the energy delivered to the higher temperature increased by 7% while the required work into the system doubled. Thus, when the demand is the greatest as the outside temperature drops, the ability of a given heat pump to meet this demand decreases. In northern latitudes, the heat pump is found to be uneconomical if it is used solely for heating, but it can and is used in southern latitudes, where there are moderate heating loads and relatively long cooling needs. However, heat pumps can still be used in northern latitudes if the evaporator is connected to a ground-based low-temperature heat source capable of providing heat at a relatively constant and moderate temperature. In such installation, U-pipes or loops are buried in the ground so an appropriate heat transfer fluid can flow through them and extract heat from the ground continuously. In southern latitudes, a single system (air conditioning and heat pump) is used for both heating and cooling, and whenever additional heating requirements are needed, electrical resistance elements in the air ducts are used to make up this additional heat requirement. Figure 10.22 shows the operation of the heat pump used for summer cooling and winter heating. Suitable valving is necessary to reverse the path of the refrigerant and to operate the system properly.

The method by which a heat pump can be switched to operate in either a cooling or a heating mode is illustrated in Figure 10.23. In the heating mode (top schematic), an expansion device drops the pressure and temperature of the refrigerant so that it may pass through the outside heat exchanger, which functions in this mode as an evaporator. There, the fluid's temperature is sufficiently low for it to pick up heat from the outside air and vaporize. The vapor is then passed through the compressor, where its temperature and pressure increase. Next, the refrigerant passes through the indoor heat exchanger (condenser), where it gives up its heat to the cooler room air and condenses. It then passes through the expansion valve to repeat the cycle.

In the cooling mode, activating the reversing valve and the appropriate expansion valve bypass reverses the flow of the refrigerant so that it completes the same cycle but with the outside heat exchanger functioning as the condenser (heat given up to the outside air) and the inside heat exchanger functioning as the evaporator (picking up heat from the room air). Switching is accomplished using an interior thermostat.

In Figure 10.22, the atmosphere is indicated to be the energy source when the heat pump is operated on the heating cycle. Unfortunately, atmospheric air is a variable-temperature fluid, and as noted earlier, as the temperature of the air decreases, requiring a greater heat output, the heat pump is less able to meet this demand. To overcome this problem, well water and the Earth itself have been used rather than atmospheric air. If well water is used, it is necessary that the supply is adequate and that the water is returned underground to prevent the local water table from lowering. When the heated water is returned, it should be at some distance from where it is initially taken from the ground, to prevent heating of the water source. If coils are buried in the Earth as explained above, it is necessary to ensure good thermal contact between the coil and the Earth.

The application of the heat pump to a specific installation or for a specific use is an economic problem. In areas where low-cost electricity is available, the heat pump finds applications in the heating and cooling of both homes and industrial buildings. Also, industrial

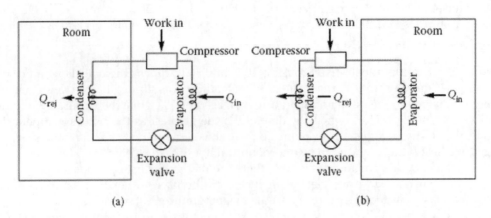

**FIGURE 10.22**
All-year heat pump operation. (a) Heating–winter. (b) Cooling–summer.

(a)

**Outside (source)**                                          **Inside conditioned space (sink)**

(b)

**Outside (sink)**                                            **Inside conditioned space (source)**

**FIGURE 10.23**
Basic heat pump operation, air-to-air heat pump. (a) Heating, (b) cooling. (From EPRI, *Heat Pump Manual*, 2nd ed., 1997. With permission.)

processes often require either heating or cooling, and the heat pump could be used for specific applications. One such application is where cooling of a process fluid is required, and the heat rejected by the heat pump could be used to heat the plant or office space. The heat pump should not be used without a prior, thorough investigation of the economics of the proposed installation being made after the engineering feasibility has been established.

## 10.7 Review

The different refrigeration systems have been introduced in this chapter. Where the objective is to keep an area cool, we should use a refrigeration system, and a heat pump to keep an area warm. For each of these desired effects, we were able to define a figure of merit called the coefficient of performance. The cycles studied were the Carnot, Rankine, Brayton, absorption, and vacuum refrigeration. In addition to studying the cycles, we also looked at the refrigerants and the practical equipment used in them. Refrigeration is an accepted part of our daily lives and the lives of most of the people in the industrialized countries of the world. Its use for food preservation, comfort, and process control has become an indispensable part of our technology. The use of the heat pump, especially when it is combined with air conditioning, is also becoming more pervasive as time goes on.

In addition to the analytical tools that were already developed, we introduced the pressure–enthalpy diagram in this chapter. This is really a modified Mollier diagram that is very useful for the analysis of refrigeration cycles. For interested students, it is suggested that the publications of the American Society of Heating, Refrigeration, and Air Conditioning Engineers (ASHRAE) be consulted for further details and information on specific system variables.

---

## Key Terms

Terms used for the first time in this chapter are as follows:

**Absorption refrigeration:** The use of a heat source to replace the compressor in a vapor- compression system.

**Bootstrap system:** A Brayton air-refrigeration cycle with heat exchangers provided with cooling air by the ram pressure of the aircraft, forcing outside air through the exchangers.

**Clearance:** The volume of gas left in the compressor after exhaust as a percentage of the volume displaced by the piston travel.

**Coefficient of performance (COP) (for heat pump):** The ratio of the energy delivered to the work input.

**Coefficient of performance (COP) (for refrigeration):** The ratio of the refrigeration effect to the work input.

**Dry compression:** Compression of vapor only in a vapor-compression cycle.

**Heat pump:** A reversed refrigeration cycle whose object is to keep an area warm.

**Refrigerant:** The working fluid in a refrigeration cycle.

**Refrigerating effect:** The removal of heat from an area by the use of mechanical energy (i.e., compressor work).

**Thermoelectric refrigerator:** The use of the reversed Seebeck–Peltier effects to keep a space at low temperature.

**Ton (of refrigeration):** 288,000 Btu/day, 12,000 Btu/h or 200 Btu/min.

**Vacuum refrigeration:** The lowering of the pressure below atmospheric pressure to lower the temperature of the refrigerant below that of the surroundings.

**Volumetric efficiency:** The volume of gas taken in by a compressor divided by the volume displaced by the piston travel.

**Wet compression:** The compression of a wet refrigerant to saturation conditions.

# Equations Developed in This Chapter

| | | |
|---|---|---|
| Coefficient of performance | $$COP = \frac{\text{refrigeration effect}}{\text{work input}}$$ | (10.1) |
| $COP_{\text{Carnot refrigeration}}$ | $$COP_{\text{Carnot refrigeration}} = \frac{T_2}{T_1 - T_2}$$ | (10.2) |
| Definition | 1 standard commercial ton = 288,000 Btu/day | |
| Rating | $$\frac{\text{horsepower}}{\text{ton of refrigeration}} = \frac{4.717}{COP}$$ | (10.4) |
| Rating | $$\frac{\text{kilowatts}}{\text{ton of refrigeration}} = \frac{3.514}{COP}$$ | (10.5) |
| Throttling—vapor-compression cycle | $h_1 = h_2$ | (10.6) |
| Cooling effect—vapor-compression cycle | $q_{\text{in}} = h_3 - h_2 = h_3 - h_1$ | (10.7) |
| Compression—vapor-compression cycle | $\text{work} = h_4 - h_3$ | (10.8) |
| Heat removal—vapor-compression cycle | $q_{\text{rej}} = h_4 - h_1$ | (10.9) |
| Coefficient of performance (vapor-compression refrigeration) | $$COP = \frac{h_3 - h_1}{h_4 - h_3}$$ | (10.11) |
| Horsepower per ton (vapor compression) | $$\text{horsepower/ton} = 4.717 \frac{h_4 - h_3}{h_3 - h_1}$$ | (10.12) |
| $COP_{\text{Brayton}}$ | $$COP_{\text{Brayton}} = \frac{\text{heat removed}}{\text{net work required}} = \frac{c_p(T_2 - T_1)}{w_{3-2} - w_{1-4}}$$ | (10.13) |
| $COP_{\text{Brayton}}$ | $$COP_{\text{Brayton}} = \frac{c_p(T_2 - T_1)}{c_p(T_3 - T_4) - c_p(T_2 - T_1)} = \frac{1}{\frac{T_3 - T_4}{T_2 - T_1} - 1}$$ | (10.14a) |
| $COP_{\text{Brayton}}$ | $$COP_{\text{Brayton}} = \frac{1}{(p_b / p_a)^{(k-1)/k-1}}$$ | (10.14b) |
| $COP_{\text{Brayton}}$ | $$COP_{\text{Brayton}} = \frac{T_2}{T_3 - T_2}$$ | (10.14c) |
| Mass balance—vacuum refrigeration | $\dot{m}_1 = \dot{m}_2 + \dot{m}_3$ | (10.16) |
| Heat balance—vacuum refrigeration | $\dot{m}_1 h_1 = \dot{m}_2 h_2 + \dot{m}_3 h_3$ | (10.17) |
| Compressor cycle work | $$W = \frac{n}{n-1} p_1 (V_1 - V_4) \left[ 1 - \left(\frac{p_2}{p_1}\right)^{\frac{n-1}{n}} \right]$$ | (10.20) |
| Compressor power | $$\dot{W} = \frac{n}{n-1} p_1 (V_1 - V_4) N \left[ 1 - \left(\frac{p_2}{p_1}\right)^{\frac{n-1}{n}} \right]$$ | (10.21) |
| Clearance | $$c = \frac{\text{clearance volume}}{\text{volume displaced}} = \frac{V_3}{V_1 - V_3}$$ | (10.22) |
| Volumetric efficiency | $$\eta_v = 1 + c - c \left(\frac{p_3}{p_1}\right)^{\frac{1}{n}}$$ | (10.24) |
| $COP_{\text{heat pump}}$ | $$COP_{\text{Carnot heat pump}} = \frac{T_1}{T_1 - T_2}$$ | (10.28) |
| $COP_{\text{heat pump}}$ | $$COP_{\text{heat pump}} = \left(1 + COP_{\text{refrig}}\right)_{\text{Carnot cycle}}$$ | (10.28a) |

## QUESTIONS

10.1    Why is it possible to reverse a Carnot cycle to obtain a refrigerating effect?

10.2    Is the COP an absolute parameter, or does it depend on the usage of cycle energy quantities?

10.3    List the five common refrigeration cycles that were discussed in this chapter.

10.4    Trace the path of the working fluid on a $T–s$ diagram, indicating each step for each refrigeration cycle.

10.5    List the desirable properties of a refrigerant.

10.6    List some commonly used refrigerants.

10.7    Sketch on a $T–s$ diagram the reversed Brayton cycle.

10.8    How does the bootstrap system differ from the simple Brayton cycle?

10.9    How does regeneration function to increase the efficiency of the simple Brayton cycle?

10.10   How is the pump replaced by an "equivalent heat source" in the absorption refrigeration cycle?

10.11   How does vacuum refrigeration work?

10.12   Thermoelectric refrigeration has been used for space applications. Why is this type of refrigeration used for this application?

10.13   Under what circumstances would you use the heat pump in northern climates?

## PROBLEMS

### Problems Involving the Carnot Cycle

10.1    A Carnot refrigeration cycle is used to keep a refrigerator at 36°F. During the summer, the room temperature in which the refrigerator is kept reaches 85°F. If 1500 Btu/min must be removed, determine (a) the COP of the cycle, (b) the power required, and (c) the rate of heat rejected to the room.

10.2    A freezer using a Carnot refrigeration cycle maintains its interior at –10°C. In the summer, it rejects heat to a room at 30°C. If 25 kW is removed as heat, determine (a) the COP of the cycle, (b) the power required, and (c) the rate of heat rejected to the room.

10.3    A refrigeration cycle operates between 15°F and 72°F. Determine the maximum COP for this cycle. If the actual COP is 3, how much more horsepower is required per ton of refrigeration by the actual cycle over the minimum possible requirement?

10.4    Derive Equation 10.4.

10.5    Derive Equation 10.5.

10.6    A reversed Carnot cycle operates between temperatures of 32°F and 86°F. Calculate the ratio of the cooling effect to the work input.

10.7    A Carnot refrigeration cycle has a COP of 4. It rejects heat to a room that is at 72°F. Determine the lowest temperature in the cycle.

10.8    A Carnot cycle is used to cool a building. If 20 kW is supplied as power and its COP is 4.2, determine the refrigerating effect in tons and $T_1/T_2$.

10.9 A Carnot refrigeration cycle takes heat from water at 33°F and discards it to a room at 72°F. Assume that the refrigeration effect is equal to 8 tons of refrigeration. (a) How much heat is discarded to the room? (b) How much power is required? (c) How much heat is removed from the water?

10.10 Determine the horsepower required for a Carnot refrigeration machine if its capacity is rated at 8 tons of refrigeration while operating between 10°F and 110°F.

10.11 If Illustrative Problem 10.6 had been a Carnot cycle operating between the same temperature limits, determine items (a), (b), and (c) required by this problem. Compare the results.

10.12 A reversed Carnot cycle is operated to remove 1000 Btu/h from the low-temperature region of the cycle, and it rejects 1500 Btu/h to the high-temperature region of the cycle. Calculate the COP for this application.

10.13 If the purpose of the unit in Problem 10.12 is to supply heat to the upper-temperature region, calculate its COP.

10.14 A Carnot refrigeration cycle operates with a COP of 4.5. The space to be cooled requires 15 MJ/h to be removed as heat. What is the minimum size motor that would be needed for this application? State your answer in horsepower.

10.15 A refrigerator operating between 0°F and 100°F removes 1000 Btu/min. If the COP is three-fourths of a Carnot refrigerator operating between the same temperatures, determine the power input and the rate of heat rejected by the actual cycle and the Carnot cycle.

10.16 A refrigerator operates between −15°C and 40°C and removes 20 kW of heat from the unit. If the COP is 80% of a Carnot refrigerator operating between the same temperatures, determine the actual power and the rate of heat rejected.

*10.17 A Carnot refrigeration unit is used to cool a space to 40°F by the transfer of heat to the surroundings at 77°F. Assume that the cooling rate (the heat removed) is equivalent to 3 tons of refrigeration. If this same unit using the same power is now used to cool the space to 10°F, estimate the cooling rate required.

## Problems Involving Vapor-Compression Cycles

10.18 A dry compression refrigeration cycle using ammonia as the working fluid operates between 80°F and 10°F. Determine the ideal COP for this cycle. Use the ammonia tables in Appendix 2.

10.19 Compare the result obtained in Problem 10.18 to an ideal Carnot cycle operating between the same temperature extremes.

10.20 Solve Problem 10.18 using the $p–h$ diagram for ammonia given in Appendix 2.

10.21 If the compressor efficiency in Problem 10.18 is 80%, determine the COP of the cycle.

10.22 Compute the number of pounds of ammonia that must be circulated per minute in a 15-ton refrigeration system operating between pressures of 15 and 140 psia, with the ammonia entering the expansion valve at saturation temperature and leaving the evaporator as dry saturated vapor.

10.23   If an ideal Carnot refrigeration cycle operates between the temperature limits shown in Problem 10.22, determine the number of pounds of ammonia that would have to be circulated per minute. How does this compare to the results of Problem 10.22?

*10.24  An ammonia vapor-compression system operates with a capacity of 100 tons. The condenser pressure is 200 psia, and the evaporator pressure is 34 psia. Determine the amount of ammonia that must be circulated per minute if the ammonia enters the expansion valve at saturation temperature and leaves the evaporator as dry saturated vapor.

10.25   An ammonia refrigeration system operates between a condenser temperature of 80°F and an evaporator temperature of −10°F. If saturated liquid leaves the condenser and dry saturated vapor leaves the evaporator, determine the pounds per minute of ammonia that is circulated in a 50-ton vapor-compression system.

10.26   A vapor-compression system uses Freon-12 as refrigerant. The vapor in the evaporator is at −10°F, and in the condenser, it is at 100°F. When the vapor leaves the evaporator, it is as a dry saturated vapor, and it enters the expansion valve as saturated liquid. Determine the refrigerating effect per pound of Freon circulated.

10.27   A 10-ton vapor-compression refrigeration system uses Freon-12 as refrigerant. The vapor in the evaporator is at −20°F and leaves as dry saturated vapor. The high-side pressure is 180 psia, with Freon-12 entering the expansion valve as a saturated liquid. Compute the Freon-12 flow.

10.28   Assuming the cycle in Problem 10.27 to be an ideal Carnot cycle, determine the Freon-12 flow. Compare your results to Problem 10.27.

*10.29  An ideal vapor-compression refrigeration cycle is used to produce 20 tons of refrigeration. If the system uses Freon-12 and has a condenser operating at 90°F and an evaporator operating at 38°F, determine the compressor power and the refrigerant flow. The vapor leaves the evaporator as a dry saturated vapor, and it enters the expansion valve as saturated liquid.

10.30A  vapor-compression system uses Freon-12. Calculate the refrigeration effect when the system is operated with dry saturated vapor leaving the evaporator at −6°F and saturated liquid entering the expansion valve at 90°F. Determine the COP for this cycle.

*10.31  A refrigeration system is operated to produce 40 tons of refrigeration. The evaporator is operated at −30°F, and the condenser is operated at 70°F. Using Freon-12 as the refrigerant, determine the theoretical COP and the weight of refrigerant circulated per minute. Dry saturated vapor leaves the evaporator, and saturated liquid enters the expansion valve.

10.32   Solve Problem 10.26 if the working refrigerant is HFC-134a.

10.33   Solve Problem 10.27 if the working refrigerant is HFC-134a.

10.34   Solve Problem 10.28 if the working refrigerant is HFC-134a.

10.35   Solve Problem 10.29 if the working refrigerant is HFC-134a.

10.36   Solve Problem 10.30 if the working refrigerant is HFC-134a.

10.37   Solve Problem 10.31 if the working refrigerant is HFC-134a.

## Problems Involving Gas Refrigeration

10.38 An ideal gas refrigeration cycle is operated so that the gas is expanded from 150 to 15 psia. Assuming that the ratio of specific heats ($k$) is 1.4 and the upper temperature of the cycle is 200°F, determine the COP and the temperature before the compression part of the cycle.

10.39 Solve for the COP in Problem 10.38 if the cycle is an ideal Carnot cycle operating between the high and low temperatures of the gas cycle. Compare this to Problem 10.38.

10.40 An ideal Brayton refrigeration cycle operates between pressures of 15 and 60 psia. Air enters the compressor at 70°F and leaves the turbine at −10°F. If the airflow is 20 $lb_m$/min, determine the COP and the horsepower required. Use $c_p = 0.24$ Btu/$lb_m$·R and $k = 1.4$.

10.41 An ideal Brayton refrigeration cycle uses air between pressures of 15 and 90 psia. The compressor inlet temperature is 78°F, and the turbine inlet temperature is 88°F. Determine the COP and the power required. Use $c_p = 0.24$ Btu/$lb_m$·R and $k = 1.4$.

10.42 If the upper temperature of a Brayton refrigeration cycle is 180°F when the lowest temperature is at −80°F and the COP is 2.7, determine the airflow per ton of refrigeration. Use $c_p = 0.24$ Btu/$lb_m$·R.

*10.43 An air turbine is used to produce power and also to cool a space. The air enters the turbine at 200 psia and 140°F and is expanded isentropically to 20 psia. If the cooled space is kept at −10°F and the turbine output is 100 hp, determine the ideal refrigeration in tons. Assume that $c_p = 0.24$ Btu/$lb_m$·R and $k = 1.4$.

10.44 An ideal Brayton refrigeration cycle uses air between pressures of 15 and 75 psia. The compressor inlet temperature is 75°F, and the turbine inlet temperature is 95°F. The machine has a rated capacity of 6 tons. Determine the power required. Use $c_p = 0.24$ Btu/$lb_m$·R and $k = 1.4$.

10.45 An ideal Brayton refrigeration cycle uses air. The pressure ratio of the compressor is 5. At the compressor inlet, the pressure is 100 kPa, and the temperature is 35°C. At the outlet of the turbine, the temperature is 0°C. Determine the work in the compressor, the work in the turbine, and the COP. Use $c_p = 1.0061$ kJ/kg·K and $k = 1.4$.

## Problems Involving Vacuum Refrigeration

*10.46 A vacuum refrigeration system cools incoming water at 80°F to 50°F. Determine the mass of water vapor that must be removed per pound of entering water.

*10.47 A vacuum refrigeration system cools incoming water from 30°C ($h_f = 125.79$ and $h_g = 2556.3$ kJ/kg) to 10°C ($h_f = 42.01$ and $h_g = 2519.8$ kJ/kg). Determine the mass of water vapor that must be removed per kilogram of entering water.

10.48 A vacuum refrigeration cycle is operated at 0.12166 psia in the vacuum chamber. If the incoming water is at 70°F, determine, per pound of incoming water, (a) pounds of cooled water, (b) pounds of vapor drawn off, (c) and refrigerating effect.

*10.49 Solve Problem 10.48 if there is a heat leak into the system equal to 10 Btu/lbm of incoming water.

*10.50  A vacuum refrigeration cycle is operated at 0.0010724 MPa in the vacuum chamber ($h_f = 33.60$ and $h_g = 2516.1$ kJ/kg). If the incoming water is at 20°C ($h_f = 83.96$ and $h_g = 2538.1$ kJ/kg) determine, per kilogram of incoming water, (a) kilograms of cooled water, (b) kilograms of vapor drawn off, and (c) refrigerating effect.

*10.51  Solve Problem 10.50 if there is a heat leak into the system equal to 6 Wh/kg of incoming water.

## Problems Involving Compressors

10.52  Calculate the volumetric efficiency of a single-cylinder compressor with a bore and stroke of $0.5 \times 0.4$ m. Clearance is 4% with a pressure ratio of 7 and $n = 1.3$.

10.53  For the compressor in Problem 10.52, find the power required to run the compressor at 200 rpm if the inlet pressure is 100 kPa.

10.54  A reciprocating compressor with a 4% clearance and a bore and stroke of $10 \times 12$ in. operates at 500 rpm. Air enters at 60°F and 14.7 psia and is discharged at 120 psia. If $n = 1.4$ for compression and expansion, find (a) volume efficiency, (b) power required, and (c) mass flow rate of air.

10.55  A double-acting reciprocating compressor runs at 200 rpm with a bore and stroke of $20 \times 25$ cm. Air is compressed isentropically from 100 to 750 kPa. Clearance is 4.5%, and the air enters at 300 K. Find (a) volumetric efficiency, (b) input power required, (c) mass flow rate of the air, (d) output air temperature, and (e) output volume flow rate.

## Problems Involving Heat Pumps

10.56  A Carnot cycle is used for both heating and cooling. If 5 hp is supplied as work and the COP for cooling is 4.0, determine the refrigerating effect and $T_1/T_2$.

10.57  A Carnot heat pump operates to keep a space cool by extracting 400 Btu/min from the space. The compressor required for this application uses 4 hp. Calculate the COP for this application.

10.58  If the purpose of the cycle in Problem 10.57 is to heat another space, calculate its COP.

10.59  If the cycle in Problem 10.10 was rated for 8 tons of equivalent heating when operating as a heat pump, determine the power required.

10.60  The device described in Illustrative Problem 10.3 is operated as a heat pump. Determine the maximum COP and the power input for the continuous supply of 1 ton of refrigeration.

10.61  A Carnot heat pump is required to maintain a house at 20°C when the outside temperature is at –10°C. Calculate the minimum horsepower required if the house loses 120 MJ/h.

10.62  A Carnot heat pump is used to maintain a house at 68°F when the outside temperature is 0°F. If 80,000 Btu/h is lost from the house, determine the horsepower required by the motor in the ideal cycle.

*10.63  Derive the expression

$$COP_{\text{heat pump}} = \left(1 + COP_{\text{refrig}}\right)_{\text{Carnot cycle}}$$

10.64 A reversed Carnot cycle is used for building heating. Assuming that 25,000 Btu/min is required to keep the building at 72°F when the outside air temperature is 40°F, determine the power required and the COP for the cycle as a heat pump.

10.65 A Carnot heat pump is used to keep a house at 70°F when the outside temperature is −10°F. If it is determined that the house requires 1.5 MJ/min in order to maintain these conditions, determine the least horsepower motor required.

10.66 A reversed Carnot cycle heats a building to a constant inside temperature of 22°C when the outside air temperature is 5°C. If 500 kW is required by the building, determine the power (kW) required and the COP for this cycle used as a heat pump.

10.67 A plant is proposed in which a heat pump will provide heat in the winter. It is estimated that the peak heating load will correspond to 50,000 Btu/h. A well will be installed that can supply water with a constant temperature of 45°F. For comfort, it is decided that the heated air will not exceed 85°F. If the cycle is operated as a Carnot heat pump, determine (a) the Carnot heat pump COP, (b) the compressor power, and (c) if an actual cycle achieves 80% of the Carnot COP, the compressor power.

*10.68 Two Carnot heat pumps operate in series, with the output of the first pump being the input of the second pump. If the lowest temperature in the cycle is 400 K and the highest temperature is 1100 K, determine the work input for each pump. Assume that both pumps have the same COP and that 500 kJ is removed from the lowest temperature sink.

*10.69 The heat transfer rate through the walls of a house is 10,000 Btu/h per degree Fahrenheit difference between the outside and the inside. Select the compressor power required for a Carnot heat pump to cool the house in the summer. Design requirements are inside, 75°F; outside, 95°F.

*10.70 For the house in Problem 10.69, select the Carnot heat pump power to heat the home in the winter, keeping the inside temperature at 70°F when the outside temperature is 30°F. What should be the heat pump rating for year-round usage?

*10.71 A hockey arena has nine new Ice Kube System heat pumps that each remove 201,600 Btu/h of heat from the ice at 10°F. The heat pumps each reject 272,100 Btu/h to the outside ground at 90°F to keep it from freezing. Determine the cost to operate all the heat pumps (real, not Carnot) for 24 h at $0.10/kWh. What would be the COP if Carnot heat pumps were used?

# 11

## Heat Transfer

### Learning Goals

After reading and studying the material in this chapter, you should be able to

1. Understand that there are three mechanisms of heat transfer—conduction, convection, and radiation—and that they can occur individually or in combination with each other.

2. Derive Fourier's law of heat conduction from some simple experiments.

3. Appreciate the fact that the thermal conductivity of solids is greater than the thermal conductivity of liquids and that, in turn, the thermal conductivity of liquids is greater than the thermal conductivity of gases.

4. Use the analogy that exists between Ohm's law and Fourier's law to solve conduction problems of thermal resistances in series and in parallel.

5. Calculate the heat transfer by conduction in cylinders.

6. Understand the need to allow for the problem of thermal contact resistance when two materials are brought into physical contact and heat flows across their interface.

7. Realize that convection heat transfer is due to the motion of a fluid relative to a body.

8. Use Newton's law of cooling to calculate the convection heat transfer, and, by comparing Fourier's equation with Newton's law of cooling, be able to write the resistance term for convection heat transfer involving the heat-transfer coefficient.

9. Understand why convection heat transfer is broken into two parts: natural convection and forced convection.

10. Use the dimensionless Nusselt, Prandtl, and Grashof numbers to calculate natural convection.

11. Use simplified forms for the calculation of natural convection for gases at atmospheric pressure.

12. Understand why we limit ourselves to the calculation of forced convection in turbulent flow inside tubes.

13. Realize that radiation heat transfer is a form of electromagnetic radiation and follows laws that are different from those controlling conduction and convection heat transfer.

14. Use the Stefan–Boltzmann law to calculate radiation heat transfer.

15. Apply the earlier portions of the chapter to calculate the performance of heat exchangers.

16. Understand that the temperatures of the fluids in a heat exchanger continuously change as the fluids go from one end of the unit to the other, leading to the concept of the logarithmic mean temperature difference.

17. Estimate the effect of fouling on the performance of heat exchangers.

18. Solve problems involving combined modes of heat transfer.

19. Apply the principles of heat transfer to the cooling of electronic equipment.

20. Understand how and how much fins increase heat transfer rates.

21. Understand how heat pipes work and how to make heat pipes more effective.

## 11.1 Introduction

In the earlier chapters of this book, we considered heat to be energy in transition, and there has been only one reference to the mechanisms that must exist for this type of energy transport to occur. The limitation that was found was from the second law of thermodynamics in the Clausius statement: "Heat cannot, of itself, pass from a lower temperature to a higher temperature." Thus, for heat transfer to occur, we can state that a temperature difference must exist between the bodies, with heat flowing from the body at the higher temperature to the body at the lower temperature. For the purposes of this book, we look at only steady-state heat transfer in the system being considered. The conditions for a steady state is that the temperatures in the system be independent of time, and as a consequence, the rate of heat transfer out of the system must equal the rate of heat transfer to the system.

The mechanisms of heat transfer that we consider in this chapter are *conduction, convection*, and *radiation*. In any industrial application, it is possible to encounter more than one of these mechanisms occurring simultaneously, and it is necessary to consider them in combination when designing or analyzing heat-transfer equipment. The formal categorization of heat transfer into separate and distinct mechanisms is a somewhat arbitrary but useful approach to complex technical problems. The usefulness of this approach will become apparent as our study progresses.

Figure 11.1 shows a modern steam generator in which specialized heat-transfer surfaces are used to achieve high heat transport or transfer efficiencies. These include conduction, convection, and radiation, the three principal modes of heat transfer.

**FIGURE 11.1**
Modern water-tube steam generator.

## 11.2 Conduction

Consider the system shown in Figure 11.2, which consists of a uniform bar of cross-sectional area $A$ perfectly insulated on all sides except at the ends; that is, heat can flow only in the $x$ direction. When one end of the bar is maintained at $t_1$ and the other at $t_2$, $\dot{Q}$ Btu/h will be transferred steadily from station ① to station ②. If the cross-sectional area of the bar is doubled while all other conditions are kept constant, it will be found that $2\dot{Q}$ is now being transferred. In other words, the rate at which heat is being transferred is directly proportional to the cross-sectional area of the bar normal to the direction of heat flow. Returning to the original bar, we now make the temperature difference $(t_1 - t_2)$ twice its original value, and again, we find the rate of heat transfer to be $2\dot{Q}$. Therefore, we conclude that the heat-transfer rate is directly proportional to the temperature difference between the ends of the bar ($\Delta t = t_1 - t_2$). Finally, returning to the original conditions, we now make the bar twice as long ($2L$), and this time, we find that only half the amount of heat is being transferred, leading to the conclusion that the rate of heat transfer is inversely proportional to the length of the bar. If we combine all these events into a mathematical statement, we have

$$\dot{Q} \propto \frac{A(t_1 - t_2)}{L} \tag{11.1}$$

or

$$\dot{Q} = \frac{-kA(t_1 - t_2)}{L} \tag{11.2}$$

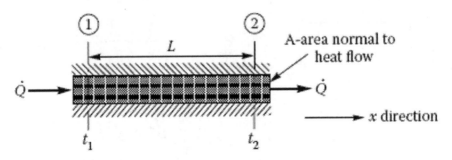

**FIGURE 11.2**
Conduction heat transfer.

where the proportionality constant, $k$, is a physical property of the material called the thermal conductivity. The negative sign has been included in Equation 11.2 to indicate a positive heat flow in the increasing $x$ direction, which is the direction of decreasing temperature. The conductivity, $k$, is usually found to be a function of temperature, but for moderate temperatures and temperature differences, it can be taken to be a constant. If we now rewrite Equation 11.2 in more general terms for one-dimensional conduction, we have

$$\dot{Q} = \frac{-kA\Delta t}{\Delta x} \tag{11.3}$$

Equation 11.3 is called *Fourier's law* of heat conduction in one dimension in honor of the noted French physicist Joseph Fourier. In this equation, the heat transfer rate $\dot{Q}$ is expressed in usual English engineering units as Btu/h, the normal area $A$ is expressed in ft.$^2$, the temperature difference $\Delta t$ is in °F, and the length $\Delta x$ is in ft., giving us $k$ in units of Btu/(h) (ft.$^2$) (°F/ft.). This unit of $k$ is often written as Btu/(h·ft.·°F). The nomenclature of Equation 11.3 is shown in Figure 11.3. Table 11.1 gives the thermal conductivities of some solids at room temperature, and Figure 11.4 shows the variation of thermal conductivity as a function of temperature for many materials. A table of the thermal conductivity of some building and insulating equipment is given in Appendix 2, Table A.6. In SI units, $k$ is W/m·°C.

The foregoing development was based on observable events in a hypothetical experiment. The conduction of heat can also be visualized as occurring as the transfer of energy

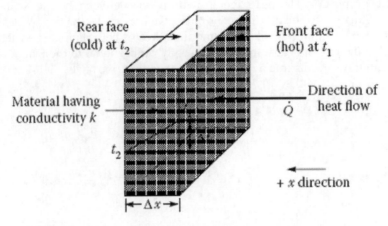

**FIGURE 11.3**
Nomenclature for Equation 11.3.

**TABLE 11.1**

Thermal Conductivities of Solids at Temperatures near 77°F (25°C)

| Material | Conductivity[a] | |
|---|---|---|
| | Btu/ (h·ft.·°F) | W/m·°C |
| Cotton wool | 0.024 | 0.041 |
| Corkboard | 0.025 | 0.043 |
| Mineral wool | 0.026 | 0.045 |
| Balsa | 0.032 | 0.055 |
| Asbestos fiber | 0.044 | 0.076 |
| White pine | 0.065 | 0.112 |
| Fir | 0.090 | 0.156 |
| Gypsum plaster | 0.30 | 0.519 |
| Common brick | 0.40 | 0.692 |
| Concrete (average house construction) | 0.80 | 1.385 |
| Porcelain | 0.87 | 1.500 |
| Mild steel | 26 | 45.0 |
| Wrought iron | 34.5 | 59.7 |
| Yellow brass | 52 | 90.0 |
| Aluminum | 118 | 204 |
| Copper | 220 | 381 |
| Silver | 242 | 419 |
| Paper | 0.029 | 0.050 |

[a] 1 Btu/(h·ft.·°F) = 1.7307 W/m·°C.

**TABLE 11.2**

Order of Magnitude of $k$ for Various Classes of Material

| Material | $k$ (W/m °C) |
|---|---|
| Gases | 0.015–0.80 |
| Insulating materials | 0.015–0.26 |
| Wood | 0.10–0.17 |
| Liquids (nonmetallic) | 0.10–0.60 |
| Brick, concrete, stone, plaster | 0.3–3.5 |
| Refractory materials | 0.80–17 |
| Metals and alloys | 8–500 |

by more active molecules at a higher temperature colliding with less active molecules at a lower temperature. Gases have longer molecular spacings and consistently exhibit much lower thermal conductivities than liquids. Because of the complex structure of solids, some have high values of $k$ while others have low values of $k$. However, for pure crystalline metals, which are good electrical conductors, there are a large number of free electrons in the lattice structure that also makes them good thermal conductors. Table 11.2 gives the order of magnitude of the thermal conductivity $k$ for various classes of materials, which may be used for rough estimates of conduction heat transfer.

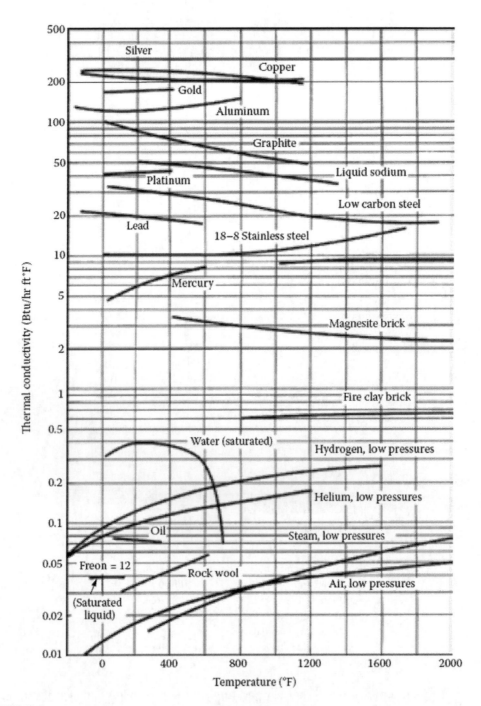

**FIGURE 11.4**
Thermal conductivity of the material. (Reprinted from W. M. Rohsenow and H. Y. Choi, *Heat, Mass and Momentum Transfer*, Englewood Cliffs, N.J.: Prentice Hall, Inc., p. 508, 1961. With permission.)

---

### ILLUSTRATIVE PROBLEM 11.1

A common brick wall 6 in. thick has one face maintained at 150°F and the other face is maintained at 80°F. Determine the heat transfer per square foot of the wall.

### SOLUTION

From Table 11.1, the thermal conductivity of the common brick is 0.40 Btu/(h·ft.·°F), and the wall is $\dfrac{6}{12}$ ft. thick. Applying Equation 11.3, we obtain

$$\frac{\dot{Q}}{A} = \frac{-k\Delta t}{\Delta x} = \frac{-(0.40)(80-150)}{6/12} = 56 \text{ Btu/h} \cdot \text{ft.}^2$$

---

### ILLUSTRATIVE PROBLEM 11.2

A brick wall 150 mm thick has one face maintained at 30°C while the other face is at 70°C. Determine the heat transfer per unit wall area.

### SOLUTION

Using $k$ of 0.692 W/m·°C, we obtain

$$\frac{\dot{Q}}{A} = q'' = \frac{-k\Delta t}{\Delta x} = \frac{-(0.692)(30-70)}{0.150} = 184.5 \text{ W/m}^2$$

For most cases of conduction heat transfer, it is satisfactory to select a value of $k$ at the mean temperature of the process.

---

The Fourier equation has a direct analog in *Ohm's law* for electrical circuits. This can be seen by rewriting Equation 11.3 in the following form:

$$\dot{Q} = \frac{\Delta t}{R_t} \tag{11.3a}$$

where $R_t = \Delta x/kA$ and is called the *thermal resistance*. Ohm's law for a direct current (dc) resistance can be expressed as

$$i = \frac{\text{Voltage Difference}}{\text{Electrical Resistance}} = \frac{\Delta E}{R_e} \tag{11.4}$$

where $\Delta E$ is the potential difference (in volts), $R_e$ is the electrical resistance (in ohms), and $i$ is the current (in amperes). Comparison of Ohm's law and the Fourier equation shows $\dot{Q}$ to be analogous to $i$, $\Delta t$ to correspond to $\Delta E$, and $R_t$ to correspond to $R_e$. Table 11.3 shows the correspondence between these systems.

**TABLE 11.3**

Thermal–Electrical Analogy

| Quantity | Thermal System | Electrical System |
|----------|----------------|-------------------|
| Potential | Temperature difference, °F | Voltage difference, volts |
| Flow | Heat transfer, Btu/h | Current, amperes |
| Resistance | Resistance, h·°F/Btu | Resistance, ohms |

---

## ILLUSTRATIVE PROBLEM 11.3

A simple resistive electrical circuit is to be used to determine the heat transfer through the wall of Illustrative Problem 11.1. If a 9-V battery is available and it is desired to have 1 A correspond to 100 Btu/h·ft.², what resistance is needed?

### SOLUTION

Referring to Figure 11.5, we have for the thermal resistance $R_t = \Delta x / kA = (6/12)/0.4 \times 1 = 1.25$ h·°F/Btu. If we now take the ratio of Equations 11.3a and 11.4,

$$\frac{\dot{Q}}{i} = \frac{\Delta t / R_t}{\Delta E / R_e}$$

but the problem statement is that $\dot{Q}/i = 100$ (numerically). Therefore,

$$100 = \frac{\Delta t}{R_t} \bigg/ \frac{\Delta E}{R_e}$$

Solving for Re yields

$$R_e = \frac{100 \Delta E R_t}{\Delta t} = \frac{100 \times 9 \times 1.25}{70} = 16.07 \text{ ohms}$$

As a check, $i = \Delta E / Re = 9/16 = 0.56$ A, which when multiplied by 100 yields $\dot{Q} = 56$ Btu/h · ft.².

**FIGURE 11.5**
Illustrative Problem 11.5.

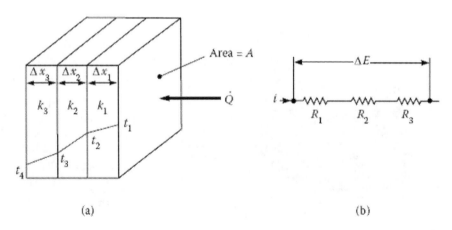

**FIGURE 11.6**
Thermal resistances in series. (a) Series wall. (b) Series electrical circuit.

The analogy between the flow of heat and the flow of electricity is very useful in both visualizing and solving heat-transfer problems. The rules applying to electrical circuits can be used to solve thermal circuits that would otherwise be quite formidable. The techniques for solving dc circuits are applicable to solving steady-state thermal conduction problems. As an application of this type of technique, let us consider a wall composed of several thermal resistances (layers) in series, as shown in Figure 11.6a, and an analogous electrical circuit, as shown in Figure 11.6b. It will be recalled that for a series dc circuit, the overall resistance ($R_{oe}$) is the sum of the individual resistances:

$$R_{oe} = R_1 + R_2 + R_3 \tag{11.5}$$

Therefore, by analogy, the overall resistance of the thermal circuit ($R_{ot}$) is

$$R_{ot} = R_1 + R_2 + R_3 = \frac{\Delta x_1}{k_1 A} + \frac{\Delta x_2}{k_2 A} + \frac{\Delta x_3}{k_3 A} \tag{11.6}$$

---

### ILLUSTRATIVE PROBLEM 11.4

A composite wall consists of 6 in. of ordinary brick ($k=0.4$ Btu/(h·ft.·°F)), 1/2 in. of concrete ($k=0.8$ Btu/(h·ft.·°F)), and 1/2 in. of plaster ($k=0.3$ Btu/(h·ft.·°F)). If the inside of the wall is maintained at 70°F when the outside is 30°F, determine the heat transfer per square foot of the wall. The overall situation is shown in Figures 11.6 and 11.7.

**SOLUTION**

For the brick,

$$R_1 = \frac{\Delta x}{kA} = \frac{6/12}{04 \times 1} = 1.25 \frac{h \cdot °F}{Btu}$$

For the concrete,

$$R_2 = \frac{\Delta x}{kA} = \frac{1/2}{12} / (08 \times 1) = 0.052 \frac{h \cdot °F}{Btu}$$

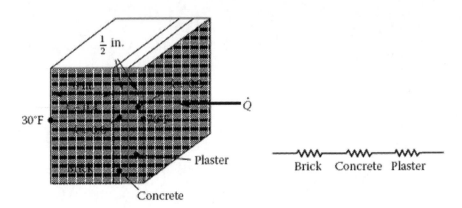

**FIGURE 11.7**
Illustrative Problem 11.4.

For the plaster,

$$R_3 = \frac{\Delta x}{kA} = \frac{1/2}{12} / (0.3 \times 1) = 0.139 \frac{h \cdot °F}{Btu}$$

The overall resistance $Ro_t = R_1 + R_2 + R_3 = 1.25 + 0.052 + 0.139 = 1.441$ h·°F/Btu, and the heat transfer $\dot{Q} = (70 - 30)/1.441 = 27.76$ Btu/h·ft.$^2$, because the area considered is 1 ft.$^2$.

---

### ILLUSTRATIVE PROBLEM 11.5

Calculate the temperature at the brick–concrete and concrete–plaster interfaces for the composite wall of Illustrative Problem 11.4.

### SOLUTION

Because each resistance in a series circuit carries the same current (heat flow), it is only necessary to successively apply Ohm's law (Fourier's equation) to each element in turn. Thus,

$$\dot{Q} = \frac{\Delta t}{R} \quad \text{or} \quad \Delta t = R\dot{Q} \quad \text{with} \quad \dot{Q} = 27.76 \frac{Btu}{h \cdot ft.^2}$$

from Illustrative Problem 11.14. Therefore, for the brick,

$$\Delta t = (12.5)(27.76) = 34.70°F$$

For the concrete,

$$\Delta t = (0.052)(27.76) = 1.44°F$$

For the plaster,

$$\Delta t = (0.139)(27.76) = \frac{3.86°F}{40.00°F}$$

Check:

$$\Delta t \text{ (overall)} = 40.00$$

The interface temperatures are

for brick–concrete: $30 + 34.70 = 64.70°F$
for concrete–plaster: $64.70 + 1.44 = 66.14°F$

---

**ILLUSTRATIVE PROBLEM 11.6**

If the composite wall of Figure 11.7 consists of 150 mm of brick ($k = 0.692$ W/m·°C) and 12 mm of concrete ($k = 1.385$ W/m·°C) with the exterior surface temperatures being 0°C and 20°C, respectively, determine the heat transfer per square meter of the wall.

**SOLUTION**

Proceeding as in Illustrative Problem 11.4, we have for the brick,

$$R = \frac{\Delta x}{kA} = \frac{0.150}{0.692 \times 1} = 0.217 \frac{°C}{W}$$

For the concrete,

$$R = \frac{\Delta x}{kA} = \frac{0.012}{1.385 \times 1} = 0.009 \frac{°C}{W}$$

For the plaster,

$$R = \frac{\Delta x}{kA} = \frac{0.012}{0.519 \times 1} = 0.023 \frac{°C}{W}$$

The total resistance is the sum of the individual resistances, which equals 0.249°C/W. The heat transfer therefore is $\frac{\dot{Q}}{A} = q'' = \frac{\Delta t}{R} = (20 - 0) / 0.249 = 80.3 \text{ W/m}^2$ for an area of 1 m².

**ILLUSTRATIVE PROBLEM 11.7**

Determine the temperature at each of the interfaces in Illustrative Problem 11.6.

**SOLUTION**

$$\Delta t = R\dot{Q}$$

for brick :        $\Delta t = 80.3 \times 0.217 = 17.43°C$

for concrete :   $\Delta t = 80.3 \times 0.009 = 0.72°C$

for plaster :      $\Delta t = 80.3 \times 0.023 = \underline{1.85°C}$

As a check, overall   $\Delta t =$                   $20.00°C$

The temperatures at the interfaces are $(20 - 17.43) = 2.57°C$, $(2.57 - 0.72) = 1.85°C$, and $(1.85 - 1.85) = 0°C$.

The electrical analogy, when applied to the case of a composite plane wall with parallel sections, yields a considerable saving and is a distinct aid in visualizing the problem. Consider the wall shown in Figure 11.8a, which consists of sections side by side. Each wall section has a different area for heat transfer, their conductivities are different, but their thicknesses are equal and the front faces are kept at $t_1$ while the back faces are kept at $t_2$. This type of wall is called a parallel wall, and we shall solve it by using the parallel electrical circuit shown in Figure 11.8b. For the parallel electrical circuit, the potential difference is the same for all elements and the total current $i$ is the sum of the branch currents. Thus,

$$i = i_1 + i_2 + i_3 \tag{11.7}$$

By analogy for the thermal circuit,

$$\dot{Q} = \dot{Q}_1 + \dot{Q}_2 + \dot{Q}_3 \tag{11.8}$$

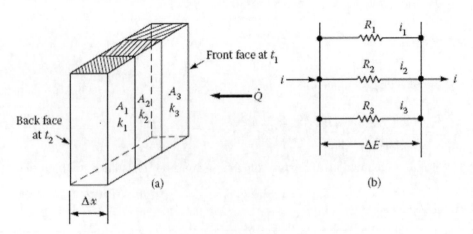

(a)                                                        (b)

**FIGURE 11.8**
Parallel circuits. (a) Parallel wall. (b) Parallel electrical circuit.

The overall resistance of the parallel electrical circuit is

$$R_{oe} = \frac{1}{(1/R_1)+(1/R_2)+(1/R_3)}$$ (11.9)

which gives us by analogy for the thermal circuit

$$R_{ot} = \frac{1}{\dfrac{1}{\Delta x/k_1 A_1} + \dfrac{1}{\Delta x/k_2 A_2} + \dfrac{1}{\Delta x/k_3 A_3}}$$ (11.10)

---

### ILLUSTRATIVE PROBLEM 11.8

A wall 7 ft. high and 6 ft. wide is 4 in. thick. Two feet of the wall is made of fir ($k=0.090$ Btu/(h·ft.·°F)), 2 ft. is made of pine ($k=0.065$ Btu/(h·ft.·°F)), and the final 2 ft. is made of corkboard ($k=0.025$ Btu/(h·ft.·°F)). Determine the heat loss from the wall if one face is maintained at 80°F while the other face is kept at 60°F.

### SOLUTION

The problem is essentially the case shown in Figure 11.8. As the first step in the solution, we shall calculate the individual thermal resistances. Thus,

$$R_{fir} = \frac{\Delta x}{kA} = \frac{4/12}{0.090 \times 7 \times 2} = 0.265 \frac{h \cdot °F}{Btu}$$

$$\frac{1}{R_{fir}} = 3.774$$

$$R_{pine} = \frac{\Delta x}{kA} = \frac{4/12}{0.065 \times 7 \times 2} = 0.366 \frac{h \cdot °F}{Btu}$$

$$\frac{1}{R_{pine}} = 2.732$$

$$R_{corkboard} = \frac{\Delta x}{kA} = \frac{4/12}{0.025 \times 7 \times 2} = 0.952 \frac{h \cdot °F}{Btu}$$

$$\frac{1}{R_{corkboard}} = 1.050$$

$$R_{overall} = \frac{1}{(1/R_1)+(1/R_2)+(1/R_3)}$$

$$= \frac{1}{3.774 + 2.732 + 1.050}$$

$$= \frac{1}{7.556} = 0.132$$

Therefore,

$$\dot{Q} = \frac{\Delta t}{R} = \frac{80 - 60}{0.132} = 151.5 \, \frac{\text{Btu}}{\text{h}}$$

As a check,

$$\dot{Q}_{fir} = \frac{\Delta t}{R} = \frac{80 - 60}{0.265} = \frac{20}{0.265} = 75.5 \, \frac{\text{Btu}}{\text{h}}$$

$$\dot{Q}_{fir} = \frac{\Delta t}{R} = \frac{80 - 60}{0.366} = \frac{20}{0.366} = 54.6 \, \frac{\text{Btu}}{\text{h}}$$

$$\dot{Q}_{corkboard} = \frac{\Delta t}{R} = \frac{80 - 60}{0.952} = \frac{20}{0.952} = 21.0 \, \frac{\text{Btu}}{\text{h}}$$

$$\dot{Q}_{total} = Q_1 + Q_2 + Q_3$$

$$= 75.5 + 54.6 + 21.0 = 151.1 \, \text{Btu/h}$$

Insulating materials, such as fiberglass batting for homes, are sold commercially with "R-values." These values relate directly to the thermal resistance $R$ of 1 ft.² of cross-sectional area. For a given material thickness $\Delta x$ and its $k$ value, the calculated $R$ is what is referred to as the "R-value" for that thickness of insulation.

## ILLUSTRATIVE PROBLEM 11.9

A 250 m² roof of a home is made up of 6 cm thick slate ($k = 1.5$ W/m °C) and mineral wool insulation ($k = 0.045$ W/m °C), supported by 3 cm thick fir panels ($k = 0.156$ W/m °C). It has been designed to allow a heat-loss rate of 1.25 kW and keep an inside temperature of 22°C when the outside temperature is –18°C. What thickness of the mineral wool is required?

## SOLUTION

The total resistance for the roof is found from the conduction equation:

$$R_{ot} = \frac{\Delta T}{Q} = \frac{40°C}{1250W} = 0.032 \, \frac{°C}{W}$$

The resistances of each component are in series and found by $R = \sum \dfrac{\Delta x}{kA}$.

$$R_{slate} = \frac{0.06 \, \text{m}}{1.5 \, \dfrac{W}{m°C} \times 250 m^2} = 0.00016 \, \frac{°C}{W}$$

$$R_{\text{fir}} = \frac{0.03\text{m}}{0.156\dfrac{\text{W}}{\text{m°C}} \times 250\text{m}^2} = 0.000769\,\frac{\text{°C}}{\text{W}}$$

Thus, $R_{\text{mineral wool}} = R_{\text{ot}} - R_{\text{slate}} - R_{\text{fir}} = 0.032 - 0.00016 - 0.000769 = 0.0311\text{°C/W}$. The mineral wool thickness required is

$$\Delta x = (Rk)_{\text{mineral wool}} \times A = 0.0311\text{°C/W} \times 0.045\ \text{W/m°C} \times 250\ \text{m}^2 = 0.35\ \text{m or }35\ \text{cm.}$$

Thus far, we have been concerned with conduction through plane walls whose heat-transfer area remained constant. The conduction of heat through pipes represents a practical problem of considerable interest in which the effective heat-transfer area is constantly changing. To solve this problem, let us consider a long, uniform, hollow cylinder whose outer surface is maintained at a temperature $t_o$ and whose inner surface is maintained at a temperature $t_i$. As is shown in Figure 11.9, we will denote the inner radius to be $r_i$, the outer radius to be $r_o$ and the radius at any portion of the cylinder to be $r$. The length of the cylinder in question will be denoted as $L$. Let us now apply the Fourier equation to a small cylinder of thickness $\Delta r$, located at a radius $r$ from the center. The surface area of this cylinder is $2\pi rL$, and its thickness is $\Delta r$. Therefore,

$$\dot{Q} = \frac{-kA\Delta t}{\Delta x} = \frac{-k2\pi rL\Delta t}{\Delta r} \tag{11.11}$$

Rearranging gives

$$\frac{\dot{Q}}{2\pi kL}\frac{\Delta r}{r} = -\Delta t \tag{11.12}$$

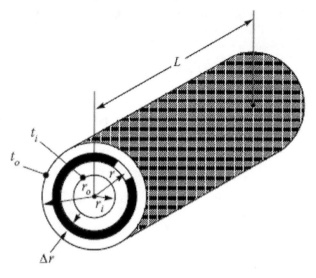

**FIGURE 11.9**
Conduction in a hollow cylinder.

Because the heat transfer $\dot{Q}$ must go through all such elements, it is a constant, and we can rewrite Equation 11.12 as

$$\frac{\dot{Q}}{2\pi kL} \sum_{r_i}^{r_o} \frac{\Delta r}{r} = -\sum_{t_i}^{t_o} \Delta t \tag{11.13}$$

where the symbol $\Sigma$ is used to denote the summation of the variable between the limits indicated. For the right-hand side of the equation, we immediately note that the sum of all the $\Delta t$'s is simply the overall temperature difference, and with the negative sign, the summation gives us $t_i - t_o$. The summation of $\Delta r/r$ is similar to the summation of $\Delta v/v$ or $\Delta t/t$, which, as we have already noted, yields the natural logarithm (ln) of the argument between its limits. Therefore, Equation 11.13 can be summed to yield

$$t_i - t_o = \frac{\dot{Q}}{2\pi kL} \ln\left(\frac{r_o}{r_i}\right) \tag{11.14a}$$

$$\dot{Q} = \frac{(t_i - t_o)2\pi kL}{\ln(r_o / r_i)} \frac{\text{Btu}}{\text{h}} \tag{11.14b}$$

and the thermal resistance

$$R_t = \frac{\ln\left(\dfrac{r_o}{r_i}\right)}{2\pi kL} \tag{11.14c}$$

---

### CALCULUS ENRICHMENT

In 1822, Joseph Fourier put forth the fundamental equation governing the conduction of heat in a solid with constant thermal conductivity. For steady-state, one-dimensional heat conduction, this equation is

$$\dot{Q} = -kA\frac{dT}{dx} \tag{a}$$

where $dT/dx$ is the temperature gradient, $\dot{Q}$ is the rate of heat flow, $A$ is the area across which the heat is flowing, and $k$ is a proportionality constant called the thermal conductivity. In the form of Equation (a), it is limited to a one-dimensional steady flow, but it can be used to analyze the steady-state heat transfer of three shapes, namely, the plane wall, the hollow cylinder, and the hollow sphere due to the symmetry of these shapes.

For the plane wall with constant thermal conductivity, Equation (a) can be integrated as follows:

$$\int_0^L \dot{Q}dx = -\int_{T_h}^{T_c} kA\,dT \tag{b}$$

where $T_h$ is the hot face temperature and $T_c$ is the cold face temperature. Integrating Equation (b) yields

$$\dot{Q}L = kA\left(T_h - T_c\right) \tag{c}$$

or

$$\dot{Q} = -\frac{kA}{L}\left(T_h - T_c\right) \tag{d}$$

Equation (d) is the same as Equation 11.3.

For the cylinder of Figure 11.9, symmetry allows us to use a one-dimensional solution. Rewriting Equation (a) for this condition in cylindrical coordinates gives us

$$\dot{Q} = -kA\frac{dT}{dr} \tag{e}$$

Because $A = 2\pi rL$,

$$\dot{Q} = -k2\pi rL\frac{dT}{dr} \tag{f}$$

Rearranging,

$$\dot{Q}\int_{r_i}^{r_o}\frac{dr}{r} = -2\pi kL\int_{T_i}^{T_o}dT \tag{g}$$

Integrating,

$$\dot{Q} = \frac{2\pi kL}{\ln\left(\dfrac{r_o}{r_i}\right)}\left(T_i - T_o\right) \tag{h}$$

Equation (h) is the same as Equation 11.14b

For the sphere, with inside radius $r_i$, outside radius $r_o$, inside temperature $T_i$, and outside temperature $T_o$, we can again use Equation (a) by symmetry. Using $A = 4\pi r^2$,

$$\dot{Q} = -k4\pi r^2\frac{dT}{dr} \tag{i}$$

Rearranging,

$$\dot{Q}\int_{r_i}^{r_o}\frac{dr}{r^2} = -4\pi k\int_{T_i}^{T_o}dT \tag{j}$$

Integrating and simplifying,

$$\dot{Q} = \frac{4\pi kr_o r_i}{\left(r_o - r_i\right)}\left(T_i - T_o\right) \tag{k}$$

The results that we have just obtained assumed that $k$ was constant. If $k$ varies linearly with temperature, as is often the case over limited temperature intervals, then

$$k = k_o\left(1 + bT\right) \tag{l}$$

If we now define an average value of $k$ to be used over the temperature interval, we find that

$$k_{avg} = k_o\left[1 + \left(b/2\right)\right]\left(T_h + T_c\right) \tag{m}$$

However, the term $k_0[1+(b/2)]$ is the arithmetic mean value of $k$ in the temperature range $T_h$ to $T_c$. The same result is obtained whether we consider the plane wall, the cylinder, or the sphere. We therefore conclude that for these cases, the average value of $k$ over the temperature range is the appropriate value to use if $k$ varies linearly with temperature.

---

### ILLUSTRATIVE PROBLEM 11.10

A bare steel pipe with an outside diameter of 3.50 in. and an inside diameter of 3.00 in. is 5 ft. long. Determine the heat loss from the pipe if the inside temperature is 240°F and its outside temperature is 120°F. Use a $k$ of 26 Btu/(h ft. °F) for the steel.

### SOLUTION

Because the ratio of $r_o/r_i$ is the same as the ratio of the corresponding diameters, we can apply Equation 11.14b directly:

$$\dot{Q} = \frac{2\pi(26)(5)(240 - 120)}{\ln(3.50/3.00)} = 635858 \; \frac{Btu}{h}$$

---

### ILLUSTRATIVE PROBLEM 11.11

A bare steel pipe with an outside diameter of 90 mm and an inside diameter of 75 mm is 2 m long. The outside temperature is 40°C, while the inside temperature is 110°C. If $k$ is 45 W/m·°C, determine the heat loss from the pipe.

### SOLUTION

Using Equation 11.14b, we obtain

$$\dot{Q} = \frac{2\pi \times 45 \times (110 - 40)}{\ln(90/75)} = 217.1 \times 10^3 \; W$$

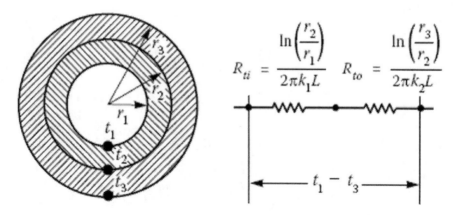

**FIGURE 11.10**
Compound cylinder.

The results of Illustrative Problems 11.10 and 11.11 show the extremely large heat loss from a bare pipe. To decrease the loss, it is usual to insulate the pipe with a material having a low thermal conductivity. When this is done, we have a situation of two resistances in series, each having variable heat-transfer areas. Referring to Figure 11.10 and applying Equation 11.14b to each of the cylinders yields the following: For the inner cylinder,

$$\dot{Q} = \frac{2\pi k_1 L(t_1 - t_2)}{\ln(r_2 / r_1)}; \quad R_{ti} = \frac{\ln(r_2 / r_1)}{2\pi k_1 L} \tag{11.14d}$$

For the outer cylinder,

$$\dot{Q} = \frac{2\pi k_2 L}{\ln(r_3 / r_1)}(t_3 - t_2); \quad R_{to} = \frac{\ln(r_3 / r_2)}{2\pi k_2 L} \tag{11.14e}$$

If at this point we apply the electrical analogy, we can write directly that the overall resistance is the sum of the individual resistances. The same result can be obtained by noting that $t_1 - t_3 = (t_1 - t_2) + (t_2 - t_3)$. Using Equations 11.14d and 11.14e, we obtain $t_1 - t_2 = \dot{Q}R_{ti}$ and $t_2 - t_3 = \dot{Q}R_{to}$. Therefore, $t_1 - t_3 = \dot{Q}(R_{ti} - R_{to})$, or

$$\dot{Q} = \frac{(t_1 - t_3)2\pi L}{(1/k_1)\ln(r_2 / r_1) + (1/k_2)\ln(r_3 / r_1)} \tag{11.15}$$

---

### ILLUSTRATIVE PROBLEM 11.12

If 1 in. of mineral wool ($k = 0.026$ Btu/(h·ft·°F)), insulation is added to the pipe of Illustrative Problem 11.10, and the temperature outside of this insulation is found to be 85°F, determine the heat loss.

### SOLUTION

Using Equation 11.15 and the data of Illustrative Problem 11.10 yields

$$R_{ti} = \frac{1}{k_1}\ln\left(\frac{r_2}{r_1}\right) = \frac{1}{k_1}\ln\left(\frac{D_2}{D_1}\right) = \frac{1}{26}\times\ln\left(\frac{3.500}{3.00}\right) = 0.006$$

$$R_{to} = \frac{1}{k_2}\ln\left(\frac{r_3}{r_2}\right) = \frac{1}{k_2}\ln\left(\frac{D_3}{D_2}\right) = \frac{1}{0.026}\times\ln\left(\frac{5.50}{3.50}\right) = 17.384$$

Therefore,

$$\dot{Q} = \frac{(t_1 - t_3)2\pi L}{R_{ti} + R_{to}} = \frac{(240-85)2\pi(5)}{0.006 + 17.384} = 280.0\frac{Btu}{h}$$

Two things are evident from the comparison of Illustrative Problems 11.10 through 11.12. The first is the decrease in heat loss of more than a thousand fold by the addition of a relatively small amount of insulation. The second is that the resistance of the steel could have been neglected, because it was only a minute fraction of the resistance of the insulation.

Before leaving the subject of conduction heat transfer, one topic must be mentioned: thermal contact resistance. When two materials are brought into physical contact and heat flows from one to the other across the corresponding interface. In such a case, it is found that there is a discontinuous and sometimes very large temperature drop at the interface plane, as shown in Figure 11.11. This temperature drop is attributed to a thermal contact resistance, which can be thought of as being caused by two parallel heat paths at the interface: solid-to-solid conduction at discrete contact points and conduction through air entrapped at the gas interface. Because of the relatively poor thermal conductivity of air (or other gases) as compared to that of metals, the larger part of the joint thermal contact resistance is due to the entrapped air. The prediction of thermal contact resistance is a complex problem that as yet has not been adequately solved, and students are cautioned that in actual practice, allowance must be made for this effect when designing industrial equipment.

**FIGURE 11.11**
Thermal contact resistance.

## 11.3 Convection

Heat transfer by convection from a body involves the motion of a fluid (liquid or gas) relative to the body. If the motion is caused by the differences in density due to the difference in temperature at various locations in the fluid, it is known as natural convection. If the motion of the fluid is caused by an external agent such as a blower, fan or compressor when using gases, or a pump when using liquids, it is called forced convection. Heat transfer from a surface whose temperature is higher than the surrounding fluid occurs in a complex fashion. We can, however, visualize it as occurring in the following sequential order. First, the particles of fluid adjacent to the walls are heated by conduction from the wall, which increases their temperature. These "hot" particles will collide with colder particles, imparting some of their energy to these colder particles. This action will occur due to particle motion as well as movement of the hotter fluid relative to the colder bulk fluid. To categorize the types of convective heat-transfer mechanisms, it is necessary to briefly discuss the mechanism of flow.

The term *laminar* (or *streamline*) *flow* is applied to a flow regime in which the flow is smooth and the fluid moves in layers or paths parallel to each other as seen in Figure 11.12a. When a fluid moves in laminar flow over a hotter surface, heat is transferred principally by molecular conduction within the fluid from layer to layer. This type of convection heat transfer yields low heat-transfer rates. In contrast to laminar flow, there is a flow regime known as *turbulent flow*. As the name implies, this type of flow is characterized by eddies that cause mixing of the layers of fluid until the layers are no longer distinguishable as seen in Figure 11.12b. The mixing of the fluid due to this turbulence causes an increase in the heat transfer, and consequently, the greater the level of turbulence, the greater the heat transfer rate.

The basic equation for convection heat transfer is known as *Newton's law of cooling* and is given by

$$\dot{Q} = hA(\Delta t)$$ (11.16)

(a)

(b)

**FIGURE 11.12**
Streamlines and velocity profiles for (a) laminar flow and (b) turbulent flow.

where

$\dot{Q}$ = heat-transfer rate (Btu/h)

$A$ = heat-transfer area (ft.$^2$)

$\Delta t$ = temperature difference between the surface and the bulk of the fluid away from the surface (°F)

$h$ = coefficient of heat transfer, film coefficient, thermal convective conductance, heat-transfer coefficient, or film heat-transfer factor (Btu/(h·ft.$^2$·°F))

By a comparison of Equation 11.16 with Equation 11.3a, we can write the thermal resistance for convective heat transfer, $R_c$, as

$$R_c = \frac{1}{h} \qquad\qquad (11.17)$$

and it is treated in the same manner as we treated the resistance concept in conduction heat transfer.

---

### ILLUSTRATIVE PROBLEM 11.13

Assume that the bare pipe in Illustrative Problem 11.10 has a heat-transfer coefficient on the outside due to natural convection to the surrounding air at 70°F of 0.9 Btu/ (h·ft.$^2$·°F). Determine the heat transfer from the pipe to the surrounding air.

### SOLUTION

The resistance of the pipe is

$$R_{pipe} = \frac{\ln\left(\dfrac{r_2}{r_1}\right)}{2\pi kL} = \frac{\ln\left(\dfrac{3.50}{3.00}\right)}{2\pi(26)5} = 0.00019 \text{ h}\cdot\text{°F/Btu}$$

$$R_{convection} = \frac{1}{hA} = \frac{1}{0.9\times(3.5)/12.5} = 0.2425 \text{ h}\cdot\text{°F/Btu}$$

$$R_{total} = 0.00019 + 0.2425 = 0.243 \text{ h}\cdot\text{°F/Btu}$$

$$\dot{Q} = \frac{\Delta t}{R} = \frac{240-70}{0.243} = 699.6 \,\frac{\text{Btu}}{\text{h}}$$

It is interesting to note that the outside film coefficient has acted in a manner similar to an insulator for this case (see Illustrative Problems 11.10 and 11.12) and that the resistance of the pipe wall is essentially negligible.

## 11.3.1 Natural Convection*

The evaluation of the *heat-transfer coefficient h* is quite difficult because it usually involves the interaction of complex physical phenomena. As noted earlier, *natural convection* heat transfer occurs as a result of density differences in the fluid caused by a body at a temperature different from the fluid exchanging heat to the fluid. These density differences cause a pumping action of the fluid relative to the body. Using the techniques of dimensional analysis, it can be shown that the parameters involved in natural convection heat transfer can be cast into the form

$$\text{Nu} = A \left( \text{Gr} \right)^a \left( \text{Pr} \right)^b \tag{11.18}$$

where

$\text{Nu}$ = Nusselt number = $hL/k$ or $hD/k$ (dimensionless)
$\text{Pr}$ = Prandtl number = $c_p \mu/k$ (dimensionless)
$\text{Gr}$ = Grashof number = $g\beta(\Delta t)L^3 \rho^2/\mu^2$ (dimensionless)
$A, a, b$ = constants depending on the system under consideration
$\beta$ = coefficient of expansion
$\rho$ = density
$\mu$ = viscosity
$g$ = acceleration of gravity
$D$ = diameter
$L$ = length
$c_p$ = specific heat at constant pressure

At this point, it should be noted that the character of the flow process must be considered. The boundary layer of the fluid will be either laminar or turbulent, and in turn, this will affect the constants in Equation 11.18. Based on experimentally determined values, it is found that when the product $(\text{Gr})(\text{Pr})$ exceeds $10^8$, there is an increase in the heat-transfer coefficient, indicating a transition from a laminar boundary layer to a turbulent boundary layer.

Equation 11.18 can be recast into the following form for Pr near unity, which is the case for many gases:

$$\text{Nu} = C_L \left( \text{Gr} \times \text{Pr} \right)^{1/4} \tag{11.19}$$

$$\text{Nu} = C_T \left( \text{Gr} \times \text{Pr} \right)^{1/3} \tag{11.20}$$

Equation 11.19 applies to laminar flow, and $C_L$ is the coefficient for laminar flow. Equation 11.20 applies to turbulent flow, and $C_T$ is the coefficient for turbulent flow. The evaluation of these equations is, at best, tedious, and in some cases, it necessitates the use of an iterative process to obtain a solution. Fortunately, the properties of air, CO, $N_2$, and $O_2$ in the temperature range 100°F to 1500°F vary in such a manner that it is possible to lump all the properties that are individually temperature-dependent into a single constant that is

---

* A portion of this material in this section is from Granet (1972).

essentially temperature-invariant to obtain the following simplified forms of Equations 11.19 and 11.20:

$$h = C_L \left( \frac{\Delta t}{L} \right)^{1/4} \tag{11.21a}$$

$$h = C_T \left( \Delta t \right)^{1/3} \tag{11.21b}$$

Equation 11.21a applies to laminar flow and Equation 11.21b to turbulent flow, $\Delta t$ is the temperature difference in °F or °C between the surface and the bulk temperature of the gas, $L$ in feet or meters is a characteristic dimension (either a length or a diameter), $h$ is the heat-transfer coefficient in Btu/(h·ft.²·°F) or W/m²·°C, and $C_L$ and $C_T$ are the constants for laminar and turbulent flow, respectively.

If the references are examined, it will be found that there is as much as 100% difference among various authors for the values of the coefficients $C_L$ and $C_T$. A consistent set of data for $C_L$ and $C_T$ that yields conservative design values is given in Equations 11.22 and 11.23.

Vertical plates or vertical cylinders up to 2 ft. in height ($L$ is the vertical height).
   English units:

$$h = 0.29 \left( \frac{\Delta t}{L} \right)^{1/4} \text{ for } 10^{-2} < \left( L^3 \Delta t \right) < 10^3 \text{ (laminar)} \tag{11.22}$$

$$h = 0.21 \left( \Delta t \right)^{1/3} \text{ for } 10^3 < \left( L^3 \Delta t \right) < 10^6 \text{ (turbulent)} \tag{11.23}$$

SI units:

$$h = 1.42 \left( \frac{\Delta t}{L} \right)^{1/4} \text{ for } 10^{-4} < \left( L^3 \Delta t \right) < 16 \text{ (laminar)}$$

$$h = 1.31 \left( \Delta t \right)^{1/3} \text{ for } 16 < \left( L^3 \Delta t \right) < 10^4 \text{ (turbulent)}$$

*Horizontal pipes.* Evidence exists to indicate that vertical pipes have higher heat-transfer coefficients than horizontal pipes, but this difference can be considered to be small and the equation given for horizontal pipes can be used for vertical pipes longer than 2 ft. high ($D$ is the pipe diameter).
   English units:

$$h = 0.25 \left( \frac{\Delta t}{D} \right)^{1/4} \text{ for } 10^{-2} < \left( D^3 \Delta t \right) < 10^3 \text{ (laminar)} \tag{11.24}$$

$$h = 0.18 \left( \Delta t \right)^{1/3} \text{ for } 10^3 < \left( D^3 \Delta t \right) < 10^6 \text{ (turbulent)} \tag{11.25}$$

SI units:

$$h = 1.32\left(\frac{\Delta t}{D}\right)^{1/4} \text{ for } 10^{-4} < \left(L^3 \Delta t\right) < 16 \text{ (laminar)}$$

$$h = 1.24\left(\Delta t\right)^{1/3} \text{ for } 16 < \left(L^3 \Delta t\right) < 10^4 \text{ (turbulent)}$$

Horizontal square plates.
  English units:

$$h = 0.27\left(\frac{\Delta t}{L}\right)^{1/4} \text{ for } 10 < \left(L^3 \Delta t\right) < 20 \text{ (laminar, hot side up)} \tag{11.26}$$

$$h = 0.22\left(\Delta t\right)^{1/3} \text{ for } 20 < \left(L^3 \Delta t\right) < 30,000 \text{ (turbulent, hot side up)} \tag{11.27}$$

$$h = 0.12\left(\frac{\Delta t}{L}\right)^{1/4} \text{ for } 0.3 < \left(L^3 \Delta t\right) < 30,000 \text{ (laminar, hot side down)} \tag{11.28}$$

SI units:

$$h = 1.32\left(\frac{\Delta t}{L}\right)^{1/4} \text{ for } 10^{-2} < \left(L^3 \Delta t\right) < 0.32 \text{ (laminar, hot side up)}$$

$$h = 1.52\left(\Delta t\right)^{1/3} \text{ for } 0.32 < \left(L^3 \Delta t\right) < 500 \text{ (turbulent, hot side up)}$$

$$h = 0.59\left(\frac{\Delta t}{L}\right)^{1/4} \text{ for } 10^{-3} < \left(L^3 \Delta t\right) < 500 \text{ (laminar, hot side down)}$$

The foregoing is presented in a graphical form for English units in Figures 11.13 through 11.15. Note that if the horizontal plate is not square, one may use the geometric mean side length, given by 2(length × width) divided by (length + width). Figure 11.13 is used first to determine whether the character of the flow is laminar or turbulent. The necessary data for entering this chart are $\Delta t$ and $L$ (or $D$). Having determined the flow character, either Figure 11.14 or 11.15 is entered. Figure 11.14 is for the laminar region, and Figure 11.15 is for the turbulent region. The use of these charts is illustrated by the following examples. Students should note that Equations 11.22 through 11.28 are readily programmable on a digital computer. It is more instructive to see trends using Figures 11.13 through 11.15.

**FIGURE 11.13**
Chart for determining character of natural convection process.

---

### ILLUSTRATIVE PROBLEM 11.14

Determine the heat-transfer coefficient from a horizontal square plate with the hot side facing down if the plate is 1 ft. × 1 ft. and $\Delta t = 100°F$.

#### SOLUTION

From Figure 11.13, the flow is found to be laminar and Figure 11.14 is to be used. Entering Figure 11.14 at $L = 1$ ft. and $\Delta t = 100°F$ and connecting the reference scale to the point marked "horizontal square—hot side down" yields $h = 0.38$ Btu/(h·ft.$^2$·°F).

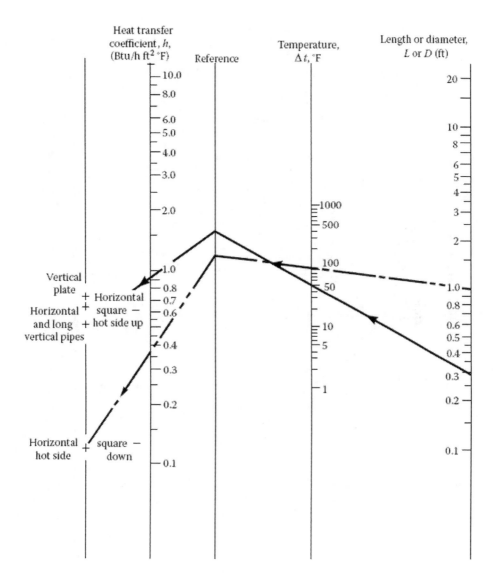

**FIGURE 11.14**
Heat-transfer coefficient of air, CO, $N_2$, and $O_2$ for natural convection, laminar flow.

## ILLUSTRATIVE PROBLEM 11.15

Determine the heat-transfer coefficient for a vertical plate if it is 10 ft. high and $\Delta t = 2°F$.

### SOLUTION

From Figure 11.14, the flow is found to be turbulent. Entering Figure 11.15 at $\Delta t = 2$ and the point for a vertical plate yields $h = 0.26$ Btu/(h·ft.$^2$·°F).

**FIGURE 11.15**
Heat-transfer coefficient of air, CO, $N_2$, and $O_2$ for natural convection, turbulent flow.

---

### ILLUSTRATIVE PROBLEM 11.16

A bare 3.50 in. outside diameter pipe has an outside temperature of 120°F and is in a room at 70°F. Determine the heat loss from a section of 5 ft. long (see Illustrative Problem 11.11) due to convection.

### SOLUTION

From Figure 11.13 at $\Delta t = 50°F$, $D = 3.5/12 = 0.292$ ft., the flow is laminar, and Figure 11.14 is to be used. From Figure 11.14, $h = 0.9$ Btu/(h·ft.²·°F), $\dot{Q} = hA\Delta = 0.9 \times \pi(3.5)/12 \times 5\,(120 - 70) = 206.2$ Btu/h.

### ILLUSTRATIVE PROBLEM 11.17

An aluminum electric box, a 15 cm cube, is placed on an insulated pad. It generates a surface temperature of 50°C when placed in 25°C air. Find the heat-loss rate.

#### SOLUTION

For purposes of convection heat transfer, the box is made up of four vertical plates (the sides) and one horizontal plate (the top). For each vertical plate, $\Delta t = 25$°C and $L = 0.15$ m, thus laminar. Likewise, the horizontal plate has laminar convection. The heat-transfer coefficient for each vertical plate is

$$h = 1.42 \left( \frac{\Delta t}{L} \right)^{1/4} \text{ or } 5.102 \text{ W/m}^2 \text{°C}$$

The heat-transfer coefficient for the horizontal plate is

$$h = 1.32 \left( \frac{\Delta t}{L} \right)^{1/4} \text{ or } 4.743 \text{ W/m}^2 \text{°C}$$

The heat-transfer rate is the sum of the values for the five plates: $Q = \Sigma hA(\Delta t)$.

$$Q = 5.102 \times 4 \times (0.15)^2 \times 25 + 4.743 \times (0.15)^2 \times 25 = 14.148 \text{ W}$$

### ILLUSTRATIVE PROBLEM 11.18

A vertical wooden wall 8 ft. high $\times$ 7 ft. wide $\times$ 3 in. thick ($k = 0.07$ Btu/(h·ft.·°F)) has warm air on one side at 80°F and cold air at 50°F on the other side. Determine the heat transfer through the wall and the wall temperature.

#### SOLUTION

This problem cannot be solved directly because the individual film resistances are functions of unknown temperature differences. Therefore, as a first approximation, let us assume that $h$ for both sides of the wall can be taken as 1/2. The wall resistance is $\Delta x/kA$, which for 1 ft.$^2$ is $(3/12)/0.07 = 3.57$. The overall series resistance therefore is $1/(1/2) + 1/(1/2) + 3.57 = 7.57$. Using 7.57, we can now obtain $\dot{Q}$ and the individual temperature differences. $\dot{Q} = (80 - 50)/7.57 = 3.96$ Btu/h·ft.$^2$, and $\Delta t$ through the "hot" air film is $3.96/(1.2) = 7.92$°F. Through the wall, $\Delta t$ is $3.57 \times 3.96 = 14.1$°F, and through the "cold" air film, $\Delta t$ is also $3.96/(1/2) = 7.92$°F. With these temperature differences, we can now enter Figures 11.13 and 11.15 to verify our approximation. From Figure 11.15, we find $h = 0.42$ Btu/(h·ft.$^2$·°F). Using $h = 0.42$, we have for the overall resistance $(1/0.42) + (1/0.42) + 3.57 = 8.33$. $Q = (80 - 50)/8.33 = 3.6$ Btu/h ft.$^2$. $\Delta t$ through both air films is $3.6/0.42 = 8.57$°F (say, 8.6°F), and through the wall, $\Delta t = 3.6 \times 3.57 = 12.85$°F. Entering Figure 11.15, we find that $h$ stays essentially 0.42, and our solution is that the heat flow is 3.6 Btu/h·ft.$^2$, the "hot" side of the wall is at $80 - 8.6$°F or 71.4°F, the "cold" side is at $50 + 8.6$°F = 58.6°F, and the temperature drop in the wall is $71.4 - 58.6 = 12.8$°F, which checks our wall $\Delta t$ calculation.

Because natural convection results from density differences in the fluid, with the warmer fluid molecules tending to rise, the orientation of the hot surface affects the rate of heat transfer. Thus, the flow from a horizontal plate becomes turbulent more readily than that from a vertical plate, because the hot fluid molecules can more easily rise from the horizontal plate. Heat transfer from a plate with its hot side down has the lowest heat-transfer coefficient, because the hot fluid molecules have the most difficult time rising from this orientation. In addition, fluid flow over hot surfaces encounters frictional forces that can slow the fluid motion. This is more of a concern in forced convection, which is discussed in Section 11.3.2. Thus, convection heat transfer is a more complex process than conduction heat transfer.

## 11.3.2 Forced Convection

*Forced convection* flow can be either laminar or turbulent, inside tubes or outside tubes, and involves changes of phase such as when a fluid is being boiled. Because of the complexity and the number of cases that would have to be studied to cover this topic, we shall limit ourselves to the situation where we have a liquid or gas flowing inside a tube in turbulent flow. For this condition, the heat-transfer coefficient can be calculated from the following type of equation:

$$\mathrm{Nu} = \mathrm{C}(\mathrm{Pr})^a (\mathrm{Re})^b \tag{11.29a}$$

where the nomenclature is the same as for Equation 11.18; that is, $a$, $b$, and $C$ are constants. Nu is the Nusselt number, Pr is the Prandtl number, and Re is the Reynolds number, which is $DV\rho/\mu$ where $D$, $V$, $\rho$, and $\mu$ are pipe diameter, fluid velocity, fluid density, and dynamic viscosity of the fluid; or $DG/\mu$, where $G$ is the mass flow rate per square foot of flow area, $\dot{m}/A$. It is to be remembered that these "numbers" are dimensionless groups, and consistent units must be used throughout. The form of Equation 11.29a that is most generally used for turbulent flow (the Reynolds number must be greater than 2100) is

$$Nu_{\text{turbulent}} = \frac{hd}{k} = 0.023\,\mathrm{Pr}^{\frac{1}{3}}\mathrm{Re}^{0.8} = 0.023 \left(\frac{c_p\mu}{k}\right)^{\frac{1}{3}} \left(\frac{DV\rho}{\mu}\right)^{0.8} \tag{11.29b}$$

with the properties of specific heat, viscosity, and thermal conductivity evaluated at the bulk temperature of the fluid. To facilitate the use of this equation for water and air flowing turbulently in tubes, Figures 11.16 through 11.20 have been developed (see Granet 1957).

Figures 11.16 and 11.17 give the viscosity of water and air and are used to check the Reynolds number to ensure that the flow is turbulent. Figures 11.18 and 11.19 yield the "basic" heat-transfer coefficient $h_1$ as a function of the weight flow $W/1000$, where $W$ is in pounds per hour. Note that in these figures, the bulk temperature of the fluid is used, not the temperature difference between the fluid and the surface. Finally, Figure 11.20 is a correction factor for the variation of the inside diameter from 1 in. The desired heat-transfer coefficient $h$ is then simply equal to $F \times h_1$.

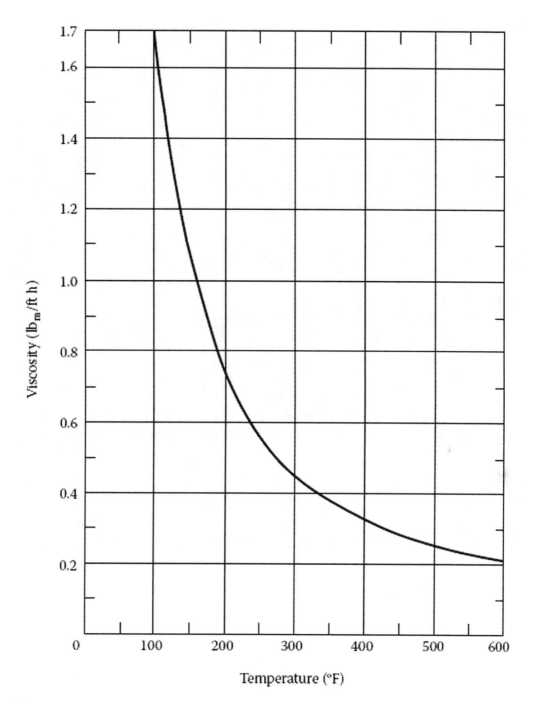

**FIGURE 11.16**
Viscosity of saturated water. (Based on data from the *Steam Tables*.)

## ILLUSTRATIVE PROBLEM 11.19

Water flows in a tube has a 1 in. outside diameter and a 0.065 in. wall (ID = 0.87 in.). If 20 lb./min of water at 400°F is flowing, calculate the heat-transfer coefficient on the inside of the tube.

### SOLUTION

The first step is to check the Reynolds number. It will be recalled that the Reynolds number is given by $DV\rho/\mu$ and is dimensionless. Therefore, we can use $D$, diameter in feet; $V$, velocity in ft./h; $\rho$, density in $lb_m/ft.^3$, and $\mu$, viscosity in $lb_m/ft.\cdot h$. Alternatively, the Reynolds number is given by $DG/\mu$, where $G$ is the mass flow rate per unit area ($lb_m/h\cdot ft.^2$). For this problem,

$$G = \frac{20 \times 60}{\dfrac{\pi (0.87)^2}{4 \times 144}} = 290680 \; lb_m/h \cdot ft.^2$$

$$= 0.33 \; lb_m/ft. \cdot h \; \left(\text{from Figure 11.16}\right)$$

Therefore, the Reynolds number is

$$Re = \frac{DG}{\mu} = \frac{(0.87/12) \times 290680}{0.33} = 63862$$

which is well into the turbulent flow regime. The next step is to enter Figure 11.18 at $W/1000$ of $20 \times 60/1000 = 1.2$ and 400°F to obtain $h_1 = 630$. From Figure 11.20, we obtain $F = 1.25$ for an inside diameter of 0.87 in. The final desired value of $h$ therefore is equal to

$$F \times h_1 = 1.25 \times 630 = 788 \; Btu/(h \cdot ft.^2 \cdot °F)$$

## ILLUSTRATIVE PROBLEM 11.20

If instead of water, atmospheric pressure air was flowing in the tube of Illustrative Problem 11.19, determine the inside film coefficient.

### SOLUTION

We first check the Reynolds number and note that $G$ is the same as for Illustrative Problem 11.19. The viscosity of air at these conditions is obtained from Figure 11.17 as 0.062 $lb_m/ft.\cdot h$. The Reynolds number, $DG/\mu$, therefore is

$$Re = \frac{DG}{\mu} = \frac{(0.87/12) \times 290680}{0.062} = 339908$$

which places the flow in the turbulent regime. Because $W/1000$ is the same as for Illustrative Problem 11.19 and equals 1.2, we now enter Figure 11.19 at 1.2 and 400°F to obtain $h_1 = 135$. Because the inside tube diameter is the same as before, $F = 1.25$. Therefore,

$$h = 135 \times 1.25 = 169 \text{ Btu/h} \cdot \text{ft.}^2 \cdot {}^\circ\text{F}$$

It is interesting to note that for equal mass flow rates, water yields a heat-transfer coefficient that is almost five times greater than air.

## 11.4 Radiation

Radiant heat transfer differs from both conduction and convection in that a medium is not required to transfer the heat. Basically, radiant heat transfer is an electromagnetic phenomenon similar to the transmission of light, x-rays, and radio waves, and all bodies radiate heat. A net interchange of heat occurs when the absorption of radiant energy by one body exceeds the energy that it is radiating. A body that absorbs all the radiation that strikes it regardless of the wavelength of the radiation is said to be a *blackbody*. Real bodies reflect as well as absorb thermal radiation, and brightly polished metals are good reflectors of

**FIGURE 11.17**
Viscosity of dry air.

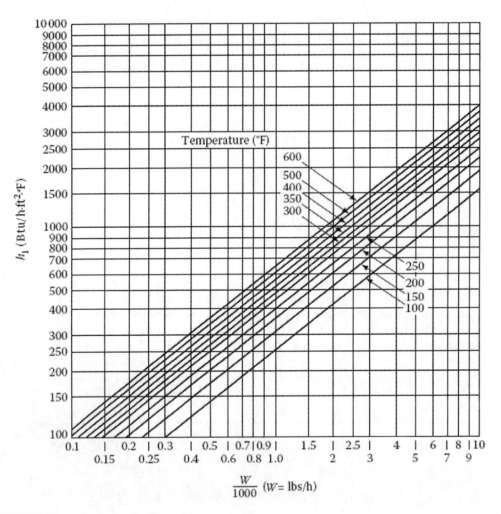

**FIGURE 11.18**
Basic heat-transfer coefficient for subsaturated water.

thermal radiation. The fraction of the incident heat that is reflected is known as the *reflec-tivity* of the body, the fraction absorbed is known as the *absorptivity*, and the effectiveness of the body as a thermal radiator at a given temperature is known as its *emissivity*. Thus, emissivity is also the ratio of the emission of heat at a given temperature to the emission of heat from a blackbody at the same temperature.

The radiation from a blackbody can be determined from the *Stefan–Boltzmann law*, which states that the radiation from a blackbody is proportional to the fourth power of the abso-lute temperature of the body. Thus,

$$\dot{Q}_r = 0.173 \times 10^{-8} A T^4 \tag{11.30}$$

where $\dot{Q}_r$ is the radiant heat transfer in Btu/h, $A$ is the radiating area in ft.$^2$, and $T$ is the absolute temperature in R (or K in the SI system). The net interchange of heat by radiation between two bodies at different temperatures can be written as

$$\dot{Q}_r = \sigma F_e F_A A \left( T_1^4 - T_2^4 \right) \tag{11.31}$$

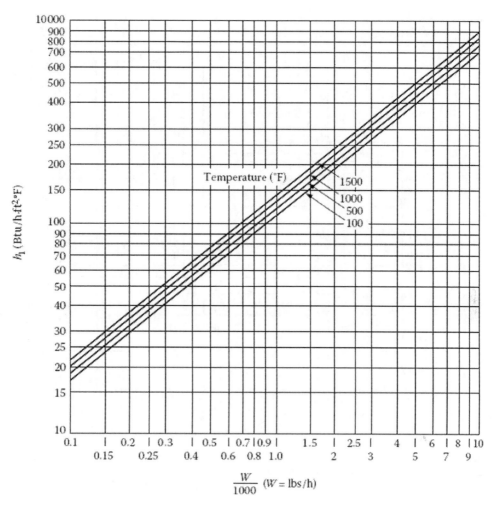

**FIGURE 11.19**
Basic heat-transfer coefficient for dry air.

where

$\sigma$ = Stefan–Boltzmann constant = $0.173 \times 10^{-8}$ Btu/(h·ft.² R⁴) (in SI, $5.669 \times 10^{-8}$ W/m²·K⁴)

$F_e$ = emissivity factor to allow for the departure of the surfaces interchanging heat from complete blackness; $F_e$ is a function of the surface emissivities and configurations

$Fa$ = geometric factor to allow for the average solid angle through which one surface "sees" the other

$A$ = area, ft.² (in SI, m²)

$T_1$, $T_2$ = absolute temperatures, R (in SI, K)

Table 11.4 and Figures 11.21 through 11.25 give the required values of $F_e$ and $F_A$ for most cases of practical interest. To evaluate $F_e$, it is necessary to know the emissivities ($e$) of the surface involved. A table of emissivity of various surfaces is given in Appendix 2 (Table A.23). In general, highly polished metals have low emissivities; the emissivity of most materials increases with temperature; most nonmetals have high emissivities; and the emissivity of a given surface will have wide variations depending on the conditions of the surface.

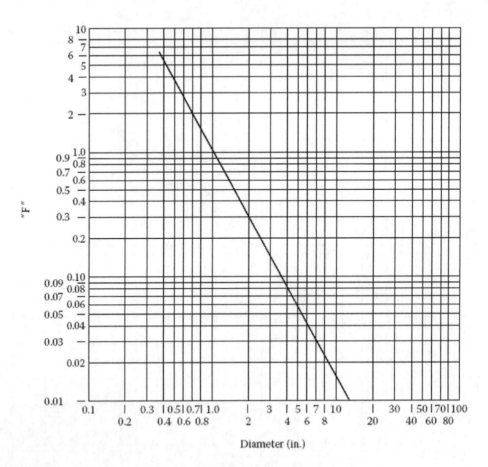

**FIGURE 11.20**
Factor $F$: multiplier of basic heat-transfer coefficient.

---

### ILLUSTRATIVE PROBLEM 11.21

A bare steel pipe of outside diameter 3.50 in. runs through a room whose walls are at 70°F (530 R) Determine the heat loss by radiation from 5 ft. of the pipe if its outside temperature is 120°F (580 R).

### SOLUTION

From Table 11.4, case 2, $F_A = 1$, $A$ is the area of the pipe, and $F_e$ is the emissivity of the steel pipe. From Appendix 2, for oxidized steel, $\varepsilon = 0.79$. Therefore,

$$\dot{Q} = 0.173 \times 10^{-8} \times 0.79 \times 1 \times \frac{\pi(3.5)}{12} \times 5 \left[ (120 + 460)^4 \right.$$

$$\left. - (70 + 460)^4 \right] = 214.5 \text{ Btu/h}$$

It should be noted that Equation 11.31 contains a $T_1^4 - T_2^4$ term, not a $(T_1 - T_2)^4$ term, a very common mistake.

**TABLE 11.4**

Radiation between Solids: Factors for Use in Equation 11.31

| Surfaces between Which Radiation Is Being Interchanged | Area, $A$ | $F_A$ | $Fe$ |
|---|---|---|---|
| 1. Infinite parallel planes. | $A_1$ or $A_2$ | 1 | $\dfrac{1}{\dfrac{1}{\varepsilon_1} + \dfrac{1}{\varepsilon_2} - 1}$ |
| 2. Completely enclosed body, small compared with enclosing body. (Subscript 1 refers to enclosed body.) | $A_1$ | 1 | $\varepsilon_1$ |
| 3. Completely enclosed body, large compared with enclosing body. (Subscript 1 refers to enclosed body.) | $A_1$ | 1 | $\dfrac{1}{\dfrac{1}{\varepsilon_1} + \dfrac{1}{\varepsilon_2} - 1}$ |
| 4. Intermediate case between 2 and 3. (Incapable of exact treatment except for special shapes. Subscript 1 refers to enclosed body.) | $A_1$ | 1 | $\varepsilon_1 > F_e > \dfrac{1}{\dfrac{1}{\varepsilon_1} + \dfrac{1}{\varepsilon_2} - 1}$ |
| 5. Concentric spheres or infinite cylinders, special case of 4. (Subscript 1 refers to enclosed body.) | $A_1$ | 1 | $\dfrac{1}{\dfrac{1}{\varepsilon_1} + \dfrac{A_1}{A_2}\dfrac{1}{\varepsilon_2} - 1}$ [a] |
| 6. Surface element $dA$ and area $A_2$. There are various special cases of 6 with results presentable in graphical form. They follow cases 7, 8, and 9. | $dA$ See special cases 7, 8, 9[b] | | $\varepsilon_1\varepsilon_2$ |
| 7. Element $dA$ and rectangular surface above and parallel to it, with one corner of rectangle contained in normal to $dA$. | $dA$ | See Figure 11.21 | $\varepsilon_1\varepsilon_2$ |
| 8. Element $dA$ and any rectangular surface above and parallel to it. Split rectangle into four having common corner at normal to $dA$ and treat as in case 7. | $dA$ | Sum of F/s determined for each rectangle as in case 7 | $\varepsilon_1\varepsilon_2$ |
| 9. Element $dA$ and circular disk in plane parallel to plane of $dA$. | $dA$ | Formula below[c] | $\varepsilon_1\varepsilon_2$ |
| 10. Two parallel and equal squares or disks of width or diameter $D$ and distance between of $L$. | $A_1$ or $A_2$ | Figure 11.22, curves 1 and 2 | $\varepsilon_1\varepsilon_2$ |
| 11. Same as case 10, except planes connected by nonconducting reradiating walls. | $A_1$ or $A_2$ | Figure 11.22, curve 3 | $\varepsilon_1\varepsilon_2$ |
| 12. Two equal rectangles in parallel planes directly opposite each other and distance $L$ between. | $A_1$ or $A_2$ | $\left(F_A F_A^{r^d}\right)^{1/2}$<br><br>Figure 11.23 | $\varepsilon_1\varepsilon_2$ or<br>$\dfrac{1}{\dfrac{1}{\varepsilon_1} + \dfrac{1}{\varepsilon_2} - 1}$ |
| 13. Two rectangles with common sides, in perpendicular planes. | $A_1$ or $A_2$ | Figure 11.24 | $\varepsilon_1\varepsilon_2$ |

*(Continued)*

**TABLE 11.4 (CONTINUED)**

Radiation between Solids: Factors for Use in Equation 11.31

| Surfaces between Which Radiation Is Being Interchanged | Area, $A$ | $F_A$ | $Fe$ |
|---|---|---|---|
| 14. Radiation from a plane to a tube bank (1 or 2 rows) above and parallel to the plane. | $A_1$ or $A_2$ | Figure 11.25 | $\varepsilon_1\,\varepsilon_2$ |

*Source:* A. I. Brown and S. M. Marco, *Introduction to Heat Transfer*, 3rd ed., New York: McGraw-Hill Book Co., 1958. With permission.

[a] This form results from assumption of completely diffuse reflection. If reflection is completely specular (mirror like), then $F_e = 1/[(1/\varepsilon_1 + 1/\varepsilon_2) - 1]$.

[b] A complete treatment of this subject, including formulas for special complicated cases and the description of a mechanical device for solving problems in radiation, is given by H. C. Hottel, *Mech. Eng.*, 52(7), 699, July 1930.

[c] Case 9, $R$ = radius of disk + distance between planes; $x$ = distance from $dA$ to normal through center of disk + distance between planes.

$$F_A = \frac{1}{2}1 - \frac{x^2 + 1 - R^2}{\sqrt{x^2 + 2(1 - R^2)x^2 + (1 + R^2)^2}}$$

[d] $F_A = F_A$ for squares equivalent to the short side of the rectangle (Figure 11.22, curve 2).
$F_A = F_A$ for squares equivalent to the short side of the rectangle (Figure 11.22, curve 2).
$F_e = \varepsilon_1\varepsilon_2$ if the areas are small compared with $L$.
$F_e = 1/[(1/\varepsilon_1 + 1/\varepsilon_1) - 1]$ if the areas are large compared with $L$.

**FIGURE 11.21**
Radiation between a surface element and a rectangle above and parallel to the surface element.

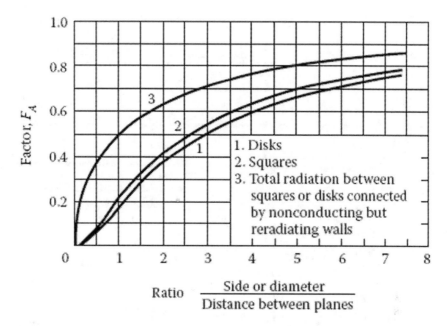

**FIGURE 11.22**

Direct radiation between equal disks or squares in parallel planes directly opposed.

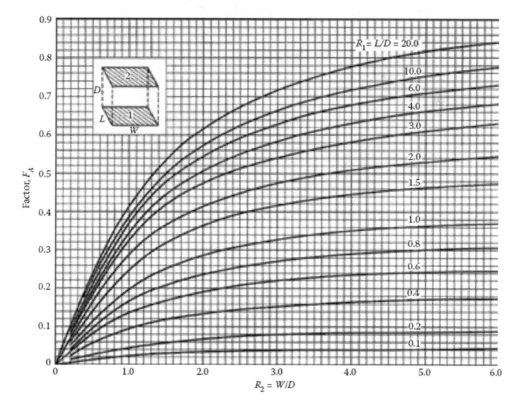

**FIGURE 11.23**

Radiation shape factor for parallel, directly opposed rectangles.

**FIGURE 11.24**
Radiation between adjacent rectangles in perpendicular planes.

In many situations where both radiation and convection occur simultaneously from a body, it is desirable to evaluate a combined heat-transfer coefficient for the process. To arrive at an effective heat-transfer coefficient for radiation, we will equate Equations 11.16 and 11.31 as follows:

$$\dot{Q}_r = h_r \left(T_1 - T_2\right) A = \sigma F_e F_A A \left(T_1^4 - T_2^4\right) \tag{11.32}$$

and

$$h_r = \frac{\sigma F_e F_A \left(T_1^4 - T_2^4\right)}{T_1 - T_2} \tag{11.33}$$

which can be rewritten as

$$h_r = F_e F_A \frac{\sigma \left(T_1^4 - T_2^4\right)}{T_1 - T_2} = F_e F_A h_r' \tag{11.34}$$

Ordinate is fraction of heat radiated from the plane to an infinite number of rows of tubes or to a plane replacing the tubes

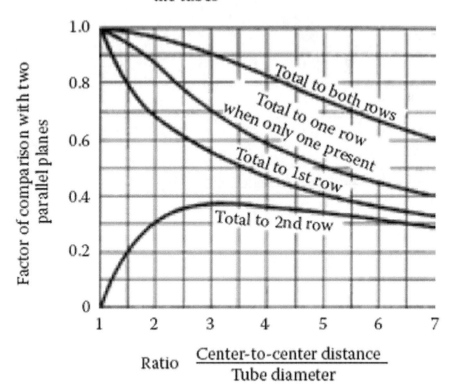

**FIGURE 11.25**
Radiation from a plane to one or two rows of tubes above and parallel to the plane.

**FIGURE 11.26**
Factor for radiation coefficient. (From I. Granet, "Coefficient of radiation heat transfer," *Design News*, September 1970.)

For small temperature differences, the term in brackets, $h_r'$, has been evaluated and plotted in terms of the upper-temperature $t_1$ in degrees Fahrenheit and the temperature difference $\Delta$ in degrees Fahrenheit between the two bodies in Figure 11.26.

---

### ILLUSTRATIVE PROBLEM 11.22

Determine the radiation heat-transfer coefficient for the pipe of Illustrative Problem 11.21.

### SOLUTION

The upper temperature is given as 120°F and the temperature difference $\Delta$ is $120 - 70 = 50$°F. Using Figure 11.26, $h_r' = 1.18$ and $h_r = F_e F_A \, h_r' = 1 \times 0.79 \times 1.18 = 0.93$. As a check, using the results of Illustrative Problem 11.21,

$$h_r = \frac{Q}{A\Delta t} = \frac{214.5}{\left[\pi(3.5/12)\right] \times 5(120-70)}$$

$$= 0.94 \text{ Btu/h} \cdot \text{ft.}^2 \cdot {}^\circ\text{F}$$

---

### ILLUSTRATIVE PROBLEM 11.23

Solve Illustrative Problem 11.16 taking into account both convection and radiation.

#### SOLUTION

Because the conditions of Illustrative Problem 11.16 are the same as for Illustrative Problems 11.21 and 11.22, we can solve this problem in two ways to obtain a check. Thus, adding the results of these problems yields $\dot{Q}_{total} = 206.2 + 214.5 = 420.7$ Btu/h. We can also approach this solution by obtaining a combined radiation and convection heat-transfer coefficient. Thus, $h_{combined} = 0.9 + 0.94 = 1.84$. $\dot{Q}_{total} = 184(\pi \times 3.5/12)5 \times (120-70) = 421.5$ Btu/h. This procedure of obtaining combined or overall heat-transfer coefficients is discussed further in Sections 11.5 and 11.6.

---

In solving radiation heat-transfer problems involving the fourth power of the absolute temperatures, the numbers can become excessively large. Because the radiation heat-transfer equation involves the Stefan–Boltzmann constant with its factor of $10^{-8}$, a useful device is to express the absolute temperature as a number times $10^2$ (e.g., 1400 is $14 \times 10^2$). When this temperature is taken to the fourth power, a number times $10^8$ results. This factor is conveniently canceled by the $10^{-8}$ of the Stefan–Boltzmann constant, leaving students with more manageable numbers.

Before proceeding, let us look at a general approach to obtaining $F_e$ for two gray bodies interchanging radiation. This method is called the network method and can be adapted to determining $F_e$ for systems in which more than two bodies are interchanging radiation.

When radiant energy is intercepted by a non-blackbody, that is, a gray body, some of the energy may be absorbed, some may be reflected, and some may be transmitted through the body, as in the case of glass. The fraction of the total incident radiation absorbed by the body is denoted by $\alpha$, the *absorptivity* of the surface; the fraction of the incident radiation that is reflected is denoted by $\rho$, the *reflectivity* of the surface; and finally, the fraction of the incident radiation that is transmitted is denoted by $\tau$, the *transmissivity* of the body. Based on these definitions, we can write

$$\alpha + \rho + \tau = 1 \tag{11.35}$$

When a body is opaque, it does not transmit any radiation, giving us $\tau = 0$, which yields the following equation for an opaque body:

$$\alpha + \rho = 1 \tag{11.35a}$$

We now denote the total incident energy per unit area to be the *irradiation*, $\dot{Q}_i$, having English units of Btu/h·ft.$^2$, and the total energy leaving the body as the *radiosity* $\dot{Q}_r$, in consistent units.

We further denote $\dot{Q}_b$ to be the amount of energy that a blackbody would emit at the same temperature as the gray surface being considered. Using these definitions, we can write

$$\dot{Q}_r = \rho\dot{Q}_i + \varepsilon\dot{Q}_b \tag{11.36}$$

Simplifying Equation 11.36 using Equation 11.35a, and assuming $\alpha = \varepsilon$ for a gray body,

$$\dot{Q}_r = (1-\alpha)\dot{Q}_i + \varepsilon\dot{Q}_b = (1-\varepsilon)\dot{Q}_i + \varepsilon\dot{Q}_b \tag{11.36a}$$

The net heat transferred to the body, $\dot{Q}_{net}$, is

$$\dot{Q}_{net} = q''_{net}A = \left(\dot{Q}_i - \dot{Q}_r\right) \tag{11.37}$$

Combining Equation 11.36a with Equation 11.37 yields

$$\dot{Q}_{net} = \underbrace{\left(\frac{\dot{Q}_r - \varepsilon\dot{Q}_b}{(1-\varepsilon)} - \dot{Q}_r\right) = \frac{\dot{Q}_r - \dot{Q}_b}{(1-\varepsilon)}}_{\varepsilon} \rightarrow q''_{net} = \underbrace{\frac{\dot{Q}_r - \dot{Q}_b}{A(1-\varepsilon)}}_{\varepsilon} \tag{11.38}$$

We can compare Equation 11.38 to Ohm's law, giving us $\dot{Q}_r - \dot{Q}_b$ as representing a potential, $(1-\varepsilon)/\varepsilon A$ representing a resistance, and $\dot{Q}_{net}$ representing a current flow. Thus, we can represent a system of two surfaces that see each other and nothing else by the circuit diagram of Figure 11.27. Each gray surface has a resistance of $(1-\varepsilon)/\varepsilon A$, and these resistances are connected by a resistance that is due to the geometry of the system, namely, $1/A_1F_{1-2}$.

In Figure 11.27, we have three resistances in series across an overall potential difference equal to $\dot{Q}_{b,1} - \dot{Q}_{b,2} = \sigma\left(T_1^4 - T_2^4\right)$. Therefore,

$$\dot{Q}_{net} = \frac{\sigma\left(T_1^4 - T_2^4\right)}{\dfrac{(1-\varepsilon_1)}{\varepsilon_1 A} + \dfrac{1}{A} + \dfrac{(1-\varepsilon_2)}{\varepsilon_2 A}} \tag{11.39}$$

**FIGURE 11.27**
Electrical analogy for two gray bodies interchanging radiation; two infinite radiating planes that see each other and nothing else.

If we factor the $A$ in the denominator and rearrange terms, we obtain

$$\dot{Q}_{net} = \frac{\sigma\left(T_1^4 - T_2^4\right)A}{\dfrac{1}{\varepsilon_1} + \dfrac{1}{\varepsilon_2} - 1}$$ (11.40)

Comparing Equation 11.40 with Equation 11.31 gives us the value of $F_e$ for this case as

$$F_e = \frac{1}{\varepsilon_1} + \frac{1}{\varepsilon_2} - 1$$ (11.41)

which is the same as case 1 in Table 11.4.

The electric analogy can be applied to many situations that would be difficult to solve in any other manner. In many cases, the complex interactions lead to a group of simultaneous equations that can be written in a condensed matrix form. A solution is then obtained by the process of inverting the matrix. The interested reader should consult the references for these procedures.

## 11.5 Heat Exchangers

When heat is transferred from one fluid to another in an industrial process without mixing, the fluids are separated by a thermally conductive barrier. Such a configuration where heat transfer takes place is known as a heat exchanger. A heat exchanger can be of varied shape and size and is usually designed to perform a specific function. The steam-generation plant uses heat exchangers as condensers, economizers, air heaters, feedwater heaters, reheaters, and so on. It is common to designate heat exchangers by their geometric shape and the relative directions of the flow of the heat-transfer fluids. For example, Figure 11.28a shows a concentric-tube (or double-pipe) unit in which the fluids would be said to be flowing parallel to each other, and the unit would be called a parallel-flow, double-pipe unit. Other common types are shown in the remainder of Figure 11.28.

The problem of calculating the heat transfer in these units differs from our previous discussion in that the temperature of one or both of the fluids varies continuously as the fluids proceed through the heat exchanger. This can be seen in Figure 11.29, where the fluid temperatures have been plotted as a function of the heat-transfer surface for the most common cases of parallel flow, counterflow, and for one fluid at a constant temperature. The subscript $h$ is used to denote the hot fluid, and the subscript $c$ denotes the cold fluid. The subscript "1" is used to denote the temperature at the entry of a fluid of the heat exchanger, and "2" denotes the temperature of the fluid at the exit of the heat exchanger. The direction of flow of each fluid through the exchanger is shown by the arrowheads on the temperature curves. The largest temperature difference between the fluids in the unit (at either inlet or outlet) is designated as $\theta_a$ and the least temperature difference between the fluids (at either inlet or outlet) is designated as $\theta_B$.

Newton's law of cooling (Equation 11.16) can be written for the heat exchangers as

$$\dot{Q} = UA\left(\Delta t\right)_m$$ (11.42)

**FIGURE 11.28**
Representative types of heat exchangers. (a) Concentric tubes counter flow. (b) Shell-and-tube counter flow heat exchanger. (c) Shell-and-tube exchanger, segmented baffles, two tube passes, one shell pass. (d) Disk and doughnut baffle. (e) Segmented baffle. (f) Crossflow tubular exchanger. (g) Plate-fin exchanger. (From W. M. Rohsenow and H. Y. Choi, *Heat, Mass and Momentum Transfer*, Englewood Cliffs, N.J.: Prentice Hall, Inc., 1961. With permission.)

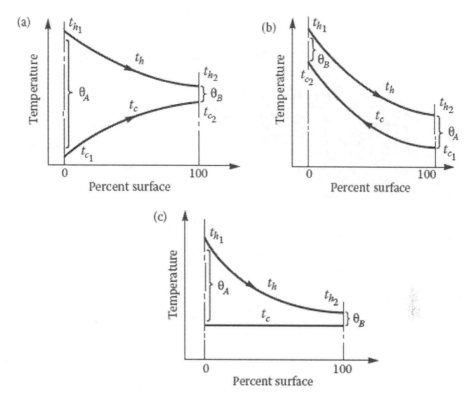

**FIGURE 11.29**
Fluid temperatures in heat exchangers. (a) Parallel flow. (b) Counter flow. (c) Constant temperature.

where $U$ is the overall conductance or overall heat-transfer coefficient having the same physical units as the convection coefficient $h$, Btu/(h·ft.$^2$·°F) or W/m$^2$°C; $A$ is the heat-transfer surface in square feet; and $(\Delta t)_m$ is an appropriate *mean temperature difference*. The *overall heat-transfer coefficient, U*, in Equation 11.42 is not usually constant for all locations in the heat exchanger, and its local value is a function of the local fluid temperatures. However, it is usual practice to evaluate the individual heat-transfer coefficients based on the arithmetic average fluid temperatures. By analogy to convection, we have 1/UA = resistance. The concept of the overall heat-transfer coefficient is best illustrated by Illustrative Problems 11.24 and 11.25.

---

### ILLUSTRATIVE PROBLEM 11.24

Determine the overall heat-transfer coefficient for the composite wall of Illustrative Problem 11.4 if $h$ on the hot side is 0.9 Btu/(h ft.$^2$·°F) and $h$ on the cold side is 1.5 Btu/(h·ft.$^2$·°F). The temperatures given are to be the respective air temperatures (see Figure 11.7).

### SOLUTION

For a plane wall, the areas are all the same, and if we use 1 ft.$^2$ of the wall surface as the reference area,

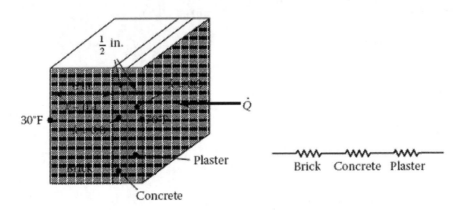

**FIGURE 11.7**
**(Repeated)** Illustrative Problem 11.24.

$$\text{brick resistance} = \frac{\Delta x}{kA} = \frac{6/12}{0.4 \times 1} = 1.25$$

$$\text{concrete resistance} = \frac{\Delta x}{kA} = \frac{\dfrac{1}{2}/12}{0.8 \times 1} = 0.052$$

$$\text{plaster resistance} = \frac{\Delta x}{kA} = \frac{\dfrac{1}{2}/12}{0.3 \times 1} = 0.139$$

$$\text{"hot film" restance} = \frac{1}{hA} = \frac{1}{0.9 \times 1} = 1.11$$

$$\text{"cold film" resistance} = \frac{1}{hA} = \frac{1}{1.5 \times 1} = 0.67$$

$$\overline{\hspace{4cm}\text{total resistance} = \hspace{2cm} 3.22}$$

The overall conductance (or overall heat-transfer coefficient) $U = 1/$(overall resistance) $= 1/3.22 = 0.31$ Btu/h·ft². In Illustrative Problem 11.24, the solution is straightforward, because the heat-transfer area is constant for all series resistances.

---

### ILLUSTRATIVE PROBLEM 11.25

A steel pipe, $k = 26$ Btu/(h·ft·°F), having an outside diameter of 3.5 in. and an inside diameter of 3.00 in., and 5 ft. long, is covered with 1 in. of mineral wool, $k = 0.026$ Btu/(h·ft·°F). If the film coefficient on the inside of the pipe is 45 Btu/(h·ft².°F) and on the outside is 0.9 Btu/(h·ft².°F), determine the overall heat-transfer coefficient. (See Figure 11.30 and Illustrative Problems 11.10 and 11.12.)

### SOLUTION

Because the same amount of heat traverses each of the paths, we can write

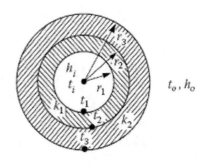

$A_i$ = Inside pipe area
$A_o$ = Outside pipe area

**FIGURE 11.30**
Illustrative Problem 11.25.

$$\dot{Q} = \dot{Q}_i = A_i h_i \left( t_i - t_1 \right) = 2\pi r_1 L h_i \left( t_i - t_1 \right)$$

$$= \frac{2\pi k_1 L}{\ln \left( r_2 / r_1 \right)} \left( t_1 - t_2 \right)$$

$$= \frac{2\pi k_2 L}{\ln \left( r_3 / r_2 \right)} \left( t_2 - t_3 \right)$$

$$= A_o h_o \left( t_o - t_o \right) = 2\pi r_3 L h_o \left( t_o - t_o \right)$$

Solving for the temperature differences and adding them yields

$$\dot{Q} = \frac{t_i - t_o}{\dfrac{1}{2\pi r_1 L h_i} + \dfrac{\ln \left( r_2 / r_1 \right)}{2\pi k_1 L} + \dfrac{\ln \left( r_3 / r_2 \right)}{2\pi k_2 L} + \dfrac{1}{2\pi r_3 L h_o}} \quad\text{(a)}$$

Although this equation looks formidable, we can simplify it and interpret it in terms of $j$ items discussed previously in this chapter. Thus,

$$\dot{Q} = \frac{t_i - t_o \, 2\pi L}{\dfrac{1}{h_i r_1} + \dfrac{\ln \left( r_2 / r_1 \right)}{k_1} + \dfrac{\ln \left( r_3 / r_2 \right)}{k_2} + \dfrac{1}{h_o r_3}} \quad\text{(b)}$$

Let us now define $U_o$ as the overall heat-transfer coefficient based on the outside pipe surface $A_o$ as

$$\dot{Q} = U_o A_o \left( t_i - t_o \right) \quad\text{(c)}$$

If we multiply the numerator and the denominator of Equation (b) by $r_3$, we obtain

$$\dot{Q} = \frac{\left( t_i - t_o \right) 2\pi L r_3}{\dfrac{r_3}{h_i r_1} + \dfrac{r_3}{k_1} \ln \left( \dfrac{r_2}{r_1} \right) + \dfrac{r_3}{k_2} \ln \left( \dfrac{r_3}{r_2} \right) + \dfrac{r_3}{h_o r_3}} \quad\text{(d)}$$

Comparison of Equations (c) and (d) yields

$$U_o = \cfrac{1}{\cfrac{1}{h_i\left(r_1 / r_3\right)} + \cfrac{r_3}{k_1}\ln\left(\cfrac{r_2}{r_1}\right) + \cfrac{r_3}{k_2}\ln\left(\cfrac{r_3}{r_2}\right) + \cfrac{1}{h_o}} \tag{e}$$

Note that $U_o$ is the overall heat-transfer coefficient based on the outside tube surface. If we had multiplied Equation (b) by $r_1/r_1$ we would obtain the overall heat-transfer coefficient $U_i$ based on the inside surface as

$$U_i = \cfrac{1}{\cfrac{1}{h_i} + \cfrac{r_1}{k_1}\ln\left(\cfrac{r_2}{r_1}\right) + \cfrac{r_1}{k_2}\ln\left(\cfrac{r_3}{r_2}\right) + \cfrac{1}{h_o\left(r_3 / r_1\right)}} \tag{f}$$

In effect, we have required that $U_oA_o = U_iA_i$. When discussing an overall heat-transfer coefficient, the reference area must be given. Proceeding with the numerical problems yields

$$\frac{1}{h_i} = \frac{1}{45} = 0.02222$$

$$\frac{r_1}{k_1}\ln\left(\frac{r_2}{r_1}\right) = \frac{3.00}{2} / \left(26 \times 12\right)\ln\left(\frac{3.50}{3.00}\right) = 0.00074$$

$$\frac{r_1}{k_1}\ln\left(\frac{r_3}{r_2}\right) = \frac{3.00}{2} / \left(0.026 \times 12\right)\ln\left(\frac{5.50}{3.50}\right) = 2.1730$$

$$\frac{1}{h_o\left(r_3 / r_1\right)} = \frac{1}{0.9\left(5.50 / 3.00\right)} = 0.6061$$

$$\Sigma = 2.8021$$

Therefore,

$$U_i = \frac{1}{2.8021} = 0.357 \; \frac{\text{Btu}}{\text{h} \cdot \text{ft.}^2 \cdot {}^\circ\text{F}}\left(\text{of inside area}\right)$$

Because $U_oA_o = U_iA_i$, $U_o = 0.357 \times A_i/A_o = 0.357 \; D_1/D_3 = (0.357)\;(3.00/5.50) = 0.195$ Btu/(h·ft.$^2$°F) (of the outside area).

The true mean temperature difference of two fluids exchanging heat in a heat exchanger cannot be determined by simply subtracting temperatures. Let us make the following assumptions in order to obtain the true temperature difference:

1. $U$ is constant over the entire heat exchanger.
2. Both fluid flows are steady, that is, constant with time.

3. The specific heat of each fluid is constant over the entire heat exchanger.

4. Heat losses are negligible.

Then $(\Delta t)_m$ in Equation 11.42 is given for parallel flow, counterflow, and constant-temperature exchangers as

$$(\Delta t)_m = \frac{\theta_A - \theta_B}{\ln(\theta_A / \theta_B)} \tag{11.43}$$

where $\theta_A$ is the greatest temperature difference between the fluids (at either inlet or outlet) and $\theta_B$ is the least temperature difference between the fluids (at either inlet or outlet; see Figure 11.29). Thus, $(\Delta t)_m$ is known as the logarithmic mean temperature difference. Because heat losses are assumed to be negligible, the heat transferred from the hot fluid must equal that received by the cold fluid. Therefore,

$$\dot{m}_c \left(c_p\right)_c \left(t_{c2} - t_{c1}\right) = \dot{m}_h \left(c_p\right)_h \left(t_{h1} - t_{h2}\right) \tag{11.44}$$

Some typical values of overall heat-transfer coefficients are given in Table 11.5 and are useful in making preliminary design calculations.

**TABLE 11.5**

Approximate Overall Coefficients for Preliminary Estimates

| Duty | Overall Coefficient, U | |
| --- | --- | --- |
|  | (Btu/(h·ft.²·°F))[a] | (W/m²°C) |
| Steam to water | | |
|   Instantaneous heater | 400–600 | 2270–3400 |
|   Storage-tank heater | 160–212 | 900–1200 |
| Steam to oil | | |
|   Heavy fuel | 10–70 | 50–400 |
|   Light fuel | 30–70 | 170–400 |
| Steam to aqueous solutions | 100–600 | 570–3400 |
| Steam condenser | 170–1050 | 1000–6000 |
| Steam to gases | 5–50 | 28–280 |
| Water to compressed air | 10–30 | 57–170 |
| Water to water, jacket water coolers | 150–300 | 850–1700 |
| Water to lubricating oil | 20–60 | 100–350 |
| Water to condensing oil vapors | 40–100 | 220–570 |
| Water to condensing alcohol | 45–120 | 255–680 |
| Water to condensing Freon-12 | 50–180 | 300–1000 |
| Water to condensing ammonia | 140–250 | 800–1400 |
| Water to organic solvents, alcohol | 40–120 | 250–700 |
| Water to boiling Freon-12 | 50–180 | 300–1000 |
| Water to gasoline | 50–180 | 300–1000 |

[a] 1 Btu/(h ft.²°F) = 5.6786 W/m²°C.

## CALCULUS ENRICHMENT

Let us consider the counterflow heat exchanger shown in the figure below. In this figure, the "hot" fluid, $h$, enters at the left and exits at the right, while the "cold" fluid, $c$, enters at the right and exits at the left. A small section of the heat exchanger having a heat-transfer area $dA$ is isolated as shown, with the temperature differences being $dt_h$ and $dt_c$, respectively. Based on making the same assumptions noted earlier, a heat balance for each fluid yields

$$d\dot{Q} = U dA \left(t_h - t_c\right) = \dot{m}_h c_{ph}\, dt_h \tag{a}$$

and

$$d\dot{Q} = U dA \left(t_h - t_c\right) = \dot{m}_c c_{pc}\, dt_c \tag{b}$$

For the entire heat exchanger,

$$\dot{Q} = UA \left(\Delta t\right)_m \tag{c}$$

where $(\Delta t)_m$ is the appropriate mean temperature difference. Also,

$$\dot{Q} = \dot{m}_h c_{ph} \left(t_{h_2} - t_{h_1}\right) = \dot{m}_c c_{pc} \left(t_{c_1} - t_{c_2}\right) \tag{d}$$

By solving Equations (a) and (b) for $dt_h$ and $dt_c$, respectively, and finding their difference,

$$d\left(t_h - t_c\right) = \frac{U dA \left(t_h - t_c\right)}{\dot{m}_h c_{ph}} - \frac{U dA \left(t_h - t_c\right)}{\dot{m}_c c_{pc}} \rightarrow \frac{d\left(t_h - t_c\right)}{\left(t_h - t_c\right)} = \left(\frac{1}{\dot{m}_h c_{ph}} - \frac{1}{\dot{m}_c c_{pc}}\right) U dA \tag{e}$$

Integrating Equation (e),

$$\int \frac{d\left(t_h - t_c\right)}{\left(t_h - t_c\right)} = \left(\frac{1}{\dot{m}_h c_{ph}} - \frac{1}{\dot{m}_c c_{pc}}\right) U \int dA \tag{f}$$

Yields,

$$\ln\left(\frac{t_{h_2} - t_{c_2}}{t_{h_1} - t_{c_1}}\right) = \left(\frac{1}{\dot{m}_h c_{ph}} - \frac{1}{\dot{m}_c c_{pc}}\right) UA \tag{g}$$

Alternatively, denoting $(t_h - t_c)$ as $\theta$ and rearranging terms yields in Equation (f), yields,

$$\int_{\theta_a}^{\theta_b} \frac{d\theta}{\theta} = \frac{\theta_a - \theta_b}{\left(\Delta t\right)_m} \tag{h}$$

Integrating Equation (h),

$$\ln\left(\frac{\theta_a}{\theta_b}\right) = \frac{\theta_a - \theta_b}{\left(\Delta t\right)_m} \tag{i}$$

and

$$\left(\Delta t\right)_m = \frac{\theta_a - \theta_b}{\ln\left(\dfrac{\theta_a}{\theta_b}\right)}$$                                                                  (j)

where $(\Delta t)_m$ is the log mean temperature difference.

Our derivation was for a counterflow heat exchanger. Using the same procedure, it can also be shown that Equation (j) is also applicable for parallel flow and for heat exchangers in which there is a change of phase in one of the fluids.

---

### ILLUSTRATIVE PROBLEM 11.26

A counterflow heat exchanger is used to cool a flow of 400 lb./min of lubricating oil. Hot oil enters at 215°F and leaves at 125°F. The specific heat of the oil is 0.85 Btu/lb.°F, and the overall coefficient of heat transfer of the unit is 40 Btu/(h·ft.²°F) (of the outside tube surface). Water enters the unit at 60°F and leaves at 90°F. Determine the outside tube surface required (A).

**SOLUTION**

From Figure 11.31, $\theta_A = 215 - 90 = 125°F$, and $\theta_B = 125 - 60 = 65°F$. Therefore,

$$\left(\Delta t\right)_m = \frac{\theta_A - \theta_B}{\ln\left(\theta_A / \theta_B\right)} = \frac{125 - 65}{\ln\left(125 / 65\right)} = 92°F$$

**FIGURE 11.31**
Illustrative Problem 11.26.

From the oil data, the heat transfer $\dot{Q} = \dot{m}c_p \Delta t = 400 \times 60 \times 0.85(215 - 125) =$ 1836000 Btu/h, and from the heat-transfer equations, $\dot{Q} = UA(\Delta t)_m = 1836000 =$ $40 \times A \times 92$. Therefore, $A = 499$ ft.$^2$ of the outside surface is required.

## ILLUSTRATIVE PROBLEM 11.27

If the best exchanger in Illustrative Problem 11.26 is operated in parallel flow, determine the outside tube surface required, A.

### SOLUTION

From Figure 11.32, $\theta_A = 215 - 60 = 155°F$, and $\theta_B = 125 - 90 = 35°F$. Therefore,

$$(\Delta t)_m = \frac{\theta_A - \theta_B}{\ln(\theta_A / \theta_B)} = \frac{155 - 35}{\ln(155 / 35)} = 80.6°F$$

Because all other conditions are the same, $\dot{Q} = 1836000$ Btu/h.

$$1836000 = 40 \times A \times 80.6$$

Therefore, $A = 569$ ft.$^2$ of the outside surface is required.

**FIGURE 11.32**
Illustrative Problem 11.27.

**TABLE 11.6**

Typical Fouling Factors

| Types of Fluid | Fouling Resistance | |
|---|---|---|
| | $(h \cdot {}^{\circ}F \cdot ft \cdot {}^2/Btu)^a$ | $(m^2 \cdot {}^{\circ}C/W)$ |
| Seawater below 50°C (122°F) | 0.0005 | 0.0001 |
| Seawater above 50°C (122°F) | 0.001 | 0.0002 |
| Treated boiler feedwater | 0.001 | 0.0002 |
| River water | 0.001–0.006 | 0.0002–0.001 |
| Fuel oil | 0.005 | 0.0009 |
| Vapor refrigerants | 0.002 | 0.0004 |
| Alcohol vapors | 0.0005 | 0.0001 |
| Steam, non-oil-bearing | 0.0005 | 0.0001 |
| Industrial air | 0.002 | 0.0004 |
| Refrigerating liquid | 0.001 | 0.0002 |

[a] Dividing the values by 5.6786 will yield $m^2 \cdot {}^{\circ}C/W$.

From the results of Illustrative Problems 11.26 and 11.27, we note that operation of a heat exchanger in parallel flow requires more surface for the same terminal temperatures than for a unit operated in counterflow. This conclusion is general because $(\Delta t)_m$ between the same terminal temperature limits is always greater for counterflow than for parallel flow. Thus, where a choice exists, it is preferable to operate a unit in counterflow, because this type of operation will give the minimum surface requirements. Another advantage of counterflow is that it is possible to raise the exit temperature of the cooling fluid closer to the inlet temperature of the hot fluid.

After a period of operation, it is found that heat exchangers cannot transfer as much heat as when they are first started and were clean. This is due to a buildup of scale, dirt, or oxide films and is known *as fouling*. This effect is taken into account in heat-exchanger design by introducing series resistances known as fouling factors in the design calculations. Table 11.6 gives typical values of fouling factors for various fluids.

---

**ILLUSTRATIVE PROBLEM 11.28**

Using Table 11.6 for the appropriate fouling factors, calculate the surface required for Illustrative Problem 11.27.

**SOLUTION**

For the oil side, a resistance (fouling factor) of 0.005 $h \cdot {}^{\circ}F \cdot ft.^2/Btu$ can be used, and for the water side, a fouling factor of 0.001 $h \cdot {}^{\circ}F ft.^2/Btu$ can be used. Because of the approximate nature of these resistances, we shall not correct them for inside or outside reference areas, and we will assume that they can be used directly with the value of $U$ of 40 $Btu/(h \cdot ft.^{2 \circ}F)$. The overall resistance and the overall heat-transfer coefficient are obtained as

$$\text{oil}, R = 0.005$$

$$\text{water}, R = 0.001$$

$$\frac{\text{clean unit, } R = 1 / 40 \, 0.025}{R_{\text{overall}} = 0.031}$$

$$U_{\text{overall}} = \frac{1}{0.031} = 32.3 \, \frac{\text{Btu}}{\text{h} \cdot \text{ft.}^2 \cdot {}^\circ\text{F}}$$

Because all other parameters are the same, the surface required will vary inversely as $U$. Therefore, $A = 569 \times 40/32.3 = 705$ ft.$^2$ or an increase in surface required of approximately 24% due to fouling. This obviously represents an important consideration in the design of industrial equipment.

In many heat exchangers, the flow paths of the fluids are not simply parallel or counterflow. In Figure 11.28c, we have an illustration of a shell-and-tube heat exchanger, which is said to have one shell pass and two tube passes. In addition, the shell-side fluid passes over segmented baffles in a sinuous path. For this type of a heat exchanger as well as others, correction factors have been developed to account for the deviation from parallel or counterflow. The correction factor is applied to the log mean temperature difference to obtain the true temperature difference. Thus,

true mean temperature difference $= F \times$ log mean temperature difference

(11.45)

$$\Delta T_{\text{true}} = F \times \Delta T_{\text{log mean}}$$

Figures 11.33 and 11.34 show the correction factor for two arrangements of flow paths. To obtain the true mean temperature difference for any of these arrangements, the log mean temperature difference *for counterflow* is multiplied by the appropriate correction factor.

Note that $t_1$ and $t_2$ refer to the fluid temperatures entering and leaving, respectively, on the tube side, and $T_1$ and $T_2$ refer to the fluid temperatures entering and leaving, respectively, on the shell side.

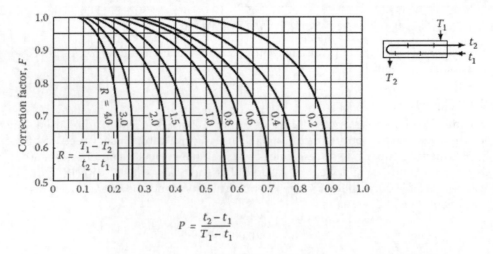

**FIGURE 11.33**
Correction factor for heat exchangers with 1 shell pass and 2, 4, or any multiple of 2 tube passes.

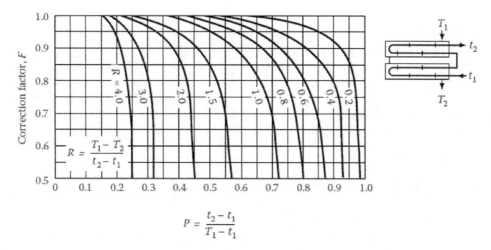

**FIGURE 11.34**
Correction factor for heat exchangers with 2 shell passes, and 4, 8, or any multiple of 4 tube passes.

---

### ILLUSTRATIVE PROBLEM 11.29

A heat exchanger has one shell pass and two tube passes similar to Figure 11.28c. Oil flows in the tube side and is cooled from 280°F to 140°F. On the shell side, water is heated from 85°F to 115°F. Determine the true mean temperature difference.

#### SOLUTION

In order to use Figure 11.33, which is applicable to this problem, we need to calculate $P$ and $R$. From the figure,

$$P = \frac{t_2 - t_1}{T_1 - t_1} = \frac{140 - 280}{85 - 280} = 0.72$$

$$R = \frac{T_1 - T_2}{t_2 - t_1} = \frac{85 - 115}{140 - 280} = 0.21$$

From Figure 11.33, $F = 0.91$. Thus,
true mean temperature difference $= 0.91 \times$ LMTD counterflow

$$\Delta T_{\text{true}} = 0.91 \frac{(280 - 115) - (140 - 85)}{\ln \dfrac{(280 - 115)}{(140 - 85)}}$$

$$\Delta T_{\text{true}} = 91°F$$

---

In selecting a heat exchanger for a particular task, one typically starts with the rate of the heat transfer needed. This depends on the fluid of interest, its specific heat, its flow rate, and the desired temperature change. The other fluid is chosen based on its properties and its compatibility with heat-exchanger materials. Its flow rate is chosen based on weighing effectiveness against pressure drop and the subsequent need for pumping power. The most common choice is the shell-and-tube exchanger with the more viscous fluid usually selected for the shell side, which offers a larger cross-sectional area to minimize the pressure drop. Other factors involved in the selection are costs, size and weight, and pumping power required. Note that the value of $U$ is usually not constant for a particular heat exchanger and so it is often overdesigned.

## 11.6 Combined Modes of Heat Transfer

Heat-transfer situations often exhibit combined heat-transfer effects of conduction, convection, and radiation. For example, an insulated steel pipe could be used to carry flowing hot water in a large room of still air. The path taken by the heat as it flows from the hot water through the pipe wall and into the room would be as follows:

1. Forced convection from the hot water to the inside wall of the pipe
2. Conduction through the pipe wall and its insulation
3. Natural convection from the outer surface of the insulation to the room air in parallel with radiation to the room

An equivalent resistance circuit is shown in Figure 11.35. An overall heat-transfer coefficient, $U$, would then be

$$U = \cfrac{1}{\cfrac{1}{h_i} + \cfrac{r_1}{k_1}\ln\left(\cfrac{r_2}{r_1}\right) + \cfrac{r_1}{k_2}\ln\left(\cfrac{r_3}{r_2}\right) + \cfrac{1}{h_0 + h_r}\left(\cfrac{r_3}{r_1}\right)} \tag{11.46}$$

Units of $U$ are typically in Btu/(h·ft.²·°F). Note that free convection and radiation occur simultaneously from the surface of the pipe's insulation. The total heat-transfer rate is thus the sum of the rates of convection and radiation.

**FIGURE 11.35**
Electrical analogy for combined heat transfer from a hot fluid flowing in a pipe to still air in a large room.

## 11.7 Cooling Electronic Equipment

The principles described in this chapter can be applied to a critical requirement of modern manufacturing: keeping electronic equipment sufficiently cool. Most electronic components have a maximum allowable operating temperature. They generate power in confined spaces, and it is important for designers to devise an adequate path for the removal of the heat developed by the dissipation of this power. Heating of parts also has the potential for enhancing corrosion and producing thermal stresses that can fracture bonds between different materials that expand at different rates with increased temperature. It is not the purpose of this text to provide an exhaustive review of the problem of cooling electronic systems; rather a brief summary of the approach will be given. For more detailed information on this application, the reader is advised to consult the references, in particular, that of Steinberg (1991) and Cengel (1998). The International Electronic Research Corporation, a manufacturer of heat sinks, also publishes a very useful "Thermal Management Guide" (International Electronic Research Corporation 1990).

Electronic equipment that generates heat during operation includes printed circuit boards (PCBs) with a variety of mounted components, power transistors, and transformers. Components are typically mounted and enclosed in a chassis, with cooling available by conduction through contact with solid heat sinks; by convection to the surrounding air, either free or forced by a small fan; and by radiation to the air and the surrounding walls. The first level of temperature control usually begins at the silicon wafer-based chip. The chip generates heat at its location, which is called the *junction*. For typical semiconductor chips, the maximum allowable junction temperature is 125°C, but lower temperatures are desirable for long-term reliability.

Manufacturers of encased chips provide a junction-to-case resistance, usually referred to as $\theta_{jc}$ in units of °C/W or °F/Btu per hour. This is where the heat transfer path begins and can include all three mechanisms of heat transfer. From the case, some heat can travel by conduction through the PCB on which the case is mounted. The PCB is often attached to a heat sink made of a highly conductive material with fins that increase the surface area. The fins transfer the heat to the surrounding air principally by convection with some radiation heat transfer in parallel. An electronics chassis often contains PCBs stacked vertically, each generating heat. Thus, the surroundings of these boards are of similar high temperatures, reducing the capability for heat removal within the chassis. Only the boards adjacent to the cooler chassis walls can lose a reasonable amount of heat by radiation. Thus, a good design mandates placing those boards with the highest power output nearest the chassis walls.

To improve heat transfer, forced convection is often employed by placing a fan in the chassis or attaching a cold plate heat exchanger to the circuit board rack. The latter is similar in principle to an automobile's radiator: two plates with multiple thin fins in the thin space between the plates are attached to the heat-generating components. Air or another fluid is passed through the fins to provide a low-resistance heat flow path from the chassis.

The additional thermal path(s) from the component case to the surrounding air is summed and identified as thermal resistance $\theta_{ca}$. Thus, the junction temperature can be determined from the temperature difference between the junction and the air when the air temperature is known. The equation is

$$\Delta T = \dot{Q}R$$

where $R = \theta_{jc} + \theta_{ca}$, and $\dot{Q}$ is the rate of heat produced, equal to the power generated. The junction temperature is thus the air temperature plus $\Delta T$.

## 11.8 Analysis of Fins

As noted in Section 11.7, cooling of equipment is facilitated by the addition of fins to the base surface. Usually the surface temperature is fixed as is the ambient temperature, so to improve heat transfer, fins provide additional surface area with a minimum of additional weight and cost. The radiator in an automobile is a good example of the usage of fins. Fins projecting from the base can be of varying geometries, but the most often used are rectangular in cross section. Such a fin, shown in Figure 11.36, will be analyzed.

The additional heat transfer provided by the fin is expressed as a fraction of the convective heat transfer that the fin's surface area would provide using the same temperature difference between the base surface and the ambient air. The fraction is given as *fin efficiency*, $\eta$, in the following equation:

$$Q_f = \eta h_c A_f \left( \Delta t \right) \tag{11.47}$$

where $Q_f$ is the added heat transfer rate, $h_c$ is the convective coefficient of the base's geometry (plate or cylinder), and $A_f$ is the fin area exposed to the air, equal to $2\,L(d + \delta/2)$. The $\delta/2$ term accounts for the convection at the tip of the fin. As fins are quite thin, $A_f$ is

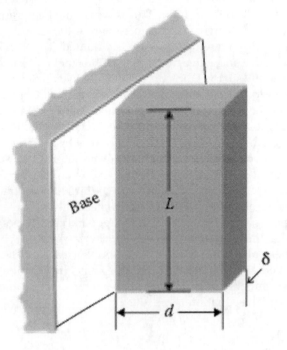

**FIGURE 11.36**
Fin with extended surface.

approximately $2Ld$, and the area of the base that is lost to the fin is negligible. The fin efficiency, between 0 and 1, is found from

$$\eta = \frac{\tanh(md)}{md} \tag{11.48}$$

where $m$ is defined as

$$m = \sqrt{\frac{Ph_c}{ka}}. \tag{11.48a}$$

with $P$ the perimeter of the fin's cross section equal to $2(L+\delta)$, $k$ the thermal conductivity of the fin material, and $a$ *the* cross-sectional area of the fin equal to $L\delta$. Tanh is the hyperbolic tangent, defined as

$$\tanh x = \frac{e^x - e^{-x}}{e^x + e^{-x}} \tag{11.48b}$$

with $e$ the base of the natural logarithm. Here, $x$ is $md$, a dimensionless variable.

The orientation of the fin is immaterial as this is accounted for in the determination of $h_c$. The effectiveness of the fin depends on its projection length $d$. While it may seem that the longer the fin, the more effective it will be, this is not so. The temperature distribution in the fin with length drops from the base temperature to the ambient temperature rather quickly. Thus, additional length does somewhat increase the heat transfer but not enough to compensate for the extra weight, space required, and cost. As a rule of thumb, the fin's length should be such that $md$ is approximately equal to 1.

The heat transfer rate can be increased by adding more fins once the proper size is selected. However, to prevent choking of the convective process, the fins should be spaced at least 0.6 in. or 1.5 cm apart.

For fins other than rectangular, the reader is referred to textbooks on heat transfer.

---

### ILLUSTRATIVE PROBLEM 11.30

To increase the heat-transfer rate from the box in Illustrative Problem 11.17, a vertical fin is added to each of the four sides (see Figure 11.37). The fins are aluminum, 15 cm high and 0.15 cm thick, and protrude 3 cm from the box. Find the new heat-loss rate.

### SOLUTION

The fins are vertical plates with the same characteristic length as the sides of the box, so the fin's heat-transfer coefficient is the same as in Problem 11.17: $h_c = 5.102$ W/m²°C.

Here $P = 0.303$ m, $k = 204.2$ W/m °C per Table 11.1, $d = 0.03$ m, $A_f = 0.0369$ m² (4 fins) and $a = 0.000225$ m². The value of $m$ in Equation 11.48a is thus

$$m = \sqrt{\frac{Ph_c}{ka}} \sqrt{\frac{0.303 \times 5.102}{204.2 \times 0.000225}} = 5.80 \text{ m}^{-1}$$

and $md = 0.174$. The tanh of $md = 0.172$ and $(\tanh md)/md = 0.99$. Thus, the fin heat-transfer rate is

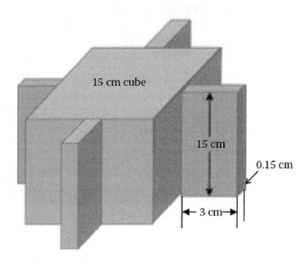

**FIGURE 11.37**
Illustrative Problem 11.30.

$$Q_f = \eta h_c A_f \left( \Delta t \right) = 0.99 \times 5.102 \times 0.0369 \times 25 = 4.66 \ W$$

This rate added to the rate for the box brings the total heat-transfer rate to 18.81 watts, which is an increase of 33% due to the usage of fins.

Note that when fins are added, the surface area of the box exposed to convection is diminished. In this example, the reduction is negligible.

## 11.9 Heat Pipes

The heat pipe is a device for enhanced thermal conduction. It can transfer large amounts of heat quickly, especially over large distances. The exterior of the pipe is sealed metal of good conductivity, such as aluminum or copper. It looks no different from a solid pipe but has an effective thermal conductivity that exceeds that of a solid pipe by many orders of magnitude. It is utilized by placing one end of the pipe in contact with the material to be cooled and the other end in a cold reservoir. It is not unusual for fins to be placed at the cold end to facilitate heat transfer.

The heat pipe is hollow and includes a working fluid and a concentric wick running the length of the pipe (see Figure 11.38). The operation of the heat pipe depends on the fluid being able to vaporize at the temperature of the material to be cooled. Thus, the fluid interior must be set at the saturation pressure that corresponds to that material's temperature. That pressure could be either above or below atmospheric. Often the fluid can be as common as water. When the pipe is placed in contact with the heat source, the fluid at that end evaporates, absorbing heat and increasing its vapor pressure. That pressure forces the vapor to the cold end of the pipe where it condenses, giving up heat to the cold reservoir. The condensed liquid returns by capillary action along the wick to the hot end where it

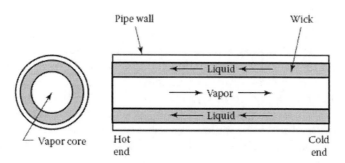

**FIGURE 11.38**
Heat pipe—section view.

re-vaporizes to repeat the heat-transfer process. The resistance to heat transfer is minimal in the boiling and condensing processes, so the net thermal resistance is only that through the thin pipe wall. To maximize the heat pipe's effectiveness, the hot end should be below the cold end so that gravity can help return the liquid to the hot end. If this is not possible, a fine wick will work better than a coarse one. For a horizontal pipe, the coarse wick usually works best. A simple heat pipe using water can have a thermal conductivity that exceeds that of copper by up to 1000 times.

The materials used in the heat pipe for the fluid, the wick, and the pipe should be chemically compatible. Otherwise, non-condensable gases may form that can degrade the heat pipe's performance. Heat pipes are usually cylindrical but can be formed in a variety of shapes and can include bends to match the shape of the material to be cooled. For example, a flat pipe can be attached to a flat electronic circuit board to keep the components cool.

## 11.10 Review

In this chapter, we have surveyed the three mechanisms of heat transfer: conduction, convection, and radiation. We studied conduction through a simple plane wall and a wall consisting of several resistances. Using the electrical analogy where Ohm's law is compared to Fourier's law, it was found that the treatment of resistances in series and parallel could be handled readily. The same reasoning was used when conduction in a hollow cylinder was considered. In convection heat transfer, there is motion of a fluid relative to a body. It therefore becomes necessary to consider the character of the flow of the fluid. That is, is the flow laminar or turbulent, and is the flow due to natural or forced convection? Because of the complexity of the subject, we could only treat several common cases in a rather simplified manner. Radiation heat transfer differs from the other modes of heat transfer in that a medium is not required to transfer the heat. Using the concept of a blackbody and the Stefan–Boltzmann equation, we were able to calculate the radiation heat transfer and to combine it with the other modes of heat transfer. We studied heat exchangers and the application of Newton's law of cooling to the calculation of their performance. When different modes of heat transfer are available, they can be combined into an overall heat transfer coefficient. Heat-transfer principles were also applied to the cooling of electronic equipment and to heat pipes. Finally, the usage of fins to enhance the rate of heat transfer was introduced.

The general subject of heat transfer is a vast field that encompasses many technical disciplines. In this chapter, we have been able to cover only briefly some of the basic concepts and topics of interest. In the past, empirical test data were used extensively, but in recent decades, great strides have been made in developing analytical and numerical procedures for the calculation of many heat-transfer problems. As examples from the space program alone, we have radiation to and from space vehicles, ablation, lightweight "super" insulation, heat transfer in high-speed flow, transpiration cooling, thermal contact resistances, heat transfer from plasmas, and compact heat exchangers. Many other examples could be cited from every branch of engineering and technology. The point to be made is that this is a truly fertile field with an almost infinite number of unsolved problems that represents great potential for original work. We hope that this chapter will serve as an introduction to the understanding and evaluation of industrial heat-transfer equipment as well as for more advanced work.

## Key Terms

Terms used for the first time in this chapter are as follows:

**Absorptivity ($\alpha$):** The fraction of the incident radiant heat that is absorbed by a body.

**Blackbody:** A body that absorbs all the radiation that strikes it.

**Coefficient of heat transfer ($h$):** The factor of proportionality in Newton's law of cooling, $\dot{Q} = hA\Delta t$.

**Conduction:** The transfer of heat from one part of a body to another part of the same body without appreciable displacement of the particles of the body. Also, it represents the flow of energy from energetic particles to less energetic particles.

**Convection:** The transfer of heat due primarily to the motion of a fluid.

**Dimensionless numbers:** The arrangement of physical parameters that form dimensionless groups. Among these are the Nusselt, Prandtl, Reynolds, and Grashof numbers.

**Emissivity:** The effectiveness of a body as a thermal radiator at a given temperature.

**Fin:** An extended surface that increases the rate of heat transfer from a body.

**Forced convection:** The motion of a fluid or gas that is produced primarily by mechanical means.

**Fourier's law:** The law of thermal conduction.

**Geometric factor ($F_A$):** A factor in the Stefan–Boltzmann equation to allow for the average solid angle through which one surface "sees" the other.

**Heat pipe:** A hollow metal pipe containing a fluid and a wick used to transfer heat faster than through pure conduction.

**Heat-transfer coefficient:** See *coefficient of heat transfer.*

**Junction:** The place where heat is maximized in an electronic component.

**Laminar flow:** A flow regime in which the fluid moves in layers or paths parallel to each other.

**Logarithmic mean temperature difference $[(\Delta t)_m]$:** The appropriate temperature difference to use for parallel flow, counterflow, and constant-temperature heat exchangers.

**Natural convection:** Motion caused by the differences in density due to the difference in temperature at various locations in the fluid.

**Ohm's law:** The basic electrical equation for direct current circuits. It is analogous to Fourier's law for heat conduction.

**Overall heat-transfer coefficient ($U$):** The proportionality term in Newton's law of cooling for heat exchangers, $\dot{Q} = UA(\Delta t)_m$.

**Radiation:** Heat that is transferred from one body to another without the need for a medium to effect the transfer; an electromagnetic phenomenon similar to the transmission of light.

**Reflectivity ($\rho$):** The fraction of the incident radiant heat that is reflected from a body, known as the reflectivity of the body.

**Stefan–Boltzmann law:** An equation governing the radiant interchange of heat between two bodies.

**Thermal conductivity ($k$):** The proportionality constant in Fourier's equation; also a physical property of a material. $\dot{Q} = -kA\Delta t/\Delta x$.

**Thermal resistance ($R_t$):** As a consequence of the analogy with Ohm's law, the thermal resistance can be written as $R_t = \Delta x/kA$.

**Transmissivity ($\tau$):** The fraction of incident radiant heat that is transmitted through the material.

**Turbulent flow:** The flow regime characterized by eddies that cause the mixing of the layers of fluid until no layers are distinguishable.

---

## Equations Developed in This Chapter

| | | |
|---|---|---|
| Fourier's equation | $\dot{Q} = \dfrac{-kA\Delta t}{\Delta x}$ | (11.3) |
| Ohm's law in heat transfer | $\dot{Q} = \dfrac{\Delta t}{R_t}$ | (11.3a) |
| Ohm's law | $i = \dfrac{\Delta E}{R_e}$ | (11.4) |
| Series thermal circuit overall resistance | $R_{oe} = R_1 + R_2 + R_3$ | (11.5) |
| Series thermal circuit overall resistance | $R_{ot} = R_1 + R_2 + R_3 = \dfrac{\Delta x_1}{k_1 A} + \dfrac{\Delta x_2}{k_2 A} + \dfrac{\Delta x_3}{k_3 A}$ | (11.6) |
| Parallel thermal circuit heat flow | $\dot{Q} = \dot{Q}_1 + \dot{Q}_2 + \dot{Q}_3$ | (11.8) |
| Parallel thermal circuit overall resistance | $R_{ot} = \dfrac{1}{\dfrac{1}{\Delta x/k_1 A_1} + \dfrac{1}{\Delta x/k_2 A_2} + \dfrac{1}{\Delta x/k_3 A_3}}$ | (11.10) |

Hollow cylinder

$$t_i - t_o = \frac{\dot{Q}}{2\pi kL}\ln\left(\frac{r_o}{r_i}\right)$$

(11.14a)

Hollow cylinder

$$\dot{Q} = \frac{(t_1 - t_o)2\pi kL}{\ln(r_o / r_i)}$$

(11.14b)

Hollow cylinder

$$R_t = \frac{\ln\left(\dfrac{r_o}{r_i}\right)}{2\pi kL}$$

(11.14c)

Compound cylinder

$$\dot{Q} = \frac{(t_1 - t_3)2\pi L}{(1/k_1)\ln(r_2 / r_1) + (1/k_2)\ln(r_3 / r_2)}$$

(11.15)

Newton's law of cooling

$$\dot{Q} = hA(\Delta t)$$

(11.16)

Convective thermal resistance

$$R_c = \frac{1}{hA}$$

(11.17)

Natural convection from gases—
   laminar flow

$$h = C_L\left(\frac{\Delta t}{L}\right)^{1/4}$$

(11.21a)

Natural convection from gases—
   turbulent flow

$$h = C_{T}{}_{,}(\Delta t)^{1/3}$$

(11.21b)

Natural convection—vertical plates
   (laminar)

$$h = 0.29\left(\frac{\Delta t}{L}\right)^{1/4}$$

(11.22)

Natural convection—vertical plates
   (turbulent)

$$h = 0.21(\Delta t)^{1/3}$$

(11.23)

Natural convection—horizontal
   pipes (laminar)

$$h = 0.25\left(\frac{\Delta t}{D}\right)^{1/4}$$

(11.24)

Natural convection—horizontal
   pipes (turbulent)

$$h = 0.18(\Delta t)^{1/3}$$

(11.25)

Natural convection—horizontal
   square plates: hot side up
   (laminar)

$$h = 0.27\left(\frac{\Delta t}{L}\right)^{1/4}$$

(11.26)

Natural convection—horizontal
   square plates: hot side up
   (turbulent)

$$h = 0.22(\Delta t)^{1/3}$$

(11.27)

Natural convection—horizontal
   square plates: hot side down

$$h = 0.12\left(\frac{\Delta t}{L}\right)^{1/4}$$

(11.28)

Forced convection inside tubes—
   turbulent flow

$$\frac{hD}{k} = 0.023\frac{C_p}{k}^{1/3}\frac{DV\rho}{}^{0.8}$$

(11.29b)

Stefan–Boltzmann equation

$$\dot{Q}_r = 0.173\times10^{-8}\,AT^4$$

(11.30)

Stefan–Boltzmann equation

$$\dot{Q}_r = \sigma F_e F_A\left(T_1^4 - T_2^4\right)$$

(11.31)

Radiation heat-transfer coefficient

$$h_r = \frac{\sigma F_e F_A\left(T_1^4 - T_2^4\right)}{T_1 - T_2}$$

(11.33)

Heat exchanger

$$\dot{Q} = UA(\Delta t)_m$$

(11.42)

Overall heat-transfer coefficient for
   a cylinder

$$U = \frac{1}{\dfrac{1}{h_i(r_1 / r_3)} + \dfrac{r_3}{k_1}\ln\left(\dfrac{r_2}{r_1}\right) + \dfrac{r_3}{k_2}\ln\left(\dfrac{r_3}{r_2}\right) + \dfrac{1}{h_o}}$$

Illustrative Problem 11.25,
Equation (e)

| Overall heat-transfer coefficient for a cylinder | $U = \dfrac{1}{\dfrac{1}{h_i} + \dfrac{r_1}{k_1}\ln\left(\dfrac{r_2}{r_1}\right) + \dfrac{r_1}{k_2}\ln\left(\dfrac{r_3}{r_2}\right) + \dfrac{1}{h_o\left(r_3/r_1\right)}}$ | Illustrative Problem 11.25, Equation (f) |
| --- | --- | --- |
| Logarithmic mean temperature difference | $(\Delta t)_m = \dfrac{\theta_A - \theta_B}{\ln\left(\theta_A/\theta_B\right)}$ | (11.43) |
| Heat-exchanger heat balance | $\dot{m}_c\left(c_p\right)_c\left(t_{c2} - t_{c1}\right) = \dot{m}_h\left(c_p\right)_h\left(t_{h1} - t_{h2}\right)$ | (11.44) |
| Fin efficiency | $\eta = \dfrac{\text{thnh}\left(md\right)}{md}$ | (11.48) |

## QUESTIONS

11.1  Explain why gases have a much lower thermal conductivity than liquids.

11.2  Compare the terms in Ohm's law and Fourier's law, and discuss their similarity.

11.3  Show how you can go from a series electrical circuit to a series thermal circuit.

11.4  Show how you can go from a parallel electrical circuit to a parallel thermal circuit.

11.5  In what way does heat transfer in a hollow cylinder differ from heat transfer in a plane wall?

11.6  What is the effect of thermal contact resistance in a series thermal circuit?

11.7  How does convection differ from conduction?

11.8  Discuss natural and forced convection. How do they differ?

11.9  What are the two flow regimes in convection heat transfer?

11.10  State Newton's law of cooling and define the thermal resistance for convective heat transfer.

11.11  How is it possible to simplify the calculation of natural convection heat transfer for gases?

11.12  Which situation did we consider when we discussed forced convection?

11.13  How does radiant heat transfer differ from conduction and convection heat transfer?

11.14  What is the temperature dependence in radiation heat transfer?

11.15  How is combined convection and radiation handled?

11.16  How does the emissivity of metals vary with the condition of the surface?

11.17  How is Newton's law of cooling written for a heat exchanger?

11.18  What is the appropriate mean temperature difference in a heat exchanger?

11.19  How is the true mean temperature difference determined in heat exchangers with differing flow paths?

11.20  Would you expect that the condensation or boiling of water in a heat exchanger would yield high or low overall heat-transfer coefficients?

11.21  Describe the effect of fouling on the performance of a heat exchanger.

11.22  How do fins increase the heat transfer rate of objects they are attached to?

## PROBLEMS

Unless indicated otherwise, use the data of Table 11.1 and Figure 11.4 for these problems.

### Problems Involving Conduction Heat Transfer

11.1   A common brick wall is 4 in. thick. Determine the rate of heat transfer through the wall if one face is at 95°F while the other is at 32°F.

11.2   Compare the insulating qualities of the wall in Problem 11.1 with that of a wooden white pine wall 2 in. thick.

11.3   Compare the insulating qualities of a conducting air gap 1/4 in. thick with the brick wall of Problem 11.1.

11.4   A furnace wall has a thermal conductivity of 0.5 Btu/(h·ft.°F). If the wall is 9 in. thick with the hot face at 2000°F and the cold face at 470°F, determine the heat loss.

11.5   A concrete wall having a thermal conductivity of 1.385 W/m°C is made 0.2 m thick. If the outer surface is kept at 0°C while the inner surface is kept at 21°C, determine the heat loss through the wall. Assume the wall is 12 m long and 2.5 m high.

11.6   A test is conducted to determine the thermal conductivity of a material. If the test specimen is 1 ft.×1 ft. and 6 in. thick and steady-state surface temperatures are found to be 100°F and 92°F, respectively, while 60 Btu/h is being applied, determine the thermal conductivity of the unknown material.

11.7   A test panel is made 3 ft.×3 ft.×1 in. thick. The panel is found to conduct 950 Btu in a test that lasts for 3 1/2 h. If the surfaces of the panel are kept at 140°F and 110°F, respectively, determine the thermal conductivity of the panel.

11.8   A wall has a thermal conductivity of 0.4 W/m°C. If the wall conducts 200 W/m² while one side is maintained at 100°C, determine the temperature of the colder surface. Assume the wall to be 2 cm thick.

11.9   A wall consists of a material that is 1 in. thick. If the surface area of the panel is 10 ft.² and the $k$ of the wall is 0.2 Btu/(h·ft.°F), determine the temperature of the "cold" side of the wall. The "hot" side of the wall is to be kept at 100°F, and there is a heat flow of 1000 Btu/h through it.

11.10  A brick wall is 100 mm thick. How much heat is transferred through the wall if one face is held at 40°C while the other is held at 0°C?

11.11  A test is conducted in which the test specimen is 1 m²×150 mm thick. Steady-state temperatures are found to be 35°C and 30°C, respectively. If 20 W is applied, determine the thermal conductivity of the material.

11.12  A furnace wall is 12 in. thick and is made of firebrick whose thermal conductivity can be taken to be 0.11 Btu/(h·ft.°F). If the heat loss is designed not to exceed 100 Btu/h·ft.², determine the outside wall temperature if the inside wall temperature is 1400°F.

11.13  A wall is designed to limit the heat transfer through it to 500 W/m². If it is of firebrick whose $k=0.22$ W/m·°C and it is 500 mm thick, what is the design temperature differential across the wall?

11.14  A corkboard wall has an area of 14 ft.² and allows a heat flow of 120 Btu/h. If the surface temperatures are 70°F and 32°F, respectively, determine the thickness of the wall.

11.15 A glass window is 3 ft.×4 ft. and 3/8 in. thick. The glass has a thermal conductivity of 0.5 Btu/(h·ft.°F). When the inner surface temperature of the glass is 20°C, the outer surface is 10°C. Determine the heat transfer due to heat conduction only.

11.16 A portion of a wall is tested to determine its heat-transfer characteristics. Assume that the wall is 6 in. thick and that 3500 Btu/h is transferred through the wall. If a temperature differential of 60°F is maintained across the wall, determine the thermal conductivity of the wall if the area is 80 ft.².

11.17 It is desired to limit the heat loss of a furnace wall to 500 Btu/h·ft.². The inner temperature of the wall is 2100°F, and the outer wall surface temperature is 470°F. If the material of the wall has a thermal conductivity of 0.5 Btu/(h·ft.°F), determine the necessary wall thickness.

11.18 A mild steel sheet is used to separate two gas streams that are at different temperatures. The steel sheet is 10 ft.×10 ft.×1/16 in. thick, and the temperatures of the faces of the sheet differ by 32°F. Calculate the heat flow through the wall.

11.19 A building has a wall composed of 3/4 in. of white pine wood, 4 in. of mineral wool insulation, and 1/2 in. of gypsum board. The value of $k$ for each material can be taken from Table 11.1. Calculate the overall heat-transfer coefficient for the wall.

11.20 If the temperature drop across the wall in Problem 11.19 is 40°F, determine the heat loss per square foot of the wall.

11.21 A composite series wall has an inside surface temperature of 125°F and an outside surface temperature of 52°F. If it consists of 3 in. of mineral wool, 1/2 in. of plaster on the outside, and 1/2 in. of corkboard on the inside, determine the heat loss per square foot of the wall.

11.22 Determine the interface temperatures in Problem 11.21.

11.23 A wall is 10 ft.×12 ft. In the wall, there is a window that is 4 ft.×4 ft. If the overall conductance of the wall is 0.28 Btu/(h·ft.²°F) and of the window is 1.13 Btu/(h·ft.²°F), determine the total heat transfer if the outside temperature is 0°F and the inside temperature is 70°F.

11.24 If a wall is 3 m×4 m and the outside temperature is –20°C while the room is maintained at 20°C, determine the heat transfer through the wall. The overall heat transfer coefficient of the wall is 0.07 W/m²·°C. Assume also that a 1 m×1 m window exists in the wall and that the window has an overall heat-transfer coefficient of 0.2 W/m²·°C.

11.25 A so-called "solid brick" construction consists of three courses of a common brick, each 4 in. thick. Comment on the insulating qualities of this wall when compared to a hollow wall filled with 3 in. of mineral wool.

11.26 A bar of steel 1 in. in diameter and 6 in. long is cut in half, and the cut ends are pressed together. If the thermal contact resistance can be thought of as an equivalent air gap of 0.001 in. thick ($k$=0.01 Btu/(h·ft.°F)), determine the percent increase in resistance to heat transfer from one end of the bar to the other due to the cutting of the bar, that is, [(cut resistance – uncut resistance)/uncut resistance]×100.

11.27 A wall is to be made of 4 in. of a common brick, 2 in. of mineral wool, and 1/2 in. of wallboard ($k$=0.04 Btu/(h·ft.°F) from Appendix 2). Determine the heat transfer from the wall if the inside wall temperature is at 74°F when the outside wall is at 0°F.

11.28    Determine the interface temperatures in Problem 11.27.

11.29    A wall consists of 100 mm of mineral wool, 12 mm of plaster, and 12 mm of cork-board. If a temperature difference of 30°C is maintained across the wall, what is the steady-state heat transfer?

11.30    What is the heat loss from a bare steel pipe ($k = 45$ W/m°C) whose inside temperature is 190°C and whose outer temperature is 100°C? The inside diameter of the pipe is 25 mm, and the outside diameter is 40 mm. Its length is 2 m.

11.31    Solve Problem 11.30 for a copper pipe ($k = 390$ W/m·°C). Compare your results.

*11.32   The equation for the heat transfer through a cylindrical wall is given by Equation 11.14b. By comparison with a plane wall, show that the mean conducting area for a cylindrical shell is $2\pi L(r_2 - r_1)/\ln(r_2/r_1)$. If $S_1$ and $S_2$ are the inner and outer surface areas, respectively, show that the mean conducting area is also $(S_2 - S_1) \ln(S_2/S_1)$.

*11.33   The heat transfer through a spherical shell of radii $r_2$ and $r_2$ is

$$Q = \frac{4\pi k r_1 r_2 \left(t_1 - t_2\right)}{r_1 - r_2}$$

Show that the mean conducting area for a spherical shell is $4\pi r_1 r_2$ or $\sqrt{S_1 S_2}$, where $S_1$ and $S_2$ are the inner and outer spherical surface areas, respectively.

11.34    A steel pipe having a 4 1/2 in. outside diameter is insulated with a molded pipe covering of diatomaceous earth ($k = 0.036$ Btu/(h·ft.·°F)) 1 1/2 in. thick. Thermocouples embedded in the insulation read 212°F on the inside and 130°F on the outside. Find the heat loss per foot length of the pipe.

11.35    Determine the heat loss from 50 ft. of 3 1/2 in. outside diameter pipe covered with 1 in. insulation ($k = 0.03$ Btu/(h·ft.·°F)) if the inner temperature of the insulation is 350°F and the outer temperature is 100°F.

11.36    A large steel pipe having an 8 in. outside diameter carries superheated steam. If the inside insulation temperature is 800°F and the pipe is insulated with 3 in. of molded pipe covering ($k = 0.036$ Btu/(h·ft.·°F)), determine the heat loss per foot of the pipe if the outside insulation temperature does not exceed 140°F.

11.37    If the steel pipe in Problem 11.36 has a wall thickness of 1/2 in. and $k_{steel} = 26$ Btu/(h·ft. ·°F), determine the heat loss per foot of the pipe. Assume that the inside of the pipe is at 800°F and the outside of the insulation remains at 140°F. Compare the results with those for Problem 11.36.

11.38    A steel boiler tube is 0.18 in. thick and has an outside diameter of 4 in. If the inside surface is at 500°F and there is a heat loss of 3800 Btu/h per foot length of the pipe, determine the outside tube temperature.

11.39    If the boiler tube in Problem 11.38 has a scale thickness of 0.1 in. built up on the inside of the tube, determine the outside tube temperature. Assume that the inside of the scale is at 500°F and that the scale has a thermal conductivity of 1.5 Btu/(h·ft.$^2$·°F).

11.40    A 3 in. thick brick wall separates 75°F air from the outside air at 25°F. What thickness of gypsum plaster overlay will keep the heat loss through the wall to 50 Btu/h·ft.$^2$?

11.41    A composite wall 8 ft. × 10 ft. is made up of 2 in. thick gypsum plaster and 1 in. thick fir. When a temperature difference of 80°F exists across the wall, what thickness of mineral wool insulation must be added to limit heat transfer through the wall to 800 Btu/h?

11.42 A 20 cm thick aluminum plate is to be insulated to keep the heat loss through it down to 100 W/m² when a $\Delta t$ of 40°C exists across it. What thickness of corkboard will accomplish this?

11.43 What is the heat-loss rate per meter of a 10 cm outside diameter×5 mm thick copper pipe covered with 1 cm of asbestos when a 50°C temperature difference exists across the insulated pipe wall?

*11.44 What thickness of mineral wool insulation is required to limit the heat-loss rate from a 6 in. outside diameter porcelain pipe, 1 in. thick, to 50 Btu/h per feet of pipe when an overall temperature gradient of 100°F exists?

*11.45 In Problem 11.44, if the outside of the insulation is at 40°F, what is the temperature at the interface of the pipe and the insulation?

**Problems Involving Convection Heat Transfer**

11.46 A bare horizontal pipe with 6 in. outside diameter has its outer surface at 140°F in a room whose air temperature is 65°F. Determine the natural convection heat-transfer coefficient on the outside of the pipe.

11.47 If instead of a pipe the surface in Problem 11.46 was a 1 ft. × 1 ft. square plate with the hot side facing up, determine the convection heat-transfer coefficient.

11.48 Determine the heat loss to the room by convection if the pipe in Problem 11.46 is 10 ft. long.

11.49 If the hot surface of the plate in Problem 11.47 faces down, determine the convection heat-transfer coefficient.

11.50 Atmospheric air flows in a 1 in. inside diameter tube at the rate of 0.2 $lb_m$/s. If the average air temperature is 150°F, determine the film coefficient on the inside of the tube.

11.51 Air flows in a 1.5 in. inside diameter tube at the rate of 0.4 $lb_m$/s and has an average temperature of 500°F. If the air is at atmospheric pressure, determine the heat-transfer coefficient on the inside of the tube.

11.52 Solve Problem 11.50 if the average air temperature is raised to 250°F. Compare your results.

11.53 Solve Problem 11.51 for a 2.0 in. inside diameter tube. Compare the results of these two problems.

11.54 If water flows in a 1 in. inside diameter tube at the rate of 1 $lb_m$/s, determine the film coefficient on the inside of the tube. The average temperature is 150°F.

11.55 Solve Problem 11.54 for a 2 in. inside diameter tube. Compare the results of these problems.

11.56 Water having an average temperature of 200°F flows in a 2 in. inside diameter pipe at the rate of 1.5 $lb_m$/s. Determine the heat-transfer coefficient on the inside of the pipe.

11.57 If the flow rate in Problem 11.56 is increased to 2.0 $lb_m$/s, determine the heat-transfer coefficient of the inside of the pipe.

11.58   A square, hollow tube is made of steel and has a thermal conductivity of 26 Btu/ (h·ft·°F). On one side, the film coefficient is 500 Btu/(h·ft.²°F), while on the other side, the film coefficient is 250 Btu/(h·ft.²°F). Estimate the overall heat-transfer coefficient if the tube is 1/2 in. thick.

11.59   A bare vertical pipe, 5 in. outside diameter has its outer surface at 120°F in a 70°F room. Find the convective heat loss per foot of the pipe.

11.60   A vertical plate, 3 ft. × 3 ft., is at 180°F on one side; the other side is insulated. Find the convective heat-loss rate to 65°F still air.

11.61   In Problem 11.51, find the heat-transfer rate per foot of the tube if the temperature of the inside of the tube is 350°F.

11.62   In Problem 11.56, find the heat-transfer rate if the pipe is 4 ft. long and its inside surface is at 150°F.

**Problems Involving Radiation Heat Transfer**

11.63   Calculate the rate of heat radiation from a blackbody per square foot of radiating area if the temperature of the body is 1000°F, 100°F, and 0°F.

11.64   Two opposed, parallel infinite black planes are at 350°F and 450°F, respectively. Determine their net heat interchange.

11.65   If the upper temperature in Problem 11.64 is made 550°F, by what percentage is the net heat transfer increased over that in Problem 11.64?

11.66   Assume that the planes in Problem 11.64 are gray with emissivities of 0.8 and 0.7, respectively. Determine the net heat interchange, and compare your answer to Problem 11.64.

11.67   Two perfectly black squares, 10 ft. × 10 ft., are spaced 20 ft. apart. What is the net rate of heat exchange between them if their temperatures are 500°F and 200°F, respectively? The planes are parallel and directly opposite each other.

11.68   Solve Problem 11.67 if the planes have emissivities of 0.8 and 0.9, respectively.

11.69   Two perfectly black rectangles, 6 ft. × 9 ft., are spaced 6 ft. apart. Their temperatures are 800°F and 200°F, respectively. If the planes are parallel and directly opposite each other, determine the net heat interchange.

11.70   Solve Problem 11.69 if both planes have emissivities of 0.8.

11.71   Solve Problem 11.69 if one plane has an emissivity of 0.85 and the other has an emissivity of 0.75.

11.72   A 3 1/2 in. outside diameter oxidized iron pipe whose outer temperature is 175°F passes through a room at 75°F. If $\varepsilon_{pipe} = 0.8$, determine $h_r$, the radiation heat-transfer coefficient.

11.73   A body whose emissivity is 0.9 is placed in a large room whose walls are at 20°C. If the body is at 250°C, determine the heat radiated to the walls per square meter of body surface. Assume that the body is small compared to the dimensions of the room.

11.74   Calculate the radiant heat interchange between two square plates having 1 ft. sides if they are separated by 1/2 ft. Assume that one square has an emissivity of 0.8 and that the other has an emissivity of 0.6. The hotter surface is at 1000°F, and the colder surface is at 100°F.

11.75 A rough metal sphere having an emissivity of 0.85 and a 12 in. diameter is hung by a nonconducting wire in a large room. Inside the sphere is an electrical heater that maintains the surface at 200°F. If the room is at 70°F, determine the wattage of the heater.

11.76 A light bulb filament may be considered to be a blackbody. If such a filament has a diameter of 0.12 mm and a length of 4 cm, determine the filament temperature. Consider radiation only, and ignore the temperature of the room or the glass envelope. The bulb is rated at 60 W.

*11.77 Consider two parallel, infinite planes having the same emissivity. The planes are at temperatures $T_1$ and $T_2$, respectively. A third plane having the same emissivity is placed between the original planes and allowed to come to its equilibrium temperature $T_3$. Show that the presence of the third plane reduces the heat transfer between the original two planes to half the value obtained when no plane (shield) is present.

11.78 What is the net radiation heat-transfer rate between the floor of a furnace at 1000°F and one of its walls at 500°F? The emissivity of each is 0.8. The floor is 6 ft. × 12 ft., and the wall is 6 ft. × 9 ft. with the 6 ft. length common to both.

11.79 Two 3 ft. diameter disks hang vertically 9 in. apart in a 75°F room. The emissivity of each is 0.7. One disk is at 200°F, and the other is at room temperature. Find the radiant heat-transfer rate between the two disks.

**Problems Involving Heat Exchangers**

11.80 A condenser operates with steam condensing at 1 psia (101.7°F). If the cooling water enters at 46°F and leaves at 90°F, determine the log mean temperature difference.

*11.81 The log mean temperature difference in a condenser operating at 1.5 psia (115.65°F) is 30°F. If the cooling water enters at 50°F, determine the exit temperature of the water.

*11.82 In an air heater operating as a counterflow heat exchanger, the air enters at 70°F and is heated to 200°F. The flue gas enters the heater at 620°F. The mass flow rate of the flue gas is 12% greater than the mass flow rate of the air. Assuming that the specific heats of the air and the 11.83 flue gas are equal, determine the log mean temperature difference in the unit.

11.84 Solve Problem 11.82 for a parallel-flow heat-transfer arrangement.

11.85 A counterflow heat exchanger operates with the hot fluid entering at 850°F and leaving at 650°F. The cold fluid enters at 150°F and leaves at 550°F. Calculate the log mean temperature difference.

11.86 Solve Problem 11.84 if the unit is operated as a parallel-flow heat exchanger.

11.87 Solve Problem 11.84 for the true temperature difference if the unit is operated with one shell pass and two tube passes with the hot fluid on the shell side.

11.88 A condenser has condensed steam on its shell side at 81.7°F. If the cooling water enters at 60°F and leaves at 70°F, determine the logarithmic mean temperature difference in the unit.

11.89 If the overall heat-transfer coefficient in Problem 11.87 is 1000 Btu/(h·ft.²°F) of the outside tube surface, determine the heat transfer per square foot of the exchanger outside tube surface.

11.90   A condenser in a power plant operates at 15°C. The cooling water enters at 5°C and leaves at 10°C. Determine the logarithmic mean temperature difference in the unit.

11.91   If the overall heat-transfer coefficient in Problem 11.89 is 250 W/m²°C of the outside tube surface, determine the heat transfer per square meter of the exchanger outside tube surface.

11.92   A counterflow heat exchanger cools oil from 175°F to 125°F. The cooling water enters at 65°F and leaves at 85°F. Calculate the logarithmic mean temperature difference in the unit.

11.93   If the heat exchanger in Problem 11.91 is operated as a parallel-flow unit, determine the logarithmic mean temperature difference.

11.94   A counterflow heat exchanger operates with oil cooled from 90°C to 50°C. The cooling water enters at 15°C and leaves at 25°C. If the overall heat-transfer coefficient based on the outside tube surface is 200 W/m²°C, determine the heat transfer per square meter of the outside tube surface.

11.95   If the heat exchanger in Problem 11.93 is operated as a parallel-flow unit, determine the heat transfer.

11.96   A condenser operates with steam condensing on the shell side at 27°C. Cooling water enters at 5°C and leaves at 10°C. If the overall heat-transfer coefficient is 5000 W/m²°C based on the outside tube surface, determine the heat transfer per square meter of the outside tube surface.

11.97   A nuclear steam generator boils water (makes steam) at 500°F on the shell side of a steam generator. Water enters the tubes at 570°F and leaves at 525°F. Determine the logarithmic mean temperature difference.

11.98   A tank-type heat exchanger is used to generate steam at 275°C. Water enters the tubes at 325°C and leaves at 300°C. Determine the logarithmic mean temperature difference.

11.99   A heat exchanger is used to transfer $1 \times 10^6$ Btu/h. At the inlet, the temperature difference between fluids is 75°F, and at the outlet, the temperature difference is 35°F. Determine the surface required if the overall heat-transfer coefficient is 175 Btu/(h·ft.²°F) based on the outside area of the tubes.

11.100  A heat exchanger is used to transfer 300 kW. At the inlet, the temperature difference between the fluids is 40°C, and at the outlet, the difference is 20°C. Determine the surface required if the overall heat-transfer coefficient is 1000 W/m²°C based on the outside area of the tubes.

11.101  How much heat is being exchanged in a heat exchanger having an overall heat-transfer coefficient of 60 Btu/(h·ft.²°F) based on the outside area of the tubes, 625 ft.² of outside tube area, and temperature differences between fluids at inlet of 62°F and at outlet of 36°F?

11.102  A heat exchanger has an overall heat-transfer coefficient of 550 W/m²°C based on the outside area of the tubes. If there is 400 m² of outside tube area and the temperature differences between fluids at the inlet is 40°C and at the outlet is 22°C, determine the total heat transfer in the unit.

*11.102 A shell-and-tube heat exchanger has a film coefficient on the inside of the tube of 100 Btu/(h·ft.²°F), a steel tube wall thickness of 0.105 in. ($k = 26$ Btu/(h·ft.°F)), a tube outside diameter of 3/4 in., and a film coefficient on the shell side (outside of the tube) of 50 Btu/(h·ft.$^2$·°F). Determine the overall film coefficient based on the inside area of the tube.

*11.103 Determine the overall film coefficient of the unit in Problem 11.102 based on the outside area of the tube.

11.104 A counterflow heat exchanger is used to cool 2000 lb$_m$/h of oil from 150°F to 100°F. Assume the $c_p$ of oil to be 0.5 Btu/lb.°F. Water ($c_p$=1 Btu/lb.°F) enters at 55°F and leaves at 75°F. $U$, based on the outside area of the tubes, is 30 Btu/(h·ft.$^2$·°F). Determine the area required.

*11.105 A shell-and-tube heat exchanger is used to condense steam on the shell side at 81.7°F. Cooling water enters the tubes at 60°F and leaves at 70°F. The tubes have a 1 in. outside diameter, a 0.902 in. inside diameter, and a conductivity of 63 Btu/(h·ft.°F). The film coefficient on the inside of the tubes is 1200 Btu/(h·ft.$^2$·°F) and on the outside is 950 Btu/(h·ft.$^2$·°F). If the unit extracts $766 \times 10^6$ Btu/h, determine the outside tube area.

11.106 In Problem 11.104, what should the cooling water flow rate be?

*11.107 A shell-and-tube heat exchanger is used to reject 400,000 kW of heat. Saturated steam enters at 40°C and leaves as 40°C saturated liquid. Cooling water (specific heat=4.19 kJ/kg K) enters at 10°C and leaves at 15°C. The tubes have a 25 mm O.D. and have an overall heat-transfer coefficient $U$=2000 W/m$^2$°C. Determine (a) the mass flow rate of cooling water required, (b) the log mean temperature difference, and (c) the length of heat-exchanger tubing required.

## Problems Involving Combined Modes of Heat Transfer

11.108 Solve Problem 11.18 assuming air films exist on both sides of the wall and that the temperatures given are the temperatures of the air on each side of the wall, respectively. Calculate the heat loss through the wall if both surface heat-transfer coefficients are equal to 1.4 Btu/(h·ft.$^2$·°F).

11.109 Solve Problem 11.15 assuming air film exists on both sides of the window and that the given temperatures are that of the air on each side, respectively. Determine the heat transfer through the window if the heat-transfer coefficients on both sides can be taken to be 1.35 Btu/(h·ft.$^2$·°F).

11.110 Solve Problem 11.21 if there are air films on both sides of the wall having heat-transfer coefficients of 1.2 Btu/(h·ft.$^2$·°F).

11.111 Determine the heat transfer in Problem 11.27 if there is an air film on each side of the wall, each having a heat-transfer coefficient of 2 Btu/(h·ft.$^2$·°F). Assume the temperatures to be the air temperatures on both sides of the wall.

11.112 If a film coefficient of 1 W/m$^2$°C exists on the inside of the wall in Problem 11.29 and the temperature differential is across the film plus the composite wall, determine the heat transfer.

11.113 A wall, 7 ft. ×7 ft., conducts 500 Btu/h. If there is a film coefficient on the inside of the wall of 1 Btu/(h·ft.$^2$·°F) and the wall has a resistance equal to 1 in. of mineral wool, determine the temperature drop across each resistance.

11.114 A composite wall consists of 1/2 in. plaster and 2 in. of fir. If the hot-side air is at 105°F, the cold-side air is at 10°F, the hot-side film coefficient is 2 Btu/(h·ft.$^2$·°F), and the cold-side film coefficient is 4 Btu/(h·ft.$^2$·°F), determine the heat transfer per square foot of the wall.

11.115   Determine the temperature drops through each resistance in Problem 11.114.

11.116   If the pipe in Problem 11.36 has an outside heat-transfer coefficient of 2.5 Btu/(h·ft.$^2$°F), and the room temperature is 140°F, determine the heat loss per foot of the pipe. Neglect the steel-wall resistance.

11.117   If in Problem 11.72 there is a convection heat-transfer coefficient of 1.5 Btu/(h·ft.$^2$°F) on the outside of the pipe, determine the combined heat-transfer coefficient and the heat loss per foot of the pipe.

11.118   In Problem 11.79, find the *total* heat-loss rate from the hotter disk if the room air is still. Consider the disk as a vertical plate with the diameter as the characteristic length.

11.119   Two 9 in.×9 in. plates hang vertically separated by a 3 in. air space. One plate is at 350°F with an emissivity of 0.8; the other is at 100°F with an e of 0.7. The air is still and at 200°F. Find the *total* heat-loss rate from the hotter plate.

*11.120  In a large kitchen, a small griddle of 1 ft.$^2$ is on a horizontal table and operates at 350°F. Its surface emissivity is 0.6. If the kitchen air is still at 70°F, how much heat does the griddle add to the room whose walls are at 75°F in its 2 h of operation?

11.121   A 4 in. outside diameter horizontal pipe ($\varepsilon = 0.8$) whose surface is at 150°F is in a large room of still air at 70°F. What percentage of its heat loss is by convection? By radiation?

11.122   A 0.03 in. nichrome wire runs horizontally through a large enclosure containing 80°F air. The temperature of the wire can reach 2250°F. Find the maximum total heat loss per foot from the wire if the emissivity of the wire is 0.3. (*Hint*: Consider the wire as a horizontal pipe.)

11.123   A 2 ft.×2 ft. plate, $\varepsilon = 0.7$, and 150°F on one side with the other side insulated, is placed in a large room at 60°F. Find its *total* heat-loss rate when it is (a) vertical; (b) horizontal, hot side up; and (c) horizontal, hot side down.

*11.124  A foot-long horizontal porcelain pipe, 5 in. outside diameter×0.5 in. thick, is in still air at 70°F. Find the overall heat-transfer coefficient, including conduction and convection, based on the outer surface area when the surface is at 200°F.

*11.125  In Problem 11.124, if the emissivity of the porcelain is 0.9, find the overall heat-transfer coefficient including radiation.

# Appendix 1: Answers to Even-Numbered Problems

## Chapter 1

| | |
|---|---|
| **1.2** | 9.78 cm |
| **1.4** | 2.894 tons |
| **1.6** | 2.95 ft. |
| **1.8** | 3900.9 g |
| **1.10** | 256.54 cm |
| **1.12** | 943.9 L/s |
| **1.14** | 0.2103 L/s |
| **1.16** | 0.00333 $m^3$/s |
| **1.18** | $5.07 \times 10^{-5}$ $m^3$/s |
| **1.20** | 102,600 L/min |
| **1.22** | 1579.9 gpm |
| **1.24** | 0.00946 $m^3$/s |
| **1.26** | 3650 L/s |
| **1.28** | 135.92 L/s |
| **1.30** | 0.0774 $m^2$ |
| **1.32** | 0.01416 $m^3$ |
| **1.34** | 0.003218 $m^3$ |
| **1.36** | 3.93 L |
| **1.38** | 29.05 $m^3$ |
| **1.40** | 3.123 $m^3$ |
| **1.42** | 0.932 mi. |
| **1.44** | 39.77 mi./h |
| **1.46** | 5339.7 N; $2.54 \times 10^{-5}$ m |
| **1.48** | 68°F, 104°F, 140°F |
| **1.50** | 4.44°C, 227°C, 87.8°C |
| **1.52** | 160°C, 320°F |
| **1.54** | −160°F, −106.7°C |
| **1.56** | −130.15°ARB |
| **1.58** | 490.5 N |
| **1.60** | (c) |
| **1.62** | 979.063 N |
| **1.64** | 10.198 kg |
| **1.66** | 61.19 kg |
| **1.68** | 0.39 $lb_f$ |
| **1.70** | 38.22 N |
| **1.72** | 10 m/$s^2$ |

| | |
|---|---|
| **1.74** | 2 kg |
| **1.76** | 2.4 N |
| **1.78** | 25 $lb_m$/ft.$^3$, 0.04 ft.$^3$/$lb_m$ |
| **1.80** | 149.6 $lb_f$/ft.$^3$, 149.6 $lb_m$/ft.$^3$ |
| **1.82** | 56.16 $lb_f$/ft.$^3$, 8829 N/m$^3$ |
| **1.84** | 2057.7 $lb_f$ |
| **1.86** | 7848 N/m$^3$, 800 kg/m$^3$, 0.00125 m$^3$/kg |
| **1.88** | 1.26 |
| **1.90** | 10.37 $lb_f$/ft.$^3$ |
| **1.92** | 8.14 N |
| **1.94** | 8.679 $lb_f$ |
| **1.96** | 435.6 N |
| **1.98** | 1.95 ft.$^3$ |
| **1.100** | 40.7 psia |
| **1.102** | 273.73 kPa |
| **1.104** | 0.87 psig |
| **1.106** | 250 kg/m$^3$, 0.25 |
| **1.108** | 39.5 psia, 7.0 ft. |
| **1.110** | 10.3 psia |
| **1.112** | 2576.4 psfa |
| **1.114** | 10.33 psf |
| **1.116** | 18.72 psia |
| **1.118** | 80,902 kPa, 3964 kN |

## Chapter 2

| | |
|---|---|
| **2.2** | 300 ft. $lb_f$, 600 ft. $lb_f$; 380 ft. $lb_f$, 600 ft. $lb_f$ |
| **2.4** | 294.3 J |
| **2.6** | 250 J |
| **2.8** | 38.86 ft. $lb_f$ |
| **2.10** | 9810 J, 6.26 m/s |
| **2.12** | 1000 ft. $lb_f$, 80.2 ft./s |
| **2.14** | 7.67 m/s |
| **2.16** | 288.86 ft. $lb_f$, 250 ft. $lb_f$, 38.86 ft. $lb_f$ |
| **2.18** | 4 kJ, 9 kJ |
| **2.20** | 2000 N/m, 62.5 J |
| **2.22** | 10.4 ft. $lb_f$ |
| **2.24** | 2500 N/m |
| **2.26** | 3.37 ft. |
| **2.28** | 4330 J |
| **2.30** | 234.38 kJ |
| **2.32** | Derivation |
| **2.34** | 13 625 kW |
| **2.36** | 50 J |
| **2.38** | 693.1 kJ/kg |

| | |
|---|---|
| **2.40** | $-1800$ ft. lb$_f$ (work in) |
| **2.42** | 14,400 ft. lb$_f$ |
| **2.44** | 161.25 kJ |
| **2.46** | 2818 ft. lb$_f$/lb$_m$ |
| **2.48** | 60.54 kJ |
| **2.50** | 1920 ft. lb$_f$ |
| **2.52** | 100 kJ |
| **2.54** | 19.31 kJ |

---

## Chapter 3

| | |
|---|---|
| **3.2** | at 1.0 MPa |
| | $h_g = 2778.1$ kJ/kg |
| | $s_g = 6.5865$ kJ/kg·K |
| | $v_g = 194.44 \times 10^{-3}$ m$^3$/kg |
| | $u_g = 2583.6$ kJ/kg |
| | at 1.1 MPa |
| | $h_g = 2781.7$ kJ/kg |
| | $s_g = 6.5536$ kJ/kg·K |
| | $v_g = 177.53 \times 10^{-3}$ m$^3$/kg |
| | $u_g = 2586.4$ kJ/kg |
| **3.4** | at 350°F |
| | $p = 134.53$ psia |
| | $v_f = 0.017988$ ft.$^3$/lb$_m$ |
| | $hf = 321.80$ Btu/lb$_m$ |
| | at 500°F |
| | $p = 680$ psia |
| | $v_f = 0.02043$ ft.$^3$/lb$_m$ |
| | $h_f = 487.7$ Btu/lb$_m$ |
| **3.6** | $h_{fg} = 826.8$ Btu/lb$_m$ |
| **3.8** | 897.44 Btu/lbm, 897.5 Btu/lbm |
| **3.10** | $h = 1143.35$ Btu/lb$_m$ |
| | $s = 1.54694$ Btu/lb$_m$·R |
| | $v = 4.213$ ft.3/lb$_m$ |
| | $u = 1065.41$ Btu/lb$_m$ |
| **3.12** | 82.3% |
| **3.14** | 2.9261 ft.$^3$/lb$_m$ |
| **3.16** | 819.71 Btu/lb$_m$ |
| **3.18** | 11,064 ft.$^3$/lb$_m$ |
| **3.20** | 0.396, 41.85 psia |
| **3.22** | 1.109 MPa |
| **3.24** | 1098.58 Btu/lb$_m$, 467.13°F (wet) |
| **3.26** | 2,004 ft.$^3$/lb$_m$ |
| **3.28** | 541.7 psia |
| **3.30** | 0.3066 m$^3$/kg |

**3.32**    1327.3 Btu/lb$_m$

**3.34**    1333.2 Btu/lb$_m$

**3.36**    66,179 ft.$^3$/lb$_m$, 1030.98 Btu/lb$_m$, 969.76 ft.$^3$/lb$_m$

**3.38**    $h = 487.5$ Btu/lbm

$v = 0.019766$ ft.$^3$/lb$_m$

$u = 472.9$ Btu/lb$_m$

$s = 0.6758$ Btu/lb$_m$·R

**3.40**    2.18 MPa

**3.42**    at 100 psia

$h_g = 1187.7$ Bm/lb$_m$

$s_g = 1.6034$ Btu/lb$_m$·R

$v_g = 4.4329$ ft.$^3$/lb$_m$

$u_g = 1105.7$ Btu/lb$_m$

at 1000 psia

$h_g = 1192.2$ Btu/lb$_m$

$s_g = 1.3902$ Btu/lbm·R

$v_g = 0.4459$ ft.$^3$/lb$_m$

$u_g = 1109.7$ Btu/lb$_m$

**3.44**    at 1.0 MPa

$h_f = 762.88$ kJ/kg

$s_f = 2.1388$ kJ/kg·K

$v_f = 0.001127$ m$^3$/kg

$u_f = 761.75$ kJ/kg

at 1.1 MPa

$h_f = 781.38$ kJ/kg

$s_f = 2.1793$ kJ/kg

$v_f = 0.001133$ m$^3$/kg

$u_f = 780.14$ kJ/kg

**3.46**    $u = 3046.2$ kJ/kg

**3.48**    $s_{fg} = 1.6529$ Btu/lb$_m$·R

**3.50**    $p = 5.6267$ kPa

$v_f = 0.001006$ m$^3$/kg

$s_f = 0.5050$ kJ/kg·K

**3.52**    $h = 2582.2$ kJ/kg

$v = 0.1571$ m$^3$/kg

$s = 6.1107$ kJ/kg·K

$u = 2406.2$ kJ/kg

**3.54**    0.3162

**3.56**    0.6723

**3.58**    0.07685 m$^3$/kg

**3.60**    $x = 0.0517$, t = 78.186°C

**3.62**    $x = 0.0909$, 0.019612 m$^3$, 0.00238 m$^3$, 2.3178 MPa

**3.64**    2.347 psia, 132.02°F, 0.01626 ft.$^3$/lb$_m$

**3.66**    212.42°C, 2422.5 kJ/kg

**3.68**    1563.7 Btu/lb$_m$

**3.70**    0.5166 ft.$^3$/lb$_m$

**3.72**    2,835 kJ/kg

**3.74**    2,811.7 kJ/kg

**3.76**    2,245.9 kJ/kg

| 3.78 | 2,620.1 kJ/kg |
| 3.80 | 487.50 Btu/lb$_m$ |
| 3.82 | 1,039.0°F, 1541.8 Btu/lb$_m$ |

---

## Chapter 4

| 4.2 | 215 kJ (into) |
| 4.4 | 70 Btu/lb$_m$ (increase) |
| 4.6 | 45 kJ/kg |
| 4.8 | −17.78 Btu/lb$_m$ |
| 4.10 | $c_p = 1$ kJ/kg·K, $\Delta V = 0.214$ m$^3$ |
| 4.12 | 120 kJ |
| 4.14 | −240 ft. lbf/lb$_m$ |
| 4.16 | −8.84 Btu/lb$_m$ |
| 4.18 | 171.5 Btu/lb$_m$ |
| 4.20 | 0.005 kJ/kg K |
| 4.22 | 1 kJ/kg·K |
| 4.24 | 75 kJ/kg |
| 4.26 | 149.4°C, 28.6 kJ/kg |
| 4.28 | 2 Btu/lb$_m$·R |
| 4.30 | 354.17 kJ/kg |
| 4.32 | −17.5 kJ (into), −24.5 kJ (decrease) |
| 4.36 | 6.47 kg |
| 4.38 | 475.8 kPa, 2,751.03 kg |
| 4.40 | 28.88 ft$^3$ |
| 4.42 | 1.21%, 247.1 psia |
| 4.44 | 36.56% |
| 4.46 | 197.6 kPa |
| 4.48 | 52.2% |
| 4.50 | 89.31 psia |
| 4.52 | 5,079.8 Btu |
| 4.54 | 412.87 psia |
| 4.56 | 2,246,400 lb$_m$/hr |
| 4.58 | $V \propto 1/D^2$, perform derivation |
| 4.60 | 254.6 m/s |
| 4.62 | 8.89 m/s |
| 4.64 | 0.927 N/s, 6.29 N/m$^3$ |
| 4.66 | 0.51 m/sec, 3.18 m/se |
| 4.68 | 5,868 ft/s |
| 4.70 | 50.3 hp |
| 4.72 | −15.26 Btu/lb$_m$; no, it is not flow work in a closed system |
| 4.74 | 1.48 Btu/lb$_m$ |
| 4.76 | 6,000 J/kg |
| 4.78 | 176.5 Btu/lb$_m$ |
| 4.80 | 808.4 kJ/kg |

| | |
|---|---|
| **4.82** | 50.46 Btu/lb$_m$ |
| **4.84** | 480 Btu/lb$_m$ |
| **4.86** | −12.55 kJ/kg |
| **4.88** | 73.1 Btu/lb$_m$ |
| **4.90** | 211.3 kJ/kg |
| **4.92** | −87.41 kW (in) |
| **4.94** | 2,619.6 ft/s |
| **4.96** | 4,151 ft/s |
| **4.98** | 715.8 m/s |
| **4.100** | 1,690.6 ft/s |
| **4.102** | 965 Btu/lb$_m$ |
| **4.104** | −16.4 Btu/lb$_m$ |
| **4.106** | −28.2 Btu/lb$_m$, in, work |
| **4.108** | 0.046 kJ/kg, 6,019 kW, 5.75 m |
| **4.110** | 312.07 °F, 92.4% |
| **4.112** | 891.31 Btu/lb$_m$ |
| **4.114** | 4.541 ft$^3$ |
| **4.116** | 74.7 Btu/lb$_m$ |

# Chapter 5

| | |
|---|---|
| **5.2** | 845.84 Btu/min, 33.4% |
| **5.4** | 1370.6 MW |
| **5.6** | requires an upper temperature of 5000 R! |
| **5.8** | 34.35% |
| **5.10** | 1 636 380 kJ/h; 40.73 L/h |
| **5.12** | 32.54% |
| **5.14** | 34.2%, 38.5%, 41.1%, lower $T_2$ |
| **5.16** | 127.27 hp, 47.27 hp, 62.86% |
| **5.18** | 115.17 Btu |
| **5.20** | 70.8%, $Q_r$=146 Btu/min, work=354 Btu/min |
| **5.22** | retest |
| **5.24** | Derivation |
| **5.26** | 30.77% |
| **5.28** | 3.37 h/day, $2.27/day |
| **5.30** | 51.5%, 39.9 Btu/min |
| **5.32** | 65.5%, 392.9 W |
| **5.34** | 0.57 hp |
| **5.36** | 7.88% |
| **5.38** | 68.86%, 217.8 kJ, 67.8 kJ |
| **5.40** | 42.1%, $T_1$=932.9 R |
| **5.42** | 1428.6 Btu/h |
| **5.44** | 1.429 MJ |
| **5.46** | 509.4 Btu/h |
| **5.48** | 0.82 hp |

| | |
|---|---|
| **5.50** | 463.8 K |
| **5.52** | 54.3 kW, 35.3 kW, 427.7 K |
| **5.54** | 85.2°F; 6.74 Btu/min·R |
| **5.56** | $\Delta s = 1.8528$ vs. 1.8526 Btu/lb$_m$·R |
| **5.58** | 0.312 Btu/lb$_m$·R |
| **5.60** | 0.1607 Btu/lb$_m$·R −0.1607 Btu/lb$_m$·R |
| **5.62** | 0.4241 kJ/kg·K |
| **5.64** | 0.2267 Btu/lb$_m$·R |
| **5.66** | 150 Btu/lb$_m$ |
| **5.68** | 0.1552 Btu/lb$_m$·R |
| **5.70** | 1.2 MJ, 4 kJ/K |
| **5.72** | 0.6390 kJ/kg·K |
| **5.74** | 530 R |
| **5.76** | 1105 R |
| **5.78** | 0 |
| **5.80** | 58.6 kJ/kg, 63.6 kJ/kg |
| **5.82** | total $\Delta s = 0.026$ Btu/lb$_m$·R |
| **5.84** | 247.1 psia, 10.72 lb$_m$ |
| **5.86** | 557,600 Btu |
| **5.88** | 1019.73 Btu/lb$_m$ |
| **5.90** | 316.9 Btu/lb$_m$ |
| **5.92** | 97.9% |
| **5.94** | 129.6 Btu/lb$_m$ |
| **5.96** | 1,058 Btu/lb$_m$ |
| **5.98** | 1,179 Btu/$_{lbm}$ |
| **5.100** | 10.3 lb$_m$/hr |
| **5.102** | 383.35 Btu/lb$_m$ |
| **5.104** | 256.5 Btu/lb$_m$ |
| **5.106** | 76.6 Btu/lb$_m$ |
| **5.108** | 4,497 ft/s |
| **5.110** | 1311.5 m/s |
| **5.112** | 1354.3 m/s |
| **5.114** | 247.05 psia, 10.72 lb$_m$ |
| **5.116** | 557,657 Btu |
| **5.118** | 1019.62 Btu/lb$_m$ |
| **5.120** | 317.0 Btu/lb$_m$ |
| **5.122** | 97.95% |
| **5.124** | 129.6 Btu/lb$_m$ |
| **5.126** | 1,058.0 Btu/lb$_m$ |
| **5.128** | 1,179 Btu/lb$_m$ |
| **5.130** | 10.3 lbm/hr |
| **5.132** | 383.4 Btu/lb$_m$ |
| **5.134** | 256.6 Btu/lb$_m$ |
| **5.136** | 76.6 Btu/lb$_m$ |
| **5.138** | 4496.4 ft/s |
| **5.140** | 1306.4 m/s |
| **5.142** | 1354.1 m/s |

## Chapter 6

| | |
|---|---|
| **6.2** | 163.8 K |
| **6.4** | 0.4 $lb_m/ft.^3$ |
| **6.6** | 13.29 $ft.^3/lb_m$ |
| **6.8** | 0.813 $lb_m/ft.^3$ |
| **6.10** | 425 psia |
| **6.12** | 1.283 kJ/kg·K, 6.48 |
| **6.14** | 0.171 $m^3$/kg |
| **6.16** | 1.06 kJ/kg·K |
| **6.18** | 70.94 kg |
| **6.20** | 0.706 $ft.^3/lb_m$ |
| **6.22** | 0.6062 $ft.^3$ |
| **6.24** | 337.9 R |
| **6.26** | 232.7 kPa |
| **6.28** | 40.93 $ft.^3$ |
| **6.30** | $c_v = 0.4725$ kJ/kg·K, $c_p = 0.6615$ kJ/kg·K |
| **6.32** | $c_v = 0.9556$ kJ/kg·K, $c_p = 1.2423$ kJ/kg·K |
| **6.34** | $R = 0.316$ kJ/kg·K, MW $= 26.31$ |
| **6.36** | MW $= 29.21$ |
| **6.38** | MW $= 27.97$, $k = 1.4$ |
| **6.40** | MW $= 28.57$, $R = 0.291$ kJ/kg·K, $c_v = 0.9091$ kJ/kg·K |
| **6.42** | 11.5 $ft.^3/lb_m$ |
| **6.44** | $-193.4$ kJ (in) |
| **6.46** | $-0.6601$ kJ/K |
| **6.48** | $-71744$ ft. $lb_f$ |
| **6.50** | 218.2 kJ |
| **6.52** | $Q = 16.9$ Btu, 0.02777 Btu/R |
| **6.54** | $v = 0.251$ $ft.^3/lb_m$, $\Delta s = -0.16928$ Btu/$lb_m$·R, $Q = -94.8$ Btu |
| **6.56** | 80 psia, 105 557 ft. $lb_f$, $\Delta u = 0$, $Q = 135.7$ Btu |
| **6.58** | 24.42 $ft.^3$, 15.0 $lb_m$ |
| **6.60** | 524.2 kJ, 0, 524.2 kJ, 1.1082 kJ/kg·K |
| **6.62** | 61.70 $lb_m$ |
| **6.64** | 33.33 psia |
| **6.66** | 0.00852 Btu/R, 1.47 |
| **6.68** | 0.11853 Btu/$lb_m$·R |
| **6.70** | 85.5 Btu/$lb_m$ |
| **6.72** | 0.17446 Btu/$lb_m$·R |
| **6.74** | 0.01760 kJ/K |
| **6.76** | 68.8 psia, 611.3°F, 0.1283 Btu/R |
| **6.78** | 641.6°F |
| **6.80** | 792.4 K, 676.8 kJ |
| **6.82** | 4480 R, 208,696 ft. $lb_f/lb_m$ |
| **6.84** | 28.8 Btu/$lb_m$ |
| **6.86** | 0.2405 Btu/$lb_m$·R |
| **6.88** | 0.2472 Btu/$lb_m$·R |
| **6.90** | 0.30525 Btu/$lb_m$·R |
| **6.92** | $+0.3529$ $m^3$, 4.0 kJ/kg·K |

| | |
|---|---|
| **6.94** | 5.83 ft.$^3$ |
| **6.96** | $-96$ Btu/lb$_m$ |
| **6.98** | work = 91.1 Btu/lb$_m$, $\Delta u = 225.7$ Btu/lb$_m$, $q = 316.0$ Btu/lb$_m$ |
| **6.100** | $q = 64.8$ Btu/lb$_m$, $\Delta s = 0.088$ Btu/lb$_m$·R |
| **6.102** | work = 5517.9 ft. lb$_f$, $V_2 = 5.76$ ft.$^3$ |
| **6.104** | 331.4 psia |
| **6.106** | 1948.5 R |
| **6.108** | 425.3 R |
| **6.110** | 395.2 R |
| **6.112** | 395.2 R |
| **6.114** | 1.37 |
| **6.116** | 1.29 |
| **6.118** | 342.4 Btu/lb$_m$ |
| **6.120** | 1.24 |
| **6.122** | 1.30 |
| **6.124** | $-85.4$ Btu/lb$_m$(in) |
| **6.126** | $-133.6$ kJ/kg (in) |
| **6.128** | 858.8 R, 104.4 psia |
| **6.130** | 0.2516 Btu/lb$_m$·R |
| **6.132** | 0.226 Btu/lb$_m$·R |
| **6.134** | 0.242 ft.$^3$/lb (ideal gas), 0.244 ft.$^3$/lb$_m$ (real) |
| **6.136** | 10 194 lb$_m$ |
| **6.138** | 7.23 lb$_m$/ft.$^3$ |
| **6.140** | 13.3°F |
| **6.142** | 320.8°F, 27.5 psia |
| **6.144** | 960.2 R |

# Chapter 7

| | |
|---|---|
| **7.2** | moles CH$_4$ = 0.06238 |
| | moles O$_2$ = 0.03125 |
| | total moles = 0.09363 |
| | $x_{CH_4} = 0.666$ |
| | $x_{O_2} = 0.334$ |
| | MW = 21.36 |
| **7.4** | CO$_2$ = 40.4% |
| | N$_2$ = 56.6% |
| | He = 2.9% |
| **7.6** | C$_2$H$_2$ = 34.53 MPa |
| | C$_4$H$_{10}$ = 15.47 MPa |
| **7.8** | MW = 29.38 |
| | $R = 52.59$ ft. lb$_f$/lb$_m$·R |
| **7.10** | $R = 210.72$ ft. lb$_f$/lb$_m$·R |
| | CO$_2$ = 8.33% |
| | He = 91.67% |
| **7.12** | 62.6 m$^3$ |

**7.14**   MW = 37.68
**7.16**   MW = 34.62 ft. $lb_f/lb_m \cdot R$
**7.18**   $CH_4 = 93.5\%$
           $CO_2 = 2.98\%$
           $N_2 = 1.90\%$
           $H_2 = 1.60\%$
           MW = 14.75
           $R = 104.75$ ft. $lb_f/lb_m \cdot R$
**7.20**   volume = 69.5 ft.$^3$
           $CH_4 = 53.5$ ft.$^3$
           $O_2 = 16.2$ ft.$^3$
           $p_{CH_4} = 57.6$ psia
           $P_{O_2} = 17.4$ psia
           $x_{CH_4} = 0.768$
           $x_{O_2} = 0.232$
           MW = 19.73
**7.22**   $p_{CO} = 44.98$ kPa
           $p_{He} = 629.80$ kPa
           $p_{total} = 674.77$ kPa
**7.24**   MW = 38.0
           $R = 40.66$ ft. $lb_f/lb_m \cdot R$
**7.26**   MW = 30.4
           $R = 0.2735$ kJ/kg·K
**7.28**   MW = 10.69
           $R = 0.778$ kJ/kg·K
           $p = 200$ kPa
**7.30**   CO = 43.1%
           $CH_4 = 12.3\%$
           $N_2 = 32.3\%$
           $O_2 = 12.3\%$
           MW = 26.02
           $R = 59.4$ ft. $lb_f/lb_m \cdot R$
**7.32**   $p_{N_2} = 40$ psia
           $p_{O_2} = 40$ psia
           $p_{CO_2}$ 20 psia
           $N_2 = 0.342$
           $O_2 = 0.390$
           $CO_2 = 0.268$
           MW = 32.81
           $R = 47.1$ ft. $lb_f/lb_m \cdot R$
**7.34**   $t = 90°F$
           $p = 167\ 4$ psia
**7.36**   $t = 42.1°C$
           $p = 1.053$ MPa
**7.38**   $t = 89.9°F$
           $p = 86.5$ psia
**7.40**   $p = 83.68$ psia
           $t = 79.05°F$

| 7.42 | 0.3801 psia, 71°F |
|------|-------------------|
| 7.44 | 51.4%, 51.5°F |
| 7.46 | 69°F, 0.3494 psia, 105 grains/$lb_m$ dry air |
| 7.48 | 59.8% |
| 7.50 | 45.9%, 57°F, 0.233 psia |
| 7.52 | 42.1%, 73°F |
| 7.54 | 30%, 26.5 Btu/$lb_m$ dry air, 46 grains/$lb_m$ dry air |
| 7.56 | 30%, 64°F |
| 7.58 | 0.0526 $lb_m$ |
| 7.60 | 46°F, 45.5 grains/$lb_m$ dry air |
| 7.62 | 4.9 Btu/$lb_m$ dry air |
| 7.64 | 6.69 Btu/$lb_m$ dry air |
| 7.66 | 62°F, 54°F, 59°F |
| 7.68 | 35.8% |
| 7.70 | 89 grains/$lb_m$ dry air, 65°F |
| 7.72 | 81%, 21 grains/$lb_m$ dry air, 6.0 Btu/$lb_m$ dry air |
| 7.74 | 20% |
| 7.76 | 63°F, 55%, 48 grains/$lb_m$ dry air |
| 7.78 | 0.9642 $lb_m$/$lb_m$ dry air, 0.0358 $lb_m$/$lb_m$ dry air |
| 7.80 | 8.19 $lb_m$/s, 359.81 $lb_m$/s |

# Chapter 8

| 8.2 | 182.5 Btu, 68.67% |
|------|-------------------|
| 8.4 | 466.7 K |
| 8.6 | Derivation |
| 8.8 | 1326 R |
| 8.10 | 200 kJ/h, 80%, 800 kJ/h |
| 8.12 | $h = 1032$ Btu/$lb_m$, $p = 5.2$ psia |
| 8.14 | $h = 1031.3$ Btu/$lb_m$, $p = 5.167$ psia |
| 8.16 | 90% |
| 8.18 | 578.26 Btu/$lb_m$, 1439.175 Btu/$lb_m$, 40.2% |
| 8.20 | 34.9% |
| 8.22 | 1.79 Btu/$lb_m$ |
| 8.24 | 1.78 Btu/$lb_m$ |
| 8.26 | 32.9% |
| 8.28 | 38.54%, 488.9 Btu/$lb_m$ |
| 8.30 | 32.91% |
| 8.32 | 32.85% |
| 8.34 | 35.4% |
| 8.36 | 35.65% |
| 8.38 | 32.8% |
| 8.40 | 36.1% |
| 8.42 | 36.86% |
| 8.44 | 86% |

| 8.46 | 78.50% |
|---|---|
| 8.48 | 73% |
| 8.50 | 59.7% |
| 8.52 | 12,365 Btu/kWh |
| 8.54 | 10.9 $lb_m$/kWh |
| 8.56 | 1170°F |
| 8.58 | 11.7% |
| 8.60 | 43.65% |
| 8.62 | 7819 Btu/kWh |
| 8.64 | 2% |
| 8.66 | 82.9% |
| 8.68 | 41.47% |
| 8.70 | 42.26% |
| 8.72 | 43.80% |
| 8.74 | 35.7% |
| 8.76 | 35.5% |
| 8.78 | 45.67% |
| 8.80 | 33.7% |
| 8.82 | 37.3% |
| 8.84 | 37.26% |
| 8.86 | 37.9% |
| 8.88 | 46.6% |
| 8.90 | 41.8% |
| 8.92 | 41.8% |
| 8.94 | 42.06% |
| 8.96 | 43.9%, a gain of 0.7% |
| 8.98 | 47.73% |
| 8.100 | Efficiency 41.9%, 23.1% |

# Chapter 9

| 9.2 | 54.6% |
|---|---|
| 9.4 | 11, 61.68% |
| 9.6 | 292.4 kJ/kg, 207.6 kJ/kg, 58.48% |
| 9.8 | 117 Btu/$lb_m$, 133 Btu/$lb_m$, 6.68 |
| 9.10 | 47.47% |
| 9.12 | 187.6 kPa |
| 9.14 | 2475.2 R, 56.47% |
| 9.16 | 579 kPa, 54.08% |
| 9.18 | 54.08%, 209 psia |
| 9.20 | 54.08%, 173.9 psia |
| 9.22 | 65.5 hp |
| 9.24 | 69.7% |
| 9.26 | 65.15% |

| | |
|---|---|
| **9.28** | Ratio% |
| | 0.01    68.7% |
| | 0.04    65.7% |
| | 0.08    62.4% |
| | 0.10    61.3% |
| **9.30** | 66.1%, 86.29 psia |
| **9.32** | 23.07, 331.1 Btu/lb$_m$, 66.22% |
| **9.34** | Derivation |
| **9.36** | 1325.78 psia, 3839 R |
| **9.38** | 36.86% |
| **9.40** | 359.04 Btu/lb$_m$, 1295 R, 48.21% |
| **9.42** | 44.8% |
| **9.44** | 880 kJ/kg, 424.2 kJ/kg, 48.21% |
| **9.46** | 893.17 K |
| **9.48** | 47.47% |
| **9.50** | 79.7% |
| **9.52** | 63.2% |
| **9.54** | 61%, 166.1 Btu/lb$_m$ |
| **9.56** | 124.8 Btu/lb$_m$ |

# Chapter 10

| | |
|---|---|
| **10.2** | 6.58, 3.8 kW, 28.8 kW |
| **10.4** | Derivation |
| **10.6** | 9.11 |
| **10.8** | 23.9 tons, 1.238 |
| **10.10** | 8.02 hp |
| **10.12** | 2 |
| **10.14** | 25.13 hp |
| **10.16** | 5.33 kW, 25.33 kW |
| **10.18** | 5.74 |
| **10.20** | 5.68 |
| **10.22** | 6.3 lb$_m$/min |
| **10.24** | 43.26 lb$_m$/min |
| **10.26** | 45.1 Btu/lb$_m$ |
| **10.28** | 46.26 lb$_m$/min |
| **10.30** | 47.91 Btu/lb$_m$, COP 3.73 |
| **10.32** | 56.6 Btu/lb$_m$ |
| **10.34** | 36.1 lb$_m$/min |
| **10.36** | 60.7 Btu/lb$_m$, COP 3.63 |
| **10.38** | 1.075, 342 R |
| **10.40** | 2.06, 4.4 hp |
| **10.42** | 9.58 lb$_m$/min/ton |
| **10.44** | 16.5 hp |
| **10.46** | 0.0282 lbm/lbm |

| | |
|---|---|
| **10.48** | 0.02808 $lb_m/lb_m$, 0.972 $lb_m/lb_m$, 29.23 Btu/$lb_m$ |
| **10.50** | 0.020286 kg/kg, 0.979714 kg/kg, 49.34 kJ/kg |
| **10.52** | 86.5% |
| **10.54** | 86.1%, 43.36 hp, 0.30 lb/s |
| **10.56** | 848.3 Btu/min, $T_1/T_2 = 1.25$ |
| **10.58** | 3.36 |
| **10.60** | 10.62, 20.8 Btu/min |
| **10.62** | 4.05 hp |
| **10.64** | 16.63, 1505 Btu/min |
| **10.66** | 28.81 kW, 17.36 |
| **10.68** | 329.1 kJ, 545.9 kJ |
| **10.70** | 30189 Btu/h |

## Chapter 11

| | |
|---|---|
| **11.2** | pine wall = 3.08 × brick wall |
| **11.4** | $\dot{Q} = 1020$ Btu/h |
| **11.6** | 3.75 Btu/h·ft·°F |
| **11.8** | 90°C |
| **11.10** | 276.8 W/m² |
| **11.12** | 491°F |
| **11.14** | 1.33 in. |
| **11.16** | 0.365 Btu/h·ft·°F |
| **11.18** | $15.98 \times 10^6$ Btu/h |
| **11.20** | 2.87 Btu/h·ft² |
| **11.22** | 114.3°F at cork/mineral wool; 52.9°F at mineral wool/plaster |
| **11.24** | 38.8 W |
| **11.26** | 43.3% |
| **11.28** | 65°F at wallboard/mineral wool; 9.6°F at mineral wool/brick |
| **11.30** | 108.3 kW |
| **11.32** | Derivation |
| **11.34** | 36.3 Btu/h·ft |
| **11.36** | 267 Btu/h·ft |
| **11.38** | 497.8°F |
| **11.40** | 1.35 in. |
| **11.42** | 17.2 mm |
| **11.44** | 1.11 in. |
| **11.46** | 0.9 Btu/h·ft²·°F |
| **11.48** | 1060 Btu/h |
| **11.50** | 85 Btu/h·ft²·°F |
| **11.52** | 86 Btu/h·ft²·°F |
| **11.54** | 900 Btu/h·ft²·°F |
| **11.56** | 435 Btu/h·ft²·°F |
| **11.58** | 147 Btu/h·ft²·°F |
| **11.60** | 1056.96 Btu/h |

| | |
|---|---|
| **11.62** | 45,553 Btu/h |
| **11.64** | 441.6 Btu/h·ft.$^2$ |
| **11.66** | 263.1 Btu/h·ft.$^2$ |
| **11.68** | 58.7 Btu/h·ft.$^2$ |
| **11.70** | 675 Btu/h·ft.$^2$ |
| **11.72** | 1.1 Btu/h·ft.$^2$·°F |
| **11.74** | 1513 Btu/h·ft.$^2$ |
| **11.76** | 2894.3 K |
| **11.78** | 41,229 Btu/h |
| **11.80** | 28.2°F |
| **11.82** | 427°F |
| **11.84** | 391.5°F |
| **11.86** | 344.5°F |
| **11.88** | 16,200 Btu/h·ft.$^2$ |
| **11.90** | 1800 W/m$^2$ |
| **11.92** | 69.2°F |
| **11.94** | 9.1 kW/m$^2$ |
| **11.96** | 43.7°F |
| **11.98** | 108.8 ft.$^2$ |
| **11.100** | 1.794×10$^6$ Btu/h |
| **11.102** | 40.5 Btu/h·ft. $^2$·°F |
| **11.104** | 28.4 ft.$^2$ |
| **11.106** | 2500 lb/h |
| **11.108** | 2241 Btu/h |
| **11.110** | 5.6 Btu/h·ft.$^2$·°F |
| **11.112** | 8.51 W/m$^2$ |
| **11.114** | 34.7 Btu/h·ft.$^2$ |
| **11.116** | 255.5 Btu/h·ft. |
| **11.118** | 1568.9 Btu/h |
| **11.120** | 1510.43 Btu |
| **11.122** | 349.54 Btu/h·ft. |
| **11.124** | 1.308 Btu/h·ft.$^2$·°F |

# Appendix 2: Supplemental Tables

| Contents of Supplemental Tables | | Page |
|---|---|---|

**TABLE A.1**

Saturation: Temperature (Steam)

| Temp. Fahr. $t$ | Press. $lb_f$ Sq. In. $p$ | Specific Volume (ft³/$lb_m$) Sat. Liquid $v_f$ | Sat. Vapor $v_g$ | Internal Energy (Btu/$lb_m$) Sat. Liquid $u_f$ | Evap. $u_{fg}$ | Sat. Vapor $u_g$ | Enthalpy (Btu/$lb_m$) Sat. Liquid $h_f$ | Evap. $h_{fg}$ | Sat. Vapor $h_g$ | Entropy (Btu/$lb_m$·R) Sat. Liquid $s_f$ | Evap. $s_{fg}$ | Sat. Vapor $s_g$ |
|---|---|---|---|---|---|---|---|---|---|---|---|---|
| 32 | .08859 | .016022 | 3305 | .01 | 1021.2 | 1021.2 | .01 | 1075.4 | 1075.4 | .00003 | 2.1870 | 2.1870 |
| 32.018 | .08866 | .016022 | 3302 | .00 | 1021.2 | 1021.2 | .01 | 1075.4 | 1075.4 | .00000 | 2.1869 | 2.1869 |
| 35 | .09992 | .016021 | 2948 | 2.99 | 1019.2 | 1022.2 | 3.00 | 1073.7 | 1076.7 | .00607 | 2.1704 | 2.1764 |
| 40 | .12166 | .016020 | 2445 | 8.02 | 1015.8 | 1023.9 | 8.02 | 1070.9 | 1078.9 | .01617 | 2.1430 | 2.1592 |
| 45 | .14748 | .016021 | 2037 | 13.04 | 1012.5 | 1025.5 | 13.04 | 1068.1 | 1081.1 | .02618 | 2.1162 | 2.1422 |
| 50 | .17803 | .016024 | 1704.2 | 18.06 | 1009.1 | 1027.2 | 18.06 | 1065.2 | 1083.3 | .03607 | 2.0899 | 2.1259 |
| 60 | .2563 | .016035 | 1206.9 | 28.08 | 1002.4 | 1030.4 | 28.08 | 1059.6 | 1087.7 | .05555 | 2.0388 | 2.0943 |
| 70 | .3632 | .016051 | 867.7 | 38.09 | 995.6 | 1033.7 | 38.09 | 1054.0 | 1092.0 | .07463 | 1.9896 | 2.0642 |
| 80 | .5073 | .016073 | 632.8 | 48.08 | 988.9 | 1037.0 | 48.09 | 1048.3 | 1096.4 | .09332 | 1.9423 | 2.0356 |
| 90 | .6988 | .016099 | 467.7 | 58.07 | 982.2 | 1040.2 | 58.07 | 1042.7 | 1100.7 | .11165 | 1.8966 | 2.0083 |
| 100 | .9503 | .016130 | 350.0 | 68.04 | 975.4 | 1043.5 | 68.05 | 1037.0 | 1105.0 | .12963 | 1.8526 | 1.9822 |
| 110 | 1.2763 | .016166 | 265.1 | 78.02 | 968.7 | 1046.7 | 78.02 | 1031.3 | 1109.3 | .14730 | 1.8101 | 1.9574 |
| 120 | 1.6945 | .016205 | 203.0 | 87.99 | 961.9 | 1049.9 | 88.00 | 1025.5 | 1113.5 | .16465 | 1.7690 | 1.9336 |
| 130 | 2.225 | .016247 | 157.17 | 97.97 | 955.1 | 1053.0 | 97.98 | 1019.8 | 1117.8 | .18172 | 1.7292 | 1.9109 |
| 140 | 2.892 | .016293 | 122.88 | 107.95 | 948.2 | 1056.2 | 107.96 | 1014.0 | 1121.9 | .19851 | 1.6907 | 1.8892 |
| 150 | 3.722 | .016343 | 96.99 | 117.95 | 941.3 | 1059.3 | 117.96 | 1008.1 | 1126.1 | .21503 | 1.6533 | 1.8684 |
| 160 | 4.745 | .016395 | 77.23 | 127.94 | 934.4 | 1062.3 | 127.96 | 1002.2 | 1130.1 | .23130 | 1.6171 | 1.8484 |
| 170 | 5.996 | .016450 | 62.02 | 137.95 | 927.4 | 1065.4 | 137.97 | 996.2 | 1134.2 | .24732 | 1.5819 | 1.8293 |
| 180 | 7.515 | .016509 | 50.20 | 147.97 | 920.4 | 1068.3 | 147.99 | 990.2 | 1138.2 | .26311 | 1.5478 | 1.8109 |
| 190 | 9.343 | .016570 | 40.95 | 158.0 | 913.3 | 1071.3 | 158.03 | 984.1 | 1142.1 | .27866 | 1.5146 | 1.7932 |
| 200 | 11.529 | .016634 | 33.63 | 168.04 | 906.2 | 1074.2 | 168.07 | 977.9 | 1145.9 | .29400 | 1.4822 | 1.7762 |
| 210 | 14.125 | .016702 | 27.82 | 178.10 | 898.9 | 1077.0 | 178.14 | 971.6 | 1149.7 | .30913 | 1.4508 | 1.7599 |
| 220 | 17.188 | .016772 | 23.15 | 188.17 | 891.7 | 1079.8 | 188.22 | 965.3 | 1153.5 | .32406 | 1.4201 | 1.7441 |
| 230 | 20.78 | .016845 | 19.386 | 198.26 | 884.3 | 1082.6 | 198.32 | 958.8 | 1157.1 | .33880 | 1.3901 | 1.7289 |
| 240 | 24.97 | .016922 | 16.327 | 208.36 | 876.9 | 1085.3 | 208.44 | 952.3 | 1160.7 | .35335 | 1.3609 | 1.7143 |
| 250 | 29.82 | .017001 | 13.826 | 218.49 | 869.4 | 1087.9 | 218.59 | 945.6 | 1164.2 | .36772 | 1.3324 | 1.7001 |
| 260 | 35.42 | .017084 | 11.768 | 228.64 | 861.8 | 1090.5 | 228.76 | 938.8 | 1167.6 | .38193 | 1.3044 | 1.6864 |

*(Continued)*

**TABLE A.1 (CONTINUED)**

Saturation: Temperature (Steam)

| Temp. Fahr. t | Press. lb/Sq. In. p | Specific Volume (ft³/lbm) | | Internal Energy (Btu/lbm) | | | Enthalpy (Btu/lbm) | | | Entropy (Btu/lbm·R) | | |
|---|---|---|---|---|---|---|---|---|---|---|---|---|
| | | Sat. Liquid $v_f$ | Sat. Vapor $v_g$ | Sat. Liquid $u_f$ | Evap. $u_{fg}$ | Sat. Vapor $u_g$ | Sat. Liquid $h_f$ | Evap. $h_{fg}$ | Sat. Vapor $h_g$ | Sat. Liquid $s_f$ | Evap. $s_{fg}$ | Sat. Vapor $s_g$ |
| 270 | 41.85 | .017170 | 10.066 | 238.82 | 854.1 | 1093.0 | 238.95 | 932.0 | 1170.9 | .39597 | 1.2771 | 1.6731 |
| 280 | 49.18 | .017259 | 8.650 | 249.02 | 846.3 | 1095.4 | 249.18 | 924.9 | 1174.1 | .40986 | 1.2504 | 1.6602 |
| 290 | 57.53 | .017352 | 7.467 | 259.25 | 838.5 | 1097.7 | 259.44 | 917.8 | 1177.2 | .42360 | 1.2241 | 1.6477 |
| 300 | 66.98 | .017448 | 6.472 | 269.52 | 830.5 | 1100.0 | 269.73 | 910.4 | 1180.2 | .47320 | 1.1984 | 1.6356 |
| 310 | 77.64 | .017548 | 5.632 | 279.81 | 822.3 | 1102.1 | 280.06 | 903.0 | 1183.0 | .45067 | 1.1731 | 1.6238 |
| 320 | 89.60 | .017652 | 4.919 | 290.14 | 814.1 | 1104.2 | 290.43 | 895.3 | 1185.8 | .46400 | 1.1483 | 1.6123 |
| 330 | 103.00 | .017760 | 4.312 | 300.51 | 805.7 | 1106.2 | 300.84 | 887.5 | 1188.4 | .47722 | 1.1238 | 1.6010 |
| 340 | 117.93 | .017872 | 3.792 | 310.91 | 797.1 | 1108.0 | 311.30 | 879.5 | 1190.8 | .49031 | 1.0997 | 1.5901 |
| 350 | 134.53 | .017988 | 3.346 | 321.35 | 788.4 | 1109.8 | 321.80 | 871.3 | 1193.1 | .50329 | 1.0760 | 1.5793 |
| 360 | 152.92 | .018108 | 2.961 | 331.84 | 779.6 | 1111.4 | 332.35 | 862.9 | 1195.2 | .51617 | 1.0526 | 1.5688 |
| 370 | 173.23 | .018233 | 2.628 | 342.37 | 770.6 | 1112.9 | 342.96 | 854.2 | 1197.2 | .52894 | 1.0295 | 1.5585 |
| 380 | 195.60 | .018363 | 2.339 | 352.95 | 761.4 | 1114.3 | 353.62 | 845.4 | 1199.0 | .54163 | 1.0067 | 1.5483 |
| 390 | 220.2 | .018498 | 2.087 | 363.58 | 752.0 | 1115.6 | 364.34 | 836.2 | 1200.6 | .55422 | .9841 | 1.5383 |
| 400 | 247.1 | .018638 | 1.8661 | 374.27 | 742.4 | 1116.6 | 375.12 | 826.8 | 1202.0 | .56672 | .9617 | 1.5284 |
| 425 | 325.6 | .019014 | 1.4249 | 401.24 | 717.4 | 1118.6 | 402.38 | 802.1 | 1204.5 | .59767 | .9066 | 1.5043 |
| 450 | 422.1 | .019433 | 1.1011 | 428.6 | 690.9 | 1119.5 | 430.2 | 775.4 | 1205.6 | .6282 | .8523 | 1.4806 |
| 475 | 539.3 | .019901 | .8594 | 456.6 | 662.6 | 1119.2 | 458.5 | 746.4 | 1204.9 | .6586 | .7985 | 1.4571 |
| 500 | 680.0 | .02043 | .6761 | 485.1 | 632.3 | 1117.4 | 487.7 | 714.8 | 1202.5 | .6888 | .7448 | 1.4335 |
| 525 | 847.1 | .02104 | .5350 | 514.5 | 599.5 | 1113.9 | 517.8 | 680.0 | 1197.8 | .7191 | .6906 | 1.4007 |
| 550 | 1044.0 | .02175 | .4249 | 544.9 | 563.7 | 1108.6 | 549.1 | 641.6 | 1190.6 | .7497 | .6354 | 1.3851 |
| 575 | 1274.0 | .02259 | .3378 | 576.5 | 524.3 | 1100.8 | 581.9 | 598.6 | 1180.4 | .7808 | .5785 | 1.3593 |
| 600 | 1541.0 | .02363 | .2677 | 609.9 | 480.1 | 1090.0 | 616.7 | 549.7 | 1166.4 | .8130 | .5187 | 1.3317 |
| 625 | 1849.7 | .02494 | .2103 | 645.7 | 429.4 | 1075.1 | 654.2 | 492.9 | 1147.0 | .8467 | .4544 | 1.3010 |
| 650 | 2205 | .02673 | .16206 | 685.0 | 368.7 | 1053.7 | 695.9 | 423.9 | 1119.8 | .8831 | .3820 | 1.2651 |
| 675 | 2616 | .02951 | .11952 | 731.0 | 289.3 | 1020.3 | 745.3 | 332.9 | 1078.2 | .9252 | .2934 | 1.2186 |
| 700 | 3090 | .03666 | .07438 | 801.7 | 145.9 | 947.7 | 822.7 | 167.5 | 990.2 | .9902 | .1444 | 1.1346 |
| 705.44 | 3204 | .05053 | .05053 | 872.6 | 0 | 872.6 | 902.5 | 0 | 902.5 | 1.0580 | 0 | 1.0580 |

**TABLE A.2**

Saturation Pressures (Steam)

| Press. Lb./Sq. In. $p$ | Temp. Fahr. $t$ | Specific Volume (ft.³/lb$_m$) Sat. Liquid $v_f$ | Sat. Vapor $v_g$ | Internal Energy (Btu/lb$_m$) Sat. Liquid $u_f$ | Evap. $u_{fg}$ | Sat. Vapor $u_g$ | Enthalpy (Btu/lb$_m$) Sat. Liquid $h_f$ | Evap. $h_{fg}$ | Sat. Vapor $h_g$ | Entropy (Btu/lb$_m \cdot$R) Sat. Liquid $s_f$ | Evap. $s_{fg}$ | Sat. Vapor $s_g$ |
|---|---|---|---|---|---|---|---|---|---|---|---|---|
| .50 | 79.56 | .016071 | 641.5 | 47.64 | 989.2 | 1036.9 | 47.65 | 1048.6 | 1096.2 | .09250 | 1.9443 | 2.0368 |
| 1.0 | 101.70 | .016136 | 333.6 | 69.74 | 974.3 | 1044.0 | 69.74 | 1036.0 | 1105.8 | .13266 | 1.8453 | 1.9779 |
| 1.5 | 115.65 | .016187 | 227.7 | 83.65 | 964.8 | 1048.5 | 83.65 | 1028.0 | 1111.7 | .15714 | 1.7867 | 1.9438 |
| 2.0 | 126.04 | .016230 | 173.75 | 94.02 | 957.8 | 1051.8 | 94.02 | 1022.1 | 1116.1 | .17499 | 1.7448 | 1.9198 |
| 3.0 | 141.43 | .016300 | 118.72 | 109.38 | 947.2 | 1056.6 | 109.39 | 1013.1 | 1122.5 | .20089 | 1.6852 | 1.8861 |
| 4.0 | 152.93 | .016358 | 90.64 | 120.88 | 939.3 | 1060.2 | 120.89 | 1006.4 | 1127.3 | .21983 | 1.6426 | 1.8624 |
| 5.0 | 162.21 | .016407 | 73.53 | 130.15 | 932.9 | 1063.0 | 130.17 | 1000.9 | 1131.0 | .23486 | 1.6093 | 1.8441 |
| 7.5 | 179.91 | .016508 | 50.30 | 147.88 | 920.4 | 1068.3 | 147.90 | 990.2 | 1138.1 | .26297 | 1.5481 | 1.8110 |
| 10 | 193.19 | .016590 | 38.42 | 161.20 | 911.0 | 1072.2 | 161.23 | 982.1 | 1143.3 | .28358 | 1.5041 | 1.7877 |
| 14.696 | 211.99 | .016715 | 26.80 | 180.10 | 897.5 | 1077.6 | 180.15 | 970.4 | 1150.5 | .31212 | 1.4446 | 1.7567 |
| 15 | 213.03 | .016723 | 26.29 | 181.14 | 896.8 | 1077.9 | 181.19 | 969.7 | 1150.9 | .31367 | 1.4414 | 1.7551 |
| 20 | 227.96 | .016830 | 20.09 | 196.19 | 885.8 | 1082.0 | 196.26 | 960.1 | 1156.4 | .33580 | 1.3962 | 1.7320 |
| 25 | 240.08 | .016922 | 16.306 | 208.44 | 867.9 | 1085.3 | 208.52 | 952.2 | 1160.7 | .35345 | 1.3607 | 1.7142 |
| 30 | 250.34 | .017004 | 13.748 | 218.84 | 869.2 | 1088.0 | 218.93 | 945.4 | 1164.3 | .36821 | 1.3314 | 1.6996 |
| 35 | 259.30 | .017078 | 11.900 | 227.93 | 862.4 | 1090.3 | 228.04 | 939.3 | 1167.4 | .38093 | 1.3064 | 1.6873 |
| 40 | 267.26 | .017146 | 10.501 | 236.03 | 856.2 | 1092.3 | 236.16 | 933.8 | 1170.0 | .39214 | 1.2845 | 1.6767 |
| 45 | 274.46 | .017209 | 9.403 | 243.37 | 850.7 | 1094.0 | 243.51 | 928.8 | 1172.3 | .40218 | 1.2651 | 1.6673 |
| 50 | 281.03 | .017269 | 8.518 | 250.08 | 845.5 | 1095.6 | 250.24 | 924.2 | 1174.4 | .41129 | 1.2476 | 1.6589 |
| 55 | 287.10 | .017325 | 7.789 | 256.28 | 840.8 | 1097.0 | 256.46 | 919.9 | 1176.3 | .41963 | 1.2317 | 1.6513 |
| 60 | 292.73 | .017378 | 7.177 | 262.06 | 836.3 | 1098.3 | 262.25 | 915.8 | 1178.0 | .42733 | 1.2170 | 1.6444 |
| 65 | 298.00 | .017429 | 6.657 | 267.46 | 832.1 | 1099.5 | 267.67 | 911.9 | 1179.6 | .43450 | 1.2035 | 1.6380 |
| 70 | 302.96 | .017478 | 6.209 | 272.56 | 828.1 | 1100.6 | 272.79 | 908.3 | 1181.0 | .44120 | 1.1909 | 1.6321 |
| 75 | 307.63 | .017524 | 5.818 | 277.37 | 824.3 | 1101.6 | 277.61 | 904.8 | 1182.4 | .44749 | 1.1790 | 1.6265 |
| 80 | 312.07 | .017570 | 5.474 | 281.95 | 820.6 | 1102.6 | 282.21 | 901.4 | 1183.6 | .45344 | 1.1679 | 1.6214 |
| 85 | 316.29 | .017613 | 5.170 | 286.30 | 817.1 | 1103.5 | 286.58 | 898.2 | 1184.8 | .45907 | 1.1574 | 1.6165 |

*(Continued)*

**TABLE A.2 (CONTINUED)**
Saturation Pressures (Steam)

| Press. Lb/Sq. In. $p$ | Temp. Fahr. $t$ | Specific Volume (ft.³/lb$_m$) | | Internal Energy (Btu/lb$_m$) | | | Enthalpy (Btu/lb$_m$) | | | Entropy (Btu/lb$_m$·R) | | |
|---|---|---|---|---|---|---|---|---|---|---|---|---|
| | | Sat. Liquid $v_f$ | Sat. Vapor $v_g$ | Sat. Liquid $u_f$ | Evap. $u_{fg}$ | Sat. Vapor $u_g$ | Sat. Liquid $h_f$ | Evap. $h_{fg}$ | Sat. Vapor $h_g$ | Sat. Liquid $s_f$ | Evap. $s_{fg}$ | Sat. Vapor $s_g$ |
| 90 | 320.31 | .017655 | 4.898 | 290.46 | 813.8 | 1104.3 | 290.76 | 895.1 | 1185.9 | .46442 | 1.1475 | 1.6119 |
| 95 | 324.16 | .017696 | 4.654 | 294.45 | 810.6 | 1105.0 | 294.76 | 892.1 | 1186.9 | .46952 | 1.1380 | 1.6076 |
| 100 | 327.86 | .017736 | 4.434 | 298.28 | 807.5 | 1105.8 | 298.61 | 889.2 | 1187.8 | .47439 | 1.1290 | 1.6034 |
| 105 | 331.41 | .017775 | 4.234 | 301.97 | 804.5 | 1106.5 | 302.31 | 886.4 | 1188.7 | .47906 | 1.1204 | 1.5995 |
| 110 | 334.82 | .017813 | 4.051 | 305.52 | 801.6 | 1107.1 | 305.88 | 883.7 | 1189.6 | .48355 | 1.1122 | 1.5957 |
| 115 | 338.12 | .017850 | 3.884 | 308.95 | 798.8 | 1107.7 | 309.33 | 881.0 | 1190.4 | .48786 | 1.1042 | 1.5921 |
| 120 | 341.30 | .017886 | 3.730 | 312.27 | 796.0 | 1108.3 | 312.67 | 878.5 | 1191.1 | .49201 | 1.0966 | 1.5886 |
| 125 | 344.39 | .017922 | 3.588 | 315.49 | 793.3 | 1108.8 | 315.90 | 875.9 | 1191.8 | .49602 | 1.0893 | 1.5853 |
| 130 | 347.37 | .017957 | 3.457 | 318.61 | 790.7 | 1109.4 | 319.04 | 873.5 | 1192.5 | .49989 | 1.0822 | 1.5821 |
| 135 | 350.27 | .017991 | 3.335 | 321.64 | 788.2 | 1109.8 | 322.08 | 871.1 | 1193.2 | .50364 | 1.0754 | 1.5790 |
| 140 | 353.08 | .018024 | 3.221 | 324.58 | 785.7 | 1110.3 | 325.05 | 868.7 | 1193.8 | .50727 | 1.0688 | 1.5761 |
| 145 | 355.82 | .018057 | 3.115 | 327.45 | 783.3 | 1110.8 | 327.93 | 866.4 | 1194.4 | .51079 | 1.0624 | 1.5732 |
| 150 | 358.48 | .018089 | 3.016 | 330.24 | 781.0 | 1111.2 | 330.75 | 864.2 | 1194.9 | .51422 | 1.0562 | 1.5704 |
| 160 | 363.60 | .018152 | 2.836 | 335.63 | 776.4 | 1112.0 | 336.16 | 859.8 | 1196.0 | .52078 | 1.0443 | 1.5651 |
| 170 | 368.47 | .018214 | 2.676 | 340.76 | 772.0 | 1112.7 | 341.33 | 855.6 | 1196.9 | .52700 | 1.0330 | 1.5600 |
| 180 | 373.13 | .018273 | 2.533 | 345.68 | 767.7 | 1113.4 | 346.29 | 851.5 | 1197.8 | .53292 | 1.0223 | 1.5553 |
| 190 | 377.59 | .018331 | 2.405 | 350.39 | 763.6 | 1114.0 | 351.04 | 847.5 | 1198.6 | .53857 | 1.0122 | 1.5507 |
| 200 | 381.86 | .018387 | 2.289 | 354.9 | 759.6 | 1114.6 | 355.6 | 843.7 | 1199.3 | .5440 | 1.0025 | 1.5464 |
| 225 | 391.87 | .018523 | 2.043 | 365.6 | 750.2 | 1115.8 | 366.3 | 834.5 | 1200.8 | .5566 | .9799 | 1.5365 |
| 250 | 401.04 | .018653 | 1.8448 | 375.4 | 741.4 | 1116.7 | 376.2 | 825.8 | 1202.1 | .5680 | .9594 | 1.5274 |
| 275 | 409.52 | .018777 | 1.6813 | 384.5 | 733.0 | 1117.5 | 385.4 | 817.6 | 1203.1 | .5786 | .9406 | 1.5192 |
| 300 | 417.43 | .018896 | 1.5442 | 393.0 | 725.1 | 1118.2 | 394.1 | 809.8 | 1203.9 | .5883 | .9232 | 1.5115 |
| 350 | 431.82 | .019124 | 1.3267 | 408.7 | 710.3 | 1119.0 | 409.9 | 795.0 | 1204.9 | .6060 | .8917 | 1.4978 |
| 400 | 444.70 | .019340 | 1.1620 | 422.8 | 696.7 | 1119.5 | 424.2 | 781.2 | 1205.5 | .6218 | .8638 | 1.4856 |
| 450 | 456.39 | .019547 | 1.0326 | 435.7 | 683.9 | 1119.6 | 437.4 | 768.2 | 1205.6 | .6360 | .8385 | 1.4746 |

*(Continued)*

**TABLE A.2 (CONTINUED)**
Saturation Pressures (Steam)

| Press. Lb/Sq. In. $p$ | Temp. Fahr. $t$ | Specific Volume (ft.³/lb$_m$) | | Internal Energy (Btu/lb$_m$) | | | Enthalpy (Btu/lb$_m$) | | | Entropy (Btu/lb$_m$·R) | | |
|---|---|---|---|---|---|---|---|---|---|---|---|---|
| | | Sat. Liquid $v_f$ | Sat. Vapor $v_g$ | Sat. Liquid $u_f$ | Evap. $u_{fg}$ | Sat. Vapor $u_g$ | Sat. Liquid $h_f$ | Evap. $h_{fg}$ | Sat. Vapor $h_g$ | Sat. Liquid $s_f$ | Evap. $s_{fg}$ | Sat. Vapor $s_g$ |
| 500 | 467.13 | .019748 | .9283 | 447.7 | 671.7 | 1119.4 | 449.5 | 755.8 | 1205.3 | .6490 | .8154 | 1.4645 |
| 550 | 477.07 | .019943 | .8423 | 458.9 | 660.2 | 1119.1 | 460.9 | 743.9 | 1204.8 | .6611 | .7941 | 1.4551 |
| 600 | 486.33 | .02013 | .7702 | 469.4 | 649.1 | 1118.6 | 471.7 | 732.4 | 1204.1 | .6723 | .7742 | 1.4464 |
| 700 | 503.23 | .02051 | .6558 | 488.9 | 628.2 | 1117.0 | 491.5 | 710.5 | 1202.0 | .6927 | .7378 | 1.4305 |
| 800 | 518.36 | .02087 | .5691 | 506.6 | 608.4 | 1115.0 | 509.7 | 689.6 | 1199.3 | .7110 | .7050 | 1.4160 |
| 900 | 532.12 | .02123 | .5009 | 523.0 | 589.6 | 1112.6 | 526.6 | 669.5 | 1196.0 | .7277 | .6750 | 1.4027 |
| 1000 | 544.75 | .02159 | .4459 | 538.4 | 571.5 | 1109.9 | 542.4 | 650.0 | 1192.4 | .7432 | .6471 | 1.3903 |
| 1250 | 572.56 | .02250 | .3454 | 573.4 | 528.3 | 1101.7 | 578.6 | 603.0 | 1181.6 | .7778 | .5841 | 1.3619 |
| 1500 | 596.39 | .02346 | .2769 | 605.0 | 486.9 | 1091.8 | 611.5 | 557.2 | 1168.7 | .8082 | .5276 | 1.3359 |
| 1750 | 617.31 | .02450 | .2268 | 634.4 | 445.9 | 1080.2 | 642.3 | 511.4 | 1153.7 | .8361 | .4748 | 1.3109 |
| 2000 | 649.20 | .02565 | .8813 | 662.4 | 404.2 | 1066.6 | 671.9 | 464.4 | 1136.3 | .8623 | .4238 | 1.2861 |
| 2250 | 652.90 | .02698 | .15692 | 689.9 | 360.7 | 1050.6 | 701.1 | 414.8 | 1115.9 | .8876 | .3728 | 1.2604 |
| 2500 | 668.31 | .02860 | .13059 | 717.7 | 313.4 | 1031.0 | 730.9 | 360.5 | 1091.4 | .9131 | .3196 | 1.2327 |
| 2750 | 682.46 | .03077 | .10717 | 747.3 | 258.6 | 1005.9 | 763.0 | 297.4 | 1060.4 | .9401 | .2604 | 1.2005 |
| 3000 | 695.52 | .03431 | .08404 | 783.4 | 185.4 | 968.8 | 802.5 | 213.0 | 1015.5 | .9732 | .1843 | 1.1575 |
| 3203.6 | 705.44 | .05053 | .05053 | 872.6 | 0 | 872.6 | 902.5 | 0 | 902.5 | 1.0580 | 0 | 1.0580 |

**TABLE A.3**

Properties of Superheated Steam

Vapor (psia)

| p (t Sat.) | 14.696 (211.99) | | | | 20 (227.96) | | | | 30 (250.34) | | | |
|---|---|---|---|---|---|---|---|---|---|---|---|---|
| t | v | u | h | s | v | u | h | s | v | u | h | s |
| Sat. | 26.80 | 1077.6 | 1150.5 | 1.7567 | 20.09 | 1082.0 | 1156.4 | 1.7320 | 13.748 | 1088.0 | 1164.3 | 1.6996 |
| 150 | 24.10 | 1054.5 | 1120.0 | 1.7090 | 17.532 | 1052.0 | 1116.9 | 1.6710 | 11.460 | 1047.3 | 1111.0 | 1.6185 |
| 160 | 24.54 | 1058.2 | 1125.0 | 1.7171 | 17.870 | 1056.0 | 1122.1 | 1.6795 | 11.701 | 1051.6 | 1116.5 | 1.6275 |
| 170 | 24.98 | 1062.0 | 1129.9 | 1.7251 | 18.204 | 1059.9 | 1127.2 | 1.6877 | 11.938 | 1055.8 | 1122.0 | 1.6364 |
| 180 | 25.42 | 1065.7 | 1134.9 | 1.7328 | 18.535 | 1063.8 | 1132.4 | 1.6957 | 12.172 | 1059.9 | 1127.5 | 1.6449 |
| 190 | 25.85 | 1069.5 | 1139.8 | 1.7405 | 18.864 | 1067.6 | 1137.4 | 1.7036 | 12.403 | 1064.0 | 1132.9 | 1.6533 |
| 200 | 26.29 | 1073.2 | 1144.7 | 1.7479 | 19.191 | 1071.4 | 1142.5 | 1.7113 | 12.631 | 1068.1 | 1138.2 | 1.6615 |
| 210 | 26.72 | 1076.9 | 1149.5 | 1.7553 | 19.515 | 1075.2 | 1147.5 | 1.7188 | 12.857 | 1072.1 | 1143.5 | 1.6694 |
| 220 | 27.15 | 1080.6 | 1154.4 | 1.7624 | 19.837 | 1079.0 | 1152.4 | 1.7262 | 13.081 | 1076.1 | 1148.7 | 1.6771 |
| 230 | 27.57 | 1084.2 | 1159.2 | 1.7695 | 20.157 | 1082.8 | 1157.4 | 1.7335 | 13.303 | 1080.0 | 1153.9 | 1.6847 |
| 240 | 28.00 | 1087.9 | 1164.0 | 1.7764 | 20.475 | 1086.5 | 1162.3 | 1.7405 | 13.523 | 1084.0 | 1159.0 | 1.6921 |
| 250 | 28.42 | 1091.5 | 1168.8 | 1.7832 | 20.79 | 1090.3 | 1167.2 | 1.7475 | 13.741 | 1087.9 | 1164.1 | 1.6994 |
| 260 | 28.85 | 1095.2 | 1173.6 | 1.7899 | 21.11 | 1094.0 | 1172.1 | 1.7543 | 13.958 | 1091.7 | 1169.2 | 1.7064 |
| 270 | 29.27 | 1098.8 | 1178.4 | 1.7965 | 21.42 | 1097.7 | 1177.0 | 1.7610 | 14.173 | 1095.6 | 1174.2 | 1.7134 |
| 280 | 29.69 | 1102.4 | 1183.1 | 1.8030 | 21.73 | 1101.4 | 1181.8 | 1.7676 | 14.387 | 1099.4 | 1179.2 | 1.7202 |
| 290 | 30.11 | 1106.0 | 1187.9 | 1.8094 | 22.05 | 1105.0 | 1186.6 | 1.7741 | 14.600 | 1103.2 | 1184.2 | 1.7269 |
| 300 | 30.52 | 1109.6 | 1192.6 | 1.8157 | 22.36 | 1108.7 | 1191.5 | 1.7805 | 14.812 | 1106.9 | 1189.2 | 1.7334 |
| 310 | 30.94 | 1113.2 | 1197.4 | 1.8219 | 22.67 | 1112.4 | 1196.3 | 1.7868 | 15.023 | 1110.7 | 1194.1 | 1.7399 |
| 320 | 31.36 | 1116.8 | 1202.1 | 1.8280 | 22.98 | 1116.0 | 1201.0 | 1.7930 | 15.233 | 1114.4 | 1199.0 | 1.7462 |
| 330 | 31.77 | 1120.4 | 1206.8 | 1.8340 | 23.28 | 1119.7 | 1205.8 | 1.7991 | 15.442 | 1118.2 | 1203.9 | 1.7525 |
| 340 | 32.19 | 1124.0 | 1211.6 | 1.8400 | 23.59 | 1123.3 | 1210.6 | 1.8051 | 15.651 | 1121.9 | 1208.8 | 1.7586 |
| 350 | 32.60 | 1127.6 | 1216.3 | 1.8458 | 23.90 | 1126.9 | 1215.4 | 1.8110 | 15.859 | 1125.6 | 1213.6 | 1.7646 |
| 360 | 33.02 | 1131.2 | 1221.0 | 1.8516 | 24.21 | 1130.6 | 1220.1 | 1.8168 | 16.067 | 1129.3 | 1218.5 | 1.7706 |
| 370 | 33.43 | 1134.8 | 1225.7 | 1.8574 | 24.51 | 1134.2 | 1224.9 | 1.8226 | 16.273 | 1133.0 | 1223.3 | 1.7765 |
| 380 | 33.84 | 1138.4 | 1230.5 | 1.8630 | 24.82 | 1137.8 | 1229.7 | 1.8283 | 16.480 | 1136.7 | 1228.1 | 1.7822 |
| 390 | 34.26 | 1142.0 | 1235.2 | 1.8686 | 25.12 | 1141.4 | 1234.4 | 1.8340 | 16.686 | 1140.3 | 1233.0 | 1.7880 |
| 400 | 34.67 | 1145.6 | 1239.9 | 1.8741 | 25.43 | 1145.1 | 1239.2 | 1.8395 | 16.891 | 1144.0 | 1237.8 | 1.7936 |

*(Continued)*

**TABLE A.3 (CONTINUED)**

Properties of Superheated Steam

| | | | | | | | | | | | | |
|---|---|---|---|---|---|---|---|---|---|---|---|---|
| 420 | 35.49 | 1152.8 | 1249.3 | 1.8850 | 26.03 | 1152.3 | 1248.7 | 1.8504 | 17.301 | 1151.4 | 1247.4 | 1.8047 |
| 440 | 36.31 | 1160.1 | 1258.8 | 1.8956 | 26.64 | 1159.6 | 1258.2 | 1.8611 | 17.709 | 1158.7 | 1257.0 | 1.8155 |
| 460 | 37.13 | 1167.3 | 1268.3 | 1.9060 | 27.25 | 1166.9 | 1267.7 | 1.8716 | 18.116 | 1166.1 | 1266.6 | 1.8260 |
| 480 | 37.95 | 1174.6 | 1277.8 | 1.9162 | 27.85 | 1174.2 | 1277.2 | 1.8819 | 18.523 | 1173.4 | 1276.2 | 1.8364 |
| 500 | 38.77 | 1181.8 | 1287.3 | 1.9263 | 28.46 | 1181.5 | 1286.8 | 1.8919 | 18.928 | 1180.8 | 1285.9 | 1.8465 |
| 520 | 39.59 | 1189.1 | 1296.8 | 1.9361 | 29.06 | 1188.8 | 1296.3 | 1.9018 | 19.333 | 1188.2 | 1295.5 | 1.8564 |
| 540 | 40.41 | 1196.5 | 1306.4 | 1.9457 | 29.66 | 1196.1 | 1305.9 | 1.9114 | 19.737 | 1195.5 | 1305.1 | 1.8661 |
| 560 | 41.22 | 1203.8 | 1315.9 | 1.9552 | 30.26 | 1203.5 | 1315.5 | 1.9210 | 20.140 | 1203.0 | 1314.8 | 1.8757 |
| 580 | 42.04 | 1211.2 | 1325.5 | 1.9645 | 30.87 | 1210.9 | 1325.2 | 1.9303 | 20.543 | 1210.4 | 1324.4 | 1.8851 |
| 600 | 42.86 | 1218.6 | 1335.2 | 1.9737 | 31.47 | 1218.4 | 1334.8 | 1.9395 | 20.95 | 1217.8 | 1334.1 | 1.8943 |
| 620 | 43.67 | 1226.1 | 1344.8 | 1.9827 | 32.07 | 1225.8 | 1344.5 | 1.9485 | 21.35 | 1225.3 | 1343.8 | 1.9034 |
| 640 | 44.49 | 1233.5 | 1354.5 | 1.9916 | 32.67 | 1233.3 | 1354.2 | 1.9575 | 21.75 | 1232.8 | 1353.6 | 1.9123 |
| 660 | 45.30 | 1241.0 | 1364.2 | 2.0004 | 33.27 | 1240.8 | 1363.9 | 1.9662 | 22.15 | 1240.4 | 1363.3 | 1.9211 |
| 680 | 46.12 | 1248.6 | 1374.0 | 2.0090 | 33.87 | 1248.3 | 1373.7 | 1.9749 | 22.55 | 1247.9 | 1373.1 | 1.9298 |
| 700 | 46.93 | 1256.1 | 1383.8 | 2.0175 | 34.47 | 1255.9 | 1383.5 | 1.9834 | 22.95 | 1255.5 | 1383.0 | 1.9384 |
| 720 | 47.75 | 1263.7 | 1393.6 | 2.0259 | 35.07 | 1263.5 | 1393.3 | 1.9918 | 23.35 | 1263.2 | 1392.8 | 1.9468 |
| 740 | 48.56 | 1271.4 | 1403.4 | 2.0342 | 35.66 | 1271.2 | 1403.2 | 2.0001 | 23.75 | 1270.8 | 1402.7 | 1.9551 |
| 760 | 49.37 | 1279.0 | 1413.3 | 2.0424 | 36.26 | 1278.8 | 1413.0 | 2.0082 | 24.15 | 1278.5 | 1412.6 | 1.9633 |
| 780 | 50.19 | 1286.7 | 1423.2 | 2.0504 | 36.86 | 1286.5 | 1423.0 | 2.0163 | 24.55 | 1286.2 | 1422.5 | 1.9714 |
| 800 | 51.00 | 1294.4 | 1433.1 | 2.0584 | 37.46 | 1294.3 | 1432.9 | 2.0243 | 24.95 | 1294.0 | 1432.5 | 1.9793 |
| 850 | 53.03 | 1313.9 | 1458.1 | 2.0778 | 38.96 | 1313.8 | 1457.9 | 2.0438 | 25.95 | 1313.5 | 1457.6 | 1.9988 |
| 900 | 55.07 | 1333.6 | 1483.4 | 2.0967 | 40.45 | 1333.5 | 1483.2 | 2.0627 | 26.95 | 1333.2 | 1482.8 | 1.0178 |
| 950 | 57.10 | 1353.5 | 1508.8 | 2.1151 | 41.94 | 1353.4 | 1508.6 | 2.0810 | 27.95 | 1353.2 | 1508.3 | 2.0362 |
| 1000 | 59.13 | 1373.7 | 1534.5 | 2.1330 | 43.44 | 1373.5 | 1534.3 | 2.0989 | 28.95 | 1373.3 | 1534.0 | 2.0541 |
| 1100 | 63.19 | 1414.6 | 1586.4 | 2.1674 | 46.42 | 1414.5 | 1586.3 | 2.1334 | 30.94 | 1414.3 | 1586.1 | 2.0886 |
| 1200 | 67.25 | 1456.5 | 1639.3 | 2.2003 | 49.41 | 1456.4 | 1639.2 | 2.1663 | 32.93 | 1456.2 | 1639.1 | 2.1215 |
| 1300 | 71.30 | 1499.3 | 1693.2 | 2.2318 | 52.39 | 1499.2 | 1693.1 | 2.1978 | 34.92 | 1499.1 | 1692.9 | 2.1530 |
| 1400 | 75.36 | 1543.0 | 1747.9 | 2.2621 | 55.37 | 1542.9 | 1747.9 | 2.2281 | 36.91 | 1542.8 | 1747.7 | 2.1833 |
| 1500 | 79.42 | 1587.6 | 1803.6 | 2.2912 | 58.35 | 1587.6 | 1803.5 | 2.2572 | 38.90 | 1587.5 | 1803.4 | 2.2125 |
| 1600 | 83.47 | 1633.2 | 1860.2 | 2.3194 | 61.33 | 1633.2 | 1860.1 | 2.2854 | 40.88 | 1633.1 | 1860.0 | 2.2407 |

*(Continued)*

**TABLE A.3 (CONTINUED)**

Properties of Superheated Steam

| t | v | u | h | s | v | u | h | s | v | u | h | s |
|---|---|---|---|---|---|---|---|---|---|---|---|---|
| 1800 | 91.58 | 1727.0 | 1976.1 | 2.3731 | 67.29 | 1727.0 | 1976.1 | 2.3391 | 44.86 | 1726.9 | 1976.0 | 2.2944 |
| 2000 | 99.69 | 1824.4 | 2095.5 | 2.4237 | 73.25 | 1824.3 | 2095.4 | 2.3897 | 48.83 | 1824.2 | 2095.3 | 2.3450 |
| 2200 | 107.80 | 1924.8 | 2218.0 | 2.4716 | 79.21 | 1924.8 | 2218.0 | 2.4376 | 52.81 | 1924.7 | 2217.9 | 2.3929 |
| 2400 | 115.91 | 2028.1 | 2343.4 | 2.5170 | 85.17 | 2028.1 | 2343.3 | 2.4830 | 56.78 | 2028.0 | 2343.3 | 2.4383 |

| p (t Sat.) | 40 (267.26) | | | | 50 (281.03) | | | | 60 (292.73) | | | |
|---|---|---|---|---|---|---|---|---|---|---|---|---|
| t | v | u | h | s | v | u | h | s | v | u | h | s |
| Sat. | 10.501 | 1092.3 | 1170.0 | 1.6767 | 8.518 | 1095.6 | 1174.4 | 1.6589 | 7.177 | 1098.3 | 1178.0 | 1.6444 |
| 200 | 9.346 | 1064.6 | 1133.8 | 1.6243 | 7.370 | 1060.9 | 1129.1 | 1.5940 | 6.047 | 1057.1 | 1124.2 | 1.5680 |
| 210 | 9.523 | 1068.8 | 1139.3 | 1.6327 | 7.519 | 1065.4 | 1135.0 | 1.6029 | 6.178 | 1061.9 | 1130.5 | 1.5774 |
| 220 | 9.699 | 1073.0 | 1144.8 | 1.6409 | 7.665 | 1069.9 | 1140.8 | 1.6115 | 6.307 | 1066.6 | 1136.6 | 1.5864 |
| 230 | 9.872 | 1077.2 | 1150.3 | 1.6488 | 7.810 | 1074.2 | 1146.5 | 1.6198 | 6.432 | 1071.2 | 1142.6 | 1.5952 |
| 240 | 10.043 | 1081.3 | 1155.6 | 1.6565 | 7.952 | 1078.6 | 1152.1 | 1.6279 | 6.556 | 1075.7 | 1148.5 | 1.6036 |
| 250 | 10.212 | 1085.4 | 1161.0 | 1.6641 | 8.092 | 1082.8 | 1157.7 | 1.6358 | 6.677 | 1080.1 | 1154.3 | 1.6118 |
| 260 | 10.380 | 1089.4 | 1166.2 | 1.6714 | 8.231 | 1087.0 | 1163.1 | 1.6434 | 6.797 | 1084.5 | 1160.0 | 1.6198 |
| 270 | 10.546 | 1093.4 | 1171.4 | 1.6786 | 8.368 | 1091.1 | 1168.5 | 1.6509 | 6.915 | 1088.8 | 1165.6 | 1.6275 |
| 280 | 10.711 | 1097.3 | 1176.6 | 1.6857 | 8.504 | 1095.2 | 1173.9 | 1.6582 | 7.031 | 1093.0 | 1171.1 | 1.6351 |
| 290 | 10.875 | 1101.2 | 1181.7 | 1.6926 | 8.639 | 1099.2 | 1179.2 | 1.6653 | 7.146 | 1097.2 | 1176.5 | 1.6424 |
| 300 | 11.038 | 1105.1 | 1186.8 | 1.6993 | 8.772 | 1103.2 | 1184.4 | 1.6722 | 7.260 | 1101.3 | 1181.9 | 1.6496 |
| 310 | 11.200 | 1109.0 | 1191.9 | 1.7059 | 8.904 | 1107.2 | 1189.6 | 1.6790 | 7.373 | 1105.4 | 1187.3 | 1.6565 |
| 320 | 11.360 | 1112.8 | 1196.9 | 1.7124 | 9.036 | 1111.2 | 1194.8 | 1.6857 | 7.485 | 1109.5 | 1192.6 | 1.6634 |
| 330 | 11.520 | 1116.6 | 1201.9 | 1.7188 | 9.166 | 1115.1 | 1199.9 | 1.6922 | 7.596 | 1113.5 | 1197.8 | 1.6700 |
| 340 | 11.680 | 1120.4 | 1206.9 | 1.7251 | 9.296 | 1119.0 | 1205.0 | 1.6986 | 7.706 | 1117.4 | 1203.0 | 1.6766 |
| 350 | 11.838 | 1124.2 | 1211.8 | 1.7312 | 9.425 | 1122.8 | 1210.0 | 1.7049 | 7.815 | 1121.4 | 1208.2 | 1.6830 |
| 360 | 11.996 | 1128.0 | 1216.8 | 1.7373 | 9.553 | 1126.7 | 1215.1 | 1.7110 | 7.924 | 1125.3 | 1213.3 | 1.6893 |
| 370 | 12.153 | 1131.7 | 1221.7 | 1.7432 | 9.681 | 1130.5 | 1220.1 | 1.7171 | 8.032 | 1129.2 | 1218.4 | 1.6955 |
| 380 | 12.310 | 1135.5 | 1226.6 | 1.7491 | 9.808 | 1134.3 | 1225.0 | 1.7231 | 8.139 | 1133.1 | 1223.5 | 1.7015 |
| 390 | 12.467 | 1139.2 | 1231.5 | 1.7549 | 9.935 | 1138.1 | 1230.0 | 1.7290 | 8.246 | 1136.9 | 1228.5 | 1.7075 |
| 400 | 12.623 | 1143.0 | 1236.4 | 1.7606 | 10.061 | 1141.9 | 1235.0 | 1.7348 | 8.353 | 1140.8 | 1233.5 | 1.7134 |
| 420 | 12.933 | 1150.4 | 1246.1 | 1.7718 | 10.312 | 1149.4 | 1244.8 | 1.7461 | 8.565 | 1148.4 | 1243.5 | 1.7249 |
| 440 | 13.243 | 1157.8 | 1255.8 | 1.7828 | 10.562 | 1156.9 | 1254.6 | 1.7572 | 8.775 | 1156.0 | 1253.4 | 1.7360 |

*(Continued)*

**TABLE A.3 (CONTINUED)**

Properties of Superheated Steam

| Temp | v | u | h | s | v | u | h | s | v | u | h | s |
|---|---|---|---|---|---|---|---|---|---|---|---|---|
| 460 | 13.551 | 1165.2 | 1265.5 | 1.7934 | 10.811 | 1164.4 | 1264.4 | 1.7679 | 8.984 | 1163.6 | 1263.3 | 1.7469 |
| 480 | 13.858 | 1172.7 | 1275.2 | 1.8038 | 11.059 | 1171.9 | 1274.2 | 1.7784 | 9.192 | 1171.1 | 1273.2 | 1.7575 |
| 500 | 14.164 | 1180.1 | 1284.9 | 1.8140 | 11.305 | 1179.4 | 1284.0 | 1.7887 | 9.399 | 1178.6 | 1283.0 | 1.7678 |
| 520 | 14.469 | 1187.5 | 1294.6 | 1.8240 | 11.551 | 1186.8 | 1293.7 | 1.7988 | 9.606 | 1186.2 | 1292.8 | 1.7780 |
| 540 | 14.774 | 1194.9 | 1304.3 | 1.8338 | 11.796 | 1194.3 | 1303.5 | 1.8086 | 9.811 | 1193.7 | 1302.6 | 1.7879 |
| 560 | 15.078 | 1202.4 | 1314.0 | 1.8434 | 12.041 | 1201.8 | 1313.2 | 1.8183 | 10.016 | 1201.2 | 1312.4 | 1.7976 |
| 580 | 15.382 | 1209.9 | 1323.7 | 1.8529 | 12.285 | 1209.3 | 1323.0 | 1.8277 | 10.221 | 1208.8 | 1322.2 | 1.8071 |
| 600 | 15.685 | 1217.3 | 1333.4 | 1.8621 | 12.529 | 1216.8 | 1332.8 | 1.8371 | 10.425 | 1216.3 | 1332.1 | 1.8165 |
| 620 | 15.988 | 1224.8 | 1343.2 | 1.8713 | 12.772 | 1224.4 | 1342.5 | 1.8462 | 10.628 | 1223.9 | 1341.9 | 1.8257 |
| 640 | 16.291 | 1232.4 | 1353.0 | 1.8802 | 13.015 | 1231.9 | 1352.4 | 1.8552 | 10.832 | 1231.5 | 1351.7 | 1.8347 |
| 660 | 16.593 | 1239.9 | 1362.8 | 1.8891 | 13.258 | 1239.5 | 1362.2 | 1.8641 | 11.035 | 1239.1 | 1361.6 | 1.8436 |
| 680 | 16.895 | 1247.5 | 1372.6 | 1.8977 | 13.500 | 1247.1 | 1372.0 | 1.8728 | 11.237 | 1246.7 | 1371.5 | 1.8523 |
| 700 | 17.196 | 1255.1 | 1382.4 | 1.9063 | 13.742 | 1254.8 | 1381.9 | 1.8814 | 11.440 | 1254.4 | 1381.4 | 1.8609 |
| 720 | 17.498 | 1262.8 | 1392.3 | 1.9147 | 13.984 | 1262.4 | 1391.8 | 1.8898 | 11.642 | 1262.0 | 1391.3 | 1.8694 |
| 740 | 17.799 | 1270.5 | 1402.2 | 1.9231 | 14.226 | 1270.1 | 1401.7 | 1.8982 | 11.844 | 1269.7 | 1401.2 | 1.8778 |
| 760 | 18.100 | 1278.2 | 1412.1 | 1.9313 | 14.467 | 1277.8 | 1411.7 | 1.9064 | 12.045 | 1277.5 | 1411.2 | 1.8860 |
| 780 | 18.401 | 1285.9 | 1422.1 | 1.9394 | 14.708 | 1285.6 | 1421.7 | 1.9145 | 12.247 | 1285.2 | 1421.2 | 1.8942 |
| 800 | 18.701 | 1293.7 | 1432.1 | 1.9474 | 14.949 | 1293.3 | 1431.7 | 1.9225 | 12.448 | 1293.0 | 1431.2 | 1.9022 |
| 850 | 19.452 | 1313.2 | 1457.2 | 1.9669 | 15.551 | 1312.9 | 1456.8 | 1.9421 | 12.951 | 1312.7 | 1456.4 | 1.9218 |
| 900 | 20.202 | 1333.0 | 1482.5 | 1.9859 | 16.152 | 1332.7 | 1482.2 | 1.9611 | 13.452 | 1332.5 | 1481.8 | 1.9408 |
| 950 | 20.951 | 1352.9 | 1508.0 | 2.0043 | 16.753 | 1352.7 | 1507.7 | 1.9796 | 13.954 | 1352.5 | 1507.4 | 1.9593 |
| 1000 | 21.700 | 1373.1 | 1533.8 | 2.0223 | 17.352 | 1372.9 | 1533.5 | 1.9975 | 14.454 | 1372.7 | 1533.2 | 1.9773 |
| 1100 | 23.20 | 1414.2 | 1585.9 | 2.0568 | 18.551 | 1414.0 | 1585.6 | 2.0321 | 15.454 | 1413.8 | 1585.4 | 2.0119 |
| 1200 | 24.69 | 1456.1 | 1638.9 | 2.0897 | 19.747 | 1456.0 | 1638.7 | 2.0650 | 16.452 | 1455.8 | 1638.5 | 2.0448 |
| 1300 | 26.18 | 1498.9 | 1692.8 | 2.1212 | 20.943 | 1498.8 | 1692.6 | 2.0966 | 17.449 | 1498.7 | 1692.4 | 2.0764 |
| 1400 | 27.68 | 1542.7 | 1747.6 | 2.1515 | 22.138 | 1542.6 | 1747.4 | 2.1269 | 18.445 | 1542.5 | 1747.3 | 2.1067 |
| 1500 | 29.17 | 1587.4 | 1803.3 | 2.1807 | 23.332 | 1587.3 | 1803.2 | 2.1561 | 19.441 | 1587.2 | 1803.0 | 2.1359 |
| 1600 | 30.66 | 1633.0 | 1859.9 | 2.2089 | 24.53 | 1632.9 | 1859.8 | 2.1843 | 20.44 | 1632.8 | 1859.7 | 2.1641 |
| 1800 | 33.64 | 1726.9 | 1975.9 | 2.2626 | 26.91 | 1726.8 | 1975.8 | 2.2380 | 22.43 | 1726.7 | 1975.7 | 2.2179 |
| 2000 | 36.62 | 1824.2 | 2095.3 | 2.3132 | 29.30 | 1824.1 | 2095.2 | 2.2886 | 24.41 | 1824.0 | 2095.1 | 2.2685 |
| 2200 | 39.61 | 1924.7 | 2217.8 | 2.3611 | 31.68 | 1924.6 | 2217.8 | 2.3365 | 26.40 | 1924.5 | 2217.7 | 2.3164 |
| 2400 | 42.59 | 2028.0 | 2343.2 | 2.4066 | 34.07 | 2027.9 | 2343.1 | 2.3820 | 28.39 | 2027.8 | 2343.1 | 2.3618 |

*(Continued)*

**TABLE A.3 (CONTINUED)**

Properties of Superheated Steam

| p (t Sat.) | 80 (312.07) | | | | 100 (327.86) | | | | 140 (353.08) | | | | 180 (373.13) | | | |
|---|---|---|---|---|---|---|---|---|---|---|---|---|---|---|---|---|
| t | v | u | h | s | v | u | h | s | v | u | h | s | v | u | h | s |
| Sat. | 5.474 | 1102.6 | 1183.6 | 1.6214 | 4.434 | 1105.8 | 1187.8 | 1.6034 | 3.221 | 1110.3 | 1193.8 | 1.5761 | 2.533 | 1113.4 | 1197.8 | 1.5553 |
| 250 | 4.902 | 1074.5 | 1147.1 | 1.5720 | | | | | | | | | | | | |
| 260 | 4.998 | 1079.3 | 1153.3 | 1.5806 | | | | | | | | | | | | |
| 270 | 5.093 | 1083.9 | 1159.3 | 1.5890 | | | | | | | | | | | | |
| 280 | 5.186 | 1088.5 | 1165.2 | 1.5970 | | | | | | | | | | | | |
| 290 | 5.277 | 1092.9 | 1171.1 | 1.6049 | | | | | | | | | | | | |
| 300 | 5.367 | 1097.4 | 1176.8 | 1.6125 | 4.228 | 1093.1 | 1171.4 | 1.5822 | | | | | | | | |
| 310 | 5.456 | 1101.7 | 1182.5 | 1.6199 | 4.303 | 1097.7 | 1177.4 | 1.5900 | | | | | | | | |
| 320 | 5.544 | 1106.0 | 1188.0 | 1.6271 | 4.377 | 1102.3 | 1183.3 | 1.5976 | 2.978 | 1089.2 | 1166.4 | 1.5414 | | | | |
| 330 | 5.631 | 1110.2 | 1193.5 | 1.6341 | 4.449 | 1106.7 | 1189.1 | 1.6050 | 3.037 | 1094.3 | 1173.0 | 1.5499 | | | | |
| 340 | 5.717 | 1114.3 | 1199.0 | 1.6409 | 4.521 | 1111.1 | 1194.8 | 1.6121 | 3.094 | 1099.3 | 1179.4 | 1.5581 | | | | |
| 350 | 5.802 | 1118.5 | 1204.3 | 1.6476 | 4.592 | 1115.4 | 1200.4 | 1.6191 | 3.150 | 1104.1 | 1185.7 | 1.5661 | 2.429 | 1101.7 | 1182.6 | 1.5368 |
| 360 | 5.886 | 1122.5 | 1209.7 | 1.6541 | 4.662 | 1119.7 | 1205.9 | 1.6259 | 3.205 | 1108.9 | 1191.9 | 1.5737 | 2.475 | 1106.9 | 1189.3 | 1.5450 |
| 370 | 5.970 | 1126.6 | 1215.0 | 1.6605 | 4.731 | 1123.9 | 1211.4 | 1.6326 | 3.259 | 1113.5 | 1198.0 | 1.5812 | 2.520 | 1111.9 | 1195.8 | 1.5529 |
| 380 | 6.053 | 1130.6 | 1220.2 | 1.6668 | 4.799 | 1128.0 | 1216.8 | 1.6391 | 3.312 | 1118.1 | 1203.9 | 1.5884 | 2.563 | 1116.7 | 1202.1 | 1.5605 |
| 390 | 6.135 | 1134.6 | 1225.4 | 1.6730 | 4.867 | 1132.1 | 1222.2 | 1.6455 | 3.364 | 1122.6 | 1209.7 | 1.5954 | 2.606 | 1121.5 | 1208.3 | 1.5678 |
| 400 | 6.217 | 1138.5 | 1230.6 | 1.6790 | 4.934 | 1136.2 | 1227.5 | 1.6517 | 3.415 | 1127.0 | 1215.5 | 1.6022 | 2.648 | 1126.2 | 1214.4 | 1.5749 |
| 410 | 6.299 | 1142.5 | 1235.7 | 1.6850 | 5.001 | 1140.3 | 1232.8 | 1.6578 | 3.466 | 1131.4 | 1221.2 | 1.6088 | 2.690 | 1130.8 | 1220.4 | 1.5818 |
| 420 | 6.380 | 1146.4 | 1240.8 | 1.6908 | 5.068 | 1144.3 | 1238.1 | 1.6638 | 3.517 | 1135.7 | 1226.8 | 1.6153 | 2.731 | 1135.3 | 1226.3 | 1.5886 |
| 430 | 6.460 | 1150.3 | 1245.9 | 1.6966 | 5.134 | 1148.3 | 1243.3 | 1.6697 | 3.567 | 1139.9 | 1232.3 | 1.6217 | 2.771 | 1139.8 | 1232.1 | 1.5951 |
| 440 | 6.541 | 1154.2 | 1251.0 | 1.7022 | 5.199 | 1152.3 | 1248.5 | 1.6755 | 3.616 | 1144.1 | 1237.8 | 1.6279 | 2.811 | 1144.2 | 1237.8 | 1.6015 |
| 450 | 6.621 | 1158.0 | 1256.0 | 1.7078 | 5.265 | 1156.2 | 1253.6 | 1.6812 | 3.665 | 1148.3 | 1243.3 | 1.6339 | 2.850 | 1148.5 | 1243.4 | 1.6078 |
| 460 | 6.700 | 1161.9 | 1261.1 | 1.7133 | 5.330 | 1160.1 | 1258.8 | 1.6868 | 3.713 | 1152.4 | 1248.6 | 1.6399 | 2.889 | 1152.8 | 1249.0 | 1.6139 |
| 470 | 6.780 | 1165.7 | 1266.1 | 1.7188 | 5.394 | 1164.1 | 1263.9 | 1.6923 | 3.762 | 1156.5 | 1254.0 | 1.6458 | 2.928 | 1157.0 | 1254.6 | 1.6199 |
| 480 | 6.859 | 1169.6 | 1271.1 | 1.7241 | 5.459 | 1168.0 | 1269.0 | 1.6978 | 3.810 | 1160.6 | 1259.3 | 1.6515 | 2.966 | 1161.3 | 1260.1 | 1.6257 |
| 490 | 6.938 | 1173.4 | 1276.1 | 1.7294 | 5.523 | 1171.8 | 1274.0 | 1.7032 | 3.857 | 1164.7 | 1264.6 | 1.6572 | 3.005 | 1165.5 | 1265.5 | 1.6315 |
| 500 | 7.017 | 1177.2 | 1281.1 | 1.7346 | 5.587 | 1175.7 | 1279.1 | 1.7085 | 3.904 | 1168.7 | 1269.9 | 1.6627 | 3.042 | 1169.6 | 1270.9 | 1.6372 |
| 510 | | | | | | | | | 3.952 | 1172.7 | 1275.1 | 1.6682 | 3.080 | 1173.7 | 1276.3 | 1.6427 |

*(Continued)*

**TABLE A.3 (CONTINUED)**
Properties of Superheated Steam

| T | v | u | h | s | v | u | h | s | v | u | h | s | v | u | h | s |
|---|---|---|---|---|---|---|---|---|---|---|---|---|---|---|---|---|
| 520 | 3.117 | 1177.8 | 1281.6 | 1.6482 | 4.045 | 1180.7 | 1285.5 | 1.6789 | 5.714 | 1183.5 | 1289.2 | 1.7189 | 7.173 | 1184.8 | 1291.0 | 1.7449 |
| 530 | 3.154 | 1181.9 | 1286.9 | 1.6536 | | | | | | | | | | | | |
| 540 | 3.191 | 1185.9 | 1292.2 | 1.6589 | 4.138 | 1188.6 | 1295.8 | 1.6893 | 5.840 | 1191.2 | 1299.2 | 1.7290 | 7.329 | 1192.4 | 1300.9 | 1.7549 |
| 550 | 3.228 | 1190.0 | 1297.5 | 1.6642 | | | | | | | | | | | | |
| 560 | 3.264 | 1194.0 | 1302.7 | 1.6693 | 4.230 | 1196.5 | 1306.0 | 1.6995 | 5.966 | 1198.9 | 1309.3 | 1.7390 | 7.485 | 1200.1 | 1310.9 | 1.7647 |
| 570 | 3.301 | 1198.0 | 1307.9 | 1.6744 | | | | | | | | | | | | |
| 580 | 3.337 | 1202.0 | 1313.1 | 1.6795 | 4.321 | 1204.3 | 1316.2 | 1.7094 | 6.091 | 1206.6 | 1319.3 | 1.7487 | 7.640 | 1207.7 | 1320.8 | 1.7743 |
| 590 | 3.373 | 1206.0 | 1318.3 | 1.6844 | | | | | | | | | | | | |
| 600 | 3.409 | 1210.0 | 1323.5 | 1.6893 | 4.412 | 1212.1 | 1326.4 | 1.7191 | 6.216 | 1214.2 | 1329.3 | 1.7582 | 7.794 | 1215.3 | 1330.7 | 1.7838 |
| 620 | 3.481 | 1217.9 | 1333.9 | 1.6990 | 4.502 | 1219.9 | 1336.6 | 1.7286 | 6.340 | 1221.9 | 1339.3 | 1.7675 | 7.948 | 1222.9 | 1340.6 | 1.7930 |
| 640 | 3.552 | 1225.8 | 1344.1 | 1.7084 | 4.592 | 1227.7 | 1346.7 | 1.7379 | 6.464 | 1229.6 | 1349.2 | 1.7767 | 8.102 | 1230.5 | 1350.5 | 1.8021 |
| 660 | 3.623 | 1233.8 | 1354.4 | 1.7177 | 4.682 | 1235.6 | 1356.8 | 1.7470 | 6.588 | 1237.3 | 1359.2 | 1.7857 | 8.255 | 1238.2 | 1360.4 | 1.8111 |
| 680 | 3.693 | 1241.7 | 1364.7 | 1.7268 | 4.771 | 1243.4 | 1367.0 | 1.7560 | 6.711 | 1245.0 | 1369.2 | 1.7946 | 8.408 | 1245.9 | 1370.4 | 1.8199 |
| 700 | 3.763 | 1249.6 | 1374.9 | 1.7357 | 4.860 | 1251.2 | 1377.1 | 1.7648 | 6.834 | 1252.8 | 1379.2 | 1.8033 | 8.561 | 1253.6 | 1380.3 | 1.8285 |
| 720 | 3.833 | 1257.5 | 1385.2 | 1.7445 | 4.949 | 1259.0 | 1387.2 | 1.7735 | 6.957 | 1260.5 | 1389.3 | 1.8118 | 8.714 | 1261.3 | 1390.3 | 1.8371 |
| 740 | 3.903 | 1265.4 | 1395.4 | 1.7531 | 5.037 | 1266.9 | 1397.4 | 1.7820 | 7.079 | 1268.3 | 1399.3 | 1.8203 | 8.866 | 1269.0 | 1400.3 | 1.8455 |
| 760 | 3.972 | 1273.4 | 1405.7 | 1.7615 | 5.125 | 1274.7 | 1407.5 | 1.7904 | 7.201 | 1276.1 | 1409.4 | 1.8286 | 9.018 | 1276.8 | 1410.3 | 1.8538 |
| 780 | 4.041 | 1281.3 | 1415.9 | 1.7699 | 5.214 | 1282.6 | 1417.7 | 1.7986 | 7.324 | 1283.9 | 1419.5 | 1.8368 | 9.170 | 1284.6 | 1420.3 | 1.8619 |
| 800 | 4.110 | 1289.3 | 1426.2 | 1.7781 | 5.301 | 1290.5 | 1427.9 | 1.8068 | 7.445 | 1291.8 | 1429.6 | 1.8449 | 9.321 | 1292.4 | 1430.4 | 1.8700 |
| 820 | 4.179 | 1297.3 | 1436.5 | 1.7862 | 5.389 | 1298.5 | 1438.1 | 1.8148 | 7.567 | 1299.7 | 1439.7 | 1.8529 | | | | |
| 840 | 4.248 | 1305.3 | 1446.7 | 1.7942 | 5.477 | 1306.4 | 1448.3 | 1.8228 | 7.689 | 1307.6 | 1449.9 | 1.8607 | | | | |
| 850 | | | | | | | | | | | | | 9.700 | 1312.1 | 1455.7 | 1.8897 |
| 860 | 4.316 | 1313.3 | 1457.1 | 1.8020 | 5.564 | 1314.4 | 1458.5 | 1.8306 | 7.810 | 1315.5 | 1460.0 | 1.8685 | | | | |
| 880 | 4.385 | 1321.3 | 1467.4 | 1.8098 | 5.651 | 1322.4 | 1468.8 | 1.8383 | 7.932 | 1323.5 | 1470.2 | 1.8762 | | | | |
| 900 | 4.453 | 1329.4 | 1477.7 | 1.8175 | 5.739 | 1330.4 | 1479.1 | 1.8459 | 8.053 | 1331.5 | 1480.5 | 1.8838 | 10.078 | 1332.0 | 1481.2 | 1.9087 |
| 920 | 4.521 | 1337.5 | 1488.1 | 1.8250 | 5.826 | 1338.5 | 1489.4 | 1.8535 | 8.174 | 1339.5 | 1490.7 | 1.8913 | | | | |
| 940 | 4.589 | 1345.6 | 1498.5 | 1.8325 | 5.913 | 1346.6 | 1499.8 | 1.8609 | 8.295 | 1347.5 | 1501.0 | 1.8987 | | | | |
| 950 | | | | | | | | | | | | | 10.455 | 1352.0 | 1506.8 | 1.9273 |
| 960 | 4.657 | 1353.8 | 1508.9 | 1.8399 | 6.000 | 1354.7 | 1510.1 | 1.8683 | 8.416 | 1355.6 | 1511.3 | 1.9060 | | | | |
| 980 | 4.725 | 1362.0 | 1519.4 | 1.8472 | 6.086 | 1362.9 | 1520.5 | 1.8755 | 8.537 | 1363.7 | 1521.7 | 1.9132 | | | | |
| 1000 | 4.793 | 1370.2 | 1529.8 | 1.8545 | 6.173 | 1371.0 | 1531.0 | 1.8827 | 8.657 | 1371.9 | 1532.1 | 1.9204 | 10.831 | 1372.3 | 1532.6 | 1.9453 |

(*Continued*)

**TABLE A.3 (CONTINUED)**

Properties of Superheated Steam

| t | v | u | h | s | v | u | h | s | v | u | h | s | v | u | h | s |
|---|---|---|---|---|---|---|---|---|---|---|---|---|---|---|---|---|
| 1020 | | | | | 8.778 | 1380.1 | 1542.5 | 1.9275 | 6.260 | 1379.3 | 1541.4 | 1.8898 | 4.861 | 1378.4 | 1540.3 | 1.8616 |
| 1040 | | | | | 8.899 | 1388.3 | 1552.9 | 1.9345 | 6.346 | 1387.5 | 1551.9 | 1.8969 | 4.928 | 1386.7 | 1550.9 | 1.8687 |
| 1060 | | | | | 9.019 | 1396.5 | 1563.4 | 1.9414 | 6.433 | 1395.8 | 1562.4 | 1.9039 | 4.996 | 1395.0 | 1561.4 | 1.8757 |
| 1080 | | | | | 9.140 | 1404.8 | 1573.9 | 1.9483 | 6.519 | 1404.1 | 1573.0 | 1.9107 | 5.063 | 1403.4 | 1572.0 | 1.8826 |
| 1100 | 11.583 | 1413.5 | 1584.9 | 1.9799 | 9.260 | 1413.1 | 1584.5 | 1.9551 | 6.605 | 1412.4 | 1583.6 | 1.9176 | 5.131 | 1411.7 | 1582.6 | 1.8894 |
| 1200 | 12.333 | 1455.5 | 1638.1 | 2.0130 | 9.861 | 1455.2 | 1637.7 | 1.9882 | 7.036 | 1454.6 | 1636.9 | 1.9507 | 5.467 | 1454.0 | 1636.1 | 1.9227 |
| 1300 | 13.081 | 1498.4 | 1692.1 | 2.0446 | 10.461 | 1498.2 | 1691.8 | 2.0198 | 7.466 | 1497.7 | 1691.1 | 1.9824 | 5.802 | 1497.2 | 1690.4 | 1.9544 |
| 1400 | 13.830 | 1542.3 | 1747.0 | 2.0749 | 11.060 | 1542.0 | 1746.7 | 2.0502 | 7.895 | 1541.6 | 1746.1 | 2.0129 | 6.137 | 1541.2 | 1745.6 | 1.9849 |
| 1500 | 14.577 | 1587.0 | 1802.8 | 2.1041 | 11.659 | 1586.8 | 1802.5 | 2.0794 | 8.324 | 1586.4 | 1802.0 | 2.0421 | 6.471 | 1586.0 | 1801.5 | 2.0142 |
| 1600 | 15.324 | 1632.6 | 1859.5 | 2.1323 | 12.257 | 1632.4 | 1859.3 | 2.1076 | 8.752 | 1632.1 | 1858.8 | 2.0704 | 6.804 | 1631.7 | 1858.4 | 2.0425 |
| 1800 | 16.818 | 1726.5 | 1975.5 | 2.1861 | 13.452 | 1726.4 | 1975.3 | 2.1614 | 9.607 | 1726.1 | 1975.0 | 2.1242 | 7.470 | 1725.8 | 1974.6 | 2.0964 |
| 2000 | 18.310 | 1823.9 | 2094.9 | 2.2367 | 14.647 | 1823.7 | 2094.8 | 2.2121 | 10.461 | 1823.5 | 2094.5 | 2.1749 | 8.135 | 1823.2 | 2094.2 | 2.1470 |
| 2200 | 19.802 | 1924.4 | 2217.5 | 2.2846 | 15.842 | 1924.3 | 2217.4 | 2.2600 | 11.315 | 1924.0 | 2217.1 | 2.2228 | 8.800 | 1923.7 | 2216.8 | 2.1950 |
| 2400 | 21.294 | 2027.7 | 2342.9 | 2.3301 | 17.036 | 2027.5 | 2342.8 | 2.3054 | 12.169 | 2027.3 | 2342.5 | 2.2682 | 9.465 | 2027.0 | 2342.2 | 2.2404 |

| p (t Sat.) | 200 (381.86) | | | | 300 (417.43) | | | | 350 (431.82) | | | | 400 (444.70) | | | |
|---|---|---|---|---|---|---|---|---|---|---|---|---|---|---|---|---|
| t | v | u | h | s | v | u | h | s | v | u | h | s | v | u | h | s |
| Sat. | 2.289 | 1114.6 | 1199.3 | 1.5464 | 1.5442 | 1118.2 | 1203.9 | 1.5115 | 1.3267 | 1119.0 | 1204.9 | 1.4978 | 1.1620 | 1119.5 | 1205.5 | 1.4856 |
| 400 | 2.361 | 1123.5 | 1210.8 | 1.5600 | 1.4915 | 1108.2 | 1191.0 | 1.4967 | | | | | | | | |
| 410 | 2.399 | 1128.2 | 1217.0 | 1.5672 | 1.5221 | 1114.0 | 1198.5 | 1.5054 | | | | | | | | |
| 420 | 2.437 | 1132.9 | 1223.1 | 1.5741 | 1.5517 | 1119.6 | 1205.7 | 1.5136 | 1.2950 | 1112.0 | 1195.9 | 1.4875 | | | | |
| 430 | 2.475 | 1137.5 | 1229.1 | 1.5809 | 1.5805 | 1125.0 | 1212.7 | 1.5216 | 1.3219 | 1117.9 | 1203.6 | 1.4962 | 1.1257 | 1110.3 | 1193.6 | 1.4723 |
| 440 | 2.511 | 1142.0 | 1234.9 | 1.5874 | 1.6086 | 1130.3 | 1219.6 | 1.5292 | 1.3480 | 1123.7 | 1211.0 | 1.5045 | 1.1506 | 1116.6 | 1201.7 | 1.4814 |
| 450 | 2.548 | 1146.4 | 1240.7 | 1.5938 | 1.6361 | 1135.4 | 1226.2 | 1.5365 | 1.3733 | 1129.2 | 1218.2 | 1.5125 | 1.1745 | 1122.6 | 1209.6 | 1.4901 |
| 460 | 2.584 | 1150.8 | 1246.5 | 1.6001 | 1.6630 | 1140.4 | 1232.7 | 1.5436 | 1.3979 | 1134.6 | 1225.2 | 1.5201 | 1.1977 | 1128.5 | 1217.1 | 1.4984 |
| 470 | 2.619 | 1155.2 | 1252.1 | 1.6062 | 1.6894 | 1145.3 | 1239.1 | 1.5505 | 1.4220 | 1139.9 | 1232.0 | 1.5275 | 1.2202 | 1134.1 | 1224.4 | 1.5063 |
| 480 | 2.654 | 1159.5 | 1257.7 | 1.6122 | 1.7154 | 1150.1 | 1245.3 | 1.5572 | 1.4455 | 1145.0 | 1238.6 | 1.5346 | 1.2421 | 1139.6 | 1231.5 | 1.5139 |
| 490 | 2.689 | 1163.7 | 1263.3 | 1.6181 | 1.7410 | 1154.8 | 1251.5 | 1.5637 | 1.4686 | 1150.0 | 1245.1 | 1.5415 | 1.2634 | 1144.9 | 1238.4 | 1.5212 |
| 500 | 2.724 | 1168.0 | 1268.8 | 1.6239 | 1.7662 | 1159.5 | 1257.5 | 1.5701 | 1.4913 | 1154.9 | 1251.5 | 1.5482 | 1.2843 | 1150.1 | 1245.2 | 1.5282 |
| 510 | 2.758 | 1172.2 | 1274.2 | 1.6295 | 1.7910 | 1164.1 | 1263.5 | 1.5763 | 1.5136 | 1159.7 | 1257.8 | 1.5546 | 1.3048 | 1155.2 | 1251.8 | 1.5351 |
| 520 | 2.792 | 1176.3 | 1279.7 | 1.6351 | 1.8156 | 1168.6 | 1269.4 | 1.5823 | 1.5356 | 1164.5 | 1263.9 | 1.5610 | 1.3249 | 1160.2 | 1258.2 | 1.5417 |

*(Continued)*

**TABLE A.3 (CONTINUED)**
Properties of Superheated Steam

| Temp | | | | | | | | | | | | | | | | | |
|---|---|---|---|---|---|---|---|---|---|---|---|---|---|---|---|---|---|
| 530 | 2.826 | 1180.5 | 1285.0 | 1.6405 | 1.8399 | 1173.1 | 1275.2 | 1.5882 | 1.5572 | 1169.1 | 1270.0 | 1.5671 | 1.3447 | 1165.0 | 1264.6 | 1.5481 |
| 540 | 2.860 | 1184.6 | 1290.4 | 1.6459 | 1.8640 | 1177.5 | 1281.0 | 1.5940 | 1.5786 | 1173.7 | 1276.0 | 1.5731 | 1.3642 | 1169.8 | 1270.8 | 1.5544 |
| 550 | 2.893 | 1188.7 | 1295.7 | 1.6512 | 1.8878 | 1181.9 | 1286.7 | 1.5997 | 1.5998 | 1178.3 | 1281.9 | 1.5790 | 1.3833 | 1174.6 | 1277.0 | 1.5605 |
| 560 | 2.926 | 1192.7 | 1301.0 | 1.6565 | 1.9114 | 1186.2 | 1292.3 | 1.6052 | 1.6207 | 1182.8 | 1287.7 | 1.5848 | 1.4023 | 1179.2 | 1283.0 | 1.5665 |
| 570 | 2.960 | 1196.8 | 1306.3 | 1.6616 | 1.9348 | 1190.5 | 1297.9 | 1.6107 | 1.6414 | 1187.2 | 1293.5 | 1.5904 | 1.4210 | 1183.8 | 1289.0 | 1.5723 |
| 580 | 2.993 | 1200.8 | 1311.6 | 1.6667 | 1.9580 | 1194.8 | 1303.5 | 1.6161 | 1.6619 | 1191.6 | 1299.3 | 1.5960 | 1.4395 | 1188.4 | 1294.9 | 1.5781 |
| 590 | 3.025 | 1204.9 | 1316.8 | 1.6717 | 1.9811 | 1199.0 | 1309.0 | 1.6214 | 1.6823 | 1196.0 | 1304.9 | 1.6014 | 1.4579 | 1192.9 | 1300.8 | 1.5837 |
| 600 | 3.058 | 1208.9 | 1322.1 | 1.6767 | 2.004 | 1203.2 | 1314.5 | 1.6266 | 1.7025 | 1200.3 | 1310.6 | 1.6068 | 1.4760 | 1197.3 | 1306.6 | 1.5892 |
| 620 | 3.123 | 1216.9 | 1332.5 | 1.6864 | 2.049 | 1211.6 | 1325.4 | 1.6368 | 1.7424 | 1208.9 | 1321.8 | 1.6172 | 1.5118 | 1206.1 | 1318.0 | 1.5999 |
| 640 | 3.188 | 1224.9 | 1342.9 | 1.6959 | 2.094 | 1220.0 | 1336.2 | 1.6467 | 1.7818 | 1217.4 | 1332.8 | 1.6274 | 1.5471 | 1214.8 | 1329.3 | 1.6103 |
| 660 | 3.252 | 1232.8 | 1353.2 | 1.7053 | 2.139 | 1228.2 | 1347.0 | 1.6564 | 1.8207 | 1225.8 | 1343.8 | 1.6372 | 1.5819 | 1223.4 | 1340.5 | 1.6203 |
| 680 | 3.316 | 1240.8 | 1363.5 | 1.7144 | 2.183 | 1236.4 | 1357.6 | 1.6658 | 1.8593 | 1234.2 | 1354.6 | 1.6469 | 1.6163 | 1231.9 | 1351.6 | 1.6301 |
| 700 | 3.379 | 1248.8 | 1373.8 | 1.7234 | 2.227 | 1244.6 | 1368.3 | 1.6751 | 1.8975 | 1242.5 | 1365.4 | 1.6562 | 1.6503 | 1240.4 | 1362.5 | 1.6397 |
| 720 | 3.442 | 1256.7 | 1384.1 | 1.7322 | 2.270 | 1252.8 | 1378.9 | 1.6841 | 1.9354 | 1250.8 | 1376.2 | 1.6654 | 1.6840 | 1248.8 | 1373.4 | 1.6490 |
| 740 | 3.505 | 1264.7 | 1394.4 | 1.7408 | 2.314 | 1261.0 | 1389.4 | 1.6930 | 1.9731 | 1259.1 | 1386.9 | 1.6744 | 1.7175 | 1257.2 | 1384.3 | 1.6581 |
| 760 | 3.568 | 1272.7 | 1404.7 | 1.7493 | 2.357 | 1269.1 | 1400.0 | 1.7017 | 2.0104 | 1267.3 | 1397.5 | 1.6832 | 1.7506 | 1265.5 | 1395.1 | 1.6670 |
| 780 | 3.631 | 1280.6 | 1415.0 | 1.7577 | 2.400 | 1277.3 | 1410.5 | 1.7103 | 2.0476 | 1275.6 | 1408.2 | 1.6919 | 1.7836 | 1273.8 | 1405.9 | 1.6758 |
| 800 | 3.693 | 1288.6 | 1425.3 | 1.7660 | 2.442 | 1285.4 | 1421.0 | 1.7187 | 2.085 | 1283.8 | 1418.8 | 1.7004 | 1.8163 | 1282.1 | 1416.6 | 1.6844 |
| 820 | 3.755 | 1296.6 | 1435.6 | 1.7741 | 2.485 | 1293.6 | 1431.5 | 1.7270 | 2.121 | 1292.0 | 1429.4 | 1.7088 | 1.8489 | 1290.5 | 1427.3 | 1.6928 |
| 840 | 3.818 | 1304.7 | 1446.0 | 1.7821 | 2.527 | 1301.7 | 1442.0 | 1.7351 | 2.158 | 1300.3 | 1440.0 | 1.7170 | 1.8813 | 1298.8 | 1438.0 | 1.7011 |
| 860 | 3.879 | 1312.7 | 1456.3 | 1.7900 | 2.569 | 1309.9 | 1452.5 | 1.7432 | 2.194 | 1308.5 | 1450.6 | 1.7251 | 1.9135 | 1307.1 | 1448.7 | 1.7093 |
| 880 | 3.941 | 1320.8 | 1466.7 | 1.7978 | 2.611 | 1318.1 | 1463.1 | 1.7511 | 2.231 | 1316.7 | 1461.2 | 1.7331 | 1.9456 | 1315.4 | 1459.4 | 1.7173 |
| 900 | 4.003 | 1328.9 | 1477.1 | 1.8055 | 2.653 | 1326.3 | 1473.6 | 1.7589 | 2.267 | 1325.0 | 1471.8 | 1.7409 | 1.9776 | 1323.7 | 1470.1 | 1.7252 |
| 920 | 4.064 | 1337.0 | 1487.5 | 1.8131 | 2.695 | 1334.5 | 1484.1 | 1.7666 | 2.303 | 1333.3 | 1482.5 | 1.7487 | 2.0094 | 1332.0 | 1480.8 | 1.7330 |
| 940 | 4.126 | 1345.2 | 1497.9 | 1.8206 | 2.736 | 1342.8 | 1494.7 | 1.7742 | 2.339 | 1341.6 | 1493.1 | 1.7563 | 2.0411 | 1340.4 | 1491.5 | 1.7407 |
| 960 | 4.187 | 1353.3 | 1508.3 | 1.8280 | 2.778 | 1351.1 | 1505.3 | 1.7817 | 2.375 | 1349.9 | 1503.7 | 1.7639 | 2.0727 | 1348.7 | 1502.2 | 1.7483 |
| 980 | 4.249 | 1361.6 | 1518.8 | 1.8353 | 2.819 | 1359.3 | 1515.8 | 1.7891 | 2.411 | 1358.2 | 1514.4 | 1.7713 | 2.1043 | 1357.1 | 1512.9 | 1.7558 |
| 1000 | 4.310 | 1369.8 | 1529.3 | 1.8425 | 2.860 | 1367.7 | 1526.5 | 1.7964 | 2.446 | 1366.6 | 1525.0 | 1.7787 | 2.136 | 1365.5 | 1523.6 | 1.7632 |
| 1020 | 4.371 | 1378.0 | 1539.8 | 1.8497 | 2.902 | 1376.0 | 1537.1 | 1.8036 | 2.482 | 1375.0 | 1535.7 | 1.7859 | 2.167 | 1373.9 | 1534.3 | 1.7705 |
| 1040 | 4.432 | 1386.3 | 1550.3 | 1.8568 | 2.943 | 1384.4 | 1547.7 | 1.8108 | 2.517 | 1383.4 | 1546.4 | 1.7931 | 2.198 | 1382.4 | 1545.1 | 1.7777 |
| 1060 | 4.493 | 1394.6 | 1560.9 | 1.8638 | 2.984 | 1392.7 | 1558.4 | 1.8178 | 2.553 | 1391.8 | 1557.1 | 1.8002 | 2.229 | 1390.8 | 1555.9 | 1.7849 |

*(Continued)*

# TABLE A.3 (CONTINUED)

## Properties of Superheated Steam

| t | v | u | h | s | v | u | h | s | v | u | h | s | v | u | h | s |
|---|---|---|---|---|---|---|---|---|---|---|---|---|---|---|---|---|
| 1080 | 4.554 | 1403.0 | 1571.5 | 1.8707 | 3.025 | 1401.2 | 1569.1 | 1.8248 | 2.588 | 1400.2 | 1567.9 | 1.8072 | 2.261 | 1399.3 | 1566.6 | 1.7919 |
| 1100 | 4.615 | 1411.4 | 1582.2 | 1.8776 | 3.066 | 1409.6 | 1579.8 | 1.8317 | 2.624 | 1408.7 | 1578.6 | 1.8142 | 2.292 | 1407.8 | 1577.4 | 1.7989 |
| 1200 | 4.918 | 1453.7 | 1635.7 | 1.9109 | 3.270 | 1452.2 | 1633.4 | 1.8653 | 2.799 | 1451.5 | 1632.8 | 1.8478 | 2.446 | 1450.7 | 1631.8 | 1.8327 |
| 1300 | 5.220 | 1496.9 | 1690.1 | 1.9427 | 3.473 | 1495.6 | 1688.4 | 1.8973 | 2.974 | 1495.0 | 1687.6 | 1.8799 | 2.599 | 1494.3 | 1686.8 | 1.8648 |
| 1400 | 5.521 | 1540.9 | 1745.3 | 1.9732 | 3.675 | 1539.8 | 1743.8 | 1.9279 | 3.148 | 1539.3 | 1743.1 | 1.9106 | 2.752 | 1538.7 | 1742.4 | 1.8956 |
| 1500 | 5.822 | 1585.8 | 1801.3 | 2.0025 | 3.877 | 1584.8 | 1800.0 | 1.9573 | 3.321 | 1584.3 | 1799.4 | 1.9401 | 2.904 | 1583.8 | 1798.8 | 1.9251 |
| 1600 | 6.123 | 1631.6 | 1858.2 | 2.0308 | 4.078 | 1630.7 | 1857.0 | 1.9857 | 3.494 | 1630.2 | 1856.5 | 1.9685 | 3.055 | 1629.8 | 1855.9 | 1.9535 |
| 1800 | 6.722 | 1725.6 | 1974.4 | 2.0847 | 4.479 | 1724.9 | 1973.5 | 2.0396 | 3.838 | 1724.5 | 1973.1 | 2.0225 | 3.357 | 1724.1 | 1972.6 | 2.0076 |
| 2000 | 7.321 | 1823.0 | 2094.0 | 2.1354 | 4.879 | 1822.3 | 2093.2 | 2.0904 | 4.182 | 1822.0 | 2092.8 | 2.0733 | 3.658 | 1821.6 | 2092.4 | 2.0584 |
| 2200 | 7.920 | 1923.6 | 2216.7 | 2.1833 | 5.280 | 1922.9 | 2216.0 | 2.1384 | 4.525 | 1922.5 | 2215.6 | 2.1212 | 3.959 | 1922.2 | 2215.2 | 2.1064 |
| 2400 | 8.518 | 2026.8 | 2342.1 | 2.2288 | 5.679 | 2026.1 | 2341.4 | 2.1838 | 4.868 | 2025.8 | 2341.1 | 2.1667 | 4.260 | 2025.4 | 2340.8 | 2.1519 |

| p (t Sat.) | 450 (456.39) | | | | 500 (467.13) | | | | 550 (477.07) | | | | 600 (486.33) | | | |
|---|---|---|---|---|---|---|---|---|---|---|---|---|---|---|---|---|
| t | v | u | h | s | v | u | h | s | v | u | h | s | v | u | h | s |
| Sat. | 1.0326 | 1119.6 | 1205.6 | 1.4746 | .9283 | 1119.4 | 1205.3 | 1.4645 | .8423 | 1119.1 | 1204.8 | 1.4551 | .7702 | 1118.6 | 1204.1 | 1.4464 |
| 450 | 1.0183 | 1115.5 | 1200.3 | 1.4687 | .9133 | 1114.7 | 1199.2 | 1.4578 | .8283 | 1114.1 | 1198.4 | 1.4483 | .7582 | 1113.9 | 1198.1 | 1.4401 |
| 460 | 1.0405 | 1121.8 | 1208.5 | 1.4777 | | | | | | | | | | | | |
| 470 | 1.0620 | 1128.0 | 1216.4 | 1.4863 | .9342 | 1121.3 | 1207.8 | 1.4671 | | | | | | | | |
| 480 | 1.0828 | 1133.8 | 1224.0 | 1.4944 | .9543 | 1127.7 | 1216.0 | 1.4759 | .8480 | 1121.1 | 1207.4 | 1.4579 | | | | |
| 490 | 1.1029 | 1139.5 | 1231.4 | 1.5022 | .9736 | 1133.8 | 1223.9 | 1.4843 | .8669 | 1127.7 | 1215.9 | 1.4669 | .7769 | 1121.1 | 1207.4 | 1.4500 |
| 500 | 1.1226 | 1145.1 | 1238.5 | 1.5097 | .9924 | 1139.7 | 1231.5 | 1.4923 | .8850 | 1134.0 | 1224.1 | 1.4755 | .7947 | 1128.0 | 1216.2 | 1.4592 |
| 510 | 1.1417 | 1150.4 | 1245.5 | 1.5170 | 1.0106 | 1145.4 | 1238.9 | 1.4999 | .9056 | 1140.1 | 1232.0 | 1.4836 | .8118 | 1134.5 | 1224.6 | 1.4679 |
| 520 | 1.1605 | 1155.7 | 1252.3 | 1.5239 | 1.0283 | 1150.9 | 1246.1 | 1.5073 | .9196 | 1146.0 | 1239.6 | 1.4914 | .8283 | 1140.7 | 1232.7 | 1.4762 |
| 530 | 1.1788 | 1160.8 | 1258.9 | 1.5307 | 1.0456 | 1156.3 | 1253.1 | 1.5144 | .9361 | 1151.7 | 1246.9 | 1.4989 | .8443 | 1146.8 | 1240.5 | 1.4841 |
| 540 | 1.1969 | 1165.8 | 1265.5 | 1.5372 | 1.0625 | 1161.6 | 1259.9 | 1.5212 | .9522 | 1157.2 | 1254.1 | 1.5061 | .8598 | 1152.6 | 1248.0 | 1.4917 |
| 550 | 1.2146 | 1170.7 | 1271.9 | 1.5436 | 1.0792 | 1166.7 | 1266.6 | 1.5279 | .9679 | 1162.6 | 1261.1 | 1.5131 | .8749 | 1158.2 | 1255.4 | 1.4990 |
| 560 | 1.2320 | 1175.6 | 1278.2 | 1.5498 | 1.0955 | 1171.8 | 1273.1 | 1.5343 | .9834 | 1167.8 | 1267.9 | 1.5198 | .8896 | 1163.7 | 1262.5 | 1.5060 |
| 570 | 1.2492 | 1180.3 | 1284.4 | 1.5559 | 1.1115 | 1176.7 | 1279.5 | 1.5406 | .9985 | 1173.0 | 1274.6 | 1.5263 | .9040 | 1169.1 | 1269.5 | 1.5128 |
| 580 | 1.2662 | 1185.0 | 1290.5 | 1.5618 | 1.1273 | 1181.6 | 1285.9 | 1.5467 | 1.0133 | 1178.0 | 1281.1 | 1.5327 | .9181 | 1174.3 | 1276.3 | 1.5194 |
| 590 | 1.2830 | 1189.7 | 1296.5 | 1.5675 | 1.1429 | 1186.4 | 1292.1 | 1.5527 | 1.0280 | 1183.0 | 1287.6 | 1.5388 | .9320 | 1179.5 | 1282.9 | 1.5253 |
| 600 | 1.2996 | 1194.3 | 1302.5 | 1.5732 | 1.1583 | 1191.1 | 1298.3 | 1.5585 | 1.0424 | 1187.9 | 1293.9 | 1.5448 | .9456 | 1184.5 | 1289.5 | 1.5320 |

*(Continued)*

**TABLE A.3 (CONTINUED)**

Properties of Superheated Steam

| | | | | | | | | | | | | | | | | |
|---|---|---|---|---|---|---|---|---|---|---|---|---|---|---|---|---|
| 610 | 1.3160 | 1198.8 | 1308.4 | 1.5788 | 1.1735 | 1195.8 | 1304.3 | 1.5642 | 1.0566 | 1192.7 | 1300.2 | 1.5507 | .9590 | 1189.5 | 1295.9 | 1.5381 |
| 620 | 1.3323 | 1203.3 | 1314.2 | 1.5842 | 1.1885 | 1200.4 | 1310.4 | 1.5698 | 1.0706 | 1197.4 | 1306.4 | 1.5565 | .9722 | 1194.4 | 1302.3 | 1.5440 |
| 630 | 1.3484 | 1207.8 | 1320.0 | 1.5896 | 1.2033 | 1205.0 | 1316.3 | 1.5753 | 1.0845 | 1202.1 | 1312.5 | 1.5621 | .9853 | 1199.2 | 1308.6 | 1.5497 |
| 640 | 1.3644 | 1212.2 | 1325.8 | 1.5948 | 1.2181 | 1209.5 | 1322.2 | 1.5807 | 1.0982 | 1206.7 | 1318.5 | 1.5676 | .9982 | 1203.9 | 1314.8 | 1.5554 |
| 650 | 1.3803 | 1216.6 | 1331.5 | 1.6000 | 1.2327 | 1214.0 | 1328.0 | 1.5860 | 1.1118 | 1211.3 | 1324.5 | 1.5730 | 1.0109 | 1208.6 | 1320.9 | 1.5609 |
| 660 | 1.3960 | 1221.0 | 1337.2 | 1.6051 | 1.2472 | 1218.4 | 1333.8 | 1.5912 | 1.1252 | 1215.9 | 1330.4 | 1.5783 | 1.0235 | 1213.3 | 1326.9 | 1.5664 |
| 670 | 1.4116 | 1225.3 | 1342.9 | 1.6101 | 1.2615 | 1222.9 | 1339.6 | 1.5963 | 1.1386 | 1220.4 | 1336.3 | 1.5836 | 1.0360 | 1217.9 | 1332.9 | 1.5717 |
| 680 | 1.4272 | 1229.6 | 1348.5 | 1.6151 | 1.2758 | 1227.3 | 1345.3 | 1.6014 | 1.1518 | 1224.9 | 1342.1 | 1.5887 | 1.0483 | 1222.5 | 1338.9 | 1.5769 |
| 690 | 1.4426 | 1233.9 | 1354.1 | 1.6199 | 1.2899 | 1231.7 | 1351.0 | 1.6063 | 1.1649 | 1229.4 | 1347.9 | 1.5938 | 1.0606 | 1227.0 | 1344.8 | 1.5821 |
| 700 | 1.4580 | 1238.2 | 1359.6 | 1.6248 | 1.3040 | 1236.0 | 1356.7 | 1.6112 | 1.1779 | 1233.8 | 1353.7 | 1.5987 | 1.0727 | 1231.5 | 1350.6 | 1.5872 |
| 720 | 1.4884 | 1246.7 | 1370.7 | 1.6342 | 1.3319 | 1244.7 | 1367.9 | 1.6208 | 1.2037 | 1242.6 | 1365.1 | 1.6085 | 1.0968 | 1240.4 | 1362.2 | 1.5971 |
| 740 | 1.5186 | 1255.2 | 1381.7 | 1.6435 | 1.3594 | 1253.3 | 1379.1 | 1.6302 | 1.2291 | 1251.3 | 1376.4 | 1.6180 | 1.1205 | 1249.3 | 1373.7 | 1.6067 |
| 760 | 1.5485 | 1263.7 | 1392.6 | 1.6525 | 1.3867 | 1261.8 | 1390.1 | 1.6394 | 1.2543 | 1260.0 | 1387.6 | 1.6273 | 1.1439 | 1258.1 | 1385.1 | 1.6161 |
| 780 | 1.5782 | 1272.1 | 1403.5 | 1.6614 | 1.4138 | 1270.3 | 1401.1 | 1.6483 | 1.2793 | 1268.6 | 1398.8 | 1.6364 | 1.1671 | 1266.8 | 1396.3 | 1.6253 |
| 800 | 1.6077 | 1280.5 | 1414.4 | 1.6701 | 1.4407 | 1278.8 | 1412.1 | 1.6571 | 1.3040 | 1277.1 | 1409.8 | 1.6452 | 1.1900 | 1275.4 | 1407.6 | 1.6343 |
| 820 | 1.6369 | 1288.9 | 1425.2 | 1.6786 | 1.4673 | 1287.3 | 1423.0 | 1.6657 | 1.3285 | 1285.7 | 1420.9 | 1.6539 | 1.2128 | 1284.1 | 1418.7 | 1.6430 |
| 840 | 1.6660 | 1297.3 | 1436.0 | 1.6870 | 1.4938 | 1295.7 | 1433.9 | 1.6742 | 1.3528 | 1294.2 | 1431.9 | 1.6625 | 1.2353 | 1292.7 | 1429.8 | 1.6517 |
| 860 | 1.6950 | 1305.6 | 1446.8 | 1.6952 | 1.5201 | 1304.2 | 1444.8 | 1.6825 | 1.3770 | 1302.7 | 1442.9 | 1.6708 | 1.2577 | 1301.2 | 1440.9 | 1.6601 |
| 880 | 1.7238 | 1314.0 | 1457.5 | 1.7033 | 1.5463 | 1312.6 | 1455.7 | 1.6906 | 1.4010 | 1311.2 | 1453.8 | 1.6791 | 1.2800 | 1309.8 | 1451.9 | 1.6684 |
| 900 | 1.7524 | 1322.4 | 1468.3 | 1.7113 | 1.5723 | 1321.0 | 1466.5 | 1.6987 | 1.4249 | 1319.7 | 1464.7 | 1.6872 | 1.3021 | 1318.4 | 1462.9 | 1.6766 |
| 920 | 1.7810 | 1330.8 | 1479.1 | 1.7191 | 1.5982 | 1329.5 | 1477.4 | 1.7066 | 1.4487 | 1328.2 | 1475.6 | 1.6951 | 1.3240 | 1326.9 | 1473.9 | 1.6846 |
| 940 | 1.8094 | 1339.2 | 1489.8 | 1.7269 | 1.6240 | 1337.9 | 1488.2 | 1.7144 | 1.4723 | 1336.7 | 1486.6 | 1.7030 | 1.3459 | 1335.5 | 1484.9 | 1.6925 |
| 960 | 1.8377 | 1347.6 | 1500.6 | 1.7345 | 1.6497 | 1346.4 | 1499.0 | 1.7221 | 1.4958 | 1345.2 | 1497.5 | 1.7107 | 1.3676 | 1344.0 | 1495.9 | 1.7003 |
| 980 | 1.8660 | 1356.0 | 1511.4 | 1.7420 | 1.6753 | 1354.9 | 1509.9 | 1.7296 | 1.5193 | 1353.7 | 1508.4 | 1.7183 | 1.3893 | 1352.6 | 1506.8 | 1.7079 |
| 1000 | 1.8941 | 1364.4 | 1522.2 | 1.7495 | 1.7008 | 1363.3 | 1520.7 | 1.7371 | 1.5426 | 1362.3 | 1519.3 | 1.7259 | 1.4108 | 1361.2 | 1517.8 | 1.7155 |
| 1020 | 1.9221 | 1372.9 | 1533.0 | 1.7568 | 1.7262 | 1371.8 | 1531.6 | 1.7445 | 1.5659 | 1370.8 | 1530.2 | 1.7333 | 1.4322 | 1369.7 | 1528.8 | 1.7230 |
| 1040 | 1.9501 | 1381.4 | 1543.8 | 1.7641 | 1.7515 | 1380.4 | 1542.4 | 1.7518 | 1.5890 | 1379.3 | 1541.1 | 1.7406 | 1.4536 | 1378.3 | 1539.7 | 1.7303 |
| 1060 | 1.9780 | 1389.9 | 1554.6 | 1.7712 | 1.7768 | 1388.9 | 1553.3 | 1.7590 | 1.6121 | 1387.9 | 1552.0 | 1.7478 | 1.4749 | 1386.9 | 1550.7 | 1.7376 |
| 1080 | 2.0058 | 1398.4 | 1565.4 | 1.7783 | 1.8020 | 1397.4 | 1564.2 | 1.7661 | 1.6352 | 1396.5 | 1562.9 | 1.7550 | 1.4961 | 1395.6 | 1561.7 | 1.7448 |
| 1100 | 2.034 | 1406.9 | 1576.3 | 1.7853 | 1.8271 | 1406.0 | 1575.1 | 1.7731 | 1.6581 | 1405.1 | 1573.9 | 1.7620 | 1.5173 | 1404.2 | 1572.7 | 1.7519 |
| 1150 | 2.103 | 1428.4 | 1603.5 | 1.8025 | 1.8896 | 1427.5 | 1602.4 | 1.7904 | 1.7152 | 1426.7 | 1601.3 | 1.7793 | 1.5699 | 1425.9 | 1600.2 | 1.7692 |

*(Continued)*

## TABLE A.3 (CONTINUED)

Properties of Superheated Steam

| t | v | u | h | s | v | u | h | s | v | u | h | s | v | u | h | s |
|---|---|---|---|---|---|---|---|---|---|---|---|---|---|---|---|---|
| 1200 | 2.172 | 1450.0 | 1630.8 | 1.8192 | 1.9518 | 1449.2 | 1629.8 | 1.8072 | 1.7720 | 1448.5 | 1628.8 | 1.7962 | 1.6222 | 1447.7 | 1627.8 | 1.7861 |
| 1250 | 2.240 | 1471.8 | 1658.3 | 1.8355 | 2.0137 | 1471.1 | 1657.4 | 1.8235 | 1.8285 | 1470.3 | 1656.5 | 1.8126 | 1.6742 | 1469.6 | 1655.5 | 1.8026 |
| 1300 | 2.308 | 1493.7 | 1685.9 | 1.8515 | 2.075 | 1493.1 | 1685.1 | 1.8395 | 1.8848 | 1492.4 | 1684.2 | 1.8286 | 1.7260 | 1491.7 | 1683.4 | 1.8186 |
| 1350 | 2.376 | 1515.8 | 1713.7 | 1.8670 | 2.137 | 1515.2 | 1712.9 | 1.8551 | 1.9409 | 1514.6 | 1712.2 | 1.8443 | 1.7775 | 1514.0 | 1711.4 | 1.8343 |
| 1400 | 2.444 | 1538.1 | 1741.7 | 1.8823 | 2.198 | 1537.6 | 1741.0 | 1.8704 | 1.9967 | 1537.0 | 1740.2 | 1.8596 | 1.8289 | 1536.5 | 1739.5 | 1.8497 |
| 1450 | 2.512 | 1560.6 | 1769.8 | 1.8972 | 2.259 | 1560.1 | 1769.2 | 1.8853 | 2.0525 | 1559.6 | 1768.5 | 1.8746 | 1.8801 | 1559.1 | 1767.8 | 1.8647 |
| 1500 | 2.580 | 1583.3 | 1798.2 | 1.9119 | 2.320 | 1582.8 | 1797.5 | 1.9000 | 2.1080 | 1582.3 | 1796.9 | 1.8892 | 1.9312 | 1581.9 | 1796.3 | 1.8794 |
| 1600 | 2.715 | 1629.3 | 1855.4 | 1.9403 | 2.442 | 1628.9 | 1854.8 | 1.9285 | 2.219 | 1628.4 | 1854.3 | 1.9178 | 2.033 | 1628.0 | 1853.7 | 1.9080 |
| 1800 | 2.983 | 1723.7 | 1972.1 | 1.9944 | 2.684 | 1723.3 | 1971.7 | 1.9827 | 2.440 | 1722.9 | 1971.2 | 1.9720 | 2.236 | 1722.6 | 1970.8 | 1.9622 |
| 2000 | 3.251 | 1821.3 | 2092.0 | 2.0453 | 2.926 | 1820.9 | 2091.6 | 2.0335 | 2.660 | 1820.6 | 2091.2 | 2.0229 | 2.438 | 1820.2 | 2090.8 | 2.0131 |
| 2200 | 3.519 | 1921.8 | 2214.9 | 2.0933 | 3.167 | 1921.5 | 2214.5 | 2.0815 | 2.879 | 1921.1 | 2214.2 | 2.0709 | 2.639 | 1920.8 | 2213.8 | 2.0612 |
| 2400 | 3.787 | 2025.1 | 2340.4 | 2.1388 | 3.408 | 2024.7 | 2340.1 | 2.1270 | 3.098 | 2024.4 | 2339.7 | 2.1164 | 2.840 | 2024.0 | 2339.4 | 2.1067 |

| p (t Sat.) | 800 (518.36) | | | | 1000 (544.75) | | | |
|---|---|---|---|---|---|---|---|---|
| t | v | u | h | s | v | u | h | s |
| Sat. | .5691 | 1115.0 | 1199.3 | 1.4160 | .4459 | 1109.9 | 1192.4 | 1.3903 |
| 500 | | | | | | | | |
| 510 | .5554 | 1107.9 | 1190.1 | 1.4066 | | | | |
| 520 | .5717 | 1116.4 | 1201.0 | 1.4178 | | | | |
| 530 | .5870 | 1124.3 | 1211.2 | 1.4282 | | | | |
| 540 | .6015 | 1131.8 | 1220.8 | 1.4378 | .4389 | 1105.2 | 1186.4 | 1.3844 |
| 550 | .6154 | 1138.8 | 1229.9 | 1.4469 | .4534 | 1114.8 | 1198.7 | 1.3966 |
| 560 | .6287 | 1145.6 | 1238.6 | 1.4555 | .4669 | 1123.6 | 1210.0 | 1.4077 |
| 570 | .6415 | 1152.0 | 1247.0 | 1.4637 | .4795 | 1131.7 | 1220.5 | 1.4179 |
| 580 | .6539 | 1158.3 | 1255.1 | 1.4714 | .4915 | 1139.4 | 1230.4 | 1.4275 |
| 590 | .6659 | 1164.3 | 1262.9 | 1.4789 | .5030 | 1146.7 | 1239.8 | 1.4365 |
| 600 | .6776 | 1170.1 | 1270.4 | 1.4861 | .5140 | 1153.7 | 1248.8 | 1.4450 |
| 610 | .6890 | 1175.8 | 1277.8 | 1.4930 | .5245 | 1160.3 | 1257.4 | 1.4531 |
| 620 | .7002 | 1181.3 | 1285.0 | 1.4997 | .5348 | 1166.7 | 1265.7 | 1.4609 |
| 630 | .7111 | 1186.7 | 1292.0 | 1.5062 | .5447 | 1172.9 | 1273.7 | 1.4682 |
| 640 | .7218 | 1192.0 | 1298.9 | 1.5125 | .5543 | 1178.9 | 1281.5 | 1.4754 |

*(Continued)*

**TABLE A.3 (CONTINUED)**
Properties of Superheated Steam

| Temp | | | | | | | | |
|---|---|---|---|---|---|---|---|---|
| 650 | .7324 | 1197.2 | 1305.6 | 1.5186 | .5637 | 1184.7 | 1289.1 | 1.4822 |
| 660 | .7428 | 1202.3 | 1312.3 | 1.5245 | .5730 | 1190.4 | 1296.5 | 1.4889 |
| 670 | .7530 | 1207.4 | 1318.8 | 1.5304 | .5820 | 1196.0 | 1303.7 | 1.4953 |
| 680 | .7631 | 1212.3 | 1325.3 | 1.5361 | .5908 | 1201.4 | 1310.8 | 1.5015 |
| 690 | .7730 | 1217.2 | 1331.7 | 1.5416 | .5995 | 1206.8 | 1317.7 | 1.5076 |
| 700 | .7829 | 1222.1 | 1338.0 | 1.5471 | .6080 | 1212.0 | 1324.6 | 1.5135 |
| 720 | .8023 | 1231.6 | 1350.4 | 1.5577 | .6247 | 1222.3 | 1337.9 | 1.5249 |
| 740 | .8212 | 1241.0 | 1362.6 | 1.5680 | .6410 | 1232.3 | 1350.9 | 1.5359 |
| 760 | .8399 | 1250.3 | 1374.6 | 1.5779 | .6569 | 1242.1 | 1363.7 | 1.5464 |
| 780 | .8583 | 1259.4 | 1386.5 | 1.5875 | .6725 | 1251.7 | 1376.2 | 1.5566 |
| 800 | .8764 | 1268.5 | 1398.2 | 1.5969 | .6878 | 1261.2 | 1388.5 | 1.5664 |
| 820 | .8943 | 1277.4 | 1409.8 | 1.6061 | .7028 | 1270.6 | 1400.6 | 1.5760 |
| 840 | .9120 | 1286.4 | 1421.4 | 1.6150 | .7176 | 1279.9 | 1412.7 | 1.5853 |
| 860 | .9295 | 1295.2 | 1432.8 | 1.6238 | .7323 | 1289.1 | 1424.6 | 1.5944 |
| 880 | .9468 | 1304.1 | 1444.2 | 1.6324 | .7467 | 1298.2 | 1436.4 | 1.6033 |
| 900 | .9640 | 1312.9 | 1455.6 | 1.6408 | .7610 | 1307.3 | 1448.1 | 1.6120 |
| 920 | .9811 | 1321.7 | 1466.9 | 1.6490 | .7752 | 1316.3 | 1459.7 | 1.6205 |
| 940 | .9980 | 1330.4 | 1478.2 | 1.6572 | .7892 | 1325.3 | 1471.3 | 1.6288 |
| 960 | 1.0149 | 1339.2 | 1489.4 | 1.6651 | .8031 | 1334.3 | 1482.9 | 1.6370 |
| 980 | 1.0316 | 1348.0 | 1500.7 | 1.6730 | .8169 | 1343.2 | 1494.4 | 1.6451 |
| 1000 | 1.0482 | 1356.7 | 1511.9 | 1.6807 | .8305 | 1352.2 | 1505.9 | 1.6530 |
| 1020 | 1.0647 | 1365.5 | 1523.1 | 1.6883 | .8441 | 1361.1 | 1517.3 | 1.6608 |
| 1040 | 1.0812 | 1374.2 | 1534.3 | 1.6959 | .8576 | 1370.0 | 1528.7 | 1.6684 |
| 1060 | 1.0975 | 1383.0 | 1545.5 | 1.7033 | .8710 | 1378.9 | 1540.1 | 1.6760 |
| 1080 | 1.1138 | 1391.7 | 1556.6 | 1.7106 | .8844 | 1387.9 | 1551.5 | 1.6834 |
| 1100 | 1.1300 | 1400.5 | 1567.8 | 1.7178 | .8976 | 1396.8 | 1562.9 | 1.6908 |
| 1120 | 1.1462 | 1409.3 | 1579.0 | 1.7249 | .9108 | 1405.7 | 1574.3 | 1.6980 |
| 1140 | 1.1623 | 1418.1 | 1590.2 | 1.7319 | .9240 | 1414.6 | 1585.6 | 1.7052 |
| 1160 | 1.1783 | 1426.9 | 1601.4 | 1.7389 | .9370 | 1423.6 | 1597.0 | 1.7122 |
| 1180 | 1.1943 | 1435.8 | 1612.6 | 1.7458 | .9500 | 1432.5 | 1608.3 | 1.7192 |
| 1200 | 1.2102 | 1444.6 | 1623.8 | 1.7526 | .9630 | 1441.5 | 1619.7 | 1.7261 |

*(Continued)*

**TABLE A.3 (CONTINUED)**

Properties of Superheated Steam

| | | | | | | | | |
|---|---|---|---|---|---|---|---|---|
| 1220 | 1.2261 | 1453.5 | 1635.9 | 1.7593 | .9759 | 1450.4 | 1631.0 | 1.7329 |
| 1240 | 1.2420 | 1462.3 | 1646.2 | 1.7659 | .9888 | 1459.4 | 1642.4 | 1.7396 |
| 1260 | 1.2578 | 1471.3 | 1657.4 | 1.7725 | 1.0016 | 1468.4 | 1653.8 | 1.7462 |
| 1280 | 1.2735 | 1480.2 | 1668.7 | 1.7790 | 1.0144 | 1477.4 | 1665.1 | 1.7528 |
| 1300 | 1.2892 | 1489.1 | 1680.0 | 1.7854 | 1.0272 | 1486.5 | 1676.5 | 1.7593 |
| 1350 | 1.3284 | 1511.6 | 1708.2 | 1.8013 | 1.0589 | 1509.1 | 1705.1 | 1.7753 |
| 1400 | 1.3674 | 1534.2 | 1736.6 | 1.8167 | 1.0905 | 1531.9 | 1733.7 | 1.7909 |
| 1450 | 1.4062 | 1556.9 | 1765.1 | 1.8319 | 1.1218 | 1554.8 | 1762.4 | 1.8061 |
| 1500 | 1.4448 | 1579.9 | 1793.7 | 1.8467 | 1.1531 | 1577.8 | 1791.2 | 1.8210 |
| 1600 | 1.5218 | 1626.2 | 1851.5 | 1.8754 | 1.2152 | 1624.4 | 1849.3 | 1.8499 |
| 1800 | 1.6749 | 1721.0 | 1969.0 | 1.9298 | 1.3384 | 1719.5 | 1967.2 | 1.9046 |
| 2000 | 1.8271 | 1818.8 | 2089.3 | 1.9808 | 1.4608 | 1817.4 | 2087.7 | 1.9557 |
| 2200 | 1.9789 | 1919.4 | 2212.4 | 2.0290 | 1.5828 | 1918.1 | 2211.0 | 2.0038 |
| 2400 | 2.1305 | 2022.7 | 2338.1 | 2.0745 | 1.7046 | 2021.3 | 2336.7 | 2.0494 |

**TABLE A.4**

Properties of Compressed Liquid (Steam)

Liquid (psia)

| p (t Sat.) | 0 | | | | 500 (467.13) | | | | 1000 (544.75) | | | |
|---|---|---|---|---|---|---|---|---|---|---|---|---|
| | v | u | h | s | v | u | h | s | v | u | h | s |
| Sat. | | | | | .019748 | 447.70 | 449.53 | .64904 | .021591 | 538.39 | 542.38 | .74320 |
| 32 | .016022 | 0.01 | 0.01 | .00003 | .015994 | .00 | 1.49 | .00000 | .015967 | .03 | 2.99 | .00005 |
| 50 | .016024 | 18.06 | 18.06 | .03607 | .015998 | 18.02 | 19.50 | .03599 | .015972 | 17.99 | 20.94 | .03592 |
| 100 | .016130 | 68.05 | 68.05 | .12963 | .016106 | 67.87 | 69.36 | .12932 | .016082 | 67.70 | 70.68 | .12901 |
| 150 | .016344 | 117.95 | 117.95 | .21504 | .016318 | 117.66 | 119.17 | .21457 | .016293 | 117.38 | 120.40 | .21410 |
| 200 | .016635 | 168.05 | 168.05 | .29402 | .016608 | 167.65 | 169.19 | .29341 | .016580 | 167.26 | 170.32 | .29281 |
| 250 | .017003 | 218.52 | 218.52 | .36777 | .016972 | 217.99 | 219.56 | .36702 | .016941 | 217.47 | 220.61 | .36628 |
| 300 | .017453 | 269.61 | 269.61 | .43732 | .017416 | 268.92 | 270.53 | .43641 | .017379 | 268.24 | 271.46 | .43352 |
| 350 | .018000 | 321.59 | 321.59 | .50359 | .017954 | 320.71 | 322.37 | .50249 | .017909 | 319.83 | 323.15 | .50140 |
| 400 | .018668 | 374.85 | 374.85 | .56740 | .018608 | 373.68 | 375.40 | .56604 | .018550 | 372.55 | 375.98 | .56472 |
| 450 | .019503 | 429.96 | 429.96 | .62970 | .019420 | 428.40 | 430.19 | .63798 | .019340 | 426.89 | 430.47 | .62632 |
| 500 | .02060 | 488.1 | 488.1 | .6919 | .02048 | 485.9 | 487.8 | .6896 | .02036 | 483.8 | 487.5 | .6874 |
| 510 | .02087 | 500.3 | 500.3 | .7046 | .02073 | 497.9 | 499.8 | .7021 | .02060 | 495.6 | 499.4 | .6997 |
| 520 | .02116 | 512.7 | 512.7 | .7173 | .02100 | 510.1 | 512.0 | .7146 | .02086 | 507.6 | 511.5 | .7121 |
| 530 | .02148 | 525.5 | 525.5 | .7303 | .02130 | 522.6 | 524.5 | .7273 | .02114 | 519.9 | 523.8 | .7245 |
| 540 | .02182 | 538.6 | 538.6 | .7434 | .02162 | 535.3 | 537.3 | .7402 | .02144 | 532.4 | 536.3 | .7372 |
| 550 | .02221 | 552.1 | 552.1 | .7569 | .02198 | 548.4 | 550.5 | .7532 | .02177 | 545.1 | 549.2 | .7499 |
| 560 | .02265 | 566.1 | 566.1 | .7707 | .02237 | 562.0 | 564.0 | .7666 | .02213 | 558.3 | 562.4 | .7630 |
| 570 | .02315 | 580.8 | 580.8 | .7851 | .02281 | 576.0 | 578.1 | .7804 | .02253 | 571.8 | 576.0 | .7763 |
| 580 | | | | | .02332 | 590.8 | 592.9 | .7946 | .02298 | 585.9 | 590.1 | .7899 |
| 590 | | | | | .02392 | 606.4 | 608.6 | .8096 | .02349 | 600.6 | 604.9 | .8041 |
| 600 | | | | | | | | | .02409 | 616.2 | 620.6 | .8189 |
| 610 | | | | | | | | | .02482 | 632.9 | 637.5 | .8348 |

*(Continued)*

**TABLE A.4 (CONTINUED)**

Properties of Compressed Liquid (Steam)

Liquid (psia)

| p (t Sat) | 4000 | | | | 18 000 | | | | 20 000 | | | |
|---|---|---|---|---|---|---|---|---|---|---|---|---|
| t | v | u | h | s | v | u | h | s | v | u | h | s |
| 32 | .015807 | .10 | 11.80 | .00005 | .015188 | 0.54 | 50.05 | .00372 | .015116 | .65 | 55.30 | .00446 |
| 50 | .015821 | 17.76 | 29.47 | .03534 | .015227 | 16.32 | 67.04 | .03021 | .015154 | 16.14 | 72.23 | .02936 |
| 100 | .015942 | 66.72 | 78.52 | .12714 | .015372 | 62.83 | 114.03 | .11817 | .015298 | 62.37 | 118.99 | .11688 |
| 150 | .016150 | 115.77 | 127.73 | .21136 | .015572 | 109.64 | 161.51 | .19941 | .015497 | 108.91 | 166.26 | .19778 |
| 200 | .016425 | 165.02 | 177.18 | .28931 | .015813 | 156.62 | 209.29 | .27474 | .015736 | 155.62 | 213.86 | .27281 |
| 250 | .016765 | 214.52 | 226.93 | .36200 | .016096 | 203.57 | 257.19 | .34472 | .016015 | 202.28 | 261.55 | .34250 |
| 300 | .017174 | 264.43 | 277.15 | .43038 | .016422 | 250.58 | 305.28 | .41021 | .016333 | 248.96 | 309.41 | .40766 |
| 350 | .017659 | 314.98 | 328.05 | .49526 | .016791 | 297.80 | 353.73 | .47197 | .016693 | 295.83 | 357.61 | .46911 |
| 400 | .018235 | 366.35 | 379.85 | .55734 | .017207 | 345.32 | 402.63 | .53057 | .017096 | 342.97 | 406.24 | .52738 |
| 450 | .018924 | 418.83 | 432.84 | .61725 | .017669 | 393.15 | 452.00 | .58639 | .017541 | 390.39 | 455.31 | .58286 |
| 500 | .019766 | 472.9 | 487.5 | .6758 | .018183 | 441.3 | 501.9 | .6397 | .018031 | 438.1 | 504.8 | .6358 |
| 520 | .020161 | 495.2 | 510.1 | .6990 | .018404 | 460.6 | 521.9 | .6605 | .018242 | 457.2 | 524.7 | .6564 |
| 540 | .020600 | 517.9 | 533.1 | .7223 | .018636 | 480.0 | 542.1 | .6808 | .018461 | 476.4 | 544.7 | .6766 |
| 560 | .021091 | 541.2 | 556.8 | .7457 | .018879 | 499.5 | 562.4 | .7009 | .018689 | 495.6 | 564.8 | .6965 |
| 580 | .021648 | 565.2 | 581.2 | .7694 | .019134 | 519.0 | 582.8 | .7207 | .018928 | 514.9 | 585.0 | .7160 |
| 600 | .02229 | 590.0 | 606.5 | .7936 | .01940 | 538.7 | 603.3 | .7403 | .01918 | 534.3 | 605.2 | .7354 |
| 620 | .02304 | 616.0 | 633.0 | .8183 | .01968 | 558.4 | 623.9 | .7596 | .01944 | 553.7 | 625.6 | .7544 |
| 640 | .02394 | 643.3 | 661.1 | .8441 | .01998 | 578.2 | 644.8 | .7787 | .01972 | 573.2 | 646.1 | .7732 |
| 660 | .02506 | 672.7 | 691.2 | .8712 | .02030 | 598.1 | 665.7 | .7976 | .02001 | 592.7 | 666.8 | .7918 |
| 680 | .02653 | 704.9 | 724.5 | .9007 | .02064 | 618.2 | 686.9 | .8163 | .02031 | 612.4 | 687.6 | .8102 |
| 700 | .02867 | 742.1 | 763.4 | .9345 | .02099 | 638.4 | 708.3 | .8349 | .02063 | 632.1 | 708.5 | .8285 |
| 710 | .03026 | 764.3 | 786.7 | .9545 | .02118 | 648.5 | 719.1 | .8442 | .02080 | 642.1 | 719.1 | .8375 |

**TABLE A.5**

Enthalpy–Entropy Diagram for Steam—A Mollier Chart for Steam

| Abs. Press. in. Hg. | Sat. Temp. °F | Abs. Press., lb. per sq. in. | Sat. Temp. °F | Abs. Press. lb. per sq. in. | Sat. Temp. °F |
|---|---|---|---|---|---|
| 0.20 | 34.56 | 1.0 | 101.74 | 120 | 341.25 |
| 0.25 | 40.23 | 2 | 126.08 | 140 | 353.02 |
| 0.30 | 44.96 | 3 | 141.18 | 160 | 363.53 |
| 0.35 | 49.06 | 4 | 152.97 | 180 | 373.06 |
| 0.40 | 52.64 | 5 | 162.24 | 200 | 381.79 |
| 0.45 | 55.87 | 6 | 170.06 | 220 | 389.86 |
| 0.50 | 58.80 | 7 | 176.83 | 240 | 397.37 |
| 0.55 | 61.48 | 8 | 182.86 | 260 | 404.42 |
| 0.60 | 63.95 | 9 | 188.28 | 280 | 411.05 |
| 0.65 | 66.26 | 10 | 193.21 | 300 | 417.33 |
| 0.70 | 68.40 | 12 | 201.96 | 400 | 444.59 |
| 0.75 | 70.43 | 14 | 209.56 | 500 | 467.01 |
| 0.80 | 72.32 | 14.696 | 212.00 | 600 | 486.21 |
| 0.85 | 74.18 | 16 | 216.32 | 700 | 503.10 |
| 0.90 | 75.84 | 18 | 222.41 | 800 | 518.23 |
| 0.95 | 77.47 | 20 | 227.96 | 1000 | 544.61 |
| 1.00 | 79.03 | 25 | 240.07 | 1200 | 567.22 |
| 1.10 | 81.95 | 30 | 250.33 | 1400 | 587.10 |
| 1.20 | 84.65 | 35 | 259.28 | 1600 | 604.90 |
| 1.30 | 87.17 | 40 | 267.25 | 1800 | 621.03 |
| 1.40 | 89.51 | 45 | 274.44 | 2000 | 635.82 |
| 1.50 | 91.72 | 50 | 281.01 | 2200 | 649.46 |
| 1.60 | 93.80 | 60 | 292.71 | 2400 | 662.12 |
| 1.70 | 95.77 | 70 | 302.92 | 2600 | 673.94 |
| 1.80 | 97.65 | 80 | 312.03 | 2800 | 684.99 |
| 1.90 | 99.43 | 90 | 320.27 | 3000 | 695.36 |
| 2.00 | 101.14 | 100 | 327.81 | 3206.2 | 705.40 |

*Source:* Modified and greatly reduced from J. H. Keenan and F. G. Keyes: *Thermodynamic Properties of Steam.* 1936. Copyright Wiley-VCH Verlag GmbH & Co. KGaA. Reproduced with permission.

**TABLE A.5**

Enthalpy–Entropy Diagram for Steam

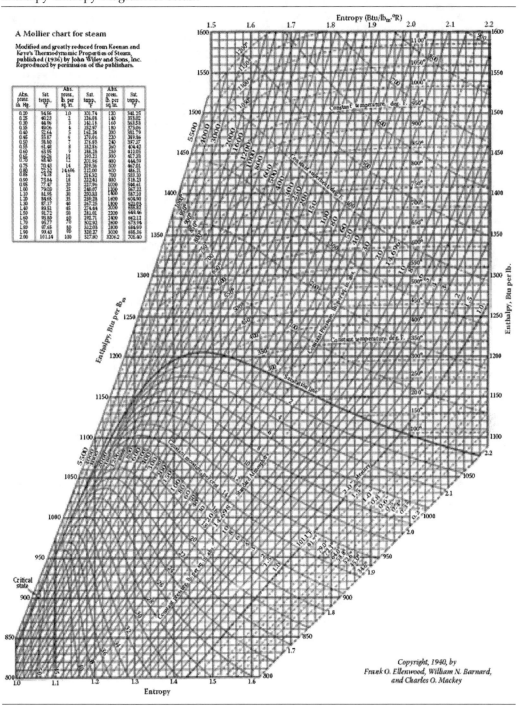

A Mollier chart for steam

Modified and greatly reduced from Keenan and Keye's Thermodynamic Properties of Steam, published (1936) by John Wiley and Sons, Inc. Reproduced by permission of the publishers.

**TABLE A.1 (SI)**

Saturation: Temperature (Steam)

| Temp. °C $T$ | Press. kPa $P$ | Specific Volume (m3kg) | | Internal Energy (kJ/kg) | | | Enthalpy (kJ/kg) | | | Entropy (kJ/kg. K) | | |
|---|---|---|---|---|---|---|---|---|---|---|---|---|
| | | Sat. Liquid $v_f$ | Sat. Vapor $v_g$ | Sat. Liquid $u_f$ | Evap. $u_{gg}$ | Sat. Vapor $u_g$ | Sat. Liquid $h_f$ | Evap. $h_{fg}$ | Sat. Vapor $h_g$ | Sat. Liquid $s_f$ | Evap. $s_{fg}$ | Sat. Vapor $s_g$ |
| 0.01 | 0.6113 | 0.001 000 | 206.14 | .00 | 2375.3 | 2375.3 | .01 | 2501.3 | 2501.4 | .0000 | 9.1562 | 9.1562 |
| 5 | 0.8721 | 0.001 000 | 147.12 | 20.97 | 2361.3 | 2382.3 | 20.98 | 2489.6 | 2510.6 | .0761 | 8.9496 | 9.0257 |
| 10 | 1.2276 | 0.001 000 | 106.38 | 42.00 | 2347.2 | 2389.2 | 42.01 | 2477.7 | 2519.8 | .1510 | 8.7498 | 8.9008 |
| 15 | 1.7051 | 0.001 001 | 77.93 | 62.99 | 2333.1 | 2396.1 | 62.99 | 2465.9 | 2528.9 | .2245 | 8.5569 | 8.7814 |
| 20 | 2.339 | 0.001 002 | 57.79 | 83.95 | 2319.0 | 2402.9 | 83.96 | 2454.1 | 2538.1 | .2966 | 8.3706 | 8.6672 |
| 25 | 3.169 | 0.001 003 | 43.36 | 104.88 | 2304.9 | 2409.8 | 104.89 | 2442.3 | 2547.2 | .3674 | 8.1905 | 8.5580 |
| 30 | 4.246 | 0.001 004 | 32.89 | 125.78 | 2290.8 | 2416.6 | 125.79 | 2430.5 | 2556.3 | .4369 | 8.0164 | 8.4533 |
| 35 | 5.628 | 0.001 006 | 25.22 | 146.67 | 2276.7 | 2423.4 | 146.68 | 2418.6 | 2565.3 | .5053 | 7.8478 | 8.3531 |
| 40 | 7.384 | 0.001 008 | 19.52 | 167.56 | 2262.6 | 2430.1 | 167.57 | 2406.7 | 2574.3 | .5725 | 7.6845 | 8.2570 |
| 45 | 9.593 | 0.001 010 | 15.26 | 188.44 | 2248.4 | 2436.8 | 188.45 | 2394.8 | 2583.2 | .6387 | 7.5261 | 8.1648 |
| 50 | 12.349 | 0.001 012 | 12.03 | 209.32 | 2234.2 | 2443.5 | 209.33 | 2382.7 | 2592.1 | .7038 | 7.3725 | 8.0763 |
| 55 | 15.758 | 0.001 015 | 9.568 | 230.21 | 2219.9 | 2450.1 | 230.23 | 2370.7 | 2600.9 | .7679 | 7.2234 | 7.9913 |
| 60 | 19.940 | 0.001 017 | 7.671 | 251.11 | 2205.5 | 2456.6 | 251.13 | 2358.5 | 2609.6 | .8312 | 7.0784 | 7.9096 |
| 65 | 25.03 | 0.001 020 | 6.197 | 272.02 | 2191.1 | 2463.1 | 272.06 | 2346.2 | 2618.3 | .8935 | 6.937.5 | 7.8310 |
| 70 | 31.19 | 0.001 023 | 5.042 | 292.95 | 2176.6 | 2469.6 | 292.98 | 2333.8 | 2626.8 | .9549 | 6.8004 | 7.7553 |
| 75 | 38.58 | 0.001 026 | 4.131 | 313.90 | 2162.0 | 2475.9 | 313.93 | 2321.4 | 2635.3 | 1.0155 | 6.6669 | 7.6824 |
| 80 | 47.39 | 0.001 029 | 3.407 | 334.86 | 2147.4 | 2482.2 | 334.91 | 2308.8 | 2643.7 | 1.0753 | 6.5369 | 7.6122 |
| 85 | 57.83 | 0.001 033 | 2.828 | 355.84 | 2132.6 | 2488.4 | 355.90 | 2296.0 | 2651.9 | 1.1343 | 6.4102 | 7.5445 |
| 90 | 70.14 | 0.001 036 | 2.361 | 376.85 | 2117.7 | 2494.5 | 376.92 | 2283.2 | 2660.1 | 1.1925 | 6.2866 | 7.4791 |
| 95 | 84.55 | 0.001 040 | 1.982 | 397.88 | 2102.7 | 2500.6 | 397.96 | 2270.2 | 2668.1 | 1.2500 | 6.1659 | 7.4159 |
| 100 | 0.101 35 | 0.001 044 | 1.6729 | 418.94 | 2087.6 | 2506.5 | 419.04 | 2257.0 | 2676.1 | 1.3069 | 6.0480 | 7.3549 |
| 105 | 0.120 82 | 0.001 048 | 1.4194 | 440.02 | 2072.3 | 2512.4 | 440.15 | 2243.7 | 2683.8 | 1.3630 | 5.9328 | 7.2958 |
| 110 | 0.143 27 | 0.001 052 | 1.2102 | 461.14 | 2057.0 | 2518.1 | 461.30 | 2230.2 | 2691.5 | 1.4185 | 5.8202 | 7.2387 |
| 115 | 0.169 06 | 0.001 056 | 1.0366 | 482.30 | 2041.4 | 2523.7 | 482.48 | 2216.5 | 2699.0 | 1.4734 | 5.7100 | 7.1833 |
| 120 | 0.198 53 | 0.001 060 | 0.8919 | 503.50 | 2025.8 | 2529.3 | 503.71 | 2202.6 | 2706.3 | 1.5276 | 5.6020 | 7.1296 |
| 125 | 0.2321 | 0.001 065 | 0.7706 | 524.74 | 2009.9 | 2534.6 | 524.99 | 2188.5 | 2713.5 | 1.5813 | 5.4962 | 7.0775 |

(*Continued*)

**TABLE A.1 (SI) (CONTINUED)**

Saturation: Temperature (Steam)

| Temp. °C T | Press. kPa P | Specific Volume (m3kg) | | Internal Energy (kJ/kg) | | | Enthalpy (kJ/kg) | | | Entropy (kJ/kg. K) | | |
|---|---|---|---|---|---|---|---|---|---|---|---|---|
| | | Sat. Liquid $v_f$ | Sat. Vapor $v_g$ | Sat. Liquid $u_f$ | Evap. $u_{gg}$ | Sat. Vapor $u_g$ | Sat. Liquid $h_f$ | Evap. $h_{fg}$ | Sat. Vapor $h_g$ | Sat. Liquid $s_f$ | Evap. $s_{fg}$ | Sat. Vapor $s_g$ |
| 130 | 0.2701 | 0.001 070 | 0.6685 | 546.02 | 1993.9 | 2539.9 | 546.31 | 2174.2 | 2720.5 | 1.6344 | 5.3925 | 7.0269 |
| 135 | 0.3130 | 0.001 075 | 0.5822 | 567.35 | 1977.7 | 2545.0 | 567.69 | 2159.6 | 2727.3 | 1.6870 | 5.2907 | 6.9777 |
| 140 | 0.3613 | 0.001 080 | 0.5089 | 588.74 | 1961.3 | 2550.0 | 589.13 | 2144.7 | 2733.9 | 1.7391 | 5.1908 | 6.9299 |
| 145 | 0.4154 | 0.001 085 | 0.4463 | 610.18 | 1944.7 | 2554.9 | 610.63 | 2129.6 | 2740.3 | 1.7907 | 5.0926 | 6.8833 |
| 150 | 0.4758 | 0.001 091 | 0.3928 | 631.68 | 1927.9 | 2559.5 | 632.20 | 2114.3 | 2746.5 | 1.8418 | 4.9960 | 6.8379 |
| 155 | 0.5431 | 0.001 096 | 0.3468 | 653.24 | 1910.8 | 2564.1 | 653.84 | 2098.6 | 2752.4 | 1.8925 | 4.9010 | 6.7935 |
| 160 | 0.6178 | 0.001102 | 0.3071 | 674.87 | 1893.5 | 2568.4 | 675.55 | 2082.6 | 2758.1 | 1.9427 | 4.8075 | 6.7502 |
| 165 | 0.7005 | 0.00108 | 0.2727 | 696.56 | 1876.0 | 2572.5 | 697.34 | 2066.2 | 2763.5 | 1.9925 | 4.7153 | 6.7078 |
| 170 | 0.7917 | 0.001114 | 0.2428 | 718.33 | 1858.1 | 2576.5 | 719.21 | 2049.5 | 2768.7 | 2.0419 | 4.6244 | 6.6663 |
| 175 | 0.8920 | 0.001121 | 0.2168 | 740.17 | 1840.0 | 2580.2 | 741.17 | 2032.4 | 2773.6 | 2.0909 | 4.5347 | 6.6256 |
| 180 | 1.0021 | 0.001127 | 0.194 05 | 762.09 | 1821.6 | 2583.7 | 763.22 | 2015.0 | 2778.2 | 2.1396 | 4.4461 | 6.5857 |
| 185 | 1.1227 | 0.001134 | 0.174 09 | 784.10 | 1802.9 | 2587.0 | 785.37 | 1997.1 | 2782.4 | 2.1879 | 4.3586 | 6.5465 |
| 190 | 1.2544 | 0.001141 | 0.156 54 | 806.19 | 1783.8 | 2590.0 | 807.62 | 1978.8 | 2786.4 | 2.2359 | 4.2720 | 6.5079 |
| 195 | 1.3978 | 0.001149 | 0.141 05 | 828.37 | 1764.4 | 2592.8 | 829.98 | 1960.0 | 2790.0 | 2.2835 | 4.1863 | 6.4698 |
| 200 | 1.5538 | 0.001157 | 0.12736 | 850.65 | 1744.7 | 2595.3 | 852.45 | 1940.7 | 2793.2 | 2.3309 | 4.1014 | 6.4323 |
| 205 | 1.7230 | 0.001164 | 0.115 21 | 873.04 | 1724.5 | 2597.5 | 875.04 | 1921.0 | 2796.0 | 2.3780 | 4.0172 | 6.3952 |
| 210 | 1.9062 | 0.001173 | 0.104 41 | 895.53 | 1703.9 | 2599.5 | 897.76 | 1900.7 | 2798.5 | 2.4248 | 3.9337 | 6.3585 |
| 215 | 2.104 | 0.001181 | 0.094 79 | 918.14 | 1682.9 | 2601.1 | 920.62 | 1879.9 | 2800.5 | 2.4714 | 3.8507 | 6.3221 |
| 220 | 2.318 | 0.001190 | 0.086 19 | 940.87 | 1661.5 | 2602.4 | 943.62 | 1858.5 | 2802.1 | 2.5178 | 3.7683 | 6.2861 |
| 225 | 2.548 | 0.001199 | 0.078 49 | 963.73 | 1639.6 | 2603.3 | 966.78 | 1836.5 | 2803.3 | 2.5639 | 3.6863 | 6.2503 |
| 230 | 2.795 | 0.001 209 | 0.071 58 | 986.74 | 1617.2 | 2603.9 | 990.12 | 1813.8 | 2804.0 | 2.6099 | 3.6047 | 6.2146 |
| 235 | 3.060 | 0.001 219 | 0.065 37 | 1009.89 | 1594.2 | 2604.1 | 1013.62 | 1790.5 | 2804.2 | 2.6558 | 3.5233 | 6.1791 |
| 240 | 3.344 | 0.001 229 | 0.059 76 | 1033.21 | 1570.8 | 2604.0 | 1037.32 | 1766.5 | 2803.8 | 2.7015 | 3.4422 | 6.1437 |
| 245 | 3.648 | 0.001 240 | 0.054 71 | 1056.71 | 1546.7 | 2603.4 | 1061.23 | 1741.7 | 2803.0 | 2.7472 | 3.3612 | 6.1083 |
| 250 | 3.973 | 0.001 251 | 0.050 13 | 1080.39 | 1522.0 | 2602.4 | 1085.36 | 1716.2 | 2801.5 | 2.7927 | 3.2802 | 6.0730 |
| 255 | 4.319 | 0.001 263 | 0.045 98 | 1104.28 | 1496.7 | 2600.9 | 1109.73 | 1689.8 | 2799.5 | 2.8383 | 3.1992 | 6.0375 |

(*Continued*)

# TABLE A.1 (SI) (CONTINUED)

Saturation: Temperature (Steam)

| Temp. °C T | Press. kPa P | Specific Volume (m3kg) | | Internal Energy (kJ/kg) | | | Enthalpy (kJ/kg) | | | Entropy (kJ/kg. K) | | |
|---|---|---|---|---|---|---|---|---|---|---|---|---|
| | | Sat. Liquid $v_f$ | Sat. Vapor $v_g$ | Sat. Liquid $u_f$ | Evap. $u_{gg}$ | Sat. Vapor $u_g$ | Sat. Liquid $h_f$ | Evap. $h_{fg}$ | Sat. Vapor $h_g$ | Sat. Liquid $s_f$ | Evap. $s_{fg}$ | Sat. Vapor $s_g$ |
| 260 | 4.688 | 0.001 276 | 0.042 21 | 1128.39 | 1470.6 | 2599.0 | 1134.37 | 1662.5 | 2796.9 | 2.8838 | 3.1181 | 6.0019 |
| 265 | 5.081 | 0.001 289 | 0.038 77 | 1152.74 | 1443.9 | 2596.6 | 1159.28 | 1634.4 | 2793.6 | 2.9294 | 3.0368 | 5.9662 |
| 270 | 5.499 | 0.001 302 | 0.035 64 | 1177.36 | 1416.3 | 2593.7 | 1184.51 | 1605.2 | 2789.7 | 2.9751 | 2.9551 | 5.9301 |
| 275 | 5.942 | 0.001 317 | 0.032 79 | 1202.25 | 1387.9 | 2590.2 | 1210.07 | 1574.9 | 2785.0 | 3.0208 | 2.8730 | 5.8938 |
| 280 | 6.412 | 0.001 332 | 0.030 17 | 1227.46 | 1358.7 | 2586.1 | 1235.99 | 1543.6 | 2779.6 | 3.0668 | 2.7903 | 5.8571 |
| 285 | 6.909 | 0.001 348 | 0.027 77 | 1253.00 | 1328.4 | 2581.4 | 1262.31 | 1511.0 | 2773.3 | 3.1130 | 2.7070 | 5.8199 |
| 290 | 7.436 | 0.001 366 | 0.025 57 | 1278.92 | 1297.1 | 2576.0 | 1289.07 | 1477.1 | 2766.2 | 3.1594 | 2.6227 | 5.7821 |
| 295 | 7.993 | 0.001 384 | 0.023 54 | 1305.2 | 1264.7 | 2569.9 | 1316.3 | 1441.8 | 2758.1 | 3.2062 | 2.5375 | 5.7437 |
| 300 | 8.581 | 0.001 404 | 0.021 67 | 1332.0 | 1231.0 | 2563.0 | 1344.0 | 1404.9 | 2749.0 | 3.2534 | 2.4511 | 5.7045 |
| 305 | 9.202 | 0.001 425 | 0.019 948 | 1359.3 | 1195.9 | 2555.2 | 1372.4 | 1366.4 | 2738.7 | 3.3010 | 2.3633 | 5.6643 |
| 310 | 9.856 | 0.001 447 | 0.018 350 | 1387.1 | 1159.4 | 2546.4 | 1401.3 | 1326.0 | 2727.3 | 3.3493 | 2.2737 | 5.6230 |
| 315 | 10.547 | 0.001 472 | 0.016 867 | 1415.5 | 1121.1 | 2536.6 | 1431.0 | 1283.5 | 2714.5 | 3.3982 | 2.1821 | 5.5804 |
| 320 | 11.274 | 0.001 499 | 0.015 488 | 1444.6 | 1080.9 | 2525.5 | 1461.5 | 1238.6 | 2700.1 | 3.4480 | 2.0882 | 5.5362 |
| 330 | 12.845 | 0.001 561 | 0.012 996 | 1505.3 | 993.7 | 2498.9 | 1525.3 | 1140.6 | 2665.9 | 3.5507 | 1.8909 | 5.4417 |
| 340 | 14.586 | 0.001 638 | 0.010 797 | 1570.3 | 894.3 | 2464.6 | 1594.2 | 1027.9 | 2622.0 | 3.6594 | 1.6763 | 5.3357 |
| 350 | 16.513 | 0.001 740 | 0.008 813 | 1641.9 | 776.6 | 2418.4 | 1670.6 | 893.4 | 2563.9 | 3.7777 | 1.4335 | 5.2112 |
| 360 | 18.651 | 0.001 893 | 0.006 945 | 1725.2 | 626.3 | 2351.5 | 1760.5 | 720.5 | 2481.0 | 3.9147 | 1.1379 | 5.0526 |
| 370 | 21.03 | 0.002 213 | 0.004 925 | 1844.0 | 384.5 | 2228.5 | 1890.5 | 441.6 | 2332.1 | 4.1106 | .6865 | 4.7971 |
| 374.14 | 22.09 | 0.003 155 | 0.003 155 | 2029.6 | 0 | 2029.6 | 2099.3 | 0 | 2099.3 | 4.4298 | 0 | 4.4298 |

**TABLE A.2 (SI)**

Saturation Pressures (Steam)

| Press. kPa $P$ | Temp. °C $T$ | Specific Volume (m³/kg) | | Internal Energy (kJ/kg) | | | Enthalpy (kJ/kg) | | | Entropy (kJ/kg. K) | | |
|---|---|---|---|---|---|---|---|---|---|---|---|---|
| | | Sat. Liquid $v_f$ | Sat. Vapor $v_g$ | Sat. Liquid $u_f$ | Evap. $u_{fg}$ | Sat. Vapor $u_g$ | Sat. Liquid $h_f$ | Evap. $h_{fg}$ | Sat. Vapor $h_g$ | Sat. Liquid $s_f$ | Evap. $s_{fg}$ | Sat. Vapor $s_g$ |
| 0.6113 | 0.01 | 0.001 000 | 206.14 | .00 | 2375.3 | 2375.3 | .01 | 2501.3 | 2501.4 | .0000 | 9.1562 | 9.1562 |
| 1.0 | 6.98 | 0.001 000 | 129.21 | 29.30 | 2355.7 | 2385.0 | 29.30 | 2484.9 | 2514.2 | .1059 | 8.8697 | 8.9756 |
| 1.5 | 13.03 | 0.001 001 | 87.98 | 54.71 | 2338.6 | 2393.3 | 54.71 | 2470.6 | 2525.3 | .1957 | 8.6322 | 8.8279 |
| 2.0 | 17.50 | 0.001 001 | 67.00 | 73.48 | 2326.0 | 2399.5 | 73.48 | 2460.0 | 2533.5 | .2607 | 8.4629 | 8.7237 |
| 2.5 | 21.08 | 0.001 002 | 54.25 | 88.48 | 2315.9 | 2404.4 | 88.49 | 2451.6 | 2540.0 | .3120 | 8.3311 | 8.6432 |
| 3.0 | 24.08 | 0.001 003 | 45.67 | 101.04 | 2307.5 | 2408.5 | 101.05 | 2444.5 | 2545.5 | .3545 | 8.2231 | 8.5776 |
| 4.0 | 28.96 | 0.001 004 | 34.80 | 121.45 | 2293.7 | 2415.2 | 121.46 | 2432.9 | 2554.4 | .4226 | 8.0520 | 8.4746 |
| 5.0 | 32.88 | 0.001 005 | 28.19 | 137.81 | 2282.7 | 2420.5 | 137.82 | 2423.7 | 2561.5 | .4764 | 7.9187 | 8.3951 |
| 7.5 | 40.29 | 0.001 008 | 19.24 | 168.78 | 2261.7 | 2430.5 | 168.79 | 2406.0 | 2574.8 | .5764 | 7.6750 | 8.2515 |
| 10 | 45.81 | 0.001 010 | 14.67 | 191.82 | 2246.1 | 2437.9 | 191.83 | 2392.8 | 2584.7 | .6493 | 7.5009 | 8.1502 |
| 15 | 53.97 | 0.001 014 | 10.02 | 225.92 | 2222.8 | 2448.7 | 225.94 | 2373.1 | 2599.1 | .7549 | 7.2536 | 8.0085 |
| 20 | 60.06 | 0.001 017 | 7.649 | 251.38 | 2205.4 | 2456.7 | 251.40 | 2358.3 | 2609.7 | .8320 | 7.0766 | 7.9085 |
| 25 | 64.97 | 0.001 020 | 6.204 | 271.90 | 2191.2 | 2463.1 | 271.93 | 2346.3 | 2618.2 | .8931 | 6.9383 | 7.8314 |
| 30 | 69.10 | 0.001 022 | 5.229 | 289.20 | 2179.2 | 2468.4 | 289.23 | 2336.1 | 2625.3 | .9439 | 6.8247 | 7.7686 |
| 40 | 75.87 | 0.001 027 | 3.993 | 317.53 | 2159.5 | 2477.0 | 317.58 | 2319.2 | 2636.8 | 1.0259 | 6.6441 | 7.6700 |
| 50 | 81.33 | 0.001 030 | 3.240 | 340.44 | 2143.4 | 2483.9 | 340.49 | 2305.4 | 2645.9 | 1.0910 | 6.5029 | 7.5939 |
| 75 | 91.78 | 0.001 037 | 2.217 | 384.31 | 2112.4 | 2496.7 | 384.39 | 2278.6 | 2663.0 | 1.2130 | 6.2434 | 7.4564 |
| **MPa** | | | | | | | | | | | | |
| 0.100 | 99.63 | 0.001 043 | 1.6940 | 417.36 | 2088.7 | 2506.1 | 417.46 | 2258.0 | 2675.5 | 1.3026 | 6.0568 | 7.3594 |
| 0.125 | 105.99 | 0.001 048 | 1.3749 | 444.19 | 2069.3 | 2513.5 | 444.32 | 2241.0 | 2685.4 | 1.3740 | 5.9104 | 7.2844 |
| 0.150 | 111.37 | 0.001 053 | 1.1593 | 466.94 | 2052.7 | 2519.7 | 467.11 | 2226.5 | 2693.6 | 1.4336 | 5.7897 | 7.2233 |
| 0.175 | 116.06 | 0.001 057 | 1.0036 | 486.80 | 2038.1 | 2524.9 | 486.99 | 2213.6 | 2700.6 | 1.4849 | 5.6868 | 7.1717 |
| 0.200 | 120.23 | 0.001 061 | 0.8857 | 504.49 | 2025.0 | 2529.5 | 504.70 | 2201.9 | 2706.7 | 1.5301 | 5.5970 | 7.1271 |
| 0.225 | 124.00 | 0.001 064 | 0.7933 | 520.47 | 2013.1 | 2533.6 | 520.72 | 2191.3 | 2712.1 | 1.5706 | 5.5173 | 7.0878 |
| 0.250 | 127.44 | 0.001 067 | 0.7187 | 535.10 | 2002.1 | 2537.2 | 535.37 | 2181.5 | 2716.9 | 1.6072 | 5.4455 | 7.0527 |
| 0.275 | 130.60 | 0.001 070 | 0.6573 | 548.59 | 1991.9 | 2540.5 | 548.89 | 2172.4 | 2721.3 | 1.6408 | 5.3801 | 7.0209 |

*(Continued)*

**TABLE A.2 (SI) (CONTINUED)**

Saturation Pressures (Steam)

| Press. kPa $P$ | Temp. °C $T$ | Specific Volume (m³/kg) Sat. Liquid $v_f$ | Specific Volume (m³/kg) Sat. Vapor $v_g$ | Internal Energy (kJ/kg) Sat. Liquid $u_f$ | Internal Energy (kJ/kg) Evap. $u_{fg}$ | Internal Energy (kJ/kg) Sat. Vapor $u_g$ | Enthalpy (kJ/kg) Sat. Liquid $h_f$ | Enthalpy (kJ/kg) Evap. $h_{fg}$ | Enthalpy (kJ/kg) Sat. Vapor $h_g$ | Entropy (kJ/kg. K) Sat. Liquid $s_f$ | Entropy (kJ/kg. K) Evap. $s_{fg}$ | Entropy (kJ/kg. K) Sat. Vapor $s_g$ |
|---|---|---|---|---|---|---|---|---|---|---|---|---|
| 0.300 | 133.55 | 0.001 073 | 0.6058 | 561.15 | 1982.4 | 2543.6 | 561.47 | 2163.8 | 2725.3 | 1.6718 | 5.3201 | 6.9919 |
| 0.325 | 136.30 | 0.001 076 | 0.5620 | 572.90 | 1973.5 | 2546.4 | 573.25 | 2155.8 | 2729.0 | 1.7006 | 5.2646 | 6.9652 |
| 0.350 | 138.88 | 0.001 079 | 0.5243 | 583.95 | 1965.0 | 2548.9 | 584.33 | 2148.1 | 2732.4 | 1.7275 | 5.2130 | 6.9405 |
| 0.375 | 141.32 | 0.001 081 | 0.4914 | 594.40 | 1956.9 | 2551.3 | 594.81 | 2140.8 | 2735.6 | 1.7528 | 5.1647 | 6.9175 |
| 0.40 | 143.63 | 0.001 084 | 0.4625 | 604.31 | 1949.3 | 2553.6 | 604.74 | 2133.8 | 2738.6 | 1.7766 | 5.1193 | 6.8959 |
| 0.45 | 147.93 | 0.001 088 | 0.4140 | 622.77 | 1934.9 | 2557.6 | 623.25 | 2120.7 | 2743.9 | 1.8207 | 5.0359 | 6.8565 |
| 0.50 | 151.86 | 0.001 093 | 0.3749 | 639.68 | 1921.6 | 2561.2 | 640.23 | 2108.5 | 2748.7 | 1.8607 | 4.9606 | 6.8213 |
| 0.55 | 155.48 | 0.001 097 | 0.3427 | 655.32 | 1909.2 | 2564.5 | 655.93 | 2097.0 | 2753.0 | 1.8973 | 4.8920 | 6.7893 |
| 0.60 | 158.85 | 0.001101 | 0.3157 | 669.90 | 1897.5 | 2567.4 | 670.56 | 2086.3 | 2756.8 | 1.9312 | 4.8288 | 6.7600 |
| 0.65 | 162.01 | 0.001104 | 0.2927 | 683.56 | 1886.5 | 2570.1 | 684.28 | 2076.0 | 2760.3 | 1.9627 | 4.7703 | 6.7331 |
| 0.70 | 164.97 | 0.001108 | 0.2729 | 696.44 | 1876.1 | 2572.5 | 697.22 | 2066.3 | 2763.5 | 1.9922 | 4.7158 | 6.7080 |
| 0.75 | 167.78 | 0.001112 | 0.2556 | 708.64 | 1866.1 | 2574.7 | 709.47 | 2057.0 | 2766.4 | 2.0200 | 4.6647 | 6.6847 |
| 0.80 | 170.43 | 0.001115 | 0.2404 | 720.22 | 1856.6 | 2576.8 | 721.11 | 2048.0 | 2769.1 | 2.0462 | 4.6166 | 6.6628 |
| 0.85 | 172.96 | 0.001118 | 0.2270 | 731.27 | 1847.4 | 2578.7 | 732.22 | 2039.4 | 2771.6 | 2.0710 | 4.5711 | 6.6421 |
| 0.90 | 175.38 | 0.001121 | 0.2150 | 741.83 | 1838.6 | 2580.5 | 742.83 | 2031.1 | 2773.9 | 2.0946 | 4.5280 | 6.6226 |
| 0.95 | 177.69 | 0.001124 | 0.2042 | 751.95 | 1830.2 | 2582.1 | 753.02 | 2023.1 | 2776.1 | 2.1172 | 4.4869 | 6.6041 |
| 1.00 | 179.91 | 0.001127 | 0.194 44 | 761.68 | 1822.0 | 2583.6 | 762.81 | 2015.3 | 2778.1 | 2.1387 | 4.4478 | 6.5865 |
| 1.10 | 184.09 | 0.001133 | 0.177 53 | 780.09 | 1806.3 | 2586.4 | 781.34 | 2000.4 | 2781.7 | 2.1792 | 4.3744 | 6.5536 |
| 1.20 | 187.99 | 0.001139 | 0.163 33 | 797.29 | 1791.5 | 2588.8 | 798.65 | 1986.2 | 2784.8 | 2.2166 | 4.3067 | 6.5233 |
| 1.30 | 191.64 | 0.001144 | 0.151 25 | 813.44 | 1777.5 | 2591.0 | 814.93 | 1972.7 | 2787.6 | 2.2515 | 4.2438 | 6.4953 |
| 1.40 | 195.07 | 0.001149 | 0.140 84 | 828.70 | 1764.1 | 2592.8 | 830.30 | 1959.7 | 2790.0 | 2.2842 | 4.1850 | 6.4693 |
| 1.50 | 198.32 | 0.001154 | 0.131 77 | 843.16 | 1751.3 | 2594.5 | 844.89 | 1947.3 | 2792.2 | 2.3150 | 4.1298 | 6.4448 |
| 1.75 | 205.76 | 0.001166 | 0.113 49 | 876.46 | 1721.4 | 2597.8 | 878.50 | 1917.9 | 2796.4 | 2.3851 | 4.0044 | 6.3896 |
| 2.00 | 212.42 | 0.001177 | 0.099 63 | 906.44 | 1693.8 | 2600.3 | 908.79 | 1890.7 | 2799.5 | 2.4474 | 3.8935 | 6.3409 |
| 2.25 | 218.45 | 0.001187 | 0.088 75 | 933.83 | 1668.2 | 2602.0 | 936.49 | 1865.2 | 2801.7 | 2.5035 | 3.7937 | 6.2972 |
| 2.5 | 223.99 | 0.001197 | 0.079 98 | 959.11 | 1644.0 | 2603.1 | 962.11 | 1841.0 | 2803.1 | 2.5547 | 3.7028 | 6.2575 |

(*Continued*)

**TABLE A.2 (SI) (CONTINUED)**

Saturation Pressures (Steam)

| Press. kPa P | Temp. °C T | Specific Volume (m³/kg) | | Internal Energy (kJ/kg) | | | Enthalpy (kJ/kg) | | | Entropy (kJ/kg, K) | | |
|---|---|---|---|---|---|---|---|---|---|---|---|---|
| | | Sat. Liquid $v_f$ | Sat. Vapor $v_g$ | Sat. Liquid $u_f$ | Evap. $u_{fg}$ | Sat. Vapor $u_g$ | Sat. Liquid $h_f$ | Evap. $h_{fg}$ | Sat. Vapor $h_g$ | Sat. Liquid $s_f$ | Evap. $s_{fg}$ | Sat. Vapor $s_g$ |
| 3.0 | 233.90 | 0.001 217 | 0.066 68 | 1004.78 | 1599.3 | 2604.1 | 1008.42 | 1795.7 | 2804.2 | 2.6457 | 3.5412 | 6.1869 |
| 3.5 | 242.60 | 0.001 235 | 0.057 07 | 1045.43 | 1558.3 | 2603.7 | 1049.75 | 1753.7 | 2803.4 | 2.7253 | 3.4000 | 6.1253 |
| 4 | 250.40 | 0.001 252 | 0.049 78 | 1082.31 | 1520.0 | 2602.3 | 1087.31 | 1714.1 | 2801.4 | 2.7964 | 3.2737 | 6.0701 |
| 5 | 263.99 | 0.001 286 | 0.039 44 | 1147.81 | 1449.3 | 2597.1 | 1154.23 | 1640.1 | 2794.3 | 2.9202 | 3.0532 | 5.9734 |
| 6 | 275.64 | 0.001 319 | 0.032 44 | 1205.44 | 1384.3 | 2589.7 | 1213.35 | 1571.0 | 2784.3 | 3.0267 | 2.8625 | 5.8892 |
| 7 | 285.88 | 0.001 351 | 0.02737 | 1257.55 | 1323.0 | 2580.5 | 1267.00 | 1505.1 | 2772.1 | 3.1211 | 2.6922 | 5.8133 |
| 8 | 295.06 | 0.001 384 | 0.023 52 | 1305.57 | 1264.2 | 2569.8 | 1316.64 | 1441.3 | 2758.0 | 3.2068 | 2.5364 | 5.7432 |
| 9 | 303.40 | 0.001 418 | 0.020 48 | 1350.51 | 1207.3 | 2557.8 | 1363.26 | 1378.9 | 2742.1 | 3.2858 | 2.3915 | 5.6772 |
| 10 | 311.06 | 0.001 452 | 0.018 026 | 1393.04 | 1151.4 | 2544.4 | 1407.56 | 1317.1 | 2724.7 | 3.3596 | 2.2544 | 5.6141 |
| 11 | 318.15 | 0.001 489 | 0.015 987 | 1433.7 | 1096.0 | 2529.8 | 1450.1 | 1255.5 | 2705.6 | 3.4295 | 2.1233 | 5.5527 |
| 12 | 324.75 | 0.001 527 | 0.014 263 | 1473.0 | 1040.7 | 2513.7 | 1491.3 | 1193.6 | 2684.9 | 3.4962 | 1.9962 | 5.4924 |
| 13 | 330.93 | 0.001 567 | 0.012 780 | 1511.1 | 985.0 | 2496.1 | 1531.5 | 1130.7 | 2662.2 | 3.5606 | 1.8718 | 5.4323 |
| 14 | 336.75 | 0.001 611 | 0.011 485 | 1548.6 | 928.2 | 2476.8 | 1571.1 | 1066.5 | 2637.6 | 3.6232 | 1.7485 | 5.3717 |
| 15 | 342.24 | 0.001 658 | 0.010 337 | 1585.6 | 869.8 | 2455.5 | 1610.5 | 1000.0 | 2610.5 | 3.6848 | 1.6249 | 5.3098 |
| 16 | 347.44 | 0.001 711 | 0.009 306 | 1622.7 | 809.0 | 2431.7 | 1650.1 | 930.6 | 2580.6 | 3.7461 | 1.4994 | 5.2455 |
| 17 | 352.37 | 0.001 770 | 0.008 364 | 1660.2 | 744.8 | 2405.0 | 1690.3 | 856.9 | 2547.2 | 3.8079 | 1.3698 | 5.1777 |
| 18 | 357.06 | 0.001 840 | 0.007489 | 1698.9 | 675.4 | 2374.3 | 1732.0 | 777.1 | 2509.1 | 3.8715 | 1.2329 | 5.1044 |
| 19 | 361.54 | 0.001 924 | 0.006 657 | 1739.9 | 598.1 | 2338.1 | 1776.5 | 688.0 | 2464.5 | 3.9388 | 1.0839 | 5.0228 |
| 20 | 365.81 | 0.002 036 | 0.005 834 | 1785.6 | 507.5 | 2293.0 | 1826.3 | 583.4 | 2409.7 | 4.0139 | .9130 | 4.9269 |
| 21 | 369.89 | 0.002 207 | 0.004 952 | 1842.1 | 388.5 | 2230.6 | 1888.4 | 446.2 | 2334.6 | 4.1075 | .6938 | 4.8013 |
| 22 | 373.80 | 0.002 742 | 0.003 568 | 1961.9 | 125.2 | 2087.1 | 2022.2 | 143.4 | 2165.6 | 4.3110 | .2216 | 4.5327 |
| 22.09 | 374.14 | 0.003 155 | 0.003 155 | 2029.6 | 0 | 2029.6 | 2099.3 | 0 | 2099.3 | 4.4298 | 0 | 4.4298 |

**TABLE A.3 (SI)**

Properties of Superheated Steam

| T | P=.010 MPa (45.81) | | | | P=.050 MPa (81.33) | | | | P=.10 MPa (99.63) | | | |
|---|---|---|---|---|---|---|---|---|---|---|---|---|
| | v | u | h | s | v | u | h | s | v | u | h | s |
| Sat. | 14.674 | 2437.9 | 2584.7 | 8.1502 | 3.240 | 2483.9 | 2645.9 | 7.5939 | 1.6940 | 2506.1 | 2675.5 | 7.3594 |
| 50 | 14.869 | 2443.9 | 2592.6 | 8.1749 | | | | | | | | |
| 100 | 17.196 | 2515.5 | 2687.5 | 8.4479 | 3.418 | 2511.6 | 2682.5 | 7.6947 | 1.6958 | 2506.7 | 2676.2 | 7.3614 |
| 150 | 19.512 | 2587.9 | 2783.0 | 8.6882 | 3.889 | 2585.6 | 2780.1 | 7.9401 | 1.9364 | 2582.8 | 2776.4 | 7.6134 |
| 200 | 21.825 | 2661.3 | 2879.5 | 8.9038 | 4.356 | 2659.9 | 2877.7 | 8.1580 | 2.172 | 2658.1 | 2875.3 | 7.8343 |
| 250 | 24.136 | 2736.0 | 2977.3 | 9.1002 | 4.820 | 2735.0 | 2976.0 | 8.3556 | 2.406 | 2733.7 | 2974.3 | 8.0333 |
| 300 | 26.445 | 2812.1 | 3076.5 | 9.2813 | 5.284 | 2811.3 | 3075.5 | 8.5373 | 2.639 | 2810.4 | 3074.3 | 8.2158 |
| 400 | 31.063 | 2968.9 | 3279.6 | 9.6077 | 6.209 | 2968.5 | 3278.9 | 8.8642 | 3.103 | 2967.9 | 3278.2 | 8.5435 |
| 500 | 35.679 | 3132.3 | 3489.1 | 9.8978 | 7.134 | 3132.0 | 3488.7 | 9.1546 | 3.565 | 3131.6 | 3488.1 | 8.8342 |
| 600 | 40.295 | 3302.5 | 3705.4 | 10.1608 | 8.057 | 3302.2 | 3705.1 | 9.4178 | 4.028 | 3301.9 | 3704.7 | 9.0976 |
| 700 | 44.91 | 3479.6 | 3928.7 | 10.4028 | 8.981 | 3479.4 | 3928.5 | 9.6599 | 4.490 | 3479.2 | 3928.2 | 9.3398 |
| 800 | 49.526 | 3663.8 | 4159.0 | 10.6281 | 9.904 | 3663.6 | 4158.9 | 9.8852 | 4.952 | 3663.5 | 4158.6 | 9.5652 |
| 900 | 54.141 | 3855.0 | 4396.4 | 10.8396 | 10.828 | 3854.9 | 4396.3 | 10.0967 | 5.414 | 3854.8 | 4396.1 | 9.7767 |
| 1000 | 58.757 | 4053.0 | 4640.6 | 11.0393 | 11.751 | 4052.9 | 4640.5 | 10.2964 | 5.875 | 4052.8 | 4640.3 | 9.9764 |
| 1100 | 63.372 | 4257.5 | 4891.2 | 11.2287 | 12.674 | 4257.4 | 4891.1 | 10.4859 | 6.337 | 4257.3 | 4891.0 | 10.1659 |
| 1200 | 67.987 | 4467.9 | 5147.8 | 11.4091 | 13.597 | 4467.8 | 5147.7 | 10.6662 | 6.799 | 4467.7 | 5147.6 | 10.3463 |
| 1300 | 72.602 | 4683.7 | 5409.7 | 11.5811 | 14.521 | 4683.6 | 5409.6 | 10.8382 | 7.260 | 4683.5 | 5409.5 | 10.5183 |

| T | P=.20 MPa (120.23) | | | | P=.30 MPa (133.55) | | | | P=.40 MPa (143.63) | | | |
|---|---|---|---|---|---|---|---|---|---|---|---|---|
| | v | u | h | s | v | u | h | s | v | u | h | s |
| Sat. | .8857 | 2529.5 | 2706.7 | 7.1272 | .6058 | 2543.6 | 2725.3 | 6.9919 | .4625 | 2553.6 | 2738.6 | 6.8959 |
| 150 | .9596 | 2576.9 | 2768.8 | 7.2795 | .6339 | 2570.8 | 2761.0 | 7.0778 | .4708 | 2564.5 | 2752.8 | 6.9299 |
| 200 | 1.0803 | 2654.4 | 2870.5 | 7.5066 | .7163 | 2650.7 | 2865.6 | 7.3115 | .5342 | 2646.8 | 2860.5 | 7.1706 |
| 250 | 1.1988 | 2731.2 | 2971.0 | 7.7086 | .7964 | 2728.7 | 2967.6 | 7.5166 | .5951 | 2726.1 | 2964.2 | 7.3789 |
| 300 | 1.3162 | 2808.6 | 3071.8 | 7.8926 | .8753 | 2806.7 | 3069.3 | 7.7022 | .6548 | 2804.8 | 3066.8 | 7.5662 |
| 400 | 1.5493 | 2966.7 | 3276.6 | 8.2218 | 1.0315 | 2965.6 | 3275.0 | 8.0330 | .7726 | 2964.4 | 3273.4 | 7.8985 |
| 500 | 1.7814 | 3130.8 | 3487.1 | 8.5133 | 1.1867 | 3130.0 | 3486.0 | 8.3251 | .8893 | 3129.2 | 3484.9 | 8.1913 |
| 600 | 2.013 | 3301.4 | 3704.0 | 8.7770 | 1.3414 | 3300.8 | 3703.2 | 8.5892 | 1.0055 | 3300.2 | 3702.4 | 8.4558 |
| 700 | 2.244 | 3478.8 | 3927.6 | 9.0194 | 1.4957 | 3478.4 | 3927.1 | 8.8319 | 1.1215 | 3477.9 | 3926.5 | 8.6987 |

*(Continued)*

**TABLE A.3 (SI) (CONTINUED)**

Properties of Superheated Steam

| T | P=.010 MPa (45.81) | | | | P=.050 MPa (81.33) | | | | P=.10 MPa (99.63) | | | |
|---|---|---|---|---|---|---|---|---|---|---|---|---|
| | $v$ | $u$ | $h$ | $s$ | $v$ | $u$ | $h$ | $s$ | $v$ | $u$ | $h$ | $s$ |
| 800 | 2.475 | 3663.1 | 4158.2 | 9.2449 | 1.6499 | 3662.9 | 4157.8 | 9.0576 | 1.2372 | 3662.4 | 4157.3 | 8.9244 |
| 900 | 2.706 | 3854.5 | 4395.8 | 9.4566 | 1.8041 | 3854.2 | 4395.4 | 9.2692 | 1.3529 | 3853.9 | 4395.1 | 9.1362 |
| 1000 | 2.937 | 4052.5 | 4640.0 | 9.6563 | 1.9581 | 4052.3 | 4639.7 | 9.4690 | 1.4685 | 4052.0 | 4639.4 | 9.3360 |
| 1100 | 3.168 | 4257.0 | 4890.7 | 9.8458 | 2.1121 | 4256.8 | 4890.4 | 9.6585 | 1.5840 | 4256.5 | 4890.2 | 9.5256 |
| 1200 | 3.399 | 4467.5 | 5147.3 | 10.0262 | 2.2661 | 4467.2 | 5147.1 | 9.8389 | 1.6996 | 4467.0 | 5146.8 | 9.7060 |
| 1300 | 3.630 | 4683.2 | 5409.3 | 10.1982 | 2.4201 | 4683.0 | 5409.0 | 10.0110 | 1.8151 | 4682.8 | 5408.8 | 9.8780 |

| T | P=.50 MPa (151.86) | | | | P=.60 MPa (158.85) | | | | P=.80 MPa (170.43) | | | |
|---|---|---|---|---|---|---|---|---|---|---|---|---|
| | $v$ | $u$ | $h$ | $s$ | $v$ | $u$ | $h$ | $s$ | $v$ | $u$ | $h$ | $s$ |
| Sat. | .3749 | 2561.2 | 2748.7 | 6.8213 | .3157 | 2567.4 | 2756.8 | 6.7600 | .2404 | 2576.8 | 2769.1 | 6.6628 |
| 200 | .4249 | 2642.9 | 2855.4 | 7.0592 | .3520 | 2638.9 | 2850.1 | 6.9665 | .2608 | 2630.6 | 2839.3 | 6.8158 |
| 250 | .4744 | 2723.5 | 2960.7 | 7.2709 | .3938 | 2720.9 | 2957.2 | 7.1816 | .2931 | 2715.5 | 2950.0 | 7.0384 |
| 300 | .5226 | 2802.9 | 3064.2 | 7.4599 | .4344 | 2801.0 | 3061.6 | 7.3724 | .3241 | 2797.2 | 3056.5 | 7.2328 |
| 350 | .5701 | 2882.6 | 3167.7 | 7.6329 | .4742 | 2881.2 | 3165.7 | 7.5464 | .3544 | 2878.2 | 3161.7 | 7.4089 |
| 400 | .6173 | 2963.2 | 3271.9 | 7.7938 | .5137 | 2962.1 | 3270.3 | 7.7079 | .3843 | 2959.7 | 3267.1 | 7.5716 |
| 500 | .7109 | 3128.4 | 3483.9 | 8.0873 | .5920 | 3127.6 | 3482.8 | 8.0021 | .4433 | 3126.0 | 3480.6 | 7.8673 |
| 600 | .8041 | 3299.6 | 3701.7 | 8.3522 | .6697 | 3299.1 | 3700.9 | 8.2674 | .5018 | 3297.9 | 3699.4 | 8.1333 |
| 700 | .8969 | 3477.5 | 3925.9 | 8.5952 | .7472 | 3477.0 | 3925.3 | 8.5107 | .5601 | 3476.2 | 3924.2 | 8.3770 |
| 800 | .9896 | 3662.1 | 4156.9 | 8.8211 | .8245 | 3661.8 | 4156.5 | 8.7367 | .6181 | 3661.1 | 4155.6 | 8.6033 |
| 900 | 1.0822 | 3853.6 | 4394.7 | 9.0329 | .9017 | 3853.4 | 4394.4 | 8.9486 | .6761 | 3852.8 | 4393.7 | 8.8153 |
| 1000 | 1.1747 | 4051.8 | 4639.1 | 9.2328 | .9788 | 4051.5 | 4638.8 | 9.1485 | .7340 | 4051.0 | 4638.2 | 9.0153 |
| 1100 | 1.2672 | 4256.3 | 4889.9 | 9.4224 | 1.0559 | 4256.1 | 4889.6 | 9.3381 | .7919 | 4255.6 | 4889.1 | 9.2050 |
| 1200 | 1.3596 | 4466.8 | 5146.6 | 9.6029 | 1.1330 | 4466.5 | 5146.3 | 9.5185 | .8497 | 4466.1 | 5145.9 | 9.3855 |
| 1300 | 1.4521 | 4682.5 | 5408.6 | 9.7749 | 1.2101 | 4682.3 | 5408.3 | 9.6906 | .9076 | 4681.8 | 5407.9 | 9.5575 |

| T | P=1.00 MPa (179.91) | | | | P=1.20 MPa (187.99) | | | | P=1.40 MPa (195.07) | | | |
|---|---|---|---|---|---|---|---|---|---|---|---|---|
| | $v$ | $u$ | $h$ | $s$ | $v$ | $u$ | $h$ | $s$ | $v$ | $u$ | $h$ | $s$ |
| Sat. | .194 44 | 2583.6 | 2778.1 | 6.5865 | .163 33 | 2588.8 | 2784.8 | 6.5233 | .140 84 | 2592.8 | 2790.0 | 6.4693 |
| 200 | .2060 | 2621.9 | 2827.9 | 6.6940 | .169 30 | 2612.8 | 2815.9 | 6.5898 | .143 02 | 2603.1 | 2803.3 | 6.4975 |
| 250 | .2327 | 2709.9 | 2942.6 | 6.9247 | .192 34 | 2704.2 | 2935.0 | 6.8294 | .163 50 | 2698.3 | 2927.2 | 6.7467 |
| 300 | .2579 | 2793.2 | 3051.2 | 7.1229 | .2138 | 2789.2 | 3045.8 | 7.0317 | .182 28 | 2785.2 | 3040.4 | 6.9534 |

*(Continued)*

# TABLE A.3 (SI) (CONTINUED)

Properties of Superheated Steam

| T | v | u | h | s | v | u | h | s | v | u | h | s |
|---|---|---|---|---|---|---|---|---|---|---|---|---|
| 350 | .2825 | 2875.2 | 3157.7 | 7.3011 | .2345 | 2872.2 | 3153.6 | 7.2121 | .2003 | 2869.2 | 3149.5 | 7.1360 |
| 400 | .3066 | 2957.3 | 3263.9 | 7.4651 | .2548 | 2954.9 | 3260.7 | 7.3374 | .2178 | 2952.5 | 3257.5 | 7.3026 |
| 500 | .3541 | 3124.4 | 3478.5 | 7.7622 | .2946 | 3122.8 | 3476.3 | 7.6759 | .2521 | 3121.1 | 3474.1 | 7.6027 |
| 600 | .4011 | 3296.8 | 3697.9 | 8.0290 | .3339 | 3295.6 | 3696.3 | 7.9435 | .2860 | 3294.4 | 3694.8 | 7.8710 |
| 700 | .4478 | 3475.3 | 3923.1 | 8.2731 | .3729 | 3474.4 | 3922.0 | 8.1881 | .3195 | 3473.6 | 3920.8 | 8.1160 |
| 800 | .4943 | 3660.4 | 4154.7 | 8.4996 | .4118 | 3659.7 | 4153.8 | 8.4148 | .3528 | 3659.0 | 4153.0 | 8.3431 |
| 900 | .5407 | 3852.2 | 4392.9 | 8.7118 | .4505 | 3851.6 | 4392.2 | 8.6272 | .3861 | 3851.1 | 4391.5 | 8.5556 |
| 1000 | .5871 | 4050.5 | 4637.6 | 8.9119 | .4892 | 4050.0 | 4637.0 | 8.8274 | .4192 | 4049.5 | 4636.4 | 8.7559 |
| 1100 | .6335 | 4255.1 | 4888.6 | 9.1017 | .5278 | 4254.6 | 4888.0 | 9.0172 | .4524 | 4254.1 | 4887.5 | 8.9457 |
| 1200 | .6798 | 4465.6 | 5145.4 | 9.2822 | .5665 | 4465.1 | 5144.9 | 9.1977 | .4855 | 4464.7 | 5144.4 | 9.1262 |
| 1300 | .7261 | 4681.3 | 5407.4 | 9.4543 | .6051 | 4680.9 | 5407.0 | 9.3698 | .5186 | 4680.4 | 5406.5 | 9.2984 |

| T | P = 1.60 MPa (201.41) | | | | P = 1.80 MPa (207.15) | | | | P = 2.00 MPa (212.42) | | | |
|---|---|---|---|---|---|---|---|---|---|---|---|---|
| | v | u | h | s | v | u | h | s | v | u | h | s |
| Sat. | .123 80 | 2596.0 | 2794.0 | 6.4218 | .11042 | 2598.4 | 2797.1 | 6.3794 | .099 63 | 2600.3 | 2799.5 | 6.3409 |
| 225 | .132 87 | 2644.7 | 2857.3 | 6.5518 | .116 73 | 2636.6 | 2846.7 | 6.4808 | .103 77 | 2628.3 | 2835.8 | 6.4147 |
| 250 | .141 84 | 2692.3 | 2919.2 | 6.6732 | .124 97 | 2686.0 | 2911.0 | 6.6066 | .111 44 | 2679.6 | 2902.5 | 6.5453 |
| 300 | .158 62 | 2781.1 | 3034.8 | 6.8844 | .140 21 | 2776.9 | 3029.2 | 6.8226 | .125 47 | 2772.6 | 3023.5 | 6.7664 |
| 350 | .174 56 | 2866.1 | 3145.4 | 7.0694 | .154 57 | 2863.0 | 3141.2 | 7.0100 | .138 57 | 2859.8 | 3137.0 | 6.9563 |
| 400 | .190 05 | 2950.1 | 3254.2 | 7.2374 | .168 47 | 2947.7 | 3250.9 | 7.1794 | .151 20 | 2945.2 | 3247.6 | 7.1271 |
| 500 | .2203 | 3119.5 | 3472.0 | 7.5390 | .195 50 | 3117.9 | 3469.8 | 7.4825 | .175 68 | 3116.2 | 3467.6 | 7.4317 |
| 600 | .2500 | 3293.3 | 3693.2 | 7.8080 | .2220 | 3292.1 | 3691.7 | 7.7523 | .199 60 | 3290.9 | 3690.1 | 7.7024 |
| 700 | .2794 | 3472.7 | 3919.7 | 8.0535 | .2482 | 3471.8 | 3918.5 | 7.9983 | .2232 | 3470.9 | 3917.4 | 7.9487 |

| T | P = 1.60 MPa (201.41) | | | | P = 1.80 MPa (207.15) | | | | P = 2.00 MPa (212.42) | | | |
|---|---|---|---|---|---|---|---|---|---|---|---|---|
| | v | u | h | s | v | u | h | s | v | u | h | s |
| 800 | .3086 | 3658.3 | 4152.1 | 8.2808 | .2742 | 3657.6 | 4151.2 | 8.2258 | .2467 | 3657.0 | 4150.3 | 8.1765 |
| 900 | .3377 | 3850.5 | 4390.8 | 8.4935 | .3001 | 3849.9 | 4390.1 | 8.4386 | .2700 | 3849.3 | 4389.4 | 8.3895 |
| 1000 | .3668 | 4049.0 | 4635.8 | 8.6938 | .3260 | 4048.5 | 4635.2 | 8.6391 | .2933 | 4048.0 | 4634.6 | 8.5901 |
| 1100 | .3958 | 4253.7 | 4887.0 | 8.8837 | .3518 | 4253.2 | 4886.4 | 8.8290 | .3166 | 4252.7 | 4885.9 | 8.7800 |
| 1200 | .4248 | 4464.2 | 5143.9 | 9.0643 | .3776 | 4463.7 | 5143.4 | 9.0096 | .3398 | 4463.3 | 5142.9 | 8.9607 |
| 1300 | .4538 | 4679.9 | 5406.0 | 9.2364 | .4034 | 4679.5 | 5405.6 | 9.1818 | .3631 | 4679.0 | 5405.1 | 9.1329 |

*(Continued)*

**TABLE A.3 (SI) (CONTINUED)**

Properties of Superheated Steam

| T | P = 2.50 Mpa (223.99) v | u | h | s | P = 3.00 Mpa (233.90) v | u | h | s | P = 3.50 MPa (242.60) v | u | h | s |
|---|---|---|---|---|---|---|---|---|---|---|---|---|
| Sat. | .079 98 | 2603.1 | 2803.1 | 6.2575 | .066 68 | 2604.1 | 2804.2 | 6.1869 | .057 07 | 2603.7 | 2803.4 | 6.1253 |
| 225 | .080 27 | 2605.6 | 2806.3 | 6.2639 | | | | | | | | |
| 250 | .087 00 | 2662.6 | 2880.1 | 6.4085 | .070 58 | 2644.0 | 2855.8 | 6.2872 | .058 72 | 2623.7 | 2829.2 | 6.1749 |
| 300 | .098 90 | 2761.6 | 3008.8 | 6.6438 | .08114 | 2750.1 | 2993.5 | 6.5390 | .068 42 | 2738.0 | 2977.5 | 6.4461 |
| 350 | .109 76 | 2851.9 | 3126.3 | 6.8403 | .090 53 | 2843.7 | 3115.3 | 6.7428 | .076 78 | 2835.3 | 3104.0 | 6.6579 |
| 400 | .120 10 | 2939.1 | 3239.3 | 7.0148 | .099 36 | 2932.8 | 3230.9 | 6.9212 | .084 53 | 2926.4 | 3222.3 | 6.8405 |
| 450 | .130 14 | 3025.5 | 3350.8 | 7.1746 | .107 87 | 3020.4 | 3344.0 | 7.0834 | .091 96 | 3015.3 | 3337.2 | 7.0052 |
| 500 | .139 98 | 3112.1 | 3462.1 | 7.3234 | .116 19 | 3108.0 | 3456.5 | 7.2338 | .099 18 | 3103.0 | 3450.9 | 7.1572 |
| 600 | .159 30 | 3288.0 | 3686.3 | 7.5960 | .132 43 | 3285.0 | 3682.3 | 7.5085 | .113 24 | 3282.1 | 3678.4 | 7.4339 |
| 700 | .178 32 | 3468.7 | 3914.5 | 7.8435 | .148 38 | 3466.5 | 3911.7 | 7.7571 | .126 99 | 3464.3 | 3908.8 | 7.6837 |
| 800 | .197 16 | 3655.3 | 4148.2 | 8.0720 | .164 14 | 3653.5 | 4145.9 | 7.9862 | .140 56 | 3651.8 | 4143.7 | 7.9134 |
| 900 | .215 90 | 3847.9 | 4387.6 | 8.2853 | .179 80 | 3846.5 | 4385.9 | 8.1999 | .154 02 | 3845.0 | 4384.1 | 8.1276 |
| 1000 | .2346 | 4046.7 | 4633.1 | 8.4861 | .195 41 | 4045.4 | 4631.6 | 8.4009 | .16743 | 4044.1 | 4630.1 | 8.3288 |
| 1100 | .2532 | 4251.5 | 4884.6 | 8.6762 | .210 98 | 4250.3 | 4883.3 | 8.5912 | .180 80 | 4249.2 | 4881.9 | 8.5192 |
| 1200 | .2718 | 4462.1 | 5141.7 | 8.8569 | .226 52 | 4460.9 | 5140.5 | 8.7720 | .194 15 | 4459.8 | 5139.3 | 8.7000 |
| 1300 | .2905 | 4677.8 | 5404.0 | 9.0291 | .242 06 | 4676.6 | 5402.8 | 8.9442 | .20749 | 4675.5 | 5401.7 | 8.8723 |

| T | P = 4.0 MPa (250.40) v | u | h | s | P = 4.5 MPa (257.49) v | u | h | s | P = 5.0 MPa (263.99) v | u | h | s |
|---|---|---|---|---|---|---|---|---|---|---|---|---|
| Sat. | .049 78 | 2602.3 | 2801.4 | 6.0701 | .044 06 | 2600.1 | 2798.3 | 6.0198 | .039 44 | 2597.1 | 2794.3 | 5.9734 |
| 275 | .054 57 | 2667.9 | 2886.2 | 6.2285 | .04730 | 2650.3 | 2863.2 | 6.1401 | .041 41 | 2631.3 | 2838.3 | 6.0544 |
| 300 | .058 84 | 2725.3 | 2960.7 | 6.3615 | .051 35 | 2712.0 | 2943.1 | 6.2828 | .045 32 | 2698.0 | 2924.5 | 6.2084 |
| 350 | .066 45 | 2826.7 | 3092.5 | 6.5821 | .058 40 | 2817.8 | 3080.6 | 6.5131 | .051 94 | 2808.7 | 3068.4 | 6.4493 |
| 400 | .073 41 | 2919.9 | 3213.6 | 6.7690 | .064 75 | 2913.3 | 3204.7 | 6.7047 | .057 81 | 2906.6 | 3195.7 | 6.6459 |
| 450 | .080 02 | 3010.2 | 3330.3 | 6.9363 | .070 74 | 3005.0 | 3323.3 | 6.8746 | .063 30 | 2999.7 | 3316.2 | 6.8186 |
| 500 | .086 43 | 3099.5 | 3445.3 | 7.0901 | .076 51 | 3095.3 | 3439.6 | 7.0301 | .068 57 | 3091.0 | 3433.8 | 6.9759 |
| 600 | .098 85 | 3279.1 | 3674.4 | 7.3688 | .087 65 | 3276.0 | 3670.5 | 7.3110 | .078 69 | 3273.0 | 3666.5 | 7.2589 |
| 700 | .11095 | 3462.1 | 3905.9 | 7.6198 | .098 47 | 3459.9 | 3903.0 | 7.5631 | .088 49 | 3457.6 | 3900.1 | 7.5122 |
| 800 | .122 87 | 3650.0 | 4141.5 | 7.8502 | .109 11 | 3648.3 | 4139.3 | 7.7942 | .098 11 | 3646.6 | 4137.1 | 7.7440 |

*(Continued)*

**TABLE A.3 (SI) (CONTINUED)**

Properties of Superheated Steam

| T | \(P = 6.0\) MPa | | | | \(P = 7.0\) MPa | | | | \(P = 8.0\) MPa | | | |
|---|---|---|---|---|---|---|---|---|---|---|---|---|
| 900 | .134 69 | 3843.6 | 4382.3 | 8.0647 | .119 65 | 3842.2 | 4380.6 | 8.0091 | .107 62 | 3840.7 | 4378.8 | 7.9593 |
| 1000 | .146 45 | 4042.9 | 4628.7 | 8.2662 | .130 13 | 4041.6 | 4627.2 | 8.2108 | .117 07 | 4040.4 | 4625.7 | 8.1612 |
| 1100 | .158 17 | 4248.0 | 4880.6 | 8.4567 | .140 56 | 4246.8 | 4879.3 | 8.4015 | .126 48 | 4245.6 | 4878.0 | 8.3520 |
| 1200 | .169 87 | 4458.6 | 5138.1 | 8.6376 | .150 98 | 4457.5 | 5136.9 | 8.5825 | .135 87 | 4456.3 | 5135.7 | 8.5331 |
| 1300 | .181 56 | 4674.3 | 5400.5 | 8.8100 | .161 39 | 4673.1 | 5399.4 | 8.7549 | .145 26 | 4672.0 | 5398.2 | 8.7055 |

| T | \(P = 6.0\) MPa (275.64) | | | | \(P = 7.0\) MPa (285.88) | | | | \(P = 8.0\) MPa (295.06) | | | |
|---|---|---|---|---|---|---|---|---|---|---|---|---|
| Sat. | .032 44 | 2589.7 | 2784.3 | 5.8892 | .02737 | 2580.5 | 2772.1 | 5.8133 | .023 52 | 2569.8 | 2758.0 | 5.7432 |
| 300 | .036 16 | 2667.2 | 2884.2 | 6.0674 | .029 47 | 2632.2 | 2838.4 | 5.9305 | .024 26 | 2590.9 | 2785.0 | 5.7906 |
| 350 | .042 23 | 2789.6 | 3043.0 | 6.3335 | .035 24 | 2769.4 | 3016.0 | 6.2283 | .029 95 | 2747.7 | 2987.3 | 6.1301 |
| 400 | .04739 | 2892.9 | 3177.2 | 6.5408 | .039 93 | 2878.6 | 3158.1 | 6.4478 | .034 32 | 2863.8 | 3138.3 | 6.3634 |
| 450 | .052 14 | 2988.9 | 3301.8 | 6.7193 | .04416 | 2978.0 | 3287.1 | 6.6327 | .038 17 | 2966.7 | 3272.0 | 6.5551 |
| 500 | .056 65 | 3082.2 | 3422.2 | 6.8803 | .048 14 | 3073.4 | 3410.3 | 6.7975 | .041 75 | 3064.3 | 3398.3 | 6.7240 |
| 550 | .061 01 | 3174.6 | 3540.6 | 7.0288 | .051 95 | 3167.2 | 3530.9 | 6.9486 | .045 16 | 3159.8 | 3521.0 | 6.8778 |
| 600 | .065 25 | 3266.9 | 3658.4 | 7.1677 | .055 65 | 3260.7 | 3650.3 | 7.0894 | .048 45 | 3254.4 | 3642.0 | 7.0206 |
| 700 | .073 52 | 3453.1 | 3894.2 | 7.4234 | .062 83 | 3448.5 | 3888.3 | 7.3476 | .054 81 | 3443.9 | 3882.4 | 7.2812 |
| 800 | .081 60 | 3643.1 | 4132.7 | 7.6566 | .069 81 | 3639.5 | 4128.2 | 7.5822 | .060 97 | 3636.0 | 4123.8 | 7.5173 |
| 900 | .089 58 | 3837.8 | 4375.3 | 7.8727 | .076 69 | 3835.0 | 4371.8 | 7.7991 | .067 02 | 3832.1 | 4368.3 | 7.7351 |
| 1000 | .097 49 | 4037.8 | 4622.7 | 8.0751 | .083 50 | 4035.3 | 4619.8 | 8.0020 | .073 01 | 4032.8 | 4616.9 | 7.9384 |
| 1100 | .105 36 | 4243.3 | 4875.4 | 8.2661 | .090 27 | 4240.9 | 4872.8 | 8.1933 | .078 96 | 4238.6 | 4870.3 | 8.1300 |
| 1200 | .113 21 | 4454.0 | 5133.3 | 8.4474 | .097 03 | 4451.7 | 5130.9 | 8.3747 | .084 89 | 4449.5 | 5128.5 | 8.3115 |
| 1300 | .121 06 | 4669.6 | 5396.0 | 8.6199 | .103 77 | 4667.3 | 5393.7 | 8.5473 | .090 80 | 4665.0 | 5391.5 | 8.4842 |

| T | \(P = 9.0\) MPa (303.40) | | | | \(P = 10.0\) MPa (311.06) | | | | \(P = 12.5\) MPa (327.89) | | | |
|---|---|---|---|---|---|---|---|---|---|---|---|---|
| Sat. | .020 48 | 2557.8 | 2742.1 | 5.6772 | .018 026 | 2544.4 | 2724.7 | 5.6141 | .013 495 | 2505.1 | 2673.8 | 5.4624 |
| 325 | .023 27 | 2646.6 | 2856.0 | 5.8712 | .019 861 | 2610.4 | 2809.1 | 5.7568 | | | | |
| 350 | .025 80 | 2724.4 | 2956.6 | 6.0361 | .022 42 | 2699.2 | 2923.4 | 5.9443 | .016 126 | 2624.6 | 2826.2 | 5.7118 |
| 400 | .029 93 | 2848.4 | 3117.8 | 6.2854 | .026 41 | 2832.4 | 3096.5 | 6.2120 | .020 00 | 2789.3 | 3039.3 | 6.0417 |
| 450 | .033 50 | 2955.2 | 3256.6 | 6.4844 | .029 75 | 2943.4 | 3240.9 | 6.4190 | .022 99 | 2912.5 | 3199.8 | 6.2719 |
| 500 | .036 77 | 3055.2 | 3386.1 | 6.6576 | .032 79 | 3045.8 | 3373.7 | 6.5966 | .025 60 | 3021.7 | 3341.8 | 6.4618 |
| 550 | .039 87 | 3152.2 | 3511.0 | 6.8142 | .035 64 | 3144.6 | 3500.9 | 6.7561 | .028 01 | 3125.0 | 3475.2 | 6.6290 |

*(Continued)*

## TABLE A.3 (SI) (CONTINUED)

Properties of Superheated Steam

| | P = 9.0 MPa (303.40) | | | | P = 10.0 MPa (311.06) | | | | P = 12.5 MPa (327.89) | | | |
|---|---|---|---|---|---|---|---|---|---|---|---|---|
| T | v | u | h | s | v | u | h | s | v | u | h | s |
| 600 | .042 85 | 3248.1 | 3633.7 | 6.9589 | .038 37 | 3241.7 | 3625.3 | 6.9029 | .030 29 | 3225.4 | 3604.0 | 6.7810 |
| 650 | .045 74 | 3343.6 | 3755.3 | 7.0943 | .041 01 | 3338.2 | 3748.2 | 7.0398 | .032 48 | 3324.4 | 3730.4 | 6.9218 |
| 700 | .048 57 | 3439.3 | 3876.5 | 7.2221 | .043 58 | 3434.7 | 3870.5 | 7.1687 | .034 60 | 3422.9 | 3855.3 | 7.0536 |
| 800 | .054 09 | 3632.5 | 4119.3 | 7.4596 | .048 59 | 3628.9 | 4114.8 | 7.4077 | .038 69 | 3620.0 | 4103.6 | 7.2965 |
| 900 | .059 50 | 3829.2 | 4364.8 | 7.6783 | .053 49 | 3826.3 | 4361.2 | 7.6272 | .042 67 | 3819.1 | 4352.5 | 7.5182 |
| 1000 | .064 85 | 4030.3 | 4614.0 | 7.8821 | .058 32 | 4027.8 | 4611.0 | 7.8315 | .046 58 | 4021.6 | 4603.8 | 7.7237 |
| 1100 | .070 16 | 4236.3 | 4867.7 | 8.0740 | .063 12 | 4234.0 | 4865.1 | 8.0237 | .050 45 | 4228.2 | 4858.8 | 7.9165 |
| 1200 | .075 44 | 4447.2 | 5126.2 | 8.2556 | .067 89 | 4444.9 | 5123.8 | 8.2055 | .054 30 | 4439.3 | 5118.0 | 8.0987 |
| 1300 | .080 72 | 4662.7 | 5389.2 | 8.4284 | .072 65 | 4660.5 | 5387.0 | 8.3783 | .058 13 | 4654.8 | 5381.4 | 8.2717 |

| | P = 15.0 MPa (342.24) | | | | P = 17.5 MPa (354.75) | | | | P = 20.0 MPa (365.81) | | | |
|---|---|---|---|---|---|---|---|---|---|---|---|---|
| T | v | u | h | s | v | u | h | s | v | u | h | s |
| Sat. | .010 337 | 2455.5 | 2610.5 | 5.3098 | .007 920 | 2390.2 | 2528.8 | 5.1419 | .005 834 | 2293.0 | 2409.7 | 4.9269 |
| 350 | .011 470 | 2520.4 | 2692.4 | 5.4421 | | | | | | | | |
| 400 | .015 649 | 2740.7 | 2975.5 | 5.8811 | .012 447 | 2685.0 | 2902.9 | 5.7213 | .009 942 | 2619.3 | 2818.1 | 5.5540 |
| 450 | .018 445 | 2879.5 | 3156.2 | 6.1404 | .015 174 | 2844.2 | 3109.7 | 6.0184 | .012 695 | 2806.2 | 3060.1 | 5.9017 |
| 500 | .020 80 | 2996.6 | 3308.6 | 6.3443 | .017358 | 2970.3 | 3274.1 | 6.2383 | .014 768 | 2942.9 | 3238.2 | 6.1401 |
| 550 | .022 93 | 3104.7 | 3448.6 | 6.5199 | .019 288 | 3083.9 | 3421.4 | 6.4230 | .016 555 | 3062.4 | 3393.5 | 6.3348 |
| 600 | .024 91 | 3208.6 | 3582.3 | 6.6776 | .021 06 | 3191.5 | 3560.1 | 6.5866 | .018 178 | 3174.0 | 3537.6 | 6.5048 |
| 650 | .026 80 | 3310.3 | 3712.3 | 6.8224 | .022 74 | 3296.0 | 3693.9 | 6.7357 | .019 693 | 3281.4 | 3675.3 | 6.6582 |
| 700 | .028 61 | 3410.9 | 3840.1 | 6.9572 | .024 34 | 3398.7 | 3824.6 | 6.8736 | .02113 | 3386.4 | 3809.0 | 6.7993 |
| 800 | .032 10 | 3610.9 | 4092.4 | 7.2040 | .02738 | 3601.8 | 4081.1 | 7.1244 | .023 85 | 3592.7 | 4069.7 | 7.0544 |
| 900 | .035 46 | 3811.9 | 4343.8 | 7.4279 | .030 31 | 3804.7 | 4335.1 | 7.3507 | .026 45 | 3797.5 | 4326.4 | 7.2830 |
| 1000 | .038 75 | 4015.4 | 4596.6 | 7.6348 | .033 16 | 4009.3 | 4589.5 | 7.5589 | .028 97 | 4003.1 | 4582.5 | 7.4925 |
| 1100 | .042 00 | 4222.6 | 4852.6 | 7.8283 | .035 97 | 4216.9 | 4846.4 | 7.7531 | .031 45 | 4211.3 | 4840.2 | 7.6874 |
| 1200 | .045 23 | 4433.8 | 5112.3 | 8.0108 | .038 76 | 4428.3 | 5106.6 | 7.9360 | .033 91 | 4422.8 | 5101.0 | 7.8707 |
| 1300 | .048 45 | 4649.1 | 5376.0 | 8.1840 | .041 54 | 4643.5 | 5370.5 | 8.1093 | .036 36 | 4638.0 | 5365.1 | 8.0442 |

| | P = 25.0 MPa | | | | P = 30.0 MPa | | | | P = 35.0 MPa | | | |
|---|---|---|---|---|---|---|---|---|---|---|---|---|
| T | v | u | h | s | v | u | h | s | v | u | h | s |
| 375 | .001 973 1 | 1798.7 | 1848.0 | 4.0320 | .001 789 2 | 1737.8 | 1791.5 | 3.9305 | .001 700 3 | 1702.9 | 1762.4 | 3.8722 |
| 400 | .006 004 | 2430.1 | 2580.2 | 5.1418 | .002 790 | 2067.4 | 2151.1 | 4.4728 | .002 100 | 1914.1 | 1987.6 | 4.2126 |

(Continued)

**TABLE A.3 (SI) (CONTINUED)**
Properties of Superheated Steam

| T | v | u | h | s | v | u | h | s | v | u | h | s |
|---|---|---|---|---|---|---|---|---|---|---|---|---|
| 425 | .007 881 | 2609.2 | 2806.3 | 5.4723 | .005 303 | 2455.1 | 2614.2 | 5.1504 | .003 428 | 2253.4 | 2373.4 | 4.7747 |
| 450 | .009 162 | 2720.7 | 2949.7 | 5.6744 | .006 735 | 2619.3 | 2821.4 | 5.4424 | .004 961 | 2498.7 | 2672.4 | 5.1962 |
| 500 | .011 123 | 2884.3 | 3162.4 | 5.9592 | .008 678 | 2820.7 | 3081.1 | 5.7905 | .006 927 | 2751.9 | 2994.4 | 5.6282 |
| 550 | .012 724 | 3017.5 | 3335.6 | 6.1765 | .010 168 | 2970.3 | 3275.4 | 6.0342 | .008 345 | 2921.0 | 3213.0 | 5.9026 |
| 600 | .014 137 | 3137.9 | 3491.4 | 6.3602 | .011 446 | 3100.5 | 3443.9 | 6.2331 | .009 527 | 3062.0 | 3395.5 | 6.1179 |
| 650 | .015 433 | 3251.6 | 3637.4 | 6.5229 | .012 596 | 3221.0 | 3598.9 | 6.4058 | .010 575 | 3189.8 | 3559.9 | 6.3010 |
| 700 | .016 646 | 3361.3 | 3777.5 | 6.6707 | .013 661 | 3335.8 | 3745.6 | 6.5606 | .011 533 | 3309.8 | 3713.5 | 6.4631 |
| 800 | .018912 | 3574.3 | 4047.1 | 6.9345 | .015 623 | 3555.5 | 4024.2 | 6.8332 | .013 278 | 3536.7 | 4001.5 | 6.7450 |
| 900 | .021 045 | 3783.0 | 4309.1 | 7.1680 | .017448 | 3768.5 | 4291.9 | 7.0718 | .014 883 | 3754.0 | 4274.9 | 6.9886 |
| 1000 | .023 10 | 3990.9 | 4568.5 | 7.3802 | .019 196 | 3978.8 | 4554.7 | 7.2867 | .016 410 | 3966.7 | 4541.1 | 7.2064 |
| 1100 | .025 12 | 4200.2 | 4828.2 | 7.5765 | .020 903 | 4189.2 | 4816.3 | 7.4845 | .017 895 | 4178.3 | 4804.6 | 7.4057 |
| 1200 | .02711 | 4412.0 | 5089.9 | 7.7605 | .022 589 | 4401.3 | 5079.0 | 7.6692 | .019 360 | 4390.7 | 5068.3 | 7.5910 |
| 1300 | .029 10 | 4626.9 | 5354.4 | 7.9342 | .024 266 | 4616.0 | 5344.0 | 7.8432 | .020 815 | 4605.1 | 5333.6 | 7.7653 |

| T | P=40.0 MPa | | | | P=50.0 MPa | | | | P=60.0 MPa | | | |
|---|---|---|---|---|---|---|---|---|---|---|---|---|
| | v | u | h | s | v | u | h | s | v | u | h | s |
| 375 | .001 640 7 | 1677.1 | 1742.8 | 3.8290 | .001 559 4 | 1638.6 | 1716.6 | 3.7639 | .001 502 8 | 1609.4 | 1699.5 | 3.7141 |
| 400 | .001 907 7 | 1854.6 | 1930.9 | 4.1135 | .001 730 9 | 1788.1 | 1874.6 | 4.0031 | .001 633 5 | 1745.4 | 1843.4 | 3.9318 |
| 425 | .002 532 | 2096.9 | 2198.1 | 4.5029 | .002 007 | 1959.7 | 2060.0 | 4.2734 | .001 816 5 | 1892.7 | 2001.7 | 4.1626 |
| 450 | .003 693 | 2365.1 | 2512.8 | 4.9459 | .002 486 | 2159.6 | 2284.0 | 4.5884 | .002 085 | 2053.9 | 2179.0 | 4.4121 |
| 500 | .005 622 | 2678.4 | 2903.3 | 5.4700 | .003 892 | 2525.5 | 2720.1 | 5.1726 | .002 956 | 2390.6 | 2567.9 | 4.9321 |
| 550 | .006 984 | 2869.7 | 3149.1 | 5.7785 | .005 118 | 2763.6 | 3019.5 | 5.5485 | .003 956 | 2658.8 | 2896.2 | 5.3441 |
| 600 | .008 094 | 3022.6 | 3346.4 | 6.0114 | .006 112 | 2942.0 | 3247.6 | 5.8178 | .004 834 | 2861.1 | 3151.2 | 5.6452 |
| 650 | .009 063 | 3158.0 | 3520.6 | 6.2054 | .006 966 | 3093.5 | 3441.8 | 6.0342 | .005 595 | 3028.8 | 3364.5 | 5.8829 |
| 700 | .009 941 | 3283.6 | 3681.2 | 6.3750 | .007 727 | 3230.5 | 3616.8 | 6.2189 | .006 272 | 3177.2 | 3553.5 | 6.0824 |
| 800 | .011 523 | 3517.8 | 3978.7 | 6.6662 | .009 076 | 3479.8 | 3933.6 | 6.5290 | .007459 | 3441.5 | 3889.1 | 6.4109 |
| 900 | .012 962 | 3739.4 | 4257.9 | 6.9150 | .010 283 | 3710.3 | 4224.4 | 6.7882 | .008 508 | 3681.0 | 4191.5 | 6.6805 |
| 1000 | .014 324 | 3954.6 | 4527.6 | 7.1356 | .011 411 | 3930.5 | 4501.1 | 7.0146 | .009 480 | 3906.4 | 4475.2 | 6.9127 |
| 1100 | .015 642 | 4167.4 | 4793.1 | 7.3364 | .012 496 | 4145.7 | 4770.5 | 7.2184 | .010 409 | 4124.1 | 4748.6 | 7.1195 |
| 1200 | .016 940 | 4380.1 | 5057.7 | 7.5224 | .013 561 | 4359.1 | 5037.2 | 7.4058 | .011 317 | 4338.2 | 5017.2 | 7.3083 |
| 1300 | .018 229 | 4594.3 | 5323.5 | 7.6969 | .014 616 | 4572.8 | 5303.6 | 7.5808 | .012 215 | 4551.4 | 5284.3 | 7.4837 |

**TABLE A.4 (SI)**

Properties of Compressed Liquid (Steam)

| T | P = 5 MPa (263.99) | | | | P = 10 MPa (311.06) | | | | P = 15 MPa (342.24) | | | |
|---|---|---|---|---|---|---|---|---|---|---|---|---|
| | v | u | h | s | v | u | h | s | v | u | h | s |
| Sat. | .001 285 9 | 1147.8 | 1154.2 | 2.9202 | .001 452 4 | 1393.0 | 1407.6 | 3.3596 | .001 658 1 | 1585.6 | 1610.5 | 3.6848 |
| 0 | .000 9977 | .04 | 5.04 | .0001 | .000 995 2 | .09 | 10.04 | .0002 | .000 992 8 | .15 | 15.05 | .0004 |
| 20 | .000 999 5 | 83.65 | 88.65 | .2956 | .000 9972 | 83.36 | 93.33 | .2945 | .000 995 0 | 83.06 | 97.99 | .2934 |
| 40 | .001 005 6 | 166.95 | 171.97 | .5705 | .001 003 4 | 166.35 | 176.38 | .5686 | .001 001 3 | 165.76 | 180.78 | .5666 |
| 60 | .001 014 9 | 250.23 | 255.30 | .8285 | .001 012 7 | 249.36 | 259.49 | .8258 | .001 010 5 | 248.51 | 263.67 | .8232 |
| 80 | .001 026 8 | 333.72 | 338.85 | 1.0720 | .001 024 5 | 332.59 | 342.83 | 1.0688 | .001 022 2 | 331.48 | 346.81 | 1.0656 |
| 100 | .001 041 0 | 417.52 | 422.72 | 1.3030 | .001 038 5 | 416.12 | 426.50 | 1.2992 | .001 036 1 | 414.74 | 430.28 | 1.2955 |
| 120 | .001 057 6 | 501.80 | 507.09 | 1.5233 | .001054 9 | 500.08 | 510.64 | 1.5189 | .001 052 2 | 498.40 | 514.19 | 1.5145 |
| 140 | .001 076 8 | 586.76 | 592.15 | 1.7343 | .001 073 7 | 584.68 | 595.42 | 1.7292 | .001 070 7 | 582.66 | 598.72 | 1.7242 |
| 160 | .001 098 8 | 672.62 | 678.12 | 1.9375 | .001 095 3 | 670.13 | 681.08 | 1.9317 | .001 091 8 | 667.71 | 684.09 | 1.9260 |
| 180 | .001 124 0 | 759.63 | 765.25 | 2.1341 | .001 119 9 | 756.65 | 767.84 | 2.1275 | .001 115 9 | 753.76 | 770.50 | 2.1210 |
| 200 | .001 153 0 | 848.1 | 853.9 | 2.3255 | .001 148 0 | 844.5 | 856.0 | 2.3178 | .001 143 3 | 841.0 | 858.2 | 2.3104 |
| 220 | .001 186 6 | 938.4 | 944.4 | 2.5128 | .001 180 5 | 934.1 | 945.9 | 2.5039 | .001 1748 | 929.9 | 947.5 | 2.4953 |
| 240 | .001 226 4 | 1031.4 | 1037.5 | 2.6979 | .001 218 7 | 1026.0 | 1038.1 | 2.6872 | .001 211 4 | 1020.8 | 1039.0 | 2.6771 |
| 260 | .001 274 9 | 1127.9 | 1134.3 | 2.8830 | .001 264 5 | 1121.1 | 1133.7 | 2.8699 | .001255 0 | 1114.6 | 1133.4 | 2.8576 |
| 280 | | | | | .001 321 6 | 1220.9 | 1234.1 | 3.0548 | .001 308 4 | 1212.5 | 1232.1 | 3.0393 |
| 300 | | | | | .001 3972 | 1328.4 | 1342.3 | 3.2469 | .001 377 0 | 1316.6 | 1337.3 | 3.2260 |
| 320 | | | | | | | | | .001 472 4 | 1431.1 | 1453.2 | 3.4247 |
| 340 | | | | | | | | | .001 631 1 | 1567.5 | 1591.9 | 3.6546 |

| T | P = 20 MPa (365.81) | | | | P = 30 MPa | | | | P = 50 MPa | | | |
|---|---|---|---|---|---|---|---|---|---|---|---|---|
| | v | u | h | s | v | u | h | s | v | u | h | s |
| Sat. | .002 036 | 1785.6 | 1826.3 | 4.0139 | | | | | | | | |
| 0 | .000 990 4 | .19 | 20.01 | .0004 | .000 985 6 | .25 | 29.82 | .0001 | .000 976 6 | .20 | 49.03 | .0014 |
| 20 | .000 992 8 | 82.77 | 102.62 | .2923 | .000 988 6 | 82.17 | 111.84 | .2899 | .000 980 4 | 81.00 | 130.02 | .2848 |
| 40 | .000 999 2 | 165.17 | 185.16 | .5646 | .000 995 1 | 164.04 | 193.89 | .5607 | .000 9872 | 161.86 | 211.21 | .5527 |
| 60 | .001 008 4 | 247.68 | 267.85 | .8206 | .001 004 2 | 246.06 | 276.19 | .8154 | .000 996 2 | 242.98 | 292.79 | .8052 |
| 80 | .001 019 9 | 330.40 | 350.80 | 1.0624 | .001 015 6 | 328.30 | 358.77 | 1.0561 | .001 0073 | 324.34 | 374.70 | 1.0440 |
| 100 | .001 033 7 | 413.39 | 434.06 | 1.2917 | .001 029 0 | 410.78 | 441.66 | 1.2844 | .001 020 1 | 405.88 | 456.89 | 1.2703 |

*(Continued)*

**TABLE A.4 (SI) (CONTINUED)**

Properties of Compressed Liquid (Steam)

| T | P=5 MPa (263.99) | | | | P=10 MPa (311.06) | | | | P=15 MPa (342.24) | | | |
|---|---|---|---|---|---|---|---|---|---|---|---|---|
| | *v* | *u* | *h* | *s* | *v* | *u* | *h* | *s* | *v* | *u* | *h* | *s* |
| 120 | .001 049 6 | 496.76 | 517.76 | 1.5102 | .001 044 5 | 493.59 | 524.93 | 1.5018 | .001 034 8 | 487.65 | 539.39 | 1.4857 |
| 140 | .001 067 8 | 580.69 | 602.04 | 1.7193 | .001 062 1 | 576.88 | 608.75 | 1.7098 | .001 051 5 | 569.77 | 622.35 | 1.6915 |
| 160 | .001 088 5 | 665.35 | 687.12 | 1.9204 | .001 082 1 | 660.82 | 693.28 | 1.9096 | .001 070 3 | 652.41 | 705.92 | 1.8891 |
| 180 | .001 112 0 | 750.95 | 773.20 | 2.1147 | .001 104 7 | 745.59 | 778.73 | 2.1024 | .001 091 2 | 735.69 | 790.25 | 2.0794 |
| 200 | .001 138 8 | 837.7 | 860.5 | 2.3031 | .001 130 2 | 831.4 | 865.3 | 2.2893 | .001 114 6 | 819.7 | 875.5 | 2.2634 |
| 220 | .001 169 3 | 925.9 | 949.3 | 2.4870 | .001 159 0 | 918.3 | 953.1 | 2.4711 | .001 140 8 | 904.7 | 961.7 | 2.4419 |
| 240 | .001 204 6 | 1016.0 | 1040.0 | 2.6674 | .001 192 0 | 1006.9 | 1042.6 | 2.6490 | .001 170 2 | 990.7 | 1049.2 | 2.6158 |
| 260 | .001 246 2 | 1108.6 | 1133.5 | 2.8459 | .001 230 3 | 1097.4 | 1134.3 | 2.8243 | .001 203 4 | 1078.1 | 1138.2 | 2.7860 |
| 280 | .001296 5 | 1204.7 | 1230.6 | 3.0248 | .001 275 5 | 1190.7 | 1229.0 | 2.9986 | .001 241 5 | 1167.2 | 1229.3 | 2.9537 |
| 300 | .001359 6 | 1306.1 | 1333.3 | 3.2071 | .001 330 4 | 1287.9 | 1327.8 | 3.1741 | .001 286 0 | 1258.7 | 1323.0 | 3.1200 |
| 320 | .001 443 7 | 1415.7 | 1444.6 | 3.3979 | .001 399 7 | 1390.7 | 1432.7 | 3.3539 | .001 338 8 | 1353.3 | 1420.2 | 3.2868 |
| 340 | .001 568 4 | 1539.7 | 1571.0 | 3.6075 | .001 492 0 | 1501.7 | 1546.5 | 3.5426 | .001 403 2 | 1452.0 | 1522.1 | 3.4557 |
| 360 | .001 822 6 | 1702.8 | 1739.3 | 3.8772 | .001 626 5 | 1626.6 | 1675.4 | 3.7494 | .001 483 8 | 1556.0 | 1630.2 | 3.6291 |
| 380 | | | | | .001 869 1 | 1781.4 | 1837.5 | 4.0012 | .001 588 4 | 1667.2 | 1746.6 | 3.8101 |

**TABLE A.5 (SI)**

Enthalpy–Entropy Diagram for Steam, SI Units

*Source:* American Society of Mechanical Engineers.

**TABLE A.6**

Thermal Conductivities of Some Building and Insulating Materials ($k$ = Btu/hft. °F)

| Material | Apparent Density $\rho$ lb$_m$/ ft.$^3$ at Room Temperature | °F | $k$ |
|---|---|---|---|
| Aerogel, silica, opacified | 8.5 | 248 | 0.013 |
| | | 554 | 0.026 |
| Asbestos-cement boards | 120 | 68 | 0.43 |
| Asbestos sheets | 55.5 | 124 | 0.096 |
| Asbestos slate | 112 | 32 | 0.087 |
| | 112 | 140 | 0.114 |
| Aluminum foil, 7 air spaces per 2.5 in. | 0.2 | 100 | 0.025 |
| | | 351 | 0.038 |
| Asphalt | 132 | 68 | 0.43 |
| Bricks: | | | |
|   Alumina (92–99% $Al_2O_3$ by weight) fused | | 801 | 1.8 |
|   Alumina (64–65% $Al_2O_3$ by weight) | | 2399 | 2.7 |
| (See also Bricks, fire clay) | 115 | 1472 | 0.62 |
| Building brickwork | | 68 | 0.4 |
| Carbon | 96.7 | | 3.0 |
|   Chrome brick (32% $Cr_2O_3$ by weight) | 200 | 392 | 0.67 |
| | 200 | 1202 | 0.85 |
| | 200 | 2399 | 1.0 |
| Diatomaceous earth, molded and fired | 38 | 399 | 0.14 |
| | 38 | 1600 | 0.18 |
| Diatomaceous earth, high burn, large pores | 37 | 392 | 0.13 |
| | 37 | 1832 | 0.34 |
| Fire clay, Missouri | | 392 | 0.58 |
| | | 1112 | 0.85 |
| | | 1832 | 0.95 |
| | | 2552 | 1.02 |
| Kaolin insulating brick | 27 | 932 | 0.15 |
| | 27 | 2102 | 0.26 |
| Kaolin insulating firebrick | 19 | 392 | 0.050 |
| | 19 | 1400 | 0.113 |
| Magnesite (86.8% MgO, 6.3% $Fe_2O_3$, 3% CaO, 2.6% $SiO_2$ by weight) | 158 | 399 | 2.2 |
| | 158 | 1202 | 1.6 |
| | 158 | 2192 | 1.1 |
| Calcium carbonate, natural | 162 | 86 | 1.3 |
|   White marble | | | 1.7 |
|   Chalk | 96 | | 0.4 |
| Calcium sulfate (4$H_2O$), artificial | 84.6 | 104 | 0.22 |
|   Plaster, artificial | 132 | 167 | 0.43 |
|   Building | 77.9 | 77 | 0.25 |
| Cardboard, corrugated | | | 0.037 |

*(Continued)*

**TABLE A.6 (CONTINUED)**

Thermal Conductivities of Some Building and Insulating Materials ($k$ = Btu/hft. °F)

| Material | Apparent Density $\rho$ lb$_m$/ft.$^3$ at Room Temperature | °F | $k$ |
|---|---|---|---|
| Celluloid | 87.3 | 86 | 0.12 |
| Charcoal flakes | 11.9 | 176 | 0.043 |
|  | 15 | 176 | 0.051 |
| Coke, petroleum |  | 212 | 3.4 |
|  |  | 932 | 2.9 |
| Coke, powdered |  | 32–212 | 0.11 |
| Concrete, cinder |  |  | 0.20 |
| 1:4 dry |  |  | 0.44 |
| Stone |  |  | 0.54 |
| Cork board | 10 | 86 | 0.025 |
| Cork, ground | 9.4 | 86 | 0.025 |
| Cotton wool | 5 | 86 | 0.024 |
| Diatomaceous earth power, coarse | 20.0 | 100 | 0.036 |
|  | 20.0 | 1600 | 0.082 |
| Fine | 17.2 | 399 | 0.040 |
|  | 17.2 | 1600 | 0.074 |
| Molded pipe covering | 26.0 | 399 | 0.051 |
|  | 26.0 | 1600 | 0.088 |
| Dolomite | 167 | 122 | 1.0 |
| Ebonite |  |  | 0.10 |
| Enamel, silicate | 38 |  | 0.5–0.75 |
| Felt, wool | 20.6 | 86 | 0.03 |
| Fiber insulating board | 14.8 | 70 | 0.028 |
| Glass |  |  | 0.2–0.73 |
| Borosilicate type | 139 | 86–167 | 0.63 |
| Soda glass |  |  | 0.3–0.44 |
| Window glass |  |  | 0.3–0.61 |
| Granite |  |  | 1.0–2.3 |
| Graphite, longitudinal |  | 68 | 95 |
| Gypsum, molded and dry | 78 | 68 | 0.25 |
| Hair felt, perpendicular to fibers | 17 | 86 | 0.021 |
| Ice | 57.5 | 32 | 1.3 |
| Kapok | 0.88 | 68 | 0.020 |
| Lampblack | 10 | 104 | 0.038 |
| Leather sole | 62.4 |  | 0.092 |
| Limestone (15.3 vol.% $H_2O$) | 103 | 75 | 0.54 |
| Magnesia, powdered | 49.7 | 117 | 0.35 |
| Magnesia, light carbonate | 19 | 70 | 0.034 |
| Magnesium oxide, compressed | 49.9 | 68 | 0.32 |

*(Continued)*

**TABLE A.6 (CONTINUED)**

Thermal Conductivities of Some Building and Insulating Materials ($k$ = Btu/hft. °F)

| Material | Apparent Density $\rho$ lb$_m$/ ft.$^3$ at Room Temperature | °F | $k$ |
|---|---|---|---|
| Marble | | | 1.2–1.7 |
| Mica, perpendicular to planes | | 122 | 0.25 |
| Mineral wool | 9.4 | 86 | 0.0225 |
| | 19.7 | 86 | 0.024 |
| Paper | | | 0.075 |
| Paraffin wax | | 32 | 0.14 |
| Porcelain | | 392 | 0.88 |
| Portland cement (*see* Concrete) | | 194 | 0.17 |
| Pumice stone | | 70–151 | 0.14 |
| Pyroxylin plastics | | | 0.075 |
| Rubber, hard | 74.8 | 32 | 0.087 |
| Para | | 70 | 0.109 |
| Soft | | 70 | 0.075–0.092 |
| Sand, dry | 94.6 | 68 | 0.19 |
| Sandstone | 140 | 104 | 1.06 |
| Sawdust | 12 | 70 | 0.03 |
| Slag, blast furnace | | 75–261 | 0.064 |
| Slag wool | 12 | 86 | 0.022 |
| Slate | | 201 | 0.86 |
| Snow | 34.7 | 32 | 0.27 |
| Sulfur, monoclinic | | 212 | 0.09–0.097 |
| Rhombic | | 70 | 0.16 |
| Wallboard, insulating type | 14.8 | 70 | 0.028 |
| Wallboard, stiff pasteboard | 43 | 86 | 0.04 |
| Wood shavings | 8.8 | 86 | 0.034 |
| Wood, across grain | | | |
| Balsa | 7–8 | 86 | 0.025–0.03 |
| Oak | 51.5 | 59 | 0.12 |
| Maple | 44.7 | 122 | 0.11 |
| Pine, white | 34.0 | 59 | 0.087 |
| Teak | 40.0 | 59 | 0.10 |
| White fir | 28.1 | 140 | 0.062 |
| Wood, parallel to grain | | | |
| Pine | 34.4 | 70 | 0.20 |
| Wool, animal | 6.9 | 86 | 0.021 |

*Source:* G. F. Babits, *Thermodynamics*, Allyn & Bacon, Inc., Boston, 1963, pp. 231–34. With permission.

**TABLE A.7**

Ammonia: Properties of Liquid and Saturated Vapor

| | Pressure | | | | Enthalpy from −40° F | | | | Entropy from −40°F | |
|---|---|---|---|---|---|---|---|---|---|---|
| Temp. °F $t$ | Abs. lb./in.$^2$ $p$ | Gauge lb./in.$^2$ $p_d$ | Volume Vapor ft.$^3$/lb. $v_g$ | Density Liquid lb./ft.$^3$ $1/v_f$ | Liquid Btu/lb. $h_f$ | Vapor Btu/ lb. $h_g$ | Latent Btu/ lb. $h_{fg}$ | Liquid Btu/lb. °F $s_f$ | Vapor Btu/lb. °F $s_g$ |
|---|---|---|---|---|---|---|---|---|---|
| −100 | 1.24 | *27.4 | 182.90 | 45.51 | −61.5 | 571.4 | 632.9 | −0.1579 | 1.6025 |
| −90 | 1.86 | *26.1 | 124.28 | 45.12 | −51.4 | 575.9 | 627.3 | −0.1309 | 1.5667 |
| −80 | 2.74 | *24.3 | 86.54 | 44.73 | −41.3 | 580.1 | 621.4 | −0.1036 | 1.5336 |
| −70 | 3.94 | *21.9 | 61.65 | 44.32 | −31.1 | 584.4 | 615.5 | −0.0771 | 1.5026 |
| −65 | 4.69 | *20.4 | 52.34 | 44.11 | −26.0 | 586.6 | 612.6 | −0.0642 | 1.4833 |
| −60 | 5.55 | *18.6 | 44.73 | 43.91 | −20.9 | 588.8 | 609.7 | −0.0514 | 1.4747 |
| −55 | 6.54 | *16.6 | 38.38 | 43.70 | −15.7 | 591.0 | 606.7 | −0.0382 | 1.4614 |
| −50 | 7.67 | *14.3 | 33.08 | 43.49 | −10.5 | 593.2 | 603.7 | −0.0254 | 1.4487 |
| −45 | 8.95 | *11.7 | 28.62 | 43.28 | −5.3 | 595.7 | 600.7 | −0.0128 | 1.4363 |
| −40 | 10.41 | *8.7 | 24.86 | 43.08 | 0.0 | 597.6 | 597.6 | 0.0000 | 1.4242 |
| −38 | 11.04 | *7.4 | 23.53 | 42.99 | 2.1 | 598.3 | 596.2 | .0051 | 1.4193 |
| −36 | 11.71 | *6.1 | 22.27 | 42.90 | 4.3 | 599.1 | 594.8 | .0101 | 1.4144 |
| −34 | 12.41 | *4.7 | 21.10 | 42.82 | 6.4 | 599.9 | 593.5 | .0151 | 1.4096 |
| −32 | 13.14 | *3.2 | 20.00 | 42.73 | 8.5 | 600.6 | 592.1 | .0201 | 1.4048 |
| −30 | 13.90 | *1.6 | 18.97 | 42.65 | 10.7 | 601.4 | 590.7 | 0.0250 | 1.4001 |
| −28 | 14.71 | 0.0 | 18.00 | 42.57 | 12.8 | 602.1 | 589.3 | .0300 | 1.3955 |
| −26 | 15.55 | 0.8 | 17.09 | 42.48 | 14.9 | 602.8 | 587.9 | .0350 | 1.3909 |
| −24 | 16.42 | 1.7 | 16.24 | 42.40 | 17.1 | 603.6 | 586.5 | .0399 | 1.3863 |
| −22 | 17.34 | 2.6 | 15.43 | 42.31 | 19.2 | 604.3 | 585.1 | .0448 | 1.3818 |
| −20 | 18.30 | 3.6 | 14.68 | 42.22 | 21.4 | 605.0 | 583.6 | 0.0497 | 1.3774 |
| −18 | 19.30 | 4.6 | 13.97 | 43.13 | 23.5 | 605.7 | 582.2 | .0545 | 1.3729 |
| −16 | 20.34 | 5.6 | 13.29 | 42.04 | 25.6 | 606.4 | 580.8 | .0594 | 1.3686 |
| −14 | 21.43 | 6.7 | 12.66 | 41.96 | 27.8 | 607.1 | 579.3 | .0642 | 1.3643 |
| −12 | 22.56 | 7.9 | 12.06 | 41.87 | 30.0 | 607.8 | 577.8 | .0690 | 1.3600 |
| −10 | 23.74 | 9.0 | 11.50 | 41.78 | 32.1 | 608.5 | 576.4 | 0.0738 | 1.3558 |
| −8 | 24.97 | 10.3 | 10.97 | 41.69 | 34.3 | 609.2 | 574.9 | .0786 | 1.3516 |

(*Continued*)

**TABLE A.7 (CONTINUED)**

Ammonia: Properties of Liquid and Saturated Vapor

| Temp. °F $t$ | Pressure Abs. lb./in.² $p$ | Pressure Gauge lb./in.² $p_d$ | Volume Vapor ft.³/lb. $v_g$ | Density Liquid lb./ft.³ $1/v_f$ | Enthalpy from −40° F Liquid Btu/lb. $h_f$ | Enthalpy from −40° F Vapor Btu/lb. $h_g$ | Enthalpy from −40° F Latent Btu/lb. $h_{fg}$ | Entropy from −40°F Liquid Btu/lb. °F $s_f$ | Entropy from −40°F Vapor Btu/lb. °F $s_g$ |
|---|---|---|---|---|---|---|---|---|---|
| −6 | 26.26 | 11.6 | 10.47 | 41.60 | 36.4 | 609.8 | 573.4 | .0833 | 1.3474 |
| −4 | 27.59 | 12.9 | 9.991 | 41.52 | 38.6 | 610.5 | 571.9 | .0880 | 1.3433 |
| −2 | 28.98 | 14.3 | 9.541 | 41.43 | 40.7 | 611.1 | 570.4 | .0928 | 1.3393 |
| 0 | 30.42 | 15.7 | 9.116 | 41.34 | 42.9 | 611.8 | 568.9 | 0.0975 | 1.3352 |
| 2 | 31.92 | 17.2 | 8.714 | 41.25 | 45.1 | 612.4 | 567.3 | .1022 | 1.3312 |
| 4 | 33.47 | 18.8 | 8.333 | 41.16 | 47.2 | 613.0 | 565.8 | .1069 | 1.3273 |
| 6 | 35.09 | 20.4 | 7.971 | 41.07 | 49.4 | 613.6 | 564.2 | .1115 | 1.3234 |
| 8 | 36.77 | 22.1 | 7.629 | 40.98 | 51.6 | 614.3 | 562.7 | .1162 | 1.3195 |
| 10 | 38.51 | 23.8 | 7.304 | 40.89 | 53.8 | 614.9 | 561.1 | 0.1208 | 1.3157 |
| 12 | 40.31 | 25.6 | 6.996 | 40.80 | 56.0 | 615.5 | 559.5 | .1254 | 1.3118 |
| 14 | 42.18 | 27.5 | 6.703 | 40.71 | 58.2 | 616.1 | 557.9 | .1300 | 1.3081 |
| 16 | 44.12 | 29.4 | 6.425 | 40.61 | 60.3 | 616.6 | 556.3 | .1346 | 1.3043 |
| 18 | 46.13 | 31.4 | 6.161 | 40.52 | 62.5 | 617.2 | 554.7 | .1392 | 1.3006 |
| 20 | 48.21 | 33.5 | 5.910 | 40.43 | 64.7 | 617.8 | 553.1 | 0.1437 | 1.2969 |
| 22 | 50.36 | 35.7 | 5.671 | 40.34 | 66.9 | 618.3 | 551.4 | .1483 | 1.2933 |
| 24 | 52.59 | 37.9 | 5.443 | 40.25 | 69.1 | 618.9 | 549.8 | .1528 | 1.2897 |
| 26 | 54.90 | 40.2 | 5.227 | 40.15 | 71.3 | 619.4 | 548.1 | .1573 | 1.2861 |
| 28 | 57.28 | 42.6 | 5.021 | 40.06 | 73.5 | 619.9 | 546.4 | .1618 | 1.2825 |
| 30 | 59.74 | 45.0 | 4.825 | 39.96 | 75.7 | 620.5 | 544.8 | 0.1663 | 1.2790 |
| 32 | 62.29 | 47.6 | 4.637 | 39.86 | 77.9 | 621.0 | 543.1 | .1708 | 1.2755 |
| 34 | 64.91 | 50.2 | 4.459 | 39.77 | 80.1 | 621.5 | 541.4 | .1753 | 1.2721 |
| 36 | 67.63 | 52.9 | 4.289 | 39.67 | 82.3 | 622.0 | 539.7 | .1797 | 1.2686 |
| 38 | 70.43 | 55.7 | 4.126 | 39.50 | 84.6 | 622.5 | 537.9 | .1841 | 1.2652 |
| 40 | 73.32 | 58.6 | 3.971 | 39.49 | 86.8 | 623.0 | 536.2 | 0.1885 | 1.2618 |
| 42 | 76.31 | 61.6 | 3.823 | 39.39 | 89.0 | 623.4 | 534.4 | .1930 | 1.2585 |
| 44 | 79.38 | 64.7 | 3.682 | 39.29 | 91.2 | 623.9 | 532.7 | .1974 | 1.2552 |

*(Continued)*

**TABLE A.7 (CONTINUED)**

Ammonia: Properties of Liquid and Saturated Vapor

| Temp. °F t | Pressure | | | Volume Vapor ft.³/lb. $v_g$ | Density Liquid lb./ft.³ $1/v_f$ | Enthalpy from –40° F | | | | Entropy from –40°F | |
|---|---|---|---|---|---|---|---|---|---|---|---|
| | Abs. lb./in.²p | Gauge lb./in.²$p_d$ | | | | Liquid Btu/lb. $h_f$ | Vapor Btu/ lb. $h_g$ | Latent Btu/ lb. $h_{fg}$ | | Liquid Btu/lb. °F $s_f$ | Vapor Btu/lb. °F $s_g$ |
| 46 | 82.55 | 67.9 | | 3.547 | 39.19 | 93.5 | 624.4 | 530.9 | | .2018 | 1.2519 |
| 48 | 85.82 | 71.1 | | 3.418 | 39.10 | 95.7 | 624.8 | 529.1 | | .2062 | 1.2486 |
| 50 | 89.19 | 74.5 | | 3.294 | 39.00 | 97.9 | 625.2 | 527.3 | | 0.2105 | 1.2453 |
| 52 | 92.66 | 78.0 | | 3.176 | 38.90 | 100.2 | 625.7 | 525.5 | | .2149 | 1.2421 |
| 54 | 96.23 | 81.5 | | 3.063 | 38.80 | 102.4 | 626.1 | 523.7 | | .2192 | 1.2389 |
| 56 | 99.91 | 85.2 | | 2.954 | 38.70 | 104.7 | 626.5 | 521.8 | | .2236 | 1.2357 |
| 58 | 103.7 | 89.0 | | 2.851 | 38.60 | 106.9 | 626.9 | 520.0 | | .2279 | 1.2325 |
| 60 | 107.6 | 92.9 | | 2.751 | 38.50 | 109.2 | 627.3 | 518.1 | | 0.2322 | 1.2294 |
| 62 | 111.6 | 96.9 | | 2.656 | 38.40 | 111.5 | 627.7 | 516.2 | | .2365 | 1.2262 |
| 64 | 115.7 | 101.0 | | 2.565 | 38.30 | 113.7 | 628.0 | 514.3 | | .2408 | 1.2231 |
| 66 | 120.0 | 105.3 | | 2.477 | 38.20 | 116.0 | 628.4 | 512.4 | | .2451 | 1.2201 |
| 68 | 124.3 | 109.6 | | 2.393 | 38.10 | 118.3 | 628.8 | 510.5 | | .2494 | 1.2170 |
| 70 | 128.8 | 114.1 | | 2.312 | 38.00 | 120.5 | 629.1 | 508.6 | | 0.2537 | 1.2140 |
| 72 | 133.4 | 118.7 | | 2.235 | 37.90 | 122.8 | 629.4 | 506.6 | | .2579 | 1.2110 |
| 74 | 138.1 | 123.4 | | 2.161 | 37.79 | 125.1 | 629.8 | 504.7 | | .2622 | 1.2080 |
| 76 | 143.0 | 128.3 | | 2.089 | 37.69 | 127.4 | 630.1 | 502.7 | | .2664 | 1.2050 |
| 78 | 147.9 | 133.2 | | 2.021 | 37.58 | 129.7 | 630.4 | 500.7 | | .2706 | 1.2020 |
| 80 | 153.0 | 138.3 | | 1.955 | 37.48 | 132.0 | 630.7 | 498.7 | | 0.2749 | 1.1991 |
| 82 | 158.3 | 143.6 | | 1.892 | 37.37 | 134.3 | 631.0 | 496.7 | | .2791 | 1.1962 |
| 84 | 163.7 | 149.0 | | 1.831 | 37.26 | 136.6 | 631.3 | 494.7 | | .2833 | 1.1933 |
| 86 | 169.2 | 154.5 | | 1.772 | 37.16 | 138.9 | 631.5 | 492.6 | | .2875 | 1.1904 |
| 88 | 174.8 | 160.1 | | 1.716 | 37.05 | 141.2 | 631.8 | 490.6 | | .2917 | 1.1875 |
| 90 | 180.6 | 165.9 | | 1.661 | 36.95 | 143.5 | 632.0 | 488.5 | | 0.2958 | 1.1846 |
| 92 | 186.6 | 171.9 | | 1.609 | 36.84 | 145.8 | 632.2 | 486.4 | | .3000 | 1.1818 |
| 94 | 192.7 | 178.0 | | 1.559 | 36.73 | 148.2 | 632.5 | 484.3 | | .3041 | 1.1789 |
| 96 | 198.9 | 184.2 | | 1.510 | 36.62 | 150.5 | 632.6 | 482.1 | | .3083 | 1.1761 |

*(Continued)*

**TABLE A.7 (CONTINUED)**

Ammonia: Properties of Liquid and Saturated Vapor

| Temp. °F t | Pressure | | Volume Vapor ft.³/lb. $v_g$ | Density Liquid lb./ft.³ 1/$v_f$ | Enthalpy from −40° F | | | | Entropy from −40°F | |
|---|---|---|---|---|---|---|---|---|---|---|
| | Abs. lb./in.² p | Gauge lb./in.² $p_d$ | | | Liquid Btu/lb. $h_f$ | Vapor Btu/ lb. $h_g$ | Latent Btu/ lb. $h_{fg}$ | Liquid Btu/lb. °F $s_f$ | Vapor Btu/lb. °F $s_g$ |
| 98 | 205.3 | 190.6 | 1.464 | 36.51 | 152.9 | 632.9 | 480.0 | .3125 | 1.1733 |
| 100 | 211.9 | 197.2 | 1.419 | 36.40 | 155.2 | 633.0 | 477.8 | 0.3166 | 1.1705 |
| 102 | 218.6 | 203.9 | 1.375 | 36.29 | 157.6 | 633.2 | 475.6 | .3207 | 1.1677 |
| 104 | 225.4 | 210.7 | 1.334 | 36.18 | 159.9 | 633.4 | 473.5 | .3248 | 1.1649 |
| 106 | 232.5 | 217.8 | 1.293 | 36.06 | 162.3 | 633.5 | 471.2 | .3289 | 1.1621 |
| 108 | 239.7 | 225.0 | 1.254 | 35.95 | 164.6 | 633.6 | 469.0 | .3330 | 1.1593 |
| 110 | 247.0 | 232.3 | 1.217 | 35.84 | 167.0 | 633.7 | 466.7 | 0.3372 | 1.1566 |
| 112 | 254.5 | 239.8 | 1.180 | 35.72 | 169.4 | 633.8 | 464.4 | .3413 | 1.1538 |
| 114 | 262.2 | 247.5 | 1.145 | 35.61 | 171.8 | 633.9 | 462.1 | .3453 | 1.1510 |
| 116 | 270.1 | 255.4 | 1.112 | 35.49 | 174.2 | 634.0 | 459.8 | .3495 | 1.1483 |
| 118 | 278.2 | 263.5 | 1.079 | 35.38 | 176.6 | 634.0 | 457.4 | .3535 | 1.1455 |
| 120 | 286.4 | 271.7 | 1.047 | 35.26 | 179.0 | 634.0 | 455.0 | 0.3576 | 1.1427 |
| 122 | 294.8 | 280.1 | 1.017 | 35.14 | 181.4 | 634.0 | 452.6 | .3618 | 1.1400 |
| 124 | 303.4 | 288.7 | 0.987 | 35.02 | 183.9 | 634.0 | 450.1 | .3659 | 1.1372 |

*Source:* Abstracted from *Tables of Thermodynamic Properties of Ammonia*, U.S. Department of Commerce, Bureau of Standards Circular 142,1945. With permission.
*Inches of mercury below one atmosphere.

**TABLE A.8**

Ammonia: Properties of Superheated Vapor

Absolute Pressure in lb./in.² (Saturation Temperature in Italics)

| Temp. °F | 5 *-63.11* | | | 10 *-41.34* | | | 15 *-27.20* | | | 20 *-16.64* | | |
|---|---|---|---|---|---|---|---|---|---|---|---|---|
| *t* | *v* | *h* | *s* | *v* | *h* | *s* | *v* | *h* | *s* | *v* | *h* | *s* |
| Sat. | 49.31 | 588.3 | 1.4857 | 25.81 | 597.1 | 1.4276 | 17.67 | 602.4 | 1.3988 | 13.50 | 606.2 | 1.3700 |
| -50 | 51.05 | 595.2 | 1.5025 | | | | | | | | | |
| -40 | 52.36 | 600.3 | .5149 | 26.58 | 603.2 | 1.4420 | | | | | | |
| -30 | 53.67 | 605.4 | .5269 | 27.26 | 608.5 | .4542 | | | | | | |
| -20 | 54.97 | 610.4 | .5385 | 27.92 | 613.7 | .4659 | 18.01 | 606.4 | 1.4031 | | | |
| -10 | 56.26 | 615.4 | .5498 | 28.58 | 618.9 | 1.4773 | 18.47 | 611.9 | .4154 | 13.74 | 610.0 | 1.3784 |
| 0 | 57.55 | 620.4 | 1.5608 | 29.24 | 624.0 | .4884 | 18.92 | 617.2 | 1.4272 | 14.09 | 615.5 | 1.3907 |
| 10 | 58.84 | 625.4 | .5716 | 29.90 | 629.1 | .4992 | 19.37 | 622.5 | .4386 | 14.44 | 621.0 | .4025 |
| 20 | 60.12 | 630.4 | .5821 | 30.55 | 634.2 | .5097 | 19.82 | 627.8 | .4497 | 14.78 | 626.4 | .4138 |
| 30 | 61.41 | 635.4 | .5925 | 31.20 | 639.3 | .5200 | 20.26 | 633.0 | .4604 | 15.11 | 631.7 | .4248 |
| 40 | 62.69 | 640.4 | .6026 | 31.85 | 644.4 | 1.5301 | 20.70 | 638.2 | .4709 | 15.45 | 637.0 | .4356 |
| 50 | 63.96 | 645.5 | 1.6125 | 32.49 | 649.5 | .5400 | 21.14 | 643.4 | 1.4812 | 15.78 | 642.3 | 1.4460 |
| 60 | 65.24 | 650.5 | .6223 | 33.14 | 654.6 | .5497 | 21.58 | 648.5 | .4912 | 16.12 | 647.5 | .4562 |
| 70 | 66.51 | 655.5 | .6319 | 33.78 | 659.7 | .5593 | 22.01 | 653.7 | .5011 | 16.45 | 652.8 | .4662 |
| 80 | 67.79 | 660.6 | .6413 | 34.42 | 664.8 | .5687 | 22.44 | 658.9 | .5108 | 16.78 | 658.0 | .4760 |
| 90 | 69.06 | 665.6 | .6506 | 35.07 | 670.0 | 1.5779 | 22.88 | 664.0 | .5203 | 17.10 | 663.2 | .4856 |
| 100 | 70.33 | 670.7 | 1.6598 | 35.71 | 675.1 | .5870 | 23.31 | 669.2 | 1.5296 | 17.43 | 668.5 | 1.4950 |
| 110 | 71.60 | 675.8 | .6689 | 36.35 | 680.3 | .5960 | 23.74 | 674.4 | .5388 | 17.76 | 673.7 | .5042 |
| 120 | 72.87 | 680.9 | .6778 | 36.99 | 685.4 | .6049 | 24.17 | 679.6 | .5478 | 18.08 | 678.9 | .5133 |
| 130 | 74.14 | 686.1 | .6865 | 37.62 | 690.6 | .6136 | 24.60 | 684.8 | .5567 | 18.41 | 684.2 | .5223 |
| 140 | 75.41 | 691.2 | .6952 | 38.26 | 695.8 | 1.6222 | 25.03 | 690.0 | .5655 | 18.73 | 689.4 | .5312 |
| 150 | 76.68 | 696.4 | 1.7038 | 38.90 | 701.1 | .6307 | 25.46 | 695.3 | 1.5742 | 19.05 | 694.7 | 1.5399 |
| 160 | 77.95 | 701.6 | .7122 | 39.54 | 706.3 | .6391 | 25.88 | 700.5 | .5827 | 19.37 | 700.0 | .5485 |
| 170 | 79.21 | 706.8 | .7206 | 40.17 | 711.6 | .6474 | 26.31 | 705.8 | .5911 | 19.70 | 705.3 | .5569 |
| 180 | 80.48 | 712.1 | .7289 | | | | 26.74 | 711.1 | .5995 | 20.02 | 710.6 | .5653 |

(*Continued*)

# TABLE A.8 (CONTINUED)

Ammonia: Properties of Superheated Vapor

Continuation rows (190–240 °F) for the lower-pressure columns carried over from the preceding page:

| Temp. °F | v | h | s | v | h | s | v | h | s |
|---|---|---|---|---|---|---|---|---|---|
| 190 | 40.81 | 716.9 | .6556 | 27.16 | 716.4 | .6077 | 20.34 | 715.9 | 1.5736 |
| 200 | 41.45 | 722.2 | 1.6637 | 27.59 | 721.7 | 1.6158 | 20.66 | 721.2 | 1.5817 |
| 220 |  |  |  | 28.44 | 732.4 | .6318 | 21.30 | 732.0 | .5978 |
| 240 |  |  |  |  |  |  | 21.94 | 742.8 | .6135 |

| Temp. °F (t) | 25 / −7.96 v | h | s | 30 / −0.57 v | h | s | 35 / 5.89 v | h | s | 40 / 11.66 v | h | s |
|---|---|---|---|---|---|---|---|---|---|---|---|---|
| Sat. | 10.96 | 609.1 | 1.3515 | 9.236 | 611.6 | 1.3364 | 7.991 | 613.6 | 1.3236 | 7.047 | 615.4 | 1.3125 |
| 0 | 11.19 | 613.8 | 1.3616 | 9.492 | 617.8 | 1.3497 | 8.078 | 616.1 | 1.3289 |  |  |  |
| 10 | 11.47 | 619.4 | .3738 | 9.731 | 623.5 | .3618 | 8.287 | 622.0 | .3413 | 7.203 | 620.4 | 1.3231 |
| 20 | 11.75 | 625.0 | .3855 | 9.966 | 629.1 | .3733 | 8.493 | 627.7 | .3532 | 7.387 | 626.3 | .3353 |
| 30 | 12.03 | 630.4 | .3967 | 10.20 | 634.6 | .3845 | 8.695 | 633.4 | .3646 | 7.568 | 632.1 | .3470 |
| 40 | 12.30 | 635.8 | .4077 | 10.43 | 640.1 | 1.3953 | 8.895 | 638.9 | 1.3756 | 7.746 | 637.8 | 1.3583 |
| 50 | 12.57 | 641.2 | 1.4183 | 10.65 | 645.5 | .4059 | 9.093 | 644.4 | .3863 | 7.922 | 643.4 | .3692 |
| 60 | 12.84 | 646.5 | .4287 | 10.88 | 650.9 | .4161 | 9.289 | 649.9 | .3967 | 8.096 | 648.9 | .3797 |
| 70 | 13.11 | 651.8 | .4388 | 11.10 | 656.2 | .4261 | 9.484 | 655.3 | .4069 | 8.268 | 654.4 | .3900 |
| 80 | 13.37 | 657.1 | .4487 | 11.33 | 661.6 | .4359 | 9.677 | 660.7 | .4168 | 8.439 | 659.9 | .4000 |
| 90 | 13.64 | 662.4 | .4584 | 11.55 | 666.9 | 1.4456 | 9.869 | 666.1 | 1.4265 | 8.609 | 665.3 | 1.4098 |
| 100 | 13.90 | 667.7 | 1.4679 | 11.77 | 672.2 | .4550 | 10.06 | 671.5 | .4360 | 8.777 | 670.7 | .4194 |
| 110 | 14.17 | 673.0 | .4772 | 11.99 | 677.5 | .4642 | 10.25 | 676.8 | .4453 | 8.945 | 676.1 | .4288 |
| 120 | 14.43 | 678.2 | .4864 | 12.21 | 682.9 | .4733 | 10.44 | 682.2 | .4545 | 9.112 | 681.5 | .4381 |
| 130 | 14.69 | 683.5 | .4954 | 12.43 | 688.2 | .4823 | 10.63 | 687.6 | .4635 | 9.278 | 686.9 | .4471 |
| 140 | 14.95 | 688.8 | .5043 | 12.65 | 693.5 | 1.4911 | 10.82 | 692.9 | 1.4724 | 9.444 | 692.3 | 1.4561 |
| 150 | 15.21 | 694.1 | 1.5131 | 12.87 | 698.8 | .4998 | 11.00 | 698.3 | .4811 | 9.609 | 697.7 | .4648 |
| 160 | 15.47 | 699.4 | .5217 | 13.08 | 704.2 | .5083 | 11.19 | 703.7 | .4897 | 9.774 | 703.1 | .4735 |
| 170 | 15.73 | 704.7 | .5303 | 13.30 | 709.6 | .5168 | 11.38 | 709.1 | .4982 | 9.938 | 708.5 | .4820 |
| 180 | 15.99 | 710.1 | .5387 | 13.52 | 714.9 | .5251 | 11.56 | 714.5 | .5066 | 10.10 | 714.0 | .4904 |
| 190 | 16.25 | 715.4 | .5470 |  |  |  |  |  |  |  |  |  |

(*Continued*)

# TABLE A.8 (CONTINUED)

Ammonia: Properties of Superheated Vapor

| Temp. °F t | 25 / −7.96 | | | 30 / −0.57 | | | 35 / 5.89 | | | 40 / 11.66 | | |
|---|---|---|---|---|---|---|---|---|---|---|---|---|
| | v | h | s | v | h | s | v | h | s | v | h | s |
| 200 | 16.50 | 720.8 | 1.5552 | 13.73 | 720.3 | 1.5334 | 11.75 | 719.9 | 1.5148 | 10.27 | 719.4 | 1.4987 |
| 220 | 17.02 | 731.6 | .5713 | 14.16 | 731.1 | .5495 | 12.12 | 730.7 | .5311 | 10.59 | 730.3 | .5150 |
| 240 | 17.53 | 742.5 | .5870 | 14.59 | 742.0 | .5653 | 12.49 | 741.7 | .5469 | 10.92 | 741.3 | .5309 |
| 260 | 18.04 | 753.4 | .6025 | 15.02 | 753.0 | .5808 | 12.86 | 752.7 | .5624 | 11.24 | 752.3 | .5465 |
| 280 | | | | 15.45 | 764.1 | .5960 | 13.23 | 763.7 | .5776 | 11.56 | 763.4 | .5617 |
| 300 | | | | | | | | | | 11.88 | 774.6 | 1.5766 |

| Temp. °F t | 50 / 21.67 | | | 60 / 30.21 | | | 70 / 37.70 | | | 80 / 44.40 | | |
|---|---|---|---|---|---|---|---|---|---|---|---|---|
| | v | h | s | v | h | s | v | h | s | v | h | s |
| Sat. | 5.710 | 618.2 | 1.2939 | 4.805 | 620.5 | 1.2787 | 4.151 | 622.4 | 1.2658 | 3.655 | 624.0 | 1.2545 |
| 30 | 5.838 | 623.4 | 1.3046 | | | | | | | | | |
| 40 | 5.988 | 629.5 | .3169 | 4.933 | 626.8 | 1.2913 | 4.177 | 623.9 | 1.2688 | | | |
| 50 | 6.135 | 635.4 | 1.3286 | 5.060 | 632.9 | 1.3035 | 4.290 | 630.4 | 1.2816 | 3.712 | 627.7 | 1.2619 |
| 60 | 6.280 | 641.2 | .3399 | 5.184 | 639.0 | .3152 | 4.401 | 636.6 | .2937 | 3.812 | 634.3 | .2745 |
| 70 | 6.423 | 646.9 | .3508 | 5.307 | 644.9 | .3265 | 4.509 | 642.7 | .3054 | 3.909 | 640.6 | .2866 |
| 80 | 6.564 | 652.6 | .3613 | 5.428 | 650.7 | .3373 | 4.615 | 648.7 | .3166 | 4.005 | 646.7 | .2981 |
| 90 | 6.704 | 658.2 | .3716 | 5.547 | 656.4 | .3479 | 4.719 | 654.6 | .3274 | 4.098 | 652.8 | .3092 |
| 100 | 6.843 | 663.7 | 1.3816 | 5.665 | 662.1 | 1.3581 | 4.822 | 660.4 | 1.3378 | 4.190 | 658.7 | 1.3199 |
| 110 | 6.980 | 669.2 | .3914 | 5.781 | 667.7 | .3681 | 4.924 | 666.1 | .3480 | 4.281 | 664.6 | .3303 |
| 120 | 7.117 | 674.7 | .4009 | 5.897 | 673.3 | .3778 | 5.025 | 671.8 | .3579 | 4.371 | 670.4 | .3404 |
| 130 | 7.252 | 680.2 | .4103 | 6.012 | 678.9 | .3873 | 5.125 | 677.5 | .3676 | 4.460 | 676.1 | .3502 |
| 140 | 7.387 | 685.7 | .4195 | 6.126 | 684.4 | .3966 | 5.224 | 683.1 | .3770 | 4.548 | 681.8 | .3598 |
| 150 | 7.521 | 691.1 | 1.4286 | 6.239 | 689.9 | 1.4058 | 5.323 | 688.7 | 1.3863 | 4.635 | 687.5 | 1.3692 |
| 160 | 7.655 | 696.6 | .4374 | 6.352 | 695.5 | .4148 | 5.420 | 694.3 | .3954 | 4.722 | 693.2 | .3784 |
| 170 | 7.788 | 702.1 | .4462 | 6.464 | 701.0 | .4236 | 5.518 | 699.9 | .4043 | 4.808 | 698.8 | .3874 |
| 180 | 7.921 | 707.5 | .4548 | 6.576 | 706.5 | .4323 | 5.615 | 705.5 | .4131 | 4.893 | 704.4 | .3963 |

*(Continued)*

# TABLE A.8 (CONTINUED)

Ammonia: Properties of Superheated Vapor

| Temp. °F | 90 50.47 | | | 100 56.05 | | | 120 66.02 | | | 140 74.79 | | |
|---|---|---|---|---|---|---|---|---|---|---|---|---|
| t | v | h | s | v | h | s | v | h | s | v | h | s |
| 190 | 8.053 | 713.0 | .4633 | 6.687 | 712.0 | .4409 | 5.711 | 711.0 | .4217 | 4.978 | 710.0 | .4050 |
| 200 | 8.185 | 718.5 | 1.4716 | 6.798 | 717.5 | 1.4493 | 5.807 | 716.6 | 1.4302 | 5.063 | 715.6 | 1.4136 |
| 210 | 8.317 | 724.0 | .4799 | 6.909 | 723.1 | .4576 | 5.902 | 722.2 | .4386 | 5.147 | 721.3 | .4220 |
| 220 | 8.448 | 729.4 | .4880 | 7.019 | 728.6 | .4658 | 5.998 | 727.7 | .4469 | 5.231 | 726.9 | .4304 |
| 240 | 8.710 | 740.5 | .5040 | 7.238 | 739.7 | .4819 | 6.187 | 738.9 | .4631 | 5.398 | 738.1 | .4467 |
| 260 | 8.970 | 751.6 | .5197 | 7.457 | 750.9 | .4976 | 6.376 | 750.1 | .4789 | 5.565 | 749.4 | .4626 |
| 280 | 9.230 | 762.7 | 1.5350 | 7.675 | 762.1 | 1.5130 | 6.563 | 761.4 | 1.4943 | 5.730 | 760.7 | 1.4781 |
| 300 | 9.489 | 774.0 | .5500 | 7.892 | 773.3 | .5281 | 6.750 | 772.7 | .5095 | 5.894 | 772.1 | .4933 |

| Temp. °F | 90 50.47 | | | 100 56.05 | | | 120 66.02 | | | 140 74.79 | | |
|---|---|---|---|---|---|---|---|---|---|---|---|---|
| t | v | h | s | v | h | s | v | h | s | v | h | s |
| Sat. | 3.266 | 625.3 | 1.2445 | 2.952 | 626.5 | 1.2356 | 2.476 | 628.4 | 1.2201 | 2.132 | 629.9 | 1.2068 |
| 50 |  |  |  |  |  |  |  |  |  |  |  |  |
| 60 | 3.353 | 631.8 | 1.2571 | 2.985 | 629.3 | 1.2409 |  |  |  |  |  |  |
| 70 | 3.442 | 638.3 | .2695 | 3.068 | 636.0 | .2539 | 2.505 | 631.3 | 1.2255 |  |  |  |
| 80 | 3.529 | 644.7 | .2814 | 3.149 | 642.6 | .2661 | 2.576 | 638.3 | .2386 | 2.166 | 633.8 | 1.2140 |
| 90 | 3.614 | 650.9 | .2928 | 3.227 | 649.0 | .2778 | 2.645 | 645.0 | .2510 | 2.228 | 640.9 | .2272 |
| 100 | 3.698 | 657.0 | 1.3038 | 3.304 | 655.2 | 1.2891 | 2.712 | 651.6 | 1.2628 | 2.288 | 647.8 | 1.2396 |
| 110 | 3.780 | 663.0 | .3144 | 3.380 | 661.3 | .2999 | 2.778 | 658.0 | .2741 | 2.347 | 654.5 | .2515 |
| 120 | 3.862 | 668.9 | .3247 | 3.454 | 667.3 | .3104 | 2.842 | 664.2 | .2850 | 2.404 | 661.1 | .2628 |

| Temp. °F | 90 50.47 | | | 100 56.05 | | | 120 66.02 | | | 140 74.59 | | |
|---|---|---|---|---|---|---|---|---|---|---|---|---|
| t | v | h | s | v | h | s | v | h | s | v | h | s |
| 130 | 3.942 | 674.7 | .3347 | 3.527 | 673.3 | .3206 | 2.905 | 670.4 | .2956 | 2.460 | 667.4 | .2738 |
| 140 | 4.021 | 680.5 | .3444 | 3.600 | 679.2 | .3305 | 2.967 | 676.5 | .3058 | 2.515 | 673.7 | .2843 |
| 150 | 4100 | 686.3 | 1.3539 | 3 672 | 685 0 | 1.3401 | 3 029 | 682 5 | 1.3157 | 2.569 | 679.9 | 1.2945 |
| 160 | 4.178 | 692.0 | .3633 | 3.743 | 690.8 | .3495 | 3.089 | 688.4 | .3254 | 2.622 | 686.0 | .3045 |
| 170 | 4.255 | 697.7 | .3724 | 3.813 | 696.6 | .3588 | 3.149 | 694.3 | .3348 | 2.675 | 692.0 | .3141 |

(Continued)

**TABLE A.8 (CONTINUED)**

Ammonia: Properties of Superheated Vapor

| t | 160 v | 160 h | 160 s | 180 v | 180 h | 180 s | 200 v | 200 h | 200 s | 220 v | 220 h | 220 s |
|---|---|---|---|---|---|---|---|---|---|---|---|---|
| 180 | 4.332 | 703.4 | .3813 | 3.883 | 702.3 | .3678 | 3.209 | 700.2 | .3441 | 2.727 | 698.0 | .3236 |
| 190 | 4.408 | 709.0 | .3901 | 3.952 | 708.0 | .3767 | 3.268 | 706.0 | .3531 | 2.779 | 704.0 | .3328 |
| 200 | 4.484 | 714.7 | 1.3988 | 4.021 | 713.7 | 1.3854 | 3.326 | 711.8 | 1.3620 | 2.830 | 709.9 | 1.3418 |
| 210 | 4.560 | 720.4 | .4073 | 4.090 | 719.4 | .3940 | 3.385 | 717.6 | .3707 | 2.880 | 715.8 | .3507 |
| 220 | 4.635 | 726.0 | .4157 | 4.158 | 725.1 | .4024 | 3.442 | 723.4 | .3793 | 2.931 | 721.6 | .3594 |
| 230 | 4.710 | 731.7 | .4239 | 4.226 | 730.8 | .4108 | 3.500 | 729.2 | .3877 | 2.981 | 727.5 | .3679 |
| 240 | 4.785 | 737.3 | .4321 | 4.294 | 736.5 | .4190 | 3.557 | 734.9 | .3960 | 3.030 | 733.3 | .3763 |
| 250 | 4.859 | 743.0 | 1.4401 | 4.361 | 742.2 | 1.4271 | 3.614 | 740.7 | 1.4042 | 3.080 | 739.2 | 1.3846 |
| 260 | 4.933 | 748.7 | .4481 | 4.428 | 747.9 | .4350 | 3.671 | 746.5 | .4123 | 3.129 | 745.0 | .3928 |
| 280 | 5.081 | 760.0 | .4637 | 4.562 | 759.4 | .4507 | 3.783 | 758.0 | .4281 | 3.227 | 756.7 | .4088 |
| 300 | 5.228 | 771.5 | .4789 | 4.695 | 770.8 | .4660 | 3.895 | 769.6 | .4435 | 3.323 | 768.3 | .4243 |

| Temp. °F | 160 / 82.64 | | | 180 / 89.78 | | | 200 / 96.34 | | | 220 / 102.42 | | |
|---|---|---|---|---|---|---|---|---|---|---|---|---|
| t | v | h | s | v | h | s | v | h | s | v | h | s |
| Sat. | 1.872 | 631.1 | 1.1952 | 1.667 | 632.0 | 1.1850 | 1.502 | 632.7 | 1.1756 | 1.367 | 633.2 | 1.1671 |
| 90 | 1.914 | 636.6 | 1.2055 | 1.668 | 632.2 | 1.1853 | | | | | | |
| 100 | 1.969 | 643.9 | 1.2186 | 1.720 | 639.9 | 1.1992 | | | | | | |
| 110 | 2.023 | 651.0 | .2311 | 1.770 | 647.3 | .2123 | 1.567 | 643.4 | 1.1947 | 1.400 | 639.4 | 1.1781 |
| 120 | 2.075 | 657.8 | .2429 | 1.818 | 654.4 | .2247 | 1.612 | 650.9 | .2077 | 1.443 | 647.3 | .1917 |
| 130 | 2.125 | 664.4 | .2542 | 1.865 | 661.3 | .2364 | 1.656 | 658.1 | .2200 | 1.485 | 654.8 | .2045 |
| 140 | 2.175 | 670.9 | .2652 | 1.910 | 668.0 | .2477 | 1.698 | 665.0 | .2317 | 1.525 | 662.0 | .2167 |
| 150 | 2.224 | 677.2 | 1.2757 | 1.955 | 674.6 | 1.2586 | 1.740 | 671.8 | 1.2429 | 1.564 | 669.0 | 1.2281 |
| 160 | 2.272 | 683.5 | .2859 | 1.999 | 681.0 | .2691 | 1.780 | 678.4 | .2537 | 1.601 | 675.8 | .2394 |
| 170 | 2.319 | 689.7 | .2958 | 2.042 | 687.3 | .2792 | 1.820 | 684.9 | .2641 | 1.638 | 682.5 | .2501 |
| 180 | 2.365 | 695.8 | .3054 | 2.084 | 693.6 | .2891 | 1.859 | 691.3 | .2742 | 1.675 | 689.1 | .2604 |
| 190 | 2.411 | 701.9 | .3148 | 2.126 | 699.8 | .2987 | 1.897 | 697.7 | .2840 | 1.710 | 695.5 | .2704 |
| 200 | 2.457 | 707.9 | 1.3240 | 2.167 | 705.9 | 1.3081 | 1.935 | 703.9 | 1.2935 | 1.745 | 701.9 | 1.2801 |
| 210 | 2.502 | 713.9 | .3331 | 2.208 | 712.0 | .3172 | 1.972 | 710.1 | .3029 | 1.780 | 708.2 | .2896 |
| 220 | 2.547 | 719.9 | .3419 | 2.248 | 718.1 | .3262 | 2.009 | 716.3 | .3120 | 1.814 | 714.4 | .2989 |
| 230 | 2.591 | 725.8 | .3506 | 2.288 | 724.1 | .3350 | 2.046 | 722.4 | .3209 | 1.848 | 720.6 | .3079 |

*(Continued)*

**TABLE A.8 (CONTINUED)**

Ammonia: Properties of Superheated Vapor

| Temp. °F t | 240 108.09 | | | 260 113.42 | | | 280 118.45 | | | 300 123.21 | | |
|---|---|---|---|---|---|---|---|---|---|---|---|---|
| | v | h | s | v | h | s | v | h | s | v | h | s |
| 240 | 2.635 | 731.7 | .3591 | 2.328 | 730.1 | .3436 | 2.082 | 728.4 | .3296 | 1.881 | 726.8 | .3168 |
| 250 | 2.679 | 737.6 | 1.3675 | 2.367 | 736.1 | 1.3521 | 2.118 | 734.5 | 1.3382 | 1.914 | 732.9 | 1.3255 |
| 260 | 2.723 | 743.5 | .3757 | 2.407 | 742.0 | .3605 | 2.154 | 740.5 | .3467 | 1.947 | 739.0 | .3340 |
| 270 | 2.766 | 749.4 | .3838 | 2.446 | 748.0 | .3687 | 2.189 | 746.5 | .3550 | 1.980 | 745.1 | .3424 |
| 280 | 2.809 | 755.3 | .3919 | 2.484 | 753.9 | .3768 | 2.225 | 752.5 | .3631 | 2.012 | 751.1 | .3507 |
| 290 | 2.852 | 761.2 | .3998 | 2.523 | 759.9 | .3847 | 2.260 | 758.5 | .3712 | 2.044 | 757.2 | .3588 |
| 300 | 2.895 | 767.1 | 1.4076 | 2.561 | 765.8 | 1.3926 | 2.295 | 764.5 | 1.3791 | 2.076 | 763.2 | 1.3668 |
| 320 | 2.980 | 778.9 | .4229 | 2.637 | 777.7 | .4081 | 2.364 | 776.5 | .3947 | 2.140 | 775.3 | .3825 |
| 340 | 3.064 | 790.7 | .4379 | 2.713 | 789.6 | .4231 | 2.432 | 788.5 | .4099 | 2.203 | 787.4 | .3978 |
| 360 | | | | | | | 2.500 | 800.5 | .4247 | 2.265 | 799.5 | .4127 |
| 380 | | | | | | | 2.568 | 812.5 | .4392 | 2.327 | 811.6 | .4273 |

| Temp. °F t | 240 108.09 | | | 260 113.42 | | | 280 118.45 | | | 300 123.21 | | |
|---|---|---|---|---|---|---|---|---|---|---|---|---|
| | v | h | s | v | h | s | v | h | s | v | h | s |
| Sat. | 1.253 | 633.6 | 1.1592 | 1.155 | 633.9 | 1.1518 | 1.072 | 634.0 | 1.1449 | 0.999 | 634.0 | 1.1383 |
| 110 | 1.261 | 635.3 | 1.1621 | | | | | | | | | |
| 120 | 1.302 | 643.5 | .1764 | 1.182 | 639.5 | 1.1617 | 1.078 | 635.4 | 1.1473 | | | |
| 130 | 1.342 | 651.3 | .1898 | 1.220 | 647.8 | .1757 | 1.115 | 644.0 | .1621 | 1.023 | 640.1 | 1.1487 |
| 140 | 1.380 | 658.8 | .2025 | 1.257 | 655.6 | .1889 | 1.151 | 652.2 | .1759 | 1.058 | 648.7 | .1632 |

| Temp. °F t | 240 108.09 | | | 260 113.42 | | | 280 118.45 | | | 300 123.21 | | |
|---|---|---|---|---|---|---|---|---|---|---|---|---|
| | v | h | s | v | h | s | v | h | s | v | h | s |
| 150 | 1.416 | 666.1 | 1.2145 | 1.292 | 663.1 | 1.2014 | 1.184 | 660.1 | 1.1888 | 1.091 | 656.9 | 1.1767 |
| 160 | 1.452 | 673.1 | .2259 | 1.326 | 670.4 | .2132 | 1.217 | 667.6 | .2011 | 1.123 | 664.7 | .1894 |
| 170 | 1.487 | 680.0 | .2369 | 1.359 | 677.5 | .2245 | 1.249 | 674.9 | .2127 | 1.153 | 672.2 | .2014 |
| 180 | 1.521 | 686.7 | .2475 | 1.391 | 684.4 | .2354 | 1.279 | 681.9 | .2239 | 1.183 | 679.5 | .2129 |
| 190 | 1.554 | 693.3 | .2577 | 1.422 | 691.1 | .2458 | 1.309 | 688.9 | .2346 | 1.211 | 686.5 | .2239 |
| 200 | 1.587 | 699.8 | 1.2677 | 1.453 | 697.7 | 1.2560 | 1.339 | 695.6 | 1.2449 | 1.239 | 693.5 | 1.2344 |

*(Continued)*

**TABLE A.8 (CONTINUED)**

Ammonia: Properties of Superheated Vapor

| | | | | | | | | | | | | |
|---|---|---|---|---|---|---|---|---|---|---|---|---|
| 210 | 1.619 | 706.2 | .2773 | 1.484 | 704.3 | .2658 | 1.367 | 702.3 | .2550 | 1.267 | 700.3 | .2447 |
| 220 | 1.651 | 712.6 | .2867 | 1.514 | 710.7 | .2754 | 1.396 | 708.8 | .2647 | 1.294 | 706.9 | .2546 |
| 230 | 1.683 | 718.9 | .2959 | 1.543 | 717.1 | .2847 | 1.424 | 715.3 | .2742 | 1.320 | 713.5 | .2642 |
| 240 | 1.714 | 725.1 | .3049 | 1.572 | 723.4 | .2938 | 1.451 | 721.8 | .2834 | 1.346 | 720.0 | .2736 |
| 250 | 1.745 | 731.3 | 1.3137 | 1.601 | 729.7 | 1.3027 | 1.478 | 728.1 | 1.2924 | 1.372 | 726.5 | 1.2827 |
| 260 | 1.775 | 737.6 | .3224 | 1.630 | 736.0 | .3115 | 1.505 | 734.4 | .3013 | 1.397 | 732.9 | .2917 |
| 270 | 1.805 | 743.6 | .3308 | 1.658 | 742.2 | .3200 | 1.532 | 740.7 | .3099 | 1.422 | 739.2 | .3004 |
| 280 | 1.835 | 749.8 | .3392 | 1.686 | 748.4 | .3285 | 1.558 | 747.0 | .3184 | 1.447 | 745.5 | .3090 |
| 290 | 1.865 | 755.9 | .3474 | 1.714 | 754.5 | .3367 | 1.584 | 753.2 | .3268 | 1.472 | 751.8 | .3175 |
| 300 | 1.895 | 762.0 | 1.3554 | 1.741 | 760.7 | 1.3449 | 1.610 | 759.4 | 1.3350 | 1.496 | 758.1 | 1.3257 |
| 320 | 1.954 | 774.1 | .3712 | 1.796 | 772.9 | .3608 | 1.661 | 771.7 | .3511 | 1.544 | 770.5 | .3419 |
| 340 | 2.012 | 786.3 | .3866 | 1.850 | 785.2 | .3763 | 1.712 | 784.0 | .3667 | 1.592 | 782.9 | .3576 |
| 360 | 2.069 | 798.4 | .4016 | 1.904 | 797.4 | .3914 | 1.762 | 796.3 | .3819 | 1.639 | 795.3 | .3729 |
| 380 | 2.126 | 810.6 | .4163 | 1.957 | 809.6 | .4062 | 1.811 | 808.7 | .3967 | 1.686 | 807.7 | .3878 |
| 400 | | | | 2.009 | 821.9 | 1.4206 | 1.861 | 821.0 | 1.4112 | 1.732 | 820.1 | 1.4024 |

*Source:* Abstracted from *Tables of Thermodynamic Properties of Ammonia*, U.S. Department of Commerce, Bureau of Standards Circular 142, 1945. With permission.

**TABLE A.9**

Pressure-Enthalpy Diagram for Ammonia

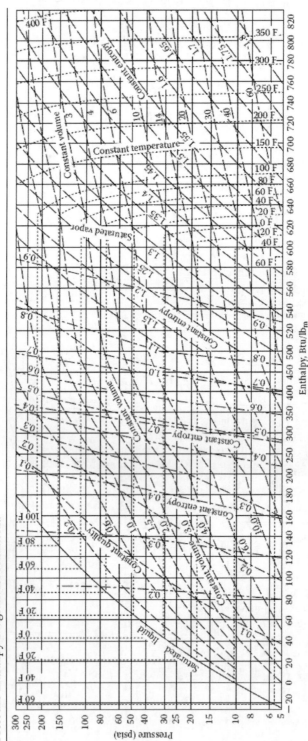

*Source:* Data from Tables of Thermodynamic Properties of Ammonia, Bureau of Standards, U.S. Department of Commerce.

**TABLE A.10**

Dichlorodifluoromethane (Freon-12): Properties of Liquid and Saturated Vapor

| Temp. °F $t$ | Pressure Abs. lb./in.² $p$ | Pressure Gauge lb./in.² $p_d$ | Volume Liquid ft.³/lb. $v_f$ | Volume Vapor ft.³/lb. $v_s$ | Density Liquid lb./ft.³ $1/v_f$ | Density Vapor lb./ft.³ $1/v_g$ | Enthalpy from −40°F Liquid Btu/lb. $h_f$ | Enthalpy from −40°F Latent Btu/lb. $h_{fg}$ | Enthalpy from −40°F Vapor Btu/lb. $h_g$ | Entropy from −40°F Liquid Btu/lb$_m$·R $s_f$ | Entropy from −40°F Vapor Btu/lbm·R $s_g$ |
|---|---|---|---|---|---|---|---|---|---|---|---|
| −152 | 0.13799 | 29.64024[a] | 0.0095673 | 197.58 | 104.52 | 0.0050614 | −23.106 | 83.734 | 60.628 | −0.063944 | 0.20818 |
| −150 | .15359 | 29.60849[a] | .0095822 | 178.65 | 104.36 | .0055976 | −22.697 | 83.534 | 60.837 | −0.062619 | .20711 |
| −145 | .19933 | 29.51537[a] | .0096198 | 139.83 | 103.95 | .0071517 | −21.674 | 83.039 | 61.365 | −0.059344 | .20452 |
| −140 | .25623 | 29.39951[a] | .0096579 | 110.46 | 103.54 | .0090533 | −20.652 | 82.548 | 61.896 | −0.056123 | .20208 |
| −135 | .32641 | 29.25663[a] | .0096966 | 88.023 | 103.13 | .011361 | −19.631 | 82.061 | 62.430 | −0.052952 | .19978 |
| −130 | .41224 | 29.08186[a] | .0097359 | 70.730 | 102.71 | .014138 | −18.609 | 81.577 | 62.968 | −0.049830 | .19760 |
| −125 | .51641 | 28.86978[a] | .0097758 | 57.283 | 102.29 | .017457 | −17.587 | 81.096 | 63.509 | −0.046754 | 0.19554 |
| −120 | .64190 | 28.61429[a] | .0098163 | 46.741 | 101.87 | .021395 | −16.565 | 80.617 | 64.052 | −0.043723 | .19359 |
| −115 | .79200 | 28.30869[a] | .0098574 | 38.410 | 101.45 | .026035 | −15.541 | 80.139 | 64.598 | −0.040734 | .19176 |
| −110 | .97034 | 27.94558[a] | .0098992 | 31.777 | 101.02 | .031470 | −14.518 | 79.663 | 65.145 | −0.037786 | .19002 |
| −105 | 1.1809 | 27.5169[a] | .0099416 | 26.458 | 100.59 | .037796 | −13.492 | 79.188 | 65.696 | −0.034877 | .18838 |
| −100 | 1.4280 | 27.0138[a] | .0099847 | 22.164 | 100.15 | .045119 | −12.466 | 78.714 | 66.248 | −0.032005 | 0.18683 |
| −95 | 1.7163 | 26.4268[a] | .010029 | 18.674 | 99.715 | .053550 | −11.438 | 78.239 | 66.801 | −0.029169 | .18536 |
| −90 | 2.0509 | 25.7456[a] | .010073 | 15.821 | 99.274 | .063207 | −10.409 | 77.764 | 67.355 | −0.026367 | .18398 |
| −85 | 2.4371 | 24.9593[a] | .010118 | 13.474 | 98.830 | .074216 | −9.3782 | 77.289 | 67.911 | −0.023599 | .18267 |
| −80 | 2.8807 | 24.0560[a] | .010164 | 11.533 | 98.382 | .086708 | −8.3451 | 76.812 | 68.467 | −0.020862 | .18143 |
| −75 | 3.3879 | 23.0234[a] | 0.010211 | 9.9184 | 97.930 | 0.10082 | −7.3101 | 76.333 | 69.023 | −0.018156 | 0.18027 |
| −70 | 3.9651 | 21.8482[a] | .010259 | 8.5687 | 97.475 | .11670 | −6.2730 | 75.853 | 69.580 | −0.015481 | .17916 |
| −65 | 4.6193 | 20.5164[a] | .010308 | 7.4347 | 97.016 | .13451 | −5.2336 | 75.371 | 70.137 | −0.012834 | .17812 |
| −60 | 5.3575 | 19.0133[a] | .010357 | 6.4774 | 96.553 | .15438 | −4.1919 | 74.885 | 70.693 | −0.010214 | .17714 |
| −55 | 6.1874 | 17.3237[a] | .010407 | 5.6656 | 96.086 | .17650 | −3.1477 | 74.397 | 71.249 | −0.007622 | .17621 |
| −50 | 7.1168 | 15.4313[a] | 0.010459 | 4.9742 | 95.616 | 0.20104 | −2.1011 | 73.906 | 71.805 | −0.005056 | 0.17533 |
| −45 | 8.1540 | 13.3196[a] | .010511 | 4.3828 | 95.141 | .22816 | −1.0519 | 73.411 | 72.359 | −0.002516 | .17451 |
| −40 | 9.3076 | 10.9709a | .010564 | 3.8750 | 94.661 | .25806 | 0 | 72.913 | 72.913 | 0 | 0.17373 |
| −38 | 9.8035 | 9.9611a | .010586 | 3.6922 | 94.469 | .27084 | 0.4215 | 72.712 | 73.134 | 0.001000 | .17343 |

*(Continued)*

## TABLE A.10 (CONTINUED)

Dichlorodifluoromethane (Freon-12): Properties of Liquid and Saturated Vapor

| Temp. °F t | Pressure Abs. lb./in.² p | Pressure Gauge lb./in.² $p_d$ | Volume Liquid ft³/lb. $v_f$ | Volume Vapor ft³/lb. $v_s$ | Density Liquid lb./ft³ $1/v_f$ | Density Vapor lb./ft³ $1/v_g$ | Enthalpy from −40°F Liquid Btu/lb. $h_f$ | Enthalpy from −40°F Latent Btu/lb. $h_{fg}$ | Enthalpy from −40°F Vapor Btu/lb. $h_g$ | Entropy from −40°F Liquid Btu/lb$_m$·R $s_f$ | Entropy from −40°F Vapor Btu/lbm·R $s_g$ |
|---|---|---|---|---|---|---|---|---|---|---|---|
| −36 | 10.320 | 8.909a | .010607 | 3.5198 | 94.275 | .28411 | .8434 | 72.511 | 73.354 | .001995 | .17313 |
| −34 | 10.858 | 7.814a | .010629 | 3.3571 | 94.081 | .29788 | 1.2659 | 72.309 | 73.575 | .002988 | .17285 |
| −32 | 11.417 | 6.675a | .010651 | 3.2035 | 93.886 | .31216 | 1.6887 | 72.106 | 73.795 | .003976 | .17257 |
| −30 | 11.999 | 5.490a | 0.010674 | 3.0585 | 93.690 | 0.32696 | 2.1120 | 71.903 | 74.015 | 0.004961 | 0.17229 |
| −28 | 12.604 | 4.259a | .010696 | 2.9214 | 93.493 | 0.34231 | 2.5358 | 71.698 | 74.234 | .005942 | .17203 |
| −26 | 13.233 | 2.979a | .010719 | 2.7917 | 93.296 | .35820 | 2.9601 | 71.494 | 74.454 | .006919 | .17177 |
| −24 | 13.886 | 1.649a | .010741 | 2.6691 | 93.098 | .37466 | 3.3848 | 71.288 | 74.673 | .007894 | .17151 |
| −22 | 14.564 | 0.270a | .010764 | 2.5529 | 92.899 | .39171 | 3.8100 | 71.081 | 74.891 | .008864 | .17126 |
| −20 | 15.267 | 0.571 | 0.010788 | 2.4429 | 92.699 | 0.40934 | 4.2357 | 70.874 | 75.110 | 0.009831 | 0.17102 |
| −18 | 15.996 | 1.300 | .010811 | 2.3387 | 92.499 | .42758 | 4.6618 | 70.666 | 75.328 | .010795 | .17078 |
| −16 | 16.753 | 2.057 | .010834 | 2.2399 | 92.298 | .44645 | 5.0885 | 70.456 | 75.545 | .011755 | .17055 |
| −14 | 17.536 | 2.840 | .010858 | 2.1461 | 92.096 | .46595 | 5.5157 | 70.246 | 75.762 | .012712 | .17032 |
| −12 | 18.348 | 3.652 | .010882 | 2.0572 | 91.893 | .48611 | 5.9434 | 70.036 | 75.979 | .013666 | .17010 |
| −10 | 19.189 | 4.493 | 0.010906 | 1.9727 | 91.689 | 0.50693 | 6.3716 | 69.824 | 76.196 | 0.014617 | 0.16989 |
| −8 | 20.059 | 5.363 | .010931 | 1.8924 | 91.485 | .52843 | 6.8003 | 69.611 | 76.411 | .015564 | .16967 |
| −6 | 20.960 | 6.264 | .010955 | 1.8161 | 91.280 | .55063 | 7.2296 | 69.397 | 76.627 | .016508 | .16947 |
| −4 | 21.891 | 7.195 | .010980 | 1.7436 | 91.074 | .57354 | 7.6594 | 69.183 | 76.842 | .017449 | .16927 |
| −2 | 22.854 | 8.158 | .011005 | 1.6745 | 90.867 | .59718 | 8.0898 | 68.967 | 77.057 | .018388 | .16907 |
| 0 | 23.849 | 9.153 | 0.011030 | 1.6089 | 90.659 | 0.62156 | 8.5207 | 68.750 | 77.271 | 0.019323 | 0.16888 |
| 2 | 24.878 | 10.182 | .011056 | 1.5463 | 90.450 | .64670 | 8.9522 | 68.533 | 77.485 | .020255 | .16869 |
| 4 | 25.939 | 11.243 | .011082 | 1.4867 | 90.240 | .67263 | 9.3843 | 68.314 | 77.698 | .021184 | .16851 |
| 5 | 26.483 | 11.787 | .011094 | 1.4580 | 90.135 | .68588 | 9.6005 | 68.204 | 77.805 | .021647 | .16842 |
| 6 | 27.036 | 12.340 | .011107 | 1.4299 | 90.030 | .69934 | 9.8169 | 68.094 | 77.911 | .022110 | .16833 |
| 8 | 28.167 | 13.471 | .011134 | 1.3758 | 89.818 | .72687 | 10.250 | 67.873 | 78.123 | .023033 | .16815 |
| 10 | 29.335 | 14.639 | 0.011160 | 1.3241 | 89.606 | 0.75523 | 10.684 | 67.651 | 78.335 | 0.023954 | 0.16798 |

*(Continued)*

# TABLE A.10 (CONTINUED)

Dichlorodifluoromethane (Freon-12): Properties of Liquid and Saturated Vapor

| Temp. °F $t$ | Pressure Abs. lb./in.² $p$ | Pressure Gauge lb./in.² $p_d$ | Volume Liquid ft.³/lb. $v_f$ | Volume Vapor ft.³/lb. $v_s$ | Density Liquid lb./ft.³ $1/v_f$ | Density Vapor lb./ft.³ $1/v_g$ | Enthalpy Liquid Btu./lb. $h_f$ | Enthalpy Latent Btu./lb. $h_{fg}$ | Enthalpy Vapor Btu./lb. $h_g$ | Entropy Liquid Btu/lb$_m$·R $s_f$ | Entropy Vapor Btu/lbm·R $s_g$ |
|---|---|---|---|---|---|---|---|---|---|---|---|
| 12 | 30.539 | 15.843 | .011187 | 1.2748 | 89.392 | .78443 | 11.118 | 67.428 | 78.546 | .024871 | .16782 |
| 14 | 31.780 | 17.084 | .011214 | 1.2278 | 89.178 | .81449 | 11.554 | 67.203 | 78.757 | .025786 | .16765 |
| 16 | 33.060 | 18.364 | .011241 | 1.1828 | 88.962 | .84544 | 11.989 | 66.977 | 78.966 | .026699 | .16750 |
| 18 | 34.378 | 19.682 | .011268 | 1.1399 | 88.746 | .87729 | 12.426 | 66.750 | 79.176 | .027608 | .16734 |
| 20 | 35.736 | 21.040 | 0.011296 | 1.0988 | 88.529 | 0.91006 | 12.863 | 66.522 | 79.385 | 0.028515 | 0.16719 |
| 22 | 37.135 | 22.439 | .011324 | 1.0596 | 88.310 | .94377 | 13.300 | 66.293 | 79.593 | .029420 | .16704 |
| 24 | 38.574 | 23.878 | .011352 | 1.0220 | 88.091 | .97843 | 13.739 | 66.061 | 79.800 | .030322 | .16690 |
| 26 | 40.056 | 25.360 | .011380 | 0.98612 | 87.870 | 1.0141 | 14.178 | 65.829 | 80.007 | .031221 | .16676 |
| 28 | 41.580 | 26.884 | .011409 | .95173 | 87.649 | 1.0507 | 14.618 | 65.596 | 80.214 | .032118 | .16662 |
| 30 | 43.148 | 28.452 | 0.011438 | 0.91880 | 87.426 | 1.0884 | 15.058 | 65.361 | 80.419 | 0.033013 | .16648 |
| 32 | 44.760 | 30.064 | .011468 | .88725 | 87.202 | 1.1271 | 15.500 | 65.124 | 80.624 | .033905 | .16635 |
| 34 | 46.417 | 31.721 | .011497 | .85702 | 86.977 | 1.1668 | 15.942 | 64.886 | 80.828 | .034796 | .16622 |
| 36 | 48.120 | 33.424 | .011527 | .82803 | 86.751 | 1.2077 | 16.384 | 64.647 | 81.031 | .035683 | .16610 |
| 38 | 49.870 | 35.174 | .011557 | .80023 | 86.524 | 1.2496 | 16.828 | 64.406 | 81.234 | .036569 | .16598 |
| 40 | 51.667 | 36.971 | .011588 | .77357 | 86.296 | 1.2927 | 17.273 | 64.163 | 81.436 | .037453 | .16586 |
| 42 | 53.513 | 38.817 | .011619 | .74798 | 86.066 | 1.3369 | 17.718 | 63.919 | 81.637 | .038334 | .16574 |
| 44 | 55.407 | 40.711 | .011650 | .72341 | 85.836 | 1.3823 | 18.164 | 63.673 | 81.837 | .039213 | .16562 |
| 46 | 57.352 | 42.656 | .011682 | .69982 | 85.604 | 1.4289 | 18.611 | 63.426 | 82.037 | .040091 | .16551 |
| 48 | 59.347 | 44.651 | .011714 | .67715 | 85.371 | 1.4768 | 19.059 | 63.177 | 82.236 | .040966 | .16540 |
| 50 | 61.394 | 46.698 | 0.011746 | 0.65537 | 85.136 | 1.5258 | 19.507 | 62.926 | 82.433 | 0.041839 | 0.16530 |
| 52 | 63.494 | 48.798 | .011779 | .63444 | 84.900 | 1.5762 | 19.957 | 62.673 | 82.630 | .042711 | .16519 |
| 54 | 65.646 | 50.950 | .011811 | .61431 | 84.663 | 1.6278 | 20.408 | 62.418 | 82.826 | .043581 | .16509 |
| 56 | 67.853 | 53.157 | .011845 | .59495 | 84.425 | 1.6808 | 20.859 | 62.162 | 83.021 | .04449 | .16499 |
| 58 | 70.115 | 55.419 | .011879 | .57632 | 84.185 | 1.7352 | 21.312 | 61.903 | 83.215 | .045316 | .16489 |
| 60 | 72.433 | 57.737 | 0.011913 | 0.55839 | 83.944 | 1.7909 | 21.766 | 61.643 | 83.409 | 0.046180 | 0.16479 |

*(Continued)*

**TABLE A.10 (CONTINUED)**

Dichlorodifluoromethane (Freon-12): Properties of Liquid and Saturated Vapor

| Temp. °F $t$ | Pressure | | Volume | | | Density | | Enthalpy from −40° F | | | Entropy from −40°F | |
|---|---|---|---|---|---|---|---|---|---|---|---|---|
| | Abs. lb./in.² $p$ | Gauge lb./in.² $p_d$ | Liquid ft.³/lb. $v_f$ | Vapor ft.³/lb. $v_s$ | | Liquid lb./ft.³ $1/v_f$ | Vapor lb./ft.³ $1/v_g$ | Liquid Btu/lb. $h_f$ | Latent Btu/lb. $h_{fg}$ | Vapor Btu/lb. $h_g$ | Liquid Btu/lb$_m$·R $s_f$ | Vapor Btu/lbm·R $s_g$ |
| 62 | 74.807 | 60.111 | .011947 | .54112 | | 83.701 | 1.8480 | 22.221 | 61.380 | 83.601 | .047044 | .16470 |
| 64 | 77.239 | 62.543 | .011982 | .52450 | | 83.457 | 1.9066 | 22.676 | 61.116 | 83.792 | .047905 | .16460 |
| 66 | 79.729 | 65.033 | .012017 | .50848 | | 83.212 | 1.9666 | 23.133 | 60.849 | 83.982 | .048765 | .16451 |
| 68 | 82.279 | 67.583 | .012053 | .49305 | | 82.965 | 2.0282 | 23.591 | 60.580 | 84.171 | .049624 | .16442 |
| 70 | 84.888 | 70.192 | 0.012089 | 0.47818 | | 82.717 | 2.0913 | 24.050 | 60.309 | 84.359 | 0.050482 | 0.16434 |
| 72 | 87.559 | 72.863 | .012126 | .46383 | | 82.467 | 2.1559 | 24.511 | 60.035 | 84.546 | .051338 | .16425 |
| 74 | 90.292 | 75.596 | .012163 | .45000 | | 82.215 | 2.2222 | 24.973 | 59.759 | 84.732 | .052193 | .16417 |
| 76 | 93.087 | 78.391 | .012201 | .43666 | | 81.962 | 2.2901 | 25.435 | 59.481 | 84.916 | .053047 | .16408 |
| 78 | 95.946 | 81.250 | .012239 | .42378 | | 81.707 | 2.3597 | 25.899 | 59.201 | 85.100 | .053900 | .16400 |
| 80 | 98.870 | 84.174 | 0.012277 | 0.41135 | | 81.450 | 2.4310 | 26.365 | 58.917 | 85.282 | 0.054751 | 0.16392 |
| 82 | 101.86 | 87.16 | .012316 | .39935 | | 81.192 | 2.5041 | 26.832 | 58.631 | 85.463 | .055602 | .16384 |
| 84 | 104.92 | 90.22 | .012356 | .38776 | | 80.932 | 2.5789 | 27.300 | 58.343 | 85.643 | .056452 | .16376 |
| 86 | 108.04 | 93.34 | .012396 | .37657 | | 80.671 | 2.6556 | 27.769 | 58.052 | 85.821 | .057301 | .16368 |
| 88 | 111.23 | 96.53 | .012437 | .36575 | | 80.407 | 2.7341 | 28.241 | 57.757 | 85.998 | .058149 | .16360 |
| 90 | 114.49 | 99.79 | 0.012478 | 0.35529 | | 80.142 | 2.8146 | 28.713 | 57.461 | 86.174 | 0.058997 | 0.16353 |
| 92 | 117.82 | 103.12 | .012520 | .34518 | | 79.874 | 2.8970 | 29.187 | 57.161 | 86.348 | .059844 | .16345 |
| 94 | 121.22 | 106.52 | .012562 | .33540 | | 79.605 | 2.9815 | 29.663 | 56.858 | 86.521 | .06090 | .16338 |
| 96 | 124.70 | 110.00 | .012605 | .32594 | | 79.334 | 3.0680 | 30.140 | 56.551 | 86.691 | .061536 | .16330 |
| 98 | 128.24 | 113.54 | .012649 | .31679 | | 79.061 | 3.1566 | 30.619 | 56.242 | 86.861 | .062381 | .16323 |
| 100 | 131.86 | 117.16 | 0.012693 | 0.30794 | | 78.785 | 3.2474 | 31.100 | 55.929 | 87.029 | 0.063227 | 0.16315 |
| 102 | 135.56 | 120.86 | .012738 | .29937 | | 78.508 | 3.3404 | 31.583 | 55.613 | 87.196 | .064072 | .16308 |
| 104 | 139.33 | 124.63 | .012783 | .29106 | | 78.228 | 3.4357 | 32.067 | 55.293 | 87.360 | .064916 | .16301 |
| 106 | 143.18 | 128.48 | .012829 | .28303 | | 77.946 | 3.5333 | 32.553 | 54.970 | 87.523 | .065761 | .16293 |
| 108 | 147.11 | 132.41 | .012876 | .27524 | | 77.662 | 3.6332 | 33.041 | 54.643 | 87.684 | .066606 | .16286 |
| 110 | 151.11 | 136.41 | 0.012924 | 0.26769 | | 77.376 | 3.7357 | 33.531 | 54.313 | 87.844 | 0.067451 | 0.16279 |

*(Continued)*

**TABLE A.10 (CONTINUED)**

Dichlorodifluoromethane (Freon-12): Properties of Liquid and Saturated Vapor

| Temp. °F $t$ | Pressure Abs. lb./in.² $p$ | Pressure Gauge lb./in.² $p_d$ | Volume Liquid ft.³/lb. $v_f$ | Volume Vapor ft.³/lb. $v_s$ | Density Liquid lb./ft.³ $1/v_f$ | Density Vapor lb./ft.³ $1/v_g$ | Enthalpy from −40° F Liquid Btu/lb. $h_f$ | Enthalpy from −40° F Latent Btu/lb. $h_{fg}$ | Enthalpy from −40° F Vapor Btu/lb. $h_g$ | Entropy from −40°F Liquid Btu/lb$_m$·R $s_f$ | Entropy from −40°F Vapor Btu/lbm·R $s_g$ |
|---|---|---|---|---|---|---|---|---|---|---|---|
| 112 | 155.19 | 140.49 | .012972 | .26037 | 77.087 | 3.8406 | 34.023 | 53.978 | 88.001 | .068296 | .16271 |
| 114 | 159.36 | 144.66 | .013022 | .25328 | 76.795 | 3.9482 | 34.517 | 53.639 | 88.156 | .069141 | .16264 |
| 116 | 163.61 | 148.91 | .013072 | .24641 | 76.501 | 4.0584 | 35.014 | 53.296 | 88.310 | .069987 | .16256 |
| 118 | 167.94 | 153.24 | .013123 | .23974 | 76.205 | 4.1713 | 35.512 | 52.949 | 88.461 | .070833 | .16249 |
| 120 | 172.35 | 157.65 | 0.013174 | 0.23326 | 75.906 | 4.2870 | 36.013 | 52.597 | 88.610 | 0.071680 | 0.16241 |
| 122 | 176.85 | 162.15 | .013227 | .22698 | 75.604 | 4.4056 | 36.516 | 52.241 | 88.757 | .072528 | .16234 |
| 124 | 181.43 | 166.73 | .013280 | .22089 | 75.299 | 4.5272 | 37.021 | 51.881 | 88.902 | .073376 | .16226 |
| 126 | 186.10 | 171.40 | .013335 | .21497 | 74.991 | 4.6518 | 37.529 | 51.515 | 89.044 | .074225 | .16218 |
| 128 | 190.86 | 176.16 | .013390 | .20922 | 74.680 | 4.7796 | 38.040 | 51.144 | 89.184 | .075075 | .16210 |
| 130 | 195.71 | 181.01 | 0.013447 | 0.20364 | 75.367 | 4.9107 | 38.553 | 50.768 | 89.321 | 0.075927 | 0.16202 |
| 132 | 200.64 | 185.94 | .013504 | .19821 | 74.050 | 5.0451 | 39.069 | 50.387 | 89.456 | .076779 | .16194 |
| 134 | 205.67 | 190.97 | .013563 | .19294 | 73.729 | 5.1829 | 39.588 | 50.000 | 89.588 | .077633 | .16185 |
| 136 | 210.79 | 196.09 | .013623 | .18782 | 73.406 | 5.3244 | 40.110 | 49.608 | 89.718 | .078489 | .16177 |
| 138 | 216.01 | 201.31 | .013684 | .18283 | 73.079 | 5.4695 | 40.634 | 49.210 | 89.844 | .079346 | .16168 |
| 140 | 221.32 | 206.62 | 0.013746 | 0.17799 | 72.748 | 5.6184 | 41.162 | 48.805 | 89.967 | 0.080205 | 0.16159 |
| 142 | 226.72 | 212.02 | .013810 | .17327 | 72.413 | 5.7713 | 41.693 | 48.394 | 90.087 | .081065 | .16150 |
| 144 | 232.22 | 217.52 | .013874 | .16868 | 72.075 | 5.9283 | 42.227 | 47.977 | 90.204 | .081928 | .16140 |
| 146 | 237.82 | 223.12 | .013941 | .16422 | 71.732 | 6.0895 | 42.765 | 47.553 | 90.318 | .082794 | .16130 |
| 148 | 243.51 | 228.81 | .014008 | .15987 | 71.386 | 6.2551 | 43.306 | 47.122 | 90.428 | .083661 | .16120 |

*Source:* R. C. Jordan and G.B. Priester, *Refrigeration and Air Conditioning*, 2nd ed., Prentice Hall, Inc, Upper Saddle River, N.J., 1956. Courtesy of E. I. Dupont de Nemours and Co.

ª Inches of mercury below one atmosphere.

**TABLE A.11**

Dichlorodifluoromethane (Freon-12): Properties of Superheated Vapor

| Temp. °F t | Abs. Pressure 0.14 lb./in.³ Gauge Pressure 29.64 in. vac. (Sat. Temp. −151.7°F) | | | Abs. Pressure 0.20 lb./in.³ Gauge Pressure 29.51 in. vac. (Sat. Temp. −144.9°F) | | | Abs. Pressure 0.40 lb./in.³ Gauge Pressure 29.11 in. vac. (Sat. Temp. −130.7°F) | | | Abs. Pressure 0.60 lb./in.³ Gauge Pressure 28.70 in. vac. (Sat. Temp. −121.6°F) | | |
|---|---|---|---|---|---|---|---|---|---|---|---|---|
| | v | h | s | v | h | s | v | h | s | v | h | s |
| Sat. | (194.91) | (60.656) | (0.20804) | (39.38) | (61.372) | (0.20449) | (72.756) | (62.897) | (0.19788) | (49.786) | (63.881) | (0.19419) |
| −150 | 196.01 | 60.840 | 0.20864 | | | | | | | | | |
| −140 | 202.37 | 61.916 | .21205 | 141.58 | 61.906 | 0.20617 | | | | | | |
| −130 | 208.73 | 63.012 | .21543 | 146.04 | 63.002 | .20955 | 72.903 | 62.970 | 0.19810 | | | |
| −120 | 215.09 | 64.127 | .21876 | 150.50 | 64.118 | .21288 | 75.139 | 64.088 | .20144 | 50.020 | 64.058 | 0.19472 |
| −110 | 221.44 | 65.261 | .22205 | 154.95 | 65.253 | .21618 | 77.374 | 65.225 | .20474 | 51.515 | 65.197 | .19802 |
| −100 | 227.80 | 66.414 | 0.22530 | 159.40 | 66.406 | 0.21943 | 79.607 | 66.381 | 0.20799 | 53.009 | 66.355 | 0.20128 |
| −90 | 234.15 | 67.585 | .22851 | 163.85 | 67.578 | .22264 | 81.839 | 67.554 | .21121 | 54.502 | 67.530 | .20451 |
| −80 | 240.50 | 68.774 | .23169 | 168.30 | 68.768 | .22582 | 84.070 | 68.746 | .21439 | 55.993 | 68.723 | .20769 |
| −70 | 246.85 | 69.981 | .23482 | 172.75 | 69.975 | .22896 | 86.299 | 69.954 | .21753 | 57.483 | 69.934 | .21084 |
| −60 | 253.20 | 71.206 | .23793 | 177.19 | 71.200 | .23206 | 88.527 | 71.181 | .22064 | 58.972 | 71.161 | .21395 |
| −50 | 259.54 | 72.447 | 0.24099 | 181.64 | 72.442 | 0.23513 | 90.755 | 72.424 | 0.22371 | 60.460 | 72.406 | 0.21702 |
| −40 | 265.89 | 73.706 | .24403 | 186.09 | 73.701 | .23816 | 92.982 | 73.684 | .22675 | 61.947 | 73.667 | .22006 |
| −30 | 272.23 | 74.980 | .24703 | 190.53 | 74.976 | .24117 | 95.207 | 74.960 | .22976 | 63.433 | 74.944 | .22307 |
| −20 | 278.58 | 76.271 | .25000 | 194.97 | 76.267 | .24414 | 97.433 | 76.252 | .23273 | 64.919 | 76.238 | .22605 |
| −10 | 284.92 | 77.578 | .25294 | 199.42 | 77.574 | .24708 | 99.657 | 77.560 | .23567 | 66.404 | 77.546 | .22899 |
| 0 | 291.27 | 78.901 | 0.25585 | 203.86 | 78.897 | 0.24998 | 101.88 | 78.884 | 0.23858 | 67.889 | 78.871 | 0.23190 |
| 10 | 297.61 | 80.239 | .25873 | 208.30 | 80.235 | .25286 | 104.11 | 80.223 | .24146 | 69.373 | 80.210 | .23479 |
| 20 | 303.95 | 81.591 | .26158 | 212.74 | 81.588 | .25571 | 106.33 | 81.576 | .24431 | 70.857 | 81.565 | .23764 |
| 30 | 310.30 | 82.959 | .26440 | 217.18 | 82.955 | .25854 | 108.55 | 82.945 | .24714 | 72.340 | 82.934 | .24046 |
| 40 | 316.64 | 84.341 | .26719 | 221.62 | 84.338 | .26133 | 110.77 | 84.327 | .24993 | 73.823 | 84.317 | .24326 |
| 50 | 322.98 | 85.737 | 0.26996 | 226.07 | 85.734 | 0.26410 | 113.00 | 85.724 | 0.25270 | 75.306 | 85.714 | 0.24603 |
| 60 | 329.32 | 87.147 | .27270 | 230.51 | 87.144 | .26684 | 115.22 | 87.135 | .25544 | 76.789 | 87.126 | .24877 |
| 70 | 335.66 | 88.570 | .27541 | 234.95 | 88.567 | .26955 | 117.44 | 88.559 | .25816 | 78.271 | 88.550 | .25149 |
| 80 | 342.01 | 90.007 | .27810 | 239.39 | 90.004 | .27224 | 119.66 | 89.996 | .26084 | 79.753 | 89.988 | .25417 |

(Continued)

**TABLE A.11 (CONTINUED)**

Dichlorodifluoromethane (Freon-12): Properties of Superheated Vapor

| t | v | h | s | v | h | s | v | h | s | v | h | s |
|---|---|---|---|---|---|---|---|---|---|---|---|---|
| 90 | 348.35 | 91.457 | .28076 | 243.82 | 91.454 | .27490 | 121.88 | 91.447 | .26351 | 81.235 | 91.439 | .25684 |
| 100 | 354.69 | 92.920 | 0.28340 | 248.26 | 92.917 | 0.27754 | 124.10 | 92.910 | 0.26614 | 82.716 | 92.902 | 0.25948 |
| 110 | 361.03 | 94.395 | .28601 | 252.70 | 94.393 | .28015 | 126.32 | 94.386 | .26876 | 84.197 | 94.378 | .26209 |
| 120 | 367.37 | 95.882 | .28860 | 257.14 | 95.880 | .28274 | 128.54 | 95.874 | .27135 | 85.679 | 95.867 | .26468 |
| 130 | 373.71 | 97.382 | .29116 | 261.58 | 97.380 | .28530 | 130.77 | 97.374 | .27391 | 87.160 | 97.367 | .26725 |
| 140 | 380.05 | 98.893 | .29370 | 266.02 | 98.891 | .28784 | 132.99 | 98.885 | .27645 | 88.641 | 98.879 | .26979 |

| Temp. °F | Abs. Pressure 0.80 lb./in.² Gauge, Pressure 28.29 in. vac. (Sat. Temp. −114.8°F) | | | Abs. Pressure 1.00 lb./in.² Gauge, Pressure 27.88 in. vac. (Sat. Temp. −109.3°F) | | | Abs. Pressure 2.0 lb./in.² Gauge, Pressure 25.85 in. vac. (Sat. Temp. −90.7°F) | | | Abs. Pressure 3.0 lb./in.² Gauge, Pressure 23.81 in. vac. (Sat. Temp. −78.8°F) | | |
|---|---|---|---|---|---|---|---|---|---|---|---|---|
| t | v | h | s | v | h | s | v | h | s | v | h | s |
| Sat. | (38.051) | (64.624) | (0.19167) | (30.896) | (65.229) | (0.18977) | (16.195) | (67.276) | (0.18417) | (11.106) | (68.604) | (0.18114) |
| −110 | 38.586 | 65.170 | 0.19324 | | | | | | | | | |
| −100 | 39.710 | 66.329 | .19651 | 31.730 | 66.303 | 0.19279 | | | | | | |
| −90 | 40.833 | 67.506 | .19974 | 32.631 | 67.482 | .19602 | 16.228 | 67.361 | 0.18440 | | | |
| −80 | 41.954 | 68.701 | .20292 | 33.531 | 68.679 | .19922 | 16.684 | 68.567 | .18762 | | | |
| −70 | 43.074 | 69.913 | .20608 | 34.429 | 69.892 | .20237 | 17.139 | 69.788 | .19079 | 11.375 | 69.683 | 0.18394 |
| −60 | 44.194 | 71.142 | .20919 | 35.327 | 71.123 | .20549 | 17.593 | 71.026 | .19393 | 11.681 | 70.928 | .18709 |
| −50 | 45.312 | 72.388 | 0.21227 | 36.223 | 72.370 | 0.20857 | 18.046 | 72.279 | 0.19703 | 11.986 | 72.188 | 0.19021 |
| −40 | 46.429 | 73.650 | .21531 | 37.119 | 73.633 | .21162 | 18.498 | 73.549 | .20009 | 12.290 | 73.463 | .19328 |
| −30 | 47.546 | 74.928 | .21832 | 38.014 | 74.913 | .21463 | 18.949 | 74.834 | .20311 | 12.594 | 74.754 | .19632 |
| −20 | 48.662 | 76.223 | .22130 | 38.908 | 76.208 | .21761 | 19.400 | 76.134 | .20611 | 12.897 | 76.059 | .19932 |
| −10 | 49.778 | 77.533 | .22424 | 39.802 | 77.519 | .22056 | 19.850 | 77.449 | .20906 | 13.199 | 77.379 | .20229 |
| 0 | 50.893 | 78.858 | 0.22716 | 40.695 | 78.845 | 0.22347 | 20.299 | 78.780 | 0.21199 | 13.500 | 78.714 | 0.20523 |
| 10 | 52.007 | 80.198 | 0.23004 | 41.587 | 80.186 | .22636 | 20.748 | 80.125 | .21488 | 13.802 | 80.063 | .20813 |
| 20 | 53.121 | 81.553 | .23290 | 42.480 | 81.542 | .22922 | 21.197 | 81.484 | .21775 | 14.102 | 81.426 | .21100 |
| 30 | 54.235 | 82.923 | .23572 | 43.372 | 82.912 | .23204 | 21.645 | 82.858 | .22058 | 14.403 | 82.803 | .21384 |
| 40 | 55.348 | 84.307 | .23852 | 44.263 | 84.297 | .23484 | 22.093 | 84.245 | .22339 | 14.703 | 84.194 | .21665 |
| 50 | 56.461 | 85.705 | 0.24129 | 45.154 | 85.695 | 0.23761 | 22.540 | 85.646 | 0.22616 | 15.002 | 85.598 | 0.21944 |
| 60 | 57.574 | 87.116 | .24403 | 46.045 | 87.107 | .24036 | 22.988 | 87.061 | .22891 | 15.302 | 87.015 | .22219 |

(Continued)

**TABLE A.11 (CONTINUED)**

Dichlorodifluoromethane (Freon-12): Properties of Superheated Vapor

| Temp. °F t | Abs. Pressure 5.0 lb./in.² Gauge Pressure 19.74 in. vac. (Sat. Temp. -62.4°F) | | | Abs. Pressure 7.5 lb./in.² Gauge Pressure 14.65 in. vac. (Sat. Temp. -48.1°F) | | | Abs. Pressure 10.0 lb./in.² Gauge Pressure 0.56 in. vac. (Sat. Temp. -37.2°F) | | | Abs. Pressure 15 lb./in.² Gauge Pressure 0.3 lb./in.³ (Sat. Temp. -20.8°F) | | |
|---|---|---|---|---|---|---|---|---|---|---|---|---|
| | $v$ | $h$ | $s$ | $v$ | $h$ | $s$ | $v$ | $h$ | $s$ | $v$ | $h$ | $s$ |
| Sat. | (6.9069) | (70.432) | (0.17759) | (4.7374) | (72.017) | (0.17501) | (3.6246) | (73.219) | (0.17331) | (2.4835) | (75.208) | (0.17111) |
| -60 | 6.9509 | 70.729 | 0.17834 | | | | | | | | | |
| -50 | 7.1378 | 72.003 | .18149 | | | | | | | | | |
| -40 | 7.3239 | 73.291 | .18459 | 4.8401 | 73.073 | 0.17755 | | | | | | |
| -30 | 7.5092 | 74.593 | .18766 | 4.9664 | 74.390 | .18065 | 3.6945 | 74.183 | 0.17557 | | | |
| -20 | 7.6938 | 75.909 | .19069 | 5.0919 | 75.719 | .18371 | 3.7906 | 75.526 | .17866 | 2.4885 | 75.131 | 0.17134 |
| -10 | 7.8777 | 77.239 | .19368 | 5.2169 | 77.061 | .18673 | 3.8861 | 76.880 | .18171 | 2.5546 | 76.512 | .17445 |
| 0 | 8.0611 | 78.582 | 0.19663 | 5.3412 | 78.415 | 0.18971 | 3.9809 | 78.246 | 0.18471 | 2.6201 | 77.902 | 0.17751 |
| 10 | 8.2441 | 79.939 | .19955 | 5.4650 | 79.782 | .19265 | 4.0753 | 79.624 | .18768 | 2.6850 | 79.302 | .18052 |
| 20 | 8.4265 | 81.309 | .20244 | 5.5884 | 81.162 | .19556 | 4.1691 | 81.014 | .19061 | 2.7494 | 80.712 | .18349 |
| 30 | 8.6086 | 82.693 | .20529 | 5.7114 | 82.555 | .19843 | 4.2626 | 82.415 | .19350 | 2.8134 | 82.131 | .18642 |
| 70 | 58.686 | 88.541 | .24675 | 46.936 | 88.533 | .24307 | 23.435 | 88.489 | .23163 | 15.601 | 88.445 | .22492 |
| 80 | 59.799 | 89.980 | .24944 | 47.826 | 89.971 | .24576 | 23.881 | 89.930 | .23433 | 15.900 | 89.889 | .22762 |
| 90 | 60.911 | 91.431 | .25210 | 48.716 | 91.423 | .24843 | 24.328 | 91.384 | .23700 | 16.198 | 91.345 | .23029 |
| 100 | 62.023 | 92.895 | 0.25474 | 49.606 | 92.887 | 0.25107 | 24.774 | 92.850 | 0.23964 | 16.497 | 92.813 | 0.23293 |
| 110 | 63.134 | 94.371 | .25736 | 50.496 | 94.364 | .25368 | 25.220 | 94.329 | .24226 | 16.795 | 94.293 | .23556 |
| 120 | 64.246 | 95.860 | .25995 | 51.386 | 95.853 | .25628 | 25.666 | 95.819 | .24485 | 17.093 | 95.785 | .23815 |
| 130 | 65.357 | 97.361 | .26251 | 52.275 | 97.354 | .25884 | 26.112 | 97.322 | .24742 | 17.391 | 97.289 | .24073 |
| 140 | 66.468 | 98.873 | .26506 | 53.165 | 98.867 | .26139 | 26.558 | 98.836 | .24997 | 17.689 | 98.805 | .24327 |
| 150 | 67.579 | 100.397 | 0.26758 | 54.054 | 100.391 | 0.26391 | 27.003 | 100.361 | 0.25249 | 17.987 | 100.332 | 0.24580 |
| 160 | 68.690 | 101.932 | .27007 | 54.943 | 101.926 | .26640 | 27.449 | 101.898 | .25499 | 18.284 | 101.869 | .24830 |
| 170 | 69.801 | 103.478 | .27255 | 55.832 | 103.472 | .26888 | 27.894 | 103.445 | .25747 | 18.582 | 103.418 | .25078 |
| 180 | 70.912 | 105.035 | .27500 | 56.721 | 105.030 | .27133 | 28.340 | 105.003 | .25992 | 18.879 | 104.977 | .25324 |
| 190 | | | | | | | 28.785 | 106.572 | .26236 | 19.176 | 106.547 | .25567 |
| 200 | | | | | | | 29.230 | 108.151 | 0.26477 | 19.473 | 108.127 | 0.25808 |

(*Continued*)

**TABLE A.11 (CONTINUED)**

Dichlorodifluoromethane (Freon-12): Properties of Superheated Vapor

| Temp. °F t | Abs. Pressure 20 lb./in.² Gauge Pressure 5.3 lb./in.² (Sat. Temp. −8.1°F) | | | Abs. Pressure 26 lb./in.² Gauge Pressure 11.3 lb./in.² (Sat. Temp. 4.1°F) | | | Abs. Pressure 32 lb./in.² Gauge Pressure 17.3 lb./in.² (Sat. Temp. 14.4°F) | | | Abs. Pressure 40 lb./in.² Gauge Pressure 25.3 lb./in.² (Sat. Temp. 25.9°F) | | |
|---|---|---|---|---|---|---|---|---|---|---|---|---|
| | v | h | s | v | h | s | v | h | s | v | h | s |
| 40 | 8.7903 | 84.090 | .20812 | 5.8340 | 83.959 | .20127 | 4.3556 | 83.828 | .19635 | 2.8770 | 83.561 | .18931 |
| 50 | 8.9717 | 85.500 | .21091 | 5.9562 | 85.377 | 0.20408 | 4.4484 | 85.252 | 0.19918 | 2.9402 | 85.001 | 0.19216 |
| 60 | 9.1528 | 86.922 | .21367 | 6.0782 | 86.806 | .20685 | 4.5408 | 86.689 | .20197 | 3.0031 | 86.451 | .19498 |
| 70 | 9.3336 | 88.358 | .21641 | 6.1999 | 88.247 | .20960 | 4.6329 | 88.136 | .20473 | 3.0657 | 87.912 | .19776 |
| 80 | 9.5142 | 89.806 | .21912 | 6.3213 | 89.701 | .21232 | 4.7248 | 89.596 | .20746 | 3.1281 | 89.383 | .20051 |
| 90 | 9.6945 | 91.266 | .22180 | 6.4425 | 91.166 | .21501 | 4.8165 | 91.067 | .21016 | 3.1902 | 90.865 | .20324 |
| 100 | 9.8747 | 92.738 | .22445 | 6.5636 | 92.643 | 0.21767 | 4.9079 | 92.548 | 0.21283 | 3.2521 | 92.357 | 0.20593 |
| 110 | 10.055 | 94.222 | .22708 | 6.6844 | 94.132 | .22031 | 4.9992 | 94.042 | .21547 | 3.3139 | 93.860 | .20859 |
| 120 | 10.234 | 95.717 | .22968 | 6.8051 | 95.632 | .22292 | 5.0903 | 95.546 | .21809 | 3.3754 | 95.373 | .21122 |
| 130 | 10.414 | 97.224 | .23226 | 6.9256 | 97.143 | .22550 | 5.1812 | 97.061 | .22068 | 3.4368 | 96.896 | .21382 |
| 140 | 10.594 | 98.743 | .23481 | 7.0459 | 98.665 | .22806 | 5.2720 | 98.586 | .22325 | 3.4981 | 98.429 | .21640 |
| 150 | 10.773 | 100.272 | .23734 | 7.1662 | 100.198 | 0.23060 | 5.3627 | 100.123 | | 3.5592 | 99.972 | 0.21895 |
| 160 | 10.952 | 101.812 | .23985 | 7.2863 | 101.741 | .23311 | 5.4533 | 101.669 | .22830 | 3.6202 | 101.525 | .22148 |
| 170 | 11.131 | 103.363 | .24233 | 7.4063 | 103.295 | .23560 | 5.5437 | 103.226 | .23080 | 3.6811 | 103.088 | .22398 |
| 180 | 11.311 | 104.925 | .24479 | 7.5262 | 104.859 | .23806 | 5.6341 | 104.793 | .23326 | 3.7419 | 104.661 | .22646 |
| 190 | 11.489 | 106.497 | .24723 | 7.6461 | 106.434 | .24050 | 5.7243 | 106.370 | .23571 | 3.8025 | 106.243 | .22891 |
| 200 | 11.668 | 108.079 | .24964 | 7.7658 | 108.018 | 0.24292 | 5.8145 | 107.957 | 0.23813 | 3.8632 | 107.835 | 0.23135 |
| 210 | 11.847 | 109.670 | .25204 | 7.8855 | 109.612 | .24532 | 5.9046 | 109.553 | .24054 | 3.9237 | 109.436 | .23375 |
| 220 | 12.026 | 111.272 | .25441 | 8.0651 | 111.215 | .24770 | 5.9946 | 111.159 | .24291 | 3.9841 | 111.046 | .23614 |
| 230 | 12.205 | 112.883 | .25677 | 8.1246 | 112.828 | .25005 | 6.0846 | 112.774 | .24527 | 4.0445 | 112.665 | .23850 |
| 240 | | | | 8.2441 | 114.451 | .25239 | 6.1745 | 114.398 | .24761 | 4.1049 | 114.292 | .24085 |
| 250 | | | | | | | 6.2643 | 116.031 | 0.24993 | 4.1651 | 115.929 | 0.24317 |

| Temp. °F t | Abs. Pressure 20 lb./in.² Gauge Pressure 5.3 lb./in.² (Sat. Temp. −8.1°F) | | | Abs. Pressure 26 lb./in.² Gauge Pressure 11.3 lb./in.² (Sat. Temp. 4.1°F) | | | Abs. Pressure 32 lb./in.² Gauge Pressure 17.3 lb./in.² (Sat. Temp. 14.4°F) | | | Abs. Pressure 40 lb./in.² Gauge Pressure 25.3 lb./in.² (Sat. Temp. 25.9°F) | | |
|---|---|---|---|---|---|---|---|---|---|---|---|---|
| | v | h | s | v | h | s | v | h | s | v | h | s |
| Sat. | (1.8977) | (76.397) | (0.16969) | (1.4835) | (77.710) | (0.16850) | (1.2198) | (78.793) | (0.16763) | (0.98743) | (80.000) | (0.16676) |
| 0 | 1.9390 | 77.550 | 0.17222 | | | | | | | | | |
| 10 | 1.9893 | 78.973 | .17528 | 1.5071 | 78.566 | 0.17033 | | | | | | |
| 20 | 2.0391 | 80.403 | .17829 | 1.5468 | 80.024 | .17340 | 1.2387 | 79.634 | 0.16939 | | | |

*(Continued)*

**TABLE A.11 (CONTINUED)**

Dichlorodifluoromethane (Freon-12): Properties of Superheated Vapor

| | | | | | | | | | | | | |
|---|---|---|---|---|---|---|---|---|---|---|---|---|
| 30 | 2.0884 | 81.842 | .18126 | 1.5861 | 81.487 | .17642 | 1.2717 | 81.123 | .17246 | 0.99865 | 80.622 | 0.16804 |
| 40 | 2.1373 | 83.289 | .18419 | 1.6248 | 82.956 | .17939 | 1.3042 | 82.616 | .17548 | 1.0258 | 82.148 | .17112 |
| 50 | 2.1858 | 84.745 | 0.18707 | 1.6632 | 84.432 | 0.18232 | 1.3363 | 84.113 | 0.17845 | 1.0526 | 83.676 | 0.17415 |
| 60 | 2.2340 | 86.210 | .18992 | 1.7013 | 85.916 | .18520 | 1.3681 | 85.616 | .18137 | 1.0789 | 85.206 | .17712 |
| 70 | 2.2819 | 87.684 | .19273 | 1.7390 | 87.407 | .18804 | 1.3995 | 87.124 | .18424 | 1.1049 | 86.739 | .18005 |
| 80 | 2.3295 | 89.168 | .19550 | 1.7765 | 88.906 | .19084 | 1.4306 | 88.639 | .18707 | 1.1306 | 88.277 | .18292 |
| 90 | 2.3769 | 90.661 | .19824 | 1.8137 | 90.413 | .19361 | 1.4615 | 90.161 | .18987 | 1.1560 | 89.819 | .18575 |
| 100 | 2.4241 | 92.164 | 0.20095 | 1.8507 | 91.929 | 0.19634 | 1.4921 | 91.690 | 0.19263 | 1.1812 | 91.367 | 0.18854 |
| 110 | 2.4711 | 93.676 | .20363 | 1.8874 | 93.453 | .19904 | 1.5225 | 93.227 | .19535 | 1.2061 | 92.920 | .19129 |
| 120 | 2.5179 | 95.198 | .20628 | 1.9240 | 94.986 | .20171 | 1.5528 | 94.771 | .19803 | 1.2309 | 94.480 | .19401 |
| 130 | 2.5645 | 96.729 | .20890 | 1.9605 | 96.527 | .20435 | 1.5828 | 96.323 | .20069 | 1.2554 | 96.047 | .19669 |
| 140 | 2.6110 | 98.270 | .21149 | 1.9967 | 98.078 | .20695 | 1.6127 | 97.883 | .20331 | 1.2798 | 97.620 | .19933 |
| 150 | 2.6573 | 99.820 | 0.21405 | 2.0329 | 99.637 | 0.20953 | 1.6425 | 99.451 | 0.20590 | 1.3041 | 99.200 | 0.20195 |
| 160 | 2.7036 | 101.380 | .21659 | 2.0689 | 101.204 | .21208 | 1.6721 | 101.027 | .20847 | 1.3282 | 100.788 | .20453 |
| 170 | 2.7497 | 102.949 | .21910 | 2.1048 | 102.781 | .21461 | 1.7017 | 102.611 | .21100 | 1.3522 | 102.383 | .20708 |
| 180 | 2.7957 | 104.528 | .22159 | 2.1406 | 104.367 | .21710 | 1.7311 | 104.204 | .21351 | 1.3761 | 103.985 | .20961 |
| 190 | 2.8416 | 106.115 | .22405 | 2.1763 | 105.961 | .21958 | 1.7604 | 105.805 | .21600 | 1.3999 | 105.595 | .21210 |
| 200 | 2.8874 | 107.712 | 0.22649 | 2.2119 | 107.563 | 0.22202 | 1.7896 | 107.414 | 0.21845 | 1.4236 | 107.212 | 0.21457 |
| 210 | 2.9332 | 109.317 | .22891 | 2.2474 | 109.174 | .22445 | 1.8187 | 109.031 | .22089 | 1.4472 | 108.837 | 0.21702 |
| 220 | 2.9789 | 110.932 | .23130 | 2.2828 | 110.794 | .22685 | 1.8478 | 110.656 | .22330 | 1.4707 | 110.469 | .21944 |
| 230 | 3.0245 | 112.555 | .23367 | 2.3182 | 112.422 | .22923 | 1.8768 | 112.289 | .22568 | 1.4942 | 112.109 | .22183 |
| 240 | 3.0700 | 114.186 | .23602 | 2.3535 | 114.058 | .23158 | 1.9057 | 113.930 | .22804 | 1.5176 | 113.757 | .22420 |
| 250 | 3.1155 | 115.826 | 0.23835 | 2.3888 | 115.703 | 0.23391 | 1.9346 | 115.579 | 0.23038 | 1.5409 | 115.412 | 0.22655 |
| 260 | 3.1609 | 117.475 | .24065 | 2.4240 | 117.355 | .23623 | 1.9634 | 117.235 | .23270 | 1.5642 | 117.074 | .22888 |
| 270 | 3.2063 | 119.131 | .24294 | 2.4592 | 119.016 | .23852 | 1.9922 | 118.900 | .23500 | 1.5874 | 118.744 | .23118 |
| 280 | 3.2517 | 120.796 | .24520 | 2.4943 | 120.684 | .24079 | 2.0209 | 120.572 | .23727 | 1.6106 | 120.421 | .23347 |
| 290 | 3.2970 | 122.469 | .24745 | 2.5293 | 122.360 | .24304 | 2.0495 | 122.251 | .23953 | 1.6337 | 122.105 | .23573 |

(*Continued*)

**TABLE A.11 (CONTINUED)**

Dichlorodifluoromethane (Freon-12): Properties of Superheated Vapor

| Temp. °F t | Abs. Pressure 50 lb./in.² Gauge, Pressure 35.3 lb./in.² (Sat. Temp. 38.2°F) | | | Abs. Pressure 60 lb./in.² Gauge, Pressure 45.3 lb./in.² (Sat. Temp. 48.6°F) | | | Abs. Pressure 80 lb./in.² Gauge, Pressure 65.3 lb./in.² (Sat. Temp. 66.2°F) | | | Abs. Pressure 100 lb./in.² Gauge, Pressure 85.3 lb./in.² (Sat. Temp. 80.8°F) | | |
|---|---|---|---|---|---|---|---|---|---|---|---|---|
| | $v$ | $h$ | $s$ | $v$ | $h$ | $s$ | $v$ | $h$ | $s$ | $v$ | $h$ | $s$ |
| *Sat.* | (0.79824) | (81.249) | (0.16597) | (0.67005) | (82.299) | (0.16537) | (0.50680) | (84.003) | (0.16450) | (0.40674) | (85.351) | (0.16389) |
| 40 | 0.80248 | 81.540 | 0.16655 | | | | | | | | | |
| 50 | 0.82502 | 83.109 | 0.16966 | 0.67272 | 82.518 | 0.16580 | | | | | | |
| 60 | .84713 | 84.676 | .17271 | .69210 | 84.126 | .16892 | | | | | | |
| 70 | .86886 | 86.243 | .17569 | .71105 | 85.729 | .17198 | 0.51269 | 84.640 | 0.16571 | | | |
| 80 | .89025 | 87.811 | .17862 | .72964 | 87.330 | .17497 | .52795 | 86.316 | .16885 | | | |
| 90 | .91134 | 89.380 | .18151 | .74790 | 88.929 | .17791 | .54281 | 87.981 | .17190 | 0.41876 | 86.964 | 0.16685 |
| 100 | .93216 | 90.953 | .18434 | .76588 | 90.528 | .18079 | .55734 | 89.640 | .17489 | 0.43138 | 88.694 | 0.16996 |
| 110 | .95275 | 92.529 | .18713 | .78360 | 92.128 | .18362 | .57158 | 91.294 | .17782 | .44365 | 90.410 | .17300 |
| 120 | .97313 | 94.110 | .18988 | .80110 | 93.731 | .18641 | .58556 | 92.945 | .18070 | .45562 | 92.116 | .17597 |
| 130 | .99332 | 95.695 | .19259 | .81840 | 95.336 | .18916 | .59931 | 94.594 | .18352 | .46733 | 93.814 | .17888 |
| 140 | 1.0133 | 97.286 | .19527 | .83551 | 96.945 | .19186 | .61286 | 96.242 | .18629 | .47881 | 95.507 | .18172 |
| 150 | 1.0332 | 98.882 | .19791 | .85247 | 98.558 | .19453 | .62623 | 97.891 | .18902 | .49009 | 97.197 | .18452 |
| 160 | 1.0529 | 100.485 | .20051 | .86928 | 100.176 | .19716 | .63943 | 99.542 | .19170 | .50118 | 98.884 | .18726 |
| 170 | 1.0725 | 102.093 | .20309 | .88596 | 101.799 | .19976 | .65250 | 101.195 | .19435 | .51212 | 100.571 | .18996 |
| 180 | 1.0920 | 103.708 | .20563 | .90252 | 103.427 | .20233 | .66543 | 102.851 | .19696 | .52291 | 102.257 | .19262 |
| 190 | 1.1114 | 105.330 | .20815 | .91896 | 105.060 | .20486 | .67824 | 104.511 | .19953 | .53358 | 103.944 | .19524 |
| 200 | 1.1307 | 106.958 | .21064 | .93531 | 106.700 | .20736 | .69095 | 106.174 | .20207 | .54413 | 105.633 | .19782 |
| 210 | 1.1499 | 108.593 | .21310 | .95157 | 108.345 | .20984 | .70356 | 107.841 | .20458 | .55457 | 107.324 | .20036 |
| 220 | 1.1690 | 110.235 | .21553 | .96775 | 109.997 | .21229 | .71609 | 109.513 | .20706 | .56492 | 109.018 | .20287 |
| 230 | 1.1880 | 111.883 | .21794 | .98385 | 111.655 | .21471 | .72853 | 111.190 | .20951 | .57519 | 110.714 | .20535 |
| 240 | 1.2070 | 113.539 | .22032 | .99988 | 113.319 | .21710 | .74090 | 112.872 | .21193 | .58538 | 112.415 | .20780 |
| 250 | 1.2259 | 115.202 | 0.22268 | 1.0159 | 114.989 | 0.21947 | 0.75320 | 114.559 | 0.21432 | 0.59549 | 114.119 | 0.21022 |
| 260 | 1.2447 | 116.871 | .22502 | 1.0318 | 116.666 | .22182 | .76544 | 116.251 | .21669 | .60554 | 115.828 | .21261 |
| 270 | 1.2636 | 118.547 | .22733 | 1.0476 | 118.350 | .22414 | .77762 | 117.949 | .21903 | .61553 | 117.540 | .21497 |

*(Continued)*

# TABLE A.11 (CONTINUED)

Dichlorodifluoromethane (Freon-12): Properties of Superheated Vapor

| Temp. °F t | Abs. Pressure 120 lb./in.² Gauge Pressure 105.3 lb./in.² (Sat. Temp. 93.3°F) | | | Abs. Pressure 140 lb./in.² Gauge Pressure 125.3 lb./in.² (Sat. Temp. 104.4°F) | | | Abs. Pressure 180 lb./in.² Gauge Pressure 165.3 lb./in.² (Sat. Temp. 123.4°F) | | | Abs. Pressure 220 lb./in.² Gauge Pressure 205.3 lb./in.² (Sat. Temp. 139.5°F) | | |
|---|---|---|---|---|---|---|---|---|---|---|---|---|
| | v | h | s | v | h | s | v | h | s | v | h | s |
| Sat. | (033886) | (86.459) | (0.16340) | (0.28964) | (87.389) | (0.16299) | (0.22276) | (88.857) | (0.16228) | (0.17917) | (89.937) | (0.16161) |
| 100 | 0.34655 | 87.675 | 0.16559 | | | | | | | | | |
| 110 | .35766 | 89.466 | .16876 | .29548 | 88.448 | 0.16486 | | | | | | |
| 120 | .36841 | 91.237 | .17184 | .30549 | 90.297 | .16808 | | | | | | |
| 130 | .37884 | 92.992 | .17484 | .31513 | 92.120 | .17120 | 0.22863 | 90.179 | 0.16454 | | | |
| 140 | .38901 | 94.736 | .17778 | .32445 | 93.923 | .17423 | .23710 | 92.136 | .16783 | 0.17957 | 90.043 | 0.16179 |
| 150 | 0.39896 | 96.471 | 0.18065 | 0.33350 | 95.709 | 0.17718 | .24519 | 94.053 | 0.17100 | 0.18746 | 92.156 | 0.16528 |
| 160 | .40870 | 98.199 | .18346 | .34232 | 97.483 | .18007 | .25297 | 95.940 | .17407 | .19487 | 94.203 | .16861 |
| 170 | .41826 | 99.922 | .18622 | .35095 | 99.247 | .18289 | .26047 | 97.803 | .17705 | .20190 | 96.199 | .17181 |
| 180 | .42766 | 101.642 | .18892 | .35939 | 101.003 | .18566 | .26775 | 99.647 | .17995 | .20861 | 98.157 | .17489 |
| 190 | .43692 | 103.359 | .19159 | .36769 | 102.754 | .18838 | .27484 | 101.475 | .18279 | .21506 | 100.084 | .17788 |
| 200 | 0.44606 | 105.076 | 0.19421 | 0.37584 | 104.501 | 0.19104 | .28176 | 103.291 | 0.18556 | 0.22130 | 101.986 | 0.18079 |
| 210 | .45508 | 106.792 | .19679 | .38387 | 106.245 | .19367 | .28852 | 105.098 | .18828 | .22735 | 103.869 | .18362 |
| 220 | .46401 | 108.509 | .19934 | .39179 | 107.987 | .19625 | .29516 | 106.896 | .19095 | .23324 | 105.735 | .18638 |
| 230 | .47284 | 110.227 | .20185 | .39961 | 109.728 | .19879 | .30168 | 108.689 | .19357 | .23900 | 107.589 | .18909 |
| 240 | .48158 | 111.948 | .20432 | .40734 | 111.470 | .20130 | .30810 | 110.478 | .19614 | .24463 | 109.432 | .19175 |
| 280 | 1.2823 | 120.231 | .22962 | 1.0634 | 120.039 | .22644 | .78975 | 119.652 | .22135 | .62546 | 119.258 | .21731 |
| 290 | 1.3010 | 121.921 | .23189 | 1.0792 | 121.736 | .22872 | .80183 | 121.361 | .22364 | .63534 | 120.980 | .21962 |
| 300 | 1.3197 | 123.618 | 0.23414 | 1.0949 | 123.438 | 0.23098 | 0.81386 | 123.075 | 0.22592 | 0.64518 | 122.707 | 0.22191 |
| 310 | 1.3383 | 125.321 | .23637 | 1.1106 | 125.147 | .23321 | .82586 | 124.795 | .22817 | .65497 | 124.439 | .22417 |
| 320 | 1.3569 | 127.032 | .23857 | 1.1262 | 126.863 | .23543 | .83781 | 126.521 | .23039 | .66472 | 126.176 | .22641 |
| 330 | 1.3754 | 128.749 | .24076 | 1.1418 | 128.585 | .23762 | .84973 | 128.253 | .23260 | .67444 | 127.917 | .22863 |
| 340 | | | | 1.1574 | 130.313 | .23980 | .86161 | 129.990 | .23479 | .68411 | 129.665 | .23083 |

(Continued)

**TABLE A.11 (CONTINUED)**

Dichlorodifluoromethane (Freon-12): Properties of Superheated Vapor

| | 0.49025 | 113.670 | 0.20677 | 0.41499 | 113.212 | 0.20377 | 0.31442 | 112.263 | 0.19868 | 0.25015 | 111.267 | 0.19435 |
|---|---|---|---|---|---|---|---|---|---|---|---|---|
| 250 | 0.49025 | 113.670 | 0.20677 | 0.41499 | 113.212 | 0.20377 | 0.31442 | 112.263 | 0.19868 | 0.25015 | 111.267 | 0.19435 |
| 260 | .49885 | 115.396 | .20918 | .42257 | 114.956 | .20621 | .32066 | 114.046 | .20117 | .25557 | 113.095 | .19691 |
| 270 | .50739 | 117.125 | .21157 | .43008 | 116.701 | .20862 | .32682 | 115.828 | .20363 | .26091 | 114.919 | .19942 |
| 280 | .51587 | 118.857 | .21393 | .43753 | 118.449 | .21100 | .33292 | 117.610 | .20605 | .26617 | 116.738 | .20190 |
| 290 | .52429 | 120.593 | .21626 | .44492 | 120.199 | .21335 | .33895 | 119.392 | .20845 | .27136 | 118.555 | .20434 |
| 300 | 0.53267 | 122.333 | 0.21856 | 0.45226 | 121.953 | 0.21567 | 0.34492 | 121.174 | 0.21081 | 0.27648 | 120.369 | 0.20674 |
| 310 | .54100 | 124.077 | .22084 | .45955 | 123.709 | .21797 | .35084 | 122.958 | .21314 | .28155 | 122.183 | .20912 |
| 320 | .54929 | 125.825 | .22310 | .46680 | 125.470 | .22024 | .35672 | 124.744 | .21545 | .28657 | 123.996 | .21146 |
| 330 | .55754 | 127.578 | .22533 | .47400 | 127.233 | .22249 | .36255 | 126.531 | .21772 | .29153 | 125.809 | .21377 |
| 340 | .56575 | 129.335 | .22754 | .48117 | 129.001 | .22471 | .36834 | 128.321 | .21998 | .29645 | 127.623 | .21605 |
| 350 | 0.57393 | 131.097 | 0.22973 | 0.48831 | 130.773 | 0.22692 | 0.37409 | 130.113 | 0.22220 | 0.30134 | 129.438 | 0.21830 |
| 360 | .58208 | 132.863 | .23190 | .49541 | 132.548 | .22910 | .37980 | 131.909 | .22441 | .30618 | 131.255 | .22053 |
| 370 | .59019 | 134.634 | .23405 | .50248 | 134.328 | .23125 | .38549 | 133.707 | .22659 | .31099 | 133.073 | .22274 |
| 380 | .59829 | 136.410 | .23618 | .50953 | 136.112 | .23339 | .39114 | 135.509 | .22875 | .31576 | 134.893 | .22492 |
| 390 | .60635 | 138.191 | .23829 | .51654 | 137.901 | .23551 | .39677 | 137.314 | .23088 | .32051 | 136.715 | .22708 |
| 400 | | | | | | | .40237 | 139.122 | 0.23300 | 0.32523 | 138.540 | 0.22921 |
| 410 | | | | | | | .40794 | 140.934 | .23509 | .32992 | 140.368 | .23133 |

*Source:* R. C. Jordan and G. B. Priester, *Refrigeration and Air Conditioning*, 2nd ed., Prentice Hall, Inc., Upper Saddle River, N.J. 1956. Courtesy of E. I. Dupont de Nemours and Co.

**TABLE A.12**

Pressure–Enthalpy Diagram for Freon-12

**TABLE A.13**

Gas Constant Values

| Gas | Chemical Formula | Molecular Weight | English Units $R$ $\frac{ft.lb.}{lb.R}$ | $c_p$ $\frac{Btu}{lb.YR}$ at 77YF | $c_p$ $\frac{Btu}{lb.YR}$ at77YF | $k,$ $\frac{c_p}{c_v}$ | SI Units $R$ $\frac{kJ}{kg \cdot °K}$ | $c_p$ $\frac{kJ}{kg \cdot °K}$ | $c_v$ $\frac{kJ}{kg \cdot °K}$ |
|---|---|---|---|---|---|---|---|---|---|
| Acetylene | $C_2H_2$ | 26.02 | 59.39 | 0.361 | 0.285 | 1.27 | 0.31955 | 1.5116 | 1.1933 |
| Air | | 28.97 | 53.34 | 0.240 | 0.171 | 1.40 | 0.28700 | 1.0052 | 0.7180 |
| Ammonia | $nh_3$ | 17.024 | 90.7 | 0.52 | 0.404 | 1.29 | 0.48802 | 2.1773 | 1.6916 |
| Argon | Ar | 39.94 | 38.68 | 0.124 | 0.075 | 1.667 | 0.20813 | 0.5207 | 0.3124 |
| Butane | $C_4H_{10}$ | 58.12 | 26.58 | 0.415 | 0.381 | 1.09 | 0.14304 | 1.7164 | 1.5734 |
| Carbon dioxide | $CO_2$ | 44.01 | 35.10 | 0.202 | 0.157 | 1.29 | 0.18892 | 0.8464 | 0.6573 |
| Carbon monoxide | CO | 28.01 | 55.16 | 0.249 | 0.178 | 1.40 | 0.29683 | 1.0411 | 0.7441 |
| Ethane | $C_2H_6$ | 30.07 | 51.38 | 0.427 | 0.361 | 1.18 | 0.27650 | 1.7662 | 1.4897 |
| Ethylene | $c_2h_4$ | 28.052 | 55.07 | 0.411 | 0.340 | 1.21 | 0.29637 | 1.5482 | 1.2518 |
| Helium | He | 4.003 | 386.0 | 1.25 | 0.753 | 1.667 | 2.07703 | 5.1926 | 3.1156 |
| Hydrogen | $h_2$ | 2.016 | 766.4 | 3.420 | 2.434 | 1.405 | 4.12418 | 14.3193 | 10.1919 |
| Methane | $ch_4$ | 16.04 | 96.35 | 0.532 | 0.403 | 1.32 | 0.51835 | 2.2537 | 1.7354 |
| Nitrogen | $n_2$ | 28.016 | 55.15 | 0.248 | 0.177 | 1.40 | 0.29680 | 1.0404 | 0.7434 |
| Octane | $C_8H_{18}$ | 114.14 | 13.53 | 0.409 | 0.392 | 1.044 | 0.07279 | 1.7113 | 1.6385 |
| Oxygen | $O_2$ | 32.000 | 48.28 | 0.219 | 0.157 | 1.39 | 0.25983 | 0.9190 | 0.6590 |
| Propane | $c_3h_8$ | 44.094 | 35.04 | 0.407 | 0.362 | 1.124 | 0.18855 | 1.6794 | 1.4909 |
| Water vapor | $h_2o$ | 18.016 | 85.76 | 0.445 | 0.335 | 1.33 | 0.46152 | 1.8649 | 1.4031 |

**TABLE A.14**

Critical Constants

| Substance | Formula | Molecular Weight | Temperature K | Temperature R | Pressure atm | Pressure lb$_f$/in.$^2$ | Volume ft.$^3$/lb. mole |
|---|---|---|---|---|---|---|---|
| Ammonia | $NH_3$ | 17.03 | 405.5 | 729.8 | 111.3 | 1636 | 1.16 |
| Argon | A | 39.944 | 151 | 272 | 48.0 | 705 | 1.20 |
| Bromine | $Br_2$ | 159.832 | 584 | 1052 | 102 | 1500 | 2.17 |
| Carbon dioxide | $CO_2$ | 44.01 | 304.2 | 547.5 | 72.9 | 1071 | 1.51 |
| Carbon monoxide | CO | 28.01 | 133 | 240 | 34.5 | 507 | 1.49 |
| Chlorine | $Cl_2$ | 70.914 | 417 | 751 | 76.1 | 1120 | 1.99 |
| Deuterium (normal) | D2 | 4.00 | 38.4 | 69.1 | 16.4 | 241 | |
| Helium | He | 4.003 | 5.3 | 9.5 | 2.26 | 33.2 | 0.926 |
| Helium$^3$ | He | 3.00 | 3.34 | 6.01 | 1.15 | 16.9 | |
| Hydrogen (normal) | $H_2$ | 2.016 | 33.3 | 59.9 | 12.8 | 188.1 | 1.04 |
| Krypton | Kr | 83.7 | 209.4 | 376.9 | 54.3 | 798 | 1.48 |
| Neon | Ne | 20.183 | 44.5 | 80.1 | 26.9 | 395 | 0.668 |
| Nitrogen | $N_2$ | 28.016 | 126.2 | 227.1 | 33.5 | 492 | 1.44 |
| Nitrous oxide | $N_2O$ | 44.02 | 309.7 | 557.4 | 71.7 | 1054 | 1.54 |
| Oxygen | $O_2$ | 32.00 | 154.8 | 278.6 | 50.1 | 736 | 1.25 |
| Sulfur dioxide | $SO_2$ | 64.06 | 430.7 | 775.2 | 77.8 | 1143 | 1.95 |
| Water | $H_2O$ | 18.016 | 647.4 | 1165.3 | 218.3 | 3208 | 0.90 |
| Xenon | Xe | 131.3 | 289.75 | 521.55 | 58.0 | 852 | 1.90 |
| Benzene | $C_6H_6$ | 78.11 | 562 | 1012 | 48.6 | 714 | 4.17 |
| *n*–Butane | $C_4H_{10}$ | 58.120 | 425.2 | 765.2 | 37.5 | 551 | 4.08 |
| Carbon tetrachloride | $CCl_4$ | 153.84 | 556.4 | 1001.5 | 45.0 | 661 | 4.42 |
| Chloroform | $CHCl_3$ | 119.39 | 536.6 | 965.8 | 54.0 | 794 | 3.85 |
| Dichlorodifluoromethane | $CCl_2F_2$ | 120.92 | 384.7 | 692.4 | 39.6 | 582 | 3.49 |
| Dichlorofluoromethane | $CHCl_2F$ | 102.93 | 451.7 | 813.0 | 51.0 | 749 | 3.16 |
| Ethane | $C_2H_5$ | 30.068 | 305.5 | 549.8 | 48.2 | 708 | 2.37 |

*(Continued)*

**TABLE A.14 (CONTINUED)**

Critical Constants

| Substance | Formula | Molecular Weight | Temperature | | Pressure | | Volume ft.³/lb. mole |
|---|---|---|---|---|---|---|---|
| | | | K | R | atm | lb$_f$/in.² | |
| Ethyl alcohol | C$_2$H$_5$OH | 46.07 | 516.0 | 929.0 | 63.0 | 926 | 2.68 |
| Ethylene | C$_2$H$_4$ | 28.052 | 282.4 | 508.3 | 50.5 | 742 | 1.99 |
| n–Hexane | C$_6$H$_{14}$ | 86.172 | 507.9 | 914.2 | 29.9 | 439 | 5.89 |
| Methane | CH$_4$ | 16.042 | 191.1 | 343.9 | 45.8 | 673 | 1.59 |
| Methyl alcohol | CH$_3$OH | 32.04 | 513.2 | 923.7 | 78.5 | 1154 | 1.89 |
| Methyl chloride | CH$_3$Cl | 50.49 | 416.3 | 749.3 | 65.9 | 968 | 2.29 |
| Propane | C$_3$H$_8$ | 44.094 | 370.0 | 665.9 | 42.0 | 617 | 3.20 |
| Propene | C$_3$H$_6$ | 42.078 | 365.0 | 656.9 | 45.6 | 670 | 2.90 |
| Propyne | C$_3$H$_4$ | 40.062 | 401 | 722 | 52.8 | 776 | |
| Trichlorofluoromethane | CCl$_3$F | 137.38 | 471.2 | 848.1 | 43.2 | 635 | 3.97 |

*Source:* G. J. Van Wylen and R. E. Sonntag, *Fundamentals of Classical Thermodynamics*, John Wiley & Sons, Inc., New York, 1965, p. 579, with permission.

**TABLE A.15**

HFC–134a Saturation Properties—Temperature Table

| Temp. °F | Pressure psia | Volume ft.³/lb. | | Density lb./ft.³ | | Enthalpy Btu/lb. | | | Entropy Btu/(lb.) (R) | | Temp. °F |
|---|---|---|---|---|---|---|---|---|---|---|---|
| | | Liquid $v_f$ | Vapor $v_g$ | Liquid $1/v_f$ | Vapor $1/v_g$ | Liquid $h_f$ | Latent $h_{fg}$ | Vapor $h_g$ | Liquid $s_f$ | Vapor $s_g$ | |
| −40 | 7.417 | 0.0113 | 5.7904 | 88.31 | 0.1727 | 0.0 | 97.2 | 97.2 | 0.0000 | 0.2316 | −40 |
| −30 | 9.851 | 0.0115 | 4.4366 | 87.31 | 0.2254 | 3.0 | 95.7 | 98.7 | 0.0070 | 0.2297 | −30 |
| −20 | 12.885 | 0.0116 | 3.4471 | 86.30 | 0.2901 | 6.0 | 94.2 | 100.2 | 0.0139 | 0.2281 | −20 |
| −15 | 14.659 | 0.0117 | 3.0525 | 85.79 | 0.3276 | 7.5 | 93.4 | 100.9 | 0.0174 | 0.2274 | −15 |
| −10 | 16.620 | 0.0117 | 2.7715 | 85.28 | 0.3688 | 9.0 | 92.6 | 101.7 | 0.0208 | 0.2267 | −10 |
| −5 | 18.784 | 0.0118 | 2.4155 | 84.76 | 0.4140 | 10.6 | 91.8 | 102.4 | 0.0241 | 0.2261 | −5 |
| 0 | 21.163 | 0.0119 | 2.1580 | 84.23 | 0.4634 | 12.1 | 91.0 | 103.1 | 0.0275 | 0.2255 | 0 |
| 5 | 23.772 | 0.0119 | 1.9327 | 83.70 | 0.5174 | 13.7 | 90.2 | 103.9 | 0.0308 | 0.2250 | 5 |
| 10 | 26.625 | 0.0120 | 1.7355 | 83.17 | 0.5762 | 15.2 | 89.4 | 104.6 | 0.0342 | 0.2244 | 10 |
| 15 | 29.739 | 0.0121 | 1.5620 | 82.63 | 0.6402 | 16.8 | 88.5 | 105.3 | 0.0375 | 0.2240 | 15 |
| 20 | 33.129 | 0.0122 | 1.4090 | 82.08 | 0.7097 | 18.4 | 87.7 | 106.0 | 0.0408 | 0.2235 | 20 |
| 25 | 36.810 | 0.0123 | 1.2737 | 81.52 | 0.7851 | 20.0 | 86.8 | 106.7 | 0.0440 | 0.2231 | 25 |
| 30 | 40.800 | 0.0124 | 1.1538 | 80.96 | 0.8667 | 21.6 | 85.9 | 107.4 | 0.0473 | 0.2227 | 30 |
| 35 | 45.115 | 0.0124 | 1.0472 | 80.40 | 0.9549 | 23.2 | 85.0 | 108.1 | 0.0505 | 0.2223 | 35 |
| 40 | 49.771 | 0.0125 | 0.9523 | 79.82 | 1.0501 | 24.8 | 84.1 | 108.8 | 0.0538 | 0.2220 | 40 |
| 45 | 54.787 | 0.0126 | 0.8675 | 79.24 | 1.1527 | 26.4 | 83.1 | 109.5 | 0.0570 | 0.2217 | 45 |
| 50 | 60.180 | 0.0127 | 0.7916 | 78.64 | 1.2633 | 28.0 | 82.1 | 110.2 | 0.0602 | 0.2214 | 50 |
| 55 | 65.963 | 0.0128 | 0.7234 | 78.04 | 1.3823 | 29.7 | 81.2 | 110.9 | 0.0634 | 0.2211 | 55 |
| 60 | 72.167 | 0.0129 | 0.6622 | 77.43 | 1.5102 | 31.4 | 80.2 | 111.5 | 0.0666 | 0.2208 | 60 |
| 65 | 78.803 | 0.0130 | 0.6069 | 76.81 | 1.6477 | 33.0 | 79.1 | 112.2 | 0.0698 | 0.2206 | 65 |
| 70 | 85.890 | 0.0131 | 0.5570 | 76.18 | 1.7952 | 34.7 | 78.1 | 112.8 | 0.0729 | 0.2203 | 70 |
| 75 | 93.447 | 0.0132 | 0.5119 | 75.54 | 1.9536 | 36.4 | 77.0 | 113.4 | 0.0761 | 0.2201 | 75 |
| 80 | 101.494 | 0.0134 | 0.4709 | 74.89 | 2.1234 | 38.1 | 75.9 | 114.0 | 0.0792 | 0.2199 | 80 |
| 85 | 110.050 | 0.0135 | 0.4337 | 74.22 | 2.3056 | 39.9 | 74.8 | 114.6 | 0.0824 | 0.2196 | 85 |
| 90 | 119.138 | 0.0136 | 0.3999 | 73.54 | 2.5009 | 41.6 | 73.6 | 115.2 | 0.0855 | 0.2194 | 90 |
| 95 | 128.782 | 0.0137 | 0.3690 | 72.84 | 2.7102 | 43.4 | 72.4 | 115.8 | 0.0886 | 0.2192 | 95 |

*(Continued)*

**TABLE A.15 (CONTINUED)**

HFC-134a Saturation Properties—Temperature Table

| Temp. °F | Pressure psia | Volume ft.³/lb. | | Density lb./ft.³ | | Enthalpy Btu/lb. | | | Entropy Btu/(lb.) (R) | | Temp. °F |
|---|---|---|---|---|---|---|---|---|---|---|---|
| | | Liquid $v_f$ | Vapor $v_g$ | Liquid $1/v_f$ | Vapor $1/v_g$ | Liquid $h_f$ | Latent $h_{fg}$ | Vapor $h_g$ | Liquid $s_f$ | Vapor $s_g$ | |
| 100 | 138.996 | 0.0139 | 0.3408 | 72.13 | 2.9347 | 45.1 | 71.2 | 116.3 | 0.0918 | 0.2190 | 100 |
| 105 | 149.804 | 0.0140 | 0.3149 | 71.40 | 3.1754 | 46.9 | 69.9 | 116.9 | 0.0949 | 0.2188 | 105 |
| 110 | 161.227 | 0.0142 | 0.2912 | 70.66 | 3.4337 | 48.7 | 68.6 | 117.4 | 0.0981 | 0.2185 | 110 |
| 115 | 173.298 | 0.0143 | 0.2695 | 69.89 | 3.7110 | 50.5 | 67.3 | 117.9 | 0.1012 | 0.2183 | 115 |
| 120 | 186.023 | 0.0145 | 0.2494 | 69.10 | 4.0089 | 52.4 | 65.9 | 118.3 | 0.1043 | 0.2181 | 120 |
| 130 | 213.572 | 0.0148 | 0.2139 | 67.46 | 4.6745 | 56.2 | 63.0 | 119.2 | 0.1106 | 0.2175 | 130 |
| 140 | 244.068 | 0.0152 | 0.1835 | 65.70 | 5.4491 | 60.0 | 59.9 | 119.9 | 0.1170 | 0.2168 | 140 |
| 150 | 277.721 | 0.0157 | 0.1573 | 63.81 | 6.3584 | 64.0 | 56.5 | 120.5 | 0.1234 | 0.2160 | 150 |
| 160 | 314.800 | 0.0162 | 0.1344 | 61.74 | 7.4390 | 68.1 | 52.7 | 120.8 | 0.1299 | 0.2150 | 160 |
| 170 | 355.547 | 0.0168 | 0.1143 | 59.45 | 8.7470 | 72.4 | 48.5 | 120.9 | 0.1366 | 0.2137 | 170 |
| 180 | 400.280 | 0.0176 | 0.0964 | 56.83 | 10.3750 | 76.9 | 43.7 | 120.7 | 0.1435 | 0.2119 | 180 |
| 190 | 449.384 | 0.0186 | 0.0800 | 53.73 | 12.4962 | 81.8 | 38.0 | 119.8 | 0.1508 | 0.2093 | 190 |
| 200 | 503.361 | 0.0201 | 0.0645 | 49.73 | 15.5155 | 87.3 | 30.7 | 118.0 | 0.1589 | 0.2054 | 200 |
| 210 | 563.037 | 0.0232 | 0.0474 | 43.19 | 21.1071 | 94.4 | 18.9 | 113.4 | 0.1693 | 0.1976 | 210 |
| 213 | 582.316 | 0.0259 | 0.0394 | 38.65 | 25.3583 | 98.4 | 11.1 | 109.5 | 0.1750 | 0.1915 | 213 |

*Source:* *Thermodynamic Properties of HFC–1Ma.* With permission of E. I. Du Pont de Nemours and Co.

**TABLE A.16**

HFC–134a Superheated Vapor–Constant Pressure Tables

**Pressure = 1.00 psia**

| Temp °F | | $v$ | $h$ | $s$ | $C_p$ | $C_p/C_v$ | $v_s$ |
|---|---|---|---|---|---|---|---|
| −97.6 | SAT LIQ | 0.01065 | −16.6 | −0.0426 | 0.2813 | 1.5295 | 3163.1 |
| −97.6 | SAT VAP | 37.87879 | 88.5 | 0.2478 | 0.1570 | 1.1488 | 447.2 |
| −90 | | 38.61004 | 89.7 | 0.2511 | 0.1588 | 1.1462 | 451.6 |
| −80 | | 38.68254 | 91.3 | 0.2554 | 0.1612 | 1.1429 | 457.3 |
| −70 | | 40.81633 | 92.9 | 0.2596 | 0.1637 | 1.1398 | 463.0 |
| −60 | | 41.84100 | 94.6 | 0.2638 | 0.1662 | 1.1370 | 468.5 |
| −50 | | 42.91845 | 96.3 | 0.2679 | 0.1686 | 1.1342 | 474.0 |
| −40 | | 44.05286 | 98.0 | 0.2720 | 0.1711 | 1.1317 | 479.3 |
| −30 | | 45.04505 | 99.7 | 0.2761 | 0.1736 | 1.1292 | 484.7 |
| −20 | | 46.08295 | 101.4 | 0.2801 | 0.1760 | 1.1269 | 489.9 |
| −10 | | 47.16981 | 103.2 | 0.2841 | 0.1785 | 1.1247 | 495.1 |
| 0 | | 48.30918 | 105.0 | 0.2880 | 0.1809 | 1.1226 | 500.2 |
| 10 | | 49.26108 | 106.8 | 0.2919 | 0.1834 | 1.1206 | 505.3 |
| 20 | | 50.25126 | 108.7 | 0.2958 | 0.1858 | 1.1187 | 510.3 |
| 30 | | 51.28205 | 110.5 | 0.2997 | 0.1882 | 1.1169 | 515.2 |
| 40 | | 52.35602 | 112.4 | 0.3035 | 0.1906 | 1.1151 | 520.1 |
| 50 | | 53.47594 | 114.4 | 0.3073 | 0.1930 | 1.1134 | 525.0 |
| 60 | | 54.64481 | 116.3 | 0.3111 | 0.1953 | 1.1118 | 529.8 |
| 70 | | 55.55556 | 118.3 | 0.3148 | 0.1977 | 1.1103 | 534.5 |
| 80 | | 56.81818 | 120.2 | 0.3186 | 0.2000 | 1.1088 | 539.3 |
| 90 | | 57.80347 | 122.3 | 0.3223 | 0.2023 | 1.1073 | 543.9 |
| 100 | | 58.82353 | 124.3 | 0.3259 | 0.2046 | 1.1059 | 548.5 |
| 110 | | 59.88024 | 126.4 | 0.3296 | 0.2069 | 1.1046 | 553.1 |

**Pressure = 5.00 psia**

| Temp °F | | $v$ | $h$ | $s$ | $C_p$ | $C_p/C_v$ | $v_s$ |
|---|---|---|---|---|---|---|---|
| −53 | SAT LIQ | 0.01116 | −3.8 | −0.0093 | 0.2929 | 1.5071 | 2768.7 |
| −53 | SAT VAP | 8.37521 | 95.2 | 0.2343 | 0.1726 | 1.1471 | 466.8 |
| −50 | | 8.44595 | 95.7 | 0.2356 | 0.1732 | 1.1458 | 468.6 |
| −40 | | 8.66551 | 97.5 | 0.2398 | 0.1752 | 1.1418 | 474.4 |
| −30 | | 8.88889 | 99.2 | 0.2439 | 0.1772 | 1.1382 | 480.0 |
| −20 | | 9.10747 | 101.0 | 0.2480 | 0.1793 | 1.1349 | 485.6 |
| −10 | | 9.32836 | 102.8 | 0.2521 | 0.1814 | 1.1318 | 491.1 |
| 0 | | 9.55110 | 104.7 | 0.2561 | 0.1835 | 1.1289 | 496.5 |
| 10 | | 9.76563 | 106.5 | 0.2601 | 0.1857 | 1.1263 | 501.8 |
| 20 | | 9.98004 | 108.4 | 0.2640 | 0.1879 | 1.1238 | 507.0 |
| 30 | | 10.20408 | 110.3 | 0.2679 | 0.1901 | 1.1215 | 512.2 |
| 40 | | 10.41667 | 112.2 | 0.2718 | 0.1923 | 1.1193 | 517.3 |
| 50 | | 10.62699 | 114.1 | 0.2756 | 0.1945 | 1.1172 | 522.3 |
| 60 | | 10.84599 | 116.1 | 0.2794 | 0.1968 | 1.1152 | 527.3 |
| 70 | | 11.06195 | 118.0 | 0.2832 | 0.1990 | 1.1134 | 532.2 |
| 80 | | 11.27396 | 120.0 | 0.2869 | 0.2012 | 1.1116 | 537.0 |
| 90 | | 11.49425 | 122.1 | 0.2906 | 0.2034 | 1.1100 | 541.8 |
| 100 | | 11.70960 | 124.1 | 0.2943 | 0.2056 | 1.1084 | 546.5 |
| 110 | | 11.91895 | 126.2 | 0.2980 | 0.2078 | 1.1068 | 551.2 |
| 120 | | 12.13592 | 128.3 | 0.3016 | 0.2100 | 1.1054 | 555.9 |
| 130 | | 12.34568 | 130.4 | 0.3052 | 0.2122 | 1.1040 | 560.5 |
| 140 | | 12.56281 | 132.5 | 0.3088 | 0.2144 | 1.1026 | 565.0 |
| 150 | | 12.77139 | 134.7 | 0.3124 | 0.2166 | 1.1013 | 569.5 |

*(Continued)*

**TABLE A.16 (CONTINUED)**

HFC–134a Superheated Vapor–Constant Pressure Tables

| Temp. °F | $v$ | $h$ | $s$ | $C_p$ | $C_p/C_v$ | $v_s$ |
|---|---|---|---|---|---|---|
| 120 | 60.97561 | 128.4 | 0.3332 | 0.2092 | 1.1033 | 557.7 |
| 130 | 62.11180 | 130.5 | 0.3368 | 0.2114 | 1.1020 | 562.2 |
| 140 | 62.89308 | 132.7 | 0.3404 | 0.2137 | 1.1008 | 566.6 |
| 150 | 64.10256 | 134.8 | 0.3439 | 0.2159 | 1.0997 | 571.1 |
| 160 | 64.93506 | 137.0 | 0.3474 | 0.2181 | 1.0985 | 575.5 |
| 170 | 66.22517 | 139.2 | 0.3509 | 0.2203 | 1.0974 | 579.8 |
| 180 | 67.11409 | 141.4 | 0.3544 | 0.2224 | 1.0964 | 584.2 |
| 190 | 68.49315 | 143.6 | 0.3579 | 0.2246 | 1.0953 | 588.5 |
| 200 | 69.44444 | 145.9 | 0.3613 | 0.2267 | 1.0943 | 592.7 |
| 210 | 70.42254 | 148.2 | 0.3648 | 0.2288 | 1.0934 | 597.0 |
| 200 | – | – | – | – | – | – |

| Temp. °F | $v$ | $h$ | $s$ | $C_p$ | $C_p/C_v$ | $v_s$ |
|---|---|---|---|---|---|---|
| 160 | 12.98701 | 136.8 | 0.3159 | 0.2187 | 1.1001 | 574.0 |
| 170 | 13.19261 | 139.0 | 0.3194 | 0.2209 | 1.0989 | 578.4 |
| 180 | 13.40483 | 141.3 | 0.3229 | 0.2230 | 1.0977 | 582.8 |
| 190 | 13.62398 | 143.5 | 0.3264 | 0.2251 | 1.0966 | 587.2 |
| 200 | 13.83126 | 145.8 | 0.3299 | 0.2272 | 1.0955 | 591.5 |
| 210 | 14.04494 | 148.0 | 0.3333 | 0.2293 | 1.0945 | 595.8 |
| 220 | 14.24501 | 150.3 | 0.3367 | 0.2313 | 1.0935 | 600.0 |
| 230 | 14.47178 | 152.7 | 0.3401 | 0.2334 | 1.0925 | 604.3 |
| 240 | 14.68429 | 155.0 | 0.3435 | 0.2354 | 1.0916 | 608.4 |
| 250 | 14.90313 | 157.4 | 0.3468 | 0.2374 | 1.0907 | 612.6 |
| 260 | – | – | – | – | – | – |

**Pressure = 10.00 psia**

| Temp. °F | | $v$ | $h$ | $s$ | $C_p$ | $C_p/C_v$ | $v_s$ |
|---|---|---|---|---|---|---|---|
| –29.5 | SAT LIQ | 0.01146 | 3.1 | 0.0074 | 0.2996 | 1.5038 | 2567.0 |
| –29.5 | SAT VAP | 4.37445 | 98.8 | 0.2297 | 0.1821 | 1.1503 | 474.4 |
| –30 | | – | – | – | – | – | – |
| –20 | | 4.48430 | 100.5 | 0.2336 | 0.1836 | 1.1457 | 480.1 |
| –10 | | 4.59770 | 102.3 | 0.2378 | 0.1852 | 1.1414 | 486.0 |
| 0 | | 4.71254 | 104.2 | 0.2419 | 0.1869 | 1.1374 | 491.8 |
| 10 | | 4.82393 | 106.1 | 0.2459 | 0.1888 | 1.1338 | 497.4 |
| 20 | | 4.93583 | 108.0 | 0.2499 | 0.1906 | 1.1306 | 502.9 |
| 30 | | 5.04796 | 109.9 | 0.2539 | 0.1926 | 1.1275 | 508.3 |
| 40 | | 5.15996 | 111.8 | 0.2578 | 0.1945 | 1.1247 | 513.7 |
| 50 | | 5.26870 | 113.8 | 0.2616 | 0.1966 | 1.1222 | 518.9 |
| 60 | | 5.37924 | 115.8 | 0.2655 | 0.1986 | 1.1197 | 524.1 |

**Pressure = 15.00 psia**

| Temp. °F | | $v$ | $h$ | $s$ | $C_p$ | $C_p/C_v$ | $v_s$ |
|---|---|---|---|---|---|---|---|
| –14.1 | SAT LIQ | 0.01167 | 7.8 | 0.0180 | 0.3043 | 1.5047 | 2437.1 |
| –14.1 | SAT VAP | 2.98686 | 101.1 | 0.2273 | 0.1888 | 1.1543 | 478.1 |
| –10 | | 3.01932 | 101.8 | 0.2290 | 0.1893 | 1.1520 | 480.7 |
| 0 | | 3.09885 | 103.7 | 0.2332 | 0.1906 | 1.1467 | 486.9 |
| 10 | | 3.17662 | 105.6 | 0.2373 | 0.1920 | 1.1420 | 492.8 |
| 20 | | 3.25415 | 107.6 | 0.2414 | 0.1935 | 1.1378 | 498.7 |
| 30 | | 3.33111 | 109.5 | 0.2454 | 0.1952 | 1.1340 | 504.4 |
| 40 | | 3.40716 | 111.5 | 0.2494 | 0.1969 | 1.1306 | 510.0 |
| 50 | | 3.48189 | 113.5 | 0.2533 | 0.1987 | 1.1274 | 515.5 |
| 60 | | 3.55745 | 115.4 | 0.2571 | 0.2005 | 1.1245 | 520.8 |
| 70 | | 3.63240 | 117.5 | 0.2610 | 0.2024 | 1.1218 | 526.1 |
| 80 | | 3.70645 | 119.5 | 0.2648 | 0.2043 | 1.1193 | 531.3 |

*(Continued)*

**TABLE A.16 (CONTINUED)**

HFC–134a Superheated Vapor–Constant Pressure Tables

**Pressure = 20.00 psia**

| Temp. °F | $v$ | $h$ | $s$ | $C_p$ | $C_p/C_v$ | $v_s$ |
|---|---|---|---|---|---|---|
| 70 | 5.48847 | 117.8 | 0.2693 | 0.2007 | 1.1175 | 529.2 |
| 80 | 5.59910 | 119.8 | 0.2731 | 0.2027 | 1.1154 | 534.2 |
| 90 | 5.70776 | 121.8 | 0.2768 | 0.2048 | 1.1134 | 539.1 |
| 100 | 5.81734 | 123.9 | 0.2805 | 0.2069 | 1.1115 | 544.0 |
| 110 | 5.92417 | 125.9 | 0.2842 | 0.2090 | 1.1097 | 548.8 |
| 120 | 6.03500 | 128.0 | 0.2879 | 0.2111 | 1.1080 | 553.6 |
| 130 | 6.14251 | 130.2 | 0.2915 | 0.2132 | 1.1064 | 558.3 |
| 140 | 6.25000 | 132.3 | 0.2951 | 0.2153 | 1.1049 | 563.0 |
| 150 | 6.35728 | 134.5 | 0.2987 | 0.2174 | 1.1034 | 567.6 |
| 160 | 6.46412 | 136.7 | 0.3022 | 0.2195 | 1.1020 | 572.1 |
| 170 | 6.57030 | 138.9 | 0.3057 | 0.2216 | 1.1007 | 576.7 |
| 180 | 6.68003 | 141.1 | 0.3093 | 0.2237 | 1.0994 | 581.1 |
| 190 | 6.78887 | 143.3 | 0.3127 | 0.2257 | 1.0982 | 585.6 |
| 200 | 6.89180 | 145.6 | 0.3162 | 0.2278 | 1.0970 | 590.0 |
| 210 | 6.99790 | 147.9 | 0.3196 | 0.2298 | 1.0959 | 594.3 |
| 220 | 7.10732 | 150.2 | 0.3231 | 0.2319 | 1.0948 | 598.6 |
| 230 | 7.21501 | 152.5 | 0.3265 | 0.2339 | 1.0938 | 602.9 |
| 240 | 7.32064 | 154.9 | 0.3298 | 0.2359 | 1.0928 | 607.2 |
| 250 | 7.42942 | 157.2 | 0.3332 | 0.2379 | 1.0918 | 611.4 |
| 260 | 7.53580 | 159.6 | 0.3366 | 0.2398 | 1.0908 | 615.6 |
| 270 | 7.63942 | 162.0 | 0.3399 | 0.2418 | 1.0899 | 619.7 |
| 280 | 7.74593 | 164.5 | 0.3432 | 0.2437 | 1.0890 | 623.8 |

| Temp. °F | | $v$ | $h$ | $s$ | $C_p$ | $C_p/C_v$ | $v_s$ |
|---|---|---|---|---|---|---|---|
| -2.4 | SAT LIQ | 0.01184 | 11.4 | 0.0259 | 0.3081 | 1.5071 | 2338.7 |
| -2.4 | SAT VAP | 2.27635 | 102.8 | 0.2258 | 0.1943 | 1.1585 | 480.2 |

**Pressure = 30.00 psia**

| Temp. °F | $v$ | $h$ | $s$ | $C_p$ | $C_p/C_v$ | $v_s$ |
|---|---|---|---|---|---|---|
| 90 | 3.78072 | 121.5 | 0.2686 | 0.2063 | 1.1169 | 536.4 |
| 100 | 3.85356 | 123.6 | 0.2723 | 0.2082 | 1.1147 | 541.4 |
| 110 | 3.92773 | 125.7 | 0.2760 | 0.2102 | 1.1127 | 546.4 |
| 120 | 4.00160 | 127.8 | 0.2797 | 0.2122 | 1.1108 | 551.3 |
| 130 | 4.07332 | 130.0 | 0.2833 | 0.2142 | 1.1090 | 556.1 |
| 140 | 4.14594 | 132.1 | 0.2869 | 0.2163 | 1.1072 | 560.9 |
| 150 | 4.21941 | 134.3 | 0.2905 | 0.2183 | 1.1056 | 565.6 |
| 160 | 4.29185 | 136.5 | 0.2941 | 0.2203 | 1.1041 | 570.3 |
| 170 | 4.36300 | 138.7 | 0.2977 | 0.2223 | 1.1026 | 574.9 |
| 180 | 4.43656 | 140.9 | 0.3012 | 0.2244 | 1.1012 | 579.4 |
| 190 | 4.50857 | 143.2 | 0.3047 | 0.2264 | 1.0999 | 584.0 |
| 200 | 4.58085 | 145.5 | 0.3081 | 0.2284 | 1.0986 | 588.4 |
| 210 | 4.65333 | 147.7 | 0.3116 | 0.2304 | 1.0974 | 592.8 |
| 220 | 4.72367 | 150.1 | 0.3150 | 0.2324 | 1.0962 | 597.2 |
| 230 | 4.79616 | 152.4 | 0.3184 | 0.2344 | 1.0951 | 601.6 |
| 240 | 4.86618 | 154.7 | 0.3218 | 0.2363 | 1.0940 | 605.9 |
| 250 | 4.93827 | 157.1 | 0.3252 | 0.2383 | 1.0929 | 610.2 |
| 260 | 5.01002 | 159.5 | 0.3285 | 0.2403 | 1.0919 | 614.4 |
| 270 | 5.08130 | 161.9 | 0.3319 | 0.2422 | 1.0910 | 618.6 |
| 280 | 5.15198 | 164.4 | 0.3352 | 0.2441 | 1.0900 | 622.8 |
| 290 | 5.22466 | 166.8 | 0.3385 | 0.2460 | 1.0891 | 626.9 |

| Temp. °F | | $v$ | $h$ | $s$ | $C_p$ | $C_p/C_v$ | $v_s$ |
|---|---|---|---|---|---|---|---|
| 15.4 | SAT LIQ | 0.01211 | 16.9 | 0.0377 | 0.3142 | 1.5132 | 2189.7 |
| 15.4 | SAT VAP | 1.54895 | 105.4 | 0.2239 | 0.2032 | 1.1670 | 482.1 |

*(Continued)*

**TABLE A.16 (CONTINUED)**

HFC–134a Superheated Vapor–Constant Pressure Tables

| T | | | | | | | | T | | | | | | |
|---|---|---|---|---|---|---|---|---|---|---|---|---|---|---|
| 0 | 2.29095 | 103.3 | 0.2268 | 0.1945 | 1.1569 | 481.8 | | 20 | 1.56863 | 106.3 | 0.2259 | 0.2033 | 1.1635 | 485.3 |
| 10 | 2.35183 | 105.2 | 0.2310 | 0.1955 | 1.1510 | 488.2 | | 30 | 1.61082 | 108.3 | 0.2301 | 0.2039 | 1.1566 | 491.9 |
| 20 | 2.41196 | 107.2 | 0.2351 | 0.1966 | 1.1457 | 494.3 | | 40 | 1.65235 | 110.4 | 0.2342 | 0.2046 | 1.1505 | 498.4 |
| 30 | 2.47158 | 109.1 | 0.2392 | 0.1979 | 1.1410 | 500.3 | | | | | | | | |
| 40 | 2.52972 | 111.1 | 0.2432 | 0.1994 | 1.1368 | 506.2 | | 50 | 1.69319 | 112.4 | 0.2383 | 0.2056 | 1.1451 | 504.6 |
| | | | | | | | | 60 | 1.73340 | 114.5 | 0.2423 | 0.2068 | 1.1403 | 510.7 |
| 50 | 2.58799 | 113.1 | 0.2472 | 0.2009 | 1.1329 | 511.9 | | 70 | 1.77336 | 116.6 | 0.2462 | 0.2080 | 1.1360 | 516.6 |
| 60 | 2.64550 | 115.1 | 0.2511 | 0.2025 | 1.1295 | 517.5 | | 80 | 1.81291 | 118.7 | 0.2501 | 0.2094 | 1.1321 | 522.4 |
| 70 | 2.70270 | 117.2 | 0.2550 | 0.2042 | 1.1263 | 523.0 | | 90 | 1.85185 | 120.8 | 0.2540 | 0.2109 | 1.1285 | 528.0 |
| 80 | 2.76014 | 119.2 | 0.2588 | 0.2060 | 1.1233 | 528.4 | | | | | | | | |
| 90 | 2.81690 | 121.3 | 0.2626 | 0.2078 | 1.1206 | 533.6 | | 100 | 1.89107 | 122.9 | 0.2578 | 0.2124 | 1.1253 | 533.5 |
| | | | | | | | | 110 | 1.92938 | 125.0 | 0.2616 | 0.2141 | 1.1223 | 538.9 |
| 100 | 2.87274 | 123.4 | 0.2664 | 0.2096 | 1.1181 | 538.8 | | 120 | 1.96773 | 127.2 | 0.2653 | 0.2157 | 1.1196 | 544.2 |
| 110 | 2.92826 | 125.5 | 0.2701 | 0.2115 | 1.1158 | 543.9 | | 130 | 2.00562 | 129.3 | 0.2690 | 0.2175 | 1.1171 | 549.4 |
| 120 | 2.98418 | 127.6 | 0.2738 | 0.2134 | 1.1136 | 549.0 | | 140 | 2.04332 | 131.5 | 0.2727 | 0.2192 | 1.1147 | 554.6 |
| 130 | 3.03951 | 129.7 | 0.2775 | 0.2153 | 1.1116 | 553.9 | | | | | | | | |
| 140 | 3.09502 | 131.9 | 0.2811 | 0.2172 | 1.1097 | 558.8 | | 150 | 2.08117 | 133.7 | 0.2764 | 0.2210 | 1.1125 | 559.6 |
| | | | | | | | | 160 | 2.11864 | 135.9 | 0.2800 | 0.2228 | 1.1105 | 564.6 |
| 150 | 3.14961 | 134.1 | 0.2847 | 0.2192 | 1.1078 | 563.6 | | 170 | 2.15610 | 138.2 | 0.2835 | 0.2247 | 1.1085 | 569.5 |
| 160 | 3.20513 | 136.3 | 0.2883 | 0.2211 | 1.1061 | 568.4 | | 180 | 2.19346 | 140.4 | 0.2871 | 0.2265 | 1.1067 | 574.3 |
| 170 | 3.26052 | 138.5 | 0.2918 | 0.2231 | 1.1045 | 573.1 | | 190 | 2.23015 | 142.7 | 0.2906 | 0.2284 | 1.1050 | 579.0 |
| 180 | 3.31455 | 140.8 | 0.2954 | 0.2251 | 1.1030 | 577.7 | | | | | | | | |
| 190 | 3.36927 | 143.0 | 0.2989 | 0.2270 | 1.1015 | 582.3 | | 200 | 2.26706 | 145.0 | 0.2941 | 0.2303 | 1.1034 | 583.7 |
| | | | | | | | | 210 | 2.30415 | 147.3 | 0.2976 | 0.2321 | 1.1019 | 588.4 |
| 200 | 3.42349 | 145.3 | 0.3024 | 0.2290 | 1.1002 | 586.9 | | 220 | 2.34082 | 149.6 | 0.3011 | 0.2340 | 1.1004 | 593.0 |
| 210 | 3.47826 | 147.6 | 0.3058 | 0.2310 | 1.0988 | 591.4 | | 230 | 2.37756 | 152.0 | 0.3045 | 0.2359 | 1.0990 | 597.5 |
| 220 | 3.53232 | 149.9 | 0.3093 | 0.2329 | 1.0976 | 595.8 | | 240 | 2.41429 | 154.4 | 0.3079 | 0.2378 | 1.0977 | 602.0 |
| 230 | 3.58680 | 152.3 | 0.3127 | 0.2349 | 1.0964 | 600.2 | | | | | | | | |
| | | | | | | | | 250 | 2.45038 | 156.7 | 0.3113 | 0.2396 | 1.0965 | 606.5 |

*(Continued)*

# TABLE A.16 (CONTINUED)

HFC-134a Superheated Vapor–Constant Pressure Tables

Continuation (from preceding page):

| Temp. °F | v | h | s | $C_p$ | $C_p/C_v$ | $v_s$ |
|---|---|---|---|---|---|---|
| 240 | 3.64033 | 154.6 | 0.3161 | 0.2368 | 1.0952 | 604.6 |
| 250 | 3.69413 | 157.0 | 0.3195 | 0.2387 | 1.0941 | 608.9 |
| 260 | 3.74813 | 159.4 | 0.3228 | 0.2407 | 1.0930 | 613.2 |
| 270 | 3.80228 | 161.8 | 0.3261 | 0.2426 | 1.0920 | 617.5 |
| 280 | 2.85654 | 164.2 | 0.3295 | 0.2445 | 1.0910 | 621.7 |
| 290 | 3.91083 | 166.7 | 0.3328 | 0.2464 | 1.0901 | 625.9 |
| 300 | 3.96354 | 169.2 | 0.3360 | 0.2482 | 1.0891 | 630.0 |

Continuation (from preceding page):

| Temp. °F | v | h | s | $C_p$ | $C_p/C_v$ | $v_s$ |
|---|---|---|---|---|---|---|
| 270 | 2.52334 | 161.6 | 0.3180 | 0.2434 | 1.0941 | 615.2 |
| 280 | 2.55951 | 164.0 | 0.3213 | 0.2452 | 1.0930 | 619.5 |
| 290 | 2.59605 | 166.5 | 0.3246 | 0.2471 | 1.0920 | 623.8 |
| 300 | 2.63227 | 169.0 | 0.3279 | 0.2489 | 1.0910 | 628.0 |
| 310 | 2.66809 | 171.5 | 0.3312 | 0.2507 | 1.0900 | 632.2 |
| 320 | 2.70416 | 174.0 | 0.3344 | 0.2526 | 1.0890 | 636.4 |

## Pressure = 40.00 psia

| Temp. °F | v | h | s | $C_p$ | $C_p/C_v$ | $v_s$ |
|---|---|---|---|---|---|---|
| 29  SAT LIQ | 0.01233 | 21.2 | 0.0467 | 0.3194 | 1.5200 | 2075.7 |
| 29  SAT VAP | 1.17606 | 107.3 | 0.2228 | 0.2106 | 1.1756 | 482.3 |
| 30 | 1.17925 | 107.5 | 0.2232 | 0.2106 | 1.1748 | 483.0 |
| 40 | 1.21242 | 109.6 | 0.2275 | 0.2105 | 1.1663 | 490.2 |
| 50 | 1.24471 | 111.7 | 0.2316 | 0.2108 | 1.1589 | 497.0 |
| 60 | 1.27665 | 113.8 | 0.2357 | 0.2114 | 1.1525 | 503.6 |
| 70 | 1.30787 | 116.0 | 0.2398 | 0.2122 | 1.1468 | 510.0 |
| 80 | 1.33887 | 118.1 | 0.2437 | 0.2131 | 1.1417 | 516.2 |
| 90 | 1.36930 | 120.2 | 0.2477 | 0.2142 | 1.1372 | 522.2 |
| 100 | 1.39958 | 122.4 | 0.2515 | 0.2155 | 1.1331 | 528.1 |
| 110 | 1.42939 | 124.5 | 0.2554 | 0.2168 | 1.1294 | 533.8 |
| 120 | 1.45900 | 126.7 | 0.2592 | 0.2183 | 1.1260 | 539.4 |
| 130 | 1.48832 | 128.9 | 0.2629 | 0.2198 | 1.1229 | 544.9 |
| 140 | 1.51745 | 131.1 | 0.2666 | 0.2213 | 1.1201 | 550.3 |
| 150 | 1.54655 | 133.3 | 0.2703 | 0.2229 | 1.1174 | 555.5 |
| 160 | 1.57505 | 135.6 | 0.2739 | 0.2246 | 1.1150 | 560.7 |

## Pressure = 50.00 psia

| Temp. °F | v | h | s | $C_p$ | $C_p/C_v$ | $v_s$ |
|---|---|---|---|---|---|---|
| 40.2  SAT LIQ | 0.01253 | 24.9 | 0.0539 | 0.3239 | 1.5271 | 1981.8 |
| 40.2  SAT VAP | 0.94805 | 108.9 | 0.2220 | 0.2172 | 1.1844 | 481.7 |
| 50 | 0.97494 | 111.0 | 0.2262 | 0.2166 | 1.1748 | 489.0 |
| 60 | 1.00180 | 113.1 | 0.2304 | 0.2165 | 1.1663 | 496.2 |
| 70 | 1.02807 | 115.3 | 0.2345 | 0.2167 | 1.1589 | 503.1 |
| 80 | 1.05396 | 117.5 | 0.2386 | 0.2172 | 1.1524 | 509.8 |
| 90 | 1.07933 | 119.7 | 0.2426 | 0.2178 | 1.1467 | 516.3 |
| 100 | 1.10436 | 121.8 | 0.2465 | 0.2187 | 1.1416 | 522.5 |
| 110 | 1.12905 | 124.0 | 0.2504 | 0.2197 | 1.1370 | 528.6 |
| 120 | 1.15354 | 126.2 | 0.2542 | 0.2209 | 1.1329 | 534.5 |
| 130 | 1.17772 | 128.5 | 0.2580 | 0.2222 | 1.1292 | 540.2 |
| 140 | 1.20178 | 130.7 | 0.2617 | 0.2235 | 1.1258 | 545.9 |
| 150 | 1.22534 | 132.9 | 0.2655 | 0.2250 | 1.1226 | 551.4 |
| 160 | 1.24891 | 135.2 | 0.2691 | 0.2265 | 1.1198 | 556.8 |
| 170 | 1.27226 | 137.5 | 0.2728 | 0.2280 | 1.1171 | 562.1 |
| 180 | 1.29550 | 139.7 | 0.2764 | 0.2296 | 1.1147 | 567.3 |
| 190 | 1.31874 | 142.0 | 0.2799 | 0.2312 | 1.1124 | 572.4 |

(Continued)

**TABLE A.16 (CONTINUED)**

HFC–134a Superheated Vapor–Constant Pressure Tables

Pressure = 60.00 psia (continued)

| Temp. °F | $v$ | $h$ | $s$ | $C_p$ | $C_p/C_v$ | $v_s$ |
|---|---|---|---|---|---|---|
| 170 | 1.60359 | 137.8 | 0.2775 | 0.2263 | 1.1127 | 565.8 |
| 180 | 1.63212 | 140.1 | 0.2811 | 0.2280 | 1.1106 | 570.8 |
| 190 | 1.66058 | 142.4 | 0.2847 | 0.2298 | 1.1086 | 575.7 |
| 200 | 1.68890 | 144.7 | 0.2882 | 0.2316 | 1.1068 | 580.6 |
| 210 | 1.71674 | 147.0 | 0.2917 | 0.2333 | 1.1050 | 585.4 |
| 220 | 1.74520 | 149.3 | 0.2952 | 0.2351 | 1.1034 | 590.1 |
| 230 | 1.77305 | 151.7 | 0.2986 | 0.2369 | 1.1018 | 594.8 |
| 240 | 1.80050 | 154.1 | 0.3020 | 0.2388 | 1.1003 | 599.4 |
| 250 | 1.82849 | 156.5 | 0.3054 | 0.2406 | 1.0989 | 604.0 |
| 260 | 1.85598 | 158.9 | 0.3088 | 0.2424 | 1.0976 | 608.5 |
| 270 | 1.88359 | 161.3 | 0.3122 | 0.2442 | 1.0963 | 612.9 |
| 280 | 1.91131 | 163.8 | 0.3155 | 0.2460 | 1.0951 | 617.4 |
| 290 | 1.93911 | 166.2 | 0.3188 | 0.2478 | 1.0939 | 621.7 |
| 300 | 1.96657 | 168.7 | 0.3221 | 0.2496 | 1.0928 | 626.1 |
| 310 | 1.99362 | 171.2 | 0.3254 | 0.2514 | 1.0917 | 630.3 |
| 320 | 2.02102 | 173.8 | 0.3286 | 0.2532 | 1.0907 | 634.6 |
| 330 | 2.04834 | 176.3 | 0.3319 | 0.2550 | 1.0897 | 638.8 |
| 340 | — | — | — | — | — | — |

Pressure = 70.00 psia (continued)

| Temp. °F | $v$ | $h$ | $s$ | $C_p$ | $C_p/C_v$ | $v_s$ |
|---|---|---|---|---|---|---|
| 200 | 1.34174 | 144.4 | 0.2835 | 0.2329 | 1.1103 | 577.4 |
| 210 | 1.36444 | 146.7 | 0.2870 | 0.2346 | 1.1083 | 582.3 |
| 220 | 1.38735 | 149.1 | 0.2905 | 0.2363 | 1.1064 | 587.2 |
| 230 | 1.41004 | 151.4 | 0.2940 | 0.2380 | 1.1046 | 592.0 |
| 240 | 1.43266 | 153.8 | 0.2974 | 0.2398 | 1.1030 | 596.8 |
| 250 | 1.45518 | 156.2 | 0.3008 | 0.2415 | 1.1014 | 601.5 |
| 260 | 1.47754 | 158.6 | 0.3042 | 0.2433 | 1.0999 | 606.1 |
| 270 | 1.50015 | 161.1 | 0.3076 | 0.2450 | 1.0985 | 610.7 |
| 280 | 1.52230 | 163.5 | 0.3109 | 0.2468 | 1.0972 | 615.2 |
| 290 | 1.54440 | 166.0 | 0.3142 | 0.2485 | 1.0959 | 619.6 |
| 300 | 1.56691 | 168.5 | 0.3176 | 0.2503 | 1.0947 | 624.1 |
| 310 | 1.58907 | 171.0 | 0.3208 | 0.2521 | 1.0935 | 628.4 |
| 320 | 1.61108 | 173.6 | 0.3241 | 0.2538 | 1.0924 | 632.8 |
| 330 | 1.63292 | 176.1 | 0.3273 | 0.2555 | 1.0913 | 637.0 |
| 340 | 1.65508 | 178.7 | 0.3306 | 0.2573 | 1.0903 | 641.3 |
| 350 | 1.67701 | 181.2 | 0.3338 | 0.2590 | 1.0893 | 645.5 |

Pressure = 60.00 psia

| Temp. °F | $v$ | $h$ | $s$ | $C_p$ | $C_p/C_v$ | $v_s$ |
|---|---|---|---|---|---|---|
| 49.8 SAT LIQ | 0.01271 | 28.0 | 0.0601 | 0.3282 | 1.5343 | 1901.3 |
| 49.8 SAT VAP | 0.79390 | 110.2 | 0.2214 | 0.2232 | 1.1934 | 480.5 |
| 50 | 0.79428 | 110.2 | 0.2215 | 0.2232 | 1.1932 | 480.6 |
| 60 | 0.81793 | 112.4 | 0.2258 | 0.2222 | 1.1820 | 488.5 |
| 70 | 0.84104 | 114.6 | 0.2300 | 0.2217 | 1.1725 | 496.0 |
| 80 | 0.86356 | 116.9 | 0.2342 | 0.2215 | 1.1643 | 503.2 |

Pressure = 70.00 psia

| Temp. °F | $v$ | $h$ | $s$ | $C_p$ | $C_p/C_v$ | $v_s$ |
|---|---|---|---|---|---|---|
| 58.3 SAT LIQ | 0.01288 | 30.8 | 0.0655 | 0.3321 | 1.5417 | 1830.2 |
| 58.3 SAT VAP | 0.68236 | 111.3 | 0.2209 | 0.2288 | 1.2026 | 478.9 |
| 60 | 0.68601 | 111.7 | 0.2217 | 0.2285 | 1.2002 | 480.3 |
| 70 | 0.70691 | 114.0 | 0.2260 | 0.2272 | 1.1880 | 488.5 |
| 80 | 0.72717 | 116.2 | 0.2302 | 0.2264 | 1.1777 | 496.3 |
| 90 | 0.74694 | 118.5 | 0.2344 | 0.2260 | 1.1688 | 503.7 |

*(Continued)*

**TABLE A.16 (CONTINUED)**

HFC–134a Superheated Vapor–Constant Pressure Tables

| T | v | | | | | | v | | | | | |
|---|---|---|---|---|---|---|---|---|---|---|---|---|
| 90  | 0.88566 | 119.1 | 0.2382 | 0.2217 | 1.1572 | 510.1 |         |       |        |        |        |       |
| 100 | 0.90728 | 121.3 | 0.2422 | 0.2222 | 1.1509 | 516.7 | 0.76628 | 120.7 | 0.2385 | 0.2259 | 1.1611 | 510.8 |
| 110 | 0.92868 | 123.5 | 0.2462 | 0.2229 | 1.1453 | 523.2 | 0.78524 | 123.0 | 0.2425 | 0.2262 | 1.1543 | 517.6 |
| 120 | 0.94967 | 125.8 | 0.2500 | 0.2237 | 1.1403 | 529.4 | 0.80386 | 125.3 | 0.2464 | 0.2267 | 1.1484 | 524.2 |
| 130 | 0.97040 | 128.0 | 0.2539 | 0.2247 | 1.1359 | 535.5 | 0.82217 | 127.5 | 0.2503 | 0.2274 | 1.1431 | 530.6 |
| 140 | 0.99098 | 130.3 | 0.2577 | 0.2258 | 1.1318 | 541.4 | 0.84034 | 129.8 | 0.2541 | 0.2283 | 1.1383 | 536.8 |
| 150 | 1.01133 | 132.5 | 0.2614 | 0.2271 | 1.1282 | 547.1 | 0.85807 | 132.1 | 0.2579 | 0.2293 | 1.1341 | 542.8 |
| 160 | 1.03135 | 134.8 | 0.2651 | 0.2284 | 1.1248 | 552.8 | 0.87573 | 134.4 | 0.2617 | 0.2304 | 1.1302 | 548.7 |
| 170 | 1.05130 | 137.1 | 0.2688 | 0.2298 | 1.1218 | 558.3 | 0.89326 | 136.7 | 0.2654 | 0.2316 | 1.1266 | 554.4 |
| 180 | 1.07101 | 139.4 | 0.2724 | 0.2312 | 1.1189 | 563.7 | 0.91058 | 139.0 | 0.2690 | 0.2329 | 1.1234 | 560.0 |
| 190 | 1.09075 | 141.7 | 0.2760 | 0.2327 | 1.1163 | 569.0 | 0.92773 | 141.4 | 0.2726 | 0.2343 | 1.1205 | 565.5 |
| 200 | 1.11012 | 144.0 | 0.2796 | 0.2343 | 1.1139 | 574.2 | 0.94473 | 143.7 | 0.2762 | 0.2357 | 1.1177 | 570.9 |
| 210 | 1.12943 | 146.4 | 0.2831 | 0.2359 | 1.1117 | 579.3 | 0.96172 | 146.1 | 0.2798 | 0.2372 | 1.1152 | 576.2 |
| 220 | 1.14903 | 148.8 | 0.2866 | 0.2375 | 1.1096 | 584.3 | 0.97847 | 148.5 | 0.2833 | 0.2387 | 1.1128 | 581.4 |
| 230 | 1.16795 | 151.1 | 0.2901 | 0.2391 | 1.1076 | 589.2 | 0.99512 | 150.9 | 0.2868 | 0.2403 | 1.1106 | 586.4 |
| 240 | 1.18723 | 153.5 | 0.2936 | 0.2408 | 1.1057 | 594.1 | 1.01184 | 153.3 | 0.2903 | 0.2419 | 1.1086 | 591.5 |
| 250 | 1.20627 | 156.0 | 0.2970 | 0.2425 | 1.1040 | 598.9 | 1.02828 | 155.7 | 0.2937 | 0.2435 | 1.1067 | 596.4 |
| 260 | 1.22519 | 158.4 | 0.3004 | 0.2442 | 1.1024 | 603.7 | 1.04482 | 158.1 | 0.2971 | 0.2451 | 1.1049 | 601.2 |
| 270 | 1.24409 | 160.8 | 0.3038 | 0.2459 | 1.1008 | 608.4 | 1.06135 | 160.6 | 0.3005 | 0.2467 | 1.1032 | 606.0 |
| 280 | 1.26295 | 163.3 | 0.3071 | 0.2476 | 1.0993 | 613.0 | 1.07747 | 163.1 | 0.3039 | 0.2484 | 1.1015 | 610.8 |
| 290 | 1.28172 | 165.8 | 0.3105 | 0.2493 | 1.0979 | 617.5 | 1.09385 | 165.6 | 0.3072 | 0.2501 | 1.1000 | 615.4 |
| 300 | 1.30039 | 168.3 | 0.3138 | 0.2510 | 1.0966 | 622.0 | 1.11012 | 168.1 | 0.3106 | 0.2517 | 1.0986 | 620.0 |
| 310 | 1.31891 | 170.8 | 0.3171 | 0.2527 | 1.0953 | 626.5 | 1.12625 | 170.6 | 0.3139 | 0.2534 | 1.0972 | 624.6 |
| 320 | 1.33761 | 173.4 | 0.3204 | 0.2544 | 1.0941 | 630.9 | 1.14233 | 173.1 | 0.3172 | 0.2551 | 1.0959 | 629.1 |
| 330 |         |       |        |        |        |       | 1.15835 | 175.7 | 0.3204 | 0.2567 | 1.0947 | 633.5 |

*(Continued)*

# TABLE A.16 (CONTINUED)

HFC–134a Superheated Vapor–Constant Pressure Tables

Continuation rows (from preceding page):

| Temp. °F | v | h | s | $C_p$ | $C_p/C_v$ | $v_s$ | | Temp. °F | v | h | s | $C_p$ | $C_p/C_v$ | $v_s$ |
|---|---|---|---|---|---|---|---|---|---|---|---|---|---|---|
| 330 | 1.35630 | 175.9 | 0.3236 | 0.2561 | 1.0930 | 635.3 | | 340 | 1.17426 | 178.3 | 0.3237 | 0.2584 | 1.0935 | 637.9 |
| 340 | 1.37457 | 178.5 | 0.3268 | 0.2578 | 1.0919 | 639.6 | | 350 | 1.19033 | 180.9 | 0.3269 | 0.2601 | 1.0923 | 642.3 |
| 350 | 1.39334 | 181.1 | 0.3301 | 0.2595 | 1.0908 | 643.9 | | 360 | 1.20642 | 183.5 | 0.3301 | 0.2618 | 1.0913 | 646.6 |
| 360 | — | — | — | — | — | — | | 370 | — | — | — | — | — | — |

Pressure = 80.00 psia

| Temp. °F | v | h | s | $C_p$ | $C_p/C_v$ | $v_s$ |
|---|---|---|---|---|---|---|
| 65.9 SAT LIQ | 0.01304 | 33.3 | 0.0703 | 0.3359 | 1.5492 | 1766.4 |
| 65.9 SAT VAP | 0.59787 | 112.3 | 0.2205 | 0.2342 | 1.2120 | 477.0 |
| 70 | 0.60580 | 113.2 | 0.2224 | 0.2333 | 1.2058 | 480.7 |
| 80 | 0.62445 | 115.6 | 0.2267 | 0.2317 | 1.1928 | 489.1 |
| 90 | 0.64255 | 117.9 | 0.2309 | 0.2306 | 1.1818 | 497.1 |
| 100 | 0.66020 | 120.2 | 0.2351 | 0.2300 | 1.1723 | 504.7 |
| 110 | 0.67741 | 122.5 | 0.2392 | 0.2298 | 1.1642 | 511.9 |
| 120 | 0.69430 | 124.8 | 0.2432 | 0.2299 | 1.1571 | 518.9 |
| 130 | 0.71083 | 127.1 | 0.2471 | 0.2303 | 1.1508 | 525.6 |
| 140 | 0.72711 | 129.4 | 0.2510 | 0.2308 | 1.1453 | 532.1 |
| 150 | 0.74322 | 131.7 | 0.2548 | 0.2316 | 1.1403 | 538.4 |
| 160 | 0.75896 | 134.0 | 0.2586 | 0.2325 | 1.1358 | 544.6 |
| 170 | 0.77465 | 136.3 | 0.2623 | 0.2335 | 1.1318 | 550.5 |
| 180 | 0.79020 | 138.7 | 0.2660 | 0.2347 | 1.1281 | 556.4 |
| 190 | 0.80548 | 141.0 | 0.2696 | 0.2359 | 1.1248 | 562.0 |
| 200 | 0.82068 | 143.4 | 0.2733 | 0.2372 | 1.1217 | 567.6 |
| 210 | 0.83577 | 145.8 | 0.2768 | 0.2386 | 1.1188 | 573.0 |
| 220 | 0.85063 | 148.2 | 0.2804 | 0.2400 | 1.1162 | 578.4 |
| 230 | 0.86558 | 150.6 | 0.2839 | 0.2414 | 1.1138 | 583.6 |

Pressure = 90.00 psia

| Temp. °F | v | h | s | $C_p$ | $C_p/C_v$ | $v_s$ |
|---|---|---|---|---|---|---|
| 72.8 SAT LIQ | 0.01319 | 35.7 | 0.0747 | 0.3396 | 1.5568 | 1708.1 |
| 72.8 SAT VAP | 0.53155 | 113.1 | 0.2202 | 0.2395 | 1.2217 | 475.0 |
| 80 | 0.54416 | 114.9 | 0.2234 | 0.2376 | 1.2100 | 481.6 |
| 90 | 0.56107 | 117.2 | 0.2278 | 0.2357 | 1.1964 | 490.2 |
| 100 | 0.57743 | 119.6 | 0.2320 | 0.2344 | 1.1849 | 498.3 |
| 110 | 0.59333 | 121.9 | 0.2361 | 0.2337 | 1.1751 | 506.0 |
| 120 | 0.60887 | 124.3 | 0.2402 | 0.2333 | 1.1666 | 513.4 |
| 130 | 0.62406 | 126.6 | 0.2442 | 0.2333 | 1.1592 | 520.5 |
| 140 | 0.63906 | 128.9 | 0.2481 | 0.2336 | 1.1527 | 527.4 |
| 150 | 0.65368 | 131.3 | 0.2520 | 0.2341 | 1.1470 | 534.0 |
| 160 | 0.66814 | 133.6 | 0.2558 | 0.2347 | 1.1418 | 540.4 |
| 170 | 0.68236 | 136.0 | 0.2596 | 0.2355 | 1.1372 | 546.6 |
| 180 | 0.69643 | 138.3 | 0.2633 | 0.2365 | 1.1331 | 552.6 |
| 190 | 0.71023 | 140.7 | 0.2670 | 0.2376 | 1.1293 | 558.5 |
| 200 | 0.72401 | 143.1 | 0.2706 | 0.2387 | 1.1258 | 564.2 |
| 210 | 0.73768 | 145.5 | 0.2742 | 0.2400 | 1.1227 | 569.9 |
| 220 | 0.75126 | 147.9 | 0.2778 | 0.2413 | 1.1197 | 575.4 |
| 230 | 0.76470 | 150.3 | 0.2813 | 0.2426 | 1.1170 | 580.8 |
| 240 | 0.77791 | 152.7 | 0.2848 | 0.2441 | 1.1145 | 586.0 |

(Continued)

**TABLE A.16 (CONTINUED)**

HFC–134a Superheated Vapor–Constant Pressure Tables

**Pressure = 80.00 psia (continued)**

| Temp. | $v$ | $h$ | $s$ | $C_p$ | $C_p/C_v$ | $v_s$ |
|---|---|---|---|---|---|---|
| 240 | 0.88020 | 153.0 | 0.2874 | 0.2429 | 1.1115 | 588.8 |
| 250 | 0.89501 | 155.4 | 0.2908 | 0.2445 | 1.1094 | 593.8 |
| 260 | 0.90950 | 157.9 | 0.2943 | 0.2460 | 1.1074 | 598.8 |
| 270 | 0.92404 | 160.4 | 0.2977 | 0.2476 | 1.1056 | 603.7 |
| 280 | 0.93853 | 162.8 | 0.3011 | 0.2492 | 1.1038 | 608.5 |
| 290 | 0.95302 | 165.3 | 0.3044 | 0.2508 | 1.1021 | 613.3 |
| 300 | 0.96721 | 167.9 | 0.3078 | 0.2525 | 1.1006 | 618.0 |
| 310 | 0.98155 | 170.4 | 0.3111 | 0.2541 | 1.0991 | 622.6 |
| 320 | 0.99582 | 172.9 | 0.3144 | 0.2557 | 1.0977 | 627.2 |
| 330 | 1.01010 | 175.5 | 0.3176 | 0.2574 | 1.0964 | 631.8 |
| 340 | 1.02417 | 178.1 | 0.3209 | 0.2590 | 1.0951 | 636.2 |
| 350 | 1.03821 | 180.7 | 0.3241 | 0.2606 | 1.0939 | 640.7 |
| 360 | 1.05230 | 183.3 | 0.3273 | 0.2623 | 1.0927 | 645.0 |
| 370 | 1.06644 | 185.9 | 0.3305 | 0.2639 | 1.0916 | 649.4 |

**Pressure = 100.00 psia**

| Temp. °F | | $v$ | $h$ | $s$ | $C_p$ | $C_p/C_v$ | $v_s$ |
|---|---|---|---|---|---|---|---|
| 79.1 | SAT LIQ | 0.01333 | 37.8 | 0.0787 | 0.3433 | 1.5646 | 1654.2 |
| 79.1 | SAT VAP | 0.47803 | 113.9 | 0.2199 | 0.2446 | 1.2317 | 472.8 |
| 80 | | 0.47952 | 114.1 | 0.2203 | 0.2442 | 1.2300 | 473.6 |
| 90 | | 0.49552 | 116.6 | 0.2248 | 0.2413 | 1.2129 | 482.9 |
| 100 | | 0.51093 | 119.0 | 0.2291 | 0.2393 | 1.1989 | 491.7 |
| 110 | | 0.52587 | 121.4 | 0.2333 | 0.2379 | 1.1871 | 499.9 |
| 120 | | 0.54037 | 123.7 | 0.2375 | 0.2370 | 1.1770 | 507.8 |
| 130 | | 0.55451 | 126.1 | 0.2415 | 0.2366 | 1.1683 | 515.3 |
| 140 | | 0.56838 | 128.5 | 0.2455 | 0.2365 | 1.1608 | 522.5 |
| 250 | | 0.79108 | 155.2 | 0.2883 | 0.2455 | 1.1122 | 591.2 |
| 260 | | 0.80431 | 157.6 | 0.2917 | 0.2470 | 1.1101 | 596.3 |
| 270 | | 0.81746 | 160.1 | 0.2951 | 0.2485 | 1.1080 | 601.4 |
| 280 | | 0.83036 | 162.6 | 0.2985 | 0.2501 | 1.1061 | 606.3 |
| 290 | | 0.84331 | 165.1 | 0.3019 | 0.2516 | 1.1043 | 611.2 |
| 300 | | 0.85616 | 167.6 | 0.3052 | 0.2532 | 1.1026 | 616.0 |
| 310 | | 0.86904 | 170.2 | 0.3086 | 0.2548 | 1.1010 | 620.7 |
| 320 | | 0.88191 | 172.7 | 0.3119 | 0.2564 | 1.0995 | 625.4 |
| 330 | | 0.89461 | 175.3 | 0.3151 | 0.2580 | 1.0981 | 630.0 |
| 340 | | 0.90728 | 177.9 | 0.3184 | 0.2596 | 1.0967 | 634.5 |
| 350 | | 0.91988 | 180.5 | 0.3216 | 0.2612 | 1.0955 | 639.0 |
| 360 | | 0.93257 | 183.1 | 0.3249 | 0.2628 | 1.0942 | 643.5 |
| 370 | | 0.94518 | 185.7 | 0.3280 | 0.2644 | 1.0930 | 647.9 |
| 380 | | 0.95767 | 188.4 | 0.3312 | 0.2660 | 1.0919 | 652.3 |

**Pressure = 120.00 psia**

| Temp. °F | | $v$ | $h$ | $s$ | $C_p$ | $C_p/C_v$ | $v_s$ |
|---|---|---|---|---|---|---|---|
| 90.5 | SAT LIQ | 0.01361 | 41.8 | 0.0858 | 0.3504 | 1.5808 | 1557.1 |
| 90.5 | SAT VAP | 0.39689 | 115.3 | 0.2194 | 0.2546 | 1.2528 | 467.9 |
| 100 | | 0.41044 | 117.7 | 0.2238 | 0.2506 | 1.2326 | 477.6 |
| 110 | | 0.42402 | 120.2 | 0.2282 | 0.2475 | 1.2153 | 487.1 |
| 120 | | 0.43710 | 122.6 | 0.2324 | 0.2453 | 1.2010 | 496.0 |
| 130 | | 0.44976 | 125.1 | 0.2366 | 0.2438 | 1.1890 | 504.4 |
| 140 | | 0.46206 | 127.5 | 0.2407 | 0.2428 | 1.1788 | 512.4 |
| 150 | | 0.47405 | 129.9 | 0.2447 | 0.2423 | 1.1700 | 520.0 |
| 160 | | 0.48577 | 132.3 | 0.2487 | 0.2421 | 1.1623 | 527.3 |

*(Continued)*

**TABLE A.16 (CONTINUED)**

HFC–134a Superheated Vapor–Constant Pressure Tables

| Temp | v | | | | | | v | | | | | |
|---|---|---|---|---|---|---|---|---|---|---|---|---|
| 150 | 0.58194 | 130.8 | 0.2494 | 0.2366 | 1.1541 | 529.4 | | | | | | |
| 160 | 0.59527 | 133.2 | 0.2533 | 0.2370 | 1.1482 | 536.1 | | | | | | |
| 170 | 0.60835 | 135.6 | 0.2570 | 0.2376 | 1.1430 | 542.5 | 0.49724 | 134.8 | 0.2525 | 0.2422 | 1.1555 | 534.3 |
| 180 | 0.62135 | 137.9 | 0.2608 | 0.2384 | 1.1383 | 548.8 | 0.50860 | 137.2 | 0.2564 | 0.2425 | 1.1495 | 541.1 |
| 190 | 0.63416 | 140.3 | 0.2645 | 0.2393 | 1.1340 | 554.9 | 0.51967 | 139.6 | 0.2601 | 0.2430 | 1.1442 | 547.6 |
| 200 | 0.64675 | 142.7 | 0.2682 | 0.2403 | 1.1301 | 560.9 | 0.53064 | 142.0 | 0.2639 | 0.2437 | 1.1394 | 554.0 |
| 210 | 0.65924 | 145.1 | 0.2718 | 0.2414 | 1.1266 | 566.7 | 0.54139 | 144.5 | 0.2675 | 0.2445 | 1.1351 | 560.1 |
| 220 | 0.67155 | 147.6 | 0.2754 | 0.2426 | 1.1234 | 572.3 | 0.55206 | 146.9 | 0.2712 | 0.2454 | 1.1311 | 566.2 |
| 230 | 0.68385 | 150.0 | 0.2789 | 0.2439 | 1.1204 | 577.9 | 0.56268 | 149.4 | 0.2747 | 0.2465 | 1.1275 | 572.0 |
| 240 | 0.69604 | 152.4 | 0.2825 | 0.2452 | 1.1177 | 583.3 | 0.57307 | 151.9 | 0.2783 | 0.2476 | 1.1243 | 577.8 |
| 250 | 0.70806 | 154.9 | 0.2859 | 0.2466 | 1.1151 | 588.6 | 0.58350 | 154.4 | 0.2818 | 0.2488 | 1.1212 | 583.4 |
| 260 | 0.72015 | 157.4 | 0.2894 | 0.2480 | 1.1128 | 593.9 | 0.59379 | 156.8 | 0.2853 | 0.2501 | 1.1184 | 588.9 |
| 270 | 0.73196 | 159.9 | 0.2928 | 0.2495 | 1.1106 | 599.0 | 0.60394 | 159.4 | 0.2888 | 0.2514 | 1.1158 | 594.2 |
| 280 | 0.74388 | 162.4 | 0.2962 | 0.2509 | 1.1085 | 604.1 | 0.61406 | 161.9 | 0.2922 | 0.2527 | 1.1134 | 599.5 |
| 290 | 0.75569 | 164.9 | 0.2996 | 0.2524 | 1.1066 | 609.0 | 0.62414 | 164.4 | 0.2956 | 0.2541 | 1.1112 | 604.7 |
| 300 | 0.76740 | 167.4 | 0.3030 | 0.2540 | 1.1047 | 613.9 | 0.63408 | 167.0 | 0.2990 | 0.2555 | 1.1091 | 609.8 |
| 310 | 0.77906 | 170.0 | 0.3063 | 0.2555 | 1.1030 | 618.7 | 0.64404 | 169.5 | 0.3023 | 0.2570 | 1.1071 | 614.8 |
| 320 | 0.79064 | 172.5 | 0.3096 | 0.2571 | 1.1014 | 623.5 | 0.65389 | 172.1 | 0.3057 | 0.2584 | 1.1052 | 619.8 |
| 330 | 0.80225 | 175.1 | 0.3129 | 0.2586 | 1.0999 | 628.2 | 0.66375 | 174.7 | 0.3090 | 0.2599 | 1.1035 | 624.6 |
| 340 | 0.81387 | 177.7 | 0.3162 | 0.2602 | 1.0984 | 632.8 | 0.67354 | 177.3 | 0.3122 | 0.2614 | 1.1019 | 629.4 |
| 350 | 0.82535 | 180.3 | 0.3194 | 0.2618 | 1.0970 | 637.4 | 0.68329 | 179.9 | 0.3155 | 0.2629 | 1.1003 | 634.1 |
| 360 | 0.83675 | 182.9 | 0.3226 | 0.2634 | 1.0957 | 641.9 | 0.69300 | 182.6 | 0.3187 | 0.2645 | 1.0988 | 638.8 |
| 370 | 0.84818 | 185.6 | 0.3258 | 0.2649 | 1.0945 | 646.4 | 0.70264 | 185.2 | 0.3220 | 0.2660 | 1.0974 | 643.4 |
| 380 | 0.85955 | 188.2 | 0.3290 | 0.2665 | 1.0933 | 650.8 | 0.71225 | 187.9 | 0.3252 | 0.2675 | 1.0961 | 648.0 |
| 390 | — | — | — | — | — | — | 0.72192 | 190.6 | 0.3283 | 0.2690 | 1.0948 | 652.5 |
| 400 | | | | | | | 0.73142 | 193.3 | 0.3315 | 0.2706 | 1.0936 | 656.9 |

*(Continued)*

**TABLE A.16 (CONTINUED)**

HFC–134a Superheated Vapor–Constant Pressure Tables

| Temp. °F | | Pressure = 140.00 psia | | | | | Temp. °F | | Pressure = 160.00 psia | | | | |
|---|---|---|---|---|---|---|---|---|---|---|---|---|---|
| | $v$ | $h$ | $s$ | $C_p$ | $C_p/C_v$ | $v_s$ | | $v$ | $h$ | $s$ | $C_p$ | $C_p/C_v$ | $v_s$ |
| 100.5 SAT LIQ | 0.01388 | 45.3 | 0.0921 | 0.3576 | 1.5980 | 1470.8 | 109.5 SAT LIQ | 0.01414 | 48.5 | 0.0977 | 0.3648 | 1.6163 | 1392.4 |
| 100.5 SAT VAP | 0.33818 | 116.4 | 0.2190 | 0.2646 | 1.2757 | 462.7 | 109.5 SAT VAP | 0.29362 | 117.3 | 0.2186 | 0.2748 | 1.3006 | 457.2 |
| 110 | 0.35042 | 118.9 | 0.2234 | 0.2593 | 1.2512 | 473.2 | 110 | 0.29426 | 117.5 | 0.2188 | 0.2744 | 1.2988 | 457.9 |
| 120 | 0.36266 | 121.4 | 0.2279 | 0.2553 | 1.2307 | 483.3 | 120 | 0.30608 | 120.2 | 0.2235 | 0.2675 | 1.2685 | 469.6 |
| 130 | 0.37438 | 124.0 | 0.2322 | 0.2523 | 1.2140 | 492.8 | 130 | 0.31726 | 122.8 | 0.2281 | 0.2624 | 1.2448 | 480.4 |
| 140 | 0.38568 | 126.5 | 0.2364 | 0.2502 | 1.2001 | 501.7 | 140 | 0.32792 | 125.4 | 0.2324 | 0.2588 | 1.2258 | 490.4 |
| 150 | 0.39662 | 129.0 | 0.2406 | 0.2487 | 1.1884 | 510.1 | 150 | 0.33817 | 128.0 | 0.2367 | 0.2561 | 1.2101 | 499.7 |
| 160 | 0.40727 | 131.5 | 0.2446 | 0.2477 | 1.1784 | 518.1 | 160 | 0.34806 | 130.5 | 0.2409 | 0.2541 | 1.1970 | 508.5 |
| 170 | 0.41764 | 133.9 | 0.2486 | 0.2472 | 1.1697 | 525.7 | 170 | 0.35767 | 133.1 | 0.2449 | 0.2528 | 1.1859 | 516.8 |
| 180 | 0.42779 | 136.4 | 0.2524 | 0.2470 | 1.1622 | 533.0 | 180 | 0.36700 | 135.6 | 0.2489 | 0.2519 | 1.1764 | 524.7 |
| 190 | 0.43777 | 138.9 | 0.2563 | 0.2470 | 1.1555 | 540.1 | 190 | 0.37611 | 138.1 | 0.2528 | 0.2515 | 1.1681 | 532.3 |
| 200 | 0.44755 | 141.3 | 0.2601 | 0.2473 | 1.1496 | 546.9 | 200 | 0.38504 | 140.6 | 0.2566 | 0.2513 | 1.1608 | 539.6 |
| 210 | 0.45714 | 143.8 | 0.2638 | 0.2478 | 1.1443 | 553.5 | 210 | 0.39378 | 143.1 | 0.2604 | 0.2514 | 1.1544 | 546.7 |
| 220 | 0.46659 | 146.3 | 0.2675 | 0.2485 | 1.1395 | 559.9 | 220 | 0.40238 | 145.7 | 0.2641 | 0.2517 | 1.1486 | 553.4 |
| 230 | 0.47596 | 148.8 | 0.2711 | 0.2492 | 1.1352 | 566.1 | 230 | 0.41085 | 148.2 | 0.2678 | 0.2522 | 1.1435 | 560.0 |
| 240 | 0.48520 | 151.3 | 0.2747 | 0.2501 | 1.1313 | 572.1 | 240 | 0.41916 | 150.7 | 0.2715 | 0.2528 | 1.1388 | 566.4 |
| 250 | 0.49432 | 153.8 | 0.2782 | 0.2511 | 1.1277 | 578.0 | 250 | 0.42746 | 153.2 | 0.2751 | 0.2536 | 1.1346 | 572.6 |
| 260 | 0.50342 | 156.3 | 0.2818 | 0.2522 | 1.1244 | 583.8 | 260 | 0.43554 | 155.8 | 0.2786 | 0.2545 | 1.1308 | 578.6 |
| 270 | 0.51235 | 158.8 | 0.2853 | 0.2534 | 1.1214 | 589.4 | 270 | 0.44364 | 158.3 | 0.2821 | 0.2555 | 1.1273 | 584.5 |
| 280 | 0.52121 | 161.4 | 0.2887 | 0.2546 | 1.1186 | 594.9 | 280 | 0.45165 | 160.9 | 0.2856 | 0.2565 | 1.1241 | 590.3 |
| 290 | 0.53011 | 163.9 | 0.2921 | 0.2558 | 1.1160 | 600.3 | 290 | 0.45954 | 163.5 | 0.2891 | 0.2577 | 1.1211 | 595.9 |
| 300 | 0.53888 | 166.5 | 0.2955 | 0.2572 | 1.1136 | 605.6 | 300 | 0.46729 | 166.0 | 0.2925 | 0.2588 | 1.1183 | 601.4 |
| 310 | 0.54750 | 169.1 | 0.2989 | 0.2585 | 1.1113 | 610.9 | 310 | 0.47508 | 168.6 | 0.2959 | 0.2601 | 1.1158 | 606.9 |
| 320 | 0.55611 | 171.7 | 0.3022 | 0.2599 | 1.1092 | 616.0 | 320 | 0.48281 | 171.2 | 0.2992 | 0.2613 | 1.1134 | 612.2 |
| 330 | 0.56478 | 174.3 | 0.3056 | 0.2613 | 1.1073 | 621.0 | 330 | 0.49053 | 173.9 | 0.3026 | 0.2627 | 1.1112 | 617.4 |
| 340 | 0.57330 | 176.9 | 0.3089 | 0.2627 | 1.1054 | 626.0 | 340 | 0.49816 | 176.5 | 0.3059 | 0.2640 | 1.1091 | 622.5 |
| 350 | 0.58184 | 179.5 | 0.3121 | 0.2641 | 1.1036 | 630.9 | 350 | 0.50566 | 179.1 | 0.3092 | 0.2654 | 1.1071 | 627.6 |

*(Continued)*

# TABLE A.16 (CONTINUED)

HFC-134a Superheated Vapor–Constant Pressure Tables

| Temp. °F | $v$ | $h$ | $s$ | $C_p$ | $C_p/C_v$ | $v_s$ | | $v$ | $h$ | $s$ | $C_p$ | $C_p/C_v$ | $v_s$ |
|---|---|---|---|---|---|---|---|---|---|---|---|---|---|
| 360 | 0.59028 | 182.2 | 0.3154 | 0.2656 | 1.1020 | 635.7 | | 0.51324 | 181.8 | 0.3124 | 0.2667 | 1.1053 | 632.5 |
| 370 | 0.59866 | 184.8 | 0.3186 | 0.2670 | 1.1004 | 640.4 | | 0.52067 | 184.5 | 0.3157 | 0.2681 | 1.1035 | 637.4 |
| 380 | 0.60716 | 187.5 | 0.3218 | 0.2685 | 1.0989 | 645.1 | | 0.52818 | 187.2 | 0.3189 | 0.2696 | 1.1019 | 642.2 |
| 390 | 0.61542 | 190.2 | 0.3250 | 0.2700 | 1.0975 | 649.7 | | 0.53565 | 189.9 | 0.3221 | 0.2710 | 1.1003 | 647.0 |
| 400 | 0.62375 | 192.9 | 0.3282 | 0.2715 | 1.0962 | 654.3 | | 0.54301 | 192.6 | 0.3253 | 0.2724 | 1.0988 | 651.7 |
| 410 | 0.63203 | 195.6 | 0.3313 | 0.2730 | 1.0949 | 658.8 | | 0.55030 | 195.3 | 0.3284 | 0.2738 | 1.0974 | 656.3 |
| 420 | | | | | | | | — | — | — | — | — | — |

## Pressure = 180.00 psia

| Temp. °F | | $v$ | $h$ | $s$ | $C_p$ | $C_p/C_v$ | $v_s$ |
|---|---|---|---|---|---|---|---|
| 117.7 | SAT LIQ | 0.01439 | 51.5 | 0.1029 | 0.3723 | 1.6360 | 1320.2 |
| 117.7 | SAT VAP | 0.25856 | 118.1 | 0.2182 | 0.2854 | 1.3280 | 451.5 |
| 120 | | 0.26125 | 118.8 | 0.2193 | 0.2831 | 1.3184 | 454.6 |
| 130 | | 0.27221 | 121.6 | 0.2241 | 0.2749 | 1.2838 | 467.0 |
| 140 | | 0.28251 | 124.3 | 0.2287 | 0.2690 | 1.2572 | 478.3 |
| 150 | | 0.29231 | 126.9 | 0.2331 | 0.2647 | 1.2361 | 488.7 |
| 160 | | 0.30170 | 129.6 | 0.2374 | 0.2615 | 1.2189 | 498.4 |
| 170 | | 0.31073 | 132.2 | 0.2415 | 0.2591 | 1.2046 | 507.5 |
| 180 | | 0.31948 | 134.8 | 0.2456 | 0.2575 | 1.1925 | 516.1 |
| 190 | | 0.32803 | 137.3 | 0.2496 | 0.2563 | 1.1822 | 524.3 |
| 200 | | 0.33629 | 139.9 | 0.2535 | 0.2556 | 1.1732 | 532.2 |
| 210 | | 0.34438 | 142.4 | 0.2573 | 0.2553 | 1.1654 | 539.7 |
| 220 | | 0.35227 | 145.0 | 0.2611 | 0.2552 | 1.1586 | 546.9 |
| 230 | | 0.36006 | 147.5 | 0.2648 | 0.2554 | 1.1524 | 553.8 |
| 240 | | 0.36776 | 150.1 | 0.2685 | 0.2557 | 1.1470 | 560.6 |
| 250 | | 0.37536 | 152.7 | 1.2722 | 0.2562 | 1.1420 | 567.1 |
| 260 | | 0.38278 | 155.2 | 0.2757 | 0.2569 | 1.1376 | 573.4 |
| 270 | | 0.39014 | 157.8 | 0.2793 | 0.2577 | 1.1335 | 579.6 |
| 280 | | 0.39739 | 160.4 | 0.2828 | 0.2586 | 1.1298 | 585.6 |
| 290 | | 0.40461 | 163.0 | 0.2863 | 0.2596 | 1.1264 | 591.5 |

## Pressure = 200.00 psia

| Temp. °F | | $v$ | $h$ | $s$ | $C_p$ | $C_p/C_v$ | $v_s$ |
|---|---|---|---|---|---|---|---|
| 125.2 | SAT LIQ | 0.01465 | 54.3 | 0.1076 | 0.3801 | 1.6574 | 1253.0 |
| 125.2 | SAT VAP | 0.23026 | 118.8 | 0.2178 | 0.2966 | 1.3582 | 445.6 |
| 130 | | 0.23544 | 120.2 | 0.2202 | 0.2908 | 1.3352 | 452.3 |
| 140 | | 0.24565 | 123.0 | 0.2250 | 0.2816 | 1.2969 | 465.3 |
| 150 | | 0.25521 | 125.8 | 0.2296 | 0.2749 | 1.2678 | 477.0 |
| 160 | | 0.26428 | 128.5 | 0.2340 | 0.2700 | 1.2449 | 487.8 |
| 170 | | 0.27294 | 131.2 | 0.2383 | 0.2664 | 1.2263 | 497.8 |
| 180 | | 0.28129 | 133.9 | 0.2425 | 0.2637 | 1.2110 | 507.2 |
| 190 | | 0.28930 | 136.5 | 0.2466 | 0.2618 | 1.1981 | 516.1 |
| 200 | | 0.29712 | 139.1 | 0.2505 | 0.2604 | 1.1872 | 524.5 |
| 210 | | 0.30477 | 141.7 | 0.2545 | 0.2595 | 1.1777 | 532.5 |
| 220 | | 0.31217 | 144.3 | 0.2583 | 0.2590 | 1.1694 | 540.2 |
| 230 | | 0.31946 | 146.9 | 0.2621 | 0.2587 | 1.1622 | 547.5 |
| 240 | | 0.32658 | 149.5 | 0.2658 | 0.2588 | 1.1557 | 554.6 |
| 250 | | 0.33360 | 152.1 | 0.2695 | 0.2590 | 1.1500 | 561.5 |
| 260 | | 0.34046 | 154.7 | 0.2731 | 0.2594 | 1.1448 | 568.1 |
| 270 | | 0.34727 | 157.3 | 0.2767 | 0.2600 | 1.1401 | 574.6 |
| 280 | | 0.35398 | 159.9 | 0.2802 | 0.2607 | 1.1359 | 580.9 |
| 290 | | 0.36063 | 162.5 | 0.2837 | 0.2615 | 1.1320 | 587.0 |

(Continued)

# TABLE A.16 (CONTINUED)

HFC–134a Superheated Vapor–Constant Pressure Tables

## Pressure = 240.00 psia

| Temp. °F | $v$ | $h$ | $s$ | $C_p$ | $C_p/C_v$ | $v_s$ |
|---|---|---|---|---|---|---|
| 138.7 SAT LIQ | 0.01517 | 59.5 | 0.1162 | 0.3972 | 1.7064 | 1130.0 |
| 138.7 SAT VAP | 0.18715 | 119.8 | 0.2169 | 0.3214 | 1.4297 | 433.3 |
| 140 | 0.18846 | 120.2 | 0.2176 | 0.3189 | 1.4201 | 435.4 |
| 150 | 0.19819 | 123.3 | 0.2227 | 0.3032 | 1.3589 | 450.9 |
| 160 | 0.20712 | 126.3 | 0.2276 | 0.2924 | 1.3154 | 464.6 |
| 170 | 0.21545 | 129.2 | 0.2322 | 0.2846 | 1.2829 | 476.9 |
| 180 | 0.22330 | 132.0 | 0.2366 | 0.2789 | 1.2575 | 488.1 |
| 190 | 0.23083 | 134.8 | 0.2409 | 0.2747 | 1.2371 | 498.6 |
| 200 | 0.23799 | 137.5 | 0.2451 | 0.2716 | 1.2204 | 508.3 |
| 300 | 0.41168 | 165.6 | 0.2897 | 0.2606 | 1.1233 | 597.2 |
| 310 | 0.41873 | 168.2 | 0.2931 | 0.2617 | 1.1204 | 602.8 |
| 320 | 0.42577 | 170.8 | 0.2965 | 0.2629 | 1.1177 | 608.3 |
| 330 | 0.43269 | 173.4 | 0.2999 | 0.2641 | 1.1152 | 613.7 |
| 340 | 0.43964 | 176.1 | 0.3032 | 0.2653 | 1.1129 | 619.0 |
| 350 | 0.44645 | 178.7 | 0.3065 | 0.2666 | 1.1107 | 624.2 |
| 360 | 0.45325 | 181.4 | 0.3098 | 0.2679 | 1.1086 | 629.3 |
| 370 | 0.46007 | 184.1 | 0.3131 | 0.2693 | 1.1067 | 634.4 |
| 380 | 0.46677 | 186.8 | 0.3163 | 0.2706 | 1.1049 | 639.3 |
| 390 | 0.47344 | 189.5 | 0.3195 | 0.2720 | 1.1032 | 644.2 |
| 400 | 0.48015 | 192.2 | 0.3227 | 0.2734 | 1.1016 | 649.0 |
| 410 | 0.48681 | 195.0 | 0.3259 | 0.2748 | 1.1000 | 653.8 |
| 420 | 0.49336 | 197.7 | 0.3290 | 0.2762 | 1.0986 | 658.5 |
| 430 | — | — | — | — | — | — |

## Pressure = 280.00 psia

| Temp. °F | $v$ | $h$ | $s$ | $C_p$ | $C_p/C_v$ | $v_s$ |
|---|---|---|---|---|---|---|
| 150.6 SAT LIQ | 0.01570 | 64.3 | 0.1238 | 0.4172 | 1.7668 | 1018.2 |
| 150.6 SAT VAP | 0.15571 | 120.5 | 0.2160 | 0.3512 | 1.5212 | 420.5 |
| 160 | 0.16463 | 123.7 | 0.2211 | 0.3275 | 1.4317 | 437.5 |
| 170 | 0.17318 | 126.8 | 0.2262 | 0.3111 | 1.3684 | 453.2 |
| 180 | 0.18101 | 129.9 | 0.2310 | 0.2998 | 1.3235 | 467.1 |
| 190 | 0.18832 | 132.9 | 0.2356 | 0.2916 | 1.2899 | 479.6 |
| 200 | 0.19519 | 135.7 | 0.2400 | 0.2857 | 1.2637 | 491.1 |
| 210 | 0.20178 | 138.6 | 0.2443 | 0.2812 | 1.2427 | 501.7 |
| 220 | 0.20809 | 141.4 | 0.2484 | 0.2778 | 1.2254 | 511.7 |
| 230 | 0.21413 | 144.1 | 0.2525 | 0.2753 | 1.2110 | 521.0 |
| 300 | 0.36716 | 165.1 | 0.2872 | 0.2624 | 1.1285 | 593.0 |
| 310 | 0.37365 | 167.7 | 0.2906 | 0.2634 | 1.1252 | 598.8 |
| 320 | 0.38008 | 170.4 | 0.2941 | 0.2645 | 1.1222 | 604.5 |
| 330 | 0.38646 | 173.0 | 0.2974 | 0.2656 | 1.1194 | 610.1 |
| 340 | 0.39282 | 175.7 | 0.3008 | 0.2667 | 1.1168 | 615.5 |
| 350 | 0.39909 | 178.3 | 0.3041 | 0.2679 | 1.1144 | 620.9 |
| 360 | 0.40532 | 181.0 | 0.3074 | 0.2692 | 1.1121 | 626.2 |
| 370 | 0.41151 | 183.7 | 0.3107 | 0.2704 | 1.1100 | 631.3 |
| 380 | 0.41762 | 186.4 | 0.3139 | 0.2717 | 1.1080 | 636.4 |
| 390 | 0.42375 | 189.2 | 0.3171 | 0.2730 | 1.1061 | 641.5 |
| 400 | 0.42989 | 191.9 | 0.3203 | 0.2743 | 1.1044 | 646.4 |
| 410 | 0.43590 | 194.7 | 0.3235 | 0.2757 | 1.1027 | 651.3 |
| 420 | 0.44191 | 197.4 | 0.3267 | 0.2770 | 1.1011 | 656.1 |
| 430 | 0.44791 | 200.2 | 0.3298 | 0.2784 | 1.0996 | 660.8 |
| 440 | — | — | — | — | — | — |

*(Continued)*

**TABLE A.16 (CONTINUED)**

HFC–134a Superheated Vapor–Constant Pressure Tables

| | | | | | | |
|---|---|---|---|---|---|---|
| 210 | 0.24495 | 140.2 | 0.2492 | 0.2692 | 1.2064 | 517.5 |
| 220 | 0.25165 | 142.9 | 0.2531 | 0.2675 | 1.1945 | 526.2 |
| 230 | 0.25818 | 145.6 | 0.2570 | 0.2664 | 1.1843 | 534.5 |
| 240 | 0.26457 | 148.2 | 0.2609 | 0.2656 | 1.1754 | 542.5 |
| 250 | 0.27079 | 150.9 | 0.2646 | 0.2652 | 1.1677 | 550.1 |
| 260 | 0.27685 | 153.5 | 0.2683 | 0.2650 | 1.1608 | 557.4 |
| 270 | 0.28289 | 156.2 | 0.2720 | 0.2651 | 1.1546 | 564.4 |
| 280 | 0.28878 | 158.8 | 0.2756 | 0.2653 | 1.1491 | 571.3 |
| 290 | 0.29459 | 161.5 | 0.2792 | 0.2658 | 1.1441 | 577.9 |
| 300 | 0.30028 | 164.1 | 0.2827 | 0.2663 | 1.1396 | 584.3 |
| 310 | 0.30593 | 166.8 | 0.2862 | 0.2670 | 1.1355 | 590.6 |
| 320 | 0.31156 | 169.5 | 0.2896 | 0.2678 | 1.1318 | 596.7 |
| 330 | 0.31709 | 172.2 | 0.2931 | 0.2687 | 1.1283 | 602.7 |
| 340 | 0.32253 | 174.8 | 0.2965 | 0.2696 | 1.1251 | 608.5 |
| 350 | 0.32794 | 177.5 | 0.2998 | 0.2706 | 1.1222 | 614.2 |
| 360 | 0.33332 | 180.3 | 0.3031 | 0.2717 | 1.1194 | 619.8 |
| 370 | 0.33866 | 183.0 | 0.3064 | 0.2728 | 1.1169 | 625.3 |
| 380 | 0.34395 | 185.7 | 0.3097 | 0.2740 | 1.1145 | 630.6 |
| 390 | 0.34921 | 188.5 | 0.3130 | 0.2751 | 1.1122 | 635.9 |
| 400 | 0.35443 | 191.2 | 0.3162 | 0.2764 | 1.1101 | 641.1 |
| 410 | 0.35962 | 194.0 | 0.3194 | 0.2776 | 1.1081 | 646.2 |
| 420 | 0.36478 | 196.8 | 0.3226 | 0.2788 | 1.1063 | 651.2 |
| 430 | 0.36990 | 199.6 | 0.3257 | 0.2801 | 1.1045 | 656.2 |
| 440 | 0.37501 | 202.4 | 0.3289 | 0.2814 | 1.1028 | 661.0 |
| 450 | – | – | – | – | – | – |

| | | | | | | |
|---|---|---|---|---|---|---|
| 240 | 0.22001 | 146.9 | 0.2564 | 0.2735 | 1.1988 | 529.8 |
| 250 | 0.22573 | 149.6 | 0.2603 | 0.2722 | 1.1882 | 538.3 |
| 260 | 0.23127 | 152.3 | 0.2641 | 0.2713 | 1.1791 | 546.3 |
| 270 | 0.23672 | 155.0 | 0.2678 | 0.2707 | 1.1711 | 554.0 |
| 280 | 0.24208 | 157.7 | 0.2715 | 0.2704 | 1.1639 | 561.5 |
| 290 | 0.24728 | 160.4 | 0.2751 | 0.2704 | 1.1576 | 568.6 |
| 300 | 0.25246 | 163.1 | 0.2787 | 0.2706 | 1.1519 | 575.6 |
| 310 | 0.25753 | 165.9 | 0.2823 | 0.2709 | 1.1468 | 582.3 |
| 320 | 0.26250 | 168.6 | 0.2858 | 0.2714 | 1.1422 | 588.9 |
| 330 | 0.26744 | 171.3 | 0.2892 | 0.2720 | 1.1379 | 595.2 |
| 340 | 0.27230 | 174.0 | 0.2927 | 0.2727 | 1.1340 | 601.4 |
| 350 | 0.27712 | 176.7 | 0.2960 | 0.2735 | 1.1305 | 607.5 |
| 360 | 0.28188 | 179.5 | 0.2994 | 0.2744 | 1.1272 | 613.4 |
| 370 | 0.28661 | 182.2 | 0.3027 | 0.2753 | 1.1241 | 619.2 |
| 380 | 0.29128 | 185.0 | 0.3060 | 0.2763 | 1.1213 | 624.8 |
| 390 | 0.29592 | 187.7 | 0.3093 | 0.2774 | 1.1186 | 630.4 |
| 400 | 0.30053 | 190.5 | 0.3126 | 0.2785 | 1.1162 | 635.8 |
| 410 | 0.30509 | 193.3 | 0.3158 | 0.2796 | 1.1138 | 641.2 |
| 420 | 0.30964 | 196.1 | 0.3190 | 0.2807 | 1.1117 | 646.4 |
| 430 | 0.31414 | 198.9 | 0.3222 | 0.2819 | 1.1096 | 651.6 |
| 440 | 0.31862 | 201.8 | 0.3253 | 0.2831 | 1.1077 | 656.7 |
| 450 | 0.32308 | 204.6 | 0.3285 | 0.2843 | 1.1059 | 661.7 |
| 460 | 0.32751 | 207.4 | 0.3316 | 0.2855 | 1.1042 | 666.6 |

*(Continued)*

**TABLE A.16 (CONTINUED)**

HFC–134a Superheated Vapor–Constant Pressure Tables

Pressure = 320.00 psia

| Temp. °F | | $v$ | $h$ | $s$ | $C_p$ | $C_p/C_v$ | $v_s$ |
|---|---|---|---|---|---|---|---|
| 161.3 | SAT LIQ | 0.01627 | 68.7 | 0.1308 | 0.4415 | 1.8437 | 914.5 |
| 161.3 | SAT VAP | 0.13160 | 120.9 | 0.2148 | 0.3889 | 1.6421 | 407.2 |
| 170 | | 0.13984 | 124.1 | 0.2200 | 0.3548 | 1.5153 | 425.6 |
| 180 | | 0.14815 | 127.5 | 0.2254 | 0.3308 | 1.4254 | 443.4 |
| 190 | | 0.15562 | 130.7 | 0.2304 | 0.3152 | 1.3655 | 458.8 |
| 200 | | 0.16251 | 133.8 | 0.2351 | 0.3043 | 1.3225 | 472.5 |
| 210 | | 0.16893 | 136.8 | 0.2396 | 0.2964 | 1.2899 | 484.9 |
| 220 | | 0.17500 | 139.7 | 0.2439 | 0.2905 | 1.2644 | 496.3 |
| 230 | | 0.18080 | 142.6 | 0.2482 | 0.2861 | 1.2438 | 506.8 |
| 240 | | 0.18638 | 145.5 | 0.2523 | 0.2828 | 1.2267 | 516.8 |
| 250 | | 0.19175 | 148.3 | 0.2563 | 0.2803 | 1.2125 | 526.1 |
| 260 | | 0.19694 | 151.1 | 0.2602 | 0.2784 | 1.2003 | 535.0 |
| 270 | | 0.20198 | 153.8 | 0.2640 | 0.2771 | 1.1898 | 543.4 |
| 280 | | 0.20692 | 156.6 | 0.2678 | 0.2762 | 1.1807 | 551.5 |
| 290 | | 0.21174 | 159.4 | 0.2715 | 0.2756 | 1.1727 | 559.3 |
| 300 | | 0.21648 | 162.1 | 0.2751 | 0.2753 | 1.1655 | 566.8 |
| 310 | | 0.22111 | 164.9 | 0.2787 | 0.2752 | 1.1592 | 574.0 |
| 320 | | 0.22567 | 167.6 | 0.2823 | 0.2753 | 1.1535 | 581.0 |
| 330 | | 0.23016 | 170.4 | 0.2858 | 0.2756 | 1.1483 | 587.7 |
| 340 | | 0.23459 | 173.1 | 0.2892 | 0.2760 | 1.1436 | 594.3 |
| 350 | | 0.23895 | 175.9 | 0.2927 | 0.2766 | 1.1393 | 600.7 |
| 360 | | 0.24327 | 178.7 | 0.2961 | 0.2772 | 1.1354 | 607.0 |
| 370 | | 0.24754 | 181.5 | 0.2994 | 0.2780 | 1.1318 | 613.1 |

Pressure = 360.00 psia

| Temp. °F | | $v$ | $h$ | $s$ | $C_p$ | $C_p/C_v$ | $v_s$ |
|---|---|---|---|---|---|---|---|
| 171 | SAT LIQ | 0.01689 | 72.9 | 0.1373 | 0.4727 | 1.9458 | 816.6 |
| 171 | SAT VAP | 0.11237 | 120.9 | 0.2135 | 0.4391 | 1.8089 | 393.7 |
| 180 | | 0.12101 | 124.6 | 0.2193 | 0.3841 | 1.6069 | 415.6 |
| 190 | | 0.12915 | 128.2 | 0.2249 | 0.3508 | 1.4840 | 435.3 |
| 200 | | 0.13633 | 131.6 | 0.2301 | 0.3303 | 1.4071 | 452.1 |
| 210 | | 0.14285 | 134.9 | 0.2350 | 0.3164 | 1.3540 | 466.8 |
| 220 | | 0.14888 | 138.0 | 0.2396 | 0.3066 | 1.3149 | 480.0 |
| 230 | | 0.15458 | 141.0 | 0.2440 | 0.2993 | 1.2847 | 492.1 |
| 240 | | 0.15998 | 144.0 | 0.2483 | 0.2939 | 1.2608 | 503.2 |
| 250 | | 0.16513 | 146.9 | 0.2524 | 0.2898 | 1.2413 | 513.6 |
| 260 | | 0.17007 | 149.8 | 0.2565 | 0.2867 | 1.2251 | 523.4 |
| 270 | | 0.17486 | 152.6 | 0.2604 | 0.2843 | 1.2114 | 532.6 |
| 280 | | 0.17950 | 155.5 | 0.2643 | 0.2825 | 1.1997 | 541.4 |
| 290 | | 0.18402 | 158.3 | 0.2680 | 0.2813 | 1.1895 | 549.8 |
| 300 | | 0.18843 | 161.1 | 0.2718 | 0.2804 | 1.1806 | 557.9 |
| 310 | | 0.19275 | 163.9 | 0.2754 | 0.2798 | 1.1728 | 565.6 |
| 320 | | 0.19698 | 166.7 | 0.2790 | 0.2795 | 1.1658 | 573.1 |
| 330 | | 0.20113 | 169.5 | 0.2826 | 0.2794 | 1.1596 | 580.3 |
| 340 | | 0.20522 | 172.3 | 0.2861 | 0.2796 | 1.1539 | 587.2 |
| 350 | | 0.20925 | 175.1 | 0.2896 | 0.2798 | 1.1488 | 594.0 |
| 360 | | 0.21322 | 177.9 | 0.2930 | 0.2803 | 1.1442 | 600.6 |
| 370 | | 0.21713 | 180.7 | 0.2964 | 0.2808 | 1.1399 | 607.0 |
| 380 | | 0.22100 | 183.5 | 0.2998 | 0.2814 | 1.1360 | 613.2 |

*(Continued)*

**TABLE A.16 (CONTINUED)**

HFC–134a Superheated Vapor–Constant Pressure Tables

**Pressure = 400.00 psia**

| Temp. °F | $v$ | $h$ | $s$ | $C_p$ | $C_p/C_v$ | $v_s$ |
|---|---|---|---|---|---|---|
| 179.9 SAT LIQ | 0.01759 | 76.9 | 0.1435 | 0.5155 | 2.0897 | 722.7 |
| 179.9 SAT VAP | 0.09649 | 120.7 | 0.2119 | 0.5106 | 2.0521 | 380.0 |
| 180 | 0.09655 | 120.7 | 0.2119 | 0.5097 | 2.0487 | 380.2 |
| 190 | 0.10650 | 125.2 | 0.2190 | 0.4139 | 1.6998 | 408.0 |
| 200 | 0.11446 | 129.1 | 0.2249 | 0.3701 | 1.5401 | 429.3 |
| 210 | 0.12136 | 132.7 | 0.2303 | 0.3445 | 1.4457 | 447.2 |
| 220 | 0.12755 | 136.0 | 0.2352 | 0.3278 | 1.3828 | 462.7 |
| 230 | 0.13325 | 139.3 | 0.2399 | 0.3160 | 1.3375 | 476.6 |
| 240 | 0.13858 | 142.4 | 0.2444 | 0.3075 | 1.3031 | 489.2 |
| 250 | 0.14362 | 145.4 | 0.2487 | 0.3011 | 1.2762 | 500.8 |
| 260 | 0.14842 | 148.4 | 0.2529 | 0.2963 | 1.2544 | 511.6 |
| 270 | 0.15302 | 151.3 | 0.2570 | 0.2926 | 1.2364 | 521.7 |
| 280 | 0.15746 | 154.3 | 0.2609 | 0.2898 | 1.2214 | 531.2 |
| 380 | 0.25176 | 184.2 | 0.3028 | 0.2788 | 1.1285 | 619.0 |
| 390 | 0.25594 | 187.0 | 0.3061 | 0.2797 | 1.1254 | 624.8 |
| 400 | 0.26008 | 189.8 | 0.3094 | 0.2806 | 1.1225 | 630.5 |
| 410 | 0.26419 | 192.6 | 0.3126 | 0.2816 | 1.1198 | 636.1 |
| 420 | 0.26827 | 195.5 | 0.3158 | 0.2827 | 1.1173 | 641.6 |
| 430 | 0.27232 | 198.3 | 0.3190 | 0.2837 | 1.1150 | 647.0 |
| 440 | 0.27633 | 201.1 | 0.3222 | 0.2848 | 1.1127 | 652.3 |
| 450 | 0.28032 | 204.0 | 0.3254 | 0.2860 | 1.1107 | 657.5 |
| 460 | 0.28428 | 206.9 | 0.3285 | 0.2871 | 1.1087 | 662.6 |
| 470 | 0.28823 | 209.7 | 0.3316 | 0.2883 | 1.1069 | 667.6 |

**Pressure = 500.00 psia**

| Temp. °F | $v$ | $h$ | $s$ | $C_p$ | $C_p/C_v$ | $v_s$ |
|---|---|---|---|---|---|---|
| 199.4 SAT LIQ | 0.02000 | 86.9 | 0.1584 | 0.7816 | 3.0254 | 496.6 |
| 199.4 SAT VAP | 0.06538 | 118.1 | 0.2057 | 0.9822 | 3.6942 | 345.3 |
| 200 | 0.06646 | 118.7 | 0.2065 | 0.9106 | 3.4381 | 348.6 |
| 210 | 0.07860 | 125.3 | 0.2164 | 0.5274 | 2.0697 | 387.4 |
| 220 | 0.08669 | 130.0 | 0.2234 | 0.4325 | 1.7319 | 413.4 |
| 230 | 0.09326 | 134.1 | 0.2294 | 0.3868 | 1.5687 | 434.1 |
| 240 | 0.09896 | 137.8 | 0.2348 | 0.3596 | 1.4704 | 451.7 |
| 250 | 0.10410 | 141.3 | 0.2397 | 0.3416 | 1.4041 | 467.2 |
| 260 | 0.10883 | 144.6 | 0.2444 | 0.3289 | 1.3561 | 481.1 |
| 270 | 0.11326 | 147.9 | 0.2489 | 0.3195 | 1.3196 | 493.8 |
| 280 | 0.11744 | 151.0 | 0.2532 | 0.3125 | 1.2909 | 505.5 |
| 290 | 0.12142 | 154.1 | 0.2573 | 0.3072 | 1.2677 | 516.4 |
| 300 | 0.12523 | 157.2 | 0.2614 | 0.3030 | 1.2486 | 526.7 |
| 390 | 0.22483 | 186.3 | 0.3031 | 0.2821 | 1.1324 | 619.3 |
| 400 | 0.22862 | 189.1 | 0.3064 | 0.2829 | 1.1291 | 625.3 |
| 410 | 0.23237 | 192.0 | 0.3097 | 0.2838 | 1.1260 | 631.1 |
| 420 | 0.23609 | 194.8 | 0.3130 | 0.2847 | 1.1231 | 636.9 |
| 430 | 0.23978 | 197.7 | 0.3162 | 0.2856 | 1.1205 | 642.5 |
| 440 | 0.24343 | 200.5 | 0.3194 | 0.2866 | 1.1180 | 648.0 |
| 450 | 0.24707 | 203.4 | 0.3226 | 0.2877 | 1.1156 | 653.4 |
| 460 | 0.25066 | 206.3 | 0.3257 | 0.2887 | 1.1134 | 658.7 |
| 470 | 0.25425 | 209.2 | 0.3288 | 0.2898 | 1.1113 | 663.9 |
| 480 | 0.25781 | 212.1 | 0.3319 | 0.2909 | 1.1093 | 669.0 |

*(Continued)*

## TABLE A.16 (CONTINUED)

HFC–134a Superheated Vapor–Constant Pressure Tables

| $T$ | $v$ | $h$ | $s$ | $C_p$ | $C_p/C_v$ | $v_s$ | $v$ | $h$ | $s$ | $C_p$ | $C_p/C_v$ | $v_s$ |
|---|---|---|---|---|---|---|---|---|---|---|---|---|
| 290 | 0.16176 | 157.1 | 0.2648 | 0.2876 | 1.2085 | 540.3 | 0.12891 | 160.2 | 0.2653 | 0.2998 | 1.2325 | 536.4 |
| 300 | 0.16594 | 160.0 | 0.2686 | 0.2860 | 1.1974 | 548.9 | 0.13248 | 163.2 | 0.2692 | 0.2974 | 1.2188 | 545.6 |
| 310 | 0.17001 | 162.9 | 0.2724 | 0.2849 | 1.1878 | 557.2 | 0.13594 | 166.1 | 0.2730 | 0.2955 | 1.2070 | 554.4 |
| 320 | 0.17399 | 165.7 | 0.2760 | 0.2841 | 1.1793 | 565.1 | 0.13930 | 169.1 | 0.2767 | 0.2940 | 1.1966 | 562.9 |
| 330 | 0.17788 | 168.5 | 0.2796 | 0.2836 | 1.1717 | 572.8 | 0.14260 | 172.0 | 0.2803 | 0.2930 | 1.1876 | 571.0 |
| 340 | 0.18171 | 171.4 | 0.2832 | 0.2833 | 1.1650 | 580.2 | 0.14582 | 174.9 | 0.2839 | 0.2923 | 1.1795 | 578.8 |
| 350 | 0.18546 | 174.2 | 0.2867 | 0.2833 | 1.1590 | 587.3 | 0.14898 | 177.9 | 0.2874 | 0.2918 | 1.1723 | 586.3 |
| 360 | 0.18916 | 177.0 | 0.2902 | 0.2834 | 1.1535 | 594.3 | 0.15208 | 180.8 | 0.2909 | 0.2916 | 1.1659 | 593.6 |
| 370 | 0.19280 | 179.9 | 0.2937 | 0.2837 | 1.1485 | 601.0 | 0.15513 | 183.7 | 0.2944 | 0.2916 | 1.1600 | 600.7 |
| 380 | 0.19639 | 182.7 | 0.2971 | 0.2842 | 1.1440 | 607.5 | 0.15813 | 186.6 | 0.2978 | 0.2917 | 1.1547 | 607.6 |
| 390 | 0.19994 | 185.6 | 0.3004 | 0.2847 | 1.1399 | 613.9 | 0.16110 | 189.5 | 0.3012 | 0.2920 | 1.1499 | 614.3 |
| 400 | 0.20345 | 188.4 | 0.3038 | 0.2853 | 1.1360 | 620.1 | 0.16402 | 192.5 | 0.3045 | 0.2923 | 1.1454 | 620.8 |
| 410 | 0.20692 | 191.3 | 0.3071 | 0.2860 | 1.1325 | 626.2 | 0.16691 | 195.4 | 0.3078 | 0.2928 | 1.1413 | 627.1 |
| 420 | 0.21034 | 194.1 | 0.3103 | 0.2868 | 1.1292 | 632.2 | 0.16976 | 198.3 | 0.3111 | 0.2934 | 1.1376 | 633.4 |
| 430 | 0.21375 | 197.0 | 0.3136 | 0.2876 | 1.1262 | 638.0 | 0.17259 | 201.2 | 0.3143 | 0.2940 | 1.1341 | 639.4 |
| 440 | 0.21712 | 199.9 | 0.3168 | 0.2885 | 1.1233 | 643.7 | 0.17538 | 204.2 | 0.3176 | 0.2947 | 1.1308 | 645.4 |
| 450 | 0.22046 | 202.8 | 0.3200 | 0.2894 | 1.1207 | 649.3 | 0.17815 | 207.1 | 0.3208 | 0.2955 | 1.1278 | 651.2 |
| 460 | 0.22377 | 205.7 | 0.3232 | 0.2904 | 1.1182 | 654.8 | 0.18090 | 210.1 | 0.3239 | 0.2963 | 1.1249 | 656.9 |
| 470 | 0.22706 | 208.6 | 0.3263 | 0.2914 | 1.1159 | 660.2 | 0.18362 | 213.1 | 0.3271 | 0.2971 | 1.1223 | 662.5 |
| 480 | 0.23033 | 211.5 | 0.3294 | 0.2924 | 1.1137 | 665.5 | 0.18632 | 216.0 | 0.3302 | 0.2981 | 1.1198 | 668.0 |
| 490 | — | — | — | — | — | — | | | | | | |

*Source:* *Thermodynamic Properties of HFC–134a.* With permission of E. I. Du Pont de Nemours and Co.

*Note:* $C_p$ = heat capacity at constant pressure in Btu/(lb.) (°F); $C_p/C_v$ = heat capacity ratio (dimensionless); $h$ = enthalpy in Btu/lb.; $s$ = entropy in Btu/(lb.) (R); $v$ = volume in ft.$^3$/lb.; $v_s$ = velocity of sound in ft./s.

**TABLE A.15 (SI)**

HFC–134a Saturation Properties—Temperature Table

| Temp. °C | Pressure kPa (abs) | Volume m³/kg Liquid $v_f$ | Vapor $v_g$ | Density kg/m³ Liquid $1/v_f$ | Vapor $1/v_g$ | Enthalpy kJ/kg Liquid $h_f$ | Latent $h_{fg}$ | Vapor $h_g$ | Entropy kJ/(kg)(°K) Liquid $s_f$ | Vapor $s_g$ | Temp. °C |
|---|---|---|---|---|---|---|---|---|---|---|---|
| −40 | 51.14 | 0.0007 | 0.3614 | 1414.6 | 2.767 | 148.4 | 225.9 | 374.3 | 0.7967 | 1.7655 | −40 |
| −35 | 66.07 | 0.0007 | 0.2843 | 1400.2 | 3.518 | 154.6 | 222.8 | 377.4 | 0.8231 | 1.7586 | −35 |
| −30 | 84.29 | 0.0007 | 0.2260 | 1385.7 | 4.424 | 160.9 | 219.6 | 380.6 | 0.8492 | 1.7525 | −30 |
| −25 | 106.32 | 0.0007 | 0.1817 | 1371.0 | 5.504 | 167.3 | 216.4 | 383.7 | 0.8750 | 1.747 | −25 |
| −20 | 132.67 | 0.0007 | 0.1474 | 1356.0 | 6.784 | 173.7 | 213.1 | 386.8 | 0.9005 | 1.7422 | −15 |
| −15 | 163.90 | 0.0007 | 0.1207 | 1340.8 | 8.288 | 180.2 | 209.7 | 389.8 | 0.9257 | 1.7379 | −15 |
| −10 | 200.60 | 0.0008 | 0.0996 | 1325.3 | 10.044 | 186.7 | 206.2 | 392.9 | 0.9507 | 1.7341 | −10 |
| −5 | 243.39 | 0.0008 | 0.0828 | 1309.4 | 12.082 | 193.3 | 202.5 | 395.9 | 0.9755 | 1.7308 | −5 |
| 0 | 292.93 | 0.0008 | 0.0693 | 1293.3 | 14.435 | 200.0 | 198.8 | 398.8 | 1.0000 | 1.7278 | 0 |
| 5 | 349.87 | 0.0008 | 0.0583 | 1276.7 | 17.140 | 206.8 | 194.9 | 401.7 | 1.0244 | 1.7252 | 5 |
| 10 | 414.92 | 0.0008 | 0.0494 | 1259.8 | 20.236 | 213.6 | 190.9 | 404.5 | 1.0485 | 1.7229 | 10 |
| 15 | 488.78 | 0.0008 | 0.0421 | 1242.3 | 23.770 | 220.5 | 186.8 | 407.3 | 1.0726 | 1.7208 | 15 |
| 20 | 572.25 | 0.0008 | 0.0360 | 1224.4 | 27.791 | 227.5 | 182.5 | 410.0 | 1.0964 | 1.7189 | 20 |
| 25 | 666.06 | 0.0008 | 0.0309 | 1205.9 | 32.359 | 234.6 | 178.0 | 412.6 | 1.1202 | 1.7171 | 25 |
| 30 | 771.02 | 0.0008 | 0.0266 | 1186.7 | 37.540 | 241.8 | 173.3 | 415.1 | 1.1439 | 1.7155 | 30 |
| 35 | 887.91 | 0.0009 | 0.0230 | 1166.8 | 43.413 | 249.2 | 168.3 | 417.5 | 1.1676 | 1.7138 | 35 |
| 40 | 1017.61 | 0.0009 | 0.0200 | 1146.1 | 50.072 | 256.6 | 163.2 | 419.8 | 1.1912 | 1.7122 | 40 |
| 45 | 1161.01 | 0.0009 | 0.0174 | 1124.5 | 57.630 | 264.2 | 157.7 | 421.9 | 1.2148 | 1.7105 | 45 |
| 50 | 1319.00 | 0.0009 | 0.0151 | 1101.8 | 66.225 | 271.9 | 151.9 | 423.8 | 1.2384 | 1.7086 | 50 |
| 55 | 1492.59 | 0.0009 | 0.0132 | 1077.9 | 76.035 | 279.8 | 145.8 | 425.6 | 1.2622 | 1.7064 | 55 |
| 60 | 1682.76 | 0.0010 | 0.0115 | 1052.5 | 87.287 | 287.9 | 139.2 | 427.1 | 1.2861 | 1.7039 | 60 |
| 65 | 1890.54 | 0.0010 | 0.0100 | 1025.3 | 100.283 | 296.2 | 132.1 | 428.3 | 1.3102 | 1.7009 | 65 |
| 70 | 2117.34 | 0.0010 | 0.0087 | 995.9 | 115.442 | 304.8 | 124.4 | 429.1 | 1.3347 | 1.6971 | 70 |
| 75 | 2364.31 | 0.0010 | 0.0075 | 963.7 | 133.373 | 313.7 | 115.8 | 429.5 | 1.3597 | 1.6924 | 75 |
| 80 | 2632.97 | 0.0011 | 0.0065 | 927.8 | 155.010 | 322.9 | 106.3 | 429.2 | 1.3854 | 1.6863 | 80 |
| 85 | 2925.11 | 0.0011 | 0.0055 | 886.7 | 181.929 | 332.8 | 95.3 | 428.1 | 1.4121 | 1.6782 | 85 |
| 90 | 3242.87 | 0.0012 | 0.0046 | 837.3 | 217.162 | 343.4 | 82.1 | 425.5 | 1.4406 | 1.6668 | 90 |
| 95 | 3589.44 | 0.0013 | 0.0037 | 772.3 | 268.255 | 355.6 | 64.9 | 420.5 | 1.4727 | 1.6489 | 95 |
| 100 | 3969.94 | 0.0015 | 0.0027 | 651.4 | 375.503 | 373.2 | 33.8 | 407.0 | 1.5187 | 1.6092 | 100 |
| 101 | 4051.35 | 0.0018 | 0.0022 | 566.4 | 457.594 | 383.0 | 13.0 | 396.0 | 1.5447 | 1.5794 | 101 |

*Source: Thermodynamic Properties of HFC–134a. With permission of E. I. Du Pont de Nemours and Co.*

**TABLE A.16 (SI)**

HFC–134a Superheated Vapor–Constant Pressure Tables

**Pressure = 40.00 kPa (abs)**

| Temp.°C | | $v$ | $h$ | $s$ | $C_p$ | $C_p/C_v$ | $v_s$ |
|---|---|---|---|---|---|---|---|
| -44.57 | SAT LIQ | 0.00070 | 142.8 | 0.7722 | 1.2311 | 1.5059 | 831.3 |
| -44.57 | SAT VAP | 0.45496 | 371.4 | 1.7725 | 0.7301 | 1.1475 | 142.8 |
| -45 | | – | – | – | – | – | – |
| -40 | | 0.46490 | 374.8 | 1.7871 | 0.7364 | 1.1440 | 144.3 |
| -35 | | 0.47574 | 378.5 | 1.8028 | 0.7437 | 1.1404 | 145.9 |
| -30 | | 0.48662 | 382.2 | 1.8183 | 0.7511 | 1.1372 | 147.4 |
| -25 | | 0.49727 | 386.0 | 1.8336 | 0.7587 | 1.1342 | 148.9 |
| -20 | | 0.50787 | 389.8 | 1.8489 | 0.7665 | 1.1314 | 150.4 |
| -15 | | 0.51867 | 393.6 | 1.8639 | 0.7745 | 1.1288 | 151.9 |
| -10 | | 0.52910 | 397.5 | 1.8789 | 0.7825 | 1.1264 | 153.4 |
| -5 | | 0.53967 | 401.5 | 1.8937 | 0.7906 | 1.1241 | 154.8 |
| 0 | | 0.55036 | 405.4 | 1.9083 | 0.7988 | 1.1219 | 156.2 |
| 5 | | 0.56085 | 409.4 | 1.9229 | 0.8070 | 1.1199 | 157.7 |
| 10 | | 0.57110 | 413.5 | 1.9374 | 0.8152 | 1.1180 | 159.0 |
| 15 | | 0.58173 | 417.6 | 1.9517 | 0.8235 | 1.1161 | 160.4 |
| 20 | | 0.59207 | 421.7 | 1.9659 | 0.8318 | 1.1144 | 161.8 |
| 25 | | 0.60241 | 425.9 | 1.9801 | 0.8401 | 1.1128 | 163.1 |
| 30 | | 0.61312 | 430.1 | 1.9941 | 0.8484 | 1.1112 | 164.4 |
| 35 | | 0.62344 | 434.4 | 2.0081 | 0.8566 | 1.1097 | 165.7 |
| 40 | | 0.63371 | 438.7 | 2.0219 | 0.8649 | 1.1082 | 167.0 |
| 45 | | 0.64392 | 443.0 | 2.0357 | 0.8731 | 1.1068 | 168.3 |
| 50 | | 0.65445 | 447.4 | 2.0494 | 0.8813 | 1.1055 | 169.6 |
| 55 | | 0.66489 | 451.9 | 2.0630 | 0.8895 | 1.1042 | 170.9 |
| 60 | | 0.67522 | 456.3 | 2.0765 | 0.8977 | 1.1030 | 172.1 |

**Pressure = 80.00 kPa (abs)**

| Temp.°C | | $v$ | $h$ | $s$ | $C_p$ | $C_p/C_v$ | $v_s$ |
|---|---|---|---|---|---|---|---|
| -31.09 | SAT LIQ | 0.00072 | 159.5 | 0.8435 | 1.2606 | 1.5038 | 768.3 |
| -31.09 | SAT VAP | 0.23747 | 379.9 | 1.7538 | 0.7718 | 1.1515 | 145.0 |
| -30 | | 0.23872 | 380.7 | 1.7572 | 0.7729 | 1.1505 | 145.4 |
| -25 | | 0.24438 | 384.6 | 1.7730 | 0.7783 | 1.1460 | 147.1 |
| -20 | | 0.24994 | 388.5 | 1.7886 | 0.7842 | 1.1420 | 148.7 |
| -15 | | 0.25549 | 392.4 | 1.8040 | 0.7904 | 1.1383 | 150.3 |
| -10 | | 0.26103 | 396.4 | 1.8192 | 0.7969 | 1.1349 | 151.9 |
| -5 | | 0.26645 | 400.4 | 1.8343 | 0.8037 | 1.1318 | 153.4 |
| 0 | | 0.27196 | 404.5 | 1.8492 | 0.8107 | 1.1290 | 154.9 |
| 5 | | 0.27732 | 408.5 | 1.8640 | 0.8179 | 1.1263 | 156.4 |
| 10 | | 0.28273 | 412.6 | 1.8786 | 0.8252 | 1.1238 | 157.8 |
| 15 | | 0.28810 | 416.8 | 1.8931 | 0.8326 | 1.1215 | 159.3 |
| 20 | | 0.29343 | 421.0 | 1.9075 | 0.8402 | 1.1193 | 160.7 |
| 25 | | 0.29878 | 425.2 | 1.9218 | 0.8478 | 1.1172 | 162.1 |
| 30 | | 0.30404 | 429.4 | 1.9359 | 0.8555 | 1.1153 | 163.5 |
| 35 | | 0.30941 | 433.7 | 1.9500 | 0.8632 | 1.1135 | 164.8 |
| 40 | | 0.31466 | 438.1 | 1.9640 | 0.8709 | 1.1118 | 166.2 |
| 45 | | 0.32000 | 442.4 | 1.9778 | 0.8787 | 1.1101 | 167.5 |
| 50 | | 0.32520 | 446.9 | 1.9916 | 0.8865 | 1.1085 | 168.8 |
| 55 | | 0.33047 | 451.3 | 2.0053 | 0.8944 | 1.1071 | 170.1 |
| 60 | | 0.33568 | 455.8 | 2.0188 | 0.9022 | 1.1056 | 171.4 |
| 65 | | 0.34095 | 460.3 | 2.0323 | 0.9100 | 1.1043 | 172.7 |
| 70 | | 0.34614 | 464.9 | 2.0457 | 0.9178 | 1.1030 | 173.9 |

*(Continued)*

## TABLE A.16 (SI) (CONTINUED)

HFC–134a Superheated Vapor–Constant Pressure Tables

| Temp. °C | v | h | s | C_p | C_p/C_v | v_s |
|---|---|---|---|---|---|---|
| 65 | 0.68540 | 460.8 | 2.0899 | 0.9058 | 1.1018 | 173.4 |
| 70 | 0.69589 | 465.4 | 2.1033 | 0.9138 | 1.1006 | 174.6 |
| 75 | 0.70621 | 470.0 | 2.1165 | 0.9219 | 1.0995 | 175.8 |
| 80 | 0.71633 | 474.6 | 2.1297 | 0.9299 | 1.0985 | 177.0 |
| 85 | 0.72674 | 479.3 | 2.1429 | 0.9378 | 1.0974 | 178.2 |
| 90 | 0.73692 | 484.0 | 2.1559 | 0.9457 | 1.0964 | 179.4 |
| 95 | 0.74738 | 488.7 | 2.1689 | 0.9535 | 1.0954 | 180.6 |
| 100 | 0.75758 | 493.5 | 2.1818 | 0.9613 | 1.0945 | 181.8 |
| 105 | 0.76805 | 498.3 | 2.1947 | 0.9691 | 1.0936 | 183.0 |
| 110 | 0.77821 | 503.2 | 2.2074 | 0.9768 | 1.0927 | 184.1 |

| Temp. °C | v | h | s | C_p | C_p/C_v | v_s |
|---|---|---|---|---|---|---|
| 75 | 0.35137 | 469.5 | 2.0591 | 0.9255 | 1.1017 | 175.2 |
| 80 | 0.35663 | 474.2 | 2.0723 | 0.9333 | 1.1005 | 176.4 |
| 85 | 0.36179 | 478.8 | 2.0855 | 0.9410 | 1.0994 | 177.6 |
| 90 | 0.36697 | 483.6 | 2.0986 | 0.9487 | 1.0982 | 178.9 |
| 95 | 0.37216 | 488.3 | 2.1116 | 0.9564 | 1.0972 | 180.1 |
| 100 | 0.37736 | 493.1 | 2.1246 | 0.9640 | 1.0961 | 181.3 |
| 105 | 0.38256 | 498.0 | 2.1375 | 0.9716 | 1.0951 | 182.5 |
| 110 | 0.38775 | 502.8 | 2.1503 | 0.9792 | 1.0942 | 183.6 |
| 115 | 0.39293 | 507.8 | 2.1630 | 0.9867 | 1.0932 | 184.8 |
| 120 | 0.39809 | 512.7 | 2.1757 | 0.9941 | 1.0923 | 186.0 |

### Pressure = 100.00 kPa (abs)

| Temp. °C | | v | h | s | C_p | C_p/C_v | v_s |
|---|---|---|---|---|---|---|---|
| −26.34 | SAT LIQ | 0.00073 | 165.6 | 0.8681 | 1.2715 | 1.5046 | 746.2 |
| −26.34 | SATVAP | 0.19246 | 382.8 | 1.7484 | 0.7876 | 1.1539 | 145.7 |
| −25 | | 0.19372 | 383.9 | 1.7527 | 0.7888 | 1.1525 | 146.1 |
| −20 | | 0.19829 | 387.9 | 1.7685 | 0.7935 | 1.1477 | 147.8 |
| −15 | | 0.20284 | 391.8 | 1.7840 | 0.7988 | 1.1434 | 149.5 |
| −10 | | 0.20734 | 395.8 | 1.7994 | 0.8045 | 1.1395 | 151.1 |
| −5 | | 0.21182 | 399.9 | 1.8146 | 0.8106 | 1.1359 | 152.7 |
| 0 | | 0.21626 | 404.0 | 1.8297 | 0.8169 | 1.1327 | 154.2 |
| 5 | | 0.22065 | 408.1 | 1.8445 | 0.8235 | 1.1296 | 155.7 |
| 10 | | 0.22502 | 412.2 | 1.8593 | 0.8304 | 1.1269 | 157.2 |
| 15 | | 0.22941 | 416.4 | 1.8739 | 0.8373 | 1.1243 | 158.7 |
| 20 | | 0.23370 | 420.6 | 1.8883 | 0.8445 | 1.1218 | 160.1 |
| 25 | | 0.23804 | 424.8 | 1.9027 | 0.8517 | 1.1196 | 161.6 |
| 30 | | 0.24231 | 429.1 | 1.9169 | 0.8591 | 1.1175 | 163.0 |

### Pressure = 150.00 kPa (abs)

| Temp. °C | | v | h | s | C_p | C_p/C_v | v_s |
|---|---|---|---|---|---|---|---|
| −17.12 | SAT LIQ | 0.00074 | 177.4 | 0.9150 | 1.2939 | 1.5081 | 703.7 |
| −17.12 | SATVAP | 0.13123 | 388.5 | 1.7397 | 0.8201 | 1.1599 | 146.5 |
| −15 | | 0.13259 | 390.3 | 1.7464 | 0.8213 | 1.1574 | 147.3 |
| −10 | | 0.13576 | 394.4 | 1.7622 | 0.8247 | 1.1519 | 149.1 |
| −5 | | 0.13889 | 398.5 | 1.7778 | 0.8287 | 1.1470 | 150.8 |
| 0 | | 0.14196 | 402.7 | 1.7931 | 0.8333 | 1.1426 | 152.4 |
| 5 | | 0.14503 | 406.9 | 1.8083 | 0.8384 | 1.1386 | 154.1 |
| 10 | | 0.14806 | 411.1 | 1.8233 | 0.8438 | 1.1350 | 155.7 |
| 15 | | 0.15106 | 415.3 | 1.8381 | 0.8496 | 1.1316 | 157.2 |
| 20 | | 0.15404 | 419.6 | 1.8528 | 0.8557 | 1.1285 | 158.7 |
| 25 | | 0.15701 | 423.9 | 1.8673 | 0.8620 | 1.1257 | 160.2 |
| 30 | | 0.15995 | 428.2 | 1.8817 | 0.8685 | 1.1231 | 161.7 |
| 35 | | 0.16289 | 432.5 | 1.8959 | 0.8752 | 1.1206 | 163.2 |
| 40 | | 0.16581 | 436.9 | 1.9101 | 0.8820 | 1.1183 | 164.6 |

*(Continued)*

**TABLE A.16 (SI) (CONTINUED)**

HFC–134a Superheated Vapor–Constant Pressure Tables

| Temp. °C | $v$ | $h$ | $s$ | $C_p$ | $C_p/C_v$ | $v_s$ |
|---|---|---|---|---|---|---|
| 35 | 0.24661 | 433.4 | 1.9310 | 0.8665 | 1.1155 | 164.3 |
| 40 | 0.25088 | 437.7 | 1.9450 | 0.8741 | 1.1136 | 165.7 |
| 45 | 0.25517 | 442.1 | 1.9589 | 0.8816 | 1.1118 | 167.1 |
| 50 | 0.25940 | 446.6 | 1.9727 | 0.8892 | 1.1101 | 168.4 |
| 55 | 0.26364 | 451.0 | 1.9864 | 0.8968 | 1.1085 | 169.7 |
| 60 | 0.26788 | 455.5 | 2.0001 | 0.9045 | 1.1070 | 171.0 |
| 65 | 0.27203 | 460.1 | 2.0136 | 0.9121 | 1.1055 | 172.3 |
| 70 | 0.27624 | 464.7 | 2.0270 | 0.9198 | 1.1042 | 173.6 |
| 75 | 0.28043 | 469.3 | 2.0404 | 0.9274 | 1.1028 | 174.9 |
| 80 | 0.28466 | 473.9 | 2.0537 | 0.9350 | 1.1016 | 176.1 |
| 85 | 0.28885 | 478.6 | 2.0669 | 0.9427 | 1.1004 | 177.4 |
| 90 | 0.29300 | 483.4 | 2.0800 | 0.9503 | 1.0992 | 178.6 |
| 95 | 0.29718 | 488.1 | 2.0930 | 0.9578 | 1.0981 | 179.8 |
| 100 | 0.30139 | 492.9 | 2.1060 | 0.9654 | 1.0970 | 181.0 |
| 105 | 0.30553 | 497.8 | 2.1189 | 0.9729 | 1.0959 | 182.2 |
| 110 | 0.30969 | 502.7 | 2.1318 | 0.9804 | 1.0949 | 183.4 |
| 115 | 0.31387 | 507.6 | 2.1445 | 0.9878 | 1.0940 | 184.6 |
| 120 | 0.31797 | 512.5 | 2.1572 | 0.9952 | 1.0930 | 185.8 |
| 125 | 0.32216 | 517.5 | 2.1698 | 1.0026 | 1.0921 | 186.9 |

| Temp. °C | $v$ | $h$ | $s$ | $C_p$ | $C_p/C_v$ | $v_s$ |
|---|---|---|---|---|---|---|
| 45 | 0.16869 | 441.4 | 1.9241 | 0.8890 | 1.1162 | 166.0 |
| 50 | 0.17156 | 445.8 | 1.9380 | 0.8960 | 1.1142 | 167.4 |
| 55 | 0.17449 | 450.3 | 1.9518 | 0.9032 | 1.1123 | 168.7 |
| 60 | 0.17734 | 454.9 | 1.9656 | 0.9103 | 1.1105 | 170.1 |
| 65 | 0.18018 | 459.4 | 1.9792 | 0.9176 | 1.1088 | 171.4 |
| 70 | 0.18305 | 464.0 | 1.9927 | 0.9249 | 1.1072 | 172.8 |
| 75 | 0.18587 | 468.7 | 2.0061 | 0.9322 | 1.1057 | 174.1 |
| 80 | 0.18875 | 473.4 | 2.0195 | 0.9395 | 1.1043 | 175.3 |
| 85 | 0.19157 | 478.1 | 2.0327 | 0.9468 | 1.1029 | 176.6 |
| 90 | 0.19440 | 482.8 | 2.0459 | 0.9541 | 1.1016 | 177.9 |
| 95 | 0.19720 | 487.6 | 2.0590 | 0.9615 | 1.1003 | 179.1 |
| 100 | 0.20000 | 492.4 | 2.0720 | 0.9688 | 1.0991 | 180.4 |
| 105 | 0.20284 | 497.3 | 2.0850 | 0.9761 | 1.0979 | 181.6 |
| 110 | 0.20563 | 502.2 | 2.0978 | 0.9834 | 1.0968 | 182.8 |
| 115 | 0.20846 | 507.1 | 2.1106 | 0.9907 | 1.0958 | 184.0 |
| 120 | 0.21124 | 512.1 | 2.1234 | 0.9979 | 1.0947 | 185.2 |
| 125 | 0.21404 | 517.1 | 2.1360 | 1.0052 | 1.0937 | 186.4 |
| 130 | 0.21683 | 522.2 | 2.1486 | 1.0123 | 1.0928 | 187.6 |
| 135 | 0.21964 | 527.2 | 2.1611 | 1.0195 | 1.0919 | 188.7 |

**Pressure = 200.00 kPa (abs)**

| Temp. °C | | $v$ | $h$ | $s$ | $C_p$ | $C_p/C_v$ | $v_s$ |
|---|---|---|---|---|---|---|---|
| −10.08 | SAT LIQ | 0.00075 | 186.6 | 0.9503 | 1.3124 | 1.5125 | 671.3 |
| −10.08 | SAT VAP | 0.09985 | 392.8 | 1.7342 | 0.8468 | 1.1661 | 146.9 |
| −10 | | 0.09989 | 392.9 | 1.7344 | 0.8468 | 1.1660 | 146.9 |
| −5 | | 0.10235 | 397.1 | 1.7504 | 0.8485 | 1.1595 | 148.8 |

**Pressure = 240.00 kPa (abs)**

| Temp. °C | | $v$ | $h$ | $s$ | $C_p$ | $C_p/C_v$ | $v_s$ |
|---|---|---|---|---|---|---|---|
| −5.37 | SAT LIQ | 0.00076 | 198.2 | 0.9736 | 1.3254 | 1.5164 | 649.7 |
| −5.37 | SAT VAP | 0.08389 | 395.6 | 1.7310 | 0.8656 | 1.1711 | 147.0 |
| −5 | | 0.08405 | 396.0 | 1.7322 | 0.8657 | 1.1706 | 147.2 |
| 0 | | 0.08615 | 400.3 | 1.7482 | 0.8663 | 1.1634 | 149.1 |

*(Continued)*

**TABLE A.16 (SI) (CONTINUED)**

HFC-134a Superheated Vapor–Constant Pressure Tables

| T (°C) | v | h | s | v | h | s |
|---|---|---|---|---|---|---|
| 0 | 0.10478 | 401.4 | 1.7661 | 0.8510 | 1.1537 | 150.6 |
| 5 | 0.10717 | 405.6 | 1.7815 | 0.8543 | 1.1485 | 152.3 |
| 10 | 0.10953 | 409.9 | 1.7968 | 0.8583 | 1.1438 | 154.0 |
| 15 | 0.11186 | 414.2 | 1.8118 | 0.8627 | 1.1396 | 155.7 |
| 20 | 0.11417 | 418.5 | 1.8267 | 0.8676 | 1.1358 | 157.3 |
| 25 | 0.11647 | 422.9 | 1.8414 | 0.8729 | 1.1323 | 158.9 |
| 30 | 0.11874 | 427.3 | 1.8560 | 0.8784 | 1.1291 | 160.4 |
| 35 | 0.12099 | 431.7 | 1.8704 | 0.8843 | 1.1261 | 162.0 |
| 40 | 0.12324 | 436.1 | 1.8847 | 0.8904 | 1.1234 | 163.4 |
| 45 | 0.12547 | 440.6 | 1.8989 | 0.8967 | 1.1208 | 164.9 |
| 50 | 0.12767 | 445.1 | 1.9129 | 0.9031 | 1.1185 | 166.3 |
| 55 | 0.12989 | 449.6 | 1.9268 | 0.9097 | 1.1163 | 167.8 |
| 60 | 0.13207 | 454.2 | 1.9406 | 0.9164 | 1.1142 | 169.2 |
| 65 | 0.13425 | 458.8 | 1.9543 | 0.9232 | 1.1122 | 170.5 |
| 70 | 0.13643 | 463.4 | 1.9679 | 0.9301 | 1.1104 | 171.9 |
| 75 | 0.13860 | 468.1 | 1.9814 | 0.9370 | 1.1087 | 173.2 |
| 80 | 0.14075 | 472.8 | 1.9948 | 0.9440 | 1.1070 | 174.6 |
| 85 | 0.14290 | 477.5 | 2.0082 | 0.9511 | 1.1055 | 175.9 |
| 90 | 0.14505 | 482.3 | 2.0214 | 0.9581 | 1.1040 | 177.2 |
| 95 | 0.14719 | 487.1 | 2.0345 | 0.9652 | 1.1026 | 178.4 |
| 100 | 0.14932 | 491.9 | 2.0476 | 0.9723 | 1.1013 | 179.7 |
| 105 | 0.15145 | 496.8 | 2.0606 | 0.9794 | 1.1000 | 181.0 |
| 110 | 0.15359 | 501.7 | 2.0735 | 0.9865 | 1.0988 | 182.2 |
| 115 | 0.15571 | 506.7 | 2.0864 | 0.9936 | 1.0976 | 183.4 |
| 120 | 0.15783 | 511.7 | 2.0991 | 1.0007 | 1.0965 | 184.7 |
| 125 | 0.15995 | 516.7 | 2.1118 | 1.0078 | 1.0954 | 185.9 |

| T (°C) | v | h | s | v | h | s |
|---|---|---|---|---|---|---|
| 5 | 0.08821 | 404.6 | 1.7639 | 0.8680 | 1.1571 | 150.9 |
| 10 | 0.09024 | 409.0 | 1.7794 | 0.8705 | 1.1515 | 152.7 |
| 15 | 0.09224 | 413.3 | 1.7947 | 0.8738 | 1.1465 | 154.4 |
| 20 | 0.09422 | 417.7 | 1.8097 | 0.8776 | 1.1420 | 156.1 |
| 25 | 0.09618 | 422.1 | 1.8246 | 0.8820 | 1.1379 | 157.8 |
| 30 | 0.09812 | 426.5 | 1.8393 | 0.8867 | 1.1341 | 159.4 |
| 35 | 0.10004 | 431.0 | 1.8539 | 0.8919 | 1.1307 | 161.0 |
| 40 | 0.10194 | 435.4 | 1.8683 | 0.8973 | 1.1276 | 162.5 |
| 45 | 0.10383 | 439.9 | 1.8825 | 0.9030 | 1.1247 | 164.0 |
| 50 | 0.10571 | 444.5 | 1.8967 | 0.9090 | 1.1220 | 165.5 |
| 55 | 0.10757 | 449.0 | 1.9107 | 0.9151 | 1.1196 | 167.0 |
| 60 | 0.10942 | 453.6 | 1.9245 | 0.9214 | 1.1172 | 168.4 |
| 65 | 0.11127 | 458.3 | 1.9383 | 0.9278 | 1.1151 | 169.8 |
| 70 | 0.11311 | 462.9 | 1.9520 | 0.9344 | 1.1130 | 171.2 |
| 75 | 0.11494 | 467.6 | 1.9656 | 0.9410 | 1.1111 | 172.6 |
| 80 | 0.11675 | 472.3 | 1.9790 | 0.9478 | 1.1093 | 173.9 |
| 85 | 0.11857 | 477.1 | 1.9924 | 0.9546 | 1.1076 | 175.3 |
| 90 | 0.12038 | 481.9 | 2.0057 | 0.9614 | 1.1060 | 176.6 |
| 95 | 0.12219 | 486.7 | 2.0189 | 0.9683 | 1.1045 | 177.9 |
| 100 | 0.12398 | 491.5 | 2.0320 | 0.9752 | 1.1030 | 179.2 |
| 105 | 0.12577 | 496.4 | 2.0450 | 0.9821 | 1.1017 | 180.5 |
| 110 | 0.12755 | 501.4 | 2.0579 | 0.9891 | 1.1004 | 181.7 |
| 115 | 0.12935 | 506.3 | 2.0708 | 0.9960 | 1.0991 | 183.0 |
| 120 | 0.13111 | 511.3 | 2.0836 | 1.0030 | 1.0979 | 184.2 |
| 125 | 0.13291 | 516.4 | 2.0963 | 1.0099 | 1.0967 | 185.4 |
| 130 | 0.13466 | 521.4 | 2.1090 | 1.0168 | 1.0956 | 186.7 |

*(Continued)*

# TABLE A.16 (SI) (CONTINUED)

HFC–134a Superheated Vapor–Constant Pressure Tables

| Temp. °C | $v$ | $h$ | $s$ | $C_p$ | $C_p/C_v$ | $v_s$ |
|---|---|---|---|---|---|---|
| 130 | 0.16207 | 521.8 | 2.1244 | 1.0148 | 1.0944 | 187.1 |
| 135 | 0.16418 | 526.8 | 2.1370 | 1.0218 | 1.0934 | 188.2 |
| 140 | 0.16628 | 532.0 | 2.1495 | 1.0288 | 1.0924 | 189.4 |
| 145 | – | – | – | – | – | – |

| Temp. °C | $v$ | $h$ | $s$ | $C_p$ | $C_p/C_v$ | $v_s$ |
|---|---|---|---|---|---|---|
| 135 | 0.13646 | 526.5 | 2.1216 | 1.0237 | 1.0946 | 187.9 |
| 140 | 0.13824 | 531.7 | 2.1341 | 1.0306 | 1.0936 | 189.0 |
| 145 | 0.13998 | 536.8 | 2.1465 | 1.0375 | 1.0926 | 190.2 |
| 150 | – | – | – | – | – | – |

Pressure = 280.00 kPa (abs)

| Temp. °C | $v$ | $h$ | $s$ | $C_p$ | $C_p/C_v$ | $v_s$ |
|---|---|---|---|---|---|---|
| −1.24 SAT LIQ | 0.00077 | 198.3 | 0.9939 | 1.3374 | 1.5204 | 630.8 |
| −1.24 SAT VAP | 0.07235 | 398.1 | 1.7285 | 0.8830 | 1.1762 | 147.0 |
| 0 | 0.07281 | 399.2 | 1.7325 | 0.8827 | 1.1741 | 147.5 |
| 5 | 0.07464 | 403.6 | 1.7485 | 0.8826 | 1.1665 | 149.5 |
| 10 | 0.07645 | 408.0 | 1.7643 | 0.8835 | 1.1598 | 151.3 |
| 15 | 0.07822 | 412.4 | 1.7797 | 0.8854 | 1.1539 | 153.2 |
| 20 | 0.07996 | 416.9 | 1.7950 | 0.8881 | 1.1486 | 154.9 |
| 25 | 0.08168 | 421.3 | 1.8100 | 0.8915 | 1.1438 | 156.7 |
| 30 | 0.08338 | 425.8 | 1.8249 | 0.8954 | 1.1395 | 158.3 |
| 35 | 0.08506 | 430.3 | 1.8396 | 0.8998 | 1.1356 | 160.0 |
| 40 | 0.08672 | 434.8 | 1.8541 | 0.9045 | 1.1320 | 161.6 |
| 45 | 0.08837 | 439.3 | 1.8685 | 0.9096 | 1.1288 | 163.1 |
| 50 | 0.09001 | 443.9 | 1.8827 | 0.9150 | 1.1258 | 164.7 |
| 55 | 0.09163 | 448.5 | 1.8968 | 0.9207 | 1.1230 | 166.2 |
| 60 | 0.09325 | 453.1 | 1.9108 | 0.9266 | 1.1204 | 167.6 |
| 65 | 0.09485 | 457.7 | 1.9246 | 0.9326 | 1.1180 | 169.1 |
| 70 | 0.09645 | 462.4 | 1.9383 | 0.9388 | 1.1158 | 170.5 |
| 75 | 0.09804 | 467.1 | 1.9520 | 0.9451 | 1.1137 | 171.9 |
| 80 | 0.09962 | 471.8 | 1.9655 | 0.9516 | 1.1117 | 173.3 |
| 85 | 0.10118 | 476.6 | 1.9789 | 0.9581 | 1.1098 | 174.7 |
| 90 | 0.10275 | 481.4 | 1.9922 | 0.9647 | 1.1081 | 176.0 |

Pressure = 300.00 kPa (abs)

| Temp. °C | $v$ | $h$ | $s$ | $C_p$ | $C_p/C_v$ | $v_s$ |
|---|---|---|---|---|---|---|
| 0.66 SAT LIQ | 0.00077 | 200.9 | 1.0032 | 1.3431 | 1.5225 | 622.1 |
| 0.66 SAT VAP | 0.06770 | 399.2 | 1.7275 | 0.8912 | 1.1787 | 147.0 |
| 5 | 0.06921 | 403.1 | 1.7415 | 0.8902 | 1.1715 | 148.7 |
| 10 | 0.07092 | 407.5 | 1.7573 | 0.8904 | 1.1642 | 150.7 |
| 15 | 0.07260 | 412.0 | 1.7729 | 0.8915 | 1.1578 | 152.5 |
| 20 | 0.07424 | 416.4 | 1.7883 | 0.8936 | 1.1521 | 154.3 |
| 25 | 0.07587 | 420.9 | 1.8034 | 0.8964 | 1.1469 | 156.1 |
| 30 | 0.07748 | 425.4 | 1.8183 | 0.8998 | 1.1423 | 157.8 |
| 35 | 0.07906 | 429.9 | 1.8331 | 0.9038 | 1.1381 | 159.5 |
| 40 | 0.08063 | 434.4 | 1.8477 | 0.9082 | 1.1343 | 161.1 |
| 45 | 0.08219 | 439.0 | 1.8621 | 0.9130 | 1.1309 | 162.7 |
| 50 | 0.08372 | 443.6 | 1.8764 | 0.9181 | 1.1277 | 164.2 |
| 55 | 0.08525 | 448.2 | 1.8905 | 0.9236 | 1.1247 | 165.8 |
| 60 | 0.08678 | 452.8 | 1.9045 | 0.9292 | 1.1220 | 167.3 |
| 65 | 0.08829 | 457.5 | 1.9184 | 0.9350 | 1.1195 | 168.7 |
| 70 | 0.08978 | 462.1 | 1.9322 | 0.9411 | 1.1171 | 170.2 |
| 75 | 0.09128 | 466.9 | 1.9458 | 0.9472 | 1.1149 | 171.6 |
| 80 | 0.09276 | 471.6 | 1.9594 | 0.9535 | 1.1129 | 173.0 |
| 85 | 0.09424 | 476.4 | 1.9728 | 0.9599 | 1.1110 | 174.4 |
| 90 | 0.09571 | 481.2 | 1.9862 | 0.9664 | 1.1091 | 175.7 |
| 95 | 0.09718 | 486.1 | 1.9994 | 0.9730 | 1.1074 | 177.1 |

(Continued)

# TABLE A.16 (SI) (CONTINUED)

HFC-134a Superheated Vapor–Constant Pressure Tables

| Temp. °C | v | h | s | $C_p$ | $C_p/C_v$ | $v_s$ |
|---|---|---|---|---|---|---|
| 95 | 0.10432 | 486.3 | 2.0055 | 0.9714 | 1.1064 | 177.3 |
| 100 | 0.10588 | 491.1 | 2.0186 | 0.9781 | 1.1049 | 178.7 |
| 105 | 0.10743 | 496.1 | 2.0317 | 0.9848 | 1.1034 | 180.0 |
| 110 | 0.10897 | 501.0 | 2.0447 | 0.9916 | 1.1020 | 181.2 |
| 115 | 0.11052 | 506.0 | 2.0576 | 0.9984 | 1.1006 | 182.5 |
| 120 | 0.11206 | 511.0 | 2.0704 | 1.0052 | 1.0993 | 183.8 |
| 125 | 0.11358 | 516.0 | 2.0831 | 1.0120 | 1.0981 | 185.0 |
| 130 | 0.11511 | 521.1 | 2.0958 | 1.0189 | 1.0969 | 186.2 |
| 135 | 0.11665 | 526.2 | 2.1084 | 1.0257 | 1.0958 | 187.5 |
| 140 | 0.11818 | 531.4 | 2.1209 | 1.0325 | 1.0947 | 188.7 |
| 145 | 0.11969 | 536.5 | 2.1334 | 1.0392 | 1.0937 | 189.9 |
| 150 | 0.12121 | 541.7 | 2.1458 | 1.0460 | 1.0927 | 191.0 |

| Temp. °C | v | h | s | $C_p$ | $C_p/C_v$ | $v_s$ |
|---|---|---|---|---|---|---|
| 100 | 0.09863 | 490.9 | 2.0126 | 0.9796 | 1.1058 | 178.4 |
| 105 | 0.10009 | 495.9 | 2.0257 | 0.9862 | 1.1042 | 179.7 |
| 110 | 0.10153 | 500.8 | 2.0387 | 0.9929 | 1.1028 | 181.0 |
| 115 | 0.10299 | 505.8 | 2.0516 | 0.9996 | 1.1014 | 182.3 |
| 120 | 0.10444 | 510.8 | 2.0645 | 1.0064 | 1.1001 | 183.5 |
| 125 | 0.10585 | 515.9 | 2.0772 | 1.0131 | 1.0988 | 184.8 |
| 130 | 0.10730 | 520.9 | 2.0899 | 1.0199 | 1.0976 | 186.0 |
| 135 | 0.10873 | 526.1 | 2.1025 | 1.0266 | 1.0964 | 187.3 |
| 140 | 0.11017 | 531.2 | 2.1151 | 1.0334 | 1.0953 | 188.5 |
| 145 | 0.11158 | 536.4 | 2.1275 | 1.0401 | 1.0943 | 189.7 |
| 150 | 0.11301 | 541.6 | 2.1399 | 1.0468 | 1.0932 | 190.9 |
| 155 | 0.11442 | 546.9 | 2.1523 | 1.0535 | 1.0922 | 192.1 |

## Pressure = 340.00 kPa (abs)

| Temp. °C | v | h | s | $C_p$ | $C_p/C_v$ | $v_s$ |
|---|---|---|---|---|---|---|
| SAT LIQ 4.18 | 0.00078 | 205.6 | 1.0204 | 1.3541 | 1.5266 | 605.9 |
| SAT VAP 4.18 | 0.05998 | 401.2 | 1.7256 | 0.9069 | 1.1838 | 146.8 |
| 5 | 0.06024 | 402.0 | 1.7283 | 0.9064 | 1.1823 | 147.2 |
| 10 | 0.06180 | 406.5 | 1.7444 | 0.9047 | 1.1736 | 149.2 |
| 15 | 0.06333 | 411.0 | 1.7602 | 0.9043 | 1.1660 | 151.2 |
| 20 | 0.06483 | 415.5 | 1.7758 | 0.9050 | 1.1594 | 153.1 |
| 25 | 0.06630 | 420.1 | 1.7911 | 0.9066 | 1.1535 | 154.9 |
| 30 | 0.06776 | 424.6 | 1.8062 | 0.9091 | 1.1482 | 156.7 |
| 35 | 0.06919 | 429.2 | 1.8211 | 0.9122 | 1.1434 | 158.4 |
| 40 | 0.07060 | 433.7 | 1.8358 | 0.9158 | 1.1391 | 160.1 |
| 45 | 0.07200 | 438.3 | 1.8504 | 0.9200 | 1.1352 | 161.8 |
| 50 | 0.07339 | 442.9 | 1.8648 | 0.9245 | 1.1316 | 163.4 |
| 55 | 0.07476 | 447.6 | 1.8790 | 0.9294 | 1.1284 | 164.9 |
| 60 | 0.07612 | 452.2 | 1.8931 | 0.9346 | 1.1254 | 166.5 |

## Pressure = 380.00 kPa (abs)

| Temp. °C | v | h | s | $C_p$ | $C_p/C_v$ | $v_s$ |
|---|---|---|---|---|---|---|
| SAT LIQ 7.4 | 0.00079 | 210.0 | 1.0360 | 1.3645 | 1.5308 | 591.1 |
| SAT VAP 7.4 | 0.05384 | 403.1 | 1.7240 | 0.9218 | 1.1890 | 146.6 |
| 10 | 0.05459 | 405.5 | 1.7325 | 0.9200 | 1.1838 | 147.8 |
| 15 | 0.05600 | 410.1 | 1.7486 | 0.9178 | 1.1750 | 149.8 |
| 20 | 0.05738 | 414.6 | 1.7644 | 0.9170 | 1.1672 | 151.8 |
| 25 | 0.05874 | 419.2 | 1.7799 | 0.9174 | 1.1604 | 153.7 |
| 30 | 0.6007 | 423.8 | 1.7952 | 0.9188 | 1.1544 | 155.6 |
| 35 | 0.06138 | 428.4 | 1.8102 | 0.9209 | 1.1490 | 157.4 |
| 40 | 0.06267 | 433.0 | 1.8251 | 0.9238 | 1.1441 | 159.1 |
| 45 | 0.06395 | 437.7 | 1.8397 | 0.9272 | 1.1398 | 160.8 |
| 50 | 0.06521 | 442.3 | 1.8542 | 0.9311 | 1.1358 | 162.5 |
| 55 | 0.06645 | 447.0 | 1.8686 | 0.9354 | 1.1321 | 164.1 |
| 60 | 0.06769 | 451.7 | 1.8827 | 0.9401 | 1.1288 | 165.7 |
| 65 | 0.06892 | 456.4 | 1.8968 | 0.9451 | 1.1258 | 167.2 |

*(Continued)*

# TABLE A.16 (SI) (CONTINUED)

HFC–134a Superheated Vapor–Constant Pressure Tables

| Temp. °C | $v$ | $h$ | $s$ | $C_p$ | $C_p/C_v$ | $v_s$ |
|---|---|---|---|---|---|---|
| 65 | 0.07747 | 456.9 | 1.9070 | 0.9400 | 1.1226 | 168.0 |
| 70 | 0.07880 | 461.6 | 1.9209 | 0.9456 | 1.1200 | 169.5 |
| 75 | 0.08013 | 466.4 | 1.9346 | 0.9515 | 1.1176 | 170.9 |
| 80 | 0.08146 | 471.1 | 1.9482 | 0.9575 | 1.1153 | 172.3 |
| 85 | 0.08280 | 475.9 | 1.9617 | 0.9636 | 1.1132 | 173.8 |
| 90 | 0.08410 | 480.8 | 1.9751 | 0.9698 | 1.1113 | 175.1 |
| 95 | 0.08540 | 485.6 | 1.9884 | 0.9761 | 1.1094 | 176.5 |
| 100 | 0.08669 | 490.5 | 2.0016 | 0.9826 | 1.1077 | 177.9 |
| 105 | 0.08800 | 495.5 | 2.0147 | 0.9890 | 1.1060 | 179.2 |
| 110 | 0.08929 | 500.4 | 2.0278 | 0.9955 | 1.1044 | 180.5 |
| 115 | 0.09058 | 505.4 | 2.0407 | 1.0021 | 1.1030 | 181.8 |
| 120 | 0.09184 | 510.4 | 2.0536 | 1.0087 | 1.1015 | 183.1 |
| 125 | 0.09314 | 515.5 | 2.0664 | 1.0153 | 1.1002 | 184.4 |
| 130 | 0.09440 | 520.6 | 2.0791 | 1.0220 | 1.0989 | 185.6 |
| 135 | 0.09568 | 525.7 | 2.0917 | 1.0286 | 1.0977 | 186.9 |
| 140 | 0.09696 | 530.9 | 2.1043 | 1.0352 | 1.0965 | 188.1 |
| 145 | 0.09822 | 536.1 | 2.1168 | 1.0419 | 1.0954 | 189.3 |
| 150 | 0.09948 | 541.3 | 2.1292 | 1.0485 | 1.0943 | 190.5 |
| 155 | 0.10074 | 546.6 | 2.1416 | 1.0551 | 1.0933 | 191.7 |
| 160 | — | — | — | — | — | — |

| Temp. °C | $v$ | $h$ | $s$ | $C_p$ | $C_p/C_v$ | $v_s$ |
|---|---|---|---|---|---|---|
| 70 | 0.07014 | 461.1 | 1.9107 | 0.9503 | 1.1229 | 168.8 |
| 75 | 0.07135 | 465.9 | 1.9245 | 0.9558 | 1.1203 | 170.2 |
| 80 | 0.07254 | 470.7 | 1.9381 | 0.9615 | 1.1179 | 171.7 |
| 85 | 0.07374 | 475.5 | 1.9517 | 0.9673 | 1.1156 | 173.1 |
| 90 | 0.07492 | 480.3 | 1.9652 | 0.9733 | 1.1135 | 174.6 |
| 95 | 0.07611 | 485.2 | 1.9785 | 0.9794 | 1.1115 | 176.0 |
| 100 | 0.07729 | 490.1 | 1.9918 | 0.9856 | 1.1096 | 177.3 |
| 105 | 0.07846 | 495.1 | 2.0049 | 0.9919 | 1.1078 | 178.7 |
| 110 | 0.07962 | 500.1 | 2.0180 | 0.9982 | 1.1061 | 180.0 |
| 115 | 0.08078 | 505.1 | 2.0310 | 1.0046 | 1.1046 | 181.3 |
| 120 | 0.08193 | 510.1 | 2.0439 | 1.0111 | 1.1031 | 182.6 |
| 125 | 0.08308 | 515.2 | 2.0567 | 1.0175 | 1.1016 | 183.9 |
| 130 | 0.08423 | 520.3 | 2.0694 | 1.0241 | 1.1003 | 185.2 |
| 135 | 0.08538 | 525.4 | 2.0821 | 1.0306 | 1.0990 | 186.5 |
| 140 | 0.08653 | 530.6 | 2.0947 | 1.0371 | 1.0977 | 187.7 |
| 145 | 0.08767 | 535.8 | 2.1072 | 1.0437 | 1.0966 | 189.0 |
| 150 | 0.08881 | 541.0 | 2.1196 | 1.0502 | 1.0954 | 190.2 |
| 155 | 0.08994 | 546.3 | 2.1320 | 1.0567 | 1.0943 | 191.4 |
| 160 | 0.09107 | 551.6 | 2.1443 | 1.0632 | 1.0933 | 192.6 |

| Pressure = 400.00 kPa (abs) | | | | | | |
|---|---|---|---|---|---|---|
| Temp. °C | $v$ | $h$ | $s$ | $C_p$ | $C_p/C_v$ | $v_s$ |
| SAT LIQ 8.91 | 0.00079 | 212.1 | 1.0433 | 1.3695 | 1.5329 | 584.1 |
| SAT VAP 8.91 | 0.05122 | 403.9 | 1.7234 | 0.9290 | 1.1916 | 146.5 |
| 10 | 0.05152 | 404.9 | 1.7269 | 0.9281 | 1.1893 | 147.0 |
| 15 | 0.05288 | 409.6 | 1.7432 | 0.9249 | 1.1797 | 149.1 |

| Pressure = 500.00 kPa (abs) | | | | | | |
|---|---|---|---|---|---|---|
| Temp. °C | $v$ | $h$ | $s$ | $C_p$ | $C_p/C_v$ | $v_s$ |
| SAT LIQ 15.71 | 0.00081 | 221.5 | 1.0759 | 1.3937 | 1.5436 | 552.8 |
| SAT VAP 15.71 | 0.04114 | 407.7 | 1.7205 | 0.9633 | 1.2049 | 145.8 |
| 20 | 0.04213 | 411.8 | 1.7347 | 0.9578 | 1.1947 | 147.8 |

*(Continued)*

## TABLE A.16 (SI) (CONTINUED)

HFC–134a Superheated Vapor–Constant Pressure Tables

| | | | | | | |
|---|---|---|---|---|---|---|
| 20 | 0.05421 | 414.2 | 1.7590 | 0.9233 | 1.1714 | 151.2 |
| 25 | 0.05552 | 418.8 | 1.7747 | 0.9230 | 1.1641 | 153.1 |
| 30 | 0.05680 | 423.4 | 1.7900 | 0.9238 | 1.1576 | 155.0 |
| 35 | 0.05806 | 428.0 | 1.8051 | 0.9254 | 1.1519 | 156.9 |
| 40 | 0.05930 | 432.7 | 1.8201 | 0.9278 | 1.1467 | 158.6 |
| 45 | 0.06052 | 437.3 | 1.8348 | 0.9309 | 1.1421 | 160.4 |
| 50 | 0.06172 | 442.0 | 1.8493 | 0.9345 | 1.1379 | 162.0 |
| 55 | 0.06293 | 446.7 | 1.8637 | 0.9385 | 1.1341 | 163.7 |
| 60 | 0.06411 | 451.4 | 1.8779 | 0.9429 | 1.1306 | 165.3 |
| 65 | 0.06529 | 456.1 | 1.8920 | 0.9477 | 1.1274 | 166.9 |
| 70 | 0.06645 | 460.8 | 1.9059 | 0.9527 | 1.1244 | 168.4 |
| 75 | 0.06761 | 465.6 | 1.9198 | 0.9580 | 1.1217 | 169.9 |
| 80 | 0.06876 | 470.4 | 1.9335 | 0.9635 | 1.1192 | 171.4 |
| 85 | 0.06989 | 475.3 | 1.9471 | 0.9692 | 1.1168 | 172.8 |
| 90 | 0.07102 | 480.1 | 1.9605 | 0.9751 | 1.1146 | 174.3 |
| 95 | 0.07215 | 485.0 | 1.9739 | 0.9810 | 1.1125 | 175.7 |
| 100 | 0.07328 | 489.9 | 1.9872 | 0.9871 | 1.1106 | 177.1 |
| 105 | 0.07440 | 494.9 | 2.0004 | 0.9933 | 1.1087 | 178.4 |
| 110 | 0.07551 | 499.9 | 2.0134 | 0.9996 | 1.1070 | 179.8 |
| 115 | 0.07661 | 504.9 | 2.0264 | 1.0059 | 1.1054 | 181.1 |
| 120 | 0.07772 | 509.9 | 2.0394 | 1.0123 | 1.1038 | 182.4 |
| 125 | 0.07881 | 515.0 | 2.0522 | 1.0187 | 1.1023 | 183.7 |
| 130 | 0.07991 | 520.1 | 2.0649 | 1.0251 | 1.1009 | 185.0 |
| 135 | 0.08101 | 525.2 | 2.0776 | 1.0316 | 1.0996 | 186.3 |
| 140 | 0.08210 | 530.4 | 2.0902 | 1.0381 | 1.0983 | 187.5 |
| 145 | 0.08318 | 535.6 | 2.1028 | 1.0445 | 1.0971 | 188.8 |

| | | | | | | |
|---|---|---|---|---|---|---|
| 25 | 0.04325 | 416.6 | 1.7508 | 0.9535 | 1.1844 | 150.0 |
| 30 | 0.04434 | 421.3 | 1.7667 | 0.9509 | 1.1755 | 152.1 |
| 35 | 0.04541 | 426.1 | 1.7822 | 0.9496 | 1.1677 | 154.1 |
| 40 | 0.04646 | 430.8 | 1.7975 | 0.9496 | 1.1609 | 156.1 |
| 45 | 0.04749 | 435.6 | 1.8125 | 0.9505 | 1.1548 | 158.0 |
| 50 | 0.04851 | 440.4 | 1.8274 | 0.9523 | 1.1493 | 159.8 |
| 55 | 0.04951 | 445.1 | 1.8420 | 0.9547 | 1.1444 | 161.5 |
| 60 | 0.05050 | 449.9 | 1.8565 | 0.9577 | 1.1400 | 163.3 |
| 65 | 0.05147 | 454.7 | 1.8708 | 0.9612 | 1.1360 | 164.9 |
| 70 | 0.05244 | 459.5 | 1.8849 | 0.9652 | 1.1323 | 166.6 |
| 75 | 0.05339 | 464.3 | 1.8989 | 0.9695 | 1.1290 | 168.2 |
| 80 | 0.05434 | 469.2 | 1.9128 | 0.9741 | 1.1259 | 169.7 |
| 85 | 0.05528 | 474.1 | 1.9265 | 0.9791 | 1.1230 | 171.3 |
| 90 | 0.05621 | 479.0 | 1.9401 | 0.9842 | 1.1204 | 172.8 |
| 95 | 0.05713 | 483.9 | 1.9536 | 0.9895 | 1.1179 | 174.2 |
| 100 | 0.05806 | 488.9 | 1.9670 | 0.9950 | 1.1156 | 175.7 |
| 105 | 0.05898 | 493.9 | 1.9803 | 1.0007 | 1.1134 | 177.1 |
| 110 | 0.05989 | 498.9 | 1.9934 | 1.0065 | 1.1114 | 178.5 |
| 115 | 0.06078 | 503.9 | 2.0065 | 1.0124 | 1.1095 | 179.9 |
| 120 | 0.06169 | 509.0 | 2.0195 | 1.0183 | 1.1077 | 181.3 |
| 125 | 0.06258 | 514.1 | 2.0324 | 1.0244 | 1.1060 | 182.6 |
| 130 | 0.06348 | 519.3 | 2.0452 | 1.0305 | 1.1044 | 184.0 |
| 135 | 0.06437 | 524.4 | 2.0580 | 1.0367 | 1.1029 | 185.3 |
| 140 | 0.06526 | 529.6 | 2.0707 | 1.0429 | 1.1015 | 186.6 |
| 145 | 0.06613 | 534.9 | 2.0832 | 1.0491 | 1.1001 | 187.8 |
| 150 | 0.06702 | 540.1 | 2.0957 | 1.0554 | 1.0988 | 189.1 |

*(Continued)*

# TABLE A.16 (SI) (CONTINUED)

HFC–134a Superheated Vapor–Constant Pressure Tables

### Pressure = 600.00 kPa (abs)

| Temp. °C | | v | h | s | $C_p$ | $C_p/C_v$ | $v_s$ |
|---|---|---|---|---|---|---|---|
| 21.54 | SAT LIQ | 0.00082 | 229.7 | 1.1038 | 1.4165 | 1.5545 | 525.8 |
| 21.54 | SAT VAP | 0.03432 | 410.8 | 1.7183 | 0.9954 | 1.2187 | 145.0 |
| 25 | | 0.03502 | 414.2 | 1.7299 | 0.9890 | 1.2090 | 146.7 |
| 30 | | 0.03600 | 419.2 | 1.7463 | 0.9819 | 1.1967 | 149.0 |
| 35 | | 0.03695 | 424.1 | 1.7623 | 0.9771 | 1.1862 | 151.3 |
| 40 | | 0.03787 | 428.9 | 1.7780 | 0.9740 | 1.1771 | 153.4 |
| 45 | | 0.03878 | 433.8 | 1.7934 | 0.9723 | 1.1692 | 155.5 |
| 50 | | 0.03967 | 438.7 | 1.8086 | 0.9718 | 1.1622 | 157.4 |
| 55 | | 0.04054 | 443.5 | 1.8235 | 0.9724 | 1.1560 | 159.3 |
| 60 | | 0.04140 | 448.4 | 1.8382 | 0.9738 | 1.1505 | 161.2 |
| 65 | | 0.04224 | 453.3 | 1.8527 | 0.9759 | 1.1455 | 163.0 |
| 70 | | 0.04308 | 458.1 | 1.8671 | 0.9786 | 1.1410 | 164.7 |
| 75 | | 0.04390 | 463.0 | 1.8813 | 0.9818 | 1.1369 | 166.4 |
| 80 | | 0.04472 | 468.0 | 1.8953 | 0.9854 | 1.1331 | 168.1 |
| 85 | | 0.04552 | 472.9 | 1.9092 | 0.9895 | 1.1297 | 169.7 |
| 90 | | 0.04633 | 477.9 | 1.9229 | 0.9938 | 1.1265 | 171.3 |
| 95 | | 0.04712 | 482.8 | 1.9365 | 0.9985 | 1.1236 | 172.8 |
| 100 | | 0.04790 | 487.8 | 1.9500 | 1.0033 | 1.1209 | 174.3 |
| 105 | | 0.04868 | 492.9 | 1.9634 | 1.0084 | 1.1184 | 175.8 |
| 110 | | 0.04945 | 497.9 | 1.9767 | 1.0137 | 1.1161 | 177.3 |
| 150 | | 0.08426 | 540.9 | 2.1152 | 1.0510 | 1.0960 | 190.0 |
| 155 | | 0.08535 | 546.1 | 2.1276 | 1.0575 | 1.0949 | 191.2 |
| 160 | | 0.08642 | 551.4 | 2.1399 | 1.0640 | 1.0938 | 192.4 |
| 165 | | – | – | – | – | – | – |

### Pressure = 700.00 kPa (abs)

| Temp. °C | | v | h | s | $C_p$ | $C_p/C_v$ | $v_s$ |
|---|---|---|---|---|---|---|---|
| 26.68 | SAT LIQ | 0.00083 | 237.0 | 1.1282 | 1.4385 | 1.5658 | 501.8 |
| 26.68 | SAT VAP | 0.02939 | 413.5 | 1.7166 | 1.0265 | 1.2332 | 144.0 |
| 30 | | 0.02999 | 416.8 | 1.7278 | 1.0182 | 1.2223 | 145.7 |
| 35 | | 0.03086 | 421.9 | 1.7444 | 1.0085 | 1.2081 | 148.2 |
| 40 | | 0.03171 | 426.9 | 1.7606 | 1.0015 | 1.1960 | 150.6 |
| 45 | | 0.03253 | 431.9 | 1.7764 | 0.9967 | 1.1857 | 152.8 |
| 50 | | 0.03333 | 436.9 | 1.7919 | 0.9935 | 1.1768 | 155.0 |
| 55 | | 0.03412 | 441.9 | 1.8071 | 0.9918 | 1.1689 | 157.0 |
| 60 | | 0.03488 | 446.8 | 1.8221 | 0.9913 | 1.1620 | 159.0 |
| 65 | | 0.03564 | 451.8 | 1.8369 | 0.9917 | 1.1559 | 160.9 |
| 70 | | 0.03638 | 456.7 | 1.8515 | 0.9929 | 1.1504 | 162.8 |
| 75 | | 0.03711 | 461.7 | 1.8658 | 0.9949 | 1.1454 | 164.6 |
| 80 | | 0.03784 | 466.7 | 1.8801 | 0.9974 | 1.1409 | 166.3 |
| 85 | | 0.03855 | 471.7 | 1.8941 | 1.0005 | 1.1369 | 168.0 |
| 90 | | 0.03925 | 476.7 | 1.9080 | 1.0040 | 1.1331 | 169.7 |
| 95 | | 0.03995 | 481.7 | 1.9217 | 1.0078 | 1.1297 | 171.3 |
| 100 | | 0.04064 | 486.8 | 1.9354 | 1.0120 | 1.1266 | 172.9 |
| 105 | | 0.04133 | 491.9 | 1.9489 | 1.0165 | 1.1236 | 174.5 |
| 110 | | 0.04201 | 496.9 | 1.9622 | 1.0212 | 1.1209 | 176.0 |
| 115 | | 0.04269 | 502.1 | 1.9755 | 1.0261 | 1.1184 | 177.5 |
| 155 | | 0.06790 | 545.4 | 2.1082 | 1.0616 | 1.0975 | 190.4 |
| 160 | | 0.06877 | 550.7 | 2.1205 | 1.0679 | 1.0964 | 191.6 |
| 165 | | 0.06964 | 556.1 | 2.1328 | 1.0742 | 1.0952 | 192.8 |
| 170 | | 0.07052 | 561.5 | 2.1450 | 1.0804 | 1.0941 | 194.0 |

*(Continued)*

# TABLE A.16 (SI) (CONTINUED)

HFC–134a Superheated Vapor–Constant Pressure Tables

| Temp. °C | $v$ | $h$ | $s$ | $C_p$ | $C_p/C_v$ | $v_s$ |
|---|---|---|---|---|---|---|
| 115 | 0.05023 | 503.0 | 1.9999 | 1.0191 | 1.1139 | 178.7 |
| 120 | 0.05100 | 508.1 | 2.0030 | 1.0247 | 1.1118 | 180.1 |
| 125 | 0.05175 | 513.3 | 2.0159 | 1.0303 | 1.1099 | 181.5 |
| 130 | 0.05252 | 518.4 | 2.0288 | 1.0361 | 1.1081 | 182.9 |
| 135 | 0.05327 | 523.6 | 2.0417 | 1.0419 | 1.1063 | 184.3 |
| 140 | 0.05402 | 528.8 | 2.0544 | 1.0478 | 1.1047 | 185.6 |
| 145 | 0.05477 | 534.1 | 2.0670 | 1.0538 | 1.1032 | 186.9 |
| 150 | 0.05552 | 539.4 | 2.0796 | 1.0598 | 1.1017 | 188.2 |
| 155 | 0.05626 | 544.7 | 2.0921 | 1.0658 | 1.1003 | 189.5 |
| 160 | 0.05700 | 550.0 | 2.1045 | 1.0719 | 1.0990 | 190.8 |
| 165 | 0.05774 | 555.4 | 2.1168 | 1.0780 | 1.0977 | 192.0 |
| 170 | 0.05848 | 560.8 | 2.1291 | 1.0840 | 1.0965 | 193.3 |
| 175 | 0.05921 | 566.3 | 2.1413 | 1.0901 | 1.0954 | 194.5 |

| Temp. °C | $v$ | $h$ | $s$ | $C_p$ | $C_p/C_v$ | $v_s$ |
|---|---|---|---|---|---|---|
| 120 | 0.04336 | 507.2 | 1.9887 | 1.0312 | 1.1161 | 179.0 |
| 125 | 0.04402 | 512.4 | 2.0018 | 1.0365 | 1.1139 | 180.4 |
| 130 | 0.04468 | 517.6 | 2.0147 | 1.0419 | 1.1118 | 181.8 |
| 135 | 0.04535 | 522.8 | 2.0276 | 1.0473 | 1.1099 | 183.2 |
| 140 | 0.04600 | 528.0 | 2.0404 | 1.0529 | 1.1081 | 184.6 |
| 145 | 0.04665 | 533.3 | 2.0531 | 1.0586 | 1.1063 | 186.0 |
| 150 | 0.04730 | 538.6 | 2.0657 | 1.0643 | 1.1047 | 187.3 |
| 155 | 0.04795 | 544.0 | 2.0782 | 1.0701 | 1.1032 | 188.7 |
| 160 | 0.04859 | 549.3 | 2.0907 | 1.0760 | 1.1017 | 190.0 |
| 165 | 0.04923 | 554.7 | 2.1031 | 1.0818 | 1.1003 | 191.3 |
| 170 | 0.04987 | 560.2 | 2.1154 | 1.0877 | 1.0990 | 192.5 |
| 175 | 0.05051 | 565.6 | 2.1276 | 1.0936 | 1.0977 | 193.8 |
| 180 | 0.05114 | 571.1 | 2.1398 | 1.0996 | 1.0965 | 195.0 |

## Pressure = 800.00 kPa (abs)

| Temp. °C | $v$ | $h$ | $s$ | $C_p$ | $C_p/C_v$ | $v_s$ |
|---|---|---|---|---|---|---|
| SAT LIQ | 0.00085 | 243.7 | 1.1500 | 1.4602 | 1.5775 | 480.2 |
| 31.29 |  |  |  |  |  |  |
| SAT VAP | 0.02565 | 415.7 | 1.7150 | 1.0569 | 1.2485 | 142.9 |
| 31.29 |  |  |  |  |  |  |
| 30 | – | – | – | – | – | – |
| 35 | 0.02626 | 419.6 | 1.7278 | 1.0453 | 1.2344 | 145.0 |
| 40 | 0.02705 | 424.8 | 1.7445 | 1.0332 | 1.2184 | 147.6 |
| 45 | 0.02782 | 430.0 | 1.7608 | 1.0243 | 1.2049 | 150.1 |
| 50 | 0.02856 | 435.1 | 1.7767 | 1.0178 | 1.1935 | 152.4 |
| 55 | 0.02928 | 440.2 | 1.7923 | 1.0133 | 1.1836 | 154.7 |
| 60 | 0.02998 | 445.2 | 1.8076 | 1.0105 | 1.1750 | 156.8 |
| 65 | 0.03067 | 450.3 | 1.8226 | 1.0089 | 1.1674 | 158.8 |
| 70 | 0.03135 | 455.3 | 1.8374 | 1.0085 | 1.1607 | 160.8 |

## Pressure = 1000.00 kPa (abs)

| Temp. °C | $v$ | $h$ | $s$ | $C_p$ | $C_p/C_v$ | $v_s$ |
|---|---|---|---|---|---|---|
| SAT LIQ | 0.00087 | 255.6 | 1.1881 | 1.5035 | 1.6025 | 442.1 |
| 39.35 |  |  |  |  |  |  |
| SAT VAP | 0.02034 | 419.5 | 1.7124 | 1.1177 | 1.2817 | 140.6 |
| 39.35 |  |  |  |  |  |  |
| 40 | 0.02044 | 420.2 | 1.7147 | 1.1144 | 1.2782 | 141.0 |
| 45 | 0.02114 | 425.7 | 1.7322 | 1.0928 | 1.2546 | 144.1 |
| 50 | 0.02181 | 431.2 | 1.7491 | 1.0766 | 1.2354 | 146.9 |
| 55 | 0.02246 | 436.5 | 1.7655 | 1.0644 | 1.2195 | 149.6 |
| 60 | 0.02308 | 441.8 | 1.7816 | 1.0552 | 1.2061 | 152.1 |
| 65 | 0.02368 | 447.1 | 1.7972 | 1.0485 | 1.1947 | 154.4 |
| 70 | 0.02427 | 452.3 | 1.8126 | 1.0438 | 1.1848 | 156.7 |
| 75 | 0.02485 | 457.5 | 1.8277 | 1.0406 | 1.1762 | 158.9 |

(Continued)

# TABLE A.16 (SI) (CONTINUED)
## HFC–134a Superheated Vapor–Constant Pressure Tables

### Pressure = 1500.00 kPa (abs)

| Temp. °C | v | h | s | $C_p$ | $C_p/C_v$ | $v_s$ |
|---|---|---|---|---|---|---|
| 55.2 SAT LIQ | 0.00093 | 280.1 | 1.2632 | 1.6205 | 1.6778 | 365.0 |
| 75 | 0.03202 | 460.4 | 1.8520 | 1.0090 | 1.1548 | 162.7 |
| 80 | 0.03267 | 465.4 | 1.8664 | 1.0103 | 1.1494 | 164.6 |
| 85 | 0.03331 | 470.5 | 1.8806 | 1.0122 | 1.1446 | 166.4 |
| 90 | 0.03395 | 475.5 | 1.8947 | 1.0147 | 1.1402 | 168.1 |
| 95 | 0.03457 | 480.6 | 1.9086 | 1.0177 | 1.1362 | 169.8 |
| 100 | 0.03520 | 485.7 | 1.9223 | 1.0212 | 1.1325 | 171.5 |
| 105 | 0.03581 | 490.8 | 1.9360 | 1.0250 | 1.1292 | 173.1 |
| 110 | 0.03642 | 495.9 | 1.9494 | 1.0291 | 1.1261 | 174.7 |
| 115 | 0.03703 | 501.1 | 1.9628 | 1.0334 | 1.1232 | 176.3 |
| 120 | 0.03763 | 506.3 | 1.9761 | 1.0380 | 1.1205 | 177.8 |
| 125 | 0.03822 | 511.5 | 1.9892 | 1.0428 | 1.1181 | 179.3 |
| 130 | 0.03881 | 516.7 | 2.0023 | 1.0478 | 1.1157 | 180.8 |
| 135 | 0.03940 | 522.0 | 2.0152 | 1.0529 | 1.1136 | 182.2 |
| 140 | 0.03998 | 527.2 | 2.0281 | 1.0582 | 1.1115 | 183.6 |
| 145 | 0.04056 | 532.5 | 2.0408 | 1.0636 | 1.1096 | 185.0 |
| 150 | 0.04114 | 537.9 | 2.0535 | 1.0690 | 1.1078 | 186.4 |
| 155 | 0.04171 | 543.2 | 2.0661 | 1.0745 | 1.1061 | 187.8 |
| 160 | 0.04229 | 548.6 | 2.0786 | 1.0801 | 1.1045 | 189.1 |
| 165 | 0.04285 | 554.0 | 2.0910 | 1.0858 | 1.1029 | 190.5 |
| 170 | 0.04342 | 559.5 | 2.1034 | 1.0915 | 1.1015 | 191.8 |
| 175 | 0.04398 | 565.0 | 2.1157 | 1.0972 | 1.1001 | 193.1 |
| 180 | 0.04455 | 570.5 | 2.1279 | 1.1030 | 1.0988 | 194.3 |
| 185 | 0.04511 | 576.0 | 2.1400 | 1.1087 | 1.0975 | 195.6 |

### Pressure 2000.00 kPa (abs)

| Temp. °C | v | h | s | $C_p$ | $C_p/C_v$ | $v_s$ |
|---|---|---|---|---|---|---|
| 67.47 SAT LIQ | 0.00099 | 300.4 | 1.3223 | 1.7690 | 1.7844 | 302.2 |
| 80 | 0.02541 | 462.7 | 1.8425 | 1.0388 | 1.1687 | 160.9 |
| 85 | 0.02596 | 467.9 | 1.8571 | 1.0381 | 1.1619 | 162.9 |
| 90 | 0.02650 | 473.1 | 1.8715 | 1.0383 | 1.1559 | 164.9 |
| 95 | 0.02703 | 478.3 | 1.8857 | 1.0393 | 1.1506 | 166.8 |
| 100 | 0.02756 | 483.5 | 1.8997 | 1.0409 | 1.1457 | 168.6 |
| 105 | 0.02807 | 488.7 | 1.9136 | 1.0432 | 1.1413 | 170.4 |
| 110 | 0.02859 | 493.9 | 1.9273 | 1.0459 | 1.1372 | 172.1 |
| 115 | 0.02909 | 499.1 | 1.9409 | 1.0490 | 1.1335 | 173.8 |
| 120 | 0.02959 | 504.4 | 1.9543 | 1.0525 | 1.1301 | 175.4 |
| 125 | 0.03008 | 509.7 | 1.9676 | 1.0563 | 1.1270 | 177.0 |
| 130 | 0.03058 | 515.0 | 1.9809 | 1.0604 | 1.1241 | 178.6 |
| 135 | 0.03106 | 520.3 | 1.9940 | 1.0647 | 1.1214 | 180.1 |
| 140 | 0.03155 | 525.6 | 2.0070 | 1.0692 | 1.1189 | 181.7 |
| 145 | 0.03203 | 531.0 | 2.0198 | 1.0739 | 1.1165 | 183.1 |
| 150 | 0.03250 | 536.3 | 2.0326 | 1.0787 | 1.1143 | 184.6 |
| 155 | 0.03298 | 541.8 | 2.0453 | 1.0837 | 1.1122 | 186.0 |
| 160 | 0.03345 | 547.2 | 2.0579 | 1.0888 | 1.1103 | 187.5 |
| 165 | 0.03392 | 552.6 | 2.0705 | 1.0940 | 1.1084 | 188.9 |
| 170 | 0.03438 | 558.1 | 2.0829 | 1.0993 | 1.1067 | 190.2 |
| 175 | 0.03485 | 563.6 | 2.0953 | 1.1046 | 1.1050 | 191.6 |
| 180 | 0.03531 | 569.2 | 2.1076 | 1.1100 | 1.1035 | 192.9 |
| 185 | 0.03577 | 574.7 | 2.1198 | 1.1154 | 1.1020 | 194.2 |
| 190 | 0.03623 | 580.3 | 2.1319 | 1.1209 | 1.1006 | 195.6 |

(Continued)

## TABLE A.16 (SI) (CONTINUED)

HFC–134a Superheated Vapor–Constant Pressure Tables

| | | | | | | | SAT VAP | | | | | | | |
|---|---|---|---|---|---|---|---|---|---|---|---|---|---|---|
| 55.2 | 0.01308 | 425.7 | 1.7063 | 1.2844 | 1.3876 | 134.2 | | 67.47 | 0.00931 | 428.8 | 1.6991 | 1.5055 | 1.5484 | 127.2 |
| 60 | 0.01363 | 431.7 | 1.7246 | 1.2358 | 1.3416 | 138.0 | | 70 | 0.00959 | 432.5 | 1.7101 | 1.4452 | 1.4943 | 129.9 |
| 65 | 0.01417 | 437.8 | 1.7427 | 1.1989 | 1.3056 | 141.7 | | | | | | | | |
| 70 | 0.01468 | 443.7 | 1.7601 | 1.1715 | 1.2776 | 145.0 | | 75 | 0.01009 | 439.5 | 1.7303 | 1.3600 | 1.4173 | 134.7 |
| 75 | 0.01516 | 449.5 | 1.7769 | 1.1507 | 1.2552 | 148.0 | | 80 | 0.01055 | 446.1 | 1.7493 | 1.3019 | 1.3638 | 138.9 |
| 80 | 0.01562 | 455.2 | 1.7932 | 1.1349 | 1.2368 | 150.9 | | 85 | 0.01097 | 452.5 | 1.7673 | 1.2601 | 1.3242 | 142.7 |
| 85 | 0.01606 | 460.9 | 1.8090 | 1.1228 | 1.2214 | 153.6 | | 90 | 0.01137 | 458.8 | 1.7845 | 1.2291 | 1.2936 | 146.2 |
| 90 | 0.01649 | 466.4 | 1.8245 | 1.1136 | 1.2084 | 156.1 | | 95 | 0.01175 | 464.8 | 1.8011 | 1.2055 | 1.2692 | 149.3 |
| 95 | 0.01690 | 472.0 | 1.8397 | 1.1067 | 1.1972 | 158.5 | | | | | | | | |
| 100 | 0.01731 | 477.5 | 1.8546 | 1.1016 | 1.1875 | 160.8 | | 100 | 0.01211 | 470.8 | 1.8173 | 1.1874 | 1.2493 | 152.3 |
| 105 | 0.01770 | 483.0 | 1.8692 | 1.0982 | 1.1790 | 163.0 | | 105 | 0.01246 | 476.7 | 1.8330 | 1.1733 | 1.2327 | 155.1 |
| 110 | 0.01809 | 488.5 | 1.8837 | 1.0959 | 1.1714 | 165.2 | | 110 | 0.01279 | 482.6 | 1.8483 | 1.1625 | 1.2186 | 157.7 |
| 115 | 0.01847 | 494.0 | 1.8978 | 1.0948 | 1.1647 | 167.2 | | 115 | 0.01312 | 488.4 | 1.8633 | 1.1542 | 1.2066 | 160.2 |
| 120 | 0.01885 | 499.4 | 1.9119 | 1.0945 | 1.1587 | 169.2 | | 120 | 0.01344 | 494.1 | 1.8781 | 1.1479 | 1.1961 | 162.6 |
| 125 | 0.01921 | 504.9 | 1.9257 | 1.0950 | 1.1532 | 171.1 | | 125 | 0.01374 | 499.8 | 1.8925 | 1.1433 | 1.1869 | 164.9 |
| 130 | 0.01958 | 510.4 | 1.9394 | 1.0962 | 1.1483 | 173.0 | | 130 | 0.01405 | 505.5 | 1.9068 | 1.1401 | 1.1789 | 167.1 |
| 135 | 0.01993 | 515.9 | 1.9529 | 1.0979 | 1.1438 | 174.8 | | 135 | 0.01434 | 511.2 | 1.9208 | 1.1380 | 1.1717 | 169.2 |
| 140 | 0.02028 | 521.4 | 1.9663 | 1.1001 | 1.1397 | 176.6 | | 140 | 0.01463 | 516.9 | 1.9347 | 1.1369 | 1.1652 | 171.3 |
| 145 | 0.02063 | 526.9 | 1.9795 | 1.1027 | 1.1360 | 178.3 | | 145 | 0.01492 | 522.6 | 1.9483 | 1.1366 | 1.1594 | 173.3 |
| 150 | 0.02098 | 532.4 | 1.9926 | 1.1057 | 1.1325 | 180.0 | | 150 | 0.01520 | 528.3 | 1.9619 | 1.1370 | 1.1541 | 175.2 |
| 155 | 0.02132 | 537.9 | 2.0056 | 1.1090 | 1.1293 | 181.6 | | 155 | 0.01547 | 534.0 | 1.9752 | 1.1380 | 1.1493 | 177.1 |
| 160 | 0.02166 | 543.5 | 2.0185 | 1.1125 | 1.1263 | 183.2 | | 160 | 0.01575 | 539.7 | 1.9884 | 1.1395 | 1.1449 | 178.9 |
| 165 | 0.02199 | 549.1 | 2.0313 | 1.1163 | 1.1235 | 184.8 | | 165 | 0.01602 | 545.4 | 2.0015 | 1.1415 | 1.1409 | 180.7 |
| 170 | 0.02233 | 554.7 | 2.0440 | 1.1203 | 1.1209 | 186.3 | | 170 | 0.01629 | 551.1 | 2.0145 | 1.1439 | 1.1372 | 182.4 |
| 175 | 0.02265 | 560.3 | 2.0566 | 1.1244 | 1.1185 | 187.9 | | 175 | 0.01655 | 556.8 | 2.0274 | 1.1466 | 1.1337 | 184.1 |
| 180 | 0.02298 | 565.9 | 2.0691 | 1.1288 | 1.1162 | 189.3 | | 180 | 0.01681 | 562.6 | 2.0401 | 1.1496 | 1.1305 | 185.7 |
| | | | | | | | | 185 | 0.01707 | 568.3 | 2.0527 | 1.1528 | 1.1276 | 187.4 |
| | | | | | | | | 190 | 0.01733 | 574.1 | 2.0653 | 1.1563 | 1.1248 | 189.0 |

*(Continued)*

# TABLE A.16 (SI) (CONTINUED)

HFC-134a Superheated Vapor–Constant Pressure Tables

| Temp. °C | $v$ | $h$ | $s$ | $C_p$ | $C_p/C_v$ | $v_s$ |
|---|---|---|---|---|---|---|
| 185 | 0.02331 | 571.6 | 2.0815 | 1.1332 | 1.1141 | 190.8 |
| 190 | 0.02363 | 577.2 | 2.0939 | 1.1378 | 1.1121 | 192.3 |
| 195 | 0.02395 | 582.9 | 2.1061 | 1.1425 | 1.1102 | 193.7 |
| 200 | 0.02427 | 588.7 | 2.1183 | 1.1472 | 1.1084 | 195.1 |
| 205 | 0.02459 | 594.4 | 2.1303 | 1.1520 | 1.1067 | 196.5 |
| 210 | 0.02491 | 600.2 | 2.1424 | 1.1569 | 1.1050 | 197.8 |
| **Pressure = 3000.00 kPa (abs)** | | | | | | |
| Temp. °C | $v$ | $h$ | $s$ | $C_p$ | $C_p/C_v$ | $v_s$ |
| 86.22 SAT LIQ | 0.00114 | 335.3 | 1.4188 | 2.3843 | 2.2770 | 195.8 |
| 86.22 SAT VAP | 0.00528 | 427.6 | 1.6758 | 2.5300 | 2.3769 | 112.1 |
| 90 | 0.00575 | 436.1 | 1.6992 | 2.0156 | 1.9317 | 119.1 |
| 95 | 0.00624 | 445.3 | 1.7245 | 1.7214 | 1.6778 | 126.1 |
| 100 | 0.00665 | 453.5 | 1.7466 | 1.5654 | 1.5427 | 131.7 |
| 105 | 0.00701 | 461.1 | 1.7667 | 1.4676 | 1.4573 | 136.6 |
| 110 | 0.00733 | 468.2 | 1.7855 | 1.4007 | 1.3977 | 140.9 |
| 115 | 0.00764 | 475.1 | 1.8034 | 1.3523 | 1.3537 | 144.8 |
| 120 | 0.00792 | 481.8 | 1.8204 | 1.3161 | 1.3197 | 148.4 |
| 125 | 0.00819 | 488.3 | 1.8369 | 1.2884 | 1.2926 | 151.7 |
| 130 | 0.00845 | 494.7 | 1.8528 | 1.2669 | 1.2704 | 154.7 |
| 135 | 0.00869 | 501.0 | 1.8683 | 1.2500 | 1.2520 | 157.6 |
| 140 | 0.00893 | 507.2 | 1.8835 | 1.2366 | 1.2363 | 160.4 |
| 145 | 0.00916 | 513.3 | 1.8983 | 1.2262 | 1.2229 | 163.0 |
| 150 | 0.00939 | 519.4 | 1.9128 | 1.2179 | 1.2113 | 165.5 |
| 155 | 0.00960 | 525.5 | 1.9271 | 1.2116 | 1.2011 | 167.9 |
| 160 | 0.00982 | 531.6 | 1.9411 | 1.2068 | 1.1921 | 170.2 |
| 165 | 0.01003 | 537.6 | 1.9549 | 1.2032 | 1.1841 | 172.4 |

| Temp. °C | $v$ | $h$ | $s$ | $C_p$ | $C_p/C_v$ | $v_s$ |
|---|---|---|---|---|---|---|
| 195 | 0.01758 | 579.9 | 2.0777 | 1.1599 | 1.1222 | 190.5 |
| 200 | 0.01784 | 585.7 | 2.0900 | 1.1638 | 1.1198 | 192.1 |
| 205 | 0.01809 | 591.5 | 2.1023 | 1.1678 | 1.1175 | 193.6 |
| 210 | 0.01834 | 597.4 | 2.1145 | 1.1719 | 1.1153 | 195.0 |
| 215 | 0.01858 | 603.2 | 2.1265 | 1.1761 | 1.1133 | 196.5 |
| 220 | 0.01883 | 609.1 | 2.1386 | 1.1804 | 1.1114 | 197.9 |
| **Pressure = 4000.00 kPa (abs)** | | | | | | |
| Temp. °C | $v$ | $h$ | $s$ | $C_p$ | $C_p/C_v$ | $v_s$ |
| 100.37 SAT LIQ | 0.00158 | 375.6 | 1.5250 | 28.1470 | 24.2211 | 95.7 |
| 100.37 SAT VAP | 0.00254 | 404.4 | 1.6022 | 42.1018 | 35.2394 | 95.0 |
| 100 | – | – | – | – | – | – |
| 105 | 0.00376 | 433.7 | 1.6804 | 3.1309 | 2.8151 | 111.6 |
| 110 | 0.00429 | 446.7 | 1.7143 | 2.2216 | 2.0540 | 120.4 |
| 115 | 0.00468 | 456.8 | 1.7406 | 1.8780 | 1.7663 | 127.1 |
| 120 | 0.00501 | 465.7 | 1.7634 | 1.6927 | 1.6106 | 132.7 |
| 125 | 0.00530 | 473.8 | 1.7840 | 1.5762 | 1.5117 | 137.5 |
| 130 | 0.00556 | 481.5 | 1.8031 | 1.4963 | 1.4429 | 141.8 |
| 135 | 0.00580 | 488.8 | 1.8212 | 1.4384 | 1.3921 | 145.8 |
| 140 | 0.00603 | 495.9 | 1.8385 | 1.3950 | 1.3530 | 149.4 |
| 145 | 0.00624 | 502.8 | 1.8550 | 1.3615 | 1.3219 | 152.8 |
| 150 | 0.00645 | 509.5 | 1.8710 | 1.3354 | 1.2965 | 155.9 |
| 155 | 0.00664 | 516.2 | 1.8866 | 1.3146 | 1.2755 | 158.9 |
| 160 | 0.00683 | 522.7 | 1.9018 | 1.2981 | 1.2577 | 161.7 |
| 165 | 0.00702 | 529.2 | 1.9166 | 1.2849 | 1.2424 | 164.4 |
| 170 | 0.00719 | 535.6 | 1.9311 | 1.2743 | 1.2293 | 167.0 |
| 175 | 0.00737 | 541.9 | 1.9454 | 1.2658 | 1.2177 | 169.4 |

*(Continued)*

## TABLE A.16 (SI) (CONTINUED)

HFC–134a Superheated Vapor–Constant Pressure Tables

| T | v | h | s | $C_p$ | $C_p/C_v$ | $v_s$ |
|---|---|---|---|---|---|---|
| 170 | 0.01023 | 543.6 | 1.9686 | 1.2007 | 1.1770 | 174.5 |
| 175 | 0.01043 | 549.6 | 1.9820 | 1.1992 | 1.1705 | 176.6 |
| 180 | 0.01063 | 555.6 | 1.9953 | 1.1984 | 1.1646 | 178.6 |
| 185 | 0.01083 | 561.6 | 2.0085 | 1.1983 | 1.1593 | 180.5 |
| 190 | 0.01102 | 567.6 | 2.0215 | 1.1988 | 1.1544 | 182.4 |
| 195 | 0.01121 | 573.6 | 2.0344 | 1.1998 | 1.1499 | 184.3 |
| 200 | 0.01139 | 579.6 | 2.0471 | 1.2012 | 1.1458 | 186.1 |
| 205 | 0.01158 | 585.6 | 2.0598 | 1.2030 | 1.1420 | 187.8 |
| 210 | 0.01176 | 591.6 | 2.0723 | 1.2051 | 1.1384 | 189.6 |
| 215 | 0.01194 | 597.6 | 2.0847 | 1.2075 | 1.1351 | 191.3 |
| 220 | 0.01212 | 603.7 | 2.0970 | 1.2102 | 1.1320 | 192.9 |
| 225 | 0.01230 | 609.7 | 2.1092 | 1.2130 | 1.1291 | 194.5 |
| 230 | 0.01248 | 615.8 | 2.1214 | 1.2161 | 1.1264 | 196.1 |
| 235 | 0.01265 | 621.9 | 2.1334 | 1.2194 | 1.1239 | 197.7 |
| 240 | 0.01283 | 628.0 | 2.1454 | 1.2227 | 1.1215 | 199.2 |

| T | v | h | s | $C_p$ | $C_p/C_v$ | $v_s$ |
|---|---|---|---|---|---|---|
| 180 | 0.00753 | 548.2 | 1.9594 | 1.2591 | 1.2076 | 171.8 |
| 185 | 0.00770 | 554.5 | 1.9732 | 1.2539 | 1.1985 | 174.1 |
| 190 | 0.00786 | 560.8 | 1.9867 | 1.2499 | 1.1904 | 176.3 |
| 195 | 0.00802 | 567.0 | 2.0001 | 1.2470 | 1.1831 | 178.4 |
| 200 | 0.00817 | 573.2 | 2.0134 | 1.2450 | 1.1765 | 180.5 |
| 205 | 0.00833 | 579.4 | 2.0265 | 1.2438 | 1.1705 | 182.5 |
| 210 | 0.00848 | 585.7 | 2.0394 | 1.2432 | 1.1650 | 184.5 |
| 215 | 0.00863 | 591.9 | 2.0522 | 1.2431 | 1.1600 | 186.4 |
| 220 | 0.00877 | 598.1 | 2.0649 | 1.2436 | 1.1554 | 188.3 |
| 225 | 0.00892 | 604.3 | 2.0774 | 1.2445 | 1.1511 | 190.1 |
| 230 | 0.00906 | 610.5 | 2.0898 | 1.2458 | 1.1471 | 191.9 |
| 235 | 0.00920 | 616.8 | 2.1022 | 1.2475 | 1.1434 | 193.6 |
| 240 | 0.00934 | 623.0 | 2.1144 | 1.2494 | 1.1400 | 195.3 |
| 245 | 0.00948 | 629.3 | 2.1265 | 1.2515 | 1.1368 | 197.0 |
| 250 | 0.00962 | 635.5 | 2.1386 | 1.2539 | 1.1337 | 198.6 |

*Source:* *Thermodynamic Properties of HFC–134a.* With permission of E. I. Du Pont de Nemours and Co.

*Note:* $C_p$ = heat capacity at constant pressure in kJ/ (kg) (°C); $C_p/C_v$ = heat capacity ratio (dimensionless); $h$ = enthalpy in kJ/kg; $s$ = entropy in kJ/ (kg) (°K); $v$ = volume in m³/kg ; $v_s$ = velocity of sound in m/s.

**TABLE A.17**

Pressure–Enthalpy Diagram for HCF–134a

*Source:* *Thermodynamic Properties of HFC-134a.* With permission of E. I. Du Pont de Nemours and Co.

**TABLE A.17 (SI)**

Pressure–Enthalpy Diagram for HCF-134a

*Source:* *Thermodynamic Properties of HFC–134a.* With permission of E. I. Du Pont de Nemours and Co.

**TABLE A.18**

Suva® 410A Saturation Properties—Temperature Table

| Temp. °F | Pressure, psia Liquid $p_f$ | Vapor $p_g$ | Volume, ft.³/lb. Liquid $v_f$ | Vapor $v_g$ | Density, lb./ft.³ Liquid $1/v_f$ | Vapor $1/v_g$ | Enthalpy, Btu/lb. Liquid $h_f$ | Latent $h_{fg}$ | Vapor $h_g$ | Entropy, Btu/(lb.)(R) Liquid $s_f$ | Vapor $s_g$ | Temp. °F |
|---|---|---|---|---|---|---|---|---|---|---|---|---|
| −150 | 0.49 | 0.49 | 0.0106 | 94.0107 | 94.39 | 0.0106 | −34.1 | 134.3 | 100.2 | −0.0946 | 0.3390 | −150 |
| −149 | 0.52 | 0.51 | 0.0106 | 89.5452 | 94.30 | 0.0112 | −33.8 | 134.1 | 100.3 | −0.0937 | 0.3380 | −149 |
| −148 | 0.55 | 0.54 | 0.0106 | 85.3254 | 94.20 | 0.0117 | −33.5 | 134.0 | 100.5 | −0.0928 | 0.3371 | −148 |
| −147 | 0.58 | 0.57 | 0.0106 | 81.3361 | 94.10 | 0.0123 | −33.2 | 133.8 | 100.6 | −0.0919 | 0.3361 | −147 |
| −146 | 0.60 | 0.60 | 0.0106 | 77.5632 | 94.01 | 0.0129 | −32.9 | 133.7 | 100.7 | −0.0910 | 0.3352 | −146 |
| −145 | 0.64 | 0.63 | 0.0106 | 73.9935 | 93.91 | 0.0135 | −32.6 | 133.5 | 100.9 | −0.0900 | 0.3343 | −145 |
| −144 | 0.67 | 0.66 | 0.0107 | 70.6147 | 93.81 | 0.0142 | −32.4 | 133.4 | 101.0 | −0.0891 | 0.3334 | −144 |
| −143 | 0.70 | 0.69 | 0.0107 | 67.4151 | 93.72 | 0.0148 | −32.1 | 133.2 | 101.2 | −0.0882 | 0.3325 | −143 |
| −142 | 0.74 | 0.73 | 0.0107 | 64.3842 | 93.62 | 0.0155 | −31.8 | 133.1 | 101.3 | −0.0873 | 0.3316 | −142 |
| −141 | 0.77 | 0.76 | 0.0107 | 61.5119 | 93.52 | 0.0163 | −31.5 | 132.9 | 101.4 | −0.0864 | 0.3307 | −141 |
| −140 | 0.81 | 0.80 | 0.0107 | 58.7888 | 93.42 | 0.0170 | −31.2 | 132.8 | 101.6 | −0.0855 | 0.3298 | −140 |
| −139 | 0.85 | 0.84 | 0.0107 | 56.2061 | 93.33 | 0.0178 | −30.9 | 132.6 | 101.7 | −0.0846 | 0.3289 | −139 |
| −138 | 0.89 | 0.88 | 0.0107 | 53.7558 | 93.23 | 0.0186 | −30.6 | 132.5 | 101.8 | −0.0838 | 0.3281 | −138 |
| −137 | 0.93 | 0.92 | 0.0107 | 51.4301 | 93.13 | 0.0194 | −30.3 | 132.3 | 102.0 | −0.0829 | 0.3272 | −137 |
| −136 | 0.98 | 0.97 | 0.0107 | 49.2219 | 93.03 | 0.0203 | −30.0 | 132.2 | 102.1 | −0.0820 | 0.3264 | −136 |
| −135 | 1.02 | 1.01 | 0.0108 | 47.1244 | 92.93 | 0.0212 | −29.7 | 132.0 | 102.3 | −0.0811 | 0.3255 | −135 |
| −134 | 1.07 | 1.06 | 0.0108 | 45.1314 | 92.84 | 0.0222 | −29.5 | 131.9 | 102.4 | −0.0802 | 0.3247 | −134 |
| −133 | 1.12 | 1.11 | 0.0108 | 43.2370 | 92.74 | 0.0231 | −29.2 | 131.7 | 102.5 | −0.0793 | 0.3239 | −133 |
| −132 | 1.17 | 1.16 | 0.0108 | 41.4356 | 92.64 | 0.0241 | −28.9 | 131.5 | 102.7 | −0.0784 | 0.3231 | −132 |
| −131 | 1.23 | 1.21 | 0.0108 | 39.7222 | 92.54 | 0.0252 | −28.6 | 131.4 | 102.8 | −0.0775 | 0.3223 | −131 |
| −130 | 1.28 | 1.27 | 0.0108 | 38.0917 | 92.44 | 0.0263 | −28.3 | 131.2 | 102.9 | −0.0766 | 0.3215 | −130 |
| −129 | 1.34 | 1.33 | 0.0108 | 36.5397 | 92.34 | 0.0274 | −28.0 | 131.1 | 103.1 | −0.0757 | 0.3207 | −129 |
| −128 | 1.40 | 1.39 | 0.0108 | 35.0619 | 92.24 | 0.0285 | −27.7 | 130.9 | 103.2 | −0.0748 | 0.3199 | −128 |
| −127 | 1.46 | 1.45 | 0.0109 | 33.6542 | 92.14 | 0.0297 | −27.4 | 130.8 | 103.4 | −0.0739 | 0.3191 | −127 |
| −126 | 1.52 | 1.51 | 0.0109 | 32.3129 | 92.04 | 0.0309 | −27.1 | 130.6 | 103.5 | −0.0730 | 0.3184 | −126 |
| −125 | 1.59 | 1.58 | 0.0109 | 31.0344 | 91.94 | 0.0322 | −26.8 | 130.4 | 103.6 | −0.0721 | 0.3176 | −125 |

*(Continued)*

**TABLE A.18 (CONTINUED)**

Suva® 410A Saturation Properties—Temperature Table

| Temp. °F | Pressure, psia | | Volume, ft.³/lb. | | Density, lb./ft.³ | | Enthalpy, Btu/lb. | | | Entropy, Btu/(lb.)(R) | | Temp. °F |
|---|---|---|---|---|---|---|---|---|---|---|---|---|
| | Liquid $p_f$ | Vapor $p_g$ | Liquid $v_f$ | Vapor $v_g$ | Liquid $1/v_f$ | Vapor $1/v_g$ | Liquid $h_f$ | Latent $h_{fg}$ | Vapor $h_g$ | Liquid $s_f$ | Vapor $s_g$ | |
| −124 | 1.66 | 1.65 | 0.0109 | 29.8154 | 91.84 | 0.0335 | −26.5 | 130.3 | 103.8 | −0.0713 | 0.3169 | −124 |
| −123 | 1.73 | 1.72 | 0.0109 | 28.6527 | 91.74 | 0.0349 | −26.2 | 130.1 | 103.9 | −0.0704 | 0.3161 | −123 |
| −122 | 1.81 | 1.79 | 0.0109 | 27.5434 | 91.64 | 0.0363 | −25.9 | 130.0 | 104.0 | −0.0695 | 0.3154 | −122 |
| −121 | 1.88 | 1.87 | 0.0109 | 26.4846 | 91.54 | 0.0378 | −25.6 | 129.8 | 104.2 | −0.0686 | 0.3147 | −121 |
| −120 | 1.96 | 1.95 | 0.0109 | 25.4738 | 91.44 | 0.0393 | −25.3 | 129.6 | 104.3 | −0.0677 | 0.3139 | −120 |
| −119 | 2.04 | 2.03 | 0.0109 | 24.5085 | 91.34 | 0.0408 | −25.0 | 129.5 | 104.5 | −0.0668 | 0.3132 | −119 |
| −118 | 2.13 | 2.12 | 0.0110 | 23.5863 | 91.24 | 0.0424 | −24.7 | 129.3 | 104.6 | −0.0660 | 0.3125 | −118 |
| −117 | 2.22 | 2.20 | 0.0110 | 22.7051 | 91.14 | 0.0440 | −24.4 | 129.1 | 104.7 | −0.0651 | 0.3118 | −117 |
| −116 | 2.31 | 2.29 | 0.0110 | 21.8628 | 91.04 | 0.0457 | −24.1 | 129.0 | 104.9 | −0.0642 | 0.3111 | −116 |
| −115 | 2.40 | 2.39 | 0.0110 | 21.0573 | 90.93 | 0.0475 | −23.8 | 128.8 | 105.0 | −0.0633 | 0.3104 | −115 |
| −114 | 2.50 | 2.48 | 0.0110 | 20.2870 | 90.83 | 0.0493 | −23.5 | 128.6 | 105.1 | −0.0624 | 0.3097 | −114 |
| −113 | 2.60 | 2.58 | 0.0110 | 19.5499 | 90.73 | 0.0512 | −23.2 | 128.5 | 105.3 | −0.0616 | 0.3091 | −113 |
| −112 | 2.70 | 2.69 | 0.0110 | 18.8446 | 90.63 | 0.0531 | −22.9 | 128.3 | 105.4 | −0.0607 | 0.3084 | −112 |
| −111 | 2.81 | 2.79 | 0.0110 | 18.1694 | 90.53 | 0.0550 | −22.6 | 128.1 | 105.5 | −0.0598 | 0.3077 | −111 |
| −110 | 2.92 | 2.91 | 0.0111 | 17.5228 | 90.42 | 0.0571 | −22.3 | 128.0 | 105.7 | −0.0589 | 0.3071 | −110 |
| −109 | 3.03 | 3.02 | 0.0111 | 16.9035 | 90.32 | 0.0592 | −22.0 | 127.8 | 105.8 | −0.0580 | 0.3064 | −109 |
| −108 | 3.15 | 3.14 | 0.0111 | 16.3101 | 90.22 | 0.0613 | −21.7 | 127.6 | 105.9 | −0.0572 | 0.3058 | −108 |
| −107 | 3.27 | 3.26 | 0.0111 | 15.7415 | 90.11 | 0.0635 | −21.4 | 127.5 | 106.1 | −0.0563 | 0.3051 | −107 |
| −106 | 3.40 | 3.38 | 0.0111 | 15.1963 | 90.01 | 0.0658 | −21.1 | 127.3 | 106.2 | −0.0554 | 0.3045 | −106 |
| −105 | 3.53 | 3.51 | 0.0111 | 14.6736 | 89.91 | 0.0681 | −20.8 | 127.1 | 106.3 | −0.0546 | 0.3039 | −105 |
| −104 | 3.66 | 3.64 | 0.0111 | 14.1722 | 89.81 | 0.0706 | −20.5 | 127.0 | 106.5 | −0.0537 | 0.3033 | −104 |
| −103 | 3.80 | 3.78 | 0.0111 | 13.6912 | 89.70 | 0.0730 | −20.2 | 126.8 | 106.6 | −0.0528 | 0.3027 | −103 |
| −102 | 3.94 | 3.92 | 0.0112 | 13.2296 | 89.60 | 0.0756 | −19.9 | 126.6 | 106.7 | −0.0520 | 0.3020 | −102 |
| −101 | 4.08 | 4.07 | 0.0112 | 12.7865 | 89.49 | 0.0782 | −19.6 | 126.4 | 106.9 | −0.0511 | 0.3014 | −101 |
| −100 | 4.23 | 4.22 | 0.0112 | 12.3611 | 89.39 | 0.0809 | −19.3 | 126.3 | 107.0 | −0.0502 | 0.3008 | −100 |
| −99 | 4.39 | 4.37 | 0.0112 | 11.9525 | 89.29 | 0.0837 | −19.0 | 126.1 | 107.1 | −0.0494 | 0.3002 | −99 |

*(Continued)*

**TABLE A.18 (CONTINUED)**

Suva® 410A Saturation Properties—Temperature Table

| Temp. °F | Pressure, psia | | Volume, ft.³/lb. | | Density, lb./ft.³ | | Enthalpy, Btu/lb. | | | Entropy, Btu/(lb.)(R) | | Temp. °F |
|---|---|---|---|---|---|---|---|---|---|---|---|---|
| | Liquid $p_f$ | Vapor $p_g$ | Liquid $v_f$ | Vapor $v_g$ | Liquid $1/v_f$ | Vapor $1/v_g$ | Liquid $h_f$ | Latent $h_{fg}$ | Vapor $h_g$ | Liquid $s_f$ | Vapor $s_g$ | |
| −98 | 4.54 | 4.53 | 0.0112 | 11.5599 | 89.18 | 0.0865 | −18.6 | 125.9 | 107.3 | −0.0485 | 0.2997 | −98 |
| −97 | 4.71 | 4.69 | 0.0112 | 11.1828 | 89.08 | 0.0894 | −18.3 | 125.7 | 107.4 | −0.0476 | 0.2991 | −97 |
| −96 | 4.88 | 4.86 | 0.0112 | 10.8203 | 88.97 | 0.0924 | −18.0 | 125.6 | 107.5 | −0.0468 | 0.2985 | −96 |
| −95 | 5.05 | 5.03 | 0.0113 | 10.4718 | 88.87 | 0.0955 | −17.7 | 125.4 | 107.7 | −0.0459 | 0.2979 | −95 |
| −94 | 5.23 | 5.21 | 0.0113 | 10.1367 | 88.76 | 0.0987 | −17.4 | 125.2 | 107.8 | −0.0450 | 0.2974 | −94 |
| −93 | 5.41 | 5.39 | 0.0113 | 9.8144 | 88.66 | 0.1019 | −17.1 | 125.0 | 107.9 | −0.0442 | 0.2968 | −93 |
| −92 | 5.60 | 5.58 | 0.0113 | 9.5043 | 88.55 | 0.1052 | −16.8 | 124.8 | 108.1 | −0.0433 | 0.2962 | −92 |
| −91 | 5.79 | 5.77 | 0.0113 | 9.2059 | 88.44 | 0.1086 | −16.5 | 124.7 | 108.2 | −0.0425 | 0.2957 | −91 |
| −90 | 5.99 | 5.97 | 0.0113 | 8.9187 | 88.34 | 0.1121 | −16.2 | 124.5 | 108.3 | −0.0416 | 0.2951 | −90 |
| −89 | 6.19 | 6.18 | 0.0113 | 8.6422 | 88.23 | 0.1157 | −15.9 | 124.3 | 108.4 | −0.0408 | 0.2946 | −89 |
| −88 | 6.40 | 6.39 | 0.0113 | 8.3760 | 88.12 | 0.1194 | −15.5 | 124.1 | 108.6 | −0.0399 | 0.2940 | −88 |
| −87 | 6.62 | 6.60 | 0.0114 | 8.1196 | 88.02 | 0.1232 | −15.2 | 123.9 | 108.7 | −0.0390 | 0.2935 | −87 |
| −86 | 6.84 | 6.82 | 0.0114 | 7.8725 | 87.91 | 0.1270 | −14.9 | 123.7 | 108.8 | −0.0382 | 0.2930 | −86 |
| −85 | 7.07 | 7.05 | 0.0114 | 7.6345 | 87.80 | 0.1310 | −14.6 | 123.6 | 109.0 | −0.0373 | 0.2925 | −85 |
| −84 | 7.30 | 7.28 | 0.0114 | 7.4050 | 87.70 | 0.1350 | −14.3 | 123.4 | 109.1 | −0.0365 | 0.2919 | −84 |
| −83 | 7.54 | 7.52 | 0.0114 | 7.1839 | 87.59 | 0.1392 | −14.0 | 123.2 | 109.2 | −0.0356 | 0.2914 | −83 |
| −82 | 7.79 | 7.77 | 0.0114 | 6.9706 | 87.48 | 0.1435 | −13.7 | 123.0 | 109.4 | −0.0348 | 0.2909 | −82 |
| −81 | 8.04 | 8.02 | 0.0114 | 6.7649 | 87.37 | 0.1478 | −13.3 | 122.8 | 109.5 | −0.0339 | 0.2904 | −81 |
| −80 | 8.30 | 8.28 | 0.0115 | 6.5665 | 87.27 | 0.1523 | −13.0 | 122.6 | 109.6 | −0.0331 | 0.2899 | −80 |
| −79 | 8.56 | 8.54 | 0.0115 | 6.3750 | 87.16 | 0.1569 | −12.7 | 122.4 | 109.7 | −0.0323 | 0.2894 | −79 |
| −78 | 8.84 | 8.82 | 0.0115 | 6.1903 | 87.05 | 0.1615 | −12.4 | 122.2 | 109.9 | −0.0314 | 0.2889 | −78 |
| −77 | 9.12 | 9.09 | 0.0115 | 6.0120 | 86.94 | 0.1663 | −12.1 | 122.1 | 110.0 | −0.0306 | 0.2884 | −77 |
| −76 | 9.40 | 9.38 | 0.0115 | 5.8398 | 86.83 | 0.1712 | −11.8 | 121.9 | 110.1 | −0.0297 | 0.2879 | −76 |
| −75 | 9.70 | 9.67 | 0.0115 | 5.6736 | 86.72 | 0.1763 | −11.4 | 121.7 | 110.2 | −0.0289 | 0.2874 | −75 |
| −74 | 10.00 | 9.97 | 0.0115 | 5.5130 | 86.61 | 0.1814 | −11.1 | 121.5 | 110.4 | −0.0280 | 0.2869 | −74 |
| −73 | 10.30 | 10.28 | 0.0116 | 5.3579 | 86.50 | 0.1866 | −10.8 | 121.3 | 110.5 | −0.0272 | 0.2865 | −73 |

*(Continued)*

**TABLE A.18 (CONTINUED)**

Suva® 410A Saturation Properties—Temperature Table

| Temp. °F | Pressure, psia | | Volume, ft.³/lb. | | Density, lb./ft.³ | | Enthalpy, Btu/lb. | | | Entropy, Btu/(lb.)(R) | | Temp. °F |
|---|---|---|---|---|---|---|---|---|---|---|---|---|
| | Liquid $p_f$ | Vapor $p_g$ | Liquid $v_f$ | Vapor $v_g$ | Liquid $1/v_f$ | Vapor $1/v_g$ | Liquid $h_f$ | Latent $h_{fg}$ | Vapor $h_g$ | Liquid $s_f$ | Vapor $s_g$ | |
| -72 | 10.62 | 10.60 | 0.0116 | 5.2080 | 86.39 | 0.1920 | -10.5 | 121.1 | 110.6 | -0.0264 | 0.2860 | -72 |
| -71 | 10.94 | 10.92 | 0.0116 | 5.0632 | 86.28 | 0.1975 | -10.2 | 120.9 | 110.7 | -0.0255 | 0.2855 | -71 |
| -70 | 11.28 | 11.25 | 0.0116 | 4.9232 | 86.17 | 0.2031 | 00 i | 120.7 | 110.9 | -0.0247 | 0.2851 | -70 |
| -69 | 11.61 | 11.59 | 0.0116 | 4.7878 | 86.06 | 0.2089 | -9.5 | 120.5 | 111.0 | -0.0238 | 0.2846 | -69 |
| -68 | 11.96 | 11.93 | 0.0116 | 4.6570 | 85.95 | 0.2147 | -9.2 | 120.3 | 111.1 | -0.0230 | 0.2841 | -68 |
| -67 | 12.32 | 12.29 | 0.0116 | 4.5304 | 85.84 | 0.2207 | -8.9 | 120.1 | 111.2 | -0.0222 | 0.2837 | -67 |
| -66 | 12.68 | 12.65 | 0.0117 | 4.4080 | 85.73 | 0.2269 | -8.5 | 119.9 | 111.4 | -0.0213 | 0.2832 | -66 |
| -65 | 13.05 | 13.02 | 0.0117 | 4.2895 | 85.62 | 0.2331 | -8.2 | 119.7 | 111.5 | -0.0205 | 0.2828 | -65 |
| -64 | 13.43 | 13.40 | 0.0117 | 4.1749 | 85.51 | 0.2395 | -7.9 | 119.5 | 111.6 | -0.0197 | 0.2823 | -64 |
| -63 | 13.82 | 13.79 | 0.0117 | 4.0639 | 85.40 | 0.2461 | -7.6 | 119.3 | 111.7 | -0.0189 | 0.2819 | -63 |
| -62 | 14.22 | 14.19 | 0.0117 | 3.9565 | 85.29 | 0.2527 | -7.3 | 119.1 | 111.8 | -0.0180 | 0.2815 | -62 |
| -61 | 14.63 | 14.60 | 0.0117 | 3.8526 | 85.17 | 0.2596 | -6.9 | 118.9 | 112.0 | -0.0172 | 0.2810 | -61 |
| -60 | 15.05 | 15.01 | 0.0118 | 3.7519 | 85.06 | 0.2665 | -6.6 | 118.7 | 112.1 | -0.0164 | 0.2806 | -60 |
| -59 | 15.47 | 15.44 | 0.0118 | 3.6543 | 84.95 | 0.2736 | -6.3 | 118.5 | 112.2 | -0.0155 | 0.2802 | -59 |
| -58 | 15.91 | 15.87 | 0.0118 | 3.5599 | 84.84 | 0.2809 | -6.0 | 118.3 | 112.3 | -0.0147 | 0.2797 | -58 |
| -57 | 16.35 | 16.32 | 0.0118 | 3.4683 | 84.72 | 0.2883 | -5.6 | 118.1 | 112.4 | -0.0139 | 0.2793 | -57 |
| -56 | 16.81 | 16.77 | 0.0118 | 3.3796 | 84.61 | 0.2959 | -5.3 | 117.9 | 112.6 | -0.0131 | 0.2789 | -56 |
| -55 | 17.28 | 17.24 | 0.0118 | 3.2937 | 84.50 | 0.3036 | -5.0 | 117.7 | 112.7 | -0.0122 | 0.2785 | -55 |
| -54 | 17.75 | 17.71 | 0.0119 | 3.2103 | 84.38 | 0.3115 | -4.6 | 117.4 | 112.8 | -0.0114 | 0.2781 | -54 |
| -53 | 18.24 | 18.19 | 0.0119 | 3.1295 | 84.27 | 0.3195 | -4.3 | 117.2 | 112.9 | -0.0106 | 0.2777 | -53 |
| -52 | 18.73 | 18.69 | 0.0119 | 3.0512 | 84.15 | 0.3277 | -4.0 | 117.0 | 113.0 | -0.0098 | 0.2773 | -52 |
| -51 | 19.24 | 19.19 | 0.0119 | 2.9752 | 84.04 | 0.3361 | -3.7 | 116.8 | 113.2 | -0.0090 | 0.2768 | -51 |
| -50 | 19.76 | 19.71 | 0.0119 | 2.9015 | 83.92 | 0.3447 | -3.3 | 116.6 | 113.3 | -0.0082 | 0.2764 | -50 |
| -49 | 20.29 | 20.24 | 0.0119 | 2.8300 | 83.81 | 0.3534 | -3.0 | 116.4 | 113.4 | -0.0073 | 0.2761 | -49 |
| -48 | 20.82 | 20.78 | 0.0119 | 2.7606 | 83.69 | 0.3622 | -2.7 | 116.2 | 113.5 | -0.0065 | 0.2757 | -48 |
| -47 | 21.38 | 21.32 | 0.0120 | 2.6933 | 83.58 | 0.3713 | -2.3 | 115.9 | 113.6 | -0.0057 | 0.2753 | -47 |

*(Continued)*

**TABLE A.18 (CONTINUED)**

Suva® 410A Saturation Properties—Temperature Table

| Temp. °F | Pressure, psia | | Volume, ft.³/lb. | | Density, lb./ft.³ | | Enthalpy, Btu/lb. | | | Entropy, Btu/(lb.)(R) | | Temp. °F |
|---|---|---|---|---|---|---|---|---|---|---|---|---|
| | Liquid $p_f$ | Vapor $p_g$ | Liquid $v_f$ | Vapor $v_g$ | Liquid $1/v_f$ | Vapor $1/v_g$ | Liquid $h_f$ | Latent $h_{fg}$ | Vapor $h_g$ | Liquid $s_f$ | Vapor $s_g$ | |
| −46 | 21.94 | 21.88 | 0.0120 | 2.6279 | 83.46 | 0.3805 | −2.0 | 115.7 | 113.7 | −0.0049 | 0.2749 | −46 |
| −45 | 22.51 | 22.46 | 0.0120 | 2.5645 | 83.35 | 0.3899 | −1.7 | 115.5 | 113.8 | −0.0041 | 0.2745 | −45 |
| −44 | 23.10 | 23.04 | 0.0120 | 2.5029 | 83.23 | 0.3995 | −1.3 | 115.3 | 114.0 | −0.0033 | 0.2741 | −44 |
| −43 | 23.69 | 23.64 | 0.0120 | 2.4430 | 83.11 | 0.4093 | −1.0 | 115.1 | 114.1 | −0.0025 | 0.2737 | −43 |
| −42 | 24.30 | 24.24 | 0.0120 | 2.3849 | 83.00 | 0.4193 | −0.7 | 114.9 | 114.2 | −0.0017 | 0.2733 | −42 |
| −41 | 24.92 | 24.86 | 0.0121 | 2.3285 | 82.88 | 0.4295 | −0.3 | 114.6 | 114.3 | −0.0008 | 0.2730 | −41 |
| −40 | 25.56 | 25.49 | 0.0121 | 2.2737 | 82.76 | 0.4398 | 0.0 | 114.4 | 114.4 | 0.0000 | 0.2726 | −40 |
| −39 | 26.20 | 26.14 | 0.0121 | 2.2204 | 82.65 | 0.4504 | 0.3 | 114.2 | 114.5 | 0.0008 | 0.2722 | −39 |
| −38 | 26.86 | 26.80 | 0.0121 | 2.1687 | 82.53 | 0.4611 | 0.7 | 114.0 | 114.6 | 0.0016 | 0.2718 | −38 |
| −37 | 27.53 | 27.47 | 0.0121 | 2.1183 | 82.41 | 0.4721 | 1.0 | 113.7 | 114.7 | 0.0024 | 0.2715 | −37 |
| −36 | 28.22 | 28.15 | 0.0122 | 2.0694 | 82.29 | 0.4832 | 1.3 | 113.5 | 114.9 | 0.0032 | 0.2711 | −36 |
| −35 | 28.92 | 28.84 | 0.0122 | 2.0219 | 82.17 | 0.4946 | 1.7 | 113.3 | 115.0 | 0.0040 | 0.2708 | −35 |
| −34 | 29.63 | 29.55 | 0.0122 | 1.9757 | 82.05 | 0.5062 | 2.0 | 113.1 | 115.1 | 0.0048 | 0.2704 | −34 |
| −33 | 30.35 | 30.28 | 0.0122 | 1.9307 | 81.93 | 0.5179 | 2.3 | 112.8 | 115.2 | 0.0056 | 0.2700 | −33 |
| −32 | 31.09 | 31.01 | 0.0122 | 1.8870 | 81.81 | 0.5300 | 2.7 | 112.6 | 115.3 | 0.0064 | 0.2697 | −32 |
| −31 | 31.85 | 31.76 | 0.0122 | 1.8444 | 81.70 | 0.5422 | 3.0 | 112.4 | 115.4 | 0.0072 | 0.2693 | −31 |
| −30 | 32.61 | 32.53 | 0.0123 | 1.8030 | 81.58 | 0.5546 | 3.4 | 112.1 | 115.5 | 0.0080 | 0.2690 | −30 |
| −29 | 33.39 | 33.31 | 0.0123 | 1.7628 | 81.45 | 0.5673 | 3.7 | 111.9 | 115.6 | 0.0088 | 0.2686 | −29 |
| −28 | 34.19 | 34.10 | 0.0123 | 1.7236 | 81.33 | 0.5802 | 4.0 | 111.7 | 115.7 | 0.0096 | 0.2683 | −28 |
| −27 | 35.00 | 34.91 | 0.0123 | 1.6854 | 81.21 | 0.5933 | 4.4 | 111.4 | 115.8 | 0.0104 | 0.2679 | −27 |
| −26 | 35.83 | 35.73 | 0.0123 | 1.6483 | 81.09 | 0.6067 | 4.7 | 111.2 | 115.9 | 0.0112 | 0.2676 | −26 |
| −25 | 36.67 | 36.57 | 0.0124 | 1.6122 | 80.97 | 0.6203 | 5.1 | 111.0 | 116.0 | 0.0120 | 0.2673 | −25 |
| −24 | 37.52 | 37.42 | 0.0124 | 1.5770 | 80.85 | 0.6341 | 5.4 | 110.7 | 116.1 | 0.0127 | 0.2669 | −24 |
| −23 | 38.39 | 38.29 | 0.0124 | 1.5427 | 80.73 | 0.6482 | 5.7 | 110.5 | 116.2 | 0.0135 | 0.2666 | −23 |
| −22 | 39.28 | 39.17 | 0.0124 | 1.5093 | 80.60 | 0.6625 | 6.1 | 110.3 | 116.3 | 0.0143 | 0.2663 | −22 |
| −21 | 40.18 | 40.07 | 0.0124 | 1.4768 | 80.48 | 0.6771 | 6.4 | 110.0 | 116.5 | 0.0151 | 0.2659 | −21 |

*(Continued)*

**TABLE A.18 (CONTINUED)**

Suva® 410A Saturation Properties—Temperature Table

| Temp. °F | Pressure, psia Liquid $p_f$ | Pressure, psia Vapor $p_g$ | Volume, ft.³/lb. Liquid $v_f$ | Volume, ft.³/lb. Vapor $v_g$ | Density, lb./ft.³ Liquid $1/v_f$ | Density, lb./ft.³ Vapor $1/v_g$ | Enthalpy, Btu/lb. Liquid $h_f$ | Enthalpy, Btu/lb. Latent $h_{fg}$ | Enthalpy, Btu/lb. Vapor $h_g$ | Entropy, Btu/(lb.)(R) Liquid $s_f$ | Entropy, Btu/(lb.)(R) Vapor $s_g$ | Temp. °F |
|---|---|---|---|---|---|---|---|---|---|---|---|---|
| −20 | 41.10 | 40.99 | 0.0124 | 1.4452 | 80.36 | 0.6920 | 6.8 | 109.8 | 116.6 | 0.0159 | 0.2656 | −20 |
| −19 | 42.03 | 41.92 | 0.0125 | 1.4143 | 80.23 | 0.7071 | 7.1 | 109.5 | 116.7 | 0.0167 | 0.2653 | −19 |
| −18 | 42.98 | 42.86 | 0.0125 | 1.3843 | 80.11 | 0.7224 | 7.5 | 109.3 | 116.8 | 0.0175 | 0.2649 | −18 |
| −17 | 43.95 | 43.83 | 0.0125 | 1.3550 | 79.99 | 0.7380 | 7.8 | 109.1 | 116.9 | 0.0183 | 0.2646 | −17 |
| −16 | 44.93 | 44.81 | 0.0125 | 1.3264 | 79.86 | 0.7539 | 8.2 | 108.8 | 117.0 | 0.0191 | 0.2643 | −16 |
| −15 | 45.94 | 45.81 | 0.0125 | 1.2986 | 79.74 | 0.7701 | 8.5 | 108.6 | 117.1 | 0.0198 | 0.2640 | −15 |
| −14 | 46.95 | 46.82 | 0.0126 | 1.2715 | 79.61 | 0.7865 | 8.8 | 108.3 | 117.2 | 0.0206 | 0.2637 | −14 |
| −13 | 47.99 | 47.85 | 0.0126 | 1.2450 | 79.49 | 0.8032 | 9.2 | 108.1 | 117.3 | 0.0214 | 0.2633 | −13 |
| −12 | 49.04 | 48.90 | 0.0126 | 1.2192 | 79.36 | 0.8202 | 9.5 | 107.8 | 117.4 | 0.0222 | 0.2630 | −12 |
| −11 | 50.11 | 49.97 | 0.0126 | 1.1941 | 79.24 | 0.8375 | 9.9 | 107.6 | 117.5 | 0.0230 | 0.2627 | −11 |
| −10 | 51.20 | 51.05 | 0.0126 | 1.1695 | 79.11 | 0.8551 | 10.2 | 107.3 | 117.6 | 0.0237 | 0.2624 | −10 |
| −9 | 52.31 | 52.16 | 0.0127 | 1.1456 | 78.98 | 0.8729 | 10.6 | 107.1 | 117.6 | 0.0245 | 0.2621 | −9 |
| −8 | 53.43 | 53.28 | 0.0127 | 1.1222 | 78.86 | 0.8911 | 10.9 | 106.8 | 117.7 | 0.0253 | 0.2618 | −8 |
| −7 | 54.57 | 54.42 | 0.0127 | 1.0995 | 78.73 | 0.9095 | 11.3 | 106.6 | 117.8 | 0.0261 | 0.2615 | −7 |
| −6 | 55.74 | 55.58 | 0.0127 | 1.0772 | 78.60 | 0.9283 | 11.6 | 106.3 | 117.9 | 0.0269 | 0.2612 | −6 |
| −5 | 56.92 | 56.75 | 0.0127 | 1.0555 | 78.47 | 0.9474 | 12.0 | 106.0 | 118.0 | 0.0276 | 0.2609 | −5 |
| −4 | 58.12 | 57.95 | 0.0128 | 1.0344 | 78.35 | 0.9668 | 12.3 | 105.8 | 118.1 | 0.0284 | 0.2606 | −4 |
| −3 | 59.34 | 59.16 | 0.0128 | 1.0137 | 78.22 | 0.9865 | 12.7 | 105.5 | 118.2 | 0.0292 | 0.2603 | −3 |
| −2 | 60.58 | 60.40 | 0.0128 | 0.9935 | 78.09 | 1.0065 | 13.0 | 105.3 | 118.3 | 0.0299 | 0.2600 | −2 |
| −1 | 61.83 | 61.65 | 0.0128 | 0.9738 | 77.96 | 1.0269 | 13.4 | 105.0 | 118.4 | 0.0307 | 0.2597 | −1 |
| 0 | 63.11 | 62.93 | 0.0128 | 0.9546 | 77.83 | 1.0475 | 13.7 | 104.7 | 118.5 | 0.0315 | 0.2594 | 0 |
| 1 | 64.41 | 64.22 | 0.0129 | 0.9358 | 77.70 | 1.0686 | 14.1 | 104.5 | 118.6 | 0.0323 | 0.2591 | 1 |
| 2 | 65.73 | 65.54 | 0.0129 | 0.9175 | 77.57 | 1.0899 | 14.5 | 104.2 | 118.7 | 0.0330 | 0.2588 | 2 |
| 3 | 67.07 | 66.87 | 0.0129 | 0.8996 | 77.44 | 1.1116 | 14.8 | 104.0 | 118.8 | 0.0338 | 0.2585 | 3 |
| 4 | 68.43 | 68.23 | 0.0129 | 0.8821 | 77.31 | 1.1337 | 15.2 | 103.7 | 118.9 | 0.0346 | 0.2582 | 4 |
| 5 | 69.81 | 69.61 | 0.0130 | 0.8650 | 77.17 | 1.1561 | 15.5 | 103.4 | 118.9 | 0.0353 | 0.2579 | 5 |
| 6 | 71.22 | 71.00 | 0.0130 | 0.8483 | 77.04 | 1.1788 | 15.9 | 103.2 | 119.0 | 0.0361 | 0.2576 | 6 |

*(Continued)*

**TABLE A.18 (CONTINUED)**

Suva® 410A Saturation Properties—Temperature Table

| Temp. °F | Pressure, psia Liquid $p_f$ | Vapor $p_g$ | Volume, ft.³/lb. Liquid $v_f$ | Vapor $v_g$ | Density, lb./ft.³ Liquid $1/v_f$ | Vapor $1/v_g$ | Enthalpy, Btu/lb. Liquid $h_f$ | Latent $h_{fg}$ | Vapor $h_g$ | Entropy, Btu/(lb.)(R) Liquid $s_f$ | Vapor $s_g$ | Temp. °F |
|---|---|---|---|---|---|---|---|---|---|---|---|---|
| 7 | 72.64 | 72.42 | 0.0130 | 0.8320 | 76.91 | 1.2020 | 16.2 | 102.9 | 119.1 | 0.0369 | 0.2573 | 7 |
| 8 | 74.09 | 73.86 | 0.0130 | 0.8160 | 76.78 | 1.2255 | 16.6 | 102.6 | 119.2 | 0.0376 | 0.2570 | 8 |
| 9 | 75.56 | 75.33 | 0.0130 | 0.8004 | 76.64 | 1.2493 | 16.9 | 102.3 | 119.3 | 0.0384 | 0.2568 | 9 |
| 10 | 77.05 | 76.81 | 0.0131 | 0.7852 | 76.51 | 1.2736 | 17.3 | 102.1 | 119.4 | 0.0392 | 0.2565 | 10 |
| 11 | 78.56 | 78.32 | 0.0131 | 0.7703 | 76.37 | 1.2982 | 17.7 | 101.8 | 119.5 | 0.0399 | 0.2562 | 11 |
| 12 | 80.09 | 79.85 | 0.0131 | 0.7557 | 76.24 | 1.3232 | 18.0 | 101.5 | 119.5 | 0.0407 | 0.2559 | 12 |
| 13 | 81.65 | 81.40 | 0.0131 | 0.7415 | 76.11 | 1.3486 | 18.4 | 101.2 | 119.6 | 0.0414 | 0.2556 | 13 |
| 14 | 83.23 | 82.98 | 0.0132 | 0.7276 | 75.97 | 1.3744 | 18.7 | 101.0 | 119.7 | 0.0422 | 0.2554 | 14 |
| 15 | 84.84 | 84.58 | 0.0132 | 0.7140 | 75.83 | 1.4006 | 19.1 | 100.7 | 119.8 | 0.0430 | 0.2551 | 15 |
| 16 | 86.46 | 86.20 | 0.0132 | 0.7007 | 75.70 | 1.4272 | 19.5 | 100.4 | 119.9 | 0.0437 | 0.2548 | 16 |
| 17 | 88.11 | 87.84 | 0.0132 | 0.6876 | 75.56 | 1.4542 | 19.8 | 100.1 | 119.9 | 0.0445 | 0.2545 | 17 |
| 18 | 89.79 | 89.51 | 0.0133 | 0.6749 | 75.42 | 1.4817 | 20.2 | 99.8 | 120.0 | 0.0452 | 0.2543 | 18 |
| 19 | 91.49 | 91.21 | 0.0133 | 0.6625 | 75.29 | 1.5095 | 20.5 | 99.6 | 120.1 | 0.0460 | 0.2540 | 19 |
| 20 | 93.21 | 92.93 | 0.0133 | 0.6503 | 75.15 | 1.5378 | 20.9 | 99.3 | 120.2 | 0.0467 | 0.2537 | 20 |
| 21 | 94.96 | 94.67 | 0.0133 | 0.6383 | 75.01 | 1.5666 | 21.3 | 99.0 | 120.3 | 0.0475 | 0.2534 | 21 |
| 22 | 96.73 | 96.43 | 0.0134 | 0.6267 | 74.87 | 1.5958 | 21.6 | 98.7 | 120.3 | 0.0483 | 0.2532 | 22 |
| 23 | 98.53 | 98.23 | 0.0134 | 0.6152 | 74.73 | 1.6254 | 22.0 | 98.4 | 120.4 | 0.0490 | 0.2529 | 23 |
| 24 | 100.36 | 100.04 | 0.0134 | 0.6041 | 74.59 | 1.6555 | 22.4 | 98.1 | 120.5 | 0.0498 | 0.2526 | 24 |
| 25 | 102.20 | 101.89 | 0.0134 | 0.5931 | 74.45 | 1.6860 | 22.7 | 97.8 | 120.6 | 0.0505 | 0.2524 | 25 |
| 26 | 104.08 | 103.76 | 0.0135 | 0.5824 | 74.31 | 1.7170 | 23.1 | 97.5 | 120.6 | 0.0513 | 0.2521 | 26 |
| 27 | 105.98 | 105.65 | 0.0135 | 0.5719 | 74.17 | 1.7485 | 23.5 | 97.2 | 120.7 | 0.0520 | 0.2518 | 27 |
| 28 | 107.91 | 107.57 | 0.0135 | 0.5616 | 74.03 | 1.7805 | 23.8 | 96.9 | 120.8 | 0.0528 | 0.2516 | 28 |
| 29 | 109.86 | 109.52 | 0.0135 | 0.5516 | 73.89 | 1.8130 | 24.2 | 96.6 | 120.9 | 0.0535 | 0.2513 | 29 |
| 30 | 111.84 | 111.49 | 0.0136 | 0.5417 | 73.74 | 1.8459 | 24.6 | 96.3 | 120.9 | 0.0543 | 0.2510 | 30 |
| 31 | 113.85 | 113.49 | 0.0136 | 0.5321 | 73.60 | 1.8794 | 24.9 | 96.0 | 121.0 | 0.0550 | 0.2508 | 31 |
| 32 | 115.88 | 115.52 | 0.0136 | 0.5227 | 73.46 | 1.9133 | 25.3 | 95.7 | 121.1 | 0.0558 | 0.2505 | 32 |

*(Continued)*

**TABLE A.18 (CONTINUED)**

Suva® 410A Saturation Properties—Temperature Table

| Temp. °F | Pressure, psia Liquid $p_f$ | Pressure, psia Vapor $p_g$ | Volume, ft.³/lb. Liquid $v_f$ | Volume, ft.³/lb. Vapor $v_g$ | Density, lb./ft.³ Liquid $1/v_f$ | Density, lb./ft.³ Vapor $1/v_g$ | Enthalpy, Btu/lb. Liquid $h_f$ | Enthalpy, Btu/lb. Latent $h_{fg}$ | Enthalpy, Btu/lb. Vapor $h_g$ | Entropy, Btu/(lb.)(R) Liquid $s_f$ | Entropy, Btu/(lb.)(R) Vapor $s_g$ | Temp. °F |
|---|---|---|---|---|---|---|---|---|---|---|---|---|
| 33 | 117.94 | 117.57 | 0.0136 | 0.5134 | 73.31 | 1.9478 | 25.7 | 95.4 | 121.1 | 0.0565 | 0.2502 | 33 |
| 34 | 120.03 | 119.66 | 0.0137 | 0.5043 | 73.17 | 1.9828 | 26.1 | 95.1 | 121.2 | 0.0573 | 0.2500 | 34 |
| 35 | 122.15 | 121.77 | 0.0137 | 0.4955 | 73.02 | 2.0183 | 26.4 | 94.8 | 121.3 | 0.0580 | 0.2497 | 35 |
| 36 | 124.30 | 123.90 | 0.0137 | 0.4868 | 72.88 | 2.0544 | 26.8 | 94.5 | 121.3 | 0.0588 | 0.2495 | 36 |
| TI | 126.47 | 126.07 | 0.0137 | 0.4782 | 72.73 | 2.0910 | 27.2 | 94.2 | 121.4 | 0.0595 | 0.2492 | 37 |
| 38 | 128.67 | 128.27 | 0.0138 | 0.4699 | 72.58 | 2.1282 | 27.5 | 93.9 | 121.5 | 0.0602 | 0.2489 | 38 |
| 39 | 130.91 | 130.49 | 0.0138 | 0.4617 | 72.43 | 2.1659 | 27.9 | 93.6 | 121.5 | 0.0610 | 0.2487 | 39 |
| 40 | 133.17 | 132.74 | 0.0138 | 0.4537 | 72.29 | 2.2042 | 28.3 | 93.3 | 121.6 | 0.0617 | 0.2484 | 40 |
| 41 | 135.46 | 135.03 | 0.0139 | 0.4458 | 72.14 | 2.2431 | 28.7 | 93.0 | 121.6 | 0.0625 | 0.2482 | 41 |
| 42 | 137.78 | 137.34 | 0.0139 | 0.4381 | 71.99 | 2.2826 | 29.1 | 92.7 | 121.7 | 0.0632 | 0.2479 | 42 |
| 43 | 140.13 | 139.68 | 0.0139 | 0.4305 | 71.84 | 2.3226 | 29.4 | 92.3 | 121.8 | 0.0640 | 0.2477 | 43 |
| 44 | 142.51 | 142.05 | 0.0139 | 0.4231 | 71.69 | 2.3633 | 29.8 | 92.0 | 121.8 | 0.0647 | 0.2474 | 44 |
| 45 | 144.92 | 144.46 | 0.0140 | 0.4159 | 71.54 | 2.4046 | 30.2 | 91.7 | 121.9 | 0.0655 | 0.2471 | 45 |
| 46 | 147.36 | 146.89 | 0.0140 | 0.4087 | 71.39 | 2.4465 | 30.6 | 91.4 | 121.9 | 0.0662 | 0.2469 | 46 |
| 47 | 149.83 | 149.35 | 0.0140 | 0.4018 | 71.23 | 2.4890 | 30.9 | 91.0 | 122.0 | 0.0669 | 0.2466 | 47 |
| 48 | 152.34 | 151.85 | 0.0141 | 0.3949 | 71.08 | 2.5322 | 31.3 | 90.7 | 122.0 | 0.0677 | 0.2464 | 48 |
| 49 | 154.87 | 154.37 | 0.0141 | 0.3882 | 70.93 | 2.5761 | 31.7 | 90.4 | 122.1 | 0.0684 | 0.2461 | 49 |
| 50 | 157.44 | 156.93 | 0.0141 | 0.3816 | 70.77 | 2.6206 | 32.1 | 90.1 | 122.2 | 0.0692 | 0.2459 | 50 |
| 51 | 160.04 | 159.52 | 0.0142 | 0.3751 | 70.62 | 2.6657 | 32.5 | 89.7 | 122.2 | 0.0699 | 0.2456 | 51 |
| 52 | 162.67 | 162.14 | 0.0142 | 0.3688 | 70.46 | 2.7116 | 32.9 | 89.4 | 122.3 | 0.0706 | 0.2454 | 52 |
| 53 | 165.33 | 164.80 | 0.0142 | 0.3626 | 70.31 | 2.7581 | 33.2 | 89.1 | 122.3 | 0.0714 | 0.2451 | 53 |
| 54 | 168.03 | 167.48 | 0.0143 | 0.3565 | 70.15 | 2.8054 | 33.6 | 88.7 | 122.3 | 0.0721 | 0.2448 | 54 |
| 55 | 170.76 | 170.20 | 0.0143 | 0.3505 | 69.99 | 2.8533 | 34.0 | 88.4 | 122.4 | 0.0729 | 0.2446 | 55 |
| 56 | 173.52 | 172.96 | 0.0143 | 0.3446 | 69.83 | 2.9020 | 34.4 | 88.0 | 122.4 | 0.0736 | 0.2443 | 56 |
| 57 | 176.32 | 175.74 | 0.0144 | 0.3388 | 69.67 | 2.9514 | 34.8 | 87.7 | 122.5 | 0.0743 | 0.2441 | 57 |
| 58 | 179.15 | 178.57 | 0.0144 | 0.3332 | 69.51 | 3.0016 | 35.2 | 87.4 | 122.5 | 0.0751 | 0.2438 | 58 |
| 59 | 182.01 | 181.42 | 0.0144 | 0.3276 | 69.35 | 3.0526 | 35.6 | 87.0 | 122.6 | 0.0758 | 0.2436 | 59 |

*(Continued)*

**TABLE A.18 (CONTINUED)**

Suva® 410A Saturation Properties—Temperature Table

| Temp. °F | Pressure, psia | | Volume, ft.³/lb. | | Density, lb./ft.³ | | Enthalpy, Btu/lb. | | | Entropy, Btu/(lb.)(R) | | Temp. °F |
|---|---|---|---|---|---|---|---|---|---|---|---|---|
| | Liquid $p_f$ | Vapor $p_g$ | Liquid $v_f$ | Vapor $v_g$ | Liquid $1/v_f$ | Vapor $1/v_g$ | Liquid $h_f$ | Latent $h_{fg}$ | Vapor $h_g$ | Liquid $s_f$ | Vapor $s_g$ | |
| 60 | 184.91 | 184.31 | 0.0145 | 0.3221 | 69.19 | 3.1043 | 36.0 | 86.7 | 122.6 | 0.0766 | 0.2433 | 60 |
| 61 | 187.84 | 187.23 | 0.0145 | 0.3168 | 69.03 | 3.1568 | 36.3 | 86.3 | 122.7 | 0.0773 | 0.2431 | 61 |
| 62 | 190.81 | 190.19 | 0.0145 | 0.3115 | 68.87 | 3.2101 | 36.7 | 86.0 | 122.7 | 0.0780 | 0.2428 | 62 |
| 63 | 193.82 | 193.18 | 0.0146 | 0.3064 | 68.70 | 3.2642 | 37.1 | 85.6 | 122.7 | 0.0788 | 0.2426 | 63 |
| 64 | 196.86 | 196.21 | 0.0146 | 0.3013 | 68.54 | 3.3191 | 37.5 | 85.3 | 122.8 | 0.0795 | 0.2423 | 64 |
| 65 | 199.93 | 199.28 | 0.0146 | 0.2963 | 68.37 | 3.3749 | 37.9 | 84.9 | 122.8 | 0.0802 | 0.2420 | 65 |
| 66 | 203.04 | 202.38 | 0.0147 | 0.2914 | 68.21 | 3.4316 | 38.3 | 84.5 | 122.8 | 0.0810 | 0.2418 | 66 |
| 67 | 206.19 | 205.51 | 0.0147 | 0.2866 | 68.04 | 3.4891 | 38.7 | 84.2 | 122.9 | 0.0817 | 0.2415 | 67 |
| 68 | 209.38 | 208.69 | 0.0147 | 0.2819 | 67.87 | 3.5475 | 39.1 | 83.8 | 122.9 | 0.0825 | 0.2413 | 68 |
| 69 | 212.60 | 211.90 | 0.0148 | 0.2773 | 67.71 | 3.6068 | 39.5 | 83.4 | 122.9 | 0.0832 | 0.2410 | 69 |
| 70 | 215.86 | 215.14 | 0.0148 | 0.2727 | 67.54 | 3.6670 | 39.9 | 83.1 | 123.0 | 0.0839 | 0.2408 | 70 |
| 71 | 219.15 | 218.43 | 0.0148 | 0.2682 | 67.37 | 3.7281 | 40.3 | 82.7 | 123.0 | 0.0847 | 0.2405 | 71 |
| 72 | 222.49 | 221.75 | 0.0149 | 0.2638 | 67.19 | 3.7902 | 40.7 | 82.3 | 123.0 | 0.0854 | 0.2402 | 72 |
| 73 | 225.86 | 225.11 | 0.0149 | 0.2595 | 67.02 | 3.8533 | 41.1 | 81.9 | 123.0 | 0.0861 | 0.2400 | 73 |
| 74 | 229.27 | 228.51 | 0.0150 | 0.2553 | 66.85 | 3.9173 | 41.5 | 81.6 | 123.1 | 0.0869 | 0.2397 | 74 |
| 75 | 232.72 | 231.94 | 0.0150 | 0.2511 | 66.68 | 3.9823 | 41.9 | 81.2 | 123.1 | 0.0876 | 0.2395 | 75 |
| 76 | 236.20 | 235.42 | 0.0150 | 0.2470 | 66.50 | 4.0484 | 42.3 | 80.8 | 123.1 | 0.0884 | 0.2392 | 76 |
| 77 | 239.73 | 238.93 | 0.0151 | 0.2430 | 66.32 | 4.1155 | 42.7 | 80.4 | 123.1 | 0.0891 | 0.2389 | 77 |
| 78 | 243.30 | 242.49 | 0.0151 | 0.2390 | 66.15 | 4.1836 | 43.1 | 80.0 | 123.1 | 0.0898 | 0.2387 | 78 |
| 79 | 246.91 | 246.08 | 0.0152 | 0.2351 | 65.97 | 4.2529 | 43.5 | 79.6 | 123.2 | 0.0906 | 0.2384 | 79 |
| 80 | 250.55 | 249.71 | 0.0152 | 0.2313 | 65.79 | 4.3232 | 43.9 | 79.2 | 123.2 | 0.0913 | 0.2381 | 80 |
| 81 | 254.24 | 253.39 | 0.0152 | 0.2276 | 65.61 | 4.3946 | 44.3 | 78.8 | 123.2 | 0.0921 | 0.2379 | 81 |
| 82 | 257.97 | 257.10 | 0.0153 | 0.2239 | 65.43 | 4.4672 | 44.8 | 78.4 | 123.2 | 0.0928 | 0.2376 | 82 |
| 83 | 261.74 | 260.85 | 0.0153 | 0.2202 | 65.25 | 4.5409 | 45.2 | 78.0 | 123.2 | 0.0936 | 0.2373 | 83 |
| 84 | 265.55 | 264.65 | 0.0154 | 0.2166 | 65.06 | 4.6159 | 45.6 | 77.6 | 123.2 | 0.0943 | 0.2371 | 84 |
| 85 | 269.40 | 268.49 | 0.0154 | 0.2131 | 64.88 | 4.6920 | 46.0 | 77.2 | 123.2 | 0.0950 | 0.2368 | 85 |

*(Continued)*

**TABLE A.18 (CONTINUED)**

Suva® 410A Saturation Properties—Temperature Table

| Temp. °F | Pressure, psia | | Volume, ft.³/lb. | | Density, lb./ft.³ | | Enthalpy, Btu/lb. | | | Entropy, Btu/(lb.)(R) | | Temp. °F |
|---|---|---|---|---|---|---|---|---|---|---|---|---|
| | Liquid $p_f$ | Vapor $p_g$ | Liquid $v_f$ | Vapor $v_g$ | Liquid $1/v_f$ | Vapor $1/v_g$ | Liquid $h_f$ | Latent $h_{fg}$ | Vapor $h_g$ | Liquid $s_f$ | Vapor $s_g$ | |
| 86 | 273.29 | 272.37 | 0.0155 | 0.2097 | 64.69 | 4.7693 | 46.4 | 76.8 | 123.2 | 0.0958 | 0.2365 | 86 |
| 87 | 277.23 | 276.29 | 0.0155 | 0.2063 | 64.51 | 4.8480 | 46.8 | 76.4 | 123.2 | 0.0965 | 0.2363 | 87 |
| 88 | 281.21 | 280.25 | 0.0155 | 0.2029 | 64.32 | 4.9278 | 47.3 | 76.0 | 123.2 | 0.0973 | 0.2360 | 88 |
| 89 | 285.23 | 284.26 | 0.0156 | 0.1996 | 64.13 | 5.0090 | 47.7 | 75.5 | 123.2 | 0.0980 | 0.2357 | 89 |
| 90 | 289.29 | 288.31 | 0.0156 | 0.1964 | 63.94 | 5.0916 | 48.1 | 75.1 | 123.2 | 0.0988 | 0.2354 | 90 |
| 91 | 293.40 | 292.40 | 0.0157 | 0.1932 | 63.74 | 5.1755 | 48.5 | 74.7 | 123.2 | 0.0995 | 0.2352 | 91 |
| 92 | 297.55 | 296.54 | 0.0157 | 0.1901 | 63.55 | 5.2608 | 48.9 | 74.3 | 123.2 | 0.1003 | 0.2349 | 92 |
| 93 | 301.75 | 300.72 | 0.0158 | 0.1870 | 63.36 | 5.3474 | 49.4 | 73.8 | 123.2 | 0.1010 | 0.2346 | 93 |
| 94 | 305.99 | 304.95 | 0.0158 | 0.1840 | 63.16 | 5.4356 | 49.8 | 73.4 | 123.2 | 0.1018 | 0.2343 | 94 |
| 95 | 310.27 | 309.22 | 0.0159 | 0.1810 | 62.96 | 5.5252 | 50.2 | 72.9 | 123.2 | 0.1025 | 0.2340 | 95 |
| 96 | 314.60 | 313.53 | 0.0159 | 0.1781 | 62.76 | 5.6164 | 50.7 | 72.5 | 123.2 | 0.1033 | 0.2338 | 96 |
| 97 | 318.98 | 317.89 | 0.0160 | 0.1752 | 62.56 | 5.7090 | 51.1 | 72.1 | 123.1 | 0.1040 | 0.2335 | 97 |
| 98 | 323.40 | 322.29 | 0.0160 | 0.1723 | 62.36 | 5.8033 | 51.5 | 71.6 | 123.1 | 0.1048 | 0.2332 | 98 |
| 99 | 327.87 | 326.75 | 0.0161 | 0.1695 | 62.16 | 5.8992 | 52.0 | 71.1 | 123.1 | 0.1056 | 0.2329 | 99 |
| 100 | 332.38 | 331.24 | 0.0161 | 0.1668 | 61.95 | 5.9967 | 52.4 | 70.7 | 123.1 | 0.1063 | 0.2326 | 100 |
| 101 | 336.94 | 335.79 | 0.0162 | 0.1640 | 61.74 | 6.0959 | 52.8 | 70.2 | 123.1 | 0.1071 | 0.2323 | 101 |
| 102 | 341.54 | 340.38 | 0.0163 | 0.1614 | 61.53 | 6.1969 | 53.3 | 69.7 | 123.0 | 0.1079 | 0.2320 | 102 |
| 103 | 346.20 | 345.02 | 0.0163 | 0.1587 | 61.32 | 6.2996 | 53.7 | 69.3 | 123.0 | 0.1086 | 0.2317 | 103 |
| 104 | 350.90 | 349.70 | 0.0164 | 0.1561 | 61.11 | 6.4041 | 54.2 | 68.8 | 123.0 | 0.1094 | 0.2314 | 104 |
| 105 | 355.65 | 354.44 | 0.0164 | 0.1536 | 60.90 | 6.5105 | 54.6 | 68.3 | 122.9 | 0.1102 | 0.2311 | 105 |
| 106 | 360.45 | 359.22 | 0.0165 | 0.1511 | 60.68 | 6.6188 | 55.1 | 67.8 | 122.9 | 0.1109 | 0.2308 | 106 |
| 107 | 365.29 | 364.05 | 0.0165 | 0.1486 | 60.46 | 6.7290 | 55.5 | 67.3 | 122.9 | 0.1117 | 0.2305 | 107 |
| 108 | 370.19 | 368.93 | 0.0166 | 0.1462 | 60.24 | 6.8413 | 56.0 | 66.8 | 122.8 | 0.1125 | 0.2302 | 108 |
| 109 | 375.13 | 373.86 | 0.0167 | 0.1438 | 60.02 | 6.9555 | 56.5 | 66.3 | 122.8 | 0.1133 | 0.2299 | 109 |
| 110 | 380.12 | 378.84 | 0.0167 | 0.1414 | 59.79 | 7.0719 | 56.9 | 65.8 | 122.7 | 0.1141 | 0.2296 | 110 |
| 111 | 385.17 | 383.86 | 0.0168 | 0.1391 | 59.57 | 7.1904 | 57.4 | 65.3 | 122.7 | 0.1148 | 0.2293 | 111 |

*(Continued)*

# TABLE A.18 (CONTINUED)

Suva® 410A Saturation Properties—Temperature Table

| Temp. °F | Pressure, psia Liquid $p_f$ | Vapor $p_g$ | Volume, ft.³/lb. Liquid $v_f$ | Vapor $v_g$ | Density, lb./ft.³ Liquid $1/v_f$ | Vapor $1/v_g$ | Enthalpy, Btu/lb. Liquid $h_f$ | Latent $h_{fg}$ | Vapor $h_g$ | Entropy, Btu/(lb.)(R) Liquid $s_f$ | Vapor $s_g$ | Temp. °F |
|---|---|---|---|---|---|---|---|---|---|---|---|---|
| 112 | 390.26 | 388.94 | 0.0169 | 0.1368 | 59.34 | 7.3112 | 57.8 | 64.8 | 122.6 | 0.1156 | 0.2289 | 112 |
| 113 | 395.40 | 394.07 | 0.0169 | 0.1345 | 59.10 | 7.4342 | 58.3 | 64.3 | 122.6 | 0.1164 | 0.2286 | 113 |
| 114 | 400.60 | 399.25 | 0.0170 | 0.1323 | 58.87 | 7.5596 | 58.8 | 63.7 | 122.5 | 0.1172 | 0.2283 | 114 |
| 115 | 405.84 | 404.49 | 0.0171 | 0.1301 | 58.63 | 7.6873 | 59.3 | 63.2 | 122.5 | 0.1180 | 0.2280 | 115 |
| 116 | 411.14 | 409.77 | 0.0171 | 0.1279 | 58.39 | 7.8176 | 59.7 | 62.6 | 122.4 | 0.1188 | 0.2276 | 116 |
| 117 | 416.49 | 415.11 | 0.0172 | 0.1258 | 58.15 | 7.9504 | 60.2 | 62.1 | 122.3 | 0.1196 | 0.2273 | 117 |
| 118 | 421.89 | 420.49 | 0.0173 | 0.1237 | 57.91 | 8.0858 | 60.7 | 61.5 | 122.3 | 0.1204 | 0.2270 | 118 |
| 119 | 427.34 | 425.94 | 0.0173 | 0.1216 | 57.66 | 8.2239 | 61.2 | 61.0 | 122.2 | 0.1213 | 0.2266 | 119 |
| 120 | 432.85 | 431.43 | 0.0174 | 0.1195 | 57.41 | 8.3648 | 61.7 | 60.4 | 122.1 | 0.1221 | 0.2263 | 120 |
| 121 | 438.41 | 436.98 | 0.0175 | 0.1175 | 57.16 | 8.5085 | 62.2 | 59.8 | 122.0 | 0.1229 | 0.2259 | 121 |
| 122 | 444.02 | 442.58 | 0.0176 | 0.1155 | 56.90 | 8.6553 | 62.7 | 59.2 | 121.9 | 0.1237 | 0.2256 | 122 |
| 123 | 449.68 | 448.24 | 0.0177 | 0.1136 | 56.64 | 8.8051 | 63.2 | 58.7 | 121.9 | 0.1246 | 0.2252 | 123 |
| 124 | 455.40 | 453.95 | 0.0177 | 0.1116 | 56.37 | 8.9581 | 63.7 | 58.1 | 121.8 | 0.1254 | 0.2249 | 124 |
| 125 | 461.18 | 459.72 | 0.0178 | 0.1097 | 56.11 | 9.1144 | 64.2 | 57.4 | 121.7 | 0.1263 | 0.2245 | 125 |
| 126 | 467.01 | 465.54 | 0.0179 | 0.1078 | 55.83 | 9.2741 | 64.8 | 56.8 | 121.6 | 0.1271 | 0.2242 | 126 |
| 127 | 472.89 | 471.42 | 0.0180 | 0.1060 | 55.56 | 9.4374 | 65.3 | 56.2 | 121.5 | 0.1280 | 0.2238 | 127 |
| 128 | 478.83 | 477.35 | 0.0181 | 0.1041 | 55.28 | 9.6043 | 65.8 | 55.6 | 121.4 | 0.1289 | 0.2234 | 128 |
| 129 | 484.83 | 483.35 | 0.0182 | 0.1023 | 54.99 | 9.7752 | 66.3 | 54.9 | 121.3 | 0.1298 | 0.2230 | 129 |
| 130 | 490.88 | 489.39 | 0.0183 | 0.1005 | 54.70 | 9.9500 | 66.9 | 54.3 | 121.2 | 0.1306 | 0.2227 | 130 |
| 131 | 496.98 | 495.50 | 0.0184 | 0.0987 | 54.41 | 10.1290 | 67.4 | 53.6 | 121.0 | 0.1315 | 0.2223 | 131 |
| 132 | 503.15 | 501.66 | 0.0185 | 0.0970 | 54.11 | 10.3125 | 68.0 | 52.9 | 120.9 | 0.1324 | 0.2219 | 132 |
| 133 | 509.37 | 507.89 | 0.0186 | 0.0952 | 53.80 | 10.5005 | 68.6 | 52.2 | 120.8 | 0.1334 | 0.2215 | 133 |
| 134 | 515.65 | 514.17 | 0.0187 | 0.0935 | 53.49 | 10.6935 | 69.1 | 51.5 | 120.7 | 0.1343 | 0.2211 | 134 |
| 135 | 521.98 | 520.51 | 0.0188 | 0.0918 | 53.18 | 10.8915 | 69.7 | 50.8 | 120.5 | 0.1352 | 0.2207 | 135 |
| 136 | 528.37 | 526.91 | 0.0189 | 0.0901 | 52.85 | 11.0950 | 70.3 | 50.1 | 120.4 | 0.1362 | 0.2203 | 136 |
| 137 | 534.83 | 533.37 | 0.0190 | 0.0885 | 52.52 | 11.3043 | 70.9 | 49.3 | 120.3 | 0.1372 | 0.2198 | 137 |

(Continued)

**TABLE A.18 (CONTINUED)**

Suva® 410A Saturation Properties—Temperature Table

| Temp. °F | Pressure, psia | | Volume, ft.³/lb. | | Density, lb./ft.³ | | Enthalpy, Btu/lb. | | | Entropy, Btu/(lb.)(R) | | Temp. °F |
|---|---|---|---|---|---|---|---|---|---|---|---|---|
| | Liquid $p_f$ | Vapor $p_g$ | Liquid $v_f$ | Vapor $v_g$ | Liquid $1/v_f$ | Vapor $1/v_g$ | Liquid $h_f$ | Latent $h_{fg}$ | Vapor $h_g$ | Liquid $s_f$ | Vapor $s_g$ | |
| 138 | 541.34 | 539.89 | 0.0192 | 0.0868 | 52.18 | 11.5197 | 71.5 | 48.6 | 120.1 | 0.1381 | 0.2194 | 138 |
| 139 | 547.90 | 546.47 | 0.0193 | 0.0852 | 51.84 | 11.7418 | 72.1 | 47.8 | 119.9 | 0.1391 | 0.2190 | 139 |
| 140 | 554.53 | 553.12 | 0.0194 | 0.0835 | 51.48 | 11.9709 | 72.8 | 47.0 | 119.8 | 0.1401 | 0.2185 | 140 |
| 141 | 561.22 | 559.82 | 0.0196 | 0.0819 | 51.12 | 12.2077 | 73.4 | 46.2 | 119.6 | 0.1412 | 0.2181 | 141 |
| 142 | 567.97 | 566.59 | 0.0197 | 0.0803 | 50.74 | 12.4528 | 74.1 | 45.4 | 119.4 | 0.1422 | 0.2176 | 142 |
| 143 | 574.77 | 573.42 | 0.0199 | 0.0787 | 50.36 | 12.7071 | 74.8 | 44.5 | 119.3 | 0.1433 | 0.2171 | 143 |
| 144 | 581.64 | 580.31 | 0.0200 | 0.0771 | 49.96 | 12.9714 | 75.4 | 43.6 | 119.1 | 0.1444 | 0.2167 | 144 |
| 145 | 588.57 | 587.27 | 0.0202 | 0.0755 | 49.55 | 13.2468 | 76.2 | 42.7 | 118.9 | 0.1455 | 0.2162 | 145 |
| 146 | 595.55 | 594.29 | 0.0204 | 0.0739 | 49.13 | 13.5347 | 76.9 | 41.8 | 118.7 | 0.1467 | 0.2157 | 146 |
| 147 | 602.60 | 601.37 | 0.0205 | 0.0723 | 48.69 | 13.8367 | 77.6 | 40.8 | 118.4 | 0.1479 | 0.2151 | 147 |
| 148 | 609.71 | 608.52 | 0.0207 | 0.0706 | 48.23 | 14.1547 | 78.4 | 39.8 | 118.2 | 0.1491 | 0.2146 | 148 |
| 149 | 616.88 | 615.74 | 0.0209 | 0.0690 | 47.76 | 14.4913 | 79.2 | 38.7 | 117.9 | 0.1504 | 0.2140 | 149 |
| 150 | 624.12 | 623.02 | 0.0212 | 0.0673 | 47.26 | 14.8496 | 80.0 | 37.6 | 117.7 | 0.1517 | 0.2135 | 150 |
| 151 | 631.41 | 630.37 | 0.0214 | 0.0656 | 46.73 | 15.2337 | 80.9 | 36.5 | 117.4 | 0.1531 | 0.2129 | 151 |
| 152 | 638.77 | 637.78 | 0.0217 | 0.0639 | 46.18 | 15.6486 | 81.8 | 35.3 | 117.1 | 0.1545 | 0.2122 | 152 |
| 153 | 646.19 | 645.26 | 0.0219 | 0.0621 | 45.59 | 16.1014 | 82.8 | 34.0 | 116.8 | 0.1560 | 0.2116 | 153 |
| 154 | 653.67 | 652.81 | 0.0222 | 0.0602 | 44.96 | 16.6006 | 83.8 | 32.7 | 116.5 | 0.1576 | 0.2109 | 154 |
| 155 | 661.22 | 660.43 | 0.0226 | 0.0583 | 44.28 | 17.1576 | 84.9 | 31.3 | 116.1 | 0.1593 | 0.2102 | 155 |
| 156 | 668.82 | 668.12 | 0.0230 | 0.0562 | 43.54 | 17.7852 | 86.0 | 29.8 | 115.8 | 0.1612 | 0.2095 | 156 |
| 157 | 676.49 | 675.87 | 0.0234 | 0.0541 | 42.71 | 18.4945 | 87.3 | 28.2 | 115.5 | 0.1632 | 0.2089 | 157 |
| 158 | 684.23 | 683.70 | 0.0239 | 0.0518 | 41.77 | 19.2867 | 88.7 | 26.5 | 115.2 | 0.1654 | 0.2083 | 158 |
| 159 | 692.02 | 691.60 | 0.0246 | 0.0496 | 40.65 | 20.1422 | 90.3 | 24.8 | 115.1 | 0.1680 | 0.2080 | 159 |
| 160 | 699.88 | 699.56 | 0.0255 | 0.0476 | 39.25 | 21.0231 | 92.3 | 23.0 | 115.2 | 0.1711 | 0.2081 | 160 |

*Source:* *Thermodynamic Properties of DuPont Suva 410A Refrigerant.* With permission of E. I. Du Pont de Nemours and Co.

**TABLE A.19**

Suva® 410A Superheated Vapor–Constant Pressure Tables (Saturated Vapor Properties in Parentheses)

| | Absolute Pressure, psia | | | | | | | | | | | | |
|---|---|---|---|---|---|---|---|---|---|---|---|---|---|
| | 1.00 | | | 2.00 | | | 3.00 | | | 4.00 | | | |
| | (−135.23°F) | | | (−119.38°F) | | | (−109.16°F) | | | (−101.46°F) | | | |
| Temp. °F | V | H | S | V | H | S | V | H | S | V | H | S | Temp. °F |
| | (47.6197) | (102.2) | (0.3257) | (24.8559) | (104.4) | (0.3135) | (17.0005) | (105.8) | (0.3065) | (12.9842) | (106.8) | (0.3017) | |
| −130 | 48.4112 | 103.0 | 0.3281 | – | – | – | – | – | – | – | – | – | −130 |
| −120 | 49.9202 | 104.5 | 0.3327 | – | – | – | – | – | – | – | – | – | −120 |
| −110 | 51.4258 | 106.1 | 0.3372 | 25.5747 | 105.9 | 0.3177 | – | – | – | – | – | – | −110 |
| −100 | 52.9283 | 107.6 | 0.3416 | 26.3382 | 107.4 | 0.3222 | 17.4741 | 107.2 | 0.3107 | 13.0415 | 107.0 | 0.3024 | −100 |
| −90 | 54.4281 | 109.2 | 0.3459 | 27.0990 | 109.1 | 0.3266 | 17.9887 | 108.9 | 0.3151 | 13.4331 | 108.7 | 0.3069 | −90 |
| −80 | 55.9256 | 110.8 | 0.3502 | 27.8575 | 110.7 | 0.3309 | 18.5009 | 110.5 | 0.3195 | 13.8223 | 110.3 | 0.3113 | −80 |
| −70 | 57.4211 | 112.5 | 0.3544 | 28.6138 | 112.3 | 0.3352 | 19.0110 | 112.2 | 0.3238 | 14.2093 | 112.0 | 0.3156 | −70 |
| −60 | 58.9148 | 114.1 | 0.3586 | 29.3684 | 114.0 | 0.3394 | 19.5192 | 113.8 | 0.3280 | 14.5944 | 113.7 | 0.3199 | −60 |
| −50 | 60.4069 | 115.8 | 0.3627 | 30.1213 | 115.7 | 0.3435 | 20.0258 | 115.5 | 0.3322 | 14.9779 | 115.4 | 0.3241 | −50 |
| −40 | 61.8976 | 117.5 | 0.3668 | 30.8728 | 117.4 | 0.3476 | 20.5310 | 117.2 | 0.3363 | 15.3599 | 117.1 | 0.3282 | −40 |
| −30 | 63.3871 | 119.2 | 0.3709 | 31.6231 | 119.1 | 0.3517 | 21.0349 | 119.0 | 0.3404 | 15.7407 | 118.9 | 0.3323 | −30 |
| −20 | 64.8755 | 120.9 | 0.3748 | 32.3723 | 120.8 | 0.3557 | 21.5378 | 120.7 | 0.3444 | 16.1204 | 120.6 | 0.3364 | −20 |
| −10 | 66.3630 | 122.7 | 0.3788 | 33.1205 | 122.6 | 0.3597 | 22.0396 | 122.5 | 0.3484 | 16.4990 | 122.4 | 0.3404 | −10 |
| 0 | 67.8496 | 124.5 | 0.3827 | 33.8679 | 124.4 | 0.3636 | 22.5406 | 124.3 | 0.3523 | 16.8768 | 124.2 | 0.3443 | 0 |
| 10 | 69.3354 | 126.2 | 0.3866 | 34.6145 | 126.2 | 0.3675 | 23.0408 | 126.1 | 0.3562 | 17.2538 | 126.0 | 0.3482 | 10 |
| 20 | 70.8206 | 128.1 | 0.3904 | 35.3604 | 128.0 | 0.3713 | 23.5403 | 127.9 | 0.3601 | 17.6302 | 127.8 | 0.3521 | 20 |
| 30 | 72.3051 | 129.9 | 0.3942 | 36.1057 | 129.8 | 0.3751 | 24.0392 | 129.8 | 0.3639 | 18.0059 | 129.7 | 0.3559 | 30 |
| 40 | 73.7891 | 131.8 | 0.3979 | 36.8505 | 131.7 | 0.3789 | 24.5376 | 131.6 | 0.3677 | 18.3811 | 131.6 | 0.3597 | 40 |
| 50 | 75.2727 | 133.6 | 0.4017 | 37.5948 | 133.6 | 0.3826 | 25.0355 | 133.5 | 0.3714 | 18.7558 | 133.5 | 0.3635 | 50 |
| 60 | 76.7558 | 135.5 | 0.4054 | 38.3387 | 135.5 | 0.3863 | 25.5329 | 135.4 | 0.3751 | 19.1300 | 135.4 | 0.3672 | 60 |
| 70 | 78.2385 | 137.5 | 0.4090 | 39.0821 | 137.4 | 0.3900 | 26.0300 | 137.4 | 0.3788 | 19.5039 | 137.3 | 0.3709 | 70 |
| 80 | 79.7209 | 139.4 | 0.4126 | 39.8253 | 139.4 | 0.3936 | 26.5267 | 139.3 | 0.3824 | 19.8774 | 139.3 | 0.3745 | 80 |
| 90 | 81.2030 | 141.4 | 0.4162 | 40.5681 | 141.3 | 0.3972 | 27.0231 | 141.3 | 0.3861 | 20.2506 | 141.2 | 0.3781 | 90 |
| 100 | 82.6848 | 143.3 | 0.4198 | 41.3106 | 143.3 | 0.4008 | 27.5192 | 143.3 | 0.3896 | 20.6235 | 143.2 | 0.3817 | 100 |

*(Continued)*

## TABLE A.19 (CONTINUED)

Suva® 410A Superheated Vapor–Constant Pressure Tables (Saturated Vapor Properties in Parentheses)

| Temp. °F | V | H | S | V | H | S | V | H | S | V | H | S | Temp. °F |
|---|---|---|---|---|---|---|---|---|---|---|---|---|---|
| 110 | 84.1663 | 145.3 | 0.4234 | 42.0529 | 145.3 | 0.4043 | 28.0151 | 145.3 | 0.3932 | 20.9962 | 145.2 | 0.3853 | 110 |
| 120 | 85.6477 | 147.4 | 0.4269 | 42.7950 | 147.3 | 0.4079 | 28.5108 | 147.3 | 0.3967 | 21.3686 | 147.2 | 0.3888 | 120 |
| 130 | 87.1288 | 149.4 | 0.4304 | 43.5369 | 149.4 | 0.4114 | 29.0062 | 149.3 | 0.4002 | 21.7409 | 149.3 | 0.3923 | 130 |
| 140 | 88.6097 | 151.5 | 0.4338 | 44.2785 | 151.4 | 0.4148 | 29.5015 | 151.4 | 0.4037 | 22.1129 | 151.4 | 0.3958 | 140 |
| 150 | 90.0905 | 153.5 | 0.4373 | 45.0201 | 153.5 | 0.4183 | 29.9966 | 153.5 | 0.4071 | 22.4848 | 153.4 | 0.3992 | 150 |
| 160 | 91.5711 | 155.6 | 0.4407 | 45.7614 | 155.6 | 0.4217 | 30.4915 | 155.6 | 0.4106 | 22.8566 | 155.5 | 0.4026 | 160 |
| 170 | 93.0516 | 157.8 | 0.4441 | 46.5026 | 157.7 | 0.4251 | 30.9863 | 157.7 | 0.4139 | 23.2282 | 157.7 | 0.4060 | 170 |
| 180 | — | — | — | 47.2437 | 159.9 | 0.4284 | 31.4810 | 159.8 | 0.4173 | 23.5996 | 159.8 | 0.4094 | 180 |
| 190 | — | — | — | 47.9847 | 162.0 | 0.4318 | 31.9756 | 162.0 | 0.4207 | 23.9710 | 162.0 | 0.4128 | 190 |
| 200 | — | — | — | — | — | — | 32.4700 | 164.2 | 0.4240 | 24.3423 | 164.2 | 0.4161 | 200 |

| | 5.00 | | | 6.00 | | | 7.00 | | | 8.00 | | | |
|---|---|---|---|---|---|---|---|---|---|---|---|---|---|
| | (−95.19°F) | | | (−89.86°F) | | | (−85.21°F) | | | (−81.08°F) | | | |
| | V | H | S | V | H | S | V | H | S | V | H | S | |
| Temp. °F | (10.5348) | (107.6) | (0.2980) | (8.8803) | (108.3) | (0.2951) | (7.6853) | (108.9) | (0.2926) | (6.7804) | (109.5) | (0.2904) | Temp. °F |
| −90 | 10.6994 | 108.5 | 0.3004 | — | — | — | — | — | — | — | — | — | −90 |
| −80 | 11.0148 | 110.2 | 0.3048 | 9.1429 | 110.0 | 0.2995 | 7.8056 | 109.8 | 0.2949 | 6.8024 | 109.7 | 0.2909 | −80 |
| −70 | 11.3280 | 111.9 | 0.3092 | 9.4070 | 111.7 | 0.3039 | 8.0346 | 111.5 | 0.2994 | 7.0052 | 111.4 | 0.2954 | −70 |
| −60 | 11.6393 | 113.6 | 0.3135 | 9.6691 | 113.4 | 0.3082 | 8.2616 | 113.3 | 0.3037 | 7.2059 | 113.1 | 0.2998 | −60 |
| −50 | 11.9490 | 115.3 | 0.3177 | 9.9295 | 115.1 | 0.3125 | 8.4870 | 115.0 | 0.3080 | 7.4050 | 114.9 | 0.3041 | −50 |
| −40 | 12.2572 | 117.0 | 0.3219 | 10.1885 | 116.9 | 0.3167 | 8.7108 | 116.8 | 0.3123 | 7.6025 | 116.6 | 0.3084 | −40 |
| −30 | 12.5641 | 118.7 | 0.3260 | 10.4462 | 118.6 | 0.3209 | 8.9334 | 118.5 | 0.3164 | 7.7987 | 118.4 | 0.3126 | −30 |
| −20 | 12.8698 | 120.5 | 0.3301 | 10.7027 | 120.4 | 0.3249 | 9.1547 | 120.3 | 0.3205 | 7.9937 | 120.2 | 0.3167 | −20 |
| −10 | 13.1746 | 122.3 | 0.3341 | 10.9583 | 122.2 | 0.3290 | 9.3751 | 122.1 | 0.3246 | 8.1877 | 122.0 | 0.3208 | −10 |
| 0 | 13.4785 | 124.1 | 0.3381 | 11.2129 | 124.0 | 0.3329 | 9.5946 | 123.9 | 0.3286 | 8.3808 | 123.8 | 0.3248 | 0 |
| 10 | 13.7816 | 125.9 | 0.3420 | 11.4668 | 125.8 | 0.3369 | 9.8133 | 125.8 | 0.3325 | 8.5731 | 125.7 | 0.3287 | 10 |
| 20 | 14.0841 | 127.8 | 0.3459 | 11.7199 | 127.7 | 0.3408 | 10.0313 | 127.6 | 0.3364 | 8.7647 | 127.5 | 0.3327 | 20 |
| 30 | 14.3859 | 129.6 | 0.3497 | 11.9725 | 129.6 | 0.3446 | 10.2486 | 129.5 | 0.3403 | 8.9557 | 129.4 | 0.3365 | 30 |
| 40 | 14.6871 | 131.5 | 0.3535 | 12.2245 | 131.4 | 0.3484 | 10.4654 | 131.4 | 0.3441 | 9.1461 | 131.3 | 0.3404 | 40 |
| 50 | 14.9879 | 133.4 | 0.3573 | 12.4760 | 133.3 | 0.3522 | 10.6817 | 133.3 | 0.3479 | 9.3360 | 133.2 | 0.3441 | 50 |
| 60 | 15.2882 | 135.3 | 0.3610 | 12.7270 | 135.3 | 0.3559 | 10.8976 | 135.2 | 0.3516 | 9.5255 | 135.1 | 0.3479 | 60 |

*(Continued)*

# TABLE A.19 (CONTINUED)

Suva® 410A Superheated Vapor–Constant Pressure Tables (Saturated Vapor Properties in Parentheses)

| Temp. °F | V | H | S | V | H | S | V | H | S | V | H | S |
|---|---|---|---|---|---|---|---|---|---|---|---|---|
| 70 | 15.5882 | 137.2 | 0.3647 | 12.9777 | 137.2 | 0.3596 | 11.1131 | 137.1 | 0.3553 | 9.7146 | 137.1 | 0.3516 |
| 80 | 15.8878 | 139.2 | 0.3683 | 13.2280 | 139.2 | 0.3633 | 11.3282 | 139.1 | 0.3590 | 9.9033 | 139.1 | 0.3553 |
| 90 | 16.1871 | 141.2 | 0.3720 | 13.4780 | 141.1 | 0.3669 | 11.5430 | 141.1 | 0.3626 | 10.0917 | 141.0 | 0.3589 |
| 100 | 16.4861 | 143.2 | 0.3755 | 13.7278 | 143.1 | 0.3705 | 11.7575 | 143.1 | 0.3662 | 10.2799 | 143.0 | 0.3625 |
| 110 | 16.7848 | 145.2 | 0.3791 | 13.9772 | 145.1 | 0.3741 | 11.9718 | 145.1 | 0.3698 | 10.4677 | 145.1 | 0.3661 |
| 120 | 17.0834 | 147.2 | 0.3826 | 14.2265 | 147.2 | 0.3776 | 12.1859 | 147.1 | 0.3733 | 10.6554 | 147.1 | 0.3696 |
| 130 | 17.3817 | 149.3 | 0.3861 | 14.4755 | 149.2 | 0.3811 | 12.3997 | 149.2 | 0.3768 | 10.8428 | 149.1 | 0.3731 |
| 140 | 17.6798 | 151.3 | 0.3896 | 14.7244 | 151.3 | 0.3846 | 12.6134 | 151.3 | 0.3803 | 11.0301 | 151.2 | 0.3766 |
| 150 | 17.9778 | 153.4 | 0.3931 | 14.9731 | 153.4 | 0.3880 | 12.8268 | 153.3 | 0.3838 | 11.2172 | 153.3 | 0.3801 |
| 160 | 18.2756 | 155.5 | 0.3965 | 15.2216 | 155.5 | 0.3915 | 13.0402 | 155.5 | 0.3872 | 11.4041 | 155.4 | 0.3835 |
| 170 | 18.5733 | 157.6 | 0.3999 | 15.4700 | 157.6 | 0.3949 | 13.2534 | 157.6 | 0.3906 | 11.5909 | 157.5 | 0.3869 |
| 180 | 18.8708 | 159.8 | 0.4033 | 15.7183 | 159.8 | 0.3983 | 13.4664 | 159.7 | 0.3940 | 11.7776 | 159.7 | 0.3903 |
| 190 | 19.1682 | 161.9 | 0.4066 | 15.9664 | 161.9 | 0.4016 | 13.6794 | 161.9 | 0.3974 | 11.9641 | 161.9 | 0.3937 |
| 200 | 19.4656 | 164.1 | 0.4100 | 16.2145 | 164.1 | 0.4049 | 13.8922 | 164.1 | 0.4007 | 12.1505 | 164.0 | 0.3970 |
| 210 | 19.7628 | 166.3 | 0.4133 | 16.4624 | 166.3 | 0.4083 | 14.1050 | 166.3 | 0.4040 | 12.3369 | 166.2 | 0.4003 |
| 220 | – | – | – | 16.7103 | 168.5 | 0.4115 | 14.3176 | 168.5 | 0.4073 | 12.5232 | 168.5 | 0.4036 |

| | 9.00 (-77.33°F) | | | 10.00 (-73.91°F) | | | 11.00 (-70.75°F) | | | 12.00 (-67.81°F) | | |
|---|---|---|---|---|---|---|---|---|---|---|---|---|
| Temp. °F | V (6.0708) | H (109.9) | S (0.2886) | V (5.4988) | H (110.4) | S (0.2869) | V (5.0277) | H (110.8) | S (0.2854) | V (4.6327) | H (111.1) | S (0.2840) |
| -70 | 6.2043 | 111.2 | 0.2919 | 5.5635 | 111.1 | 0.2887 | 5.0391 | 110.9 | 0.2857 | – | – | – |
| -60 | 6.3847 | 113.0 | 0.2963 | 5.7276 | 112.8 | 0.2931 | 5.1899 | 112.7 | 0.2903 | 4.7417 | 112.5 | 0.2876 |
| -50 | 6.5633 | 114.7 | 0.3007 | 5.8899 | 114.6 | 0.2975 | 5.3388 | 114.5 | 0.2947 | 4.8795 | 114.3 | 0.2920 |
| -40 | 6.7403 | 116.5 | 0.3050 | 6.0506 | 116.4 | 0.3018 | 5.4861 | 116.3 | 0.2990 | 5.0157 | 116.1 | 0.2964 |
| -30 | 6.9161 | 118.3 | 0.3092 | 6.2099 | 118.2 | 0.3061 | 5.6321 | 118.1 | 0.3033 | 5.1505 | 118.0 | 0.3007 |
| -20 | 7.0906 | 120.1 | 0.3133 | 6.3681 | 120.0 | 0.3102 | 5.7769 | 119.9 | 0.3075 | 5.2842 | 119.8 | 0.3049 |
| -10 | 7.2641 | 121.9 | 0.3174 | 6.5252 | 121.8 | 0.3143 | 5.9206 | 121.7 | 0.3116 | 5.4167 | 121.6 | 0.3090 |
| 0 | 7.4367 | 123.8 | 0.3214 | 6.6814 | 123.7 | 0.3184 | 6.0634 | 123.6 | 0.3156 | 5.5484 | 123.5 | 0.3131 |
| 10 | 7.6085 | 125.6 | 0.3254 | 6.8368 | 125.5 | 0.3224 | 6.2054 | 125.4 | 0.3196 | 5.6792 | 125.4 | 0.3171 |
| 20 | 7.7796 | 127.5 | 0.3293 | 6.9915 | 127.4 | 0.3263 | 6.3466 | 127.3 | 0.3236 | 5.8092 | 127.2 | 0.3211 |

(Continued)

**TABLE A.19 (CONTINUED)**

Suva® 410A Superheated Vapor–Constant Pressure Tables (Saturated Vapor Properties in Parentheses)

| Temp. °F | 13.00 (−65.06°F) V | H | S | 14.00 (−62.48°F) V | H | S | 14.696 (−60.76°F) V | H | S | 15.00 (−60.03°F) V | H | S | Temp. °F |
|---|---|---|---|---|---|---|---|---|---|---|---|---|---|
| | (4.2967) | (111.5) | (0.2828) | (4.0070) | (111.8) | (0.2817) | (3.8280) | (112.0) | (0.2809) | (3.7548) | (112.1) | (0.2806) | |
| −60 | 4.3623 | 112.4 | 0.2851 | 4.0371 | 112.2 | 0.2828 | 3.8368 | 112.1 | 0.2813 | 3.7552 | 112.1 | 0.2806 | −60 |
| −50 | 4.4908 | 114.2 | 0.2896 | 4.1575 | 114.1 | 0.2873 | 3.9523 | 114.0 | 0.2858 | 3.8687 | 113.9 | 0.2852 | −50 |
| −40 | 4.6176 | 116.0 | 0.2940 | 4.2763 | 115.9 | 0.2917 | 4.0662 | 115.8 | 0.2902 | 3.9805 | 115.8 | 0.2896 | −40 |
| −30 | 4.7430 | 117.8 | 0.2983 | 4.3937 | 117.7 | 0.2961 | 4.1786 | 117.7 | 0.2946 | 4.0909 | 117.6 | 0.2940 | −30 |
| 30 | 7.9501 | 129.3 | 0.3332 | 7.1455 | 129.3 | 0.3302 | 6.4873 | 129.2 | 0.3275 | 5.9387 | 129.1 | 0.3250 | 30 |
| 40 | 8.1199 | 131.2 | 0.3370 | 7.2990 | 131.2 | 0.3341 | 6.6273 | 131.1 | 0.3313 | 6.0676 | 131.0 | 0.3289 | 40 |
| 50 | 8.2893 | 133.2 | 0.3408 | 7.4520 | 133.1 | 0.3379 | 6.7669 | 133.0 | 0.3352 | 6.1959 | 133.0 | 0.3327 | 50 |
| 60 | 8.4583 | 135.1 | 0.3446 | 7.6045 | 135.0 | 0.3416 | 6.9060 | 135.0 | 0.3389 | 6.3238 | 134.9 | 0.3365 | 60 |
| 70 | 8.6269 | 137.0 | 0.3483 | 7.7567 | 137.0 | 0.3453 | 7.0447 | 136.9 | 0.3427 | 6.4514 | 136.9 | 0.3402 | 70 |
| 80 | 8.7951 | 139.0 | 0.3520 | 7.9085 | 138.9 | 0.3490 | 7.1830 | 138.9 | 0.3463 | 6.5785 | 138.8 | 0.3439 | 80 |
| 90 | 8.9630 | 141.0 | 0.3556 | 8.0599 | 140.9 | 0.3527 | 7.3211 | 140.9 | 0.3500 | 6.7053 | 140.8 | 0.3476 | 90 |
| 100 | 9.1306 | 143.0 | 0.3592 | 8.2111 | 142.9 | 0.3563 | 7.4588 | 142.9 | 0.3536 | 6.8319 | 142.9 | 0.3512 | 100 |
| 110 | 9.2979 | 145.0 | 0.3628 | 8.3620 | 145.0 | 0.3599 | 7.5963 | 144.9 | 0.3572 | 6.9582 | 144.9 | 0.3548 | 110 |
| 120 | 9.4650 | 147.0 | 0.3664 | 8.5127 | 147.0 | 0.3634 | 7.7335 | 147.0 | 0.3608 | 7.0842 | 146.9 | 0.3583 | 120 |
| 130 | 9.6319 | 149.1 | 0.3699 | 8.6632 | 149.1 | 0.3669 | 7.8706 | 149.0 | 0.3643 | 7.2101 | 149.0 | 0.3619 | 130 |
| 140 | 9.7986 | 151.2 | 0.3734 | 8.8135 | 151.1 | 0.3704 | 8.0074 | 151.1 | 0.3678 | 7.3357 | 151.1 | 0.3654 | 140 |
| 150 | 9.9652 | 153.3 | 0.3768 | 8.9636 | 153.2 | 0.3739 | 8.1441 | 153.2 | 0.3713 | 7.4612 | 153.2 | 0.3688 | 150 |
| 160 | 10.1316 | 155.4 | 0.3803 | 9.1136 | 155.4 | 0.3773 | 8.2807 | 155.3 | 0.3747 | 7.5866 | 155.3 | 0.3723 | 160 |
| 170 | 10.2978 | 157.5 | 0.3837 | 9.2634 | 157.5 | 0.3808 | 8.4171 | 157.5 | 0.3781 | 7.7118 | 157.4 | 0.3757 | 170 |
| 180 | 10.4640 | 159.7 | 0.3871 | 9.4131 | 159.6 | 0.3841 | 8.5533 | 159.6 | 0.3815 | 7.8368 | 159.6 | 0.3791 | 180 |
| 190 | 10.6300 | 161.8 | 0.3904 | 9.5627 | 161.8 | 0.3875 | 8.6895 | 161.8 | 0.3849 | 7.9618 | 161.7 | 0.3825 | 190 |
| 200 | 10.7959 | 164.0 | 0.3938 | 9.7122 | 164.0 | 0.3908 | 8.8255 | 164.0 | 0.3882 | 8.0866 | 163.9 | 0.3858 | 200 |
| 210 | 10.9617 | 166.2 | 0.3971 | 9.8616 | 166.2 | 0.3942 | 8.9615 | 166.2 | 0.3915 | 8.2114 | 166.1 | 0.3891 | 210 |
| 220 | 11.1275 | 168.4 | 0.4004 | 10.0109 | 168.4 | 0.3975 | 9.0973 | 168.4 | 0.3948 | 8.3360 | 168.4 | 0.3924 | 220 |
| 230 | 11.2931 | 170.7 | 0.4036 | 10.1601 | 170.7 | 0.4007 | 9.2331 | 170.6 | 0.3981 | 8.4606 | 170.6 | 0.3957 | 230 |
| 240 | — | — | — | — | — | — | — | — | — | 8.5851 | 172.9 | 0.3989 | 240 |

*(Continued)*

# TABLE A.19 (CONTINUED)

Suva® 410A Superheated Vapor–Constant Pressure Tables (Saturated Vapor Properties in Parentheses)

| | | | | | | | | | | | | | |
|---|---|---|---|---|---|---|---|---|---|---|---|---|---|
| -20 | 4.8672 | 119.7 | 0.3025 | 4.5098 | 119.6 | 0.3003 | 4.2897 | 119.5 | 0.2989 | 4.2000 | 119.5 | 0.2982 | -20 |
| -10 | 4.9903 | 121.5 | 0.3067 | 4.6248 | 121.4 | 0.3045 | 4.3998 | 121.4 | 0.3030 | 4.3080 | 121.3 | 0.3024 | -10 |
| 0 | 5.1125 | 123.4 | 0.3108 | 4.7389 | 123.3 | 0.3086 | 4.5089 | 123.2 | 0.3072 | 4.4151 | 123.2 | 0.3066 | 0 |
| 10 | 5.2339 | 125.3 | 0.3148 | 4.8522 | 125.2 | 0.3126 | 4.6172 | 125.1 | 0.3112 | 4.5214 | 125.1 | 0.3106 | 10 |
| 20 | 5.3545 | 127.2 | 0.3188 | 4.9647 | 127.1 | 0.3166 | 4.7247 | 127.0 | 0.3152 | 4.6269 | 127.0 | 0.3146 | 20 |
| 30 | 5.4745 | 129.1 | 0.3227 | 5.0766 | 129.0 | 0.3206 | 4.8316 | 128.9 | 0.3192 | 4.7317 | 128.9 | 0.3186 | 30 |
| 40 | 5.5939 | 131.0 | 0.3266 | 5.1879 | 130.9 | 0.3244 | 4.9379 | 130.9 | 0.3230 | 4.8360 | 130.8 | 0.3225 | 40 |
| 50 | 5.7128 | 132.9 | 0.3304 | 5.2987 | 132.8 | 0.3283 | 5.0437 | 132.8 | 0.3269 | 4.9398 | 132.8 | 0.3263 | 50 |
| 60 | 5.8313 | 134.9 | 0.3342 | 5.4090 | 134.8 | 0.3321 | 5.1491 | 134.8 | 0.3307 | 5.0431 | 134.7 | 0.3301 | 60 |
| 70 | 5.9493 | 136.8 | 0.3379 | 5.5190 | 136.8 | 0.3358 | 5.2540 | 136.7 | 0.3344 | 5.1460 | 136.7 | 0.3339 | 70 |
| 80 | 6.0670 | 138.8 | 0.3416 | 5.6285 | 138.7 | 0.3395 | 5.3586 | 138.7 | 0.3382 | 5.2485 | 138.7 | 0.3376 | 80 |
| 90 | 6.1843 | 140.8 | 0.3453 | 5.7378 | 140.7 | 0.3432 | 5.4628 | 140.7 | 0.3418 | 5.3507 | 140.7 | 0.3413 | 90 |
| 100 | 6.3014 | 142.8 | 0.3489 | 5.8467 | 142.8 | 0.3468 | 5.5668 | 142.7 | 0.3455 | 5.4527 | 142.7 | 0.3449 | 100 |
| 110 | 6.4182 | 144.8 | 0.3525 | 5.9554 | 144.8 | 0.3504 | 5.6705 | 144.8 | 0.3491 | 5.5543 | 144.8 | 0.3485 | 110 |
| 120 | 6.5348 | 146.9 | 0.3561 | 6.0639 | 146.8 | 0.3540 | 5.7739 | 146.8 | 0.3527 | 5.6557 | 146.8 | 0.3521 | 120 |
| 130 | 6.6512 | 149.0 | 0.3596 | 6.1721 | 148.9 | 0.3576 | 5.8772 | 148.9 | 0.3562 | 5.7570 | 148.9 | 0.3556 | 130 |
| 140 | 6.7674 | 151.0 | 0.3631 | 6.2802 | 151.0 | 0.3611 | 5.9803 | 151.0 | 0.3597 | 5.8580 | 151.0 | 0.3591 | 140 |
| 150 | 6.8834 | 153.1 | 0.3666 | 6.3881 | 153.1 | 0.3645 | 6.0832 | 153.1 | 0.3632 | 5.9588 | 153.1 | 0.3626 | 150 |
| 160 | 6.9993 | 155.3 | 0.3701 | 6.4958 | 155.2 | 0.3680 | 6.1859 | 155.2 | 0.3666 | 6.0595 | 155.2 | 0.3661 | 160 |
| 170 | 7.1150 | 157.4 | 0.3735 | 6.6034 | 157.4 | 0.3714 | 6.2885 | 157.3 | 0.3701 | 6.1601 | 157.3 | 0.3695 | 170 |
| 180 | 7.2306 | 159.5 | 0.3769 | 6.7109 | 159.5 | 0.3748 | 6.3910 | 159.5 | 0.3735 | 6.2605 | 159.5 | 0.3729 | 180 |
| 190 | 7.3460 | 161.7 | 0.3802 | 6.8183 | 161.7 | 0.3782 | 6.4933 | 161.7 | 0.3768 | 6.3608 | 161.7 | 0.3763 | 190 |
| 200 | 7.4614 | 163.9 | 0.3836 | 6.9255 | 163.9 | 0.3815 | 6.5956 | 163.9 | 0.3802 | 6.4611 | 163.9 | 0.3796 | 200 |
| 210 | 7.5767 | 166.1 | 0.3869 | 7.0327 | 166.1 | 0.3848 | 6.6977 | 166.1 | 0.3835 | 6.5612 | 166.1 | 0.3829 | 210 |
| 220 | 7.6919 | 168.3 | 0.3902 | 7.1397 | 168.3 | 0.3881 | 6.7998 | 168.3 | 0.3868 | 6.6612 | 168.3 | 0.3862 | 220 |
| 230 | 7.8070 | 170.6 | 0.3935 | 7.2467 | 170.6 | 0.3914 | 6.9018 | 170.5 | 0.3901 | 6.7611 | 170.5 | 0.3895 | 230 |
| 240 | 7.9220 | 172.8 | 0.3967 | 7.3536 | 172.8 | 0.3947 | 7.0037 | 172.8 | 0.3933 | 6.8610 | 172.8 | 0.3928 | 240 |

(Continued)

**TABLE A.19 (CONTINUED)**

Suva® 410A Superheated Vapor–Constant Pressure Tables (Saturated Vapor Properties in Parentheses)

| Temp. °F | 16.00 (−57.71°F) | | | 17.00 (−55.50°F) | | | 18.00 (−53.40°F) | | | 19.00 (−51.38°F) | | | Temp. °F |
|---|---|---|---|---|---|---|---|---|---|---|---|---|---|
| | V | H | S | V | H | S | V | H | S | V | H | S | |
| | (3.5331) | (112.4) | (0.2796) | (3.3367) | (112.6) | (0.2787) | (3.1614) | (112.9) | (0.2778) | (3.0039) | (113.1) | (0.2770) | |
| −50 | 3.6158 | 113.8 | 0.2831 | 3.3927 | 113.6 | 0.2812 | 3.1943 | 113.5 | 0.2794 | 3.0167 | 113.4 | 0.2776 | −50 |
| −40 | 3.7216 | 115.6 | 0.2876 | 3.4931 | 115.5 | 0.2857 | 3.2899 | 115.4 | 0.2839 | 3.1081 | 115.3 | 0.2822 | −40 |
| −30 | 3.8259 | 117.5 | 0.2920 | 3.5920 | 117.4 | 0.2901 | 3.3841 | 117.3 | 0.2884 | 3.1980 | 117.2 | 0.2867 | −30 |
| −20 | 3.9289 | 119.4 | 0.2963 | 3.6897 | 119.3 | 0.2944 | 3.4770 | 119.2 | 0.2927 | 3.2866 | 119.0 | 0.2910 | −20 |
| −10 | 4.0308 | 121.2 | 0.3005 | 3.7862 | 121.1 | 0.2987 | 3.5687 | 121.0 | 0.2969 | 3.3741 | 120.9 | 0.2953 | −10 |
| 0 | 4.1318 | 123.1 | 0.3046 | 3.8818 | 123.0 | 0.3028 | 3.6595 | 122.9 | 0.3011 | 3.4606 | 122.9 | 0.2995 | 0 |
| 10 | 4.2319 | 125.0 | 0.3087 | 3.9765 | 124.9 | 0.3069 | 3.7494 | 124.9 | 0.3052 | 3.5462 | 124.8 | 0.3036 | 10 |
| 20 | 4.3313 | 126.9 | 0.3127 | 4.0704 | 126.8 | 0.3109 | 3.8385 | 126.8 | 0.3093 | 3.6310 | 126.7 | 0.3077 | 20 |
| 30 | 4.4300 | 128.8 | 0.3167 | 4.1637 | 128.8 | 0.3149 | 3.9270 | 128.7 | 0.3132 | 3.7152 | 128.6 | 0.3116 | 30 |
| 40 | 4.5281 | 130.8 | 0.3206 | 4.2564 | 130.7 | 0.3188 | 4.0149 | 130.6 | 0.3172 | 3.7988 | 130.6 | 0.3156 | 40 |
| 50 | 4.6257 | 132.7 | 0.3244 | 4.3486 | 132.7 | 0.3227 | 4.1023 | 132.6 | 0.3210 | 3.8819 | 132.5 | 0.3195 | 50 |
| 60 | 4.7229 | 134.7 | 0.3283 | 4.4403 | 134.6 | 0.3265 | 4.1892 | 134.6 | 0.3249 | 3.9645 | 134.5 | 0.3233 | 60 |
| 70 | 4.8196 | 136.7 | 0.3320 | 4.5317 | 136.6 | 0.3303 | 4.2757 | 136.5 | 0.3286 | 4.0467 | 136.5 | 0.3271 | 70 |
| 80 | 4.9160 | 138.6 | 0.3357 | 4.6226 | 138.6 | 0.3340 | 4.3618 | 138.5 | 0.3324 | 4.1285 | 138.5 | 0.3308 | 80 |
| 90 | 5.0121 | 140.6 | 0.3394 | 4.7132 | 140.6 | 0.3377 | 4.4476 | 140.6 | 0.3361 | 4.2100 | 140.5 | 0.3345 | 90 |
| 100 | 5.1078 | 142.7 | 0.3431 | 4.8036 | 142.6 | 0.3414 | 4.5331 | 142.6 | 0.3397 | 4.2911 | 142.5 | 0.3382 | 100 |
| 110 | 5.2033 | 144.7 | 0.3467 | 4.8937 | 144.7 | 0.3450 | 4.6184 | 144.6 | 0.3433 | 4.3721 | 144.6 | 0.3418 | 110 |
| 120 | 5.2986 | 146.8 | 0.3503 | 4.9835 | 146.7 | 0.3485 | 4.7034 | 146.7 | 0.3469 | 4.4528 | 146.6 | 0.3454 | 120 |
| 130 | 5.3937 | 148.8 | 0.3538 | 5.0731 | 148.8 | 0.3521 | 4.7882 | 148.8 | 0.3505 | 4.5332 | 148.7 | 0.3490 | 130 |
| 140 | 5.4885 | 150.9 | 0.3573 | 5.1626 | 150.9 | 0.3556 | 4.8728 | 150.9 | 0.3540 | 4.6135 | 150.8 | 0.3525 | 140 |
| 150 | 5.5832 | 153.0 | 0.3608 | 5.2518 | 153.0 | 0.3591 | 4.9572 | 153.0 | 0.3575 | 4.6936 | 152.9 | 0.3560 | 150 |
| 160 | 5.6778 | 155.2 | 0.3643 | 5.3409 | 155.1 | 0.3626 | 5.0415 | 155.1 | 0.3610 | 4.7736 | 155.1 | 0.3594 | 160 |
| 170 | 5.7722 | 157.3 | 0.3677 | 5.4299 | 157.3 | 0.3660 | 5.1256 | 157.2 | 0.3644 | 4.8534 | 157.2 | 0.3629 | 170 |
| 180 | 5.8664 | 159.5 | 0.3711 | 5.5187 | 159.4 | 0.3694 | 5.2096 | 159.4 | 0.3678 | 4.9331 | 159.4 | 0.3663 | 180 |
| 190 | 5.9606 | 161.6 | 0.3745 | 5.6075 | 161.6 | 0.3728 | 5.2935 | 161.6 | 0.3712 | 5.0127 | 161.6 | 0.3697 | 190 |
| 200 | 6.0547 | 163.8 | 0.3778 | 5.6961 | 163.8 | 0.3761 | 5.3773 | 163.8 | 0.3745 | 5.0921 | 163.7 | 0.3730 | 200 |

*(Continued)*

# TABLE A.19 (CONTINUED)

Suva® 410A Superheated Vapor–Constant Pressure Tables (Saturated Vapor Properties in Parentheses)

| Temp. °F | V | H | S | V | H | S | V | H | S | V | H | S | Temp. °F |
|---|---|---|---|---|---|---|---|---|---|---|---|---|---|
| 210 | 6.1486 | 166.0 | 0.3811 | 5.7846 | 166.0 | 0.3795 | 5.4610 | 166.0 | 0.3779 | 5.1715 | 166.0 | 0.3764 | 210 |
| 220 | 6.2425 | 168.3 | 0.3844 | 5.8730 | 168.2 | 0.3828 | 5.5446 | 168.2 | 0.3812 | 5.2508 | 168.2 | 0.3797 | 220 |
| 230 | 6.3363 | 170.5 | 0.3877 | 5.9614 | 170.5 | 0.3860 | 5.6281 | 170.5 | 0.3845 | 5.3300 | 170.4 | 0.3829 | 230 |
| 240 | 6.4300 | 172.8 | 0.3910 | 6.0497 | 172.8 | 0.3893 | 5.7116 | 172.7 | 0.3877 | 5.4091 | 172.7 | 0.3862 | 240 |
| 250 | 6.5236 | 175.1 | 0.3942 | 6.1379 | 175.0 | 0.3925 | 5.7950 | 175.0 | 0.3910 | 5.4882 | 175.0 | 0.3894 | 250 |

| | 20.00 | | | 21.00 | | | 22.00 | | | 23.00 | | | |
| | M9.45°F | | | (−47.59°F) | | | (−45.80°F) | | | (−44.07°F) | | | |
| | V | H | S | V | H | S | V | H | S | V | H | S | |
| Temp. °F | (2.8617) | (113.3) | (0.2762) | (2.7326) | (113.5) | (0.2755) | (2.6148) | (113.8) | (0.2748) | (2.5070) | (114.0) | (0.2741) | Temp. °F |
| −40 | 2.9445 | 115.1 | 0.2806 | 2.7963 | 115.0 | 0.2790 | 2.6616 | 114.9 | 0.2775 | 2.5386 | 114.7 | 0.2760 | −40 |
| −30 | 3.0305 | 117.0 | 0.2850 | 2.8790 | 116.9 | 0.2835 | 2.7412 | 116.8 | 0.2820 | 2.6153 | 116.7 | 0.2806 | −30 |
| −20 | 3.1153 | 118.9 | 0.2894 | 2.9603 | 118.8 | 0.2879 | 2.8193 | 118.7 | 0.2864 | 2.6906 | 118.6 | 0.2850 | −20 |
| −10 | 3.1989 | 120.8 | 0.2937 | 3.0404 | 120.7 | 0.2922 | 2.8963 | 120.6 | 0.2908 | 2.7647 | 120.5 | 0.2894 | −10 |
| 0 | 3.2816 | 122.8 | 0.2979 | 3.1196 | 122.7 | 0.2964 | 2.9723 | 122.6 | 0.2950 | 2.8378 | 122.5 | 0.2936 | 0 |
| 10 | 3.3633 | 124.7 | 0.3021 | 3.1978 | 124.6 | 0.3006 | 3.0474 | 124.5 | 0.2992 | 2.9100A | 124.4 | 0.2978 | 10 |
| 20 | 3.4443 | 126.6 | 0.3061 | 3.2753 | 126.5 | 0.3047 | 3.1217 | 126.5 | 0.3033 | 2.9814 | 126.4 | 0.3019 | 20 |
| 30 | 3.5246 | 128.6 | 0.3101 | 3.3521 | 128.5 | 0.3087 | 3.1953 | 128.4 | 0.3073 | 3.0521 | 128.3 | 0.3060 | 30 |
| 40 | 3.6043 | 130.5 | 0.3141 | 3.4283 | 130.4 | 0.3126 | 3.2683 | 130.4 | 0.3113 | 3.1222 | 130.3 | 0.3099 | 40 |
| 50 | 3.6835 | 132.5 | 0.3180 | 3.5040 | 132.4 | 0.3165 | 3.3408 | 132.3 | 0.3152 | 3.1918 | 132.3 | 0.3139 | 50 |
| 60 | 3.7622 | 134.4 | 0.3218 | 3.5792 | 134.4 | 0.3204 | 3.4128 | 134.3 | 0.3190 | 3.2609 | 134.3 | 0.3177 | 60 |
| 70 | 3.8405 | 136.4 | 0.3256 | 3.6540 | 136.4 | 0.3242 | 3.4844 | 136.3 | 0.3228 | 3.3296 | 136.3 | 0.3215 | 70 |
| 80 | 3.9184 | 138.4 | 0.3294 | 3.7284 | 138.4 | 0.3279 | 3.5557 | 138.3 | 0.3266 | 3.3979 | 138.3 | 0.3253 | 80 |
| 90 | 3.9960 | 140.5 | 0.3331 | 3.8025 | 140.4 | 0.3317 | 3.6266 | 140.4 | 0.3303 | 3.4659 | 140.3 | 0.3290 | 90 |
| 100 | 4.0734 | 142.5 | 0.3367 | 3.8763 | 142.4 | 0.3353 | 3.6972 | 142.4 | 0.3340 | 3.5336 | 142.4 | 0.3327 | 100 |
| 110 | 4.1504 | 144.5 | 0.3404 | 3.9498 | 144.5 | 0.3390 | 3.7675 | 144.5 | 0.3376 | 3.6010 | 144.4 | 0.3364 | 110 |
| 120 | 4.2272 | 146.6 | 0.3439 | 4.0231 | 146.6 | 0.3426 | 3.8376 | 146.5 | 0.3412 | 3.6682 | 146.5 | 0.3400 | 120 |
| 130 | 4.3038 | 148.7 | 0.3475 | 4.0962 | 148.6 | 0.3461 | 3.9075 | 148.6 | 0.3448 | 3.7351 | 148.6 | 0.3435 | 130 |
| 140 | 4.3802 | 150.8 | 0.3510 | 4.1691 | 150.7 | 0.3497 | 3.9771 | 150.7 | 0.3483 | 3.8019 | 150.7 | 0.3471 | 140 |
| 150 | 4.4564 | 152.9 | 0.3545 | 4.2418 | 152.9 | 0.3532 | 4.0467 | 152.8 | 0.3518 | 3.8685 | 152.8 | 0.3506 | 150 |

(Continued)

**TABLE A.19 (CONTINUED)**

Suva® 410A Superheated Vapor–Constant Pressure Tables (Saturated Vapor Properties in Parentheses)

| Temp. °F | 24.00 (−42.40°F) | | | 25.00 (−40.78°F) | | | 26.00 (−39.21°F) | | | 27.00 (−37.69°F) | | | Temp. °F |
|---|---|---|---|---|---|---|---|---|---|---|---|---|---|
| | V | H | S | V | H | S | V | H | S | V | H | S | |
| | (2.4078) | (114.1) | (0.2735) | (2.3163) | (114.3) | (0.2729) | (2.2316) | (114.5) | (0.2723) | (2.1530) | (114.7) | (0.2717) | |
| −40 | 2.4258 | 114.6 | 0.2746 | 2.3220 | 114.5 | 0.2732 | – | – | – | – | – | – | −40 |
| −30 | 2.4999 | 116.6 | 0.2792 | 2.3937 | 116.4 | 0.2779 | 2.2956 | 116.3 | 0.2766 | 2.2048 | 116.2 | 0.2753 | −30 |
| −20 | 2.5726 | 118.5 | 0.2837 | 2.4639 | 118.4 | 0.2824 | 2.3637 | 118.3 | 0.2811 | 2.2708 | 118.2 | 0.2799 | −20 |
| −10 | 2.6440 | 120.4 | 0.2880 | 2.5330 | 120.3 | 0.2868 | 2.4305 | 120.2 | 0.2855 | 2.3356 | 120.1 | 0.2843 | −10 |
| 0 | 2.7145 | 122.4 | 0.2923 | 2.6010 | 122.3 | 0.2911 | 2.4963 | 122.2 | 0.2898 | 2.3993 | 122.1 | 0.2886 | 0 |
| 10 | 2.7840 | 124.3 | 0.2965 | 2.6681 | 124.3 | 0.2953 | 2.5612 | 124.2 | 0.2941 | 2.4621 | 124.1 | 0.2929 | 10 |
| 20 | 2.8528 | 126.3 | 0.3006 | 2.7345 | 126.2 | 0.2994 | 2.6252 | 126.1 | 0.2982 | 2.5241 | 126.1 | 0.2970 | 20 |
| 30 | 2.9209 | 128.3 | 0.3047 | 2.8001 | 128.2 | 0.3035 | 2.6886 | 128.1 | 0.3023 | 2.5854 | 128.0 | 0.3011 | 30 |
| 40 | 2.9883 | 130.2 | 0.3087 | 2.8651 | 130.2 | 0.3075 | 2.7513 | 130.1 | 0.3063 | 2.6460 | 130.0 | 0.3051 | 40 |
| 50 | 3.0552 | 132.2 | 0.3126 | 2.9296 | 132.2 | 0.3114 | 2.8136 | 132.1 | 0.3102 | 2.7061 | 132.0 | 0.3091 | 50 |
| 60 | 3.1217 | 134.2 | 0.3165 | 2.9936 | 134.1 | 0.3153 | 2.8753 | 134.1 | 0.3141 | 2.7658 | 134.0 | 0.3130 | 60 |
| 70 | 3.1877 | 136.2 | 0.3203 | 3.0571 | 136.2 | 0.3191 | 2.9366 | 136.1 | 0.3180 | 2.8250 | 136.0 | 0.3168 | 70 |
| 80 | 3.2533 | 138.2 | 0.3241 | 3.1203 | 138.2 | 0.3229 | 2.9975 | 138.1 | 0.3217 | 2.8838 | 138.1 | 0.3206 | 80 |
| 90 | 3.3186 | 140.3 | 0.3278 | 3.1832 | 140.2 | 0.3266 | 3.0581 | 140.2 | 0.3255 | 2.9423 | 140.1 | 0.3244 | 90 |
| 160 | 4.5325 | 155.0 | 0.3580 | 4.3143 | 155.0 | 0.3566 | 4.1160 | 155.0 | 0.3553 | 3.9349 | 154.9 | 0.3541 | 160 |
| 170 | 4.6084 | 157.2 | 0.3614 | 4.3867 | 157.1 | 0.3601 | 4.1852 | 157.1 | 0.3588 | 4.0012 | 157.1 | 0.3575 | 170 |
| 180 | 4.6842 | 159.3 | 0.3648 | 4.4590 | 159.3 | 0.3635 | 4.2543 | 159.3 | 0.3622 | 4.0674 | 159.3 | 0.3609 | 180 |
| 190 | 4.7599 | 161.5 | 0.3682 | 4.5312 | 161.5 | 0.3669 | 4.3233 | 161.5 | 0.3656 | 4.1334 | 161.4 | 0.3643 | 190 |
| 200 | 4.8355 | 163.7 | 0.3716 | 4.6032 | 163.7 | 0.3702 | 4.3921 | 163.7 | 0.3689 | 4.1994 | 163.6 | 0.3677 | 200 |
| 210 | 4.9109 | 165.9 | 0.3749 | 4.6752 | 165.9 | 0.3736 | 4.4609 | 165.9 | 0.3723 | 4.2652 | 165.9 | 0.3710 | 210 |
| 220 | 4.9863 | 168.2 | 0.3782 | 4.7471 | 168.1 | 0.3769 | 4.5295 | 168.1 | 0.3756 | 4.3309 | 168.1 | 0.3743 | 220 |
| 230 | 5.0616 | 170.4 | 0.3815 | 4.8188 | 170.4 | 0.3802 | 4.5981 | 170.4 | 0.3789 | 4.3966 | 170.3 | 0.3776 | 230 |
| 240 | 5.1369 | 172.7 | 0.3848 | 4.8906 | 172.7 | 0.3834 | 4.6666 | 172.6 | 0.3821 | 4.4622 | 172.6 | 0.3809 | 240 |
| 250 | 5.2120 | 175.0 | 0.3880 | 4.9622 | 174.9 | 0.3867 | 4.7351 | 174.9 | 0.3854 | 4.5277 | 174.9 | 0.3841 | 250 |
| 260 | 5.2871 | 177.3 | 0.3912 | 5.0338 | 177.3 | 0.3899 | 4.8035 | 177.2 | 0.3886 | 4.5932 | 177.2 | 0.3874 | 260 |

*(Continued)*

# TABLE A.19 (CONTINUED)

Suva® 410A Superheated Vapor–Constant Pressure Tables (Saturated Vapor Properties in Parentheses)

| Temp. °F | 28.00 (−36.21°F) | | | 29.00 (−34.78°F) | | | 30.00 (−33.38°F) | | | 31.00 (−32.02°F) | | | Temp. °F |
|---|---|---|---|---|---|---|---|---|---|---|---|---|---|
| | V (2.0798) | H (114.8) | S (0.2712) | V (2.0115) | H (115.0) | S (0.2707) | V (1.9476) | H (115.1) | S (0.2702) | V (1.8877) | H (115.3) | S (0.2697) | |
| −30 | 2.1204 | 116.1 | 0.2741 | 2.0418 | 115.9 | 0.2729 | 1.9685 | 115.8 | 0.2718 | 1.8998 | 115.7 | 0.2706 | −30 |
| −20 | 2.1845 | 118.1 | 0.2787 | 2.1042 | 117.9 | 0.2775 | 2.0292 | 117.8 | 0.2764 | 1.9590 | 117.7 | 0.2753 | −20 |
| −10 | 2.2474 | 120.0 | 0.2831 | 2.1653 | 119.9 | 0.2820 | 2.0887 | 119.8 | 0.2809 | 2.0169 | 119.7 | 0.2798 | −10 |
| 0 | 2.3092 | 122.0 | 0.2875 | 2.2253 | 121.9 | 0.2864 | 2.1470 | 121.8 | 0.2853 | 2.0738 | 121.7 | 0.2842 | 0 |
| 10 | 2.3701 | 124.0 | 0.2917 | 2.2844 | 123.9 | 0.2906 | 2.2044 | 123.8 | 0.2896 | 2.1296 | 123.7 | 0.2885 | 10 |
| 20 | 2.4301 | 126.0 | 0.2959 | 2.3427 | 125.9 | 0.2948 | 2.2610 | 125.8 | 0.2938 | 2.1846 | 125.7 | 0.2928 | 20 |
| 30 | 2.4895 | 128.0 | 0.3000 | 2.4002 | 127.9 | 0.2989 | 2.3169 | 127.8 | 0.2979 | 2.2389 | 127.7 | 0.2969 | 30 |
| 100 | 3.3837 | 142.3 | 0.3315 | 3.2457 | 142.3 | 0.3303 | 3.1184 | 142.2 | 0.3292 | 3.0005 | 142.2 | 0.3281 | 100 |
| 110 | 3.4484 | 144.4 | 0.3351 | 3.3080 | 144.3 | 0.3340 | 3.1784 | 144.3 | 0.3328 | 3.0584 | 144.2 | 0.3318 | 110 |
| 120 | 3.5129 | 146.4 | 0.3388 | 3.3700 | 146.4 | 0.3376 | 3.2381 | 146.4 | 0.3365 | 3.1160 | 146.3 | 0.3354 | 120 |
| 130 | 3.5772 | 148.5 | 0.3423 | 3.4318 | 148.5 | 0.3412 | 3.2977 | 148.5 | 0.3400 | 3.1735 | 148.4 | 0.3390 | 130 |
| 140 | 3.6413 | 150.6 | 0.3459 | 3.4935 | 150.6 | 0.3447 | 3.3571 | 150.6 | 0.3436 | 3.2307 | 150.5 | 0.3425 | 140 |
| 150 | 3.7052 | 152.8 | 0.3494 | 3.5549 | 152.7 | 0.3482 | 3.4162 | 152.7 | 0.3471 | 3.2878 | 152.7 | 0.3460 | 150 |
| 160 | 3.7689 | 154.9 | 0.3529 | 3.6162 | 154.9 | 0.3517 | 3.4753 | 154.8 | 0.3506 | 3.3447 | 154.8 | 0.3495 | 160 |
| 170 | 3.8325 | 157.1 | 0.3563 | 3.6774 | 157.0 | 0.3552 | 3.5341 | 157.0 | 0.3540 | 3.4015 | 157.0 | 0.3530 | 170 |
| 180 | 3.8960 | 159.2 | 0.3597 | 3.7384 | 159.2 | 0.3586 | 3.5929 | 159.2 | 0.3575 | 3.4581 | 159.1 | 0.3564 | 180 |
| 190 | 3.9594 | 161.4 | 0.3631 | 3.7993 | 161.4 | 0.3620 | 3.6515 | 161.4 | 0.3609 | 3.5147 | 161.3 | 0.3598 | 190 |
| 200 | 4.0227 | 163.6 | 0.3665 | 3.8601 | 163.6 | 0.3653 | 3.7100 | 163.6 | 0.3642 | 3.5711 | 163.5 | 0.3632 | 200 |
| 210 | 4.0858 | 165.8 | 0.3698 | 3.9208 | 165.8 | 0.3687 | 3.7685 | 165.8 | 0.3676 | 3.6274 | 165.8 | 0.3665 | 210 |
| 220 | 4.1489 | 168.1 | 0.3731 | 3.9814 | 168.0 | 0.3720 | 3.8268 | 168.0 | 0.3709 | 3.6836 | 168.0 | 0.3698 | 220 |
| 230 | 4.2119 | 170.3 | 0.3764 | 4.0419 | 170.3 | 0.3753 | 3.8850 | 170.3 | 0.3742 | 3.7398 | 170.3 | 0.3731 | 230 |
| 240 | 4.2748 | 172.6 | 0.3797 | 4.1024 | 172.6 | 0.3786 | 3.9432 | 172.5 | 0.3775 | 3.7959 | 172.5 | 0.3764 | 240 |
| 250 | 4.3376 | 174.9 | 0.3829 | 4.1628 | 174.9 | 0.3818 | 4.0013 | 174.8 | 0.3807 | 3.8519 | 174.8 | 0.3797 | 250 |
| 260 | 4.4004 | 177.2 | 0.3862 | 4.2231 | 177.2 | 0.3850 | 4.0594 | 177.1 | 0.3839 | 3.9078 | 177.1 | 0.3829 | 260 |
| 270 | – | – | – | – | – | – | 4.1174 | 179.5 | 0.3871 | 3.9637 | 179.4 | 0.3861 | 270 |

(Continued)

# TABLE A.19 (CONTINUED)

Suva® 410A Superheated Vapor–Constant Pressure Tables (Saturated Vapor Properties in Parentheses)

| Temp. °F | 32.00 (−30.69°F) V | H | S | 33.00 (−29.39°F) V | H | S | 34.00 (−28.13°F) V | H | S | 35.00 (−26.89°F) V | H | S | Temp. °F |
|---|---|---|---|---|---|---|---|---|---|---|---|---|---|
| | (1.8314) | (115.4) | (0.2692) | (1.7784) | (115.6) | (0.2688) | (1.7284) | (115.7) | (0.2683) | (1.6812) | (115.8) | (0.2679) | |
| 40 | 2.5482 | 130.0 | 0.3040 | 2.4571 | 129.9 | 0.3030 | 2.3721 | 129.8 | 0.3020 | 2.2926 | 129.8 | 0.3009 | 40 |
| 50 | 2.6064 | 132.0 | 0.3080 | 2.5135 | 131.9 | 0.3070 | 2.4268 | 131.8 | 0.3059 | 2.3457 | 131.8 | 0.3049 | 50 |
| 60 | 2.6641 | 134.0 | 0.3119 | 2.5694 | 133.9 | 0.3109 | 2.4810 | 133.9 | 0.3099 | 2.3983 | 133.8 | 0.3089 | 60 |
| 70 | 2.7213 | 136.0 | 0.3158 | 2.6248 | 135.9 | 0.3147 | 2.5348 | 135.9 | 0.3137 | 2.4505 | 135.8 | 0.3128 | 70 |
| 80 | 2.7782 | 138.0 | 0.3196 | 2.6799 | 138.0 | 0.3185 | 2.5881 | 137.9 | 0.3175 | 2.5023 | 137.9 | 0.3166 | 80 |
| 90 | 2.8347 | 140.1 | 0.3233 | 2.7346 | 140.0 | 0.3223 | 2.6412 | 140.0 | 0.3213 | 2.5537 | 139.9 | 0.3203 | 90 |
| 100 | 2.8910 | 142.1 | 0.3270 | 2.7890 | 142.1 | 0.3260 | 2.6939 | 142.0 | 0.3250 | 2.6049 | 142.0 | 0.3241 | 100 |
| 110 | 2.9469 | 144.2 | 0.3307 | 2.8432 | 144.2 | 0.3297 | 2.7463 | 144.1 | 0.3287 | 2.6557 | 144.1 | 0.3277 | 110 |
| 120 | 3.0026 | 146.3 | 0.3343 | 2.8971 | 146.2 | 0.3333 | 2.7985 | 146.2 | 0.3323 | 2.7063 | 146.2 | 0.3314 | 120 |
| 130 | 3.0581 | 148.4 | 0.3379 | 2.9507 | 148.3 | 0.3369 | 2.8505 | 148.3 | 0.3359 | 2.7567 | 148.3 | 0.3350 | 130 |
| 140 | 3.1134 | 150.5 | 0.3415 | 3.0042 | 150.5 | 0.3405 | 2.9023 | 150.4 | 0.3395 | 2.8069 | 150.4 | 0.3386 | 140 |
| 150 | 3.1686 | 152.6 | 0.3450 | 3.0575 | 152.6 | 0.3440 | 2.9539 | 152.6 | 0.3430 | 2.8570 | 152.5 | 0.3421 | 150 |
| 160 | 3.2235 | 154.8 | 0.3485 | 3.1107 | 154.7 | 0.3475 | 3.0054 | 154.7 | 0.3465 | 2.9068 | 154.7 | 0.3456 | 160 |
| 170 | 3.2783 | 156.9 | 0.3519 | 3.1637 | 156.9 | 0.3509 | 3.0567 | 156.9 | 0.3500 | 2.9565 | 156.8 | 0.3491 | 170 |
| 180 | 3.3330 | 159.1 | 0.3554 | 3.2166 | 159.1 | 0.3544 | 3.1078 | 159.0 | 0.3534 | 3.0061 | 159.0 | 0.3525 | 180 |
| 190 | 3.3876 | 161.3 | 0.3588 | 3.2693 | 161.3 | 0.3578 | 3.1589 | 161.2 | 0.3568 | 3.0556 | 161.2 | 0.3559 | 190 |
| 200 | 3.4421 | 163.5 | 0.3621 | 3.3219 | 163.5 | 0.3612 | 3.2098 | 163.4 | 0.3602 | 3.1050 | 163.4 | 0.3593 | 200 |
| 210 | 3.4964 | 165.7 | 0.3655 | 3.3745 | 165.7 | 0.3645 | 3.2607 | 165.7 | 0.3635 | 3.1542 | 165.7 | 0.3626 | 210 |
| 220 | 3.5507 | 168.0 | 0.3688 | 3.4269 | 167.9 | 0.3678 | 3.3114 | 167.9 | 0.3669 | 3.2034 | 167.9 | 0.3660 | 220 |
| 230 | 3.6049 | 170.2 | 0.3721 | 3.4793 | 170.2 | 0.3711 | 3.3621 | 170.2 | 0.3702 | 3.2525 | 170.2 | 0.3693 | 230 |
| 240 | 3.6590 | 172.5 | 0.3754 | 3.5316 | 172.5 | 0.3744 | 3.4127 | 172.5 | 0.3735 | 3.3015 | 172.4 | 0.3725 | 240 |
| 250 | 3.7131 | 174.8 | 0.3786 | 3.5839 | 174.8 | 0.3777 | 3.4632 | 174.7 | 0.3767 | 3.3504 | 174.7 | 0.3758 | 250 |
| 260 | 3.7671 | 177.1 | 0.3819 | 3.6360 | 177.1 | 0.3809 | 3.5137 | 177.1 | 0.3799 | 3.3993 | 177.0 | 0.3790 | 260 |
| 270 | 3.8210 | 179.4 | 0.3851 | 3.6881 | 179.4 | 0.3841 | 3.5641 | 179.4 | 0.3831 | 3.4481 | 179.4 | 0.3822 | 270 |

| Temp. °F | 32.00 (−30.69°F) V | H | S | 33.00 (−29.39°F) V | H | S | 34.00 (−28.13°F) V | H | S | 35.00 (−26.89°F) V | H | S | Temp. °F |
|---|---|---|---|---|---|---|---|---|---|---|---|---|---|
| | (1.8314) | (115.4) | (0.2692) | (1.7784) | (115.6) | (0.2688) | (1.7284) | (115.7) | (0.2683) | (1.6812) | (115.8) | (0.2679) | |
| −30 | 1.8354 | 115.6 | 0.2695 | – | – | – | – | – | – | – | – | – | −30 |
| −20 | 1.8932 | 117.6 | 0.2742 | 1.8314 | 117.5 | 0.2732 | 1.7731 | 117.4 | 0.2722 | 1.7182 | 117.3 | 0.2712 | −20 |

(Continued)

**TABLE A.19 (CONTINUED)**

Suva® 410A Superheated Vapor–Constant Pressure Tables (Saturated Vapor Properties in Parentheses)

| Temp | | | | | | | | | | | | | Temp |
|---|---|---|---|---|---|---|---|---|---|---|---|---|---|
| -10 | 1.9497 | 119.6 | 0.2788 | 1.8865 | 119.5 | 0.2777 | 1.8270 | 119.4 | 0.2767 | 1.7709 | 119.3 | 0.2758 | -10 |
| 0 | 2.0051 | 121.6 | 0.2832 | 1.9405 | 121.5 | 0.2822 | 1.8797 | 121.4 | 0.2812 | 1.8224 | 121.3 | 0.2803 | 0 |
| 10 | 2.0595 | 123.6 | 0.2875 | 1.9935 | 123.6 | 0.2865 | 1.9315 | 123.5 | 0.2856 | 1.8730 | 123.4 | 0.2846 | 10 |
| 20 | 2.1130 | 125.7 | 0.2918 | 2.0457 | 125.6 | 0.2908 | 1.9824 | 125.5 | 0.2898 | 1.9227 | 125.4 | 0.2889 | 20 |
| 30 | 2.1658 | 127.7 | 0.2959 | 2.0972 | 127.6 | 0.2949 | 2.0325 | 127.5 | 0.2940 | 1.9716 | 127.4 | 0.2931 | 30 |
| 40 | 2.2180 | 129.7 | 0.3000 | 2.1480 | 129.6 | 0.2990 | 2.0821 | 129.5 | 0.2981 | 2.0199 | 129.5 | 0.2972 | 40 |
| 50 | 2.2697 | 131.7 | 0.3040 | 2.1982 | 131.6 | 0.3030 | 2.1310 | 131.6 | 0.3021 | 2.0676 | 131.5 | 0.3012 | 50 |
| 60 | 2.3208 | 133.7 | 0.3079 | 2.2480 | 133.7 | 0.3070 | 2.1794 | 133.6 | 0.3061 | 2.1148 | 133.6 | 0.3052 | 60 |
| 70 | 2.3715 | 135.8 | 0.3118 | 2.2973 | 135.7 | 0.3109 | 2.2274 | 135.7 | 0.3100 | 2.1616 | 135.6 | 0.3091 | 70 |
| 80 | 2.4218 | 137.8 | 0.3156 | 2.3462 | 137.8 | 0.3147 | 2.2750 | 137.7 | 0.3138 | 2.2079 | 137.7 | 0.3130 | 80 |
| 90 | 2.4718 | 139.9 | 0.3194 | 2.3948 | 139.8 | 0.3185 | 2.3223 | 139.8 | 0.3176 | 2.2540 | 139.7 | 0.3168 | 90 |
| 100 | 2.5214 | 141.9 | 0.3231 | 2.4430 | 141.9 | 0.3222 | 2.3692 | 141.8 | 0.3214 | 2.2997 | 141.8 | 0.3205 | 100 |
| 110 | 2.5708 | 144.0 | 0.3268 | 2.4910 | 144.0 | 0.3259 | 2.4159 | 143.9 | 0.3251 | 2.3451 | 143.9 | 0.3242 | 110 |
| 120 | 2.6199 | 146.1 | 0.3305 | 2.5387 | 146.1 | 0.3296 | 2.4623 | 146.0 | 0.3287 | 2.3903 | 146.0 | 0.3279 | 120 |
| 130 | 2.6688 | 148.2 | 0.3341 | 2.5863 | 148.2 | 0.3332 | 2.5085 | 148.1 | 0.3323 | 2.4352 | 148.1 | 0.3315 | 130 |
| 140 | 2.7175 | 150.3 | 0.3376 | 2.6336 | 150.3 | 0.3368 | 2.5545 | 150.3 | 0.3359 | 2.4800 | 150.2 | 0.3351 | 140 |
| 150 | 2.7661 | 152.5 | 0.3412 | 2.6807 | 152.4 | 0.3403 | 2.6004 | 152.4 | 0.3394 | 2.5246 | 152.4 | 0.3386 | 150 |
| 160 | 2.8145 | 154.6 | 0.3447 | 2.7277 | 154.6 | 0.3438 | 2.6460 | 154.6 | 0.3429 | 2.5690 | 154.5 | 0.3421 | 160 |
| 170 | 2.8627 | 156.8 | 0.3481 | 2.7745 | 156.8 | 0.3473 | 2.6915 | 156.7 | 0.3464 | 2.6133 | 156.7 | 0.3456 | 170 |
| 180 | 2.9108 | 159.0 | 0.3516 | 2.8212 | 159.0 | 0.3507 | 2.7369 | 158.9 | 0.3499 | 2.6574 | 158.9 | 0.3490 | 180 |
| 190 | 2.9588 | 161.2 | 0.3550 | 2.8678 | 161.2 | 0.3541 | 2.7822 | 161.1 | 0.3533 | 2.7014 | 161.1 | 0.3524 | 190 |
| 200 | 3.0066 | 163.4 | 0.3584 | 2.9143 | 163.4 | 0.3575 | 2.8273 | 163.3 | 0.3567 | 2.7454 | 163.3 | 0.3558 | 200 |
| 210 | 3.0544 | 165.6 | 0.3617 | 2.9606 | 165.6 | 0.3609 | 2.8724 | 165.6 | 0.3600 | 2.7892 | 165.5 | 0.3592 | 210 |
| 220 | 3.1021 | 167.9 | 0.3651 | 3.0069 | 167.8 | 0.3642 | 2.9173 | 167.8 | 0.3633 | 2.8329 | 167.8 | 0.3625 | 220 |
| 230 | 3.1497 | 170.1 | 0.3684 | 3.0531 | 170.1 | 0.3675 | 2.9622 | 170.1 | 0.3667 | 2.8765 | 170.1 | 0.3658 | 230 |
| 240 | 3.1972 | 172.4 | 0.3716 | 3.0992 | 172.4 | 0.3708 | 3.0070 | 172.4 | 0.3699 | 2.9201 | 172.3 | 0.3691 | 240 |
| 250 | 3.2446 | 174.7 | 0.3749 | 3.1453 | 174.7 | 0.3740 | 3.0518 | 174.7 | 0.3732 | 2.9636 | 174.6 | 0.3724 | 250 |
| 260 | 3.2920 | 177.0 | 0.3781 | 3.1913 | 177.0 | 0.3773 | 3.0964 | 177.0 | 0.3764 | 3.0070 | 176.9 | 0.3756 | 260 |
| 270 | 3.3394 | 179.3 | 0.3813 | 3.2372 | 179.3 | 0.3805 | 3.1411 | 179.3 | 0.3796 | 3.0504 | 179.3 | 0.3788 | 270 |
| 280 | — | — | — | 3.2831 | 181.7 | 0.3837 | 3.1856 | 181.6 | 0.3828 | 3.0937 | 181.6 | 0.3820 | 280 |

*(Continued)*

# TABLE A.19 (CONTINUED)
Suva® 410A Superheated Vapor–Constant Pressure Tables (Saturated Vapor Properties in Parentheses)

| Temp. °F | 36.00 (−25.67°F) | | | 37.00 (−24.49°F) | | | 38.00 (−23.22°F) | | | 39.00 (−22.19°F) | | | Temp. °F |
|---|---|---|---|---|---|---|---|---|---|---|---|---|---|
| | V (1.6365) | H (116.0) | S (0.2675) | V (1.5941) | H (116.1) | S (0.2671) | V (1.5539) | H (116.2) | S (0.2667) | V (1.5157) | H (116.3) | S (0.2663) | |
| −20 | 1.6663 | 117.1 | 0.2702 | 1.6172 | 117.0 | 0.2692 | 1.5707 | 116.9 | 0.2683 | 1.5265 | 116.8 | 0.2674 | −20 |
| −10 | 1.7179 | 119.2 | 0.2748 | 1.6678 | 119.1 | 0.2739 | 1.6202 | 119.0 | 0.2730 | 1.5751 | 118.9 | 0.2721 | −10 |
| 0 | 1.7683 | 121.3 | 0.2793 | 1.7171 | 121.2 | 0.2784 | 1.6686 | 121.1 | 0.2775 | 1.6225 | 121.0 | 0.2766 | 0 |
| 10 | 1.8177 | 123.3 | 0.2837 | 1.7654 | 123.2 | 0.2828 | 1.7159 | 123.1 | 0.2819 | 1.6688 | 123.0 | 0.2811 | 10 |
| 20 | 1.8662 | 125.3 | 0.2880 | 1.8129 | 125.2 | 0.2871 | 1.7623 | 125.2 | 0.2863 | 1.7143 | 125.1 | 0.2854 | 20 |
| 30 | 1.9140 | 127.4 | 0.2922 | 1.8596 | 127.3 | 0.2913 | 1.8080 | 127.2 | 0.2905 | 1.7590 | 127.1 | 0.2897 | 30 |
| 40 | 1.9612 | 129.4 | 0.2963 | 1.9056 | 129.3 | 0.2955 | 1.8530 | 129.3 | 0.2946 | 1.8030 | 129.2 | 0.2938 | 40 |
| 50 | 2.0077 | 131.4 | 0.3004 | 1.9511 | 131.4 | 0.2995 | 1.8974 | 131.3 | 0.2987 | 1.8465 | 131.2 | 0.2979 | 50 |
| 60 | 2.0538 | 133.5 | 0.3043 | 1.9960 | 133.4 | 0.3035 | 1.9413 | 133.4 | 0.3027 | 1.8894 | 133.3 | 0.3019 | 60 |
| 70 | 2.0994 | 135.5 | 0.3083 | 2.0405 | 135.5 | 0.3074 | 1.9847 | 135.4 | 0.3066 | 1.9318 | 135.4 | 0.3058 | 70 |
| 80 | 2.1446 | 137.6 | 0.3121 | 2.0846 | 137.6 | 0.3113 | 2.0278 | 137.5 | 0.3105 | 1.9739 | 137.4 | 0.3097 | 80 |
| 90 | 2.1894 | 139.7 | 0.3159 | 2.1284 | 139.6 | 0.3151 | 2.0705 | 139.6 | 0.3143 | 2.0156 | 139.5 | 0.3135 | 90 |
| 100 | 2.2340 | 141.8 | 0.3197 | 2.1718 | 141.7 | 0.3189 | 2.1129 | 141.7 | 0.3181 | 2.0570 | 141.6 | 0.3173 | 100 |
| 110 | 2.2782 | 143.8 | 0.3234 | 2.2150 | 143.8 | 0.3226 | 2.1550 | 143.8 | 0.3218 | 2.0982 | 143.7 | 0.3210 | 110 |
| 120 | 2.3222 | 146.0 | 0.3270 | 2.2579 | 145.9 | 0.3262 | 2.1969 | 145.9 | 0.3255 | 2.1390 | 145.8 | 0.3247 | 120 |
| 130 | 2.3660 | 148.1 | 0.3307 | 2.3006 | 148.0 | 0.3299 | 2.2385 | 148.0 | 0.3291 | 2.1797 | 148.0 | 0.3283 | 130 |
| 140 | 2.4096 | 150.2 | 0.3342 | 2.3430 | 150.2 | 0.3334 | 2.2800 | 150.1 | 0.3327 | 2.2201 | 150.1 | 0.3319 | 140 |
| 150 | 2.4530 | 152.3 | 0.3378 | 2.3853 | 152.3 | 0.3370 | 2.3212 | 152.3 | 0.3362 | 2.2604 | 152.2 | 0.3355 | 150 |
| 160 | 2.4963 | 154.5 | 0.3413 | 2.4275 | 154.5 | 0.3405 | 2.3623 | 154.4 | 0.3397 | 2.3005 | 154.4 | 0.3390 | 160 |
| 170 | 2.5394 | 156.7 | 0.3448 | 2.4695 | 156.6 | 0.3440 | 2.4032 | 156.6 | 0.3432 | 2.3404 | 156.6 | 0.3425 | 170 |
| 180 | 2.5823 | 158.9 | 0.3482 | 2.5113 | 158.8 | 0.3474 | 2.4441 | 158.8 | 0.3467 | 2.3802 | 158.8 | 0.3459 | 180 |
| 190 | 2.6252 | 161.1 | 0.3516 | 2.5531 | 161.0 | 0.3509 | 2.4848 | 161.0 | 0.3501 | 2.4199 | 161.0 | 0.3494 | 190 |
| 200 | 2.6679 | 163.3 | 0.3550 | 2.5947 | 163.3 | 0.3543 | 2.5253 | 163.2 | 0.3535 | 2.4595 | 163.2 | 0.3528 | 200 |
| 210 | 2.7106 | 165.5 | 0.3584 | 2.6362 | 165.5 | 0.3576 | 2.5658 | 165.5 | 0.3569 | 2.4990 | 165.4 | 0.3561 | 210 |
| 220 | 2.7531 | 167.8 | 0.3617 | 2.6777 | 167.7 | 0.3610 | 2.6062 | 167.7 | 0.3602 | 2.5384 | 167.7 | 0.3595 | 220 |
| 230 | 2.7956 | 170.0 | 0.3650 | 2.7190 | 170.0 | 0.3643 | 2.6465 | 170.0 | 0.3635 | 2.5777 | 170.0 | 0.3628 | 230 |

*(Continued)*

# TABLE A.19 (CONTINUED)

Suva® 410A Superheated Vapor–Constant Pressure Tables (Saturated Vapor Properties in Parentheses)

| Temp. °F | 40.00 (−21.08°F) | | | 41.00 (−19.98°F) | | | 42.00 (−18.91°F) | | | 43.00 (−17.86°F) | | | Temp. °F |
|---|---|---|---|---|---|---|---|---|---|---|---|---|---|
| | V | H | S | V | H | S | V | H | S | V | H | S | |
| | (1.4793) | (116.4) | (0.2659) | (1.4447) | (116.6) | (0.2656) | (1.4116) | (116.7) | (0.2652) | (1.3801) | (116.8) | (0.2649) | |
| −20 | 1.4846 | 116.7 | 0.2665 | — | — | — | — | — | — | — | — | — | −20 |
| −10 | 1.5323 | 118.8 | 0.2712 | 1.4915 | 118.7 | 0.2703 | 1.4526 | 118.6 | 0.2695 | 1.4155 | 118.4 | 0.2687 | −10 |
| 0 | 1.5787 | 120.9 | 0.2758 | 1.5371 | 120.8 | 0.2749 | 1.4974 | 120.7 | 0.2741 | 1.4596 | 120.6 | 0.2733 | 0 |
| 10 | 1.6242 | 122.9 | 0.2802 | 1.5817 | 122.8 | 0.2794 | 1.5412 | 122.8 | 0.2786 | 1.5025 | 122.7 | 0.2778 | 10 |
| 20 | 1.6687 | 125.0 | 0.2846 | 1.6253 | 124.9 | 0.2838 | 1.5840 | 124.8 | 0.2830 | 1.5446 | 124.8 | 0.2822 | 20 |
| 30 | 1.7125 | 127.1 | 0.2889 | 1.6682 | 127.0 | 0.2881 | 1.6261 | 126.9 | 0.2873 | 1.5858 | 126.8 | 0.2865 | 30 |
| 40 | 1.7556 | 129.1 | 0.2930 | 1.7104 | 129.1 | 0.2922 | 1.6674 | 129.0 | 0.2915 | 1.6264 | 128.9 | 0.2907 | 40 |
| 50 | 1.7981 | 131.2 | 0.2971 | 1.7520 | 131.1 | 0.2963 | 1.7082 | 131.1 | 0.2956 | 1.6664 | 131.0 | 0.2948 | 50 |
| 60 | 1.8401 | 133.2 | 0.3011 | 1.7931 | 133.2 | 0.3003 | 1.7485 | 133.1 | 0.2996 | 1.7058 | 133.1 | 0.2989 | 60 |
| 70 | 1.8816 | 135.3 | 0.3051 | 1.8338 | 135.3 | 0.3043 | 1.7882 | 135.2 | 0.3036 | 1.7448 | 135.1 | 0.3028 | 70 |
| 80 | 1.9227 | 137.4 | 0.3089 | 1.8740 | 137.3 | 0.3082 | 1.8276 | 137.3 | 0.3075 | 1.7834 | 137.2 | 0.3067 | 80 |
| 90 | 1.9635 | 139.5 | 0.3128 | 1.9139 | 139.4 | 0.3120 | 1.8667 | 139.4 | 0.3113 | 1.8216 | 139.3 | 0.3106 | 90 |
| 100 | 2.0040 | 141.6 | 0.3165 | 1.9535 | 141.5 | 0.3158 | 1.9054 | 141.5 | 0.3151 | 1.8595 | 141.4 | 0.3144 | 100 |
| 110 | 2.0441 | 143.7 | 0.3203 | 1.9927 | 143.6 | 0.3195 | 1.9438 | 143.6 | 0.3188 | 1.8971 | 143.5 | 0.3181 | 110 |
| 120 | 2.0841 | 145.8 | 0.3239 | 2.0318 | 145.7 | 0.3232 | 1.9820 | 145.7 | 0.3225 | 1.9345 | 145.7 | 0.3218 | 120 |
| 130 | 2.1238 | 147.9 | 0.3276 | 2.0706 | 147.9 | 0.3269 | 2.0199 | 147.8 | 0.3261 | 1.9716 | 147.8 | 0.3255 | 130 |
| 140 | 2.1633 | 150.1 | 0.3312 | 2.1092 | 150.0 | 0.3305 | 2.0577 | 150.0 | 0.3298 | 2.0086 | 149.9 | 0.3291 | 140 |
| 150 | 2.2026 | 152.2 | 0.3347 | 2.1476 | 152.2 | 0.3340 | 2.0952 | 152.1 | 0.3333 | 2.0453 | 152.1 | 0.3326 | 150 |
| 160 | 2.2417 | 154.4 | 0.3383 | 2.1858 | 154.3 | 0.3375 | 2.1326 | 154.3 | 0.3368 | 2.0819 | 154.3 | 0.3362 | 160 |
| 170 | 2.2807 | 156.6 | 0.3417 | 2.2239 | 156.5 | 0.3410 | 2.1699 | 156.5 | 0.3403 | 2.1183 | 156.5 | 0.3397 | 170 |
| 180 | 2.3196 | 158.7 | 0.3452 | 2.2619 | 158.7 | 0.3445 | 2.2070 | 158.7 | 0.3438 | 2.1546 | 158.7 | 0.3431 | 180 |
| 240 | 2.8380 | 172.3 | 0.3683 | 2.7603 | 172.3 | 0.3675 | 2.6867 | 172.3 | 0.3668 | 2.6169 | 172.2 | 0.3661 | 240 |
| 250 | 2.8803 | 174.6 | 0.3716 | 2.8015 | 174.6 | 0.3708 | 2.7269 | 174.6 | 0.3701 | 2.6561 | 174.5 | 0.3693 | 250 |
| 260 | 2.9226 | 176.9 | 0.3748 | 2.8427 | 176.9 | 0.3740 | 2.7670 | 176.9 | 0.3733 | 2.6952 | 176.9 | 0.3726 | 260 |
| 270 | 2.9648 | 179.3 | 0.3780 | 2.8838 | 179.2 | 0.3773 | 2.8070 | 179.2 | 0.3765 | 2.7342 | 179.2 | 0.3758 | 270 |
| 280 | 3.0069 | 181.6 | 0.3812 | 2.9248 | 181.6 | 0.3805 | 2.8470 | 181.6 | 0.3797 | 2.7732 | 181.5 | 0.3790 | 280 |

(Continued)

# TABLE A.19 (CONTINUED)

Suva® 410A Superheated Vapor–Constant Pressure Tables (Saturated Vapor Properties in Parentheses)

| Temp. °F | 44.00 (−16.82°F) | | | 45.00 (−15.81°F) | | | 46.00 (−14.81°F) | | | 47.00 (−13.82°F) | | |
|---|---|---|---|---|---|---|---|---|---|---|---|---|
| | V | H | S | V | H | S | V | H | S | V | H | S |
| | (1.3499) | (116.9) | (0.2646) | (1.3210) | (117.0) | (0.2642) | (1.2933) | (117.1) | (0.2639) | (1.2668) | (117.2) | (0.2636) |
| −10 | 1.3801 | 118.3 | 0.2678 | 1.3463 | 118.2 | 0.2670 | 1.3139 | 118.1 | 0.2662 | 1.2829 | 118.0 | 0.2655 |
| 0 | 1.4234 | 120.5 | 0.2725 | 1.3889 | 120.4 | 0.2717 | 1.3559 | 120.3 | 0.2710 | 1.3242 | 120.2 | 0.2702 |
| 10 | 1.4657 | 122.6 | 0.2770 | 1.4304 | 122.5 | 0.2763 | 1.3967 | 122.4 | 0.2755 | 1.3644 | 122.3 | 0.2748 |
| 20 | 1.5070 | 124.7 | 0.2815 | 1.4710 | 124.6 | 0.2807 | 1.4366 | 124.5 | 0.2800 | 1.4037 | 124.4 | 0.2792 |
| 30 | 1.5475 | 126.8 | 0.2858 | 1.5108 | 126.7 | 0.2850 | 1.4757 | 126.6 | 0.2843 | 1.4421 | 126.5 | 0.2836 |
| 40 | 1.5873 | 128.8 | 0.2900 | 1.5499 | 128.8 | 0.2893 | 1.5141 | 128.7 | 0.2885 | 1.4798 | 128.6 | 0.2878 |
| 50 | 1.6265 | 130.9 | 0.2941 | 1.5883 | 130.9 | 0.2934 | 1.5519 | 130.8 | 0.2927 | 1.5169 | 130.7 | 0.2920 |
| 60 | 1.6652 | 133.0 | 0.2981 | 1.6263 | 132.9 | 0.2974 | 1.5891 | 132.9 | 0.2968 | 1.5535 | 132.8 | 0.2961 |
| 70 | 1.7034 | 135.1 | 0.3021 | 1.6638 | 135.0 | 0.3014 | 1.6259 | 135.0 | 0.3007 | 1.5896 | 134.9 | 0.3001 |
| 80 | 1.7412 | 137.2 | 0.3060 | 1.7008 | 137.1 | 0.3053 | 1.6622 | 137.1 | 0.3047 | 1.6253 | 137.0 | 0.3040 |
| 90 | 1.7786 | 139.3 | 0.3099 | 1.7375 | 139.2 | 0.3092 | 1.6982 | 139.2 | 0.3085 | 1.6606 | 139.1 | 0.3079 |
| 100 | 1.8158 | 141.4 | 0.3137 | 1.7739 | 141.3 | 0.3130 | 1.7339 | 141.3 | 0.3123 | 1.6956 | 141.2 | 0.3117 |
| 110 | 1.8526 | 143.5 | 0.3174 | 1.8100 | 143.5 | 0.3168 | 1.7693 | 143.4 | 0.3161 | 1.7303 | 143.4 | 0.3155 |
| 120 | 1.8892 | 145.6 | 0.3211 | 1.8459 | 145.6 | 0.3205 | 1.8044 | 145.5 | 0.3198 | 1.7648 | 145.5 | 0.3192 |
| 130 | 1.9255 | 147.8 | 0.3248 | 1.8815 | 147.7 | 0.3241 | 1.8393 | 147.7 | 0.3235 | 1.7990 | 147.6 | 0.3228 |
| 190 | 2.3583 | 161.0 | 0.3486 | 2.2998 | 160.9 | 0.3479 | 2.2440 | 160.9 | 0.3472 | 2.1908 | 160.9 | 0.3466 |
| 200 | 2.3970 | 163.2 | 0.3520 | 2.3375 | 163.1 | 0.3513 | 2.2809 | 163.1 | 0.3506 | 2.2268 | 163.1 | 0.3500 |
| 210 | 2.4355 | 165.4 | 0.3554 | 2.3751 | 165.4 | 0.3547 | 2.3176 | 165.4 | 0.3540 | 2.2628 | 165.3 | 0.3533 |
| 220 | 2.4740 | 167.7 | 0.3587 | 2.4127 | 167.6 | 0.3580 | 2.3543 | 167.6 | 0.3573 | 2.2987 | 167.6 | 0.3567 |
| 230 | 2.5123 | 169.9 | 0.3621 | 2.4502 | 169.9 | 0.3614 | 2.3909 | 169.9 | 0.3607 | 2.3345 | 169.9 | 0.3600 |
| 240 | 2.5506 | 172.2 | 0.3653 | 2.4875 | 172.2 | 0.3646 | 2.4275 | 172.2 | 0.3640 | 2.3702 | 172.2 | 0.3633 |
| 250 | 2.5888 | 174.5 | 0.3686 | 2.5249 | 174.5 | 0.3679 | 2.4639 | 174.5 | 0.3672 | 2.4058 | 174.5 | 0.3666 |
| 260 | 2.6270 | 176.8 | 0.3719 | 2.5621 | 176.8 | 0.3712 | 2.5003 | 176.8 | 0.3705 | 2.4414 | 176.8 | 0.3698 |
| 270 | 2.6651 | 179.2 | 0.3751 | 2.5993 | 179.2 | 0.3744 | 2.5367 | 179.1 | 0.3737 | 2.4769 | 179.1 | 0.3730 |
| 280 | 2.7031 | 181.5 | 0.3783 | 2.6364 | 181.5 | 0.3776 | 2.5729 | 181.5 | 0.3769 | 2.5124 | 181.5 | 0.3762 |
| 290 | 2.6735 | 183.9 | 0.3807 | 2.6092 | 183.8 | 0.3801 | 2.5478 | 183.8 | 0.3794 | – | – | – |

(Continued)

**TABLE A.19 (CONTINUED)**

Suva® 410A Superheated Vapor–Constant Pressure Tables (Saturated Vapor Properties in Parentheses)

| Temp. °F | 48.00 (−12.86°F) | | | 49.00 (−11.91°F) | | | 50.00 (−10.97°F) | | | 55.00 (−6.50°F) | | | Temp. °F |
|---|---|---|---|---|---|---|---|---|---|---|---|---|---|
| | V (1.2413) | H (117.3) | S (0.2633) | V (1.2169) | H (117.4) | S (0.2630) | V (1.1933) | H (117.5) | S (0.2627) | V (1.0882) | H (117.9) | S (0.2613) | |
| −10 | 1.2532 | 117.9 | 0.2647 | 1.2247 | 117.8 | 0.2639 | 1.1972 | 117.7 | 0.2632 | – | – | – | −10 |
| 0 | 1.2939 | 120.1 | 0.2694 | 1.2648 | 120.0 | 0.2687 | 1.2368 | 119.9 | 0.2680 | 1.1121 | 119.3 | 0.2645 | 0 |
| 10 | 1.3335 | 122.2 | 0.2741 | 1.3038 | 122.1 | 0.2733 | 1.2752 | 122.0 | 0.2726 | 1.1481 | 121.5 | 0.2692 | 10 |
| 20 | 1.3721 | 124.3 | 0.2785 | 1.3418 | 124.2 | 0.2778 | 1.3127 | 124.2 | 0.2771 | 1.1830 | 123.7 | 0.2738 | 20 |
| 30 | 1.4099 | 126.4 | 0.2829 | 1.3790 | 126.4 | 0.2822 | 1.3493 | 126.3 | 0.2815 | 1.2171 | 125.9 | 0.2783 | 30 |
| 40 | 1.4470 | 128.6 | 0.2872 | 1.4155 | 128.5 | 0.2865 | 1.3852 | 128.4 | 0.2858 | 1.2504 | 128.0 | 0.2827 | 40 |
| 50 | 1.4834 | 130.7 | 0.2913 | 1.4513 | 130.6 | 0.2907 | 1.4205 | 130.5 | 0.2900 | 1.2830 | 130.2 | 0.2869 | 50 |
| 60 | 1.5194 | 132.8 | 0.2954 | 1.4866 | 132.7 | 0.2948 | 1.4552 | 132.6 | 0.2941 | 1.3151 | 132.3 | 0.2910 | 60 |
| 70 | 1.5548 | 134.9 | 0.2994 | 1.5215 | 134.8 | 0.2988 | 1.4894 | 134.7 | 0.2981 | 1.3468 | 134.5 | 0.2951 | 70 |
| 80 | 1.5899 | 137.0 | 0.3034 | 1.5559 | 136.9 | 0.3027 | 1.5233 | 136.9 | 0.3021 | 1.3779 | 136.6 | 0.2991 | 80 |
| 140 | 1.9617 | 149.9 | 0.3284 | 1.9169 | 149.9 | 0.3277 | 1.8740 | 149.8 | 0.3271 | 1.8330 | 149.8 | 0.3264 | 140 |
| 150 | 1.9976 | 152.1 | 0.3320 | 1.9521 | 152.0 | 0.3313 | 1.9085 | 152.0 | 0.3307 | 1.8668 | 152.0 | 0.3300 | 150 |
| 160 | 2.0334 | 154.2 | 0.3355 | 1.9872 | 154.2 | 0.3348 | 1.9429 | 154.2 | 0.3342 | 1.9005 | 154.1 | 0.3336 | 160 |
| 170 | 2.0691 | 156.4 | 0.3390 | 2.0221 | 156.4 | 0.3383 | 1.9771 | 156.4 | 0.3377 | 1.9340 | 156.3 | 0.3371 | 170 |
| 180 | 2.1046 | 158.6 | 0.3425 | 2.0568 | 158.6 | 0.3418 | 2.0111 | 158.6 | 0.3412 | 1.9674 | 158.5 | 0.3406 | 180 |
| 190 | 2.1400 | 160.8 | 0.3459 | 2.0915 | 160.8 | 0.3452 | 2.0451 | 160.8 | 0.3446 | 2.0006 | 160.8 | 0.3440 | 190 |
| 200 | 2.1753 | 163.1 | 0.3493 | 2.1260 | 163.0 | 0.3487 | 2.0789 | 163.0 | 0.3480 | 2.0338 | 163.0 | 0.3474 | 200 |
| 210 | 2.2105 | 165.3 | 0.3527 | 2.1605 | 165.3 | 0.3520 | 2.1126 | 165.3 | 0.3514 | 2.0668 | 165.2 | 0.3508 | 210 |
| 220 | 2.2456 | 167.6 | 0.3560 | 2.1948 | 167.5 | 0.3554 | 2.1463 | 167.5 | 0.3548 | 2.0998 | 167.5 | 0.3541 | 220 |
| 230 | 2.2806 | 169.8 | 0.3593 | 2.2291 | 169.8 | 0.3587 | 2.1798 | 169.8 | 0.3581 | 2.1326 | 169.8 | 0.3575 | 230 |
| 240 | 2.3155 | 172.1 | 0.3626 | 2.2633 | 172.1 | 0.3620 | 2.2133 | 172.1 | 0.3614 | 2.1654 | 172.1 | 0.3608 | 240 |
| 250 | 2.3504 | 174.4 | 0.3659 | 2.2974 | 174.4 | 0.3653 | 2.2467 | 174.4 | 0.3646 | 2.1981 | 174.4 | 0.3640 | 250 |
| 260 | 2.3852 | 176.8 | 0.3692 | 2.3314 | 176.7 | 0.3685 | 2.2800 | 176.7 | 0.3679 | 2.2308 | 176.7 | 0.3673 | 260 |
| 270 | 2.4199 | 179.1 | 0.3724 | 2.3654 | 179.1 | 0.3717 | 2.3133 | 179.0 | 0.3711 | 2.2634 | 179.0 | 0.3705 | 270 |
| 280 | 2.4546 | 181.4 | 0.3756 | 2.3993 | 181.4 | 0.3749 | 2.3465 | 181.4 | 0.3743 | 2.2959 | 181.4 | 0.3737 | 280 |
| 290 | 2.4892 | 183.8 | 0.3788 | 2.4332 | 183.8 | 0.3781 | 2.3797 | 183.8 | 0.3775 | 2.3284 | 183.8 | 0.3769 | 290 |

*(Continued)*

**TABLE A.19 (CONTINUED)**

Suva® 410A Superheated Vapor–Constant Pressure Tables (Saturated Vapor Properties in Parentheses)

| Temp. °F | V | H | S | V | H | S | V | H | S | V | H | S | Temp. °F |
|---|---|---|---|---|---|---|---|---|---|---|---|---|---|
| 90 | 1.6245 | 139.1 | 0.3072 | 1.5899 | 139.0 | 0.3066 | 1.5567 | 139.0 | 0.3060 | 1.4088 | 138.7 | 0.3030 | 90 |
| 100 | 1.6589 | 141.2 | 0.3111 | 1.6237 | 141.1 | 0.3104 | 1.5899 | 141.1 | 0.3098 | 1.4392 | 140.9 | 0.3069 | 100 |
| 110 | 1.6929 | 143.3 | 0.3148 | 1.6571 | 143.3 | 0.3142 | 1.6227 | 143.2 | 0.3136 | 1.4694 | 143.0 | 0.3107 | 110 |
| 120 | 1.7267 | 145.5 | 0.3185 | 1.6903 | 145.4 | 0.3179 | 1.6553 | 145.4 | 0.3173 | 1.4993 | 145.2 | 0.3144 | 120 |
| 130 | 1.7603 | 147.6 | 0.3222 | 1.7232 | 147.6 | 0.3216 | 1.6876 | 147.5 | 0.3210 | 1.5290 | 147.3 | 0.3181 | 130 |
| 140 | 1.7937 | 149.8 | 0.3258 | 1.7560 | 149.7 | 0.3252 | 1.7198 | 149.7 | 0.3246 | 1.5585 | 149.5 | 0.3218 | 140 |
| 150 | 1.8269 | 151.9 | 0.3294 | 1.7885 | 151.9 | 0.3288 | 1.7517 | 151.9 | 0.3282 | 1.5877 | 151.7 | 0.3254 | 150 |
| 160 | 1.8599 | 154.1 | 0.3330 | 1.8209 | 154.1 | 0.3324 | 1.7835 | 154.0 | 0.3318 | 1.6168 | 153.9 | 0.3290 | 160 |
| 170 | 1.8927 | 156.3 | 0.3365 | 1.8531 | 156.3 | 0.3359 | 1.8151 | 156.2 | 0.3353 | 1.6458 | 156.1 | 0.3325 | 170 |
| 180 | 1.9254 | 158.5 | 0.3399 | 1.8852 | 158.5 | 0.3393 | 1.8466 | 158.4 | 0.3388 | 1.6746 | 158.3 | 0.3360 | 180 |
| 190 | 1.9580 | 160.7 | 0.3434 | 1.9172 | 160.7 | 0.3428 | 1.8780 | 160.7 | 0.3422 | 1.7033 | 160.5 | 0.3394 | 190 |
| 200 | 1.9905 | 163.0 | 0.3468 | 1.9491 | 162.9 | 0.3462 | 1.9092 | 162.9 | 0.3456 | 1.7319 | 162.8 | 0.3429 | 200 |
| 210 | 2.0229 | 165.2 | 0.3502 | 1.9808 | 165.2 | 0.3496 | 1.9404 | 165.2 | 0.3490 | 1.7603 | 165.0 | 0.3463 | 210 |
| 220 | 2.0552 | 167.5 | 0.3535 | 2.0125 | 167.4 | 0.3529 | 1.9715 | 167.4 | 0.3524 | 1.7887 | 167.3 | 0.3496 | 220 |
| 230 | 2.0874 | 169.7 | 0.3569 | 2.0441 | 169.7 | 0.3563 | 2.0024 | 169.7 | 0.3557 | 1.8170 | 169.6 | 0.3530 | 230 |
| 240 | 2.1196 | 172.0 | 0.3602 | 2.0756 | 172.0 | 0.3596 | 2.0334 | 172.0 | 0.3590 | 1.8452 | 171.9 | 0.3563 | 240 |
| 250 | 2.1516 | 174.3 | 0.3634 | 2.1070 | 174.3 | 0.3629 | 2.0642 | 174.3 | 0.3623 | 1.8734 | 174.2 | 0.3596 | 250 |
| 260 | 2.1836 | 176.7 | 0.3667 | 2.1384 | 176.6 | 0.3661 | 2.0950 | 176.6 | 0.3655 | 1.9015 | 176.5 | 0.3628 | 260 |
| 270 | 2.2156 | 179.0 | 0.3699 | 2.1697 | 179.0 | 0.3693 | 2.1257 | 179.0 | 0.3688 | 1.9295 | 178.9 | 0.3660 | 270 |
| 280 | 2.2475 | 181.4 | 0.3731 | 2.2010 | 181.3 | 0.3725 | 2.1563 | 181.3 | 0.3720 | 1.9575 | 181.2 | 0.3693 | 280 |
| 290 | 2.2793 | 183.7 | 0.3763 | 2.2322 | 183.7 | 0.3757 | 2.1869 | 183.7 | 0.3752 | 1.9854 | 183.6 | 0.3724 | 290 |
| 300 | — | — | — | — | — | — | — | — | — | 2.0133 | 186.0 | 0.3756 | 300 |

| | 60.00 | | | 65.00 | | | 70.00 | | | 75.00 | | | |
| | (−2.32°F) | | | (1.59°F) | | | (5.28°F) | | | (8.78°F) | | | |
| Temp. °F | V | H | S | V | H | S | V | H | S | V | H | S | Temp. °F |
|---|---|---|---|---|---|---|---|---|---|---|---|---|---|
| | (1.0000) | (118.3) | (0.2601) | (0.9249) | (118.6) | (0.2589) | (0.8602) | (119.0) | (0.2578) | (0.8039) | (119.3) | (0.2568) | |
| 0 | 1.0080 | 118.8 | 0.2612 | — | — | — | — | — | — | — | — | — | 0 |
| 10 | 1.0419 | 121.1 | 0.2661 | 0.9519 | 120.6 | 0.2631 | 0.8746 | 120.1 | 0.2602 | 0.8074 | 119.6 | 0.2574 | 10 |
| 20 | 1.0748 | 123.3 | 0.2708 | 0.9831 | 122.8 | 0.2678 | 0.9043 | 122.4 | 0.2651 | 0.8360 | 121.9 | 0.2624 | 20 |
| 30 | 1.1067 | 125.5 | 0.2753 | 1.0133 | 125.1 | 0.2725 | 0.9331 | 124.7 | 0.2698 | 0.8634 | 124.2 | 0.2672 | 30 |

*(Continued)*

**TABLE A.19 (CONTINUED)**

Suva® 410A Superheated Vapor–Constant Pressure Tables (Saturated Vapor Properties in Parentheses)

| Temp | v | h | s | v | h | s | v | h | s | v | h | s |
|---|---|---|---|---|---|---|---|---|---|---|---|---|
| 40 | 0.8900 | 126.5 | 0.2718 | 0.9610 | 126.9 | 0.2743 | 1.0427 | 127.3 | 0.2769 | 1.1379 | 127.7 | 0.2797 |
| 50 | 0.9159 | 128.8 | 0.2763 | 0.9881 | 129.1 | 0.2787 | 1.0714 | 129.5 | 0.2813 | 1.1684 | 129.8 | 0.2840 |
| 60 | 0.9411 | 131.0 | 0.2807 | 1.0147 | 131.4 | 0.2830 | 1.0995 | 131.7 | 0.2855 | 1.1984 | 132.0 | 0.2882 |
| 70 | 0.9658 | 133.3 | 0.2849 | 1.0407 | 133.6 | 0.2872 | 1.1271 | 133.9 | 0.2897 | 1.2278 | 134.2 | 0.2923 |
| 80 | 0.9900 | 135.5 | 0.2891 | 1.0663 | 135.8 | 0.2913 | 1.1542 | 136.0 | 0.2938 | 1.2568 | 136.3 | 0.2963 |
| 90 | 1.0138 | 137.7 | 0.2931 | 1.0915 | 137.9 | 0.2954 | 1.1810 | 138.2 | 0.2978 | 1.2854 | 138.5 | 0.3003 |
| 100 | 1.0373 | 139.9 | 0.2971 | 1.1163 | 140.1 | 0.2993 | 1.2074 | 140.4 | 0.3017 | 1.3137 | 140.6 | 0.3042 |
| 110 | 1.0604 | 142.1 | 0.3010 | 1.1408 | 142.3 | 0.3032 | 1.2335 | 142.6 | 0.3055 | 1.3416 | 142.8 | 0.3080 |
| 120 | 1.0832 | 144.3 | 0.3049 | 1.1650 | 144.5 | 0.3070 | 1.2593 | 144.7 | 0.3093 | 1.3693 | 145.0 | 0.3118 |
| 130 | 1.1058 | 146.5 | 0.3086 | 1.1890 | 146.7 | 0.3108 | 1.2849 | 146.9 | 0.3131 | 1.3968 | 147.1 | 0.3155 |
| 140 | 1.1282 | 148.7 | 0.3124 | 1.2127 | 148.9 | 0.3145 | 1.3102 | 149.1 | 0.3167 | 1.4240 | 149.3 | 0.3192 |
| 150 | 1.1503 | 151.0 | 0.3160 | 1.2363 | 151.1 | 0.3181 | 1.3354 | 151.3 | 0.3204 | 1.4511 | 151.5 | 0.3228 |
| 160 | 1.1723 | 153.2 | 0.3197 | 1.2597 | 153.4 | 0.3218 | 1.3604 | 153.5 | 0.3240 | 1.4779 | 153.7 | 0.3264 |
| 170 | 1.1942 | 155.4 | 0.3233 | 1.2829 | 155.6 | 0.3253 | 1.3853 | 155.8 | 0.3275 | 1.5047 | 155.9 | 0.3299 |
| 180 | 1.2158 | 157.7 | 0.3268 | 1.3060 | 157.8 | 0.3289 | 1.4100 | 158.0 | 0.3311 | 1.5313 | 158.1 | 0.3334 |
| 190 | 1.2374 | 159.9 | 0.3303 | 1.3289 | 160.1 | 0.3323 | 1.4345 | 160.2 | 0.3345 | 1.5577 | 160.4 | 0.3369 |
| 200 | 1.2588 | 162.2 | 0.3338 | 1.3518 | 162.3 | 0.3358 | 1.4590 | 162.5 | 0.3380 | 1.5841 | 162.6 | 0.3403 |
| 210 | 1.2802 | 164.5 | 0.3372 | 1.3745 | 164.6 | 0.3392 | 1.4833 | 164.8 | 0.3414 | 1.6103 | 164.9 | 0.3437 |
| 220 | 1.3014 | 166.8 | 0.3406 | 1.3971 | 166.9 | 0.3426 | 1.5076 | 167.0 | 0.3448 | 1.6364 | 167.2 | 0.3471 |
| 230 | 1.3226 | 169.1 | 0.3440 | 1.4197 | 169.2 | 0.3460 | 1.5318 | 169.3 | 0.3481 | 1.6625 | 169.5 | 0.3505 |
| 240 | 1.3436 | 171.4 | 0.3473 | 1.4422 | 171.5 | 0.3493 | 1.5559 | 171.6 | 0.3515 | 1.6885 | 171.8 | 0.3538 |
| 250 | 1.3646 | 173.7 | 0.3506 | 1.4646 | 173.8 | 0.3526 | 1.5799 | 174.0 | 0.3548 | 1.7144 | 174.1 | 0.3571 |
| 260 | 1.3855 | 176.1 | 0.3539 | 1.4869 | 176.2 | 0.3559 | 1.6038 | 176.3 | 0.3580 | 1.7403 | 176.4 | 0.3603 |
| 270 | 1.4064 | 178.4 | 0.3571 | 1.5092 | 178.5 | 0.3591 | 1.6277 | 178.7 | 0.3613 | 1.7660 | 178.8 | 0.3636 |
| 280 | 1.4272 | 180.8 | 0.3604 | 1.5314 | 180.9 | 0.3624 | 1.6516 | 181.0 | 0.3645 | 1.7918 | 181.1 | 0.3668 |
| 290 | 1.4480 | 183.2 | 0.3636 | 1.5535 | 183.3 | 0.3656 | 1.6754 | 183.4 | 0.3677 | 1.8175 | 183.5 | 0.3700 |
| 300 | 1.4687 | 185.6 | 0.3668 | 1.5757 | 185.7 | 0.3687 | 1.6991 | 185.8 | 0.3709 | 1.8431 | 185.9 | 0.3731 |
| 310 | 1.4894 | 188.0 | 0.3699 | 1.5977 | 188.1 | 0.3719 | 1.7228 | 188.2 | 0.3740 | — | — | — |

*(Continued)*

**TABLE A.19 (CONTINUED)**

Suva® 410A Superheated Vapor–Constant Pressure Tables (Saturated Vapor Properties in Parentheses)

| Temp. °F | 80.00 (12.10°F) | | | 85.00 (15.26°F) | | | 90.00 (18.29°F) | | | 95.00 (21.19°F) | | | Temp. °F |
|---|---|---|---|---|---|---|---|---|---|---|---|---|---|
| | V (0.7543) | H (119.5) | S (0.2599) | V (0.7105) | H (119.8) | S (0.2550) | V (0.6713) | H (120.1) | S (0.2542) | V (0.6361) | H (120.3) | S (0.2534) | |
| 20 | 0.7760 | 121.5 | 0.2599 | 0.7229 | 121.0 | 0.2574 | 0.6756 | 120.5 | 0.2551 | – | – | – | 20 |
| 30 | 0.8024 | 123.8 | 0.2648 | 0.7484 | 123.4 | 0.2624 | 0.7004 | 122.9 | 0.2601 | 0.6572 | 122.5 | 0.2579 | 30 |
| 40 | 0.8279 | 126.1 | 0.2695 | 0.7730 | 125.7 | 0.2672 | 0.7241 | 125.3 | 0.2650 | 0.6803 | 124.9 | 0.2629 | 40 |
| 50 | 0.8526 | 128.4 | 0.2740 | 0.7967 | 128.1 | 0.2718 | 0.7470 | 127.7 | 0.2697 | 0.7024 | 127.3 | 0.2676 | 50 |
| 60 | 0.8767 | 130.7 | 0.2784 | 0.8198 | 130.4 | 0.2763 | 0.7692 | 130.0 | 0.2742 | 0.7238 | 129.7 | 0.2722 | 60 |
| 70 | 0.9002 | 133.0 | 0.2827 | 0.8423 | 132.6 | 0.2806 | 0.7908 | 132.3 | 0.2786 | 0.7446 | 132.0 | 0.2767 | 70 |
| 80 | 0.9233 | 135.2 | 0.2869 | 0.8643 | 134.9 | 0.2848 | 0.8119 | 134.6 | 0.2829 | 0.7649 | 134.3 | 0.2810 | 80 |
| 90 | 0.9459 | 137.4 | 0.2910 | 0.8859 | 137.2 | 0.2890 | 0.8325 | 136.9 | 0.2870 | 0.7848 | 136.6 | 0.2852 | 90 |
| 100 | 0.9681 | 139.6 | 0.2950 | 0.9071 | 139.4 | 0.2930 | 0.8528 | 139.2 | 0.2911 | 0.8042 | 138.9 | 0.2893 | 100 |
| 110 | 0.9900 | 141.9 | 0.2989 | 0.9279 | 141.6 | 0.2970 | 0.8727 | 141.4 | 0.2951 | 0.8233 | 141.2 | 0.2933 | 110 |
| 120 | 1.0117 | 144.1 | 0.3028 | 0.9485 | 143.9 | 0.3009 | 0.8923 | 143.7 | 0.2990 | 0.8421 | 143.4 | 0.2973 | 120 |
| 130 | 1.0330 | 146.3 | 0.3066 | 0.9688 | 146.1 | 0.3047 | 0.9117 | 145.9 | 0.3029 | 0.8606 | 145.7 | 0.3011 | 130 |
| 140 | 1.0542 | 148.5 | 0.3104 | 0.9889 | 148.4 | 0.3085 | 0.9308 | 148.2 | 0.3067 | 0.8789 | 148.0 | 0.3049 | 140 |
| 150 | 1.0751 | 150.8 | 0.3141 | 1.0088 | 150.6 | 0.3122 | 0.9498 | 150.4 | 0.3104 | 0.8970 | 150.2 | 0.3087 | 150 |
| 160 | 1.0959 | 153.0 | 0.3177 | 1.0285 | 152.8 | 0.3158 | 0.9685 | 152.7 | 0.3141 | 0.9149 | 152.5 | 0.3124 | 160 |
| 170 | 1.1165 | 155.3 | 0.3213 | 1.0480 | 155.1 | 0.3194 | 0.9871 | 154.9 | 0.3177 | 0.9326 | 154.8 | 0.3160 | 170 |
| 180 | 1.1370 | 157.5 | 0.3249 | 1.0674 | 157.4 | 0.3230 | 1.0055 | 157.2 | 0.3213 | 0.9502 | 157.1 | 0.3196 | 180 |
| 190 | 1.1573 | 159.8 | 0.3284 | 1.0866 | 159.6 | 0.3265 | 1.0238 | 159.5 | 0.3248 | 0.9676 | 159.3 | 0.3232 | 190 |
| 200 | 1.1775 | 162.1 | 0.3318 | 1.1058 | 161.9 | 0.3300 | 1.0420 | 161.8 | 0.3283 | 0.9849 | 161.6 | 0.3267 | 200 |
| 210 | 1.1976 | 164.4 | 0.3353 | 1.1248 | 164.2 | 0.3335 | 1.0601 | 164.1 | 0.3318 | 1.0021 | 163.9 | 0.3301 | 210 |
| 220 | 1.2176 | 166.6 | 0.3387 | 1.1437 | 166.5 | 0.3369 | 1.0780 | 166.4 | 0.3352 | 1.0192 | 166.3 | 0.3336 | 220 |
| 230 | 1.2376 | 169.0 | 0.3421 | 1.1626 | 168.8 | 0.3403 | 1.0959 | 168.7 | 0.3386 | 1.0362 | 168.6 | 0.3370 | 230 |
| 240 | 1.2574 | 171.3 | 0.3454 | 1.1813 | 171.2 | 0.3436 | 1.1137 | 171.0 | 0.3419 | 1.0532 | 170.9 | 0.3403 | 240 |
| 250 | 1.2772 | 173.6 | 0.3487 | 1.2000 | 173.5 | 0.3470 | 1.1314 | 173.4 | 0.3453 | 1.0700 | 173.3 | 0.3437 | 250 |
| 260 | 1.2969 | 176.0 | 0.3520 | 1.2186 | 175.9 | 0.3502 | 1.1491 | 175.8 | 0.3486 | 1.0868 | 175.6 | 0.3470 | 260 |
| 270 | 1.3165 | 178.3 | 0.3553 | 1.2372 | 178.2 | 0.3535 | 1.1667 | 178.1 | 0.3518 | 1.1036 | 178.0 | 0.3503 | 270 |

*(Continued)*

# TABLE A.19 (CONTINUED)

Suva® 410A Superheated Vapor–Constant Pressure Tables (Saturated Vapor Properties in Parentheses)

| Temp. °F | 100.00 (23.98°F) | | | 110.00 (29.25°F) | | | 120.00 (34.16°F) | | | 130.00 (38.78°F) | | |
|---|---|---|---|---|---|---|---|---|---|---|---|---|
| | V (0.6043) | H (120.5) | S (0.2526) | V (0.5491) | H (120.9) | S (0.2512) | V (0.5029) | H (121.2) | S (0.2499) | V (0.4635) | H (121.5) | S (0.2487) |
| 30 | 0.6183 | 122.0 | 0.2558 | 0.5508 | 121.1 | 0.2516 | – | – | – | – | – | – |
| 40 | 0.6407 | 124.5 | 0.2608 | 0.5722 | 123.7 | 0.2569 | 0.5148 | 122.8 | 0.2531 | 0.4659 | 121.8 | 0.2494 |
| 50 | 0.6622 | 126.9 | 0.2656 | 0.5926 | 126.2 | 0.2618 | 0.5344 | 125.4 | 0.2582 | 0.4848 | 124.5 | 0.2548 |
| 60 | 0.6830 | 129.3 | 0.2703 | 0.6122 | 128.6 | 0.2666 | 0.5531 | 127.9 | 0.2632 | 0.5028 | 127.2 | 0.2598 |
| 70 | 0.7031 | 131.7 | 0.2748 | 0.6311 | 131.0 | 0.2712 | 0.5710 | 130.4 | 0.2679 | 0.5200 | 129.7 | 0.2647 |
| 80 | 0.7226 | 134.0 | 0.2792 | 0.6495 | 133.4 | 0.2757 | 0.5884 | 132.8 | 0.2724 | 0.5366 | 132.2 | 0.2694 |
| 90 | 0.7417 | 136.3 | 0.2834 | 0.6673 | 135.8 | 0.2800 | 0.6053 | 135.2 | 0.2769 | 0.5526 | 134.7 | 0.2739 |
| 100 | 0.7604 | 138.6 | 0.2875 | 0.6848 | 138.1 | 0.2842 | 0.6217 | 137.6 | 0.2812 | 0.5682 | 137.1 | 0.2782 |
| 110 | 0.7788 | 140.9 | 0.2916 | 0.7019 | 140.5 | 0.2884 | 0.6377 | 140.0 | 0.2853 | 0.5833 | 139.5 | 0.2825 |
| 120 | 0.7968 | 143.2 | 0.2956 | 0.7186 | 142.8 | 0.2924 | 0.6534 | 142.3 | 0.2894 | 0.5981 | 141.9 | 0.2866 |
| 130 | 0.8146 | 145.5 | 0.2995 | 0.7351 | 145.1 | 0.2963 | 0.6688 | 144.6 | 0.2934 | 0.6126 | 144.2 | 0.2907 |
| 140 | 0.8321 | 147.8 | 0.3033 | 0.7513 | 147.4 | 0.3002 | 0.6839 | 147.0 | 0.2973 | 0.6269 | 146.6 | 0.2946 |
| 150 | 0.8494 | 150.0 | 0.3071 | 0.7673 | 149.7 | 0.3040 | 0.6988 | 149.3 | 0.3012 | 0.6409 | 148.9 | 0.2985 |
| 160 | 0.8666 | 152.3 | 0.3108 | 0.7831 | 152.0 | 0.3077 | 0.7135 | 151.6 | 0.3049 | 0.6547 | 151.3 | 0.3023 |
| 170 | 0.8835 | 154.6 | 0.3144 | 0.7988 | 154.3 | 0.3114 | 0.7281 | 153.9 | 0.3087 | 0.6683 | 153.6 | 0.3061 |
| 180 | 0.9003 | 156.9 | 0.3180 | 0.8142 | 156.6 | 0.3151 | 0.7425 | 156.3 | 0.3123 | 0.6817 | 155.9 | 0.3098 |
| 190 | 0.9170 | 159.2 | 0.3216 | 0.8296 | 158.9 | 0.3186 | 0.7567 | 158.6 | 0.3159 | 0.6950 | 158.3 | 0.3134 |
| 200 | 0.9335 | 161.5 | 0.3251 | 0.8448 | 161.2 | 0.3222 | 0.7708 | 160.9 | 0.3195 | 0.7082 | 160.6 | 0.3170 |
| 210 | 0.9500 | 163.8 | 0.3286 | 0.8599 | 163.5 | 0.3257 | 0.7848 | 163.3 | 0.3230 | 0.7213 | 163.0 | 0.3205 |
| 280 | 1.3361 | 180.7 | 0.3585 | 1.2557 | 180.6 | 0.3568 | 1.1842 | 180.5 | 0.3551 | 1.1202 | 180.4 | 0.3535 |
| 290 | 1.3556 | 183.1 | 0.3617 | 1.2741 | 183.0 | 0.3600 | 1.2017 | 182.9 | 0.3583 | 1.1369 | 182.8 | 0.3567 |
| 300 | 1.3751 | 185.5 | 0.3649 | 1.2925 | 185.4 | 0.3632 | 1.2191 | 185.3 | 0.3615 | 1.1534 | 185.2 | 0.3599 |
| 310 | 1.3945 | 187.9 | 0.3681 | 1.3109 | 187.8 | 0.3663 | 1.2365 | 187.7 | 0.3647 | 1.1699 | 187.7 | 0.3631 |
| 320 | 1.4139 | 190.4 | 0.3712 | 1.3292 | 190.3 | 0.3695 | 1.2538 | 190.2 | 0.3678 | 1.1864 | 190.1 | 0.3663 |
| 330 | 1.2029 | 192.6 | 0.3694 | – | – | – | – | – | – | – | – | – |

*(Continued)*

# TABLE A.19 (CONTINUED)

Suva® 410A Superheated Vapor–Constant Pressure Tables (Saturated Vapor Properties in Parentheses)

| Temp. °F | 140.00 (43.14°F) | | | 150.00 (47.26°F) | | | 160.00 (51.18°F) | | | 170.00 (54.93°F) | | | Temp. °F |
|---|---|---|---|---|---|---|---|---|---|---|---|---|---|
| | V | H | S | V | H | S | V | H | S | V | H | S | |
| 220 | 0.9663 | 166.1 | 0.3320 | 0.8749 | 165.9 | 0.3291 | 0.7987 | 165.6 | 0.3265 | 0.7343 | 165.3 | 0.3240 | 220 |
| 230 | 0.9826 | 168.5 | 0.3354 | 0.8898 | 168.2 | 0.3326 | 0.8125 | 168.0 | 0.3299 | 0.7471 | 167.7 | 0.3275 | 230 |
| 240 | 0.9987 | 170.8 | 0.3388 | 0.9046 | 170.6 | 0.3360 | 0.8262 | 170.3 | 0.3333 | 0.7599 | 170.1 | 0.3309 | 240 |
| 250 | 1.0148 | 173.2 | 0.3422 | 0.9194 | 172.9 | 0.3393 | 0.8399 | 172.7 | 0.3367 | 0.7726 | 172.5 | 0.3343 | 250 |
| 260 | 1.0308 | 175.5 | 0.3455 | 0.9341 | 175.3 | 0.3426 | 0.8534 | 175.1 | 0.3400 | 0.7852 | 174.9 | 0.3376 | 260 |
| 270 | 1.0468 | 177.9 | 0.3488 | 0.9487 | 177.7 | 0.3459 | 0.8669 | 177.5 | 0.3433 | 0.7978 | 177.3 | 0.3409 | 270 |
| 280 | 1.0627 | 180.3 | 0.3520 | 0.9632 | 180.1 | 0.3492 | 0.8804 | 179.9 | 0.3466 | 0.8103 | 179.7 | 0.3442 | 280 |
| 290 | 1.0785 | 182.7 | 0.3552 | 0.9777 | 182.5 | 0.3524 | 0.8938 | 182.3 | 0.3499 | 0.8227 | 182.1 | 0.3475 | 290 |
| 300 | 1.0943 | 185.1 | 0.3584 | 0.9922 | 184.9 | 0.3557 | 0.9071 | 184.7 | 0.3531 | 0.8351 | 184.5 | 0.3507 | 300 |
| 310 | 1.1101 | 187.6 | 0.3616 | 1.0066 | (187.40) | 0.3588 | 0.9204 | 187.2 | 0.3563 | 0.8475 | 187.0 | 0.3539 | 310 |
| 320 | 1.1258 | 190.0 | 0.3648 | 1.0210 | 189.8 | 0.3620 | 0.9337 | 189.6 | 0.3595 | 0.8598 | 189.5 | 0.3571 | 320 |
| 330 | 1.1414 | 192.5 | 0.3679 | 1.0353 | 192.3 | 0.3652 | 0.9469 | 192.1 | 0.3626 | 0.8720 | 191.9 | 0.3603 | 330 |
| 340 | – | – | – | – | – | – | 0.9600 | 194.6 | 0.3657 | 0.8843 | 194.4 | 0.3634 | 340 |
| | (0.4295) | (121.8) | (0.2476) | (0.4000) | (122.0) | (0.2466) | (0.3740) | (122.2) | (0.2456) | (0.3509) | (122.4) | (0.2446) | |
| 50 | 0.4421 | 123.7 | 0.2514 | 0.4048 | 122.8 | 0.2481 | – | – | – | – | – | – | 50 |
| 60 | 0.4596 | 126.4 | 0.2566 | 0.4219 | 125.6 | 0.2535 | 0.3887 | 124.8 | 0.2505 | 0.3592 | 123.9 | 0.2475 | 60 |
| 70 | 0.4762 | 129.0 | 0.2616 | 0.4380 | 128.3 | 0.2587 | 0.4044 | 127.5 | 0.2558 | 0.3747 | 126.8 | 0.2530 | 70 |
| 80 | 0.4921 | 131.6 | 0.2664 | 0.4534 | 130.9 | 0.2636 | 0.4194 | 130.2 | 0.2608 | 0.3893 | 129.5 | 0.2582 | 80 |
| 90 | 0.5074 | 134.1 | 0.2710 | 0.4681 | 133.5 | 0.2683 | 0.4336 | 132.9 | 0.2657 | 0.4031 | 132.2 | 0.2631 | 90 |
| 100 | 0.5222 | 136.5 | 0.2755 | 0.4823 | 136.0 | 0.2728 | 0.4474 | 135.4 | 0.2703 | 0.4164 | 134.8 | 0.2678 | 100 |
| 110 | 0.5366 | 139.0 | 0.2798 | 0.4961 | 138.5 | 0.2772 | 0.4606 | 137.9 | 0.2747 | 0.4292 | 137.4 | 0.2724 | 110 |
| 120 | 0.5507 | 141.4 | 0.2840 | 0.5096 | 140.9 | 0.2815 | 0.4735 | 140.4 | 0.2791 | 0.4416 | 139.9 | 0.2768 | 120 |
| 130 | 0.5644 | 143.8 | 0.2881 | 0.5226 | 143.3 | 0.2856 | 0.4860 | 142.9 | 0.2833 | 0.4537 | 142.4 | 0.2810 | 130 |
| 140 | 0.5779 | 146.2 | 0.2921 | 0.5355 | 145.7 | 0.2897 | 0.4983 | 145.3 | 0.2874 | 0.4654 | 144.9 | 0.2852 | 140 |
| 150 | 0.5911 | 148.5 | 0.2960 | 0.5480 | 148.1 | 0.2936 | 0.5103 | 147.8 | 0.2914 | 0.4769 | 147.4 | 0.2892 | 150 |
| 160 | 0.6042 | 150.9 | 0.2999 | 0.5604 | 150.5 | 0.2975 | 0.5220 | 150.2 | 0.2953 | 0.4882 | 149.8 | 0.2932 | 160 |

*(Continued)*

**TABLE A.19 (CONTINUED)**

Suva® 410A Superheated Vapor–Constant Pressure Tables (Saturated Vapor Properties in Parentheses)

| Temp. °F | V | H | S | V | H | S | V | H | S | V | H | S | Temp. °F |
|---|---|---|---|---|---|---|---|---|---|---|---|---|---|
| 170 | 0.6170 | 153.3 | 0.3036 | 0.5725 | 152.9 | 0.3013 | 0.5336 | 152.6 | 0.2992 | 0.4992 | 152.2 | 0.2971 | 170 |
| 180 | 0.6297 | 155.6 | 0.3074 | 0.5845 | 155.3 | 0.3051 | 0.5450 | 155.0 | 0.3029 | 0.5101 | 154.6 | 0.3009 | 180 |
| 190 | 0.6422 | 158.0 | 0.3110 | 0.5963 | 157.7 | 0.3088 | 0.5562 | 157.4 | 0.3067 | 0.5208 | 157.0 | 0.3046 | 190 |
| 200 | 0.6546 | 160.3 | 0.3146 | 0.6080 | 160.0 | 0.3124 | 0.5673 | 159.7 | 0.3103 | 0.5314 | 159.5 | 0.3083 | 200 |
| 210 | 0.6668 | 162.7 | 0.3182 | 0.6196 | 162.4 | 0.3160 | 0.5783 | 162.1 | 0.3139 | 0.5418 | 161.9 | 0.3119 | 210 |
| 220 | 0.6790 | 165.1 | 0.3217 | 0.6311 | 164.8 | 0.3195 | 0.5892 | 164.5 | 0.3175 | 0.5522 | 164.3 | 0.3155 | 220 |
| 230 | 0.6910 | 167.5 | 0.3252 | 0.6424 | 167.2 | 0.3230 | 0.5999 | 166.9 | 0.3210 | 0.5624 | 166.7 | 0.3190 | 230 |
| 240 | 0.7030 | 169.8 | 0.3286 | 0.6537 | 169.6 | 0.3265 | 0.6106 | 169.4 | 0.3244 | 0.5725 | 169.1 | 0.3225 | 240 |
| 250 | 0.7149 | 172.2 | 0.3320 | 0.6649 | 172.0 | 0.3299 | 0.6212 | 171.8 | 0.3279 | 0.5826 | 171.5 | 0.3260 | 250 |
| 260 | 0.7267 | 174.6 | 0.3354 | 0.6761 | 174.4 | 0.3333 | 0.6317 | 174.2 | 0.3313 | 0.5926 | 174.0 | 0.3294 | 260 |
| 270 | 0.7385 | 177.1 | 0.3387 | 0.6871 | 176.8 | 0.3366 | 0.6421 | 176.6 | 0.3346 | 0.6025 | 176.4 | 0.3327 | 270 |
| 280 | 0.7502 | 179.5 | 0.3420 | 0.6981 | 179.3 | 0.3399 | 0.6525 | 179.1 | 0.3379 | 0.6123 | 178.8 | 0.3361 | 280 |
| 290 | 0.7618 | 181.9 | 0.3453 | 0.7090 | 181.7 | 0.3432 | 0.6629 | 181.5 | 0.3412 | 0.6221 | 181.3 | 0.3394 | 290 |
| 300 | 0.7734 | 184.4 | 0.3485 | 0.7199 | 184.2 | 0.3464 | 0.6731 | 184.0 | 0.3445 | 0.6318 | 183.8 | 0.3426 | 300 |
| 310 | 0.7850 | 186.8 | 0.3517 | 0.7308 | 186.6 | 0.3497 | 0.6834 | 186.4 | 0.3477 | 0.6415 | 186.2 | 0.3459 | 310 |
| 320 | 0.7964 | 189.3 | 0.3549 | 0.7416 | 189.1 | 0.3528 | 0.6935 | 188.9 | 0.3509 | 0.6512 | 188.7 | 0.3491 | 320 |
| 330 | 0.8079 | 191.8 | 0.3581 | 0.7523 | 191.6 | 0.3560 | 0.7037 | 191.4 | 0.3541 | 0.6608 | 191.2 | 0.3523 | 330 |
| 340 | 0.8193 | 194.3 | 0.3612 | 0.7630 | 194.1 | 0.3592 | 0.7138 | 193.9 | 0.3572 | 0.6703 | 193.7 | 0.3554 | 340 |
| 350 | 0.8307 | 196.8 | 0.3643 | 0.7737 | 196.6 | 0.3623 | 0.7239 | 196.4 | 0.3604 | 0.6799 | 196.3 | 0.3586 | 350 |
| 360 | – | – | – | – | – | – | 0.7339 | 199.0 | 0.3635 | 0.6894 | 198.8 | 0.3617 | 360 |

| | 180.00 | | | 190.00 | | | 200.00 | | | 220.00 | | | |
| | (58.50°F) | | | (61.94°F) | | | (65.23°F) | | | (71.48°F) | | | |
| | V | H | S | V | H | S | V | H | S | V | H | S | |
| Temp. °F | (0.3303) | (122.6) | (0.2437) | (0.3119) | (122.7) | (0.2428) | (0.2951) | (122.8) | (0.2420) | (0.2661) | (123.0) | (0.2404) | Temp. °F |
| 60 | 0.3327 | 123.0 | 0.2446 | – | – | – | 0.3022 | 124.3 | 0.2448 | – | – | – | 60 |
| 70 | 0.3480 | 126.0 | 0.2503 | 0.3240 | 125.2 | 0.2475 | 0.3162 | 127.3 | 0.2505 | – | – | – | 70 |
| 80 | 0.3624 | 128.8 | 0.2556 | 0.3382 | 128.1 | 0.2530 | 0.3293 | 130.2 | 0.2558 | 0.2780 | 125.8 | 0.2455 | 80 |
| 90 | 0.3759 | 131.6 | 0.2606 | 0.3515 | 130.9 | 0.2582 | 0.3417 | 133.0 | 0.2609 | 0.2908 | 128.8 | 0.2512 | 90 |
| 100 | 0.3888 | 134.3 | 0.2655 | 0.3641 | 133.7 | 0.2631 | 0.3536 | 135.8 | 0.2657 | 0.3029 | 131.8 | 0.2565 | 100 |
| 110 | 0.4013 | 136.9 | 0.2701 | 0.3762 | 136.3 | 0.2679 | | | | 0.3143 | 134.6 | 0.2615 | 110 |

*(Continued)*

**TABLE A.19 (CONTINUED)**

Suva® 410A Superheated Vapor–Constant Pressure Tables (Saturated Vapor Properties in Parentheses)

| | | | | | | | | | | | | | |
|---|---|---|---|---|---|---|---|---|---|---|---|---|---|
| 120 | 0.4133 | 139.4 | 0.2746 | 0.3878 | 138.9 | 0.2724 | 0.3649 | 138.4 | 0.2703 | 0.3251 | 137.4 | 0.2663 | 120 |
| 130 | 0.4249 | 142.0 | 0.2789 | 0.3991 | 141.5 | 0.2768 | 0.3758 | 141.0 | 0.2748 | 0.3355 | 140.1 | 0.2709 | 130 |
| 140 | 0.4362 | 144.5 | 0.2831 | 0.4100 | 144.0 | 0.2811 | 0.3864 | 143.6 | 0.2791 | 0.3456 | 142.7 | 0.2753 | 140 |
| 150 | 0.4472 | 147.0 | 0.2872 | 0.4207 | 146.5 | 0.2852 | 0.3967 | 146.1 | 0.2833 | 0.3553 | 145.3 | 0.2796 | 150 |
| 160 | 0.4581 | 149.4 | 0.2912 | 0.4311 | 149.0 | 0.2892 | 0.4068 | 148.6 | 0.2874 | 0.3648 | 147.9 | 0.2838 | 160 |
| 170 | 0.4686 | 151.9 | 0.2951 | 0.4413 | 151.5 | 0.2932 | 0.4166 | 151.1 | 0.2914 | 0.3740 | 150.4 | 0.2879 | 170 |
| 180 | 0.4791 | 154.3 | 0.2989 | 0.4513 | 154.0 | 0.2971 | 0.4263 | 153.6 | 0.2953 | 0.3830 | 152.9 | 0.2919 | 180 |
| 190 | 0.4893 | 156.7 | 0.3027 | 0.4611 | 156.4 | 0.3009 | 0.4357 | 156.1 | 0.2991 | 0.3919 | 155.4 | 0.2957 | 190 |
| 200 | 0.4994 | 159.1 | 0.3064 | 0.4708 | 158.8 | 0.3046 | 0.4451 | 158.5 | 0.3029 | 0.4006 | 157.9 | 0.2996 | 200 |
| 210 | 0.5094 | 161.6 | 0.3101 | 0.4804 | 161.3 | 0.3083 | 0.4543 | 161.0 | 0.3065 | 0.4091 | 160.4 | 0.3033 | 210 |
| 220 | 0.5193 | 164.0 | 0.3137 | 0.4898 | 163.7 | 0.3119 | 0.4633 | 163.4 | 0.3102 | 0.4175 | 162.9 | 0.3070 | 220 |
| 230 | 0.5290 | 166.4 | 0.3172 | 0.4992 | 166.2 | 0.3154 | 0.4723 | 165.9 | 0.3138 | 0.4258 | 165.4 | 0.3106 | 230 |
| 240 | 0.5387 | 168.9 | 0.3207 | 0.5084 | 168.6 | 0.3190 | 0.4811 | 168.4 | 0.3173 | 0.4341 | 167.8 | 0.3142 | 240 |
| 250 | 0.5483 | 171.3 | 0.3242 | 0.5176 | 171.1 | 0.3224 | 0.4899 | 170.8 | 0.3208 | 0.4422 | 170.3 | 0.3177 | 250 |
| 260 | 0.5578 | 173.7 | 0.3276 | 0.5266 | 173.5 | 0.3259 | 0.4986 | 173.3 | 0.3242 | 0.4502 | 172.8 | 0.3211 | 260 |
| 270 | 0.5672 | 176.2 | 0.3309 | 0.5356 | 176.0 | 0.3293 | 0.5072 | 175.7 | 0.3276 | 0.4582 | 175.3 | 0.3246 | 270 |
| 280 | 0.5766 | 178.6 | 0.3343 | 0.5446 | 178.4 | 0.3326 | 0.5158 | 178.2 | 0.3310 | 0.4661 | 177.8 | 0.3280 | 280 |
| 290 | 0.5859 | 181.1 | 0.3376 | 0.5535 | 180.9 | 0.3359 | 0.5243 | 180.7 | 0.3343 | 0.4739 | 180.3 | 0.3313 | 290 |
| 300 | 0.5951 | 183.6 | 0.3409 | 0.5623 | 183.4 | 0.3392 | 0.5328 | 183.2 | 0.3376 | 0.4817 | 182.8 | 0.3346 | 300 |
| 310 | 0.6044 | 186.1 | 0.3441 | 0.5711 | 185.9 | 0.3425 | 0.5412 | 185.7 | 0.3409 | 0.4894 | 185.3 | 0.3379 | 310 |
| 320 | 0.6135 | 188.6 | 0.3473 | 0.5798 | 188.4 | 0.3457 | 0.5495 | 188.2 | 0.3441 | 0.4971 | 187.8 | 0.3412 | 320 |
| 330 | 0.6226 | 191.1 | 0.3505 | 0.5885 | 190.9 | 0.3489 | 0.5578 | 190.7 | 0.3473 | 0.5048 | 190.3 | 0.3444 | 330 |
| 340 | 0.6317 | 193.6 | 0.3537 | 0.5972 | 193.4 | 0.3521 | 0.5661 | 193.2 | 0.3505 | 0.5124 | 192.9 | 0.3476 | 340 |
| 350 | 0.6408 | 196.1 | 0.3569 | 0.6058 | 195.9 | 0.3552 | 0.5743 | 195.8 | 0.3537 | 0.5199 | 195.4 | 0.3508 | 350 |
| 360 | 0.6498 | 198.6 | 0.3600 | 0.6144 | 198.5 | 0.3583 | 0.5825 | 198.3 | 0.3568 | 0.5275 | 198.0 | 0.3539 | 360 |
| 370 | 0.6229 | 201.0 | 0.3614 | 0.5907 | 200.9 | 0.3599 | 0.5350 | 200.6 | 0.3570 | – | – | – | 370 |
| 380 | – | – | – | – | – | – | – | – | – | 0.5424 | 203.1 | 0.3601 | 380 |

*(Continued)*

**TABLE A.19 (CONTINUED)**

Suva® 410A Superheated Vapor–Constant Pressure Tables (Saturated Vapor Properties in Parentheses)

| Temp. °F | 240.00 (77.30°F) | | | 260.00 (82.77°F) | | | 280.00 (87.94°F) | | | 300.00 (92.83°F) | | | Temp. °F |
|---|---|---|---|---|---|---|---|---|---|---|---|---|---|
| | V | H | S | V | H | S | V | H | S | V | H | S | |
| | (0.2418) | (123.1) | (0.2389) | (0.2210) | (123.2) | (0.2374) | (0.2031) | (123.2) | (0.2360) | (0.1875) | (123.2) | (0.2347) | |
| 80 | 0.2455 | 124.1 | 0.2406 | – | – | – | – | – | – | – | – | – | 80 |
| 90 | 0.2583 | 127.3 | 0.2466 | 0.2303 | 125.7 | 0.2420 | 0.2058 | 124.0 | 0.2374 | – | – | – | 90 |
| 100 | 0.2702 | 130.4 | 0.2522 | 0.2422 | 129.0 | 0.2480 | 0.2178 | 127.5 | 0.2438 | 0.1962 | 125.9 | 0.2395 | 100 |
| 110 | 0.2813 | 133.4 | 0.2575 | 0.2531 | 132.1 | 0.2535 | 0.2287 | 130.8 | 0.2496 | 0.2073 | 129.4 | 0.2457 | 110 |
| 120 | 0.2918 | 136.3 | 0.2624 | 0.2634 | 135.1 | 0.2587 | 0.2389 | 133.9 | 0.2550 | 0.2174 | 132.7 | 0.2514 | 120 |
| 130 | 0.3018 | 139.1 | 0.2672 | 0.2731 | 138.0 | 0.2636 | 0.2484 | 136.9 | 0.2602 | 0.2268 | 135.8 | 0.2568 | 130 |
| 140 | 0.3114 | 141.8 | 0.2718 | 0.2824 | 140.8 | 0.2683 | 0.2574 | 139.8 | 0.2650 | 0.2356 | 138.8 | 0.2618 | 140 |
| 150 | 0.3207 | 144.4 | 0.2762 | 0.2913 | 143.6 | 0.2729 | 0.2660 | 142.6 | 0.2697 | 0.2440 | 141.7 | 0.2666 | 150 |
| 160 | 0.3297 | 147.1 | 0.2804 | 0.2999 | 146.2 | 0.2772 | 0.2743 | 145.4 | 0.2742 | 0.2521 | 144.5 | 0.2712 | 160 |
| 170 | 0.3384 | 149.7 | 0.2846 | 0.3083 | 148.9 | 0.2815 | 0.2823 | 148.1 | 0.2785 | 0.2598 | 147.3 | 0.2756 | 170 |
| 180 | 0.3469 | 152.2 | 0.2886 | 0.3164 | 151.5 | 0.2856 | 0.2901 | 150.8 | 0.2827 | 0.2673 | 150.0 | 0.2799 | 180 |
| 190 | 0.3553 | 154.8 | 0.2926 | 0.3243 | 154.1 | 0.2896 | 0.2976 | 153.4 | 0.2868 | 0.2745 | 152.7 | 0.2841 | 190 |
| 200 | 0.3634 | 157.3 | 0.2965 | 0.3320 | 156.7 | 0.2936 | 0.3050 | 156.0 | 0.2908 | 0.2816 | 155.4 | 0.2881 | 200 |
| 210 | 0.3715 | 159.8 | 0.3003 | 0.3396 | 159.2 | 0.2974 | 0.3122 | 158.6 | 0.2947 | 0.2885 | 158.0 | 0.2921 | 210 |
| 220 | 0.3794 | 162.3 | 0.3040 | 0.3470 | 161.8 | 0.3012 | 0.3193 | 161.2 | 0.2985 | 0.2953 | 160.6 | 0.2960 | 220 |
| 230 | 0.3871 | 164.8 | 0.3076 | 0.3544 | 164.3 | 0.3049 | 0.3262 | 163.7 | 0.3022 | 0.3019 | 163.2 | 0.2998 | 230 |
| 240 | 0.3948 | 167.3 | 0.3112 | 0.3616 | 166.8 | 0.3085 | 0.3331 | 166.3 | 0.3059 | 0.3084 | 165.8 | 0.3035 | 240 |
| 250 | 0.4024 | 169.8 | 0.3148 | 0.3687 | 169.3 | 0.3121 | 0.3398 | 168.8 | 0.3095 | 0.3148 | 168.3 | 0.3071 | 250 |
| 260 | 0.4099 | 172.3 | 0.3183 | 0.3757 | 171.9 | 0.3156 | 0.3465 | 171.4 | 0.3131 | 0.3211 | 170.9 | 0.3107 | 260 |
| 270 | 0.4173 | 174.8 | 0.3217 | 0.3827 | 174.4 | 0.3191 | 0.3530 | 173.9 | 0.3166 | 0.3273 | 173.5 | 0.3143 | 270 |
| 280 | 0.4247 | 177.4 | 0.3252 | 0.3896 | 176.9 | 0.3225 | 0.3595 | 176.5 | 0.3201 | 0.3335 | 176.0 | 0.3177 | 280 |
| 290 | 0.4319 | 179.9 | 0.3285 | 0.3964 | 179.4 | 0.3259 | 0.3660 | 179.0 | 0.3235 | 0.3396 | 178.6 | 0.3212 | 290 |
| 300 | 0.4392 | 182.4 | 0.3319 | 0.4032 | 182.0 | 0.3293 | 0.3723 | 181.6 | 0.3269 | 0.3456 | 181.2 | 0.3246 | 300 |
| 310 | 0.4464 | 184.9 | 0.3352 | 0.4099 | 184.5 | 0.3326 | 0.3787 | 184.1 | 0.3302 | 0.3516 | 183.7 | 0.3279 | 310 |
| 320 | 0.4535 | 187.4 | 0.3384 | 0.4166 | 187.1 | 0.3359 | 0.3849 | 186.7 | 0.3335 | 0.3575 | 186.3 | 0.3313 | 320 |
| 330 | 0.4606 | 190.0 | 0.3417 | 0.4232 | 189.6 | 0.3391 | 0.3911 | 189.3 | 0.3368 | 0.3634 | 188.9 | 0.3346 | 330 |

*(Continued)*

# TABLE A.19 (CONTINUED)

Suva® 410A Superheated Vapor–Constant Pressure Tables (Saturated Vapor Properties in Parentheses)

| Temp. °F | 320.00 (97.48°F) | | | 340.00 (101.92°F) | | | 360.00 (106.16°F) | | | 380.00 (110.23°F) | | |
|---|---|---|---|---|---|---|---|---|---|---|---|---|
| | V (0.1738) | H (123.1) | S (0.2333) | V (0.1616) | H (123.0) | S (0.2320) | V (0.1507) | H (122.9) | S (0.2308) | V (0.1409) | H (122.7) | S (0.2295) |
| 100 | 0.1769 | 124.1 | 0.2351 | — | — | — | — | — | — | — | — | — |
| 110 | 0.1882 | 127.9 | 0.2418 | 0.1710 | 126.3 | 0.2378 | 0.1552 | 124.5 | 0.2337 | — | — | — |
| 120 | 0.1984 | 131.4 | 0.2478 | 0.1813 | 130.0 | 0.2442 | 0.1659 | 128.5 | 0.2406 | 0.1517 | 126.9 | 0.2368 |
| 130 | 0.2077 | 134.6 | 0.2534 | 0.1907 | 133.4 | 0.2501 | 0.1754 | 132.1 | 0.2468 | 0.1614 | 130.8 | 0.2434 |
| 140 | 0.2164 | 137.7 | 0.2587 | 0.1994 | 136.6 | 0.2555 | 0.1841 | 135.5 | 0.2524 | 0.1702 | 134.3 | 0.2494 |
| 150 | 0.2247 | 140.7 | 0.2636 | 0.2075 | 139.7 | 0.2607 | 0.1921 | 138.7 | 0.2578 | 0.1783 | 137.6 | 0.2549 |
| 160 | 0.2325 | 143.7 | 0.2683 | 0.2152 | 142.7 | 0.2655 | 0.1997 | 141.8 | 0.2628 | 0.1858 | 140.8 | 0.2601 |
| 170 | 0.2400 | 146.5 | 0.2729 | 0.2225 | 145.6 | 0.2702 | 0.2069 | 144.8 | 0.2676 | 0.1929 | 143.9 | 0.2650 |
| 180 | 0.2473 | 149.3 | 0.2773 | 0.2296 | 148.5 | 0.2747 | 0.2138 | 147.7 | 0.2721 | 0.1996 | 146.9 | 0.2697 |
| 190 | 0.2543 | 152.0 | 0.2815 | 0.2364 | 151.3 | 0.2790 | 0.2204 | 150.5 | 0.2765 | 0.2061 | 149.8 | 0.2742 |
| 200 | 0.2611 | 154.7 | 0.2856 | 0.2430 | 154.0 | 0.2832 | 0.2268 | 153.3 | 0.2808 | 0.2123 | 152.6 | 0.2785 |
| 210 | 0.2677 | 157.4 | 0.2896 | 0.2494 | 156.7 | 0.2872 | 0.2330 | 156.1 | 0.2849 | 0.2184 | 155.4 | 0.2827 |
| 220 | 0.2742 | 160.0 | 0.2935 | 0.2556 | 159.4 | 0.2912 | 0.2390 | 158.8 | 0.2890 | 0.2242 | 158.2 | 0.2868 |
| 230 | 0.2805 | 162.6 | 0.2974 | 0.2617 | 162.1 | 0.2951 | 0.2449 | 161.5 | 0.2929 | 0.2299 | 160.9 | 0.2908 |
| 240 | 0.2868 | 165.2 | 0.3011 | 0.2677 | 164.7 | 0.2989 | 0.2507 | 164.2 | 0.2967 | 0.2355 | 163.6 | 0.2947 |
| 250 | 0.2929 | 167.8 | 0.3048 | 0.2735 | 167.3 | 0.3026 | 0.2563 | 166.8 | 0.3005 | 0.2409 | 166.3 | 0.2985 |
| 260 | 0.2989 | 170.4 | 0.3084 | 0.2793 | 169.9 | 0.3063 | 0.2619 | 169.4 | 0.3042 | 0.2463 | 168.9 | 0.3022 |
| 270 | 0.3048 | 173.0 | 0.3120 | 0.2850 | 172.5 | 0.3099 | 0.2673 | 172.1 | 0.3078 | 0.2515 | 171.6 | 0.3059 |
| 280 | 0.3107 | 175.6 | 0.3155 | 0.2906 | 175.2 | 0.3134 | 0.2727 | 174.7 | 0.3114 | 0.2567 | 174.2 | 0.3095 |
| 340 | 0.4676 | 192.5 | 0.3449 | 0.4298 | 192.2 | 0.3424 | 0.3973 | 191.8 | 0.3400 | 0.3692 | 191.5 | 0.3378 |
| 350 | 0.4746 | 195.1 | 0.3481 | 0.4363 | 194.8 | 0.3456 | 0.4035 | 194.4 | 0.3432 | 0.3750 | 194.1 | 0.3410 |
| 360 | 0.4816 | 197.7 | 0.3512 | 0.4428 | 197.3 | 0.3487 | 0.4096 | 197.0 | 0.3464 | 0.3808 | 196.7 | 0.3442 |
| 370 | 0.4885 | 200.2 | 0.3543 | 0.4493 | 199.9 | 0.3519 | 0.4156 | 199.6 | 0.3496 | 0.3865 | 199.3 | 0.3474 |
| 380 | 0.4955 | 202.8 | 0.3575 | 0.4557 | 202.5 | 0.3550 | 0.4217 | 202.2 | 0.3527 | 0.3922 | 201.9 | 0.3505 |
| 390 | — | — | — | 0.4621 | 205.1 | 0.3581 | 0.4277 | 204.8 | 0.3558 | 0.3978 | 204.5 | 0.3536 |
| 400 | — | — | — | — | — | — | — | — | — | 0.4035 | 207.2 | 0.3567 |

(Continued)

# TABLE A.19 (CONTINUED)

Suva® 410A Superheated Vapor–Constant Pressure Tables (Saturated Vapor Properties in Parentheses)

| Temp. °F | 400.00 (114.14°F) | | | 450.00 (123.31°F) | | | 500.00 (131.73°F) | | | 550.00 (139.53°F) | | | Temp. °F |
|---|---|---|---|---|---|---|---|---|---|---|---|---|---|
| | V (0.1320) | H (122.5) | S (0.2283) | V (0.1130) | H (121.8) | S (0.2251) | V (0.0974) | H (121.0) | S (0.2220) | V (0.0843) | H (119.9) | S (0.2187) | |
| 120 | 0.1386 | 125.2 | 0.2329 | – | – | – | – | – | – | – | – | – | 120 |
| 130 | 0.1487 | 129.3 | 0.2399 | 0.1205 | 125.2 | 0.2308 | – | – | – | – | – | – | 130 |
| 140 | 0.1576 | 133.0 | 0.2462 | 0.1301 | 129.6 | 0.2382 | 0.1066 | 125.4 | 0.2295 | 0.0849 | 120.2 | 0.2193 | 140 |
| 150 | 0.1656 | 136.5 | 0.2520 | 0.1385 | 133.5 | 0.2448 | 0.1158 | 130.1 | 0.2372 | 0.0958 | 126.0 | 0.2289 | 150 |
| 160 | 0.1731 | 139.8 | 0.2574 | 0.1461 | 137.2 | 0.2507 | 0.1237 | 134.2 | 0.2439 | 0.1046 | 130.8 | 0.2367 | 160 |
| 170 | 0.1802 | 143.0 | 0.2624 | 0.1530 | 140.6 | 0.2562 | 0.1309 | 138.0 | 0.2499 | 0.1121 | 135.1 | 0.2435 | 170 |
| 180 | 0.1868 | 146.0 | 0.2672 | 0.1596 | 143.9 | 0.2613 | 0.1374 | 141.5 | 0.2555 | 0.1188 | 138.9 | 0.2497 | 180 |
| 190 | 0.1932 | 149.0 | 0.2718 | 0.1657 | 147.0 | 0.2662 | 0.1435 | 144.9 | 0.2607 | 0.1250 | 142.6 | 0.2553 | 190 |
| 200 | 0.1993 | 151.9 | 0.2763 | 0.1716 | 150.1 | 0.2709 | 0.1492 | 148.1 | 0.2657 | 0.1307 | 146.0 | 0.2605 | 200 |
| 210 | 0.2051 | 154.8 | 0.2806 | 0.1772 | 153.0 | 0.2753 | 0.1546 | 151.2 | 0.2704 | 0.1360 | 149.3 | 0.2655 | 210 |
| 220 | 0.2108 | 157.5 | 0.2847 | 0.1826 | 155.9 | 0.2797 | 0.1598 | 154.3 | 0.2749 | 0.1411 | 152.5 | 0.2702 | 220 |
| 290 | 0.3165 | 178.2 | 0.3190 | 0.2961 | 177.8 | 0.3169 | 0.2780 | 177.3 | 0.3149 | 0.2618 | 176.9 | 0.3130 | 290 |
| 300 | 0.3222 | 180.8 | 0.3224 | 0.3016 | 180.3 | 0.3204 | 0.2832 | 179.9 | 0.3184 | 0.2668 | 179.5 | 0.3165 | 300 |
| 310 | 0.3279 | 183.3 | 0.3258 | 0.3070 | 182.9 | 0.3238 | 0.2884 | 182.6 | 0.3218 | 0.2718 | 182.2 | 0.3199 | 310 |
| 320 | 0.3335 | 185.9 | 0.3291 | 0.3124 | 185.6 | 0.3271 | 0.2935 | 185.2 | 0.3252 | 0.2767 | 184.8 | 0.3233 | 320 |
| 330 | 0.3391 | 188.5 | 0.3324 | 0.3177 | 188.2 | 0.3304 | 0.2986 | 187.8 | 0.3285 | 0.2816 | 187.4 | 0.3267 | 330 |
| 340 | 0.3446 | 191.1 | 0.3357 | 0.3229 | 190.8 | 0.3337 | 0.3036 | 190.4 | 0.3318 | 0.2864 | 190.1 | 0.3300 | 340 |
| 350 | 0.3501 | 193.7 | 0.3389 | 0.3281 | 193.4 | 0.3370 | 0.3086 | 193.0 | 0.3351 | 0.2912 | 192.7 | 0.3333 | 350 |
| 360 | 0.3556 | 196.3 | 0.3422 | 0.3333 | 196.0 | 0.3402 | 0.3136 | 195.7 | 0.3383 | 0.2959 | 195.3 | 0.3365 | 360 |
| 370 | 0.3610 | 199.0 | 0.3453 | 0.3385 | 198.6 | 0.3434 | 0.3185 | 198.3 | 0.3415 | 0.3006 | 198.0 | 0.3398 | 370 |
| 380 | 0.3664 | 201.6 | 0.3485 | 0.3436 | 201.3 | 0.3465 | 0.3234 | 201.0 | 0.3447 | 0.3053 | 200.6 | 0.3429 | 380 |
| 390 | 0.3717 | 204.2 | 0.3516 | 0.3487 | 203.9 | 0.3497 | 0.3282 | 203.6 | 0.3478 | 0.3099 | 203.3 | 0.3461 | 390 |
| 400 | 0.3770 | 206.9 | 0.3547 | 0.3537 | 206.6 | 0.3528 | 0.3330 | 206.3 | 0.3510 | 0.3145 | 206.0 | 0.3492 | 400 |
| 410 | – | – | – | 0.3588 | 209.2 | 0.3559 | 0.3378 | 209.0 | 0.3541 | 0.3191 | 208.7 | 0.3523 | 410 |
| 420 | – | – | – | – | – | – | – | – | – | 0.3237 | 211.4 | 0.3554 | 420 |

(Continued)

**TABLE A.19 (CONTINUED)**

Suva® 410A Superheated Vapor–Constant Pressure Tables (Saturated Vapor Properties in Parentheses)

| Temp | V | H | S | V | H | S | V | H | S | V | H | S | Temp |
|---|---|---|---|---|---|---|---|---|---|---|---|---|---|
| 230 | 0.2164 | 160.3 | 0.2887 | 0.1878 | 158.8 | 0.2838 | 0.1648 | 157.2 | 0.2792 | 0.1459 | 155.6 | 0.2748 | 230 |
| 240 | 0.2218 | 163.0 | 0.2927 | 0.1928 | 161.6 | 0.2879 | 0.1696 | 160.1 | 0.2834 | 0.1505 | 158.6 | 0.2791 | 240 |
| 250 | 0.2271 | 165.8 | 0.2965 | 0.1978 | 164.4 | 0.2919 | 0.1743 | 163.0 | 0.2875 | 0.1550 | 161.6 | 0.2833 | 250 |
| 260 | 0.2322 | 168.4 | 0.3003 | 0.2026 | 167.2 | 0.2957 | 0.1788 | 165.9 | 0.2914 | 0.1593 | 164.5 | 0.2874 | 260 |
| 270 | 0.2373 | 171.1 | 0.3040 | 0.2073 | 169.9 | 0.2995 | 0.1832 | 168.7 | 0.2953 | 0.1635 | 167.4 | 0.2914 | 270 |
| 280 | 0.2423 | 173.8 | 0.3076 | 0.2119 | 172.6 | 0.3032 | 0.1876 | 171.4 | 0.2991 | 0.1676 | 170.2 | 0.2953 | 280 |
| 290 | 0.2472 | 176.4 | 0.3112 | 0.2164 | 175.3 | 0.3068 | 0.1918 | 174.2 | 0.3028 | 0.1716 | 173.1 | 0.2991 | 290 |
| 300 | 0.2521 | 179.1 | 0.3147 | 0.2209 | 178.0 | 0.3104 | 0.1960 | 177.0 | 0.3065 | 0.1756 | 175.9 | 0.3028 | 300 |
| 310 | 0.2569 | 181.7 | 0.3182 | 0.2253 | 180.7 | 0.3139 | 0.2001 | 179.7 | 0.3100 | 0.1794 | 178.7 | 0.3064 | 310 |
| 320 | 0.2616 | 184.4 | 0.3216 | 0.2296 | 183.4 | 0.3174 | 0.2041 | 182.4 | 0.3136 | 0.1832 | 181.4 | 0.3100 | 320 |
| 330 | 0.2663 | 187.0 | 0.3249 | 0.2339 | 186.1 | 0.3208 | 0.2081 | 185.2 | 0.3170 | 0.1869 | 184.2 | 0.3135 | 330 |
| 340 | 0.2709 | 189.7 | 0.3283 | 0.2381 | 188.8 | 0.3242 | 0.2120 | 187.9 | 0.3205 | 0.1906 | 186.9 | 0.3170 | 340 |
| 350 | 0.2755 | 192.3 | 0.3316 | 0.2423 | 191.5 | 0.3275 | 0.2158 | 190.6 | 0.3238 | 0.1942 | 189.7 | 0.3204 | 350 |
| 360 | 0.2800 | 195.0 | 0.3348 | 0.2465 | 194.2 | 0.3308 | 0.2197 | 193.3 | 0.3272 | 0.1978 | 192.4 | 0.3238 | 360 |
| 370 | 0.2845 | 197.7 | 0.3381 | 0.2506 | 196.8 | 0.3341 | 0.2235 | 196.0 | 0.3305 | 0.2013 | 195.2 | 0.3271 | 370 |
| 380 | 0.2890 | 200.3 | 0.3413 | 0.2547 | 199.5 | 0.3373 | 0.2272 | 198.7 | 0.3337 | 0.2048 | 197.9 | 0.3304 | 380 |
| 390 | 0.2934 | 203.0 | 0.3444 | 0.2587 | 202.2 | 0.3405 | 0.2309 | 201.5 | 0.3370 | 0.2083 | 200.7 | 0.3337 | 390 |
| 400 | 0.2979 | 205.7 | 0.3476 | 0.2627 | 204.9 | 0.3437 | 0.2346 | 204.2 | 0.3401 | 0.2117 | 203.4 | 0.3369 | 400 |
| 410 | 0.3022 | 208.4 | 0.3507 | 0.2667 | 207.7 | 0.3468 | 0.2383 | 206.9 | 0.3433 | 0.2151 | 206.2 | 0.3401 | 410 |
| 420 | 0.3066 | 211.1 | 0.3538 | 0.2706 | 210.4 | 0.3499 | 0.2419 | 209.7 | 0.3464 | 0.2184 | 209.0 | 0.3432 | 420 |
| 430 | – | – | – | 0.2746 | 213.1 | 0.3530 | 0.2455 | 212.4 | 0.3495 | 0.2218 | 211.7 | 0.3463 | 430 |
| 440 | – | – | – | – | – | – | 0.2491 | 215.2 | 0.3526 | 0.2251 | 214.5 | 0.3494 | 440 |

*Source:* *Thermodynamic Properties of DuPont Suva 410A Refrigerant.* With permission of E. I. Du Pont de Nemours and Co.

*Note:* $V$ = volume in ft.$^3$/lb.; $H$ = enthalpy in Btu/lb; $S$ = entropy in Btu/(lb.) (R).

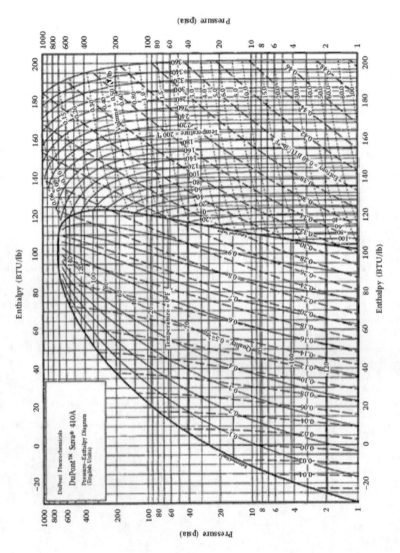

Source:    *Thermodynamic Properties of DuPont Suva 410A Refrigerant.* With permission of E. I. Du Pont de Nemours and Co.

## TABLE A.18 (SI)

Suva® 410A Saturation Properties—Temperature Table

| Temp. °C | Pressure, kPa | | Volume, m³/kg | | Density, kg/m³ | | Enthalpy, kJ/kg | | | Entropy, kJ/(kg)(K) | | Temp. °C |
|---|---|---|---|---|---|---|---|---|---|---|---|---|
| | Liquid $p_f$ | Vapor $p_g$ | Liquid $v_f$ | Vapor $v_g$ | Liquid $1/v_f$ | Vapor $1/v_g$ | Liquid $h_f$ | Latent $h_{fg}$ | Vapor $h_g$ | Liquid $s_f$ | Vapor $s_g$ | |
| -100 | 3.8 | 3.7 | 0.0007 | 5.3267 | 1509.0 | 0.188 | 63.3 | 311.4 | 374.7 | 0.3789 | 2.1774 | -100 |
| -99 | 4.1 | 4.1 | 0.0007 | 4.8882 | 1506.2 | 0.205 | 64.5 | 310.8 | 375.3 | 0.3857 | 2.1703 | -99 |
| -98 | 4.5 | 4.4 | 0.0007 | 4.4913 | 1503.4 | 0.223 | 65.7 | 310.2 | 375.8 | 0.3925 | 2.1633 | -98 |
| -97 | 4.9 | 4.9 | 0.0007 | 4.1317 | 1500.6 | 0.242 | 66.9 | 309.5 | 376.4 | 0.3993 | 2.1565 | -97 |
| -96 | 5.4 | 5.3 | 0.0007 | 3.8053 | 1497.8 | 0.263 | 68.1 | 308.9 | 377.0 | 0.4061 | 2.1498 | -96 |
| -95 | 5.9 | 5.8 | 0.0007 | 3.5088 | 1494.9 | 0.285 | 69.3 | 308.3 | 377.6 | 0.4128 | 2.1432 | -95 |
| -94 | 6.4 | 6.3 | 0.0007 | 3.2391 | 1492.1 | 0.309 | 70.5 | 307.6 | 378.2 | 0.4196 | 2.1367 | -94 |
| -93 | 6.9 | 6.8 | 0.0007 | 2.9935 | 1489.3 | 0.334 | 71.7 | 307.0 | 378.7 | 0.4263 | 2.1304 | -93 |
| -92 | 7.5 | 7.4 | 0.0007 | 2.7694 | 1486.4 | 0.361 | 73.0 | 306.3 | 379.3 | 0.4330 | 2.1241 | -92 |
| -91 | 8.2 | 8.1 | 0.0007 | 2.5649 | 1483.6 | 0.390 | 74.2 | 305.7 | 379.9 | 0.4398 | 2.1180 | -91 |
| -90 | 8.8 | 8.8 | 0.0007 | 2.3780 | 1480.7 | 0.421 | 75.4 | 305.0 | 380.5 | 0.4465 | 2.1120 | -90 |
| -89 | 9.6 | 9.5 | 0.0007 | 2.2069 | 1477.9 | 0.453 | 76.6 | 304.4 | 381.0 | 0.4532 | 2.1061 | -89 |
| -88 | 10.3 | 10.2 | 0.0007 | 2.0502 | 1475.0 | 0.488 | 77.9 | 303.7 | 381.6 | 0.4599 | 2.1003 | -88 |
| -87 | 11.2 | 11.1 | 0.0007 | 1.9065 | 1472.1 | 0.525 | 79.1 | 303.0 | 382.2 | 0.4666 | 2.0945 | -87 |
| -86 | 12.0 | 12.0 | 0.0007 | 1.7746 | 1469.2 | 0.564 | 80.4 | 302.4 | 382.7 | 0.4732 | 2.0889 | -86 |
| -85 | 13.0 | 12.9 | 0.0007 | 1.6534 | 1466.3 | 0.605 | 81.6 | 301.7 | 383.3 | 0.4799 | 2.0834 | -85 |
| -84 | 14.0 | 13.9 | 0.0007 | 1.5419 | 1463.4 | 0.649 | 82.9 | 301.0 | 383.9 | 0.4866 | 2.0780 | -84 |
| -83 | 15.0 | 14.9 | 0.0007 | 1.4391 | 1460.5 | 0.695 | 84.1 | 300.3 | 384.4 | 0.4932 | 2.0727 | -83 |
| -82 | 16.2 | 16.1 | 0.0007 | 1.3445 | 1457.6 | 0.744 | 85.4 | 299.7 | 385.0 | 0.4998 | 2.0674 | -82 |
| -81 | 17.4 | 17.3 | 0.0007 | 1.2571 | 1454.7 | 0.795 | 86.6 | 299.0 | 385.6 | 0.5064 | 2.0623 | -81 |
| -80 | 18.6 | 18.5 | 0.0007 | 1.1764 | 1451.7 | 0.850 | 87.9 | 298.3 | 386.1 | 0.5130 | 2.0572 | -80 |
| -79 | 20.0 | 19.9 | 0.0007 | 1.1019 | 1448.8 | 0.908 | 89.2 | 297.6 | 386.7 | 0.5196 | 2.0523 | -79 |
| -78 | 21.4 | 21.3 | 0.0007 | 1.0328 | 1445.8 | 0.968 | 90.4 | 296.9 | 387.3 | 0.5262 | 2.0474 | -78 |
| -77 | 22.9 | 22.8 | 0.0007 | 0.9689 | 1442.8 | 1.032 | 91.7 | 296.1 | 387.8 | 0.5328 | 2.0426 | -77 |
| -76 | 24.5 | 24.4 | 0.0007 | 0.9097 | 1439.9 | 1.099 | 93.0 | 295.4 | 388.4 | 0.5393 | 2.0378 | -76 |
| -75 | 26.2 | 26.1 | 0.0007 | 0.8547 | 1436.9 | 1.170 | 94.2 | 294.7 | 389.0 | 0.5459 | 2.0332 | -75 |
| -74 | 27.9 | 27.8 | 0.0007 | 0.8037 | 1433.9 | 1.244 | 95.5 | 294.0 | 389.5 | 0.5524 | 2.0286 | -74 |

*(Continued)*

**TABLE A.18 (SI) (CONTINUED)**

Suva® 410A Saturation Properties—Temperature Table

| Temp. °C | Pressure, kPa Liquid $p_f$ | Pressure, kPa Vapor $p_g$ | Volume, m³/kg Liquid $v_f$ | Volume, m³/kg Vapor $v_g$ | Density, kg/m³ Liquid $1/v_f$ | Density, kg/m³ Vapor $1/v_g$ | Enthalpy, kJ/kg Liquid $h_f$ | Enthalpy, kJ/kg Latent $h_{fg}$ | Enthalpy, kJ/kg Vapor $h_g$ | Entropy, kJ/(kg)(K) Liquid $s_f$ | Entropy, kJ/(kg)(K) Vapor $s_g$ | Temp. °C |
|---|---|---|---|---|---|---|---|---|---|---|---|---|
| -73 | 29.8 | 29.7 | 0.0007 | 0.7563 | 1430.9 | 1.322 | 96.8 | 293.2 | 390.1 | 0.5589 | 2.0241 | -73 |
| -72 | 31.8 | 31.7 | 0.0007 | 0.7121 | 1427.9 | 1.404 | 98.1 | 292.5 | 390.6 | 0.5654 | 2.0196 | -72 |
| -71 | 33.9 | 33.7 | 0.0007 | 0.6711 | 1424.8 | 1.490 | 99.4 | 291.8 | 391.2 | 0.5719 | 2.0153 | -71 |
| -70 | 36.0 | 35.9 | 0.0007 | 0.6328 | 1421.8 | 1.580 | 100.7 | 291.0 | 391.7 | 0.5784 | 2.0110 | -70 |
| -69 | 38.3 | 38.2 | 0.0007 | 0.5972 | 1418.8 | 1.675 | 102.0 | 290.3 | 392.3 | 0.5849 | 2.0067 | -69 |
| -68 | 40.7 | 40.6 | 0.0007 | 0.5639 | 1415.7 | 1.773 | 103.3 | 289.5 | 392.8 | 0.5913 | 2.0026 | -68 |
| -67 | 43.3 | 43.2 | 0.0007 | 0.5328 | 1412.7 | 1.877 | 104.6 | 288.8 | 393.4 | 0.5978 | 1.9985 | -67 |
| -66 | 45.9 | 45.8 | 0.0007 | 0.5038 | 1409.6 | 1.985 | 105.9 | 288.0 | 393.9 | 0.6042 | 1.9944 | -66 |
| -65 | 48.7 | 48.6 | 0.0007 | 0.4766 | 1406.5 | 2.098 | 107.2 | 287.2 | 394.4 | 0.6106 | 1.9904 | -65 |
| -64 | 51.7 | 51.5 | 0.0007 | 0.4512 | 1403.4 | 2.216 | 108.5 | 286.4 | 395.0 | 0.6170 | 1.9865 | -64 |
| -63 | 54.7 | 54.6 | 0.0007 | 0.4274 | 1400.3 | 2.340 | 109.9 | 285.7 | 395.5 | 0.6234 | 1.9827 | -63 |
| -62 | 57.9 | 57.8 | 0.0007 | 0.4051 | 1397.2 | 2.469 | 111.2 | 284.9 | 396.0 | 0.6298 | 1.9788 | -62 |
| -61 | 61.3 | 61.2 | 0.0007 | 0.3842 | 1394.1 | 2.603 | 112.5 | 284.1 | 396.6 | 0.6361 | 1.9751 | -61 |
| -60 | 64.8 | 64.7 | 0.0007 | 0.3646 | 1390.9 | 2.743 | 113.8 | 283.3 | 397.1 | 0.6425 | 1.9714 | -60 |
| -59 | 68.5 | 68.3 | 0.0007 | 0.3461 | 1387.8 | 2.889 | 115.2 | 282.5 | 397.6 | 0.6488 | 1.9678 | -59 |
| -58 | 72.3 | 72.2 | 0.0007 | 0.3288 | 1384.6 | 3.041 | 116.5 | 281.6 | 398.2 | 0.6551 | 1.9642 | -58 |
| -57 | 76.4 | 76.2 | 0.0007 | 0.3126 | 1381.4 | 3.199 | 117.9 | 280.8 | 398.7 | 0.6614 | 1.9606 | -57 |
| -56 | 80.6 | 80.4 | 0.0007 | 0.2972 | 1378.3 | 3.364 | 119.2 | 280.0 | 399.2 | 0.6677 | 1.9571 | -56 |
| -55 | 84.9 | 84.7 | 0.0007 | 0.2828 | 1375.1 | 3.536 | 120.5 | 279.2 | 399.7 | 0.6740 | 1.9537 | -55 |
| -54 | 89.5 | 89.3 | 0.0007 | 0.2692 | 1371.9 | 3.714 | 121.9 | 278.3 | 400.2 | 0.6803 | 1.9503 | -54 |
| -53 | 94.2 | 94.0 | 0.0007 | 0.2565 | 1368.7 | 3.899 | 123.2 | 277.5 | 400.7 | 0.6865 | 1.9470 | -53 |
| -52 | 99.2 | 99.0 | 0.0007 | 0.2444 | 1365.4 | 4.092 | 124.6 | 276.6 | 401.2 | 0.6928 | 1.9437 | -52 |
| -51 | 104.3 | 104.1 | 0.0007 | 0.2330 | 1362.2 | 4.292 | 126.0 | 275.8 | 401.7 | 0.6990 | 1.9404 | -51 |
| -50 | 109.7 | 109.4 | 0.0007 | 0.2222 | 1358.9 | 4.500 | 127.3 | 274.9 | 402.2 | 0.7052 | 1.9372 | -50 |
| -49 | 115.3 | 115.0 | 0.0007 | 0.2121 | 1355.7 | 4.715 | 128.7 | 274.1 | 402.7 | 0.7114 | 1.9340 | -49 |
| -48 | 121.1 | 120.8 | 0.0007 | 0.2025 | 1352.4 | 4.939 | 130.1 | 273.2 | 403.2 | 0.7176 | 1.9309 | -48 |
| -47 | 127.1 | 126.8 | 0.0007 | 0.1934 | 1349.1 | 5.171 | 131.4 | 272.3 | 403.7 | 0.7238 | 1.9278 | -47 |

(Continued)

**TABLE A.18 (SI) (CONTINUED)**

Suva® 410A Saturation Properties—Temperature Table

| Temp. °C | Pressure, kPa | | Volume, m³/kg | | Density, kg/m³ | | Enthalpy, kJ/kg | | | Entropy, kJ/(kg)(K) | | Temp. °C |
|---|---|---|---|---|---|---|---|---|---|---|---|---|
| | Liquid $p_f$ | Vapor $p_g$ | Liquid $v_f$ | Vapor $v_g$ | Liquid $1/v_f$ | Vapor $1/v_g$ | Liquid $h_f$ | Latent $h_{fg}$ | Vapor $h_g$ | Liquid $s_f$ | Vapor $s_g$ | |
| -46 | 133.4 | 133.0 | 0.0007 | 0.1848 | 1345.8 | 5.411 | 132.8 | 271.4 | 404.2 | 0.7299 | 1.9248 | -46 |
| -45 | 139.9 | 139.5 | 0.0007 | 0.1767 | 1342.5 | 5.660 | 134.2 | 270.5 | 404.7 | | 1.9217 | -45 |
| -44 | 146.6 | 146.3 | 0.0008 | 0.1690 | 1339.2 | 5.918 | 135.6 | 269.6 | 405.2 | | 1.9188 | -44 |
| -43 | 153.6 | 153.2 | 0.0008 | 0.1617 | 1335.8 | 6.186 | 137.0 | 268.7 | 405.7 | | 1.9158 | -43 |
| -42 | 160.9 | 160.5 | 0.0008 | 0.1547 | 1332.5 | 6.462 | 138.4 | 267.8 | 406.1 | | 1.9129 | -42 |
| -41 | 168.4 | 168.0 | 0.0008 | 0.1482 | 1329.1 | 6.749 | 139.7 | 266.9 | 406.6 | | 1.9101 | -41 |
| -40 | 176.2 | 175.8 | 0.0008 | 0.1419 | 1325.7 | 7.045 | 141.1 | 265.9 | 407.1 | | 1.9072 | -40 |
| -39 | 184.3 | 183.8 | 0.0008 | 0.1360 | 1322.3 | 7.352 | 142.5 | 265.0 | 407.6 | | 1.9045 | -39 |
| -38 | 192.7 | 192.2 | 0.0008 | 0.1304 | 1318.9 | 7.669 | 144.0 | 264.1 | 408.0 | | 1.9017 | -38 |
| -37 | 201.3 | 200.8 | 0.0008 | 0.1251 | 1315.5 | 7.996 | 145.4 | 263.1 | 408.5 | | 1.8990 | -37 |
| -36 | 210.3 | 209.8 | 0.0008 | 0.1200 | 1312.1 | 8.335 | 146.8 | 262.2 | 408.9 | | 1.8963 | -36 |
| -35 | 219.6 | 219.0 | 0.0008 | 0.1151 | 1308.6 | 8.685 | 148.2 | 261.2 | 409.4 | | 1.8936 | -35 |
| -34 | 229.2 | 228.6 | 0.0008 | 0.1105 | 1305.2 | 9.046 | 149.6 | 260.2 | 409.8 | | 1.8910 | -34 |
| -33 | 239.1 | 238.4 | 0.0008 | 0.1062 | 1301.7 | 9.419 | 151.0 | 259.3 | 410.3 | | 1.8884 | -33 |
| -32 | 249.3 | 248.6 | 0.0008 | 0.1020 | 1298.2 | 9.805 | 152.4 | 258.3 | 410.7 | | 1.8858 | -32 |
| -31 | 259.9 | 259.2 | 0.0008 | 0.0980 | 1294.7 | 10.202 | 153.9 | 257.3 | 411.2 | | 1.8832 | -31 |
| -30 | 270.8 | 270.1 | 0.0008 | 0.0942 | 1291.2 | 10.613 | 155.3 | 256.3 | 411.6 | | 1.8807 | -30 |
| -29 | 282.1 | 281.3 | 0.0008 | 0.0906 | 1287.6 | 11.036 | 156.7 | 255.3 | 412.0 | | 1.8782 | -29 |
| -28 | 293.7 | 292.9 | 0.0008 | 0.0872 | 1284.1 | 11.473 | 158.2 | 254.3 | 412.5 | | 1.8757 | -28 |
| -27 | 305.7 | 304.9 | 0.0008 | 0.0839 | 1280.5 | 11.923 | 159.6 | 253.3 | 412.9 | | 1.8733 | -27 |
| -26 | 318.1 | 317.2 | 0.0008 | 0.0807 | 1276.9 | 12.388 | 161.1 | 252.2 | 413.3 | | 1.8709 | -26 |
| -25 | 330.9 | 329.9 | 0.0008 | 0.0777 | 1273.3 | 12.866 | 162.5 | 251.2 | 413.7 | | 1.8685 | -25 |
| -24 | 344.0 | 343.0 | 0.0008 | 0.0749 | 1269.7 | 13.360 | 164.0 | 250.1 | 414.1 | | 1.8661 | -24 |
| -23 | 357.6 | 356.6 | 0.0008 | 0.0721 | 1266.0 | 13.868 | 165.4 | 249.1 | 414.5 | | 1.8638 | -23 |
| -22 | 371.5 | 370.5 | 0.0008 | 0.0695 | 1262.3 | 14.391 | 166.9 | 248.0 | 414.9 | | 1.8614 | -22 |
| -21 | 385.9 | 384.8 | 0.0008 | 0.0670 | 1258.7 | 14.931 | 168.4 | 247.0 | 415.3 | | 1.8591 | -21 |
| -20 | 400.7 | 399.5 | 0.0008 | 0.0646 | 1255.0 | 15.486 | 169.8 | 245.9 | 415.7 | | 1.8569 | -20 |

*(Continued)*

**TABLE A.18 (SI) (CONTINUED)**

Suva® 410A Saturation Properties—Temperature Table

| Temp. °C | Pressure, kPa Liquid $p_f$ | Pressure, kPa Vapor $p_g$ | Volume, m³/kg Liquid $v_f$ | Volume, m³/kg Vapor $v_g$ | Density, kg/m³ Liquid $1/v_f$ | Density, kg/m³ Vapor $1/v_g$ | Enthalpy, kJ/kg Liquid $h_f$ | Enthalpy, kJ/kg Latent $h_{fg}$ | Enthalpy, kJ/kg Vapor $h_g$ | Entropy, kJ/(kg)(K) Liquid $s_f$ | Entropy, kJ/(kg)(K) Vapor $s_g$ | Temp. °C |
|---|---|---|---|---|---|---|---|---|---|---|---|---|
| -19 | 415.9 | 414.7 | 0.0008 | 0.0623 | 1251.3 | 16.058 | 171.3 | 244.8 | 416.1 | | 1.8546 | -19 |
| -18 | 431.6 | 430.3 | 0.0008 | 0.0601 | 1247.5 | 16.647 | 172.8 | 243.7 | 416.5 | | 1.8523 | -18 |
| -17 | 447.7 | 446.4 | 0.0008 | 0.0580 | 1243.8 | 17.253 | 174.3 | 242.6 | 416.9 | | 1.8501 | -17 |
| -16 | 464.3 | 462.9 | 0.0008 | 0.0559 | 1240.0 | 17.877 | 175.7 | 241.5 | 417.2 | | 1.8479 | -16 |
| -15 | 481.3 | 479.9 | 0.0008 | 0.0540 | 1236.2 | 18.519 | 177.2 | 240.4 | 417.6 | | 1.8457 | -15 |
| -14 | 498.9 | 497.4 | 0.0008 | 0.0521 | 1232.4 | 19.179 | 178.7 | 239.3 | 418.0 | | 1.8436 | -14 |
| -13 | 516.9 | 515.3 | 0.0008 | 0.0504 | 1228.6 | 19.859 | 180.2 | 238.1 | 418.3 | 0.9260 | 1.8414 | -13 |
| -12 | 535.4 | 533.7 | 0.0008 | 0.0486 | 1224.7 | 20.558 | 181.7 | 237.0 | 418.7 | 0.9318 | 1.8393 | -12 |
| -11 | 554.4 | 552.7 | 0.0008 | 0.0470 | 1220.8 | 21.276 | 183.2 | 235.8 | 419.1 | 0.9375 | 1.8372 | -11 |
| -10 | 573.9 | 572.1 | 0.0008 | 0.0454 | 1216.9 | 22.016 | 184.7 | 234.7 | 419.4 | 0.9432 | 1.8351 | -10 |
| -9 | 593.9 | 592.1 | 0.0008 | 0.0439 | 1213.0 | 22.776 | 186.2 | 233.5 | 419.7 | 0.9489 | 1.8330 | -9 |
| -8 | 614.4 | 612.6 | 0.0008 | 0.0425 | 1209.1 | 23.558 | 187.7 | 232.3 | 420.1 | 0.9547 | 1.8309 | -8 |
| -7 | 635.5 | 633.6 | 0.0008 | 0.0411 | 1205.1 | 24.361 | 189.3 | 231.1 | 420.4 | 0.9604 | 1.8288 | -7 |
| -6 | 657.2 | 655.1 | 0.0008 | 0.0397 | 1201.1 | 25.187 | 190.8 | 229.9 | 420.7 | 0.9660 | 1.8268 | -6 |
| -5 | 679.3 | 677.3 | 0.0008 | 0.0384 | 1197.1 | 26.036 | 192.3 | 228.7 | 421.0 | 0.9717 | 1.8247 | -5 |
| -4 | 702.1 | 699.9 | 0.0008 | 0.0372 | 1193.1 | 26.909 | 193.8 | 227.5 | 421.4 | 0.9774 | 1.8227 | -4 |
| -3 | 725.4 | 723.2 | 0.0008 | 0.0360 | 1189.0 | 27.806 | 195.4 | 226.3 | 421.7 | 0.9830 | 1.8207 | -3 |
| -2 | 749.3 | 747.0 | 0.0008 | 0.0348 | 1184.9 | 28.728 | 196.9 | 225.1 | 422.0 | 0.9887 | 1.8187 | -2 |
| -1 | 773.9 | 771.4 | 0.0009 | 0.0337 | 1180.8 | 29.675 | 198.5 | 223.8 | 422.3 | 0.9943 | 1.8167 | -1 |
| 0 | 799.0 | 796.5 | 0.0009 | 0.0326 | 1176.7 | 30.649 | 200.0 | 222.5 | 422.5 | 1.0000 | 1.8147 | 0 |
| 1 | 824.7 | 822.1 | 0.0009 | 0.0316 | 1172.5 | 31.649 | 201.6 | 221.3 | 422.8 | 1.0056 | 1.8128 | 1 |
| 2 | 851.0 | 848.4 | 0.0009 | 0.0306 | 1168.3 | 32.676 | 203.1 | 220.0 | 423.1 | 1.0112 | 1.8108 | 2 |
| 3 | 878.0 | 875.3 | 0.0009 | 0.0297 | 1164.1 | 33.732 | 204.7 | 218.7 | 423.4 | 1.0168 | 1.8088 | 3 |
| 4 | 905.6 | 902.8 | 0.0009 | 0.0287 | 1159.8 | 34.817 | 206.2 | 217.4 | 423.6 | 1.0225 | 1.8069 | 4 |
| 5 | 933.9 | 931.0 | 0.0009 | 0.0278 | 1155.5 | 35.931 | 207.8 | 216.1 | 423.9 | 1.0281 | 1.8049 | 5 |
| 6 | 962.9 | 959.8 | 0.0009 | 0.0270 | 1151.2 | 37.076 | 209.4 | 214.8 | 424.1 | 1.0337 | 1.8030 | 6 |
| 7 | 992.5 | 989.3 | 0.0009 | 0.0261 | 1146.9 | 38.252 | 211.0 | 213.4 | 424.4 | 1.0392 | 1.8011 | 7 |

(*Continued*)

**TABLE A.18 (SI) (CONTINUED)**

Suva® 410A Saturation Properties—Temperature Table

| Temp. °C | Pressure, kPa Liquid $p_f$ | Pressure, kPa Vapor $p_g$ | Volume, m³/kg Liquid $v_f$ | Volume, m³/kg Vapor $v_g$ | Density, kg/m³ Liquid $1/v_f$ | Density, kg/m³ Vapor $1/v_g$ | Enthalpy, kJ/kg Liquid $h_f$ | Enthalpy, kJ/kg Latent $h_{fg}$ | Enthalpy, kJ/kg Vapor $h_g$ | Entropy, kJ/(kg)(K) Liquid $s_f$ | Entropy, kJ/(kg)(K) Vapor $s_g$ | Temp. °C |
|---|---|---|---|---|---|---|---|---|---|---|---|---|
| 8 | 1022.8 | 1019.5 | 0.0009 | 0.0253 | 1142.5 | 39.461 | 212.6 | 212.1 | 424.6 | 1.0448 | 1.7991 | 8 |
| 9 | 1053.8 | 1050.4 | 0.0009 | 0.0246 | 1138.1 | 40.702 | 214.1 | 210.7 | 424.9 | 1.0504 | 1.7972 | 9 |
| 10 | 1085.5 | 1082.0 | 0.0009 | 0.0238 | 1133.7 | 41.977 | 215.7 | 209.3 | 425.1 | 1.0560 | 1.7953 | 10 |
| 11 | 1117.9 | 1114.3 | 0.0009 | 0.0231 | 1129.2 | 43.288 | 217.3 | 207.9 | 425.3 | 1.0616 | 1.7934 | 11 |
| 12 | 1151.0 | 1147.3 | 0.0009 | 0.0224 | 1124.7 | 44.634 | 219.0 | 206.5 | 425.5 | 1.0671 | 1.7914 | 12 |
| 13 | 1184.9 | 1181.1 | 0.0009 | 0.0217 | 1120.1 | 46.017 | 220.6 | 205.1 | 425.7 | 1.0727 | 1.7895 | 13 |
| 14 | 1219.5 | 1215.6 | 0.0009 | 0.0211 | 1115.6 | 47.437 | 222.2 | 203.7 | 425.9 | 1.0783 | 1.7876 | 14 |
| 15 | 1254.9 | 1250.8 | 0.0009 | 0.0205 | 1110.9 | 48.897 | 223.8 | 202.2 | 426.1 | 1.0838 | 1.7857 | 15 |
| 16 | 1291.0 | 1286.9 | 0.0009 | 0.0198 | 1106.3 | 50.398 | 225.4 | 200.8 | 426.2 | 1.0894 | 1.7838 | 16 |
| 17 | 1328.0 | 1323.7 | 0.0009 | 0.0193 | 1101.6 | 51.939 | 227.1 | 199.3 | 426.4 | 1.0949 | 1.7818 | 17 |
| 18 | 1365.7 | 1361.3 | 0.0009 | 0.0187 | 1096.9 | 53.523 | 228.7 | 197.8 | 426.5 | 1.1005 | 1.7799 | 18 |
| 19 | 1404.2 | 1399.6 | 0.0009 | 0.0181 | 1092.1 | 55.152 | 230.4 | 196.3 | 426.7 | 1.1060 | 1.7780 | 19 |
| 20 | 1443.6 | 1438.8 | 0.0009 | 0.0176 | 1087.2 | 56.825 | 232.0 | 194.8 | 426.8 | 1.1116 | 1.7760 | 20 |
| 21 | 1483.7 | 1478.9 | 0.0009 | 0.0171 | 1082.4 | 58.545 | 233.7 | 193.2 | 426.9 | 1.1172 | 1.7741 | 21 |
| 22 | 1524.7 | 1519.7 | 0.0009 | 0.0166 | 1077.5 | 60.314 | 235.4 | 191.7 | 427.1 | 1.1227 | 1.7721 | 22 |
| 23 | 1566.6 | 1561.4 | 0.0009 | 0.0161 | 1072.5 | 62.132 | 237.1 | 190.1 | 427.2 | 1.1283 | 1.7702 | 23 |
| 24 | 1609.3 | 1604.0 | 0.0009 | 0.0156 | 1067.5 | 64.001 | 238.7 | 188.5 | 427.3 | 1.1338 | 1.7682 | 24 |
| 25 | 1652.9 | 1647.4 | 0.0009 | 0.0152 | 1062.4 | 65.924 | 240.4 | 186.9 | 427.3 | 1.1394 | 1.7662 | 25 |
| 26 | 1697.3 | 1691.7 | 0.0010 | 0.0147 | 1057.3 | 67.901 | 242.1 | 185.3 | 427.4 | 1.1450 | 1.7643 | 26 |
| 27 | 1742.7 | 1736.9 | 0.0010 | 0.0143 | 1052.1 | 69.935 | 243.9 | 183.6 | 427.5 | 1.1506 | 1.7623 | 27 |
| 28 | 1788.9 | 1783.0 | 0.0010 | 0.0139 | 1046.9 | 72.028 | 245.6 | 181.9 | 427.5 | 1.1562 | 1.7603 | 28 |
| 29 | 1836.1 | 1830.0 | 0.0010 | 0.0135 | 1041.6 | 74.181 | 247.3 | 180.2 | 427.5 | 1.1618 | 1.7582 | 29 |
| 30 | 1884.2 | 1877.9 | 0.0010 | 0.0131 | 1036.3 | 76.398 | 249.1 | 178.5 | 427.6 | 1.1674 | 1.7562 | 30 |
| 31 | 1933.3 | 1926.8 | 0.0010 | 0.0127 | 1030.9 | 78.679 | 250.8 | 176.8 | 427.6 | 1.1730 | 1.7541 | 31 |
| 32 | 1983.3 | 1976.6 | 0.0010 | 0.0123 | 1025.4 | 81.028 | 252.6 | 175.0 | 427.6 | 1.1786 | 1.7521 | 32 |
| 33 | 2034.3 | 2027.4 | 0.0010 | 0.0120 | 1019.9 | 83.447 | 254.3 | 173.2 | 427.5 | 1.1843 | 1.7500 | 33 |
| 34 | 2086.3 | 2079.2 | 0.0010 | 0.0116 | 1014.2 | 85.939 | 256.1 | 171.4 | 427.5 | 1.1899 | 1.7479 | 34 |

*(Continued)*

**TABLE A.18 (SI) (CONTINUED)**

Suva® 410A Saturation Properties—Temperature Table

| Temp. °C | Pressure, kPa | | Volume, m³/kg | | Density, kg/m³ | | Enthalpy, kJ/kg | | | Entropy, kJ/(kg)(K) | | Temp. °C |
|---|---|---|---|---|---|---|---|---|---|---|---|---|
| | Liquid $p_f$ | Vapor $p_g$ | Liquid $v_f$ | Vapor $v_g$ | Liquid $1/v_f$ | Vapor $1/v_g$ | Liquid $h_f$ | Latent $h_{fg}$ | Vapor $h_g$ | Liquid $s_f$ | Vapor $s_g$ | |
| 35 | 2139.2 | 2132.0 | 0.0010 | 0.0113 | 1008.6 | 88.506 | 257.9 | 169.5 | 427.5 | 1.1956 | 1.7458 | 35 |
| 36 | 2193.2 | 2185.7 | 0.0010 | 0.0110 | 1002.8 | 91.151 | 259.7 | 167.7 | 427.4 | 1.2013 | 1.7436 | 36 |
| 37 | 2248.1 | 2240.5 | 0.0010 | 0.0107 | 996.9 | 93.879 | 261.5 | 165.8 | 427.3 | 1.2070 | 1.7414 | 37 |
| 38 | 2304.2 | 2296.3 | 0.0010 | 0.0103 | 991.0 | 96.691 | 263.4 | 163.8 | 427.2 | 1.2127 | 1.7392 | 38 |
| 39 | 2361.2 | 2353.2 | 0.0010 | 0.0100 | 985.0 | 99.592 | 265.2 | 161.9 | 427.1 | 1.2185 | 1.7370 | 39 |
| 40 | 2419.3 | 2411.1 | 0.0010 | 0.0098 | 978.9 | 102.585 | 267.1 | 159.9 | 427.0 | 1.2243 | 1.7348 | 40 |
| 41 | 2478.5 | 2470.1 | 0.0010 | 0.0095 | 972.7 | 105.674 | 269.0 | 157.8 | 426.8 | 1.2301 | 1.7325 | 41 |
| 42 | 2538.8 | 2530.2 | 0.0010 | 0.0092 | 966.4 | 108.864 | 270.9 | 155.8 | 426.7 | 1.2359 | 1.7302 | 42 |
| 43 | 2600.1 | 2591.3 | 0.0010 | 0.0089 | 960.0 | 112.159 | 272.8 | 153.7 | 426.5 | 1.2418 | 1.7278 | 43 |
| 44 | 2662.6 | 2653.6 | 0.0011 | 0.0087 | 953.4 | 115.564 | 274.8 | 151.5 | 426.3 | 1.2477 | 1.7255 | 44 |
| 45 | 2726.1 | 2717.0 | 0.0011 | 0.0084 | 946.8 | 119.085 | 276.7 | 149.3 | 426.0 | 1.2537 | 1.7230 | 45 |
| 46 | 2790.9 | 2781.6 | 0.0011 | 0.0082 | 940.0 | 122.727 | 278.7 | 147.1 | 425.8 | 1.2597 | 1.7206 | 46 |
| 47 | 2856.7 | 2847.3 | 0.0011 | 0.0079 | 933.1 | 126.497 | 280.7 | 144.8 | 425.5 | 1.2658 | 1.7181 | 47 |
| 48 | 2923.8 | 2914.2 | 0.0011 | 0.0077 | 926.0 | 130.402 | 282.7 | 142.5 | 425.2 | 1.2719 | 1.7156 | 48 |
| 49 | 2991.9 | 2982.2 | 0.0011 | 0.0074 | 918.8 | 134.448 | 284.8 | 140.1 | 424.9 | 1.2781 | 1.7130 | 49 |
| 50 | 3061.3 | 3051.5 | 0.0011 | 0.0072 | 911.4 | 138.645 | 286.9 | 137.7 | 424.6 | 1.2843 | 1.7104 | 50 |
| 51 | 3131.9 | 3122.0 | 0.0011 | 0.0070 | 903.9 | 143.001 | 289.0 | 135.2 | 424.2 | 1.2906 | 1.7077 | 51 |
| 52 | 3203.7 | 3193.7 | 0.0011 | 0.0068 | 896.1 | 147.527 | 291.2 | 132.6 | 423.8 | 1.2971 | 1.7050 | 52 |
| 53 | 3276.7 | 3266.6 | 0.0011 | 0.0066 | 888.2 | 152.235 | 293.4 | 130.0 | 423.4 | 1.3036 | 1.7022 | 53 |
| 54 | 3351.0 | 3340.9 | 0.0011 | 0.0064 | 880.0 | 157.139 | 295.6 | 127.3 | 423.0 | 1.3102 | 1.6994 | 54 |
| 55 | 3426.5 | 3416.3 | 0.0012 | 0.0062 | 871.5 | 162.252 | 297.9 | 124.6 | 422.5 | 1.3169 | 1.6965 | 55 |
| 56 | 3503.3 | 3493.1 | 0.0012 | 0.0060 | 862.8 | 167.594 | 300.3 | 121.7 | 422.0 | 1.3238 | 1.6935 | 56 |
| 57 | 3581.3 | 3571.2 | 0.0012 | 0.0058 | 853.8 | 173.187 | 302.7 | 118.7 | 421.4 | 1.3308 | 1.6904 | 57 |
| 58 | 3660.7 | 3650.7 | 0.0012 | 0.0056 | 844.5 | 179.056 | 305.1 | 115.7 | 420.8 | 1.3380 | 1.6873 | 58 |
| 59 | 3741.3 | 3731.5 | 0.0012 | 0.0054 | 834.8 | 185.232 | 307.7 | 112.5 | 420.2 | 1.3453 | 1.6841 | 59 |
| 60 | 3823.3 | 3813.6 | 0.0012 | 0.0052 | 824.7 | 191.757 | 310.3 | 109.2 | 419.5 | 1.3529 | 1.6808 | 60 |
| 61 | 3906.6 | 3897.1 | 0.0012 | 0.0050 | 814.1 | 198.680 | 313.0 | 105.8 | 418.8 | 1.3608 | 1.6773 | 61 |

*(Continued)*

**TABLE A.18 (SI) (CONTINUED)**

Suva® 410A Saturation Properties—Temperature Table

| Temp. °C | Pressure, kPa | | Volume, m³/kg | | Density, kg/m³ | | Enthalpy, kJ/kg | | | Entropy, kJ/(kg)(K) | | Temp. °C |
|---|---|---|---|---|---|---|---|---|---|---|---|---|
| | Liquid $p_f$ | Vapor $p_g$ | Liquid $v_f$ | Vapor $v_g$ | Liquid $1/v_f$ | Vapor $1/v_g$ | Liquid $h_f$ | Latent $h_{fg}$ | Vapor $h_g$ | Liquid $s_f$ | Vapor $s_g$ | |
| 62 | 3991.2 | 3982.0 | 0.0013 | 0.0049 | 802.9 | 206.069 | 315.9 | 102.2 | 418.1 | 1.3689 | 1.6738 | 62 |
| 63 | 4077.2 | 4068.4 | 0.0013 | 0.0047 | 791.1 | 214.014 | 318.8 | 98.4 | 417.2 | 1.3774 | 1.6700 | 63 |
| 64 | 4164.5 | 4156.1 | 0.0013 | 0.0045 | 778.5 | 222.641 | 322.0 | 94.3 | 416.3 | 1.3863 | 1.6661 | 64 |
| 65 | 4253.2 | 4245.4 | 0.0013 | 0.0043 | 765.0 | 232.131 | 325.3 | 90.0 | 415.3 | 1.3958 | 1.6620 | 65 |
| 66 | 4343.3 | 4336.1 | 0.0013 | 0.0041 | 750.3 | 242.755 | 328.8 | 85.3 | 414.2 | 1.4059 | 1.6575 | 66 |
| 67 | 4434.7 | 4428.2 | 0.0014 | 0.0039 | 734.2 | 254.940 | 332.7 | 80.3 | 412.9 | 1.4168 | 1.6527 | 67 |
| 68 | 4527.6 | 4521.9 | 0.0014 | 0.0037 | 716.0 | 269.366 | 336.9 | 74.6 | 411.5 | 1.4289 | 1.6476 | 68 |
| 69 | 4621.8 | 4617.2 | 0.0014 | 0.0035 | 694.9 | 287.059 | 341.7 | 68.4 | 410.1 | 1.4425 | 1.6424 | 69 |
| 70 | 4717.5 | 4713.9 | 0.0015 | 0.0032 | 669.1 | 308.947 | 347.3 | 61.6 | 408.9 | 1.4586 | 1.6380 | 70 |

**TABLE A.19 (SI)**

Suva® 410A Superheated Vapor–Constant Pressure Tables (Saturated Vapor Properties in Parentheses)

Absolute Pressure, kPa

| Temp. °C | 10.0 (−88.31°C) | | | 20.0 (−78.91°C) | | | 30.0 (−72.84°C) | | | 40.0 (−68.26°C) | | | Temp. °C |
|---|---|---|---|---|---|---|---|---|---|---|---|---|---|
| | V (2.0972) | H (381.4) | S (2.1020) | V (1.0953) | H (386.8) | S (2.0518) | V (0.7491) | H (390.2) | S (2.0234) | V (0.5721) | H (392.7) | S (2.0036) | |
| −85 | 2.1363 | 383.5 | 2.1134 | – | – | – | – | – | – | – | – | – | −85 |
| −80 | 2.1951 | 386.8 | 2.1304 | – | – | – | – | – | – | – | – | – | −80 |
| −75 | 2.2538 | 390.1 | 2.1472 | 1.1188 | 389.4 | 2.0651 | – | – | – | – | – | – | −75 |
| −70 | 2.3123 | 393.4 | 2.1637 | 1.1487 | 392.7 | 2.0819 | 0.7607 | 392.1 | 2.0331 | – | – | – | −70 |
| −65 | 2.3707 | 396.7 | 2.1800 | 1.1784 | 396.2 | 2.0985 | 0.7810 | 395.6 | 2.0499 | 0.5822 | 395.0 | 2.0147 | −65 |
| −60 | 2.4290 | 400.1 | 2.1962 | 1.2081 | 399.6 | 2.1148 | 0.8011 | 399.0 | 2.0664 | 0.5976 | 398.5 | 2.0315 | −60 |
| −55 | 2.4872 | 403.6 | 2.2121 | 1.2377 | 403.1 | 2.1309 | 0.8211 | 402.6 | 2.0827 | 0.6129 | 402.0 | 2.0479 | −55 |
| −50 | 2.5453 | 407.0 | 2.2278 | 1.2672 | 406.6 | 2.1468 | 0.8411 | 406.1 | 2.0988 | 0.6280 | 405.6 | 2.0642 | −50 |
| −45 | 2.6033 | 410.5 | 2.2434 | 1.2966 | 410.1 | 2.1625 | 0.8609 | 409.7 | 2.1146 | 0.6431 | 409.2 | 2.0802 | −45 |
| −40 | 2.6613 | 414.1 | 2.2588 | 1.3259 | 413.7 | 2.1780 | 0.8807 | 413.3 | 2.1303 | 0.6581 | 412.9 | 2.0960 | −40 |
| −35 | 2.7192 | 417.7 | 2.2740 | 1.3551 | 417.3 | 2.1934 | 0.9005 | 416.9 | 2.1458 | 0.6731 | 416.6 | 2.1116 | −35 |
| −30 | 2.7770 | 421.3 | 2.2891 | 1.3844 | 421.0 | 2.2086 | 0.9201 | 420.6 | 2.1610 | 0.6880 | 420.3 | 2.1270 | −30 |
| −25 | 2.8348 | 425.0 | 2.3040 | 1.4135 | 424.7 | 2.2236 | 0.9397 | 424.3 | 2.1762 | 0.7028 | 424.0 | 2.1422 | −25 |
| −20 | 2.8926 | 428.7 | 2.3188 | 1.4426 | 428.4 | 2.2385 | 0.9593 | 428.1 | 2.1911 | 0.7176 | 427.8 | 2.1573 | −20 |
| −15 | 2.9503 | 432.4 | 2.3334 | 1.4717 | 432.2 | 2.2532 | 0.9788 | 431.9 | 2.2059 | 0.7324 | 431.6 | 2.1721 | −15 |
| −10 | 3.0079 | 436.2 | 2.3479 | 1.5007 | 436.0 | 2.2678 | 0.9983 | 435.7 | 2.2206 | 0.7471 | 435.4 | 2.1869 | −10 |
| −5 | 3.0656 | 440.0 | 2.3623 | 1.5297 | 439.8 | 2.2822 | 1.0178 | 439.5 | 2.2351 | 0.7618 | 439.3 | 2.2014 | −5 |
| 0 | 3.1232 | 443.9 | 2.3766 | 1.5587 | 443.7 | 2.2966 | 1.0372 | 443.4 | 2.2495 | 0.7764 | 443.2 | 2.2159 | 0 |
| 5 | 3.1808 | 447.8 | 2.3908 | 1.5876 | 447.6 | 2.3108 | 1.0566 | 447.4 | 2.2637 | 0.7911 | 447.1 | 2.2302 | 5 |
| 10 | 3.2383 | 451.7 | 2.4048 | 1.6166 | 451.5 | 2.3248 | 1.0760 | 451.3 | 2.2779 | 0.8057 | 451.1 | 2.2444 | 10 |
| 15 | 3.2958 | 455.7 | 2.4187 | 1.6455 | 455.5 | 2.3388 | 1.0953 | 455.3 | 2.2919 | 0.8203 | 455.1 | 2.2584 | 15 |
| 20 | 3.3534 | 459.7 | 2.4325 | 1.6743 | 459.5 | 2.3527 | 1.1147 | 459.4 | 2.3057 | 0.8348 | 459.2 | 2.2723 | 20 |
| 25 | 3.4109 | 463.8 | 2.4462 | 1.7032 | 463.6 | 2.3664 | 1.1340 | 463.4 | 2.3195 | 0.8494 | 463.3 | 2.2861 | 25 |
| 30 | 3.4683 | 467.9 | 2.4598 | 1.7320 | 467.7 | 2.3800 | 1.1533 | 467.5 | 2.3332 | 0.8639 | 467.4 | 2.2998 | 30 |

(*Continued*)

# TABLE A.19 (SI) (CONTINUED)

Suva® 410A Superheated Vapor–Constant Pressure Tables (Saturated Vapor Properties in Parentheses)

Continuation rows (temperatures 35–85 °C; pressure-column headers appear on the preceding page):

| Temp. °C | V | H | S | V | H | S | V | H | S | V | H | S |
|---|---|---|---|---|---|---|---|---|---|---|---|---|
| 35 | 3.5258 | 472.0 | 2.4733 | 1.7609 | 471.8 | 2.3936 | 1.1726 | 471.7 | 2.3468 | 0.8784 | 471.5 | 2.3134 |
| 40 | 3.5833 | 476.2 | 2.4868 | 1.7897 | 476.0 | 2.4070 | 1.1918 | 475.9 | 2.3602 | 0.8929 | 475.7 | 2.3269 |
| 45 | 3.6407 | 480.4 | 2.5001 | 1.8185 | 480.2 | 2.4204 | 1.2111 | 480.1 | 2.3736 | 0.9074 | 480.0 | 2.3403 |
| 50 | 3.6981 | 484.6 | 2.5133 | 1.8473 | 484.5 | 2.4336 | 1.2303 | 484.3 | 2.3869 | 0.9219 | 484.2 | 2.3536 |
| 55 | 3.7555 | 488.9 | 2.5264 | 1.8761 | 488.8 | 2.4468 | 1.2496 | 488.6 | 2.4000 | 0.9363 | 488.5 | 2.3668 |
| 60 | 3.8130 | 493.2 | 2.5395 | 1.9048 | 493.1 | 2.4598 | 1.2688 | 493.0 | 2.4131 | 0.9508 | 492.8 | 2.3799 |
| 65 | 3.8704 | 497.6 | 2.5524 | 1.9336 | 497.4 | 2.4728 | 1.2880 | 497.3 | 2.4261 | 0.9652 | 497.2 | 2.3929 |
| 70 | – | – | – | 1.9624 | 501.8 | 2.4857 | 1.3072 | 501.7 | 2.4390 | 0.9797 | 501.6 | 2.4058 |
| 75 | – | – | – | 1.9911 | 506.3 | 2.4985 | 1.3264 | 506.2 | 2.4518 | 0.9941 | 506.1 | 2.4187 |
| 80 | – | – | – | – | – | – | 1.3456 | 510.6 | 2.4646 | 1.0085 | 510.5 | 2.4314 |
| 85 | – | – | – | – | – | – | – | – | – | 1.0229 | 515.0 | 2.4441 |

| | 50.0 (−64.52°C) | | | 60.0 (−61.34°C) | | | 70.0 (−58.56°C) | | | 80.0 (−56.09°C) | | | |
|---|---|---|---|---|---|---|---|---|---|---|---|---|---|
| Temp. °C | V (0.4641) | H (394.7) | S (1.9885) | V (0.3911) | H (396.4) | S (1.9764) | V (0.3384) | H (397.9) | S (1.9662) | V (0.2985) | H (399.1) | S (1.9574) | Temp. °C |
| −60 | 0.4754 | 397.9 | 2.0039 | 0.3940 | 397.4 | 1.9810 | – | – | – | – | – | – | −60 |
| −55 | 0.4879 | 401.5 | 2.0206 | 0.4045 | 401.0 | 1.9979 | 0.3449 | 400.5 | 1.9783 | 0.3003 | 400.0 | 1.9612 | −55 |
| −50 | 0.5002 | 405.2 | 2.0370 | 0.4149 | 404.7 | 2.0145 | 0.3540 | 404.2 | 1.9951 | 0.3083 | 403.7 | 1.9782 | −50 |
| −45 | 0.5124 | 408.8 | 2.0532 | 0.4253 | 408.4 | 2.0308 | 0.3630 | 407.9 | 2.0116 | 0.3163 | 407.5 | 1.9948 | −45 |
| −40 | 0.5246 | 412.5 | 2.0691 | 0.4355 | 412.1 | 2.0469 | 0.3719 | 411.7 | 2.0279 | 0.3242 | 411.2 | 2.0112 | −40 |
| −35 | 0.5367 | 416.2 | 2.0848 | 0.4457 | 415.8 | 2.0627 | 0.3807 | 415.4 | 2.0438 | 0.3320 | 415.0 | 2.0273 | −35 |
| −30 | 0.5487 | 419.9 | 2.1003 | 0.4558 | 419.6 | 2.0783 | 0.3895 | 419.2 | 2.0596 | 0.3397 | 418.9 | 2.0431 | −30 |
| −25 | 0.5607 | 423.7 | 2.1156 | 0.4659 | 423.4 | 2.0937 | 0.3982 | 423.0 | 2.0751 | 0.3474 | 422.7 | 2.0588 | −25 |
| −20 | 0.5726 | 427.5 | 2.1308 | 0.4759 | 427.2 | 2.1090 | 0.4069 | 426.9 | 2.0904 | 0.3551 | 426.5 | 2.0742 | −20 |
| −15 | 0.5845 | 431.3 | 2.1457 | 0.4859 | 431.0 | 2.1240 | 0.4155 | 430.7 | 2.1055 | 0.3627 | 430.4 | 2.0894 | −15 |
| −10 | 0.5964 | 435.2 | 2.1605 | 0.4959 | 434.9 | 2.1389 | 0.4241 | 434.6 | 2.1205 | 0.3703 | 434.3 | 2.1044 | −10 |
| −5 | 0.6082 | 439.0 | 2.1752 | 0.5058 | 438.8 | 2.1536 | 0.4327 | 438.5 | 2.1352 | 0.3778 | 438.3 | 2.1192 | −5 |
| 0 | 0.6200 | 443.0 | 2.1897 | 0.5157 | 442.7 | 2.1681 | 0.4412 | 442.5 | 2.1498 | 0.3853 | 442.3 | 2.1339 | 0 |
| 5 | 0.6318 | 446.9 | 2.2040 | 0.5255 | 446.7 | 2.1825 | 0.4497 | 446.5 | 2.1643 | 0.3928 | 446.3 | 2.1484 | 5 |

*(Continued)*

**TABLE A.19 (SI) (CONTINUED)**

Suva® 410A Superheated Vapor–Constant Pressure Tables (Saturated Vapor Properties in Parentheses)

| Temp. °C | 90.0 (−53.84°C) | | | 100.0 (−51.79°C) | | | 101.325 (−51.53°C) | | | 110.0 (−49.90°C) | | | Temp. °C |
|---|---|---|---|---|---|---|---|---|---|---|---|---|---|
| | V | H | S | V | H | S | V | H | S | V | H | S | |
| (Sat.) | (0.2672) | (400.3) | (1.9498) | (0.2420) | (401.3) | (1.9430) | (0.2390) | (401.5) | (1.9421) | (0.2212) | (402.3) | (1.9369) | |
| −50 | 0.2728 | 403.2 | 1.9630 | 0.2443 | 402.7 | 1.9492 | 0.2410 | 402.7 | 1.9474 | – | – | – | −50 |
| −45 | 0.2800 | 407.0 | 1.9798 | 0.2509 | 406.6 | 1.9662 | 0.2475 | 406.5 | 1.9645 | 0.2271 | 406.1 | 1.9537 | −45 |
| −40 | 0.2870 | 410.8 | 1.9963 | 0.2573 | 410.4 | 1.9828 | 0.2538 | 410.3 | 1.9811 | 0.2330 | 410.0 | 1.9705 | −40 |
| −35 | 0.2941 | 414.6 | 2.0125 | 0.2637 | 414.3 | 1.9992 | 0.2602 | 414.2 | 1.9975 | 0.2389 | 413.9 | 1.9870 | −35 |
| −30 | 0.3010 | 418.5 | 2.0285 | 0.2701 | 418.1 | 2.0153 | 0.2664 | 418.1 | 2.0136 | 0.2447 | 417.8 | 2.0032 | −30 |
| −25 | 0.3079 | 422.4 | 2.0442 | 0.2763 | 422.0 | 2.0311 | 0.2726 | 422.0 | 2.0295 | 0.2505 | 421.7 | 2.0192 | −25 |
| −20 | 0.3148 | 426.2 | 2.0597 | 0.2826 | 425.9 | 2.0467 | 0.2788 | 425.9 | 2.0451 | 0.2562 | 425.6 | 2.0349 | −20 |
| −15 | 0.3216 | 430.1 | 2.0750 | 0.2887 | 429.9 | 2.0621 | 0.2849 | 429.8 | 2.0605 | 0.2618 | 429.6 | 2.0503 | −15 |
| 10 | 0.6435 | 450.9 | 2.2182 | 0.5354 | 450.7 | 2.1968 | 0.4581 | 450.5 | 2.1786 | 0.4002 | 450.3 | 2.1628 | 10 |
| 15 | 0.6552 | 454.9 | 2.2323 | 0.5452 | 454.7 | 2.2109 | 0.4666 | 454.5 | 2.1928 | 0.4076 | 454.4 | 2.1770 | 15 |
| 20 | 0.6669 | 459.0 | 2.2463 | 0.5550 | 458.8 | 2.2250 | 0.4750 | 458.6 | 2.2068 | 0.4150 | 458.4 | 2.1911 | 20 |
| 25 | 0.6786 | 463.1 | 2.2601 | 0.5647 | 462.9 | 2.2388 | 0.4834 | 462.7 | 2.2207 | 0.4224 | 462.6 | 2.2050 | 25 |
| 30 | 0.6903 | 467.2 | 2.2739 | 0.5745 | 467.1 | 2.2526 | 0.4918 | 466.9 | 2.2345 | 0.4298 | 466.7 | 2.2188 | 30 |
| 35 | 0.7019 | 471.4 | 2.2875 | 0.5842 | 471.2 | 2.2662 | 0.5002 | 471.1 | 2.2482 | 0.4371 | 470.9 | 2.2325 | 35 |
| 40 | 0.7135 | 475.6 | 2.3010 | 0.5940 | 475.4 | 2.2798 | 0.5085 | 475.3 | 2.2618 | 0.4445 | 475.1 | 2.2461 | 40 |
| 45 | 0.7252 | 479.8 | 2.3144 | 0.6037 | 479.7 | 2.2932 | 0.5169 | 479.5 | 2.2752 | 0.4518 | 479.4 | 2.2596 | 45 |
| 50 | 0.7368 | 484.1 | 2.3277 | 0.6134 | 483.9 | 2.3065 | 0.5252 | 483.8 | 2.2886 | 0.4591 | 483.7 | 2.2730 | 50 |
| 55 | 0.7484 | 488.4 | 2.3409 | 0.6231 | 488.3 | 2.3198 | 0.5336 | 488.1 | 2.3018 | 0.4664 | 488.0 | 2.2863 | 55 |
| 60 | 0.7600 | 492.7 | 2.3541 | 0.6327 | 492.6 | 2.3329 | 0.5419 | 492.5 | 2.3150 | 0.4737 | 492.4 | 2.2994 | 60 |
| 65 | 0.7715 | 497.1 | 2.3671 | 0.6424 | 497.0 | 2.3460 | 0.5502 | 496.9 | 2.3281 | 0.4810 | 496.7 | 2.3125 | 65 |
| 70 | 0.7831 | 501.5 | 2.3800 | 0.6521 | 501.4 | 2.3589 | 0.5585 | 501.3 | 2.3410 | 0.4883 | 501.2 | 2.3255 | 70 |
| 75 | 0.7947 | 505.9 | 2.3929 | 0.6617 | 505.8 | 2.3718 | 0.5668 | 505.7 | 2.3539 | 0.4956 | 505.6 | 2.3384 | 75 |
| 80 | 0.8062 | 510.4 | 2.4057 | 0.6714 | 510.3 | 2.3846 | 0.5751 | 510.2 | 2.3667 | 0.5028 | 510.1 | 2.3512 | 80 |
| 85 | 0.8178 | 514.9 | 2.4183 | 0.6810 | 514.8 | 2.3973 | 0.5834 | 514.7 | 2.3794 | 0.5101 | 514.6 | 2.3639 | 85 |
| 90 | 0.8294 | 519.5 | 2.4309 | 0.6907 | 519.4 | 2.4099 | 0.5916 | 519.3 | 2.3920 | 0.5174 | 519.2 | 2.3766 | 90 |
| 95 | – | – | – | – | – | – | 0.5999 | 523.9 | 2.4046 | 0.5246 | 523.8 | 2.3891 | 95 |

*(Continued)*

**TABLE A.19 (SI) (CONTINUED)**

Suva® 410A Superheated Vapor–Constant Pressure Tables (Saturated Vapor Properties in Parentheses)

| Temp. °C | 120.0 (−48.13°C) V | H | S | 130.0 (−46.48°C) V | H | S | 140.0 (−44.93°C) V | H | S | 150.0 (−43.46°C) V | H | S |
|---|---|---|---|---|---|---|---|---|---|---|---|---|
| | (0.2037) | (403.2) | (1.9313) | (0.1889) | (404.0) | (1.9262) | (0.1761) | (404.7) | (1.9215) | (0.1650) | (405.5) | (1.9172) |
| −10 | 0.3284 | 434.1 | 2.0901 | 0.2949 | 433.8 | 2.0773 | 0.2909 | 433.8 | 2.0756 | 0.2675 | 433.5 | 2.0655 |
| −5 | 0.3351 | 438.0 | 2.1050 | 0.3010 | 437.8 | 2.0922 | 0.2970 | 437.7 | 2.0906 | 0.2730 | 437.5 | 2.0806 |
| 0 | 0.3418 | 442.0 | 2.1197 | 0.3071 | 441.8 | 2.1070 | 0.3030 | 441.8 | 2.1054 | 0.2786 | 441.5 | 2.0954 |
| 5 | 0.3485 | 446.0 | 2.1343 | 0.3131 | 445.8 | 2.1216 | 0.3089 | 445.8 | 2.1200 | 0.2841 | 445.6 | 2.1101 |
| 10 | 0.3552 | 450.1 | 2.1487 | 0.3191 | 449.9 | 2.1361 | 0.3149 | 449.8 | 2.1345 | 0.2896 | 449.7 | 2.1246 |
| 15 | 0.3618 | 454.2 | 2.1630 | 0.3251 | 454.0 | 2.1504 | 0.3208 | 453.9 | 2.1488 | 0.2951 | 453.8 | 2.1390 |
| 20 | 0.3684 | 458.3 | 2.1771 | 0.3311 | 458.1 | 2.1646 | 0.3267 | 458.0 | 2.1630 | 0.3005 | 457.9 | 2.1532 |
| 25 | 0.3750 | 462.4 | 2.1911 | 0.3370 | 462.2 | 2.1786 | 0.3326 | 462.2 | 2.1770 | 0.3060 | 462.0 | 2.1672 |
| 30 | 0.3816 | 466.6 | 2.2049 | 0.3430 | 466.4 | 2.1925 | 0.3384 | 466.4 | 2.1909 | 0.3114 | 466.2 | 2.1811 |
| 35 | 0.3881 | 470.8 | 2.2187 | 0.3489 | 470.6 | 2.2062 | 0.3443 | 470.6 | 2.2047 | 0.3168 | 470.4 | 2.1949 |
| 40 | 0.3947 | 475.0 | 2.2323 | 0.3548 | 474.8 | 2.2199 | 0.3501 | 474.8 | 2.2183 | 0.3222 | 474.7 | 2.2086 |
| 45 | 0.4012 | 479.3 | 2.2458 | 0.3607 | 479.1 | 2.2334 | 0.3559 | 479.1 | 2.2319 | 0.3276 | 479.0 | 2.2222 |
| 50 | 0.4077 | 483.5 | 2.2592 | 0.3666 | 483.4 | 2.2468 | 0.3617 | 483.4 | 2.2453 | 0.3329 | 483.3 | 2.2356 |
| 55 | 0.4142 | 487.9 | 2.2725 | 0.3725 | 487.7 | 2.2601 | 0.3675 | 487.7 | 2.2586 | 0.3383 | 487.6 | 2.2489 |
| 60 | 0.4207 | 492.2 | 2.2857 | 0.3783 | 492.1 | 2.2733 | 0.3733 | 492.1 | 2.2718 | 0.3436 | 492.0 | 2.2622 |
| 65 | 0.4272 | 496.6 | 2.2988 | 0.3842 | 496.5 | 2.2865 | 0.3791 | 496.5 | 2.2849 | 0.3490 | 496.4 | 2.2753 |
| 70 | 0.4337 | 501.1 | 2.3118 | 0.3900 | 501.0 | 2.2995 | 0.3849 | 500.9 | 2.2979 | 0.3543 | 500.8 | 2.2883 |
| 75 | 0.4402 | 505.5 | 2.3247 | 0.3959 | 505.4 | 2.3124 | 0.3907 | 505.4 | 2.3109 | 0.3596 | 505.3 | 2.3013 |
| 80 | 0.4467 | 510.0 | 2.3375 | 0.4017 | 509.9 | 2.3252 | 0.3964 | 509.9 | 2.3237 | 0.3649 | 509.8 | 2.3141 |
| 85 | 0.4531 | 514.5 | 2.3502 | 0.4075 | 514.4 | 2.3380 | 0.4022 | 514.4 | 2.3364 | 0.3702 | 514.4 | 2.3268 |
| 90 | 0.4596 | 519.1 | 2.3629 | 0.4134 | 519.0 | 2.3506 | 0.4079 | 519.0 | 2.3491 | 0.3755 | 518.9 | 2.3395 |
| 95 | 0.4660 | 523.7 | 2.3754 | 0.4192 | 523.6 | 2.3632 | 0.4137 | 523.6 | 2.3617 | 0.3808 | 523.5 | 2.3521 |
| 100 | 0.4725 | 528.3 | 2.3879 | 0.4250 | 528.2 | 2.3757 | 0.4194 | 528.2 | 2.3742 | 0.3861 | 528.2 | 2.3646 |
| 105 | – | – | – | – | – | – | – | – | – | 0.3914 | 532.8 | 2.3770 |

| Temp. °C | 120.0 (−48.13°C) V | H | S | 130.0 (−46.48°C) V | H | S | 140.0 (−44.93°C) V | H | S | 150.0 (−43.46°C) V | H | S | Temp. °C |
|---|---|---|---|---|---|---|---|---|---|---|---|---|---|
| | (0.2037) | (403.2) | (1.9313) | (0.1889) | (404.0) | (1.9262) | (0.1761) | (404.7) | (1.9215) | (0.1650) | (405.5) | (1.9172) | |
| −45 | 0.2072 | 405.6 | 1.9421 | 0.1904 | 405.2 | 1.9314 | – | – | – | – | – | – | −45 |
| −40 | 0.2128 | 409.5 | 1.9591 | 0.1956 | 409.1 | 1.9485 | 0.1809 | 408.7 | 1.9386 | 0.1681 | 408.2 | 1.9292 | −40 |

*(Continued)*

**TABLE A.19 (SI) (CONTINUED)**

Suva® 410A Superheated Vapor–Constant Pressure Tables (Saturated Vapor Properties in Parentheses)

| T | | | | | | | | | | | | | T |
|---|---|---|---|---|---|---|---|---|---|---|---|---|---|
| -35 | 0.2182 | 413.5 | 1.9758 | 0.2007 | 413.1 | 1.9653 | 0.1857 | 412.7 | 1.9555 | 0.1727 | 412.3 | 1.9463 | -35 |
| -30 | 0.2236 | 417.4 | 1.9921 | 0.2057 | 417.0 | 1.9818 | 0.1904 | 416.7 | 1.9721 | 0.1771 | 416.3 | 1.9630 | -30 |
| -25 | 0.2289 | 421.3 | 2.0082 | 0.2107 | 421.0 | 1.9979 | 0.1950 | 420.7 | 1.9884 | 0.1815 | 420.3 | 1.9794 | -25 |
| -20 | 0.2342 | 425.3 | 2.0239 | 0.2156 | 425.0 | 2.0138 | 0.1996 | 424.7 | 2.0044 | 0.1858 | 424.3 | 1.9955 | -20 |
| -15 | 0.2394 | 429.3 | 2.0395 | 0.2205 | 429.0 | 2.0294 | 0.2042 | 428.7 | 2.0201 | 0.1901 | 428.4 | 2.0113 | -15 |
| -10 | 0.2446 | 433.3 | 2.0548 | 0.2253 | 433.0 | 2.0448 | 0.2087 | 432.7 | 2.0355 | 0.1943 | 432.4 | 2.0268 | -10 |
| -5 | 0.2498 | 437.3 | 2.0699 | 0.2301 | 437.0 | 2.0600 | 0.2132 | 436.8 | 2.0508 | 0.1985 | 436.5 | 2.0421 | -5 |
| 0 | 0.2549 | 441.3 | 2.0848 | 0.2348 | 441.1 | 2.0750 | 0.2176 | 440.8 | 2.0658 | 0.2027 | 440.6 | 2.0572 | 0 |
| 5 | 0.2600 | 445.4 | 2.0995 | 0.2395 | 445.1 | 2.0898 | 0.2220 | 444.9 | 2.0807 | 0.2068 | 444.7 | 2.0721 | 5 |
| 10 | 0.2650 | 449.5 | 2.1141 | 0.2442 | 449.2 | 2.1044 | 0.2264 | 449.0 | 2.0953 | 0.2110 | 448.8 | 2.0868 | 10 |
| 15 | 0.2701 | 453.6 | 2.1285 | 0.2489 | 453.4 | 2.1188 | 0.2308 | 453.2 | 2.1098 | 0.2151 | 453.0 | 2.1014 | 15 |
| 20 | 0.2751 | 457.7 | 2.1427 | 0.2536 | 457.5 | 2.1331 | 0.2351 | 457.3 | 2.1241 | 0.2191 | 457.1 | 2.1157 | 20 |
| 25 | 0.2801 | 461.9 | 2.1568 | 0.2582 | 461.7 | 2.1472 | 0.2394 | 461.5 | 2.1383 | 0.2232 | 461.3 | 2.1299 | 25 |
| 30 | 0.2851 | 466.1 | 2.1708 | 0.2628 | 465.9 | 2.1612 | 0.2437 | 465.7 | 2.1523 | 0.2272 | 465.6 | 2.1440 | 30 |
| 35 | 0.2901 | 470.3 | 2.1846 | 0.2674 | 470.1 | 2.1750 | 0.2480 | 470.0 | 2.1662 | 0.2312 | 469.8 | 2.1579 | 35 |
| 40 | 0.2950 | 474.5 | 2.1983 | 0.2720 | 474.4 | 2.1888 | 0.2523 | 474.2 | 2.1799 | 0.2352 | 474.1 | 2.1717 | 40 |
| 45 | 0.3000 | 478.8 | 2.2119 | 0.2766 | 478.7 | 2.2024 | 0.2566 | 478.5 | 2.1936 | 0.2392 | 478.4 | 2.1853 | 45 |
| 50 | 0.3049 | 483.1 | 2.2253 | 0.2812 | 483.0 | 2.2159 | 0.2608 | 482.9 | 2.2071 | 0.2432 | 482.7 | 2.1988 | 50 |
| 55 | 0.3098 | 487.5 | 2.2387 | 0.2857 | 487.4 | 2.2292 | 0.2651 | 487.2 | 2.2205 | 0.2472 | 487.1 | 2.2123 | 55 |
| 60 | 0.3147 | 491.9 | 2.2519 | 0.2903 | 491.8 | 2.2425 | 0.2693 | 491.6 | 2.2337 | 0.2511 | 491.5 | 2.2256 | 60 |
| 65 | 0.3196 | 496.3 | 2.2651 | 0.2948 | 496.2 | 2.2556 | 0.2735 | 496.1 | 2.2469 | 0.2551 | 495.9 | 2.2388 | 65 |
| 70 | 0.3245 | 500.7 | 2.2781 | 0.2993 | 500.6 | 2.2687 | 0.2777 | 500.5 | 2.2600 | 0.2590 | 500.4 | 2.2518 | 70 |
| 75 | 0.3294 | 505.2 | 2.2911 | 0.3038 | 505.1 | 2.2817 | 0.2819 | 505.0 | 2.2730 | 0.2629 | 504.9 | 2.2648 | 75 |
| 80 | 0.3343 | 509.7 | 2.3039 | 0.3083 | 509.6 | 2.2945 | 0.2861 | 509.5 | 2.2858 | 0.2669 | 509.4 | 2.2777 | 80 |
| 85 | 0.3392 | 514.3 | 2.3167 | 0.3129 | 514.2 | 2.3073 | 0.2903 | 514.1 | 2.2986 | 0.2708 | 514.0 | 2.2905 | 85 |
| 90 | 0.3440 | 518.8 | 2.3294 | 0.3174 | 518.7 | 2.3200 | 0.2945 | 518.6 | 2.3113 | 0.2747 | 518.5 | 2.3032 | 90 |
| 95 | 0.3489 | 523.4 | 2.3420 | 0.3219 | 523.3 | 2.3326 | 0.2987 | 523.3 | 2.3240 | 0.2786 | 523.2 | 2.3159 | 95 |
| 100 | 0.3537 | 528.1 | 2.3545 | 0.3263 | 528.0 | 2.3451 | 0.3029 | 527.9 | 2.3365 | 0.2825 | 527.8 | 2.3284 | 100 |
| 105 | 0.3586 | 532.7 | 2.3669 | 0.3308 | 532.7 | 2.3576 | 0.3070 | 532.6 | 2.3489 | 0.2864 | 532.5 | 2.3409 | 105 |
| 110 | — | — | — | — | — | — | 0.3112 | 537.3 | 2.3613 | 0.2903 | 537.2 | 2.3533 | 110 |

*(Continued)*

**TABLE A.19 (SI) (CONTINUED)**

Suva® 410A Superheated Vapor–Constant Pressure Tables (Saturated Vapor Properties in Parentheses)

| Temp. °C | 160.0 (−42.07°C) | | | 170.0 (−40.74°C) | | | 180.0 (−39.47°C) | | | 190.0 (−38.26°C) | | | Temp. °C |
|---|---|---|---|---|---|---|---|---|---|---|---|---|---|
| | V (0.1552) | H (406.1) | S (1.9131) | V (0.1465) | H (406.7) | S (1.9093) | V (0.1388) | H (407.3) | S (1.9058) | V (0.1318) | H (407.9) | S (1.9024) | |
| −40 | 0.1570 | 407.8 | 1.9204 | 0.1471 | 407.4 | 1.9119 | — | — | — | — | — | — | −40 |
| −35 | 0.1613 | 411.9 | 1.9376 | 0.1512 | 411.4 | 1.9293 | 0.1423 | 411.0 | 1.9214 | 0.1342 | 410.6 | 1.9139 | −35 |
| −30 | 0.1655 | 415.9 | 1.9544 | 0.1552 | 415.5 | 1.9463 | 0.1461 | 415.1 | 1.9385 | 0.1379 | 414.8 | 1.9311 | −30 |
| −25 | 0.1696 | 420.0 | 1.9709 | 0.1591 | 419.6 | 1.9629 | 0.1498 | 419.3 | 1.9553 | 0.1415 | 418.9 | 1.9480 | −25 |
| −20 | 0.1737 | 424.0 | 1.9871 | 0.1630 | 423.7 | 1.9792 | 0.1535 | 423.4 | 1.9716 | 0.1450 | 423.0 | 1.9645 | −20 |
| −15 | 0.1778 | 428.1 | 2.0030 | 0.1669 | 427.8 | 1.9952 | 0.1572 | 427.5 | 1.9877 | 0.1485 | 427.2 | 1.9806 | −15 |
| −10 | 0.1818 | 432.2 | 2.0186 | 0.1707 | 431.9 | 2.0109 | 0.1608 | 431.6 | 2.0035 | 0.1520 | 431.3 | 1.9965 | −10 |
| −5 | 0.1857 | 436.2 | 2.0340 | 0.1744 | 436.0 | 2.0263 | 0.1644 | 435.7 | 2.0190 | 0.1554 | 435.5 | 2.0121 | −5 |
| 0 | 0.1897 | 440.3 | 2.0492 | 0.1781 | 440.1 | 2.0416 | 0.1679 | 439.9 | 2.0343 | 0.1588 | 439.6 | 2.0274 | 0 |
| 5 | 0.1936 | 444.5 | 2.0641 | 0.1818 | 444.2 | 2.0566 | 0.1714 | 444.0 | 2.0494 | 0.1621 | 443.8 | 2.0426 | 5 |
| 10 | 0.1974 | 448.6 | 2.0789 | 0.1855 | 448.4 | 2.0714 | 0.1749 | 448.2 | 2.0642 | 0.1654 | 448.0 | 2.0575 | 10 |
| 15 | 0.2013 | 452.8 | 2.0934 | 0.1892 | 452.6 | 2.0860 | 0.1784 | 452.4 | 2.0789 | 0.1687 | 452.2 | 2.0722 | 15 |
| 20 | 0.2051 | 456.9 | 2.1078 | 0.1928 | 456.8 | 2.1004 | 0.1818 | 456.6 | 2.0934 | 0.1720 | 456.4 | 2.0867 | 20 |
| 25 | 0.2089 | 461.2 | 2.1221 | 0.1964 | 461.0 | 2.1147 | 0.1852 | 460.8 | 2.1077 | 0.1752 | 460.6 | 2.1011 | 25 |
| 30 | 0.2127 | 465.4 | 2.1362 | 0.2000 | 465.2 | 2.1288 | 0.1886 | 465.1 | 2.1219 | 0.1784 | 464.9 | 2.1152 | 30 |
| 35 | 0.2165 | 469.7 | 2.1501 | 0.2035 | 469.5 | 2.1428 | 0.1920 | 469.3 | 2.1359 | 0.1817 | 469.2 | 2.1293 | 35 |
| 40 | 0.2203 | 473.9 | 2.1639 | 0.2071 | 473.8 | 2.1566 | 0.1954 | 473.6 | 2.1497 | 0.1849 | 473.5 | 2.1432 | 40 |
| 45 | 0.2240 | 478.3 | 2.1776 | 0.2106 | 478.1 | 2.1703 | 0.1987 | 478.0 | 2.1634 | 0.1880 | 477.8 | 2.1569 | 45 |
| 50 | 0.2278 | 482.6 | 2.1911 | 0.2141 | 482.5 | 2.1839 | 0.2020 | 482.3 | 2.1770 | 0.1912 | 482.2 | 2.1705 | 50 |
| 55 | 0.2315 | 487.0 | 2.2046 | 0.2177 | 486.9 | 2.1973 | 0.2054 | 486.7 | 2.1905 | 0.1944 | 486.6 | 2.1840 | 55 |
| 60 | 0.2352 | 491.4 | 2.2179 | 0.2212 | 491.3 | 2.2107 | 0.2087 | 491.1 | 2.2039 | 0.1975 | 491.0 | 2.1974 | 60 |
| 65 | 0.2389 | 495.8 | 2.2311 | 0.2247 | 495.7 | 2.2239 | 0.2120 | 495.6 | 2.2171 | 0.2007 | 495.5 | 2.2107 | 65 |
| 70 | 0.2426 | 500.3 | 2.2442 | 0.2282 | 500.2 | 2.2370 | 0.2153 | 500.1 | 2.2303 | 0.2038 | 499.9 | 2.2238 | 70 |
| 75 | 0.2463 | 504.8 | 2.2572 | 0.2317 | 504.7 | 2.2501 | 0.2186 | 504.6 | 2.2433 | 0.2070 | 504.5 | 2.2369 | 75 |
| 80 | 0.2500 | 509.3 | 2.2701 | 0.2351 | 509.2 | 2.2630 | 0.2219 | 509.1 | 2.2562 | 0.2101 | 509.0 | 2.2498 | 80 |
| 85 | 0.2537 | 513.9 | 2.2829 | 0.2386 | 513.8 | 2.2758 | 0.2252 | 513.7 | 2.2691 | 0.2132 | 513.6 | 2.2627 | 85 |

(Continued)

# TABLE A.19 (SI) (CONTINUED)

Suva® 410A Superheated Vapor–Constant Pressure Tables (Saturated Vapor Properties in Parentheses)

| Temp. °C | 200.0 (−37.09°C) | | | 210.0 (−35.97°C) | | | 220.0 (−34.89°C) | | | 230.0 (−33.85°C) | | | Temp. °C |
|---|---|---|---|---|---|---|---|---|---|---|---|---|---|
| | V (0.1255) | H (408.4) | S (1.8992) | V (0.1198) | H (409.0) | S (1.8962) | V (0.1146) | H (409.4) | S (1.8933) | V (0.1099) | H (409.9) | S (1.8906) | |
| −35 | 0.1270 | 410.2 | 1.9066 | 0.1205 | 409.8 | 1.8997 | — | — | — | — | — | — | −35 |
| −30 | 0.1306 | 414.4 | 1.9240 | 0.1239 | 414.0 | 1.9172 | 0.1178 | 413.6 | 1.9106 | 0.1123 | 413.2 | 1.9043 | −30 |
| −25 | 0.1340 | 418.5 | 1.9410 | 0.1272 | 418.2 | 1.9343 | 0.1210 | 417.8 | 1.9278 | 0.1154 | 417.5 | 1.9216 | −25 |
| −20 | 0.1374 | 422.7 | 1.9576 | 0.1305 | 422.4 | 1.9510 | 0.1242 | 422.0 | 1.9446 | 0.1184 | 421.7 | 1.9385 | −20 |
| −15 | 0.1407 | 426.9 | 1.9738 | 0.1337 | 426.6 | 1.9673 | 0.1273 | 426.3 | 1.9611 | 0.1214 | 425.9 | 1.9551 | −15 |
| −10 | 0.1440 | 431.0 | 1.9898 | 0.1368 | 430.7 | 1.9834 | 0.1303 | 430.5 | 1.9772 | 0.1243 | 430.2 | 1.9713 | −10 |
| −5 | 0.1473 | 435.2 | 2.0055 | 0.1400 | 434.9 | 1.9991 | 0.1333 | 434.7 | 1.9930 | 0.1272 | 434.4 | 1.9872 | −5 |
| 0 | 0.1505 | 439.4 | 2.0209 | 0.1431 | 439.1 | 2.0146 | 0.1363 | 438.9 | 2.0086 | 0.1301 | 438.6 | 2.0028 | 0 |
| 5 | 0.1537 | 443.5 | 2.0361 | 0.1461 | 443.3 | 2.0298 | 0.1392 | 443.1 | 2.0239 | 0.1329 | 442.8 | 2.0182 | 5 |
| 10 | 0.1569 | 447.7 | 2.0510 | 0.1491 | 447.5 | 2.0449 | 0.1421 | 447.3 | 2.0390 | 0.1357 | 447.1 | 2.0333 | 10 |
| 15 | 0.1600 | 452.0 | 2.0658 | 0.1521 | 451.8 | 2.0597 | 0.1450 | 451.6 | 2.0538 | 0.1385 | 451.3 | 2.0482 | 15 |
| 20 | 0.1631 | 456.2 | 2.0803 | 0.1551 | 456.0 | 2.0743 | 0.1479 | 455.8 | 2.0685 | 0.1412 | 455.6 | 2.0629 | 20 |
| 25 | 0.1662 | 460.4 | 2.0947 | 0.1581 | 460.3 | 2.0887 | 0.1507 | 460.1 | 2.0829 | 0.1439 | 459.9 | 2.0774 | 25 |
| 30 | 0.1693 | 464.7 | 2.1090 | 0.1610 | 464.6 | 2.1030 | 0.1535 | 464.4 | 2.0972 | 0.1466 | 464.2 | 2.0917 | 30 |
| 35 | 0.1724 | 469.0 | 2.1230 | 0.1640 | 468.9 | 2.1170 | 0.1563 | 468.7 | 2.1113 | 0.1493 | 468.5 | 2.1058 | 35 |
| 40 | 0.1754 | 473.3 | 2.1369 | 0.1669 | 473.2 | 2.1310 | 0.1591 | 473.0 | 2.1253 | 0.1520 | 472.9 | 2.1198 | 40 |
| 45 | 0.1785 | 477.7 | 2.1507 | 0.1698 | 477.5 | 2.1448 | 0.1619 | 477.4 | 2.1391 | 0.1547 | 477.3 | 2.1337 | 45 |
| 50 | 0.1815 | 482.1 | 2.1644 | 0.1727 | 481.9 | 2.1585 | 0.1647 | 481.8 | 2.1528 | 0.1573 | 481.7 | 2.1474 | 50 |
| 55 | 0.1845 | 486.5 | 2.1779 | 0.1755 | 486.3 | 2.1720 | 0.1674 | 486.2 | 2.1664 | 0.1600 | 486.1 | 2.1610 | 55 |
| 90 | 0.2574 | 518.5 | 2.2957 | 0.2421 | 518.4 | 2.2885 | 0.2285 | 518.3 | 2.2818 | 0.2163 | 518.2 | 2.2754 | 90 |
| 95 | 0.2610 | 523.1 | 2.3083 | 0.2455 | 523.0 | 2.3012 | 0.2317 | 522.9 | 2.2945 | 0.2194 | 522.8 | 2.2881 | 95 |
| 100 | 0.2647 | 527.7 | 2.3209 | 0.2490 | 527.6 | 2.3137 | 0.2350 | 527.5 | 2.3070 | 0.2225 | 527.5 | 2.3007 | 100 |
| 105 | 0.2684 | 532.4 | 2.3333 | 0.2524 | 532.3 | 2.3262 | 0.2383 | 532.2 | 2.3195 | 0.2256 | 532.2 | 2.3132 | 105 |
| 110 | 0.2720 | 537.1 | 2.3457 | 0.2559 | 537.0 | 2.3386 | 0.2415 | 537.0 | 2.3319 | 0.2287 | 536.9 | 2.3256 | 110 |
| 115 | — | — | — | — | — | — | 0.2448 | 541.7 | 2.3442 | 0.2318 | 541.6 | 2.3379 | 115 |

(Continued)

**TABLE A.19 (SI) (CONTINUED)**

Suva® 410A Superheated Vapor–Constant Pressure Tables (Saturated Vapor Properties in Parentheses)

| Temp. °C | 240.0 (−32.84°C) | | | 250.0 (−31.87°C) | | | 260.0 (−30.92°C) | | | 270.0 (−30.01°C) | | |
|---|---|---|---|---|---|---|---|---|---|---|---|---|
| | V (0.1055) | H (410.4) | S (1.8880) | V (0.1015) | H (410.8) | S (1.8854) | V (0.0977) | H (411.2) | S (1.8830) | V (0.0943) | H (411.6) | S (1.8807) |
| 60 | 0.1875 | 490.9 | 2.1913 | 0.1784 | 490.8 | 2.1854 | 0.1701 | 490.6 | 2.1798 | 0.1626 | 490.5 | 2.1744 |
| 65 | 0.1905 | 495.3 | 2.2045 | 0.1813 | 495.2 | 2.1987 | 0.1729 | 495.1 | 2.1931 | 0.1652 | 495.0 | 2.1878 |
| 70 | 0.1935 | 499.8 | 2.2177 | 0.1841 | 499.7 | 2.2119 | 0.1756 | 499.6 | 2.2063 | 0.1678 | 499.5 | 2.2010 |
| 75 | 0.1965 | 504.3 | 2.2308 | 0.1870 | 504.2 | 2.2250 | 0.1783 | 504.1 | 2.2194 | 0.1704 | 504.0 | 2.2141 |
| 80 | 0.1994 | 508.9 | 2.2437 | 0.1898 | 508.8 | 2.2379 | 0.1810 | 508.7 | 2.2324 | 0.1730 | 508.6 | 2.2271 |
| 85 | 0.2024 | 513.5 | 2.2566 | 0.1926 | 513.4 | 2.2508 | 0.1837 | 513.3 | 2.2453 | 0.1756 | 513.2 | 2.2400 |
| 90 | 0.2054 | 518.1 | 2.2694 | 0.1955 | 518.0 | 2.2636 | 0.1864 | 517.9 | 2.2581 | 0.1782 | 517.8 | 2.2528 |
| 95 | 0.2083 | 522.7 | 2.2820 | 0.1983 | 522.6 | 2.2763 | 0.1891 | 522.5 | 2.2708 | 0.1808 | 522.4 | 2.2655 |
| 100 | 0.2113 | 527.4 | 2.2946 | 0.2011 | 527.3 | 2.2889 | 0.1918 | 527.2 | 2.2834 | 0.1834 | 527.1 | 2.2781 |
| 105 | 0.2142 | 532.1 | 2.3071 | 0.2039 | 532.0 | 2.3014 | 0.1945 | 531.9 | 2.2959 | 0.1860 | 531.8 | 2.2906 |
| 110 | 0.2171 | 536.8 | 2.3195 | 0.2067 | 536.7 | 2.3138 | 0.1972 | 536.6 | 2.3083 | 0.1885 | 536.6 | 2.3031 |
| 115 | 0.2201 | 541.6 | 2.3319 | 0.2095 | 541.5 | 2.3261 | 0.1999 | 541.4 | 2.3207 | 0.1911 | 541.3 | 2.3154 |
| 120 | — | — | — | — | — | — | 0.2026 | 546.2 | 2.3329 | 0.1936 | 546.1 | 2.3277 |

| Temp. °C | 240.0 (−32.84°C) | | | 250.0 (−31.87°C) | | | 260.0 (−30.92°C) | | | 270.0 (−30.01°C) | | |
|---|---|---|---|---|---|---|---|---|---|---|---|---|
| | V (0.1055) | H (410.4) | S (1.8880) | V (0.1015) | H (410.8) | S (1.8854) | V (0.0977) | H (411.2) | S (1.8830) | V (0.0943) | H (411.6) | S (1.8807) |
| −30 | 0.1072 | 412.8 | 1.8981 | 0.1026 | 412.4 | 1.8922 | 0.0983 | 412.0 | 1.8864 | 0.0943 | 411.6 | 1.8807 |
| −25 | 0.1102 | 417.1 | 1.9156 | 0.1055 | 416.7 | 1.9098 | 0.1011 | 416.4 | 1.9041 | 0.0970 | 416.0 | 1.8986 |
| −20 | 0.1132 | 421.4 | 1.9326 | 0.1083 | 421.0 | 1.9269 | 0.1038 | 420.7 | 1.9214 | 0.0997 | 420.4 | 1.9160 |
| −15 | 0.1160 | 425.6 | 1.9493 | 0.1111 | 425.3 | 1.9437 | 0.1065 | 425.0 | 1.9382 | 0.1023 | 424.7 | 1.9330 |
| −10 | 0.1189 | 429.9 | 1.9656 | 0.1138 | 429.6 | 1.9601 | 0.1092 | 429.3 | 1.9547 | 0.1049 | 429.0 | 1.9495 |
| −5 | 0.1216 | 434.1 | 1.9815 | 0.1165 | 433.9 | 1.9761 | 0.1118 | 433.6 | 1.9708 | 0.1074 | 433.3 | 1.9657 |
| 0 | 0.1244 | 438.4 | 1.9972 | 0.1192 | 438.1 | 1.9919 | 0.1144 | 437.9 | 1.9867 | 0.1099 | 437.6 | 1.9816 |
| 5 | 0.1271 | 442.6 | 2.0126 | 0.1218 | 442.4 | 2.0073 | 0.1169 | 442.1 | 2.0022 | 0.1123 | 441.9 | 1.9972 |
| 10 | 0.1298 | 446.9 | 2.0278 | 0.1244 | 446.7 | 2.0226 | 0.1194 | 446.4 | 2.0175 | 0.1148 | 446.2 | 2.0126 |
| 15 | 0.1325 | 451.1 | 2.0428 | 0.1270 | 450.9 | 2.0376 | 0.1219 | 450.7 | 2.0325 | 0.1172 | 450.5 | 2.0277 |
| 20 | 0.1351 | 455.4 | 2.0575 | 0.1295 | 455.2 | 2.0523 | 0.1244 | 455.0 | 2.0473 | 0.1196 | 454.8 | 2.0425 |
| 25 | 0.1377 | 459.7 | 2.0720 | 0.1321 | 459.5 | 2.0669 | 0.1268 | 459.4 | 2.0620 | 0.1219 | 459.2 | 2.0572 |
| 30 | 0.1404 | 464.0 | 2.0864 | 0.1346 | 463.9 | 2.0813 | 0.1292 | 463.7 | 2.0764 | 0.1243 | 463.5 | 2.0716 |

(*Continued*)

## TABLE A.19 (SI) (CONTINUED)

Suva® 410A Superheated Vapor–Constant Pressure Tables (Saturated Vapor Properties in Parentheses)

| Temp. °C | 280.0 (−29.12°C) | | | 290.0 (−28.25°C) | | | 300.0 (−27.40°C) | | | 310.0 (−26.58°C) | | | Temp. °C |
|---|---|---|---|---|---|---|---|---|---|---|---|---|---|
| | V | H | S | V | H | S | V | H | S | V | H | S | |
| | (0.0910) | (412.0) | (1.8785) | (0.0880) | (412.4) | (1.8764) | (0.0852) | (412.7) | (1.8743) | (0.0825) | (413.1) | (1.8723) | |
| −25 | 0.0932 | 415.6 | 1.8933 | 0.0897 | 415.2 | 1.8881 | 0.0864 | 414.9 | 1.8830 | 0.0833 | 414.5 | 1.8781 | −25 |
| −20 | 0.0958 | 420.0 | 1.9108 | 0.0922 | 419.7 | 1.9057 | 0.0889 | 419.3 | 1.9008 | 0.0857 | 419.0 | 1.8959 | −20 |
| −15 | 0.0984 | 424.4 | 1.9279 | 0.0947 | 424.0 | 1.9229 | 0.0913 | 423.7 | 1.9180 | 0.0881 | 423.4 | 1.9133 | −15 |
| −10 | 0.1009 | 428.7 | 1.9445 | 0.0971 | 428.4 | 1.9396 | 0.0937 | 428.1 | 1.9349 | 0.0904 | 427.8 | 1.9302 | −10 |
| −5 | 0.1033 | 433.0 | 1.9608 | 0.0995 | 432.8 | 1.9560 | 0.0960 | 432.5 | 1.9513 | 0.0927 | 432.2 | 1.9468 | −5 |
| 0 | 0.1057 | 437.4 | 1.9768 | 0.1019 | 437.1 | 1.9720 | 0.0983 | 436.8 | 1.9674 | 0.0949 | 436.6 | 1.9629 | 0 |
| 5 | 0.1081 | 441.7 | 1.9924 | 0.1042 | 441.4 | 1.9877 | 0.1005 | 441.2 | 1.9832 | 0.0971 | 441.0 | 1.9788 | 5 |
| 35 | 0.1429 | 468.4 | 2.1006 | 0.1371 | 468.2 | 2.0955 | 0.1316 | 468.1 | 2.0906 | 0.1266 | 467.9 | 2.0859 | 35 |
| 40 | 0.1455 | 472.7 | 2.1146 | 0.1395 | 472.6 | 2.1096 | 0.1340 | 472.4 | 2.1047 | 0.1289 | 472.3 | 2.1000 | 40 |
| 45 | 0.1481 | 477.1 | 2.1285 | 0.1420 | 477.0 | 2.1235 | 0.1364 | 476.8 | 2.1186 | 0.1312 | 476.7 | 2.1140 | 45 |
| 50 | 0.1506 | 481.5 | 2.1422 | 0.1445 | 481.4 | 2.1372 | 0.1388 | 481.2 | 2.1324 | 0.1335 | 481.1 | 2.1278 | 50 |
| 55 | 0.1532 | 485.9 | 2.1558 | 0.1469 | 485.8 | 2.1509 | 0.1411 | 485.7 | 2.1461 | 0.1358 | 485.6 | 2.1415 | 55 |
| 60 | 0.1557 | 490.4 | 2.1693 | 0.1493 | 490.3 | 2.1643 | 0.1435 | 490.1 | 2.1596 | 0.1380 | 490.0 | 2.1550 | 60 |
| 65 | 0.1582 | 494.9 | 2.1826 | 0.1517 | 494.8 | 2.1777 | 0.1458 | 494.6 | 2.1730 | 0.1403 | 494.5 | 2.1684 | 65 |
| 70 | 0.1607 | 499.4 | 2.1959 | 0.1542 | 499.3 | 2.1910 | 0.1481 | 499.2 | 2.1862 | 0.1425 | 499.0 | 2.1817 | 70 |
| 75 | 0.1632 | 503.9 | 2.2090 | 0.1566 | 503.8 | 2.2041 | 0.1504 | 503.7 | 2.1994 | 0.1448 | 503.6 | 2.1948 | 75 |
| 80 | 0.1657 | 508.5 | 2.2220 | 0.1590 | 508.4 | 2.2171 | 0.1527 | 508.3 | 2.2124 | 0.1470 | 508.2 | 2.2079 | 80 |
| 85 | 0.1682 | 513.1 | 2.2349 | 0.1614 | 513.0 | 2.2300 | 0.1551 | 512.9 | 2.2253 | 0.1492 | 512.8 | 2.2208 | 85 |
| 90 | 0.1707 | 517.7 | 2.2477 | 0.1638 | 517.6 | 2.2429 | 0.1574 | 517.5 | 2.2382 | 0.1514 | 517.4 | 2.2337 | 90 |
| 95 | 0.1732 | 522.3 | 2.2604 | 0.1661 | 522.3 | 2.2556 | 0.1596 | 522.2 | 2.2509 | 0.1536 | 522.1 | 2.2464 | 95 |
| 100 | 0.1756 | 527.0 | 2.2731 | 0.1685 | 526.9 | 2.2682 | 0.1619 | 526.8 | 2.2636 | 0.1558 | 526.8 | 2.2591 | 100 |
| 105 | 0.1781 | 531.7 | 2.2856 | 0.1709 | 531.6 | 2.2808 | 0.1642 | 531.6 | 2.2761 | 0.1581 | 531.5 | 2.2716 | 105 |
| 110 | 0.1806 | 536.5 | 2.2980 | 0.1733 | 536.4 | 2.2932 | 0.1665 | 536.3 | 2.2886 | 0.1603 | 536.2 | 2.2841 | 110 |
| 115 | 0.1830 | 541.2 | 2.3104 | 0.1756 | 541.2 | 2.3056 | 0.1688 | 541.1 | 2.3010 | 0.1624 | 541.0 | 2.2965 | 115 |
| 120 | 0.1855 | 546.0 | 2.3227 | 0.1780 | 546.0 | 2.3179 | 0.1711 | 545.9 | 2.3133 | 0.1646 | 545.8 | 2.3088 | 120 |

*(Continued)*

# TABLE A.19 (SI) (CONTINUED)

Suva® 410A Superheated Vapor–Constant Pressure Tables (Saturated Vapor Properties in Parentheses)

| Temp. °C | 320.0 (−25.78°C) | | | 330.0 (−24.99°C) | | | 340.0 (−24.23°C) | | | 350.0 (−23.48°C) | | | Temp. °C |
|---|---|---|---|---|---|---|---|---|---|---|---|---|---|
| | V | H | S | V | H | S | V | H | S | V | H | S | |
| | (0.0800) | (413.4) | (1.8703) | (0.0777) | (413.7) | (1.8685) | (0.0755) | (414.0) | (1.8667) | (0.0734) | (414.3) | (1.8649) | |
| 10 | 0.1105 | 446.0 | 2.0078 | 0.1065 | 445.8 | 2.0032 | 0.1027 | 445.5 | 1.9987 | 0.0992 | 445.3 | 1.9944 | 10 |
| 15 | 0.1128 | 450.3 | 2.0229 | 0.1087 | 450.1 | 2.0184 | 0.1049 | 449.9 | 2.0140 | 0.1014 | 449.7 | 2.0096 | 15 |
| 20 | 0.1151 | 454.6 | 2.0379 | 0.1110 | 454.5 | 2.0333 | 0.1071 | 454.3 | 2.0289 | 0.1035 | 454.1 | 2.0247 | 20 |
| 25 | 0.1174 | 459.0 | 2.0526 | 0.1132 | 458.8 | 2.0481 | 0.1093 | 458.6 | 2.0437 | 0.1056 | 458.4 | 2.0395 | 25 |
| 30 | 0.1197 | 463.4 | 2.0670 | 0.1154 | 463.2 | 2.0626 | 0.1114 | 463.0 | 2.0583 | 0.1077 | 462.8 | 2.0541 | 30 |
| 35 | 0.1219 | 467.7 | 2.0814 | 0.1176 | 467.6 | 2.0769 | 0.1135 | 467.4 | 2.0727 | 0.1097 | 467.2 | 2.0685 | 35 |
| 40 | 0.1242 | 472.1 | 2.0955 | 0.1197 | 472.0 | 2.0911 | 0.1156 | 471.8 | 2.0869 | 0.1117 | 471.7 | 2.0827 | 40 |
| 45 | 0.1264 | 476.5 | 2.1095 | 0.1219 | 476.4 | 2.1051 | 0.1177 | 476.2 | 2.1009 | 0.1138 | 476.1 | 2.0968 | 45 |
| 50 | 0.1286 | 481.0 | 2.1233 | 0.1240 | 480.8 | 2.1190 | 0.1198 | 480.7 | 2.1148 | 0.1158 | 480.6 | 2.1107 | 50 |
| 55 | 0.1308 | 485.4 | 2.1370 | 0.1261 | 485.3 | 2.1327 | 0.1218 | 485.2 | 2.1285 | 0.1178 | 485.0 | 2.1244 | 55 |
| 60 | 0.1330 | 489.9 | 2.1505 | 0.1283 | 489.8 | 2.1462 | 0.1239 | 489.7 | 2.1421 | 0.1198 | 489.5 | 2.1381 | 60 |
| 65 | 0.1351 | 494.4 | 2.1640 | 0.1304 | 494.3 | 2.1597 | 0.1259 | 494.2 | 2.1555 | 0.1218 | 494.0 | 2.1515 | 65 |
| 70 | 0.1373 | 498.9 | 2.1772 | 0.1325 | 498.8 | 2.1730 | 0.1280 | 498.7 | 2.1689 | 0.1237 | 498.6 | 2.1649 | 70 |
| 75 | 0.1395 | 503.5 | 2.1904 | 0.1346 | 503.4 | 2.1862 | 0.1300 | 503.3 | 2.1821 | 0.1257 | 503.2 | 2.1781 | 75 |
| 80 | 0.1416 | 508.1 | 2.2035 | 0.1366 | 508.0 | 2.1993 | 0.1320 | 507.9 | 2.1952 | 0.1276 | 507.8 | 2.1912 | 80 |
| 85 | 0.1438 | 512.7 | 2.2164 | 0.1387 | 512.6 | 2.2122 | 0.1340 | 512.5 | 2.2081 | 0.1296 | 512.4 | 2.2042 | 85 |
| 90 | 0.1459 | 517.3 | 2.2293 | 0.1408 | 517.2 | 2.2251 | 0.1360 | 517.1 | 2.2210 | 0.1315 | 517.0 | 2.2171 | 90 |
| 95 | 0.1481 | 522.0 | 2.2421 | 0.1429 | 521.9 | 2.2379 | 0.1380 | 521.8 | 2.2338 | 0.1335 | 521.7 | 2.2299 | 95 |
| 100 | 0.1502 | 526.7 | 2.2547 | 0.1449 | 526.6 | 2.2505 | 0.1400 | 526.5 | 2.2465 | 0.1354 | 526.4 | 2.2426 | 100 |
| 105 | 0.1523 | 531.4 | 2.2673 | 0.1470 | 531.3 | 2.2631 | 0.1420 | 531.2 | 2.2591 | 0.1373 | 531.1 | 2.2552 | 105 |
| 110 | 0.1544 | 536.1 | 2.2798 | 0.1490 | 536.1 | 2.2756 | 0.1440 | 536.0 | 2.2716 | 0.1393 | 535.9 | 2.2677 | 110 |
| 115 | 0.1566 | 540.9 | 2.2922 | 0.1511 | 540.8 | 2.2880 | 0.1460 | 540.8 | 2.2840 | 0.1412 | 540.7 | 2.2801 | 115 |
| 120 | 0.1587 | 545.7 | 2.3045 | 0.1531 | 545.7 | 2.3003 | 0.1480 | 545.6 | 2.2963 | 0.1431 | 545.5 | 2.2924 | 120 |
| 125 | 0.1608 | 550.6 | 2.3167 | 0.1552 | 550.5 | 2.3126 | 0.1499 | 550.4 | 2.3086 | 0.1450 | 550.4 | 2.3047 | 125 |
| −25 | 0.0804 | 414.1 | 1.8732 | – | – | – | – | – | – | – | – | – | −25 |
| −20 | 0.0828 | 418.6 | 1.8912 | 0.0800 | 418.3 | 1.8866 | 0.0774 | 417.9 | 1.8821 | 0.0750 | 417.5 | 1.8776 | −20 |

*(Continued)*

**TABLE A.19 (SI) (CONTINUED)**

Suva® 410A Superheated Vapor–Constant Pressure Tables (Saturated Vapor Properties in Parentheses)

| Temp | v | h | s | v | h | s | v | h | s | v | h | s |
|---|---|---|---|---|---|---|---|---|---|---|---|---|
| -15 | 0.0851 | 423.1 | 1.9087 | 0.0823 | 422.8 | 1.9042 | 0.0796 | 422.4 | 1.8998 | 0.0771 | 422.1 | 1.8955 |
| -10 | 0.0873 | 427.5 | 1.9257 | 0.0845 | 427.2 | 1.9213 | 0.0818 | 426.9 | 1.9170 | 0.0792 | 426.6 | 1.9128 |
| -5 | 0.0896 | 431.9 | 1.9423 | 0.0866 | 431.7 | 1.9380 | 0.0839 | 431.4 | 1.9338 | 0.0813 | 431.1 | 1.9297 |
| 0 | 0.0917 | 436.3 | 1.9586 | 0.0888 | 436.1 | 1.9543 | 0.0860 | 435.8 | 1.9502 | 0.0833 | 435.5 | 1.9461 |
| 5 | 0.0939 | 440.7 | 1.9745 | 0.0908 | 440.5 | 1.9703 | 0.0880 | 440.2 | 1.9662 | 0.0853 | 440.0 | 1.9622 |
| 10 | 0.0960 | 445.1 | 1.9901 | 0.0929 | 444.9 | 1.9860 | 0.0900 | 444.6 | 1.9820 | 0.0873 | 444.4 | 1.9780 |
| 15 | 0.0980 | 449.5 | 2.0055 | 0.0949 | 449.3 | 2.0014 | 0.0920 | 449.1 | 1.9974 | 0.0892 | 448.8 | 1.9935 |
| 20 | 0.1001 | 453.9 | 2.0205 | 0.0969 | 453.7 | 2.0165 | 0.0939 | 453.5 | 2.0126 | 0.0911 | 453.3 | 2.0088 |
| 25 | 0.1021 | 458.3 | 2.0354 | 0.0989 | 458.1 | 2.0314 | 0.0958 | 457.9 | 2.0275 | 0.0930 | 457.7 | 2.0237 |
| 30 | 0.1041 | 462.7 | 2.0500 | 0.1009 | 462.5 | 2.0461 | 0.0978 | 462.3 | 2.0422 | 0.0948 | 462.1 | 2.0385 |
| 35 | 0.1061 | 467.1 | 2.0645 | 0.1028 | 466.9 | 2.0606 | 0.0996 | 466.7 | 2.0567 | 0.0967 | 466.6 | 2.0530 |
| 40 | 0.1081 | 471.5 | 2.0787 | 0.1047 | 471.3 | 2.0749 | 0.1015 | 471.2 | 2.0711 | 0.0985 | 471.0 | 2.0674 |
| 45 | 0.1101 | 475.9 | 2.0928 | 0.1066 | 475.8 | 2.0890 | 0.1034 | 475.7 | 2.0852 | 0.1003 | 475.5 | 2.0815 |
| 50 | 0.1120 | 480.4 | 2.1067 | 0.1085 | 480.3 | 2.1029 | 0.1052 | 480.1 | 2.0992 | 0.1021 | 480.0 | 2.0955 |
| 55 | 0.1140 | 484.9 | 2.1205 | 0.1104 | 484.8 | 2.1167 | 0.1071 | 484.6 | 2.1130 | 0.1039 | 484.5 | 2.1094 |
| 60 | 0.1159 | 489.4 | 2.1341 | 0.1123 | 489.3 | 2.1303 | 0.1089 | 489.1 | 2.1266 | 0.1057 | 489.0 | 2.1230 |
| 65 | 0.1178 | 493.9 | 2.1476 | 0.1142 | 493.8 | 2.1438 | 0.1107 | 493.7 | 2.1402 | 0.1075 | 493.6 | 2.1366 |
| 70 | 0.1198 | 498.5 | 2.1610 | 0.1160 | 498.4 | 2.1572 | 0.1125 | 498.3 | 2.1536 | 0.1092 | 498.1 | 2.1500 |
| 75 | 0.1217 | 503.1 | 2.1742 | 0.1179 | 502.9 | 2.1705 | 0.1143 | 502.8 | 2.1668 | 0.1110 | 502.7 | 2.1633 |
| 80 | 0.1236 | 507.7 | 2.1873 | 0.1197 | 507.5 | 2.1836 | 0.1161 | 507.4 | 2.1800 | 0.1127 | 507.3 | 2.1764 |
| 85 | 0.1255 | 512.3 | 2.2003 | 0.1216 | 512.2 | 2.1966 | 0.1179 | 512.1 | 2.1930 | 0.1145 | 512.0 | 2.1895 |
| 90 | 0.1273 | 516.9 | 2.2132 | 0.1234 | 516.8 | 2.2095 | 0.1197 | 516.7 | 2.2059 | 0.1162 | 516.6 | 2.2024 |
| 95 | 0.1292 | 521.6 | 2.2260 | 0.1252 | 521.5 | 2.2223 | 0.1215 | 521.4 | 2.2187 | 0.1179 | 521.3 | 2.2152 |
| 100 | 0.1311 | 526.3 | 2.2387 | 0.1271 | 526.2 | 2.2351 | 0.1233 | 526.1 | 2.2315 | 0.1197 | 526.1 | 2.2280 |
| 105 | 0.1330 | 531.1 | 2.2514 | 0.1289 | 531.0 | 2.2477 | 0.1250 | 530.9 | 2.2441 | 0.1214 | 530.8 | 2.2406 |
| 110 | 0.1349 | 535.8 | 2.2639 | 0.1307 | 535.7 | 2.2602 | 0.1268 | 535.7 | 2.2566 | 0.1231 | 535.6 | 2.2531 |
| 115 | 0.1367 | 540.6 | 2.2763 | 0.1325 | 540.5 | 2.2726 | 0.1285 | 540.5 | 2.2691 | 0.1248 | 540.4 | 2.2656 |
| 120 | 0.1386 | 545.4 | 2.2886 | 0.1343 | 545.4 | 2.2850 | 0.1303 | 545.3 | 2.2814 | 0.1265 | 545.2 | 2.2780 |
| 125 | 0.1404 | 550.3 | 2.3009 | 0.1361 | 550.2 | 2.2972 | 0.1321 | 550.1 | 2.2937 | 0.1282 | 550.1 | 2.2902 |
| 130 | 0.1379 | 555.1 | 2.3094 | 0.1338 | 555.0 | 2.3059 | 0.1299 | 554.9 | 2.3024 | – | – | – |

*(Continued)*

## TABLE A.19 (SI) (CONTINUED)
Suva® 410A Superheated Vapor–Constant Pressure Tables (Saturated Vapor Properties in Parentheses)

| Temp. °C | 360.0 (−22.75°C) | | | 370.0 (−22.03°C) | | | 380.0 (−21.33°C) | | | 390.0 (−20.64°C) | | | Temp. °C |
|---|---|---|---|---|---|---|---|---|---|---|---|---|---|
| | V (0.0714) | H (414.6) | S (1.8632) | V (0.0696) | H (414.9) | S (1.8615) | V (0.0678) | H (415.2) | S (1.8599) | V (0.0661) | H (415.5) | S (1.8583) | |
| −20 | 0.0726 | 417.2 | 1.8733 | 0.0704 | 416.8 | 1.8690 | 0.0683 | 416.4 | 1.8648 | 0.0664 | 416.1 | 1.8607 | −20 |
| −15 | 0.0748 | 421.8 | 1.8912 | 0.0725 | 421.4 | 1.8871 | 0.0704 | 421.1 | 1.8830 | 0.0684 | 420.8 | 1.8790 | −15 |
| −10 | 0.0768 | 426.3 | 1.9087 | 0.0745 | 426.0 | 1.9046 | 0.0724 | 425.7 | 1.9006 | 0.0703 | 425.4 | 1.8968 | −10 |
| −5 | 0.0788 | 430.8 | 1.9256 | 0.0765 | 430.5 | 1.9217 | 0.0743 | 430.2 | 1.9178 | 0.0722 | 430.0 | 1.9140 | −5 |
| 0 | 0.0808 | 435.3 | 1.9422 | 0.0785 | 435.0 | 1.9383 | 0.0762 | 434.8 | 1.9345 | 0.0741 | 434.5 | 1.9308 | 0 |
| 5 | 0.0828 | 439.7 | 1.9583 | 0.0804 | 439.5 | 1.9545 | 0.0781 | 439.3 | 1.9508 | 0.0759 | 439.0 | 1.9472 | 5 |
| 10 | 0.0847 | 444.2 | 1.9742 | 0.0822 | 444.0 | 1.9704 | 0.0799 | 443.7 | 1.9668 | 0.0777 | 443.5 | 1.9632 | 10 |
| 15 | 0.0866 | 448.6 | 1.9897 | 0.0841 | 448.4 | 1.9860 | 0.0817 | 448.2 | 1.9824 | 0.0795 | 448.0 | 1.9789 | 15 |
| 20 | 0.0884 | 453.1 | 2.0050 | 0.0859 | 452.9 | 2.0014 | 0.0835 | 452.7 | 1.9978 | 0.0812 | 452.5 | 1.9943 | 20 |
| 25 | 0.0903 | 457.5 | 2.0200 | 0.0877 | 457.3 | 2.0164 | 0.0853 | 457.1 | 2.0129 | 0.0829 | 457.0 | 2.0094 | 25 |
| 30 | 0.0921 | 462.0 | 2.0348 | 0.0895 | 461.8 | 2.0313 | 0.0870 | 461.6 | 2.0278 | 0.0846 | 461.4 | 2.0243 | 30 |
| 35 | 0.0939 | 466.4 | 2.0494 | 0.0912 | 466.2 | 2.0459 | 0.0887 | 466.1 | 2.0424 | 0.0863 | 465.9 | 2.0390 | 35 |
| 40 | 0.0957 | 470.9 | 2.0638 | 0.0930 | 470.7 | 2.0603 | 0.0904 | 470.6 | 2.0568 | 0.0880 | 470.4 | 2.0535 | 40 |
| 45 | 0.0974 | 475.4 | 2.0780 | 0.0947 | 475.2 | 2.0745 | 0.0921 | 475.1 | 2.0711 | 0.0896 | 474.9 | 2.0678 | 45 |
| 50 | 0.0992 | 479.9 | 2.0920 | 0.0964 | 479.7 | 2.0885 | 0.0938 | 479.6 | 2.0851 | 0.0913 | 479.4 | 2.0818 | 50 |
| 55 | 0.1009 | 484.4 | 2.1058 | 0.0981 | 484.2 | 2.1024 | 0.0954 | 484.1 | 2.0990 | 0.0929 | 484.0 | 2.0958 | 55 |
| 60 | 0.1027 | 488.9 | 2.1195 | 0.0998 | 488.8 | 2.1161 | 0.0971 | 488.6 | 2.1128 | 0.0945 | 488.5 | 2.1095 | 60 |
| 65 | 0.1044 | 493.4 | 2.1331 | 0.1015 | 493.3 | 2.1297 | 0.0987 | 493.2 | 2.1264 | 0.0961 | 493.1 | 2.1231 | 65 |
| 70 | 0.1061 | 498.0 | 2.1465 | 0.1032 | 497.9 | 2.1431 | 0.1004 | 497.8 | 2.1398 | 0.0977 | 497.7 | 2.1366 | 70 |
| 75 | 0.1078 | 502.6 | 2.1598 | 0.1048 | 502.5 | 2.1564 | 0.1020 | 502.4 | 2.1532 | 0.0993 | 502.3 | 2.1500 | 75 |
| 80 | 0.1095 | 507.2 | 2.1730 | 0.1065 | 507.1 | 2.1696 | 0.1036 | 507.0 | 2.1664 | 0.1009 | 506.9 | 2.1632 | 80 |
| 85 | 0.1112 | 511.9 | 2.1860 | 0.1081 | 511.8 | 2.1827 | 0.1052 | 511.7 | 2.1794 | 0.1024 | 511.6 | 2.1763 | 85 |
| 90 | 0.1129 | 516.5 | 2.1990 | 0.1098 | 516.5 | 2.1957 | 0.1068 | 516.4 | 2.1924 | 0.1040 | 516.3 | 2.1892 | 90 |
| 95 | 0.1146 | 521.2 | 2.2118 | 0.1114 | 521.2 | 2.2085 | 0.1084 | 521.1 | 2.2053 | 0.1056 | 521.0 | 2.2021 | 95 |
| 100 | 0.1163 | 526.0 | 2.2246 | 0.1131 | 525.9 | 2.2213 | 0.1100 | 525.8 | 2.2180 | 0.1071 | 525.7 | 2.2149 | 100 |
| 105 | 0.1179 | 530.7 | 2.2372 | 0.1147 | 530.6 | 2.2339 | 0.1116 | 530.5 | 2.2307 | 0.1087 | 530.5 | 2.2276 | 105 |

*(Continued)*

# TABLE A.19 (SI) (CONTINUED)

Suva® 410A Superheated Vapor–Constant Pressure Tables (Saturated Vapor Properties in Parentheses)

| Temp. °C | 400.0 V | H | S | 425.0 V | H | S | 450.0 V | H | S | 475.0 V | H | S | Temp. °C |
|---|---|---|---|---|---|---|---|---|---|---|---|---|---|
| 110 | 0.1196 | 535.5 | 2.2498 | 0.1163 | 535.4 | 2.2465 | 0.1132 | 535.3 | 2.2433 | 0.1102 | 535.2 | 2.2401 | 110 |
| 115 | 0.1213 | 540.3 | 2.2622 | 0.1179 | 540.2 | 2.2589 | 0.1148 | 540.1 | 2.2557 | 0.1118 | 540.1 | 2.2526 | 115 |
| 120 | 0.1229 | 545.1 | 2.2746 | 0.1196 | 545.0 | 2.2713 | 0.1164 | 545.0 | 2.2681 | 0.1133 | 544.9 | 2.2650 | 120 |
| 125 | 0.1246 | 550.0 | 2.2869 | 0.1212 | 549.9 | 2.2836 | 0.1179 | 549.8 | 2.2804 | 0.1149 | 549.8 | 2.2773 | 125 |
| 130 | 0.1263 | 554.9 | 2.2991 | 0.1228 | 554.8 | 2.2958 | 0.1195 | 554.7 | 2.2926 | 0.1164 | 554.7 | 2.2895 | 130 |

| Temp. °C | 400.0 (−19.97°C) V | H | S | 425.0 (−18.34°C) V | H | S | 450.0 (−16.78°C) V | H | S | 475.0 (−15.29°C) V | H | S | Temp. °C |
|---|---|---|---|---|---|---|---|---|---|---|---|---|---|
| | (0.0645) | (415.7) | (1.8568) | (0.0608) | (416.4) | (1.8531) | (0.0575) | (417.0) | (1.8496) | (0.0545) | (417.5) | (1.8464) | |
| −15 | 0.0665 | 420.4 | 1.8751 | 0.0621 | 419.6 | 1.8655 | 0.0582 | 418.7 | 1.8563 | 0.0546 | 417.8 | 1.8474 | −15 |
| −10 | 0.0684 | 425.1 | 1.8929 | 0.0639 | 424.3 | 1.8837 | 0.0599 | 423.5 | 1.8747 | 0.0564 | 422.7 | 1.8661 | −10 |
| −5 | 0.0703 | 429.7 | 1.9103 | 0.0657 | 428.9 | 1.9012 | 0.0617 | 428.2 | 1.8925 | 0.0580 | 427.5 | 1.8842 | −5 |
| 0 | 0.0721 | 434.2 | 1.9271 | 0.0675 | 433.6 | 1.9183 | 0.0633 | 432.9 | 1.9098 | 0.0597 | 432.2 | 1.9017 | 0 |
| 5 | 0.0739 | 438.8 | 1.9436 | 0.0692 | 438.1 | 1.9349 | 0.0650 | 437.5 | 1.9266 | 0.0612 | 436.9 | 1.9187 | 5 |
| 10 | 0.0756 | 443.3 | 1.9597 | 0.0708 | 442.7 | 1.9512 | 0.0666 | 442.1 | 1.9430 | 0.0628 | 441.5 | 1.9352 | 10 |
| 15 | 0.0774 | 447.8 | 1.9754 | 0.0725 | 447.2 | 1.9670 | 0.0682 | 446.7 | 1.9591 | 0.0643 | 446.1 | 1.9514 | 15 |
| 20 | 0.0791 | 452.3 | 1.9909 | 0.0741 | 451.8 | 1.9826 | 0.0697 | 451.3 | 1.9748 | 0.0658 | 450.7 | 1.9672 | 20 |
| 25 | 0.0807 | 456.8 | 2.0061 | 0.0757 | 456.3 | 1.9979 | 0.0712 | 455.8 | 1.9902 | 0.0672 | 455.3 | 1.9828 | 25 |
| 30 | 0.0824 | 461.3 | 2.0210 | 0.0773 | 460.8 | 2.0130 | 0.0727 | 460.4 | 2.0053 | 0.0687 | 459.9 | 1.9980 | 30 |
| 35 | 0.0841 | 465.8 | 2.0357 | 0.0789 | 465.3 | 2.0278 | 0.0742 | 464.9 | 2.0202 | 0.0701 | 464.5 | 2.0130 | 35 |
| 40 | 0.0857 | 470.3 | 2.0502 | 0.0804 | 469.9 | 2.0423 | 0.0757 | 469.5 | 2.0348 | 0.0715 | 469.1 | 2.0277 | 40 |
| 45 | 0.0873 | 474.8 | 2.0645 | 0.0819 | 474.4 | 2.0567 | 0.0772 | 474.0 | 2.0493 | 0.0729 | 473.6 | 2.0422 | 45 |
| 50 | 0.0889 | 479.3 | 2.0786 | 0.0834 | 478.9 | 2.0709 | 0.0786 | 478.6 | 2.0635 | 0.0743 | 478.2 | 2.0565 | 50 |
| 55 | 0.0905 | 483.8 | 2.0926 | 0.0849 | 483.5 | 2.0849 | 0.0800 | 483.2 | 2.0776 | 0.0756 | 482.8 | 2.0706 | 55 |
| 60 | 0.0921 | 488.4 | 2.1064 | 0.0864 | 488.1 | 2.0987 | 0.0814 | 487.8 | 2.0915 | 0.0770 | 487.4 | 2.0846 | 60 |
| 65 | 0.0936 | 493.0 | 2.1200 | 0.0879 | 492.7 | 2.1124 | 0.0829 | 492.4 | 2.1052 | 0.0783 | 492.1 | 2.0983 | 65 |
| 70 | 0.0952 | 497.6 | 2.1335 | 0.0894 | 497.3 | 2.1259 | 0.0843 | 497.0 | 2.1188 | 0.0797 | 496.7 | 2.1119 | 70 |
| 75 | 0.0967 | 502.2 | 2.1468 | 0.0909 | 501.9 | 2.1393 | 0.0856 | 501.6 | 2.1322 | 0.0810 | 501.4 | 2.1254 | 75 |
| 80 | 0.0983 | 506.8 | 2.1601 | 0.0923 | 506.6 | 2.1526 | 0.0870 | 506.3 | 2.1455 | 0.0823 | 506.0 | 2.1387 | 80 |

(Continued)

## TABLE A.19 (SI) (CONTINUED)

Suva® 410A Superheated Vapor–Constant Pressure Tables (Saturated Vapor Properties in Parentheses)

| Temp. °C | 500.0 (−13.85°C) | | | 525.0 (−12.47°C) | | | 550.0 (−11.14°C) | | | 575.0 (−9.85°C) | | | Temp. °C |
|---|---|---|---|---|---|---|---|---|---|---|---|---|---|
| | V | H | S | V | H | S | V | H | S | V | H | S | |
| | (0.0519) | (418.0) | (1.8432) | (0.0494) | (418.5) | (1.8403) | (0.0472) | (419.0) | (1.8375) | (0.0452) | (419.4) | (1.8347) | |
| −10 | 0.0532 | 421.8 | 1.8578 | 0.0502 | 421.0 | 1.8497 | 0.0476 | 420.2 | 1.8419 | – | – | – | −10 |
| −5 | 0.0548 | 426.7 | 1.8761 | 0.0518 | 425.9 | 1.8683 | 0.0491 | 425.2 | 1.8607 | 0.0466 | 424.4 | 1.8533 | −5 |
| 0 | 0.0563 | 431.5 | 1.8938 | 0.0533 | 430.8 | 1.8862 | 0.0506 | 430.1 | 1.8789 | 0.0481 | 429.4 | 1.8717 | 0 |
| 5 | 0.0579 | 436.2 | 1.9110 | 0.0548 | 435.6 | 1.9036 | 0.0520 | 434.9 | 1.8965 | 0.0495 | 434.3 | 1.8895 | 5 |
| 10 | 0.0593 | 440.9 | 1.9277 | 0.0562 | 440.3 | 1.9205 | 0.0534 | 439.7 | 1.9135 | 0.0508 | 439.1 | 1.9068 | 10 |
| 15 | 0.0608 | 445.6 | 1.9441 | 0.0576 | 445.0 | 1.9370 | 0.0548 | 444.5 | 1.9302 | 0.0521 | 443.9 | 1.9236 | 15 |
| 20 | 0.0622 | 450.2 | 1.9600 | 0.0590 | 449.7 | 1.9531 | 0.0561 | 449.2 | 1.9464 | 0.0534 | 448.7 | 1.9399 | 20 |
| 25 | 0.0636 | 454.9 | 1.9757 | 0.0604 | 454.4 | 1.9688 | 0.0574 | 453.9 | 1.9623 | 0.0547 | 453.4 | 1.9559 | 25 |
| 30 | 0.0650 | 459.5 | 1.9910 | 0.0617 | 459.0 | 1.9843 | 0.0587 | 458.5 | 1.9778 | 0.0559 | 458.1 | 1.9715 | 30 |
| 35 | 0.0664 | 464.1 | 2.0060 | 0.0630 | 463.6 | 1.9994 | 0.0599 | 463.2 | 1.9930 | 0.0571 | 462.8 | 1.9869 | 35 |
| 40 | 0.0677 | 468.7 | 2.0209 | 0.0643 | 468.3 | 2.0143 | 0.0612 | 467.9 | 2.0080 | 0.0583 | 467.4 | 2.0019 | 40 |
| 45 | 0.0690 | 473.3 | 2.0354 | 0.0656 | 472.9 | 2.0290 | 0.0624 | 472.5 | 2.0227 | 0.0595 | 472.1 | 2.0167 | 45 |
| 50 | 0.0704 | 477.9 | 2.0498 | 0.0668 | 477.5 | 2.0434 | 0.0636 | 477.2 | 2.0372 | 0.0607 | 476.8 | 2.0313 | 50 |
| 55 | 0.0717 | 482.5 | 2.0640 | 0.0681 | 482.1 | 2.0576 | 0.0648 | 481.8 | 2.0515 | 0.0618 | 481.5 | 2.0456 | 55 |
| 85 | 0.0998 | 511.5 | 2.1732 | 0.0938 | 511.2 | 2.1657 | 0.0884 | 511.0 | 2.1587 | 0.0836 | 510.7 | 2.1519 | 85 |
| 90 | 0.1013 | 516.2 | 2.1862 | 0.0952 | 515.9 | 2.1787 | 0.0898 | 515.7 | 2.1717 | 0.0849 | 515.4 | 2.1650 | 90 |
| 95 | 0.1029 | 520.9 | 2.1990 | 0.0967 | 520.6 | 2.1916 | 0.0912 | 520.4 | 2.1846 | 0.0862 | 520.2 | 2.1780 | 95 |
| 100 | 0.1044 | 525.6 | 2.2118 | 0.0981 | 525.4 | 2.2044 | 0.0925 | 525.2 | 2.1975 | 0.0875 | 524.9 | 2.1908 | 100 |
| 105 | 0.1059 | 530.4 | 2.2245 | 0.0995 | 530.2 | 2.2171 | 0.0939 | 529.9 | 2.2102 | 0.0888 | 529.7 | 2.2036 | 105 |
| 110 | 0.1074 | 535.2 | 2.2371 | 0.1010 | 535.0 | 2.2297 | 0.0952 | 534.7 | 2.2228 | 0.0901 | 534.5 | 2.2162 | 110 |
| 115 | 0.1089 | 540.0 | 2.2496 | 0.1024 | 539.8 | 2.2422 | 0.0966 | 539.6 | 2.2353 | 0.0914 | 539.4 | 2.2288 | 115 |
| 120 | 0.1104 | 544.8 | 2.2620 | 0.1038 | 544.6 | 2.2547 | 0.0979 | 544.4 | 2.2478 | 0.0926 | 544.2 | 2.2412 | 120 |
| 125 | 0.1119 | 549.7 | 2.2743 | 0.1052 | 549.5 | 2.2670 | 0.0993 | 549.3 | 2.2601 | 0.0939 | 549.1 | 2.2536 | 125 |
| 130 | 0.1134 | 554.6 | 2.2865 | 0.1066 | 554.4 | 2.2792 | 0.1006 | 554.2 | 2.2724 | 0.0952 | 554.0 | 2.2658 | 130 |
| 135 | 0.1149 | 559.5 | 2.2986 | 0.1080 | 559.3 | 2.2914 | 0.1019 | 559.2 | 2.2845 | 0.0965 | 559.0 | 2.2780 | 135 |

*(Continued)*

## TABLE A.19 (SI) (CONTINUED)

Suva® 410A Superheated Vapor–Constant Pressure Tables (Saturated Vapor Properties in Parentheses)

| Temp. °C | 600.0 (−8.61°C) V | H | S | 625.0 (−7.40°C) V | H | S | 650.0 (−6.24°C) V | H | S | 675.0 (−5.10°C) V | H | S |
|---|---|---|---|---|---|---|---|---|---|---|---|---|
| | (0.0433) | (419.9) | (1.8322) | (0.0416) | (420.3) | (1.8297) | (0.0400) | (420.6) | (1.8273) | (0.0385) | (421.0) | (1.8249) |
| −5 | 0.0444 | 423.6 | 1.8461 | 0.0423 | 422.8 | 1.8390 | 0.0404 | 422.0 | 1.8321 | 0.0386 | 421.1 | 1.8253 |
| 0 | 0.0458 | 428.6 | 1.8648 | 0.0437 | 427.9 | 1.8580 | 0.0417 | 427.1 | 1.8513 | 0.0399 | 426.4 | 1.8448 |
| 5 | 0.0471 | 433.6 | 1.8828 | 0.0450 | 432.9 | 1.8762 | 0.0430 | 432.2 | 1.8698 | 0.0411 | 431.5 | 1.8635 |
| 10 | 0.0484 | 438.5 | 1.9002 | 0.0463 | 437.9 | 1.8938 | 0.0442 | 437.2 | 1.8876 | 0.0424 | 436.6 | 1.8815 |
| 15 | 0.0497 | 443.3 | 1.9172 | 0.0475 | 442.8 | 1.9109 | 0.0454 | 442.2 | 1.9049 | 0.0435 | 441.6 | 1.8990 |
| 20 | 0.0510 | 448.1 | 1.9337 | 0.0487 | 447.6 | 1.9276 | 0.0466 | 447.0 | 1.9217 | 0.0447 | 446.5 | 1.9159 |
| 25 | 0.0522 | 452.9 | 1.9498 | 0.0499 | 452.4 | 1.9438 | 0.0478 | 451.9 | 1.9380 | 0.0458 | 451.4 | 1.9324 |
| 60 | 0.0730 | 487.1 | 2.0780 | 0.0693 | 486.8 | 2.0717 | 0.0660 | 486.5 | 2.0656 | 0.0630 | 486.1 | 2.0598 |
| 65 | 0.0742 | 491.8 | 2.0918 | 0.0705 | 491.4 | 2.0855 | 0.0672 | 491.1 | 2.0795 | 0.0641 | 490.8 | 2.0738 |
| 70 | 0.0755 | 496.4 | 2.1054 | 0.0718 | 496.1 | 2.0992 | 0.0684 | 495.8 | 2.0933 | 0.0653 | 495.5 | 2.0876 |
| 75 | 0.0768 | 501.1 | 2.1190 | 0.0730 | 500.8 | 2.1128 | 0.0695 | 500.5 | 2.1069 | 0.0664 | 500.2 | 2.1012 |
| 80 | 0.0780 | 505.8 | 2.1323 | 0.0742 | 505.5 | 2.1262 | 0.0707 | 505.2 | 2.1203 | 0.0675 | 505.0 | 2.1147 |
| 85 | 0.0793 | 510.5 | 2.1456 | 0.0754 | 510.2 | 2.1394 | 0.0718 | 510.0 | 2.1336 | 0.0686 | 509.7 | 2.1280 |
| 90 | 0.0805 | 515.2 | 2.1587 | 0.0766 | 515.0 | 2.1526 | 0.0730 | 514.7 | 2.1468 | 0.0697 | 514.5 | 2.1412 |
| 95 | 0.0818 | 519.9 | 2.1717 | 0.0778 | 519.7 | 2.1656 | 0.0741 | 519.5 | 2.1598 | 0.0708 | 519.2 | 2.1543 |
| 100 | 0.0830 | 524.7 | 2.1845 | 0.0789 | 524.5 | 2.1785 | 0.0752 | 524.3 | 2.1727 | 0.0719 | 524.0 | 2.1672 |
| 105 | 0.0842 | 529.5 | 2.1973 | 0.0801 | 529.3 | 2.1913 | 0.0764 | 529.1 | 2.1855 | 0.0729 | 528.9 | 2.1800 |
| 110 | 0.0855 | 534.3 | 2.2100 | 0.0813 | 534.1 | 2.2040 | 0.0775 | 533.9 | 2.1983 | 0.0740 | 533.7 | 2.1928 |
| 115 | 0.0867 | 539.2 | 2.2225 | 0.0825 | 539.0 | 2.2166 | 0.0786 | 538.8 | 2.2109 | 0.0751 | 538.6 | 2.2054 |
| 120 | 0.0879 | 544.0 | 2.2350 | 0.0836 | 543.9 | 2.2290 | 0.0797 | 543.7 | 2.2234 | 0.0762 | 543.5 | 2.2179 |
| 125 | 0.0891 | 548.9 | 2.2474 | 0.0848 | 548.8 | 2.2414 | 0.0808 | 548.6 | 2.2358 | 0.0772 | 548.4 | 2.2303 |
| 130 | 0.0903 | 553.9 | 2.2596 | 0.0859 | 553.7 | 2.2537 | 0.0819 | 553.5 | 2.2481 | 0.0783 | 553.3 | 2.2427 |
| 135 | 0.0915 | 558.8 | 2.2718 | 0.0871 | 558.6 | 2.2660 | 0.0830 | 558.5 | 2.2603 | 0.0793 | 558.3 | 2.2549 |
| 140 | 0.0927 | 563.8 | 2.2840 | 0.0882 | 563.6 | 2.2781 | 0.0841 | 563.5 | 2.2725 | 0.0804 | 563.3 | 2.2671 |
| 145 | – | – | – | 0.0815 | 568.3 | 2.2791 | – | – | – | – | – | – |

(Continued)

**TABLE A.19 (SI) (CONTINUED)**

Suva® 410A Superheated Vapor–Constant Pressure Tables (Saturated Vapor Properties in Parentheses)

| Temp. °C | 700.0 (−4.00°C) | | | 725.0 (−2.92°C) | | | 750.0 (−1.88°C) | | | 800.0 (0.14°C) | | | Temp. °C |
|---|---|---|---|---|---|---|---|---|---|---|---|---|---|
| | V | H | S | V | H | S | V | H | S | V | H | S | |
| | (0.0372) | (421.4) | (1.8227) | (0.0359) | (421.7) | (1.8206) | (0.0347) | (422.0) | (1.8185) | (0.0325) | (422.6) | (1.8145) | |
| 30 | 0.0534 | 457.6 | 1.9655 | 0.0511 | 457.2 | 1.9597 | 0.0489 | 456.7 | 1.9540 | 0.0469 | 456.2 | 1.9485 | 30 |
| 35 | 0.0546 | 462.3 | 1.9809 | 0.0522 | 461.9 | 1.9752 | 0.0500 | 461.5 | 1.9696 | 0.0480 | 461.0 | 1.9642 | 35 |
| 40 | 0.0557 | 467.0 | 1.9961 | 0.0533 | 466.6 | 1.9904 | 0.0511 | 466.2 | 1.9849 | 0.0491 | 465.8 | 1.9796 | 40 |
| 45 | 0.0569 | 471.7 | 2.0110 | 0.0544 | 471.3 | 2.0054 | 0.0522 | 471.0 | 2.0000 | 0.0501 | 470.6 | 1.9947 | 45 |
| 50 | 0.0580 | 476.4 | 2.0256 | 0.0555 | 476.1 | 2.0201 | 0.0532 | 475.7 | 2.0147 | 0.0511 | 475.3 | 2.0095 | 50 |
| 55 | 0.0591 | 481.1 | 2.0400 | 0.0566 | 480.8 | 2.0345 | 0.0543 | 480.4 | 2.0293 | 0.0521 | 480.1 | 2.0241 | 55 |
| 60 | 0.0602 | 485.8 | 2.0542 | 0.0577 | 485.5 | 2.0488 | 0.0553 | 485.2 | 2.0436 | 0.0531 | 484.8 | 2.0385 | 60 |
| 65 | 0.0613 | 490.5 | 2.0682 | 0.0587 | 490.2 | 2.0629 | 0.0563 | 489.9 | 2.0577 | 0.0541 | 489.6 | 2.0527 | 65 |
| 70 | 0.0624 | 495.2 | 2.0820 | 0.0598 | 494.9 | 2.0767 | 0.0574 | 494.6 | 2.0716 | 0.0551 | 494.3 | 2.0666 | 70 |
| 75 | 0.0635 | 500.0 | 2.0957 | 0.0608 | 499.7 | 2.0904 | 0.0584 | 499.4 | 2.0854 | 0.0561 | 499.1 | 2.0804 | 75 |
| 80 | 0.0645 | 504.7 | 2.1092 | 0.0618 | 504.4 | 2.1040 | 0.0594 | 504.2 | 2.0989 | 0.0570 | 503.9 | 2.0941 | 80 |
| 85 | 0.0656 | 509.4 | 2.1226 | 0.0629 | 509.2 | 2.1174 | 0.0603 | 508.9 | 2.1124 | 0.0580 | 508.7 | 2.1075 | 85 |
| 90 | 0.0667 | 514.2 | 2.1358 | 0.0639 | 514.0 | 2.1307 | 0.0613 | 513.7 | 2.1257 | 0.0590 | 513.5 | 2.1208 | 90 |
| 95 | 0.0677 | 519.0 | 2.1489 | 0.0649 | 518.8 | 2.1438 | 0.0623 | 518.5 | 2.1388 | 0.0599 | 518.3 | 2.1340 | 95 |
| 100 | 0.0688 | 523.8 | 2.1619 | 0.0659 | 523.6 | 2.1568 | 0.0633 | 523.4 | 2.1518 | 0.0608 | 523.1 | 2.1471 | 100 |
| 105 | 0.0698 | 528.6 | 2.1748 | 0.0669 | 528.4 | 2.1697 | 0.0642 | 528.2 | 2.1647 | 0.0618 | 528.0 | 2.1600 | 105 |
| 110 | 0.0708 | 533.5 | 2.1875 | 0.0679 | 533.3 | 2.1824 | 0.0652 | 533.1 | 2.1775 | 0.0627 | 532.9 | 2.1728 | 110 |
| 115 | 0.0719 | 538.4 | 2.2001 | 0.0689 | 538.2 | 2.1951 | 0.0662 | 538.0 | 2.1902 | 0.0636 | 537.8 | 2.1855 | 115 |
| 120 | 0.0729 | 543.3 | 2.2127 | 0.0699 | 543.1 | 2.2076 | 0.0671 | 542.9 | 2.2028 | 0.0646 | 542.7 | 2.1981 | 120 |
| 125 | 0.0739 | 548.2 | 2.2251 | 0.0709 | 548.0 | 2.2201 | 0.0681 | 547.8 | 2.2153 | 0.0655 | 547.6 | 2.2106 | 125 |
| 130 | 0.0749 | 553.1 | 2.2375 | 0.0719 | 553.0 | 2.2325 | 0.0690 | 552.8 | 2.2277 | 0.0664 | 552.6 | 2.2230 | 130 |
| 135 | 0.0760 | 558.1 | 2.2497 | 0.0728 | 557.9 | 2.2447 | 0.0700 | 557.8 | 2.2399 | 0.0673 | 557.6 | 2.2353 | 135 |
| 140 | 0.0770 | 563.1 | 2.2619 | 0.0738 | 562.9 | 2.2569 | 0.0709 | 562.8 | 2.2521 | 0.0682 | 562.6 | 2.2475 | 140 |
| 145 | 0.0780 | 568.1 | 2.2740 | 0.0748 | 568.0 | 2.2690 | 0.0718 | 567.8 | 2.2642 | 0.0691 | 567.6 | 2.2596 | 145 |

| Temp. °C | 700.0 (−4.00°C) | | | 725.0 (−2.92°C) | | | 750.0 (−1.88°C) | | | 800.0 (0.14°C) | | | Temp. °C |
|---|---|---|---|---|---|---|---|---|---|---|---|---|---|
| | V | H | S | V | H | S | V | H | S | V | H | S | |
| | (0.0372) | (421.4) | (1.8227) | (0.0359) | (421.7) | (1.8206) | (0.0347) | (422.0) | (1.8185) | (0.0325) | (422.6) | (1.8145) | |
| 0 | 0.0382 | 425.6 | 1.8384 | 0.0366 | 424.8 | 1.8322 | 0.0351 | 424.1 | 1.8260 | 0 | | | 0 |
| 5 | 0.0394 | 430.8 | 1.8574 | 0.0378 | 430.1 | 1.8514 | 0.0363 | 429.4 | 1.8454 | 0.0336 | 427.9 | 1.8339 | 5 |

*(Continued)*

**TABLE A.19 (SI) (CONTINUED)**

Suva® 410A Superheated Vapor–Constant Pressure Tables (Saturated Vapor Properties in Parentheses)

| Temp | v | h | s | v | h | s | v | h | s | v | h | s | Temp |
|---|---|---|---|---|---|---|---|---|---|---|---|---|---|
| 10 | 0.0406 | 436.0 | 1.8756 | 0.0390 | 435.3 | 1.8698 | 0.0375 | 434.6 | 1.8641 | 0.0347 | 433.3 | 1.8530 | 10 |
| 15 | 0.0418 | 441.0 | 1.8932 | 0.0401 | 440.4 | 1.8876 | 0.0386 | 439.8 | 1.8821 | 0.0358 | 438.6 | 1.8714 | 15 |
| 20 | 0.0429 | 446.0 | 1.9103 | 0.0412 | 445.4 | 1.9048 | 0.0397 | 444.8 | 1.8995 | 0.0368 | 443.7 | 1.8891 | 20 |
| 25 | 0.0440 | 450.9 | 1.9269 | 0.0423 | 450.3 | 1.9216 | 0.0407 | 449.8 | 1.9164 | 0.0378 | 448.8 | 1.9062 | 25 |
| 30 | 0.0451 | 455.7 | 1.9431 | 0.0433 | 455.3 | 1.9379 | 0.0417 | 454.8 | 1.9328 | 0.0388 | 453.8 | 1.9229 | 30 |
| 35 | 0.0461 | 460.6 | 1.9589 | 0.0444 | 460.1 | 1.9538 | 0.0427 | 459.7 | 1.9488 | 0.0398 | 458.8 | 1.9392 | 35 |
| 40 | 0.0471 | 465.4 | 1.9744 | 0.0454 | 465.0 | 1.9694 | 0.0437 | 464.5 | 1.9645 | 0.0407 | 463.7 | 1.9550 | 40 |
| 45 | 0.0482 | 470.2 | 1.9896 | 0.0464 | 469.8 | 1.9847 | 0.0447 | 469.4 | 1.9798 | 0.0416 | 468.6 | 1.9705 | 45 |
| 50 | 0.0492 | 475.0 | 2.0045 | 0.0473 | 474.6 | 1.9996 | 0.0456 | 474.2 | 1.9949 | 0.0425 | 473.5 | 1.9857 | 50 |
| 55 | 0.0501 | 479.7 | 2.0192 | 0.0483 | 479.4 | 2.0144 | 0.0465 | 479.0 | 2.0097 | 0.0434 | 478.3 | 2.0007 | 55 |
| 60 | 0.0511 | 484.5 | 2.0336 | 0.0492 | 484.2 | 2.0288 | 0.0475 | 483.8 | 2.0242 | 0.0443 | 483.2 | 2.0153 | 60 |
| 65 | 0.0521 | 489.3 | 2.0478 | 0.0502 | 489.0 | 2.0431 | 0.0484 | 488.6 | 2.0385 | 0.0451 | 488.0 | 2.0297 | 65 |
| 70 | 0.0530 | 494.0 | 2.0618 | 0.0511 | 493.7 | 2.0572 | 0.0493 | 493.4 | 2.0526 | 0.0460 | 492.8 | 2.0439 | 70 |
| 75 | 0.0540 | 498.8 | 2.0757 | 0.0520 | 498.5 | 2.0710 | 0.0502 | 498.3 | 2.0666 | 0.0468 | 497.7 | 2.0579 | 75 |
| 80 | 0.0549 | 503.6 | 2.0893 | 0.0529 | 503.3 | 2.0847 | 0.0510 | 503.1 | 2.0803 | 0.0477 | 502.5 | 2.0718 | 80 |
| 85 | 0.0558 | 508.4 | 2.1028 | 0.0538 | 508.2 | 2.0983 | 0.0519 | 507.9 | 2.0939 | 0.0485 | 507.4 | 2.0854 | 85 |
| 90 | 0.0568 | 513.2 | 2.1162 | 0.0547 | 513.0 | 2.1117 | 0.0528 | 512.7 | 2.1073 | 0.0493 | 512.2 | 2.0989 | 90 |
| 95 | 0.0577 | 518.1 | 2.1294 | 0.0556 | 517.8 | 2.1249 | 0.0536 | 517.6 | 2.1205 | 0.0501 | 517.1 | 2.1122 | 95 |
| 100 | 0.0586 | 522.9 | 2.1425 | 0.0565 | 522.7 | 2.1380 | 0.0545 | 522.5 | 2.1337 | 0.0509 | 522.0 | 2.1254 | 100 |
| 105 | 0.0595 | 527.8 | 2.1554 | 0.0573 | 527.6 | 2.1510 | 0.0554 | 527.3 | 2.1467 | 0.0517 | 526.9 | 2.1384 | 105 |
| 110 | 0.0604 | 532.7 | 2.1683 | 0.0582 | 532.5 | 2.1638 | 0.0562 | 532.2 | 2.1596 | 0.0525 | 531.8 | 2.1514 | 110 |
| 115 | 0.0613 | 537.6 | 2.1810 | 0.0591 | 537.4 | 2.1766 | 0.0570 | 537.2 | 2.1723 | 0.0533 | 536.8 | 2.1642 | 115 |
| 120 | 0.0622 | 542.5 | 2.1936 | 0.0600 | 542.3 | 2.1892 | 0.0579 | 542.1 | 2.1850 | 0.0541 | 541.7 | 2.1768 | 120 |
| 125 | 0.0631 | 547.4 | 2.2061 | 0.0608 | 547.3 | 2.2017 | 0.0587 | 547.1 | 2.1975 | 0.0549 | 546.7 | 2.1894 | 125 |
| 130 | 0.0639 | 552.4 | 2.2185 | 0.0617 | 552.2 | 2.2142 | 0.0595 | 552.0 | 2.2100 | 0.0557 | 551.7 | 2.2019 | 130 |
| 135 | 0.0648 | 557.4 | 2.2308 | 0.0625 | 557.2 | 2.2265 | 0.0604 | 557.1 | 2.2223 | 0.0565 | 556.7 | 2.2143 | 135 |
| 140 | 0.0657 | 562.4 | 2.2431 | 0.0634 | 562.3 | 2.2387 | 0.0612 | 562.1 | 2.2345 | 0.0573 | 561.7 | 2.2266 | 140 |
| 145 | 0.0666 | 567.5 | 2.2552 | 0.0642 | 567.3 | 2.2509 | 0.0620 | 567.1 | 2.2467 | 0.0580 | 566.8 | 2.2387 | 145 |
| 150 | 0.0675 | 572.5 | 2.2672 | 0.0651 | 572.4 | 2.2629 | 0.0628 | 572.2 | 2.2588 | 0.0588 | 571.9 | 2.2508 | 150 |
| 155 | – | – | – | – | – | – | – | – | – | 0.0596 | 577.0 | 2.2629 | 155 |

*(Continued)*

# TABLE A.19 (SI) (CONTINUED)

Suva® 410A Superheated Vapor–Constant Pressure Tables (Saturated Vapor Properties in Parentheses)

| Temp. °C | 850.0 (2.06°C) | | | 900.0 (3.90°C) | | | 950.0 (5.66°C) | | | 1000.0 (7.36°C) | | | Temp. °C |
|---|---|---|---|---|---|---|---|---|---|---|---|---|---|
| | V (0.0305) | H (423.1) | S (1.8107) | V (0.0288) | H (423.6) | S (1.8071) | V (0.0273) | H (424.1) | S (1.8036) | V (0.0259) | H (424.5) | S (1.8004) | |
| 5 | 0.0312 | 426.4 | 1.8226 | 0.0291 | 424.9 | 1.8116 | – | – | – | – | – | – | 5 |
| 10 | 0.0323 | 431.9 | 1.8422 | 0.0301 | 430.5 | 1.8318 | 0.0282 | 429.1 | 1.8215 | 0.0264 | 427.6 | 1.8115 | 10 |
| 15 | 0.0333 | 437.3 | 1.8610 | 0.0311 | 436.0 | 1.8510 | 0.0291 | 434.7 | 1.8412 | 0.0274 | 433.4 | 1.8316 | 15 |
| 20 | 0.0343 | 442.6 | 1.8791 | 0.0321 | 441.4 | 1.8694 | 0.0301 | 440.2 | 1.8600 | 0.0283 | 438.9 | 1.8508 | 20 |
| 25 | 0.0353 | 447.7 | 1.8965 | 0.0330 | 446.6 | 1.8872 | 0.0310 | 445.5 | 1.8781 | 0.0292 | 444.4 | 1.8693 | 25 |
| 30 | 0.0362 | 452.8 | 1.9135 | 0.0339 | 451.8 | 1.9044 | 0.0319 | 450.8 | 1.8956 | 0.0300 | 449.7 | 1.8870 | 30 |
| 35 | 0.0372 | 457.8 | 1.9299 | 0.0348 | 456.9 | 1.9211 | 0.0327 | 455.9 | 1.9125 | 0.0309 | 455.0 | 1.9042 | 35 |
| 40 | 0.0380 | 462.8 | 1.9460 | 0.0357 | 461.9 | 1.9373 | 0.0336 | 461.1 | 1.9290 | 0.0317 | 460.2 | 1.9209 | 40 |
| 45 | 0.0389 | 467.8 | 1.9617 | 0.0365 | 467.0 | 1.9532 | 0.0344 | 466.1 | 1.9450 | 0.0324 | 465.3 | 1.9371 | 45 |
| 50 | 0.0398 | 472.7 | 1.9770 | 0.0374 | 471.9 | 1.9687 | 0.0352 | 471.1 | 1.9607 | 0.0332 | 470.4 | 1.9529 | 50 |
| 55 | 0.0406 | 477.6 | 1.9921 | 0.0382 | 476.9 | 1.9839 | 0.0360 | 476.1 | 1.9760 | 0.0340 | 475.4 | 1.9684 | 55 |
| 60 | 0.0415 | 482.5 | 2.0068 | 0.0390 | 481.8 | 1.9988 | 0.0367 | 481.1 | 1.9910 | 0.0347 | 480.4 | 1.9836 | 60 |
| 65 | 0.0423 | 487.4 | 2.0214 | 0.0397 | 486.7 | 2.0134 | 0.0375 | 486.1 | 2.0058 | 0.0354 | 485.4 | 1.9984 | 65 |
| 70 | 0.0431 | 492.2 | 2.0357 | 0.0405 | 491.6 | 2.0278 | 0.0382 | 491.0 | 2.0203 | 0.0361 | 490.4 | 2.0130 | 70 |
| 75 | 0.0439 | 497.1 | 2.0498 | 0.0413 | 496.5 | 2.0420 | 0.0389 | 495.9 | 2.0345 | 0.0368 | 495.3 | 2.0274 | 75 |
| 80 | 0.0447 | 502.0 | 2.0637 | 0.0420 | 501.4 | 2.0560 | 0.0397 | 500.9 | 2.0486 | 0.0375 | 500.3 | 2.0415 | 80 |
| 85 | 0.0455 | 506.8 | 2.0774 | 0.0428 | 506.3 | 2.0697 | 0.0404 | 505.8 | 2.0625 | 0.0382 | 505.3 | 2.0555 | 85 |
| 90 | 0.0463 | 511.7 | 2.0909 | 0.0435 | 511.2 | 2.0834 | 0.0411 | 510.7 | 2.0761 | 0.0389 | 510.2 | 2.0692 | 90 |
| 95 | 0.0470 | 516.6 | 2.1043 | 0.0443 | 516.1 | 2.0968 | 0.0418 | 515.7 | 2.0896 | 0.0396 | 515.2 | 2.0828 | 95 |
| 100 | 0.0478 | 521.5 | 2.1175 | 0.0450 | 521.1 | 2.1101 | 0.0425 | 520.6 | 2.1030 | 0.0402 | 520.1 | 2.0962 | 100 |
| 105 | 0.0486 | 526.5 | 2.1306 | 0.0457 | 526.0 | 2.1232 | 0.0432 | 525.6 | 2.1162 | 0.0409 | 525.1 | 2.1094 | 105 |
| 110 | 0.0493 | 531.4 | 2.1436 | 0.0464 | 531.0 | 2.1362 | 0.0439 | 530.5 | 2.1292 | 0.0416 | 530.1 | 2.1225 | 110 |
| 115 | 0.0501 | 536.3 | 2.1565 | 0.0472 | 535.9 | 2.1491 | 0.0446 | 535.5 | 2.1422 | 0.0422 | 535.1 | 2.1355 | 115 |
| 120 | 0.0508 | 541.3 | 2.1692 | 0.0479 | 540.9 | 2.1619 | 0.0452 | 540.5 | 2.1550 | 0.0429 | 540.1 | 2.1484 | 120 |
| 125 | 0.0516 | 546.3 | 2.1818 | 0.0486 | 545.9 | 2.1745 | 0.0459 | 545.5 | 2.1677 | 0.0435 | 545.2 | 2.1611 | 125 |
| 130 | 0.0523 | 551.3 | 2.1943 | 0.0493 | 550.9 | 2.1871 | 0.0466 | 550.6 | 2.1802 | 0.0441 | 550.2 | 2.1737 | 130 |

*(Continued)*

**TABLE A.19 (SI) (CONTINUED)**

Suva® 410A Superheated Vapor–Constant Pressure Tables (Saturated Vapor Properties in Parentheses)

| Temp. °C | 1100.0 (10.56°C) | | | 1200.0 (13.55°C) | | | 1300.0 (16.36°C) | | | 1400.0 (19.01°C) | | | Temp. °C |
|---|---|---|---|---|---|---|---|---|---|---|---|---|---|
| | V | H | S | V | H | S | V | H | S | V | H | S | |
| | (0.0234) | (425.2) | (1.7942) | (0.0214) | (425.8) | (1.7885) | (0.0196) | (426.3) | (1.7831) | (0.0181) | (426.7) | (1.7780) | |
| 15 | 0.0243 | 430.6 | 1.8130 | 0.0216 | 427.6 | 1.7948 | – | – | – | – | – | – | 15 |
| 20 | 0.0251 | 436.4 | 1.8331 | 0.0225 | 433.8 | 1.8159 | 0.0203 | 431.0 | 1.7991 | 0.0183 | 428.0 | 1.7825 | 20 |
| 25 | 0.0260 | 442.1 | 1.8523 | 0.0233 | 439.7 | 1.8359 | 0.0211 | 437.2 | 1.8201 | 0.0191 | 434.5 | 1.8045 | 25 |
| 30 | 0.0268 | 447.6 | 1.8707 | 0.0241 | 445.4 | 1.8550 | 0.0218 | 443.1 | 1.8399 | 0.0198 | 440.7 | 1.8251 | 30 |
| 35 | 0.0276 | 453.0 | 1.8884 | 0.0249 | 451.0 | 1.8733 | 0.0226 | 448.9 | 1.8588 | 0.0206 | 446.7 | 1.8447 | 35 |
| 40 | 0.0284 | 458.3 | 1.9055 | 0.0256 | 456.5 | 1.8909 | 0.0233 | 454.5 | 1.8769 | 0.0212 | 452.5 | 1.8635 | 40 |
| 45 | 0.0291 | 463.6 | 1.9221 | 0.0263 | 461.8 | 1.9079 | 0.0239 | 460.0 | 1.8944 | 0.0219 | 458.2 | 1.8814 | 45 |
| 50 | 0.0298 | 468.8 | 1.9383 | 0.0270 | 467.1 | 1.9245 | 0.0246 | 465.5 | 1.9113 | 0.0225 | 463.8 | 1.8988 | 50 |
| 55 | 0.0305 | 473.9 | 1.9541 | 0.0276 | 472.4 | 1.9406 | 0.0252 | 470.8 | 1.9278 | 0.0231 | 469.2 | 1.9156 | 55 |
| 60 | 0.0312 | 479.0 | 1.9695 | 0.0283 | 477.6 | 1.9563 | 0.0258 | 476.1 | 1.9438 | 0.0237 | 474.6 | 1.9319 | 60 |
| 65 | 0.0319 | 484.1 | 1.9846 | 0.0289 | 482.7 | 1.9716 | 0.0264 | 481.3 | 1.9594 | 0.0243 | 480.0 | 1.9477 | 65 |
| 70 | 0.0325 | 489.1 | 1.9994 | 0.0296 | 487.8 | 1.9866 | 0.0270 | 486.5 | 1.9746 | 0.0248 | 485.2 | 1.9632 | 70 |
| 75 | 0.0332 | 494.1 | 2.0139 | 0.0302 | 492.9 | 2.0014 | 0.0276 | 491.7 | 1.9896 | 0.0254 | 490.5 | 1.9784 | 75 |
| 80 | 0.0338 | 499.2 | 2.0282 | 0.0308 | 498.0 | 2.0159 | 0.0282 | 496.9 | 2.0042 | 0.0259 | 495.7 | 1.9933 | 80 |
| 85 | 0.0345 | 504.2 | 2.0423 | 0.0314 | 503.1 | 2.0301 | 0.0287 | 502.0 | 2.0186 | 0.0265 | 500.9 | 2.0078 | 85 |
| 90 | 0.0351 | 509.2 | 2.0562 | 0.0320 | 508.1 | 2.0441 | 0.0293 | 507.1 | 2.0328 | 0.0270 | 506.0 | 2.0222 | 90 |
| 95 | 0.0357 | 514.2 | 2.0699 | 0.0325 | 513.2 | 2.0580 | 0.0298 | 512.2 | 2.0468 | 0.0275 | 511.2 | 2.0362 | 95 |
| 100 | 0.0363 | 519.2 | 2.0834 | 0.0331 | 518.2 | 2.0716 | 0.0304 | 517.3 | 2.0605 | 0.0280 | 516.3 | 2.0501 | 100 |
| 105 | 0.0370 | 524.2 | 2.0968 | 0.0337 | 523.3 | 2.0850 | 0.0309 | 522.4 | 2.0741 | 0.0285 | 521.5 | 2.0638 | 105 |
| 135 | 0.0530 | 556.3 | 2.2067 | 0.0500 | 556.0 | 2.1995 | 0.0472 | 555.6 | 2.1927 | 0.0448 | 555.3 | 2.1862 | 135 |
| 140 | 0.0538 | 561.4 | 2.2190 | 0.0507 | 561.1 | 2.2119 | 0.0479 | 560.7 | 2.2051 | 0.0454 | 560.4 | 2.1986 | 140 |
| 145 | 0.0545 | 566.5 | 2.2312 | 0.0514 | 566.1 | 2.2241 | 0.0486 | 565.8 | 2.2173 | 0.0461 | 565.5 | 2.2109 | 145 |
| 150 | 0.0552 | 571.6 | 2.2433 | 0.0521 | 571.3 | 2.2362 | 0.0492 | 570.9 | 2.2295 | 0.0467 | 570.6 | 2.2231 | 150 |
| 155 | 0.0560 | 576.7 | 2.2554 | 0.0528 | 576.4 | 2.2483 | 0.0499 | 576.1 | 2.2416 | 0.0473 | 575.8 | 2.2352 | 155 |
| 160 | 0.0505 | 581.2 | 2.2536 | 0.0479 | 580.9 | 2.2472 | – | – | – | – | – | – | 160 |

*(Continued)*

## TABLE A.19 (SI) (CONTINUED)

Suva® 410A Superheated Vapor–Constant Pressure Tables (Saturated Vapor Properties in Parentheses)

| Temp. °C | 1500.0 (21.52°C) | | | 1600.0 (23.91°C) | | | 1700.0 (26.19°C) | | | 1800.0 (28.36°C) | | |
|---|---|---|---|---|---|---|---|---|---|---|---|---|
| | V (0.0168) | H (427.0) | S (1.7731) | V (0.0157) | H (427.2) | S (1.7684) | V (0.0146) | H (427.4) | S (1.7639) | V (0.0137) | H (427.5) | S (1.7595) |
| 110 | 0.0376 | 529.2 | 2.1100 | 0.0342 | 528.4 | 2.0983 | 0.0314 | 527.5 | 2.0875 | 0.0290 | 526.6 | 2.0773 |
| 115 | 0.0382 | 534.3 | 2.1230 | 0.0348 | 533.4 | 2.1115 | 0.0319 | 532.6 | 2.1007 | 0.0295 | 531.7 | 2.0906 |
| 120 | 0.0388 | 539.3 | 2.1360 | 0.0354 | 538.5 | 2.1245 | 0.0325 | 537.7 | 2.1138 | 0.0300 | 536.9 | 2.1038 |
| 125 | 0.0394 | 544.4 | 2.1488 | 0.0359 | 543.6 | 2.1374 | 0.0330 | 542.8 | 2.1268 | 0.0305 | 542.0 | 2.1168 |
| 130 | 0.0399 | 549.5 | 2.1614 | 0.0364 | 548.7 | 2.1501 | 0.0335 | 548.0 | 2.1396 | 0.0309 | 547.2 | 2.1297 |
| 135 | 0.0405 | 554.6 | 2.1740 | 0.0370 | 553.8 | 2.1627 | 0.0340 | 553.1 | 2.1523 | 0.0314 | 552.4 | 2.1425 |
| 140 | 0.0411 | 559.7 | 2.1864 | 0.0375 | 559.0 | 2.1752 | 0.0345 | 558.3 | 2.1648 | 0.0319 | 557.6 | 2.1551 |
| 145 | 0.0417 | 564.8 | 2.1988 | 0.0381 | 564.1 | 2.1876 | 0.0350 | 563.4 | 2.1773 | 0.0324 | 562.8 | 2.1676 |
| 150 | 0.0423 | 569.9 | 2.2110 | 0.0386 | 569.3 | 2.1999 | 0.0355 | 568.6 | 2.1896 | 0.0328 | 568.0 | 2.1800 |
| 155 | 0.0429 | 575.1 | 2.2232 | 0.0391 | 574.5 | 2.2121 | 0.0360 | 573.8 | 2.2019 | 0.0333 | 573.2 | 2.1923 |
| 160 | 0.0434 | 580.3 | 2.2352 | 0.0397 | 579.7 | 2.2242 | 0.0365 | 579.1 | 2.2140 | 0.0338 | 578.5 | 2.2045 |
| 165 | 0.0440 | 585.5 | 2.2472 | 0.0402 | 584.9 | 2.2362 | 0.0370 | 584.3 | 2.2261 | 0.0342 | 583.7 | 2.2166 |
| 170 | 0.0375 | 589.6 | 2.2380 | 0.0347 | 589.0 | 2.2286 | – | – | – | – | – | – |
| 25 | 0.0174 | 431.7 | 1.7891 | 0.0158 | 428.8 | 1.7736 | – | – | – | – | – | – |
| 30 | 0.0181 | 438.2 | 1.8107 | 0.0166 | 435.6 | 1.7963 | 0.0152 | 432.9 | 1.7820 | 0.0140 | 430.0 | 1.7676 |
| 35 | 0.0188 | 444.5 | 1.8310 | 0.0173 | 442.1 | 1.8176 | 0.0159 | 439.7 | 1.8042 | 0.0147 | 437.1 | 1.7910 |
| 40 | 0.0195 | 450.5 | 1.8504 | 0.0179 | 448.3 | 1.8376 | 0.0165 | 446.1 | 1.8251 | 0.0153 | 443.8 | 1.8126 |
| 45 | 0.0201 | 456.3 | 1.8689 | 0.0185 | 454.4 | 1.8567 | 0.0172 | 452.3 | 1.8447 | 0.0159 | 450.3 | 1.8330 |
| 50 | 0.0207 | 462.0 | 1.8867 | 0.0191 | 460.2 | 1.8749 | 0.0177 | 458.4 | 1.8635 | 0.0165 | 456.5 | 1.8523 |
| 55 | 0.0213 | 467.6 | 1.9038 | 0.0197 | 465.9 | 1.8925 | 0.0183 | 464.2 | 1.8815 | 0.0170 | 462.5 | 1.8708 |
| 60 | 0.0219 | 473.1 | 1.9205 | 0.0202 | 471.5 | 1.9095 | 0.0188 | 470.0 | 1.8989 | 0.0175 | 468.3 | 1.8886 |
| 65 | 0.0224 | 478.5 | 1.9366 | 0.0208 | 477.1 | 1.9260 | 0.0193 | 475.6 | 1.9157 | 0.0180 | 474.1 | 1.9057 |
| 70 | 0.0229 | 483.9 | 1.9524 | 0.0213 | 482.5 | 1.9420 | 0.0198 | 481.1 | 1.9320 | 0.0185 | 479.7 | 1.9223 |
| 75 | 0.0235 | 489.2 | 1.9678 | 0.0218 | 487.9 | 1.9576 | 0.0203 | 486.6 | 1.9478 | 0.0190 | 485.3 | 1.9384 |
| 80 | 0.0240 | 494.5 | 1.9828 | 0.0223 | 493.3 | 1.9728 | 0.0208 | 492.0 | 1.9633 | 0.0195 | 490.8 | 1.9541 |

*(Continued)*

# TABLE A.19 (SI) (CONTINUED)

Suva® 410A Superheated Vapor–Constant Pressure Tables (Saturated Vapor Properties in Parentheses)

| Temp. °C | 1900.0 (30.45 °C) | | | 2000.0 (32.46 °C) | | | 2200.0 (36.26 °C) | | | 2400.0 (39.81 °C) | | | Temp. °C |
|---|---|---|---|---|---|---|---|---|---|---|---|---|---|
| | V | H | S | V | H | S | V | H | S | V | H | S | |
| | (0.0129) | (427.6) | (1.7553) | (0.0122) | (427.6) | (1.7511) | (0.0109) | (427.4) | (1.7430) | (0.0098) | (427.0) | (1.7352) | |
| 35 | 0.0136 | 434.4 | 1.7776 | 0.0125 | 431.5 | 1.7641 | – | – | – | – | – | – | 35 |
| 40 | 0.0142 | 441.4 | 1.8003 | 0.0132 | 438.9 | 1.7879 | 0.0114 | 433.5 | 1.7627 | 0.0098 | 427.3 | 1.7363 | 40 |
| 45 | 0.0148 | 448.1 | 1.8214 | 0.0138 | 445.9 | 1.8099 | 0.0120 | 441.1 | 1.7868 | 0.0105 | 435.8 | 1.7632 | 45 |
| 50 | 0.0154 | 454.5 | 1.8414 | 0.0143 | 452.5 | 1.8305 | 0.0126 | 448.2 | 1.8090 | 0.0110 | 443.6 | 1.7874 | 50 |
| 55 | 0.0159 | 460.7 | 1.8603 | 0.0149 | 458.8 | 1.8500 | 0.0131 | 455.0 | 1.8297 | 0.0116 | 450.8 | 1.8096 | 55 |
| 85 | 0.0245 | 499.7 | 1.9976 | 0.0228 | 498.6 | 1.9878 | 0.0213 | 497.4 | 1.9784 | 0.0199 | 496.3 | 1.9694 | 85 |
| 90 | 0.0250 | 505.0 | 2.0120 | 0.0232 | 503.9 | 2.0024 | 0.0217 | 502.8 | 1.9932 | 0.0203 | 501.7 | 1.9844 | 90 |
| 95 | 0.0255 | 510.2 | 2.0263 | 0.0237 | 509.1 | 2.0168 | 0.0222 | 508.1 | 2.0077 | 0.0208 | 507.0 | 1.9990 | 95 |
| 100 | 0.0260 | 515.3 | 2.0403 | 0.0242 | 514.4 | 2.0309 | 0.0226 | 513.4 | 2.0220 | 0.0212 | 512.4 | 2.0134 | 100 |
| 105 | 0.0264 | 520.5 | 2.0541 | 0.0246 | 519.6 | 2.0448 | 0.0230 | 518.6 | 2.0360 | 0.0216 | 517.7 | 2.0276 | 105 |
| 110 | 0.0269 | 525.7 | 2.0677 | 0.0251 | 524.8 | 2.0586 | 0.0235 | 523.9 | 2.0499 | 0.0220 | 523.0 | 2.0415 | 110 |
| 115 | 0.0274 | 530.9 | 2.0811 | 0.0255 | 530.0 | 2.0721 | 0.0239 | 529.2 | 2.0635 | 0.0224 | 528.3 | 2.0553 | 115 |
| 120 | 0.0278 | 536.1 | 2.0944 | 0.0260 | 535.2 | 2.0854 | 0.0243 | 534.4 | 2.0769 | 0.0228 | 533.6 | 2.0688 | 120 |
| 125 | 0.0283 | 541.2 | 2.1075 | 0.0264 | 540.5 | 2.0986 | 0.0247 | 539.6 | 2.0902 | 0.0232 | 538.8 | 2.0821 | 125 |
| 130 | 0.0287 | 546.4 | 2.1204 | 0.0268 | 545.7 | 2.1116 | 0.0251 | 544.9 | 2.1033 | 0.0236 | 544.1 | 2.0953 | 130 |
| 135 | 0.0292 | 551.6 | 2.1332 | 0.0272 | 550.9 | 2.1245 | 0.0255 | 550.2 | 2.1162 | 0.0240 | 549.4 | 2.1083 | 135 |
| 140 | 0.0296 | 556.8 | 2.1459 | 0.0277 | 556.1 | 2.1373 | 0.0259 | 555.4 | 2.1290 | 0.0244 | 554.7 | 2.1212 | 140 |
| 145 | 0.0301 | 562.1 | 2.1585 | 0.0281 | 561.4 | 2.1499 | 0.0263 | 560.7 | 2.1417 | 0.0248 | 560.0 | 2.1340 | 145 |
| 150 | 0.0305 | 567.3 | 2.1709 | 0.0285 | 566.6 | 2.1624 | 0.0267 | 566.0 | 2.1543 | 0.0251 | 565.3 | 2.1466 | 150 |
| 155 | 0.0310 | 572.6 | 2.1833 | 0.0289 | 571.9 | 2.1748 | 0.0271 | 571.3 | 2.1667 | 0.0255 | 570.6 | 2.1591 | 155 |
| 160 | 0.0314 | 577.8 | 2.1955 | 0.0293 | 577.2 | 2.1871 | 0.0275 | 576.6 | 2.1791 | 0.0259 | 575.9 | 2.1714 | 160 |
| 165 | 0.0318 | 583.1 | 2.2077 | 0.0297 | 582.5 | 2.1992 | 0.0279 | 581.9 | 2.1913 | 0.0263 | 581.3 | 2.1837 | 165 |
| 170 | 0.0323 | 588.4 | 2.2197 | 0.0302 | 587.8 | 2.2113 | 0.0283 | 587.2 | 2.2034 | 0.0266 | 586.6 | 2.1959 | 170 |
| 175 | 0.0327 | 593.7 | 2.2316 | 0.0306 | 593.2 | 2.2233 | 0.0287 | 592.6 | 2.2154 | 0.0270 | 592.0 | 2.2079 | 175 |
| 180 | – | – | – | – | – | – | 0.0291 | 598.0 | 2.2274 | 0.0274 | 597.7 | 2.2199 | 180 |

*(Continued)*

**TABLE A.19 (SI) (CONTINUED)**

Suva® 410A Superheated Vapor–Constant Pressure Tables (Saturated Vapor Properties in Parentheses)

| | | | | | | | | | | | | | |
|---|---|---|---|---|---|---|---|---|---|---|---|---|---|
| 60 | 0.0164 | 466.7 | 1.8785 | 0.0154 | 465.0 | 1.8686 | 0.0136 | 461.4 | 1.8493 | 0.0120 | 457.7 | 1.8305 | 60 |
| 65 | 0.0169 | 472.5 | 1.8960 | 0.0158 | 471.0 | 1.8865 | 0.0140 | 467.7 | 1.8680 | 0.0125 | 464.3 | 1.8501 | 65 |
| 70 | 0.0174 | 478.3 | 1.9128 | 0.0163 | 476.8 | 1.9037 | 0.0145 | 473.8 | 1.8859 | 0.0129 | 470.7 | 1.8688 | 70 |
| 75 | 0.0178 | 484.0 | 1.9292 | 0.0167 | 482.6 | 1.9203 | 0.0149 | 479.8 | 1.9032 | 0.0134 | 476.9 | 1.8867 | 75 |
| 80 | 0.0183 | 489.5 | 1.9451 | 0.0172 | 488.3 | 1.9365 | 0.0153 | 485.6 | 1.9199 | 0.0137 | 482.9 | 1.9040 | 80 |
| 85 | 0.0187 | 495.1 | 1.9607 | 0.0176 | 493.9 | 1.9522 | 0.0157 | 491.4 | 1.9360 | 0.0141 | 488.8 | 1.9206 | 85 |
| 90 | 0.0191 | 500.5 | 1.9758 | 0.0180 | 499.4 | 1.9676 | 0.0161 | 497.1 | 1.9518 | 0.0145 | 494.7 | 1.9368 | 90 |
| 95 | 0.0195 | 506.0 | 1.9907 | 0.0184 | 504.9 | 1.9826 | 0.0165 | 502.7 | 1.9671 | 0.0149 | 500.4 | 1.9525 | 95 |
| 100 | 0.0199 | 511.3 | 2.0052 | 0.0188 | 510.3 | 1.9973 | 0.0169 | 508.2 | 1.9821 | 0.0152 | 506.1 | 1.9679 | 100 |
| 105 | 0.0203 | 516.7 | 2.0195 | 0.0192 | 515.7 | 2.0117 | 0.0172 | 513.8 | 1.9968 | 0.0156 | 511.7 | 1.9828 | 105 |
| 110 | 0.0207 | 522.1 | 2.0335 | 0.0196 | 521.1 | 2.0259 | 0.0176 | 519.2 | 2.0112 | 0.0159 | 517.3 | 1.9975 | 110 |
| 115 | 0.0211 | 527.4 | 2.0474 | 0.0199 | 526.5 | 2.0398 | 0.0179 | 524.7 | 2.0254 | 0.0162 | 522.9 | 2.0119 | 115 |
| 120 | 0.0215 | 532.7 | 2.0610 | 0.0203 | 531.9 | 2.0535 | 0.0183 | 530.1 | 2.0393 | 0.0166 | 528.4 | 2.0260 | 120 |
| 125 | 0.0219 | 538.0 | 2.0744 | 0.0207 | 537.2 | 2.0670 | 0.0186 | 535.6 | 2.0530 | 0.0169 | 533.9 | 2.0399 | 125 |
| 130 | 0.0223 | 543.3 | 2.0877 | 0.0210 | 542.6 | 2.0804 | 0.0189 | 541.0 | 2.0665 | 0.0172 | 539.4 | 2.0536 | 130 |
| 135 | 0.0226 | 548.7 | 2.1008 | 0.0214 | 547.9 | 2.0935 | 0.0193 | 546.4 | 2.0799 | 0.0175 | 544.8 | 2.0671 | 135 |
| 140 | 0.0230 | 554.0 | 2.1137 | 0.0218 | 553.2 | 2.1066 | 0.0196 | 551.8 | 2.0930 | 0.0178 | 550.3 | 2.0804 | 140 |
| 145 | 0.0234 | 559.3 | 2.1265 | 0.0221 | 558.6 | 2.1194 | 0.0199 | 557.2 | 2.1060 | 0.0181 | 555.7 | 2.0935 | 145 |
| 150 | 0.0237 | 564.6 | 2.1392 | 0.0225 | 563.9 | 2.1321 | 0.0202 | 562.6 | 2.1188 | 0.0184 | 561.2 | 2.1065 | 150 |
| 155 | 0.0241 | 570.0 | 2.1517 | 0.0228 | 569.3 | 2.1447 | 0.0206 | 568.0 | 2.1315 | 0.0187 | 566.6 | 2.1193 | 155 |
| 160 | 0.0244 | 575.3 | 2.1642 | 0.0231 | 574.7 | 2.1572 | 0.0209 | 573.4 | 2.1441 | 0.0190 | 572.1 | 2.1319 | 160 |
| 165 | 0.0248 | 580.7 | 2.1765 | 0.0235 | 580.1 | 2.1696 | 0.0212 | 578.8 | 2.1566 | 0.0193 | 577.6 | 2.1445 | 165 |
| 170 | 0.0251 | 586.0 | 2.1887 | 0.0238 | 585.5 | 2.1818 | 0.0215 | 584.2 | 2.1689 | 0.0196 | 583.0 | 2.1569 | 170 |
| 175 | 0.0255 | 591.4 | 2.2008 | 0.0242 | 590.9 | 2.1939 | 0.0218 | 589.7 | 2.1811 | 0.0199 | 588.5 | 2.1692 | 175 |
| 180 | 0.0258 | 596.9 | 2.2128 | 0.0245 | 596.3 | 2.2060 | 0.0221 | 595.2 | 2.1932 | 0.0202 | 594.0 | 2.1814 | 180 |
| 185 | 0.0262 | 602.3 | 2.2247 | 0.0248 | 601.7 | 2.2179 | 0.0224 | 600.6 | 2.2052 | 0.0205 | 599.5 | 2.1935 | 185 |
| 190 | – | – | – | – | – | – | 0.0227 | 606.1 | 2.2171 | 0.0207 | 605.0 | 2.2054 | 190 |

*(Continued)*

**TABLE A.19 (SI) (CONTINUED)**

Suva® 410A Superheated Vapor–Constant Pressure Tables (Saturated Vapor Properties in Parentheses)

| Temp. °C | 2600.0 (43.14°C) | | | 2800.0 (46.28°C) | | | 3000.0 (49.26°C) | | | 3200.0 (52.09°C) | | | Temp. °C |
|---|---|---|---|---|---|---|---|---|---|---|---|---|---|
| | V (0.0089) | H (426.4) | S (1.7275) | V (0.0081) | H (425.7) | S (1.7199) | V (0.0074) | H (424.8) | S (1.7123) | V (0.0068) | H (423.8) | S (1.7047) | |
| 45 | 0.0091 | 429.9 | 1.7384 | – | – | – | – | – | – | – | – | – | 45 |
| 50 | 0.0097 | 438.5 | 1.7653 | 0.0086 | 432.9 | 1.7422 | 0.0075 | 426.4 | 1.7172 | – | – | – | 50 |
| 55 | 0.0103 | 446.4 | 1.7894 | 0.0091 | 441.5 | 1.7688 | 0.0081 | 436.2 | 1.7472 | 0.0071 | 430.1 | 1.7242 | 55 |
| 60 | 0.0107 | 453.7 | 1.8117 | 0.0096 | 449.5 | 1.7928 | 0.0086 | 444.9 | 1.7735 | 0.0077 | 439.8 | 1.7534 | 60 |
| 65 | 0.0112 | 460.7 | 1.8325 | 0.0101 | 456.9 | 1.8149 | 0.0091 | 452.8 | 1.7972 | 0.0082 | 448.4 | 1.7791 | 65 |
| 70 | 0.0116 | 467.4 | 1.8521 | 0.0105 | 463.9 | 1.8355 | 0.0095 | 460.3 | 1.8191 | 0.0086 | 456.4 | 1.8025 | 70 |
| 75 | 0.0120 | 473.8 | 1.8707 | 0.0109 | 470.7 | 1.8551 | 0.0099 | 467.4 | 1.8396 | 0.0090 | 463.9 | 1.8241 | 75 |
| 80 | 0.0124 | 480.1 | 1.8886 | 0.0113 | 477.2 | 1.8736 | 0.0103 | 474.1 | 1.8589 | 0.0094 | 471.0 | 1.8444 | 80 |
| 85 | 0.0128 | 486.2 | 1.9058 | 0.0116 | 483.5 | 1.8914 | 0.0106 | 480.7 | 1.8774 | 0.0097 | 477.8 | 1.8636 | 85 |
| 90 | 0.0131 | 492.2 | 1.9224 | 0.0120 | 489.7 | 1.9085 | 0.0110 | 487.1 | 1.8951 | 0.0101 | 484.4 | 1.8818 | 90 |
| 95 | 0.0135 | 498.1 | 1.9386 | 0.0123 | 495.7 | 1.9251 | 0.0113 | 493.3 | 1.9121 | 0.0104 | 490.8 | 1.8994 | 95 |
| 100 | 0.0138 | 503.9 | 1.9542 | 0.0126 | 501.7 | 1.9412 | 0.0116 | 499.4 | 1.9286 | 0.0107 | 497.1 | 1.9163 | 100 |
| 105 | 0.0142 | 509.7 | 1.9695 | 0.0130 | 507.6 | 1.9568 | 0.0119 | 505.4 | 1.9445 | 0.0110 | 503.2 | 1.9327 | 105 |
| 110 | 0.0145 | 515.4 | 1.9845 | 0.0133 | 513.4 | 1.9720 | 0.0122 | 511.3 | 1.9601 | 0.0113 | 509.3 | 1.9485 | 110 |
| 115 | 0.0148 | 521.0 | 1.9991 | 0.0136 | 519.1 | 1.9869 | 0.0125 | 517.2 | 1.9752 | 0.0116 | 515.2 | 1.9640 | 115 |
| 120 | 0.0151 | 526.6 | 2.0135 | 0.0139 | 524.8 | 2.0015 | 0.0128 | 523.0 | 1.9901 | 0.0118 | 521.1 | 1.9791 | 120 |
| 125 | 0.0154 | 532.2 | 2.0276 | 0.0142 | 530.5 | 2.0158 | 0.0131 | 528.7 | 2.0046 | 0.0121 | 526.9 | 1.9938 | 125 |
| 130 | 0.0157 | 537.7 | 2.0414 | 0.0144 | 536.1 | 2.0299 | 0.0133 | 534.4 | 2.0188 | 0.0124 | 532.7 | 2.0083 | 130 |
| 135 | 0.0160 | 543.3 | 2.0551 | 0.0147 | 541.7 | 2.0437 | 0.0136 | 540.1 | 2.0328 | 0.0126 | 538.5 | 2.0224 | 135 |
| 140 | 0.0163 | 548.8 | 2.0685 | 0.0150 | 547.3 | 2.0573 | 0.0139 | 545.7 | 2.0466 | 0.0129 | 544.2 | 2.0364 | 140 |
| 145 | 0.0166 | 554.3 | 2.0818 | 0.0153 | 552.8 | 2.0707 | 0.0141 | 551.4 | 2.0601 | 0.0131 | 549.9 | 2.0500 | 145 |
| 150 | 0.0169 | 559.8 | 2.0948 | 0.0155 | 558.4 | 2.0839 | 0.0144 | 557.0 | 2.0735 | 0.0134 | 555.6 | 2.0635 | 150 |
| 155 | 0.0171 | 565.3 | 2.1078 | 0.0158 | 564.0 | 2.0969 | 0.0146 | 562.6 | 2.0866 | 0.0136 | 561.2 | 2.0768 | 155 |
| 160 | 0.0174 | 570.8 | 2.1205 | 0.0161 | 569.5 | 2.1098 | 0.0149 | 568.2 | 2.0996 | 0.0139 | 566.9 | 2.0899 | 160 |
| 165 | 0.0177 | 576.3 | 2.1332 | 0.0163 | 575.1 | 2.1225 | 0.0151 | 573.8 | 2.1125 | 0.0141 | 572.5 | 2.1029 | 165 |
| 170 | 0.0180 | 581.8 | 2.1457 | 0.0166 | 580.6 | 2.1351 | 0.0154 | 579.4 | 2.1252 | 0.0143 | 578.1 | 2.1157 | 170 |

*(Continued)*

## TABLE A.19 (SI) (CONTINUED)

Suva® 410A Superheated Vapor–Constant Pressure Tables (Saturated Vapor Properties in Parentheses)

| Temp. °C | 3400.0 (54.78°C) V | H | S | 3600.0 (57.36°C) V | H | S | 3800.0 (59.84°C) V | H | S | 4000.0 (62.21°C) V | H | S |
|---|---|---|---|---|---|---|---|---|---|---|---|---|
| | (0.0062) | (422.6) | (1.6971) | (0.0057) | (421.2) | (1.6893) | (0.0052) | (419.7) | (1.6813) | (0.0048) | (417.9) | (1.6730) |
| 55 | 0.0062 | 423.1 | 1.6987 | – | – | – | – | – | – | – | – | – |
| 60 | 0.0068 | 434.1 | 1.7321 | 0.0060 | 427.7 | 1.7089 | 0.0053 | 420.1 | 1.6827 | – | – | – |
| 65 | 0.0074 | 443.7 | 1.7604 | 0.0066 | 438.4 | 1.7408 | 0.0059 | 432.5 | 1.7195 | 0.0052 | 425.6 | 1.6959 |
| 70 | 0.0078 | 452.2 | 1.7856 | 0.0071 | 447.8 | 1.7682 | 0.0064 | 442.8 | 1.7500 | 0.0058 | 437.4 | 1.7305 |
| 75 | 0.0082 | 460.2 | 1.8086 | 0.0075 | 456.2 | 1.7928 | 0.0068 | 452.0 | 1.7766 | 0.0062 | 447.4 | 1.7596 |
| 80 | 0.0086 | 467.6 | 1.8299 | 0.0079 | 464.1 | 1.8153 | 0.0072 | 460.4 | 1.8005 | 0.0066 | 456.5 | 1.7853 |
| 85 | 0.0089 | 474.8 | 1.8499 | 0.0082 | 471.6 | 1.8362 | 0.0076 | 468.3 | 1.8225 | 0.0070 | 464.7 | 1.8086 |
| 90 | 0.0093 | 481.6 | 1.8688 | 0.0086 | 478.7 | 1.8559 | 0.0079 | 475.7 | 1.8431 | 0.0073 | 472.5 | 1.8302 |
| 95 | 0.0096 | 488.2 | 1.8869 | 0.0089 | 485.5 | 1.8747 | 0.0082 | 482.8 | 1.8625 | 0.0076 | 479.9 | 1.8504 |
| 100 | 0.0099 | 494.7 | 1.9043 | 0.0092 | 492.2 | 1.8926 | 0.0085 | 489.6 | 1.8810 | 0.0079 | 487.0 | 1.8695 |
| 105 | 0.0102 | 501.0 | 1.9211 | 0.0095 | 498.6 | 1.9098 | 0.0088 | 496.3 | 1.8987 | 0.0082 | 493.8 | 1.8877 |
| 110 | 0.0105 | 507.1 | 1.9373 | 0.0097 | 505.0 | 1.9264 | 0.0091 | 502.7 | 1.9157 | 0.0085 | 500.5 | 1.9051 |
| 115 | 0.0107 | 513.2 | 1.9531 | 0.0100 | 511.2 | 1.9425 | 0.0093 | 509.1 | 1.9321 | 0.0087 | 506.9 | 1.9220 |
| 120 | 0.0110 | 519.2 | 1.9684 | 0.0103 | 517.3 | 1.9581 | 0.0096 | 515.3 | 1.9481 | 0.0090 | 513.3 | 1.9382 |
| 125 | 0.0113 | 525.1 | 1.9834 | 0.0105 | 523.3 | 1.9734 | 0.0098 | 521.4 | 1.9636 | 0.0092 | 519.5 | 1.9540 |
| 130 | 0.0115 | 531.0 | 1.9981 | 0.0108 | 529.3 | 1.9882 | 0.0101 | 527.5 | 1.9787 | 0.0095 | 525.7 | 1.9694 |
| 135 | 0.0118 | 536.8 | 2.0124 | 0.0110 | 535.2 | 2.0028 | 0.0103 | 533.5 | 1.9935 | 0.0097 | 531.8 | 1.9844 |
| 140 | 0.0120 | 542.6 | 2.0265 | 0.0112 | 541.0 | 2.0171 | 0.0105 | 539.4 | 2.0079 | 0.0099 | 537.8 | 1.9990 |
| 175 | 0.0182 | 587.3 | 2.1581 | 0.0168 | 586.2 | 2.1476 | 0.0156 | 585.0 | 2.1377 | 0.0145 | 583.8 | 2.1283 |
| 180 | 0.0185 | 592.9 | 2.1703 | 0.0171 | 591.7 | 2.1599 | 0.0159 | 590.6 | 2.1501 | 0.0148 | 589.4 | 2.1408 |
| 185 | 0.0188 | 598.4 | 2.1825 | 0.0173 | 597.3 | 2.1722 | 0.0161 | 596.2 | 2.1624 | 0.0150 | 595.0 | 2.1532 |
| 190 | 0.0190 | 604.0 | 2.1945 | 0.0176 | 602.9 | 2.1843 | 0.0163 | 601.8 | 2.1746 | 0.0152 | 600.7 | 2.1655 |
| 195 | 0.0193 | 609.5 | 2.2065 | 0.0178 | 608.5 | 2.1963 | 0.0166 | 607.4 | 2.1867 | 0.0155 | 606.3 | 2.1776 |
| 200 | – | – | – | 0.0181 | 614.1 | 2.2082 | 0.0168 | 613.0 | 2.1987 | 0.0157 | 612.0 | 2.1896 |
| 205 | – | – | – | – | – | – | – | – | – | 0.0159 | 617.7 | 2.2016 |

*(Continued)*

**TABLE A.19 (SI) (CONTINUED)**

Suva® 410A Superheated Vapor–Constant Pressure Tables (Saturated Vapor Properties in Parentheses)

| T | V | H | S | V | H | S | V | H | S | V | H | S | T |
|---|---|---|---|---|---|---|---|---|---|---|---|---|---|
| 145 | 0.0122 | 548.4 | 2.0404 | 0.0115 | 546.9 | 2.0311 | 0.0108 | 545.3 | 2.0221 | 0.0101 | 543.8 | 2.0134 | 145 |
| 150 | 0.0125 | 554.1 | 2.0540 | 0.0117 | 552.7 | 2.0449 | 0.0110 | 551.2 | 2.0360 | 0.0103 | 549.7 | 2.0275 | 150 |
| 155 | 0.0127 | 559.8 | 2.0674 | 0.0119 | 558.4 | 2.0584 | 0.0112 | 557.0 | 2.0497 | 0.0106 | 555.6 | 2.0413 | 155 |
| 160 | 0.0129 | 565.5 | 2.0807 | 0.0121 | 564.2 | 2.0718 | 0.0114 | 562.8 | 2.0632 | 0.0108 | 561.4 | 2.0549 | 160 |
| 165 | 0.0132 | 571.2 | 2.0937 | 0.0123 | 569.9 | 2.0849 | 0.0116 | 568.6 | 2.0765 | 0.0110 | 567.3 | 2.0683 | 165 |
| 170 | 0.0134 | 576.9 | 2.1066 | 0.0126 | 575.6 | 2.0979 | 0.0118 | 574.4 | 2.0896 | 0.0112 | 573.1 | 2.0815 | 170 |
| 175 | 0.0136 | 582.6 | 2.1193 | 0.0128 | 581.4 | 2.1108 | 0.0120 | 580.1 | 2.1025 | 0.0114 | 578.9 | 2.0946 | 175 |
| 180 | 0.0138 | 588.2 | 2.1319 | 0.0130 | 587.1 | 2.1234 | 0.0122 | 585.9 | 2.1153 | 0.0116 | 584.7 | 2.1074 | 180 |
| 185 | 0.0140 | 593.9 | 2.1444 | 0.0132 | 592.8 | 2.1360 | 0.0124 | 591.6 | 2.1279 | 0.0117 | 590.5 | 2.1201 | 185 |
| 190 | 0.0143 | 599.6 | 2.1567 | 0.0134 | 598.5 | 2.1484 | 0.0126 | 597.4 | 2.1404 | 0.0119 | 596.3 | 2.1327 | 190 |
| 195 | 0.0145 | 605.3 | 2.1689 | 0.0136 | 604.2 | 2.1607 | 0.0128 | 603.1 | 2.1527 | 0.0121 | 602.1 | 2.1451 | 195 |
| 200 | 0.0147 | 611.0 | 2.1810 | 0.0138 | 609.9 | 2.1728 | 0.0130 | 608.9 | 2.1649 | 0.0123 | 607.8 | 2.1574 | 200 |
| 205 | 0.0149 | 616.7 | 2.1930 | 0.0140 | 615.7 | 2.1849 | 0.0132 | 614.6 | 2.1771 | 0.0125 | 613.6 | 2.1696 | 205 |
| 210 | – | – | – | 0.0142 | 621.4 | 2.1968 | 0.0134 | 620.4 | 2.1891 | 0.0127 | 619.4 | 2.1816 | 210 |
| 215 | – | – | – | – | – | – | – | – | – | 0.0129 | 625.2 | 2.1936 | 215 |

*Note:* $H$ = enthalpy in kJ/kg; $S$ = entropy in kJ/(kg) (K); $V$ = volume in $m^3$/kg.

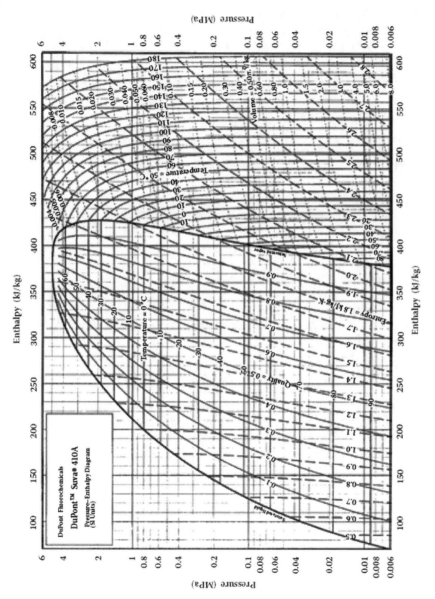

*Source:*    *Thermodynamic Properties of DuPont Suva 410A Refrigerant.* With permission of E. I. Du Pont de Nemours and Co.

**TABLE A.20**

Thermodynamic Properties of Air at Low Pressure

| T, R | $h$, Btu/lb | $p_r$ | $u$, Btu/lb | $v_r$ | $\phi$ Btu/lb. °R |
|---|---|---|---|---|---|
| 200 | 47.67 | 0.04320 | 33.96 | 1714.9 | 0.36303 |
| 220 | 52.46 | 0.06026 | 37.38 | 1352.5 | 0.38584 |
| 240 | 57.25 | 0.08165 | 40.80 | 1088.8 | 0.40666 |
| 260 | 62.03 | 0.10797 | 44.21 | 892.0 | 0.42582 |
| 280 | 66.82 | 0.13986 | 47.63 | 741.6 | 0.44356 |
| 300 | 71.61 | 0.17795 | 51.04 | 624.5 | 0.46007 |
| 320 | 76.40 | 0.22290 | 54.46 | 531.8 | 0.47550 |
| 340 | 81.18 | 0.27545 | 57.87 | 457.2 | 0.49002 |
| 360 | 85.97 | 0.3363 | 61.29 | 396.6 | 0.50369 |
| 380 | 90.75 | 0.4061 | 64.70 | 346.6 | 0.51663 |
| 400 | 95.53 | 0.4858 | 68.11 | 305.0 | 0.52890 |
| 420 | 100.32 | 0.5760 | 71.52 | 270.1 | 0.54058 |
| 440 | 105.11 | 0.6776 | 74.93 | 240.6 | 0.55172 |
| 460 | 109.90 | 0.7913 | 78.36 | 215.33 | 0.56235 |
| 480 | 114.69 | 0.9182 | 81.77 | 193.65 | 0.57255 |
| 500 | 119.48 | 1.0590 | 85.20 | 174.90 | 0.58233 |
| 520 | 124.27 | 1.2147 | 88.62 | 158.58 | 0.59173 |
| 540 | 129.06 | 1.3860 | 92.04 | 144.32 | 0.60078 |
| 560 | 133.86 | 1.5742 | 95.47 | 131.78 | 0.60950 |
| 580 | 138.66 | 1.7800 | 98.90 | 120.70 | 0.61793 |
| 600 | 143.47 | 2.005 | 102.34 | 110.88 | 0.62607 |
| 620 | 148.28 | 2.249 | 105.78 | 102.12 | 0.63395 |
| 640 | 153.09 | 2.514 | 109.21 | 94.30 | 0.64159 |
| 660 | 157.92 | 2.801 | 112.67 | 87.27 | 0.64902 |
| 680 | 162.73 | 3.111 | 116.12 | 80.96 | 0.65621 |
| 700 | 167.56 | 3.446 | 119.58 | 75.25 | 0.66321 |
| 720 | 172.39 | 3.806 | 123.04 | 70.07 | 0.67002 |
| 740 | 177.23 | 4.193 | 126.51 | 65.38 | 0.67665 |
| 760 | 182.08 | 4.607 | 129.99 | 61.10 | 0.68312 |
| 780 | 186.94 | 5.051 | 133.47 | 57.20 | 0.68942 |
| 800 | 191.81 | 5.526 | 136.97 | 53.63 | 0.69558 |
| 820 | 196.69 | 6.033 | 140.47 | 50.35 | 0.70160 |
| 840 | 201.56 | 6.573 | 143.98 | 47.34 | 0.70747 |
| 860 | 206.46 | 7.149 | 147.50 | 44.57 | 0.71323 |
| 880 | 211.35 | 7.761 | 151.02 | 42.01 | 0.71886 |
| 900 | 216.26 | 8.411 | 154.57 | 39.64 | 0.72438 |
| 920 | 221.18 | 9.102 | 158.12 | 37.44 | 0.72979 |
| 940 | 226.11 | 9.834 | 161.68 | 35.41 | 0.73509 |
| 960 | 231.06 | 10.610 | 165.26 | 33.52 | 0.74030 |
| 980 | 236.02 | 11.430 | 168.83 | 31.76 | 0.74540 |
| 1000 | 240.98 | 12.298 | 172.43 | 30.12 | 0.75042 |
| 1020 | 245.97 | 13.215 | 176.04 | 28.59 | 0.75536 |
| 1040 | 250.95 | 14.182 | 179.66 | 27.17 | 0.76019 |

*(Continued)*

**TABLE A.20 (CONTINUED)**

Thermodynamic Properties of Air at Low Pressure

| T, R | h, Btu/lb | $p_r$ | u, Btu/lb | $v_r$ | φ Btu/lb. °R |
|---|---|---|---|---|---|
| 1060 | 255.96 | 15.203 | 183.29 | 25.82 | 0.76496 |
| 1080 | 260.97 | 16.278 | 186.93 | 24.58 | 0.76964 |
| 1100 | 265.99 | 17.413 | 190.58 | 23.40 | 0.77426 |
| 1120 | 271.03 | 18.604 | 194.25 | 22.30 | 0.77880 |
| 1140 | 276.08 | 19.858 | 197.94 | 21.27 | 0.78326 |
| 1160 | 281.14 | 21.18 | 201.63 | 20.293 | 0.78767 |
| 1180 | 286.21 | 22.56 | 205.33 | 19.377 | 0.79201 |
| 1200 | 291.30 | 24.01 | 209.05 | 18.514 | 0.79628 |
| 1220 | 296.41 | 25.53 | 212.78 | 17.700 | 0.80050 |
| 1240 | 301.52 | 27.13 | 216.53 | 16.932 | 0.80466 |
| 1260 | 306.65 | 28.80 | 220.28 | 16.205 | 0.80876 |
| 1280 | 311.79 | 30.55 | 224.05 | 15.518 | 0.81280 |
| 1300 | 316.94 | 32.39 | 227.83 | 14.868 | 0.81680 |
| 1320 | 322.11 | 34.31 | 231.63 | 14.253 | 0.82075 |
| 1340 | 327.29 | 36.31 | 235.43 | 13.670 | 0.82464 |
| 1360 | 332.48 | 38.41 | 239.25 | 13.118 | 0.82848 |
| 1380 | 337.68 | 40.59 | 243.08 | 12.593 | 0.83229 |
| 1400 | 342.90 | 42.88 | 246.93 | 12.095 | 0.83604 |
| 1420 | 348.14 | 45.26 | 250.79 | 11.622 | 0.83975 |
| 1440 | 353.37 | 47.75 | 254.66 | 11.172 | 0.84341 |
| 1460 | 358.63 | 50.34 | 258.54 | 10.743 | 0.84704 |
| 1480 | 363.89 | 53.04 | 262.44 | 10.336 | 0.85062 |
| 1500 | 369.17 | 55.86 | 266.34 | 9.948 | 0.85416 |
| 1520 | 374.47 | 58.78 | 270.26 | 9.578 | 0.85767 |
| 1540 | 379.77 | 61.83 | 274.20 | 9.226 | 0.86113 |
| 1560 | 385.08 | 65.00 | 278.13 | 8.890 | 0.86456 |
| 1580 | 390.40 | 68.30 | 282.09 | 8.569 | 0.86794 |
| 1600 | 395.74 | 71.73 | 286.06 | 8.263 | 0.87130 |
| 1620 | 401.09 | 75.29 | 290.04 | 7.971 | 0.87462 |
| 1640 | 406.45 | 78.99 | 294.03 | 7.691 | 0.87791 |
| 1660 | 411.82 | 82.83 | 298.02 | 7.424 | 0.88116 |
| 1680 | 417.20 | 86.82 | 302.04 | 7.168 | 0.88439 |
| 1700 | 422.59 | 90.95 | 306.06 | 6.924 | 0.88758 |
| 1720 | 428.00 | 95.24 | 310.09 | 6.690 | 0.89074 |
| 1740 | 433.41 | 99.69 | 314.13 | 6.465 | 0.89387 |
| 1760 | 438.83 | 104.30 | 318.18 | 6.251 | 0.89697 |
| 1780 | 444.26 | 109.08 | 322.24 | 6.045 | 0.90003 |
| 1800 | 449.71 | 114.03 | 326.32 | 5.847 | 0.90308 |
| 1820 | 455.17 | 119.16 | 330.40 | 5.658 | 0.90609 |
| 1840 | 460.63 | 124.47 | 334.50 | 5.476 | 0.90908 |
| 1860 | 466.12 | 129.95 | 338.61 | 5.302 | 0.91203 |
| 1880 | 471.60 | 135.64 | 342.73 | 5.134 | 0.91497 |
| 1900 | 477.09 | 141.51 | 346.85 | 4.974 | 0.91788 |

*(Continued)*

**TABLE A.20 (CONTINUED)**

Thermodynamic Properties of Air at Low Pressure

| T, R | h, Btu/lb | $p_r$ | u, Btu/lb | $v_r$ | φ Btu/lb. °R |
|------|-----------|-------|-----------|-------|--------------|
| 1920 | 482.60 | 147.59 | 350.98 | 4.819 | 0.92076 |
| 1940 | 488.12 | 153.87 | 355.12 | 4.670 | 0.92362 |
| 1960 | 493.64 | 160.37 | 359.28 | 4.527 | 0.92645 |
| 1980 | 499.17 | 167.07 | 363.43 | 4.390 | 0.92926 |
| 2000 | 504.71 | 174.00 | 367.61 | 4.258 | 0.93205 |
| 2020 | 510.26 | 181.16 | 371.79 | 4.130 | 0.93481 |
| 2040 | 515.82 | 188.54 | 375.98 | 4.008 | 0.93756 |
| 2060 | 521.39 | 196.16 | 380.18 | 3.890 | 0.94026 |
| 2080 | 526.97 | 204.02 | 384.39 | 3.777 | 0.94296 |
| 2100 | 532.55 | 212.1 | 388.60 | 3.667 | 0.94564 |
| 2120 | 538.15 | 220.5 | 392.83 | 3.561 | 0.94829 |
| 2140 | 543.74 | 229.1 | 397.05 | 3.460 | 0.95092 |
| 2160 | 549.35 | 238.0 | 401.29 | 3.362 | 0.95352 |
| 2180 | 554.97 | 247.2 | 405.53 | 3.267 | 0.95611 |
| 2200 | 560.59 | 256.6 | 409.78 | 3.176 | 0.95868 |
| 2220 | 566.23 | 266.3 | 414.05 | 3.088 | 0.96123 |
| 2240 | 571.86 | 276.3 | 418.31 | 3.003 | 0.96376 |
| 2260 | 577.51 | 286.6 | 422.59 | 2.921 | 0.96626 |
| 2280 | 583.16 | 297.2 | 426.87 | 2.841 | 0.96876 |
| 2300 | 588.82 | 308.1 | 431.16 | 2.765 | 0.97123 |
| 2320 | 594.49 | 319.4 | 435.46 | 2.691 | 0.97369 |
| 2340 | 600.16 | 330.9 | 439.76 | 2.619 | 0.97611 |
| 2360 | 605.84 | 342.8 | 444.07 | 2.550 | 0.97853 |
| 2380 | 611.53 | 355.0 | 448.38 | 2.483 | 0.98092 |
| 2400 | 617.22 | 367.6 | 452.70 | 2.419 | 0.98331 |

*Source:* Abridged from Table 1 in Joseph H. Keenan and Joseph Kaye: *Gas Tables.* 1948. Copyright Wiley–VCH Verlag GmbH & Co. KGaA. Reproduced with permission.

## TABLE A.21

Properties of Some Gases at Low Pressure

| Temp R | Products of Combustion, 400% Theoretical Air | | Products of Combustion, 200% Theoretical Air | | Nitrogen | | Oxygen | |
|---|---|---|---|---|---|---|---|---|
| | $\bar{h}$ | $\bar{\phi}$ | $\bar{h}$ | $\bar{\phi}$ | $\bar{h}$ | $\bar{\phi}$ | $\bar{h}$ | $\bar{\phi}$ |
| 537 | 3746.8 | 46.318 | 3774.9 | 46.300 | 3729.5 | 45.755 | 3725.1 | 48.986 |
| 600 | 4191.9 | 47.101 | 4226.3 | 47.094 | 4167.9 | 46.514 | 4168.3 | 49.762 |
| 700 | 4901.7 | 48.195 | 4947.7 | 48.207 | 4864.9 | 47.588 | 4879.3 | 50.858 |
| 800 | 5617.5 | 49.150 | 5676.3 | 49.179 | 5564.4 | 48.522 | 5602.0 | 51.821 |
| 900 | 6340.3 | 50.002 | 6413.0 | 50.047 | 6268.1 | 49.352 | 6337.9 | 52.688 |
| 1000 | 7072.1 | 50.773 | 7159.8 | 50.833 | 6977.9 | 50.099 | 7087.5 | 53.477 |
| 1100 | 7812.9 | 51.479 | 7916.4 | 51.555 | 7695.0 | 50.783 | 7850.4 | 54.204 |
| 1200 | 8563.4 | 52.132 | 8683.6 | 52.222 | 8420.0 | 51.413 | 8625.8 | 54.879 |
| 1300 | 9324.1 | 52.741 | 9461.7 | 52.845 | 9153.9 | 52.001 | 9412.9 | 55.508 |
| 1400 | 10095.0 | 53.312 | 10250.7 | 53.430 | 9896.9 | 52.551 | 10210.4 | 56.099 |
| 1500 | 10875.6 | 53.851 | 11050.2 | 53.981 | 10648.9 | 53.071 | 11017.1 | 56.656 |
| 1600 | 11665.6 | 54.360 | 11859.6 | 54.504 | 11409.7 | 53.561 | 11832.5 | 57.182 |
| 1700 | 12464.3 | 54.844 | 12678.6 | 55.000 | 12178.9 | 54.028 | 12655.6 | 57.680 |
| 1800 | 13271.7 | 55.306 | 13507.0 | 55.473 | 12956.3 | 54.472 | 13485.8 | 58.155 |
| 1900 | 14087.2 | 55.747 | 14344.1 | 55.926 | 13741.6 | 54.896 | 14322.1 | 58.607 |
| 2000 | 14910.3 | 56.169 | 15189.3 | 56.360 | 14534.4 | 55.303 | 15164.0 | 59.039 |
| 2100 | 15740.5 | 56.574 | 16042.4 | 56.777 | 15334.0 | 55.694 | 16010.9 | 59.451 |
| 2200 | 16577.1 | 56.964 | 16902.5 | 57.177 | 16139.8 | 56.068 | 16862.6 | 59.848 |
| 2300 | 17419.8 | 57.338 | 17769.3 | 57.562 | 16951.2 | 56.429 | 17718.8 | 60.228 |
| 2400 | 18268.0 | 57.699 | 18642.1 | 57.933 | 17767.9 | 56.777 | 18579.2 | 60.594 |
| 2500 | 19121.4 | 58.048 | 19520.7 | 58.292 | 18589.5 | 57.112 | 19443.4 | 60.946 |
| 2600 | 19979.7 | 58.384 | 20404.6 | 58.639 | 19415.8 | 57.436 | 20311.4 | 61.287 |
| 2700 | 20842.8 | 58.710 | 21293.8 | 58.974 | 20246.4 | 57.750 | 21182.9 | 61.616 |
| 2800 | 21709.8 | 59.026 | 22187.5 | 59.300 | 21081.1 | 58.053 | 22057.8 | 61.934 |
| 2900 | 22581.4 | 59.331 | 23086.0 | 59.615 | 21919.5 | 58.348 | 22936.1 | 62.242 |
| 3000 | 23456.6 | 59.628 | 23988.5 | 59.921 | 22761.5 | 58.632 | 23817.7 | 62.540 |
| 3100 | 24335.5 | 59.916 | 24895.3 | 60.218 | 23606.8 | 58.910 | 24702.5 | 62.831 |
| 3200 | 25217.8 | 60.196 | 25805.6 | 60.507 | 24455.0 | 59.179 | 25590.5 | 63.113 |
| 3300 | 26102.9 | 60.469 | 26719.2 | 60.789 | 25306.0 | 59.442 | 26481.6 | 63.386 |
| 3400 | 26991.4 | 60.734 | 27636.4 | 61.063 | 26159.7 | 59.697 | 27375.9 | 63.654 |
| 3500 | | | 28556.8 | 61.329 | 27015.9 | 59.944 | 28273.3 | 63.914 |
| 3600 | | | 29479.9 | 61.590 | 27874.4 | 60.186 | 29173.9 | 64.168 |
| 3700 | | | 30406.0 | 61.843 | 28735.1 | 60.422 | 30077.5 | 64.415 |
| 3800 | | | 31334.8 | 62.091 | 29597.9 | 60.652 | 30984.1 | 64.657 |
| 3900 | | | 32266.2 | 62.333 | 30462.8 | 60.877 | 31893.6 | 64.893 |
| 4000 | | | | | 31329.4 | 61.097 | 32806.1 | 65.123 |
| 4100 | | | | | 32198.0 | 61.310 | 33721.6 | 65.350 |
| 4200 | | | | | 33068.1 | 61.520 | 34639.9 | 65.571 |
| 4300 | | | | | 33939.9 | 61.726 | 35561.1 | 65.788 |
| 4400 | | | | | 34813.1 | 61.927 | 36485.0 | 66.000 |
| 4500 | | | | | 35687.8 | 62.123 | 37411.8 | 66.208 |
| 4600 | | | | | 36563.8 | 62.316 | 38341.4 | 66.413 |

*(Continued)*

**TABLE A.21 (CONTINUED)**

Properties of Some Gases at Low Pressure

| Temp R | Products of Combustion, 400% Theoretical Air | | Products of Combustion, 200% Theoretical Air | | Nitrogen | | Oxygen | |
|---|---|---|---|---|---|---|---|---|
| | $\bar{h}$ | $\phi$ | $\bar{h}$ | $\phi$ | $\bar{h}$ | $\phi$ | $\bar{h}$ | $\phi$ |
| 4700 | | | | | 37441.1 | 62.504 | 39273.6 | 66.613 |
| 4800 | | | | | 38319.5 | 62.689 | 40208.6 | 66.809 |
| 4900 | | | | | 39199.1 | 62.870 | 41146.1 | 67.003 |
| 5000 | | | | | 40079.8 | 63.049 | 42086.3 | 67.193 |
| 5100 | | | | | 40961.6 | 63.223 | 43029.1 | 67.380 |
| 5200 | | | | | 41844.4 | 63.395 | 43974.3 | 67.562 |
| 5300 | | | | | 42728.3 | 63.563 | 44922.2 | 67.743 |

*Source:* Abridged from Tables 4, 7, 11, 13, 15, 17, 19, and 21 in Joseph H. Keenan and Joseph Kaye: *Gas Tables.* 1948. Copyright Wiley-VCH Verlag GmbH & Co. KGaA. Reproduced with permission.

*Note:* $\bar{h}$ = enthalpy Btu/lbmole; $\phi = \int_{T-0}^{T} c_{P0} \dfrac{dT}{T}$, Btu/lb. mole R; $\phi$ is essentially equal to the absolute entropy at 1 atm pressure, Btu/lb. mole R.

**TABLE A.21 (SI)**

Properties of Some Gases at Low Pressure

| | Water Vapor | | Carbon Dioxide | | Hydrogen | | Carbon Monoxide | |
|---|---|---|---|---|---|---|---|---|
| Temp R | $\bar{h}$ | $\bar{\phi}$ | $\bar{h}$ | $\bar{\phi}$ | $\bar{h}$ | $\bar{\phi}$ | $\bar{h}$ | $\bar{\phi}$ |
| 537 | 4258.3 | 45.079 | 4030.2 | 51.032 | 3640.3 | 31.194 | 3729.5 | 47.272 |
| 600 | 4764.7 | 45.970 | 4600.9 | 52.038 | 4075.6 | 31.959 | 3168.0 | 48.044 |
| 700 | 5575.4 | 47.219 | 5552.0 | 53.503 | 4770.2 | 33.031 | 4866.0 | 49.120 |
| 800 | 6396.9 | 48.316 | 6552.9 | 54.839 | 5467.1 | 33.961 | 5568.2 | 50.058 |
| 900 | 7230.9 | 49.298 | 7597.6 | 56.070 | 6165.3 | 34.784 | 6276.4 | 50.892 |
| 1000 | 8078.9 | 50.191 | 8682.1 | 57.212 | 6864.5 | 35.520 | 6992.2 | 51.646 |
| 1100 | 8942.0 | 51.013 | 9802.6 | 58.281 | 7564.6 | 36.188 | 7716.8 | 52.337 |
| 1200 | 9820.4 | 51.777 | 10955.3 | 59.283 | 8265.8 | 36.798 | 8450.8 | 52.976 |
| 1300 | 10714.5 | 52.494 | 12136.9 | 60.229 | 8968.7 | 37.360 | 9194.6 | 53.571 |
| 1400 | 11624.8 | 53.168 | 13344.7 | 61.124 | 9673.8 | 37.883 | 9948.1 | 54.129 |
| 1500 | 12551.4 | 53.808 | 14576.0 | 61.974 | 10381.5 | 38.372 | 10711.1 | 54.655 |
| 1600 | 13494.9 | 54.418 | 15829.0 | 62.783 | 11092.5 | 38.830 | 11483.4 | 55.154 |
| 1700 | 14455.4 | 54.999 | 17101.4 | 63.555 | 11807.4 | 39.264 | 12264.3 | 55.628 |
| 1800 | 15433.0 | 55.559 | 18391.5 | 64.292 | 12526.8 | 39.675 | 13053.2 | 56.078 |
| 1900 | 16427.5 | 56.097 | 19697.8 | 64.999 | 13250.9 | 40.067 | 13849.8 | 56.509 |
| 2000 | 17439.0 | 56.617 | 21018.7 | 65.676 | 13980.1 | 40.441 | 14653.2 | 56.922 |
| 2100 | 18466.9 | 57.119 | 22352.7 | 66.327 | 14714.5 | 40.799 | 15463.3 | 57.317 |
| 2200 | 19510.8 | 57.605 | 23699.0 | 66.953 | 15454.4 | 41.143 | 16279.4 | 57.696 |
| 2300 | 20570.6 | 58.077 | 25056.3 | 67.557 | 16199.8 | 41.475 | 17101.0 | 58.062 |
| 2400 | 21645.7 | 58.535 | 26424.0 | 68.139 | 16950.6 | 41.794 | 17927.4 | 58.414 |
| 2500 | 22735.4 | 58.980 | 27801.2 | 68.702 | 17707.3 | 42.104 | 18758.8 | 58.754 |
| 2600 | 23839.5 | 59.414 | 29187.1 | 69.245 | 18469.7 | 42.403 | 19594.3 | 59.081 |
| 2700 | 24957.2 | 59.837 | 30581.2 | 69.771 | 19237.8 | 42.692 | 20434.0 | 59.398 |
| 2800 | 26088.0 | 60.248 | 31982.8 | 70.282 | 20011.8 | 42.973 | 21277.2 | 59.705 |
| 2900 | 27231.2 | 60.650 | 33391.5 | 70.776 | 20791.5 | 43.247 | 22123.8 | 60.002 |
| 3000 | 28386.3 | 61.043 | 34806.6 | 71.255 | 21576.9 | 43.514 | 22973.4 | 60.290 |
| 3100 | 29552.8 | 61.426 | 36227.9 | 71.722 | 22367.7 | 43.773 | 23826.0 | 60.569 |
| 3200 | 30730.2 | 61.801 | 37654.7 | 72.175 | 23164.1 | 44.026 | 24681.2 | 60.841 |
| 3300 | 31918.2 | 62.167 | 39086.7 | 72.616 | 23965.5 | 44.273 | 25539.0 | 61.105 |
| 3400 | 33116.0 | 62.526 | 40523.6 | 73.045 | 24771.9 | 44.513 | 26399.3 | 61.362 |
| 3500 | 34323.5 | 62.876 | 41965.2 | 73.462 | 25582.9 | 44.748 | 27261.8 | 61.612 |
| 3600 | 35540.1 | 63.221 | 43411.0 | 73.870 | 26398.5 | 44.978 | 28126.6 | 61.855 |
| 3700 | 36765.4 | 63.557 | 44860.6 | 74.267 | 27218.5 | 45.203 | 28993.5 | 62.093 |
| 3800 | 37998.9 | 63.887 | 46314.0 | 74.655 | 28042.8 | 45.423 | 29862.3 | 62.325 |
| 3900 | 39240.2 | 64.210 | 47771.0 | 75.033 | 28871.1 | 45.638 | 30732.9 | 62.551 |
| 4000 | 40489.1 | 64.528 | 49231.4 | 75.404 | 29703.5 | 45.849 | 31605.2 | 62.772 |
| 4100 | 41745.4 | 64.839 | 50695.1 | 75.765 | 30539.8 | 46.056 | 32479.1 | 62.988 |
| 4200 | 43008.4 | 65.144 | 52162.0 | 76.119 | 31379.8 | 46.257 | 33354.4 | 63.198 |
| 4300 | 44278.0 | 65.444 | 53632.1 | 76.464 | 32223.5 | 46.456 | 34231.2 | 63.405 |

(*Continued*)

**TABLE A.21 (SI) (CONTINUED)**

Properties of Some Gases at Low Pressure

| Temp $R$ | Water Vapor | | Carbon Dioxide | | Hydrogen | | Carbon Monoxide | |
|---|---|---|---|---|---|---|---|---|
| | $\bar{h}$ | $\bar{\phi}$ | $\bar{h}$ | $\bar{\phi}$ | $\bar{h}$ | $\bar{\phi}$ | $\bar{h}$ | $\bar{\phi}$ |
| 4400 | 45553.9 | 65.738 | 55105.1 | 76.803 | 33070.9 | 46.651 | 35109.2 | 63.607 |
| 4500 | 46835.9 | 66.028 | 56581.0 | 77.135 | 33921.6 | 46.842 | 35988.6 | 63.805 |
| 4600 | 48123.6 | 66.312 | 58059.7 | 77.460 | 34775.7 | 47.030 | 36869.3 | 63.998 |
| 4700 | 49416.9 | 66.591 | 59541.1 | 77.779 | 35633.0 | 47.215 | 37751.0 | 64.188 |
| 4800 | 50715.5 | 66.866 | 61024.9 | 78.091 | 36493.4 | 47.396 | 38633.9 | 64.374 |
| 4900 | 52019.0 | 67.135 | 62511.3 | 78.398 | 37356.9 | 47.574 | 39517.8 | 64.556 |
| 5000 | 53327.4 | 67.401 | 64000.0 | 78.698 | 38223.3 | 47.749 | 40402.7 | 64.735 |
| 5100 | 54640.3 | 67.662 | 65490.9 | 78.994 | 39092.8 | 47.921 | 41288.6 | 64.910 |
| 5200 | 55957.4 | 67.918 | 66984.0 | 79.284 | 39965.1 | 48.090 | 42175.5 | 65.082 |
| 5300 | 57278.7 | 68.172 | 68479.1 | 79.569 | 40840.2 | 48.257 | 43063.2 | 65.252 |

**TABLE A.22**

Normal Total Emissivity of Various Surfaces

| Surfaces | °F | $\epsilon$ |
|---|---|---|
| **a. Metals and Their Oxides** | | |
| Aluminum: | | |
| Highly polished plate, 98.3% pure | 440, 1070 | 0.039, 0.057 |
| Polished plate | 73 | 0.040 |
| Rough plate | 78 | 0.055 |
| Oxidized at 1110°F | 390, 1110 | 0.11, 0.19 |
| Al-surfaced roofing | 110 | 0.216 |
| Al-treated surfaces, heated at 1110°F | | |
| Copper | 390, 1110 | 0.18, 0.19 |
| Steel | 390, 1110 | 0.52, 0.57 |
| Brass: | | |
| Highly polished: | | |
| 73.2% Cu, 26.7% Zn, by weight | 476, 674 | 0.028, 0.031 |
| 62.4% Cu, 36.8% Zn, 0.4% Pb, 0.3% Al, by weight | 494, 710 | 0.0388, 0.037 |
| 82.9% Cu, 17.0% Zn, by weight | 530 | 0.030 |
| Hard-rolled, polished, but direction of polishing visible | 70 | 0.038 |
| But somewhat attacked | 73 | 0.043 |
| But traces of stearin from polish left on | 75 | 0.053 |
| Polished: | 100, 600 | 0.096, 0.096 |
| Rolled plate: | | |
| Natural surface | 72 | 0.06 |
| Rubbed with coarse emery | 72 | 0.20 |
| Dull plate | 120, 660 | 0.22 |
| Oxidized by heating at 1110°F | 390, 1110 | 0.61, 0.59 |
| Chromium: | | |
| *See* Nickel alloys for Ni–Cr steels | | |
| Copper: | | |
| Carefully polished electrolytic Cu | 176 | 0.018 |
| Commercial, emeried, polished, but pits remaining | 66 | 0.030 |
| Scraped shiny, but not mirrorlike | 72 | 0.072 |
| Polished | 242 | 0.023 |
| Plate heated at 1110°F | 390, 1110 | 0.57, 0.57 |
| Cuprous oxide | 1470, 2010 | 0.66, 0.54 |
| Plate, heated for a long time, covered with thick oxide layer | 77 | 0.78 |
| Molten copper | 1970, 2300 | 0.16, 0.13 |
| Gold: | | |
| Pure, highly polished | 440, 1160 | 0.018, 0.035 |
| Iron and steel: | | |
| Metallic surfaces (or very thin oxide layer): | | |
| Electrolytic iron, highly polished | 350, 440 | 0.052, 0.074 |
| Polished iron | 800, 1880 | 0.144, 0.377 |
| Iron freshly emeried | 68 | 0.242 |
| Cast iron, polished | 392 | 0.21 |
| Wrought iron, highly polished | 100, 480 | 0.28 |
| Cast iron, newly turned | 72 | 0.435 |

*(Continued)*

**TABLE A.22 (CONTINUED)**

Normal Total Emissivity of Various Surfaces

| Surfaces | °F | $\epsilon$ |
|---|---|---|
| Polished steel casting | 1420, 1900 | 0.52, 0.56 |
| Ground sheet steel | 1720, 2010 | 0.55, 0.61 |
| Smooth sheet iron | 1650, 1900 | 0.55, 0.60 |
| Cast iron, turned on lathe | 1620, 1810 | 0.60, 0.70 |
| Oxidized surfaces: | | |
| Iron plate, pickled, then rusted red | 68 | 0.612 |
| Then completely rusted | 67 | 0.685 |
| Rolled sheet steel | 70 | 0.657 |
| Oxidized iron | 212 | 0.736 |
| Cast iron, oxidized at 1100°F | 390, 1110 | 0.64, 0.78 |
| Steel oxidized at 1100°F | 390, 1110 | 0.79, 0.79 |
| Smooth, oxidized electrolytic iron | 260, 980 | 0.78, 0.82 |
| Iron oxide | 930, 2190 | 0.85, 0.89 |
| Rough ingot iron | 1700, 2040 | 0.87, 0.95 |
| Sheet steel, strong rough oxide layer | 75 | 0.80 |
| Dense shiny oxide layer | 75 | 0.82 |
| Cast plate: | | |
| Smooth | 73 | 0.80 |
| Rough | 73 | 0.82 |
| Cast iron, rough, strongly oxidized | 100, 480 | 0.95 |
| Wrought iron, dull-oxidized | 70, 680 | 0.94 |
| Steel plate, rough | 100, 700 | 0.94, 0.97 |
| High-temperature alloy steels; *see* Nickel alloys | | |
| Molten metals: | | |
| Molten cast iron | 2370, 2550 | 0.29, 0.29 |
| Molten mild steel | 2910, 3270 | 0.28, 0.28 |
| Lead: | | |
| Pure (99.96%) unoxidized | 260, 440 | 0.057, 0.075 |
| Gray oxidized | 75 | 0.281 |
| Oxidized at 390°F | 390 | 0.63 |
| Mercury, pure clean | 32, 212 | 0.09, 0.12 |
| Molybdenum filament | 1340, 4700 | 0.096, 0.292 |
| Ni-Cu alloy, oxidized at 1110°F | 390, 1110 | 0.41, 0.46 |
| Nickel: | | |
| Electroplated on polished iron, then polished | 74 | 0.045 |
| Technically pure (98.9% Ni by weight, +Mn), polished | 440, 710 | 0.07, 0.087 |
| Electroplated on pickled iron, not polished | 68 | 0.11 |
| Wire | 368, 1844 | 0.096, 0.186 |
| Plate, oxidized by heating at 1110°F | 390, 1110 | 0.37, 0.48 |
| Nickel oxide | 1200, 2290 | 0.59, 0.86 |
| Nickel alloys: | | |
| Cr-Ni alloy | 125, 1894 | 0.64, 0.76 |
| (18–32% Ni, 55–68% Cu, 20% Zn by weight), gray oxidized | 70 | 0.262 |
| Alloy steel (8% Ni, 18% Cr); light silvery, rough, brown after heating | 420, 914 | 0.44, 0.36 |

*(Continued)*

**TABLE A.22 (CONTINUED)**

Normal Total Emissivity of Various Surfaces

| Surfaces | °F | $\epsilon$ |
|---|---|---|
| Same, after 24 h heating at 980°F | 420, 980 | 0.62, 0.73 |
| Alloy (20% Ni, 25% Cr), brown, splotched, oxidized from service | 420, 980 | 0.90, 0.97 |
| Alloy (60% Ni, 12% Cr), smooth, black, firm adhesive oxide coat from service | 520, 1045 | 0.89, 0.82 |
| Platinum: | | |
| Pure, polished plate | 440, 1160 | 0.054, 0.104 |
| Strip | 1700, 2960 | 0.12, 0.17 |
| Filament | 80, 2240 | 0.036, 0.192 |
| Wire | 440, 2510 | 0.073, 0.182 |
| Silver: | | |
| Polished, pure | 440, 1160 | 0.0198, 0.0324 |
| Polished | 100, 700 | 0.0221, 0.0312 |
| Steel, *see* Iron | | |
| Tantalum filament | 2420, 4580 | 0.193, 0.31 |
| Tin, bright, tinned iron sheet | 76 | 0.043, 0.064 |
| Tungsten: | | |
| Filament, aged | 80, 6000 | 0.032, 0.35 |
| Filament | 6000 | 0.39 |
| Zinc: | | |
| Commercial, 99.1% pure, polished | 440, 620 | 0.045, 0.053 |
| Oxidized by heating at 750°F | 750 | 0.11 |
| Galvanized sheet iron: | | |
| Fairly bright | 82 | 0.228 |
| Gray, oxidized | 75 | 0.276 |
| **b. Refractories, Building Materials, Paints, and Miscellaneous** | | |
| Asbestos board | 74 | 0.96 |
| Asbestos paper | 100, 700 | 0.93, 0.945 |
| Brick: | | |
| Red, rough, but no gross irregularities | 70 | 0.93 |
| Silica unglazed, rough | 1832 | 0.80 |
| Silica glazed, rough | 2012 | 0.85 |
| Grog brick, glazed | 2012 | 0.75 |
| *See* Refractory materials, below | | |
| Carbon: | | |
| T-carbon, 0.9% ash | 260, 1160 | 0.81, 0.79 |
| Carbon filament | 1900, 2560 | 0.526 |
| Candle soot | 206, 520 | 0.952 |
| Lampblack: | | |
| Water-glass coating | 209, 362 | 0.959, 0.947 |
| Water-glass coating | 260, 440 | 0.957, 0.952 |
| Thin layer on iron plate | 69 | 0.927 |
| Thick coat | 68 | 0.967 |
| 0.003 in. or thicker | 100, 700 | 0.945 |
| Enamel, white, fused on iron | 66 | 0.897 |
| Glass, smooth | 72 | 0.937 |
| Gypsum, 0.02 in. thick or smooth on blackened plate | 70 | 0.903 |
| Marble, light gray, polished | 72 | 0.931 |

(*Continued*)

**TABLE A.22 (CONTINUED)**

Normal Total Emissivity of Various Surfaces

| Surfaces | °F | ϵ |
|---|---|---|
| Oak, planed | 70 | 0.895 |
| Oil layers on polished nickel (lubricating oil): | | |
| Polished surface alone | 68 | 0.045 |
| +0.001 in. oil | 68 | 0.27 |
| +0.002 in. oil | 68 | 0.46 |
| +0.005 in. oil | 68 | 0.72 |
| + ∞ | 68 | 0.82 |
| Oil layers on aluminum foil (linseed oil): | | |
| Aluminum foil | 212 | 0.087 |
| +1 coat oil | 212 | 0.561 |
| +2 coats oil | 212 | 0.574 |
| Paints, lacquers, varnishes: | | |
| Snow-white enamel varnish on rough iron plate | 73 | 0.906 |
| Black shiny lacquer, sprayed on iron | 76 | 0.875 |
| Black shiny shellac on tinned iron sheet | 70 | 0.821 |
| Black-matte shellac | 170, 295 | 0.91 |
| Black lacquer | 100, 200 | 0.80, 0.95 |
| Flat black lacquer | 100, 200 | 0.96, 0.98 |
| White lacquer | 100, 200 | 0.80, 0.95 |
| Oil paints, 16 different, all colors | 212 | 0.92, 0.96 |
| Aluminum paints and lacquers: | | |
| 10% Al, 22% lacquer body, on rough or smooth surface | 212 | 0.52 |
| 26% Al, 27% lacquer body, on rough or smooth surface | 212 | 0.30 |
| Other aluminum paints, varying age and Al content | 212 | 0.27, 0.67 |
| Aluminum lacquer, varnish binder, on rough plate | 70 | 0.39 |
| Aluminum paint, after heating to 620°F | 300, 600 | 0.35 |
| Paper, thin: | | |
| Pasted on tinned iron plate | 66 | 0.924 |
| Pasted on rough iron plate | 66 | 0.929 |
| Pasted on black lacquered plate | 66 | 0.944 |
| Plaster, rough, lime | 50, 190 | 0.91 |
| Porcelain, glazed | 72 | 0.924 |
| Quartz, rough, fused | 70 | 0.932 |
| Refractory materials, 40 different | 1110, 1830 | |
| Poor radiat.ors | | 0.65, 0.75 0.70 |
| Good radiators | | 0.80, 0.85 0.85, 0.90 |
| Roofing paper | 69 | 0.91 |
| Rubber: | | |
| Hard, glossy plate | 74 | 0.945 |
| Soft, gray, rough (reclaimed) | 76 | 0.859 |
| Serpentine, polished | 74 | 0.900 |
| Water | 32, 212 | 0.95, 0.963 |

*Source:* A. I. Brown and S. M. Marco, *Introduction to Heat Transfer*, 3rd ed., McGraw-Hill Book Company, New York, 1958, pp. 54–58. With permission.

**TABLE A.23**

Psychrometric Chart for Air–Water Vapor Mixtures

*Source:* Courtesy of General Electric.

**TABLE A.23 (SI)**

Psychrometric Chart for Air–Water Vapor Mixtures

*Source:* *ASHRAE Handbook, Fundamentals,* American Society of Heating, Refrigerating and Air-Conditioning Engineers Inc., 2005. With permission.

# Bibliography

American Iron and Steel Institute. *AISI Metric Practice Guide: SI Units and Conversion Factors for the Steel Industry*. Washington, DC: AISI, 1975.

American Society of Heating, Refrigerating, and Air-Conditioning Engineers. *ASHRAE Handbook*, New York: ASHRAE, 2016.

American Society of Mechanical Engineers. *ASME Orientation and Guide for Use of SI (Metric) Units*, 5th ed. New York: ASME, 1974.

American Society of Mechanical Engineers. *ASME Text Booklet: SI Units in Strength of Materials*. New York: ASME, 1975.

American Society of Mechanical Engineers. *ASME Steam Tables*, 6th ed. New York: ASME, 1993.

American Solar Energy Society, 2400 Central Avenue, Suite G-l, Boulder, CO, 80301.

American Wind Energy Association, 122 C Street, Suite 380, Washington, DC, 20001.

Annamalai, K., I. Puri, and M. Jog. *Advanced Engineering Thermodynamics*, 2nd ed. Boca Raton, FL: CRC Press, 2011.

Baker, H. D., E. A. Ryder, and H. N. Baker. *Temperature Measurement in Engineering*. New York: John Wiley & Sons, Inc., 1961.

Benedict, R. P. *Fundamentals of Temperature, Pressure, and Flow Measurements*. New York: John Wiley & Sons, Inc., 1969.

Bluestein, M. "Nature's Carnot Engine: The Hurricane," *Proceedings of ASME International Mechanical Engineering Conference, IMECE2006–13359*, November, 2006.

Cengel, Y. A. *Heat Transfer: A Practical Approach*. New York: McGraw-Hill Companies, 1998.

Clifford, G. E. *Modern Heating and Ventilating Systems Design*. Upper Saddle River, NJ: Regents/Prentice Hall, 1993.

Dodge, B. F. *Chemical Engineering Thermodynamics*. New York: McGraw-Hill Book Company, 1944.

Durham, F. P. *Thermodynamics*, 2nd ed. Upper Saddle River, NJ: Prentice Hall, Inc., 1959, p. 43.

Electric Power Research Institute, 3412 Hillview Avenue, Palo Alto, CA, 95160.

Fagenbaum, J. "The Stirling: A 'Hot' Engine for the 80's?" *Mechanical Engineering*, Vol. 105, No. 5, May 1983, pp. 18–29.

Fermi, E. *Thermodynamics*. New York: Dover Publications, 1956, p. 11.

Ferris, E. A. *Modern Steam Generating Equipment*. New York: Combustion Engineering Inc., 1960.

Granet, I. "Heat Transfer Performance Curves," *Chemical Engineering*, March 1955, Vol. 62, No. 3, pp. 187–190.

Granet, I. "Heat Transfer to Subsaturated Water, Dry Air and Hydrogen in Turbulent Flow Inside a Tube," *Journal of the American Society of Naval Engineers*, Vol. 69, No. 4, November 1957, pp. 787–794.

Granet, I. "The Coefficient of Radiant Heat Transfer," *Design News*, September 29, 1970.

Granet, I. "Natural Convection Heat Transfer Aids," *Design News*, March 20, 1972.

Hatsopoulos, G. N., and J. H. Keenan. "A Single Axiom for Classical Thermodynamics," ASME Paper 61-WA-100, 1961.

Howell, J. R., and R. O. Buckius. *Fundamentals of Engineering Thermodynamics*, 2nd ed. New York: McGraw-Hill Book Company, 1987.

Howell, J. R., and R. O. Buckius. *Fundamentals of Engineering Thermodynamics*, 2nd ed. New York: McGraw-Hill Book Company, 1992, p. 284.

Huang, F. F. *Engineering Thermodynamics: Fundamentals and Applications*. New York: Macmillan Publishing Co., Inc., 1992.

International Electronic Research Corporation. *Thermal Management Guide*. Burbank, CA: IERC, 1990.

Irvine, T. E., and J. P. Hartnett. *Steam and Gas Tables with Computer Equations*. New York: Hemisphere Publishing Corp., 1984.

Janna, W. S. *Introduction to Fluid Mechanics*, 3rd ed. Boston, MA: PWS-Kent Publishing Company, 1993.

Jones, C. "A Survey of Curtiss–Wright's 1958–1971 Rotary Combustion Engine Technological Developments," Paper 720468. Detroit, MI: Society of Automotive Engineers National Automobile Engineering Meeting, May 1972.

Jordan, R. C., and G. B. Priester. *Refrigeration and Air Conditioning*, 2nd ed. Upper Saddle River, NJ: Prentice Hall, Inc., 1956.

Keenan, J. H. *Thermodynamics*. New York: John Wiley & Sons, Inc., 1941, p. 58.

Keenan, J. H., and J. Kaye. *Gas Tables*. New York: John Wiley & Sons, Inc., 1948.

Keenan, J. H., F. G. Keyes, P. G. Hill, and J. G. Moore. *Steam Tables-Thermodynamic Properties of Water Including Vapor, Liquid, and Solid Phases*. New York: John Wiley & Sons, Inc., 1969. Also by the same authors and under the same title (*SI Units*), 1978.

Kreith, F., and R. Bezdek. "Can Industry Afford Solar Energy?" *Mechanical Engineering*, Vol. 105, No. 3, March 1983, pp. 35–41.

Kreith, F., and M. S. Bohn. *Principles of Heat Transfer*, 5th ed. St. Paul, MN: West Publishing Co., 1993.

Liley, P. E. *2000 Solved Problems in Mechanical Engineering Thermodynamics*. New York: McGraw-Hill Book Co., 1989.

McAdams, W. H. *Heat Transmission*, 3rd ed. New York: McGraw-Hill Book Company, 1954.

McQuiston, F. C., and J. D. Parker. *Heating, Ventilating, and Air-Conditioning*, 4th ed. New York: John Wiley & Sons, Inc., 1994.

Moebius Research, Inc. *Heat Pump Manual*, 2nd ed. Palo Alto, CA: Electric Power Research Institute, 1997.

Moran, M. J., and H. N. Shapiro. *Fundamentals of Engineering Thermodynamics*, 2nd ed. New York: John Wiley & Sons, Inc., 1992.

Naef, F. E., and D. N. Burwell. "Mini-OTEC Results," *7th Energy Technology Conference.*, 1980, Washington, DC.

Obert, E. F. *Concepts of Thermodynamics*. New York: McGraw-Hill Book Company, 1960, p. 59.

Pita, E. G. *Air Conditioning Principles and Systems*, 3rd ed. Upper Saddle River, NJ: Pearson Education, Inc., 1998.

Potter, M. C., and C. W. Sommerton. *Theory and Problems of Engineering Thermodynamics*. New York: McGraw-Hill, Inc., 1993.

*Power* magazine special reports: B. G. A. Skrotzki, "Steam Turbines," June 1962; B. G. A. Skrotzki, "Gas Turbines," December 1963; R. J. Bender, "Steam Generation," June 1964; R. K. Evans, "Nuclear Power Reactors," March 1965; R. G. Schweiger, "Heat Exchangers," June 1970.

Rathakrishnan, E. *Elements of Heat Transfer*. Boca Raton, FL: CRC Press, 2012.

Renewal Products Co. *How an Internal Combustion Engine Works*. Fairless Hills, PA: The Company, 1960.

Reynolds, W. C. *Thermodynamic Properties in SI (Graphs, Tables and Computational Equations for 40 Substances)*. Stanford, CA: Department of Mechanical Engineering, Stanford University, 1979.

Rohsenow, W. M., and H. Y. Choi. *Heat, Mass and Momentum Transfer*. Upper Saddle River, NJ: Prentice Hall, Inc., 1961.

Severns, W. H., and J. F. Fellows. *Air Conditioning and Refrigeration*. New York: John Wiley & Sons, Inc., 1958.

Shapiro, A. H. *The Dynamics and Thermodynamics of Fluid Flow*, Vol. 1. New York: The Ronald Press Company, 1953.

Shavit, A., and C. Gutfinger. *Thermodynamics: From Concepts to Applications*, 2nd ed. Boca Raton, FL: CRC Press, 2009.

Steinberg, D. S. *Cooling Techniques for Electronic Equipment*, 2nd ed. New York: John Wiley & Sons, Inc., 1991.

Stoecker, W. F., and J. W. Jones. *Refrigeration and Air Conditioning*, 2nd ed. New York: McGraw-Hill Book Company, 1982.

Stultz, S. C., and J. B. Kitto, eds. *Steam: Its Generation and Use*. Barberton, OH: The Babcock and Wilcox Company, 1992.

Todd, J. P., and H. B. Ellis. *An Introduction to Thermodynamics for Engineering Technologists*. New York: John Wiley & Sons, Inc., 1981.

U.S. Atomic Energy Commission. *Power Reactors*. Washington, DC: USAEC. Technical Information Service, May 1958.

Van Wylen, G. J., and R. E. Sonntag. *Fundamentals of Classical Thermodynamics*, 2nd ed. New York: John Wiley & Sons, Inc., 1973; SI version, 1976.

Waxberg, H. *Economic Study of Advanced Steam Conditions*. New York: Ebasco Services, Inc., June 1981.

Wolfe, H. C. ed. *Temperature, Its Measurement and Control in Science and Industry*. New York: Reinhold Publishing, 1955.

Wood, D. B. *Applications of Thermodynamics*, 2nd ed. Reading, MA: Addison-Wesley Publishing Co., Inc., 1982.

Zemansky, M. W. *Heat and Thermodynamics*, 4th ed. New York: McGraw-Hill Book Company, 1957, p. 59.

# Index